FOR THE
IB DIPLOMA PROGRAMME

THIRD EDITION
Biology

C. J. Clegg
Andrew Davis
Christopher Talbot

This book marks 60 years since the marriage of my mother and father, Mary and Brian Davis, on 10 August 1963. It is dedicated to them both.

Andrew Davis
2023

IB advisers: The Publishers would like to thank the following for their advice and support in the development of this project: Marcela Rodriguez.

The Publishers would also like to thank the International Baccalaureate Organization for permission to re-use their past examination questions in the online materials.

Although every effort has been made to ensure that website addresses are correct at time of going to press, Hodder Education cannot be held responsible for the content of any website mentioned in this book. It is sometimes possible to find a relocated web page by typing in the address of the home page for a website in the URL window of your browser.

Hachette UK's policy is to use papers that are natural, renewable and recyclable products and made from wood grown in well-managed forests and other controlled sources. The logging and manufacturing processes are expected to conform to the environmental regulations of the country of origin.

Orders: please contact Hachette UK Distribution, Hely Hutchinson Centre, Milton Road, Didcot, Oxfordshire, OX11 7HH. Telephone: +44 (0)1235 827827. Email: education@hachette.com Lines are open from 9 a.m. to 5 p.m., Monday to Friday. You can also order through our website: www.hoddereducation.com

ISBN: 978 1 3983 6424 0

© C. J. Clegg and Andrew Davis 2023
First published in 2007
Second edition published in 2014
This edition published in 2023 by

Hodder Education,
An Hachette UK Company
Carmelite House
50 Victoria Embankment
London EC4Y 0DZ

www.hoddereducation.com

The authorised representative in the EEA is Hachette Ireland, 8 Castlecourt Centre, Dublin 15, D15 XTP3, Ireland (email: info@hbgi.ie)

Impression number   10 9 8 7 6 5

Year   2027 2026 2025

All rights reserved. Apart from any use permitted under UK copyright law, no part of this publication may be reproduced or transmitted in any form or by any means, electronic or mechanical, including photocopying and recording, or held within any information storage and retrieval system, without permission in writing from the publisher or under licence from the Copyright Licensing Agency Limited. Further details of such licences (for reprographic reproduction) may be obtained from the Copyright Licensing Agency Limited, www.cla.co.uk

Cover art © Thierry – stock.adobe.com

Illustrations by Barking Dog, Aptara, Inc., and Oxford Designers & Illustrators

Typeset in Berkeley Oldstyle 10/14pt by DC Graphic Design Limited, Hextable, Kent

Printed in Great Britain by Bell & Bain Ltd, Glasgow

A catalogue record for this title is available from the British Library.

Contents

> Each of the four themes covered in this book is broken down into four levels of organization. These four levels are colour coded as follows:
> **1 Molecules**
> **2 Cells**
> **3 Organisms**
> **4 Ecosystems**

Introduction v

How to use this book vi

Tools and Inquiry viii

A Unity and diversity

A1.1 Water 2
A1.2 Nucleic acids 14
A2.1 Origins of cells (HL only) 34
A2.2 Cell structure 49
A2.3 Viruses (HL only) 88
A3.1 Diversity of organisms 102
A3.2 Classification and cladistics (HL only) 125
A4.1 Evolution and speciation 138
A4.2 Conservation of biodiversity 156

B Form and function

B1.1 Carbohydrates and lipids 181
B1.2 Proteins 205
B2.1 Membranes and membrane transport 223
B2.2 Organelles and compartmentalization 244
B2.3 Cell specialization 256
B3.1 Gas exchange 273
B3.2 Transport 295
B3.3 Muscle and motility (HL only) 326
B4.1 Adaptation to environment 341
B4.2 Ecological niches 363

C Interaction and interdependence

- **C1.1** Enzymes and metabolism 380
- **C1.2** Cell respiration 405
- **C1.3** Photosynthesis 425
- **C2.1** Chemical signalling (HL only) 450
- **C2.2** Neural signalling 467
- **C3.1** Integration of body systems 490
- **C3.2** Defence against disease 517
- **C4.1** Populations and communities 548
- **C4.2** Transfers of energy and matter 578

D Continuity and change

- **D1.1** DNA replication 602
- **D1.2** Protein synthesis 615
- **D1.3** Mutation and gene editing 637
- **D2.1** Cell and nuclear division 648
- **D2.2** Gene expression (HL only) 670
- **D2.3** Water potential 681
- **D3.1** Reproduction 693
- **D3.2** Inheritance 725
- **D3.3** Homeostasis 760
- **D4.1** Natural selection 779
- **D4.2** Stability and change 798
- **D4.3** Climate change 820

Acknowledgements 838

Index 841

Free online content

Go to our website www.hoddereducation.co.uk/ib-extras for free access to the following:
- Practice exam-style questions for each chapter
- Glossary
- Answers to self-assessment questions and practice exam-style questions
- Answers to linking questions
- Tools and Inquiries reference guide
- Internal Assessment – the scientific investigation

Introduction

Welcome to *Biology for the IB Diploma Third Edition*, updated and designed to meet the criteria of the new International Baccalaureate (IB) Diploma Programme Biology Guide. This coursebook provides complete coverage of the new IB Biology Diploma syllabus, with first teaching from 2023. Differentiated content for SL and HL students is clearly identified throughout.

The aim of this syllabus is to integrate concepts, topic content and the nature of science through inquiry. This book comprises four main themes, each made up of two broad integrating concepts:

- **Theme A**: Unity and diversity
- **Theme B**: Form and function
- **Theme C**: Interaction and interdependence
- **Theme D**: Continuity and change

Each theme is then further divided into four levels of biological organization. In this coursebook, each level is colour coded as follows:

1 Molecules   **2 Cells**   **3 Organisms**   **4 Ecosystems**

About the authors

Chris Clegg is an experienced teacher and examiner of biology and has written many internationally respected textbooks for pre-university courses. He was encouraged to write by his colleague and mentor at his school, textbook writer and teacher D.G. Mackean, in the 1970s and became his co-author on numerous books. He eventually took over the biology coursebook mantle from Don in the 1980s.

Andrew Davis has taught biology for over 20 years. He is the author of several IB textbooks, including *Biology for the IB Diploma Study and Revision Guide, IB Diploma: Internal assessment for Biology: Skills for success*, and *Biology for the MYP 4 & 5: By Concept*. He is also author of online teaching and learning resources: *Biology for the IB Diploma Teaching and Learning* and *Biology for the IB MYP 4 & 5 Dynamic Learning*.

IB advisors

Chris Talbot graduated in Biochemistry from the University of Sussex in the United Kingdom. He has Masters Degrees in Life Sciences (Chemistry) and in Science Education from the National Technological University in the Republic of Singapore. He has taught IB Chemistry, IB Biology and Theory of Knowledge (TOK) in a number of local and international schools in Singapore. He is the author of numerous science textbooks, including *Chemistry for the MYP 4&5: By Concept*.

John Sprague has been teaching TOK for 20 years, in the UK, Switzerland and Singapore. Previously Director of IB at Sevenoaks School in the UK, he now teaches philosophy and TOK at Tanglin Trust School, Singapore.

 The 'In cooperation with IB' logo signifies that this coursebook has been rigorously reviewed by the IB to ensure it fully aligns with the current IB curriculum and offers high-quality guidance and support for IB teaching and learning.

How to use this book

The following features of this book will help you to consolidate and develop your understanding of biology, through concept-based learning:

Guiding questions

- There are two guiding questions at the start of every chapter, as signposts for inquiry.
- These questions will help you to view the content of the syllabus through the conceptual lenses of both the themes and the levels of biological organization.

SYLLABUS CONTENT

▶ This coursebook follows the order of the contents of the IB Biology Diploma syllabus.
▶ At the beginning of each chapter is a list of the content to be covered, with all subsections clearly linked to the content statements and showing the breadth and depth of understanding required.

Key terms
◆ Definitions appear throughout the margins to provide context and to help you understand the language of biology. There is also a glossary of all key terms at www.hoddereducation.co.uk/ib-extras.

Concepts

The four themes that underpin the IB Biology Diploma course (A Unity and diversity, B Form and function, C Interaction and interdependence, and D Continuity and change) are integrated into the conceptual understandings of all the units to ensure that a conceptual thread is woven throughout the course.

Conceptual understanding therefore enhances your overall understanding of the course, making the subject more meaningful. This understanding assists you in developing clear evidence of synthesis and evaluation in your responses to questions asked in the assessment, and helps you make connections across the course.

Concepts are explored in context and can be found throughout the chapter.

● Common mistake
These detail some common misunderstandings and typical errors made by students, so that you can avoid making the same mistakes yourself.

Tools

The Tools features explore the skills and techniques that you require and are integrated into the biology content to be practised in context. The skills in the study of biology can be assessed through internal and external assessment.

Inquiry

The application and development of the Inquiry process is supported throughout this coursebook, in close association with the Tools.

● Top tips!
This feature includes advice relating to the content being discussed and tips to help you retain the knowledge you need.

WORKED EXAMPLES

These provide a step-by-step guide showing you how to answer the kind of quantitative questions that you might encounter in your studies and in the assessment.

TOK

Links to Theory of Knowledge (TOK) allow you to develop critical-thinking skills and deepen scientific understanding by discussing the subject beyond the scope of the curriculum.

Links

Due to the conceptual nature of biology, many topics are connected. The Links feature states where relevant material is covered elsewhere in the coursebook. They may also help you to start creating your own linking questions.

Nature of science

Nature of science (NOS) is an overarching theme in the biology course that seeks to explore conceptual understandings related to the purpose, features and impact of scientific knowledge. It can be examined in biology papers. NOS explores the scientific process itself, and how science is represented and understood by the general public. It covers 11 aspects: Observations, Patterns and trends, Hypotheses, Experiments, Measurements, Models, Evidence, Theories, Falsification, Science as a shared endeavour and the Global impact of science. It also examines the way in which science is the basis for technological developments and how these new technologies, in turn, drive developments in science.

ATL ACTIVITY

Approaches to learning (ATL), including learning through inquiry, are integral to IB pedagogy. These activities are designed to get you to think about real-world applications of biology.

Going further

Written for students interested in further study, this optional feature contains material that goes beyond the IB Diploma Biology Guide.

LINKING QUESTIONS

These questions are listed at the end of each chapter and are for all students to attempt (apart from those in HL-only chapters). They are designed to strengthen your understanding by making connections across the themes. The linking questions encourage you to apply broad, integrated and discipline-specific concepts from one topic to another, ideally networking your knowledge. Practise answering the linking questions first, on your own or in groups. Sample answers and structures are provided online at www.hoddereducation.co.uk/ib-extras. The list in this coursebook is not exhaustive; you may encounter other connections between concepts, leading you to create your own linking questions.

Self-assessment questions appear throughout the chapters, phrased to assist comprehension and recall, but also to help familiarize you with the assessment implications of the command terms. These command terms are defined in the online glossary. Practice exam-style questions for each chapter allow you to check your understanding and prepare for the assessments. The questions are in the style of those in the examination so that you get practise seeing the command terms and the weight of the answers with the mark scheme. Practice exam-style questions and their answers, together with self-assessment answers, are on the accompanying website, IB Extras: www.hoddereducation.co.uk/ib-extras

Skills are highlighted with this icon. Students are expected to be able to show these skills in the examination, so we have explicitly pointed these out when they are mentioned in the Guide.

International mindedness is indicated with this icon. It explores how the exchange of information and ideas across national boundaries has been essential to the progress of science and illustrates the international aspects of biology.

The **IB learner profile** icon indicates material that is particularly useful to help you towards developing the following attributes: to be inquirers, knowledgeable, thinkers, communicators, principled, open-minded, caring, risk-takers, balanced and reflective. When you see the icon, think about what learner profile attribute you might be demonstrating – it could be more than one.

How to use this book

Tools and Inquiry

Skills in the study of biology

The skills and techniques you must experience through this biology course are encompassed within the tools. These support the application and development of the inquiry process in the delivery of the course.

■ Tools
- **Tool 1**: Experimental techniques
- **Tool 2**: Technology
- **Tool 3**: Mathematics

■ Inquiry process
- **Inquiry 1**: Exploring and designing
- **Inquiry 2**: Collecting and processing data
- **Inquiry 3**: Concluding and evaluating

Throughout the programme, you will be given opportunities to encounter and practise the skills; and instead of stand-alone topics, they will be integrated into the teaching of the syllabus when they are relevant to the topics being covered.

You can see what the Tools and Inquiry boxes look like in the *How to use this book* section on page vi.

The skills in the study of biology can be assessed through internal and external assessment.

The approaches to learning provide the framework for the development of these skills.

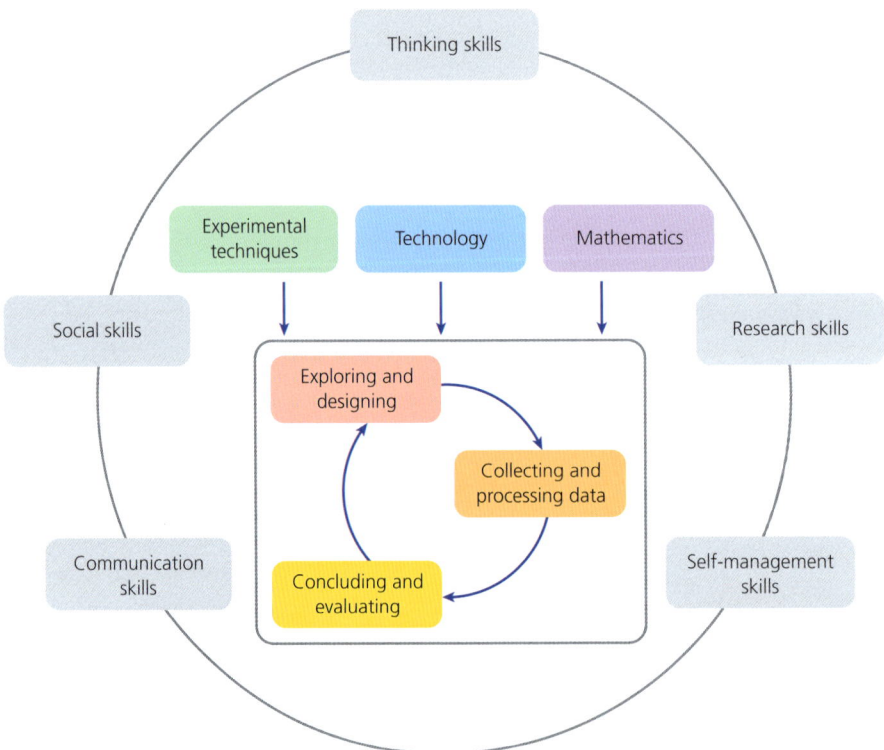

■ Skills for biology   *From IB Diploma Programme Biology Guide, page 29*

Visit the following website to view the online Tools and Inquiries reference guide:
www.hoddereducation.co.uk/ib-extras

Tool 1: Experimental techniques

Skill	Description
Addressing safety of self, others and the environment	Recognize and address relevant safety, ethical or environmental issues in an investigation.
Measuring variables	Understand how to accurately measure the following to an appropriate level of precision: • mass • volume • time • temperature • length. Make careful observations, including the following: • counts • drawing annotated diagrams from observation • making appropriate qualitative observations • classifying.
Applying techniques	Show awareness of the purpose and practice of: • paper or thin layer chromatography • colorimetry or spectrophotometry • serial dilutions • physical and digital molecular modelling • a light microscope and eyepiece graticule • preparation of temporary mounts • identifying and classifying organisms • using a variety of sampling techniques/using random and systematic sampling • karyotyping and karyograms • cladogram analysis.

Tool 2: Technology

Skill	Description
Applying technology to collect data	• Use sensors. • Identify and extract data from databases. • Generate data from models and simulations.
Applying technology to process data	• Use spreadsheets to manipulate data. • Represent data in a graphical form. • Use computer modelling. • Carry out image analysis.

Tool 3: Mathematics

Skill	Description
Applying general mathematics	• Use basic arithmetic and algebraic calculations to solve problems. • Carry out calculations involving: decimals, fractions, percentages, ratios, proportions, frequencies (including allele frequencies), densities, approximations and reciprocals. • Calculate measures of central tendency: mean, median and mode. • Apply measures of dispersion: range, standard deviation (SD), standard error (SE), interquartile range (IQR). • Use and interpret scientific notation (for example, 3.5×10^6). • Use approximation and estimation. • Calculate scales of magnification. • Calculate rates of change from graphical or tabulated data. • Understand direct and inverse proportionality between variables, as well as positive and negative correlations between variables. • Calculate and interpret percentage change and percentage difference. • Distinguish between continuous and discrete variables. • Calculate the actual size from a micrograph that has a scale bar. • Apply Simpson's reciprocal index. • Apply the Lincoln index. • Apply the chi-squared test. • Apply the *t*-test.
Using units, symbols and numerical values	• Apply and use SI prefixes and units or non-SI metric units. • Express quantities and uncertainties to an appropriate number of decimal places.
Processing uncertainties	• Understand the significance of uncertainties in raw and processed data. • Record uncertainties in measurements as a range (±) to an appropriate precision. • Express ranges, degrees of precision, standard error or standard deviations as error bars. • Express measurement and processed uncertainties to an appropriate number of decimal places or level of precision. • Apply the coefficient of determination (R^2) to evaluate the fit of a trend line. • Interpret values of the correlation coefficient and identify correlations as positive or negative. • Apply and interpret appropriate tests of statistical significance (for example, chi-squared test).
Graphing	• Sketch graphs, with labelled but unscaled axes, to qualitatively describe trends. • Construct and interpret tables, charts and graphs for raw and processed data including bar charts, histograms, scatter graphs, line and curve graphs, logarithmic graphs, pie charts and box-and-whisker plots. • Plot linear and non-linear graphs showing the relationship between two variables with appropriate scales and axes. • Draw lines or curves of best fit. • Interpret features of graphs including gradient, changes in gradient, intercepts, maxima and minima. • Draw and interpret uncertainty/error bars. • Extrapolate and interpolate graphs. • Design dichotomous keys. • Represent energy flow in the form of food chains, food webs and pyramids of energy. • Represent familial genetic relationships using pedigree charts.

Inquiry process

■ Inquiry 1: Exploring and designing

Skill	Description
Exploring	• Demonstrate independent thinking, initiative and insight. • Consult a variety of sources. • Select sufficient and relevant sources of information. • Formulate research questions and hypotheses. • State and explain predictions using scientific understanding.
Designing	• Demonstrate creativity in the designing, implementation and presentation of the investigation. • Develop investigations that involve hands-on laboratory experiments, databases, simulations, modelling and surveys. • Identify and justify the choice of dependent, independent and control variables. • Justify the range and quantity of measurements. • Design and explain a valid methodology. • Pilot methodologies.
Controlling variables	Appreciate when and how to: • calibrate measuring apparatus • maintain constant environmental conditions of systems • choose representative random samples and minimize sampling errors • set up a control run where appropriate.

■ Inquiry 2: Collecting and processing data

Skill	Description
Collecting data	• Identify and record relevant qualitative observations. • Collect and record sufficient relevant quantitative data. • Identify and address issues that arise during data collection.
Processing data	• Carry out relevant and accurate data processing.
Interpreting results	• Interpret qualitative and quantitative data. • Interpret diagrams, graphs and charts. • Identify, describe and explain patterns, trends and relationships. • Identify and justify the removal or inclusion of outliers in data (no mathematical processing is required). • Assess accuracy, precision, reliability and validity.

■ Inquiry 3: Concluding and evaluating

Skill	Description
Concluding	• Interpret processed data and analysis to draw and justify conclusions. • Compare the outcomes of an investigation to the accepted scientific context. • Relate the outcomes of an investigation to the stated research question or hypothesis. • Discuss the impact of uncertainties on the conclusions.
Evaluating	• Evaluate hypotheses. • Identify and discuss sources and impacts of random and systematic errors. • Evaluate the implications of methodological weaknesses, limitations and assumptions on conclusions. • Explain realistic and relevant improvements to an investigation.

Tables from IB Diploma Programme Biology Guide, pages 29–33

Water

> **Guiding questions**
> - What physical and chemical properties of water make it essential for life?
> - What are the challenges and opportunities of water as a habitat?

Concept: Unity and diversity

Common ancestry has given living organisms many shared features while evolution has resulted in the rich biodiversity of life on Earth.

SYLLABUS CONTENT

This chapter covers the following syllabus content:
- ▶ A1.1.1 Water as the medium for life
- ▶ A1.1.2 Hydrogen bonds as a consequence of the polar covalent bonds within water molecules
- ▶ A1.1.3 Cohesion of water molecules due to hydrogen bonding and consequences for organisms
- ▶ A1.1.4 Adhesion of water to materials that are polar or charged and impacts for organisms
- ▶ A1.1.5 Solvent properties of water linked to its role as a medium for metabolism and for transport in plants and animals
- ▶ A1.1.6 Physical properties of water and the consequences for animals in aquatic habitats
- ▶ A1.1.7 Extraplanetary origin of water on Earth and reasons for its retention (HL only)
- ▶ A1.1.8 The relationship between the search for extraterrestrial life and the presence of water (HL only)

Concept: Unity

All living organisms require water to exist. Enzymes – biological molecules that increase the rate of chemical reactions – need to be dissolved in water to work. Water provides a chemically stable medium for life processes to operate.

Water: the medium for life

The Earth is covered mainly by water and so appears a mostly blue planet when viewed from space. Approximately 71% of our planet's surface is water, with 97% found in oceans and only 3% as fresh water. Evidence from the geological record indicates that water has existed on Earth for 3.8 billion years. The Earth formed an estimated 4.5 billion years ago, so water has existed on its surface for most of its history. The first cells originated in water, where the oceans blocked harmful ultraviolet radiation from the Sun, allowing the first life to evolve. Water remains the medium in which most processes of life occur.

Water forms a large proportion of living organisms – between 65% and 95% by mass of most multicellular plants and animals (about 80% of a human cell consists of water). Despite this, and the fact that water has some unusual properties, water is a substance that is often taken for granted. As we will see in this chapter, the properties of water allow life to exist at a range of scales – from the smallest bacteria to the tallest tree – and without water life would not exist on Earth.

> **ATL A1.1A**
>
> Freshwater is a limited resource globally. Work in a group to produce an informative poster on the threats to freshwater sources and the solutions available for providing sufficient, clean drinking water for all.

♦ **Covalent bond**: a bond between atoms in which pairs of electrons are shared.

♦ **Polar molecule**: a molecule where there is an unequal distribution of electrical charge: one end is slightly positive and the other end is slightly negative.

♦ **Hydrogen bond**: a weak attractive intermolecular force; a hydrogen atom in a molecule is attracted to an electronegative atom, such as oxygen, in a different molecule.

Hydrogen bonds

The water molecule consists of one atom of oxygen and two atoms of hydrogen combined by sharing pairs of electrons (**covalent bonding**). However, the molecule is V-shaped rather than linear. The nucleus of the oxygen atom draws electrons (negatively charged) away from the hydrogen nuclei (positively charged) with an interesting consequence. Although overall the water molecule is electrically neutral, there is a net negative charge on the oxygen atom and a net positive charge on the hydrogen atoms. The water molecule therefore carries an unequal distribution of electrical charge within it. This arrangement is known as a **polar molecule** (Figure A1.1.1).

With water molecules, the positively charged hydrogen atoms of one molecule are attracted to negatively charged oxygen atoms of nearby water molecules, causing attractive forces called **hydrogen bonds** (Figure A1.1.1). These intermolecular forces are weak compared to covalent bonds, yet they are strong enough to hold water molecules together and to attract water molecules to charged particles or to a charged surface. Hydrogen bonds largely account for the unique properties of water. We will examine these properties next.

1 Distinguish between ionic and covalent bonding.

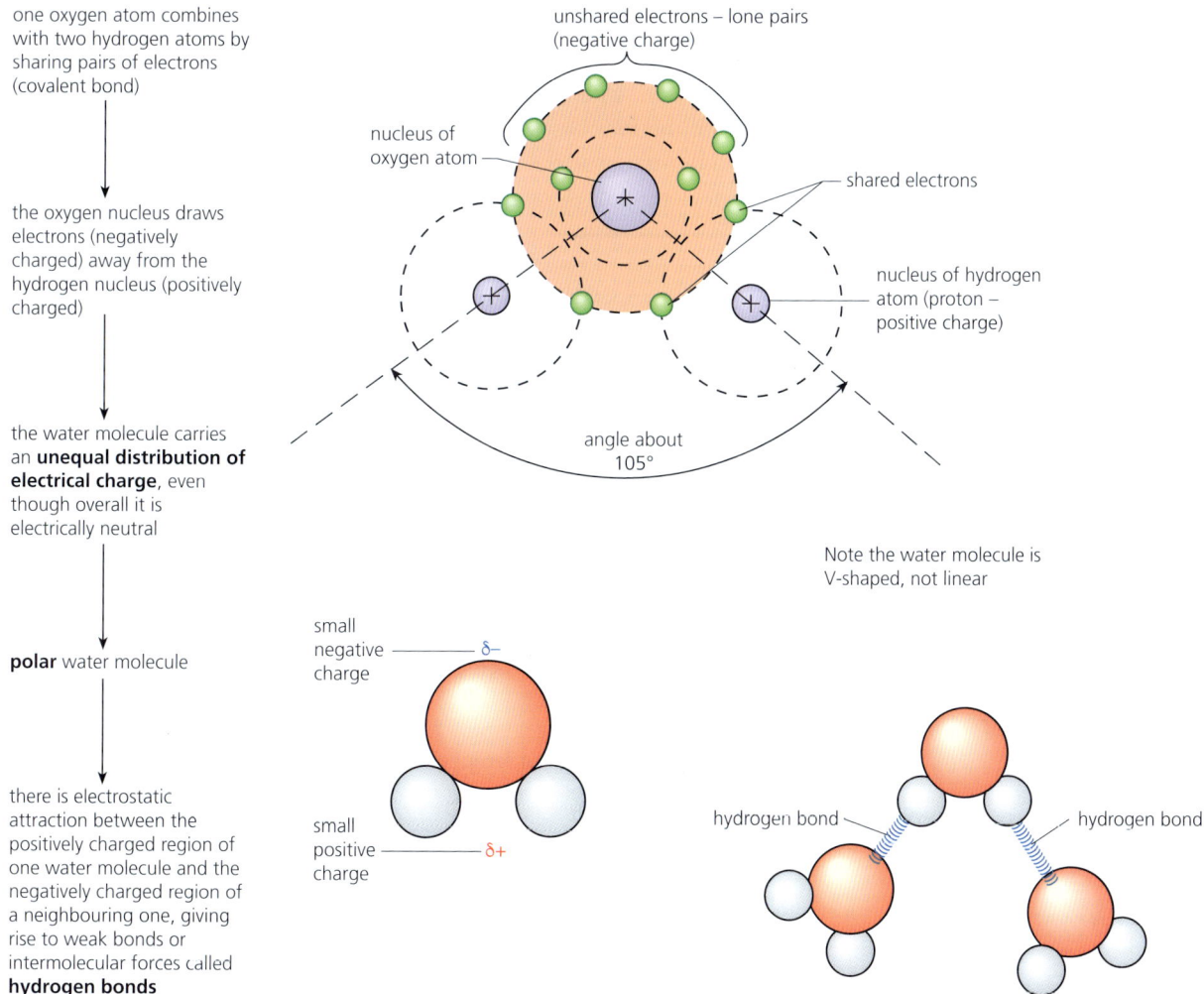

■ Figure A1.1.1 The water molecule and the hydrogen bonds it forms

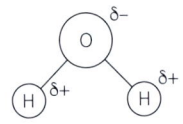

■ Figure A1.1.2 The polarity of water

Figure A1.1.2 (left) shows how to indicate polarity in a water molecule.

Top tip!

You need to be able to represent two or more water molecules and hydrogen bonds between them. Delta (δ) symbols indicate a small charge.

■ Figure A1.1.3 Hydrogen bonds between water molecules; the dashed line between the oxygen and hydrogen atoms represents a hydrogen bond

Common mistake

A common mistake is suggesting that hydrogen bonding occurs within water molecules. Do not confuse intra- (within) and inter- (between) molecular bonding. Covalent bonding acts **within** a water molecule; hydrogen bonds are formed **between** water molecules.

2 **List** the important properties of water that are due to its polar nature.

Tool 2: Technology

Using computer modelling

Computer modelling allows scientists to explore how the structure of water is essential for maintaining its properties and, therefore, in maintaining life.

Ruth Lynden-Bell and co-workers at Queen's University Belfast used computer simulations to model changes in water's properties. The bond angle in water molecules is 104.5°: they found that if this was changed to 90°, or if the hydrogen bonds were about 15% weaker, the three-dimensional network of hydrogen bonds – crucial to the liquid's unique properties – would be severely disrupted or fall apart.

TOK

The central principle of homeopathy is that water can retain a 'memory' of substances previously dissolved in it, even after any number of serial dilutions. Such claims about the 'memory of water' are categorized as 'pseudoscientific', meaning that while the theories or ideas might look as if they follow the scientific method as normally applied by expert scientists, they do not.

What are the criteria that can be used to distinguish scientific claims from pseudoscientific claims?

The scientific method uses hypothesis, observations and falsification to develop new scientific ideas. This means that scientists set out to challenge hypotheses and look for evidence that might prove them false. If researchers only seek more and more *confirmation* of their ideas, rather than trying to find how their ideas might be false, it is possible that their results could be biased. The results may appear well established, but really the research is either irrelevant or ignores false results. One characteristic of 'pseudoscience' is that it only looks for evidence that supports its claims.

Confirmation bias

Confirmation bias refers to the tendency to search for, interpret and favour information or data in a way that confirms your pre-existing beliefs or hypotheses. You may be guilty of this when you use an internet search engine to settle an argument and only look for results that confirm what you already think.

The concept of water having 'memory' of what it has previously encountered contradicts current scientific understanding of physical chemistry. Another characteristic of pseudoscientific theories is that they are at odds with well-established scientific findings; they are wildly surprising. The responsible scientific approach is therefore to replicate the tests to see whether the same results are found. With the cooperation of Benveniste's own team, a group from *Nature* tried to repeat Benveniste's findings but failed, ultimately showing that there was no evidence that water had any sort of chemical 'memory'. Subsequent investigations did not support Benveniste's findings. Given the scientific evidence, then (as opposed to anecdotal evidence), there is no reason to believe that water has a chemical memory.

◆ **Cohesion**: force by which individual molecules of the same type attract and associate ('stick together').

◆ **Surface tension**: property of the surface of a liquid that allows it to resist an external force, due to the cohesion between water molecules.

Cohesion of water molecules and the consequences for organisms

Cohesion is the force by which individual molecules of the same type attract and associate (stick together). Water molecules stick together because of hydrogen bonding. These bonds continually break and reform with surrounding water molecules, although at any one moment a large number are held together by their hydrogen bonds. Cohesive forces allow water molecules to be drawn up xylem vessels in plants by the evaporative loss of water from the leaves (Figure A1.1.4). Compared with other liquids, water has extremely strong cohesive properties that prevent it 'breaking' under tension.

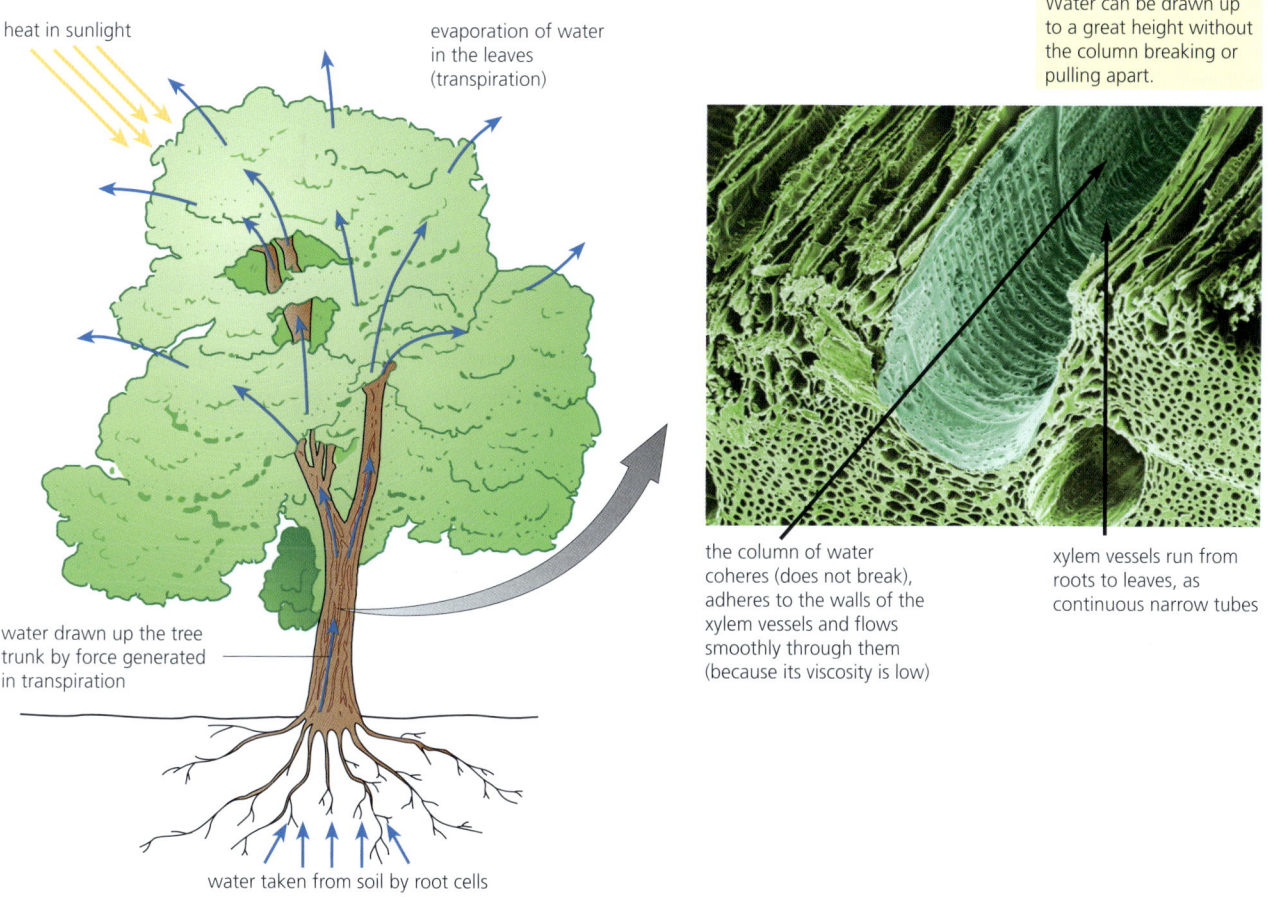

■ **Figure A1.1.4** Water is drawn up a tree trunk through xylem vessels: cohesive forces stop the water column from breaking and help draw water up the tree

Link
The transport of water from roots to leaves during transpiration is covered in Chapter B3.2, page 303.

Common mistake
A common mistake is suggesting that hydrogen bonds are strong – this is not the case. A single hydrogen bond is a weak interaction. It is only because there are many hydrogen bonds in water that they collectively exert large cohesive forces.

Related to the property of cohesion is the property of **surface tension**. The outermost molecules of water form hydrogen bonds with the water molecules below them. This gives water a very high surface tension (Figure A1.1.5), higher than any other liquid except mercury. The water molecules on the surface have no neighbouring water molecules above and therefore exhibit stronger attractive forces upon their nearest neighbours on and below the surface. Water's strong surface tension allows it to form almost completely spherical droplets.

A1.1 Water

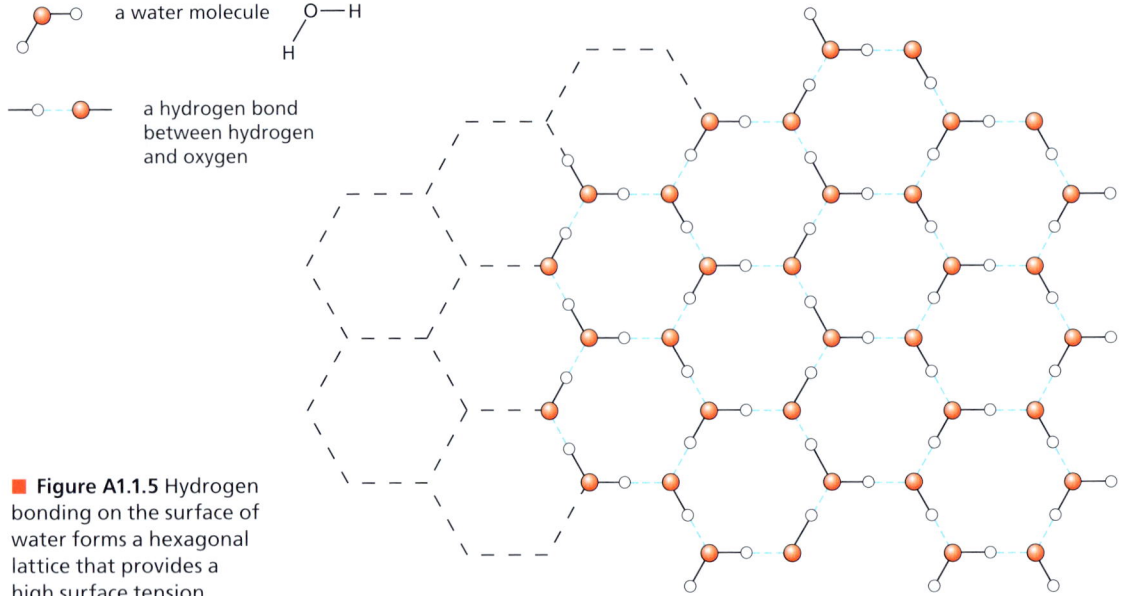

■ **Figure A1.1.5** Hydrogen bonding on the surface of water forms a hexagonal lattice that provides a high surface tension

Within a body of liquid, there is no net force on a molecule because the cohesive forces exerted by the neighbouring molecules all cancel out (see Figure A1.1.5). However, for a molecule on the surface of the liquid, there is a net inward cohesive force since there is no attractive force acting from above. This inward net force causes the molecules on the surface to contract and to resist being stretched or broken. Thus, the surface is under tension, hence the name 'surface tension'.

The surface tension of water is exploited by insects that 'surface skate' (Figure A1.1.6). The insect's waxy cuticle prevents the wetting of its body, and the mass of the insect is not great enough to break the surface tension.

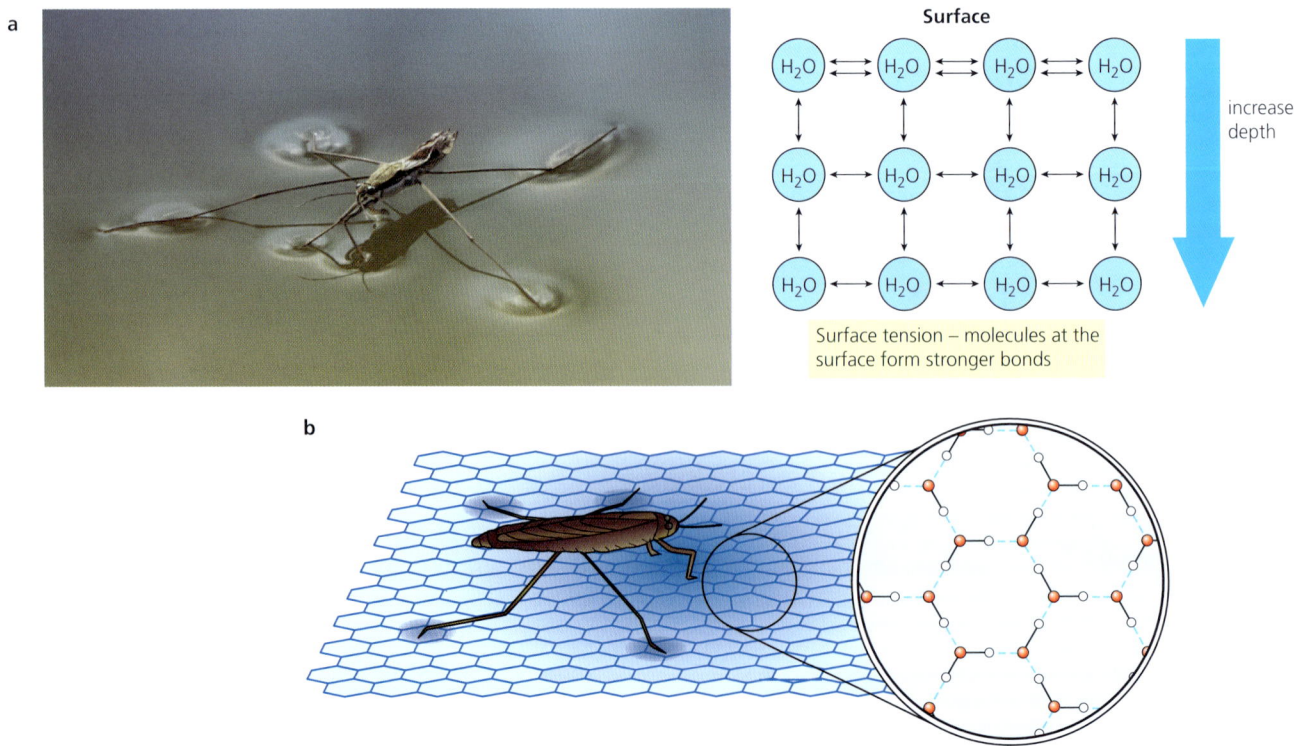

■ **Figure A1.1.6** a) A pond skater moving over the water surface; b) the surface tension supports the pond skater – the surface is depressed but the hydrogen bonds hold it together

Theme A: Unity and diversity – Molecules

ATL A1.1B

What other examples of surface tension are there? How does the knowledge of surface tension help you understand everyday phenomena and experiences? For example:
- Why are droplets of water pulled into a spherical shape?
- Why is it better to wash in hot water rather than cold water?
- Why are soaps and detergents used to clean clothes?

You could use the website below or other sources to research other examples of surface tension and why the property is useful to know about.

www.usgs.gov/special-topics/water-science-school/science/surface-tension-and-water#overview

◆ **Viscosity**: a measure of a fluid's resistance to flow.

Below the surface, water molecules slide past each other very easily. This property is described as low **viscosity**. Consequently, water flows readily through narrow capillaries, tiny gaps and pores.

Top tip!

The diffusion of molecules through a solvent, such as water, is inversely proportional to the viscosity of the solvent. Temperature affects the viscosity of liquids, for example, the viscosity of water at 25°C is approximately half that than when the temperature is 4°C. As we will see in Theme B2.1, the diffusion rate of molecules is extremely important for the processes that are needed to sustain life.

Inquiry 1: Exploring and designing

Designing

Surface tension is one of water's most important properties. It causes water to collect in drops.

Design an investigation to show the properties of water's surface tension using a paper clip.

Use the following equipment:
- drinking glass
- water
- liquid dishwashing detergent
- paper clips
- piece of paper towel.

◆ **Adhesion**: the force by which individual molecules stick to surrounding materials and surfaces.

◆ **Hydrophilic**: attracted to water; e.g. hydrogen bonds are readily formed between a molecule and water.

Adhesion of water and the impacts for organisms

Adhesion is the force by which individual molecules cling to surrounding materials and surfaces. Materials and substances with an affinity for water are described as **hydrophilic** (page 8). Water adheres strongly to most surfaces and can be drawn up long columns, for example through narrow tubes such as the xylem vessels of plant stems, without danger of the water column breaking (Figure A1.1.4). It should be noted that cohesion is a far more significant force in xylem transport and explains how tensions can be resisted. Adhesion is only significant when air-filled xylem vessels refill with aqueous sap under positive pressures, which is something that happens only rarely (no more than once a year). Figure A1.1.7 shows both adhesive and cohesive forces at work in a xylem vessel.

Common mistake

The terms 'cohesion' and 'adhesion' are sometimes treated as if their meanings are interchangeable, but this is not the case. If they were, we would have one word for these forces rather than two! Cohesion ('co' means 'together') is attraction between water molecules, while adhesion ('ad' means 'toward') is attraction to a surface.

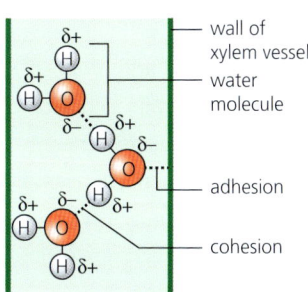

■ Figure A1.1.7 Adhesive and cohesive forces supporting a column of water in a xylem vessel

A1.1 Water

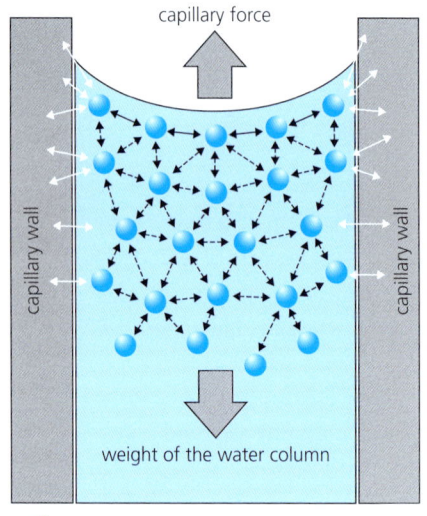

- ● water molecules
- ←---→ cohesion between water molecules
- ←——→ cohesion between water molecules on the surface
- ⬛ adhesion between water molecules and capillary wall

■ **Figure A1.1.8** Channels in soils and spaces between cellulose fibres in the cell wall act as capillary tubes, drawing water through the plant

■ Capillary action in soils and plant cell walls

Soil contains many vertical, thin channels known as **capillary tubes**, in which plant roots are located. When water enters capillary tubes, adhesion between the water molecules and the wall of the capillary draws water up the small tube: this is called **capillary action**. In this way, plants bring water up from the water table to the roots when the ground becomes dry.

The cell walls of plants are made from a fibrous material called cellulose (see page 193). Cellulose is polar/hydrophilic to a certain degree. Fibrous materials can act like wicks, drawing water up into the material by capillary action (see Figure A1.1.8). Cell walls can draw water by capillary action from nearby xylem vessels, keeping water flowing through plant tissue. Cells that are directly exposed to the air, such as those found in the spongy mesophyll tissue of leaves (page 283), remain constantly wetted by capillary action into these cells. Water evaporates from the moist, blotting-paper-like cell walls of the mesophyll and then diffuses out of leaves through pores on the surface of the leaf (stomata), enabling water to be transported up the plant.

Solvent properties of water

Hydrogen bonds pull water molecules very close to each other because the potential energy of the hydrogen bonds is greater than the kinetic energies of the water molecules up to 100 °C (at atmospheric pressure). This is why water is a liquid at the temperatures and pressure that exist over much of the Earth's surface. As a result, we have a liquid medium with distinctive thermal and solvent properties.

♦ **Capillary tubes**: channels with a very small internal diameter.

♦ **Capillary action**: the tendency of a liquid to move up against gravity when confined within a narrow tube (capillary). Also known as capillarity.

♦ **Solute**: dissolved molecule or ion in a solution.

♦ **Solvent**: a liquid in which another substance can be dissolved.

♦ **Hydrophobic**: repelled by water.

Water is a powerful solvent for polar substances such as ionic substances like sodium chloride (Na^+ and Cl^-). All cations (positively charged ions) and anions (negatively charged ions) become surrounded by a layer of orientated water molecules (Figure A1.1.9).

There is a diverse range of hydrophilic molecules that dissolve in water, such as carbon-containing (organic) molecules with ionized groups (for example, amino acids have a negatively charged carboxyl group, $-COO^-$, and a positively charged amino group, $-NH_3^+$); soluble organic molecules like sugars dissolve in water due to the formation of hydrogen bonds with their slightly charged hydroxyl groups ($-OH$). Once they have dissolved, molecules or ions (the **solute**) are free to move around in water (the **solvent**) by diffusion and, as a result, are more chemically reactive than when in the undissolved solid.

On the other hand, non-polar substances are repelled by water, as in the case of oil on the surface of water. Non-polar substances are **hydrophobic**. The functions of some molecules in cells depends on them being hydrophobic and insoluble. For example, the cell membrane is made from phospholipids, the tails of which are hydrophobic and form the internal structure of the membrane.

Of the common gases, carbon dioxide (CO_2), oxygen (O_2) and nitrogen (N_2), only carbon dioxide is particularly soluble in water; nitrogen and oxygen are only slightly soluble in water. Carbon dioxide is moderately soluble in water because a proportion of it undergoes a chemical reaction to form carbonic acid (H_2CO_3 (aq)), which immediately ionizes or dissociates to form hydrogen ions, H^+ (aq), and hydrogencarbonate ions, HCO_3^- (aq).

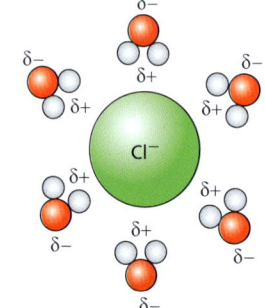

■ **Figure A1.1.9** The hydration of sodium and chloride ions

> **Link**
> The structure of cell membranes is covered in Chapter B2.1, page 223.

> **Link**
> For more on enzymes, see Chapter C1.1, page 380.

> **3** In an aqueous solution of glucose, **state** which component is the solvent and which is the solute.

> ◆ **Buoyancy**: the ability of any fluid to provide a vertical upwards force on an object placed in or on it.
>
> ◆ **Thermal conductivity (k)**: the measure of how easily heat flows through a specific type of material.
>
> ◆ **Specific heat capacity**: the amount of energy required to raise the temperature of 1 kg of a substance by 1°C.

Oxygen and nitrogen have low solubility in water because they are non-polar and do not form hydrogen bonds with water. In addition, they do not undergo dissociation or ionisation. One consequence of the poor solubility of oxygen in water is the evolution of respiratory pigments, for example haemoglobin, which greatly increase the oxygen-carrying capacity of blood relative to that of pure water.

Most enzymes catalyse reactions in aqueous solution. Enzymes require a certain level of water in their structures to maintain enzyme shape and stability, enabling them to function effectively. Most naturally occurring enzymes cannot form their active forms without being immersed in water. Hydrogen bonds often act as bridges between enzyme binding sites and their substrates. Although most enzymes act in aqueous solutions, sometimes an enzyme may be in a fixed position, such as within a cell membrane. In these cases, the location of the enzyme allows reactions to be localized to particular sites.

Physical properties of water and the consequences for animals in aquatic habitats

The physical properties of water depend on the hydrogen bonding between water molecules and include **buoyancy**, **viscosity**, **thermal conductivity** and **specific heat capacity**.

Buoyancy is the ability of any fluid (liquid or gas) to provide a vertical upwards force on an object placed in or on it. Objects float in water when their average density is less than water and sink when they are denser. The density of a substance is its mass per unit volume.

> ● **Common mistake**
>
> Do not confuse the terms 'heat capacity' with 'specific heat capacity'. Heat capacity is the amount of heat required to change the temperature of a body by one degree. The amount of heat energy **per unit mass** is needed to calculate the specific heat capacity. Unlike heat capacity, the specific heat capacity is therefore independent of mass or volume.

The differences between these properties in air and water are shown in Table A1.1.1.

■ **Table A1.1.1** Physical properties of water and air at 20°C and 1 atm pressure (for air, molar mass is a weighted average, since molar masses can only be calculated for pure substances)

	Water	Air
density, ρ (kg m^{-3})	998.21	1.204
thermal conductivity, k (W m K^{-1})	0.598	0.02154
specific heat capacity, c_p (J kg^{-1} °C^{-1})	4184	1007
dynamic viscosity, η (kg m^{-1} s^{-1})	1.002×10^{-3}	1.825×10^{-5}

For example, at sea level, air is 784 times less dense than water, and a volume of air at sea level has 0.13% of the density of the same volume of water.

The importance of these factors in relation to life can be illustrated by looking at two animals that live in water as well as in the air or on land, such as the black-throated loon (*Gavia arctica*) and the ringed seal (*Pusa hispida*), as shown in Figure A1.1.10.

■ **Figure A1.1.10** The black-throated loon, *Gavia arctica*, (left) and the ringed seal, *Pusa hispida*, (right)

> **Top tip!**
>
> When referring to an organism, either the common name, e.g. black-throated loon, or the scientific name, for this example *Gavia arctica*, is acceptable.

The black-throated loon (*Gavia arctica*) is a diving bird species, catching its prey (mostly fish) underwater. It breeds in the vicinity of deep freshwater lakes throughout northern Europe, the west coast of Alaska, and Asia. From August, it migrates south to areas around the Black Sea and the Mediterranean Sea, and to north-east Atlantic coasts and the eastern and western Pacific Ocean. It returns to its breeding grounds in early April when sea ice in those areas has melted.

Ringed seals (*Pusa hispida*) live in the Arctic and sub-arctic regions of the North Pole. They live on packs of ice, but also spend much of their time in the sea, under the ice. They are quite small seals, usually less than 1.5 m in length, and have a distinct pattern of dark spots surrounded by light grey rings on its fur – explaining its common name.

■ Specific heat capacity and the temperature of water

A relatively large amount of energy is required to raise the temperature of water, because a lot of energy is needed to break the large number of hydrogen bonds that restrict the movement of water molecules. This property of water is its specific heat capacity. Consequently, aquatic environments (rivers, ponds, lakes and seas) are very slow to change temperature when the surrounding air temperature changes. Aquatic environments have relatively more stable temperatures than terrestrial (land) environments. As organisms, and the cells from which they are made, are largely composed of water, water's ability to absorb and lose heat without undergoing a large temperature change also provides thermal cushioning within the organisms themselves, protecting cells and organisms from large fluctuations in temperature.

a water molecule

○—○ a hydrogen bond

■ **Figure A1.1.11** In ice the water molecules are hydrogen bonded in an open tetrahedral lattice, which makes ice less dense than liquid water

The relatively stable sea temperatures enable seals to live and feed throughout the year. The specific heat capacity of air is lower than water, so air temperature tends to fluctuate more. In winter, the very low air temperatures cause surface water to freeze. One of the interesting properties of water is that, unlike many other substances, it floats when it freezes because the density of ice is lower than that of liquid water. This is due to the behaviour of hydrogen bonds and how they make water molecules interact (Figure A1.1.11). Water has its highest density at 4 °C. The ice forms a platform on which seals can live. Ringed seals have claws to dig through ice to produce holes so that they can emerge from their aquatic habitat to breathe. This enables them to live under and on the ice throughout the year.

■ Thermal conductivity

> **Top tip!**
>
> Restricting convection is as important as restricting conduction in maintaining the body temperature of birds.

Water has a higher thermal conductivity than air, with water conducting heat 28 times better than air. By trapping air in its feathers, the black-throated loon forms an effective insulating layer between its skin and the outside air. Feathers also restrict convection currents by trapping a thin layer of air that is not able to move easily, which also helps to maintain the body temperature of the bird. In contrast, the seal relies on thick blubber to insulate its body. Layers of ice also have insulating properties because ice's thermal conductivity is low, like the thermal conductivity of air, which stops heat being transferred into the surroundings, even when the temperature is very low. The ice traps thermal energy in the water beneath the ice, increasing sea temperatures.

Buoyancy

The black-throated loon can swim large distances underwater. However, bird anatomy is adapted for life in the air and on land, with hollow bones to decrease weight and air trapped between feathers to provide insulation. These adaptations can be problematic in water, as buoyancy needs to be overcome to catch underwater prey species. The loon has solid bones to increase its weight and to compress air from its lungs and feathers to decrease buoyancy and enable successful diving.

Fat is stored in animals as adipose tissue, usually under the skin (subcutaneous fat). Aquatic diving mammals, such as seals, have a great deal of subcutaneous fat, which is known as blubber. Blubber acts as a buoyancy aid, as well as providing thermal insulation.

Viscosity

Viscosity is the resistance to flow. Water is more viscous than air (see Table A1.1.1). Bird plumage is adapted to hold and deflect air to make lift easier to achieve flight. When the black-throated loon flies, the light feathers can move through the air easily and with minimum friction.

Interactions between water molecules at the surface of water form surface tension (page 21). Below the surface, however, water molecules slide past each other very easily. This property is described as low viscosity. The hydrodynamic shapes of the loon and seal enable both animals to move through the water easily. Both animals need to produce resistance against the water to achieve movement. The seal uses its flippers (modified arms) to propel itself through the water. The black-throated loon has webbed feet that provide a large surface area to push against water. The feet are located laterally and towards the back end of the body to allow maximum propulsion; it also avoids the formation of turbulent eddies in the water and therefore reduces drag.

4 **Explain** the properties of water using examples of two animals that live in water as well as in the air or on land, such as the black-throated loon (*Gavia arctica*) and the ringed seal (*Pusa hispida*).

> **ATL A1.1C**
>
> Find out about another two animals that live in water – one should be a mammal that also lives on land, and the other should be a bird. Research the adaptations that help them to survive in these environments, using the same factors that are covered in this section: buoyancy, viscosity, thermal conductivity and specific heat capacity.

Extraplanetary origin of water on Earth

The Earth formed approximately 4.5 billion years ago, in an environment too hot for water to condense into liquid. This means that the Earth's water must have an extraplanetary origin. As the distance from the Sun increases, water vapour can condense directly into water ice. It is in these regions that water could first have formed, thereby providing the origin of Earth's water.

Researchers examining the composition of asteroids, and the meteorites that form by breaking off from them, have hypothesised that asteroids are most likely to be the source of Earth's water. Such asteroids still contain ice and organic material (amino acids), and so could have delivered water and organic molecules to Earth, which are both critical for the possible evolution of life (see page 36 for further discussion of the origin of life). A group of meteorites, known as **carbonaceous chondrites** – some of the oldest meteorites in the solar system – can be up to 28% water and have a water composition similar to ocean water. The water molecules are incorporated in the crystal structures of minerals. The composition of water, and its possible origin, can be assessed using isotopes of hydrogen (Figure A1.1.12) and the relative proportions in which they appear: deuterium (hydrogen-2) has a nucleus with one proton and one neutron, while protium (hydrogen-1) has just one proton in its nucleus. With hydrogen isotopes that closely match Earth's seawater, the water in these meteorites could have been the source of the Earth's oceans.

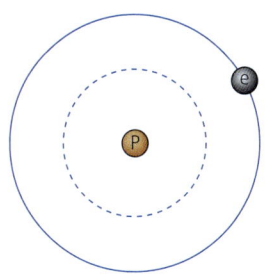

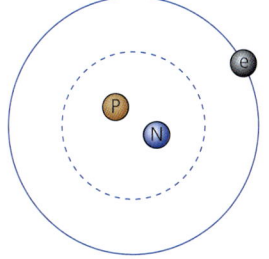

■ Figure A1.1.12 Isotopes of hydrogen

HL ONLY

Top tip!

There are many hypotheses for how water first arrived on the Earth, including being carried on icy comets and the creation of water beneath the planet's surface itself. For this syllabus we are only considering the asteroid hypothesis.

◆ **Goldilocks zone**: also known as the 'habitable zone'; the area around a star where it is not too hot or too cold for liquid water to exist on the surface of surrounding planets.

Two 4.5-billion-year-old meteorites containing liquid water, found on Earth, support this hypothesis. When meteorites heat up, such as during an impact with our planet, they release their water as gas, which is then trapped by the Earth's gravitational attraction.

The Earth's current deuterium to protium ratio also matches ancient eucrite achondrites, one type of meteorite, which originates from a large asteroid known as Vesta, located in the outer asteroid belt (the asteroid belt is located between the orbits of Mars and Jupiter). Carbonaceous chondrites and eucrite achondrites are therefore hypothesised to have delivered water to Earth.

Once present on Earth, the temperatures were cool enough to allow water vapour to condense into liquid water. Gravity enabled the water to be retained on the Earth's surface, rather than being dispersed into space.

The relationship between the search for extraterrestrial life and the presence of water

Given life's dependence on water, any planet where life as we know it is to exist must also have water present. Planets where water can exist must be at the right distance from their nearest star – too close and the water boils and evaporates, too far away and the water is frozen. The distance from the star where liquid water, and therefore life, can exist is called the **Goldilocks zone**, from the nineteenth-century British fairy tale *Goldilocks and the three bears*.

5 **State** what is meant by the 'Goldilocks zone'.

■ **Figure A1.1.13** The 'Goldilocks zone' for our solar system; Mars is also in the habitable zone along with Earth but it is too small to keep an atmosphere, which is needed to sustain life

Theme A: Unity and diversity – Molecules

ATL A1.1D

Scientists at NASA are planning to send astronauts to Mars (www.nasa.gov/topics/moon-to-mars). Carrying water from Earth to Mars is impractical – it is far too heavy to carry all the water required for a mission in a rocket. The plan would be to collect water from Mars itself: to do that, scientists would need to know if and where water is located.

What evidence is there for water on Mars, both now and in the past? What would the presence of water tell us about the possibility of finding life on Mars? What type of organisms could we expect to find on Mars?

To establish whether a distant planet may contain water, scientists use a technique called 'transit spectroscopy'. As a planet passes in front of its nearest star ('transits'), light passes through the planet's atmosphere; this light is analysed to see which wavelengths are being absorbed or deflected. This analysis shows which elements and molecules, such as water, are present in the atmosphere. In this way, planets outside of our solar system (exoplanets) may be said to have a 'water signature'. Scientists are looking for exoplanets that have a water signature, that are the right distance from their nearest star and are also the right size for life to exist. Discoveries such as the planet Kepler-186f, which is Earth-size and located in a Goldilocks zone, are likely contenders for life existing in other solar systems.

ATL A1.1E

Working in small groups, make a presentation about the conditions needed for life and how scientists are looking for such exoplanets. What techniques are there for finding water on other planets? What other exoplanets have been found? Could these planets ever be a potential home for humanity? The presentations should clearly explain the science behind 'Goldilocks planets' and how scientists go about looking for them.

LINKING QUESTIONS

1. How do the various intermolecular forces of attraction affect biological systems?
2. Which biological processes only happen at or near surfaces?

A1.2 Nucleic acids

> ## Guiding questions
> - How does the structure of nucleic acids allow hereditary information to be stored?
> - How does the structure of DNA facilitate accurate replication?

SYLLABUS CONTENT

This chapter covers the following syllabus content:
- ▶ A1.2.1 DNA as the genetic material of all living organisms
- ▶ A1.2.2 Components of a nucleotide
- ▶ A1.2.3 Sugar–phosphate bonding and the sugar–phosphate 'backbone' of DNA and RNA
- ▶ A1.2.4 Bases in each nucleic acid that form the basis of a code
- ▶ A1.2.5 RNA as a polymer formed by condensation of nucleotide monomers
- ▶ A1.2.6 DNA as a double helix made of two antiparallel strands of nucleotides with two strands linked by hydrogen bonding between complementary base pairs
- ▶ A1.2.7 Differences between DNA and RNA
- ▶ A1.2.8 Role of complementary base pairing in allowing genetic information to be replicated and expressed
- ▶ A1.2.9 Diversity of possible DNA base sequences and the limitless capacity of DNA for storing information
- ▶ A1.2.10 Conservation of the genetic code across all life forms as evidence of universal common ancestry
- ▶ A1.2.11 Directionality of RNA and DNA (HL only)
- ▶ A1.2.12 Purine-to-pyrimidine bonding as a component of DNA helix stability (HL only)
- ▶ A1.2.13 Structure of a nucleosome (HL only)
- ▶ A1.2.14 Evidence from the Hershey–Chase experiment for DNA as the genetic material (HL only)
- ▶ A1.2.15 Chargaff's data on the relative amounts of pyrimidine and purine bases across diverse life forms (HL only)

◆ **Nucleic acid**: polynucleotide chain of one of two types, deoxyribonucleic acid (DNA) or ribonucleic acid (RNA).

◆ **Genetic code**: the order of bases in DNA (of a chromosome) that determines the sequence of amino acids in a protein.

◆ **Deoxyribonucleic acid (DNA)**: a form of nucleic acid consisting of two complementary chains of deoxyribonucleotide subunits, and containing the bases adenine, thymine, guanine and cytosine.

◆ **Ribonucleic acid (RNA)**: a form of nucleic acid containing the pentose sugar ribose, and the organic bases adenine, guanine, uracil and cytosine.

Link
Viruses and their life cycles are covered in more detail in Chapter A2.3 (HL only), page 88.

DNA as the genetic material of all living organisms

DNA is a **nucleic acid**. Nucleic acids are the information molecules of cells (and also of viruses – see below, and page 88) found throughout the living world. The code containing the information in nucleic acids, known as the **genetic code**, is universal. This means that it makes sense in all organisms. It is not specific to a few organisms or to just one group, but to all groups and species.

There are two types of nucleic acid found in living cells: **deoxyribonucleic acid (DNA)** and **ribonucleic acid (RNA)**. DNA is the genetic material and occurs in the chromosomes of the nucleus (and also certain cell organelles, chloroplasts and mitochondria – see page 69). While some RNA also occurs in the nucleus, most is found in the cytoplasm – particularly in the ribosomes.

Some viruses use RNA as their genetic material, such as SARS-CoV-2, the virus that causes COVID-19. Other diseases caused by RNA viruses include the common cold, influenza, Dengue fever, hepatitis C, rabies, Ebola, polio, mumps and measles. Viruses depend on the cells of living organisms to survive and replicate, and so are not considered to be living; thus it is true to say that DNA is the genetic material of all living organisms.

Theme A: Unity and diversity – Molecules

◆ **Protein**: a long sequence of amino acid residues combined together (primary structure), and taking up a particular shape (secondary and tertiary structure).

● Top tip!

The presence of genetic material in a structure does not necessarily indicate life. Viruses, which are usually considered to be non-living, contain genetic material. In addition, DNA is chemically stable so can persist in dead organic matter and some fossils.

◆ **Nucleotide**: phosphate ester of a nucleoside – an organic base combined with pentose sugar and phosphate (Pi).

◆ **Cytosine**: a pyrimidine nitrogenous base found in nucleic acids (DNA and RNA) that pairs with guanine.

◆ **Guanine**: a purine nitrogenous base found in nucleic acids (DNA and RNA) that pairs with cytosine.

◆ **Adenine**: a purine nitrogenous base, found in the coenzymes ATP and NADP and in nucleic acids (DNA and RNA), that pairs with thymine.

◆ **Thymine**: a pyrimidine nitrogenous base found in DNA that pairs with adenine.

◆ **Pentose**: a 5-carbon monosaccharide sugar.

◆ **Condensation**: formation of larger molecules involving the removal of water from smaller component molecules.

Both DNA and RNA have roles in the day-to-day control of cells and organisms, as we shall see shortly. First, we will look into the structure of nucleotides and the way they are built up (synthesized) to form the unique DNA double helix.

Concept: Unity

All living organisms and viruses contain nucleic acids. This universality of the genetic code indicates the inter-connectivity of life on Earth and explains how viruses can take over and use the biological machinery of cells.

ATL A1.2A

DNA has long been known as a major chemical in the nucleus. DNA is associated with **proteins** in the nucleus and, during the early 1900s, proteins were considered better candidates as molecules able to transmit large amounts of hereditary information from generation to generation rather than DNA.

Find out about the early work on DNA by Friedrich Miescher and Phoebus Levene. What did they conclude about the structure and role of proteins and DNA in the nucleus, and to what extent were their hypotheses falsified by subsequent work? Use this site to find out more: www.dnaftb.org/15/animation.html

Inquiry 1: Exploring and designing

Exploring; designing

DNA can be extracted from any tissue. Using the following statements about DNA, plan a method to extract DNA from animal or plant tissue:

- Cell walls can be broken up by heating and mashing.
- DNA is soluble in water.
- DNA is not soluble in ethanol.
- DNA is found in the nuclei of cells.
- Nuclei have a membrane around them made of lipids (fats).
- Cell membranes are made of lipids (fats).
- Detergents (such as washing up liquid) dissolve fats.
- Salty water causes DNA to clump together to make larger molecules in solution.
- Precipitates form more easily in cold liquids.

Write out a set of instructions for DNA extraction and explain why you need to take each step.

This site has a methodology you can follow: https://learn.genetics.utah.edu/content/labs/extraction/howto

Components of a nucleotide

A **nucleotide** consists of three substances combined through covalent chemical bonding. These are:
- a **nitrogenous base** – the four bases of DNA are **cytosine** (C), **guanine** (G), **adenine** (A) and **thymine** (T)
- a **pentose** sugar – deoxyribose occurs in DNA and ribose in RNA
- a **phosphate** group (phosphate diester).

These components are combined by an enzyme-controlled **condensation** reaction to form a nucleotide. Condensation reactions occur when two molecules combine, producing water as a by-product. Enzymes are biological catalysts that speed up and control biological reactions. Since any one of the four bases can be incorporated, four different types of nucleotide can be found in DNA.

How these components are combined is shown in Figure A1.2.1, together with the diagrammatic way the components are represented to illustrate their spatial arrangement. Simple shapes are used rather than complex structural formulas, and these shapes are all that are required here. (You need to be able to draw simple diagrams, using these symbols.)

> ● **Top tip!**
>
> In diagrams of nucleotides, use circles, pentagons and rectangles to represent phosphates, pentose sugars and bases. The positions of the components relative to each other need to be accurately represented.

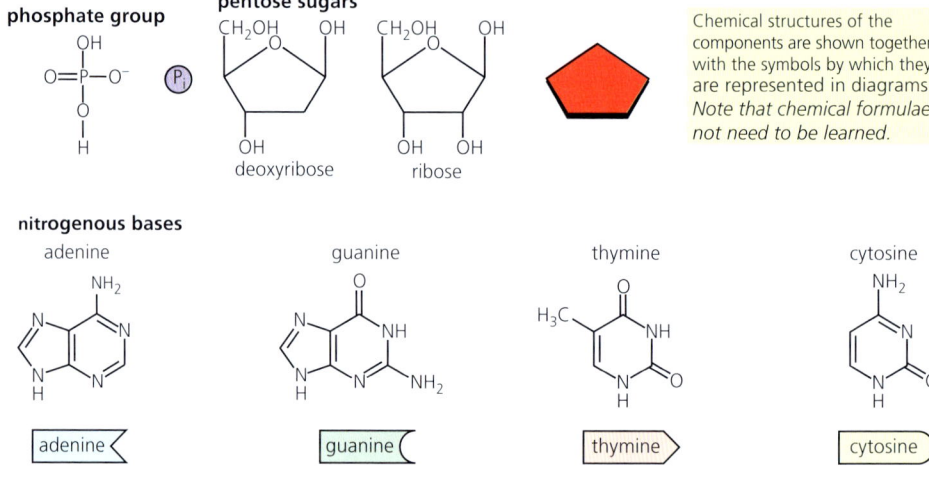

■ **Figure A1.2.1** The components of nucleotides

The bases can be divided into two groups: the **purines** (adenine and guanine) and the **pyrimidines** (cytosine and thymine), based on their molecular structure (see Figure A1.2.1).

The sugar–phosphate 'backbone' of DNA and RNA

■ Sugar–phosphate bonding

Nucleotides may chemically combine, one nucleotide at a time, by condensation reactions to form large molecules (with high values of molar mass) called nucleic acids or **polynucleotides** (Figure A1.2.2). So, nucleic acids are very long, thread-like (linear) macromolecules with alternating sugar and phosphate molecules forming the 'backbone'. This part of the nucleic acid molecule is uniform and unvarying. Sugar–phosphate bonding creates a continuous chain of covalently bonded atoms in each strand of DNA (and also RNA) nucleotides, which forms a strong backbone to the molecule.

1 **Distinguish** between a nitrogenous base and a base found in inorganic chemistry.

◆ **Purine**: one of two types of chemical compound used to make nucleotides, the building blocks of DNA and RNA. Examples are adenine and guanine.

◆ **Pyrimidine**: one of two types of chemical compound used to make nucleotides. Examples are cytosine, thymine and uracil. Cytosine and thymine are used to make DNA; cytosine and uracil are used to make RNA.

◆ **Polynucleotide**: a long, unbranched chain of nucleotides, as found in DNA and RNA.

Theme A: Unity and diversity – Molecules

Nucleotides become chemically combined together, phosphate to pentose sugar, by covalent bonds, with a sequence of bases attached to the sugar residues. Up to 5 million nucleotides condense together in this way, forming a polynucleotide (nucleic acid).

Along the strand, each base is attached to a pentose sugar molecule. The bases project sideways (Figure A1.2.2). Since the bases vary, they represent a unique sequence that carries the coded information held by the nucleic acid.

Bases in each nucleic acid form the basis of a code

Information in DNA lies in the sequence of the nitrogenous bases – cytosine, guanine, adenine and thymine – forming the genetic code. This sequence dictates the order in which specific amino acids are assembled and combined to synthesize a protein. The code lies in the sequence in one of the DNA strands, the coding strand. The other strand is complementary to it (see Figure A1.2.6, page 19). The coding strand is always read in the same direction, by enzymes.

The code is a three-letter or triplet code, meaning that each sequence of three bases stands for one of the 20 amino acids, and is called a **codon**. With a four-letter alphabet (C, G, A, T), there are 64 possible different triplet combinations ($4 \times 4 \times 4$).

Nucleic acids code for the production of proteins in cells. Proteins make up about two-thirds of the total dry mass of a cell. They differ from carbohydrates and lipids in that they contain the element nitrogen and sometimes the element sulfur, as well as carbon, hydrogen and oxygen.

Links

For more on proteins see Chapter B1.2, page 205, and Chapter D1.2, page 615. The codons for the 20 amino acids found in proteins are in Chapter D1.2, page 617.

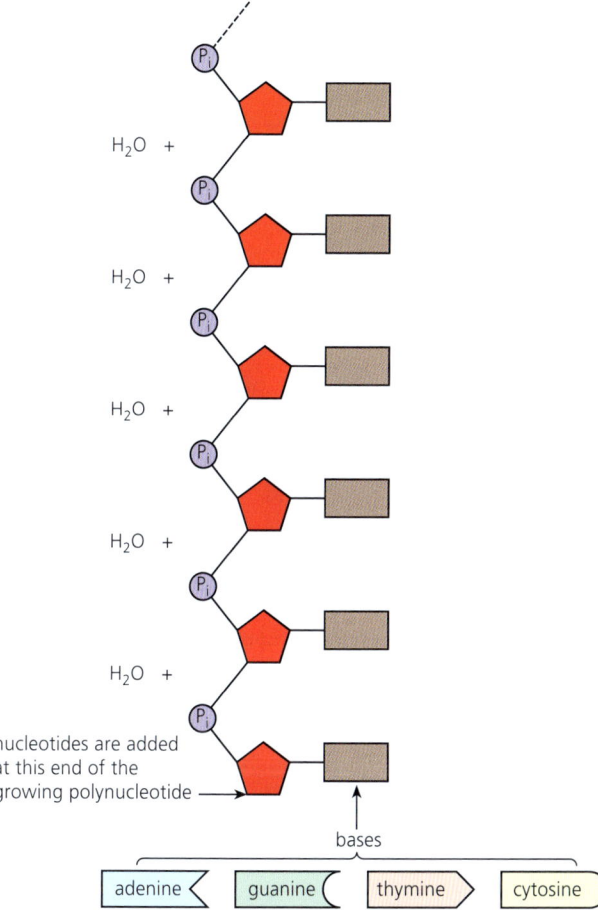

■ **Figure A1.2.2** How nucleotides make up nucleic acid

2 **Distinguish** between a nitrogenous base, a nucleotide and a nucleic acid.

◆ **Codon**: three consecutive bases in DNA (or RNA) which specify an amino acid.

◆ **Polymer**: large organic molecules made up of repeating subunits (monomers).

◆ **Monomer**: a molecule that chemically combines with other monomers, via covalent bond formation, to form a polymer.

◆ **Uracil**: a pyrimidine nitrogenous base found in RNA (not DNA); it pairs with adenine.

RNA polymers

RNA molecules are relatively short in length, compared with DNA. In fact, RNA molecules tend to be from a hundred to thousands of nucleotides long, depending on their particular role.

The RNA molecule is a **polymer**. In messenger RNA (mRNA) it is a single strand of polynucleotide in which the sugar **monomer** is ribose (see pages 18, 22). The bases found in RNA (Figure A1.2.3) are cytosine, guanine, adenine and **uracil** (which replaces thymine of DNA).

The carbon atoms in organic molecules such as ribose can be numbered (Figure A1.2.4). The numbering runs from right to left, clockwise. This enables the bonds between adjacent sugars and their phosphate neighbours to be identified, along with the direction in which the polynucleotide is orientated.

A1.2 Nucleic acids

Top tip!

Make sure you can draw and recognize diagrams of the structure of single nucleotides and RNA polymers.

◆ **Messenger RNA (mRNA)**: single-stranded ribonucleic acid, formed by the process of transcription of the genetic code in the nucleus, that then moves to ribosomes in the cytoplasm.

◆ **Transfer RNA (tRNA)**: short lengths of RNA that combine with specific amino acids prior to protein synthesis.

◆ **Ribosomal RNA (rRNA)**: molecule that forms part of the protein-synthesizing organelle known as a ribosome.

◆ **Phosphodiester bond**: the linkage between the 3' carbon atom of one sugar molecule and the 5' carbon atom of another (deoxyribose in DNA and ribose in RNA) in nucleic acids.

3 **Draw** a labelled diagram of:
   a an RNA single nucleotide
   b an RNA polymer.

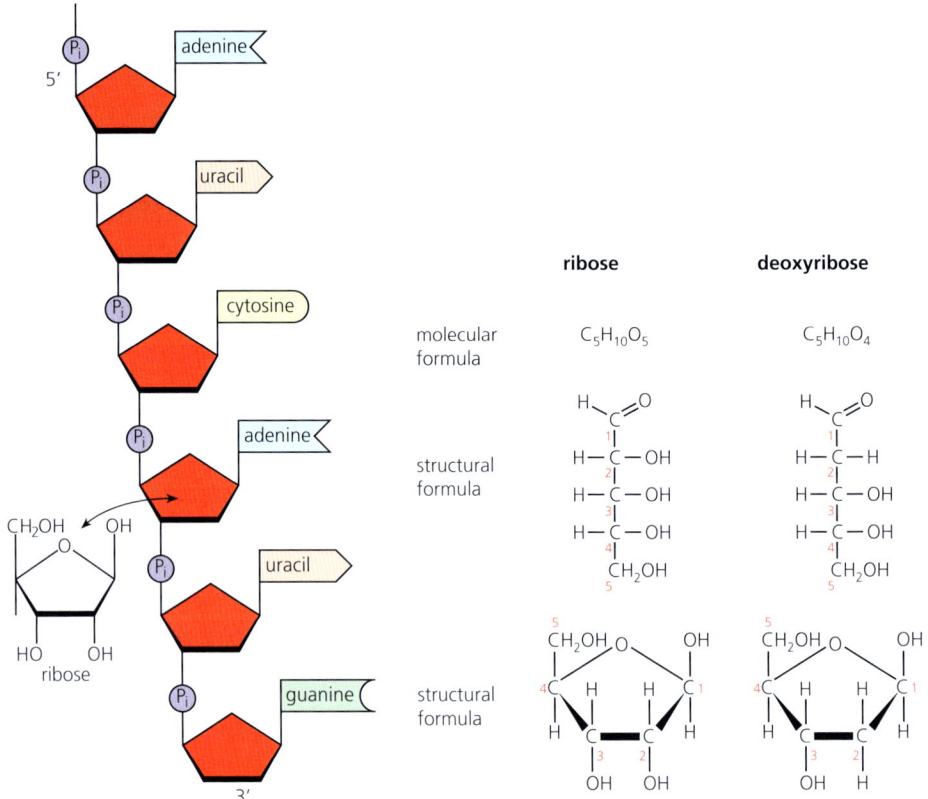

■ **Figure A1.2.3** An RNA polymer

■ **Figure A1.2.4** The numbering of carbon atoms in ribose and deoxyribose (in their straight chain (linear) and cyclic forms)

It is the convention to refer to the first carbon as 1' carbon, the second as 2' carbon, and so on.

There are three functional types of RNA: **messenger RNA (mRNA)**, **transfer RNA (tRNA)** and **ribosomal RNA (rRNA)**. mRNA is formed in the nucleus and is transported out through nuclear pores to the ribosomes in the cytoplasm. tRNA and rRNA are also made in the nucleus and occur in the cytoplasm.

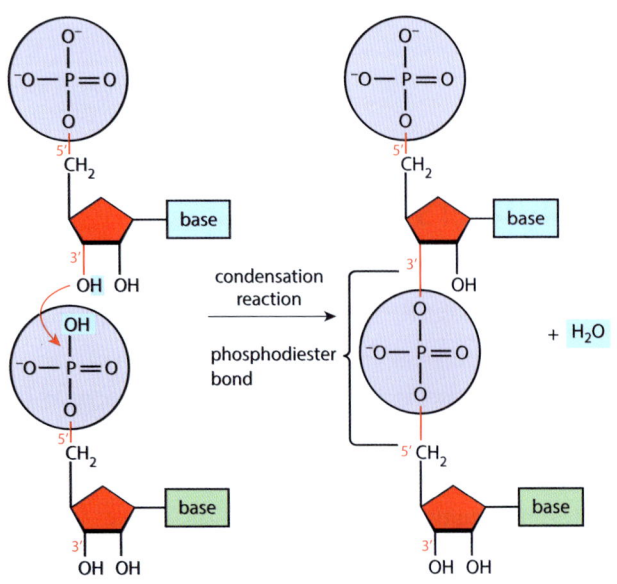

■ **Figure A1.2.5** RNA is formed by the condensation of nucleotide monomers

RNA is formed by the condensation of many nucleotide monomers. These condensation reactions link the pentose sugar and phosphate groups of adjacent nucleotides, so forming the new strands (Figure A1.2.5). The bond formed between adjacent nucleotides is called a 3'–5' **phosphodiester bond**.

♦ **Double helix**: two interlocking helices joined by hydrogen bonds between the pairs of purine–pyrimidine bases (A pairs with T and G with C). The helical structure makes a 360° twist after each 10 nucleotides, i.e. every 3.4 nm.

4 **Explain** what is meant by *antiparallel strands*.

The DNA double helix

The DNA molecule consists of two antiparallel polynucleotide strands, paired together, and held by hydrogen bonds. The two strands take the shape of a **double helix** (Figure A1.2.6).

The two strands are termed 'antiparallel' because one runs from a 5' carbon to a 3' carbon, and the other from a 3' carbon to a 5' carbon.

> ● **Top tip!**
>
> Note: when drawing DNA's structure, two antiparallel strands should be drawn, but the helical shape is not required. Adenine (A) should be shown paired with thymine (T), and guanine (G) paired with cytosine (C). For completion, Figure A1.2.6 shows the numbers of hydrogen bonds between adjacent bases – this detail does not need to be included when drawing the structure of DNA in exams. Only a small section of the DNA need be included to illustrate the way in which the nucleotides are arranged.

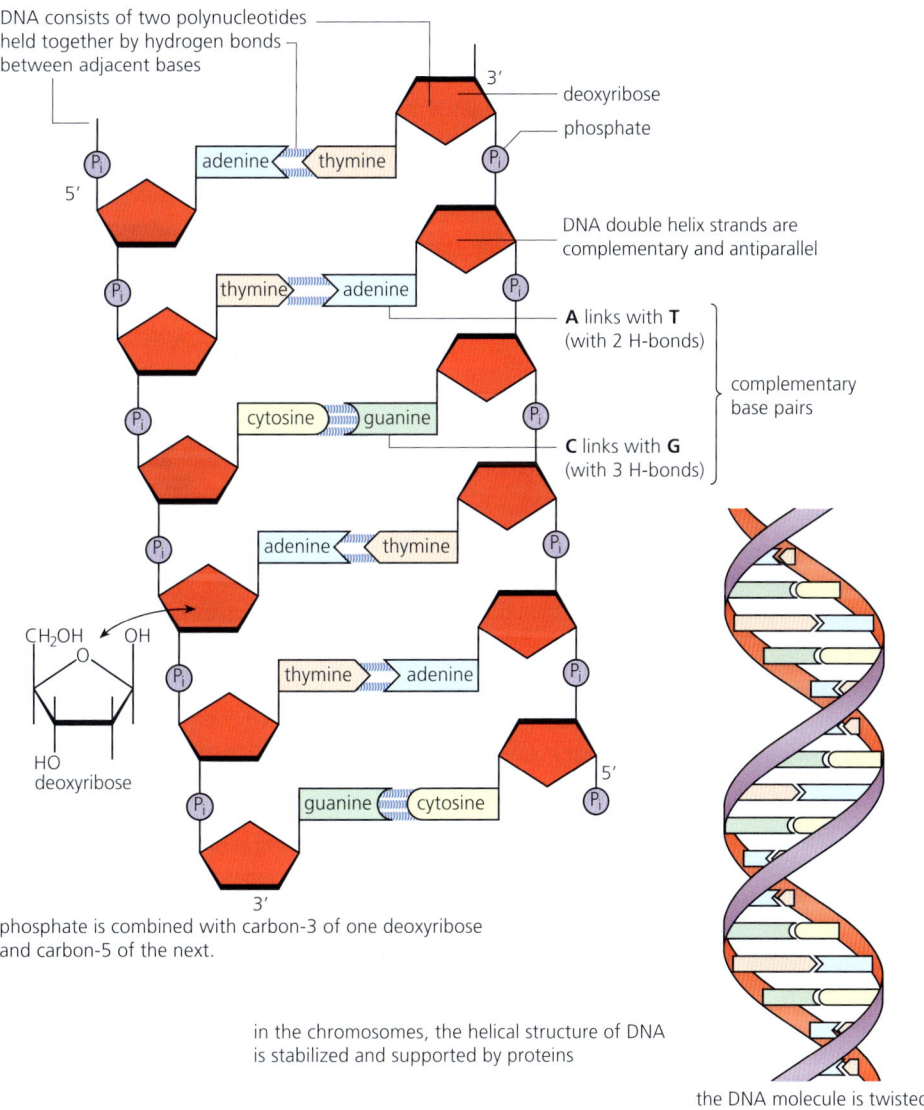

Figure A1.2.6 The DNA double helix

A1.2 Nucleic acids

Nature of science: Models

Creating the model of DNA

A model is a simplified description of a biological system, a concept or biological process. Models are used to describe and explain phenomena that cannot be experienced directly or to simplify complex systems such as ecosystems. The first model of the structure of DNA was developed in 1953 by Francis Crick and James Watson (see Figure A1.2.11 on page 26). Bringing together experimental results from other scientists, Crick and Watson used evidence to deduce the likely structure of the DNA molecule. Models are useful because they allow biologists to mimic the real world, make calculations and test predictions. Biological models include mathematical models that can be simulated on computer software. Models, like theories, may be modified as new experimental data are discovered.

Common mistake

A common mistake when drawing DNA is to link nitrogenous bases to phosphates rather than the pentose sugars. Another common error is to link the phosphate groups to the oxygen in the sugar ring, rather than to C4 via C5. Pay careful attention to the detail contained in the DNA molecule, and how the different parts of the nucleotide connect together, and make sure you can accurately recreate this from memory. Practise drawing molecular diagrams.

TOK

What are the implications of having, or not having, knowledge?

The 'object' shown in Figure A1.2.7 is an X-ray diffraction image of a DNA fibre, taken by Raymond Gosling in May 1952 at King's College London. Calculations by James Watson and Francis Crick from the photograph gave crucial dimensions for the double helix model they built.

The publication in 1953 in *Nature* by Watson and Crick of DNA's structure and a mechanism for replication of DNA marked a paradigm shift from classical genetics to molecular genetics.

The initial implications of this new scientific knowledge to Watson, Crick and other biologists (the knowers) were that scientists could now develop new molecular knowledge about the genetic code and protein synthesis based on new forms of experiments.

At a later stage, knowledge of DNA structure and complementary base pairing led to recombinant DNA research, genetic engineering, human molecular genetics, monoclonal antibodies and, most recently, CRISPR (gene editing). This new DNA-based knowledge has led to more effective medical treatments, new drugs and better disease diagnosis (for patients). However, there are new ethical implications arising from potential CRISPR editing of the inherited germline ('designer babies') for non-therapeutic and enhancement reasons.

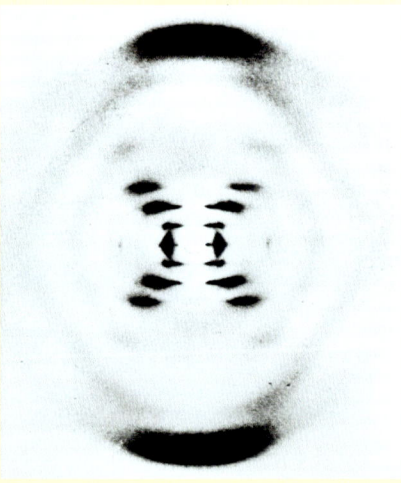

■ Figure A1.2.7 X-ray diffraction image of DNA

Genetic fingerprinting is perhaps the most well-known DNA technology to the general public (another group of knowers) and is a 'tool' used to track down relatives, establish paternity and identify dead bodies. However, it has also led to the conviction of many criminals and to the freeing from prison of many individuals who were wrongly convicted.

Possible social implications of these new advances in genetic knowledge are that genetic fingerprinting may act as a deterrent to certain kinds of crime and lead to greater satisfaction with the criminal justice system. There will also be economic costs associated with the adoption of the forensic use of DNA technology, and another implication might be the unauthorised disclosure or misuse of the data, e.g., typing by insurance companies.

Hence, the development and application of scientific knowledge has led to ethical decisions that can be justified through utilitarian reasoning: the idea that whether actions are ethical or not depends on their effects. Utilitarianism 'maximises utility', and favours technology that produces the largest amount of 'good' or 'happiness'.

However, some social scientists fear that this type of research into the human genetic code will encourage some people to argue that social problems, such as violence and drug abuse, are explainable in terms of human genes, and therefore deterministic abnormalities. They are worried that social interventions such as counselling, affirmative action and educational opportunities will lose funding.

ATL A1.2B

Crick and Watson brought together the experimental results of many other scientists, and from this evidence they deduced the likely structure of the DNA molecule. Rosalind Franklin (Figure A1.2.8) is often called 'the forgotten woman of DNA'. What role did she play in developing the first model of DNA? Why has she not been recognized in the same way as Watson and Crick in this pivotal scientific breakthrough?

■ Figure A1.2.8 Rosalind Franklin produced the key X-ray diffraction pattern of DNA at King's College London

◆ **Gene**: heritable factor that consists of a length of DNA that codes for a protein.

◆ **Chromosome**: length of DNA that carries specific genes in a linear sequence.

◆ **Locus**: the particular position of a gene on homologous chromosomes.

◆ **Allele**: different versions of the same gene.

Link
The concept of the gene pool is covered in more detail in Chapter D4.1, page 790.

■ Genes

Within the DNA molecule, there are sections that code for proteins – these sections are called **genes**. A gene is a heritable factor that influences a specific character. By 'character' we mean some feature of an organism, such as 'height' in the garden pea plant or 'blood group' in humans. 'Heritable' means genes are factors that pass from parent to offspring during reproduction.

■ Chromosomes

Genes are located on **chromosomes**. Each gene occupies a specific position on a chromosome; therefore each chromosome is a linear series of genes. Furthermore, the gene for a particular characteristic is always found at the same position or **locus** (plural, loci) on a particular chromosome. For example, the gene controlling height in the garden pea plant is always present in the exact same position on one particular chromosome of that plant. However, that gene for height may code for 'tall' or it may code for 'dwarf', as we shall see shortly. In other words, there are different forms of genes. In fact, each gene has two or more forms and these are called **alleles**. The word 'allele' just means 'alternative form'. For a given gene, many alleles may exist in the **gene pool** of the species.

◆ **Homologous chromosomes**: pairs of chromosomes, one from each parent, that carry the same sequence of genes (but not necessarily the same alleles of those genes).

Now, the chromosomes of eukaryotic cells occur in pairs called **homologous chromosomes**. ('Homologous' means 'similar in structure'.) One of each pair came originally from one parent, and the other one of the pair came from the other parent. So, for example, humans have 46 chromosomes, 23 coming originally from each parent in the process of sexual reproduction. Homologous chromosomes resemble each other in structure and they contain the same sequence of genes.

In Figure A1.2.9, we see some of the genes and their alleles in place on a homologous pair of chromosomes.

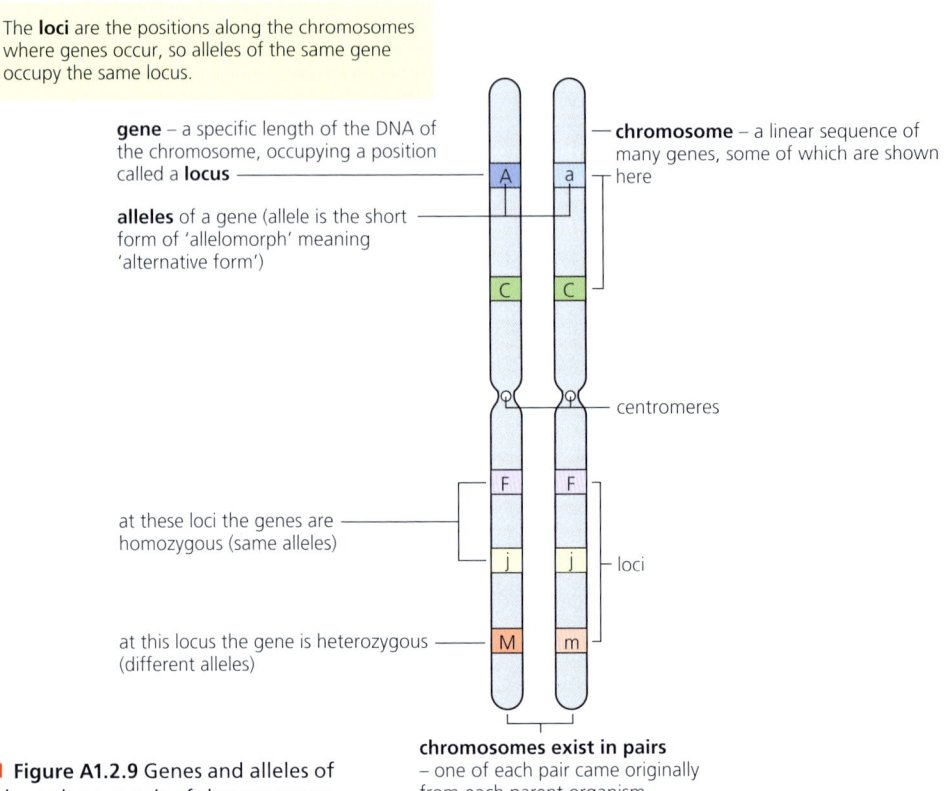

■ Figure A1.2.9 Genes and alleles of a homologous pair of chromosomes

> ● **Top tip!**
> You should be able to sketch the distinction between ribose and deoxyribose.

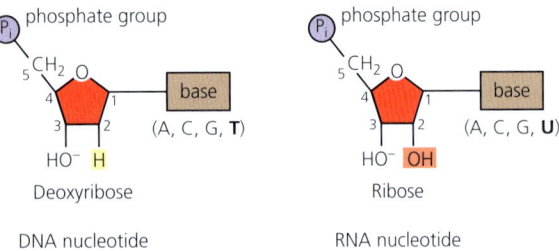

■ Figure A1.2.10 DNA and RNA nucleotides

Differences between DNA and RNA

Both DNA and RNA are made of nucleotides although they have different structures. One main difference is that DNA has the pentose sugar deoxyribose while RNA contains ribose (Figure A1.2.10).

5 **Construct** a table to **distinguish** between DNA and RNA.

 **Top tip!**

When distinguishing between DNA and RNA, you need to refer to the number of strands present, the types of nitrogenous bases and the type of pentose sugar.

◆ **Complementary base pairing**: this describes how the nitrogenous bases of nucleic acids align with each other in a specific way, i.e. adenine pairs with thymine (or uracil in RNA) and cytosine with guanine; complementary bases are held together by hydrogen bonds.

Link

The process of DNA replication is covered in more detail in Chapter D1.1, page 602.

Role of complementary base pairing

The pairing of bases is between adenine (A) and thymine (T), and between cytosine (C) and guanine (G), simply because these are the only combinations that fit together along the helix. This pairing, known as **complementary base pairing**, also makes possible the very precise way that DNA is copied in a process called replication. From the model of DNA in Figure A1.2.6, we can also see that when A pairs with T, they are held together by two hydrogen bonds; when C pairs with G, they are held by three hydrogen bonds. Only these pairs can form hydrogen bonds. Due to base pairing and the formation of specific hydrogen bonds, the sequence of bases in one strand of the helix determines the sequence of bases in the other. Complementary base pairing allows genetic information to be replicated and expressed.

 **Top tip!**

Complementarity is based on hydrogen bonding. The bases pair up in the way they do due to the hydrogen bonds that form between the pairs of bases.

ATL A1.2C

Watson and Crick developed their model of DNA in 1953. The first model they developed was not successful, however, as the various components did not fit together correctly. They developed their final and correct model once all the parts of nucleotides were accurately represented. By building a model, they were able to understand how DNA can replicate (copy) itself, thereby passing on genetic information from generation to generation.

Can you think of a way of representing DNA using material you have in the lab or in your home? How can you represent the different components and how they are connected together?

6 In base pairing, organic bases are held together (A–T, C–G) by hydrogen bonds. **State** which parts of these organic molecules form the hydrogen bonds.

Diversity of possible DNA base sequences

Although the genetic code is comprised of only four bases – A, C, T and G – the order in which they can be combined is immeasurable. The diversity of possible DNA base sequences means that DNA has a limitless capacity for storing information, with diversity by any length of DNA molecule and any base sequence possible.

An indication of the storage capacity of DNA is the number of genes that can be contained within it. Species vary in the number of genes they have – some have many more than others. Table A1.2.1 lists the numbers of genes present in a range of common organisms. Notice that the list includes one bacterium, as well as certain plants and animals, and that the water flea has more genes than a human, but the fruit fly has fewer.

A1.2 Nucleic acids

> **Concept: Diversity**
>
> The four-letter genetic code (A, C, T and G, coding for A, C, U and G in RNA) allows a huge variety of different proteins to be coded for. The human body contains thousands of different proteins, each with a specific function. The evolution of DNA over millions of years (see Chapter A4.1, page 141) has led to the vast diversity of life seen on Earth, from microscopic bacteria to large organisms such as the blue whale and redwood trees.

■ Table A1.2.1 Estimated approximate numbers of protein-coding genes

Species (animals and protists)	Number of genes	Species (plants, fungi, prokaryotes)	Number of genes
Daphnia (water flea)	31 000	*Oryza sativa* (rice)	41 500
Homo sapiens (human)	20 000	*Vitis vinifera* (grape)	30 450
Canis familiaris (domestic dog)	19 000	*Arabidopsis thaliana* (rockcress)	27 000
Drosophila melanogaster (fruit fly)	14 000	*Saccharomyces cerevisiae* (yeast)	6 000
Plasmodium (malarial parasite)	5 000	*Escherichia coli* (bacterium)	4 300

Another indication of the data-storing capability of DNA is the number of base pairs that DNA contains (Table A1.2.2).

■ Table A1.2.2 A comparison of genome size

Species	Total number of base pairs (bp)
T2 phage (a virus specific to a bacterium)	3 569 (3.5 kb)
Escherichia coli (bacterium)	4 600 000 (4.6 Mb)
Drosophila melanogaster (fruit fly)	123 000 000 (123 Mb)
Oryza sativa (rice)	430 000 000 (430 Mb)
Homo sapiens (human)	3 200 000 000 (3.2 Gb)
Paris japonica (canopy plant)	150 000 000 000 (150 Gb)

Genome size refers to the amount of DNA contained in a genome (which is the genetic code in one complete set of chromosomes), where 1 Mb = 1 000 000 bp and 1 Gb = 1000 Mb. Table A1.2.2 expresses the number of base pairs in terminology familiar from computer data storage (Mb and Gb), although here the units are different: gigabases (Gb) and megabases (Mb) rather than gigabytes and megabytes.

In a human cell, the DNA held in the nucleus measures about 2 m in total length. This length contains 3.2 Gb of 'data' – a phenomenal quantity of genetic code. Within this DNA, it is estimated that humans have between 20 000 and 25 000 protein-coding genes. These figures are an indication of how DNA offers an enormous capacity for storing data with great economy.

Conservation of the genetic code

We now know that the 64 codons in the genetic code of DNA have the same meaning and code for the same amino acids in nearly all organisms. This supports the idea of a common origin of life on Earth; that the very first DNA has sustained an unbroken chain of life from the first cells on Earth to all cells in organisms alive today. Only the most minor variations in the genetic code have arisen in the evolution and expansion of life since it originated 3.5 billion years ago.

Over many generations, changes in the sequence of bases in the **genome**, and therefore in the mRNA and order of amino acids that they assemble, can occur due to **mutations**. Many sequences, both in areas which code for proteins (so-called 'coding sequences') and those that do not (known as **non-coding sequences**), persist unchanged, however, or with only minor modifications over many generations: these are known as **conserved sequences**. It is possible that highly conserved sequences have a functional value, although the reasons for non-coding sequences are unclear. Even if the base sequences in coding areas of DNA change, the sequences of amino acids they code for may not, because each amino acid has several different mRNA codes, and so mutations in a coding sequence do not necessarily affect the amino acid sequence of its protein product (these are called synonymous mutations).

◆ **Genome**: the whole of the genetic information of an organism or cell.

◆ **Mutation**: a change in the amount or the chemical structure (i.e. base sequence) of DNA of a chromosome.

TOK

Highly repetitive DNA sequences were once described as 'junk DNA', the label 'junk' showing a degree of confidence that those sequences had no role. To what extent do you think the labels and categories used in the pursuit of knowledge affect the knowledge that we obtain?

The most highly conserved genes are those that can be found in all organisms. These include proteins required for transcription and translation, and those found in ribosomes. The fact that such genes exist indicate that all life is interlinked, with universal ancestry for all life on Earth. Histone proteins, which help to package DNA within nuclei (see page 26), are also highly conserved in terms of sequence and structure, again suggesting universal ancestry for all species.

Common mistake

A common mistake is to equate 'genome', or genomic size (which is the size of the genome), to the total number of genes in an organism, rather than the correct definition – the total amount of DNA.

Directionality of RNA and DNA

We can identify direction or polarity in the DNA double helix. The phosphate groups along each strand are bridges between carbon-3 of one sugar molecule and carbon-5 of the next, and one chain runs from 5' to 3' while the other runs from 3' to 5' (see Figure A1.2.6). (Remember, the carbon atoms of organic molecules can be numbered, page 18.) That is, the two chains of DNA are antiparallel, as illustrated in Figure A1.2.6. The existence of direction in DNA strands becomes important in DNA replication (when DNA is copied), when the genetic code is transcribed into mRNA (a process called **transcription**), and when the message encoded in the mRNA is read to form proteins (a process called **translation**).

Information in the DNA lies in the sequence of the bases: cytosine (C), guanine (G), adenine (A) and thymine (T). This sequence dictates the order in which specific amino acids are assembled and combined together. The code lies in the sequence in one of the strands, the coding strand; the other strand is complementary to it. It is the coding strand that becomes the template for transcription. The coding strand is always read in the same direction (in the 3' to 5' direction). A single-stranded molecule of RNA is formed by complementary base pairing (the RNA strand is synthesized in the 5' to 3' direction). The mRNA are translated in the 5' to 3' direction into amino acids by a ribosome to produce a **polypeptide** chain. The details of these processes will be explored in subsequent chapters (see page 619).

Link
Translation and transcription are discussed further in Chapter D1.2, page 615.

♦ **Polypeptide**: a chain of amino acid residues linked by peptide linkages.

Common mistake

DNA has a role beyond coding for proteins. Although the genes within chromosomes code for polypeptides, some regions of DNA do not code for proteins but have other important functions. Some regions of DNA regulate the expression of genes, and other sections code for the RNA that attaches to amino acids and also play a role in the formation of proteins at ribosomes (tRNA), for example.

Purine-to-pyrimidine bonding

When Watson and Crick assembled the first model of DNA in 1953 (see page 23), they used cardboard cut-outs to represent the different bases and other nucleotide subunits. Their first attempts used molecular shapes for thymine and guanine that were incorrect, and they arranged the different atoms of different elements from which the bases were made in the wrong configuration. This meant that the DNA model did not fit together correctly, as the lengths of the base pairings were incorrect. Following suggestions from the American scientist Jerry Donohue, in which the correct shapes for the bases were proposed, Watson made new cardboard cut-outs of the two bases, and found that the

A1.2 Nucleic acids

complementary bases now fitted together perfectly (i.e., A with T and C with G), with each pair held together by hydrogen bonds (Figure A1.2.6). The structure also matched Chargaff's rules (see page 31, later in this section).

> ## Tool 1: Experimental techniques
>
> ### Physical molecular modelling
>
> Physical model making helped Watson and Crick to establish the structure of DNA in a number of ways:
>
> - it allowed them to combine what was known about the chemical content of DNA with information from X-ray diffraction studies
> - by building scale models of the components of DNA, they were able to attempt to fit them together in a way that agreed with the data from other sources, such as Chargaff's rules
> - they made several arrangements of the scale model until they found the best one that fitted all the data.

■ Figure A1.2.11 Watson and Crick with their demonstration model of DNA

Watson and Crick discovered that the base pairings, A to T and C to G, are of equal length. This means that whatever the base sequence, the DNA helix has the same three-dimensional structure. The hydrogen bonding between complementary bases also confers stability (Figure A1.2.12), making DNA the ideal molecule for the storage of information in cells.

Nucleosome structure

In cells with a true nucleus (eukaryotes – see page 65), DNA occurs in the chromosomes in the nucleus, along with protein. More than 50% of a chromosome contains protein. While some of the proteins of the chromosome are enzymes involved in copying and repair reactions of DNA, the bulk of chromosome protein has a support and packaging role for DNA.

Why is packaging necessary?

Take the case of human DNA. In the nucleus, the total length of the DNA of the chromosomes is over 2 m. We know this is shared out between 46 chromosomes, and that each chromosome contains one very long DNA molecule. Chromosomes are different lengths, depending on the number of genes they contain, but we can estimate that within a typical chromosome of 5 μm length (where 1 μm = 1/1000 mm), there is a DNA molecule approximately 5 cm long. This means that about 50 000 μm of DNA is packed into 5 μm of chromosome.

This phenomenal packaging is achieved by coiling the DNA double helix and looping it around protein beads called **nucleosomes**, as illustrated in Figure A1.2.13.

The packaging proteins of the nucleosome, called **histones**, are a basic (positively charged) protein containing a high concentration of amino acid residues with additional basic groups ($-NH_2$), such as lysine and arginine (see also page 211, Theme B). In nucleosomes, eight histone molecules combine to make a single bead. Around each bead, the DNA double helix is wrapped in a double loop.

TOK

Crick and Watson had a distinctive method of working, including reinterpreting already-published data and developing others' studies, leading to the building of models (Figure A1.2.11).

To what extent were their achievements the product of both cooperation and competition?

In researching this, remember to consult a variety of relevant sources of information.

♦ **Nucleosome**: a sequence of DNA wound around eight histone protein cores – a repeating unit of eukaryotic chromatin.

♦ **Histone**: protein (rich in the amino acids arginine and lysine) that forms the scaffolding of chromosomes and is used in chromosome condensation to form nucleosomes.

Theme A: Unity and diversity – Molecules

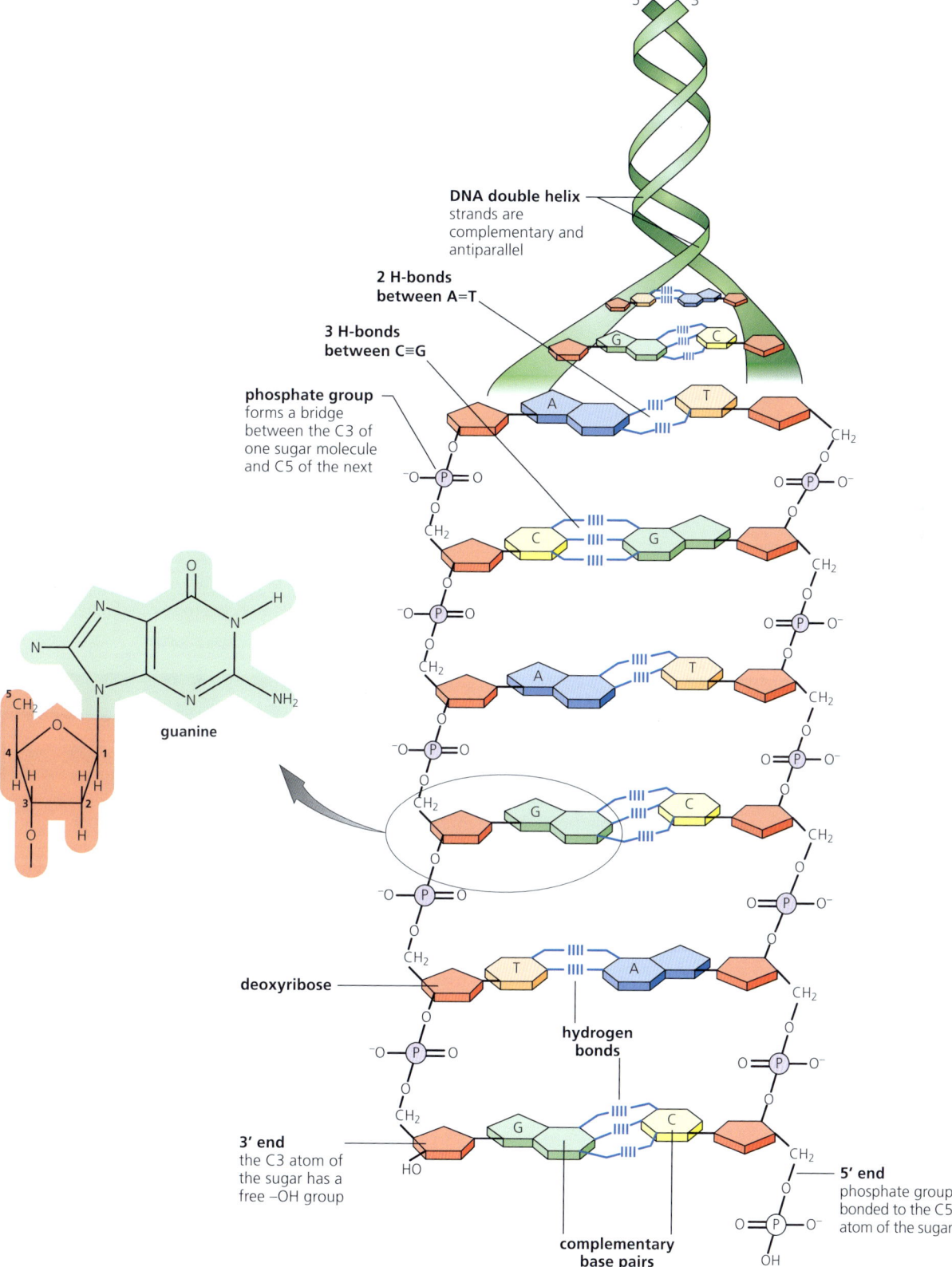

■ Figure A1.2.12 Direction, base pairing and hydrogen bonding between purine and pyrimidine bases in the DNA double helix

A1.2 Nucleic acids

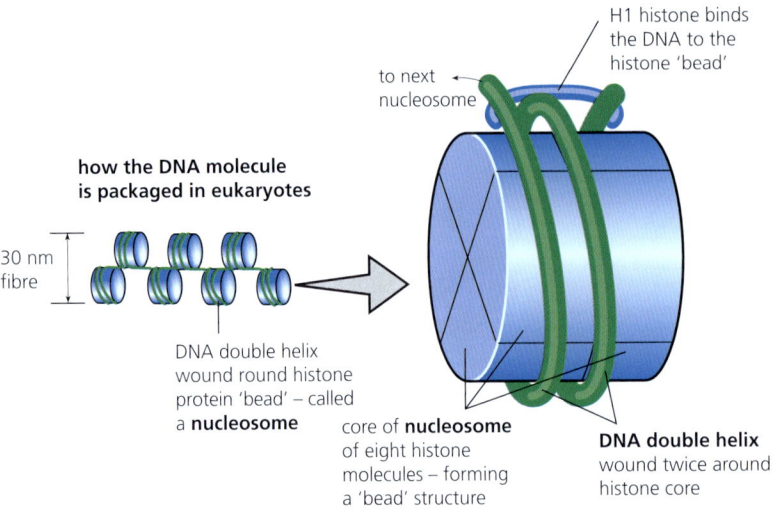
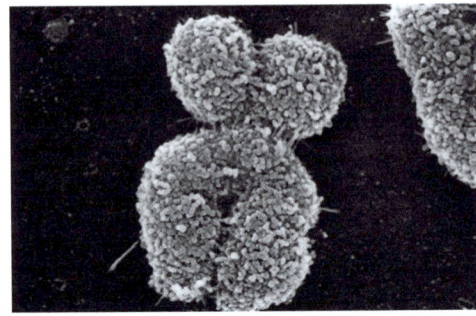

■ Figure A1.2.13 The nucleosome and supercoiling of DNA

♦ **Non-histone chromosomal protein**: proteins that remain in the chromatin once histone proteins have been removed; they play a key role in the regulation of gene expression.

At times of cell division, when the nucleus divides, the whole beaded thread is coiled up, forming the chromatin fibre. The chromatin fibre is again coiled, and the coils are looped around a 'scaffold' protein fibre made of a **non-histone chromosomal protein**. This whole structure is folded (supercoiled) into the much-condensed chromosome (Figure A1.2.13).

Clearly, the nucleosomes are the key structures in the safe storage of these phenomenal lengths of DNA that are packed in the nuclei. However, nucleosomes also allow access to selected lengths of the DNA (particular genes) during transcription – a process we will discuss shortly.

7 **Explain** a main advantage of chromosomes being 'supercoiled' during the process of cell division.

Tool 1: Experimental techniques

Digital molecular modelling

A molecular visualization of DNA was created for the 50th anniversary of the discovery of the double helix. The dynamics and molecular shapes were based on X-ray crystallographic models and other data. You can observe an animation of the packaging of the DNA molecule in nucleosomes at:

www.lindenbiomedical.com/animation

We can conclude that the much smaller genomes of prokaryotes (organisms without a true nucleus, i.e. bacteria) do not require this packaging, as protein is absent from the circular chromosomes of bacteria. Here, the DNA is described as 'naked'.

● Common mistake

A group of bacteria called eubacteria have DNA that is not associated with histone proteins – this is termed 'naked' DNA. Some students incorrectly use this term to describe DNA that is not enclosed in a nuclear membrane, as is the case in all bacteria. The term 'naked DNA' should be reserved for DNA that is not associated with histone proteins.

■ Use of molecular visualization software

Molecular visualization software can be used to study the association between the proteins and DNA within a nucleosome. This page shows DNA wrapped around a nucleosome:
www.wehi.edu.au/wehi-tv/nucleosomes

Theme A: Unity and diversity – Molecules

Search for your own image of a nucleosome using the Protein Data Bank (PDB):

Access the PDB: **www.rcsb.org/pdb/home/home.do**

1. At the top of the page, search for 'nucleosomes' – this will take you to a list of images. Select one – this will take you to a page that has an image of the nucleosome and information about it. At the top of the page, select '3D view'. This will show you an image that you can use your mouse to drag, rotate and zoom in and out of the structure. Alternatively, you can 'Select Orientation' from the menu on the right.
2. Rotate the nucleosome so that you can see the two copies of each histone protein, with DNA wrapped around each. Each protein has a tail that extends out from the core. DNA is wrapped nearly twice around the octamer core.
3. You can alter the image by selecting a different style of molecular visualization. The default is 'Mol* (Javascript)' but other options can be accessed on the menu at the bottom right of the screen ('Select a different viewer'). If 'JSmol' is selected, the structure can be seen using a variety of different styles and colours.
4. Access the JSmol viewer. Select 'colour by amino acid'. What role do the positively charged amino acids play in the association of the protein core with the negatively charged DNA?

This site also has a molecular visualization of a nucleosome:
https://proteopedia.org/wiki/index.php/Nucleosomes

It is possibly to modify the image using the selection underneath the visualization. For example, 'Show protein as cartoons'. The original view can be restored by clicking on 'Restore original view'.

Count how many times the DNA is wound around the histones (to make this easier, you may want to 'Hide protein'). Count the number of histone proteins (H2A, H2B, H4 and H3). Note the tails coming from the histone core. The N-terminal tail that projects from the histone core for each protein is used in regulating gene expression through chemical modification.

Another site allows you to download free software to view the three-dimensional structure of molecules: **https://pymol.org/2/**

This programme enables you to upload PDB files and allows you to zoom in and rotate molecules.

The Hershey and Chase DNA experiment

Since about 50% of a chromosome consists of protein, it is not surprising that scientists once speculated that the protein of chromosomes might be the information substance of the cell. For example, there is more chemical 'variety' within a protein than in nucleic acid. However, this idea proved incorrect. We now know that the DNA of the chromosomes holds the information that codes for the sequence of amino acids from which the proteins of the cell cytoplasm are synthesized.

How was this established?

The evidence for the unique importance of DNA was proved by an experiment carried out by two experimental scientists, Martha Chase and Alfred Hershey, with a bacteriophage virus. A **bacteriophage** (or **phage**) is a virus that parasitizes a bacterium. A virus particle consists of a protein coat (capsid) surrounding a nucleic acid core. Once a virus has gained entry to a host cell, it may take over the cell's metabolism, switching it to the production of new viruses. Eventually, the remains of the host cell break down (lysis) and the new virus particles escape – now able to repeat the infection in new host cells. The life cycle of a bacteriophage, a virus with a complex 'head' and 'tail' structure, is shown in Figure A1.2.14.

◆ **Bacteriophage**: a virus that parasitizes bacteria (also known as a phage).

Link
The life cycle of a virus is covered more fully in Chapter A2.3, page 88.

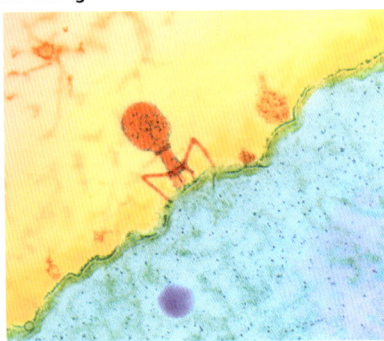

electron micrograph of bacteriophage infecting a bacterium

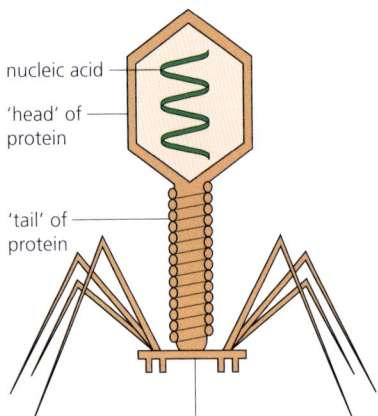

structure of the phage
- nucleic acid
- 'head' of protein
- 'tail' of protein
- baseplate of protein

steps to replication of the phage

1 The phage attaches to the bacterial wall and then injects the virus DNA.

2 Virus DNA takes over the host's synthesis machinery.

3 New viruses are assembled and then escape to repeat the infection cycle.

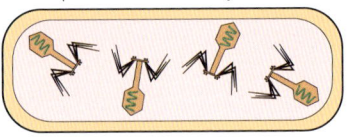

■ Figure A1.2.14 The life cycle of a bacteriophage

In 1952, Chase and Hershey used a bacteriophage that parasitizes the bacterium *Escherichia coli* to answer the question of whether genetic information lies in the protein coat (capsid) or the DNA (core) (Figure A1.2.15).

Two batches of the bacteriophage were produced, one with radioactive phosphorus atoms (^{32}P) built into the DNA core (so here the DNA was labelled) and one with radioactive sulfur atoms (^{35}S) built into the protein coat (here the protein was labelled). Note that sulfur occurs in protein, but there is no sulfur in DNA. Likewise, phosphorus occurs in DNA, but there is no phosphorus in protein. So, we can be sure the radioactive labels were specific.

Is it the **protein coat** or the **DNA** of a bacteriophage that enters the host cell and takes over the cell's machinery, so causing new viruses to be produced?

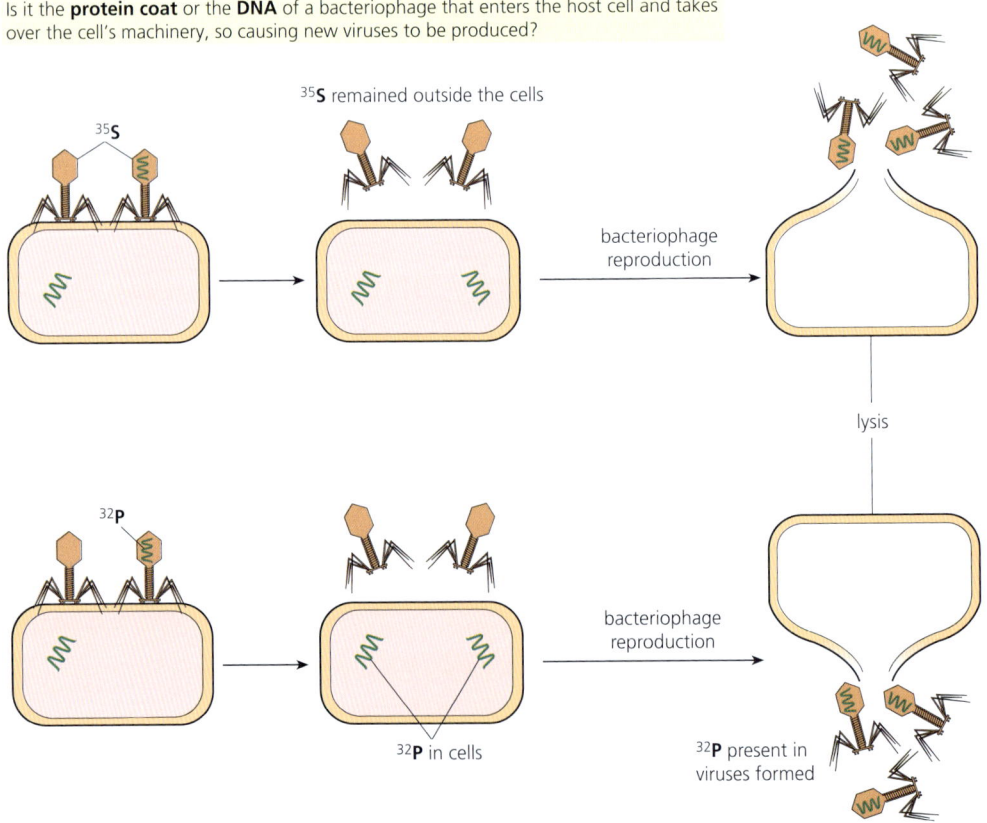

■ Figure A1.2.15 The Hershey–Chase experiment

Only the DNA part of the virus got into the host cell (and radioactively labelled DNA was present in the new viruses formed). It was the virus DNA that controlled the formation of new viruses in the host, so Hershey and Chase concluded that **DNA carries the genetic message**.

Two identical cultures of *E. coli* were infected, one with the ^{32}P-labelled virus and one with the ^{35}S-labelled virus. Subsequently, radioactively labelled viruses were obtained only from the bacteria infected with virus labelled with ^{32}P. In fact, the ^{35}S label did not enter the host cell at all. Chase and Hershey's experiment clearly demonstrated that it is the DNA part of the virus which enters the host cell and carries the genetic information for the production of new viruses.

8 **Deduce** what would have been the outcome of the Hershey–Chase experiment (Figure A1.2.15) if protein had been the carrier of genetic information.

ATL A1.2D

Find out more about Alfred Hershey and his work with Martha Chase using this site: www.dnaftb.org/18/animation.html Select 'animation'; and click on the icon 'jump to' to move to the fourth section where Alfred Hershey discusses his research on bacteriophage genetics.

Nature of science: Global impact of science

The Hershey–Chase experiment illustrates how technological developments can open up new possibilities for experiments. When radioisotopes were made available to scientists as research tools, the Hershey–Chase experiment became possible.

After the nuclear explosions over Japan at the end of the Second World War, the US government publicised the peacetime benefits of nuclear knowledge through nuclear power and the medical uses of radioisotopes. The availability of reactor-produced radioisotopes equipped virus researchers (and biochemists more generally) with a valuable new research tool. Isotopes are especially suitable for studying the dynamics of chemical transformation over time, through metabolic pathways or life cycles.

Chargaff's data on the relative amounts of pyrimidine and purine bases

The discovery of the principle of base pairing by Watson and Crick was the result of their interpretation of the work of Erwin Chargaff. In 1935, Chargaff had analysed the composition of DNA from a range of organisms and found rather remarkable patterns. Apparently, the significance of these patterns was not immediately obvious to Chargaff, though.

His discoveries were:
- the numbers of purine bases (adenine and guanine) always equalled the number of pyrimidine bases (cytosine and thymine)
- the number of adenine bases equalled the number of thymine bases, and the number of guanine bases equalled the number of cytosine bases.

What does this mean?

The organic bases found in DNA are of two distinct types with contrasting shapes:
- cytosine and thymine are pyrimidines or single-ring bases
- adenine and guanine are purines or double-ring bases.

Only a purine will fit with a pyrimidine between the sugar–phosphate backbones, when base pairing occurs (Figure A1.2.6, page 41). So, in DNA adenine must pair with thymine, and cytosine must pair with guanine.

Nature of science: Falsification

The problem of induction

Biology is not a body of unchanging facts, but a process of generating new biological knowledge, theories and laws, using the scientific method. There is no single agreed scientific method, but a number of variations, all of which can be used to generate new scientific knowledge. The scientific method you will be familiar with from your practical work (investigations) is known as the Baconian or **inductive** scientific method.

The inductive scientific method begins with observations and the collection of raw data. The data are analysed and a hypothesis is generated. An investigation is then designed to test the validity of the hypothesis. A general theory may then be generated from the specific data and the hypothesis.

Biologists use both inductive and deductive reasoning to study biological problems.

Inductive reasoning (Figure A1.2.16) is sometimes termed the 'bottom up' approach. When inductive reasoning is used, specific observations and measurements may show a general pattern. This may then lead to a hypothesis that can be further explored and may also lead to the drawing of some general conclusions.

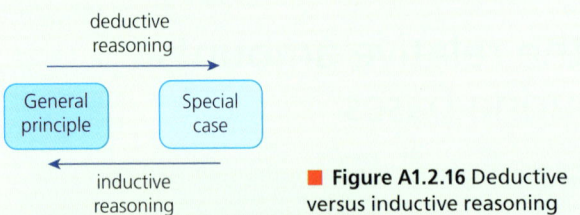

■ **Figure A1.2.16** Deductive versus inductive reasoning

In this case, one might construct an inductive argument along the following lines:

Organisms A, B and C all have characteristic X.

Therefore, all items in the same class as A, B and C probably also have X.

For example:
- This sawfly stung me. It is a hymenopteran.
- This wasp stung me. It is a hymenopteran.
- This fire ant stung me. It is a hymenopteran.

There is a pattern here: it might seem that all hymenopterans (a large order of insects) have stingers.

One potential issue here is the **problem of induction**. Using data from many specific observations discovered in the past to create general observations about what will always happen in the future, is to assume; namely that the future will be just the way it was when you gathered your data. But you may be as yet unaware of something in your past observations which means the generalisation is not true. For example, many hymenopterans (stingless bees and ants, male honeybees (drones), etc.) do not have stingers. Perhaps your previous observations only *happen* to be of those hymenopterans that did have a stinger, you just never saw all the others that did not! (You might not discover this unless you test every single hymenopteran species for stinging capability, and this is simply impractical.)

Inductive reasoning involves forming generalisations from specific examples. Biology uses inductive reasoning – generalisations based on empirical evidence – as the basis of its justification for knowledge. A reliable scientific conclusion will be based on a large number of repeated investigations.

However, inductive reasoning can never give certainty. We also cannot be sure that the generalizations made in the past will continue to hold in the future. The impossibility of reaching certainty through induction is known as the 'problem of induction'. Inductive generalisations (biological theories) may, therefore, be shown to be wrong by new data and should only be thought of as 'tentative'. A single counter-example falsifies an inductive conclusion. (However, some theories, such as cell theory or the 'theory of gravity', are so well confirmed that there is little room for rationally doubting them.)

As a counter to the view that scientists are simply looking for further data to confirm their hypotheses (which are always only tentative anyway), the philosopher Karl Popper rejected the idea that science creates new knowledge by inductive steps. He suggested that scientists may work intuitively and creatively to generate a hypothesis before collecting data. This guides the scientist to plan and carry out investigations to collect data to test the hypothesis. The data will then either support the hypothesis or falsify (i.e. disprove) the hypothesis, but the falsifying data are actually more helpful in developing knowledge. Popper suggested that **falsification** is an important part of the scientific process because a hypothesis which is confirmed after one or many experiments, may yet for some unknown reason be falsified later. However, if a hypothesis is shown to be false, then genuine knowledge is gained: the hypothesis is *not* true.

The Russian-American biochemist Phoebus Levene (1869–1940), who discovered ribose sugar in 1909 and deoxyribose sugar in 1929, suggested (incorrectly, with hindsight) the structure of nucleic acid as a repeating tetramer. He called the phosphate–sugar–base unit a nucleotide.

◆ **Falsification**: a process used by scientists in which a hypothesis is tested by trying to show that it is false. Where a hypothesis cannot be shown to be false after repeated experiments conducted by different groups of scientists, it is considered a strong hypothesis.

Levene did not recognize that the compositions of nucleic acids were organism-specific, and he did not recognize that in organisms the four nucleotides are not present in equal amounts. This was due to the inaccuracy of the analytical techniques available at that time, which did not allow a reliable determination of the relative amounts of nucleotides in nucleic acids.

It was only in the second half of the 1940s that Erwin Chargaff established the organism-specificity of nucleic acids and the special relationships among the amounts of nucleotides in any organism. The tetranucleotide hypothesis became obsolete after the structure of DNA was determined, since it was realized that a structure in which a four-member unit is being repeated could not carry the genetic information that must be involved in heredity.

The tetranucleotide hypothesis, and Chargaff's falsification, is an example of how the problem of induction can be addressed by the certainty of falsification. In this case, Chargaff's data falsified the tetranucleotide hypothesis that there was a repeating sequence of the four bases in DNA.

LINKING QUESTIONS

1. What makes RNA more likely to have been the first genetic material, rather than DNA?
2. How can polymerization result in emergent properties?

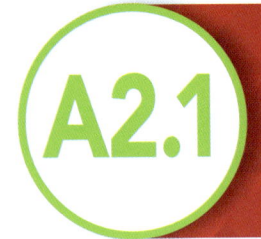

Origins of cells

Guiding questions

- What plausible hypothesis could account for the origin of life?
- What intermediate stages could there have been between non-living matter and the first living cells?

SYLLABUS CONTENT

This chapter covers the following syllabus content:
- ▶ A2.1.1 Conditions on early Earth and the pre-biotic formation of carbon compounds (HL only)
- ▶ A2.1.2 Cells as the smallest units of self-sustaining life (HL only)
- ▶ A2.1.3 Challenge of explaining the spontaneous origin of cells (HL only)
- ▶ A2.1.4 Evidence for the origin of carbon compounds (HL only)
- ▶ A2.1.5 Spontaneous formation of vesicles by coalescence of fatty acids into spherical bilayers (HL only)
- ▶ A2.1.6 RNA as a presumed first genetic material (HL only)
- ▶ A2.1.7 Evidence for a last universal common ancestor (HL only)
- ▶ A2.1.8 Approaches used to estimate dates of the first living cells and the last universal common ancestor (HL only)
- ▶ A2.1.9 Evidence for the evolution of the last universal common ancestor in the vicinity of hydrothermal vents (HL only)

◆ **Greenhouse gas**: the heating caused by the atmosphere on Earth's surface because certain atmospheric gases absorb and emit infrared radiation.

◆ **Greenhouse effect**: process in which greenhouse gases trap outgoing long-wave radiation from the Earth, causing the planet to be warmer than it would otherwise be.

1 **Compare and contrast** the atmospheric conditions of the early Earth with the atmosphere of today.

Conditions on early Earth and the pre-biotic formation of carbon compounds

Around 4.3 billion years ago, much of the surface of the Earth was molten rock. This time is known as the Hadean eon (from Greek mythology where 'Hades' is the God of the Underworld, and the term is associated with 'Hell', indicating the conditions on Earth at the time). As Earth cooled, gases released by volcanic activity formed the atmosphere. The atmosphere included ammonia (NH_3), nitrogen, methane, water and significantly higher levels of carbon dioxide compared to today's atmosphere (see Figure A2.1.1). Both methane and carbon dioxide are **greenhouse gases** – this means that they absorb and react with infrared radiation emitted from the surface of the planet, causing the surface of the Earth to heat up, resulting in higher temperatures (a process known as the **greenhouse effect**).

The atmosphere today has relatively high levels of oxygen, essential for sustaining life. Also important for the preservation of life on the Earth's surface is the presence of ozone in the stratosphere (Figure A2.1.2). Ozone (O_3) is formed naturally through the interaction of solar ultraviolet (UV) radiation with molecular oxygen (O_2). On the early Earth there was a lack of free oxygen and, therefore, ozone in the atmosphere, resulting in ultraviolet light penetration and high levels of UV light at the surface of the planet. UV radiation 3.7 billion years ago was 100 times more intense than today. UV causes damage to DNA and causes it to mutate – lower levels of UV allow life to exist on the surface of the Earth today.

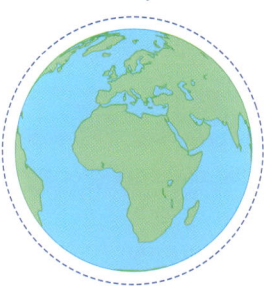

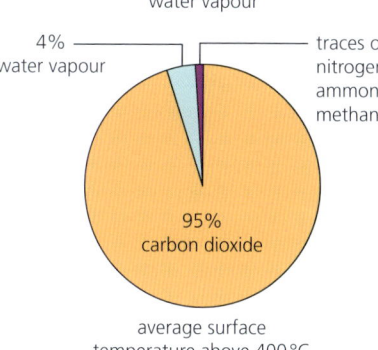

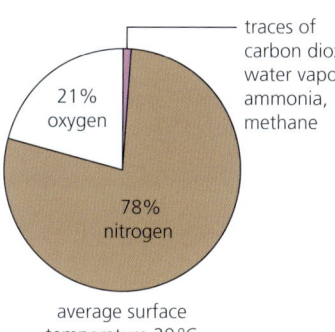

■ Figure A2.1.1 The atmosphere of the early Earth and the atmosphere today

The conditions on the early Earth may have caused a variety of carbon compounds to form spontaneously by chemical processes that do not occur now. If energy was added to the gases that made up Earth's early atmosphere, the building blocks of life (for example, amino acids, peptides, ribose, nucleobases, fatty acids, nucleotides and oligonucleotides) could have been created. The idea that high energy or UV radiation led to the formation of the first biological molecules is called the 'primordial (or pre-biotic) soup' hypothesis and was originally proposed independently by Alexander Oparin in 1924 and JBS Haldane in 1929.

However, high-energy UV light is destructive of the chemistry of early life. When a molecule is destroyed, it is broken into smaller, very reactive pieces that undergo additional reactions, eventually recombining to form larger high-energy molecules. For example, pyruvic acid, a molecule that is central to key metabolic pathways in cells, forms larger molecules when dissolved in water and illuminated with UV light.

2 **List** the gases which cause the greenhouse effect.

Research has shown that the bases of nucleotides are extremely efficient at reducing the harmful effects of UV radiation, thereby protecting the pentose sugar and phosphate components of nucleic acids. In the presence of strong UV light, RNA was shown to be more likely to form chains than other molecules. It is therefore possible that the high UV levels on primordial Earth, rather than being an obstacle to the origin of life, may have acted as a selection factor that drove the process forwards.

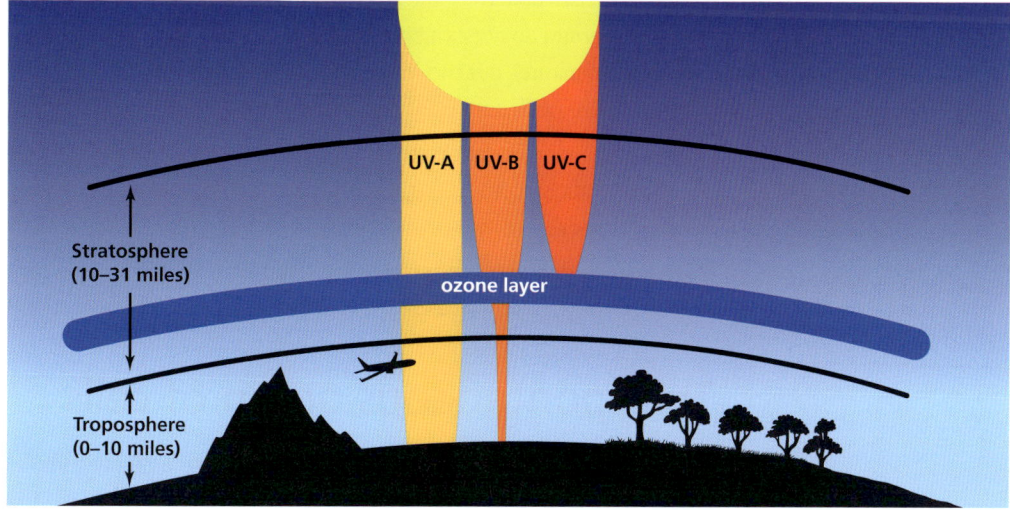

■ Figure A2.1.2 Ozone absorbs the most energetic frequencies of ultraviolet radiation, known as UV-C and UV-B, which are frequencies that harm living organisms

A2.1 Origins of cells

Cells as the smallest units of self-sustaining life

Cells are the fundamental self-sustaining units of life that reproduce via cell division. All present-day cells are believed to have evolved from a common ancestral cell that existed around 4 billion years ago.

All cells are enclosed by a plasma membrane that separates the inside of the cell from its environment. All cells contain DNA to store genetic information and use it as a template for the synthesis of RNA molecules and proteins.

The identification of cellular life processes, e.g., respiration, nutrition, reproduction and so on (see also page 73), does not define life and it does not explain how life appeared. The three principles below can be used to define life:

- Living organisms must be able to evolve through natural selection. This requires them to be able to reproduce and have a hereditary system that must show genetic variation.
- Life forms are contained and separate from, but in communication with, their surrounding environment, like a cell.
- Life forms are chemical and physical 'machines' that receive and respond to information.

There is also a need to take a systems approach and not just to view life as simply a large set of biochemical reactions at the molecular level. Systems biologists view living organisms as complex systems that process information about themselves and their environment.

NASA's working definition of life is: 'life is a self-sustaining chemical system capable of Darwinian evolution'. This definition of 'life' would include viruses, but many of the problems in attempting to define life are because there is only one example – life on Earth – where all the organisms have the same basic biochemistry. It is therefore difficult to distinguish which properties of life on Earth are unique and which are needed in a general sense to qualify as 'life'. The most promising place for life in our solar system is probably Mars since it may have sub-surface pockets of water.

However, some scientists have proposed that life could evolve on other planets in the universe using a liquid other than water, e.g., ammonia. If these hypothetical organisms existed, they might be described as 'weird life' because they would have to be fundamentally different from terrestrial life.

Viruses are regarded as non-living. Viruses are non-cellular and therefore lack organelles to carry out metabolism independently, e.g., releasing energy in the form of ATP (see page 67) and protein synthesis. They can only replicate inside living cells using their cellular components.

However, viruses do have some features of living organisms. They do not respond to stimuli from their surroundings and do not exhibit homeostasis (the keeping of internal conditions within narrow limits, such as temperature), but they do respond to external stimuli, such as evading immunity, adapting to drug treatment (with antivirals) or changing host range by mutation (especially RNA viruses).

Viruses contain hereditary material in the form of nucleic acids (DNA or RNA) with genes that code for specific structures, e.g. capsid coat proteins, and reverse transcriptase and integrase (for retroviruses).

The genomes of viruses are prone to rapid mutation and are one factor in the evolution of viruses by natural selection. Many viruses can engage in a form of recombination known as genetic shift.

The spontaneous origin of cells

Cells are consistent across all biological systems and are highly complex structures that can currently only be produced by the division of pre-existing cells. However, the first cells must have had non-biotic (non-biological) origins, and one of the most important questions in biology is how did these first cells evolve? Answers may be found by looking at the features of cells and then considering how they first evolved and appeared.

> **Link**
> Plasma membranes are covered in detail in Chapter B2.1, page 223.

> **Link**
> Natural selection is covered in Chapters A4.1, page 140, and D4.1, page 779.

> **Link**
> For more on Darwinian evolution, see Chapter A4.1, page 140.

> **Link**
> Viruses are covered in detail in Chapter A2.3, page 88.

All cells have three common features:

1. a stable, partially permeable membrane that surrounds cell components
2. genetic material that can be passed on when new cells are formed, and which controls the function and behaviour of cells
3. metabolic processes that allow energy generation, enabling growth, self-maintenance and reproduction.

In biology, the term 'evolution' specifically means the processes that have transformed life on Earth from its earliest beginnings to the diversity of forms we know about today, living and extinct. It is an organizing principle of modern biology. It helps us make sense of the ways living things are related to each other, for example.

The evolution of life in geological time has involved major steps – none more so than the origin of the first cells. Unless these first cells arrived here from somewhere else in the universe, they must have arisen from non-living materials, starting from the components of the Earth's atmosphere at the time. We can only speculate about these very first steps.

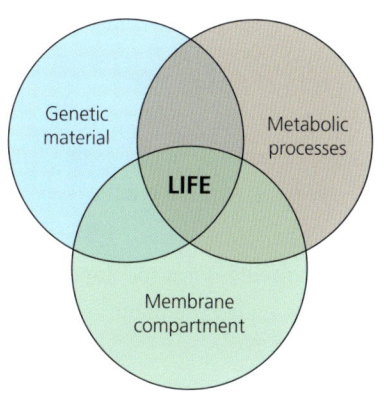

■ **Figure A2.1.3** Three factors essential for sustaining life

Concept: Diversity

The common features of all cells have allowed a large diversity of forms to evolve over a period of some 4 billion years.

Following on from the discussion above, the formation of living cells from non-living materials would have required the following steps:
- the synthesis of simple organic molecules, such as sugars and amino acids
- the assembly of these molecules into polymers (page 39)
- the development of self-replicating molecules, such as the nucleic acids
- the retention of these molecules within membranous sacs, so that an internal chemistry developed, different from the surrounding environment.

As well as the features shown in Figure A2.1.3, and the processes listed above, these elements must have spontaneously self-assembled to form the first cells.

Nature of science: Hypotheses

The origin and the evolution of the earliest cells are among the most intriguing topics being debated in the scientific community. Traditionally, two approaches have been used to understand how life on the Earth originated. The bottom-up approach, favoured by chemists, for example Miller's experiment (see Figure A2.1.4, page 39), attempts to reconstruct the conditions of primitive Earth. The top-down approach is favoured by biologists, who study modern organisms to find the relics of their ancestors to reconstruct ancient metabolic pathways and molecular processes.

However, knowledge about evolutionary history is not restricted to perfectly replicating a point in the geological past or finding fossils. Biologists interested in chemical evolution or the emergence of the first protocell can carry out experiments to test the mechanisms upon which a theory rests – and can do so in laboratory conditions that match what scientists do know about conditions that likely existed somewhere on the pre-biotic Earth. However, one of the problems of testing hypotheses in this way is that the exact conditions on pre-biotic Earth cannot be replicated. This approach can be illustrated for the three main competing theories for the origin of life.

3 **List** the three common features shared by all cells.

4 **Outline** the steps that would be needed for the formation of living cells from non-living materials.

A2.1 Origins of cells

The three competing theories for the origin of life

1 Protocell-first

A cell-like compartment that had a basic metabolism but lacked a fully developed genetic system; it arose spontaneously with the ability to grow and then divide into daughters that tended to resemble the mother cell. These **protocells** evolved adaptively until they eventually acquired a genetic system (likely RNA, then later DNA).

Sample prediction: cell-like units capable of growing and dividing without genetic molecules could be engineered or, better still, be seen to arise spontaneously in the laboratory under controlled conditions. The exact nature of the earliest cells would be difficult to prove, however, because the first protocells did not fossilize.

2 Gene-first

A genetic molecule (thought to be RNA) or a small set of genetic molecules arose spontaneously, capable of replication. The replicators evolved adaptively by natural selection, forming genetic variants that could assemble a cell membrane and start metabolising.

Sample predictions: a context could be found in which complex RNA molecules form spontaneously; spontaneously arising RNAs can show collective self-replication and open-ended evolution.

3 Metabolism-first

A self-sustaining system of simple reactions, capable of feeding on nutrients and energy arose spontaneously, perhaps adsorbed on to a mineral surface. The chemical mixture evolved adaptively, eventually evolving cells (perhaps via selection for dispersal) and genetic systems (perhaps via selection for catalysis of metabolic reactions).

Sample predictions: chemical mixtures given a flux of nutrients/energy in the laboratory should sometimes demonstrate autocatalysis and show evidence of adaptive evolution; autocatalytic reaction systems can produce lipids and genetic polymers.

Many scientists favour the metabolism-first theory because rapid growth, replication and division, which are essential processes for protocells' evolution, all require significant amounts of energy.

◆ **Protocell**: pre-cellular or cell-like entity, e.g. a lipid droplet with a few molecules inside.

Top tip!

Claims in science, including hypotheses and theories, must be testable. In some cases, scientists struggle with hypotheses that are difficult to test. In this case, the exact conditions on pre-biotic Earth cannot be replicated and the first protocells did not fossilize.

TOK

Knowledge that is beyond the capability of science is perhaps knowledge that we do not have the technology to discover at this current moment in time. Although it is possible to investigate the origin of the first cells, this is not something that can be verified as it is not possible to check whether the hypotheses are true. The validity of hypotheses regarding the origin of life can only be tested by accumulating evidence that supports a particular theory. What knowledge, if any, is likely to always remain beyond the capabilities of science to investigate or verify?

Evidence for the origin of carbon compounds

The molecules that make up living things are built mainly from carbon, hydrogen and oxygen, with some nitrogen, phosphorus and sulfur; a small number of other elements are also present (metals and their ions are very important in living organisms). Today, living things make these molecules by the action of enzymes in their cells, but for life to originate from non-living material, the first step was the non-living synthesis of simple organic molecules.

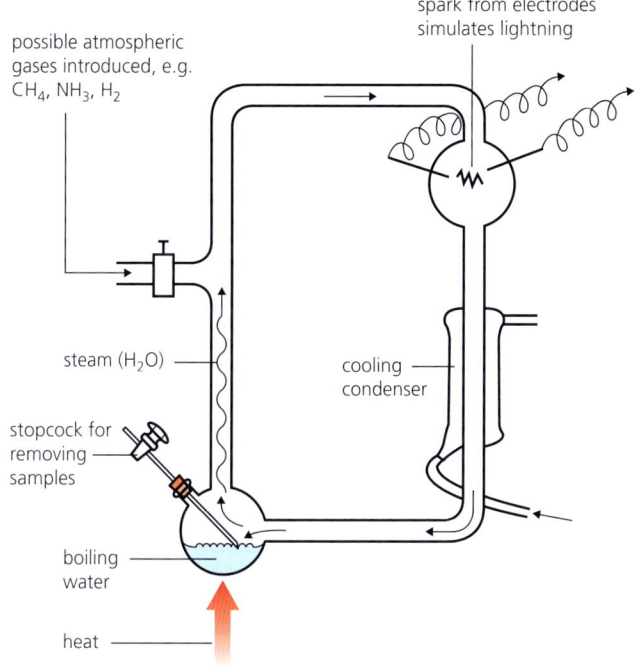

Apparatus like this has been used with various gases to investigate the organic molecules that may be synthesized.

■ Figure A2.1.4 Apparatus for simulating early chemical evolution

SL Miller and HC Urey (1953) investigated how simple organic molecules might have arisen from the chemicals present on Earth before there was life. They used a reaction vessel in which specific environmental conditions could be reproduced (Figure A2.1.4). For example, strong electric sparks (simulating lightning) were passed through mixtures of methane, ammonia, hydrogen and water vapour for a period of time. They discovered that amino acids (some known components of cell proteins) were formed naturally, as well as other compounds.

This approach confirmed that organic molecules can be synthesized outside cells, in the absence of oxygen. The experiment has subsequently been repeated, sometimes using different gaseous mixtures and other sources of energy (ultraviolet radiation, in particular), in similar apparatus. The products have included amino acids, fatty acids and sugars such as glucose. In addition, nucleotide bases have been formed and, in some cases, simple polymers of all these molecules have been found. So, we can see how it is possible that a wide range of organic compounds could have formed on the pre-biotic Earth, including some of the building blocks of the cells of organisms.

TOK

To what extent can you argue that Miller and Urey's experimental response to a seemingly insoluble issue was a uniquely scientific response?

ATL A2.1A

Carry out further research into Stanley Miller and Harold Urey's work on synthesizing organic molecules in a pre-biotic world: www.dnaftb.org/26/animation.html

What first led Urey and Miller to study the origin of life on Earth? What research came before them, and what work followed their discoveries?

■ Assembly of the polymers of living organisms

For polymers to be assembled in the absence of cells and enzymes would have required a concentration of biologically important molecules such as monosaccharides (simple sugars – the building blocks for polysaccharides), amino acids (building blocks for proteins) and fatty acids (for lipid synthesis). They would need to come together in 'pockets' where further chemical reactions between them were possible. Clays have been shown to be important in the **polymerization** of monomers – they promote phosphodiester bond formation (see page 18) by binding and concentrating nucleotides. Microscopic layers of clay may have played a similar role in the formation of the first polyribonucleotides. This might have happened in water close to lava flows from volcanoes or at the vents of submarine volcanoes where the environment is hot, the pressure is high and the gases being vented are often rich in sulfur compounds (e.g. H_2S) and other compounds. There is some evidence for the latter (see page 47).

◆ **Polymerization**: process by which relatively small molecules, called monomers, combine chemically to produce a larger molecule called a polymer.

A2.1 Origins of cells

Evaluating the Miller–Urey experiment

◆ **Evaluation**: make an appraisal by weighing up the strengths and limitations.

◆ **Accuracy**: how close to the true value a result is.

◆ **Precision**: describes the reproducibility of repeated measurements of the same quantity and how close they are to each other.

> ### Inquiry 3: Concluding and evaluating
>
> #### Evaluating
>
> An **evaluation** is an important procedure towards the end of a scientific investigation. Did you demonstrate your hypothesis? What were the limitations and strengths of the investigation? How could you improve the experiment? What else could you measure or change?
>
> When commenting on limitations, consider the procedures, the equipment, the use of equipment, the quality of the data (for example, their **accuracy** and **precision**) and the relevance of the data. To what extent may the limitations have affected the results? Propose realistic improvements that address the limitations.

The conditions of the Miller–Urey experiment were believed at the time (1953) to simulate the atmosphere of early Earth, which was assumed to be reducing (hydrogen rich) and rich in methane. However, current thinking is that methane was in low abundance in the early atmosphere (except perhaps for brief periods), with carbon largely in the form of carbon dioxide (oxygen rich). Also, the Miller–Urey experiment used electrical discharges rather than UV light to simulate high-energy input into the early Earth system. However, organic molecules such as amino acids and bases are generated when carbon dioxide, nitrogen and water are subjected to ionizing (nuclear) radiation and ultraviolet light, as well as electrical discharges.

Despite the success of the experiments of Miller and Urey, efforts to reproduce the conditions of pre-biotic chemistry have not until recently succeeded in generating nucleotides. Nucleotides have now been chemically synthesized via a new approach involving four simple organic molecules – cyanamide, cyanoacetylene, glycolaldehyde and glyceraldehyde – that are readily produced under reasonable pre-biotic conditions.

Spontaneous formation of vesicles

All cells are made from membranes that separate genetic material and chemical reactants in metabolic processes from the external environment. It is likely that the earliest cells, or protocells, were formed from basic membranes.

But how did these protocells form?

Fatty acids are likely to have formed the components of protocell membranes because they are **amphipathic**, which means that they have a polar end that is attracted to water and a non-polar end that is repelled by it. Scientists have shown that if a few lipid molecules are in water, they form a monolayer on the surface of water and, with more lipid present, bilayers form. Lengths of these bilayers are likely to have formed **microspheres** (Figure A2.1.5) or very small **vesicles**. Vesicles therefore form spontaneously (i.e. without an external cause or stimulus) by coalescence of fatty acids into spherical bilayers.

◆ **Amphipathic**: a molecule that has two different affinities – a polar end that is attracted to water and a non-polar end that is repelled by it.

◆ **Microsphere**: a microscopic hollow sphere made from a lipid bilayer.

◆ **Vesicle**: membrane-bound sac.

Perhaps simple microspheres, surrounding a portion of a pre-biotic 'soup' of polymers and monomers, were the forerunners of cells. These may have formed membrane systems with a distinctive internal chemistry, as they developed a chemical environment different from their surroundings.

> **Link**
>
> Phospholipids and cell membranes are covered in detail in Chapters B1.1, page 202, and B2.1, page 223.

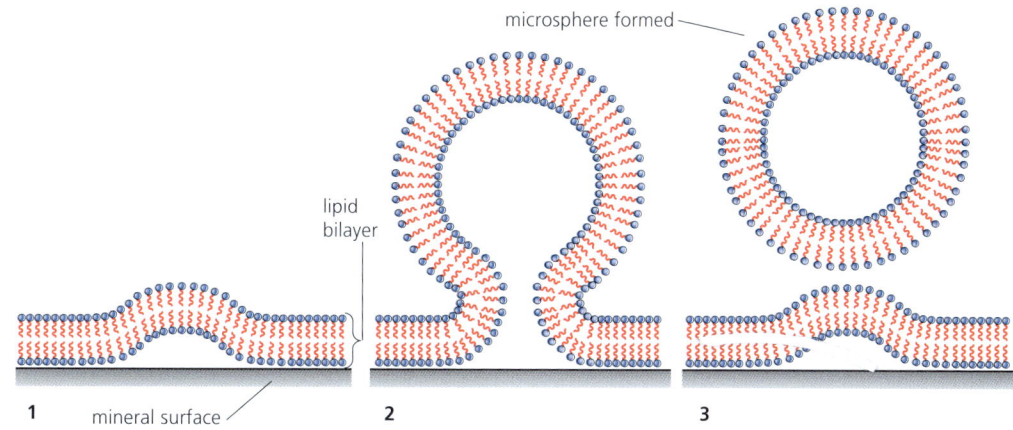

■ **Figure A2.1.5** Steps in the formation of microspheres

Cell membranes are made from modified lipids, called phospholipids, which have structures with hydrophobic tails and hydrophilic heads. It is likely that fatty acids rather than phospholipids formed the first membranes as they are chemically simpler than phospholipids.

The evolution of cell membranes, from primitive to more modern, may therefore have followed the following pathway:

- Protocells formed from fatty acids. Fatty acids are extremely stable compounds and may have accumulated to significant levels on the early Earth.
- Condensation of fatty acids with glycerol to form triglycerides, a highly stabilizing membrane component.
- Phosphorylation (the addition of a phosphate to the triglyceride) forms the simplest phospholipid.

RNA as a presumed first genetic material

For the evolution of life from a mixture of polymers and their monomers, two special situations need to emerge:

- a 'self-replication' system
- an ability to catalyse chemical change.

Today in living cells these essential situations are achieved by DNA, the home of the genetic code, and enzymes, which are typically large, globular proteins (see Chapter C1.1, page 380). However, neither of these have been synthesized in any experiments that repeat Miller and Urey's demonstration of how biologically important molecules might have been synthesized in the pre-biotic world.

So what may have filled the roles of DNA and enzymes in the origin of life?

A likely answer came as a by-product of a genetic engineering experiment, investigating the enzymes needed to join short lengths of RNA. It was discovered that RNA, as well as being information molecules, may also function as enzymes. Perhaps short lengths of RNA combined the roles of information molecules and enzymes in the evolution of life itself.

In Chapter A1.2, we explored the structure of RNA and its similarities and differences to DNA. We have also seen how hydrogen bonds between adjacent bases confer stability in DNA molecules. Messenger RNA (mRNA) is clearly a simpler molecule than DNA (it is single stranded rather than double stranded), and hydrogen bonds can occur between nucleotides in the same chain, causing RNA to fold up in a unique way, determined by its nucleotide sequence. This folding can confer enzymatic properties on the RNA – a property that would have been needed in the earliest forms of life. This is known as the **RNA world** hypothesis.

> ◆ **RNA world**: hypothesis that proposes that the earliest life forms (protocells) may have used RNA alone for the storage of genetic material.

> **Concept: Unity**
>
> As nucleic acid is in all living organisms, this suggests a means by which the first cells arose.

A2.1 Origins of cells

◆ **Central dogma**: the idea that the transfer of genetic information from DNA of the chromosome to mRNA to protein (amino acid sequence) is irreversible.

Top tip!

The idea that information always flows in this direction (DNA to RNA to protein) in cells was called the central dogma of cell biology, implying it was always the case. However, in retroviruses such as HIV (see Chapters A2.3 and C3.2), the information in RNA in the cytoplasm is translated into DNA within a host cell and then becomes attached to the DNA of a chromosome in the host's nucleus.

ATL A2.1B

Crick's original formulation of the central dogma was: 'Once information has got into a protein it cannot get out again.'

What did he mean by this statement and how does it allow for reverse transcription, retroviruses and other biological phenomena?

Read about Crick's ideas here: www.ncbi.nlm.nih.gov/pmc/articles/PMC5602739

◆ **Ribozyme**: RNA molecule capable of acting as an enzyme.

The RNA world

The **central dogma** of molecular biology is that DNA makes RNA, which makes protein (Figure A2.1.6). Nucleic acids are required for protein synthesis, but proteins are required to synthesize nucleic acids. This makes it difficult to see how this interdependent system could have evolved by natural selection.

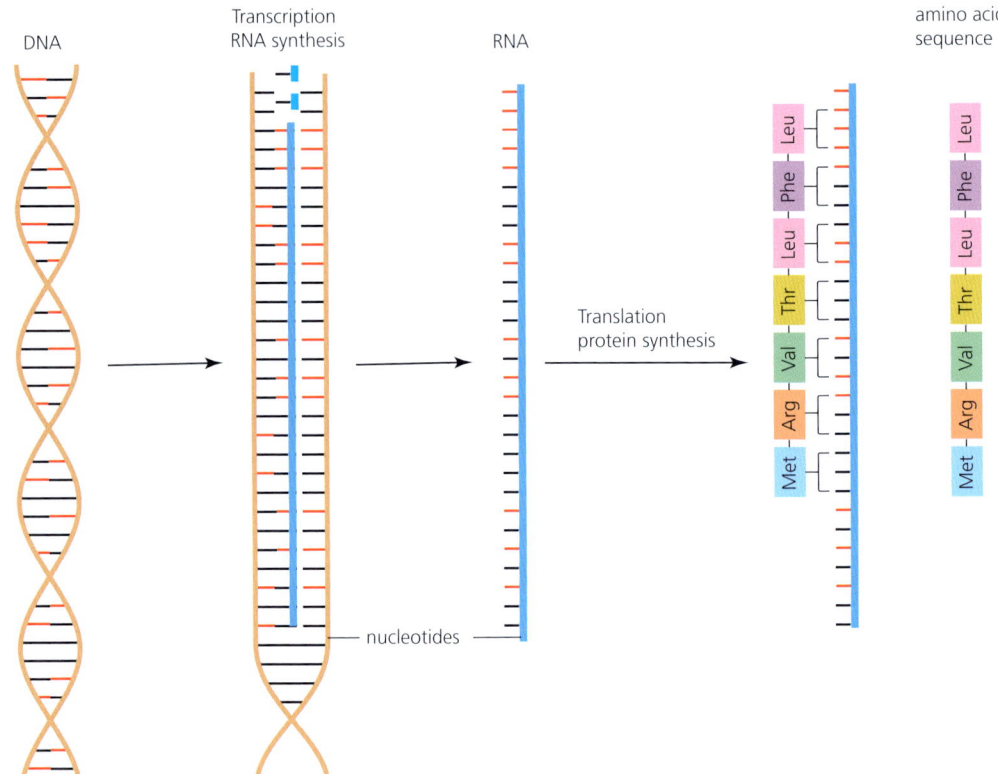

■ Figure A2.1.6 The central dogma showing flow of information from DNA to mRNA to proteins

One view is that an RNA world existed in protocells before modern cells containing DNA and proteins. According to the RNA world hypothesis, RNA is now an intermediate between genes and proteins, but it stored genetic information and catalysed chemical reactions in protocells. Only later in evolutionary time did DNA take over as the genetic material and proteins become the major catalysts and structural components of modern cells.

However, there still appear to be relics of the RNA world in modern cells. RNA primers (short nucleic acid sequences) are used in eukaryotic DNA replication, and ribosomal RNA appears to be involved in the catalysis of peptide bond formation in the ribosome. RNA molecules that have catalytic properties are known as **ribozymes**. These molecules have a unique folded three-dimensional shape that acts as an active site. In 1989, Thomas Cech and Sidney Altman were awarded the Nobel Prize in Chemistry for the discovery of catalytic RNA.

The unique potential of RNA molecules to act both as carriers of genetic information and as catalysts is thought to have enabled them to have a central role in the origin of life. Although self-replicating systems of RNA molecules have not been found in nature, scientists are attempting to synthesize them in the laboratory.

Evidence that RNA arose before DNA in evolution can be found in the chemical differences between them in their pentose sugars. Ribose present in RNA is readily formed from methanal (H_2CO), which is one of the principal products of the Miller–Urey experiment (page 39). In modern cells, deoxyribose is produced from ribose in a reaction catalysed by a protein-based enzyme.

> **Top tip!**
>
> Catalysis, self-replication of molecules, self-assembly and the emergence of compartmentalization were necessary requirements for the evolution of the first cells.

The other differences between RNA and DNA, the stable double helix of DNA and the use of the base thymine rather than uracil, further increase DNA's chemical stability by making the molecule easier to repair by enzymes.

The basic chemical reaction of the ribosome – joining together amino acids from an RNA template – is ultimately catalysed by RNA. This is perhaps even stronger evidence for the RNA world than the more generalized ability of RNA to act as a catalyst of various reactions, as it is consistent with the idea that protein synthesis could have first been developed in a pre-protein world, using an RNA enzyme (the early ribosome) to make the proteins from an RNA template.

The RNA world hypothesis has been an important paradigm shift in the scientific study of life's origins. Although this concept does not fully explain how life originated, it has helped to guide scientific thinking and has served to focus experimental efforts.

> **TOK**
>
> The concept that RNA can have both informational and functional roles, and that ribozymes can act as catalysts for chemical reactions between other RNA molecules, represented a paradigm shift in how scientists viewed the evolution of early life. The RNA world hypothesis provided a means by which the first nucleic acids could have developed and the role they played in the first cells.
>
> What role do paradigm shifts play in the progression of scientific knowledge?

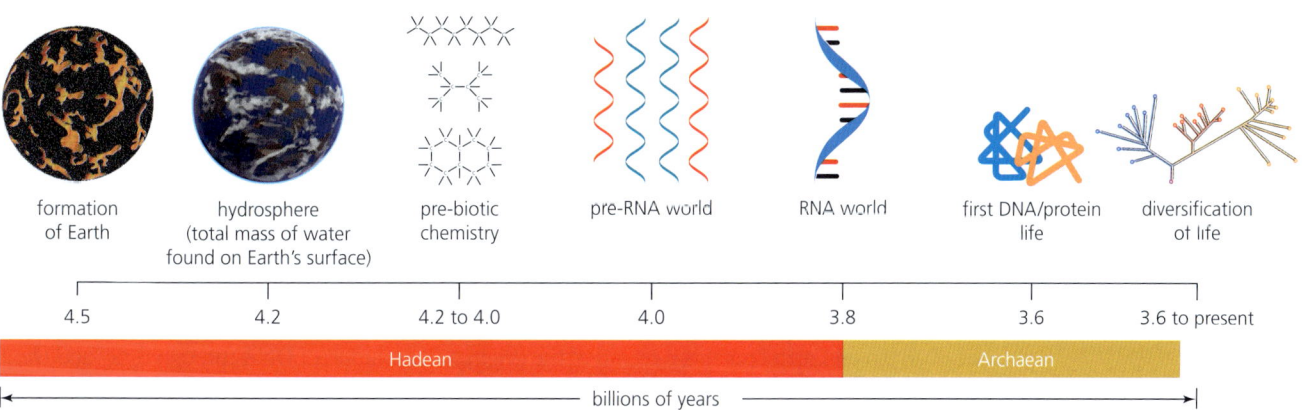

■ **Figure A2.1.7** Geological timeline for the early Earth and the appearance of life; the sequence of events shows the formation of Earth through the Hadean eon and into the Archaean eon, and the corresponding events in the origin of life according to the RNA world hypothesis

5 Describe two properties of RNA which may have contributed to the origin of life.

> **Top tip!**
>
> RNA can be replicated and also has some catalytic activity so may have acted initially as both the genetic material and the enzymes of the earliest cells.

Evidence for a last universal common ancestor

◆ **Common ancestor**: the most recent species from which two or more different species have evolved.

Fossil remains and other evidence (such as anatomical and biochemical similarities) provide evidence about how species are related. The skeletons of the apes, for example, show that the gibbon, gorilla, chimpanzee, orang-utan and human are related and share a **common ancestor** (an ancestor species they all share), and suggest how evolution by natural selection has allowed them to adapt to different environments and lifestyles.

A2.1 Origins of cells

DNA can be used to determine similarities and differences between species. Species with very similar genes will be closely related, whereas those with very different DNA will be only distantly related. As we have already discussed, all life on Earth is related to each other and, ultimately, we all share a common ancestor, which is believed to have existed some 4 billion years ago. This organism is therefore the evolutionary link between the abiotic phase of Earth's history and the biotic phase. This organism is known as **LUCA**, which stands for '**L**ast **U**niversal **C**ommon **A**ncestor'. If all life on Earth is represented as a tree, LUCA is the organism at the base of the tree (Figure A2.1.8). There is an unbroken line of descent from us to LUCA. All organisms share the same biochemistry, the same bases in DNA and the same shared amino acids. The shared genetic code of (most) organisms – the code for translating RNA to protein – is very good evidence for common ancestry, since multiple codes would be possible if different lineages somehow independently evolved protein synthesis.

> **Concept: Unity**
>
> All living organisms can be related to a universal common ancestor, from which the tree of life arose.

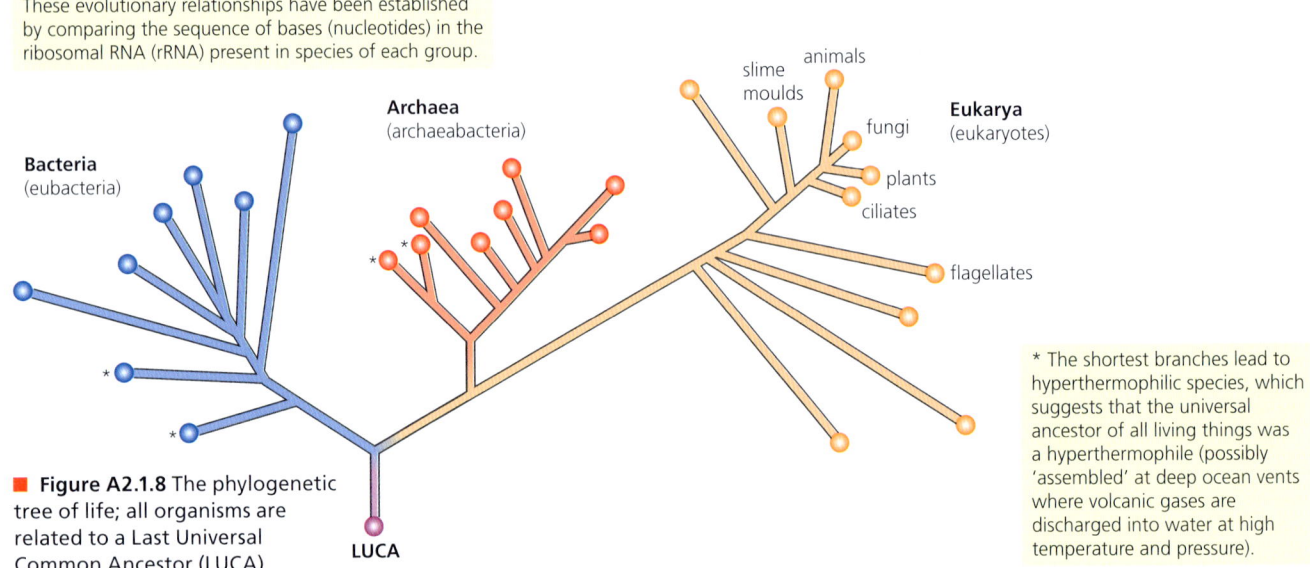

■ Figure A2.1.8 The phylogenetic tree of life; all organisms are related to a Last Universal Common Ancestor (LUCA)

* The shortest branches lead to hyperthermophilic species, which suggests that the universal ancestor of all living things was a hyperthermophile (possibly 'assembled' at deep ocean vents where volcanic gases are discharged into water at high temperature and pressure).

These evolutionary relationships have been established by comparing the sequence of bases (nucleotides) in the ribosomal RNA (rRNA) present in species of each group.

> **Common mistake**
>
> LUCA should not be referred to as a 'protocell'. The term protocell refers to a pre-cellular or cell-like entity (e.g. a lipid droplet with a few molecules inside; something much less elaborate than, for example, the ancestral prokaryote).

We have already discussed how the genetic code is universal (see Theme A1.2, page 14). The genetic code of all life on Earth contains a record of the origin and evolution of DNA through time, and the proteins for which it forms the 'blueprint'. Scientists have analysed DNA from modern bacteria (the eubacteria) and from two archaea (a separate group of extremophile prokaryotes, which can live in extreme environments such as deep ocean vents, in high-temperature habitats such as geysers, in salt pans, in acidic conditions and in polar environments) and searched for genes shared by them as an indicator of genes that were inherited from a common ancestor. Researchers searched DNA databanks, analysing the genomes of 2000 modern microbes sequenced over two decades: 355 gene families were discovered that were widespread among the bacteria from six million total genes, which means they were likely to be genes passed down from LUCA. Care was taken by researchers to eliminate genes that could have been transferred between bacteria laterally (through processes such as conjugation – the process by which one bacterium transfers genetic material to another through direct contact – see page 119) rather than by descent. Genomic analysis was then used to predict the likely structure and function of LUCA (given that DNA codes for proteins, and that proteins determine the structure and function of organisms). This analysis is discussed below (page 47).

It is likely that other forms of life evolved at the same time as LUCA but then became extinct by competing for common resources, making LUCA the surviving organism from which all species evolved. It is also likely that descendants of LUCA also competed with species that subsequently became extinct, shaping the tree of life that we see today.

ATL A2.1C

Figure A2.1.8 shows the three-domain 'model' of life. Recent research has suggested that a 'two-domain' model may better reflect the evolution of life on Earth. This is illustrated in this journal paper: www.nature.com/articles/nmicrobiol2016116

Find out about the two-domain tree of life. How does it relate to what you know about the origin of the first cells, for example the endosymbiont theory, page 83, Chapter A2.2?

This 2017 article: *Looking for LUCA, the Last Universal Common Ancestor* is a good starting point: https://astrobiology.nasa.gov/news/looking-for-luca-the-last-universal-common-ancestor/

TOK

What is the role of imagination and intuition in the creation of hypotheses in the natural sciences?

Science is creative in a similar way to art, music or literature. Scientists must use their imagination to formulate a hypothesis – that is, a testable scientific explanation.

Although imagination, faith and intuition (guiding a scientist in one particular direction) may be used in developing hypotheses and theories about the origin of cells, the validity of scientific arguments must eventually be tested by experimentation or, if that is not possible, simulation.

Biologists may often disagree about the choice of methodology and the value and importance of specific data, or about the appropriateness of particular assumptions and simplifications that are made – and therefore disagree about what conclusions are justified. However, they tend to agree about the principles of logical reasoning that connect evidence (data) and assumptions with conclusions.

What differentiates the natural sciences from other knowledge is subjecting hypotheses to empirical testing by observing whether predictions derived from a hypothesis are confirmed from relevant observations and, if possible, experiments. According to Karl Popper, a hypothesis is scientific if there is possibility of falsification.

The central hypothesis regarding the origin of the cell is that organic molecules self-assembled within a vesicle to form the first protocell (Figure A2.1.9) approximately 4 billion years ago.

Biologists have focused on four critical processes: the formation of organic molecules such as amino acids and nucleic acids (especially RNA), the polymerization of these molecules, the formation of membranes, and the development of metabolic pathways for energy transfer.

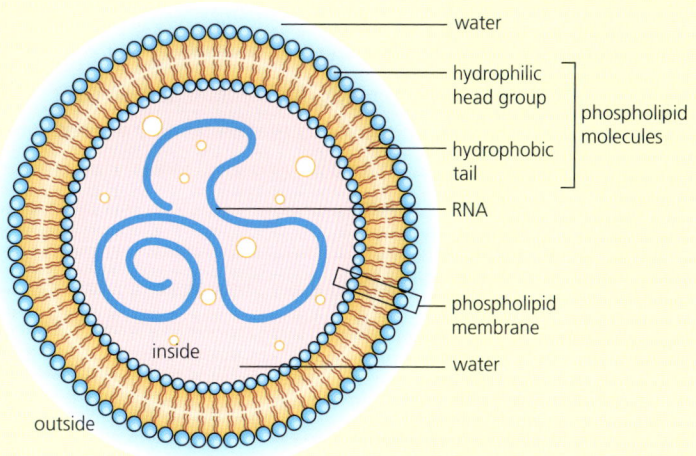

■ **Figure A2.1.9** A possible protocell consisting of self-replicating RNA and proteins within a lipid vesicle

A2.1 Origins of cells

Competing theories exist for how each of these processes evolved, and in what sequence. Although there is a scientific consensus that organic molecules came first via chemical evolution, views differ between scientists over whether metabolism, polymerized molecules (nucleic acids and proteins) or membranes then evolved.

For example, the 'iron–sulfur world' model was initially proposed as a hypothesis without experimental testing. The hypothesis postulated that the first steps in polymerization and protocell formation would take place in a hot and high-pressure (several kilometres under the sea) iron–sulfur hydrothermal environment under the sea.

The hypothesis gained support after the discovery of sub-ocean hydrothermal vents, but experimental designs to test it have been hampered by the challenges of simulating high pressures and temperatures in the laboratory.

We still do not know how life originated on the Earth, and we will possibly never know, since scientists are studying a historical problem for which critical evidence and data may have completely disappeared. The lack of sources forces scientists to imagine and then reconstruct and test the events that could have happened.

Estimating the dates of the first living cells and LUCA

Life has been evolving on Earth over an immense length of time. LUCA may have existed some 4 billion years ago, some 560 million years after the creation of the Earth. But how have scientists determined the age of the first life?

We learn something about the history of life from the evidence found in fossils. Fossilization is an extremely rare, chance event. Predators, scavengers and bacterial action normally break down dead plant and animal structures before they can be fossilized. Of the relatively few fossils formed, most remain buried or, if they do become exposed, are overlooked or accidentally destroyed. Bacteria are known to have been fossilized in rock.

Nevertheless, numerous fossils have been found and more continue to be discovered all the time. If the fossil, or the rock that surrounds it, can be accurately dated (using radiometric dating techniques), we have good evidence of the history of life. Radiometric dating measures the amounts of naturally occurring radioactive substances such carbon-14 (in relation to the amount of carbon-12), or the ratio of potassium-40 to argon-40.

Older rocks will contain more ancient groups of organisms. It can be expected that the oldest rocks on the planet (providing they allowed for fossilization to take place and biological remains were not destroyed by geological forces, for example in metamorphic rock where great pressure and heat create extreme conditions where fossils cannot survive) will contain evidence of the oldest life on Earth.

Genomic analysis also offers techniques to establish the age of ancient organisms. Changes occur in DNA over time. These gene mutations provide the key to estimating dates of the first living cells and the last universal common ancestor. By estimating the average time for mutations to take place, and then extrapolating this back through time, the dates when organisms shared a common ancestor can be estimated. Similarly, the amino acid composition of proteins can be used in a similar way, because change to the genetic code leads to alteration of protein composition and structure.

Biochemical changes, like those discussed above, may occur at a constant rate and, if so, may be used as a 'molecular clock'. If the rate of change can be reliably estimated, it does record the time that has passed between the separation of evolutionary lines. By examining the changes in specific genes common to first life and species alive today, the molecular clock can be used to estimate when these genes converged in a common ancestor, and the rate of biochemical change used to estimate the time over which these changes would have occurred, giving a date for the earliest life on Earth.

Top tip!
Fossils in a rock layer (stratum) that has been accurately dated give us clues to the community of organisms living at a particular time in the past (although this is an incomplete picture). The fossil record may also suggest the sequence in which groups of species evolved and the timing of the appearance of the major groups.

Link
The molecular clock is discussed further on page 130.

Evidence for the evolution of LUCA

Given that LUCA probably evolved in deep-sea hydrothermal vents, rocks formed by ancient seafloor hydrothermal vent precipitates are the most likely to contain fossils of first life.

Scientists have found fossilized evidence of bacteria from ancient seafloor hydrothermal vent precipitates from the Nuvvuagittuq Greenstone Belt in Quebec, Canada. Rock was cut into hundredth-of-a-millimetre slices and examined under a light microscope. These sedimentary rocks contained structures similar to those produced by modern bacteria found at hydrothermal vents. The fossils have been dated to at least 3.77 billion years old, but could be up to 4.28 billion years old, making them the oldest fossil remains to be found to date, indicating that these are some of the first cells to have existed on the Earth.

◆ **Chemosynthesis**: inorganic molecules are oxidized to release energy; this energy is used to synthesize glucose.

◆ **Extremophile**: an organism that lives in conditions of extreme temperature, acidity, alkalinity, salinity, pressure or chemical concentration.

The fossil structures are very small, around half the width of human hair. They are tubes made of haematite, the mineral form of iron(III) oxide (Figure A2.1.10). They are like filamentous microbes from modern hydrothermal vent precipitates. The Nuvvuagittuq rocks also contain carbonate and carbonaceous material, which provides supporting evidence of oxidation and biological activity. Iron-oxidizing microbial communities, which take iron atoms out of the water and remove electrons from it for energy transfer within metabolism, can currently be found associated with widespread hydrothermal vents at the ocean floor. It is likely that the bacteria found in the Nuvvuagittuq rock formation had a similar biochemistry. While different to the metabolism indicated for LUCA, the environment found at deep-sea hydrothermal vents provides many opportunities for diverse forms of energy generation using **chemosynthetic** pathways.

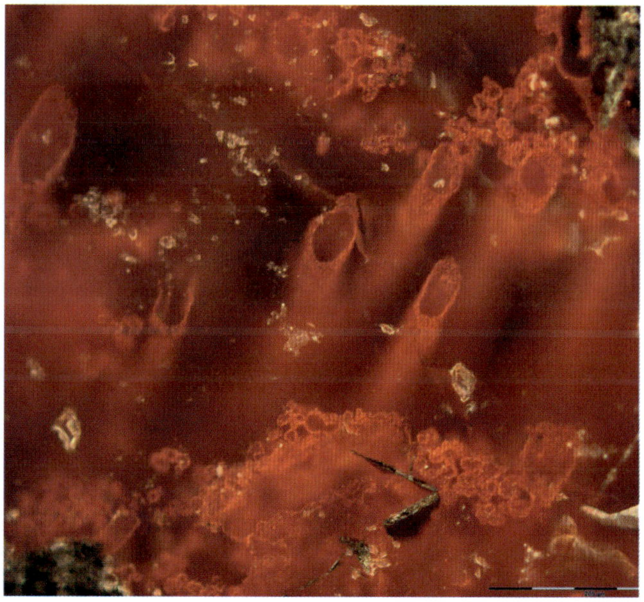

LUCA would have contained conserved genes that are present in all cells. Genetic sequences for conserved genes typically involve proteins associated with ribosomes, where the genetic code is translated into proteins. Because protein synthesis is the most energy-intensive activity of a cell (around 75% of a cell's ATP is used for protein synthesis) these conserved genes tell us that LUCA released and used energy. These universal conserved sequences from genomic analysis do not tell scientists the metabolic nature of this first ancestor, and so other techniques are needed.

The 355 protein families determined from the genetic analysis discussed above are not distributed throughout all organisms, and so can be used to suggest the likely physiology of a possible LUCA. For example, LUCA contained a gene for making a protein called 'reverse gyrase' (an enzyme that helps maintain DNA's structure and stability), which is found today in **extremophiles** existing in high-temperature environments including hydrothermal vents.

■ **Figure A2.1.10** Haematite tubes found in Nuvvuagittuq – the remains of ancient bacteria, at least 3.77 billion years old

The properties and functions of these proteins indicate that LUCA had the following characteristics:
- anaerobic (survived without oxygen)
- CO_2-fixing (converted carbon dioxide into glucose)
- H_2-dependent (used molecular hydrogen as an energy source, rather than sunlight)
- N_2-fixing (converted nitrogen into ammonia, for subsequent synthesis of amino acids)
- thermophilic (survived in areas of very high temperature – up to 122 °C).

A2.1 Origins of cells

The analysis indicated that modern-day microbes with similar physiologies include *Clostridia* (an anaerobic bacteria found in soil and the intestines of humans and other animals) and methanogens (anaerobic bacteria that produce methane as a waste product).

The genes identified by scientists as being passed down from LUCA were those of an **autotrophic** (i.e. it can synthesize glucose) extremophile organism that probably lived in hydrothermal vents: areas where seawater and magma meet on the ocean floor (Figure A2.1.11). LUCA inhabited a geochemically active environment rich in hydrogen, carbon dioxide and iron. As discussed above, similar prokaryotic organisms still live in these environments, among the toxic plumes of sulfides and metals. Given the genetic analysis and recent fossil evidence, many researchers believe this is where life first began.

◆ **Autotrophic**: synthesizing glucose from simple inorganic substances using an external source of energy.

6 **Explain** why LUCA (the Last Universal Common Ancestor) is thought to be the evolutionary link between the abiotic phase of Earth's history and the biotic phase.

■ **Figure A2.1.11** A hydrothermal vent

ATL A2.1D

Tardigrades are an example of an extremophile organism that can survive in the vacuum of space. Research this organism and find out about the range of conditions it can survive in. What physiological adaptations does it have to survive in such extreme environments? Start your reading here: https://serc.carleton.edu/microbelife/topics/tardigrade/index.html

LINKING QUESTIONS

1. For what reasons is heredity an essential feature of living things?
2. What is needed for structures to be able to evolve by natural selection?

A2.2 Cell structure

> **Guiding questions**
> - What are the features common to all cells and the features that differ?
> - How is microscopy used to investigate cell structure?

SYLLABUS CONTENT

This chapter covers the following syllabus content:
- ▶ A2.2.1 Cells as the basic structural unit of all living organisms
- ▶ A2.2.2 Microscopy skills
- ▶ A2.2.3 Developments in microscopy
- ▶ A2.2.4 Structures common to cells in all living organisms
- ▶ A2.2.5 Prokaryote cell structure
- ▶ A2.2.6 Eukaryote cell structure
- ▶ A2.2.7 Processes of life in unicellular organisms
- ▶ A2.2.8 Differences in eukaryotic cell structure between animals, fungi and plants
- ▶ A2.2.9 Atypical cell structure in eukaryotes
- ▶ A2.2.10 Cell types and cell structures viewed in light and electron micrographs
- ▶ A2.2.11 Drawing and annotation based on electron micrographs
- ▶ A2.2.12 Origin of eukaryotic cells by endosymbiosis (HL only)
- ▶ A2.2.13 Cell differentiation as the process for developing specialized tissues in multicellular organisms (HL only)
- ▶ A2.2.14 Evolution of multicellularity (HL only)

Introduction to cells

The cell is the basic structural unit of all living organisms – it is the smallest part of an organism that we can say is alive. It is cells that carry out the essential processes of life. We think of them as self-contained units of structure and function.

Cells are extremely small – most are only visible as distinct structures when we use a microscope (although a few types of cell are just large enough to be seen by the naked eye).

Observations of cells were first reported over 300 years ago, following the early development of microscopes. You may have already used a light microscope to view living cells, such as the single-celled organism *Amoeba*, shown in Figure A2.2.1.

> **Concept: Unity**
>
> All living organisms are made from cells. Some organisms are made of single cells (such as protists and bacteria) and others are multicellular (animals, plants and most fungi). In multicellular organisms, cells are the building blocks for tissues.

● **Common mistake**

Students sometimes use the terms 'cell' and 'tissue' as if they are synonymous (i.e. are the same thing) – this is not the case. Cells are the basic structural unit of all living organisms, whereas tissues are a collection of cells of similar structure and function.

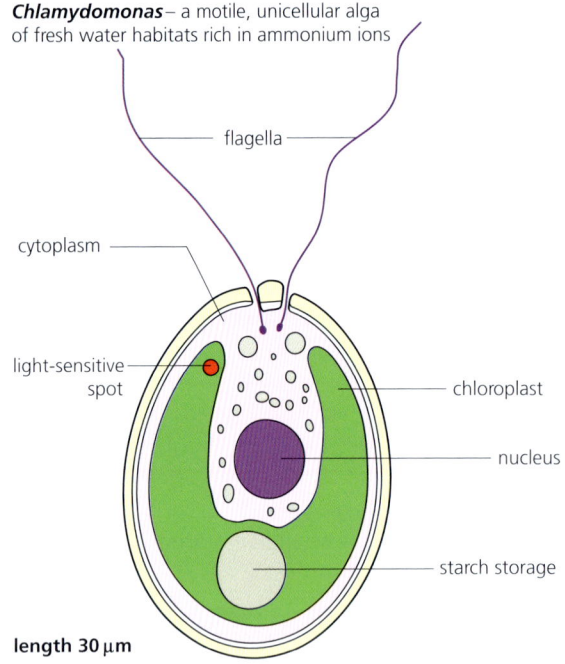
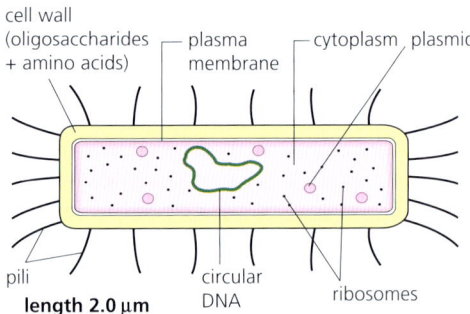

■ Figure A2.2.1 Introducing unicellular organization

◆ **Unicellular**: consisting of a single cell (e.g. prokaryotes, protists and some fungi).
◆ **Protists**: eukaryotes consisting of single-celled organisms.
◆ **Multicellular**: consisting of many cells (e.g. animals, plants and most fungi).

■ Unicellular and multicellular organisms

Some organisms are made of a single cell and are known as **unicellular**. Examples of unicellular organisms are introduced in Figure A2.2.1. There are vast numbers of different unicellular organisms in the living world, many with very long evolutionary histories. One type of unicellular organism is in a kingdom called the Protoctista (such as *Chlamydomonas* and *Amoeba* in Figure A2.2.1 – organisms in this kingdom are referred to as **protists**) and another are the bacteria (*Escherichia coli*, Figure A2.2.1).

Other organisms are made of many cells and are known as **multicellular** organisms. Examples of multicellular organisms are mammals and flowering plants.

● Top tip!

Much of the biology in this book is about multicellular organisms and the processes that go on in these organisms. But remember, single-celled organisms carry out all the essential functions of life too, within a single cell.

■ Cell theory

All organisms are composed of one or more cells. **Cell theory** includes the idea that cells are the unit of structure and function in living organisms. The cell theory states that:
- cells can only arise from pre-existing cells
- living organisms are composed of cells, which are the smallest unit of life
- organisms consisting of only one cell carry out all functions of life in that cell; cells perform life functions at some point in their existence.

Although most organisms conform to cell theory, there are exceptions (see page 78).

Many biologists contributed to the development of the cell theory. This concept evolved gradually in western Europe during the nineteenth century because of the steadily accelerating pace of developments in microscopy and biochemistry.

■ **Table A2.2.1** Units of length used in microscopy

1 metre (m) = 1000 millimetres (mm)
1 mm (10^{-3} m) = 1000 micrometres (µm) (or microns)
1 µm (10^{-6} m) = 1000 nanometres (nm)

Cell size

Since cells are so small, we need appropriate units to measure them. The **metre** (symbol **m**) is the standard unit of length used in science (it is an internationally agreed unit, or **SI unit**).

Look at Table A2.2.1, showing the subdivisions of the metre that are used to measure cells and their contents.

These units are listed in descending order of size. You will see that each subdivision is one thousandth of the unit above it. The smallest units are probably quite new to you; they may take some getting used to.

So, the dimensions of cells are expressed in the unit called a **micrometre** or micron (**µm**). Notice this unit is one thousandth (10^{-3}) of a millimetre. This gives us a clear idea about how small cells are when compared to the millimetre, which you can see on a standard ruler.

Bacteria are really small, typically 0.1–2 µm in size, whereas the cells of plants and animals are often in the range of 50–150 µm or larger. In fact, the lengths of the unicellular organisms shown in Figure A2.2.1 are approximately:

Escherichia coli   2 µm

Chlamydomonas   30 µm

Amoeba   400 µm (but its shape and, therefore, length varies greatly).

Cell size determines the rate of diffusion of substances across the plasma membrane; by being small, this rate is maximized (see Chapter B2.3, page 260).

Table A2.2.2 shows the average sizes of cells and their components in decreasing size. The organelles and other structures that are found in cells are discussed in detail later in this chapter (pages 63–73).

■ **Table A2.2.2** The size of cells and their components

Cell and component	Diameter	Cell component	Diameter
plant cell	40 µm (average)	lysosome	0.2–0.5 µm
animal cell	20 µm (average)	centriole	0.15 µm
nucleus	10–20 µm	microtubule	24 nm
chloroplast	5–10 µm	ribosome	20 nm
bacterium	1 µm	microfilament	7 nm
mitochondrion	0.5–1.5 µm	DNA molecule	2 nm

1 **Calculate** how many cells of 100 µm diameter will fit side by side along a millimetre.

● Nature of science: Theories

Collecting and analysing biological observations can lead to important conclusions based on **inductive reasoning**. Induction involves formulating generalisations from many related specific observations (see page 32).

The claim of cell theory that 'all organisms will consist of one or more cells' is derived from inductive reasoning. This was based on microscopic observations by many biologists in a wide range of organisms. In inductive reasoning, biologists begin with specific observations and measurements and then detect patterns or common features. They may then formulate a hypothesis that can be tested, and finally develop a theory.

A related process known as **deductive reasoning** is used to test the theories, such as cell theory, produced by induction. Deductive reasoning proceeds from the more general to the more specific. This ultimately leads biologists to test the hypotheses with specific data that either support or falsify the theory.

The philosopher Karl Popper rejected the use of inductive reasoning in science, claiming that for induction to be true, every example of its inference must be true. Biologist have found that there are a small number of cells and organisms that are exceptions to the cell theory (page 78). However, cell theory remains an important and unifying concept in biology, necessary for inductive science.

Induction and deduction are important types of logical reasoning and both contribute to the construction of scientific knowledge (Figure A2.2.2). Inductive reasoning is a form of logic directly opposite to that of deductive reasoning. Inductive reasoning is covered in Chapter A1.2, page 32.

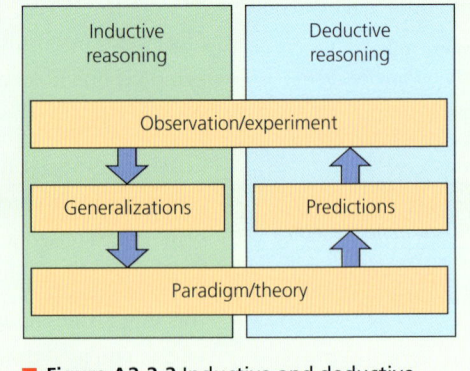

■ **Figure A2.2.2** Inductive and deductive reasoning in the scientific method

> ### Top tip!
> You should have experience of making temporary mounts of cells and tissues, staining, measuring sizes using an eyepiece graticule, focusing with coarse and fine adjustments, calculating actual size and magnification, producing a scale bar and taking photographs. Adaptors are available that allow cameras or smartphones to be attached to the eyepiece lens to enable photos to be taken. See more on this in the Tools section on page 56.

Microscopy skills

■ Examining cells and recording structure and size

We use microscopes to magnify the cells of biological specimens in order to view them. Figure A2.2.3 shows two types of light microscope.

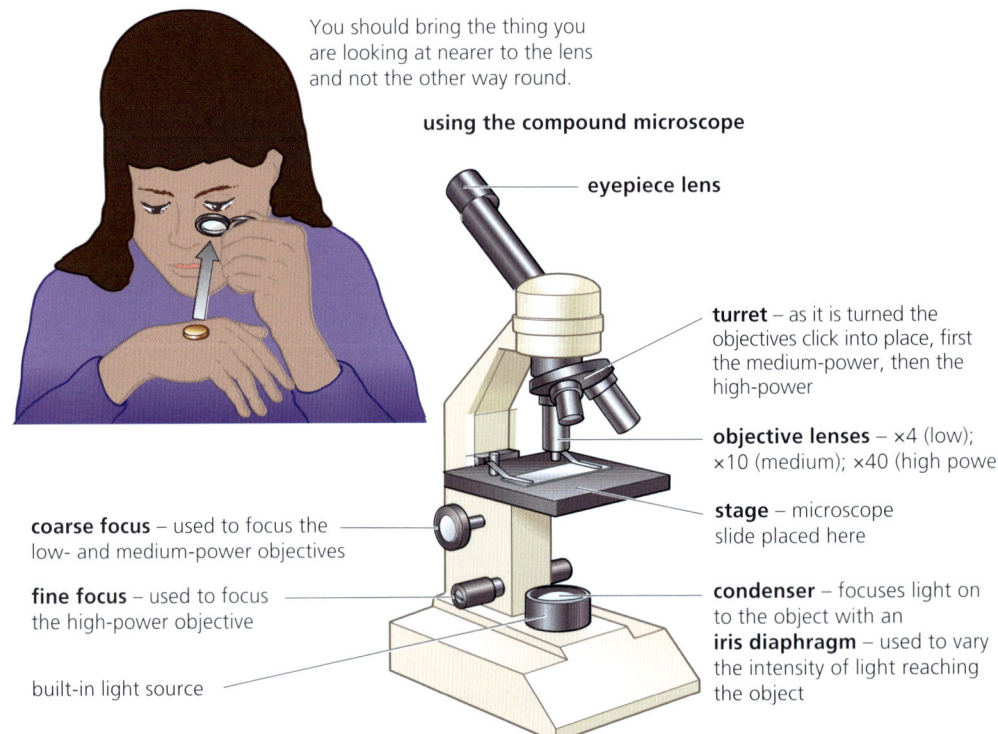

■ **Figure A2.2.3** Light microscopy

In the simple microscope (**hand lens**), the instrument can be held very close to the eye and is today mostly used to observe external structure. Some of the earliest hand lenses detailed observations of living cells.

Light (compound) microscopes have been instrumental in understanding cell structure and function, and revealing organisms that are not visible with the naked eye. The **objective lens** forms an image (in the microscope tube) that is then further magnified by the **eyepiece lens**, producing a greatly enlarged image. Initially, coarse focus enables you to locate and see the specimen under low power. Fine focus can be used to resolve the image under higher powers.

> ### Tool 1: Experimental techniques
>
> #### The light microscope
> 1 Select a low-power lens. Make sure the lens clicks into position.
> 2 Examine the prepared slide without the microscope and note the position, colour and rough size of specimen.
> 3 Place the slide on the stage, coverslip uppermost, viewing it from the side. Position it with stage adjustment controls so that the specimen is lit up.
> 4 Focus using first the coarse and then the fine focusing controls. Use **both hands** to alter the focusing controls; this helps keep the controls working properly and to not go out of alignment.
> Note: The image will be reversed and upside down when seen by viewing the slide directly. When focusing, always move the stage down, away from the objective lens, to avoid moving the slide on to the objective lens (which could damage the lens and break the slide).
> 5 For higher magnifications, swing in the relevant objective lens carefully, checking there is space for it. Adjust the focus using the fine control only. If the object is in the centre of the field of view with the ×10 objective, it should remain in view with the ×40 objective.
> 6 When you have finished using the microscope:
>   - turn the objective lens back to ×10 and then lower the stage
>   - remove the last slide and return it to the correct section in the tray
>   - clean the stage if necessary and check eyepiece lenses and objective lenses are clean
>   - unplug the cable and store tidily, replacing the dust cover.

> ### Common mistake
>
> When using a compound light microscope:
> - Never force any of the controls.
> - Never touch any of the glass surfaces with anything other than a clean, dry lens tissue.
> - Do not hold the microscope with one hand. When moving the microscope, hold the stand above the stage with one hand and rest the base of the stand on your other hand.
> - Do not tilt the microscope. Always keep the microscope vertical (or the eyepiece may fall out).
> - Do not touch the surface of lenses with your fingers.
> - Do not allow any solvent to touch a lens.
> - When focusing, move the stage down, away from the objective lens, to avoid moving the slide on to the objective lens.

Biological material to be examined by compound microscopy must be sufficiently transparent for light rays to pass through. When bulky tissues and parts of organs are to be examined, thin sections are cut. Thin sections are largely colourless.

■ Table A2.2.3 The skills of light microscopy

You need to master and be able to demonstrate these aspects of good practice
Knowledge of the parts of your microscope and care of the instrument – its light source, lenses and focusing mechanisms.
Use in low-power magnification first, using prepared slides and temporary mounts.
Switching to high-power magnification, maintaining focus and examining different parts of the image.
Types of microscope slides and the preparation of temporary mounts, both stained and unstained.

Tool 1: Experimental techniques

Preparation of a temporary mount

Cells can be mounted on slides so they can be viewed under a microscope. These can be disposed of once the cells have been seen and studied. Stains can be used to view cell features more clearly. Techniques describing how temporary mounts can be made are outlined here.

Living cells are not only very small but also transparent. In light microscopy it is common practice to add dyes or stains to introduce sufficient contrast and so differentiate structure. Dyes and stains that are taken up by cells are especially useful.

Observing the nucleus, cytoplasm and cell membrane in human cheek cells

Take a smear from the inside lining of your cheek using a fresh, unused cotton bud you remove from the pack. Touch the materials removed by the cotton bud on to the centre of a microscope slide and add a cover slip (see Figure A2.2.4). Dispose of the cotton bud safely and hygienically. Handle the microscope slide yourself and, at the end of the observation, immerse the slide in 1% sodium hypochlorite solution to sterilize the slide and cover slip. To observe the structure of human cheek cells, irrigate the slide with a drop of methylene blue stain (following the procedure shown in Figure A2.2.4) and examine some of the individual cells with medium- and high-power magnification.

Making a temporary mount

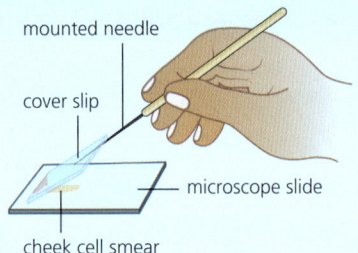

Irrigating a temporary mount

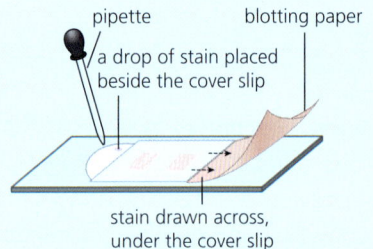

■ Figure A2.2.4 Preparing living cells for light microscopy

You could also observe chloroplasts in moss leaf cells, or the nucleus, cell wall and vacuole in an onion epidermis cell.

 TOK

Living tissues prepared for examination under the microscope are typically cut into thin sections and stained. Both processes may alter the appearance of cells. Is our knowledge acquired with the aid of technology fundamentally different from that which we acquire from our unaided sense? If so, what may be done about this, in practical terms?

Tool 1: Experimental techniques

The eyepiece graticule

The size of a cell can be measured under the microscope. A transparent scale, called a **graticule**, is mounted in the eyepiece at the focal plane (there is a ledge for it to rest on). In this position, when the object under observation is in focus, so too is the scale. The size (for example, length or diameter) of the object may then be recorded in arbitrary units. Next, the graticule scale is calibrated using a **stage micrometer** – in effect, a tiny, transparent ruler, which is placed on the microscope stage in place of the slide and then observed. With the eyepiece and stage micrometer scales superimposed, the true dimensions of the object can be estimated in micrometres. Figure A2.2.5 shows how this is done.

Theme A: Unity and diversity – Cells

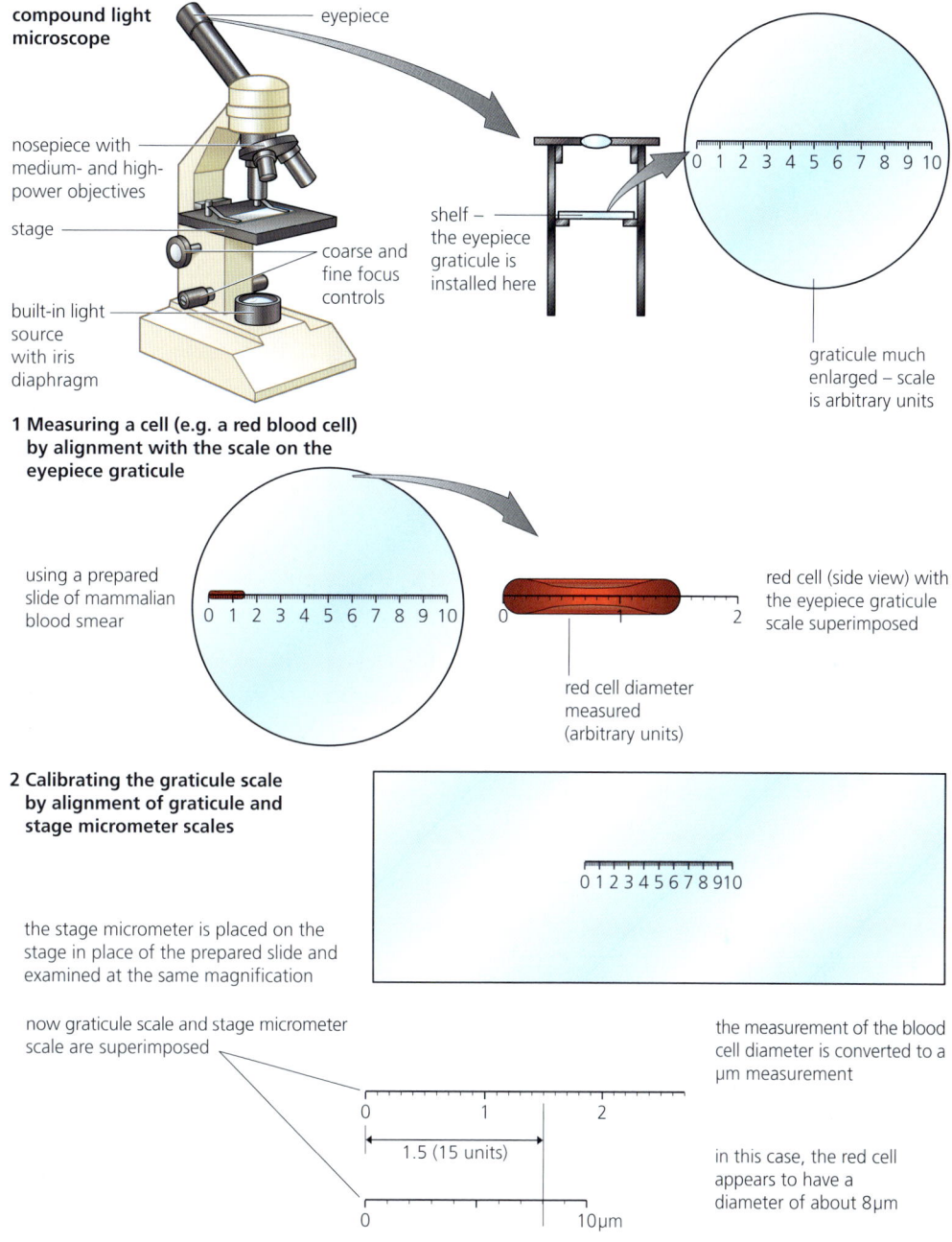

■ Figure A2.2.5 Measuring the size of cells

Once the size of a cell has been measured, a scale bar line may be added to a micrograph or drawing to record the actual size of the structure, as illustrated in Figure A2.2.6.

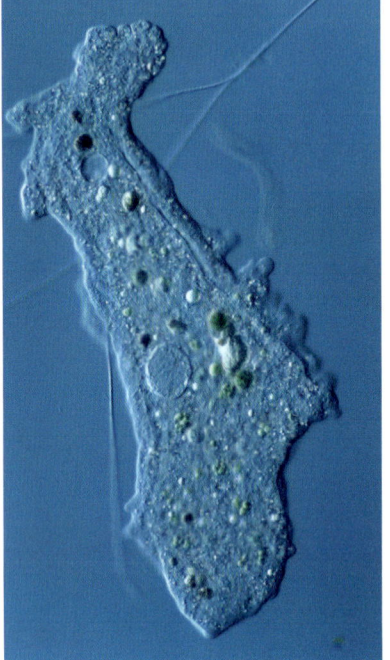

photomicrograph of *Amoeba proteus* (living specimen) – phase contrast microscopy

interpretive drawing
- cell surface membrane
- small food vacuoles
- pseudopodia
- nucleus
- large food vacuole
- cytoplasm – outer, clear (ectoplasm) and inner, granular (endoplasm)
- contractile vacuole
- scale bar 0.1 mm

■ Figure A2.2.6 Recording size by means of scale bars

ATL A2.2A

Converting smartphones into 'CellCams'

Photographs of microscope images (photomicrographs) have traditionally been taken using expensive camera-mounted apparatus. However, smartphones can take high-quality images without the need for such specialized apparatus. It is difficult to take photographs using hand-held smartphones, because slight variations in the angle of the phone make the image invisible or partially obscured. Simple methods can be used to attach the smartphone to the eyepiece of the microscope, enabling images to be taken. Such adaptors enable you to create your own 'CellCam'.

Take a toilet-paper cardboard tube (4.5 cm diameter and 4 cm in length) and add an adhesive foam weather-strip (1–2 cm wide) to the inside so that the cardboard tube fits securely around the eyepiece of the microscope. For a mid-mounted smartphone camera, attach the tube to the camera using ice-cream sticks secured either side of the tube using elastic bands (Figure A2.2.7a). For a corner-mounted camera, cut a slit in one edge of the tube and slot the camera inside (Figure A2.2.7b). Put the cardboard tube down over the eyepiece of the microscope. The CellCam can then be used to capture images or video of specimens.

Try making a CellCam using your smartphone and the methodology above and taking photos of specimens under the microscope. Remember to be careful when putting the cardboard tube over the eyepiece of the microscope.

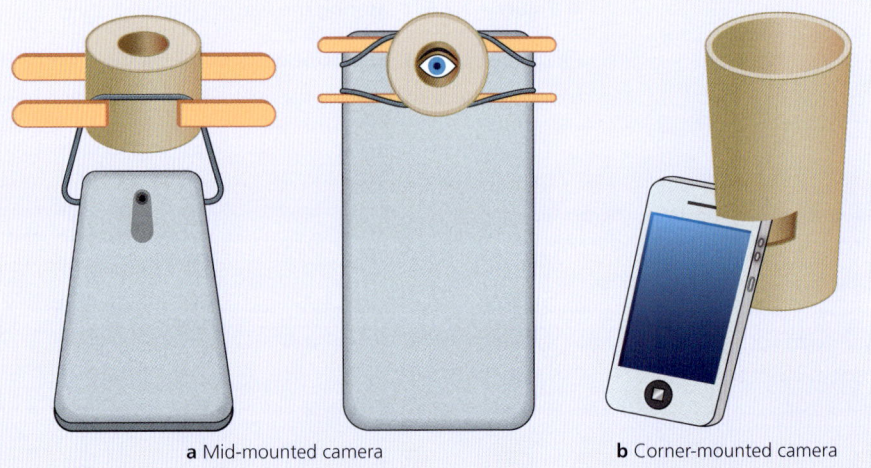

■ Figure A2.2.7 Converting a smartphone into a 'CellCam': a) mid-mounted camera b) corner-mounted camera

a Mid-mounted camera   **b** Corner-mounted camera

Magnification of an image

◆ **Magnification**: how many more times larger an object appears.

Magnification is the number of times larger an image is than the specimen. The magnification obtained with a compound microscope depends on which of the lenses you use. For example, using a ×10 eyepiece and a ×10 objective lens (medium power), the image is magnified ×100 (10 × 10). When you switch to the ×40 objective (high power) with the same eyepiece lens, then the magnification becomes ×400 (10 × 40). These are the most usual orders of magnification you will use in your laboratory work.

2 **Calculate** what magnification occurs with a ×6 eyepiece and a ×10 objective.

There is actually **no limit to magnification**. For example, if a magnified image is photographed, then further enlargement can be made photographically. This is what may happen with photomicrographs shown in books and articles.

When an image from a light microscope is magnified photographically **the detail will be no greater**, see Figure A2.2.9, page 59.

> ### Tool 3: Mathematics
>
> ### Calculating scales of magnification
>
> Make a temporary mount of a plant cell using the techniques outlined on page 54. Pondweed is a good plant to use as the leaves are thin and easily removed from the plant. Draw one plant cell using the method shown on page 56. Use an eyepiece graticule to estimate the size (either the length or width) of a plant cell using the technique in Figure A2.2.5. Now calculate the magnification of the image you have drawn.
>
> Magnification is given by the formula:
>
> $$\text{magnification} = \frac{\text{size of image}}{\text{size of specimen}}$$
>
> So, for a particular plant cell of 150 µm diameter, photographed with a microscope and then enlarged in a drawing or photographically, the magnification in a print showing the cell at 15 cm diameter (150 000 µm) is:
>
> $$\frac{150\,000}{150} = \times 1000$$
>
> If a further enlargement is made to show the same cell at 30 cm diameter (300 000 µm), then the magnification is
>
> $$\frac{300\,000}{150} = \times 2000$$

3 Using the scale bar given in Figure A2.2.6, **calculate** the maximum observed length of the *Amoeba* cell.

4 **Calculate** the magnification of the image of *Escherichia coli* in Figure A2.2.1 on page 50.

Scale bars

To add a scale bar to your drawing of a biological specimen:

1 Use the stage micrometer and eyepiece graticule to work out the distance between two markings on the eyepiece graticule, i.e. the number of micrometres equivalent to one unit (or division) on the eyepiece graticule (Figure A2.2.5).

2 Remove the stage micrometer and place the specimen on the stage.

3 Measure the length of the specimen using the eyepiece graticule. The measurement will be in graticule units (which will depend on the magnification you are using).

4 Determine the length of the specimen in micrometres by multiplying the number of graticule units by the length represented by one unit. For example, if the length of the specimen is 20 graticule units, and the length of each unit represents 10 µm, the total length of the specimen will be 20 × 10 = 200 µm.

● **Top tip!**

Scale bars can be used as a way of indicating the actual sizes in drawings and micrographs, and can be used to calculate magnification. Magnification is calculated by dividing the actual length of the scale bar by the length indicated on the scale bar.

Ideally your scale bar needs to be around 20% of the length of the specimen. If the specimen is 200 μm then the scale bar would be 20% of 200 = 40 μm. Now draw the specimen on a piece of paper or in your lab notebook and measure the length of your drawing. If the length of the drawing is 100 mm, then the scale bar needs to be drawn as 20% of this length = 20 mm. Draw a line next to your drawing 20 mm in length and mark on this the actual length it represents (40 μm).

The actual length represented by the scale bar should be a whole number, and so the measurement taken from the specimen may need to be rounded. For example, if the specimen is 52.5 μm, 20% of this number is 10.5 μm. This can be rounded down to give 10 μm – ratios can then be used to establish the length of the scale bar. 10/52.5 gives one ratio (the length of the scale bar to the actual length of the organism). If the length of the drawing is 96 mm, the second ratio is $x/96$ where x is the length of the scale bar. When the ratios are resolved:

$10/52.5 = x/96$

$x = (10 \times 96)/52.5 = 18.3 \text{ mm}$

The scale bar is then drawn to the length 18.3 mm, and the actual length that this represents recorded as 10 μm.

Top tip!

The size of cells, or components of cells, can be calculated given the amount of magnification and a scale drawing of the object. Simple equations can be used to calculate the magnification or actual size of the specimen.
- **I** = size of image (drawing of an object on paper)
- **A** = actual size of the object being measured
- **M** = magnification (the size of an object compared to its actual size, i.e. the number of times larger an image is than the specimen)

So, M = I/A; A = I/M and I = A × M.

A memory diagram can be used, showing how to calculate the magnification, actual size or image size of an object (Figure A2.2.8).

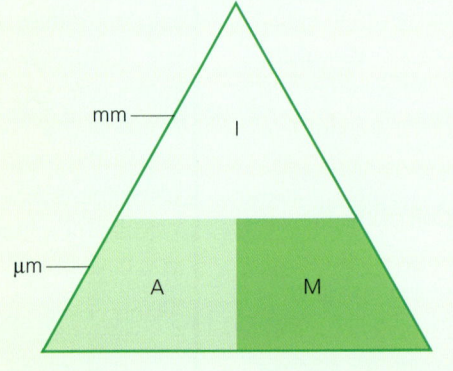

■ **Figure A2.2.8** Memory diagram showing how to calculate the magnification, actual size or image size of an object

Remember the equation as AIM or IAM, and remember to convert units so that they are the same for both I and A.

Top tip!

Remember to convert all values to the same unit of measurement: if you do not do this, your results will be incorrect by a factor of 100, 1000 or even 1 000 000.
- Convert mm into μm by multiplying by 1000.
- Convert μm into mm by dividing by 1000.

5 A highly magnified electron micrograph of the bacterium *Escherichia coli* was accompanied by a scale bar of length 23 mm and labelled 1 μm. The following features were measured. Complete the following table. Express their actual size in appropriate units.

Feature	Measurement on scale bar (mm)	Actual size
thickness of the wall	1	
length of a flagellum	32	
width of the cell	24	
length of the cell	5	

Resolution of an image

◆ **Resolution**: the amount of detail that can be seen.

The **resolution** (resolving power) of a microscope is its ability to separate visually small objects that are very close together. If two separate objects cannot be resolved, they are seen as one object. Merely enlarging them does not separate them. Resolution is a property of lenses that is quite different from their magnification – and is more important.

Resolution is determined by the wavelength of light. Light is composed of relatively long wavelengths, whereas shorter wavelengths give better resolution. For the light microscope, the limit of resolution is about 0.2 μm. As a result, two objects less than 0.2 μm apart may be seen as one object, which means that the image is blurred at low resolution. Figure A2.2.9 shows the importance of resolution over magnification. In this figure, an image of a chloroplast at the same magnification is shown at low (left-hand image) and high resolution (right-hand image). The high-resolution image has been taken using an electron microscope, and the low-resolution image with a light microscope. Details of the internal and external membrane structure cannot be seen under low resolution, however much the image is enlarged. This is because the membrane is too thin to be seen and resolved as separate structures using the light microscope.

6 **Distinguish** between resolution and magnification.

◆ **Quantitative data**: numerical measurements with units (and associated random uncertainty), which are often digitally recorded and processed.

◆ **Qualitative data**: descriptive information used to record the conditions in which data are recorded, or to describe the features or properties of an object.

a chloroplast enlarged (×6000) from a photomicrograph obtained by light microscopy

b chloroplast from a transmission electron micrograph

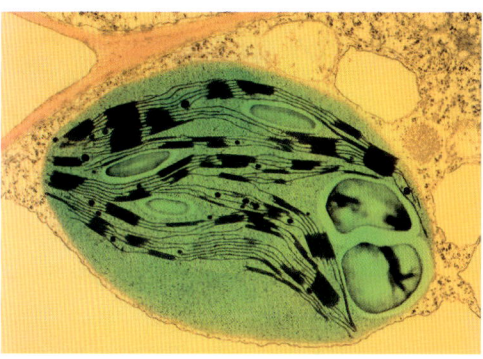

■ **Figure A2.2.9** Magnification a) without and b) with resolution; the micrograph in b) has been colourized

Nature of science: Measurements

Using instruments is a form of quantitative observation

Biology – like all natural sciences – is an observation-based science, perhaps more so than chemistry or physics. Living organisms are very diverse and the first task of any biological investigation is the observation and recording of biological phenomena.

Observations in biology are varied and may include counting, drawing, photographing, video recording (of animal behaviour) or measuring changes in concentration or pH during biochemical reactions. **Quantitative observations** are ones that involve numerical data, for example the recording values for a dependent variable. **Qualitative observations** reinforce quantitative data: they record observations relevant to the study.

Observations and measurements are the basis for the development of new biological hypotheses. In biology, the observations often arise from questions of the form, 'How does variable X affect variable Y?' Knowledge acquired by the senses is known as **empirical data**.

Biologists often 'enhance' their senses by using technology with a variety of analytical instruments. This enables them to study phenomena that are beyond or outside the direct limits of our human senses.

For example, the electron microscope has been used to study the ultrastructure of cells and the technique of NMR (nuclear magnetic resonance spectroscopy) has been used to study the folding and unfolding of proteins. In this technique, molecules are studied by recording the interaction of high-frequency radio waves with the nuclei of molecules placed in a strong magnetic field. The effect of ultrasound has been studied on organisms such as bats, dolphins and porpoises which use it for 'echo location'. Humans are not able to directly perceive electromagnetic radiation or sound waves with these frequencies (energies).

Developments in microscopy

The technology used to view cells and their internal structure has developed through time, allowing greater detail to be discovered, which has revealed increasing knowledge of how cells function.

■ Electron microscopy – the discovery of cell ultrastructure

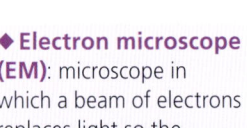

Microscopes were invented simultaneously in different parts of the world at a time when information travelled slowly. Modern-day advances in microscopy and communications have allowed for improvements in the ability to investigate and collaborate, enriching scientific endeavour.

The **electron microscope (EM)** uses electrons to make a magnified image in much the same way as the optical microscope uses light. Electrons can show particle and wave properties, and need to travel through a vacuum. This means the biological material must be dead. However, because an electron beam has a much shorter wavelength, its resolving power is much greater. When the electron microscope is used with biological materials, the limit of resolution is about 5 nm. (The size of nanometres is given in Table A2.2.1, page 51.)

Only with the electron microscope can the detailed structures of cell organelles be observed. This is why the electron microscope is used to resolve the fine detail of the contents of cells, the organelles and cell membranes, collectively known as cell **ultrastructure**. It is difficult to exaggerate the importance of electron microscopy in providing our detailed knowledge of cells.

In the electron microscope, the electron beam is generated by an **electron gun**, and focusing is by **electromagnets**, rather than by glass lenses. We cannot see electrons, so the electron beam is focused on to a **fluorescent screen** for viewing, or on to a **photographic plate** for permanent recording (Figure A2.2.10).

In **transmission electron microscopy (TEM)**, the electron beam is passed through an extremely thin section of fixed, biological material. Membranes and other structures are stained with heavy metal ions, making them electron-opaque so they stand out as dark areas in the image.

In **scanning electron microscopy (SEM)**, a narrow electron beam is scanned back and forth across the surface of the specimen. Electrons that are reflected or emitted from this surface are detected and converted into a three-dimensional image (Figure A2.2.11).

◆ **Electron microscope (EM)**: microscope in which a beam of electrons replaces light so the powers of magnification and resolution are correspondingly much greater.

◆ **Ultrastructure**: fine structure of cells, determined by electron microscopy.

Electron microscopes have a greater resolving power than light microscopes. Their application to biology has established the presence and structure of all the cell organelles.

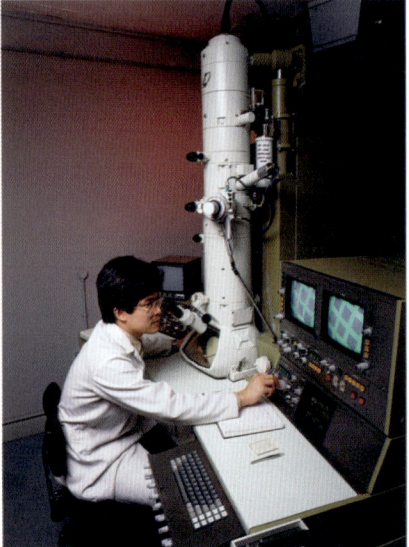

■ **Figure A2.2.10** Using the transmission electron microscope

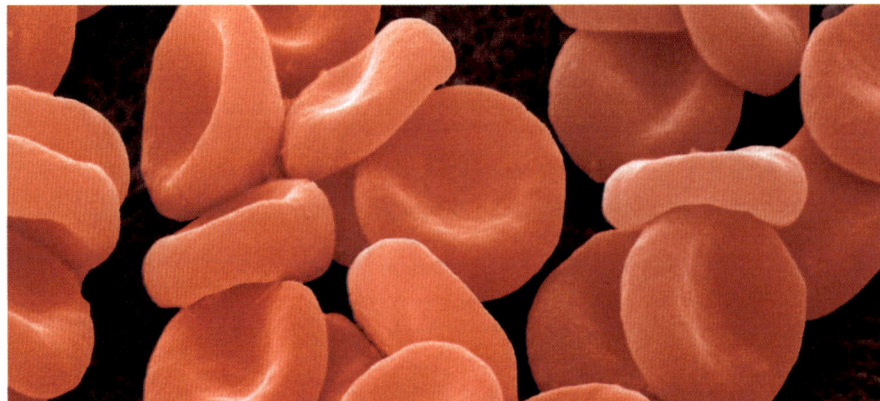

■ **Figure A2.2.11** A scanning electron micrograph of red blood cells (5.7 µm in diameter); the image from an electron microscope is in black and white – this image has been coloured

Theme A: Unity and diversity – Cells

7 **List** the features of cells that can be observed by electron microscopy that are not visible by light microscopy.

8 **State** two problems that arise in electron microscopy because of the nature of an electron in relation to the living cell.

TOK

The investigation of cell structures by observation of electron micrographs of very thin sections of tissue (after dehydration and staining) raises the issue of whether the structures observed are actually present (or are artefacts). The solution to this problem, described above, is an example of how scientific knowledge may require multiple observations assisted by technology.

The impact of electron microscopy on cell biology: the presence and structure of organelles

The nucleus is the largest substructure (organelle) of a cell and may be observed with a light microscope. However, most organelles cannot be viewed by light microscopy and none are large enough for internal details to be seen. It is by means of the electron microscope that we have learnt about the fine details of cell structure. Because electron microscopes have much greater resolving power than light microscopes, they can reveal structures that are not visible by light microscopy, such as organelles made from membranes.

ATL A2.2B

What do you think are the strengths and limitations of light microscopy compared to electron microscopy? Discuss in a group and draw up a comparison table.

Cryogenic electron microscopy

A more recent development in electron microscopy is a technique that involves flash-freezing solutions of proteins or other biomolecules and then exposing them to electrons to produce very high-resolution images of individual molecules. This technique is called **cryogenic electron microscopy**, because of the cryogenic temperatures used (Figure A2.2.12). The images are used to reconstruct the three-dimensional shape of the molecule, which in turn reveals its function. Cryogenic electron microscopy is used to reveal how proteins work, how they malfunction in disease and how to target them with drugs. The most recent cryogenic electron microscopes are able to locate individual atoms within a protein. In 2017, the Nobel Prize in chemistry was awarded to three scientists (Jacques Dubochet, Joachim Frank and Richard Henderson) for developing cryogenic electron microscopy.

Traditional electron microscopes place the specimen in a chamber that is kept under vacuum. This allows a pure beam of electrons to interact with the sample, without interference from particles that would be present in air. However, some materials, such as biological molecules, are not compatible with the high-vacuum conditions and intense electron beams used in traditional TEMs. Under the conditions of a vacuum, the water that surrounds the molecules evaporates, and the high-energy electrons destroy the molecules. Because cryogenic electron microscopes use frozen samples, less-intense electron beams can be used and the evaporation of water is no longer a problem.

Prior to cryogenic electron microscopy, scientists relied on a technique called X-ray crystallography to view biomolecules, which involves crystallizing molecules, exposing them to X-rays and then reconstructing their shape from the patterns of diffracted X-rays. This technique was used to reveal the structure of DNA (see page 20).

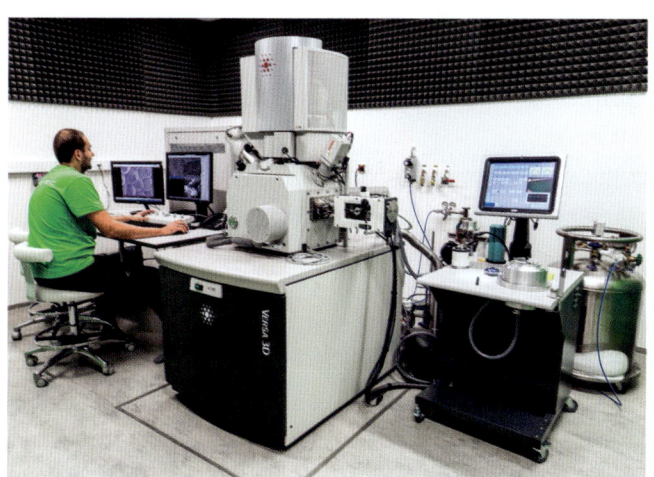

■ Figure A2.2.12 Cryogenic electron microscope

Freeze fracture

In an **alternative method of preparation**, biological material is *instantly* frozen solid in liquid nitrogen. At atmospheric pressure this liquid is at −196 °C. At this temperature living materials do not change shape as the water present in them solidifies instantly.

This solidified tissue is then broken up in a vacuum, and the exposed surfaces are allowed to lose some of their ice; the surface is described as 'etched'.

Finally, a carbon replica (a form of 'mask') of this exposed surface is made and coated with heavy metal, such as gold, to strengthen it. The mask of the surface is then examined in the electron microscope. The resulting electron micrograph is described as being produced by **freeze etching**.

A comparison of a cell nucleus observed by both transmission electron microscopy and by freeze etching is shown in Figure A2.2.13.

Look at these images carefully.

The picture we get of nucleus structure is consistent; therefore, we can be confident that our views of cell structure obtained by electron microscopy are realistic.

◆ **Freeze etching**: preparation of specimens for electron microscope examination by freezing, fracturing along natural structural lines and preparing a replica.

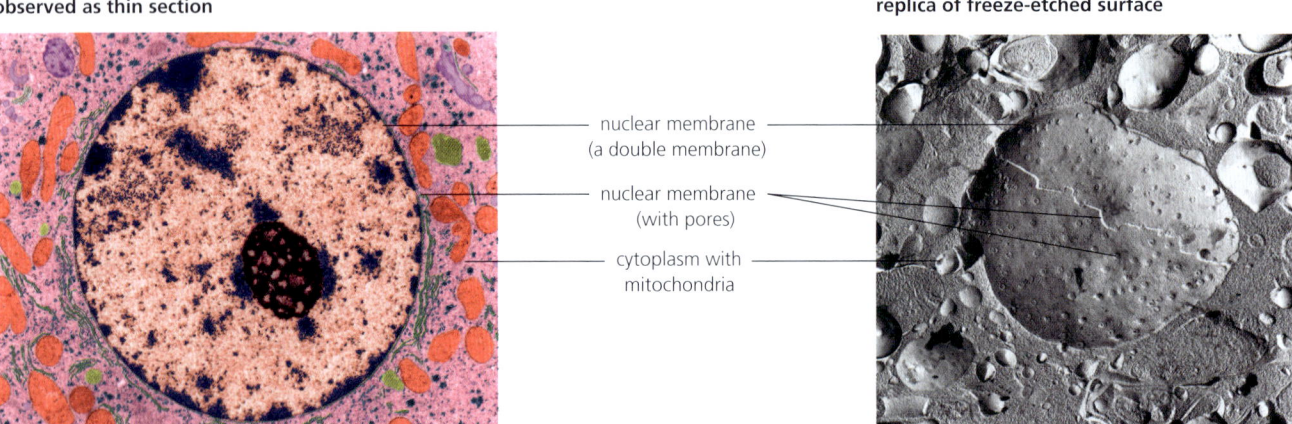

■ **Figure A2.2.13** Electron micrographs from thin-sectioned and freeze-etched material showing the nucleus of a liver cell

Fluorescence microscopy

Fluorescent dyes absorb light at one wavelength and emit it at another longer wavelength. Some such dyes bind specifically to target molecules in cells, e.g., DNA, and can reveal their cellular location when examined with a fluorescence microscope.

A fluorescence microscope is used to detect cells stained with fluorescent dyes. This is similar to an ordinary light microscope except that the illuminating light is passed through two sets of filters. The first set filters the light before it reaches the specimen, passing only those wavelengths that excite the specifically chosen fluorescent dye. The second filter blocks out this light and passes only those wavelengths emitted when the dye fluoresces. Dyed objects show up in bright colour on a dark background (Figure A2.2.14).

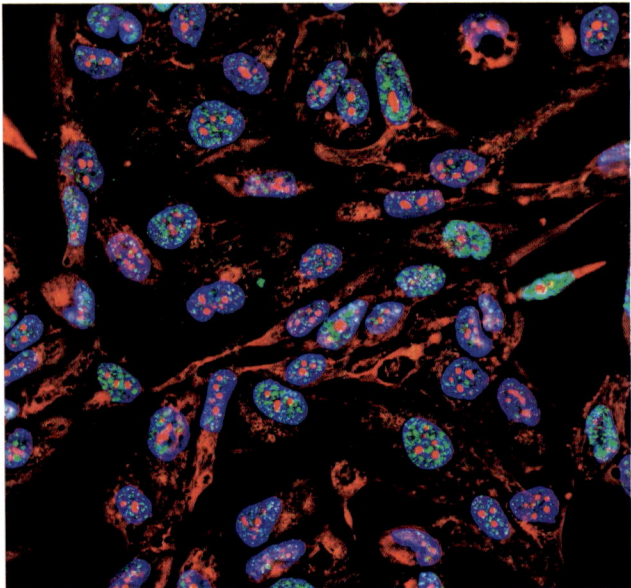

■ **Figure A2.2.14** Fluorescent imaging: immunofluorescence of cancer cells with nuclei in blue, cytoplasm in red and damaged DNA in green

Theme A: Unity and diversity – Cells

- ◆ **Immunofluorescence**: a technique for determining the location of target cells by reaction with an antibody labelled with a fluorescent dye.
- ◆ **Nucleus**: largest organelle of eukaryotic cells; controls and directs the activity of the cell.
- ◆ **Cytoplasm**: fluid that fills each cell and is enclosed by the plasma membrane.
- ◆ **Plasma membrane**: the membrane of lipid and protein that forms the surface of cells; constructed as a fluid mosaic membrane.
- ◆ **Differentiation**: process by which cells become specialized, when some genes and not others are expressed in a cell's genome.

Other dyes can be coupled to antibody molecules, which then act as highly specific and versatile staining reagents that bind selectively to targeted biomolecules (for example, proteins or DNA), showing their distribution in the cell. Such techniques were used to explain the structure of cell plasma membranes, and the mechanism by which muscles contract.

Immunofluorescence uses antibodies chemically labelled with fluorescent dyes to visualize molecules under a light microscope. The antibodies attach to affected cells and the dyes indicate which cells have been marked in this way. Fluorescence microscopy has allowed doctors to make a diagnosis on whether cancer is present based on small samples of tissue (Figure A2.2.14).

Structures common to cells in all living organisms

■ The features of cells

A cell consists of a **nucleus** surrounded by **cytoplasm**, composed mainly of water, contained within the cell membrane made of lipids. The nucleus is the structure that controls and directs the activities of the cell. It contains DNA, which is the hereditary material of all living organisms (see page 44). The cytoplasm provides the medium in which many of the metabolic reactions of the cell occur, with water being the essential molecule needed for enzymes to function at their optimum (see page 210). It is the site of the enzyme-controlled chemical reactions of life, which we call 'metabolism'. The cell membrane, known as the **plasma membrane**, is the barrier controlling entry to and exit from the cytoplasm.

Newly formed cells grow and enlarge. A growing cell can normally divide into two cells. Cell division is very often restricted to unspecialized cells before they become modified for a particular task – a process known as **differentiation**.

> **Concept: Diversity**
>
> Because of specialization, cells show great variety in shape, structure and function. This variety in structure reflects the evolutionary adaptations of cells to different environments and to different specialized functions – for example, within multicellular organisms.

■ Animal and plant cells

No 'typical' cell exists – there is a very great deal of variety among cells. However, we shall see that most cells have features in common. Viewed using a compound microscope, the appearance of a plant cell and animal cell are shown in Figure A2.2.15. Both are bound by a membrane and contain a nucleus. The plant cell has additional organelles that are visible (chloroplasts and a permanent vacuole).

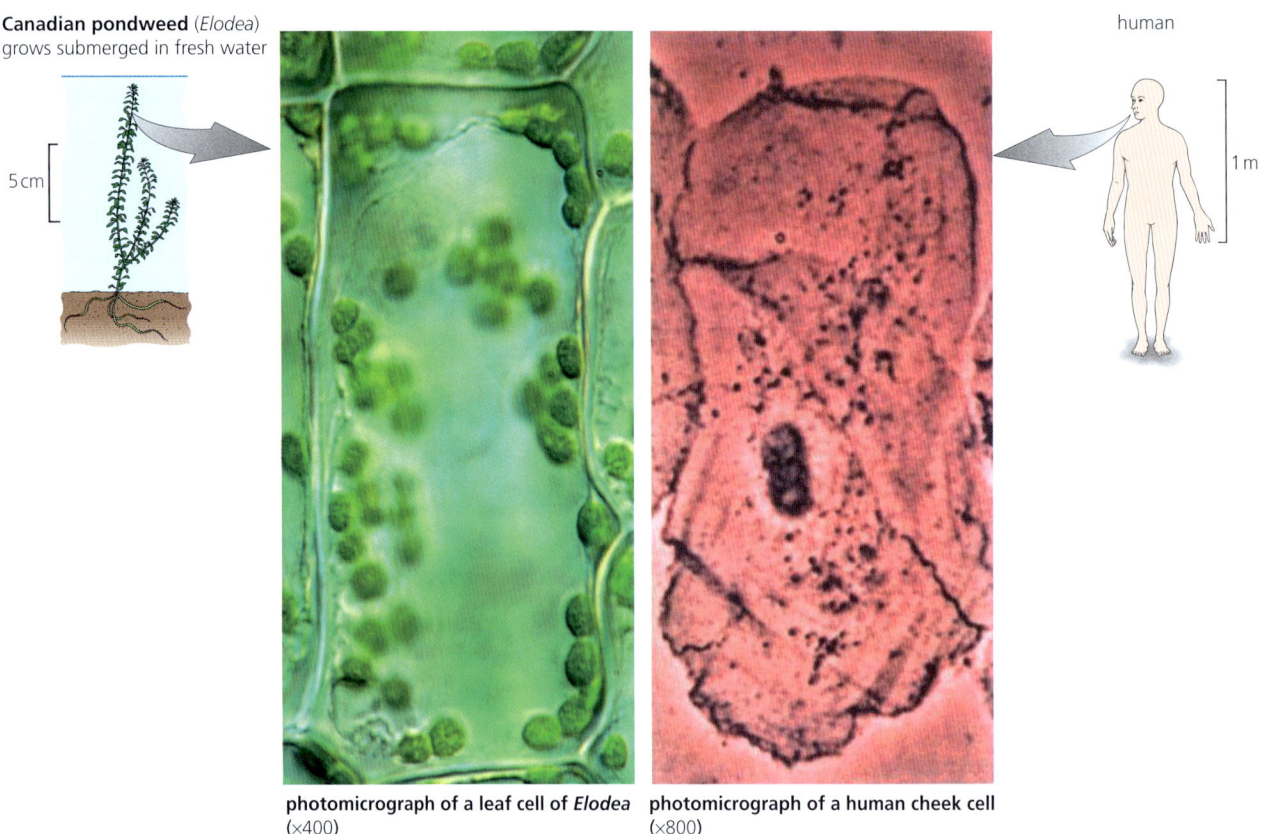

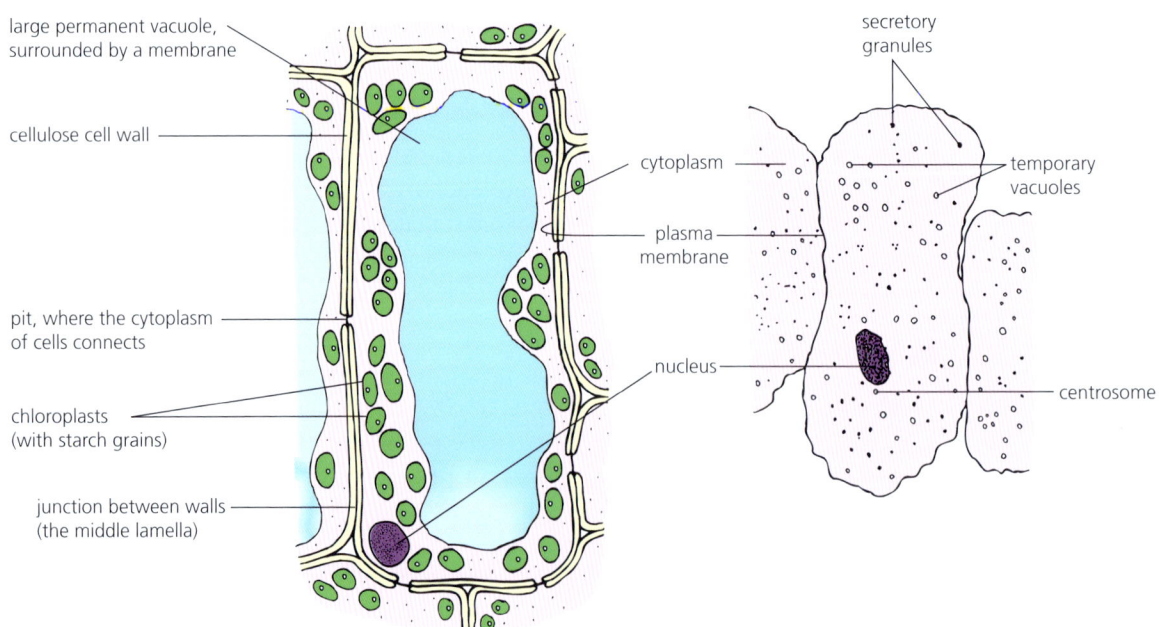

■ **Figure A2.2.15** Animal and plant cells with interpretive drawings to indicate the structure and function of organelles

◆ **Organelle**: a unit of cell substructure.

◆ **Eukaryote**: group of organisms that have a nuclear membrane surrounding their genetic material, i.e., they have a true nucleus.

◆ **Prokaryote**: unicellular organism without a true nucleus; they have a loop of naked DNA.

Animal and plant cells have at least three structures in common. These are their cytoplasm with its nucleus, surrounded by a plasma membrane. In addition, there are many tiny structures in the cytoplasm, called **organelles**, most of them common to both animal and plant cells. An organelle is a discrete structure within a cell, having a specific function. Some organelles are too small or thin to be seen at this magnification. We will learn about the structure of organelles using the electron microscope (page 60).

There are some important basic differences between plant and animal cells, which will be explored later (see page 68 in this section).

Cells of plants, animals, fungi and protoctista have cells with a large, obvious nucleus (although there are exceptions to this – see page 78). The surrounding cytoplasm contains many different membranous organelles. These types of cells are called **eukaryotic** cells (meaning a 'true nucleus').

On the other hand, bacteria contain no true nucleus and their cytoplasm does not have the organelles of eukaryotes. These are called **prokaryotic** cells (meaning 'before the nucleus').

This distinction between prokaryotic and eukaryotic cells is a fundamental division and is more significant than the differences between plants and animals. We will first examine the detailed structure of the prokaryotic cell.

> ### Concept: Unity
>
> Although there are a large variety of different cell types, all cells can be divided into two distinctive groups: eukaryotes (cell that have a nucleus) and prokaryotes (cells that do not have a nucleus). Each cell type has distinctive features within the group; there are also similarities that all cells share (e.g. DNA, a plasma membrane, cytoplasm and ribosomes).

Prokaryote cell structure

We have seen that the use of the electron microscope in biology led to the discovery of eukaryotic and prokaryotic cell structure (page 60). Bacteria and cyanobacteria (a group of bacteria that can photosynthesize) are prokaryotes.

The generalized structure of a bacterium is shown in Figure A2.2.16. The prokaryote cell structure varies (as shown in the figure), but the distinctive features of the prokaryotes are:
- they are exceedingly small – about the size of individual organelles found in the cells of eukaryotes
- they contain no true nucleus but have a single loop of DNA in the cytoplasm, referred to as a nucleoid
- their cytoplasm does not have the membrane-bound organelles of eukaryotes.

Link
In HL Biology, you need to know the differences between the two domains of bacteria. See Chapter A3.2, page 137.

Top tip!

There are two broad groups, or domains, of prokaryotes – eubacteria and archaea. In this section, and when the term 'bacteria' is used, the group being referred to is the eubacteria.

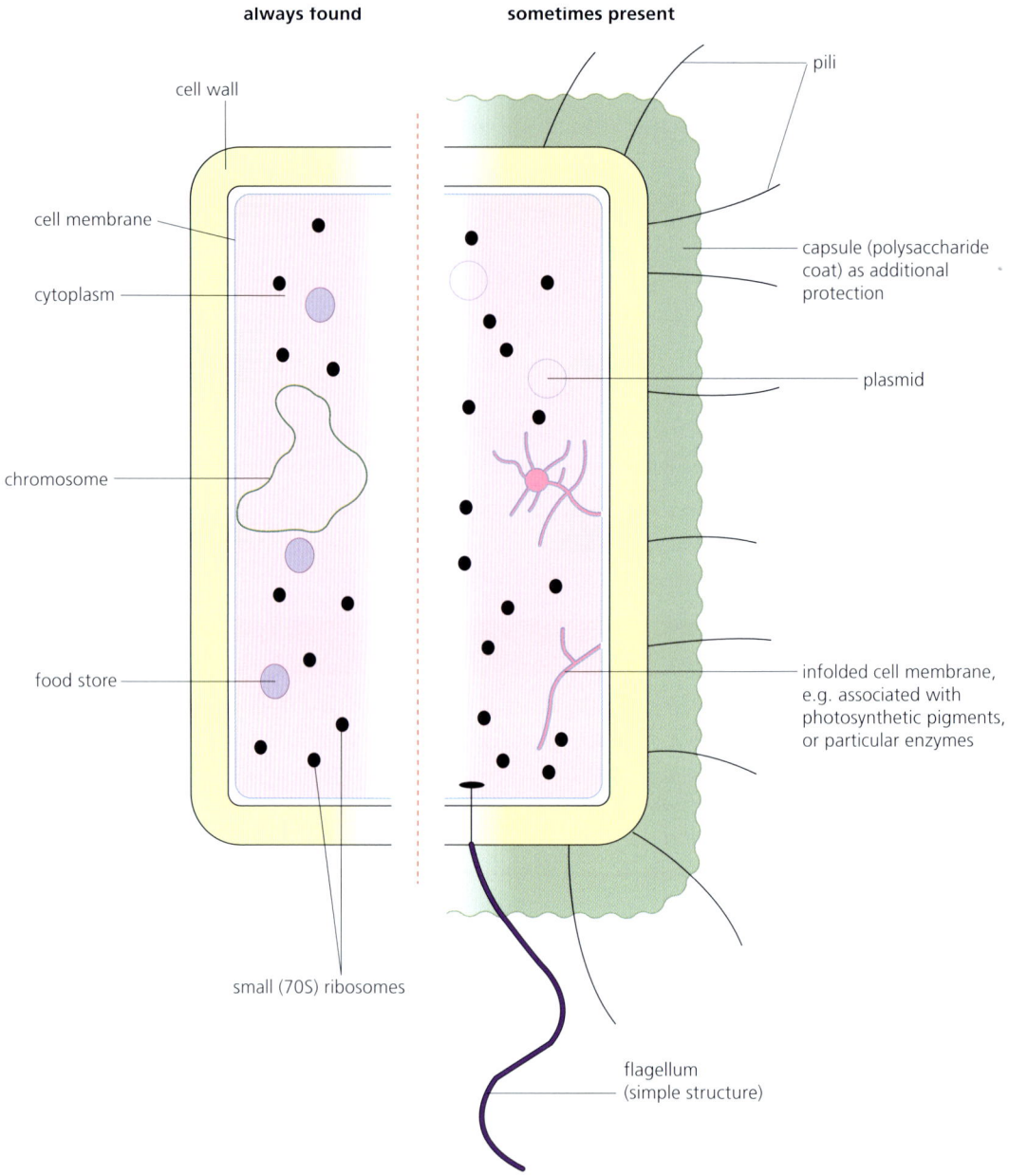

■ **Figure A2.2.16** Generalized structure of a bacterium

In Figure A2.2.17, the ultrastructure of *Escherichia coli* is shown. *E. coli* is a common bacterium that occurs in huge numbers in the lower intestine of humans and other endothermic (once known as 'warm-blooded') vertebrates, such as mammals. It is a major component of the faeces of these animals.

This tiny organism was named by a bacteriologist, Professor T. Escherich, in 1885. Notice the scale bar in Figure A2.2.17. This bacterium is typically about 1–3 μm in length – about the size of a mitochondrion in a eukaryotic cell.

All bacteria contain the following cell components:
- a cell wall (made from peptidoglycan – a polymer made of polysaccharide and peptide chains)
- plasma membrane
- cytoplasm
- naked DNA (i.e. not associated with histone proteins – see page 26) in a loop
- 70S ribosomes.

9 **Calculate** the approximate magnification of the image of *E. coli* in Figure A2.2.17.

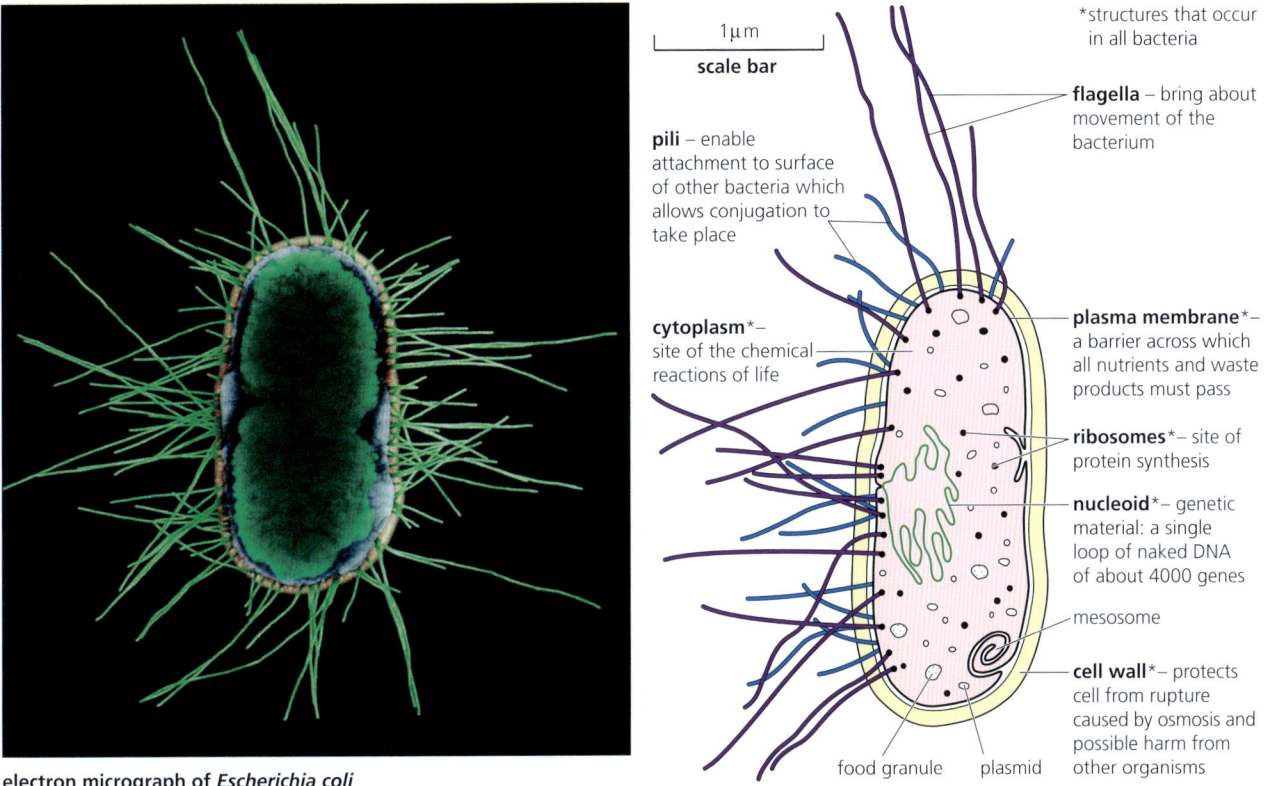

Figure A2.2.17 The structure of *Escherichia coli*, together with an interpretive drawing

Ribosomes are organelles where protein synthesis takes place. There are two different sizes of ribosomes found in cells – smaller 70S ribosomes are found in bacteria and larger 80S ribosomes are found in eukaryotic cells (see page 69).

Unlike eukaryotic chromosomes, which are linear (page 110) with many contained in each cell, prokaryotic chromosomes are singular and form a loop within the nucleoid region of the cell (Figures A2.2.16 and A2.2.17).

> ### ATL A2.2C
>
> Species of bacteria can be divided into two broad groups – gram-positive and gram-negative bacteria. Find out the differences between these two groups. How are these differences determined, and why is it important for scientists to know the differences?
>
> Gram-positive eubacteria include *Bacillus* and *Staphylococcus*. Find out the structure of these two species. How do they differ from the bacterium shown in Figure A2.2.17?

Eukaryote cell structure

The chemicals in the cytoplasm are substances formed and used in the chemical reactions of life. All the reactions of life are known collectively as **metabolism**, and the chemicals are known as **metabolites**.

Cytoplasm and organelles are contained within the plasma membrane. This membrane is clearly a barrier of sorts. It must be crossed by all the metabolites that move between the cytoplasm and the environment of the cell. We will return to the structure of the cell membrane and how molecules enter and leave cells in Theme B2.

◆ **Metabolism**: integrated network of all the biochemical reactions of life.

◆ **Metabolite**: a chemical substance involved in metabolism.

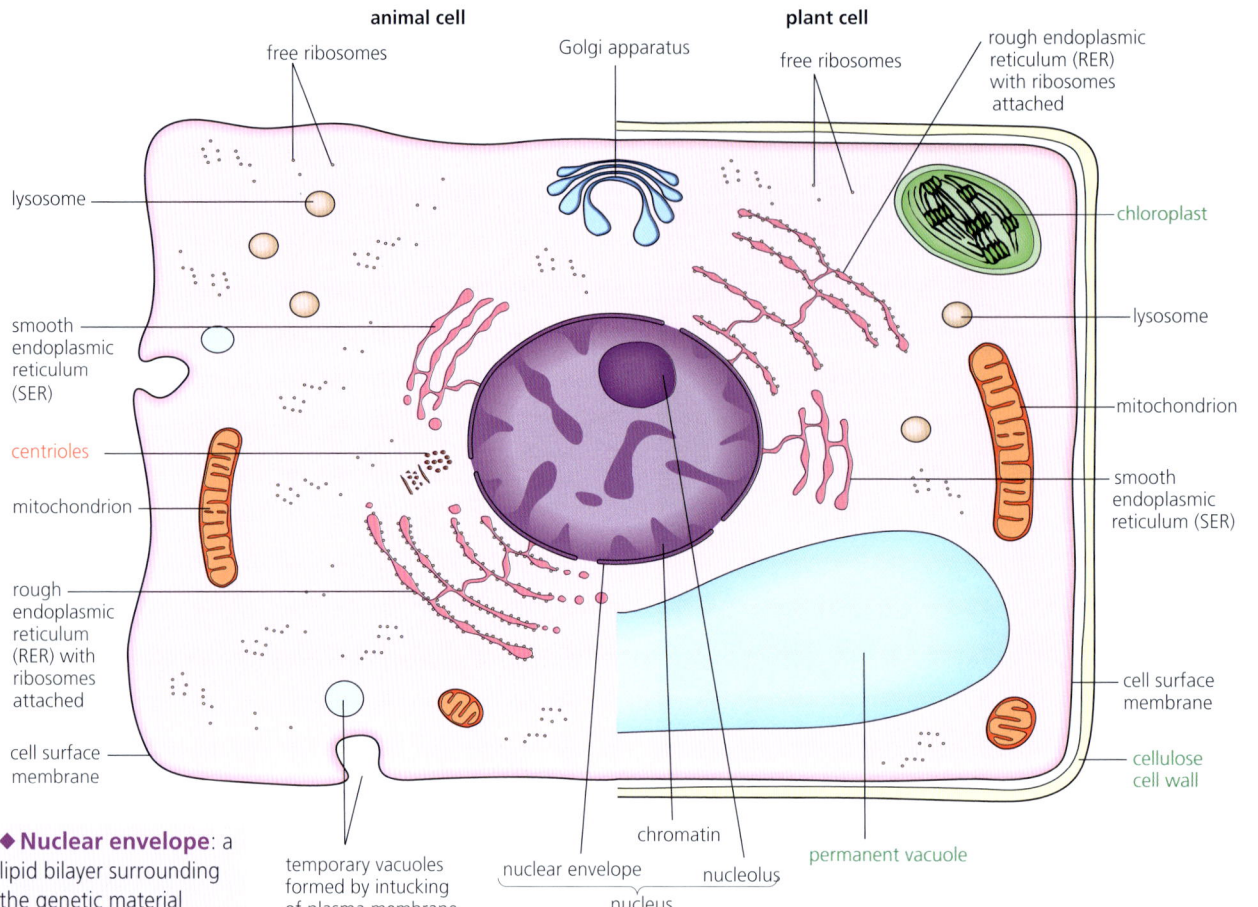

■ **Figure A2.2.18** The ultrastructure of the eukaryotic animal and plant cell

◆ **Nuclear envelope**: a lipid bilayer surrounding the genetic material of the cell, containing nuclear pores that control the movement of molecules between the inside of the nucleus and the cytoplasm.

◆ **Nuclear pore**: a protein-lined channel in the nuclear envelope; regulates the transport of molecules between the nucleus and the cytoplasm.

◆ **Chromosome**: length of DNA that carries specific genes in a linear sequence.

◆ **Chromatin**: nuclear material comprised of DNA and histone proteins in the nucleus of eukaryotic cells at interphase; forms into chromosomes during mitosis and meiosis.

We next consider the structure and function of the organelles.

Our knowledge of organelles has been built up by examining electron micrographs of many different cells. The outcome, a detailed picture of the ultrastructure of animal and plant cells, is represented diagrammatically in a generalized cell in Figure A2.2.18.

■ Introducing the organelles present in all eukaryotic cells

Nucleus

The appearance of the **nucleus** in electron micrographs is shown in Figure A2.2.13 (page 62). The nucleus is the largest organelle in the eukaryotic cell, typically 10–20 μm in diameter. It is surrounded by a double-layered membrane, the **nuclear envelope**. This contains many **nuclear pores** formed by specific proteins. These pores are tiny, about 100 nm in diameter. However, the pores are so numerous that they make up about one-third of the nuclear membrane's surface area. This suggests that communications between nucleus and cytoplasm are important.

The nucleus contains the **chromosomes**. These thread-like structures are visible at the time the nucleus divides (page 651). Chromosomes are made of DNA bound to **histone** proteins (page 26). Histones help to pack the DNA into condensed structures so that they can be moved within the cell during division. At other times, the chromosomes appear as a diffuse network called **chromatin**.

Theme A: Unity and diversity – Cells

Common mistake

Do not confuse the terms 'nucleus' and 'nucleolus'. The nucleolus is a sub-organelle located inside the nucleus, whereas the nucleus is a membrane-bound organelle in the cell. The nucleolus produces ribosomes, and the nucleus contains the genetic material of the cell.

◆ **Nucleolus**: compact region of nucleus where ribosomes are synthesized.

◆ **Mitochondrion (plural, mitochondria)**: organelle in eukaryotic cells, site of Krebs cycle and the electron-transport pathway.

◆ **Cristae**: folds in the inner membrane of mitochondria.

◆ **Matrix**: the fluid that is surrounded by the inner membrane of a mitochondrion, containing enzymes, ribosomes and mitochondrial DNA.

Common mistake

A common misconception is to regard a chromosome solely as a condensed structure, visible with a light microscope, implying that chromosomes are not present unless a cell is carrying out nuclear division. Chromosomes are continuously present in the nucleus of a eukaryotic cell; the uncondensed state should be seen as normal as this form persists longest and is when the genes can be transcribed into proteins.

A **nucleolus** (plural, nucleoli) is a tiny, rounded, darkly-staining body. One or more **nucleoli** are present in the nucleus. It is the site where the sub-units of ribosomes (see below) are synthesized. Chromatin, chromosomes and the nucleolus are visible only if stained with certain dyes. The everyday role of the nucleus in cell management, and its behaviour when the cell divides, are the subject of Chapter D2.1 (page 648).

Most cells contain one nucleus but there are interesting exceptions. For example, both the red blood cells of mammals (page 78) and the sieve tube element of the phloem of flowering plants (page 79) do not have a nucleus. Both lose their nucleus as they mature.

Mitochondria

Mitochondria appear mostly as rod-shaped or cylindrical organelles in electron micrographs (Figure A2.2.19). Occasionally their shape is more variable. They are relatively large organelles, typically 0.5–1.5 µm wide and 3.0–10.0 µm long. Mitochondria are found in all cells and are usually present in very large numbers. Very metabolically active cells contain thousands of them in their cytoplasm, for example, muscle fibres and hormone-secreting cells. Human liver cells can individually have 2000 mitochondria.

The mitochondrion also has a double membrane. The outer membrane is a smooth boundary, while the inner membrane is folded to form **cristae**. The interior of the mitochondrion contains an aqueous solution of metabolites and enzymes. This is called the **matrix**. The mitochondrion is the site of the aerobic stages of respiration (see Chapter C1.2, page 416).

a mitochondrion, cut open to show the inner membrane and cristae

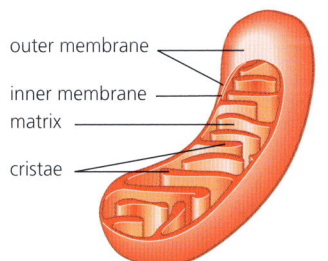

outer membrane
inner membrane
matrix
cristae

In the mitochondrion many of the enzymes of respiration are housed, and the 'energy currency' molecules adenosine triphosphate (ATP) are formed.

TEM of a thin section of a mitochondrion

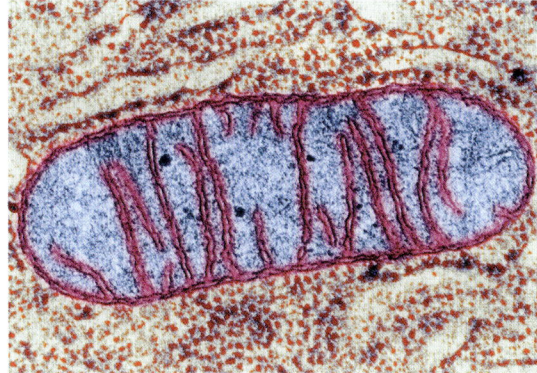

■ Figure A2.2.19 The mitochondrion

Ribosomes

Ribosomes are tiny structures, approximately 25 nm in diameter. They are built of two subunits and do not have membranes as part of their structures. Chemically, they consist of protein and the nucleic acid RNA. Ribosomes are found free in the cytoplasm and bound to endoplasmic reticulum (rough endoplasmic reticulum – RER, see below). They also occur within mitochondria and in chloroplasts. The sizes of tiny objects like ribosomes are recorded in Svedberg units (S). This is a measure of their rate of sedimentation in centrifugation, rather than of their actual size.

small subunit

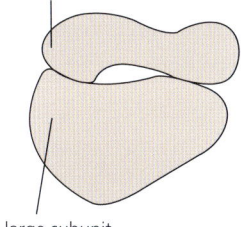

large subunit

■ Figure A2.2.20 The ribosome

10 **Explain** why the nucleus in a human cheek cell (Figure A2.2.15, page 64) may be viewed by light microscopy in an appropriately stained cell, but the ribosomes cannot.

Ribosomes of mitochondria and chloroplasts are slightly smaller (70S) than those in the rest of the cell (80S). As we have already seen, prokaryotes contain 70S ribosomes (page 66).

Ribosomes are the sites where proteins are made in cells. The structure of a ribosome is shown in Figure A2.2.20. Many different types of cell contain vast numbers of ribosomes. Some of the cell proteins produced in the ribosomes have structural roles. Collagen is an example (Chapter B1.2, page 219). Most cell proteins are enzymes. These are biological catalysts: they cause the reactions of metabolism to occur more quickly under the conditions found within the cytoplasm.

Endoplasmic reticulum

The **endoplasmic reticulum** consists of a network of folded membranes formed into sheets, tubes or sacs that are extensively interconnected. Endoplasmic reticulum is contiguous with (connected to) the membrane of the nuclear envelope (Figure A2.2.21). The cytoplasm of metabolically active cells is commonly packed with endoplasmic reticulum. In Figure A2.2.21 we can see there are two distinct types of endoplasmic reticulum.

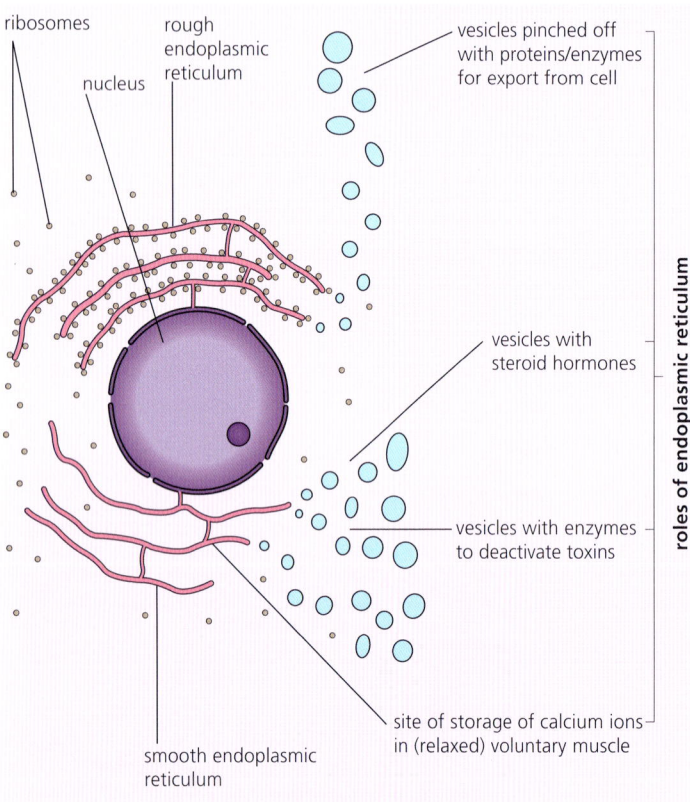
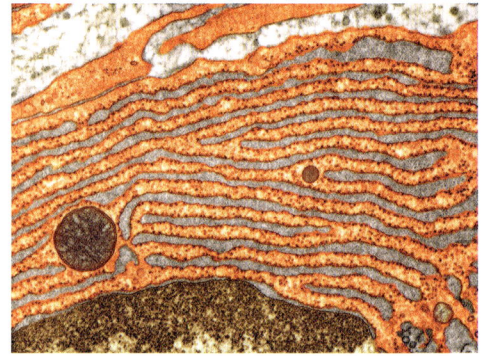
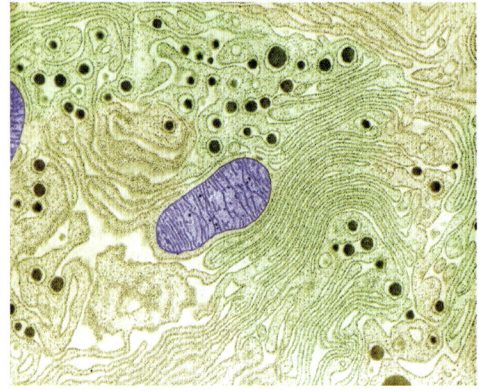

■ Figure A2.2.21 Endoplasmic reticulum, rough (RER) and smooth (SER)

◆ **Endoplasmic reticulum**: system of branching membranes in the cytoplasm of eukaryotic cells, existing as rough ER (with ribosomes) or as smooth ER (without ribosomes).

- **Rough endoplasmic reticulum (RER)** has ribosomes attached. At its margin, vesicles are formed from swellings. A vesicle is a small, spherical organelle bounded by a single membrane, which becomes pinched off as it separates. These tiny sacs are then used to store and transport substances around the cell. For example, the RER is the site of synthesis of proteins that are 'packaged' in the vesicles and then typically discharged from the cell by a process called exocytosis. Digestive enzymes are discharged in this way.
- **Smooth endoplasmic reticulum (SER)** has no ribosomes. SER is the site of synthesis of substances needed by cells. For example, the SER is important in the manufacture of lipids. In the cytoplasm of voluntary muscle fibres, a special form of SER is the site of storage of calcium ions, which have an important role in the contraction of muscle fibres.

Golgi apparatus

◆ **Golgi apparatus**: a stack of flattened membranes in the cytoplasm involved in the processing, modifying and packaging of molecules.

The **Golgi apparatus** consists of a stack-like collection of flattened membranous sacs (Figure A2.2.22). One side of the stack of membranes is formed by the fusion of membranes of vesicles from the SER and RER. At the opposite side of the stack, vesicles are formed from swellings at the margins that, again, become pinched off.

The Golgi apparatus occurs in all cells, but it is especially prominent in metabolically active cells – for example, secretory cells. It is the site of synthesis of specific biochemicals, such as hormones and enzymes. These are then packaged into vesicles.

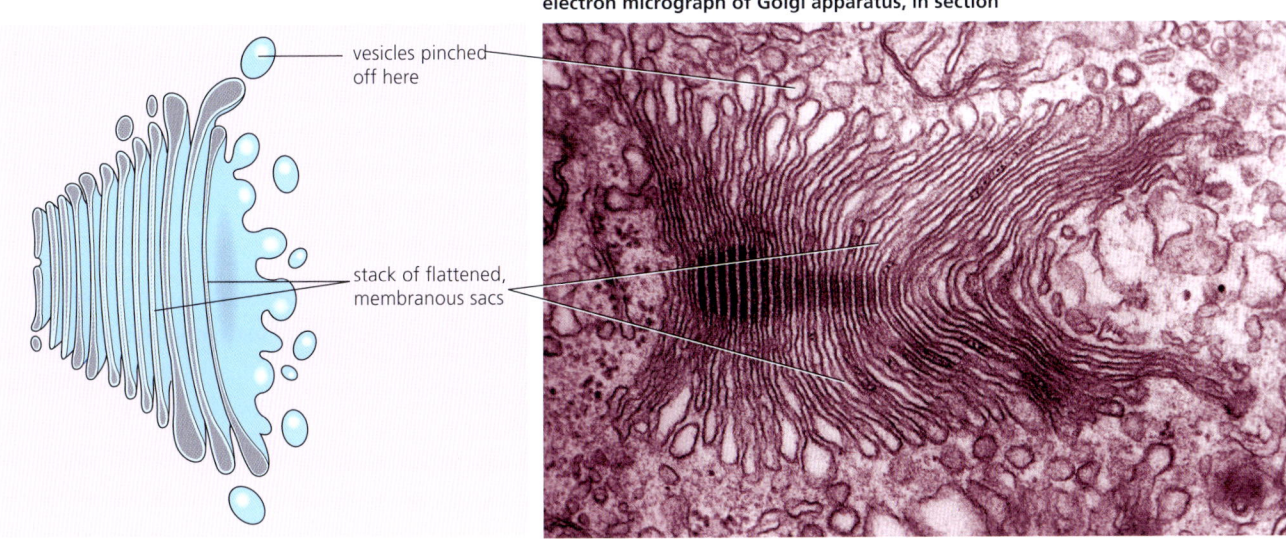

■ Figure A2.2.22 The Golgi apparatus

Lysosomes

◆ **Lysosomes**: membrane-bound vesicles, common in the cytoplasm, containing digestive enzymes.

◆ **Non-permanent vacuoles**: small vacuoles in animal cells used to temporarily store materials or to transport substances.

The Golgi apparatus produces a variety of different vesicles. In animal cells these vesicles may form **lysosomes**. Some vesicles develop into **non-permanent vacuoles**, which are small, temporary vacuoles that help sequester waste products.

Lysosomes are tiny spherical vesicles bound by a single membrane (Figure A2.2.23). They contain a concentrated mixture of digestive enzymes. These are correctly known as hydrolytic enzymes. They are produced in the Golgi apparatus or by the RER.

Lysosomes are involved in the breakdown of the contents of 'food' vacuoles. For example, harmful bacteria that invade the body are taken up into tiny vacuoles (they are engulfed) by special white blood cells called macrophages. Macrophages are part of the body's defence system (Chapter C3.2, page 526).

Any foreign matter or food particles taken up into these vacuoles are then broken down. This occurs when lysosomes fuse with the vacuole. The products of digestion then escape into the liquid of the cytoplasm. Lysosomes will also destroy damaged organelles in this way.

When an organism dies, it is the hydrolytic enzymes in the lysosomes of the cells that escape into the cytoplasm and cause self-digestion, known as autolysis. Lysosomes are also involved in a process called apoptosis, or 'programmed cell death', which occurs in cells damaged by infection or mutation.

11 **Outline** the roles of lysosomes, the Golgi apparatus and the endoplasmic reticulum in the eukaryotic cell.

12 **Outline** how the electron microscope has increased our knowledge of cell structure.

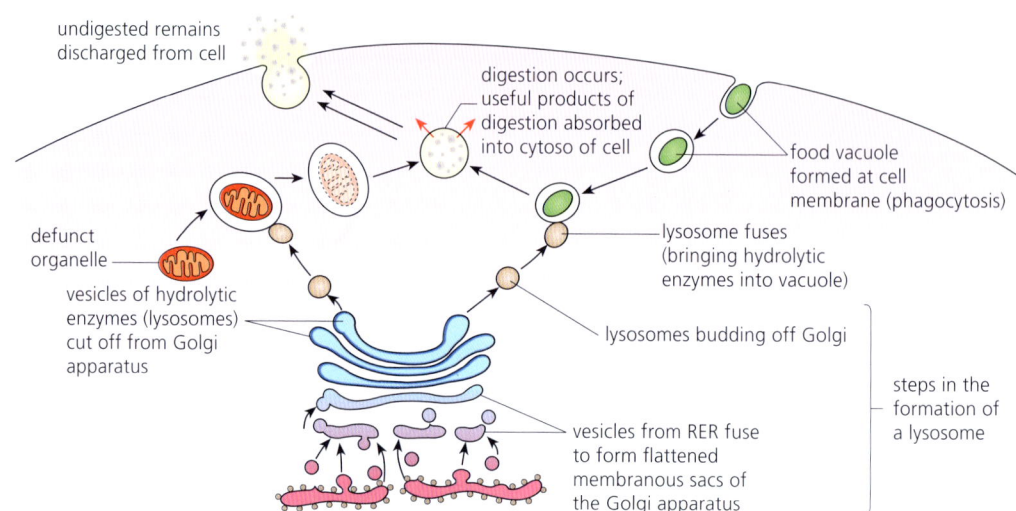

■ Figure A2.2.23 Lysosomes

Plasma membrane – the cell surface membrane

The plasma membrane is an extremely thin structure – 7 nm thick. It consists of a lipid bilayer in which proteins are embedded. This membrane has a number of roles. Firstly, it retains the cytoplasm. The cell surface membrane also forms the barrier across which all substances entering and leaving the cell must pass. In addition, it is where the cell is identified by surrounding cells.

The detailed structure and function of the cell surface membrane is the subject of Chapter B2.1 (page 223).

Cytoskeleton

The **cytoskeleton** is a network of fibres extending throughout the cytoplasm. It organises the structures and activities within the cell. There are three main types of fibre in the cytoskeleton: microtubules, microfilaments and intermediate filaments.

Microtubules are the thickest class of the cytoskeletal fibres and are straight, unbranched, hollow cylindrical fibres (Figure A2.2.24) about 25 nm in diameter. The cells of all eukaryotes, whether plants or animals, have a well-organized system of these microtubules that shape and support the cytoplasm. The network of microtubules is made of a helically arranged globular protein called **tubulin**. This is built up and broken down in the cell as the microtubule framework is required in different places for different tasks.

◆ **Cytoskeleton**: a microscopic network of protein filaments and tubules in the cytoplasm that give cells shape and are used for internal organization.

◆ **Microtubule**: tiny, hollow protein tube in cytoplasm (e.g. a component of the spindle).

What are the functions of microtubules?

The movement of chromosomes during cell division is achieved by the lengthening and shortening of microtubules. Microtubules are involved in movement of other cell components within the cytoplasm too, acting to guide and direct them.

The movement of organelles is facilitated by motor proteins (Figure A2.2.25) carrying organelles along the microtubules to their destination. For example, microtubules guide secretory vesicles from the Golgi apparatus to the plasma membrane.

It provides mechanical support and maintains the shape of the cell. It is also dynamic and can be quickly dismantled and reassembled at another part of the cell, changing the shape of the cell.

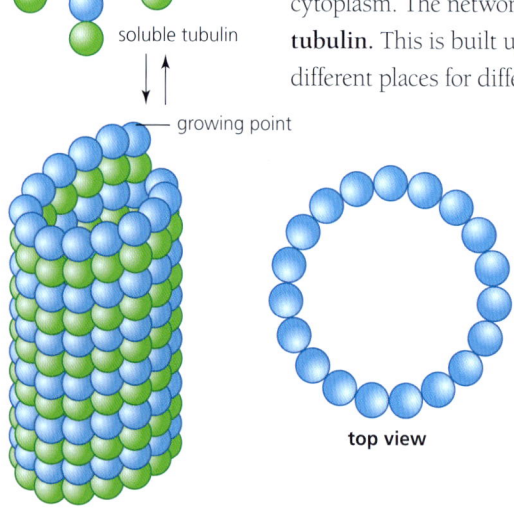

■ Figure A2.2.24 Structure of microtubules

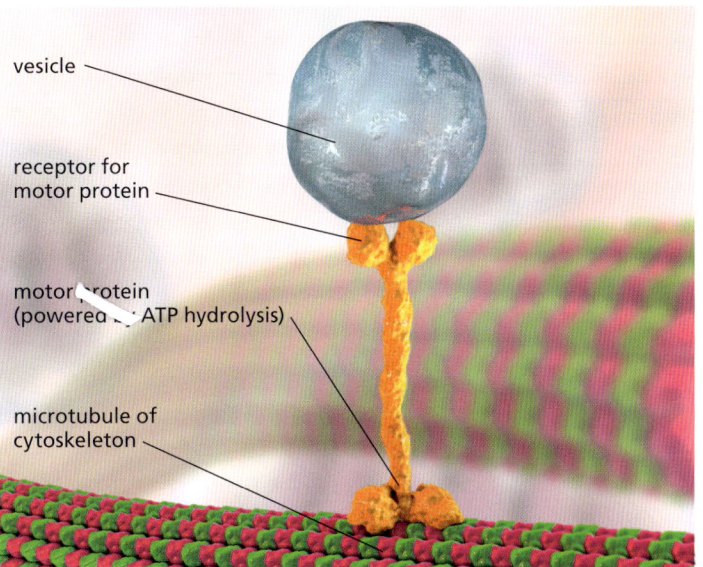

■ **Figure A2.2.25** Computer illustration of a microtubule with motor proteins

◆ **Microfilament**: the thinnest class of the cytoskeletal fibres, made of solid rods of globular protein called actin.

Microfilaments (Figure A2.2.26) are the thinnest class of the cytoskeletal fibres, about 7 nm in diameter, and made up of solid rods of globular protein called actin. Each filament consists of a twisted double chain of actin molecules. Microfilaments are designed to resist tension.

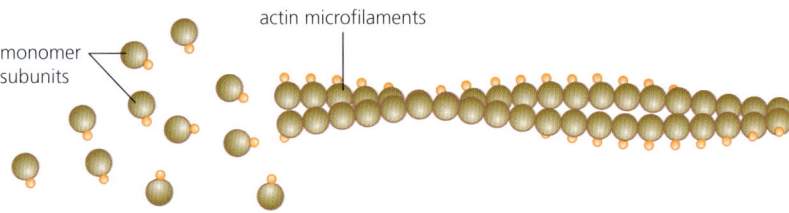

■ **Figure A2.2.26** Structure and assembly of microfilaments

The functions of microfilaments are:
- They are involved in cleavage furrow formation to divide the cell during cell division.
- They play a part in cell motility, particularly in muscle contraction.
- They maintain and change cell shape.

● **Top tip!**

In plant cells, the flow of cytoplasm over actin filaments is important for the distribution of chemical substances within the cell. This cytoplasmic streaming is due to actin–myosin interactions.

13 Distinguish between the structure of prokaryotic and eukaryotic cells.

14 Construct a table to **distinguish** between the chromosomes of prokaryotic and eukaryotic cells.

Processes of life in unicellular organisms

Unicellular and multicellular organisms share the same functions of life. The functions of life are:

- **homeostasis**: maintaining a constant, stable internal environment
- **metabolism**: the complex network of interdependent and interacting chemical reactions occurring in living organisms
- **nutrition**: process by which organisms take in and/or make use of food//nutrients
- **movement**: moving from one place to another
- **excretion**: elimination of waste created by metabolic processes inside cells
- **growth**: increase in size, mass or number of cells within an organism
- **response to stimuli (sensitivity)**: reaction to changes in the environment
- **reproduction**: formation of new individuals by sexual or asexual means.

Concept: Unity

All organisms, whether unicellular and multicellular, share the same life processes.

The functions of life can be demonstrated in unicellular organisms as shown in Table A2.2.4.

■ **Table A2.2.4** Life processes in unicellular organisms

Paramecium – a large protozoan (about 600 μm), common in freshwater ponds		*Chlorella* – a small alga (about 20 μm), abundant in freshwater ponds where its presence colours the water green
■ Figure A2.2.27a *Paramecium*		■ Figure A2.2.27b *Chlorella*
Contractile vacuoles are used to expel water from the cell that has entered by osmosis. This maintains water levels within closely controlled limits.	**Homeostasis**	Water enters the cell by osmosis and is collected in contractile vacuoles. These vacuoles expel the water through the plasma membrane so that internal water levels are kept within acceptable limits.
Obtains the biochemicals it requires for metabolism by digestion of food particles. Energy transferred by respiration makes this possible. Respires aerobically, transferring energy to maintain cell functions.	**Metabolism (including respiration)**	Manufactures all biochemicals it requires for metabolism using sugars from photosynthesis and ions (such as nitrates) from the surrounding water. Energy transferred by respiration makes this possible. Respires aerobically, transferring energy to maintain cell functions.
A 'particle feeder', it takes in small floating unicellular organisms into food vacuoles in the cytoplasm where the contents are digested and the products absorbed.	**Nutrition**	Synthesises sugars by photosynthesis in the light, using carbon dioxide and water (in a way that is almost identical to photosynthesis in flowering plants).
Loss of waste products (mainly CO_2 and NH_3) over the entire cell surface.	**Excretion**	Loss of waste products (mainly CO_2) over the entire cell surface.
Commonly, reproduction occurs by nuclear division followed by a transverse constriction of the cytoplasm.	**Reproduction**	Periodically the cell contents divide into four autospores that each form a cell wall around themselves. Eventually these are released by breakdown of the mother-cell wall.
Moves rapidly through the water, rotating as it goes. Food vacuoles within can be seen being carried around the cytoplasm.	**Movement**	A stationary cell. Cytoplasm streaming allows redistribution of biological molecules within the cell.
Typically detects favourable food particles in the water and moves towards them.	**Sensitivity**	Typically responds to the absence of light by nuclear division followed by cell division.
Small cells grow to full size prior to cell division (dividing into two cells).	**Growth/development**	Small cells grow to full size prior to cell division into autospores.

> **Concept: Diversity**
>
> Although some cellular structures are common to all eukaryotic cells, there are structures that are only found in particular types of cell, which relate to functional differences between different groups of organisms (kingdoms).

> **ATL A2.2D**
>
> 1. Put a drop of pond water on to a slide. Look at micro-organisms (microscopic living unicellular organisms) under a microscope.
> 2. Find *Euglena*. If it is moving too fast, use a slide with a dimple in it (a dip in the middle) – this will restrict its movement to one part of the slide.
> 3. If you don't have access to pond water containing *Euglena*, access the following website: **https://sketchfab.com/3d-models/euglena-17b2a7d702834a759c42f4927109d74a** or use the search terms: pond life, microscope.
> 4. What organelles (structures in the cell) can you see? Do the features belong to the animal kingdom? Do the features belong to the plant kingdom? How would you classify this organism? Use scientific reasoning to explain what you have observed.
> 5. Draw and label a diagram of *Euglena*, using your interpretation of the organism you have seen under the microscope. Use scientific reasoning to deduce the functions of the structures you have observed, and then annotate the diagram to explain these features.

Differences in eukaryotic cell structure between animals, fungi and plants

As well as the organelles that occur in all eukaryotic cells, there are structures that are only found in particular types of cell.

■ Structures typically found in animal cells (and some protists)

Animal cells have organelles not present in plant cells. These will be explored now.

Centrioles

A **centriole** is a tiny organelle consisting of nine sets of three **microtubules** each (see page 72), arranged in a short, hollow cylinder. In animal cells, two centrioles occur at right-angles, just outside the nucleus, forming the **centrosome** (which can be seen in Figure A2.2.28). Before an animal cell divides, the centrioles replicate, and their role is to grow the spindle fibres – the **spindle** is the structure responsible for the movement of chromosomes during nuclear division.

◆ **Centriole**: a small cylindrical organelle which occurs in pairs; located near the nucleus in animal cells; involved in the production of spindle fibres during cell division.

◆ **Centrosome**: organelle situated near the nucleus in animal cells, involved in the formation of the spindle prior to nuclear division.

◆ **Spindle**: structure formed from microtubules, associated with the movements of chromosomes in mitosis and meiosis.

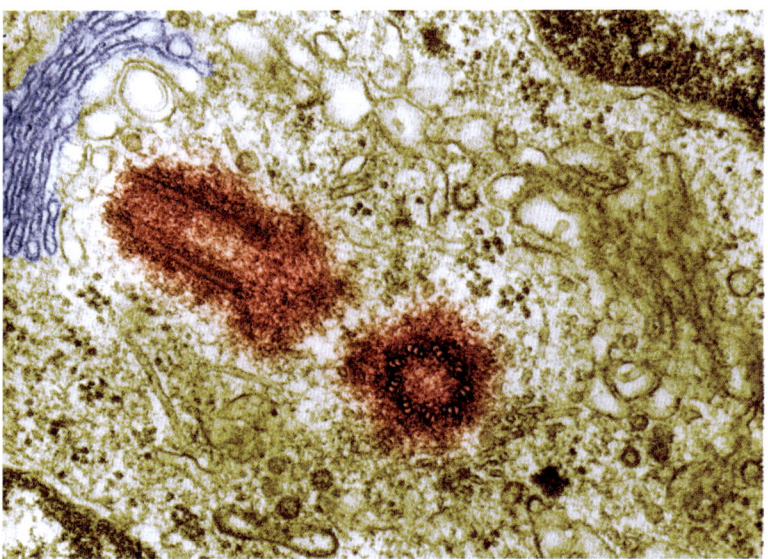

■ Figure A2.2.28 Electron micrograph of a cell showing two centrioles (in red) at right angles to each other, forming a centrosome

A2.2 Cell structure

◆ **Cilium (plural, cilia)**: motile, hair-like outgrowth from the surface of certain eukaryotic cells, which move rhythmically to propel objects such as mucus in the trachea and eggs in oviducts.

◆ **Flagellum (plural, flagella)**: a long, thin structure occurring singly or in groups on some cells and tissues; used to propel unicellular organisms and to move liquids past anchored cells (flagella of prokaryotes and eukaryotes have a different internal structure).

◆ **Chloroplast**: organelle that is the site of photosynthesis and contains chlorophyll.

◆ **Thylakoid**: membrane system of chloroplast.

◆ **Granum (plural, grana)**: stacked discs of membranes found within the chloroplast, containing the photosynthetic pigments, and the site of the light-dependent reaction of photosynthesis.

◆ **Stroma**: colourless fluid contents of the chloroplast; site of the light-independent reaction of photosynthesis.

◆ **Plastid**: an organelle containing pigments (e.g. chloroplast).

◆ **Vacuole**: fluid-filled space in the cytoplasm, especially large and permanent in plant cells.

◆ **Tonoplast**: membrane around the plant cell vacuole.

Link

For more on photosynthesis, see Chapter C1.3, page 425.

Cilia and flagella

Cilia (singular, cilium) and **flagella** (singular, flagellum) are made up of microtubules (see Figure A2.2.24) and are almost structurally identical, except that cilia are usually shorter and more numerous than flagella. They are organelles that project from the surface of certain cells and both can move.

Cilia occur in large numbers on certain cells, such as the ciliated lining (epithelium) of the air tubes serving the lungs (bronchi), where they cause the movement of mucus across the cell surface. It is the cilia of this 'bronchial tree' that cigarette smoke destroys over time. They are also found inside the oviduct where they help move the egg (ovum) from the ovary to the uterus. Cilia are also found in some protists (e.g. *Euglena* and *Paramecium*). Flagella occur singly, typically on small, motile cells, such as sperm, or they may occur in pairs.

A cilium is about 10 µm in length and 0.2 µm in diameter. A flagellum is about 100 µm in length and 0.2 µm in diameter. Both contain a ring of nine microtubule doublets surrounding a central pair of microtubules (9 + 2 arrangement). Both contain a basal body at the base of each cilium or flagellum. Basal bodies are identical in structure to centrioles and help anchor the cilia and flagella to the cell.

■ Structures typically found in plant cells (and some protists)

Just as animal cells have structures not present in plant cells, plants contain organelles that are not present in animal cells.

Chloroplasts

Chloroplasts are large organelles, typically biconvex in shape, about 4–10 µm long and 2–3 µm wide. They occur in green plants, where most occur in the mesophyll cells of leaves. A mesophyll cell may be packed with 50 or more chloroplasts. Photosynthesis is the process that occurs in chloroplasts.

Look at the chloroplasts in the electron micrograph in Figure A2.2.29. In each chloroplast, two continuous membranes form the envelope. A third system of membranes is tucked to form a system of branching membranes called lamellae or **thylakoids**. The thylakoids are arranged in flattened circular piles called **grana** (singular, granum). These look a little like a stack of coins. It is here that the chlorophylls and other pigments are located. Most chloroplasts incorporate several grana, and the number of membranes present in each 'stack' is larger in leaves grown at lower light levels. Between them, the branching membranes are very loosely arranged in an aqueous matrix, usually containing small starch grains. This part of the chloroplast is called the **stroma**.

Chloroplasts are one of a larger group of organelles called **plastids**. Plastids are found in many plant cells but never in animals. The other members of the plastid family are **leucoplasts** (colourless plastids) in which starch is stored, and **chromoplasts** (coloured plastids), containing non-photosynthetic pigments such as carotene, and occurring in flower petals and the root tissue of carrots, and anthocyanin which shades and protects the photosynthetic apparatus by absorbing excess visible and UV light.

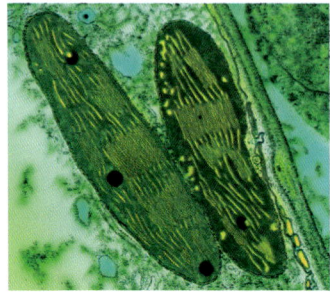

■ Figure A2.2.29 The chloroplast

Permanent vacuole

A **vacuole** is a fluid-filled space within the cytoplasm (see Figures A2.2.15 and A2.2.18, pages 64 and 68), surrounded by a single membrane (called a **tonoplast**). The tonoplast separates the contents of the vacuole from the cell's cytoplasm. The main functions of vacuoles include maintaining cell turgor pressure and regulating waste. As the vacuole fills with water, it pushes the cytoplasm against the cell wall, creating turgor pressure in the cells which maintains the rigidity of plant tissue.

15 **Discuss**, in tabular form, the location of membranes and their function within plant and animal cells.

16 **Outline** the functions of the cell wall in a plant cell.

17 **Describe** how a typical plant cell differs from a typical animal cell.

● Common mistake

Cell walls are not only found in plant cells – prokaryote cell walls exist as well, and cell walls are also part of fungal cell structure.

◆ **Glycoprotein**: membrane protein with a carbohydrate chain attached.

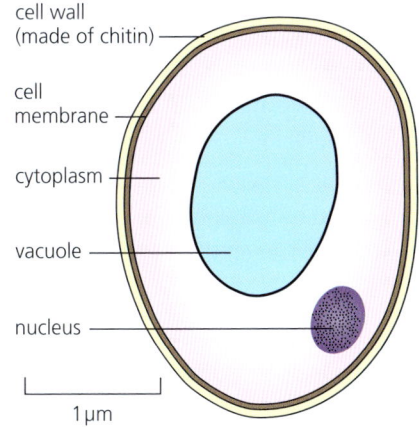

■ **Figure A2.2.30** Diagram of a yeast cell; yeast is a single-celled fungus

18 **Compare and contrast** the structure of plant, animal and fungal cells.

Ions in the sap within the vacuole provide the negative osmotic pressure (or *potential*) that draws water molecules into the vacuole (see Chapter D2.3, page 692). Chemicals can be permanently stored in the vacuole, such as pigments (for example, betalains, the red pigment in beetroot). Carbohydrates, proteins and fats are stored in the vacuoles of storage cells in seeds. In addition, the vacuole functions in a similar way to lysosomes in animal cells – waste material is moved there for disposal. The environment inside a vacuole is slightly acid (about pH 5.0), allowing hydrolase enzymes to break down large molecules. The products of the breakdown process are retained within the vacuole.

● Top tip!

Plant cells frequently have a large, permanent vacuole present. By contrast, animal cells may have small vacuoles, but these are mostly temporary.

■ Extracellular components

We have noted that the contents of cells are contained within the plasma membrane. However, cells may secrete material outside the plasma membrane; for example, plant cells have an external wall, while many animal cells secrete glycoproteins.

The cell wall

The plant cell differs from an animal cell in that it is surrounded by a wall. This wall is completely external to the cell; it is not an organelle. Plant cell walls are primarily constructed of cellulose – a polysaccharide and an extremely strong material. Cellulose molecules are very long and are arranged in bundles called microfibrils (discussed in Chapter B1.1, page 193).

Cell walls make the boundaries of plant cells easy to see when plant tissues are examined by microscopy. The presence of this strong structure allows the plant cell to develop high internal pressure due to water uptake, without danger of the cell bursting. This is a major difference between the cell water relations of plants and animals.

Cell walls are also found in fungi (see more below) and prokaryotic cells, although the chemicals used in the construction of the walls are different in each case.

Extracellular glycoproteins around animal cells

Many animal cells can adhere to one other. This property enables cells to form compact tissues and organs. Other animal cells occur in simple sheets or layers, attached to a basement membrane below them. These cases of adhesion are brought about by **glycoproteins** that the cells have secreted. Glycoproteins are large molecules of protein to which a small number of sugar molecules bonded together (called oligosaccharides) are attached by covalent bonding.

■ Fungal cells

Fungi are a large group of eukaryotic organisms that obtain their food by absorbing nutrients from their external environment. They cannot photosynthesize as they do not contain chloroplasts. Fungal cells have a cell wall but are made of a different material compared to plants and bacteria – chitin. The majority are multicellular, although some, such as yeast, are unicellular.

The cell structure of yeast is shown in Figure A2.2.30.

> **Top tip!**
> Some bacteria have flagella for movement, although the structure is different to those in animals.

◆ **Multinucleate**: a cell that has two or more nuclei.

◆ **Hypha**: the tubular filament 'plant' body of a fungus, which in certain species is divided by cross walls into either multicellular or unicellular compartments.

Atypical cell structure in eukaryotes

Although most organisms conform to the cell theory (page 50), there are exceptions. In addition to the familiar unicellular and multicellular organization of living things, there are a few **multinucleate** organs and organisms that are not divided into separate cells. These are known as atypical cells. An example of an atypical organism is the mould *Mucor*, in which the main body of the organism consists of fine, thread-like structures called **hyphae** (Figure A2.2.31). An example of an atypical organ is the striped muscle fibres that make up the skeletal muscles of mammals (Figure A2.2.32).

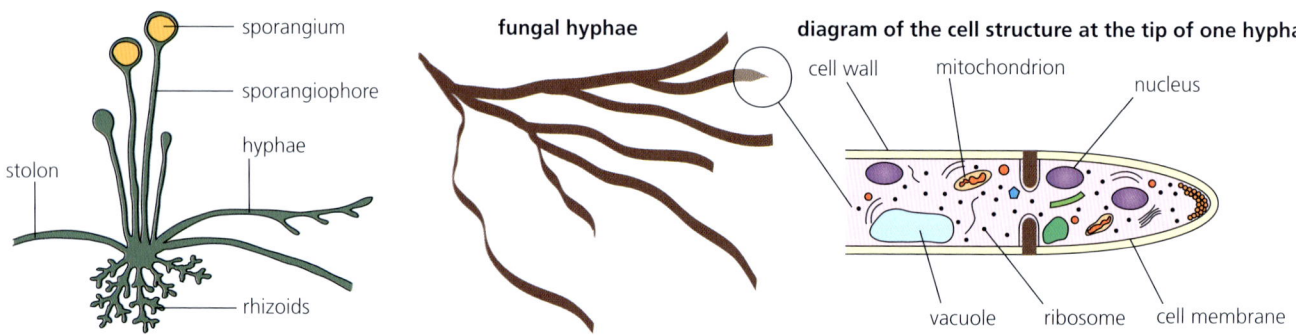

■ Figure A2.2.31 Atypical structure in *Mucor*

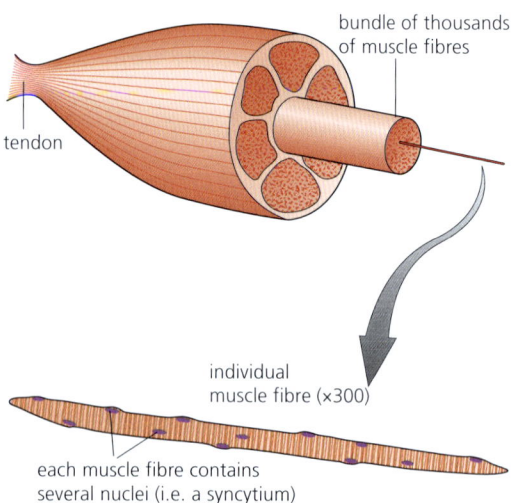

■ Figure A2.2.32 Atypical structure in skeletal muscle fibres of mammals

Another example of atypical cell structure is demonstrated by red blood cells (Figure A2.2.11, page 60). These cells have no nucleus. This adaptation allows more room for haemoglobin – the red pigment that carries oxygen in the blood and gives the cell its red colour. The lack of nucleus also creates a biconcave shape for the cell, which gives a greater surface area for the diffusion of oxygen into and out from the red blood cell.

Phloem tissue transports sugars as sucrose in plants. Plant tissues will be explored in more detail in Theme B3, but for now we will focus on one cell type in the phloem – the sieve tube elements. Phloem tissue (Figure A2.2.33) consists of sieve tubes and companion cells. The sieve tube elements are another example of an atypical cell structure in eukaryotes.

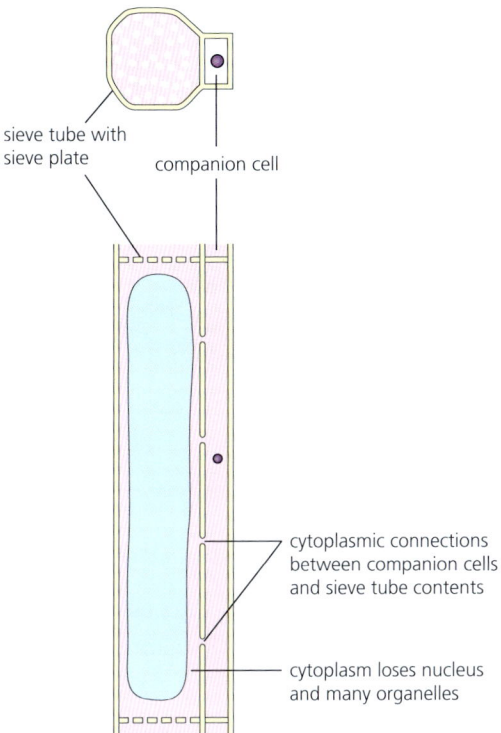
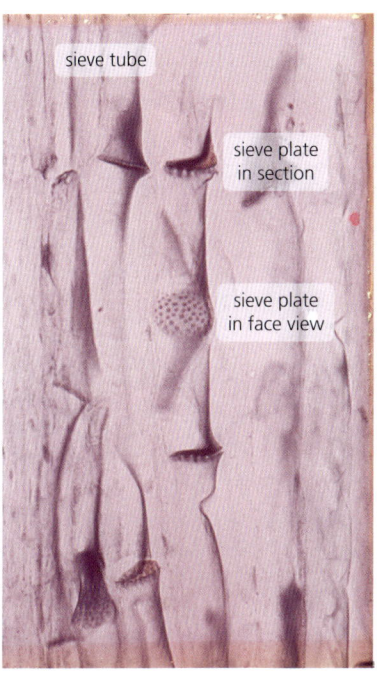

■ **Figure A2.2.33** The structure of phloem tissue

Sieve tubes are narrow, elongated elements, connected end to end to form tubes. There is no nucleus or many other organelles in the cytoplasm of a mature sieve tube. The end walls, known as sieve plates, are perforated by pores and each sieve tube is connected to a companion cell by strands of cytoplasm, called plasmodesmata, that pass through narrow gaps (called pits) in the walls. The companion cells are believed to service and maintain the cytoplasm of the sieve tube, because the sieve tubes have no nucleus.

> ● **Top tip!**
>
> You may be asked to state the limitations of cell theory. These include:
> - striated muscle cells contain many nuclei whereas most eukaryotic cells have one nucleus
> - red blood cells have no nucleus whereas most eukaryotic cells have one nucleus
> - multicellular fungi, such as *Mucor*, have multicellular hyphal cells
> - sieve tube elements in phloem tissue do not contain a nucleus, and rely on companion cells to provide metabolic support
> - viruses have some characteristics of living organisms but are not cells.
>
> If all cells come from pre-existing cells, where did the first one come from?

Cell types and structures viewed in light and electron micrographs

You should be able to identify cells in light or electron micrographs as prokaryote, plant or animal. In electron micrographs, you should be able to identify these structures: nucleoid region (see Figure A2.2.18, page 68), prokaryotic cell wall, nucleus, mitochondrion, chloroplast, sap vacuole, Golgi apparatus, rough and smooth endoplasmic reticulum, chromosomes, ribosomes, cell wall, plasma membrane and microvilli.

A2.2 Cell structure

19 **Examine** the electron micrographs of specialized animal and plant cells in Figure A2.2.34.

a **List** the organelles common to the animal and plant cells illustrated in Figure A2.2.34. **Annotate** your list by recording the principal role or function of these structures.

b **List** separately any organelles you observe to be present only in the plant cell.

ATL A2.2E

Identify the features of structure in the cell in Figure A.2.2.35 – its shape, size and the organelles present. Based on these observations, deduce the specialized role of the cell, giving your reasons.

electron micrograph of an exocrine gland cell of the mammalian pancreas

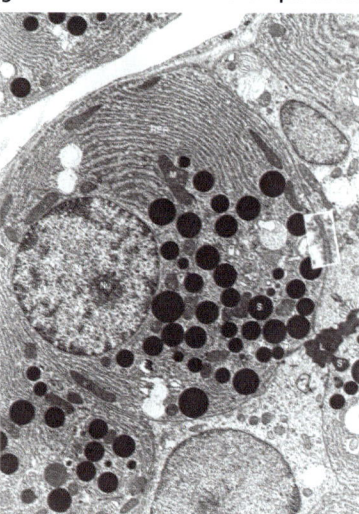

electron micrograph of a spongy mesophyll cell

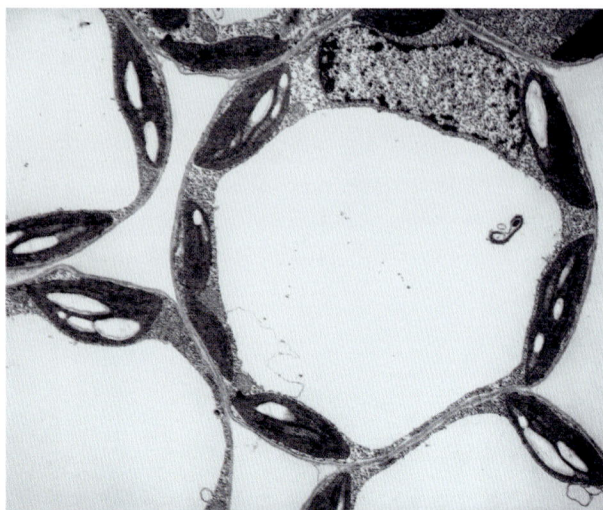

■ **Figure A2.2.34** Electron micrographs of named animal and plant cells

Electron micrographs can be used to show structures not visible by light microscopy, such as microvilli present on the surface of epithelial cells in the small intestine (Figure A2.2.35).

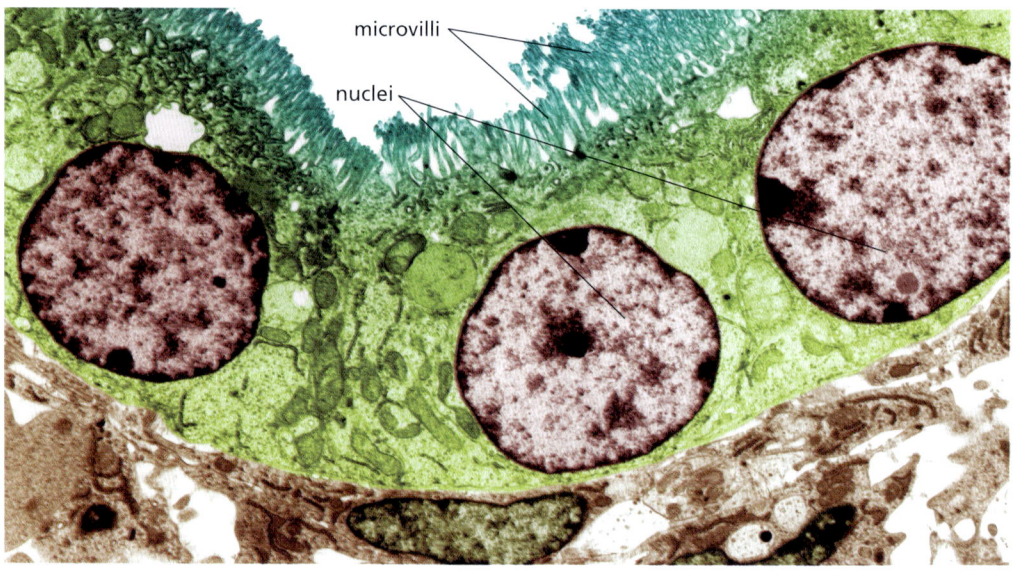

■ **Figure A2.2.35** Electron micrograph of a specialized cell in tissue lining the small intestine (×4500)

Drawing and annotating based on electron micrographs

You should be able to draw and annotate diagrams of organelles (nucleus, mitochondria, chloroplasts, sap vacuole, Golgi apparatus, RER and SER, and chromosomes) as well as other cell structures (cell wall, plasma membrane, secretory vesicles and microvilli) shown in electron micrographs.

Tool 1: Experimental techniques

Drawing annotated diagrams from observation

For a clear, simple drawing:
- use a sharp HB pencil and a clean eraser
- use unlined paper and a separate sheet for each specimen you record
- draw clear, sharp outlines, avoiding shading or colouring (density of structures may be represented by degrees of stippling)
- label each drawing with appropriate information, such as the species, conditions (living or stained; if stained, note which stain was used) and type of section (transverse section, TS, or longitudinal section, LS)
- label your drawing fully, with labels well clear of the structures shown, remembering that label lines should not cross
- annotate (add notes about function, role and development) if appropriate
- include a statement of the magnification under which the specimen has been observed.

Figure A2.2.36 shows an interpretive drawing of a liver cell and the original TEM photo on which it is based.

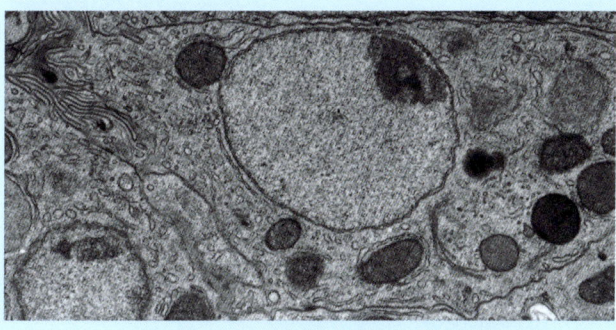

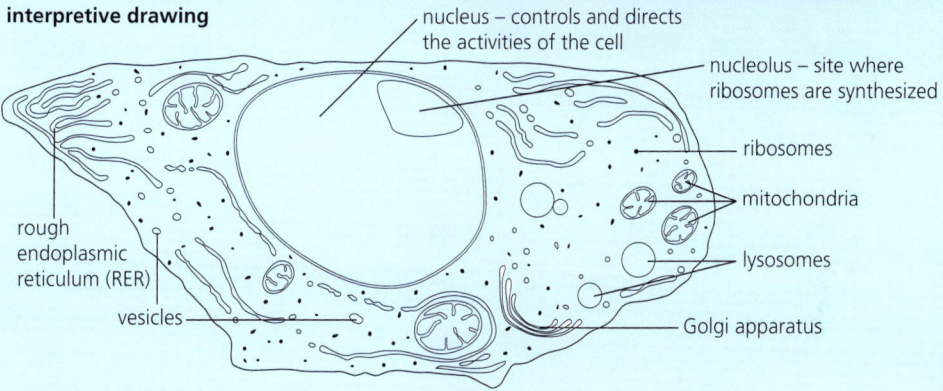

■ Figure A2.2.36 Transmission electron micrograph of a mammalian liver cell (×15 000) with interpretive drawing

● Top tip!
When drawing organelles and other cell structures, you need to include the functions in your annotations.

● Top tip!
When drawing eukaryotic cells, follow this guidance:
- Cell walls should be shown with two continuous lines to indicate the thickness.
- Cell membranes should be shown as a single continuous line.
- Chromosomes can either be drawn as single lines or 'X' shaped, indicating two chromatids joined by a centromere in a cell about to divide.
- When drawing a plant cell or fungal cell, do not forget to add a single line to represent the membrane directly beneath the cell wall.
- Nuclear membranes should be shown with a double membrane and nuclear pores.
- Vacuole membranes should be shown as a single continuous line.
- Chloroplasts should be shown with a double line to indicate the envelope and internal structure (thylakoids/grana).
- Mitochondria should be shown with a double membrane (indicating cristae).
- Avoid overlapping, multiple or discontinuous lines for structures that have a single continuous edge.

● Common mistake
Many students mistakenly think that the organelles (e.g. chloroplasts and mitochondria) are cells with cell membranes, cell walls or even nuclei. Look carefully at electron micrographs and the information provided before you answer a question on organelles.

Images of cells and tissues may be magnified, displayed, projected and saved for printing by the technique of **digital microscopy**. A digital microscope is used or, alternatively, an appropriate video camera is connected by microscope coupler or eyepiece adaptor that replaces the standard microscope eyepiece. Images are displayed via video recorder, TV monitor or computer, and may be printed out by the latter.

> 20 **Draw** and **annotate** a representation of the electron micrograph of the spongy mesophyll cell in Figure A2.2.34, using the interpretive drawing of an animal cell in Figure A2.2.36 as a model, and following the guidelines on biological drawing on page 81.

Inquiry 2: Collecting and processing data

Collecting data

Draw, label and annotate the cell shown in Figure A2.2.37.

What kingdom is the cell from? What evidence did you use to reach this conclusion?

Notice that the magnification of this electron micrograph is recorded. Using this information, it is possible to estimate the actual length of the cell, expressed in μm.

Draw a table with two columns and ten rows.

- In the first column, list the organelles whose structure and roles are explained in this section, pages 68–77.
- Then record the presence of each type of organelle that you can confidently identify in the cell by means of a tick.

Remember, a section through a cell is likely to expose some, but not necessarily all, of the organelles present, especially if their numbers are limited. Illustrations such as Figure A2.2.18 on page 68 are idealized representations of the complete range of organelles of eukaryotic cells.

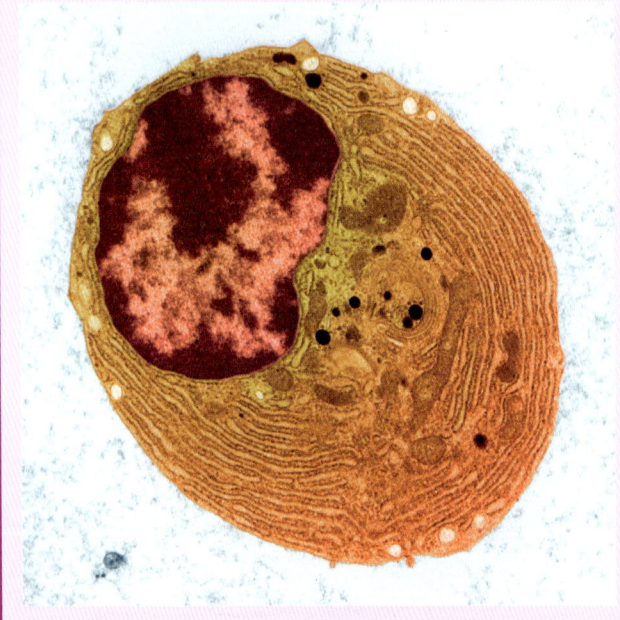

■ **Figure A2.2.37** Electron micrograph of a cell, ×70 000, colourized

Origin of eukaryotic cells by endosymbiosis

Prokaryote cells differ from microspheres (see Figure A2.1.5 on page 41) in a number of ways. For example, attached to the plasma membrane in the prokaryote cell is a circular chromosome of DNA. Also, a cell wall of complex chemistry is secreted outside the membrane barrier. However, the first prokaryotes could have survived nutritionally on the organic molecules of the pre-biotic 'soup'. In this early environment, with its abundance of organic molecules surrounding simple cells, 'digestion' and 'respiration' would have demanded limited enzymatic machinery. These biochemical sophistications would have to evolve with time – if life originated in this manner.

Present-day prokaryotes appear similar to fossil prokaryotes, some of which are 3500 million years old. By comparison, the earliest eukaryote cells date back only 1000 million years.

◆ **Endosymbiotic theory**: theory that suggests that some organelles in eukaryotic cells, e.g. mitochondria and chloroplasts, were once free-living prokaryotic microbes that were incorporated into an ancestral host cell.

21 **Explain** why we can expect that, of all the fossils found in sedimentary rock, eukaryotic fossils in the lowest strata may bear the least resemblance to present-day forms.

Common mistake

A common error in explaining endosymbiosis is to say that the original engulfing process was carried out by a eukaryote, rather than a prokaryote. The theory of endosymbiosis states that a larger prokaryotic cell engulfed smaller prokaryotic cells, ultimately forming the first eukaryotic cell.

How did eukaryotic cells arise?

The origin of eukaryotic cells can be explained by the **endosymbiotic theory** (Figure A2.2.38) proposed by Lynn Margulis in 1967. Evidence suggests that all eukaryotes evolved from a common unicellular ancestor that had a nucleus and reproduced sexually. The eukaryotic cell may have formed from large prokaryote cells that came to contain their chromosome in a sac of infolded plasma membrane. If so, a distinct nucleus was now present. But how did the other organelles originate? Remember, membranous organelles are a feature of eukaryotes, additional to their discrete nucleus.

ATL A2.2F

Research has shown that some genes or DNA sequences in mitochondria are similar to those found in a photosynthetic bacterium. What is the biochemistry of this bacterium? Why was photosynthesis lost and respiration retained in these primitive mitochondria (proto-mitochondria)?

Use this paper as a starting point for your research: www.ncbi.nlm.nih.gov/pmc/articles/PMC1634775

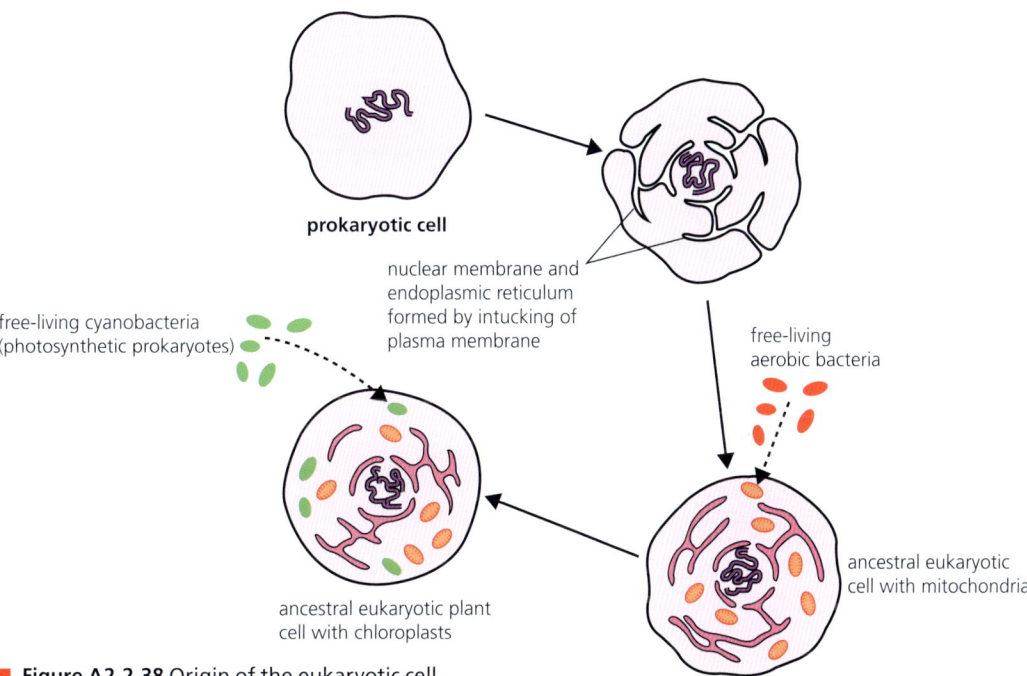

■ **Figure A2.2.38** Origin of the eukaryotic cell

In the evolution of the eukaryotic cell, prokaryotic cells (which had been taken up into food vacuoles for digestion) may have survived as organelles inside the host cell, rather than becoming food items. The double membrane of the mitochondrion is the result of endocytosis (the outer membrane originating from the food vacuole and the inner membrane of the mitochondrion from the engulfed cell). The engulfed prokaryotic cells were not destroyed or digested and would have become integrated by natural selection into the biochemistry of their 'host' cell over time. An 'endosymbiotic' (or mutualistic) relationship developed.

Top tip!

Endocytosis would have allowed a free-living prokaryote to enter and remain in a host prokaryotic cell. Because the prokaryote made itself useful, for example providing energy in the case of an aerobic prokaryote, there was a selective advantage in developing a mutualistic relationship rather than digesting the engulfed cell.

■ Table A2.2.5 Evidence for the endosymbiotic theory

Unicellular organisms are known to inhabit larger eukaryotic cells. For example, green algae (*Chlorella* – see page 74) are known to form mutualistic endosymbiotic relationship with species of *Paramecium* (page 74).
Chloroplasts and mitochondria reproduce by binary fission, just as prokaryotes do.
Chloroplasts and mitochondria contain circular DNA (not associated with histone proteins), like that of prokaryotes.
Chloroplasts and mitochondria contain 70S ribosomes, also found in prokaryotes.
Chloroplasts and mitochondria transcribe mRNA from their DNA and synthesize specific proteins in their ribosomes, as prokaryotes do.
Chloroplasts and mitochondria are similar in size to prokaryotes.
Chloroplasts and mitochondria have double membranes: double membrane suggests engulfing by endocytosis.

This theory explains why mitochondria (and chloroplasts) contain a loop of naked DNA, together with 70S ribosomes, just like a bacterial cell. It is these features that suggest these organelles are descendants of free-living prokaryotic organisms that came to inhabit larger cells.

The nature of the engulfed cell would have determined their role in the cell:

- If the engulfed prokaryotic cell respired aerobically, consuming oxygen, it would have provided the host cell with energy in the form of ATP, and formed mitochondria.
- If the engulfed cell could photosynthesize, it would have formed chloroplasts.

This concept is known as the **endosymbiotic origin of eukaryotes** (Table A2.2.5).

Top tip!

Evidence that mitochondria and chloroplasts evolved by endosymbiosis include the presence of 70S ribosomes in mitochondria and chloroplasts, naked circular DNA and the ability to replicate.

22 **Explain** what is meant by the 'endosymbiont theory'.

Top tip!

You should recognize that the strength of a theory comes from the observations the theory explains and the predictions it supports. A wide range of observations are accounted for by the theory of endosymbiosis.

Nature of science: Theories

Factors which determine the strength of a theory

A scientific theory not only explains known observations and data; it also allows scientists to make predictions of what they should observe. Scientific theories must be testable. New observations should support the theory and, if they do not, then the theory must be modified or rejected. The longer the central features of a theory hold, the more observations and data it predicts, the more tests it passes, the more it explains, then the stronger the theory.

The evolutionary origins of eukaryotes can be grouped into **autogenous theories** and **endosymbiotic theories**. Although both theories involve natural selection, they provide different explanations and novel predictions.

In autogenous theories, it is suggested that all the structures and functions of eukaryotes evolved gradually from a single ancestor of prokaryotes. One common feature of autogenous theories is the proposed infolding of regions of the cell membrane forming internal vesicles, which subsequently evolved into the various organelles.

However, the presence of DNA in the mitochondria and chloroplast cannot be explained by most autogenous theories. The presence of extra-nuclear DNA in mitochondria and chloroplasts was discovered indirectly in the 1960s and 1970s. American evolutionary biologist Margulis carried out experiments providing indirect evidence for DNA in the cytoplasm of algae (i.e. in the chloroplast).

In endosymbiotic theories it is believed that certain eukaryotic organelles evolved from prokaryotic organisms, which entered into symbiosis with an ancestor of eukaryotic cells (the proto-eukaryote).

The most widely accepted and supported variant of the endosymbiosis theory hypothesizes that the mitochondrion and chloroplast evolved from bacteria. There is a broad range of biochemical, molecular and cell structural data to strongly support this theory of endosymbiosis, especially for the chloroplast evolving from cyanobacteria.

However, debate continues about the exact nature of the bacteria that were involved in the evolution of the mitochondrion in the proto-eukaryote. Margulis also proposed that the eukaryotic flagellum originated by a process of endosymbiosis, but very little evidence supports this proposal.

The discovery of DNA in mitochondria and chloroplasts was supportive of the endosymbiont hypothesis, but it was the later evolutionary analysis of the mitochondrial and chloroplast genomes themselves that provided the strongest evidence for the theory.

The DNA sequences of chloroplasts were most similar to those of free-living photosynthetic cyanobacteria (as one would predict based on their shared photosynthetic properties, pigmentation, etc.), while the mitochondrial DNA was clearly bacterial but did not initially have a strong connection to a specific type of bacteria.

Around the 1980s it started to become clear that mitochondrial DNA was specifically linked to alphaproteobacteria (a large and diverse taxon of bacteria) and Rickettsiales (Figure A2.2.39) (non-motile, gram-negative bacteria associated with disease) in particular, and this result has stood the test of time.

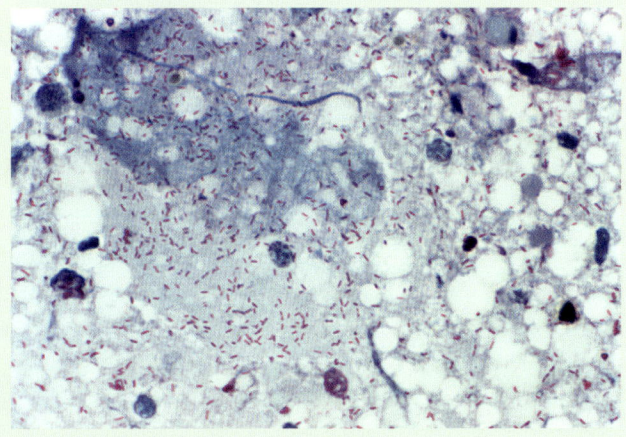

■ Figure A2.2.39 *Rickettsia rickettsii*

Cell differentiation and the development of specialized tissues in multicellular organisms

We have seen that unicellular organisms, although structurally simple, carry out all the functions and activities of life within a single cell. The cell takes in nutrients, respires, excretes, is sensitive to internal and external conditions (and may respond to them), may move, and eventually divides or reproduces (page 73).

By contrast, most multicellular organisms – all animals and plants, the majority of fungi and seaweeds in the Protoctista – are made of cells, most of which are highly **specialized** to perform a particular role or function (Figure A2.2.40). Specialized cells are organized into tissues and organs. A **tissue** is a group of similar cells that are specialized to perform a particular function, such as the heart muscle tissue of a mammal or phloem tissue in plants (see page 79). An **organ** is a collection of different tissues which performs a specialized and coordinated function, such as the heart of a mammal or the leaf of plants. So, the tissues, organs and organ systems of multicellular organisms consist of specialized cells.

◆ **Tissue**: collection of cells of similar structure and function.

◆ **Organ**: a part of an organism, consisting of a collection of tissues, having a definite form and structure, and performing one or more specialized functions.

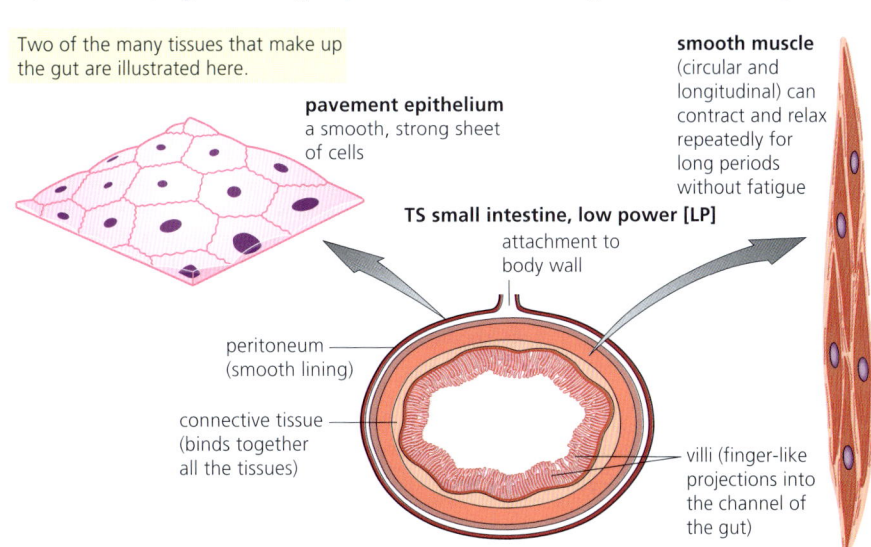

■ Figure A2.2.40 Tissues of part of the mammalian gut

A2.2 Cell structure

Cells in a multicellular organism, though they all contain the same DNA, are differentiated to carry out specialized functions. They express ('switch on') different sets of genes according to their developmental history and to the signals they receive from their environment. These changes in gene expression are often triggered by changes in the environment of the cell.

Control of cell specialization

We have seen that the nucleus of each cell is the structure that controls and directs the activities of the cell. The information required for this exists in the form of a nucleic acid, DNA. The nucleus of a cell contains the DNA in thread-like chromosomes, which are linear sequences of **genes** (Chapter A1.2, page 14). Genes control the development of each cell within the mature organism. We can define a gene in different ways, including:

- a specific region of a chromosome which is capable of determining the development of a specific characteristic of an organism
- a specific length of the DNA double helix (hundreds or thousands of base pairs long) which codes for a protein.

When a cell is becoming specialized, we say the cell is differentiating: some of its genes are being activated and expressed. These genes determine how the cell develops. In Chapter D2.2 (page 670) we explore both what happens during gene expression and the mechanism by which a cell's chemical reactions are controlled. For now, we can just note that the nucleus of each cell contains all the information required to make each type of cell present within the whole organism; only a selected part of that information is needed in any one cell and tissue. How a cell specializes, and which genes are activated, are controlled by the immediate environment of the differentiating cell and the cell's position in the developing organism.

Differences between the proteomes of cells within a multicellular organism

The **proteome** of an organism is the totality of proteins expressed within a cell, tissue or organism at a certain time. Because the genome – the whole of the genetic information of an organism – is unique to each individual, the proteome it expresses is also unique.

The genome is the same in all the cells of an organism because all cells in a multicellular organism are ultimately derived from one original cell, by cell division. The genome instructs the expression of proteins. Because individual tissues have specific structures relating to their function, different proteins are expressed in each specific tissue cell type. Specific genes are expressed (turned on or off) in different cells, according to a required function. As the proteome is all the proteins produced by a cell, the proteome varies with the function, location or environmental conditions of the cell through the process of cell differentiation.

The cost of specialization

Specialized cells are efficient at carrying out their particular function, such as transport, support or protection. We say the resulting differences between cells are due to **division of labour**. By specialization, increased efficiency is achieved, but at a price. The specialized cells are now totally dependent on the activities of other cells. For example, in animals, nerve cells are adapted for the transport of nerve impulses but depend on red blood cells for oxygen, and on heart muscle cells to pump the blood. This modification of cell structure to support differing functions is another reason why no 'typical' cell really exists.

> ● **Common mistake**
>
> Differentiation of cells is often dealt with superficially: if asked about this in a question you need to refer to some genes being expressed and others suppressed, and not just say that 'cells specialize'. The strongest answers explain how cells use genes selectively and give specific examples of specialized tissue and their functions.

◆ **Proteome**: the entire set of proteins expressed by a genome.

◆ **Division of labour**: the carrying out of specialized functions by different types of cell in a multicellular organism.

Evolution of multicellularity

Many fungi and eukaryotic algae and all plants and animals are multicellular, indicating that multicellularity has evolved repeatedly. The broad distribution of multicellular taxa across the eukaryotic tree of life suggests that multicellularity is relatively 'easy' to evolve with few barriers. The various mechanisms needed for multicellularity (e.g. cell adhesion, communication between cells and differentiation) found across taxa show that there are many ways it can evolve.

Multicellularity has the advantages of allowing larger body size and cell specialization. The frequent occurrence of multicellularity is most probably due to strong selective pressures favouring it, the few genetic changes necessary to enable the change, or a combination of these factors.

LINKING QUESTIONS
1. What explains the use of certain molecular building blocks in all living cells?
2. What are the features of a compelling theory?

A2.3 Viruses

> **Guiding questions**
> - How can viruses exist with so few genes?
> - In what way do viruses vary?

> **SYLLABUS CONTENT**
>
> This chapter covers the following syllabus content:
> ▶ A2.3.1 Structural features common to viruses (HL only)
> ▶ A2.3.2 Diversity of structure in viruses (HL only)
> ▶ A2.3.3 Lytic cycle of a virus (HL only)
> ▶ A2.3.4 Lysogenic cycle of a virus (HL only)
> ▶ A2.3.5 Evidence for several origins of viruses from other organisms (HL only)
> ▶ A2.3.6 Rapid evolution in viruses (HL only)

◆ **Virus**: non-cellular parasite of animals, plants and bacteria that consists of nucleic acid surrounded by a protein coat (a capsid).

◆ **Parasite**: an organism that lives on or in another organism (its host) for most of its life cycle, deriving nutrients from its host.

Introduction to viruses

All living organisms (eukaryotes and prokaryotes) have nucleic acids that make up their genomes. **Viruses** contain nucleic acids but require living cells to reproduce. This makes viruses obligate **parasites** as they can only reproduce within a host cell, since they do not have the ribosomes and metabolic enzymes necessary for protein synthesis. Viruses are known to infect all cellular organisms and are highly diverse in their shape and structure. They are classified according to their genome and life cycle (Figure A2.3.1).

This topic will look at their common features, the diversity of their structures, the life cycles of viruses and the evidence of their origin and rapid evolution.

> ● **Common mistake**
>
> A common mistake is to say that a virus is a cell (cellular organism). In fact, a virus is considered 'non-cellular', meaning that no part of it is cellular.

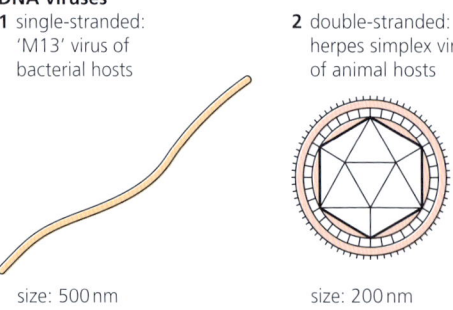

DNA viruses
1 single-stranded: 'M13' virus of bacterial hosts
size: 500 nm

2 double-stranded: herpes simplex virus of animal hosts
size: 200 nm

RNA viruses
1 single-stranded: poliovirus of animal hosts
size: 25 nm

human immunodeficiency virus, HIV (retrovirus)
size: 100 nm

2 double-stranded: reovirus of animal hosts
size: 80 nm

■ **Figure A2.3.1** A range of different viruses classified according to their nucleic acids

> ● **Top tip!**
>
> Viruses are not part of the Linnaean system of classification or the three domain system.

Unlike LUCA for cellular organisms (discussed on page 47), there is no presumed single common ancestor for viruses. One method of classifying viruses is based on the genome present in the virus. The nucleic acid present will be DNA or RNA, which will either be single or double stranded and with positive or negative polarity.

Theme A: Unity and diversity – Cells

◆ **Micro-organism**: a living organism that cannot be seen by the naked eye, such as bacteria, protists and single-celled fungi.

◆ **Capsid**: the protein coat of a virus.

1 **State** the differences in structure between a bacterial cell and a virus.

Some infections are caused by **micro-organisms**, such as bacteria. Other infections are caused by viruses. However, bacteria and viruses have very different structures and life cycles. Table A2.3.1 summarises the differences between bacteria (unicellular) and viruses (non-cellular).

■ Table A2.3.1 Summary of the differences between viruses and bacteria

Characteristic	Viruses	Bacteria
Cellular	no	yes
Approximate diameter (μm)	0.02–0.2	1–5
Nucleic acid	either DNA or RNA; never both	both DNA and RNA
Ribosomes	rarely present	70S
Nature of outer surface	protein **capsid** and lipoprotein envelope (for some)	cell wall containing peptidoglycan
Motility (movement)	none	some (via rotating flagella)
Method of replication	viral replication (protein synthesis followed by self-assembly) inside cells	binary fission (generally outside cells)

Nature of science: Observations and science as a shared endeavour

Predicting the existence of viruses

Viruses were first observed in 1931 when electron microscopes were developed, but scientists had already hypothesized their existence based on careful observations. Researchers used special, fine filters to remove bacteria from tissues that were infected. If bacteria were causing the infection, the filtered tissues should no longer be able to make other organisms sick. However, the filtered tissues remained infective. This meant an infectious micro-organism much smaller than bacteria was causing the infection. In 1915, English bacteriologist Frederick Twort discovered **bacteriophages**, the group of viruses whose host cells are bacteria. He observed tiny clear spots within bacterial colonies and hypothesized that an organism was killing the bacteria.

Edward Jenner and Louis Pasteur developed the first vaccines, even though they did not know viruses existed as they were too small to detect with a compound light microscope. Jenner developed the first vaccine against smallpox in 1798. At the time, many people who got smallpox died. However, those who had earlier contracted cowpox never died. (Workers who handled cows typically caught cowpox at some stage.) Jenner saw the significance of the protection the patients had acquired. He extracted fluid from a cowpox pustule on an infected milkmaid, and injected it into himself and into the arm of an eight-year-old boy. The child got a mild cowpox infection but, when exposed to smallpox, remained healthy. Jenner named this technique vaccination (after the Latin 'vacca', which means cow). Of course, he did not understand the cause of smallpox; the chemical nature of viruses was not reported until 1935.

Almost 90 years after Jenner's smallpox vaccine, Pasteur developed the first vaccine against rabies. Pasteur was looking for what caused rabies and thought it could be a pathogen too small to be detected with a microscope (at the time). Knowledge about viruses therefore goes back before 1931, before viruses were first observed, even if scientists did not know what viruses were.

The discovery of viruses led to a whole new research field – one that is a shared international endeavour. Today, the International Committee on Taxonomy of Viruses (ICTV), formed in 1966, coordinates the development, refinement and maintenance of virus taxonomy. The ICTV has developed a universal taxonomic scheme for viruses, aiming to accurately describe, name and classify every virus that infects living organisms. It uses a slightly modified version of the standard Linnaean classification system to classify viruses – starting at order and going down to species. For example, HIV is in the Retroviridae family because it is a retrovirus and contains RNA and reverse transcriptase. It is in the subfamily Orthoretrovirinae because it is spherical and a tumour-causing virus in vertebrates. The ICTV maintains a database detailing the current knowledge and provides the database and other virus-related information to the public (see **http://ictv.global**).

Structural features common to viruses

Concept: Unity

All viruses have common structural features: small, fixed size; nucleic acid (DNA or RNA) as genetic material; a capsid made of protein; no cytoplasm; and few or no enzymes.

> **Top tip!**
> No virus has been found to contain ribosomes (except the Arenaviruses, but the ribosomes are not required for virus replication and their presence may simply be an inadvertent result of the packaging process).

There are relatively few features shared by all viruses. However, all viruses have the following features in common:
- they are extremely small and have a fixed size range of 20–400 nm in diameter
- they have nucleic acid (DNA or RNA) as genetic material
- a capsid made of protein surrounds the nucleic acid
- no cytoplasm
- few or even no enzymes.

> **Common mistake**
> Viruses do not have a nucleus or nucleoid. These terms apply only to eukaryotes and prokaryotes. There is a viral genome, which consists of RNA or DNA.

Diversity of structure in viruses

Viruses differ in many aspects from host range (the number of species and cell types that they can infect), to the type of genome, their structure and their method of replication. They are highly diverse in their shape and structure, as seen in Figure A2.3.2.

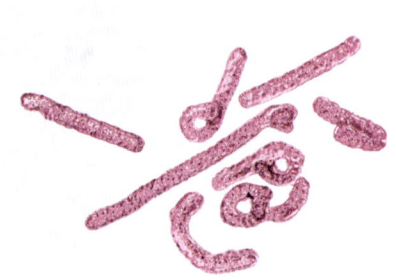
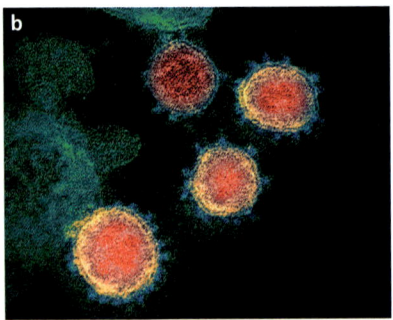
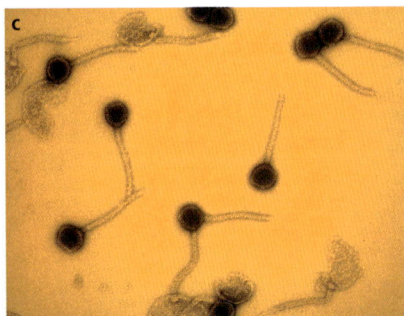

■ Figure A2.3.2 The diverse shapes of three different viruses: a) TEM of Marburg virus, b) coloured TEM of COVID-19 coronavirus particles, c) coloured TEM of bacteriophage lambda

> **Concept: Diversity**
> Viruses are highly diverse in their shape and structure. Genetic material may be RNA or DNA, which can be either single- or double-stranded. Some viruses are enveloped in host cell membrane and others are not enveloped.

> **ATL A2.3A**
> 1. The Biointeractive Virus explorer website www.biointeractive.org/classroom-resources/virus-explorer has three-dimensional models of ten different viruses including coronavirus, influenza A, HIV, tobacco mosaic virus (TMV) and adenovirus. Click on 'start interactive' to view the viruses. You can then click on each virus image to rotate them, view from different angles and see in cross-section. You can also explore diagrams of the viruses' replication cycles. Suggest possible groupings for the different viruses based on their structure. What classification system would you use for these viruses?
> 2. The Baltimore classification (first defined in 1971) is a classification system that places viruses into one of seven groups depending on a combination of their nucleic acid (DNA or RNA), whether they are single-stranded or double-stranded, and their method of replication. Use the internet to find out about the Baltimore classification of viruses and produce an annotated diagram as a poster outlining this approach to virus classification.

The viral genome

DNA viruses include adenoviruses (one of the causes of eye diseases) and most bacteriophages (a virus that parasitizes bacteria).

◆ **Coronavirus**: a large and diverse family of RNA viruses that cause respiratory diseases in animals and humans.

The genetic code is shared between viruses and living organisms, meaning that the host cells transcribe their DNA into mRNA and then translate it into protein. RNA viruses include HIV and **coronaviruses**.

> ● **Common mistake**
>
> A coronavirus came to public attention in February 2020 when the World Health Organization (WHO) announced COVID-19 as the name of a new respiratory disease that was spreading rapidly. Many thought that the coronavirus itself was a new virus. However, coronaviruses were around prior to the 2019 pandemic; human coronavirus was first discovered in the 1960s. SARS-CoV-2, the virus which causes COVID-19 disease, is thought to have transferred from an animal host to humans in late 2019.

◆ **SARS-CoV-2**: severe acute respiratory syndrome coronavirus 2, the virus that causes COVID-19.

◆ **Positive-sense RNA**: viral RNA that has the same base sequence as mRNA, which allows it to function as a template for protein synthesis during viral replication.

◆ **Negative-sense RNA**: viral RNA that is complementary to mRNA (i.e. the antisense strand of the viral mRNA) and so it cannot directly encode for protein synthesis; it must be replicated to mRNA before protein production can begin.

The RNA viruses can be divided again into two groups, depending on how their RNA is reproduced and used in the host cell. In some RNA viruses, e.g. **SARS-CoV-2**, nucleic acid replication occurs entirely in the cytoplasm, but influenza viruses replicate their RNA in the nucleus. In some viruses, e.g. HIV, their RNA is used as a template for synthesizing viral DNA in the cytoplasm, using a viral enzyme called reverse transcriptase.

For viruses with an RNA genome, they may possess either **positive-sense RNA** (i.e. identical to viral mRNA and thus can be immediately translated into proteins by the host cell ribosomes) as SARS-CoV-2 does, or **negative-sense RNA** (i.e. complementary to viral mRNA, so must be converted to positive-sense RNA by RNA polymerase before translation) as in influenza viruses.

The viral genome is either a linear molecule of nucleic acid, for example as in coronaviruses, or a circular molecule of nucleic acid, for example as in hepatitis B. In some viruses, the nucleic acid is **single-stranded**, as seen in the hepatitis E virus and coronaviruses. In others, however, it is **double-stranded**, for example as in HIV and SV-40. Some kinds of viruses may have more than one copy of the genome, for example HIV and influenza viruses.

The genomes of some simple viruses, such as that of the bacteriophage Qbeta, are only a few thousand nucleotides in length and contain only four genes. Other viruses, particularly those with a complex structure, contain more genes. Bacteriophage T4 (see Figure A2.3.3) is an example of a virus having 289 genes. These extra genes are largely involved in the formation of the complex capsid.

Viruses have a small genome, which makes them highly adapted for virus replication, in which each infected host cell produces many copies of the viral genome from a single DNA or RNA template. Rapid replication is favoured by a small genome size: the smaller the genome, the faster it can replicate.

The size of viral genomes also depends on the type of host cell. Viruses with prokaryotic host cells tend to replicate rapidly due to the high rate of binary fission. This is reflected in the compact nature of the genome with overlapping genes of many bacteriophages, leading to the minimum genome size.

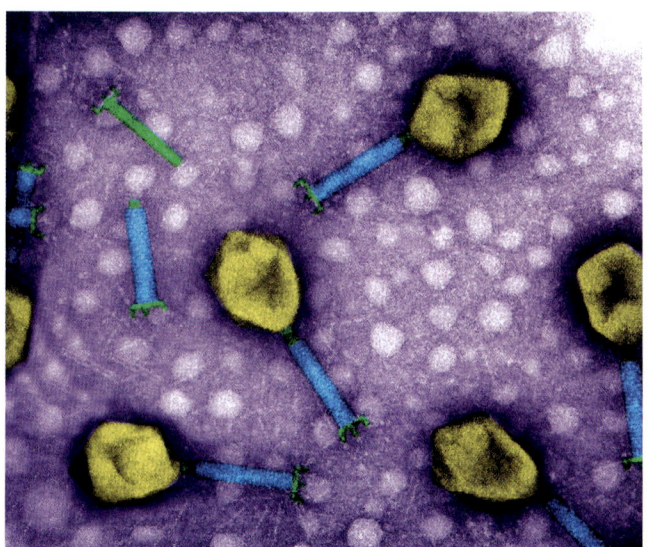

■ **Figure A2.3.3** Coloured TEM of T4 bacteriophage

> ● **Top tip!**
>
> There is a roughly inverse relationship between genome size and dependence on the host for essential functions in DNA replication.

A2.3 Viruses

Going further

A gene is usually defined as having a unique position on a chromosome. However, individual genes can overlap or share one or more nucleotides with adjacent genes – something initially discovered in viruses and mitochondria but now also known from bacteria and higher organisms.

◼ Capsid

The capsid (Figure A2.3.4) is a protein coat enclosing the viral genome. There are a variety of shapes of capsid, including helical (for example the tobacco mosaic virus (TMV)), conical (for example, HIV) and polyhedral (for example, adenovirus). Capsids are synthesized from many protein subunits called **capsomeres**. Some viruses carry viral enzyme molecules within their capsids.

◆ **Capsomere**: one of the individual proteins that make up a viral capsid.

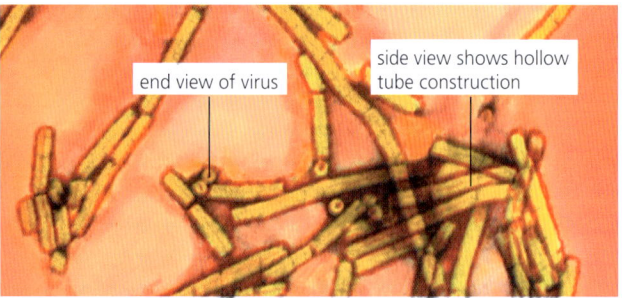

transmission electron micrograph of TMV (×40 000) negatively stained

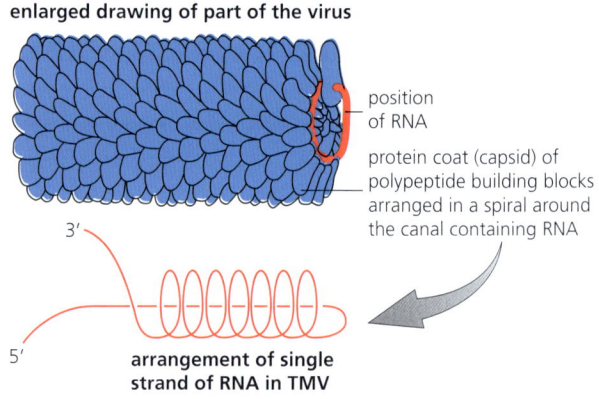

healthy leaves

infected leaf

■ Figure A2.3.4 The tobacco mosaic virus (TMV), which infects the tobacco plant

Some of the most complex capsids are found among bacteriophages (phages). The capsids of phages have elongated icosahedral heads enclosing their genome. These are made of equilateral triangles fused together in a spherical shape, forming a 20-sided structure, which is the optimal way to form a closed shell using identical protein sub-units (Figure A2.3.5). The TEM image on page 91 shows the icosahedral head of a Phage T4 (Figure A2.3.3). Joined to the head is a tail shaft with fibres that the phages use to attach to a bacterial cell wall. Capsids protect the viral genome, but they are often a target of the immune system.

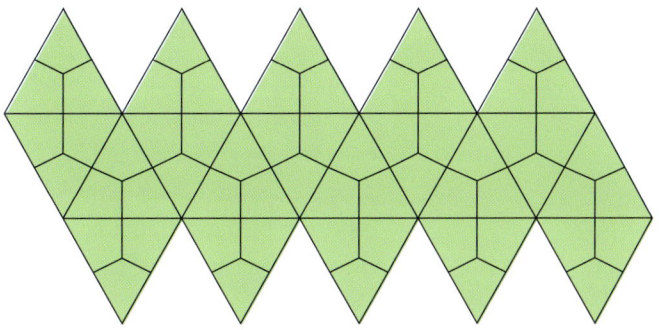 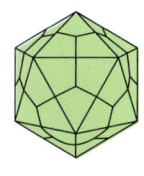

■ **Figure A2.3.5** Unfolded two-dimensional (left) and folded three-dimensional (right) icosahedron

Envelope

Viruses can be classified as **enveloped** (for example, HIV, influenza viruses and coronaviruses) or **non-enveloped** (bacteriophages and poliovirus). The viral **envelope** encloses the **nucleocapsids** of many viruses that infect animals, for example HIV and coronaviruses.

Enveloped viruses contain host cell phospholipids, usually from the plasma membrane of the host cell embedded with virally-encoded spike glycoproteins. However, the coronaviruses take their envelope from internal membranes of the host cell rather than the plasma membrane. The viral envelope protects the **virion** from enzymes and other chemicals, giving them an advantage over virions with only a capsid.

Glycoproteins on viral envelopes help viruses enter host cells by binding to receptor molecules on the cell surface. Virions recognize and bind to specific host proteins, resulting in possible uptake of virions into the cell.

◆ **Envelope**: a membrane typically from the host cell plasma membrane with viral glycoproteins.

◆ **Nucleocapsid**: the capsid of a virus with an envelope.

◆ **Virion**: an isolated but infectious virus particle found outside the host cell.

Link

Glycoproteins and phospholipids are covered in more detail in B1.1, pages 195–203, and B2.1, page 223 and page 234.

Nature of science: Patterns and trends

Measuring the infectivity of viruses

For many years viruses could be studied only by observing the effects of virus-infected extracts on living animals. The effects of infection with virus are commonly the production of disease symptoms and pathological changes, such as tissue damage. An approximate measure of the infectivity of the virus-infected extract could be estimated by observing how far it could be diluted yet remain capable of causing infection in living animals.

Common mistake

A common misconception is that a virus can move independently. Viruses cannot move by themselves; they can only move passively through phloem, blood, other body fluids or the air. When they reach a host cell, they bind to receptors on the cell's surface, which causes the virus particle to be drawn into the cell.

Viral reproductive life cycles

In this section we will look at two reproductive strategies, the lytic and lysogenic life cycles, using the example of bacteriophage lambda.

The stages of viral reproduction are generally:
- attachment (or adsorption) to host cells
- entry (or penetration)
- replication of viral genome
- transcription of virus mRNA and protein synthesis using host cell ribosomes
- maturation and assembly of viruses and release from host cell.

Virus particles cannot carry out life processes independently, and so use the mechanisms available within the host cell to carry out their functions. They rely on the host cell for energy supply, nutrition, protein synthesis and other life functions.

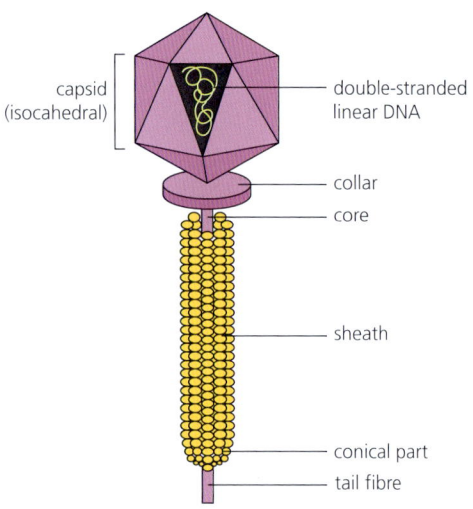

■ Figure A2.3.6 Structure of bacteriophage lambda

■ Bacteriophage lambda

Bacteriophages (or phages for short) are viruses that infect bacteria. They played a central role in the development of molecular biology, especially in our understanding of gene structure and expression. They are also important in the laboratory, providing a way in which genes can be transferred from one bacterium to another. With the development of gene cloning, phages were used as vectors for cloned DNA.

Bacteriophages are very specific in the bacteria they infect. Bacteriophage lambda (λ phage) (Figure A2.3.6) only infects *E. coli*. The wild type of this virus has a temperate life cycle that allows it to be in the bacterial genome through lysogeny or enter a lytic cycle during which it kills the bacterial cell. We will now look at the differences between lytic and lysogenic life cycles.

■ Lytic life cycle

♦ **Lytic**: a phage life cycle where a phage attaches to a host bacterium, injects its DNA which undergoes replication to form new virions, which then lyse the cell.

♦ **Lysis**: breakdown, typically of cells.

♦ **Virulent**: the ability of a virus (or bacterium) to cause rapid and severe disease.

The **lytic cycle** is one of the life cycles viruses can undergo. At the start of the reproductive life cycle, the tail fibre of the bacteriophage lambda binds to specific receptors on the surface of *E. coli* (a process called attachment) before injecting its genome (double-stranded DNA) into the host cell (Figure A2.3.7). The protein capsid is left outside.

The entry of the viral DNA into the bacterium is known as penetration. The viral DNA remains as a separate molecule within the bacterial cell, rather than becoming incorporated with the host DNA, and replicates separately from the host bacterial DNA. The bacterial DNA is broken down via DNA hydrolysis, while the viral DNA is replicated and used to form new virus particles (a process called biosynthesis). Viral genes are expressed using host enzymes to produce viral proteins, which self-assemble to form mature bacteriophage lambda virions (maturation process). The release of bacteriophage lambda from its host *E. coli* cell involves the **lysis** of the bacterial cell wall, resulting in the death of the cell.

The lytic cycle results in the destruction of the infected cell and its membrane. Bacteriophages that only use the lytic cycle are called **virulent** phages.

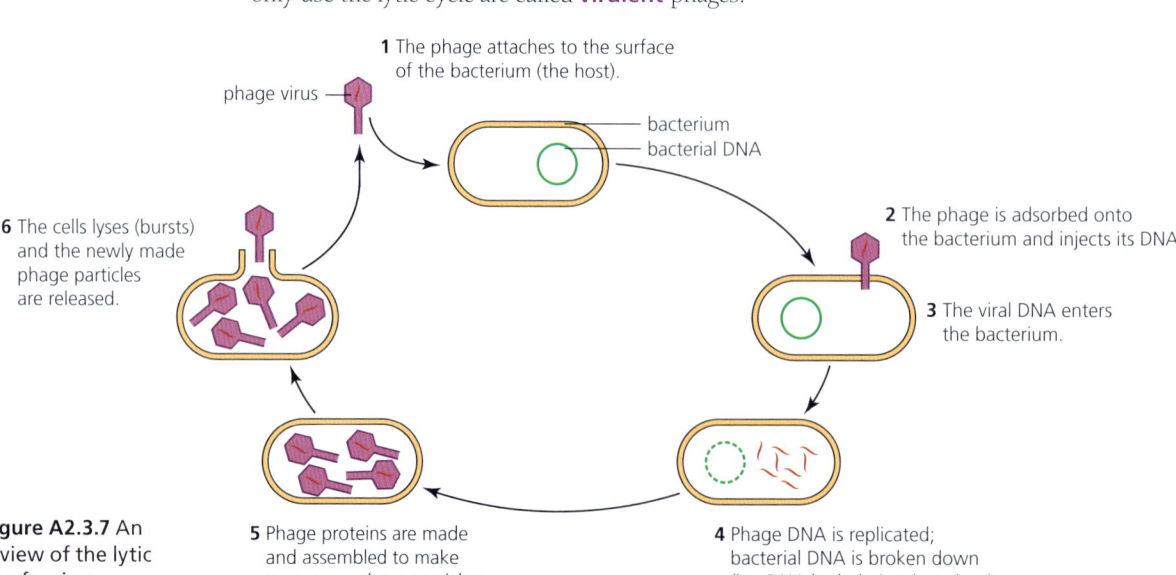

■ Figure A2.3.7 An overview of the lytic cycle of a virus

◆ **Lysogenic**: a phage life cycle where a phage attaches to a bacterium and injects its genome, but it does not undergo a full replication cycle; instead it becomes resident within the bacterial host where it is maintained in a dormant state.

◆ **Prophage**: a bacteriophage in an inactive state in which the genome is typically integrated into the chromosome of the bacterial host.

Lysogenic life cycle

In the **lysogenic** life cycle, the phage infects the bacteria cell by injecting its DNA into the cell (Figure A2.3.8). Viral DNA is incorporated into the host cell's DNA to form a **prophage**. This is different to the lytic life cycle, where the viral DNA remains separate. A bacterial host with a prophage is called a **lysogen**, which is where the term lysogenic comes from. In the lysogenic life cycle, the bacteriophage's viral DNA replicates together with the DNA of the *E. coli* host cell when it replicates, but since the prophage contains genes, it can give novel properties to *E. coli*.

Healthy bacteria grown in rich medium contain high levels of proteases (enzymes that break down proteins) and when infected are more likely to support lytic replication. The role of these enzymes is to catalyse the splitting of specific bonds in viral precursor protein or in cellular proteins – this is essential for the completion of the viral infectious cycle. In contrast, poorly growing bacteria have lower levels of proteases and so will encourage establishment of lysogeny.

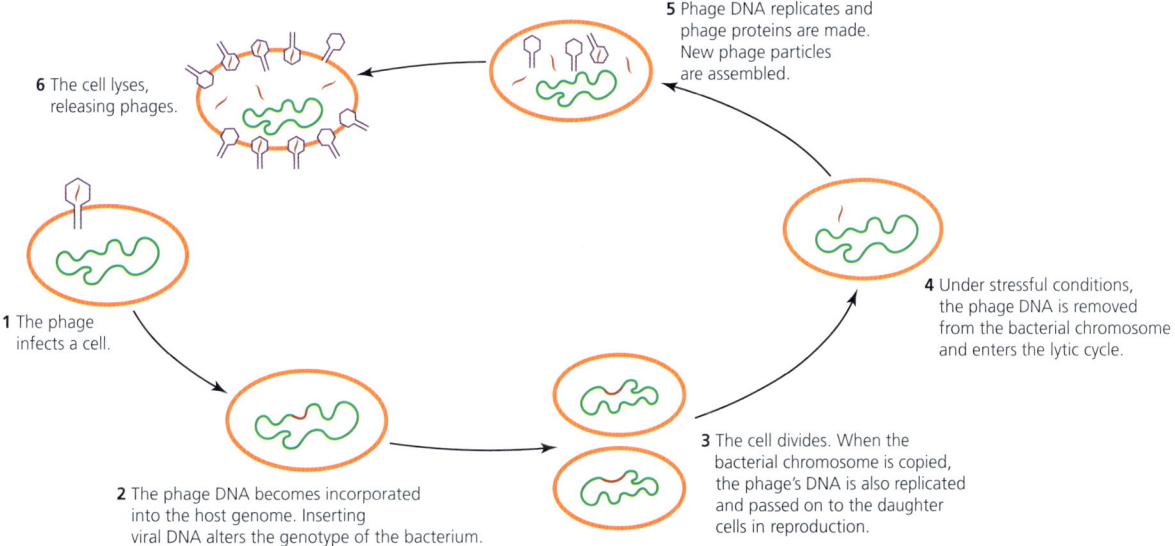

■ **Figure A2.3.8** An overview of the lysogenic cycle of a virus; note that the last two stages (5 and 6) are part of the lytic cycle

Top tip!

Bacteriophages that can adopt either a lytic or a lysogenic replication cycle are termed **temperate phages**.

2 Distinguish between the lytic and lysogenic life cycles of a virus.

Nature of science: Experiments

Determining the concentration of bacteriophage particles

If a bacteriophage infects a liquid culture of bacteria, it may result in completely clearing the culture. However, if the bacteria are spread on the surface of an agar plate, the phages released from an individual infected cell will only be able to infect neighbouring bacteria, which will result in localized clearing of the bacterial lawn, referred to as plaques. This allows a biologist to determine the concentration of bacteriophage particles in a suspension (the phage titre), using the assumption that the number of plaques corresponds to the number of bacteriophage particles present in the sample.

HL ONLY

> ### Tool 3: Mathematics
>
> #### Applying general mathematics
>
> The multiplicity of infection (MOI) is one factor that determines whether a virus enters the lytic cycle or lysogenic (dormant) phase.
>
> $$\text{MOI} = \frac{\text{number of infectious virus particles}}{\text{number of target cells present}}$$
>
> A scientist needed to use an MOI of 0.5 for an investigation.
>
> The virus particles were at a concentration of 2×10^9 cm^{-3} and the bacteria were at a concentration of 8×10^8 cm^{-3}.
>
> Calculate the volume of virus particles that should be added to 0.25 cm^3 of bacteria.

3 **Discuss** why viruses are considered non-living and why they are obligate parasites.

4 **Describe** the role of the structural components of viruses.

ATL A2.3B

Using the features in Table A2.3.2, research and create summary tables for SARS-CoV-2 (COVID-19), influenza and HIV.

■ Table A2.3.2 Summary of bacteriophage lambda

Feature	Description
Viral genome	linear DNA (with sticky ends)
Capsid	complex capsid comprising of an icosahedral head, a tail sheath and one tail fibre
Envelope	naked virus (no envelope)
Host	bacteria *E. coli* (gram-negative)
Cycle	lytic and lysogenic cycles
Attachment	tail fibre attaches to receptor on surface of host cell
Entry/penetration	(non-contractile) tail sheath inserts viral genome into host cell; capsid is left outside the host cell
Integration into host chromosome	integrated into bacterial host chromosome as prophage in the lysogenic cycle
Degradation of host cell DNA	DNA of host cell is only degraded during the lytic stage
Replication of viral genome	lysogenic: prophage replicates with the host chromosomes as host cell undergoes binary fission
	lytic: host cell DNA polymerase used to replicate phage DNA
Synthesis of viral proteins	transcription and translation by host cell machinery
Release	when lytic cycle is induced, prophage is removed from bacterial host chromosome; osmotic lysis occurs and the death of host cell releases phage particles

Evidence for several origins of viruses from other organisms

There are many different types of virus depending on the cells that they infect. It seems unlikely that all currently known viruses have a common ancestor. Many researchers hypothesize that viruses have probably arisen numerous times in the past by one or more mechanisms.

Viruses are therefore considered to be **polyphyletic** in origin, meaning they have a few independent origins – almost certainly at different times from cellular organisms. Three main hypotheses have been suggested to explain the evolutionary origin of viruses.

The **virus-first** hypothesis suggests that viruses evolved before or co-evolved with their current host cells. Viruses may have come from bits of RNA that had self-complementary sequences, allowing it to fold itself like a protein. Gradually the bits of RNA became more complex and could later self-replicate. They then began infecting other cells – becoming viruses.

The **escaped genes** (progressive) hypothesis (Figure A2.3.9) suggests that viruses arose from genetic elements that gained the ability to move between cells. Transposons are repetitive DNA sequences that can move between chromosomes in eukaryotes.

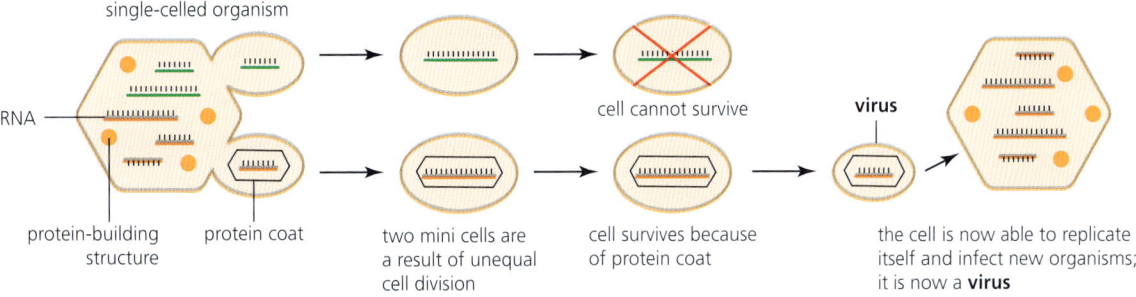

■ Figure A2.3.9 The escaped genes (progressive) hypothesis for the origin of viruses

The **regressive** (or reduction) hypothesis (Figure A2.3.10) suggests that viruses are remnants of cellular organisms (cells).

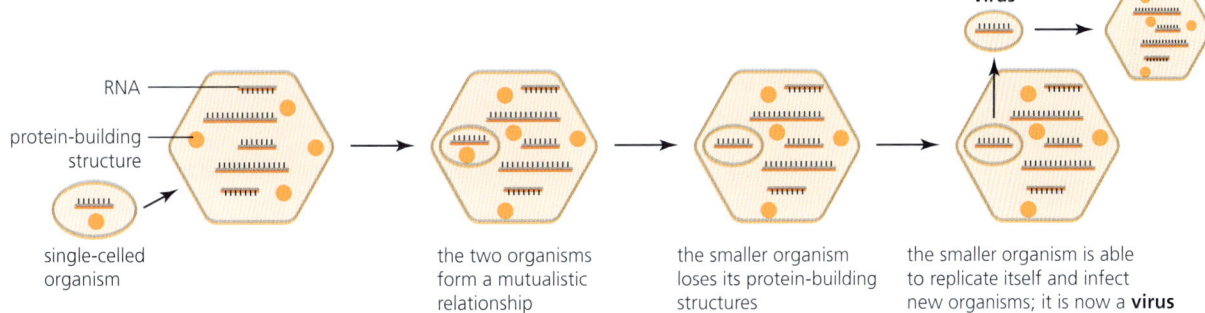

■ Figure A2.3.10 The regressive (reduction) hypothesis for the origin of viruses

Both the escaped genes and regressive hypotheses assume that cells existed before viruses, but some researchers have proposed that viruses may have been the first replicating entities (appearing before LUCA) involved in the origin of life on Earth.

There is a huge range of diversity in viruses (with no gene shared by all viruses) and it has been argued that precursors of a group of viruses known as **nucleocytoplasmic large DNA viruses (NCLDV)** were involved in the emergence of eukaryotic cells. This group of newly discovered viruses has a large genome and the presence of genes involved in DNA repair, DNA replication, transcription and translation. They replicate in the host cell's nucleus and cytoplasm. It is hypothesised that the nucleus in eukaryotic cells may have arisen from an endosymbiosis in which a complex, enveloped DNA virus became incorporated into an ancestral eukaryotic cell (see Chapter A2.2, page 82).

Some researchers hypothesise that NCLDV viruses evolved through a regressive process, during which their ancestral cells lost key genes and adopted an obligate parasite replication strategy. Their dependence on parasitism is likely to have caused the loss of genes that enabled them to survive outside a cell. The irreversible succession of gene losses may have been the dominant force in viral evolution.

A2.3 Viruses

Link

Convergent evolution and analogous structures are covered in more detail in Chapter A4.1, page 145.

 Top tip!

Viruses are evolutionarily dynamic and respond to changes in the macroenvironment and microenvironment (the body and immune system).

 Top tip!

The term 'proofreading' is used in genetics to refer to the error-correcting processes.

Link

Evolution and Darwinism are covered in more detail in Chapter A4.1, page 138, and Chapter D4.1, page 779.

 Top tip!

A variant describes a subtype or strain of a virus, e.g., delta and omicron variants of SARS-CoV-2, that is genetically distinct from the main strain, but not sufficiently different to be termed a distinct strain (e.g., influenza A, B, C and D).

It is possible that no single hypothesis explains the origin of viruses. Certain viruses, such as the retroviruses, probably arose through a progressive process, when mobile genetic elements gained the ability to travel between cells, becoming infectious agents. On the other hand, large DNA viruses probably arose through a regressive process, whereby once-independent entities lost essential genes over time and adopted a parasitic replication strategy. It is also possible that all viruses arose via a mechanism that has yet to be discovered.

Since all viruses share an extreme form of obligate parasitism on host cells, the structural features they have in common could be regarded as a form of **convergent evolution**. Convergent evolution occurs when species occupy similar ecological niches and adapt in similar ways in response to similar selective pressures. Traits that arise through convergent evolution are referred to as 'analogous structures'.

Rapid evolution of viruses

The evolution rate in viruses can be defined as the number of nucleotide substitutions per nucleotide site, per year. Various factors affect the rate including mutations, recombination, large population size and a rapid life cycle.

Mutation in viruses

Mutations are a natural by-product of viral replication. The mutations occur at a frequency determined by the error rate of the replication enzymes (and are then adjusted by the effect of any proofreading in the virus if it occurs).

The success of the mutant after generation is then determined by whether it has any form of advantage in a Darwinian sense (i.e. if it has a selective advantage that enables the virus to survive).

The mutations that give **variants** a selective advantage regarding viral replication, transmission or evading the immune system of the host organism will increase in frequency, while those that reduce viral fitness will be removed from the population of circulating viruses.

The mutation of existing viruses is a major source of new diseases. RNA viruses tend to have an unusually higher rate of mutation compared to DNA viruses because errors in replicating their RNA genomes are not corrected by proofreading. Coronaviruses, however, have a lower mutation rate than most RNA viruses because they produce an RNA polymerase that corrects some of the errors made during replication. However, the viruses are evolving faster than their host cells, which gives the virus an advantage in overcoming the immune response.

Case study: Influenza

The three types of influenza (flu) viruses are A, B and C. It is an enveloped virus with a genome made up of negative sense, single-stranded, segmented RNA. Influenza viruses infect the epithelial cells of the respiratory tract (and in birds the intestinal tract). The virions recognize their target cells by complementary binding of haemagglutinin (spike protein) in the viral envelope and specific sialic acid receptors on the cell plasma membrane.

Some mutations enable existing viruses to evolve into new strains that can cause disease in individuals who have developed immunity to the ancestral virus. Flu epidemics, for example, are caused by new strains of influenza virus genetically different enough from earlier strains that people have little immunity to them.

◆ **Antigenic drift**: small changes (caused by mutations) in viral genes that can lead to changes in the surface proteins of a virus, HA (haemagglutinin) and NA (neuraminidase).

◆ **Antigenic shift**: an abrupt, major change in an influenza A virus, resulting in new HA and/or new HA and NA proteins. Happens when recombination occurs.

◆ **Recombination**: occurs when viruses of two different parent strains co-infect the same host cell and interact during replication to generate virus progeny that have some genes from both parent strains.

Antigenic drift consists of mutations in the genes of influenza viruses that can lead to changes in the surface proteins of the virus: haemagglutinin (HA) and neuraminidase (NA). The surface proteins of influenza virus particles are antigens. These proteins are recognized by the immune system and can trigger an immune response, including the production of antibodies that stop the development of disease. The changes associated with antigenic drift happen continually over time as influenza viruses replicate, leading to surface antigens that are no longer recognized by the immune system. This in turn leads to the development of disease and the signs and symptoms of an illness.

Antigenic shift (Figure A2.3.11) is a sudden major change in an influenza A virus, resulting in new haemagglutinin and neuraminidase proteins. Antigenic shift can result in a new influenza A subtype. This process results in **recombination**, where DNA from different strains of the virus are recombined to produce new combinations of alleles.

If two different influenza viruses are present in the same cell, the newly produced viruses can take segments from either of the infecting viruses to make a new type of influenza virus.

Antigenic shift can happen if an influenza virus from an animal population gains the ability to infect humans. Such animal-origin viruses can contain haemagglutinin and neuraminidase protein combinations that are different enough from human viruses that most people will not have immunity to the novel virus.

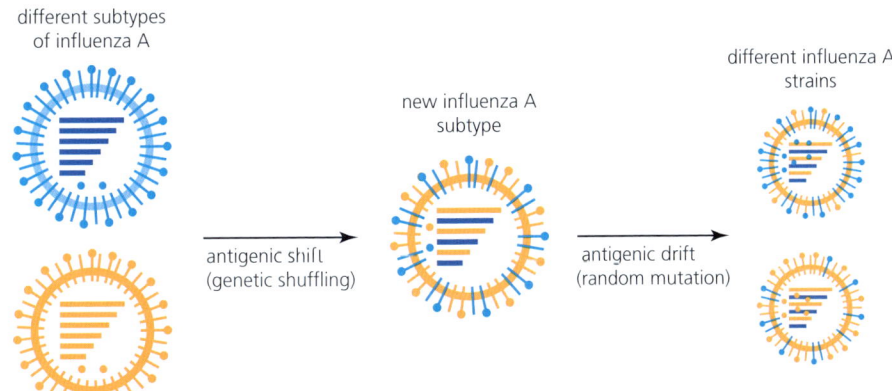

■ **Figure A2.3.11** Conceptual illustration of antigenic drift and antigenic shift

Link

HIV is covered in more detail in Chapter C3.2, page 529.

◆ **Retrovirus**: a virus that has RNA as its nucleic acid and uses the enzyme reverse transcriptase to copy its genome into the DNA of the host cell's chromosomes.

◆ **Reverse transcriptase**: an enzyme found in HIV (and other retroviruses). HIV uses reverse transcriptase to convert its RNA into viral DNA, a process called reverse transcription.

◆ **Provirus**: a virus that has integrated into a host's genome and is replicated when the cell replicates its DNA.

■ Case study: HIV

Acquired Immune Deficiency Syndrome (AIDS) is a condition caused by the human immunodeficiency virus (HIV). HIV (Figure A2.3.12) belongs to a small group of viruses known as **retroviruses**, which have a unique life cycle and mechanism of viral replication. A retrovirus is a type of virus that uses RNA as its genetic material.

Having entering the cytoplasm of a host cell, the virus RNA is converted into DNA under the control of an enzyme called **reverse transcriptase**. The retrovirus then integrates its viral DNA into the DNA of the host cell's nucleus to form a **provirus**.

HIV has a high mutation rate in its genome due to errors by reverse transcriptase, which lacks a proofreading mechanism. AIDS patients often carry many different genetic variants of HIV that are distinct from the original HIV virus that infected them.

This is a significant problem in treating the infection. Antiviral drugs that block essential viral enzymes work only temporarily, because new strains of the virus resistant to these drugs arise rapidly by mutation.

However, combinations of drugs that target different virus proteins (typically three drugs with different targets) are highly effective at inhibiting the virus and preventing replication.

The population size of HIV is the total number of infectious proviruses integrated into the cellular DNA of an individual at a given time. The high mutation rates of viruses such as in HIV, combined with short generation times and large population sizes, allows viruses to evolve rapidly and adapt to the host.

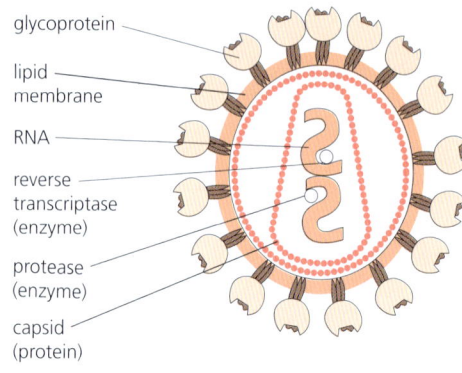

■ Figure A2.3.12 The human immunodeficiency virus (HIV)

 Top tip!

Emerging diseases, such as HIV and COVID-19, are often virulent as they are not adapted to their new host species.

Nature of science: Falsification

Discovery of reverse transcriptase in viruses

The central dogma of molecular biology proposes that genetic information flows unidirectionally from DNA to RNA to protein. However, in 1970 the enzyme reverse transcriptase was discovered in retroviruses by David Baltimore, thereby falsifying the first step of the central dogma. Its discovery laid the foundation for the discovery of the HIV virus in 1984 and led to the development of effective therapy for treating HIV infection. The transfer of genetic information from RNA to DNA has since been recognized as a general principle of evolution operating not just in viruses but throughout biology.

 ATL A2.3C

Choose a virus (that causes a well-known infectious disease) to write about, such as rabies or Ebola. You could create a poster showing the structure of its capsid and genome, its life cycle in a named host cell, its evolutionary history, its effects on human physiology and current treatments.

Consequences of rapid viral evolution

The rapid evolution of viruses has consequences for treating the diseases caused by these pathogens. Yearly flu vaccines are developed to ensure their efficacy (the percentage reduction in a disease in a group of people who received a vaccination in a clinical trial). In the most recent COVID-19 pandemic, many variants evolved, sometimes with different symptoms. The evolution of new variants leads to difficulties in treating the virus, with the vaccines first used in the pandemic becoming less efficacious as new variants emerged.

Tool 2: Technology

Using computer modelling

Compartmental models are a very general modelling technique often used in the mathematical modelling of infectious diseases, including COVID-19. The population is classified into compartments with labels, for example, S, I or R, (susceptible, infectious or recovered).

People in the population may move from S to I to R. Figure A2.3.13 below shows the SIR model for the transition between these states. The SIR model is accurate for describing infectious diseases that are only transmitted by human to human and where recovery gives long-lasting resistance.

parameters, such as the reproductive number, R_0, where R_0 is the expected number of secondary cases produced by a single infection in a population that is completely susceptible. The models can show how different interventions, such as a lockdown or vaccination, may affect the outcome of an epidemic.

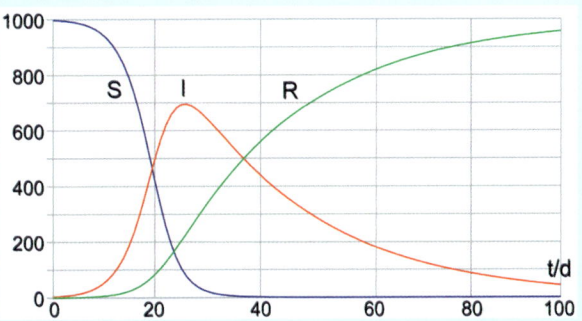

■ **Figure A2.3.14** SIR model showing numbers of susceptible, infectious or recovered in a population

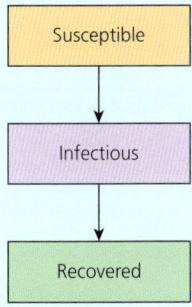

■ **Figure A2.3.13** SIR model showing the transition between susceptible, infectious and recovered states

Models try to predict how a viral or bacterial disease spreads, the total number infected, or the time period of an epidemic, and estimate various epidemiological

Figure A2.3.14 shows an SIR model for an outbreak of an infectious disease. The y-axis is the number of people and the x-axis is the time in days. S stands for 'numbers of susceptible', i.e. members of the population who can catch the infection; I for 'infectious', or those in the population who have the infection and can spread it; and R for 'recovered', those who have had the disease but are now better.

---

5 **Describe** the graph model in Figure A2.3.14 and then **deduce** how S, I and R interact with each other.

ATL A2.3D

What factors determine the spread of a virus? Use the following site to explore various scenarios: www.washingtonpost.com/graphics/2020/world/corona-simulator/

What lessons have you learned about the most effective strategies for limiting the spread of rapidly evolving viruses?

LINKING QUESTIONS

1. What mechanisms contribute to convergent evolution?
2. To what extent is the history of life characterized by increasing complexity or simplicity?

A2.3 Viruses

A3.1 Diversity of organisms

Guiding questions

- What is a species?
- What patterns are seen in the diversity of genomes within and between species?

SYLLABUS CONTENT

This chapter covers the following syllabus content:
- ▶ A3.1.1 Variation between organisms as a defining feature of life
- ▶ A3.1.2 Species as groups of organisms with shared traits
- ▶ A3.1.3 Binomial system for naming organisms
- ▶ A3.1.4 Biological species concept
- ▶ A3.1.5 Difficulties distinguishing between populations and species due to divergence of non-interbreeding populations during speciation
- ▶ A3.1.6 Diversity in chromosome numbers of plant and animal species
- ▶ A3.1.7 Karyotyping and karyograms
- ▶ A3.1.8 Unity and diversity of genomes within species
- ▶ A3.1.9 Diversity of eukaryote genomes
- ▶ A3.1.10 Comparison of genome sizes
- ▶ A3.1.11 Current and potential future uses of whole genome sequencing
- ▶ A3.1.12 Difficulties applying the biological species concept to asexually reproducing species and to bacteria that have horizontal gene transfer (HL only)
- ▶ A3.1.13 Chromosome number as a shared trait within a species (HL only)
- ▶ A3.1.14 Engagement with local plant or animal species to develop a dichotomous key (HL only)
- ▶ A3.1.15 Identification of species from environmental DNA in a habitat using barcodes (HL only)

◆ **Species**: a group of individuals of common ancestry that closely resemble each other and that are normally capable of interbreeding to produce fertile offspring.

◆ **Variation**: differences between individuals of a species. May be caused by environmental and/or genetic factors.

Link
The theory of evolution by natural selection is explored more fully in Chapter D4.1, page 779.

Concept: Diversity
All species show variation. This is one of the pillars on which evolution by natural selection is based.

Variation between organisms

In 1831, scientist Charles Darwin was employed as a naturalist on the ship *HMS Beagle*. The journey lasted five years and provided the fieldwork for Darwin's famous book *On the Origin of Species by Means of Natural Selection*, which was published in 1859. During his travels, and particularly during his visit to the Galápagos Islands, Charles Darwin noted that all **species** showed large **variation**: each individual was slightly different from others in the population (see Figure A3.1.1). This observation formed a central pillar of his theory of evolution by natural selection – an explanation of how all species on Earth arose. Darwin had also observed that humans can select for different breeds of animals (for example pigeons – see page 141), and that if this 'artificial selection' could lead to new varieties then there was no reason why, given enough time, nature could not select for new varieties and, ultimately, new species (hence 'natural selection').

■ **Figure A3.1.1** Variation in one species of beetle; the wing cases of fourteen-spot ladybirds (*Propylea 14-punctata*) show variation in colouration patterns between these individuals, which were all collected from the same nettle plant

Theme A: Unity and diversity – Organisms

◆ **Continuous variation**: where there is a continuum of variation from one phenotype to another, e.g. human height.

◆ **Polygenic inheritance**: inheritance of phenotypic characteristics (such as height in humans) that are determined by the collective effects of several different genes.

◆ **Discontinuous variation**: arises when the characteristic concerned is one of two or more discrete types with no intermediate forms. Examples include the garden pea plant (tall or dwarf) and human ABO blood grouping (group A, B, AB or O).

◆ **Allele**: an alternative form of a gene occupying a specific locus on a chromosome.

◆ **Homeotic mutation**: alteration in genes that determine the type or location of a body part during an organism's development.

◆ **Morphology**: form and structure of an organism.

■ Continuous variation

As Darwin noted, all species show variation. Some of this variation can be at any value of measurement for a specific characteristic, such as height, and is called **continuous variation**. Many features of humans are controlled by **polygenes** (i.e. several different genes), including body weight and height. A graph of the height of a population of 400 people, Figure A3.1.2, shows continuous variation between the shortest person at 160 cm and the tallest person at 185 cm, with a mean height of 173 cm.

Human height is determined genetically by interactions of the alleles of several genes, probably located at loci on different chromosomes.

Variation in the height of adult humans
The results cluster around a mean value and show a normal distribution. For the purpose of the graph, the heights are collected into arbitrary groups, each of a height range of 2 cm.

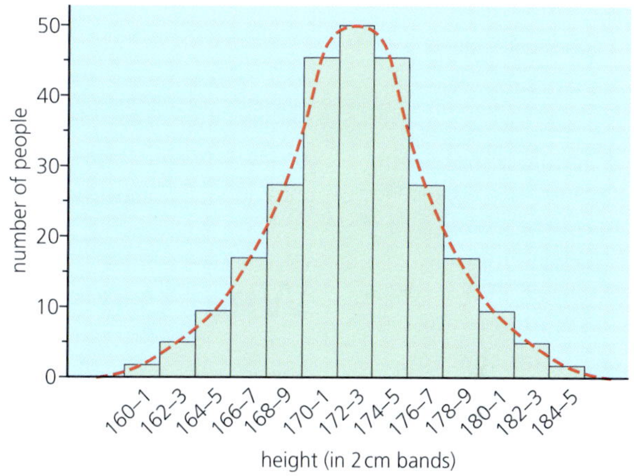

■ **Figure A3.1.2** Human height as a case of polygenic inheritance

'Normal distribution' means that when the frequency of measurement classes is plotted against the measurement classes, a symmetrical bell-shaped curve is obtained, with values grouped symmetrically around a central value (Figure A3.1.2).

Another factor that may affect the appearance of the phenotype (the physical appearance of an organism and its internal physiology) and produce variation between organisms is the effect of environmental conditions. This can affect the continuous variation of a species by altering the physiology of individuals. For example, a tall plant may appear almost dwarf if it has been consistently deprived of adequate essential mineral ions. Similarly, the physique of humans may be greatly affected by the levels of nourishment received, particularly as children. So, the phenotype of an organism is the product of both its genotype and the influences of the environment.

■ Discontinuous variation

Some variation in species is not continuous but discrete, such as blood type, and so is known as **discontinuous variation**. Discontinuous variation is, therefore, variation that has distinct groups that organisms belong to. There is no intermediate form and no overlap between the two phenotypes. Characteristics that demonstrate discontinuous variation are controlled by a single gene, usually with two **alleles** (the human blood group system has three alleles – see Chapter D3.2, page 734).

Discontinuous variation is not subject to environmental factors and is determined solely by the genotype (genetic composition) of the species.

In this way, species can show both continuous and discontinuous patterns of variation.

■ Distinguishing between different species

Figure A3.1.1 shows that there is morphological variation within a species, where **morphology** refers to the 'form and structure' of an organism. The question then arises, how can one species be distinguished from another species that also shows significant variation? Some individuals of one species, within the variation shown, may resemble individuals of another species.

● **Top tip!**

Discontinuous variation can be due to threshold effects in development or mutations of large effect, including **homeotic mutations** that cause discrete changes in form from a single mutation (e.g., loss of one pair of wings in insects; conversion of petals to stamens in plants).

A3.1 Diversity of organisms

Link
For more on classifying organisms see Chapter A3.2, page 125.

1 **Distinguish** between continuous and discontinuous variation.

♦ **Sexual dimorphism**: differences in appearance between males and females of the same species, such as in colour, shape, size and structure.

♦ **Species concept**: a working definition of a species and a methodology for determining whether two organisms are members of the same species.

♦ **Morphological species concept**: species are groups of organisms with shared traits.

♦ **Taxonomy**: the science of classification.

♦ **Taxon**: a classificatory grouping.

Historically, **taxonomists** – scientists who define and classify groups of biological organisms based on shared characteristics – over many years have learnt to distinguish one species from another. Many species show **sexual dimorphism**, where males appear different from females (e.g. lions), which also leads to morphological variation within a species. By collecting and studying a large number of individuals within a species, taxonomists can distinguish one species from another irrespective of the complex variation shown. This is the basis for naming and classifying organisms.

Most natural history museums, where the process of classification and storage of the Earth's biodiversity takes place, have molecular laboratories for the DNA profiling of species within their collection. This has allowed taxonomists to check on the current classification of species, make necessary changes, and to analyse new species to ensure they are placed correctly within the classification system (see page 135). DNA profiling (Chapter D1.1, page 608) allows for genetic variation within a species and ensures that clear parameters can be set when deciding whether an individual belongs to one species or another.

What is a species?

Species were originally defined by their appearance – the **species concept** – based on what they looked like (shape, colour and other distinctive features). This is known as the **morphological species concept**. This concept therefore defines a species in terms of its body shape and other structural features (or 'traits'). Two organisms that had different morphologies were placed in different species. Species have been defined in this way since before modern science. This original concept of 'species' was used by Swedish scientist Carl Linnaeus (see more below) in the 1700s.

Although much criticised, the concept of morphological species is still a widely used species concept in everyday life, and still retains an important place within the biological sciences, particularly in the case of plants.

Advantages:
- can be applied to asexual and sexual organisms
- does not require any information on the extent of gene flow
- can be applied to extinct and fossilized species
- easiest and fastest concept to apply in the field because it is based only on the appearance of the organism.

Limitations:
- relies on subjective criteria and researchers may disagree on which structural features distinguish a species
- different individuals in a species may appear very different, such as males and females.

The binomial system for naming organisms

Classification is essential to biology because there are too many different living organisms to sort and compare unless they are organized into manageable categories. There are currently approximately 1.8 million described species, and so a process is needed to divide organisms into groups of similar species. Biological classification schemes are the invention of biologists and are based upon the best available evidence at the time. With an effective classification system in use, it is easier to organize our ideas about organisms and make generalizations.

The science of classification is called **taxonomy**. The word comes from **taxa** (singular, taxon), which is the general name for groups or categories within a classification system.

The scheme of classification must be flexible, allowing newly discovered living organisms to be added where they fit best. It should also include fossils since we believe living and extinct species are related.

The process of classification involves:
- giving every organism an agreed name
- imposing a scheme upon the diversity of living things.

◆ **Binomial system**: double names for organisms, in Latin, with the generic name preceding the specific name.

◆ **Genus**: a group of similar and closely related species.

Carl Linnaeus (Figure A3.1.3) was a Swedish botanist, physician and zoologist, who in the 1730s developed the method of classifying organisms that is still used today: binomial nomenclature or the **binomial system** (meaning 'two-part name'). Linnaeus is known as the father of modern taxonomy and is also considered one of the fathers of modern ecology. Linnaeus published *Systema Naturae* (The Natural World) in 1734 (Figure A3.1.4), in which he divided flowering plants into classes determined by the structure of their sexual organs. In 1749 he introduced the binomial nomenclature for which he is now famous. Each plant was given a Latin noun (the **genus**) followed by an adjective (the **species**). The generic name comes first and begins with a capital letter, followed by the species name. Traditionally, a species name is written in italics (or is underlined). Humans, for example, have the species name *Homo sapiens*, where '*Homo*' is the genus and '*sapiens*' is the species. There can be several different species in a genus – we share the *Homo* genus with several other (now extinct) species, such as the Neanderthals (*Homo neanderthalensis*) – see Figure A3.1.5.

Species in the same genus share similar traits. For example, the skulls from genus *Homo* shown in Figure A3.1.5 have a larger brain capacity and smaller face compared to other species of the family Hominidae (the great apes). The similar traits relate to similar life histories and adaptations to the environment (Chapter B4.2).

■ **Figure A3.1.3** Carl Linnaeus

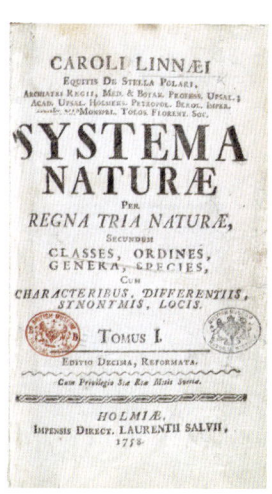

■ **Figure A3.1.4** Carl Linnaeus' most famous work; *Systema Naturae* showed for the first time how life could be classified

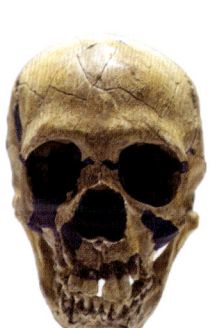 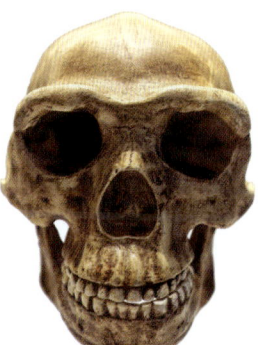  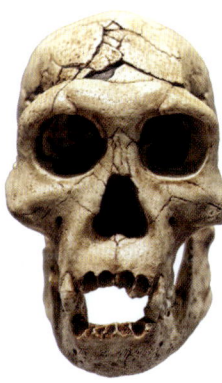

Homo neanderthalensis   *Homo antecessor*   *Homo sapiens*   *Homo erectus*

■ **Figure A3.1.5** Skulls of species in the genus *Homo*

As shown in Figure A3.1.6, closely related organisms have the same generic name; only their species names differ. You will see that, when organisms are referred to frequently, the full name is given the first time, but after the generic name is shortened to the first (capital) letter. Thus, in continuing references to humans in an article or scientific paper, *Homo sapiens* would become *H. sapiens*.

● **Top tip!**

When writing species names, the convention is to start the genus with a capital letter and the species with a lower-case letter. The species name should be either written in italics or underlined.

A3.1 Diversity of organisms

2 **Outline** the process by which species are named.

Panthera leo (lion)

Panthera tigris (tiger)

> **Concept: Unity**
>
> The binomial system provides a universal means of classifying organisms that can be understood worldwide.

■ **Figure A3.1.6** Naming organisms by the binomial system

● Common mistake

Students often begin the genus name with a lower-case letter, for example *homo sapiens*. This is incorrect. The genus should start with a capital, e.g. *Homo*, and the full species name should be underlined when written by hand, e.g. Homo sapiens.

● Nature of science: Science as a shared endeavour

Why is a standardized, universal method for naming species needed?

Many organisms have local common names, but these often differ around the world. They do not allow observers to be confident that they are all talking about the same organism. For example, the name 'magpie' represents entirely different birds commonly seen in Europe, in Asia and in Sri Lanka (Figure A3.1.7). Instead, scientists use the international binomial system, so that everyone everywhere knows exactly which organism is being referred to. International cooperation and collaboration continue to develop and apply a common set of species names for use throughout the world. This agreed convention and the common terminology for species names used globally helps prevent ambiguous communication.

European magpie
(*Pica pica*)

Asian magpie
(*Platysmurus leucopterus*)

Sri Lankan magpie
(*Urocissa ornata*)

3 Scientific names of organisms are often difficult to pronounce or remember. **State** why they are used.

■ **Figure A3.1.7** 'Magpie' species of the world

The biological species concept

♦ **Biological species concept**: a species is a group of organisms that can breed and produce fertile offspring.

The most widely used definition of species in modern science was developed by zoologist Ernst Mayr in 1942. According to his **biological species concept**, a species is a group of organisms that can breed and produce fertile offspring.

There are limitations to the biological species concept:
- It does not apply to organisms that reproduce asexually all or most of the time, e.g. prokaryotes.
- It does not apply to organisms that are extinct.
- It overemphasizes gene flow and downplays the role of natural selection, as natural selection can cause many pairs of species that are morphologically and ecologically distinct to remain distinct and yet have gene flow between them.

■ Alternative species concepts

♦ **Ecological species concept**: a species is a group of organisms that is adapted to a particular set of resources (niche); this concept explains differences in form and behaviour between species as adaptations to resource availability.

♦ **Evolutionary species concept**: a species is a single lineage of populations descending from a common ancestor, which maintains its identity from other such lineages and which has its own selection pressures and evolutionary outcome.

♦ **Population**: interacting groups of organisms of the same species living in an area.

An alternative to the original morphological species concept is the **ecological species concept**. Ecology is the study of organisms in relation to their environment, and so this concept uses ecological factors to define species. Each species has a particular role in the environment, known as its 'niche'. The niche of a species is everything about where and how it lives – what it eats, how it reproduces, its habitat, and so on. Each species has therefore, by definition, a unique niche. The ecological species concept defines a species in terms of its ecological niche, i.e. the sum of how members of the species interact with the non-living (abiotic) and living (biotic) aspects of their environment. For example, two species of finches on the Galápagos Islands may be similar in appearance but distinguishable based on what they feed on.

Advantages of the ecological species concept:
- It can accommodate asexual as well as sexual species.
- It emphasizes the role of disruptive natural selection as organisms adapt to different environmental conditions.

Limitations of the ecological species concept:
- Niches are generally difficult to identify because a niche involves the interaction between the organism and its environment (both biotic and abiotic).

Today, genetic techniques allow scientists to compare genetic sequences from different organisms to compare similarities and potential ancestry. These techniques have allowed the **evolutionary species concept** to be developed, which gives scientists a detailed overview not only of what defines a species, but also the interrelationship between species over geological time.

Table A3.1.1 compares the four different species concepts.

■ **Table A3.1.1** Comparing different species concepts

Species concept	Approach and methodology
Morphological species concept	relies on morphological data and emphasizes groups of physical traits that are unique to each species
Biological species concept	relies on behavioural information and emphasizes reproductive isolation between groups
Ecological species concept	relies on detailed information about how organisms interact with their biotic (biological) and abiotic (non-living) environment
Evolutionary species concept	relies on genetic data and emphasizes distinct evolutionary links between groups

4 **Distinguish** between the biological and evolutionary species concepts.

Difficulties distinguishing between a population and a species

In nature, organisms are members of local **populations**. A population is a group of individuals of the same species, living close enough to be able to interbreed (see more in Chapter C4.1, page 548). Species typically exist in localized populations, although the boundaries of a local population can be hard to define.

A3.1 Diversity of organisms

Look at Figure A3.1.8 below.

You can see that a population of garden snails might occupy a small part of a garden, perhaps around a compost heap. A population of thrushes (snail-eating birds) might occupy some gardens and surrounding fields. So, the area occupied by a population depends on the size of the organism and on how mobile it is, for example, as well as on environmental factors (such as food supply and predation).

The boundaries of a population may be hard to define. Some populations are completely 'open', with individuals moving in from, or out to, other nearby populations. Alternatively, some populations are 'closed'; they are isolated populations almost completely cut off from neighbours of the same species. The snails in the traffic island flower bed (Figure A3.1.8) are a good example of a closed population.

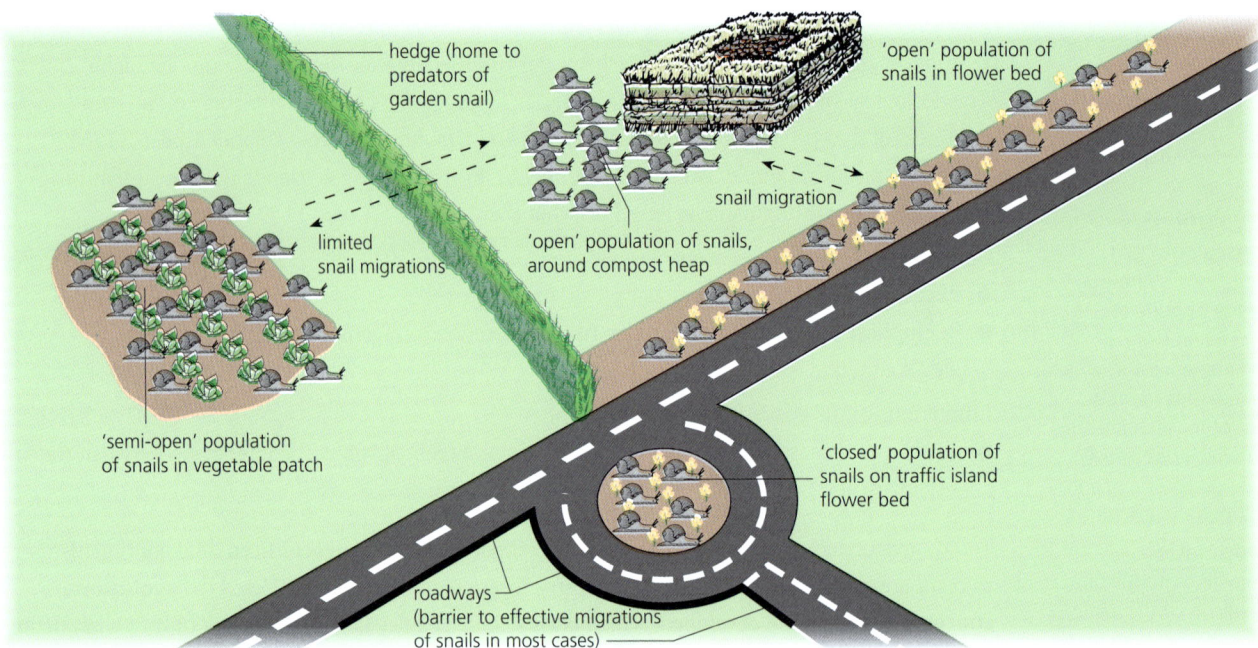

■ **Figure A3.1.8** The concept of population

Link
The topic of speciation by splitting of pre-existing species is explored further in Chapter A4.1, page 147.

Gene pools are discussed in Chapter D4.1, page 790.

◆ **Speciation**: the process by which new species form, where one species is split into two or more species.

Individuals in local populations tend to resemble each other. They may become quite different from members of other populations. Local populations are very important because they are a potential starting point for **speciation**. Speciation is the name we give to the process by which one species splits into two or more species. Present-day flora and fauna have arisen by change from pre-existing forms of life. The term 'speciation' emphasizes the fact that species change. The fossil record provides evidence for this process (see page 161).

Development of barriers within local populations is a possible cause. Before separation, individuals share a common **gene pool** (See Chapter D4.1, page 790) but, after isolation, processes such as mutation can cause change in one population but not in the other. Alternatively, a new population may form from a tiny sample that became separated from a much larger population. While the number of individuals in the new population may rapidly increase, the gene pool from which they formed may have been totally unrepresentative of the original, with many alleles lost altogether. In this way, populations of the same species can change, through accumulation of gradual changes in the genotype through time, to eventually form new, genetically distinct species.

Going further

The founder effect and genetic drift

The **founder effect** occurs when a few individuals become geographically isolated from a larger population, and this smaller group may establish a new population whose gene pool differs from the source population. The isolation mechanism indiscriminately chooses some individuals, but not others, from the source population.

Genetic drift describes the random fluctuations in the numbers of alleles in a population. Genetic drift takes place when the occurrence of alleles increases or decreases by chance over time. These variations in the presence of alleles are measured as changes in allele frequencies.

Inquiry 1: Exploring and designing

Exploring

Think about the definitions of 'species' and 'population', and the way in which separate populations of one species can evolve into distinct species. Is there a problem here?

Isolated populations will show variation both within and between populations. At what point do variations between isolated populations show sufficient difference to allow separate groups to be called new species?

Research problems regarding species concepts. Are there ways of deciding when speciation has occurred? Make sure you consult a variety of relevant sources of information.

Speciation usually happens gradually rather than by a single act, with populations becoming more and more different in their traits. It can therefore be an arbitrary decision whether two populations are regarded as the same or different species. A further difficulty arises because of the definition of 'species'. According to the biological species concept, a species is a group of organisms that can breed and produce fertile offspring. If populations have diverged and can no longer interbreed (i.e. the individuals of one population do not reproduce with those of another), how can it be determined that the isolated populations have evolved sufficiently to form new species? It is possible that separate populations can still, in theory, interbreed, and are still therefore members of the same species, but this cannot be proven since they are isolated from one another and so cannot physically meet up and interbreed.

 ### Common mistake

The smallest unit of evolution is a population. One common misconception about evolution is that individual organisms evolve during their lifetimes. Natural selection (Chapters A4.1 and D4.1) acts on individuals, but it is populations that evolve. Individual organisms cannot evolve. But a population can have genetic variation in traits and can undergo Darwinian evolution, resulting in all the changes in genotypes and phenotypes associated with evolutionary change.

The diversity in chromosome numbers of plant and animal species

Concept: Diversity

Chromosomes vary in number and shape among organisms. Bacteria have circular chromosomes whereas animals and plants have linear chromosomes. Variety in chromosome number can be related to evolutionary origin and functional differences between organisms.

Concept: Unity

The number of chromosomes in the cells of different species varies, but in any one species the number of chromosomes per cell is normally constant (Table A3.1.2). A fruit fly, for example, has 8 chromosomes, while a rice plant has 24.

The variety, or diversity, of organisms is demonstrated by the varying number of chromosomes seen in animal and plant species (Table A3.1.2). There are five features of the chromosomes of eukaryotic organisms that are helpful to remember.

■ 1 The number of chromosomes per species is fixed

The number of chromosomes in the cells of different species varies, but in any one species the number of chromosomes per cell is normally constant (Table A3.1.2). For example, the mouse has 40 chromosomes per cell, the onion has 16, humans have 46 and the sunflower has 34. Each species has a characteristic chromosome number. Note, these are all *even* numbers.

■ **Table A3.1.2** Diploid chromosome numbers compared

Species	Number of chromosomes per cell	Species	Number of chromosomes per cell
Parascaris equorum (roundworm)	2	*Mus musculus* (mouse)	40
Drosophila melanogaster (fruit fly)	8	*Homo sapiens* (human)	46
Oryza sativa (rice)	24	*Pan troglodytes* (chimpanzee)	48
Helianthus annuus (sunflower)	34	*Canis familiaris* (dog)	78

> **Top tip!**
> You need to know, as an example of diversity in chromosome numbers, that humans have 46 chromosomes and chimpanzees have 48.

■ 2 The shape of a chromosome is characteristic

Chromosomes are long, thin structures of a fixed length. Somewhere along the length of the chromosome is a narrow region called the **centromere**. Centromeres may occur anywhere along the chromosome, but they are always in the same position on any given chromosome. The position of the centromere and the length of chromosome (as well as the banding patterns after staining) on each side enable scientists to identify particular chromosomes in photomicrographs.

■ 3 The chromosomes of a cell occur in pairs called homologous pairs

We have seen that the chromosomes of a cell occur in pairs, called **homologous pairs** (Figure A1.2.9). One of each pair came originally from the male parent and one from the female parent. Cells in which the chromosomes are in homologous pairs are described as having a **diploid** nucleus. We describe this as $2n$ where the symbol 'n' represents one set of chromosomes. A cell that has one chromosome of each pair has a **haploid** nucleus. We represent this as n. A sex cell has a haploid nucleus – formed as a result of the nuclear division known as meiosis (page 271).

◆ **Centromere**: constriction of the chromosome, the region that becomes attached to the spindle fibres during nuclear division.

◆ **Diploid**: cells having two sets of chromosomes (one from each parent organism).

◆ **Haploid**: cells having one of set of chromosomes.

Link

The terms 'locus' and 'allele' were introduced in Chapter A1.2, page 21.

◆ **Homozygous**: having two identical alleles of a gene.

◆ **Heterozygous**: having two different alleles of a gene.

5 **Define** the terms:
  a gene
  b allele.

6 **Explain** how genes and alleles differ.

■ 4 Genes occur at specific loci

We have seen that chromosomes carry genes in a linear sequence. The position of a gene is called a **locus** (plural, loci), and each gene has two or more forms, called **alleles** (Figure A1.2.9). The two alleles may carry exactly the *same* 'message' – the same sequence of bases coding for an identical protein. A diploid organism that has the same allele of a gene at the gene's locus on both copies of the homologous chromosomes in its cells is described as **homozygous**.

Alternatively, the two alleles may be *different*. A diploid organism that has different alleles of a gene at the gene's locus on both copies of the homologous pair is **heterozygous**.

■ 5 Chromosomes are copied precisely

Between nuclear divisions, while the chromosomes are uncoiled (in the form of chromatin) and cannot be seen, each chromosome is copied. It is said to **replicate**.

Replication occurs in the cell cycle, during interphase (page 651). The two identical structures formed are called **chromatids** (Figure A3.1.9). The chromatids remain attached by their centromeres until they are separated during nuclear division. After division of the centromeres, the chromatids are recognized as chromosomes again.

Of course, when chromosomes are copied, the critical event is the copying of the DNA double helix that runs the length of the chromosome. Replication occurs in a very precise way, brought about by specific enzymes, as will be discussed in Chapter D1.1 (page 602).

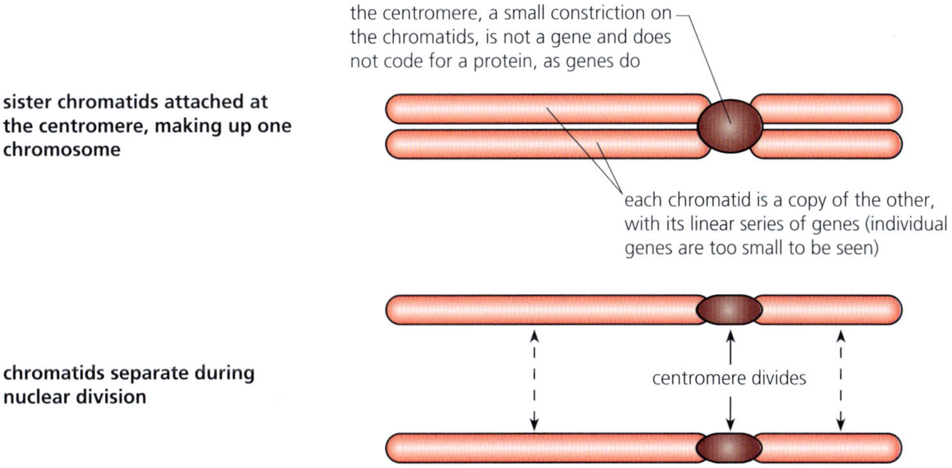

■ Figure A3.1.9 One chromosome as two chromatids

Karyotyping and karyograms

◆ **Karyotype**: the number and type of chromosomes present in an organism.

The number and type of chromosomes in the nucleus is known as the **karyotype**. In Figure A3.1.10 on the left-hand side, the karyotype of a diploid human male cell is shown, much enlarged.

These chromosomes are seen at an early stage of the nuclear division called mitosis (page 653). You can see that at this stage each chromosome is present as two chromatids, held together by its centromere.

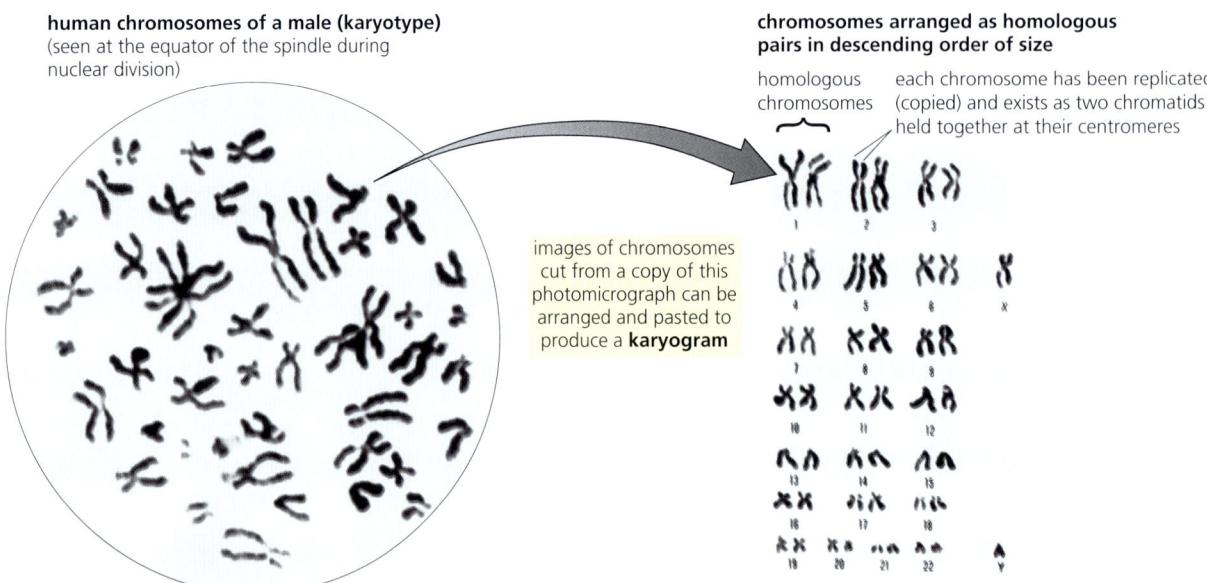

■ Figure A3.1.10 Chromosomes as homologous pairs, seen during nuclear division

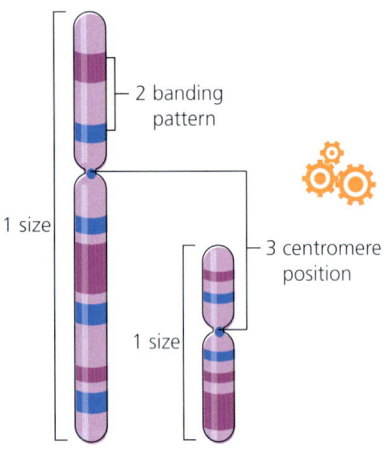

■ Figure A3.1.11 Three features can be used to identify chromosomes

◆ **Karyogram**: a diagram or photograph showing the chromosomes of a cell, arranged in homologous pairs in descending order of size.

◆ **Sex chromosome**: a chromosome which determines sex rather than other body (somatic) characteristics.

Link
For more on mitosis and cell division see Chapter D2.1, page 653.

For the image on the right in Figure A3.1.10, the individual chromosomes were cut out from a copy of the original photograph. These were then arranged in homologous pairs in descending order of size, and numbered. A photograph of this type is called a **karyogram**.

A karyotype, as well as being the number and appearance of chromosomes, also includes their length, banding pattern and centromere position. These differences can be seen in a karyotype stained to show **banding patterns** (Figure A3.1.11). This is the easiest way to tell two different chromosomes apart. Banding patterns are shown when chromosomes are stained with Giemsa dye, or fluorochromes when chromosomes are to be studied with a fluorescent microscope (page 62). The light and dark bands describe the position of genes on the chromosome – the size and location of bands on chromosomes make each chromosome pair unique. Centromeres are regions in chromosomes that appear as a constriction (Chapter D2.1) and play a role in the separation of chromosomes into daughter cells during cell division (mitosis and meiosis). Again, the position of the centromere varies (Figure A3.1.11) and can be used to distinguish different chromosomes.

Sex chromosomes

In Figure A3.1.10 you can see that two chromosomes are not numbered on the karyogram. Rather, they are labelled X (next to 6) and Y (next to 22). These are known as the **sex chromosomes**; they decide the sex of the individual – a male in this case. They are often shown as the final pair of chromosomes in the karyogram. All the other chromosomes (pairs numbered 1 to 22) are called **autosomes**. Karyograms are used by genetic counsellors to detect the presence of (rare) chromosomal abnormalities, such as Down syndrome.

Link
Chromosomal abnormalities are discussed further in Chapter D2.1, page 658.

■ Investigating the differences in human and chimpanzee chromosome numbers

Humans (*Homo sapiens*) have 23 pairs of chromosomes. However, every other living species in the family Hominidae, including our closest relatives the bonobo (*Pan paniscus*) and chimpanzee (*Pan troglodytes*), have 24 chromosome pairs. DNA sequencing has shown that less than 3% of human DNA is different from the chimpanzee. What might have happened to change the number of chromosomes in humans? Have a look at Figure A3.1.12, which compares the karyograms of chimpanzees and humans. What do you notice?

An alignment of the human and chimpanzee chromosomes shows that the centromere of human chromosome 2 lines up almost exactly with that of chimpanzee chromosome 2A. The hypothesis has been suggested that two of the ape chromosomes must have fused to form one large one. Let's evaluate the evidence in the Inquiry.

7 **Outline** the evidence that chromosome 2 in humans arose from the fusion of chromosomes in a shared ancestor.

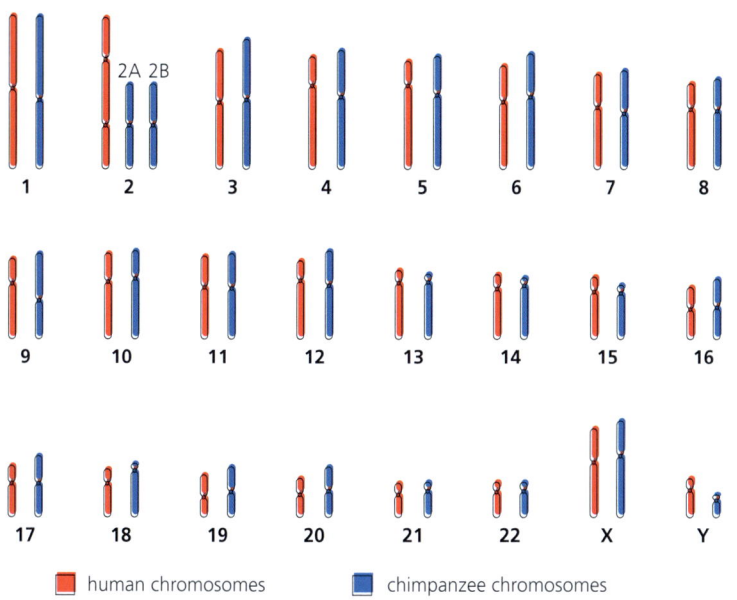

■ Figure A3.1.12 A comparison of human and chimpanzee chromosomes

Inquiry 3: Concluding and evaluating

Concluding

1 Scientists compared human and chimpanzee chromosomes and found remarkable similarity in banding between human chromosome 2 and chimp chromosomes that were originally numbered 12 and 13.
2 Human chromosome 2 has two sets of centromeres: usually a chromosome has just one centromere, but in the human chromosome 2 there are remnants of a second centromere (Figure A3.1.13).
3 There is a segment of the long arm of human chromosome 2, close to its centromere, that has significant similarity to chimpanzee telomeric DNA (see Figure A3.1.13). The detailed sequence in this region is exactly what you would expect to see if the two chimpanzee chromosomes had fused end-to-end.

The evidence supports the hypothesis of an ancestral fusion event between the chromosomes originally numbered 12 and 13 in chimpanzees. This 'new' human chromosome is, in fact, the second largest chromosome as viewed under a microscope, and it is therefore named chromosome 2. The chimpanzee chromosomes involved in the fusion (12 and 13) have been renamed as 2A and 2B in recognition of this fusion event (Figure A3.1.13).

The exact mechanism through which these two chimpanzee chromosomes fused is not yet completely understood, although studies show that no genes from the ends of chimpanzee chromosomes 2A and 2B appear to have been lost.

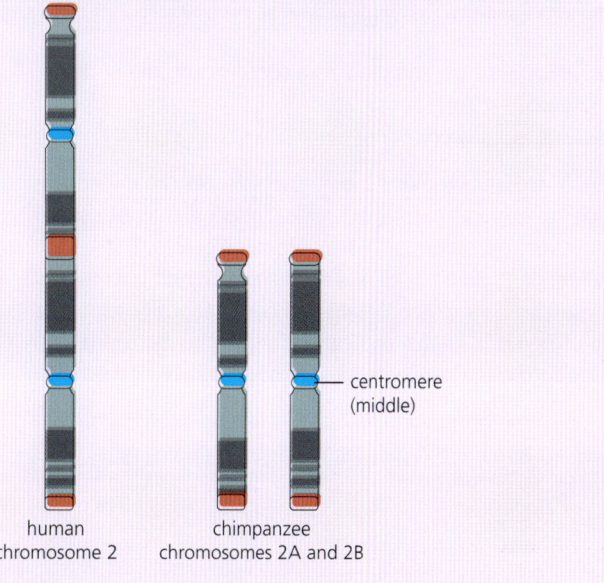

■ Figure A3.1.13 Evolution of human chromosome 2

Tool 2: Technology

Identifying and extracting data from databases

Genes are located at a specific position on a chromosome (its locus – see Chapter A1.2, page 21). Online databases can be used to locate the locus of specific genes in the human genome. The database also provides information about the function of the gene (e.g. its polypeptide product).

Procedure
1. Access the Online Mendelian Inheritance in Man website: **http://omim.org**
2. Go to the Gene Map search engine: **www.omim.org/search/advanced/entry**
3. Enter the name of a gene into the search engine (access a list of genes here: **http://en.wikipedia.org/wiki/List_of_human_genes**). A box will appear with information about the gene, including the chromosome it is found on ('Location', the number before the colon), its locus and details of its function.
4. Alternatively, you can enter the number or name of a chromosome – autosomal (1–22) or the sex chromosomes (X or Y) – and a complete sequence of gene loci for the chromosome will be displayed.

In small groups, prepare a poster or presentation about one gene. Include the following information:
- Which chromosome is it found on?
- What is its exact locus?
- What is the role of the gene in the body?
- How did you locate the gene and information about it?

Nature of science: Hypotheses

Distinguishing between testable hypotheses and non-testable statements

A good scientific hypothesis has the following qualities:
- It is testable via experiments and verifiable through observations.
- It is predictive and predicts a specific outcome under particular conditions.
- It is explanatory.

There are limits to what the natural sciences can investigate. Science explores natural or physical phenomena but cannot investigate paranormal and supernatural events. These limitations are largely due to falsifiability, which means that hypotheses should be testable via observations or experiments. If an idea is falsifiable and can potentially be disproved, then it is scientific and may be evaluated by the scientific method.

Many areas of human knowledge are 'non-scientific' when viewed through this perspective. Science also cannot investigate questions in ethics, morality, aesthetics (principles concerning the nature and appreciation of beauty), philosophy and metaphysics (which concerns existence and the nature of entities that exist).

Falsifiability provides a characteristic of scientific knowledge, but in some areas of scientific research falsifiability is not always possible. Some of the scientific ideas that are being developed are so complex or so new that it is not possible to falsify them using experiments and observations. Disciplines such as evolutionary biology, geology and astronomy contain ideas that are scientific but not falsifiable.

The origin of chromosome 2 is a testable hypothesis because observations made of the chromosomes in our nearest relative agree with genomic analysis which suggests a common ancestor between humans and chimpanzees some 4 million years ago. A variety of scientific evidence can therefore be used to develop a credible explanation for the origin of chromosome 2 in *Homo sapiens*.

> **Concept: Unity**
>
> Organisms in the same species share most of their genome.

Unity and diversity of genomes within species

The **genome** is all the genetic information of an organism (see also Chapter A1.2, page 24). Organisms in the same species share most of their genome but variations occur, which give some diversity. The most common form of genetic variation among humans is called **single-nucleotide polymorphism (SNP)**.

Single-nucleotide polymorphisms

◆ **Genome**: the whole of the genetic information of an organism or cell.

◆ **Single-nucleotide polymorphism (SNP)**: polymorphism involving variation of a single base pair.

An SNP represents a difference in a single nucleotide. For example, an SNP may replace the nucleotide guanine (G) with the nucleotide adenine (A). SNPs occur throughout the genome and occur, on average, approximately once in every 1000 nucleotides in humans, which means there may be between 4 and 5 million SNPs in the genome of one person.

SNPs may be unique or occur in many individuals, and scientists have found more than 100 million SNPs in populations around the world. These variations are usually found in the non-coding parts of DNA found between genes. In this way, SNPs contribute to human diversity.

> **Concept: Diversity**
>
> Variations such as single-nucleotide polymorphisms give some diversity in species' genomes.

SNPs can act as biological markers, helping scientists locate genes that are associated with disease. When SNPs occur within a gene or in a regulatory region near a gene, they may play a more direct role in disease by affecting the gene's function.

Scientists have shown that certain SNPs may help to predict an individual's response to specific drugs and risk of developing particular diseases. SNPs can also be used to trace the inheritance of genetic diseases within families.

The diversity of eukaryote genomes

Genomes vary in overall size, which is determined by the total amount of DNA. Genomes also vary in base sequence. Variation between species is much larger than variation within a species.

ATL A3.1A

Nucleotide sequences of specific genes can be compared from different species. The genomes from different species can be identified and then compared using two online resources:
- GenBank – a genetic database of DNA sequences. It can be used to identify the DNA sequence for a gene in a number of different species: www.ncbi.nlm.nih.gov/genbank
- Clustal Omega – a programme that compares DNA sequences: www.ebi.ac.uk/Tools/msa/clustalo

Complete the following steps:
1. Access GenBank. Change the search parameter from 'nucleotide' to 'gene'.
2. Type in the name of the gene of interest, for example 'cytochrome oxidase 1' or 'haemoglobin beta'.
3. Choose the species of interest (right side of screen – Results by taxon) and click on the link (under 'Name / Gene ID').
4. Scroll to the 'Genomic regions, transcripts and products' section and click on the 'FASTA' link.
5. Access Clustal Omega. Change the input sequence type to 'DNA' and paste the relevant FASTA sequences (from GenBank) into the space provided (Step 1). Alternatively, sequences can be saved as a document in plain text format (.txt) and then uploaded.
6. Before the sequence, designate a species name preceded by a forward arrow, e.g. '>Mouse' or '>Human'.
7. Repeat with a second sequence from GenBank. Paste the sequence into the same 'Step 1' box, underneath the first sequence. Do not forget to name the sequence.
8. When all sequences have been included, click 'submit' under 'Step 3 – submit your job'.
9. The programme will compare the genome of the two species selected, showing similarities and differences between DNA sequences.

When you compare nucleotide sequences of two species think about:
- To what extent are the base sequences similar?
- What do the differences tell you about the two species and their evolution?

Common mistake

Genome or genome size refers to the total amount of DNA measured in base pairs, not the number of genes.

Comparing genome sizes

Not all DNA codes for proteins. Genome size refers to the amount of DNA in a set of chromosomes in a species, measured in millions of base pairs (bp).

Genome size varies between organisms (see Chapter A1.2, Table A1.2.2, page 24). However, larger genomes do not necessary confer greater complexity. Some organisms, in particular plants, form new species by combining whole sets of chromosomes (a process called polyploidy – see Chapter A4.1, page 154). This results in extremely large genome sizes, which have nothing to do with the complexity of the organism.

Common mistake

DNA in chromosomes and in mitochondria are considered part of the genome, as well as chloroplast DNA in plants, but not mRNA or tRNA.

Tool 2: Technology

Identifying and extracting data from databases

You should be able to extract information about genome size for different taxonomic groups from a database to compare genome size to organism complexity. Various databases exist to compare genome sizes of different organisms:

- Animal Genome Size Database:
  www.genomesize.com/index.php
  Search for the genome size of species using common or scientific names. Data give chromosome number and genome size, measured by 'C-value'. The C-value is the mass of DNA in picograms (1 pg ≈ 1 billion base pairs or 1000 Mb of DNA) in a haploid set of chromosomes (often measured from gametes).

- The genomes of plants can be researched using: https://cvalues.science.kew.org

- Ensembl: www.ensembl.org/info/about/species.html
  Select an organism from the drop-down list then click on 'View karyotype' (Figure A3.1.14) to see image of all chromosomes and genome size.

Comparing genome size with organism complexity

1. Select two organisms and find their genome size.
2. Find out about, or use your knowledge of, the size and complexity of these organisms, and relate this to the size of their genome.
3. Is there a correlation?

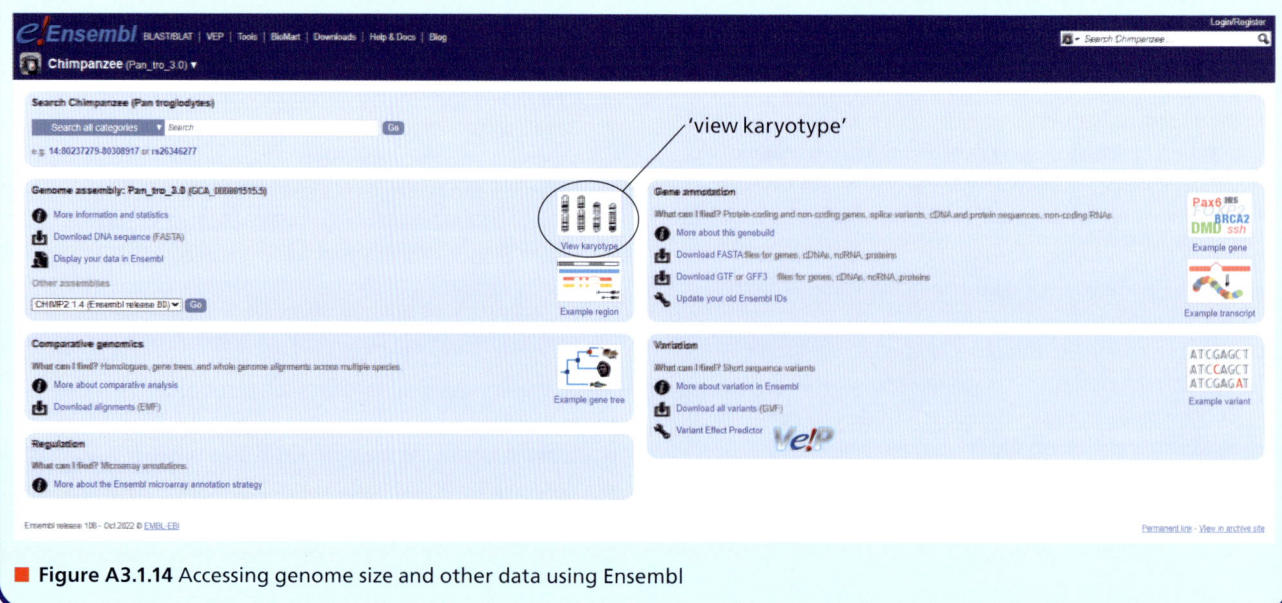

■ **Figure A3.1.14** Accessing genome size and other data using Ensembl

Current and potential future uses of whole genome sequencing

The Human Genome Project

The Human Genome Project (HGP), an initiative to map the entire human genome, was a publicly funded project that was launched in 1990. The ultimate objective of the HGP was to discover the base sequence of the entire human genome. The work was shared among more than 200 laboratories around the world, avoiding duplication of effort. On 26 June 2000, an announcement established that the sequencing of the human genome had been achieved.

At the same time as the HGP was underway, teams of scientists set about the sequencing of the DNA of other organisms. Initially, this included the common human gut bacterium (*E. coli*), the fruit fly, the mouse, bakers' yeast and *Arabidopsis* (a weed related to important crops such as broccoli and oilseed rape). Since then, scientists have sequenced the genomes of about 3500 species of complex life with around 100 being sequenced at 'reference quality', which is used for in-depth research.

■ **Figure A3.1.15** Here the order of the nucleotide bases is being determined in a molecule of DNA by a largely automated process

Developments in technology have meant that the speed of genome sequencing has increased while the cost has decreased.

Ultimately, most, if not all, aspects of biological investigation will benefit from the results of the HGP.

1 How many individual genes do we have and how do they work?

Genes may be located in base-sequence data by the detection of sequences that are uninterrupted by 'start' and 'stop' codons. Such regions are more likely to code for a protein. The 3×10^9 bases that make up the human genome represent about 20 000 genes, far fewer than was expected. *Drosophila* (the fruit fly) has almost half our number of genes, and a rice plant has more than 40 000 genes.

Promoters and enhancers are associated with many genes and may determine which cells express which genes, when they are expressed and at what level.

The genetic code of DNA is not altered by the environment (except in cases of mutation) – information flows out of genes and not back into them. However, while our genes are largely immune to direct outside influence, our experience may regulate the expression of particular genes. For example, genes are switched on and off by promoters, and this may occur in response to external factors. If so, our genes may be responding to our actions, as well as causing them (see Epigenesis on page 127).

2 Locating the cause of genetic disorders

Another outcome of the HGP is the ability to locate genes that are responsible for human genetic disorders. More than half of all genetic disorders are due to a mutation of a single gene that is commonly recessive. To prove a gene is associated with a disease, it must be shown that patients have a mutation in that gene, but that unaffected individuals do not. The outcome of the mutation in people is the loss of the ability to form the normal product of the gene. Common genetic diseases include cystic fibrosis (page 751), sickle-cell disease (page 623) and haemophilia (page 739).

> **Top tip!**
>
> Bioinformatics is the storage, manipulation and analysis of biological information via computer science. At the centre of this development are the creation and maintenance of databases concerning nucleic acid sequences and the proteins derived from them.

3 Development of the new discipline of bioinformatics

The genomes of many prokaryotes and eukaryotes have been sequenced already, as well as that of humans. This huge volume of data requires organization, storage and indexing to make practical use of the subsequent analyses. These tasks involve applied mathematics, informatics, statistics and computer science.

■ Whole genome sequencing: current and potential future uses

Currently, genome sequencing is used to research evolutionary relationships (page 115). In the future, with advancing technology, potential applications could include more widespread use of personalized medicine, where an individual's genetic profile is used to guide decisions regarding the prevention, diagnosis and treatment of disease.

Malaria, the most significant insect-borne disease, poses a threat to 40% of the world's population. Most malaria cases are found in Africa, south of the Sahara, and it is here that 90% of the fatalities due to this disease occur. Malaria is caused by *Plasmodium*, a protist, which is transmitted from one infected person to another by bloodsucking mosquitoes of the genus *Anopheles*. Of the four species of *Plasmodium*, only one (*P. falciparum*) causes severe illness. Vaccine research began in the 1980s but was at first unsuccessful. Now the situation is more hopeful. For example, the entire genome sequence of *P. falciparum* has been analysed. It consists of 14 chromosomes encoding approximately 5300 genes. This development is a major boost to the attempts to design a successful vaccine. The sequencing of the genome of *P. falciparum* could help lead to the development of vaccines by identifying potential antigens to which antibodies can bind. However, the complex parasitic lifestyle of *Plasmodium* makes it difficult to eradicate. *Plasmodium* has quickly developed drug resistance against each currently effective drug. The development of an effective malaria vaccine remains a major international challenge.

The difficulties applying the biological species concept to asexually reproducing species and bacteria

The biological species concept, discussed on page 106, does not work well for groups of organisms that do not reproduce sexually or where genes can be transferred from one species to another, for example in prokaryotes.

■ Genetic transfer in prokaryotes

♦ **Binary fission**: when a cell divides into two daughter cells, typically in asexual reproduction of prokaryotes.

Even though bacteria reproduce asexually via **binary fission**, their populations show great genetic diversity. Within a given bacterial species, the term **strain** refers to a lineage that has genetic differences compared to another strain. For example, one strain of *E. coli* may be resistant to an antibiotic while another strain may be sensitive to the same antibiotic.

The genetic diversity in bacteria comes primarily from two sources. The first way is through mutation. A mutation can occur that alters the DNA sequence of the bacterial genome and affects the traits of bacterial cells. For example, a mutation may give rise to a bacterial strain that requires a specific amino acid from an outside source for growth, while other strains of the same species can make this amino acid.

The second way that genetic diversity can be generated is by genetic transfer/horizontal gene transfer, in which genetic material is transferred from one bacterial cell to another. Genetic transfer can occur in three very different ways: **transformation**, **transduction** and **conjugation**.

Transformation

Transformation is the alteration of a bacterial cell's genotype by the uptake of naked, foreign DNA from the surrounding environment (Figure A3.1.16). Usually, this DNA is released into the environment when another bacterium has lysed (broken down). Many bacteria possess cell-surface proteins that recognize and transport DNA from closely related species into the cell. This foreign DNA can then be incorporated into the genome, either by insertion or recombination via crossing over at homologous regions.

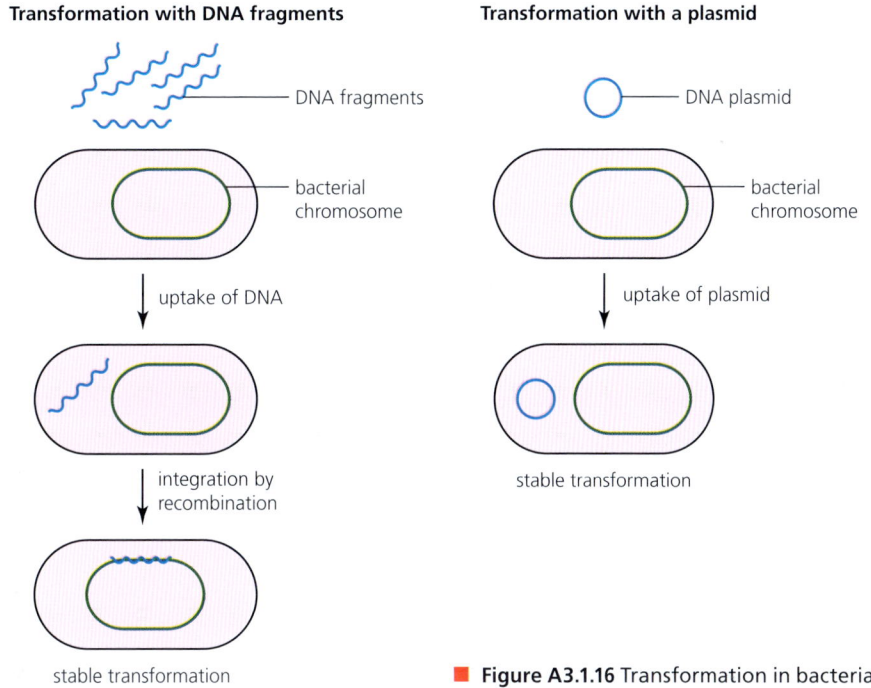

Figure A3.1.16 Transformation in bacteria

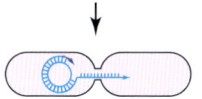

a Conjugation tube forms between a donor and a recipient. An enzyme cuts the plasmid.

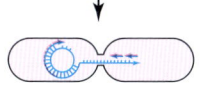

b Plasmid DNA replication starts. The free DNA strand starts moving through the tube.

c In the recipient cell, replication starts on the transferred DNA.

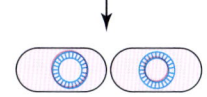

d The cells move apart and the plasmid in each forms a circle.

Figure A3.1.17 Conjugation in bacteria

Transduction

Transduction occurs when a bacteriophage (a virus that infects a bacterium – see Chapter A2.3, page 94) infects a bacterial cell and then transfers some of the cell's DNA to another bacterium. The mechanism of transduction is actually an error in a phage lytic cycle. During the synthesis of phage DNA and proteins, the bacterial chromosome is degraded into small pieces. When the new viruses are assembled, coat proteins occasionally surround a piece of bacterial DNA instead of phage genetic material. This creates an abnormal phage carrying bacterial chromosomal DNA.

Conjugation

Conjugation involves a direct physical interaction between two bacterial cells and the transfer of genetic material from a donor bacterium to a recipient bacterium (Figure A3.1.17).

The DNA transfer is one-way, i.e. one bacterium donates DNA, and the other bacterium receives the DNA. The donor sex pili (singular, pilus) attaches to the recipient. The ability to form sex pili and donate DNA during conjugation is due to the presence of an F factor (F = fertility). The F plasmid consists of several genes that are required to produce sex pili and also may carry genes that confer a growth advantage for the bacterium. After contacting a recipient cell, a sex pilus draws the donor and recipient cells closer together. A temporary cytoplasmic mating bridge or tube then forms between the two cells, providing an avenue for DNA transfer.

A3.1 Diversity of organisms

■ A species concept for prokaryotes and other asexually reproducing species

As DNA can be exchanged between bacteria, it is difficult to apply the biological species concept to them. There are variable rates of genetic transfer between populations and the genotypes of bacteria are constantly changing. The requirement in the biological species concept definition to 'breed and produce fertile offspring' does not apply to bacteria and other asexually reproducing organisms. Therefore, species definitions with an evolutionary basis (so called 'phylogenetic' ones, as mentioned on page 107) may be more appropriate for prokaryotes.

Chromosome number within a species

All individuals within a species have the same number of chromosomes (with the exception of those with chromosome non-disjunction – see page 658). Chromosomes occur in *even* numbers. This is because chromosomes occur in pairs (these pairs are called homologous chromosomes), one from each parent. To create gametes (sex cells), each set of homologous chromosomes separates into a different daughter cell (the gamete) so that each gamete has only one set of chromosomes (see page 656). At fertilization, when one male gamete fuses with one female gamete to create a genetically unique individual, the diploid (i.e. two of each chromosome) number is restored. Homologous chromosomes divide into separate gametes during sexual cell division (meiosis) so that the diploid number can be conserved in offspring. If the chromosomes did not separate in this way, the number of chromosomes would double in each generation. Within a species, the diploid and haploid (i.e. one set of chromosomes) number is constant, allowing gametes to form and fertilization to restore the diploid number. Breeding between different species is not possible because a mismatch in the number of chromosomes in gametes leads to an inability of chromosomes to pair up following fertilization. Anatomical factors can also make individuals from different species reproductively incompatible.

Cross-breeding between closely related species is unlikely to produce fertile offspring if the parent chromosome numbers are different. If the offspring produced by hybridization of different but closely related species leads to an uneven number of chromosomes, such as in the case of the horse and donkey where a mule is produced (Figure A3.1.18), the chromosomes of the offspring will be unable to divide equally into gametes. Mules have 63 chromosomes: a number that cannot be divided by two, so mules are not fertile (cannot form functional gametes) for this reason.

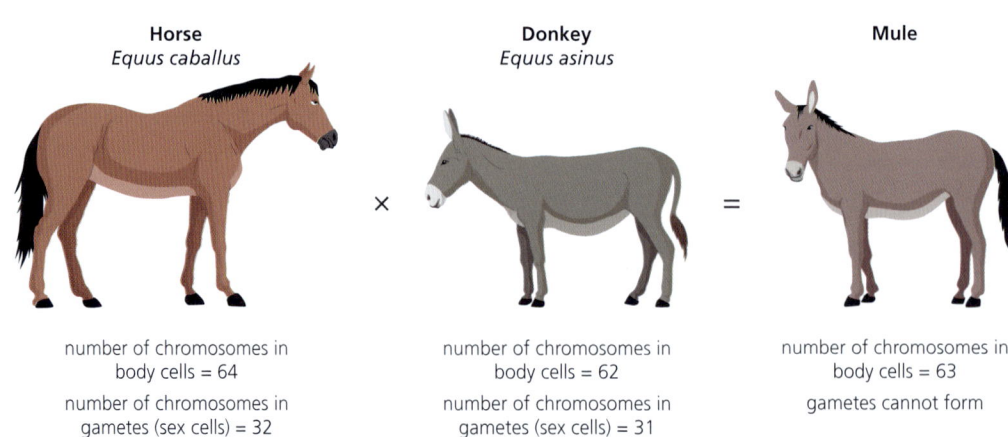

■ **Figure A3.1.18** Cross-breeding between closely related species is unlikely to produce fertile offspring if parent chromosome numbers are different

◆ **Dichotomous key**: identification key in which a group of organisms is progressively divided into two groups of smaller size.

Developing a dichotomous key

The process of identifying unknown organisms – for example, in ecological field work – is important but time-consuming. We often attempt this by making comparisons, using identification books that are illustrated with drawings and photographs, and provide information on habitat and habits, to give us clues as to the identity of organisms.

Alternatively, the use of keys may assist us in the identification of unknown organisms, for example a **dichotomous key**. The advantage of using keys is that it requires careful observation. We learn a great deal about the structural features of organisms and get some understanding of how different organisms may be related.

The steps in the construction of a dichotomous key are illustrated first (Figures A3.1.19–A3.1.22). *Follow the steps in Tool 3, then put them into practice yourself.*

Tool 3: Mathematics

Designing dichotomous keys

The steps in key construction are illustrated using eight different tree leaves, as shown in Figure A3.1.19. When selecting a leaf, care must be taken to ensure that it is entirely representative of most of the leaves of that particular tree.

First, each leaf is carefully examined, and the most significant structural features are listed in a matrix, where their presence (or absence) is recorded against each specimen, as shown in Figure A3.1.20.

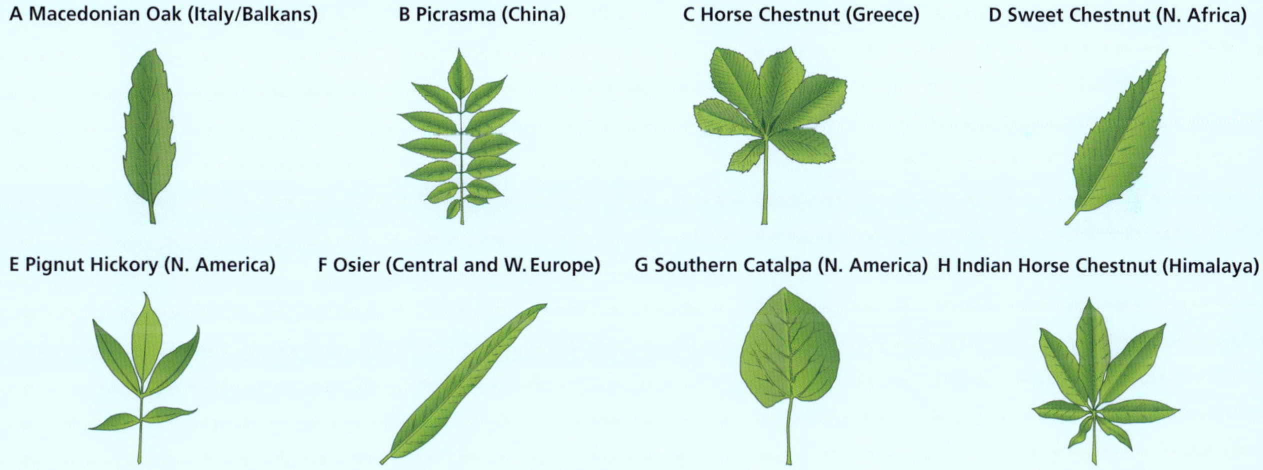

A Macedonian Oak (Italy/Balkans) **B** Picrasma (China) **C** Horse Chestnut (Greece) **D** Sweet Chestnut (N. Africa)

E Pignut Hickory (N. America) **F** Osier (Central and W. Europe) **G** Southern Catalpa (N. America) **H** Indian Horse Chestnut (Himalaya)

■ **Figure A3.1.19** Collection of leaves for the construction of a dichotomous key

A3.1 Diversity of organisms

HL ONLY

Feature of leaf: present (✓) or absent (–)	Tree leaves, identified by number							
	A	B	C	D	E	F	G	H
	Macedonian Oak	Picrasma	Horse Chestnut	Sweet Chestnut	Pignut Hickory	Osier	Southern Catalpa	Indian Horse Chestnut
leaf not divided into leaflets (leaf entire)	✓	–	–	✓	–	✓	✓	–
leaf consists of leaflets	–	✓	✓	–	✓	–	–	✓
leaf blade narrow, with almost parallel sides	✓	–	–	–	–	✓	–	–
leaf blade broad	–	–	–	✓	–	–	✓	–
leaflets radiate from one point (palmate)	–	–	✓	–	–	–	–	✓
leaflets arranged in two rows along stalk	–	✓	–	–	✓	–	–	–
leaf / leaflet margin smooth	–	–	–	–	✓	✓	✓	✓
leaf / leaflet margin toothed, like a saw	✓	✓	✓	✓	–	–	–	–
leaf blade heart-shaped	–	–	–	–	–	–	✓	–
leaf blade boat-shaped (widest in middle)	–	–	–	✓	–	–	✓	–
leaflet paddle-shaped, widest near one end	–	–	✓	–	–	–	–	–
leaflets 5 in number or less	–	–	–	–	✓	–	–	–
leaflets between 6 and 10 in number	–	–	✓	–	–	–	–	✓
leaflets more than 10 in number	–	✓	–	–	–	–	–	–

■ **Figure A3.1.20** Matrix of characteristics shown by one or more of the sample

■ **Figure A3.1.21** Dichotomous flow diagram of leaf characteristics

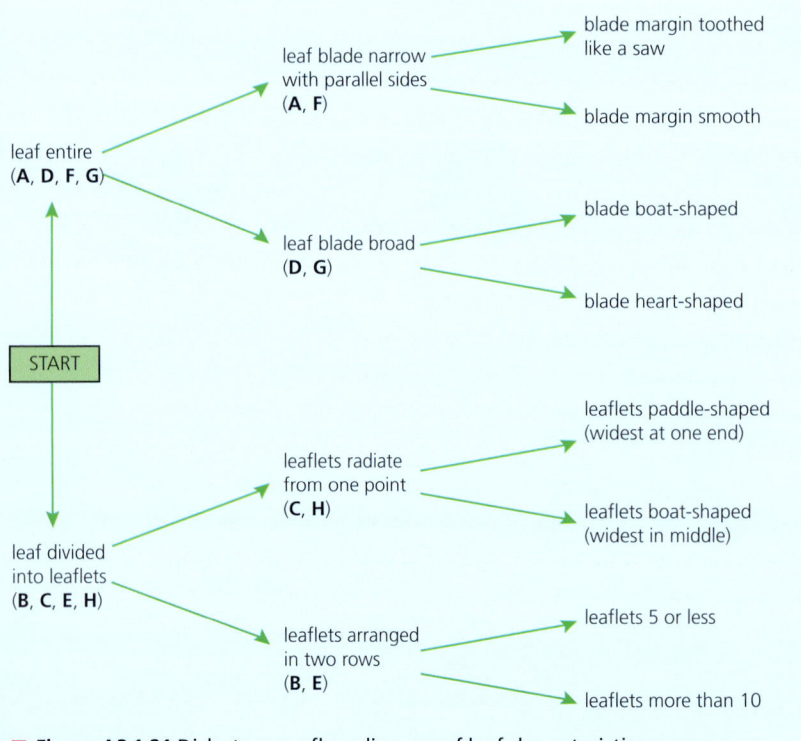

From the matrix, a characteristic shown by half (or around half) of the leaves is selected. This divides the specimens into two groups. A dichotomous flow diagram is constructed, progressively dividing the specimens into smaller groups. Each division point is labelled with the critical diagnostic feature(s), as shown in Figure A3.1.21.

Finally, a dichotomous key is constructed, reducing the dichotomy points in the flow diagram to alternative statements to which the answer is either 'yes' or 'no'. Each alternative leaf is given a number to which the reader must refer to carry on the identification, until all eight leaves have been identified (Figure A3.1.22).

		Go to ...
1	Leaves entire, not divided into leaflets	2
	Leaf blade divided into leaflets	5
2	Leaf blade narrow, with almost parallel sides	3
	Leaf blade broad rather than narrow	4
3	Blade margin toothed like a saw	Macedonian Oak
	Blade margin smooth	Osier
4	Blade boat-shaped	Sweet Chestnut
	Blade heart-shaped	Southern Catalpa
5	Leaflets radiate from one point	6
	Leaflets arranged in two rows	7
6	Leaflets paddle-shaped – widest at one end	Horse Chestnut
	Leaflets boat-shaped – widest in the middle	Indian Horse Chestnut
7	Leaflets 5 (or less) in number	Pignut Hickory
	Leaflets more than 10 in number	Picrasma

■ **Figure A3.1.22** Dichotomous key to the sample of eight tree leaves

Top tip!

Limitations of keys include:
- The organism might not be in the key.
- Individuals of one species may vary in morphology, e.g. show sexual dimorphism.
- Terminology can be difficult for non-experts.
- There might not be a key available for the organisms under investigation.
- Some features cannot be easily established in the field – for example, whether an animal has a placenta or not, or whether an animal is endothermic or ectothermic.

Inquiry 1: Exploring and designing

Exploring; designing

Design a method of classifying animals and plants that are commonly found in gardens or farmland that might be useful to an enthusiastic gardener or farmer. Select animals or plants that are easy to identify from their morphological (i.e. physical) characteristics. Consider and address safety, ethical and environmental issues when planning how to collect your organisms.

Inquiry 2: Collecting and processing data

Collecting data; processing data

Using your methodology created in Inquiry 1, sample insects or plants in your local area. What qualitative features can you use to separate different species?

Tool 3: Mathematics

Designing dichotomous keys

Use the steps laid out in the Tool 3 box starting on page 121 and design a dichotomous key for your specimens. Make sure that the key enables each of the species to be identified using sets of paired questions.

A3.1 Diversity of organisms

Identifying species from environmental DNA in a habitat using barcodes

As mentioned earlier, approximately 1.8 million species have been described. Some scientists estimate that there are between 8 and 10 million species on Earth, although this number may be higher.

Whichever estimate is used, there is clearly a large number of species that still await identification and classification. The current system of taxonomy requires expertise and time – both of which are currently lacking as many species face extinction (see Chapter A4.1, page 149).

'Barcoding' is a method to identify, track and manage items for sale. Every item you buy in the shops is identified by a unique barcode. A barcode has lines and spaces of varying thicknesses printed in different combinations. Similarly, the DNA profile produced by gel electrophoresis (see page 605) is unique to the individual it comes from, and is made from bars and gaps, forming specific patterns. Although individuals may have slightly different genetic profiles, overall 'barcodes' can be developed which represent the genetic makeup of the species. **DNA barcoding** is a method of species identification using a short section of DNA from a specific gene or genes. The genes chosen have less intraspecific (within species) variation than interspecific (between species) variation. For most animal groups, biologists use the sequence obtained from the mitochondrial cytochrome c oxidase gene. Small samples of DNA can be amplified using a technique called polymerase chain reaction (PCR) – see page 605.

Traditional taxonomy is a laborious process that requires specialist training. Natural history museums, where such work is carried out, have limited staff and capacity to identify species at a sufficient speed. In contrast, DNA barcoding offers a fast and accurate technique to identify species, avoiding the limitations of traditional methods (for example, inability to identify damaged or partial specimens, and the problem of morphological variation within species).

Collection data, relating to the location and environment from which the specimens were sampled, along with photographs and genetic data, can be made available on the internet so that scientists from around the world can access information about the species (Figure A3.1.23).

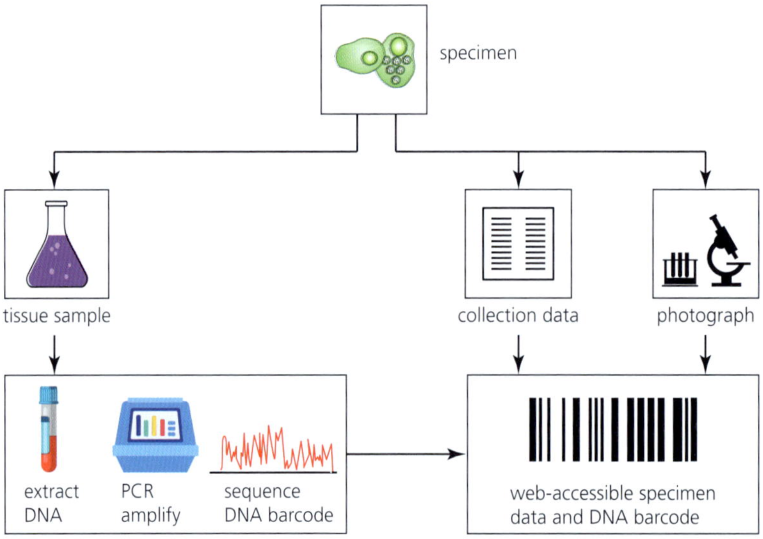

◆ **DNA barcoding**: species identification using short sections of DNA.

ATL A3.1B

Find out about an area where DNA barcoding has been used to establish the taxonomy of species in the environment. How can this information be used to carry out surveys rapidly to establish the biodiversity of sample sites? How could this information be used to establish the effect of human disturbance on natural systems?

LINKING QUESTIONS

1. What might cause a species to persist or go extinct?
2. How do species exemplify both continuous and discontinuous patterns of variation?

■ **Figure A3.1.23** Species identification data available on the internet

Once barcodes for a sufficient number of species have been established, environments can be sampled to obtain DNA from a variety of organisms, which allows the biodiversity of habitats to be investigated rapidly.

A3.2 Classification and cladistics

Guiding questions
- What tools are used to classify organisms into taxonomic groups?
- How do cladistics methods differ from traditional taxonomic methods?

SYLLABUS CONTENT

This chapter covers the following syllabus content:
- ▶ A3.2.1 Need for classification of organisms (HL only)
- ▶ A3.2.2 Difficulties classifying organisms into the traditional hierarchy of taxa (HL only)
- ▶ A3.2.3 Advantages of classification corresponding to evolutionary relationships (HL only)
- ▶ A3.2.4 Clades as groups of organisms with common ancestry and shared characteristics (HL only)
- ▶ A3.2.5 Gradual accumulation of sequence differences as the basis for estimates of when clades diverged from a common ancestor (HL only)
- ▶ A3.2.6 Base sequences of genes or amino acid sequences of proteins as the basis for constructing cladograms (HL only)
- ▶ A3.2.7 Analysing cladograms (HL only)
- ▶ A3.2.8 Using cladistics to investigate whether the classification of groups corresponds to evolutionary relationships (HL only)
- ▶ A3.2.9 Classification of all organisms into three domains using evidence from rRNA base sequences (HL only)

The need to classify organisms

Concept: Diversity

Given the great diversity of life, there is a need to classify organisms in order to provide a framework for biological knowledge. This in turn reveals functional and structural similarities and differences between organisms.

Given the great variety of life on Earth, and the immense number of species, there is a need to group organisms together based on shared similarities. In chemistry, the periodic table is the central organizing framework for all elements. Similarly, classification gives biology an organizational structure that facilitates the study of the subject. Rather than a series of unrelated facts, taxonomy gives biology a rationale for grouping organisms based on shared characteristics, which in turn reveals functional and structural similarities and differences. After classification is completed, a broad-range of further study is carried out, which would not be possible without first classifying the organisms.

Concept: Unity

The system of classification is universal (i.e. everyone uses it). By using the same methodology and language to describe and group species, collaboration is possible between scientists around the world.

The quickest way to classify living things is on their immediate and obvious similarities and differences. For example, we might classify together animals that fly, simply because the essential organ, wings, are so easily seen. This would include almost all birds and many insects, as well as the bats and certain fossil dinosaurs. However, resemblances between the wings of the bird and the insect are superficial. Both are aerofoils (structures that generate 'lift' when moved through the air) but they are built from different tissues and have different origins in the body. This will be returned to in Chapter A4.1.

Difficulties classifying organisms into the traditional hierarchy of taxa

The scheme of classification

In classification, the aim is to use as many characteristics as possible when placing similar organisms together and separating dissimilar ones. Just as similar species are grouped together into the same **genus** (plural, genera), similar genera are grouped together into families.

This approach is extended from families to orders, then classes, phyla and kingdoms. This is the hierarchical scheme of classification, with each successive group containing more and more different kinds of organism. In a natural classification, the genus and accompanying higher taxa consist of all the species that have evolved from one common ancestral species.

The taxa used in taxonomy are given in Figure A3.2.1.

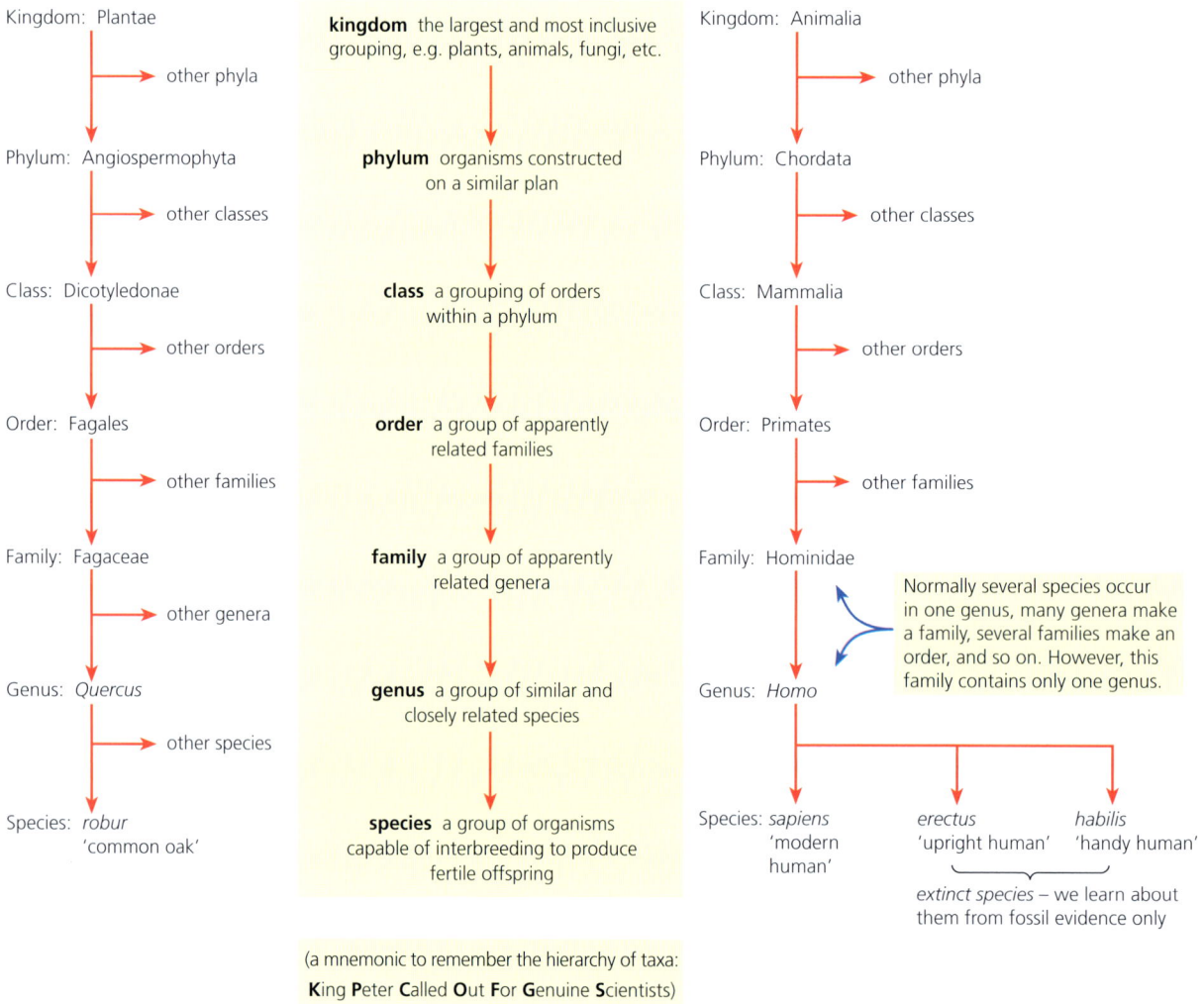

■ **Figure A3.2.1** The taxa used in taxonomy, applied to genera from two different kingdoms, Plantae and Animalia

Domains and kingdoms

◆ **Kingdom**: second highest taxonomic rank, below domain.

At one time the living world seemed to divide naturally into two **kingdoms** consisting of the plants (with autotrophic nutrition) and the animals (with heterotrophic nutrition). These two kingdoms grew from the original disciplines of biology, namely botany, the study of plants, and zoology, the study of animals. Fungi and micro-organisms were conveniently 'added' to botany.

Initially there was only one problem: fungi possessed the typically 'animal' heterotrophic nutrition but were superficially 'plant-like' in appearance. Then, with the use of the electron microscope, came the discovery of the two types of cell structure, namely prokaryotic and eukaryotic (page 79). As a result bacteria, which have prokaryotic cells, could no longer be 'plants' since plants have eukaryotic cells. The division of living things into kingdoms needed changing. This led to the division of living things into five kingdoms rather than two (Table A3.2.1).

Today, taxonomists sometimes reclassify groups of species when new evidence (usually molecular based) shows that a previous taxon contains species that have evolved from different ancestral species.

■ **Table A3.2.1** The five kingdom model of classification

Prokaryotae	the prokaryote kingdom, the bacteria and cyanobacteria (a group of photosynthetic bacteria), predominately unicellular organisms
Protoctista	the protoctistan kingdom (eukaryotes), predominately unicellular, and seen as resembling the ancestors of the fungi, plants and animals
Fungi	the fungal kingdom (eukaryotes), predominately multicellular organisms, non-motile and with heterotrophic nutrition
Plantae	the plant kingdom (eukaryotes), multicellular organisms, non-motile, with autotrophic nutrition
Animalia	the animal kingdom (eukaryotes), multicellular organisms, motile, with heterotrophic nutrition

Note that viruses are not classified as living organisms.

◆ **Domain**: the highest taxonomic rank in the hierarchical biological classification system, above the kingdom level. There are three domains of life: archaea, eubacteria and eukarya.

When RNA sequencing became possible, American microbiologist and biophysicist Carl Woese theorized that since all living organisms contained ribosomes, the ribosomal RNA sequence could be used to determine the relatedness of all organisms.

His work revealed that the prokaryotes comprise two distinct groups – the bacteria and the archaea – that diverged early in the history of life on Earth. The living world therefore has three major **domains**: eubacteria, archaea and eukarya.

The five-kingdom method of classifying species as shown in Table A3.2.1 was modified in the 1970s following these new discoveries in the areas of biochemistry and genetics, leading to the introduction of a classification group even larger than the kingdoms – domains. The prokaryotae were divided into two domains, eubacteria and archaea, with the remaining four eukaryote kingdoms forming the eukarya domain (see more details later in this chapter on page 137).

It can be difficult to determine which similarities between species are most relevant when grouping species.

- It is important to distinguish similarities that are based on shared ancestry from those that have evolved independently but under similar selective pressures and so appear the same or similar, for example human and octopus eyes (known as analogous structures).
- Species that are similar in appearance may not be closely related – their resemblance is due to analogous adaptations to very similar environments (for example, both bats and dragonflies have wings for flight, see Chapter A4.1, page 146). In contrast, structures that are homologous may look very different, such as forelimbs in vertebrates, but represent common evolutionary origins.
- Morphology (form and structure) of organisms can lead to mistakes in classification, due to misinterpreting whether structures are analogous or homologous. Base or amino acid sequences are more accurate ways of determining members of a clade (see below) because they represent true homology.

1 In Figure A3.2.1, one plant and one animal species are classified from kingdom to species level. **Suggest** how these flow diagrams could be modified to show their classification from domain level.

Link
Analogous and homologous structures will be covered in detail in Chapter A4.1, page 145.

A3.2 Classification and cladistics

◆ **Cladistics**: a classification system used to construct evolutionary trees. Organisms are categorized based on shared derived characteristics that can be traced to a group's most recent common ancestor and are not present in more distant ancestors. Characteristics can be anatomical, physiological, behavioural, or genetic and protein sequences.

◆ **Clade**: a group of organisms that have evolved from a common ancestor.

◆ **Cladogram**: a diagram used in cladistics that shows evolutionary relations among organisms.

Nature of science: Theories

A paradigm shift in classification

Charles Darwin noted that for each species a 'number of intermediate forms must have existed, linking together all species in each group by gradations as fine as our existing varieties'. A fixed ranking of taxa (kingdom, phylum and so on) is arbitrary because it does not reflect the gradation of variation. The traditional hierarchy does not always match patterns of divergence caused by evolution. An alternative classification system is **cladistics**, using unranked clades (see below). Cladistics is a very powerful methodology for classification. It is entirely based on the observation of shared derived characteristics known as **synapomorphies**. It categorizes organisms based on these shared derived characteristics that can be traced to a group's most recent common ancestor and are not present in more distant ancestors. Relationships are shown as evolutionary trees. This alternative classification system is an example of the paradigm shift that sometimes occurs in scientific theories. Paradigm shifts take place when a new theory replaces an old one.

Advantages of classification corresponding to evolutionary relationships

Ideally, a classification system should show the evolutionary relationships between organisms. A phylogenetic classification system is based on evolutionary relationships and not just similarities in physical traits that may or may not have evolutionary significance. Through the process of evolution, organisms have changed into groups with a common ancestry and common characteristics.

By grouping organisms using biochemical information, such as that provided by DNA and proteins, changes in species though time can be mapped on to a 'tree of life', showing the interconnections and true evolutionary relationships between species (see page 135 later in this chapter, and Chapter A4.1, page 141). The characteristics of organisms within such a group can be predicted because they are shared within a **clade**.

Classification systems that use morphological characteristics do not necessarily show evolutionary relationships because analogous structures can evolve separately in groups of organisms that do not share a recent common ancestor.

2 **Distinguish** between the two diagrams in Figure A3.2.2.

Clades

When classifying living things using evolutionary relationships, taxonomists display these relationships on phylogenetic trees called **cladograms**. Figure A3.2.2 is an example of a cladogram.

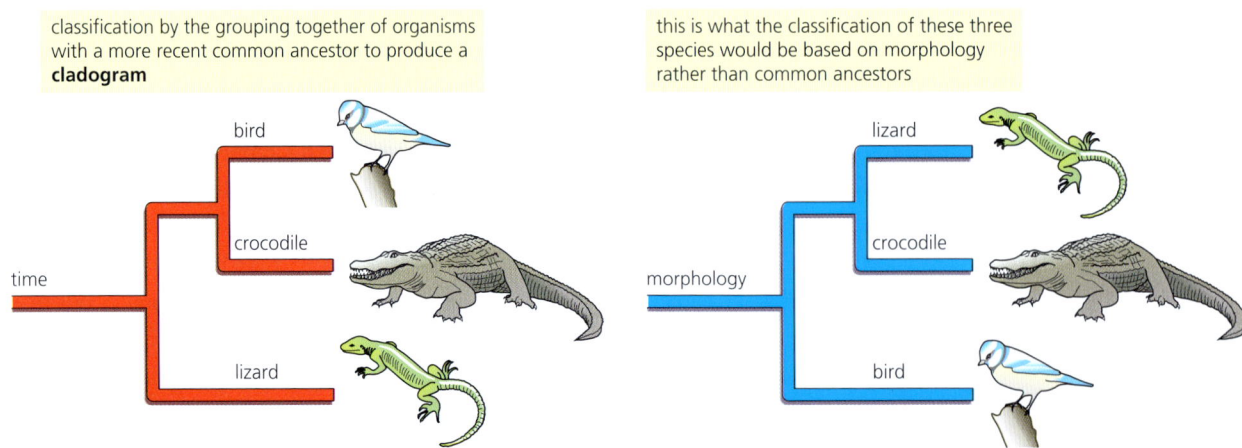

■ **Figure A3.2.2** Cladistics (left diagram) – one of two ways of classifying

In **cladistics**, classification is based upon an analysis of relatedness, and the product is a cladogram. **A clade is a group of organisms that have evolved from a common ancestor.** Figure A3.2.2 shows that a bird, crocodile and lizard all have a common ancestor, but that the bird and crocodile share a more recent ancestor. A clade is **monophyletic**, that is, one that includes all the descendants of one, but only one, ancestor.

Cladograms have two important features:
- **branch points** in the tree – representing the **time** at which a divide between two taxa occurred
- the **degree of divergence** between branches – representing the **differences** that have developed between the two taxa since they diverged.

The most objective evidence for placing organisms in the same clade comes from base sequences of genes or amino acid sequences of proteins. Morphological traits can be used to assign organisms to clades, although care must be taken to ensure that these characteristics are homologous and not analogous.

In summary, a cladogram:
- shows patterns of shared characteristics
- is a diagram that shows the evolutionary relationships among a group of organisms and classifies organisms according to the order in time at which branches arise along their phylogenetic tree.

Taxonomists ponder the question: which features are more significant in a phylogenetic taxonomy, and so should receive the greater emphasis in devising a scheme? Differences in DNA and protein sequences provide a system that accurately maps the relationship between species, whereas morphological differences, while being the only option for many organisms (for example, those only with fossil remains), can lead to problems where convergent evolution has taken place (see Chapter A4.1, page 145).

Cladistics was developed from a body of taxonomic theory known as phylogenetic systematics and is organised around three principles: (i) monophyletic taxa are natural, (ii) organisms are related through evolutionary descent, and (iii) evolutionary modifications uniquely shared by organisms are evidence of their unique phylogenetic history.

Evidence of evolutionary relationships comes from a range of studies (Table A3.2.2).

■ **Table A3.2.2** Evidence of evolutionary relationships

Palaeontology (study of fossils)	Fossils tell us about organisms in the geological past. Sometimes they represent intermediate forms – links between groups of organisms, such as *Archaeopteryx*, a possible 'missing link' between non-avian feathered dinosaurs and birds.
Comparative biochemistry	The composition of nucleic acids and cell proteins establish degrees of relatedness – closely related organisms show fewer differences in the composition of specific nucleic acids and cell proteins.
Comparative embryology	Study of the development of an organism from zygote to adult may reveal its evolution. For example, the arrangement of arteries and development of the heart in early vertebrate embryos follow similar patterns.
Comparative anatomy	Studies of comparative anatomy show that many structural features are basically similar.

> **Top tip!**
> The difficulty that cladistics creates is that there is no obvious way in which to create such a hierarchical taxonomy (and therefore a naming system) from a phylogeny (the history of the evolution of a species or group). There are, for practical purposes, a finite number of levels in a taxonomic hierarchy but an indefinite number of ancestors in a phylogeny.

■ Limitations of cladistics

While cladistics is a powerful tool for testing phylogenetic hypotheses, it is not without conceptual and practical problems. Cladistics makes the implicit assumption that these shared derived characters are the result of linear or vertical genetic transfer: vertical in the sense of a pedigree, with genetic transfer only from parent to offspring. This is a reasonable assumption in most eukaryotes.

Lateral or horizontal genetic transfer, e.g., transduction or conjugation (see page 119), however, disregards this assumption. Cladistics does not really have a mechanism for dealing with this problem so it has been of little use for taxonomy of prokaryotes.

Originally bacterial taxonomists used a technique called numerical taxonomy, where as many physical and biochemical characteristics as possible (sometimes as many as 200) would be identified in a group of related bacteria. A matrix would then be created that would group those isolates that had the most shared characteristics together.

This was a completely morphological method (that is, based solely on phenotype). It was an expensive and arduous process, but it seemed to result in robust groupings. Unfortunately, it does not give any insight into how the identified groups were related to each other in an evolutionary sense.

There is significant genetic variation in bacterial species. There can be as much as 30% DNA sequence variation between two *E. coli* genomes and, unlike eukaryotes, much of that difference is due to the presence or absence of genes, not just sequence variations between different versions of the same gene. In other words, one *E. coli* strain can have hundreds of genes that another strain of *E. coli* completely lacks.

So, the approach that bacterial taxonomists currently use is what is called the **phylo-morphological species concept**. This is uses a combination of ribosomal phylogenetic information and phenotypic characteristics to form groups of similar organisms, and then using a somewhat arbitrary threshold of 70% overall DNA similarity to delineate species boundaries.

Sequence differences as the basis for clades divergence

Today, the similarities and differences in the biochemistry of organisms have become extremely important in establishing the evolutionary links between different species and the point at which common ancestors existed. For example, the β chain of haemoglobin, which is built from 146 amino acid residues, shows variation in the sequence of amino acids in different species. Haemoglobin structure is determined by inherited genes, so the more closely related species are, the more likely their amino acid sequence is to match (Table A3.2.3). Variations are thought to have arisen by mutation of an 'ancestral' gene for haemoglobin. If so, the longer ago a species diverged from a common ancestor, the more likely it is that differences may have arisen.

◆ **Molecular clock**: a measure of evolutionary change over time based on the mutation rate of DNA sequences or the proteins they encode; can be used for estimating how long ago two related organisms diverged from a common ancestor.

■ Table A3.2.3 Number of amino acid differences in β chain of haemoglobin compared to human haemoglobin

Species	Differences	Species	Differences
human	0	kangaroo	38
gorilla	1	chicken	45
gibbon	2	frog	67
rhesus monkey	8	lamprey	125
mouse	27	sea slug (mollusc)	127

Similar studies have been made of the differences in the polypeptide chains of other protein molecules, including ones common to all eukaryotes and prokaryotes. One such is the universally occurring electron transport carrier, cytochrome c (see page 124).

■ Biochemical variation used as an evolutionary or molecular clock

Biochemical changes like those discussed above may occur at a constant rate and, if so, may be used as a **molecular clock**. If the rate of change can be reliably estimated, it records the time that has passed between the separation of evolutionary lines.

Top tip!

The molecular clock can only give estimates because mutation rates are affected by the length of the generation time, the size of a population, the intensity of selective pressure and other factors.

The construction of cladograms

DNA and protein comparisons are done by aligning the sequences from different species and using computer software to identify the best phylogenetic tree joining the sequences, and thus inferring the phylogeny of the species.

■ Using amino acid sequences to build cladograms

Cladograms can be constructed using amino acid sequences of proteins. An example comes from immunological studies – a means of detecting differences in specific proteins of species and, therefore (indirectly), their genetic relatedness.

Serum is a liquid produced from blood samples from which blood cells and fibrinogen have been removed. Protein molecules present in the serum act as antigens if the serum is injected into an animal with an immune system that lacks these proteins (page 195).

Typically, a rabbit is used when investigating relatedness to humans. In the rabbit, the injected serum causes the production of antibodies against the human proteins. Then, serum produced from the rabbit's blood (now containing antibodies against human proteins) can be tested against serum from a range of animals. The more closely related a test animal is to humans, the greater the reaction of rabbit antibodies with human-like antigens (bringing about observable precipitation, Figure A3.2.3).

The precipitation produced by the reaction of treated rabbit serum with human serum is taken as 100%. For each species tested, the greater the precipitation, the more recently the species shared a common ancestor with humans. This technique, called **comparative serology**, has been used by taxonomists to establish phylogenetic links in mammals and in invertebrates.

> **Immunological studies** are a means of detecting differences in specific proteins of species and, therefore (indirectly), their **relatedness**.

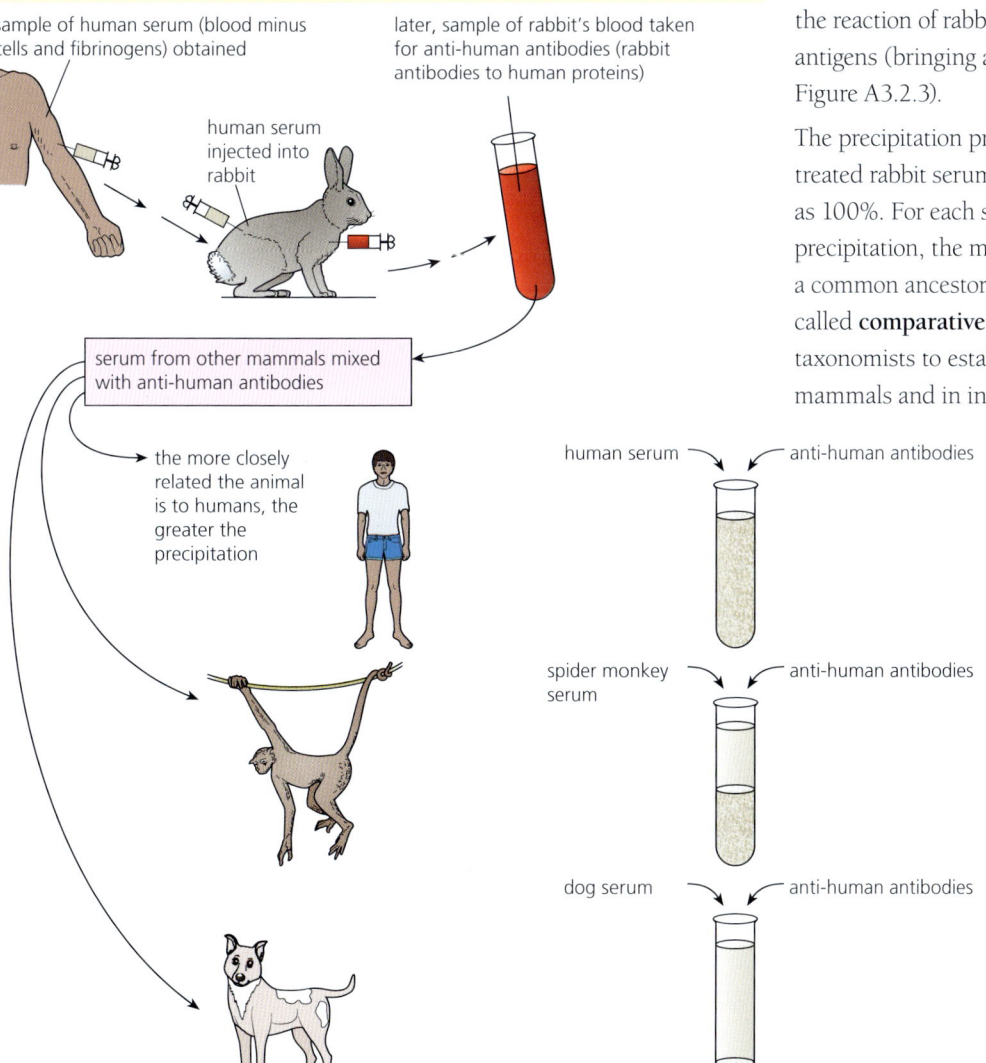

■ Figure A3.2.3 Investigating evolutionary relationships by the immune reaction

A3.2 Classification and cladistics

■ Table A3.2.4 Relatedness investigated via the immune reaction

1	2	3	4	5
Species	Precipitation/%	Difference from human/%	Difference to common ancestor (half difference from human)	Postulated time since common ancestor/mya
human	100	–	–	–
chimpanzee	95	5	2.5	4
gorilla	95	5	2.5	4
orang-utan	85	15	7.5	13
gibbon	82	18	9	15
baboon	73	27	13.5	23
spider monkey	60	40	20	34
lemur	35	65	32.5	55
dog	25	75	37.5	64
kangaroo	8	92	46	79

The list of animals tested in this way is given in Table A3.2.4. Of course, we do not know the common ancestor to these animals and the blood of that ancestor is not available to test. However, if the sequence of the 584 amino acids that make up the blood protein albumin changes at a constant rate, the percentage immunological 'distance' between humans and any of these animals will be a sum of the distance 'back' to the common ancestor plus the difference 'forward' to the listed animal. Hence, the difference between a listed animal and human can be halved to estimate the difference between a modern form and the common ancestor.

The divergent evolution of the primates is known from geological (fossil) evidence. So, the forward rate of change since the lemur gives us the rate of the molecular clock – namely 35% in 60 million years, or 0.6% every million years.

This calculation has been applied to all the data (Table A3.2.4, column 5).

We can construct a cladogram based on the biochemical data in Table A3.2.4, as shown in Figure A3.2.4.

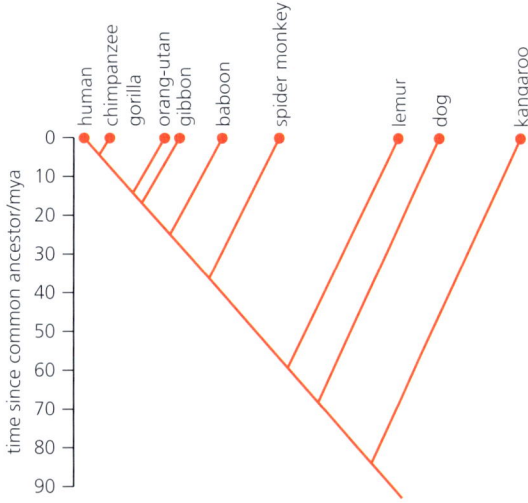

■ Figure A3.2.4 A cladogram based on immunological evidence in Table A3.2.4

Nature of science: Hypotheses

Different criteria for judgement can lead to different hypotheses. Parsimony is the concept that the simplest explanation for the data is preferred. In the analysis of phylogeny, parsimony means that a hypothesis of relationships that requires the smallest number of character changes is most likely to be correct. Parsimony analysis is used to select the most probable cladogram in which observed sequence variation between clades is accounted for with the smallest number of sequence changes.

The term 'Occam's razor' has the same meaning as parsimony and is the term more often used in everyday life. It states that the simplest explanation is usually the best one and advises not to make more assumptions than you absolutely need to.

DNA hybridization is a technique that involves matching the DNA of different species to discover how closely they are related.

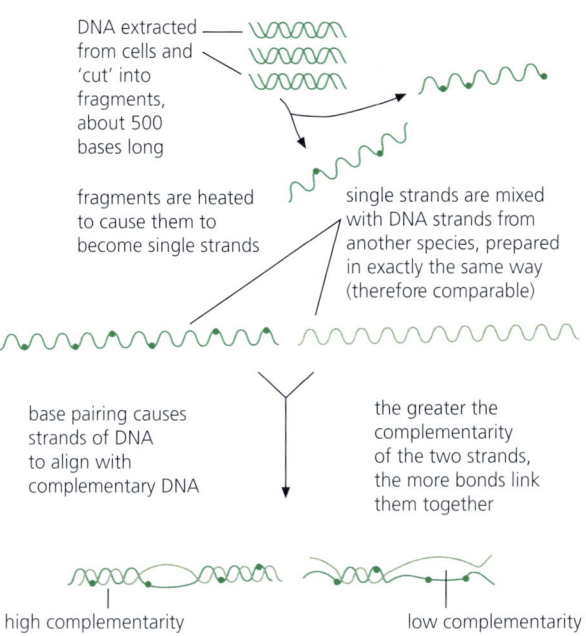

The closeness of the two DNAs is measured by finding the temperature at which they separate – the fewer bonds formed, the lower the temperature required.

The degree of relatedness of the DNA of **primate species** can be correlated with the estimated number of years since they shared a common ancestor.

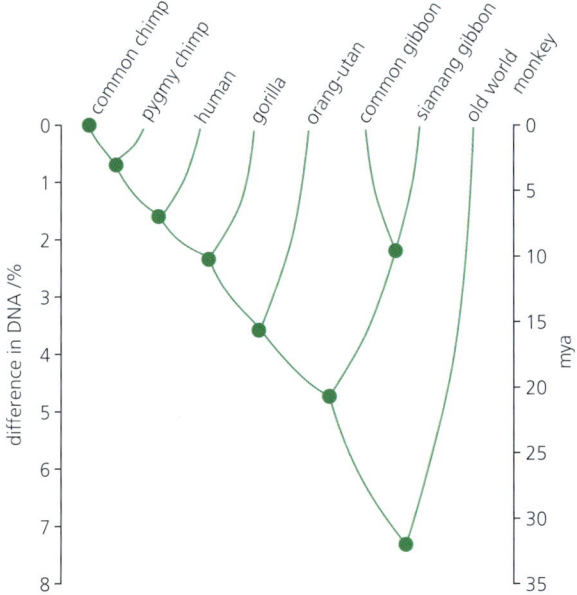

■ Figure A3.2.5 Genetic difference between DNA samples

Relatedness measured from DNA samples

Cladograms can also be constructed using base sequences of genes.

It is possible to measure the relatedness of different groups of organisms by the amount of difference between specific molecules, such as differences in the base sequence of genes in DNA. The genetic differences between the DNA of various organisms give us data on degrees of divergence. *Look at Figure A3.2.5*. Here, the degree of relatedness of the DNA of primate species suggests the number of years that have elapsed since the various primates shared a common ancestor.

In cladograms, the more derived characteristics two organisms share, the closer their evolutionary relationship (i.e. the more recently their common ancestor lived).

- Points at which two branches form are called **nodes**, which represent speciation events (see Figure A3.2.6 overleaf).
- Close relationships are shown by a recent fork – the closer the fork in the branch between two organisms, the closer their relationship.
- **Terminal branches** lead to **terminal nodes** that represent organisms for which data exist (genetic or amino acid sequences); the organisms may be either still living (extant) or extinct.
- The **root** is the central trunk of a cladogram that indicates the ancestor common to all groups branching from it.

Cladograms provide strong evidence for evolutionary relationships, although they cannot be regarded as absolute proof. This is because they are constructed based on the assumption that the smallest number of mutations possible account for differences between species (see below). If such assumptions are incorrect, errors may occur in cladograms. By using several different cladograms that have been derived independently using different data, such errors can be avoided.

Figure A3.2.7 overleaf shows that the lion and jaguar (*Panthera leo* and *Panthera onca*) are the most closely related cat species.

◆ **Node**: a branching point from the ancestral population in a cladogram. It represents a speciation event.
◆ **Terminal branch**: indicates both extinct and extant species in a cladogram, leading to a terminal node.
◆ **Terminal node**: represents the hypothetical last common ancestral interbreeding population of the taxon labelled at a tip of a cladogram.
◆ **Root**: the central trunk of a cladogram, indicating the common ancestor to all groups that branch from it.

A3.2 Classification and cladistics

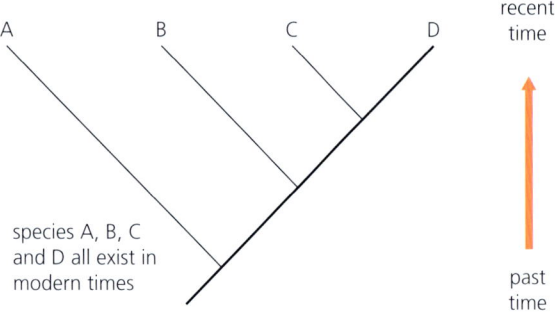

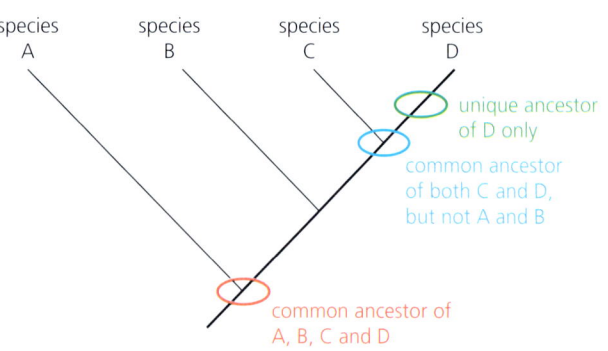

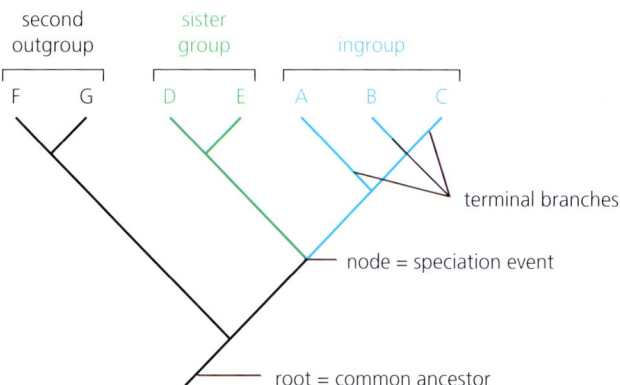

■ **Figure A3.2.6** A cladogram shows evolutionary relationships among a group of organisms

Common mistake

Do not look at the end positions of species on a cladogram to determine relatedness, look at the nodes on the cladogram.

Tool 1: Experimental techniques

Classifying and cladogram analysis

Cladograms can be used to ensure that the classification of species reflects evolutionary relationships. Analyse the cladogram in Figure A3.2.7. Which species diverged from the ancestor first? Which species is most closely related to the snow leopard? Which species do leopards share a common ancestor with? **Comment** on the evolution of species in the genus *Panthera*.

Common mistake

The branching off points of a cladogram are often poorly explained. The points where a cladogram branches (the nodes) represent common ancestors to the species that come after the branch point. All branch tips arising from a given branching point are descendants of the common ancestor at that branching point.

Top tip!

You should be able to deduce evolutionary relationships, common ancestors and clades from a cladogram. You should understand the terms 'root', 'node' and 'terminal branch', and also that a node represents a hypothetical common ancestor.

3 **Discuss** how cladograms can be used in classification of organisms.

TOK

Think about the different classification systems that have been used – for example the fixed ranking of taxa versus classification corresponding to evolutionary relationships. How has each system resulted in different conclusions about the relationship between species and the way in which life is interrelated? To what extent do the classification systems we use in the pursuit of knowledge affect the conclusions that we reach?

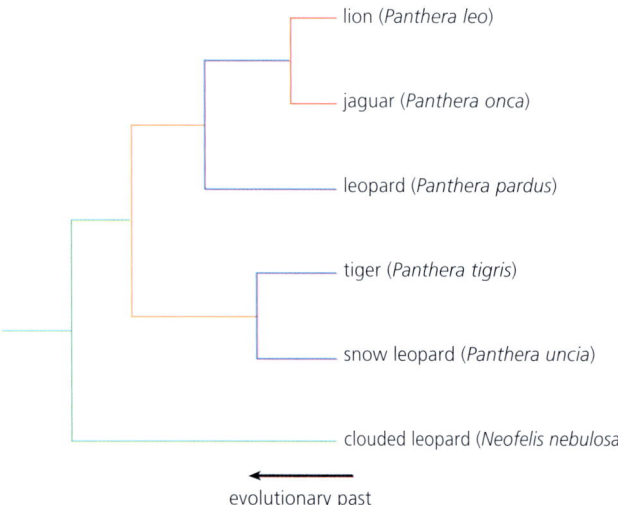

■ **Figure A3.2.7** Part of the phylogenetic tree for the Felidae

Theme A: Unity and diversity – Organisms

ATL A3.2A

Access this website: https://tidal.northwestern.edu/blog/bat

In this interactive activity you create cladograms and add distinguishing features that determine branches in the diagram. The programme will give you a rating to determine how successful you have been.

1. Click on the 'list' option on the right-hand side of the screen and select the game level you want to work on (there are levels 1–7). Each level has a different selection of organisms and get progressively harder.
2. Drag the images of the organisms into the central part of the screen. Rearrange them to create the cladogram that best resembles their relationships.
3. Drag the distinguishing features (each has a different coloured dot, e.g. Level 1 orange – cells with nuclei) from the left-hand side of the screen on to the correct node, which defines the shared features of a clade.
4. At the end of the activity, you will be given a score. You may need to work with a partner in the later stages and discuss the harder levels in groups. Can you complete all seven levels?

Investigating whether the classification of groups corresponds to evolutionary relationships

Cladistics can be used to investigate whether the established classification of groups corresponds to evolutionary relationships.

■ Reclassification of the figwort family (Scrophulariaceae)

The Scrophulariaceae (figwort family) – one of the groups of flowering plants that has long been recognized by plant taxonomists – was a large and varied family. Typically, genera of this family had irregular (zygomorphic) flowers, like those of the snapdragon and the foxglove, for example. The flowering plant families were designated before most biochemical studies were applied to plant taxonomy. Nevertheless, many families appear to be (largely) natural classifications, for example the rose family (Rosaceae), which includes many fruit plants, and the Crucifereae, which includes many economically important plants.

Today, flowering plant classification is being revisited. (Indeed, all classification is being revisited.) Most evidence for plant evolutionary relationships now comes from a comparison of DNA sequences in only one to three genes (these are conserved genes – see page 25) found in the chloroplasts of the plant cells. These genes are of modest length, each about 1000 nucleotides long. However, they have provided more secure information regarding evolutionary history than the differences in anatomy and morphology on which the traditional classifications were based.

In the reorganization of the figwort family, these three genes have been compared from many species of plants, including some formerly classified in the Scrophulariceae. The outcome has been the repositioning of several genera into other families (and the reassignment of others, previously in other families, to the newly defined Scrophulariaceae). Examine the cladogram that sums up the evidence and identifies some major changes (Figure A3.2.8 overleaf) and the illustrations of representative plants involved (Figure A3.2.9).

Top tip!

This case study is an example of how plant species have been transferred between families and is useful to develop understanding of the processes involved. Note: you are not required to memorize the details of the case study.

HL ONLY

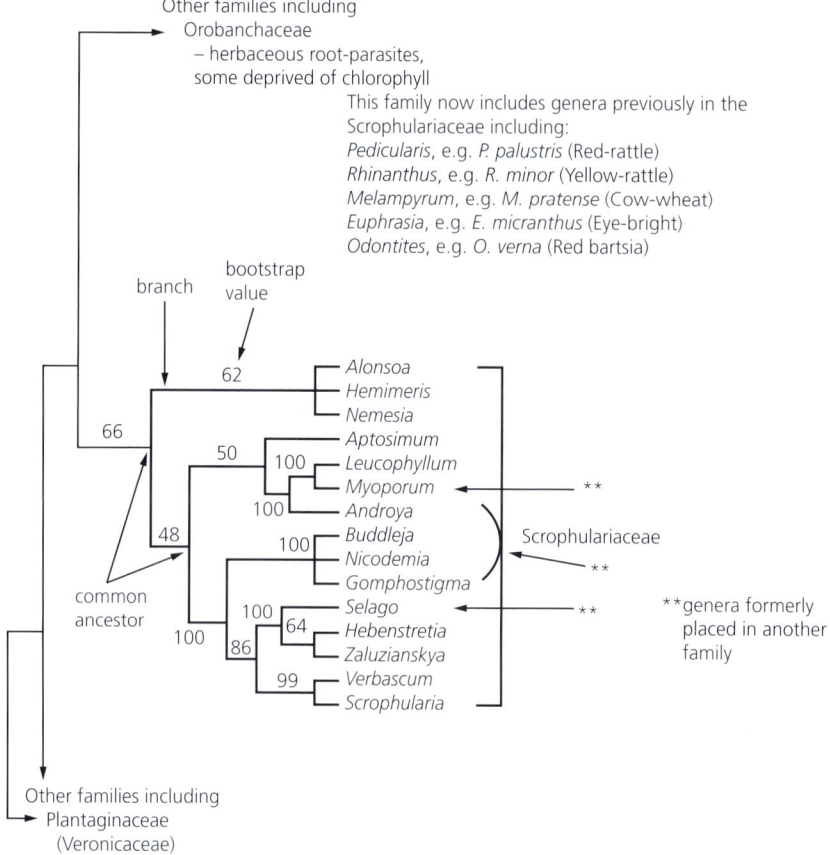

Other families including
Orobanchaceae
– herbaceous root-parasites,
some deprived of chlorophyll

This family now includes genera previously in the Scrophulariaceae including:
Pedicularis, e.g. *P. palustris* (Red-rattle)
Rhinanthus, e.g. *R. minor* (Yellow-rattle)
Melampyrum, e.g. *M. pratense* (Cow-wheat)
Euphrasia, e.g. *E. micranthus* (Eye-bright)
Odontites, e.g. *O. verna* (Red bartsia)

**genera formerly placed in another family

Other families including
Plantaginaceae
(Veronicaceae)

This family now includes genera previously in the Scrophulariaceae including:
Antirrhinum, e.g. *A. majus* (Snapdragon)
Digitalis, e.g. *D. purpurea* (Foxglove)
Veronica, e.g. *V. officinalis* (Speedwell)

■ **Figure A3.2.8** Cladogram summarizing changes to the family Scrophulariaceae: 'bootstrap values' provide a measure of accuracy of sample estimates. On a scale of 0–100 it is a measure of support for individual branches in the cladogram

Plants of the family Orobanchaceae, previously in the large, diverse family Scrophulariaceae

Yellow-rattle
(*Rhinanthus minor*)

Eye-bright
(*Euphrasia micranthus*)

Plants of the modern family Scrophulariaceae

Mullein
(*Verbascum lychnitis*)

Figwort
(*Scrophularia nodosa*)

Plants of the family Plantaginaceae, previously in the family Scrophulariaceae

Antirrhinum
(*Antirrhinum majus*)

Foxglove
(*Digitalis purpurea*)

■ **Figure A3.2.9** Plants associated with the Scrophulariaceae, now and previously

A note of caution

Despite these additional, largely biochemical, sources of evidence, the evolutionary relationships of organisms are still only partly understood. Consequently, current taxonomy is only partly a phylogenetic classification.

4 **Suggest** why molecular and computing techniques have made the construction of cladograms more accurate.

Nature of science: Falsification

You should appreciate that theories and other scientific knowledge claims may eventually be falsified. In this example, similarities in morphology between the flowering plants was due to convergent evolution rather than common ancestry and suggested a classification that was shown to be false by cladistics.

Theme A: Unity and diversity – Organisms

> **TOK**
>
> A major advance in the study of bacteria was the recognition, in 1977, by Carl Woese that archaea have a separate line of evolutionary descent to bacteria. Famous scientists, including Luria and Mayr, objected to his division of the prokaryotes. This poses the question: *to what extent is conservatism in science desirable?*

> **Top tip!**
>
> The archaea have been found in many extreme environments, such as deep ocean vents, geysers, salt pans and polar environments. However, their habitats are broader than this: some species occur only in anaerobic conditions, such as in the guts of termites and cattle, others at the bottom of ponds and swamps among the rotting plants.

Classification of all organisms into three domains

The realization that prokaryotes are composed of two very different groups followed the discovery of the distinctive biochemistry of the bacteria found in extremely hostile environments (the **extremophiles**), such as the 'heat-loving' bacteria found in hot springs at about 70 °C (see Theme A2.1, page 47). This led on to the new scheme of classification. These micro-organisms of extreme habitats have cells that we can identify as prokaryotic. However, one group of prokaryotes, now identified as the domain archaea (see below), contains rRNA that is unique and has molecular regions distinctly different from the rRNA of other prokaryotes and eukaryotes. The fact that RNA molecules present in the ribosomes of extremophiles are different from those of previously known bacteria is the key piece of evidence that has led to the reclassification of life.

Further analyses of their biochemistry in comparison with that of other groups has suggested new evolutionary relationships (Figure A2.1.8, page 44).

As a result, we now recognize three major forms of life, called domains. The organisms of each domain share a distinctive, unique pattern of ribosomal RNA and there are other differences which establish their evolutionary relationships (Table A3.2.5). These domains are:

- the **archaea** (the extremophile prokaryotes)
- the **eubacteria** (the true bacteria)
- the **eukarya** (all eukaryotic cells – the protoctista, fungi, plants and animals).

 Table A3.2.5 Biochemical differences between the domains

Biochemical features	Domains		
	Archaea	Eubacteria	Eukarya
DNA of chromosome(s)	circular genome	circular genome	linear chromosomes
Bound protein (histone) present in DNA	present	absent	present
Introns in genes	typically absent	typically absent	frequent
Cell wall	present – not made of peptidoglycan	present – made of peptidoglycan	sometimes present – never made of peptidoglycan
Lipids of cell membrane bilayer	archaeal membranes contain lipids that differ from those of eubacteria and eukaryotes (Figure A3.2.10)		

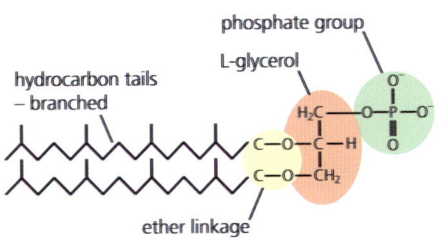
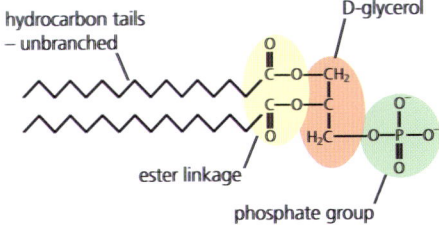

■ **Figure A3.2.10** Lipid structure of cell membranes in the three domains

> **LINKING QUESTIONS**
>
> 1. How can similarities between distantly related organisms be explained?
> 2. What are some examples of ideas over which biologists disagree?

A3.2 Classification and cladistics

Evolution and speciation

> **Guiding questions**
> - What is the evidence for evolution?
> - How do analogous and homologous structures exemplify commonality and diversity?

SYLLABUS CONTENT

This chapter covers the following syllabus content:
- ▶ A4.1.1 Evolution as change in the heritable characteristics of a population
- ▶ A4.1.2 Evidence for evolution from base sequences in DNA or RNA and amino acid sequences in proteins
- ▶ A4.1.3 Evidence for evolution from selective breeding of domesticated animals and crop plants
- ▶ A4.1.4 Evidence for evolution from homologous structures
- ▶ A4.1.5 Convergent evolution as the origin of analogous structures
- ▶ A4.1.6 Speciation by splitting of pre-existing species
- ▶ A4.1.7 Roles of reproductive isolation and differential selection in speciation
- ▶ A4.1.8 Differences and similarities between sympatric and allopatric speciation (HL only)
- ▶ A4.1.9 Adaptive radiation as a source of biodiversity (HL only)
- ▶ A4.1.10 Barriers to hybridization and sterility of interspecific hybrids as mechanisms for preventing the mixing of alleles between species (HL only)
- ▶ A4.1.11 Abrupt speciation in plants by hybridization and polyploidy (HL only)

Evolution as change in the heritable characteristics of a population

◆ **Evolution**: cumulative change in the heritable characteristics of a population.

Today, it is generally accepted that present-day flora and fauna have arisen by change ('descent with modification'), probably very gradual change, from pre-existing forms of life. By **evolution** we mean the development of life, from its earliest beginnings to the diversity of organisms we know about today, living and extinct. Evolution has occurred over geological time, but modern evidence shows that evolution can happen over much more rapid timescales (such as the evolution of bacteria – see page 534).

> **ATL A4.1A**
>
> Explore the diversity of life on Earth using the Linnaean Society's interactive tree of life, OneZoom: **www.onezoom.org/linnean**, which connects all species together in a spiralling structure that shows evolutionary history. Each leaf on the OneZoom tree represents a species and the branches show how they are connected through evolution.
>
> Think of a species that is of interest to you (for example, the blue whale). Can you find this species on the tree of life? To which organisms is it most closely related?

Concept: Diversity

Evolution is a process that explains the great variety of life seen on Earth. Changes in the heritable characteristics of a population, combined with selection pressures, have resulted in diversity across all forms of organisms.

Link

Mass extinctions are covered in Chapter A4.2, page 162.

Scientists have long appreciated that species have changed over time. Evidence from the fossil record, for example, makes it clear that life has changed substantially over geological history, becoming more complex over time but with many species now extinct. Mass extinctions have occurred in the past, where over 75% of species have been wiped out. In 1809, French naturalist Jean-Baptiste Lamarck proposed an explanation for change in species over time. His theory proposed that all the physical changes occurring in an individual during its lifetime can be inherited by its offspring. The theory influenced evolutionary thought throughout the first half of the nineteenth century.

■ Lamarck's theory of evolution

- **Change through use and disuse**: The organs that are used frequently by the organism develop and the characteristics that are used seldom are lost in the succeeding generations. For example, as a giraffe stretches its neck to eat leaves the neck elongates, and this is then passed on to the next generation. The organs that the organisms have stopped using would shrink with time. As the organisms adapted to their surroundings, they became increasingly complex compared to the simpler forms. Lamarck proposed that complexity arises due to usage or disuse of particular traits.
- **Inheritance of acquired characters**: An individual acquires certain characteristics during its lifetime. These characters are inherited by their offspring as well. Lamarck explained this with an example of a blacksmith. A blacksmith has strong arms due to the nature of their work. He proposed that any children a blacksmith conceives will inherit the development of strong muscles.
- **Effect of environment and new needs**: The environment influences all organisms. A slight change in the environment brings about changes in organisms. This gives rise to new needs, which in turn produces new structures and changes the habits of organisms.

Supposed examples of evolution by acquired characters (Lamarckism)

- The ancestors of the giraffe appeared similar to modern horses, with smaller necks and forelimbs compared to modern giraffes. Lamarckism states that the ancestors of giraffes, striving to reach the leaves of trees as an alternative source of food, were able to lengthen their necks. If such characters are inherited, the next generation of giraffes would be born with longer necks and would then stretch them further (Figure A4.1.1).
- Weightlifters acquire better-developed muscles, which can then be passed on to their offspring.

In contrast, the theory of evolution proposed by Charles Darwin (page 140 and Chapter D4.1) proposed that variation within a population leads to selection pressures, which enables organisms that are better adapted to the environment to survive and pass on this advantage to future generations (with reference to giraffes, see Figure A4.1.1).

Lamarck

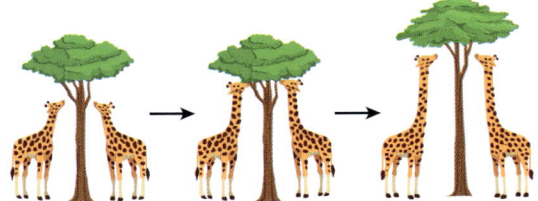

Darwin

| Ancestral giraffes stretched their necks to reach higher vegetation. | The offspring inherited the stretched necks. | Over generations, offspring repeatedly inherited longer and longer stretched necks. | In populations of ancestral giraffes, some individuals had longer necks than others. | Giraffes with longer necks were better adapted to the environment and had more offspring, also with longer necks. | Over generations, the population evolved to have longer necks. |

 Figure A4.1.1 Comparing Lamarck's and Darwin's theories of evolution

A4.1 Evolution and speciation

1 Compare and contrast Lamarck and Darwin's theories of evolution.

● Top tip!

The key element of Darwinism, and not of Lamarckism, is that evolution is driven by differences between individuals in the population. In Lamarckism all the changes could be occurring in all individuals simultaneously.

◆ **Epigenetics**: the study of heritable changes in gene activity that are not caused by changes in the DNA base sequences.

Concept: Unity

Darwinian evolution explains the origin of all life on Earth. This theory unifies biology by providing a comprehensive explanation for the pattern of similarities and differences that exist in all living organisms.

Many years after the lives of both Lamarck and Darwin, knowledge of DNA structure and function led to a revolution in science's understanding of inheritance. DNA, as the biological molecule of heredity, shows how genetics underpins the mechanism by which evolution takes place. It confirms Darwin's theory of evolution by natural selection, where variation in populations plays a key role. As we now know (but unknown to Darwin), variation is caused by random mutations in DNA (page 642) leading to alterations in the genetic makeup of members of a species (new alleles) which may confer a selective advantage. If modification of DNA code occurs in reproductive cells, these changes can be passed on to the next generation.

Modern genetics shows that acquired biological characters cannot be inherited, unless they cause some change to the organism's DNA or the way in which it is 'read'. Changes in the neck of a giraffe during its lifetime cannot be passed on as the changes do not affect the DNA that codes for neck length. The giraffe evolved its long neck because offspring inherited alleles producing a slightly longer neck, which allowed them to reproduce faster than those who carried the alleles for shorter necks.

Interestingly, **epigenetics** (see page 672) shows how environmental factors that affect the survivorship of parents can alter the environment in which genes operate, and that these changes can be inherited by offspring, leading to alteration of gene expression in the next generation. These changes are not, however, Lamarckian, as the changes in gene expression in offspring, although affected by the environment experienced by the parents, do not lead to long-term or *evolutionary* changes. Epigenetic inheritance may last for only a few generations, so it is not a stable basis for evolutionary change.

Evolution is therefore the development of new types of living organism from pre-existing types by the accumulation of genetic differences over many generations through the process of natural selection of chance variations. The changes that an organism acquires in its lifetime are not inherited and are not transmitted to its offspring. So, evolution is the process of cumulative change in the heritable characteristics of a population. This is an organizing principle of modern biology.

● Top tip!

The modern definition of evolution is 'change in the heritable characteristics of a population'. This definition helps to distinguish Darwinian evolution from Lamarckism. Acquired changes that are not genetic in origin are not regarded as evolution.

● Nature of science: Theories

Explaining evolution by natural selection

The theory of evolution by natural selection predicts and explains a broad range of observations and is unlikely ever to be falsified. However, the nature of science makes it impossible to formally prove that it is true by correspondence. It is a pragmatic truth and is therefore referred to as a theory, despite all the supporting evidence.

The theory of evolution is an empirical description of reality, rather than being a mathematical theory that can be shown to be true through formal logic. In this way, the theory of evolution has the same status as, for example, the atomic theory in chemistry. It is not necessarily true, but it accurately describes the world. Given this, it is a mistake to think that some general scientific statements are scientific theories while other general scientific statements are scientific facts. All are just empirically true, based on measurement and observation.

> **Concept: Unity**
>
> All living organisms are descended from a shared common ancestor: something that is demonstrated by the universality of the genetic code and similarities in cell structure.

Evidence for evolution

Evidence for evolution comes from many sources, including from the study of fossils, artificial selection in the production of domesticated breeds, studies of the comparative anatomy of groups of related organisms and from molecular information. We will look at some of these sources now.

■ Evidence from base sequences in DNA or RNA and amino acid sequences in proteins

All living things have DNA as their genetic material, with a genetic code that is virtually universal. The processes of 'reading' the code and protein synthesis, using RNA and translation in ribosomes, are very similar in prokaryotes and eukaryotes, too. Processes such as respiration involve the same types of steps and similar or identical intermediates and biochemical reactions, similarly catalysed. ATP is the universal energy currency. Also, among the autotrophic organisms, the biochemistry of photosynthesis is virtually identical.

This biochemical commonality suggests a common origin for life, as the biochemical differences between the living things of today are limited. Some of the earliest events in the evolution of life must have been biochemical, and the results have been inherited widely. However, large molecules such as nucleic acids and the proteins they may code for are subjected to changes with time, but this change may be an aid to the study of evolution and genetic relatedness. It is possible to measure the relatedness of different groups of organism by the amount of difference between specific molecules such as DNA, proteins and enzyme systems. This relatedness is a function of time since particular organisms shared a common ancestor. In this way, sequence data give powerful evidence of common ancestry.

> **Link**
>
> Selective breeding (artificial selection) is further covered in Chapter D4.1, page 797.

■ Evidence from selective breeding of domesticated animals and crop plants

Selective breeding, or **artificial selection**, is caused by humans. It is usually a deliberate and planned activity. It involves identifying the largest, the best or the most useful of the progeny, and using them as the next generation of parents. Continuous removal of progeny showing less-desired features, generation by generation, leads to deliberate genetic change. Indeed, the genetic constitution of the population may change rapidly. Although both artificial selection and natural selection have different selection mechanisms (one by nature and one by humans), the generation of the variation on which artificial selection acts is no different from the generation of variation in natural populations (i.e. random, by chance mutations).

Charles Darwin started breeding pigeons because of his interest in variation in organisms (Figure A4.1.2). In the *Origin of Species*, he noted there were more than a dozen varieties of pigeon which, had they been presented as wild birds to an expert, would have been recognized as separate species. All these pigeons were descendants of the rock dove (*Columba livia*), a common wild bird.

♦ **Artificial selection**: selection in breeding, carried out deliberately by humans to alter populations.

2 **Explain** the key differences between natural and artificial selection.

> From his breeding of pigeons, Darwin noted that there were more than a dozen varieties that, had they been presented to an ornithologist as wild birds, would have been classsified as separate species.

■ Figure A4.1.2 Charles Darwin's observation of pigeon breeding

A4.1 Evolution and speciation

> **ATL A4.1B**
>
> Explore artificial selection in pigeons using this site:
> https://learn.genetics.utah.edu/content/pigeons/pigeonetics
> In the scenarios presented in the game, can you successfully breed pigeons with the desired characteristics?

Darwin argued that, if so much change can be induced in so few generations, species must be able to evolve into other species by the gradual accumulation of small genetic changes as environmental conditions change – selecting some progeny and not others. By artificial selection, the plants and animals used by humans (such as in agriculture, transport, companionship and leisure) have been bred from wild organisms. The origins of artificial selection go back to the earliest developments of agriculture, although at that stage successful practice no doubt evolved by accident (Figures A4.1.3 and A4.1.4).

Wild sheep or European mouflon (*Ovis aries musimon*) occur today on Sardinia and Corsica

What domestication of wild animals involves.

- The identification of a population of a 'species' as a useful source of hides, meat, etc. and learning how to tell these animals apart from related species. Herd animals (e.g. sheep and cattle) are naturally sociable, and lend themselves to this.
- The selective killing (culling) of the least suitable members of this herd in order to meet immediate needs for food and materials for living.
- Encouraging breeding among the docile, well-endowed members of the herd, and providing protection against predators and disease.
- Selecting from the progeny individuals with the most useful features, and making them the future breeding stock.
- Maintenance of the breeding stock during unfavourable seasons.
- Ultimately, the establishment of a domesticated herd, dependent on the herdspeople rather than living wild, leading to the possibility of trading individuals of a breeding stock with neighbouring herdspeople's stocks.

Soay sheep of the Outer Hebrides suggest to us what the earliest domesticated sheep looked like

Modern selective breeding has produced shorter animals with a woolly fleece in place of coarse hair and with muscle of higher fat content; many breeds have lost their horns

■ **Figure A4.1.3** From wild to domesticated species, and the origins of selective breeding skills

■ **Figure A4.1.4** Selective breeding of wild mustard to create different types of vegetable

Variation between different domesticated animal breeds and varieties of crop plant, and between them and the original wild species, shows how rapidly evolutionary changes can occur.

Examples of crop improvement by selective breeding

Cereal grains are highly significant components of the human diet. Plant matter forms the bulk of human food intake in both developed and less-developed countries, but it is the plants of one family, the grasses, that we and most of our livestock depend on (Figure A4.1.5). This family includes cereals, which are the fruits (grain) of cultivated grass species. They are relatively easy to grow and the mature grains they yield are comparatively easy to store. Grains contain significant quantities of protein, as well as starch, as we shall see shortly.

■ **Figure A4.1.5** Crops (cereal grasses) that have been selectively bred from wild grasses

The cereals wheat, barley and oats and rice are the most important arable crops grown by agricultural communities in the Northern Hemisphere. They are examples of cultivated grass plants adapted to temperate climates with moderate amounts of rainfall. For example, wheat grows best in cool springs with moderate moisture for early growth, followed by sunny summer months that are dry for harvesting.

Bread wheat (*Triticum aestivum*) arose by natural crossings of various wild wheats that occurred around 8000 years ago in the Fertile Crescent of the Middle East (an area that today includes parts of Israel, Jordan, Lebanon, Syria, Egypt, northern Iraq, southern Turkey and Iran). Today, 'winter' wheat varieties are planted in the autumn and produce side shoots before lower temperatures during the winter months suspend growth. Winter wheat matures in early summer. 'Spring' wheat varieties are sown after winter and are adapted to a shorter growing season (but give lower yields).

> **Top tip!**
>
> Today's growers and consumers seek the following maize characteristics:
> - plants that grow vigorously in the local environment (climate, soil) but are resistant to diseases and result in high-yielding cobs
> - crops of fairly standard size (height) and with cobs that are ready at the same time, to facilitate harvesting.

There are two classes of wheat varieties. Hard wheats are higher in proteins, are grown in areas of lower rainfall, and are used to produce bread that can be kept for longer periods. Soft wheat varieties are starchier and are used to make pasta and 'French bread'. They are grown in more humid conditions.

Maize is the second most important food crop. It was probably originally domesticated in an area we now call Mexico. Today maize is grown all over the world, but most intensively in the USA. Modern varieties are hybrids – selected from the original form of wild maize (teosinte), which can still be found in the wild.

Plants originally grown in the wild would have had some of these characteristics, but not all. Selective breeding has been undertaken over the years to achieve the quality and consistency required.

> ### Concept: Diversity
>
> Many modern food crops have been created by artificial selection from wild ancestral populations. Selective breeding has enabled humanity to diversify its sources of food and provide sufficient yield to feed a growing population.

> ### Inquiry 1: Exploring and designing
>
> **Designing**
>
> Develop a laboratory experiment that investigates selective breeding in plants. Which species will you select and how will you ensure viable repeated generations? You will need a species that shows genetic variability and traits that can be easily measured. The following site will give you some ideas:
>
> https://fastplants.org/2018/03/15/observing-variation-fast-plants
>
> How will you observe variation and measure selection?

> 3 Charles Darwin argued that the great variety of breeds that we have produced in domestication supports the concept of evolution. **Outline** how this is so.
>
> 4 Dogs of the breeds known as Alsatians, Pekinese and Dachshunds are different in appearance yet are all classified as members of the same species. **Explain** how this is justified.

◆ **Homologous structures**: similar structures due to common ancestry.

◆ **Natural classification**: organisms grouped by as many common features as possible, and therefore likely to reflect evolutionary relationships.

◆ **Phylogenetic classification**: a classification based on evolutionary relationships (rather than on appearances).

■ Evidence from homologous structures

The body structures of some organisms appear fundamentally similar. For example, the limbs of vertebrates seem to conform to a common plan – called the pentadactyl limb (meaning 'five fingered'). Scientists describe these limbs as **homologous structures** as they occupy similar positions in an organism, have a common underlying basic structure, but may have evolved different functions (Figure A4.1.6). The fact that limbs of vertebrates conform but show modification suggests these organisms share a common ancestry. From this common origin, the tetrapod vertebrates have diverged over a long period of time. This process is called **adaptive radiation** (see also page 152). Classification based on homologous structures is a **natural classification**, or **phylogenetic classification**, because it is based on similarities and differences that are due to close relationships between organisms because they share common ancestors, reflecting evolutionary relationships.

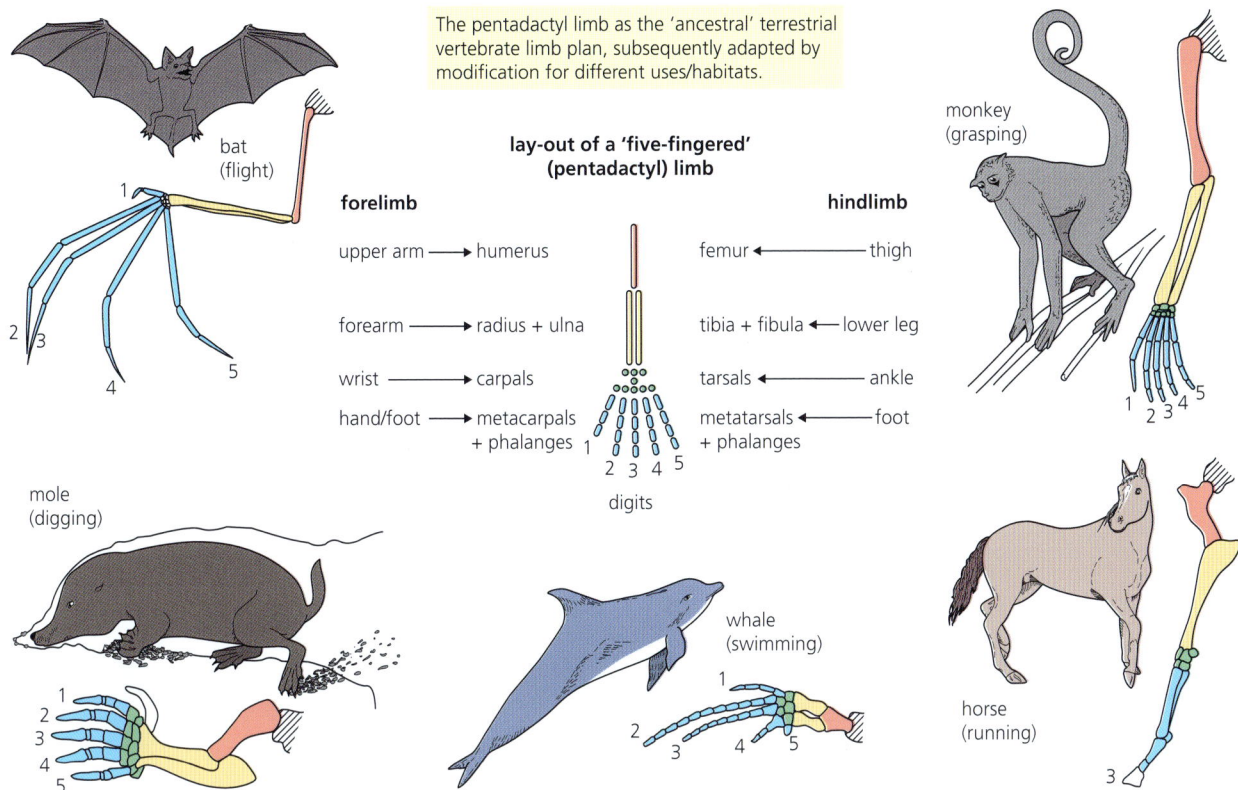

■ **Figure A4.1.6** Homologous structures show adaptive radiation

Convergent evolution

◆ **Analogous structure**: a feature with a similar function and superficial structural similarity, but different fundamental structure and evolutionary origin.

◆ **Artificial classification**: classifying organisms on the basis of few, self-evident features.

◆ **Convergent evolution**: the process by which distantly related organisms independently evolve analogous traits due to similar selection pressures.

We can contrast **homologous structures** with other structures of organisms that have similar functions but fundamentally different origins. Their resemblances are superficial; these are described as **analogous structures**. The wings of an insect and a bat are analogous, for example, as they look similar but were derived through separate evolutionary pathways. Classification based on analogous structures is an **artificial classification**. Table A4.1.1 compares analogous and homologous structures.

■ **Table A4.1.1** Analogous and homologous structures

Analogous structures	Homologous structures
resemble each other in function	are similar in fundamental structure
differ in their fundamental structure	are similar in position and development, but not necessarily in function
demonstrate only superficial resemblances	are similar because of common ancestry
for example, wings of birds and insects	for example, limbs of vertebrates

Convergent evolution occurs when different species evolve similar biological adaptations in response to similar selective pressures. This happens when species occupy similar ecological niches. Traits that arise through convergent evolution are referred to as analogous structures. Analogous features are features that have similar functions in different organisms but have different evolutionary origins. The similarity of function makes them look similar – such as the fins in dolphins and sharks, or bird and insect wings – but they are not similar either in terms of anatomy or origin (Figure A4.1.7).

Link
Galápagos finches are also an example of divergent evolution, see Figure A4.1.15, page 153.

◆ **Divergent evolution**: occurs when an ancestral species splits into two reproductively isolated groups, causing each group to develop different traits due to their respective selective pressures and natural selection.

Convergent evolution contrasts with evolution from a common ancestor, which is known as **divergent evolution** (Figure A4.1.7). The vertebrates and invertebrates, for example, share a common ancestor. Homologous structures are ones that may appear different but show common evolutionary origins from a common ancestor, such as the forelimb of a bat and the forelimb of a whale (Figure A4.1.7).

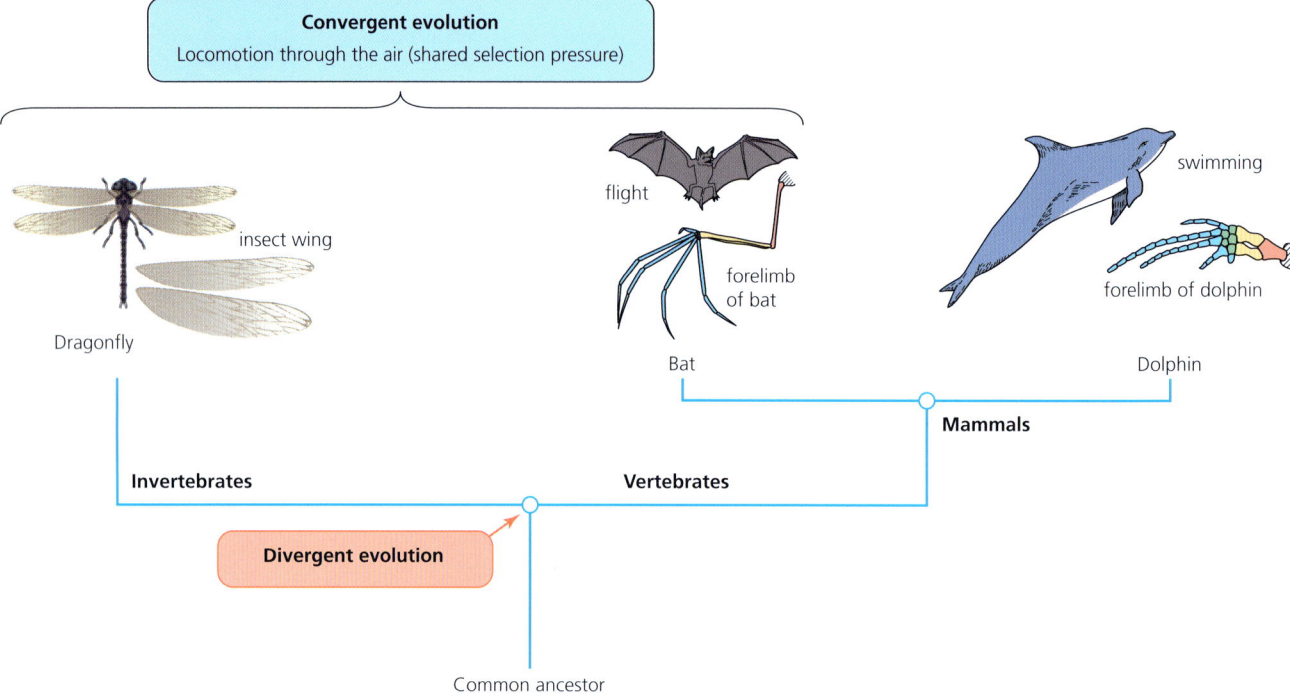

■ Figure A4.1.7 Convergent and divergent evolution

Common mistake

Do not confuse the terms 'homologous' and 'analogous'. Analogous structures are ones that appear the same between different, unrelated species due to similar selection pressures, whereas homologous structures are ones that may appear different but show common evolutionary origins from a common ancestor. Classification is based on homologous not analogous structures.

Convergent evolution can be observed at both the phenotypic and molecular levels (amino acid and DNA sequences). It is now thought that morphology and physiology often converge owing to the evolution of similar molecular mechanisms in independent lineages (page 146).

Most mammals have a placenta, which supports the offspring until birth. A separate group of mammals, the marsupials, do not have a placenta and so give birth to undeveloped offspring that they then need to support outside the body. Today, the marsupials are largely found in Australia. The placental mammals and Australian marsupials show many examples of convergent evolution. Animals with comparable appearances occur in both groups, due to similar selection pressures.

ATL A4.1C

Look at the following pictures of a dolphin and a dogfish (Figure A4.1.8). Are these animals closely related? Which features may make you think that they are closely related?

Then think about what you know about the groups of animals they are in – are they in fact closely related? Which of the two would be more closely related to a human?

■ **Figure A4.1.8** a) A dolphin and b) a dogfish

Use these sites to help you find out the features of both animals:

Dolphins: **www.dolphins-world.com**

Dogfish: **http://britishseafishing.co.uk/dogfish**

Analyse and evaluate the information you have found out about each animal to explain the implications of using only physical characteristics to classify organisms.
- Why should features that indicate common evolutionary origin be used (such as characteristics that group all mammals together compared with characteristics that group all fish) rather than superficial physical characteristics (e.g. the similarity in appearance between dogfish and dolphins)? Justify your answer using scientifically supported judgements.
- Explain the way in which science is applied and used to address the problem of classifying organisms. Why is it important to use the correct method of classification?
- Explain why it is important to use the same method of classification throughout the scientific community.

TOK

In the study of evolutionary history, do experiments have any part to play in establishing knowledge? If an experimental approach has a limited role, is the study of evolution a 'science'?

◆ **Speciation**: the process by which new species form, where one species is split into two or more species.

Speciation by splitting of pre-existing species

For a new species to form – a process called **speciation** – populations of an ancestral species need to become separated so that different environmental factors will act on them.

Look at the following diagrams in Figure A4.1.9.

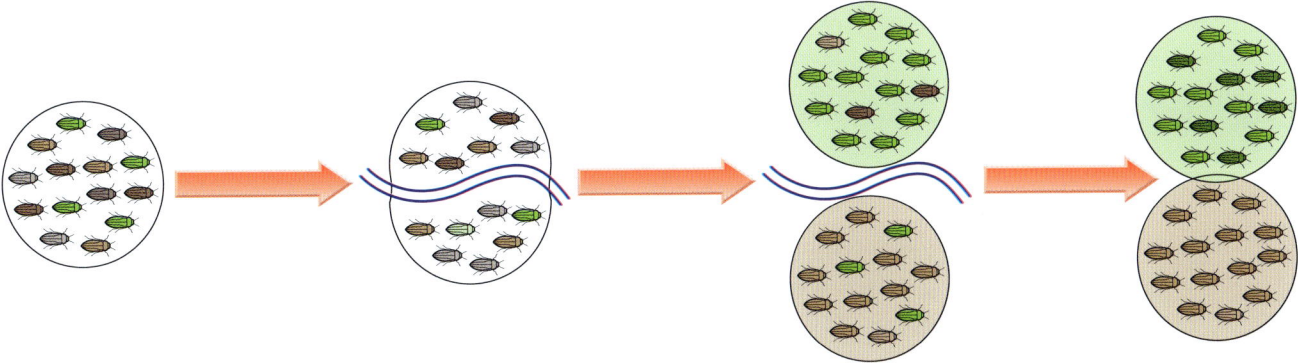

■ **Figure A4.1.9** Beetle speciation

A4.1 Evolution and speciation

> **Top tip!**
>
> Without a barrier, two populations would continue to mix and exchange genes; this would maintain the variety present in the initial population but not allow new species to evolve.

Something has happened to separate the initial population of beetles into two isolated populations – in this case, a river has formed over thousands of years, but the isolation of the populations could equally be due to the creation of a mountain range through movement of the Earth's plates, or some other barrier. The environments on either side of the barrier are different, and so there are different selection pressures. Again, in this example, individuals to the north of the river are given an advantage by having a green colouration, matching their habitat which is covered in lush vegetation; individuals to the south of the river are given an advantage by having a brown colour as their habitat is arid with more exposed earth (Figure A4.1.10).

> **Going further**
>
> Today, evolutionary biologists think that a complete barrier is not necessary for speciation to occur. Populations in different environments can diverge, even to the point of becoming different species, with some exchange of genes. What is needed is that selection is strong enough (e.g. the two environments are different enough) to overcome the homogenising (i.e. making uniform or similar) effect of gene exchange. An example is the evolution of apple maggot flies from ancestors that infested hawthorn, following the introduction of apple trees to North America.

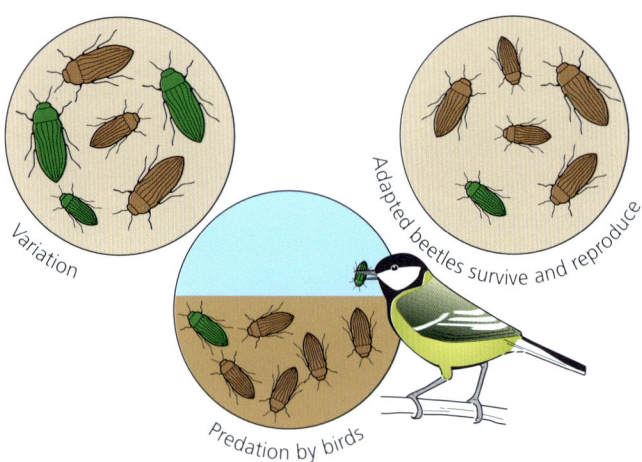

■ **Figure A4.1.10** Predation pressures on beetles

Progressive evolutionary change in a species is not speciation, as individuals within the population maintain the same gene pool and interbreed with one another. The barrier, by separating populations, isolates the gene pools of two populations so they can no longer interbreed and exchange genes (they become reproductively isolated). This means that over time, through the process of natural selection (see Chapter D4.1), adaptive genes are selected and others are lost, and so the two gene pools change and evolve. Eventually the two gene pools are so different that even if the two populations come back together (if the river dries up, for example) the individuals from each population cannot interbreed – the two populations have become different species.

Through the process of speciation, the number of species increases (as opposed to periods of extinction when the number of species decreases, see page 162). Speciation is the only way new species have appeared. There are different mechanisms for speciation and these will be explored below.

5 **Discuss** the factors that contribute to speciation.

> ● **Common mistake**
>
> It is incorrect to say that speciation is 'evolutionary change over time in a species'. Speciation is the splitting of a species into two or more separate species.

> ● **Common mistake**
>
> Do not confuse the term 'speciation' for species classification and taxonomy. Speciation is a process leading to the evolution of new species, whereas classification and taxonomy refer to the way in which species are divided into groups based on shared evolutionary origin (see page 126).

Going further

The Red Queen hypothesis

The biologist Leigh Van Valen proposed a 'law of extinction' based on the constant probability (as opposed to rate) of extinction in families of related organisms. He analysed data compiled from the duration of tens of thousands of genera throughout the fossil record. Figure A4.1.11 shows a linear relationship between survival time and the logarithm of the number of genera, suggesting that the probability of extinction is constant over time.

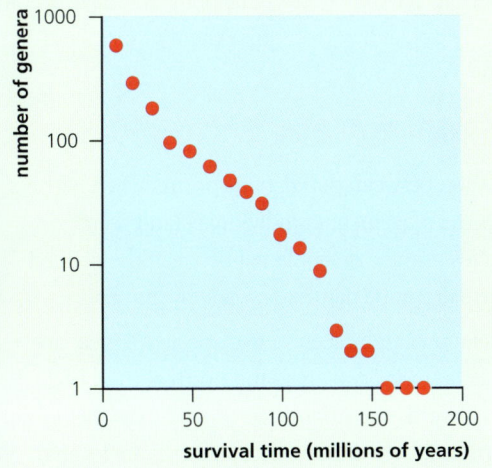

■ **Figure A4.1.11** The law of extinction ('genera', on y-axis, is the plural form of genus)

To explain his law of extinction, Van Valen proposed the **Red Queen hypothesis**, which describes how the coevolution of competing species creates a dynamic equilibrium in which the probability of extinction remains fairly constant over time.

In an ecosystem, a species interacts not with one species of competitor or parasite but with many. As each species evolves by natural selection, it changes the environment of many other species. The environment of each species changes as other organisms evolve.

Van Valen proposed that, even in a constant physical environment, evolutionary change will continue indefinitely as each species evolves to match other species' adaptations. This is an evolutionary 'arms race': as one species evolves adaptations that make it more competitive, its competitors experience selection pressures that force them to evolve to match it. Species that lag too far behind will become extinct.

In terms of speciation, the barrier (for example, a river) need not separate two physically different environments. Even in the same environment, two isolated populations will go on evolving due to the Red Queen process and get more and more different, to the point where they cannot interbreed.

The hypothesis is named after a remark made by the Red Queen in Lewis Carroll's *Through the Looking Glass*, where she told Alice that it took all the running she could do to keep in the same place.

Roles of reproductive isolation and differential selection in speciation

A first step to speciation may be when a local population (particularly a small local population) becomes completely cut off in some way, such as in colonizing an island. Even then, many generations may elapse before the composition of the gene pool has changed sufficiently to allow us to define a different species. However, changes in local gene pools are an early indication of speciation.

Occasionally, a population is suddenly divided into two isolated populations by the appearance of a barrier (Figure A4.1.12 and A4.1.9). Before separation, individuals shared a common gene pool, but after isolation disturbing processes such as natural selection, mutation and random genetic drift (where there is random fluctuations in the frequency of a particular allele in a population) may occur independently in both populations, causing them to diverge in their features and characteristics.

Reproductive isolation

By definition, different species cannot interbreed and have fertile offspring; gene flow between them is prevented. When members of related populations have evolved to this point and have become fully reproductively isolated, we recognise them as members of different species.

Geographical isolation

Geographic isolation between populations occurs when barriers arise and restrict the movement of individuals (and their spores and gametes in the case of plants) between the divided populations. Barriers can be natural or made by humans.

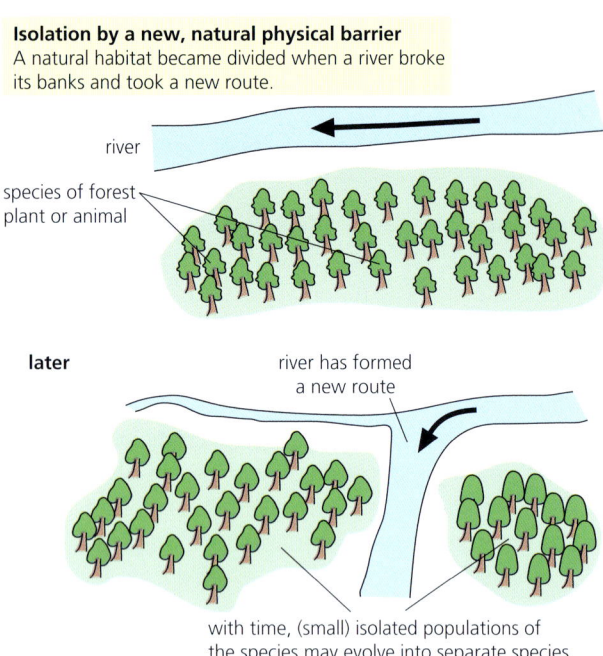

■ Figure A4.1.12 A geographical barrier

An example of speciation driven by geographical isolation can be seen in two species of great ape found in western and central Africa. Chimpanzees (*Pan troglodytes*) are found north of the Congo River (in two groups – central African chimpanzees towards the Atlantic coast, and a distinct group further east, see Figure A4.1.13). A closely related species, the bonobos (*Pan paniscus*) are located south of the river (Figure A4.1.13). The Congo River forms a barrier between the two different species, so that there is geographical isolation between them.

Although chimpanzees and bonobos appear superficially similar, there are significant differences in the morphology and behaviour of both species (Table A4.1.2).

■ Table A4.1.2 Differences between chimpanzees and bonobos

Chimpanzee (*Pan troglodytes*)	Bonobo (*Pan paniscus*)
larger	smaller
male-dominated societies	female-dominated societies
display aggressive behaviour	display peaceful behaviour
strong and sturdy body	long legs, narrow shoulders, small head
location: across west and central Africa; north of Congo River	location: Democratic Republic of the Congo; south of the Congo River

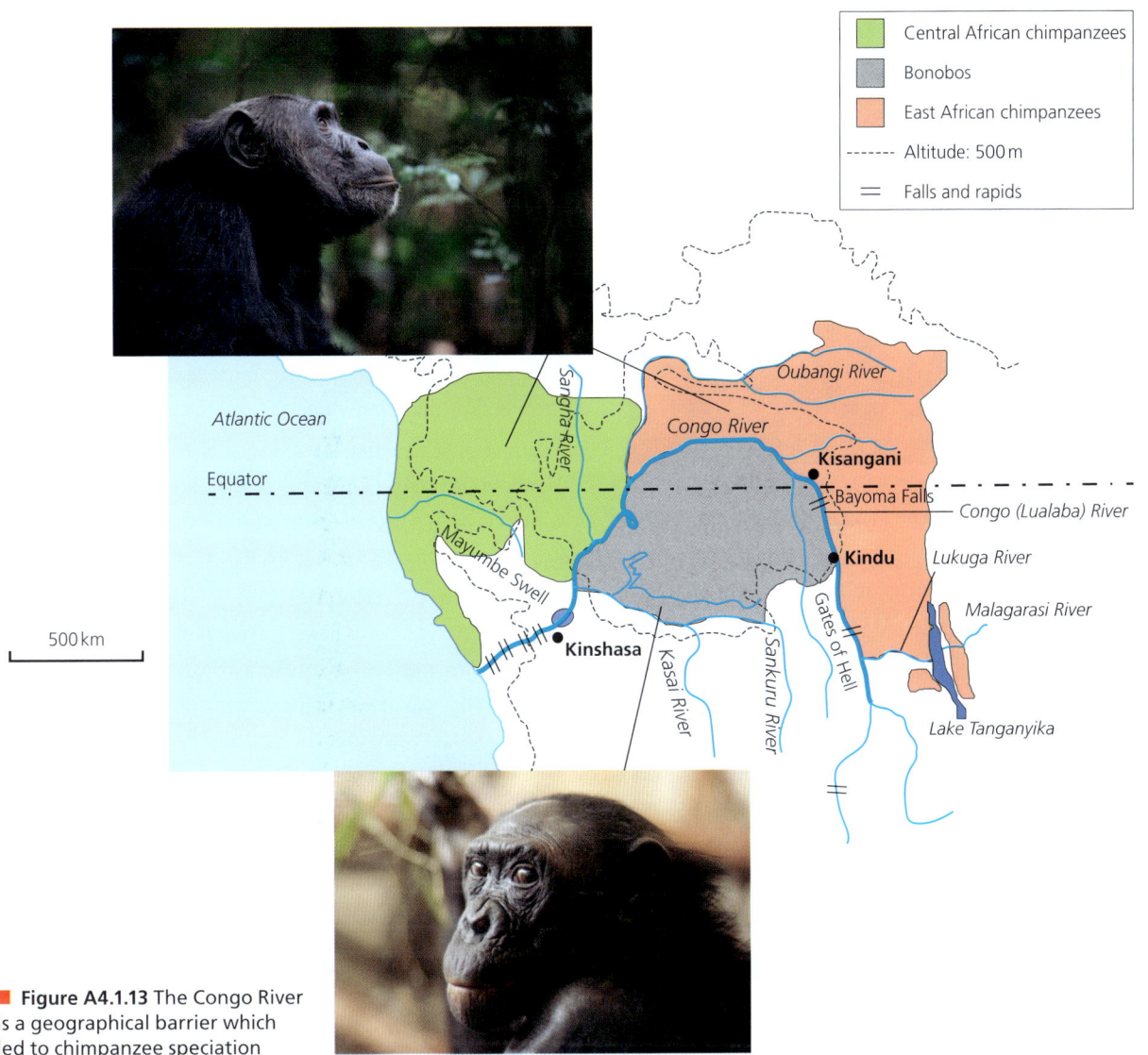

■ **Figure A4.1.13** The Congo River is a geographical barrier which led to chimpanzee speciation

Scientists agree that chimpanzees and bonobos diverged from a common ancestor between 1 million and 2 million years ago. For many years, the established view was that the Congo River appeared in the Pleistocene epoch (which began around 1.5 million years ago) and separated the common ancestor of chimpanzees and bonobos. Different environmental conditions on either side of the river led to them evolving into two species. Evidence from 2015, however, suggests that the Congo River has acted as a geographical barrier for far longer than previously thought, at least 34 million years. Therefore, scientists have hypothesized that when the Congo River first formed, the ancestor of the bonobo did not occupy the range the species does today, but that during times when the Congo River's level decreased during the Pleistocene, one or more founder populations of ancestral bonobos crossed the river, forming the populations from which the current day bonobo evolved.

Differences and similarities between sympatric and allopatric speciation

Apart from cases of instant speciation by polyploidy (see page 154), species do not generally evolve in a rapid way. The process is usually gradual, taking place over a long period of time. In fact, speciation may occur over several thousand years and, in all cases, requires 'isolation'.

A4.1 Evolution and speciation

■ Figure A4.1.14 A male Bird of Paradise (bottom) displays to a potential female mate (top) who looks on

As we have already seen, reproductive isolation occurs when two potentially compatible populations are prevented from interbreeding. As well as geographical isolation, this can also be due to the following types of isolation:
- **Temporal isolation** – this occurs when organisms produce gametes at different times or seasons. An example in animals is rainbow trout (spring) and brown trout (autumn). Another type of temporal isolation is when populations may be active at different times of the day or night.
- **Behavioural isolation** – organisms acquire distinctive behaviour routines, such as in courtship or mating, not matched by other individuals of the species.

An example of behavioural isolation is seen in the Birds of Paradise (Figure A4.1.14). Here, bright, glittering and prominently posing males seek to secure the attention of females. Elaborate and attractive display traits have evolved because they apparently satisfy some innate preference in the female birds. Competition between males has led to changes in display and plumage. The divergence of populations to display in different ways, and to have different preferences, may begin with very slight preference differences that get exaggerated over time. The resulting progressive elaboration of plumage and performance may be at the expense of flight ability, vulnerability to the attention of predators and, possibly, thermoregulation. Despite this, the genetic constitution of the next generation is strongly influenced by the few sexually successful males in each generation. It is the critical female selection that follows in response that leads to isolation of populations and, ultimately, reproductive isolation. Courtship behaviour of this type prevents hybridization in animal species.

◆ **Allopatric speciation**: speciation that occurs when two groups of organisms are spatially separated by a physical or geographic barrier, e.g. mountain ranges and large rivers.

◆ **Sympatric speciation**: speciation that occurs without geographical separation. Groups from the same ancestral population evolve into separate species due to temporal or behavioural isolation that prevents individuals of one species from mating with another.

◆ **Adaptive radiation**: the diversification of an ancestral species into new species, characterized by great ecological and morphological diversity, filling different ecological niches.

> ### Going further
>
> Temporal and behavioural isolation are examples of 'intrinsic isolation' due to evolved features of organisms, rather than imposed from the outside (extrinsic isolation). The question is, was geographical isolation required for the evolution of these intrinsic barriers? The current view is that geographic isolation may facilitate this evolution but it is not essential.

We have noted above some examples of the ways populations become isolated. Reviewing these, we see they fall into two groups, depending on the way isolation is brought about:
- Isolating mechanisms that involve spatial separation, such as geographical isolation, are known as **allopatric speciation** (meaning literally 'different country').
- Isolating mechanisms that occur within the same location, such as temporal and behavioural isolation, are known as **sympatric speciation** (meaning literally 'same country').

6 **Distinguish** between allopatric and sympatric speciation.

Adaptive radiation as a source of biodiversity

The concept of **adaptive radiation** was touched on when we discussed how the pentadactyl limb in vertebrates is evidence for evolution (see page 145 earlier in this chapter). Another example of adaptive radiation is illustrated by the beaks of the finches of the Galápagos Islands (Figure A4.1.15). These birds mostly failed to catch Charles Darwin's attention on his visits to the islands – the detailed study of these birds was undertaken later by the ornithologist David Lack.

The variation in finch beak morphology is a genetically controlled characteristic. It reflects differences in feeding habits. The evidence the finches provided for Darwin's theory of evolution by natural selection so impressed Lack that he coined the name 'Darwin's finches'. It has stayed with them, misleading though it is.

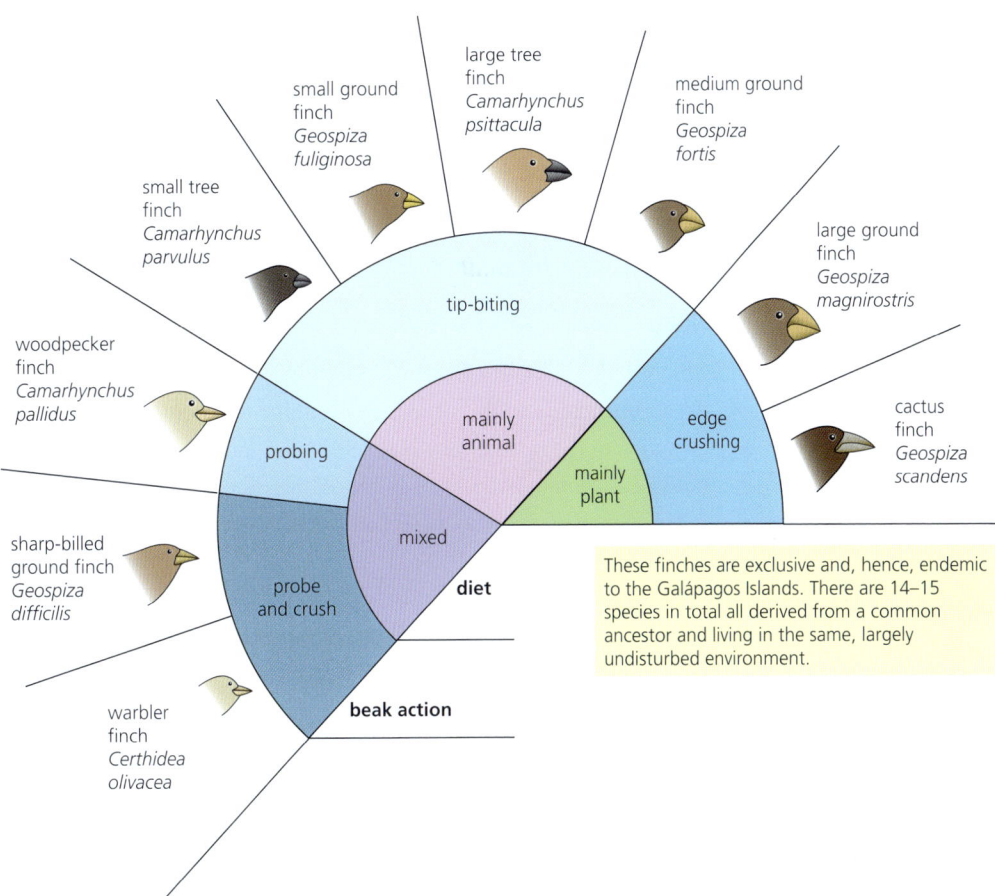

■ Figure A4.1.15 Adaptive radiation in Galápagos finches

Concept: Diversity

Adaptive radiation leads to the diversification of an ancestral species into new niches through the process of colonization and adaptation to local environments.

The numerous, closely related species of finch can coexist without competing and reducing the number of species because of the differences in their beak morphology. The Galápagos Islands provided a large variety of different opportunities for the ancestral finch species landing on the islands. These different roles, or 'niches' (see page 156), provided a range of different selection pressures which, over time, led to the variety of finch species seen on the Galápagos Islands today. For example, the warbler finch (*Certhidea olivacea*) has a different diet to the cactus finch (*Geospiza scandens*), meaning they do not compete and can coexist, thus increasing biodiversity.

Barriers to hybridization and sterility of interspecific hybrids

We have already seen how courtship displays can lead to sympatric speciation (page 152). Such displays can prevent hybridization in animal species. We have also seen how differences in chromosome number lead to hybrid sterility (as in the example of the mule, Chapter A3.1, page 120).

The barriers that prevent interbreeding between closely related species occur either before fertilization can be attempted (pre-zygotic isolation) or after fertilization has occurred (post-zygotic isolation). Table A4.1.3 shows a summary of the underlying isolating mechanisms discussed in this section.

■ Table A4.1.3 Summary of reproductive isolating mechanisms

Pre-zygotic reproductive isolation	Post-zygotic reproductive isolation
prevention of mating due to: • habitat differences that prevent meeting • behavioural differences, such as different mating rituals • temporal differences, such as being fertile at different times or seasons	hybrids formed are not viable and die prematurely hybrids formed are infertile because the chromosomes cannot pair up in meiosis to produce haploid gametes hybrids are less fertile: with each successive generation fewer survive, leading to them eventually all dying out

Abrupt speciation in plants by hybridization and polyploidy

In the examples considered up to now, speciation has been a gradual process. However, speciation can occur abruptly too, as we will examine next.

An abrupt change in the structure or number of chromosomes may lead to a new species. Such an alteration in the number of whole sets of chromosomes is known as **polyploidy**. Normally, adult body cells contain two sets of chromosomes (the 'diploid' condition – see page 110). Polyploidization is the gaining of one or more sets of chromosomes over and above the number found in diploid cells. Polyploids are largely restricted to plants and, indeed, polyploidy is thought to be one of the most important evolutionary processes in plants, especially the angiosperms (flowering plants). Some animals that reproduce asexually are polyploids too, but the sex determination mechanism of vertebrates prevents polyploidy in these animals.

◆ **Polyploidy**: a process that results in more than two sets of chromosomes in the nuclear genome; also known as whole genome duplication (WGD).

There are two mechanisms by which polyploids form:
- a diploid gamete fuses with a haploid one, resulting in **triploid** offspring ('triploid bridge'), i.e. forming offspring with three sets of chromosomes
- a diploid gamete fuses with another diploid gamete, resulting in **tetraploid** embryos (with four sets of chromosomes).

Diploid gametes can form when the spindle fails to form during meiosis (see page 656 for further detail of meiosis – sexual cell division), resulting in an additional set of chromosomes being present in the gamete.

Polyploid organisms can be divided into two categories (see Figure A4.1.16):
- An **autopolyploid** is the result of doubling a diploid genome.
- An **allopolyploid** is the result of two distinct genomes combining; this occurs by interspecific hybridization followed by either chromosome doubling or fusing of gametes, or by interspecific hybridization of two tetraploids.

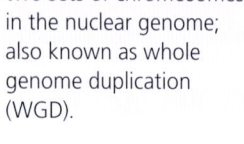

Top tip!

Polyploidy causes speciation most frequently because a new tetraploid produces only infertile triploid offspring when it mates with either diploid ancestor. Triploids are infertile because the chromosomes cannot pair properly in meiosis.

Many crop species are polyploids, such as *Triticum aestivum* (wheat), *Arachis hypogaea* (peanut), and *Fragaria ananassa* (strawberry). Invasive species (which are found in areas outside their normal area of distribution – see page 172) are often polyploids too.

ATL A4.1D

Why would mapping the genome of wheat be important in food production? Carry out research to find out about how mapping can be carried out (e.g. see https://learn.genetics.utah.edu/content/cotton/genome) and how could this be used, e.g. finding the location of genes beneficial to increasing yield, protecting plants against disease, and so on, to help with food production.

Figure A4.1.16 Comparing autopolyploidy and allopolyploidy

■ **Figure A4.1.17** Pale smartweed (*Persicaria lapathifolia*) flowering; a plant of the family Polygonaceae

The speed of invasive plant evolution, allowing them to spread and establish themselves throughout the world, can in part be explained by polyploidization. Interspecific hybridization (allopolyploidy) results in new genetic combinations that can be acted upon by natural selection and results in rapid evolution. Hybrids often show improved biological function compared to either parent species ('hybrid vigour') and are generally larger in size, more fertile and more invasive compared to the parent species. Scientists have shown that invasive species are more likely to be polyploid than native species, and the proportion of polyploids increases with more advanced invasion.

The plant genus *Persicaria* (in the family Polygonaceae) shows many instances of allopolyploid speciation. This genus includes many important weeds. Fifteen species appear to be allopolyploids. One of the diploid species, *Persicaria lapathifolia* (pale smartweed, Figure A4.1.17), has been involved in at least six cases of allopolyploid speciation. This species is geographically and ecologically widespread, and also bears numerous conspicuous flowers, which may explain its ability to hybridize with other species.

> **LINKING QUESTIONS**
>
> 1 How does the theory of evolution by natural selection predict and explain the unity and diversity of life on Earth?
> 2 What counts as strong evidence in biology?

A4.2 Conservation of biodiversity

Guiding questions
- What factors are causing the sixth mass extinction of species?
- How can conservationists minimize the loss of biodiversity?

SYLLABUS CONTENT

This chapter covers the following syllabus content:
- ▶ A4.2.1 Biodiversity as the variety of life in all its forms, levels and combinations
- ▶ A4.2.2 Comparisons between current number of species on Earth and past levels of biodiversity
- ▶ A4.2.3 Causes of anthropogenic species extinction
- ▶ A4.2.4 Causes of ecosystem loss
- ▶ A4.2.5 Evidence for a biodiversity crisis
- ▶ A4.2.6 Causes of the current biodiversity crisis
- ▶ A4.2.7 Need for several approaches to conservation of biodiversity
- ▶ A4.2.8 Selection of evolutionarily distinct and globally endangered species for conservation prioritization in the EDGE of Existence programme

Note: There is no higher-level only content in A4.2.

Concept: Diversity

Biodiversity is the variety of life in all its forms, levels and combinations, and includes ecosystem diversity, species diversity and genetic diversity.

◆ **Biodiversity**: the amount of biological or living diversity per unit area. It includes the concepts of species diversity, habitat diversity and genetic diversity.

Common mistake

Some students think that biodiversity is only seen in animals. This is incorrect – it is a term that can be applied to all living organisms.

Biodiversity as the variety of life in all its forms, levels and combinations

There are vast numbers of different types of living organism in the world – almost unlimited diversity, in fact. We refer to this as **biodiversity**, which is a contraction of the words 'biological' and 'diversity'. By biodiversity we mean 'the total number of different species living in a defined area or ecosystem'. It is also applied to the incredible abundance of different types of species on Earth.

Biodiversity is the amount of biological or living diversity per unit area. It includes the concepts of species diversity, habitat diversity and genetic diversity:

- **Ecosystem diversity** is the range of different habitats or number of ecological niches per unit area in an ecosystem. For example, a woodland may contain many different habitats (e.g. river, soil, trees) and so have a high ecosystem diversity, whereas a desert has few (e.g. sand, occasional vegetation) and so has a low diversity. Conservation of ecosystem diversity usually leads to the conservation of species and genetic diversity.
- **Species diversity** is the variety of species per unit area. This includes both the number of species present and their relative abundance.
- **Genetic diversity** is the range of genetic material present in a gene pool or population of a species. A large gene pool leads to high genetic diversity and a small gene pool to low genetic diversity. Although the term normally refers to the diversity within one species, it can also be used to refer to the diversity of genes in all species within an area.

Richness and **evenness** are components of species diversity (see below). Richness is a term that refers to the number of a species in an area, and evenness refers to the relative abundance of each species (Figure A4.2.1). A community with high evenness is one that has a similar abundance of all species – this implies a complex ecosystem where there are many different niches that support a wide range of different species. In contrast, low evenness refers to a community where one or a handful of species dominate – this suggests lower complexity and a smaller number of potential niches, where a few species can dominate.

■ **Figure A4.2.1** Species richness versus species diversity. The species richness is the same in both ecosystems (ecosystem A and ecosystem B have three species of beetle in each). In ecosystem B, diversity is higher than ecosystem A because evenness is higher. One species dominates in ecosystem A, indicating a less-complex ecosystem

Common mistake

'Species richness' is not the same as 'diversity'. Richness is the number of species in an area, whereas the diversity is a measure of the number of species and their relative abundance.

Inquiry 2: Collecting and processing data

Collecting data; processing data; interpreting results

Collect your own data to investigate the richness and evenness of the species sampled in your local environment. What qualitative data can you record? What issues will arise during data collection and how will you address these (i.e. consider safety, ethical and environmental issues)? How will you assess the accuracy, precision, reliability and validity of your data?

■ Simpson's reciprocal index

Diversity indices, such as **Simpson's reciprocal index**, can be used to describe and compare communities. Diversity indices can be used to assess whether the impact of human development on ecosystems is sustainable or not. When comparing communities, consider the following:
- low diversity could indicate pollution, eutrophication or recent colonization of a site
- the number of species present in an area is often indicative of general patterns of biodiversity, although disturbed sites often have artificially increased species richness due to the mixing of habitats that are usually spatially separate in undisturbed sites.

A4.2 Conservation of biodiversity

Common mistake

Do not confuse Simpson's reciprocal index with the Simpson's index – they use different equations and are interpreted in different ways. This IB Biology course uses the Simpson's reciprocal index (equation and working shown on this page).

Top tip!

Diverse communities are ones that have high evenness – each species is approximately equally abundant, indicating a complex ecosystem with lots of niches.

Tool 3: Mathematics

Applying Simpson's reciprocal index

Species diversity refers to the number of species and their relative abundance. It can be calculated using diversity indices.

Species diversity can be calculated using Simpson's reciprocal index, using the equation:

$$D = \frac{N(N-1)}{\sum n(n-1)}$$

where:

D = diversity index, N = total number of organisms of all species found, and n = number of individuals of a particular species.

Index values are relative to each other and not absolute, unlike measures of, say, temperature, which are on a fixed scale. This means that two different areas can be compared using the index, but a value on its own is not useful.

Other important points to note about Simpson's reciprocal index are:

- Comparisons must be made between areas containing the same type of organism in the same ecosystem.
- A high value of D suggests a stable and ancient site, where all species have similar abundance (or 'evenness', see page 157).
- A low value of D could suggest disturbance through, say, logging, pollution, recent colonization or agricultural management, where one species may dominate.

Analysis of the biodiversity of two communities

Below, an analysis of two habitats is carried out using Simpson's reciprocal index of diversity. Table A4.2.1 contains data from two different habitats. The total number of species ('species richness') and total number of individuals is the same for each habitat.

WORKED EXAMPLE

Calculate the diversity of habitats X and Y, and comment on the differences between them.

 Table A4.2.1 The number of species and their abundance in two habitats

Species found	Number found in habitat X	Number found in habitat Y
A	10	3
B	10	5
C	10	2
D	10	36
E	10	4
Number of species	5	5
Number of individuals	50	50

The Simpson's reciprocal index must be calculated for each habitat. This can be done using a table to calculate components of the index.

■ Table A4.2.2 Calculating the Simpson's reciprocal index for two habitats

Species	Number (n) found in habitat X	n(n – 1)	Number (n) found in habitat Y	n(n – 1)
A	10	10(9) = 90	3	3(2) = 6
B	10	10(9) = 90	5	5(4) = 20
C	10	10(9) = 90	2	2(1) = 2
D	10	10(9) = 90	36	36(35) = 1260
E	10	10(9) = 90	4	4(3) = 12
	$\Sigma n(n-1)$	= 450	$\Sigma n(n-1)$	= 1300

Species diversity for each habitat:

habitat X: $D = \dfrac{50(49)}{450} = \dfrac{2450}{450} = 5.44$

habitat Y: $D = \dfrac{50(49)}{1300} = \dfrac{2450}{1300} = 1.88$

What do these values say about each habitat?
- There is greater 'evenness' between species in habitat X. This implies more opportunity for niches in this habitat, due to greater resources and space.
- There is less competition due to non-overlapping niches in habitat X.
- One species does not dominate in X, reflecting greater habitat complexity.
- Habitat Y is less complex with fewer, or overlapping, niches. One species can dominate, leading to lower diversity.

Tool 3: Mathematics

Applying Simpson's reciprocal index

Calculate the Simpson's reciprocal index for the data you gathered in your biodiversity investigation as part of your Inquiry on page 157. Ideally you will have sampled at least two different areas so that comparisons can be made.

Inquiry 3: Concluding and evaluating

Concluding; evaluating

Use the calculation from your biodiversity study to draw conclusions about the outcome of your investigation. Compare your results to other similar studies – does your study show similar patterns and conclusions? Link the outcomes of your investigation to your original research question – did you confirm or disprove your hypothesis? What were the limitations of the experiment and what improvements could you make next time?

1 **Calculate** Simpson's reciprocal index for habitat B using data from the table below.

■ Table A4.2.3 Comparing the species diversity of five habitats

Species found	Number of individuals found in each habitat				
	Habitat A	Habitat B	Habitat C	Habitat D	Habitat E
Species 1	25	50	80	97	100
Species 2	25	30	10	1	0
Species 3	25	15	5	1	0
Species 4	25	5	5	1	0
Simpson's reciprocal index (D)	4.13		1.37	1.04	1.00

2 **Describe** and **explain** the differences between the habitats shown in the table.

3 **Suggest** reasons for the different values of Simpson's reciprocal index recorded in the different habitats.

Comparisons between current number of species on Earth and past levels of biodiversity

Scientists do not know how many different types of organisms exist. Millions of species have been discovered, named and described, but there are many more species to be discovered. Estimates of the number of species on Earth range from 5 to 100 million, with the scientific consensus currently being around 9 million species. This estimate is broken down as follows:

- animals: 7.77 million (12% of which are described)
- fungi: 0.61 million (7% of which are described)
- plants: 0.30 million (70% of which are described)
- other species: 0.07 million.

We might have expected that all different species have been discovered in the countries that pioneered the systematic study of plants and animals. However, this is not the case; previously unknown organisms are frequently found in all countries.

Figure A4.2.2 is a representation of the relative numbers of known species that are estimated to exist in many of the major divisions of living things.

Of the 1.8 million described species, more than 50% are insects; the higher plants, mostly flowering plants, are the next largest group. By contrast, only 4000 species of mammals are known, about 0.25% of all known species.

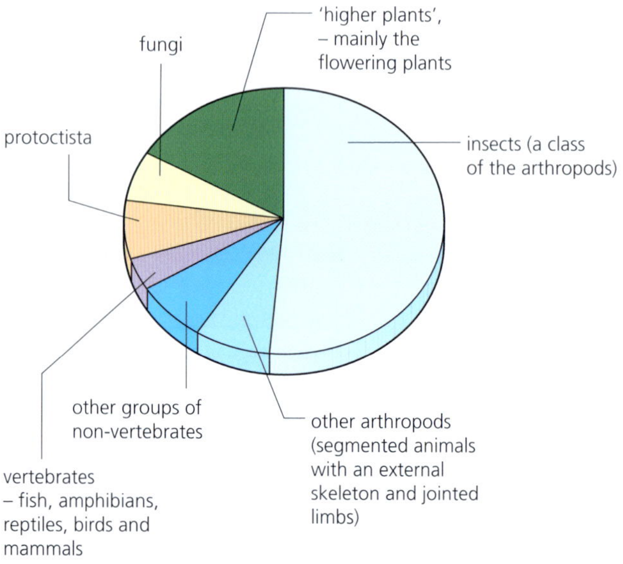

■ Figure A4.2.2 The relative number of animal and plant groups

Theme A: Unity and diversity – Ecosystems

■ Evidence from fossils

We learn something about the history of life from the evidence of fossils. Fossilization is an extremely rare, chance event. Predators, scavengers and bacterial action normally break down dead plant and animal structures before they can be fossilized. Of the relatively few fossils formed, most remain buried or, if they do become exposed, are overlooked or accidentally destroyed.

Nevertheless, numerous fossils have been found – and more continue to be discovered all the time. The various types of fossil and the steps in fossil formation by petrification are illustrated in Figure A4.2.3. Where it is the case that the fossil, or the rock that surrounds it, can be accurately dated (using radiometric dating techniques), we have good evidence of the history of life. Radiometric dating measures the amounts of naturally occurring radioactive substances such as carbon-14 (in relation to the amount of carbon-12), or the ratio of potassium-40 to argon-40.

The fossils in a rock layer (stratum) that has been accurately dated give us clues about the community of organisms living at a particular time in the past, although this is necessarily an incomplete picture. The fossil record may also suggest the sequence in which groups of species evolved, and the timing of the appearance of the major phyla. Evidence from fossils suggests that there are currently more species alive on Earth today than at any time in the past.

Fossil forms

petrification – organic matter of the dead organism is replaced by mineral ions

mould – the organic matter decays, but the space left becomes a mould, filled by mineral matter

trace – an impression of a form, such as a leaf or a footprint, made in layers that then harden

preservation – of the intact whole organism; for example, in amber (resin exuded from a conifer, which then solidified) or in anaerobic, acidic peat

Steps of fossil formation by petrification

1. dead remains of organisms may fall into a lake or sea and become buried in silt or sand, in anaerobic, low-temperature conditions
2. hard parts of skeleton or lignified plant tissues may persist and become impregnated by silica or carbonate ions, hardening them
3. remains hardened in this way become compressed in layers of sedimentary rock
4. after millions of years, upthrust may bring rocks to the surface and erosion of these rocks commences
5. land movements may expose some fossils and a few are discovered by chance but, of the relatively few organisms fossilized, very few will ever be found by humans

■ Figure A4.2.3 The process of fossilization

● TOK

What knowledge is likely to always remain beyond the capabilities of science to investigate or verify?

Some scientists propose that evidence from fossils suggests that there are currently a greater number of species alive on Earth today than at any time in the past. Others argue that this is debateable. Due to the paucity of evidence from the fossil record, particularly in rocks from deep time, we cannot be certain whether species diversity was ever higher in the past than it is now. There have been five global extinction events (see page 149). Following a mass extinction, there seems consistently to have been very little biodiversity at the beginning of each geological era; this has subsequently evolved. In addition, not all species fossilize well, for example organisms with soft bodies such as jellyfish, so there would be no fossil record of these extinct species. Scientists are also unlikely to find all the fossils that are hidden in the Earth. Therefore, some common ancestors will remain beyond the ability of science to verify.

ATL A4.2A

The dinosaurs were the dominant group of organisms on the Earth for around 180 million years. Then, 66 million years ago, the majority were wiped out, with only one group surviving that evolved into the birds. What caused the dinosaurs to go extinct? Watch the following film and summarize the main points in an article or poster:

www.biointeractive.org/classroom-resources/day-mesozoic-died

Nature of science: Observations

As classification tools have evolved in taxonomy, as seen throughout Theme A, the determination of what a species (or genus, or kingdom) is has changed.

In bacterial taxonomy there has been a tug of war between what are often referred to as the 'lumpers' and the 'splitters', with the former tending to lump together similar organisms into the same species, while the latter splits off subgroups within species into new species.

Let's take the example of *Salmonella*, a genus in the family Enterobacteriaceae, which are Gram-negative. They are motile facultative anaerobes (meaning they are bacteria that can move and carry out aerobic respiration but can switch to anaerobic respiration in the absence of oxygen) and can cause a wide range of illnesses in both human and animals.

Originally there were over 200 species of *Salmonella* based on the host range, the clinical expression of infection, biochemical reactions and antigens of the bacterium. Then, the results from DNA–DNA hybridization techniques (a molecular biology technique that measures the degree of genetic similarity between DNA sequences) suggested all *Salmonella* species were highly related to one another and should thus be designated as one species, *S. enterica*, with some subspecies. However, the debate about the taxonomy continues with lumpers versus splitters.

There is some debate about whether it even makes sense to talk about species in bacteria at all. Species are fundamentally clusters of phylogeny, physical traits, gene variants, and so on. However, diversity in bacteria does not always cluster and, when it does, it can do so at many different levels.

Top tip!

In this section we will focus on anthropogenic causes of the current mass extinction rather than the non-anthropogenic causes of previous mass extinctions.

◆ **Anthropogenic**: relating to human activity.

Causes of anthropogenic species extinction

On Earth, there have been five mass extinctions in the past in which more than 75% of all species have gone extinct over an extended period (see page 139). The causes for these extinctions were natural – most ultimately linked to climate change, often caused by volcanic activity. The last mass extinction occurred 66 million years ago and was caused by the impact of an asteroid, where change to Earth happened instantaneously and led to the extinction of the dinosaurs. Today, scientists believe we are in a sixth mass extinction, this time caused by human actions rather than natural causes. There are many different human-related (**anthropogenic**) causes of species extinction, many of which are covered in the case studies below.

> **Top tip!**
>
> Remember, you can use either the common name or the scientific name when referring to an organism.

Case study 1: North Island giant moas

The North Island giant moa (*Dinornis novaezealandiae*) is one of three extinct moa in the genus *Dinornis* that were endemic to New Zealand (Figure A4.2.4). They were a group of flightless birds and were the second tallest of the nine moa species, measuring up to 2 metres from the ground to their back and up to 3 metres tall including their necks. The North Island giant moa also showed sexual dimorphism (differences between females and males) with adult females being much larger than adult males.

■ Figure A4.2.4 The North Island giant moa (*novaezealandiae*)

This species of moa lived on New Zealand's North Island in the lowlands, shrublands, grasslands, dune lands and forests. It is an example of terrestrial megafauna, which are the large or giant animals of an area, habitat or geological period, extinct and/or extant (living). The population size of the giant moa remained stable over the past 40 000 years until the arrival of humans, the Māori, in New Zealand around 1280.

The giant moa, along with other moa genera, were hunted for food. All taxa in this genus were extinct by 1500 in New Zealand. The most important contributing factor was probably farming, however, since the forests were cut and burned down and the ground was turned into arable land.

Studies of ancient DNA from the bones of giant extinct New Zealand birds demonstrate that significant climate and geological environmental changes did not have a major impact on their populations. Ancient DNA, radiocarbon dating and stable dietary isotope analysis have shown that, before humans arrived, moa adapted to the effects of climate change on their species by tracking their preferred habitat as it expanded, contracted and shifted during warming and cooling events.

Moa were very large and as so played a major role in shaping the structure and composition of vegetation communities. The extinction of moa could have affected New Zealand's ecosystems through altering vegetation composition and structure, regeneration patterns and fire frequency.

> **Link**
>
> Food webs are covered in detail in Chapter C4.2, page 581; predator–prey relationships are discussed in Chapter C4.1, page 574.

Case study 2: Caribbean monk seals

The Caribbean monk seal (*Neomonachus tropicalis*) was declared extinct in 2008 and is an example of the loss of a marine species (Figure A4.2.5). At their peak abundance, this species had a widespread distribution throughout the Caribbean Sea, Gulf of Mexico and western Atlantic Ocean, and their habitat included areas around the east coast of Central America and north coast of South America.

The Caribbean monk seal was part of a group called Pinnipedia (pinnipeds), which includes sea lions and walruses. They are the only pinniped species to have become extinct. Reasons for the seal's extinction included being hunted for fur and meat, and oil from its blubber; being captured for display in museums and zoos; and overfishing activities that disturbed their habitat and reduced their prey species (mainly fish). They were easy to kill because of their tame and non-aggressive behaviour, which meant that hunters could get close to them. NOAA (National Oceanic and Atmospheric Administration) Fisheries has stated it was the first species of seal to become extinct because of human causes.

The extinction of the monk seal had a huge knock-on effect across the Caribbean's food web. The monk seal was a top predator, feeding on a variety of fish and invertebrates, so its disappearance allowed some species of fish to expand at the expense of others, significantly altering the

■ **Figure A4.2.5** The Caribbean monk seal (*Neomonachus tropicalis*)

biodiversity of the areas where the seal had been found. At one point there were an estimated 233 000–338 000 monk seals distributed among 13 colonies across the Caribbean. Such an extensive population could only survive because there was a large quantity of food available: each adult seal would eat approximately 245 kg of fish per year. The loss of historically dense monk seal colonies, and their consumption rates, severely impacted other species in the Caribbean coral reefs.

■ Case study 3: Falkland Islands wolf

The Falkland Islands wolf (Figure A4.2.6) was the only native land mammal of the Falkland Islands. These remote South Atlantic islands, about 480 kilometres northeast of the southern tip of South America, were first sighted by Europeans in 1692. In 1833, Charles Darwin visited the islands and described the wolf as 'common and tame'.

The Falkland Islands wolf is said to have lived in burrows. As there were no native rodents on the islands (the usual prey), it is probable that its diet consisted of ground-nesting birds (such as geese and penguins), grubs, insects and some seashore scavenging.

■ **Figure A4.2.6** The Falkland Islands wolf (*Dusicyon australis*)

ATL A4.2B

The now-extinct Falkland Islands wolf was the only native land mammal on the Falkland Islands. Even Charles Darwin was confused by the wolf: what was a large predator doing on a small set of islands nearly 500 kilometres away from South America?

Find out about the evolution of the Falkland Islands wolf and why such a tame predator was found on these islands. Here is a good starting point:
www.nationalgeographic.com/science/article/the-origin-of-the-friendly-wolf-that-confused-darwin

The many settlers of the islands (mainly Scottish inhabitants but also some French and English) considered the Falkland Islands wolf a threat to their sheep. A huge-scale operation of poisoning and shooting began with the aim of leading the wolf to extinction. The operation was successful very rapidly, assisted by the lack of forests and the tameness of the animal (due to the absence of predators the animal trusted humans, who would lure it with a piece of meat and then kill it).

The Falkland Islands wolf was not particularly threatening nor was it a significant predator, although the removal of a top predator would have had an impact on the rest of the food chain, for example increasing the population size of its prey.

Causes of ecosystem loss

◆ **Ecosystem**: a community of organisms and the physical environment with which it interacts.

◆ **Community**: a group of different species living in an area.

◆ **Abiotic**: the non-living components of an ecosystem.

An **ecosystem** is defined as a community of organisms and their surroundings, the environment in which they live and with which they interact. A **community** (i.e. a group of different species living in an area) forms an ecosystem by its interactions with the **abiotic** (non-living) environment (see Chapter C4.1, page 579).

Disturbance and loss of ecosystems can be natural (for example, due to typhoons or volcanic activity), but currently is predominantly due to anthropogenic causes.

Top tip!

We will focus on causes of ecosystem loss that are directly or indirectly anthropogenic (caused by humans). Remember to include case studies in your answers to strengthen them.

■ Case study: The loss of mixed dipterocarp forests in South East Asia

Tropical rainforests cover only 6% of the Earth's land surface but may contain up to 50% of all species. They are found in South America, Africa, India, South East Asia and Australia, close to the equator. The climate is warm and stable, with temperatures varying from 20 °C at night to 35 °C at midday, and high rainfall, with up to 2500 mm per year. The constant warm temperatures, high insolation (sunlight) and high rainfall lead to high levels of photosynthesis and high productivity. High productivity leads to high amounts of biological matter and food, which in turn leads to ecosystem complexity, abundant resources (such as food) and niche (the role an organism plays in its community) diversity. Abundant niches lead to high species richness and high biodiversity.

Link
Niches are covered in detail in Chapter B4.2, page 363.

Tropical rainforests are vulnerable to disturbance. As they have high biodiversity, many species are affected when the rainforests are disturbed. Deforestation and forest degradation are caused principally by demands for wood (Figure A4.2.7), for land for cattle to provide beef, and for plantation crops including soya and biofuels (such as palm oil in South East Asia). Tropical rainforests grow on nutrient-poor soils that are thin and easily eroded once the forest has been cleared, which means that once forest is cleared and the soil eroded, it is difficult to re-establish a forest cover.

Dipterocarp trees are the main hardwood species in the tropical rainforests of South East Asia. More than 270 species of dipterocarp trees have been identified so far in the island of Borneo, of which 155 are endemic.

In South East Asia, large areas of dipterocarp forest have been cleared to grow oil palms (*Elaeis guineensis*), see Figure A4.2.8. In Malaysia, for example, 20% of land area is now covered by oil palm plantations, compared with 1% in 1974. Palm oil is used in food production (for example, margarine, cooking oil, ice cream, ready meals, biscuits and cakes), domestic products such as detergents and cosmetics, and to provide biofuel. The global production of palm oil exceeds 35 million tonnes per year.

■ Figure A4.2.7 Logging activities in Sabah, Malaysian Borneo

■ Figure A4.2.8 Oil palm trees have replaced tropical rainforest over large areas of Borneo

Once the original forest has been removed, natural nutrient recycling is also lost. As the soils are generally nutrient poor, oil palm trees require fertilizer to be applied to produce yields that return a reasonable profit. Fertilizers can have various negative environmental impacts (page 808). Fertilizer application in oil palm plantations has also been linked to increases in ground-level ozone, which can have detrimental effects on both humans and wildlife, and also reduces plant productivity.

The causes and effects of the loss of mixed dipterocarp forest in Borneo in South East Asia are summarized in Table A4.2.4.

■ Table A4.2.4 The loss of dipterocarp forest in South East Asia has severe ecological impacts

Description	Borneo is the third-largest island in the world and has historically been covered by tropical rainforest – it is one of the oldest rainforests in the world at approximately 130 million years old.
	The rainforest is dominated by dipterocarp trees: long-lived, tall, hardwood trees that are valuable timber species.
	The island has high species diversity, with 15 000 plant species, 220 mammal species and 420 bird species recorded.
	Around 20% of the mammal species are endemic to Borneo.
	300 species of tree can be found in 1 hectare of forest.
	Many species are on the International Union for Conservation of Nature (IUCN) Red List – for example orang-utan, sun bear and Asian elephant.
Human threats	Borneo's rainforests have been logged commercially for export market since the 1970s, with logging accelerating in the 1980s and 1990s.
	In 1974, forest cover was estimated to be 6.4 million ha (88% of the total land area); only 4.5 million ha remained in 1985 (i.e. a 30% reduction in 11 years).
	The deforestation rate (in 2000–5) was 3.9% per annum.
	At the peak of logging operations, trees were removed in large numbers (up to 100 m^3 ha^{-1} in volume).
	Conventional logging methods are not selective and cause damage to the remaining forest.
	More recently, selective logging (reduced-impact logging – RIL) methods have been used. These cause less damage and allow faster regeneration of forest.
	Since the 1990s, oil palm has been planted on cleared land (Figure A4.2.8).
	One hectare of oil palm yields up to 5000 kilograms of crude palm oil.
	Most oil palm plantations are owned by the state or by transnational corporations.
Consequences of disturbance	Damage to the remaining forest is proportional to the amount of timber removed.
	Heavily-logged forest using conventional methods can remove 94% more timber than in RIL sites.
	Conventional logging methods severely damage forest structure, leading to an increase in light-loving, riverine (i.e., living or situated on the banks of a river) species such as *Macaranga* trees.
	Changes in forest structure reduce biodiversity.
	Oil palm plantations are monocultures with low species diversity.
	Oil palm plantations fragment rainforest, block migration routes and remove habitats for animals.
	Insecticides and herbicides are used to control insect pests and weeds, and so reduce biodiversity.
	Animals such as Asian elephants and orang-utans that stray into the plantations can be illegally killed.
	Without forest cover, soil is eroded, making it difficult or impossible for the original vegetation to regrow.
	Loss of forest reduces transpiration from leaves, which affects local weather patterns, leading to drier areas that are more prone to fire.
	The valuable role that the ecosystem provides through biodiversity and controlling weather patterns has been reduced.
	The role of the rainforest as an economic resource (for example though ecotourism) has been reduced.

■ Figure A4.2.9 A satellite image of the Great Barrier Reef from 2004; the coast of Queensland is on the left, with the reef systems clearly visible on the right

■ Case study: The loss of the Great Barrier Reef

The Great Barrier Reef Marine Park is 345 000 km^2. It is in the Coral Sea, off the coast of northeast Australia (Figure A4.2.9). It is an important part of Indigenous Australian culture and spirituality. It is also a very popular destination for tourists, especially in the Cairns region, where it is economically significant. Fishing occurs in the region as well, generating AU$1 billion per year.

Links

The ecology of coral reefs is covered in more detail in Chapter B4.1, page 353, while the threats to coral reefs are covered in Chapter D4.3, page 832.

Coral reefs, like rainforests, are amazingly diverse (and for similar reasons – such as their location, complexity and high productivity). The Great Barrier Reef stretches 2300 kilometres along the Queensland coastline of northern Australia and is made up of 3000 individual reef systems. It is home to 1500 species of fish, 359 types of hard coral, a third of the world's soft corals, 6 of the world's 7 species of threatened marine turtle and more than 30 species of marine mammals including vulnerable dugongs (sea cows). In addition, there are 5000 to 8000 molluscs; thousands of different sponges, worms and crustaceans; 800 species of echinoderms (starfish, sea urchins) and 215 bird species, of which 29 are seabirds (e.g. reef herons, ospreys, pelicans, frigate birds and shearwaters).

The threats to this ecosystem are many and varied; the causes and effects of the loss of the coral in the Great Barrier Reef are summarized in Table A4.2.5.

■ Table A4.2.5 The causes and effects of the loss of the Great Barrier Reef

Description	The world's largest coral reef ecosystem; with its diversity of species and habitats, it is one of the richest and most complex natural ecosystems on the planet.
	The area the Great Barrier Reef inhabits has been exposed to many glacial cycles and first began to grow about 18 million years ago (although the current reef is built upon older reef platforms).
	Ecological, sociopolitical and economic pressures are causing the degradation of the coral reef and, as a consequence, are threatening the biodiversity of the area.
	The Great Barrier Reef was made a World Heritage Site in 1981.
Human threats	The fragile coral is easily damaged by divers' fins and boat anchors.
	Although it is illegal to remove coral pieces from the country of origin, tourists still break off bits as souvenirs.
	Over-fishing disrupts the balance of species in the food chain, and there may also be accidental damage from anchors and pollution from boats.
	Seafloor trawling for prawns is still permitted in over half of the marine park, resulting in the unintentional capture of other species and the destruction of the seafloor.
	Changes in land use in Australia, from low-level subsistence agriculture to large-scale farming, means the plantations need heavy input of fertilizers and pesticides. Run-off from the soils into the sea has caused inorganic nitrogen pollution to increase by 3000%.
	Combined with sewage and pollution from coastal settlements such as Cairns, there are excessive nutrients in the water and algal blooms occur.
	Sedimentation (leading to mud pollution) has increased by 800% due to deforestation of mangroves to make space for tourist developments, housing and farming.
	Climate change is also affecting the reef: increases in sea temperature cause coral bleaching (plant and algal life on the reef dies, so the reef loses colour) – aerial surveys in 2002 showed that almost 60 per cent of reefs were bleached to some degree.
Consequences of disturbance	Available habitats for sea turtles (coral reefs and seagrass beds) are being damaged by sedimentation, nutrient run-off and tourist development.
	Destructive fishing techniques and climate change cause reduction in population numbers.
	Mass coral bleaching events occurred in 1998, 2002, 2016, 2017 and 2020 caused by increased sea temperatures due to climate change.
	Increases in sea level and changes to sea temperatures may have a permanent effect on the Great Barrier Reef, causing loss in the biodiversity and ecological value of the area.
	Climate change may be causing some fish species to move away from the reef to seek waters at their preferred temperature. This leads to increased mortality in seabirds that prey on the fish.

Evidence for a biodiversity crisis

Evidence for the current biodiversity crisis can be drawn from reports and other sources from the Intergovernmental Science-Policy Platform on Biodiversity and Ecosystem Services (IPBES).

IPBES (**https://ipbes.net**) is an independent, intergovernmental body founded in 2012 by 94 countries to strengthen the link between scientific information and policy decisions relating to conservation and the sustainable use of biodiversity, long-term human well-being and sustainable development. Although it is not a United Nations (UN) organization, the United Nations Environment Programme (UNEP) provides help to IPBES.

Results from reliable surveys of biodiversity in a wide a range of habitats around the world are required to establish current levels of biodiversity and to allow estimates of species extinction rates to be calculated. Such surveys may use the Simpson's reciprocal index to compare communities in undisturbed and disturbed habitats (see page 157 earlier this chapter). Surveys need to be repeated to provide evidence of change in species richness and evenness. There are opportunities for contributions from both expert scientists and citizen scientists. Citizen science is scientific research carried out, in whole or in part, by amateur (i.e. non-professional) scientists.

There are four common features of citizen science:
- anyone can participate
- participants use the same protocol so data can be combined and be high quality
- data can help professional scientists come to reliable conclusions
- a wide community of scientists and volunteers work together and share data to which the public, as well as scientists, have access.

Citizen science allows large quantities of data to be collected – more than would be possible by professional scientists alone. Given the scale and imminent threats posed by the current biodiversity crisis (the sixth mass extinction, see page 162), citizen science offers a rapid and global mechanism for gathering data that can be used to support the conservation of species.

Nature of science: Evidence

The role of peer review and 'citizen science'

To be verifiable, evidence usually comes from a published source that has been peer-reviewed, which allows the methodology to be checked. Scientists must adopt sceptical attitudes to claims and evaluate them using evidence.

A peer review is the evaluation of work by other experts in the same field as the producers of the work (peers). The purpose is to check the quality, validity and credibility of scientific work; it is a form of self-regulation within the scientific community to maintain the scientific integrity of published papers. Independent verification of data analyses and scientific conclusions gives objectivity to a scientific paper.

However, studies show that even after peer review, some articles still contain inaccuracies and demonstrate that most rejected papers will go on to be published somewhere else.

This review process is usually 'blind': the identity of the reviewers is not revealed to the authors by the paper's editorial board, and the names of the authors may not be revealed to the reviewers to prevent bias. However, scientists sometimes choose to bypass the peer-review process and publish their work on non-peer-reviewed platforms, such as websites and magazines.

Data can also be recorded by citizens rather than scientists. As mentioned above, this is known as 'citizen science' and can be done by anyone from the public with an interest in the projects being studied.

Citizen science brings benefits but also unique methodological concerns. The people carrying out the research are not experts and so there is a danger that data are inaccurately recorded. However, members of the public who carry out the work are supported by experts and training is available, as well as access to scientific protocols (procedures) so that data can be combined from many sources and be of high quality.

Read more about citizen science here:
https://scistarter.org/citizen-science

Causes of the current biodiversity crisis

■ Human population growth

For most of Earth's history, natural processes have influenced and shaped life, such as plate tectonic movements, ocean and atmospheric currents, and volcanic activity. More recently, one species – *Homo sapiens* – has been the dominant influence on Earth's ecosystems. Early humans lived in balance with nature, as hunter-gatherers, and had little impact on their environment. Populations were low in number and people lived off the land.

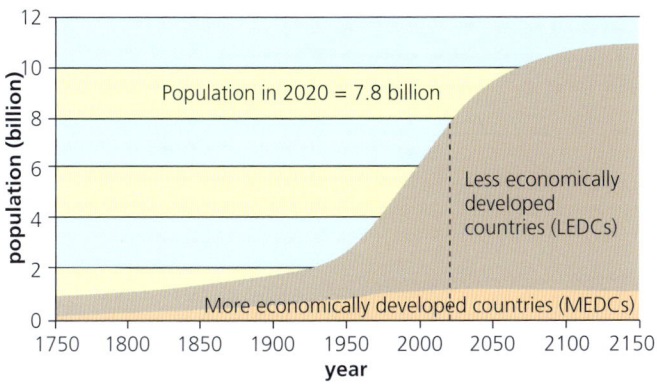

■ Figure A4.2.10 Human population increase over 400 years

As humanity spread out from Africa, eventually becoming farmers and clearing land to grow crops, the impact of *Homo sapiens* on the planet grew. The development of settled agriculture represents one of the most significant changes in human history and enabled human populations to start growing. This period, known as the Neolithic ('new stone age') revolution, began in the 'fertile crescent' in the Middle East about 10 000 years ago, and forever changed the way that humanity interacts with the environment.

In recent times, humanity's tremendous growth in population, from around 2 million in the early Neolithic period to over 7 billion (7×10^9) today has put even more pressure on the Earth's natural systems.

 Following the development of agriculture, human populations became settled in growing communities. Over the past 10 000 years, the domestication of many species of plants and animals (see page 142) has occurred independently and in different ways in different parts of the world. Production of staple foods enabled the human population to grow. During industrial revolutions, increased access to energy further fuelled population growth. The world's population doubled between 1804 and 1922, 1922 and 1959, 1959 and 1974. It is, therefore, taking less and less time for the population to double. In the twentieth century, population growth became exponential (exponential means increasingly rapid growth).

Other factors have contributed to an increase in human population:
- better healthcare
- more nutritious food
- cleaner water
- better sanitation.

The biggest increase in population is in less economically developed countries (LEDCs) rather than in more economically developed countries (MEDCs). High infant death rates increase the pressure on women to have more children, and in some agricultural societies parents have larger families to provide labour for the farm and as security for the parents in old age. Lack of access to contraception, through education or medical services, also leads to increased birth rates.

The impact of exponential growth is that enormous amounts of extra resources are needed to support growing populations (for food, housing and clothing). Human population growth is an overarching cause of biodiversity loss, together with other specific causes such as hunting and other forms of over-exploitation; urbanization; deforestation and clearance of land for agriculture with consequent loss of natural habitats; pollution and spread of pest and diseases; and alien invasive species due to global transport.

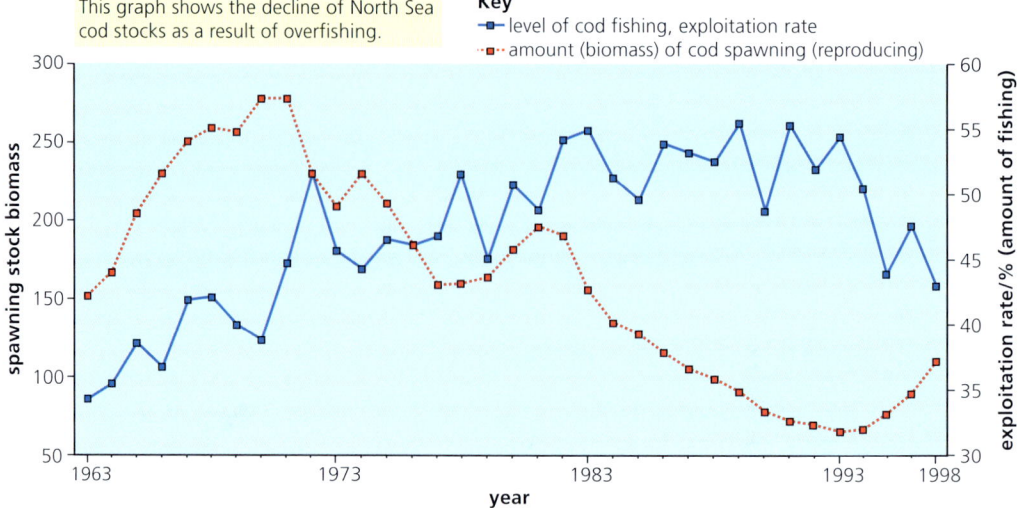

This graph shows the decline of North Sea cod stocks as a result of overfishing.

As the amount of fishing has been reduced (since about 1990), the decline in spawning fish has been reversed. With time, this trend may allow stocks to recover completely.

■ **Figure A4.2.11** Overfishing of North Atlantic cod in European waters

Hunting and other forms of over-exploitation

Overharvesting and hunting have led to a significant reduction in population size of many species. Animals are hunted for food, medicines, souvenirs, fashion and to supply the exotic pet trade.

Overharvesting of North Atlantic cod in the 1960s and 1970s led to significant reduction in population numbers (Figure A4.2.11).

Urbanization

Today, some 4.2 billion (4.2×10^9) people (55% of the world's population) live in cities. It is predicted that by 2050, 68% of the world's population will live in urban areas.

Urbanization refers to the increase in the proportion of people living in towns and cities. Living areas are built on land that was once covered by natural habitats. For example, New York was built on land that was a natural wetland environment: these areas were drained and filled in during the development of the city. With an increasing global population and the increase in the numbers of people living in urban centres, the trend of increasing urbanization and loss of natural ecosystems (and therefore biodiversity) can be expected to continue. The density of people living in cities is very high compared to more rural areas, so it could be argued that increasing urbanization overall provides an effective and efficient way of housing a rapidly growing population, compared to less-dense housing solutions.

Deforestation and clearance of land for agriculture

Deforestation

Habitat loss due to deforestation is one of the major causes of biodiversity loss. This is especially true when biodiverse ecosystems, such as tropical rainforests, are cleared (Figure A4.2.12, and see page 799).

Rainforests cover only 6% of the Earth's surface, yet may be home to 50% of all species. The climate in equatorial areas provides optimal conditions for photosynthesis and, therefore, the production of food for consumer organisms (see pages 356 and 592).

Tropical rainforests are rich in natural resources such as timber and so are vulnerable to exploitation, with an average of 1.5 hectares (equivalent to a football pitch) lost every 6 seconds in 2019.

Theme A: Unity and diversity – Ecosystems

Conversion of land for agriculture and mining

The increase in the world's human population from around 3 billion (3×10^9) people in the 1950s to over 7 billion people in 2023 (Figure A4.2.10) has led to increased demand for food. Conversion of land to agriculture to produce food (Figure A4.2.13) has led to further habitat loss. Nearly 40% of the Earth's land surface is used for agriculture (Figure A4.2.14), with an area approximately the size of South America used for crop production, and even more land ($3.2–3.6 \times 10^9$ hectares) being used to raise livestock such as cattle.

■ Figure A4.2.12 Bulldozer making a logging road in rainforest, Sabah, Borneo

■ Figure A4.2.13 A range of crops being grown in the UK

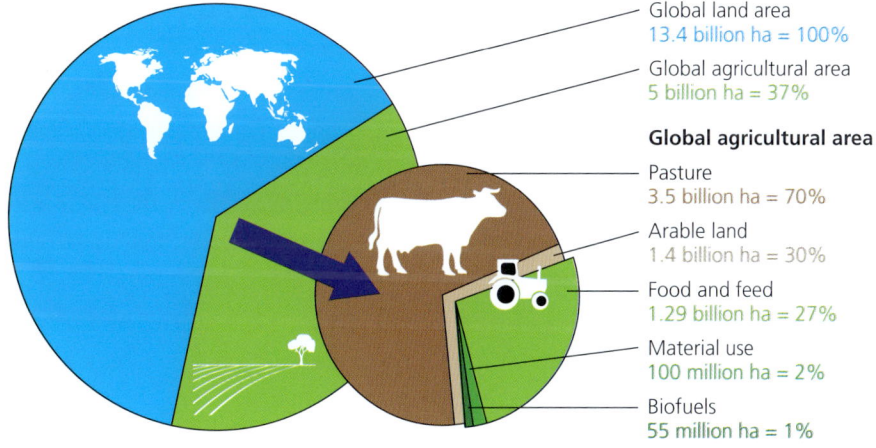

■ Figure A4.2.14 Nearly 40% of the Earth's surface is used for agriculture. This pie graph shows how agricultural land is used for different purposes

Agricultural practices have led to the destruction of native habitats and replaced them with monocultures (i.e. crops of only one species). Monocultures represent a large loss of diversity compared to the native ecosystems. However, increasing awareness of this has led to the re-establishment of hedgerows and undisturbed corridors that encourage more natural communities to return.

Habitats can also be lost through mining activities. Mobile phones contain an essential metallic element (tantalum) which is obtained by mining coltan (a metallic ore that contains the elements niobium and tantalum). Coltan is found mainly in the eastern regions of the Democratic Republic of Congo – mining activities in these areas have led to extensive habitat destruction of forests that contain gorillas and other endangered animals.

Natural habitats have also been cleared to make way for plantation crops. Sugar cane plantations have replaced tropical forest ecosystems, such as mangrove in Australia, and oil palm plantations throughout South East Asia have led to the widespread loss of tropical forests (page 165).

Some parts of the world are more intensively farmed than others (Figure A4.2.15).

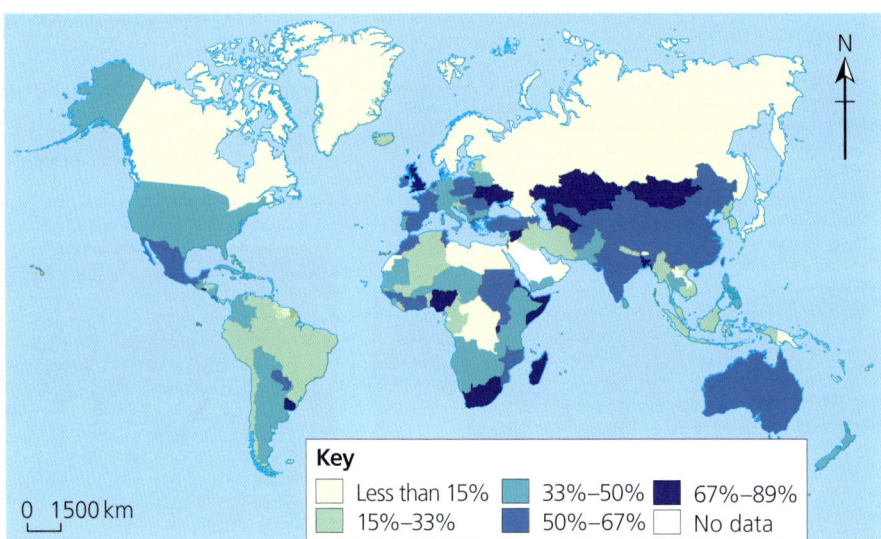

■ **Figure A4.2.15** Agricultural land as a percentage of total land area

◼ Pollution and the spread of disease

Pollution is the addition to an environment of a substance or agent (such as heat) by human activity, at a rate greater than that at which it can be rendered harmless by the environment, and which has an appreciable effect on the organisms in that environment. Pollution, including chemicals, litter, nets, plastic bags and oil spills, damages habitats and kills animals and plants, leading to the loss of life and reduction in population numbers of species.

Human effects on the environment cause the spread of diseases. The golden toad (*Incilius periglenes*) was a small, shiny, bright toad once common in a small region of high-altitude, cloud-covered tropical forests, 0.5 km by 8 km in area, above the city of Monteverde in Costa Rica. This species was first discovered in 1966 but was recorded as extinct by the IUCN in 2004. The last recorded sighting of the toad was in 1989. Possible reasons for its extinction include airborne pollution or infection by fungus or parasites spread by increasing global temperatures.

Small populations are particularly prone to being affected by diseases. Reduced population size leads to a reduced gene pool, leaving a species more prone to disease and inbreeding (for example, the critically endangered Asiatic cheetah in Iran).

◼ Invasive species

When humans started trading with one another between continents, many species were removed from their native habitats and transported vast distances into new environments where they had not existed before and where they would not naturally have reached. In all countries around the world, there are now many species of flora and fauna that are non-native to that region. Some non-native species were deliberately introduced, for example in the UK the rabbit was introduced from Normandy, France, and the buddleia (a flowering plant endemic to Asia, Africa and the Americas) from China. Other species may arrive in a country by accident, as unwanted colonizers. These **alien species** in many instances became **invasive species** when they adversely affected endemic (native) species by competing with them, leading to a reduction in the population of endemic species.

> **Link**
> The effects of pollution are explored in depth in Chapter D4.2, page 810; how fertilizer use in agriculture leads to pollution is on page 808 and plastic pollution is on page 812. Atmospheric pollution (climate change) is discussed further in Chapter D4.3, page 820.

◆ **Alien species**: species that are introduced into an area by human activity.

◆ **Invasive species**: an alien species that has increased rapidly in number, having a negative effect on the environment and on native species.

● Common mistake

The term 'alien species' is not synonymous with 'invasive species'. 'Alien' refers to a species that has been introduced outside its natural distribution. 'Invasive' means that a species expands into and modifies ecosystems to which it has been introduced. A species may be alien without being invasive.

Invasive species case study: Lionfish

Over the past decade the red lionfish (*Pterois volitans*), a reef fish native to the Indo-Pacific ocean, has become increasingly abundant in the Atlantic and Caribbean oceans, where it has not historically been found (Figure A4.2.16). Current distribution includes the Atlantic coast of the USA, the Caribbean coasts of Central and South America, the Gulf of Mexico, the Greater Antilles and the Leeward Islands. Scientists believe that the fish escaped from aquaria in Florida into US coastal waters and has since expanded in numbers due to a lack of competition and predation, along with abundant food supplies.

■ Figure A4.2.16 A lionfish on a coral reef

Lionfish are venomous and aggressive marine fish. They belong to the family Scorpaenidae; the genus to which they belong (*Pterois*) is characterized by red, white and black stripes (used to put off predators by indicating toxicity, or by breaking up body form), and elaborate pectoral and dorsal fins. This makes them inedible to the predators in the Atlantic and Caribbean. Lionfish have been shown to overpopulate reef areas and force native species to move to areas where conditions may be less favourable for them. They therefore pose a major potential threat to reef ecological systems on the east coast of the USA and the Caribbean.

■ Figure A4.2.17 Water hyacinth in a Louisiana swamp

Invasive species case study: Water hyacinth

Water hyacinth (*Eichhornia crassipes*, Figure A4.2.17) was introduced as an ornamental plant to the USA from South America in the 1880s. By the early 1960s, the floating water hyacinth covered more than 50 000 hectares of public lakes and navigable rivers. The plant grows rapidly, forming dense mats that can spread across water surfaces, eventually covering entire bodies of water. This reduces the native algae and plankton in the water, which are food for the native species of fish and other wildlife. The water hyacinth is therefore destroying native wetlands and waterways, reproducing very quickly and crowding out native species.

Invasive species can lead to the extinction of the native species. For example, in Hawaii many species of endemic snail have been wiped out by habitat loss, the introduction of rats to the islands, and also following the deliberate introduction in 1955 of the carnivorous snail *Euglandia* to control the alien African snail. *Euglandia* ate not only the African snail but also the endemic snail species.

If an alien species does not have the usual limiting factors that control it in its home environment, such as predation, disease and competitors, it will be able to reproduce rapidly and increase in numbers, putting pressure on native species.

ATL A4.2C

Find out about an invasive species in your local area. Where did the species come from and how was it introduced? How is it affecting native species? Is it possible for the invasive species to be removed or managed?

Approaches to conservation of biodiversity

Conservation, in ecological terms, means striving to 'keep what we have'. Conservation biology is the scientific study of Earth's biodiversity with the aim of protecting habitats and ecosystems, and therefore species, from human-made disturbances, such as deforestation and pollution.

Conservation activities aim to slow the rate of extinction caused by the knock-on effects of unsustainable exploitation of natural resources and to maintain biotic interactions between species.

In situ conservation

◆ **In situ conservation**: the conservation of species in their natural habitat.

In situ conservation is the conservation of species in their natural habitat. This means that endangered species, for example, are conserved in their native habitat. Not only are the endangered animals protected, but also the habitat and ecosystem in which they live, leading to the preservation of many other species. In situ conservation works within the boundaries of conservation areas or nature reserves.

In situ conservation may require active management of **nature reserves** or **national parks**. This may mean active clearing of overgrowth, limiting predators, controlling poaching, controlling access, reintroducing species that have become locally extinct and removing alien species. In addition to such measures, successfully protected areas also:

- provide vital habitat for indigenous species; this can include habitat and food for migrating species such as birds
- create community support for the area
- receive adequate funding and resources
- carry out relevant ecological research and monitoring
- play an important role in education
- are protected by legislation
- have policing and guarding policies
- give the site economic value.

The effect of biogeographic factors

Biogeographic factors affect species diversity and so need to be considered when planning nature reserves.

Island biogeography theory predicts that smaller islands of habitat will contain fewer species than larger islands. Therefore, it is inevitable that protected areas will have lost some of the diversity seen in the original undisturbed ecosystem. The principles of island biogeography can be applied to the design of reserves (Figure A4.2.18).

■ **Table A4.2.6** Features of protected areas that are better or worse for conservation (see Figure A4.2.18)

	Better	Worse
A	single large area	single small area
B	single large area	several small areas of the same total size
C	intact habitat	fragmented and disturbed habitat
D	areas connected by corridors	separated areas
E	round (= fewer edge effects)	not round (= more edge effects)

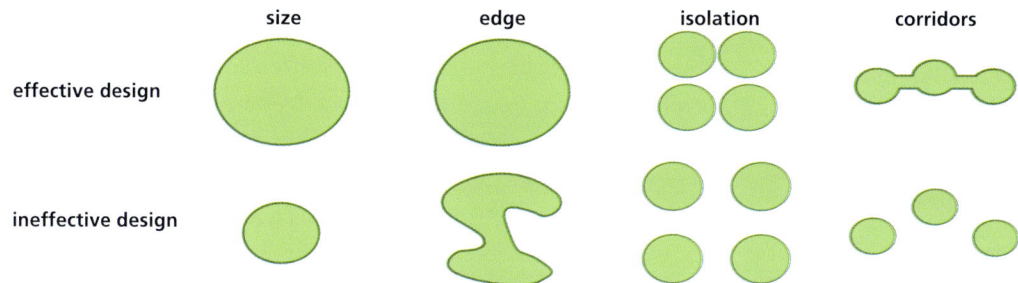

■ Figure A4.2.18 Features of effective nature reserves

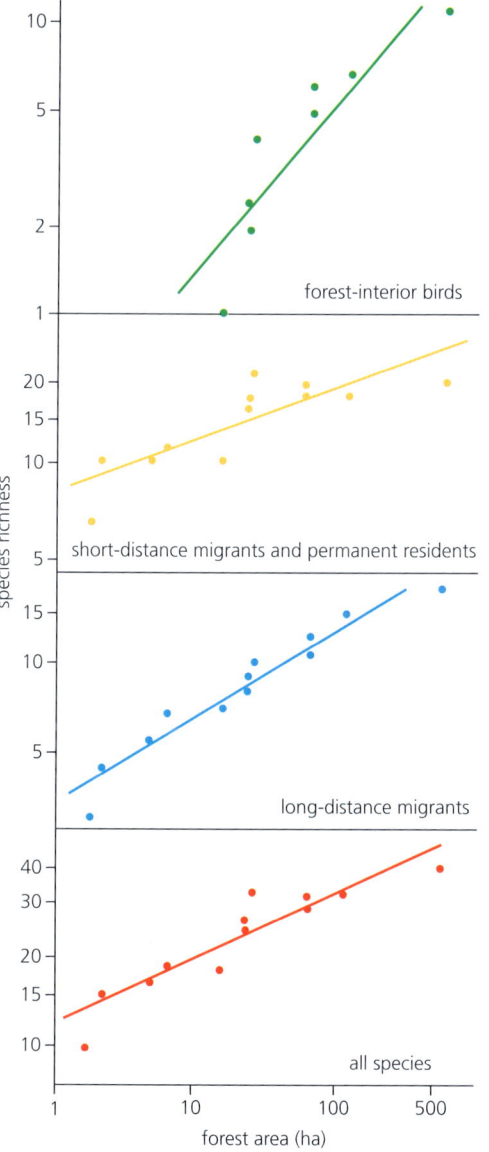

■ Figure A4.2.19 The theory of island biogeography explains species diversity

Nature reserves that are better for conservation have the following features.
- They are large, so that:
  ○ they support a greater range of habitats and, therefore, greater species diversity
  ○ there are higher population numbers of each species
  ○ there is greater productivity at each trophic level, leading to longer food chains and greater stability
  ○ they can maintain top carnivores and large mammals.
- They have a low perimeter to area ratio to reduce edge effects. Fewer edge effects mean more of the area is undisturbed. Edge conditions are very different to those of the interior habitat (hotter, less humid and windier), and so flora and fauna that are interior specialists cannot survive in edge conditions. The best shape for a reserve is a circle as this has the lowest edge to area ratio – it is better than an extended strip of land, or one that has an undulating edge, even if the total area is the same.
- If areas are divided, then fragmented areas need to be in close proximity to allow animals and plants to move between fragments, or there needs to be corridors to join fragments.
  ○ Gene flow between fragmented reserves is maintained by enabling movement along corridors.
  ○ Movement of large mammals and top carnivores between fragments is maintained by corridors.

Figure A4.2.19 shows how variables associated with island biogeography theory correlate with species richness.

A4.2 Conservation of biodiversity

♦ **Rewilding**: a form of environmental conservation that seeks to reinstate natural processes and, where appropriate, missing species, allowing the complex interactions that exist in an ecosystem to be reinstated.

Rewilding and reclamation of degraded ecosystems

Other cases of in situ conservation involve 'rewilding' and reclamation of degraded ecosystems. **Rewilding** aims to restore ecosystems and reverse declines in biodiversity by allowing wildlife and natural processes to reclaim areas no longer under human management. It is a form of environmental conservation that can significantly increase biodiversity in an area. Rewilding reintroduces lost animal species to natural environments, such as top predators that have a significant effect on the food web and trophic levels in an ecosystem. For example, the US National Park Service began to reintroduce the grey wolf (*Canis lupus*) into Yellowstone National Park in the mid-1990s. This lowered the local elk population (*Cervus canadensis*) population and their overgrazing of plants.

Rewilding has three main principles:

1. Core habitat areas that support biodiversity are established.
2. Connectivity: corridors of habitat connect the core areas, allowing movement of biodiversity between different parts of the landscape.
3. Carnivores: top predators are needed to maintain the ecological balance of communities. Carnivores represent the apex predators within ecosystems whose presence and reintroduction within a landscape enables ecosystem restoration.

The benefits of rewilding include:
- Increasing storage of carbon from the atmosphere: for example, in the UK it is estimated that by restoring and protecting native woodland, peat bogs, heaths and grasslands (a total area of over 6 million hectares), 47 million tonnes of CO_2 per year could be captured and stored. This figure is more than a tenth of current UK greenhouse gas emissions.
- Helping wildlife adapt to climate change: connections between different wildlife reserves allow animals to move and habitats to adapt as climate change leads to the northward shift of ecosystems (as the temperature in northern latitudes, on average, warms). This has the potential to save a significant number of species from climate-driven decline or extinction.
- Reversing biodiversity loss: rewilding allows a reversal of biodiversity decline and extinction.
- Supporting economic opportunities for local people: rewilding has the potential to help communities grow and thrive through nature-based enterprises, production and employment opportunities.
- Improving health and well-being: reclaimed and restored areas provide clean water, healthy soils, flood defences, food and unpolluted air, all of which support human health and well-being. It's important that everyone has access to natural environments, especially people who live in urban areas.

4 Evaluate the particular challenges in management of a nature reserve that is maintained near your home, school or college.

Ecosystem restoration means preventing, halting and reversing the damage caused to degraded ecosystems. Efforts to restore ecosystems and species can involve active management of the land, for example replanting native trees to restore forest cover (Figure A4.2.20). Rewilding may involve reorganizing and regenerating wildness in an ecologically degraded landscape. This may modify the biodiversity that would have originally been present, perhaps to manage the landscape in the long term with maximum sustainability in a changing world (e.g. through climate change). Restoration ecology aims to return an ecosystem to as close to its former state as possible after a major disturbance.

■ **Figure A4.2.20** Active management of the land in Scotland, UK, where exotic pine tree species are removed and replaced with native broadleaf trees

◆ **Ex situ conservation**: the preservation of species outside their natural habitats.

◆ **Germ plasm**: cells that are capable of, or are produced by, sexual cell division or fertilization, such as sperm, pollen and seeds (i.e. they contain the genetic material that is passed down from one generation to the next).

5 **Outline** the strengths and weaknesses of ex situ conservation.

Ex situ conservation

Ex situ conservation is the preservation of species outside their natural habitats. For animals, this usually takes place in zoos, which carry out captive breeding and reintroduction programmes:
- a small population is obtained from the wild or from other zoos
- enclosures for animals are made as similar to their natural habitat as possible
- breeding can be assisted through artificial insemination.

Botanical gardens have a role in the ex situ conservation of plants, where both living collections and seed banks are used to store genetic diversity. The storage of **germ plasm** represents a live information source for all the genes in a species of plant, which can be conserved for long periods and regenerated whenever it is required in the future. Germ plasm conservation is the most successful method to conserve the genetic characteristics of endangered and commercially valuable species.

Tissue banks store body tissue and can be used to provide material from which researchers can extract genomic DNA, potentially to restore endangered or even extinct organisms.

Case study: The golden lion tamarin

The golden lion tamarin (*Leontopithecus rosalia*, Figure A4.2.21) was critically endangered, but its conservation status has improved due to human intervention. Found in the tropical rainforests of Brazil, 90% of the golden lion tamarin's original habitat has been cut down and the remaining forest is small and fragmented. The species is only found in one small area of Brazil and so is especially vulnerable to extinction. The monkey was thought by some to carry human diseases, such as yellow fever and malaria, and so was hunted to near extinction. The loss of the species would have affected other species such as insects and small lizards, which the monkey ate, causing those species to become more numerous. Larger animals that preyed on the monkey would decrease in number. Overall, the food chains of the rainforest would become shorter, producing changes in other trophic levels and imbalances in the forest food web.

■ **Figure A4.2.21** A golden lion tamarin monkey (*Leontopithecus rosalia*)

Captive breeding programmes (ex situ conservation) in places such as Bristol Zoo in the UK have increased numbers, allowing release into the wild. There are also efforts to preserve the native forests of the monkey in Brazil (in situ conservation), for example at the Reserva Biológica de Poço das Antas, near Rio de Janeiro. Now, the numbers in the wild have increased from a low of 400 in the 1970s to about 1000 today.

The case of the golden lion tamarin indicates how both in situ and ex situ conservation is needed to conserve species effectively: combining both in situ (e.g. protected areas) and ex situ (e.g. zoos and captive breeding) methods can be the best solution for species conservation in many instances.

Case study: The Bengal tiger

The Bengal tiger (*Panthera tigris tigris*, Figure A4.2.22) is found mainly in India, with smaller populations in Bangladesh, Nepal, Bhutan and China. Although it is one of the most numerous subspecies of tiger, with more than 2500 left in the wild, the subspecies is under threat from habitat loss and poaching. Tiger reserves established throughout India in the 1970s helped to stabilize numbers and are helping to protect the species. Mangrove forests in Bangladesh and India, known as the Sundarbans, are the only mangrove forests where tigers are found. The Sundarbans are threatened by increasing sea level caused by climate change. The Bengal tiger is helping to attract tourists to the Sundarbans and other areas, thus raising money for conservation.

■ Figure A4.2.22 The Bengal tiger (*Panthera tigris tigris*)

'Project Tiger' was launched in India in 1972, with the joint aim of preserving areas where the tiger is found as well as ensuring the ongoing survival of a healthy population of tigers. Captive breeding of Bengal tigers is helping to maintain the genetic diversity of the species. A Bengal tiger studbook keeps a record of which Bengal tigers are kept in captivity and their breeding history. By avoiding inbreeding within populations and cross-breeding with other subspecies of tiger, the genetic integrity of the Bengal tiger can be maintained.

Again, a combination of habitat protection and captive breeding is helping to conserve this important animal. No single approach by itself is sufficient, and different species require different measures. Threats to the Bengal tiger are different to those of the golden lion tamarin. The golden lion tamarin is endangered largely due to its restricted distribution within a small area of Brazil, while the Bengal tiger is endangered due to habitat loss and poaching. One species is a small primate and the other a big carnivore that requires large areas to hunt for prey. Each has specific ecological requirements that need to be protected if they are to be conserved.

EDGE of Existence programme

■ International conventions on biodiversity

International conventions have shaped attitudes towards sustainability. The UN Conference on the Human Environment (Stockholm, 1972) was the first time that the international community met to consider global environment and development needs together. It led to the Stockholm Declaration, which played an essential role in setting targets and triggering action at both local and international levels.

In 1992, the UN Rio Earth Summit resulted in the Rio Declaration and Agenda 21:
- The Earth Summit was attended by 172 governments and set the agenda for the sustainable development of the Earth's resources.
- The Earth Summit led to agreement on two legally binding conventions: the UN Convention on Biological Diversity (CBD) and the UN Framework Convention on Climate Change (UNFCCC).
- Both the UNCBD and UNFCCC are governed by the Conference of the Parties (CoP), which meets either annually or biennially to assess the success and future directions of the Convention. For example, CoP 15 of the CBD was held between 25 April and 8 May 2022, in Kunming, China.

The CBD developed the 'Vision of the Strategic Plan for Biodiversity' that ran between 2010 and 2020. Its stated aim was: 'By 2050, biodiversity is valued, conserved, restored and wisely used, maintaining ecosystem services, sustaining a healthy planet and delivering benefits essential for all people.'

A 'post-2020 Framework' has now been put in place which seeks to 'bring about a transformation in society's relationship with biodiversity'. Both plans highlight the need for 'transformative change' to better appreciate the value of biodiversity and to avert its loss.

Phylogenetic diversity and conservation

Phylogenetic diversity is an important aspect of biodiversity, as it measures the evolutionary history represented by a set of species. The extinction of a species that represents an entire branch of the 'tree of life' means a significant loss of diversity, as well as the loss of an evolutionarily distinct species.

A practical methodology for applying the concept of phylogenetic diversity to conservation is being used in the 'EDGE lists' produced by the Zoological Society of London (ZSL). EDGE (Evolutionarily Distinct and Globally Endangered) species are those which disproportionately represent threatened phylogenetic diversity.

EDGE species are those that have an above-median Evolutionary Distinct (ED) score and are also threatened with extinction (Critically Endangered, Endangered or Vulnerable on the IUCN Red List). There are currently over 550 EDGE mammal species (around 10% of all species) and over 900 EDGE amphibian species (around 13% of all species). Potential EDGE species are those with high ED scores but whose conservation status is unclear.

EDGE species identified so far include mammals, amphibians, birds, corals, reptiles, gymnosperms (a group of plants which include conifers) and elasmobranchs (sharks, rays and skates).

EDGE species represent an opportunity to stop the loss of phylogenetic diversity. In 2012, the IUCN adopted a resolution that recognized the importance of conserving threatened evolutionarily distinct lineages.

ATL A4.2D

You need to be able to understand the rationale behind focusing conservation efforts on evolutionarily distinct and globally endangered species (EDGE).

Find out about the EDGE programme here: www.edgeofexistence.org

Using the EDGE programme website, research the answers to the following questions:
1. How are priority EDGE species identified?
2. What is meant by 'Evolutionarily Distinct' (ED) species?
3. What is the IUCN Red List and how is this used to assess the conservation status of plant and animal species?

Nature of science: Global impact of science

Complex issues associated with conservation efforts

There are many arguments for preserving species and habitats. These include:

Ethical	Every species has a right to survive; we have a responsibility to safeguard resources for future generations.
	These are very broad and can include the intrinsic value of the species or the utilitarian value.
Cultural	The culture of a community can determine how it interacts with the natural world; for example, the indigenous Khmer Dauem community in Cambodia have, through their customs and traditions, been protecting the critically endangered Siamese crocodile and endangered Asian elephant.
Economic	Value of genetic resources for humans (e.g. improved crops).
	Commercial considerations of the natural capital (e.g. new medicines).
	Value of ecotourism (which benefits from higher levels of diversity).
Political	Governments are responsible for putting in place local conservation strategies, so decisions depend on which political party is in place (in democratic countries).
	Strategies that are appealing to the public are more likely to be put forward as these are more likely to get the party that proposes them elected.
	Political tensions arise when more than one appropriate management strategy exists.
Ecological or environmental	Conserving rare habitats (e.g. endemic species require specific habitats).
	Ecosystems with high levels of diversity are generally more stable: healthy ecosystems are more likely to provide ecological services (e.g. pollination, food production).
	Species diversity should be preserved as it can have knock-on effects on the rest of the food web.
Social	Loss of natural ecosystems can lead to loss of people's homes, sources of livelihood and culture.
	Areas of high biodiversity provide income for local people through tourism, for example, and so support social cohesion and cultural services.

Each of these issues will affect decisions on which species should be prioritized for conservation efforts. Because the issues are complex, they need to be debated to ensure the most effective conservation strategies are implemented.

LINKING QUESTIONS

1. In what ways is diversity a property of life at all levels of biological organization?
2. How does variation contribute to the stability of ecological communities?

B1.1 Carbohydrates and lipids

Concept: Form and function

Adaptations are forms that correspond to function. These adaptations persist from generation to generation because they increase the chances of survival.

Guiding questions

- In what ways do variations in form allow diversity of function in carbohydrates and lipids?
- How do carbohydrates and lipids compare as energy storage compounds?

SYLLABUS CONTENT

This chapter covers the following syllabus content:
- B1.1.1 Chemical properties of a carbon atom allowing for the formation of diverse compounds upon which life is based
- B1.1.2 Production of macromolecules by condensation reactions that link monomers to form a polymer
- B1.1.3 Digestion of polymers into monomers by hydrolysis reactions
- B1.1.4 Form and function of monosaccharides
- B1.1.5 Polysaccharides as energy storage compounds
- B1.1.6 Structure of cellulose related to its function as a structural polysaccharide in plants
- B1.1.7 Role of glycoproteins in cell–cell recognition
- B1.1.8 Hydrophobic properties of lipids
- B1.1.9 Formation of triglycerides and phospholipids by condensation reactions
- B1.1.10 Difference between saturated, monounsaturated and polyunsaturated fatty acids
- B1.1.11 Triglycerides in adipose tissues for energy storage and thermal insulation
- B1.1.12 Formation of phospholipid bilayers as a consequence of the hydrophobic and hydrophilic regions
- B1.1.13 Ability of non-polar steroids to pass through the phospholipid bilayer

Note: There is no higher-level only content in B1.1.

Chemical properties of the carbon atom

◆ **Macromolecule**: very large organic molecule (e.g. protein, nucleic acid or polysaccharide).

Living organisms contain lots of different molecules. Many are large **macromolecules** made from smaller molecules. The most common atoms present in these molecules are carbon (C), hydrogen (H), oxygen (O) and nitrogen (N). Some biomolecules contain sulfur (S) and phosphorus (P).

Carbon has unique properties that make life possible. These are outlined below.

1. Atoms combine (or 'bond') to form molecules in ways that produce a stable arrangement of electrons in the outer shells of each atom. Atoms are most stable when their outer shell of electrons is complete. The first electron shell of an atom can hold up to two electrons and then it is full. The second shell can hold a maximum of eight electrons. Carbon is a relatively small atom. It has four electrons in its second shell and can form four strong, stable **covalent** bonds. A single covalent bond is formed when two atoms share a pair of outer electrons. For example, carbon atoms form four single bonds in the methane molecule (CH_4), a hydrocarbon, by sharing their four outer electrons (Figure B1.1.1) with four hydrogen atoms.

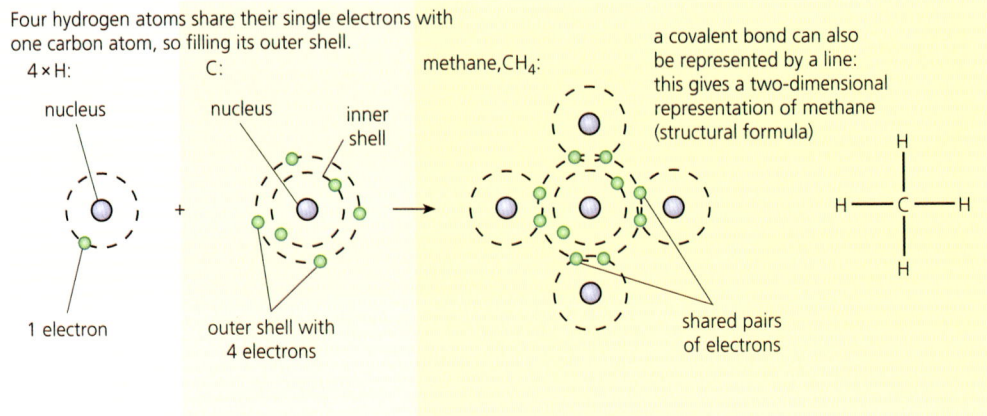

■ **Figure B1.1.1** Methane, the simplest organic compound

> **Top tip!**
> The 'skeletal' formulae in Figure B1.1.2 show carbon atoms but not hydrogen atoms. Full skeletal diagrams have carbon atoms that are implied when two lines intersect – the hydrogen atoms are at the end of each line.

2. Covalent bonds are the strongest bonds found in biological molecules. This means they need the greatest input of energy to break them. So, covalent bonds provide great stability to biological molecules, many of which are very large. Carbon atoms are also able to form covalent bonds with atoms of oxygen, nitrogen and sulfur, forming different groups of organic molecules with distinctive properties.

3. The covalent bonds are only broken during specific chemical reactions with other molecules. Molecules are attracted to each other by much weaker intermolecular forces, which can be easily broken and reformed.

4. Carbon has a critical role in the cell because of its ability to form four strong covalent bonds with other carbon atoms to give a stable chain, branched chains and cyclic structures ('rings') (Figure B1.1.2).

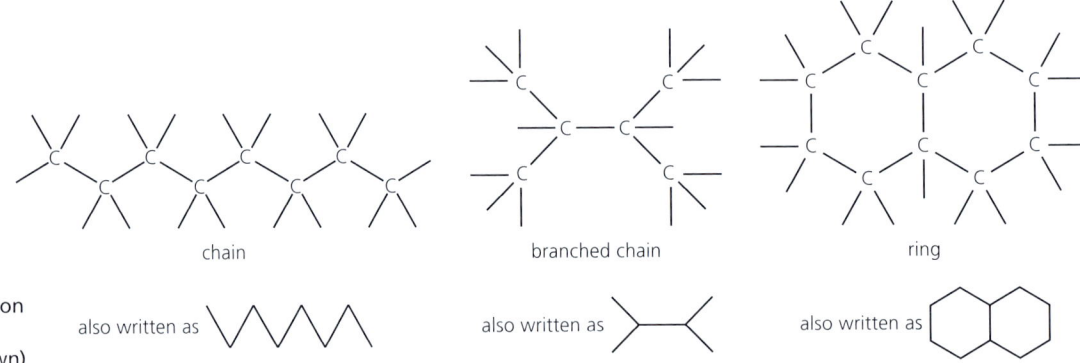

■ **Figure B1.1.2** Carbon 'skeletons' (where hydrogen is not shown)

At least 2.5 million organic compounds exist – more than the total of known compounds of all the other elements, in fact. Organic compounds include amino acids, which form proteins; fatty acids, a component of lipids; and carbohydrates such as glucose (Figure B1.1.3).

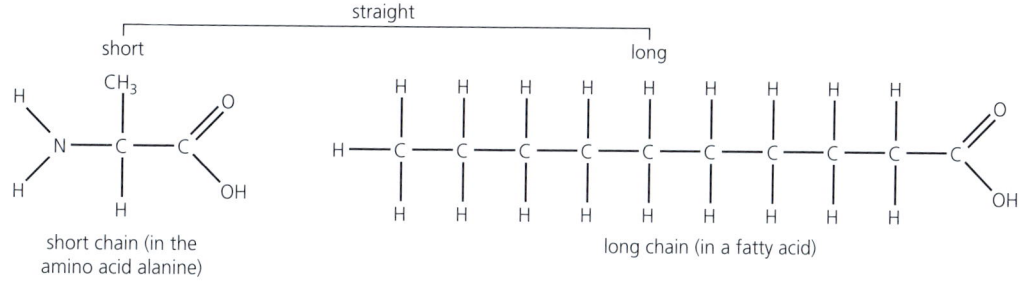

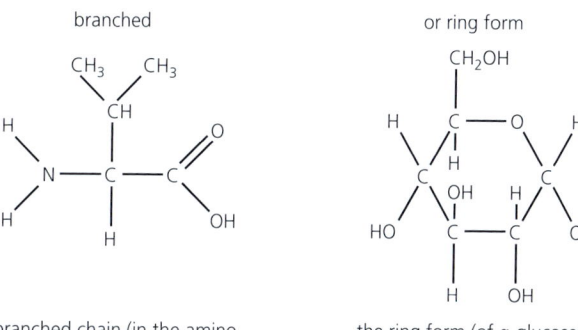

Figure B1.1.3 Examples of different structures of organic compounds

> **Top tip!**
> The diversity of carbon compounds includes molecules with branched or unbranched chains and single or multiple rings, e.g. see Figures B1.1.2 and B1.1.3.

5 The four covalent bonds of carbon atoms point to the corners of a regular tetrahedron (a pyramid with a triangular base, Figure B1.1.4a). This is because the four pairs of electrons repel each other and so position themselves as far away from each other as possible. If there are different groups attached to each of the four bonds around a carbon atom, there are two different ways of arranging the groups (Figure B1.1.4e). This can lead to forms of molecules that are mirror images of each other. Carbon atoms with four different atoms or groups attached are said to be asymmetric. This is another cause of variety among organic molecules.

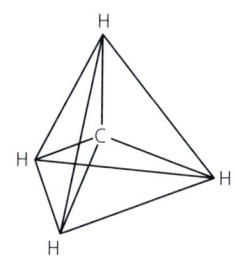

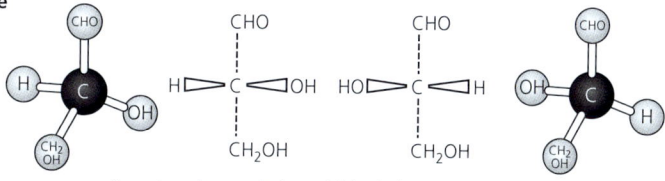

these two forms of glyceraldehyde have very similar chemical properties, but cell enzymes tell them apart – and will only react with one of them

> **Top tip!**
> Carbon atoms can form up to four single bonds or a combination of single and double bonds with other carbon atoms or atoms of other non-metallic elements.

> ◆ **Double bond**: a chemical bond in which two pairs of electrons are shared between two atoms. This type of bond involves four bonding electrons between atoms, rather than the usual two bonding electrons involved in a single bond.

Figure B1.1.4 The tetrahedral carbon atom

6 Carbon atoms can form more than one bond between them (Figure B1.1.5). For example, carbon atoms may share two pairs of electrons to form a **double bond**. Nitrogen and oxygen also form double bonds. Carbon compounds that contain carbon=carbon double bonds are known to chemists as **unsaturated**. This introduces yet more variety to the range of carbon compounds that make up cells. We will meet unsaturated fats that are biologically important shortly.

B1.1 Carbohydrates and lipids

a double bond is formed when two pairs of electrons are shared, for example C═C in ethene (ethene is a plant growth regulator)

other double bonds, common in naturally occurring compounds

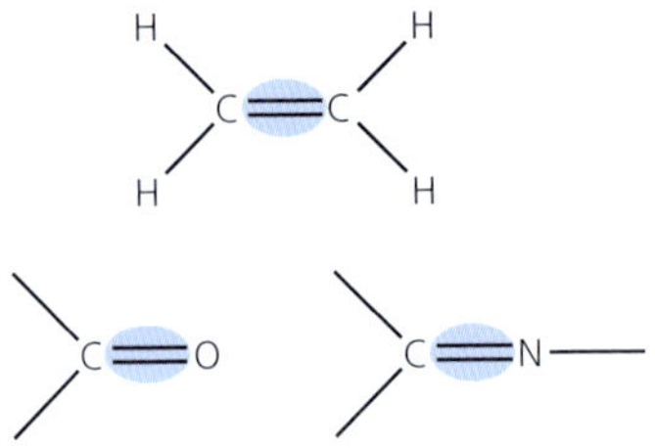

■ **Figure B1.1.5** Carbon double bonds

◆ **Functional group**: the chemically active part of a member of a series of organic molecules.

7 Many biomolecules will have **functional groups** (Figure B1.1.6) attached to a carbon skeleton. These functional groups make the molecule reactive and able to react with other molecules to form larger molecules.

Many biological compounds have functional groups containing carbon bonded to oxygen (Figure B1.1.6). Each functional group gives the molecule unique physical and chemical properties. Carbohydrates contain the hydroxyl group, fatty acids contain the carboxyl group and lipids contain the ester group.

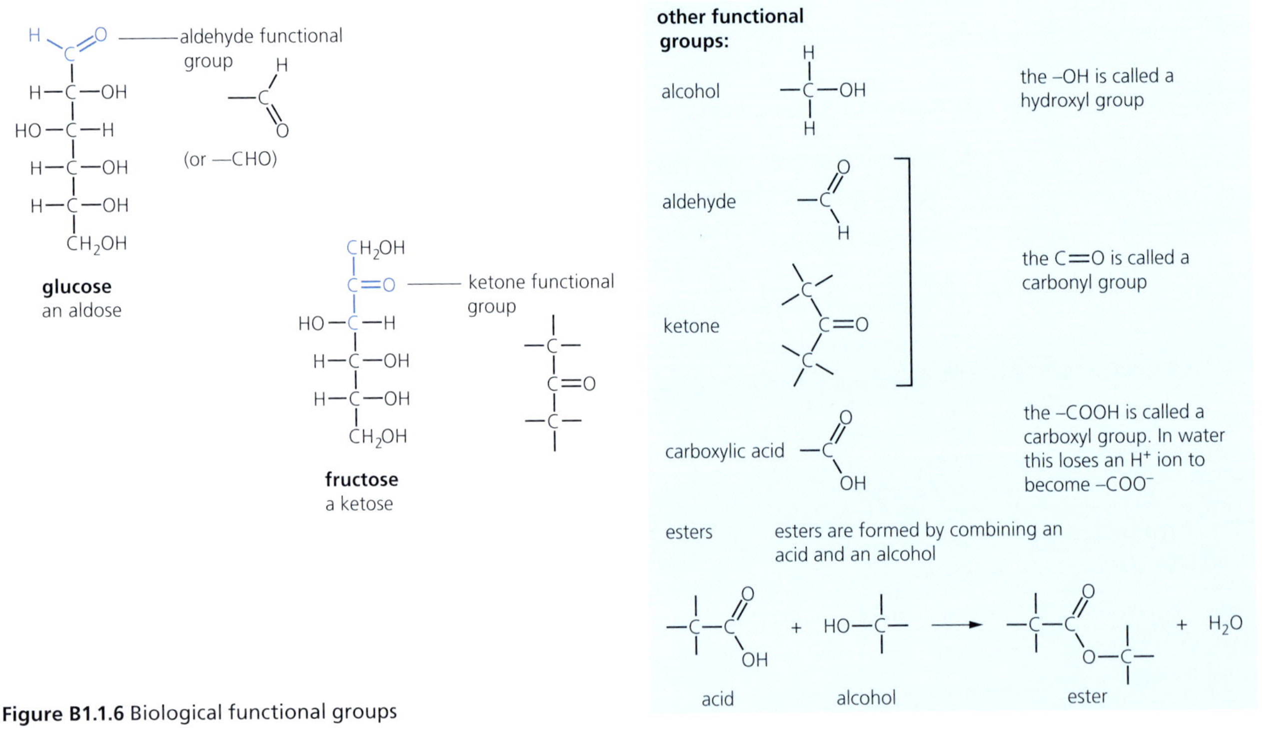

■ **Figure B1.1.6** Biological functional groups

> ● **Top tip!**
>
> There are four major classes of biomolecules that are synthesized by cells: nucleic acids (DNA and RNA), proteins, lipids (fats and oils) and polysaccharides (carbohydrates).

All carbon compounds therefore fit into a relatively small numbers of 'families' of compounds (homologous series), with the families identified by the functional group, which gives them their characteristic chemical properties. The remainder of the organic molecule, apart from the functional group, has little or no effect on the chemical properties of the functional group, and is referred to as the **R-group**.

Due to these properties, molecules containing carbon exist in vast numbers. We now know that living organisms control their composition and functions by a complex web of chemical reactions. One outcome is the presence in cells of a huge range of organic compounds. These will be introduced in following sections.

Theme B: Form and function – Molecules

TOK

Regarding the four major classes of biomolecules, how does the way that we organize or classify knowledge affect what we know?

■ Organic and inorganic compounds

Organic molecules are molecules that are made of carbon and hydrogen, such as urea, $CO(NH_2)_2$, proteins, sugars and lipids, which are found in living organisms. Hydrocarbons such as methane, CH_4, are organic, although not found in living organisms. Some simple carbon compounds are not considered organic, for example carbon dioxide (CO_2), carbonic acid (H_2CO_3) and hydrogen carbonate ions (HCO_3^-). Inorganic compounds are all other compounds, such as water (H_2O) and phosphoric(V) acid (H_3PO_4).

Nature of science: Science as a shared endeavour

Scientific conventions – SI metric units

Since the eighteenth century, scientists have sought to establish common systems of measurements to enable international collaboration across science disciplines and ensure replication and comparability of experiments. Measurements are based on the **International System of Units** (*Système International d'Unités*), abbreviated to **SI**, and metric systems. There are two types of quantities:

1 fundamental quantities, such as length (m), time (s), temperature (K) and mass (kg)
2 derived quantities, such as volume (m^3) and density ($kg\,m^{-3}$).

The metric multipliers (see Table B1.1.1) are used with very large numbers (for example, in ecology) and very small numbers (often used in microscopy and biochemistry).

■ Table B1.1.1 Metric multipliers

Power of 10	Prefix name	Symbol
10^{-9}	nano	n
10^{-6}	micro	μ
10^{-3}	milli	m
10^{-2}	centi	c
10^{3}	kilo	k
10^{6}	mega	M
10^{9}	giga	G

Tool 3: Mathematics

Applying and using SI prefixes and units

SI units are based on the units of six base quantities (see Table B1.1.2). Although it is not formally an SI unit, the degree Celsius (°C) is often used as a measure of temperature.

■ Table B1.1.2 Selected SI base units

Property measured	Name of unit	Abbreviated unit
time	seconds	s
mass	kilogram	kg
length	metre	m
temperature	kelvin	K
electric current	ampere	A
amount of substance	mole	mol

Units indicate what a given quantity is measured in. A measured quantity without units is meaningless, although some derived quantities in biology do not have units, for example pH and absorbance (because they are based on logarithmic functions).

Very large or small numbers must be expressed in scientific notation, for example the number of units in one mole of any substance ($6.02 \times 10^{23}\,mol^{-1}$) and the charge of an electron ($1.60 \times 10^{-19}\,C$).

Converting between different multiples is a matter of multiplying by the appropriate factor. When converting a quantity q from a factor 10^a to a factor 10^b, the quantity needs to be multiplied by a factor 10^{a-b}.

TOK

In Theory of Knowledge, you might explore the importance of having a standardized system of measurements in the pursuit of scientific knowledge. What effect on the knowledge produced would this sort of standardizing have? Is it possible to imagine different ways of measuring features of the world such as mass or temperature? By choosing to describe certain features using this standardized system, is it a challenge to identify other features of the world?

Condensation and hydrolysis reactions

Condensation and **hydrolysis** (Figure B1.1.7) are two of the most important reactions that take place in metabolism. They are responsible for the formation of proteins, nucleic acids (DNA and RNA – see Theme A1.2, page 18), triglycerides and polysaccharides such as starch and cellulose.

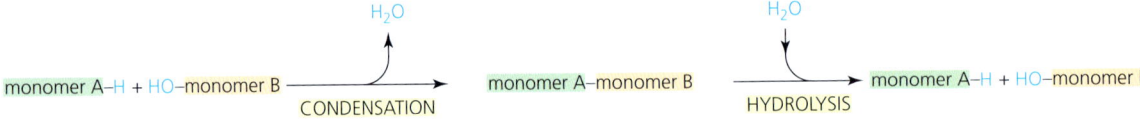

■ **Figure B1.1.7** Condensation versus hydrolysis

◆ **Condensation reaction**: reaction that combines two molecules while removing a small molecule (usually water).

◆ **Hydrolysis reaction**: reaction where hydrogen and hydroxide ions from water are added to a large molecule causing it to split into smaller molecules.

◆ **Monosaccharide**: any of the class of sugars (e.g. glucose) that cannot be hydrolysed to give a simpler sugar.

◆ **Disaccharide**: a sugar that is a condensation product of two monosaccharides (e.g. maltose).

◆ **Glycosidic linkage**: a covalent bond between monosaccharide residues in disaccharides and polysaccharides.

1 **List** three monosaccharides.

During condensation a new covalent bond is formed and a molecule of water is produced. The process is under enzyme control and requires energy. Hydrolysis is the reverse of condensation and involves water molecules acting as a reactant and breaking a covalent bond. Hydrolysis means 'splitting by water'. The water molecule is split to provide the –H and –OH groups to produce the monomers.

For example, when two **monosaccharide** molecules are combined to form a **disaccharide**, a molecule of water is also formed as a product, so this type of reaction is a condensation reaction. The linkage between monosaccharide residues after the removal of H–O–H between them is called a **glycosidic linkage** (Figure B1.1.8). These are strong covalent bonds. The condensation reaction happens in the absence of enzymes but very slowly. Catalysis (the use of an enzyme) results in a large increase in the rate of condensation.

In the reverse process, disaccharides are 'digested' to their component monosaccharides in a hydrolysis reaction. It is catalysed by an enzyme, too, but it is a different enzyme from the one that brings about the condensation reaction.

sucrose + water $\underset{\text{condensation}}{\overset{\text{hydrolysis}}{\rightleftharpoons}}$ glucose + fructose

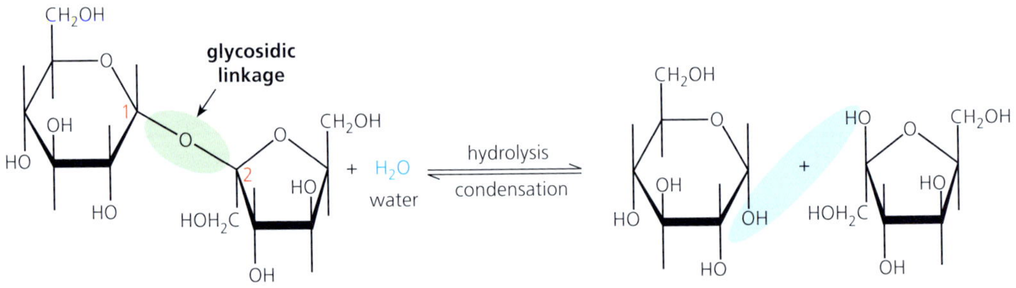

This structural formula shows us how the glycosidic linkage forms/breaks, but the structural formulae of disaccharides does not need to be memorized.

maltose + water $\underset{\text{condensation}}{\overset{\text{hydrolysis}}{\rightleftharpoons}}$ glucose + glucose

lactose + water $\underset{\text{condensation}}{\overset{\text{hydrolysis}}{\rightleftharpoons}}$ galactose + glucose

■ **Figure B1.1.8** Disaccharides and the monosaccharides that form them through hydrolysis and condensation reactions

Apart from sucrose, other disaccharide sugars produced by cells and used in metabolism include:

- maltose, formed by condensation reaction of two molecules of glucose
- lactose, formed by condensation reaction of galactose and glucose.

> **Top tip!**
> Polysaccharides, polypeptides and nucleic acids (macromolecules) are formed by condensation reactions.

◆ **Polysaccharide**: very high molecular mass carbohydrate formed by condensation of large numbers of monosaccharide units, with the removal of water.

◆ **Polymer**: large organic molecule made up of repeating subunits (monomers).

2 **Define**, with an example, *condensation reaction*.

Link
Enzymes and metabolism are covered in more detail in Chapter C1.1, page 380.

■ Production of macromolecules by condensation reactions

Formation of polysaccharides

Polysaccharides are built from many monosaccharide molecules condensed together, linked by glycosidic bonds. 'Poly' means many, with often thousands of monosaccharide molecules making up a polysaccharide. A polysaccharide is an example of a giant molecule called a **polymer**. Normally, each polysaccharide contains only one type of monomer: for example, starch is built from the monomer glucose. The structure and function of the polysaccharides (starch, glycogen and cellulose) are covered later in the chapter (see page 191).

Formation of polypeptides

Amino acids are the building blocks of polypeptides (as discussed previously in Theme A1.2, page 39, and see Chapter B1.2). Polypeptides are formed by condensation reactions between many amino acids. These reactions occur at ribosomes in the cytoplasm and at ribosomes on the surface of the rough endoplasmic reticulum (Chapter A2.2, page 69).

> **ATL B1.1A**
>
> The length of the C–N bond in the peptide bond in proteins is 0.133 nm. Express this in micrometres, picometres and Angstroms (Å) – 1 Å = 10^{-10} m – a non-SI unit that is commonly used to describe protein structures.

Formation of nucleic acids

The formation of nucleic acids (RNA and DNA) by condensation reactions is covered in detail in Chapter A1.2, page 18.

■ Digestion of polymers into monomers by hydrolysis reactions

Digestion of polysaccharides

To release the energy-rich monomers contained in starch and glycogen polymers, the macromolecules must be broken down (or 'digested') into their component parts. The process of digestion is brought about by hydrolysis reactions, where the glycosidic bonds in polysaccharides are broken with the addition of water (Figure B1.1.9). The rate of these reactions is increased by enzyme action. Amylase, an enzyme found in the salivary glands and pancreas of mammals, is secreted on to food to facilitate the hydrolysis of polysaccharides.

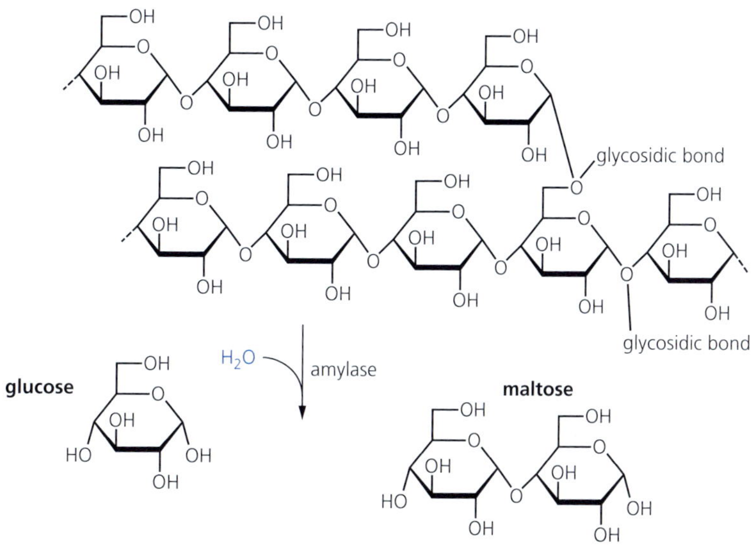

■ Figure B1.1.9 Hydrolysis of starch into smaller sugar molecules

B1.1 Carbohydrates and lipids

Digestion of polypeptides

Proteins (polypeptides) are similarly digested into shorter chain peptides and, ultimately, amino acids through hydrolysis reactions (Figure B1.1.10). In animals, protease enzymes catalyse these reactions.

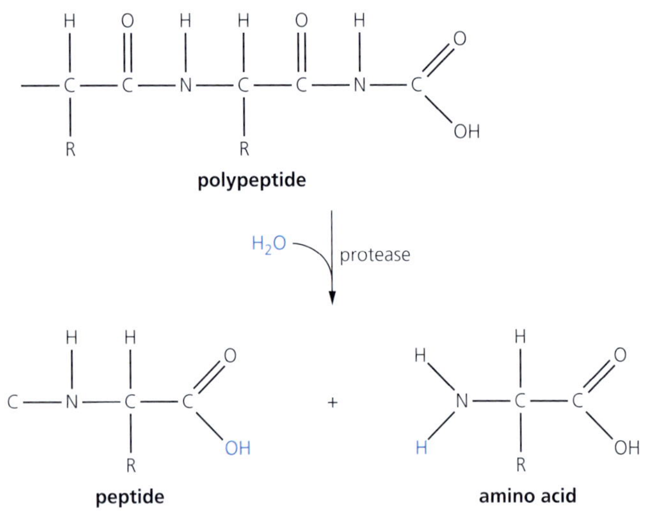

Digestion of nucleic acids

In nucleotides, the phosphodiester bond is the linkage between the third carbon atom (3' carbon) of one sugar molecule and the fifth carbon atom (5' carbon) of another sugar molecule.

The hydrolysis of the phosphodiester bond is as follows in Figure B1.1.11.

In the reaction, the hydroxyl (−OH) group is at the 3' carbon and the phosphate group is at the 5' carbon of the deoxyribose sugar. Thus, when the phosphodiester bond is hydrolysed, the phosphodiester bond breaks and forms $3-OH-deoxyribose-5-PO_4^{3-}$.

■ Figure B1.1.10 Hydrolysis of a peptide bond

3 Distinguish between condensation and hydrolysis reactions. Give an example of each.

● Top tip!

Water molecules are split to provide the −H and −OH groups that are incorporated to produce monomers, which explains the name of 'hydrolysis' reactions.

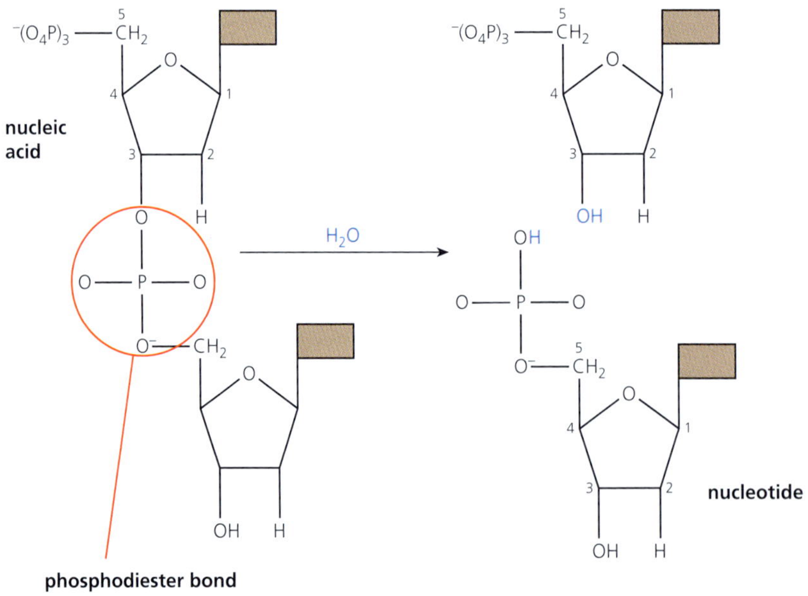

■ Figure B1.1.11 The hydrolysis of a nucleic acid

Form and function of monosaccharides

Introduction to carbohydrates

Link

The prefix D- (e.g. D-fructose) indicates the molecule is an optical isomer. Optical isomers are covered further overleaf.

Carbohydrates are the largest group of organic compounds found in living things and include sugars, cellulose and starch. Carbohydrates have the general formula $C_x(H_2O)_y$, where $x = 3, 4, 5$ (pentoses) and 6 (hexoses). There are some heptoses, $x = 7$, for example mannoheptulose in avocados. Examples of carbohydrates include D-fructose ($C_6H_{12}O_6$), D-ribose ($C_5H_{10}O_5$) and D-maltose ($C_{12}H_{22}O_{11}$): these three are **cyclic** molecules and have a number of hydroxyl groups (–OH) bonded to the carbon atoms in the ring (Figure B1.1.13).

Table B1.1.3 is a summary of the three types of carbohydrates commonly found in living things.

■ **Table B1.1.3** Carbohydrates of cells and organisms

Monosaccharides	Disaccharides	Polysaccharides
simple sugars (e.g. glucose and fructose with six carbon atoms, ribose with five carbon atoms)	two simple sugars condensed together (e.g. sucrose, lactose, maltose)	very many simple sugars condensed together (e.g. starch, glycogen, cellulose)

Monosaccharides – the simple sugars

Monosaccharides are carbohydrates with relatively small molecules. They taste sweet and are soluble in water.

Glucose

In biology, **glucose** is an especially important monosaccharide because:
- all green leaves synthesize glucose in chloroplasts using light energy
- our bodies transport glucose in the blood
- all cells use glucose in respiration – we call it one of the respiratory substrates
- in cells and organisms, glucose is the building block for many larger molecules, such as cellulose and starch in plants and glycogen in animals.

Link

Hydrogen bonds are covered in detail in Chapter A1.1, page 3.

Glucose is a polar molecule that dissolves in water due to the formation of hydrogen bonds with their slightly charged hydroxyl groups (–OH). The negatively charged oxygen ions in water attract and surround the positively charged hydroxyl groups in glucose (the O in the –OH is ☐– and the H is ☐+). The solubility of glucose in water means it can be transported in blood plasma.

The structure of glucose gives it stability. The carbon atoms in the glucose ring each have four covalent bonds. The best, or optimum, angle between all these bonds is 109.5° – an angle that results in a perfect tetrahedron (see Figure B1.1.4), which is a very stable structure. If, for any reason, these bonds are forced into greater or smaller angles, then the molecule will be less stable.

Top tip!

Pay attention to the properties of monosaccharides such as glucose (e.g., its solubility, transportability, chemical stability and the yield of energy) as these properties affect how the monosaccharide is used by organisms.

Glucose is the main substrate used in respiration in cells. It can be completely broken down via aerobic respiration to release large amounts of energy (in the form of ATP – see page 412), releasing carbon dioxide and water as waste products. The metabolic processes involved use glucose directly, whereas other substrates (derived from lipids and proteins) need to be processed first (Chapter C1.2, page 407). Glucose metabolism is directly linked to the production of ATP.

The structure of glucose

Glucose has the chemical or **molecular formula** $C_6H_{12}O_6$. This type of formula tells us what the component atoms are and the number of each in the molecule. Thus, glucose is a 6-carbon sugar, or hexose. However, the molecular formula does not tell us the structure of the molecule.

B1.1 Carbohydrates and lipids

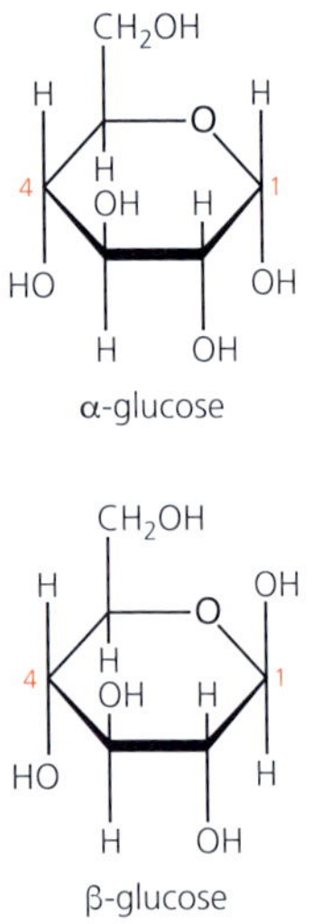

Glucose can be folded, taking a ring (cyclic) form. It is written on paper as a linear molecule and in solution can exist in this form in equilibrium with the cyclic structure. Figure B1.1.12 shows the structural formula of glucose.

The carbon atoms of an organic molecule may be numbered. This allows us to identify which atoms are affected when the molecule reacts and changes shape. For example, as the glucose ring forms, the oxygen on carbon-5 attaches itself to carbon-1. Note that the glucose ring contains five carbon atoms and an oxygen atom (Figure B1.1.12).

Isomers

Compounds that have the same component atoms in their molecules, but which differ in the arrangement of the atoms are known as **isomers**. Many organic compounds exist in isomeric forms. For example, in the ring structure of glucose the positions of the –H and –OH groups that are attached to carbon atom 1 may interchange, giving rise to two isomers known as **α-glucose** and **β-glucose** (Figure B1.1.12). The alpha and beta forms are known as **anomers**: a type of geometric variation found at certain atoms in carbohydrate molecules. These small differences make only minor changes to the properties of the sugar, but are recognized by enzymes and hence can have major effects in the construction of polymers (see Figure B1.1.13).

Two organic compounds can be identical except that they are mirror images of each other. They have the same chemical properties but, in solution, they rotate the plane of plane-polarized light in opposite directions, and so they are **optical isomers**. One form rotates the plane of polarized light to the right (represented in the name by **D-**), and the other rotates the plane of polarized light to the left (represented by **L-**).

■ **Figure B1.1.12** The differences between α- and β-glucose

◆ **Isomer**: chemical compounds of the same chemical formula but different structural formulae.

● Top tip!

You need to be able to recognize pentoses (such as ribose) and hexoses (such as glucose) as monosaccharides from molecular diagrams showing them in ring forms.

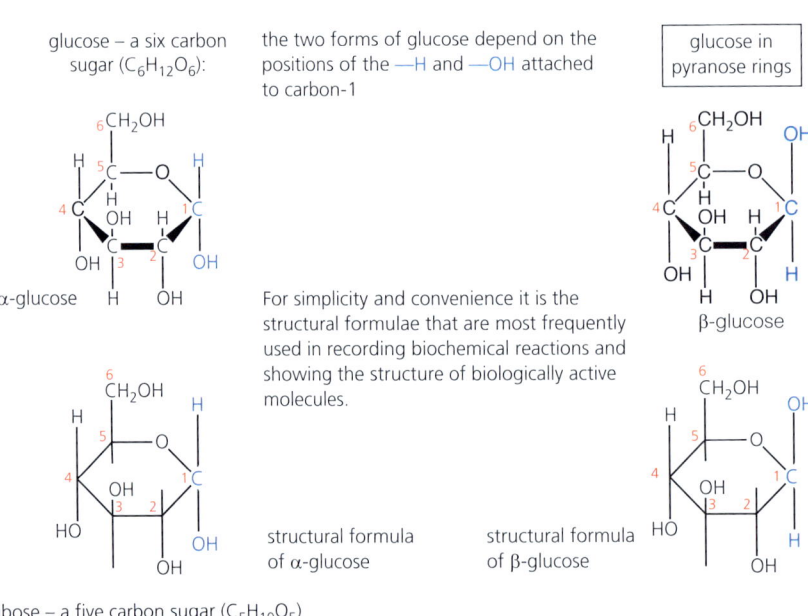

■ **Figure B1.1.13** Molecular structures of D-glucose and D-ribose

> **Concept: Form**
>
> Glucose exists in different forms, which determine the properties of the molecule.

In the following pages you will learn about the roles of α-D-glucose and β-D-glucose in cell chemistry. The names of organic compounds – and of the linkages between compounds when they combine – can be complex because their names give this information. When we compare the structure of the polysaccharide cellulose with that of starch and glycogen, the significance of this difference will become apparent.

As well as monosaccharides, carbohydrates are also found as disaccharides (two carbohydrate monomers joined together) and polysaccharides (as we have already seen, where many monomers join together).

Polysaccharides and energy storage

Starch

> **Top tip!**
>
> Starch is a useful store of energy in plant cells because the insoluble starch has no osmotic effects (i.e. water is not drawn into the cells, which would happen if glucose were dissolved in water within the cytoplasm).

Starch is a mixture of two polysaccharides:
- **Amylose** is an unbranched chain of several thousand 1,4 linked **α-glucose** units.
- **Amylopectin** has shorter chains of 1,4 linked **α-glucose** units but, in addition, there are branch points of α-1,6 glycosidic links along its chains (Figure B1.1.14).

In starch, the bonds between glucose residues bring the molecules together as a **helix**. The whole starch molecule is stabilized by many hydrogen bonds between parts of the component glucose molecules.

Starch is the major storage carbohydrate of most plants. It is laid down as compact grains in **plastids** (membrane-bound organelles found in the cells of plants, such as the chloroplast – Chapter A2.2, page 76). Starch is an important energy source in the diet of many animals, too. Its usefulness lies in the compactness and insolubility of its molecule. It is readily hydrolysed to form sugar when required. We sometimes see 'soluble starch' as an ingredient of manufactured foods. Here the starch molecules have been broken down into short lengths, making them more easily dissolved.

> **Concept: Function**
>
> The structure of glucose determines its function. In plants, α-glucose units combine to form amylose and amylopectin.

We test for starch by adding a solution of iodine in potassium iodide. Iodine molecules fit neatly into the centre of the starch helix, creating an intense blue–black colour.

> **Top tip!**
>
> The compact nature of starch in plants is due to the coiling and branching during polymerization, making it an ideal energy store, as it takes up little space in the cell.

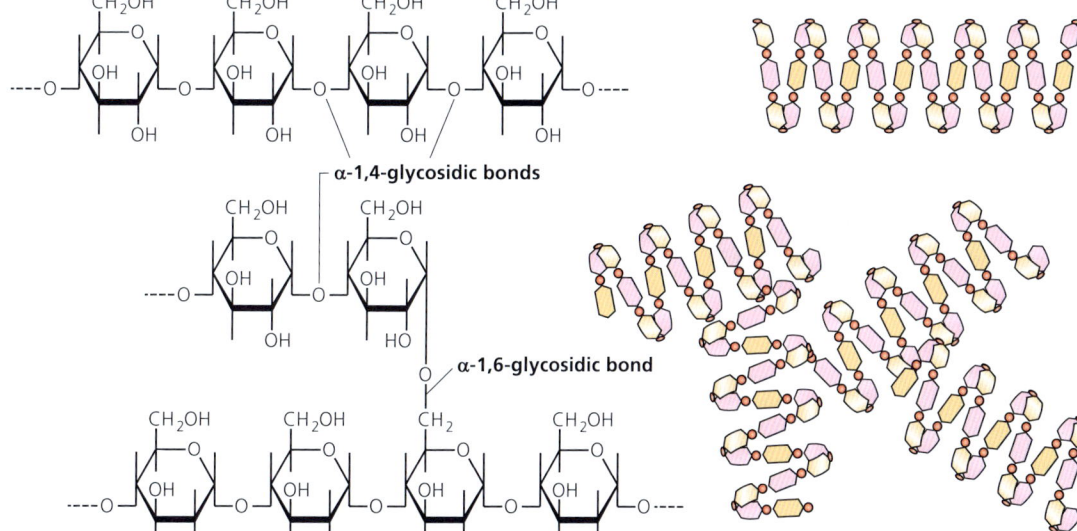

■ Figure B1.1.14 Starch

Inquiry 2: Collecting and processing data

Collecting data; interpreting results

Chemical tests can be carried out that identify biological molecules such as starch, disaccharides and monosaccharides. Lab-made solutions of the chemical compounds can be tested as well as samples from actual food. Reagents can be used to test for different food groups.

- Starch can be tested for with **iodine**: the polysaccharide forms a blue–black starch–polyiodide complex.
- Monosaccharides and disaccharides can be tested for using the **Benedict's test**. Benedict's reagent is an aqueous solution of copper(II) sulfate (copper(II) ions), sodium carbonate and sodium citrate. All monosaccharides and most disaccharides (except sucrose) will reduce copper(II) sulfate, producing an orange/red precipitate of copper(I) oxide on heating, so they are called **reducing sugars**. The colour and density of the precipitate gives an indication of the amount of reducing sugar present, so this is a semi-quantitative test.

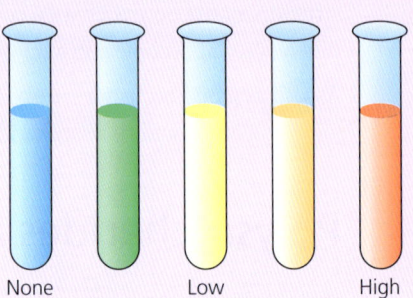

■ Figure B1.1.15 Colour changes using Benedict's test indicate the approximate quantity of monosaccharides present

The sugar content of foods such as fruit is affected by the conditions in which it is kept, and the age of the fruit. Select a variety of different fruit of different ages.

Carry out the Benedict's and starch test on your selection of different fruit. What can you interpret about the quantity of different food groups in your samples? Assess the accuracy and precision of your investigation.

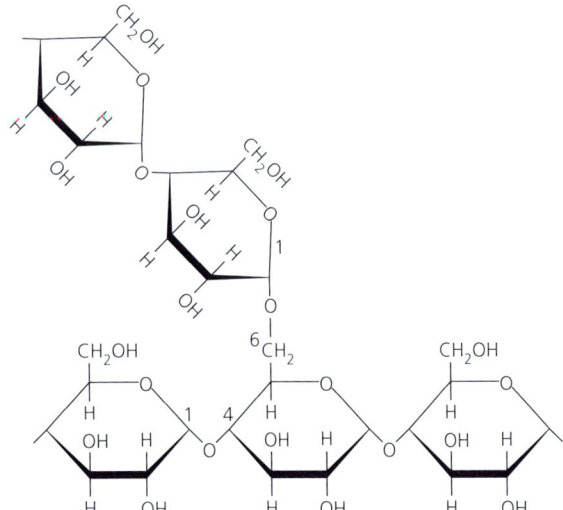

■ Figure B1.1.16 The structure of glycogen, showing the branched chain of α-glucose subunits

■ Glycogen

Glycogen is a polymer of **α-glucose**, chemically very similar to amylopectin, although larger and more highly branched (Figure B1.1.16). Glycogen is formed by condensation reactions between monomers of α-glucose. Glycogen is one of our body's energy reserves and is hydrolysed and respired as needed.

> **Top tip!**
>
> As well as being made and used by animals as an energy reserve, glycogen is also made by some fungi and bacteria.

> **Concept: Function**
>
> The function of glycogen is the same as that of starch – to store energy in cells. Glycogen is chemically very similar to amylopectin, although larger and more highly branched.

4 Deduce why glucose is stored in muscle and liver in the form of glycogen, not as individual glucose molecules.

Glycogen is a useful store of energy in cells where there is a large amount of glucose because, as with starch, the insoluble glycogen has no osmotic effects. The presence of extensive branching in glycogen means it has a compact structure that allows for more glucose molecules to be stored within a small volume. The presence of many non-reducing ends due to the many branches allows for more rapid enzyme-controlled hydrolysis of glycogen to release stored glucose when there is an increase in energy demand.

5 **Explain** how the structure of glycogen enables it to perform its function in animals.

> **Top tip!**
> Granules of glycogen are seen in liver cells, muscle fibres and throughout the tissues of the human body. The one exception is in the cells of the brain. Here, virtually no energy reserves are stored. Instead, brain cells require a constant supply of glucose from the blood circulation.

> **Top tip!**
> Glycogen and starch make good storage compounds because of their relative insolubility (due to their large molecular sizes) and the relative ease of adding or removing α-glucose monomers by condensation and hydrolysis to build or mobilize the energy stores.

> **Top tip!**
> Make sure you understand the distinction between cellulose molecules, fibrils and fibres. Molecules of β-glucose join together to form cellulose molecules, which join together via hydrogen bonding to form cellulose fibrils. Many fibrils join together to form cellulose fibres.

The structure of cellulose

Cellulose is by far the most abundant carbohydrate – it makes up more than 50% of all organic carbon. (Remember that the gas carbon dioxide, CO_2, and the mineral calcium carbonate, $CaCO_3$, are examples of 'inorganic' carbon.) The cell walls of green plants and the debris of plants in and on the soil are where most cellulose is found.

Cellulose molecules are a polymer of **β-glucose** molecules combined together by glycosidic bonds between the carbon-4 of one β-glucose molecule and the carbon-1 of the next. Cellulose molecules are straight and uncoiled. Successive glucose units are linked at 180° to each other (Figure B1.1.17). A β-glucose molecule must be rotated in orientation to the preceding molecules because a glycosidic bond is formed by removing water from adjacent –OH groups (Figure B1.1.17): to ensure that the two –OH groups are pointing in the same direction requires one of the two β-glucoses to rotate relative to the other. This structure is stabilized and strengthened by hydrogen bonds between adjacent glucose units in the same strand, in fibrils of cellulose and between parallel strands. In plant cell walls additional strength comes from the cellulose fibres being laid down in layers that run in different directions. Many cellulose fibrils together form a cellulose fibre.

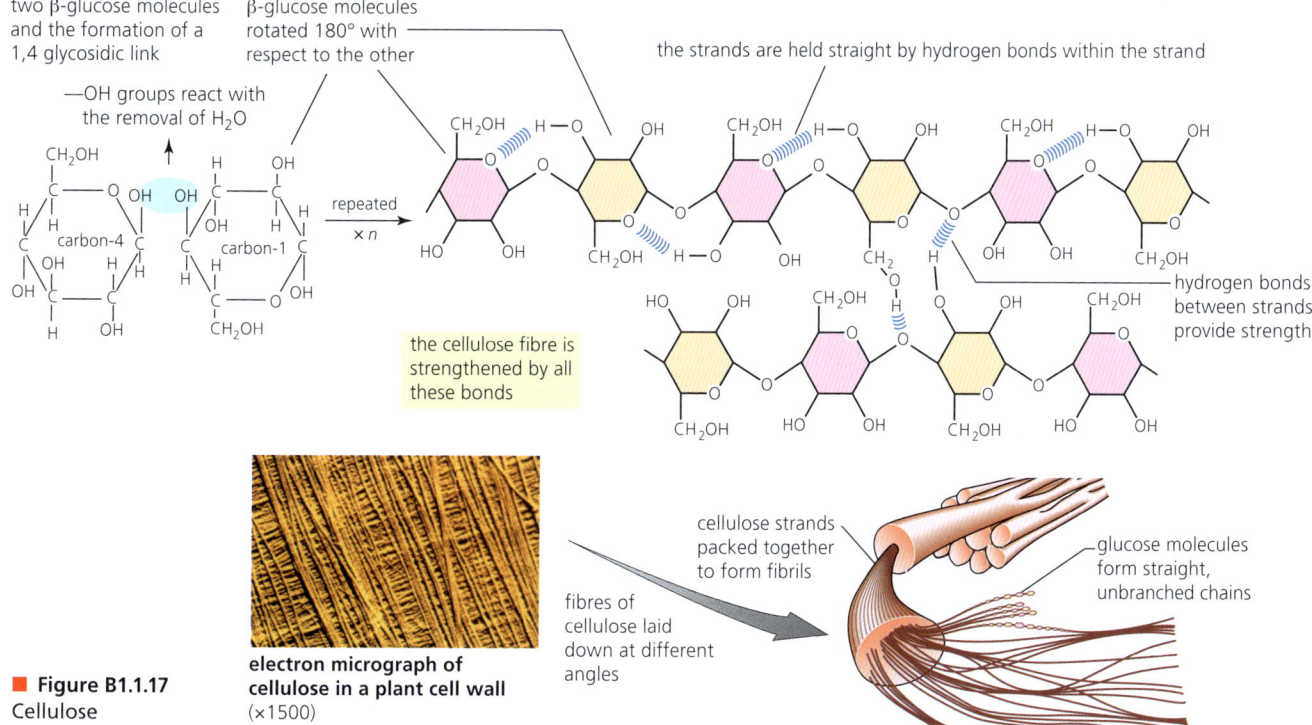

■ Figure B1.1.17 Cellulose

B1.1 Carbohydrates and lipids

> **Concept: Form**
>
> Cellulose is made from β-glucose molecules and, because these molecules join together differently to α-glucose, cellulose has a different form to starch and glycogen.

Due to the structure of cellulose, it is an extremely strong material – insoluble, tough and durable, and slightly elastic. When cellulose fibres are extracted from plants, the fibres have many industrial uses. We use cellulose fibres as cotton and manufacture them into paper, rayon fibres for clothes, nitrocellulose for explosives, cellulose acetate for fibres of multiple uses, and cellophane for packaging.

Common mistake

Make sure you know the differences between the two isomers of glucose: α-glucose is used in the synthesis of starch and glycogen, and β-glucose in the synthesis of cellulose. Do not confuse the two, and make sure you refer to the correct specific isomer when you answer questions about them.

> **Concept: Form and function**
>
> Because the structure of cellulose is different from starch and glycogen, the function of cellulose is also different. Starch and glycogen are used for energy storage, whereas cellulose is used primarily as structural support and mechanical strength in plant cell walls. The orientation of β-glucose molecules in cellulose allows hydrogen bonds to form between parallel strands, and between adjacent glucose units in the same strand, which strengthens the cellulose polymer and enables it to carry out its function in the cell wall of plants.

6 Starch is a powdery material whereas cellulose is a strong, fibrous substance, yet both are made of glucose. **Identify** the features of the cellulose molecule that account for its strength.

7 **Compare and contrast** the structure and function of starch, glycogen and cellulose.

Figure B1.1.18 shows the differences between starch and cellulose, and the role the different forms of glucose play in orientating the monomers, which confers different properties on each polysaccharide.

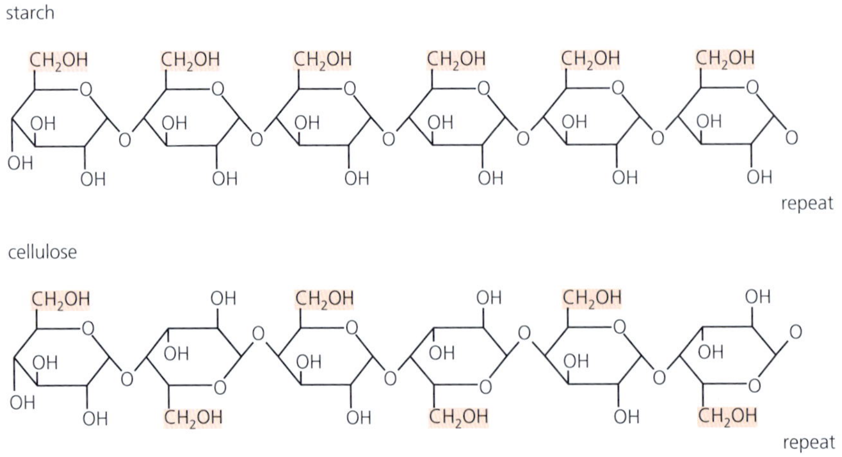

8 **Outline** the roles of monosaccharides, disaccharides and polysaccharides in animals and plants. Use an example for each.

■ Figure B1.1.18 Comparing starch and cellulose polymers; both molecules are shown as linear chains so that the differences in the orientation of glucose monomers is made clear

The role of glycoproteins in cell–cell recognition

The plasma membrane has three major components: phospholipids (see page 198), proteins and carbohydrates. The carbohydrate molecules of the cell surface membrane are relatively short chain polysaccharides, some attached to the proteins (**glycoproteins**) and some to the lipids (**glycolipids**). Glycoproteins and glycolipids are only found on the outer surface of the plasma membrane. These carbohydrate components on the cell's exterior surface are referred to as the **glycocalyx** (meaning 'sugar coating').

Along with peripheral proteins (proteins on the outside of the plasma membrane), carbohydrates allow cells to recognize each other because of the unique shapes that these macromolecules have. This recognition function allows the immune system to differentiate between body cells ('self') and foreign cells or tissues ('non-self'). It is important that the cells of an organism are recognized so that the immune system does not attack and destroy them. Similarly, it is important for foreign cells or viruses to be identified and dealt with by the immune system. The glycoprotein or glycolipid types on the surfaces of viruses may change frequently, preventing immune cells from recognizing and attacking them.

The glycocalyx is important for cell identification and self/non-self-determination, as well as for embryonic development and cell-to-cell attachments to form tissues. The roles of the glycocalyx are:
- cell–cell recognition
- as receptor sites for chemical signals, such as hormone messengers
- to assist in the binding together of cells to form tissues.

■ Blood transfusions can cause an immune reaction

Antigens are substances that can cause an immune response. The glycocalyx can act in this way because the formation of the carbohydrates on the outer surface of the plasma membrane is genetically determined – the shape of the macromolecules on the membrane surface can be used to identify cells that belong to the body and those that do not. The specificity of antigens is the result of the variety of amino acid sequences that are possible, allowing for responses that are customized to specific pathogens. In this way, the glycocalyx allows body cells to be recognized and distinguished from 'non-self' (for example, bacteria or viruses).

Antibodies are proteins that are produced by one group of white blood cells (lymphocytes) to attach to foreign (non-self) material and allow the immune system to remove it. Specific antibodies recognize and bind to specific antigens (see Chapter C3.2, page 524). Table B1.1.4 summarizes the composition and roles of antigens and antibodies.

■ Table B1.1.4 Antigens and antibodies and their role in the immune response

	Antigen	Antibody
Description	substance that can induce an immune response	proteins that recognize and bind to antigens
Type of molecule	glycoprotein	protein
Origin	within the body ('self') or externally ('non-self')	within the body

Human blood cells carry antigens on their plasma membrane, of which the **ABO system** is one of the most important as far as blood transfusions are concerned. You may already know your own blood grouping; each of us carries a particular combination and this must be determined before we can receive a transfusion of blood.

Link

Glycoproteins and glycolipids are discussed further along with the plasma membrane and its components in Chapter B2.1, page 234.

◆ **Glycoprotein**: membrane protein with a glycocalyx attached.

◆ **Glycolipid**: lipids with a glycocalyx attached.

◆ **Glycocalyx**: long carbohydrate molecules attached to membrane proteins and membrane lipids.

◆ **Antigen**: a substance (usually glycoproteins or other protein) capable of binding specifically to an antibody. It is recognized by the body as foreign (non-self) and stimulates an immune response.

◆ **Antibody**: a protein produced by blood plasma cells derived from B lymphocytes when in the presence of a specific antigen, which then binds with the antigen, aiding its destruction.

Link

Antigens as recognition molecules that trigger antibody production are covered in Chapter C3.2, page 525.

◆ **ABO system**: a system of four basic blood types (A, AB, B and O), based on the presence or absence of specific inherited antigens.

◆ **Agglutination**: process in which red blood cells are clumped together by an antibody.

Link

ABO blood groups are an example of multiple alleles; they are covered in further detail along with Mendelian laws in Chapter D3.2, page 734.

Blood groups demonstrate a special example of the antigen–antibody reaction. Looking at the detail of the ABO system in Table B1.1.5, you will notice that people tend to have an antibody in the plasma against whichever antigen they lack. This is always present, even though the blood has not been in contact with the relevant antigen. So, for example, if a person of blood group A accidentally receives a transfusion of group B blood, then the anti-B antibodies in the recipient's plasma make the 'foreign' B cells clump together (the term is **agglutinate**).

The clumped blood cells block smaller vessels and capillaries, which may be fatal. Consequently, you can see why blood of group O is so useful for transfusion purposes – it has neither A nor B antigens on the red blood cells.

The inheritance of ABO blood groups is controlled by multiple alleles and is inherited according to Mendelian laws (Chapter D3.2, page 734).

Top tip!

The ABO system is based on the fact that different antigens are exposed on the surface of red blood cells. In blood types A, B and AB, the antigens are different sugars; O-type red blood cells do not have A or B antigens, which makes them different from other blood types.

ATL B1.1B

Along with the ABO blood group system, there is another system called the Rhesus (Rh) blood group system. What is it and why was it developed? How does it differ from the ABO system? What effect can pregnancy have on the Rh system and how can this affect subsequent pregnancies and the unborn child?

Top tip!

Blood group O is useful for transfusion purposes because it has neither A nor B antigens on the red blood cells, and so agglutination cannot take place.

■ **Table B1.1.5** Blood group and transfusion possibilities

ABO system	Blood group A	Blood group B	Blood group AB	Blood group O
Red blood cell surface	A antigens	B antigens	A + B antigens	neither
Plasma	anti-B antibodies	anti-A antibodies	neither	both anti-A and anti-B antibodies
Blood groups that may be used for transfusion	A, O	B, O	A, B, AB, O	O

*Note: blood group O is the **universal donor**, blood group AB is the **universal recipient***

9 **Distinguish** between the terms 'antigen' and 'antibody'.

Top tip!

If a small quantity of blood is given in a transfusion, the type of antibodies in the plasma received does not matter because of dilution by the plasma of the recipient's blood. For a large transfusion, however, the match of antigens and antibodies needs to be perfect, for example, A antigens with anti-B antibodies, so that agglutination cannot take place.

Hydrophobic properties of lipids

Introduction to lipids

♦ **Triglyceride**: an ester made from glycerol and three fatty acid groups.

♦ **Unsaturated fat**: fat with double bond(s) between carbons in the hydrocarbon chain.

♦ **Saturated fat**: fat with a fully hydrogenated carbon backbone (i.e. no double bonds present).

Lipids are fats (semi-solids) and oils (liquids). They are poorly soluble in water. In fact, they generally behave as 'water-repelling' molecules, a property described as **hydrophobic**. However, lipids can be dissolved in less-polar organic solvents such as ethanol (e.g. alcohol) and propanone (acetone). Lipids only contain carbon, hydrogen and oxygen, but have relatively fewer oxygen atoms than carbohydrates.

The simplest lipids are **triglycerides** and are formed from the reaction between a trihydric alcohol (with three hydroxyl groups) called glycerol (propane-1,2,3-triol) and fatty acids – long chain carboxylic acids with long hydrocarbon 'tails'. There are many different types of fatty acid molecules. Some have one or more carbon–carbon double bonds in their hydrocarbon tail and are **unsaturated** (Figure B1.1.19). Fatty acid molecules with no carbon–carbon double bonds are **saturated**. Saturated and unsaturated fatty acids will be discussed further on page 199.

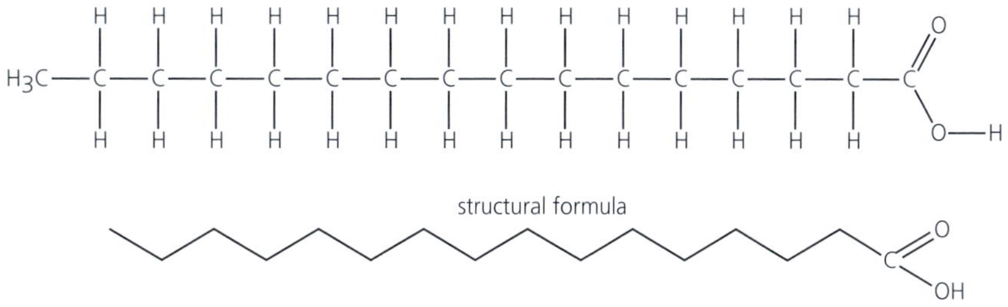

■ **Figure B1.1.19** A saturated fatty acid, H_3C-$(CH_2)_{14}$-$COOH$

● Common mistake

Glycerol is not a fatty acid. Students sometimes make this mistake because glycerol is part of a triglyceride together with fatty acids. It is also not a sugar, because the ratio of elements carbon : hydrogen : oxygen needed to be considered a sugar (1 : 2 : 1) is incorrect (the ratio in glycerol is 3 : 8 : 3). Glycerol in fact belongs to the alcohol family of organic compounds.

Lipids are present as animal fats and plant oils, and as the phospholipids of cell membranes. Fats and oils seem rather different substances, but their only difference is that at about 20 °C (room temperature) oils are liquid and fats are solid.

The structure of a fatty acid commonly found in cells and the structure of glycerol are shown in Figure B1.1.20.

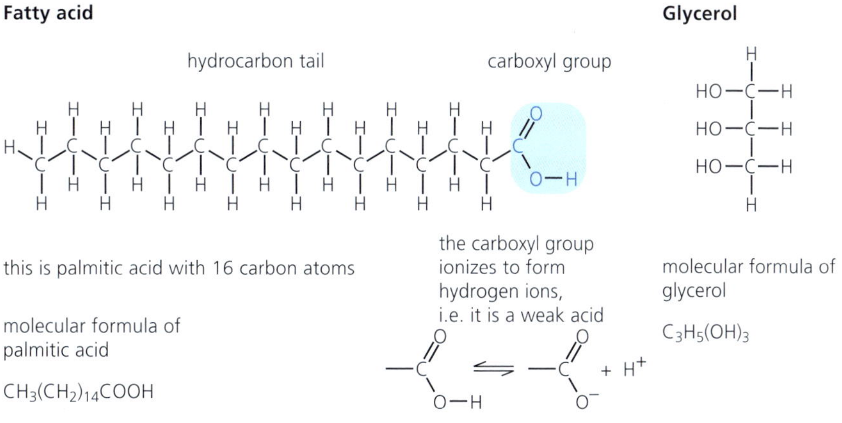

Fatty acid

hydrocarbon tail   carboxyl group

this is palmitic acid with 16 carbon atoms

the carboxyl group ionizes to form hydrogen ions, i.e. it is a weak acid

molecular formula of palmitic acid

$CH_3(CH_2)_{14}COOH$

Glycerol

molecular formula of glycerol

$C_3H_5(OH)_3$

■ **Figure B1.1.20** Fatty acids and glycerol, the building blocks of lipids

B1.1 Carbohydrates and lipids

> **Top tip!**
>
> Lipids are substances in living organisms that dissolve in non-polar solvents but are only sparingly soluble in aqueous solvents. Lipids include fats, oils, waxes and steroids.

> **Common mistake**
>
> Are lipids monomers or a polymer? They are not considered to be polymers. Polymers are made of the same repeating units, whereas lipids have different kinds of monomers.

The long hydrocarbon tails present in fats and oils are typically of about 16–18 carbon atoms long but may be any even number between 14 and 22. The hydrophobic properties of triglycerides are due to these hydrocarbon tails. A molecule of triglyceride is quite large, but relatively small when compared to polymers such as starch and cellulose. It is only because of their hydrophobic properties that triglyceride molecules clump together (aggregate) into huge globules in the presence of water, giving them the *appearance* of polymers.

We describe fatty acid molecules as 'acids' because in aqueous solution their functional group (–COOH) tends to ionize (slightly) to produce hydrogen ions, which is the property of an acid:

$$-COOH(aq) \rightleftharpoons -COO^-(aq) + H^+(aq)$$

It is this functional group of each of the three organic acids that reacts with the three hydroxyl (–OH) functional groups of glycerol to form a triglyceride (Figure B1.1.21). The bonds formed in this case are known as **ester bonds**.

Formation of triglycerides and phospholipids by condensation reactions

Triglycerides are formed by condensation reactions between fatty acids and glycerol, in which water is removed. Three fatty acids combine with one glycerol to form a triglyceride molecule. The steps to triglyceride formation are shown in Figure B1.1.21. In cells, enzymes catalyse the formation of triglyceride molecules by condensation and also the breakdown of glycerides by hydrolysis.

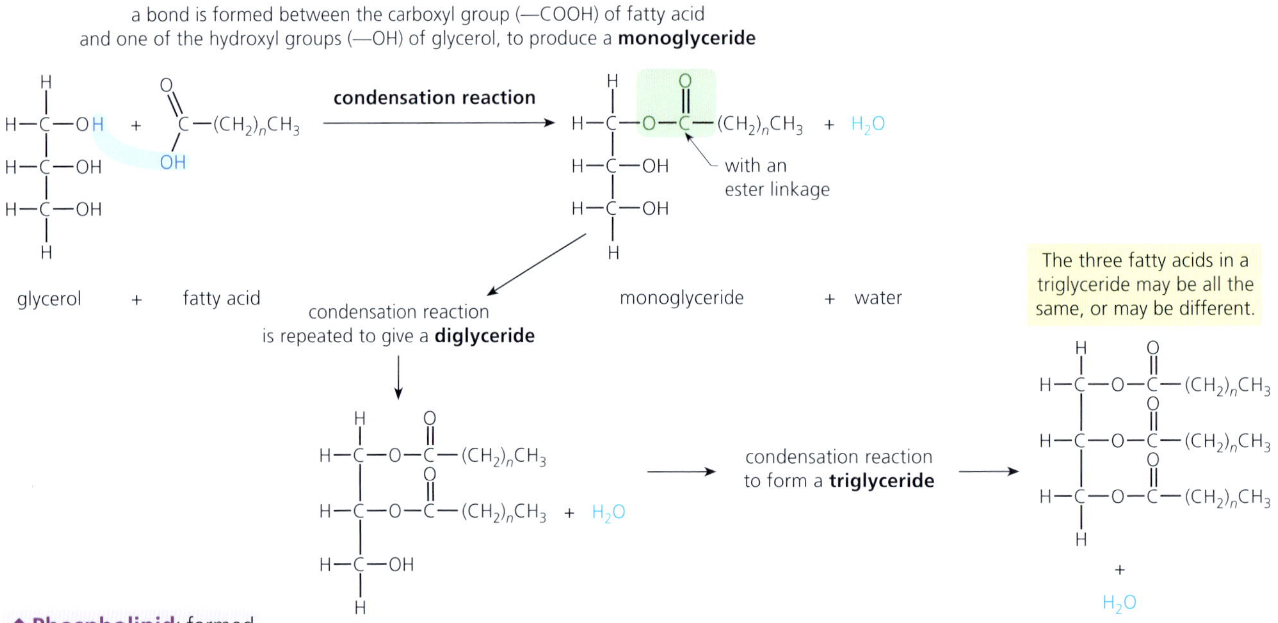

■ **Figure B1.1.21** Triglycerides are formed by condensation from three fatty acids and one glycerol

> ◆ **Phospholipid**: formed from a triacylglycerol in which one of the fatty acid groups is replaced by an ionized phosphate group.

Glycerol can also bond with phosphate, along with two fatty acids. The molecule formed is called a **phospholipid**. Figure B1.1.22 shows a triglyceride and Figure B1.1.23, for comparison, a phospholipid. Phospholipids play an important role in plasma membranes.

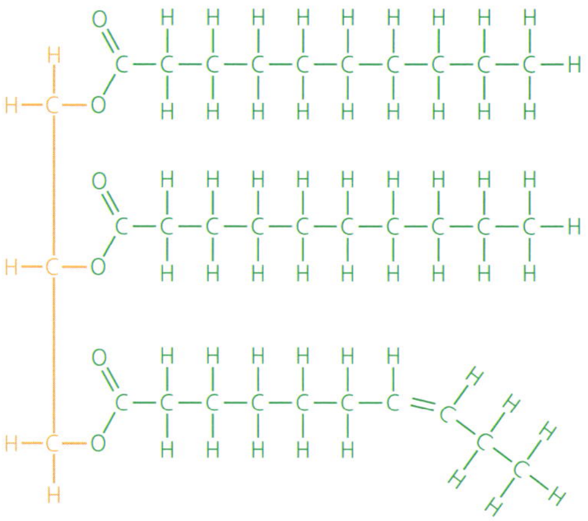

■ **Figure B1.1.22** A triglyceride molecule: a glycerol molecule (orange) combined with three fatty acids (green). The top two fatty acids are saturated (i.e. no double bonds between a pair of carbon atoms) and the bottom one is unsaturated (one double carbon–carbon bond shown)

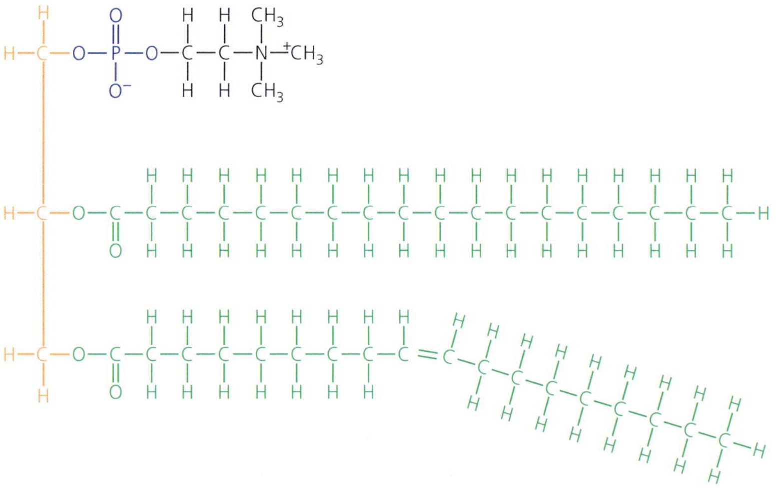

■ **Figure B1.1.23** A phospholipid molecule: a glycerol molecule (orange) combined with two fatty acids (green) and a phosphate group (blue). The top fatty acid is saturated and the bottom one unsaturated

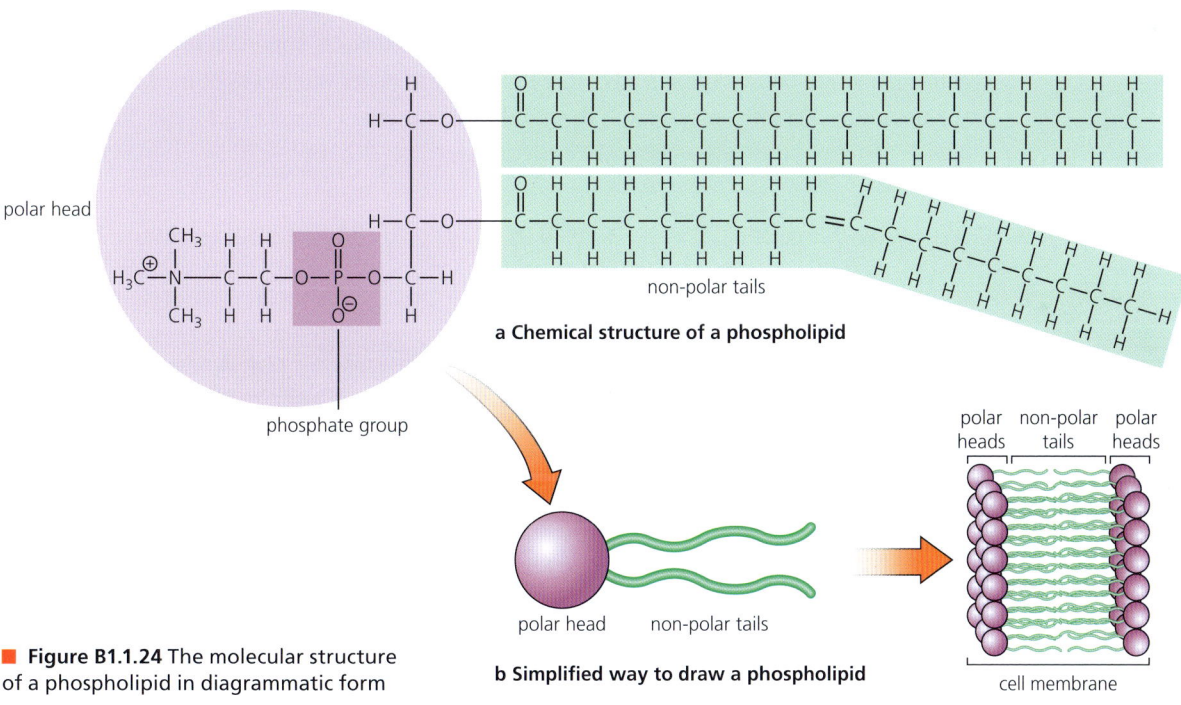

■ **Figure B1.1.24** The molecular structure of a phospholipid in diagrammatic form

a Chemical structure of a phospholipid

b Simplified way to draw a phospholipid

Differences between saturated, monounsaturated and polyunsaturated fatty acids

> **Top tip!**
>
> One glycerol molecule can link three fatty acid molecules or two fatty acid molecules and one phosphate group.

We have seen that the fatty acids combined in a triglyceride may vary in length. In fact, the fatty acids present in dietary lipids (the lipids we commonly eat) vary in another, more important way, too. To understand this difference we need remember that carbon atoms, combined together in chains, may contain one or more **double bonds**. A double bond is formed when adjacent carbon atoms share *two pairs* of electrons, rather than the single electron pair shared in a single bond (Figure B1.1.1, page 182).

B1.1 Carbohydrates and lipids

Differences between the melting points of the fats is related to the number of C=C double bonds. Double bonds cause a bend in the carbon chain, which prevents the chains from coming near each other and interacting strongly. In turn, the weak bonds between the molecules make for a lower melting point.

We have already seen that fatty acids can be saturated or unsaturated (see structures in phospholipids in Figure B1.1.23). This difference is of special importance in the fatty acids that are components of our dietary lipids (Figure B1.1.25):

- Lipids built exclusively from saturated fatty acids are known as **saturated fats**. Saturated fatty acids are major constituents of butter and cocoa butter.
- Lipids built from one or more unsaturated fatty acid are referred to as **unsaturated fats** by dieticians. These occur in significant quantities in many common fats and oils – they make up about 70% of the lipids present in olive oil. Where there is a single double bond in the carbon chain of a fatty acid, the compound is referred to as a **monounsaturated** fatty acid. However, it is possible and common for there to be two or more double bonds in the carbon chain. Lipids with two (and sometimes three) double bonds occur in large amounts in vegetable seed oils, such as maize, soya and sunflower seed oils. These are examples of **polyunsaturated fatty acids**. Fats with unsaturated fatty acids melt at a lower temperature than those with saturated fatty acids, because their unsaturated hydrocarbon tails do not pack so closely together as those of saturated fats. This difference between saturated and polyunsaturated fats is important in the manufacture of margarine and butter-type spreads, since these 'spread better, straight from the fridge'. Polyunsaturated fats are important to the health of our arteries (Chapter B3.2, page 301).

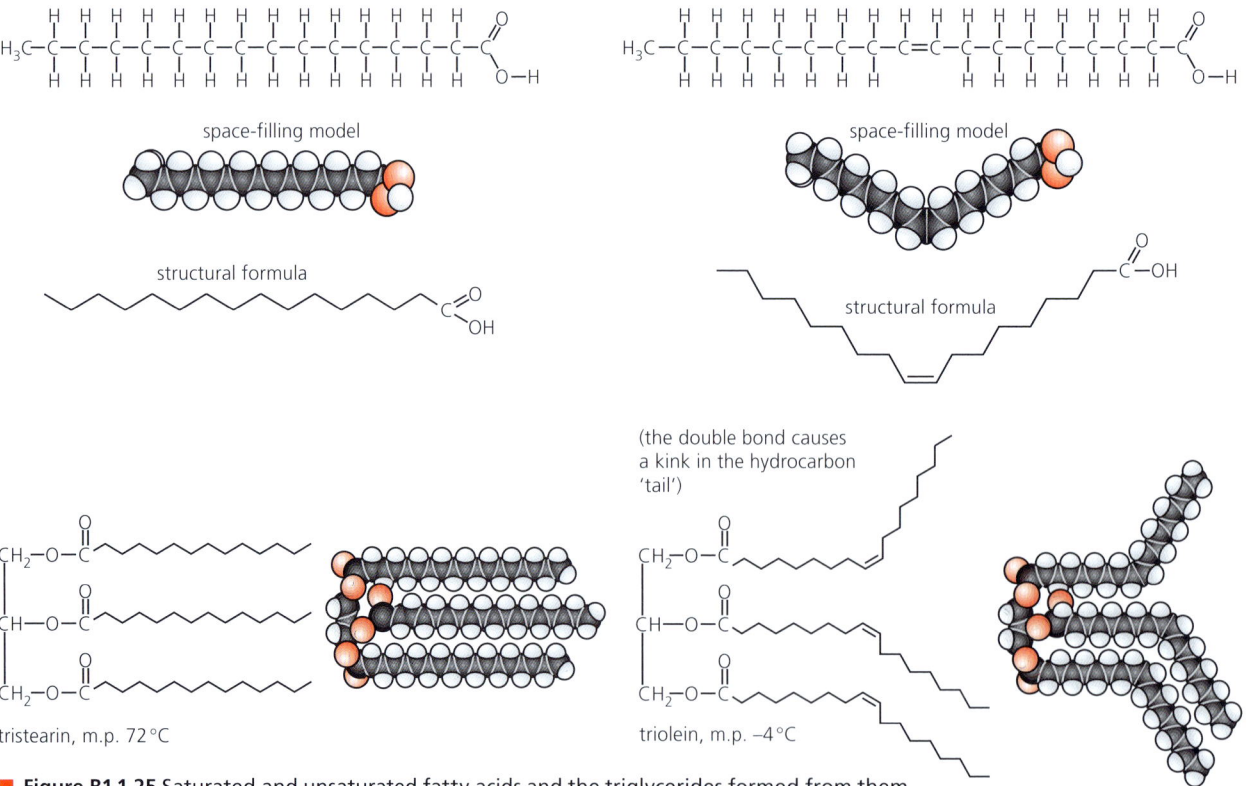

■ **Figure B1.1.25** Saturated and unsaturated fatty acids and the triglycerides formed from them

Unsaturated fatty acids are used for energy storage in plants and saturated fatty acids in endotherms (warm-blooded animals). Plants and cold-blooded animals store unsaturated fatty acids in their cells because they can efficiently utilize energy with less oxygen. Cold-blooded animals use a prevalence of unsaturated fatty acids (including in the cell membrane) because they have lower melting points than saturated fats, and so remain liquid at temperatures lower (usually 5 °C or lower) than those rich in saturated fatty acids. Warm-blooded animals utilize saturated fatty acids because they become liquid at higher temperatures.

■ Health consequences of lipid content of diets

Diets with an excess of lipids and fatty acids provide more energy-rich items than the body requires. People in such affluent situations are in danger of becoming overweight and then obese because of the storage of excess fat in the fat cells that make up the adipose tissue stored around the body organs and under the skin. People who are chronically overweight have enhanced likelihoods of acquiring type 2 diabetes (page 764) and high blood pressure (hypertension, page 297).

Diets that are consistently very low in energy-rich foods, including lipids and carbohydrates, also cause major health risks. Such diets do not contain sufficient fatty acids, so people typically respire the amino acids derived from protein digestion, rather than using them to build and maintain tissues. On a continuing low-energy diet, muscle proteins are broken down and the body wastes away. For the nutritionally deprived members of communities in less economically developed countries this is a constant danger.

> **TOK**
>
> It is said there are conflicting views as to the harms and benefits of fats in diets. To what extent is this true in your country? Why?

ATL B1.1C

Studies have been carried out on people who have a high daily intake of omega-3 fatty acids, such as Greenland Inuit. Here, diets are exceptionally rich in oily fish meat. Associated with this diet (and lifestyle), there is observed a very much reduced likelihood of heart attacks compared with other human groups. Other studies (referred to as 'randomized control trials'), however, suggest it may be difficult to show the clear benefits of omega-3 fatty acids.

Why do you think that one study indicated a link between omega-3 fatty acids intake and prevention of heart disease and other studies say that it is difficult to show the clear benefits?

Find out about omega-3 fatty acids. What is the scientific basis for claims that it may help to prevent heart disease?

◆ **Adipose tissue**: a tissue found beneath the skin layer, containing fat cells.

> **Link**
>
> For an example about fat stores in marine mammals, see the case study of the ringed seal in Chapter A1.1, page 10.

Triglycerides in adipose tissue

■ Fat as a buoyancy aid and thermal insulator

Because fats and oils are insoluble in water, they can be safely stored in cells and tissues without osmotic consequences. Normally, fats are stored in the bodies of animals in fat cells (adipocytes). Fat is stored in animals as **adipose tissue**, typically under the skin where it is known as subcutaneous fat. Aquatic diving mammals have so much it is identified as blubber. This gives buoyancy to the body, because fat is not as dense as muscle or bone. If fat reserves like these have a restricted blood supply and the heat of the body is not distributed to the fat under the skin (as is commonly the case), then the subcutaneous fat also functions as a heat insulation layer. The huge stores of fat that build up in the bodies of marine mammals may be seen as an insulation against body heat loss into the surrounding intensely cold waters.

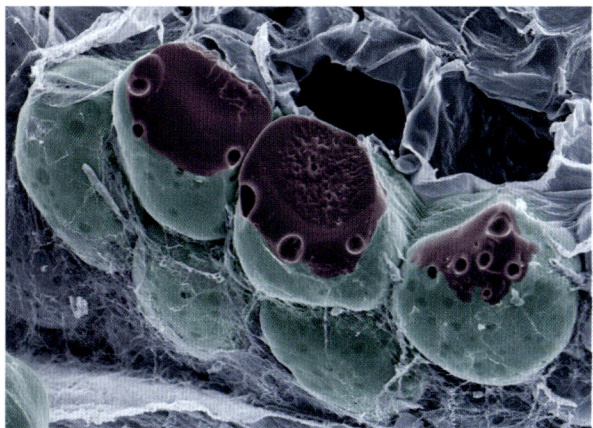

■ Figure B1.1.26 Adipose tissue

B1.1 Carbohydrates and lipids

Energy source and metabolic water source

When triglycerides are oxidised in respiration, a lot of energy is transferred to make ATP (see page 424). Mass for mass, fats and oils release more than twice as much energy as carbohydrates do when they are respired. This is because fats are more reduced (i.e. have more hydrogen atoms relative to carbon and oxygen atoms) than carbohydrates. More of the oxygen in the respiration of fats comes from the atmosphere. In the oxidation of carbohydrates, more oxygen is present in the carbohydrate molecule itself. Fat therefore forms a concentrated, insoluble energy store.

Fat layers are typical of animals that endure long, unfavourable seasons in which they survive by using the concentrated reserves of fat stored in their bodies. Oils are often a major energy store in plants, their seeds and fruits, and it is common for fruits and seeds to be used commercially as a source of edible oils for humans, including maize, olives and sunflower.

Complete oxidation of fats and oils produces a large amount of water, far more than when the same mass of carbohydrate is respired. Desert animals like the camel and the desert rat retain much of this 'metabolic water' within their bodies, helping them survive when there is no liquid water for drinking. Bird and reptile embryos while developing in their shells also benefit from metabolic water formed by the oxidation of the stored fat in the yolk of their eggs. Whales are mammals surrounded by salt water, which they cannot drink, so they rely solely on metabolic water.

> **Top tip!**
> Carbohydrates and lipids are both sources of energy. Carbohydrates are used for short-term storage and lipids for long-term storage. When comparing carbohydrates to lipids for the amount of energy storage, it is essential to compare them based on the same given mass.

Formation of phospholipid bilayers

Phospholipids are the major components of cell membranes. In phospholipid molecules, two of the –OH groups in glycerol are bonded to fatty acids, while the third to phosphoric acid, which is bonded to another small group (Figure B1.1.23, page 199).

A phospholipid has a similar chemical structure to triglyceride, but one of the fatty acid groups is replaced by phosphate. This phosphate is ionized and is therefore water soluble. So, phospholipids combine the hydrophobic properties of the hydrocarbon tails with hydrophilic properties of the phosphate.

Phospholipid molecules form monolayers and bilayers in water (Figure B1.1.27). A phospholipid bilayer is a major component of the plasma membrane of cells.

10 Explain why lipids are more suitable for long-term energy storage in humans than carbohydrates.

11 Distinguish between lipids and carbohydrates as energy stores.

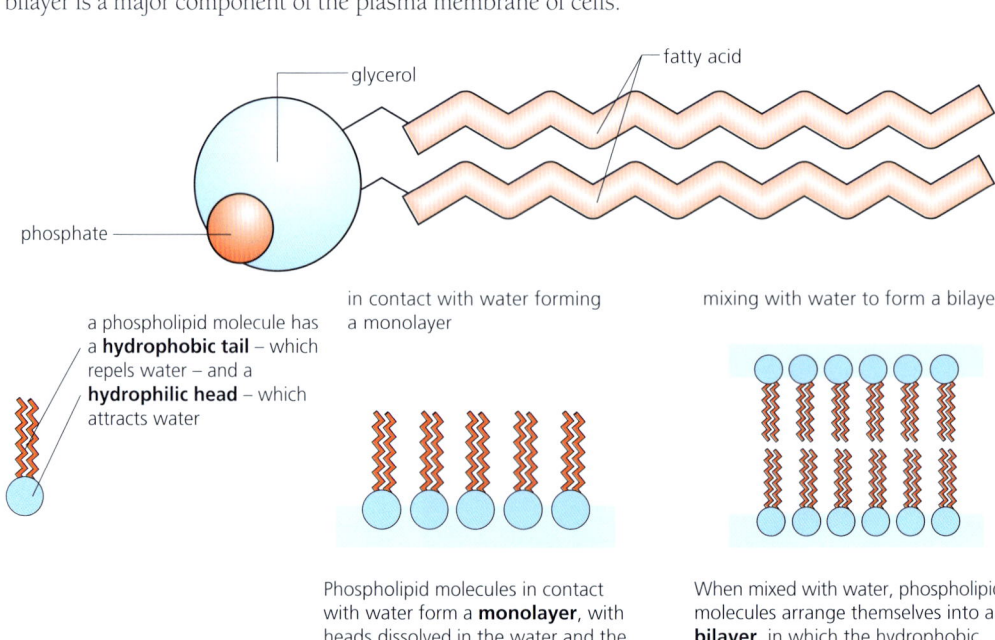

■ Figure B1.1.27 The phospholipid molecule and its response when added to water (the formation of monolayers and bilayers)

◆ **Amphipathic**: having hydrophobic and hydrophilic parts on the same molecule.

Top tip!
Make sure you understand and can use the term 'amphipathic'.

Link
Cholesterol's role in the lipid bilayer is covered in Chapter B2.1, page 237.

12 List two properties of phospholipids.

Link
Osmosis and the movement into and out of cells is covered in detail in Chapter D2.3, page 683.

The phospholipid has a 'head' composed of a glycerol group, to which is attached one ionized phosphate group. This latter part of the molecule has hydrophilic properties (see Chapter A1.1, page 7). Hydrogen bonds readily form between the phosphate head and water molecules (page 3).

The remainder of the phospholipid consists of two long fatty acid residues made up of hydrocarbon chains. These have hydrophobic properties and are repelled by water. So, phospholipid is unusual in being partly hydrophilic and partly hydrophobic. This is referred to as **amphipathic**.

■ What are the consequences of the amphipathic nature of phospholipid?

A small quantity of phospholipid in contact with a solid surface (a clean glass plate is suitable) remains as a discrete bubble; the phospholipid molecules do not spread out. However, when a similar tiny drop of phospholipid is added to water, it instantly spreads over the entire surface. The molecules float with their hydrophilic 'heads' in contact with the water molecules, and with their hydrocarbon tails exposed above and away from the water, forming a monolayer of phospholipid molecules (Figure B1.1.27). When more phospholipid is added, the molecules arrange themselves as a bilayer, with the hydrocarbon tails facing together. This is how they are arranged in the plasma membrane.

Additional attractions in the lipid bilayer – between the hydrophobic hydrocarbon tails on the inside, and between the hydrophilic glycerol/phosphate heads and the surrounding water on the outside – make a stable, strong barrier.

Cholesterol also forms part of the plasma membrane (see Chapter B2.1, page 237). Cholesterol is a steroid (see page 204), with a hydroxyl (–OH) group and hydrocarbon chain on either side of the carbon ring structure. Table B1.1.6 compares and contrasts triglycerides, phospholipids and cholesterol.

■ **Table B1.1.6** Comparing and contrasting triglycerides, phospholipids and cholesterol

Feature	Triglyceride	Phospholipid	Cholesterol
Components	three fatty acid molecules, one glycerol (propane-1,2,3-triol) molecule	two fatty acid molecules, one glycerol molecule, one phosphate group nitrogen-containing base, choline	carbon skeleton consisting of four fused carbon rings, a hydrocarbon tail and one hydroxyl group (–OH)
Bonds between components	ester bonds formed by condensation reactions (under enzyme control)	two ester bonds one phosphoester bond formed by a condensation reaction (under enzyme control)	mainly carbon-carbon single bonds, carbon-hydrogen single bonds
Properties	non-polar (not amphipathic) insoluble in water (exerts no osmotic effect on cells) soluble in organic solvents more compact (than carbohydrates)	amphipathic soluble in both water and oil more compact than carbohydrates	virtually non-polar almost insoluble in water soluble in organic solvents more compact than carbohydrates
Function	energy storage thermal insulation protection buoyancy	basic structure of cell membrane by forming the phospholipid bilayer association with oligosaccharides to form glycolipids, which help in cell–cell recognition and cell–cell adhesion	component of cell membrane regulates membrane fluidity maintains mechanical stability prevents leakage of small polar molecules precursor for synthesis of steroid hormones

Ability of non-polar steroids to pass through the phospholipid bilayer

◆ **Steroid**: a group of four-ring hydrocarbons, of which cholesterol, oestradiol and testosterone are examples.

◆ **Oestradiol**: a steroid hormone; a sex hormone of female mammals.

◆ **Testosterone**: a steroid hormone; the main sex hormone of male mammals.

The lipid bilayer of the plasma membrane is permeable to non-polar substances, including **steroids**. This is because non-polar molecules can pass through the fatty acid interior of the cell membrane, which is also non-polar in nature. This includes the hormones **oestradiol** and **testosterone**. Oestradiol is involved with coordination of the menstrual cycle and development of secondary sexual characteristics in females, and testosterone in male secondary sexual development. These hormones are chemically very similar molecules; they are manufactured from the steroid cholesterol, which is made in the liver and also absorbed as part of the diet. (Steroids are a form of lipid, page 197.)

Perhaps the most noticeable effect of the increased secretion of the sex hormones is the stimulation of muscle protein formation and bone growth. Because of this effect, testosterone and oestradiol (and also the hormone progesterone) are known as anabolic steroids (anabolic means 'build up'). The effects of the female sex hormones are less marked in this respect than those of testosterone in the male.

Figure B1.1.28 shows the molecular diagrams of testosterone and oestradiol. Note the distinctive ringed structure of steroid molecules, and the differences to other lipids (e.g. those shown in Figures B1.1.22 and B1.1.23). The two steroid molecules are made from four rings joined together, and only differ in two functional groups (highlighted in red in Figure B1.1.28). As a result of these small alterations in structure, each hormone has different effects in female and male vertebrates.

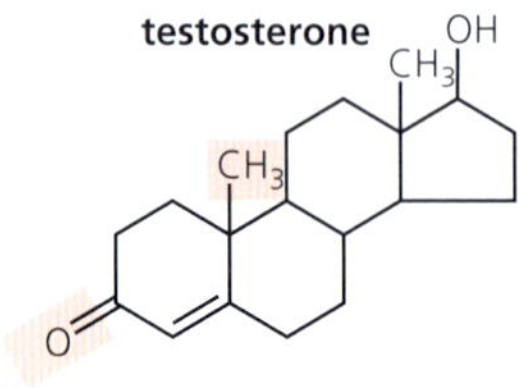

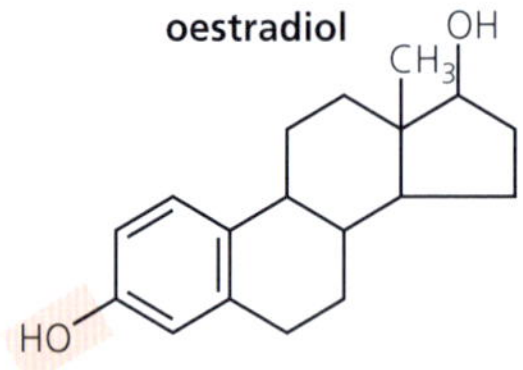

■ **Figure B1.1.28** Molecular structure of oestradiol and testosterone

> ● **Top tip!**
>
> You need to know that oestradiol and testosterone are steroid molecules, and be able to identify compounds as steroids from molecular diagrams. Look closely at Figure B1.1.28 and pick out the identifying features of steroids.

LINKING QUESTIONS

1. How can compounds synthesized by living organisms accumulate and become carbon sinks?
2. What are the roles of oxidation and reduction in biological systems?

B1.2 Proteins

Guiding questions

- What is the relationship between amino acid sequence and the diversity in form and function of proteins?
- How are protein molecules affected by their chemical and physical environments?

SYLLABUS CONTENT

This chapter covers the following syllabus content:
- ▶ B1.2.1 Generalized structure of an amino acid
- ▶ B1.2.2 Condensation reactions forming dipeptides and longer chains of amino acids
- ▶ B1.2.3 Dietary requirements for amino acids
- ▶ B1.2.4 Infinite variety of possible peptide chains
- ▶ B1.2.5 Effect of pH and temperature on protein structure
- ▶ B1.2.6 Chemical diversity in the R-groups of amino acids as a basis for the immense diversity in protein form and function (HL only)
- ▶ B1.2.7 Impact of primary structure on the conformation of proteins (HL only)
- ▶ B1.2.8 Pleating and coiling of secondary structure of proteins (HL only)
- ▶ B1.2.9 Dependence of tertiary structure on hydrogen bonds, ionic bonds, disulfide covalent bonds and hydrophobic interactions (HL only)
- ▶ B1.2.10 Effect of polar and non-polar amino acids on tertiary structure of proteins (HL only)
- ▶ B1.2.11 Quaternary structure of non-conjugated and conjugated proteins (HL only)
- ▶ B1.2.12 Relationship of form and function in globular and fibrous proteins (HL only)

Amino acid generalized structure

Proteins make up about two-thirds of the total dry mass of a cell. They differ from carbohydrates and lipids in that they contain atoms of the element nitrogen and usually the element sulfur, in addition to carbon, hydrogen and oxygen.

Proteins are crucial biomolecules for basic life processes. They are responsible for transport throughout a cell or organism, for maintaining cellular structures and for basic metabolism among other processes. Essential functions such as transport of oxygen throughout blood and its storage in muscle cells are carried out by proteins such as haemoglobin and myoglobin. Proteins form channels and pumps that transport ions, water and other molecules through the cell membrane. Proteins carry molecules from the cell membrane to intracellular compartments and organelles. Many proteins have enzymatic activity.

Top tip!

The term 'amino acid residue' refers to amino acids bonded in a polypeptide chain.

Amino acids are the molecules from which peptides and proteins are built – typically several hundred or even thousands of amino acid molecules are combined together to form a protein. Notice that the terms 'polypeptide' and 'protein' can be used interchangeably but, when a polypeptide is about 20 amino acid residues long, it is generally agreed to have become a protein.

Amino acids – the building blocks of peptides

Figure B1.2.1 shows the structure of an amino acid. As their name implies, amino acids carry two functional groups:
- an amino group (–NH_2)
- an organic acid group (carboxyl group, –COOH).

These groups are attached to the same alpha (α) carbon atom in the amino acids, which get built up into proteins. Also attached here is a side-chain part of the molecule, called an **R-group**.

> **Top tip!**
> You need to be able to draw a diagram of a generalized amino acid showing the α carbon atom with amine, carboxyl, R-group and hydrogen attached.

> **Top tip!**
> Amino acids can be represented by three letter codes or one letter abbreviations, e.g., from Figure B1.2.1, glycine is represented by gly and G, alanine is represented by ala and A.

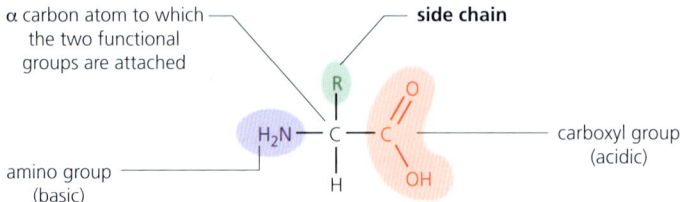

The 20 different amino acids that make up proteins in cells and organisms differ in their side chains. Below are three illustrations but *details of R-groups are not required*.

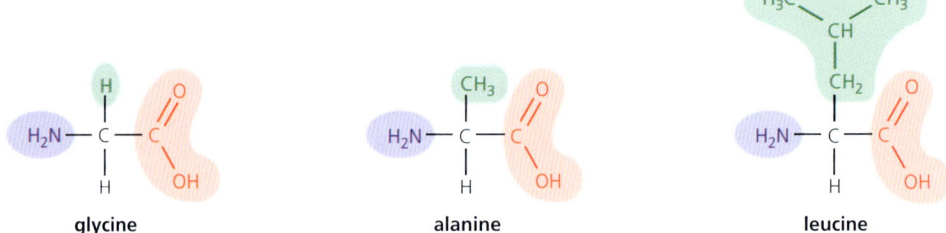

■ **Figure B1.2.1** The structure of amino acids

Proteins of living things are built from just 20 different amino acids, in differing proportions. All we need to note is that the R-groups of these amino acids are all very different and, consequently, amino acids (and the proteins containing them) have different chemical characteristics.

Going further

Enantiomers of amino acids

All the naturally occurring amino acids found in proteins have the L-configuration and not the D-configuration (see earlier discussion about optical isomers and glucose on page 190). D- and L-enantiomers refer to the configurational stereochemistry of the molecule. The central α carbon atom in the amino acid alanine and the other amino acids is a chiral centre because it has four different atoms or groups bonded to it. Two different enantiomers can exist and are mirror image forms (Figure B1.2.2). The critical point here is that cells use only the L forms to synthesize proteins, and this leads to highly specific three-dimensional shapes.

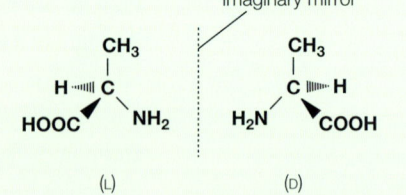

■ **Figure B1.2.2** The enantiomers of alanine are mirror images of each other

Forming dipeptides and longer chains of amino acids

◆ **Peptide linkage**: a covalent bonding of the amino group of one amino acid to the carboxyl group of another (with the loss of a molecule of water).

◆ **Tripeptide**: peptide of three amino acid residues.

Two amino acids combine together with the loss of water to form a dipeptide. This is another example of a catalysed **condensation reaction**. The condensation reaction (in ribosomes) occurs between the amino group (–NH_2) of one amino acid and the carboxyl group (–COOH) of another amino acid, forming a dipeptide (plus a molecule of water) that is held together by a peptide bond (Figure B1.2.3) called a **peptide linkage**.

amino acids combine and bond together the amino group of one with the carboxyl group of the other

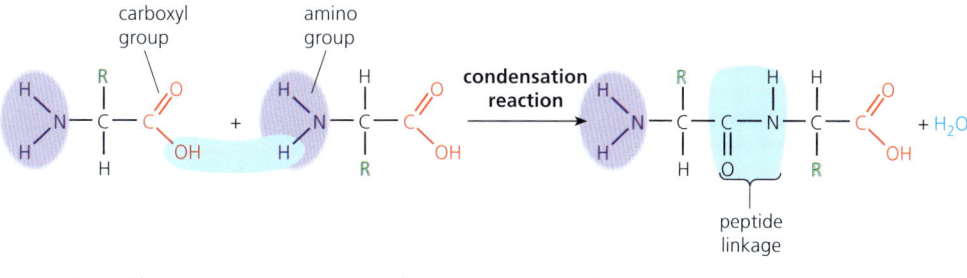

When a further amino acid residue is attached by condensation reaction, a tripeptide is formed. In this way, long strings of amino acid residues are assembled to form polypeptides and proteins.

■ **Figure B1.2.3** Peptide linkage formation

The start of an amino acid chain is referred to as the N-terminus, and the end is the C-terminus.

A further condensation reaction between the dipeptide and another amino acid results in a **tripeptide**. In this way, long chains (linear sequences) of amino acid residues joined by peptide linkages are formed. Thus, peptides or protein chains are assembled one amino acid at a time in the presence of a specific enzyme (peptidyl transferase, in the ribosome – page 253). Multiple amino acids can be joined together to form a polypeptide chain, which folds to form a protein.

Proteins differ in the variety, number and order of their constituent amino acids. The order of amino acids in the polypeptide chain is controlled by the sequence of bases in the DNA of a chromosome (Chapter A1.2, page 25). Changing one amino acid in the sequence of a protein may alter its properties completely.

Each protein has a specific shape, and folds into a particular conformation, depending on its amino acid sequence. The shape of a protein determines its function. Some proteins, such as haemoglobin, consist of one or more polypeptides linked together (see page 212).

● **Top tip!**
You should be able to write the word equation for the condensation reaction forming a dipeptide and longer chains of amino acids, and be able to draw a generalized dipeptide.

1 **Calculate** how many different tripeptide molecules composed of three amino acids linked together by a peptide bond can be made from a set of 20 naturally occurring amino acids. (This is a combination problem.)

2 **Outline** the structure of proteins.

> **Tool 1: Experimental techniques**
>
> ### Physical and digital molecular modelling
>
> Molecular models can be used to model biochemical reactions, such as condensation reactions. These help the reactions to be understood and remembered. Computer software gives a much more accurate representation of biological structures (such as DNA and proteins) and assessment of theoretical predictions. This website has a list of molecular visualization software you could use:
> **www.rcsb.org/docs/additional-resources/molecular-graphics-software**

Dietary requirements for amino acids

■ Table B1.2.1 Essential and non-essential nutrients

Essential nutrients	some amino acids some unsaturated fatty acids some minerals vitamins water
Non-essential nutrients	carbohydrates (such as starch and glucose) as respiratory substrates; fatty acids from lipids are alternatives

A **nutrient** is a chemical substance found in foods that is used in the human body. Some of these nutrients are essential components of our diets but others are non-essential.

An **essential nutrient** is one that cannot be synthesized by the body and therefore has to be included in the diet. On the other hand, **non-essential nutrients** are those that are made in the body or that have a replacement nutrient that can fulfil the same dietary purpose (Table B1.2.1).

◆ **Nutrient**: a chemical substance found in foods that is used in the human body. Any substance used or required by an organism as food.

Essential amino acids

Proteins in our diet are first digested – broken down – by proteases into their constituent amino acids. These amino acids are then absorbed into the body and contribute to the pool of amino acids from which new proteins are built (Figure B1.2.4). We have seen how proteins are built up by condensation reactions between amino acids, taking place within ribosomes (page 207).

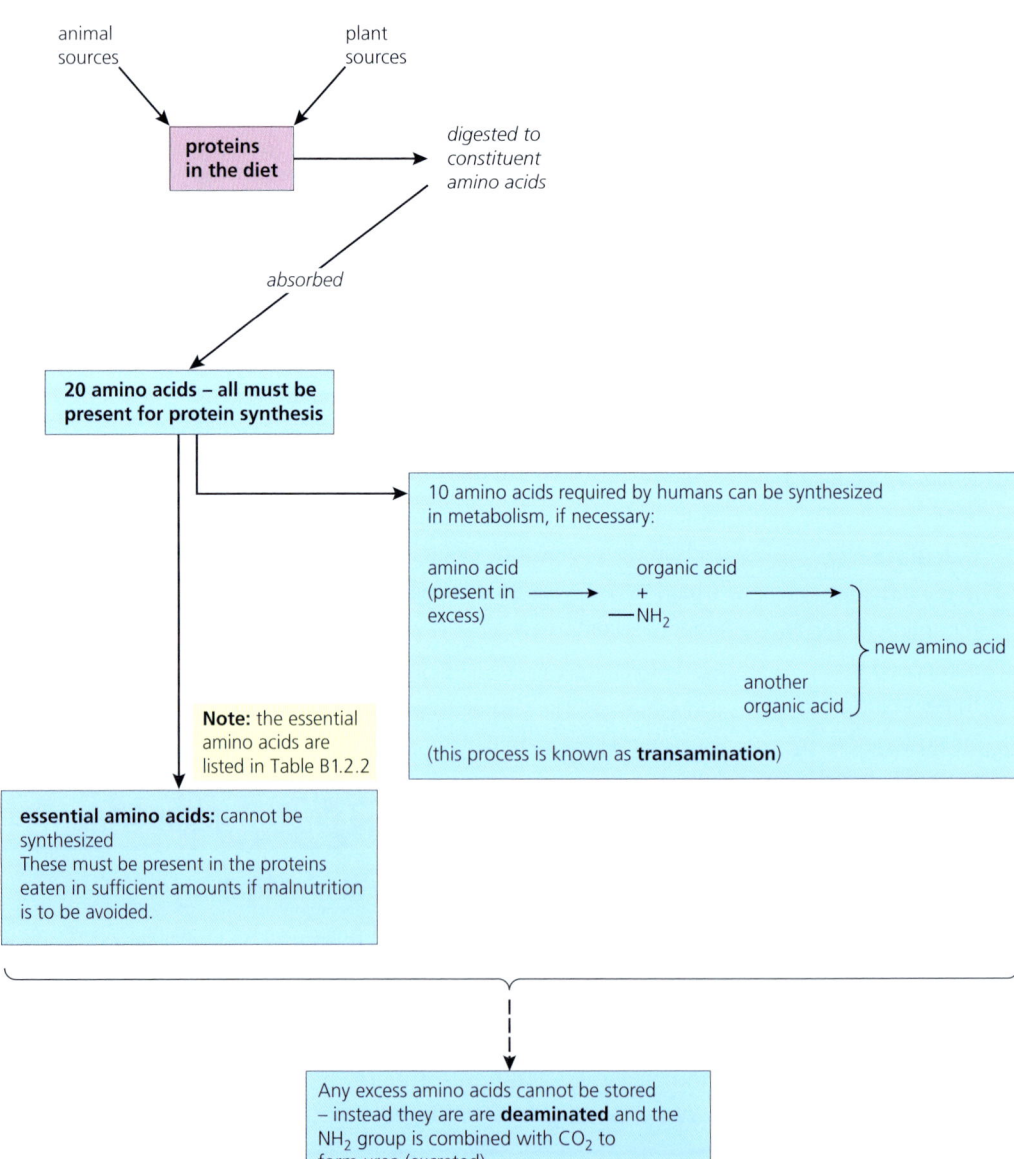

■ Figure B1.2.4 The supply of amino acids in human nutrition

◆ **Essential amino acids**: amino acids that cannot be synthesized and must be obtained from food.

◆ **Non-essential amino acids**: amino acids that can be made from other amino acids.

◆ **Proteome**: the entire set of proteins expressed by a genome.

About 20 amino acids are necessary components of the range of protein molecules made in our bodies. Of these, half cannot be synthesized in the body, at least at some stage of life, and are therefore **essential amino acids** (Table B1.2.2). At any time when one or more of these essential amino acids is in short supply in the food eaten, the body cannot make sufficient of the proteins it requires – a condition known as **protein deficiency malnutrition**.

Non-essential amino acids are made from other amino acids. Humans can only synthesize 10 of the 20 standard amino acids. Tyrosine and cysteine, for example, are synthesized from the essential amino acids phenylalanine and methionine, respectively. The liver has enzymes such as transaminases and is responsible for non-essential amino acid synthesis through a process called transamination. There are many different pathways for the production of non-essential amino acids, many involving processes in glycolysis and the Krebs cycle (stages in cellular respiration – see Chapter C1.2).

■ Table B1.2.2 Essential amino acids

Essential always	histidine, isoleucine, leucine, lysine, methionine, phenylalanine, tryptophan and valine
Essential in the diet of infants	arginine
Essential if the amino acid phenylalanine is absent	threonine

Vegan diets require attention to ensure essential amino acids are consumed. There are a few plant-based sources that contain all essential amino acids, including quinoa, buckwheat, hemp seeds and soy.

Top tip!

Examples of essential and non-essential amino acids are given here if you want to investigate them further, but you will not be required to recall this detail. However, you do need to know the difference between the two types of amino acid (essential and non-essential) and why these differences are important.

The infinite variety of possible peptide chains

Each type of protein has a unique sequence of amino acids and there are many thousands of proteins in cells. The **proteome** is the entire set of proteins found in an individual organism (see Chapter A2.2, page 86).

The 20 amino acids used to make proteins are coded for in the genetic code. A gene is a sequence of DNA that encodes a polypeptide sequence or protein molecule. Typically, one gene codes for one protein. Peptide chains can have any number of amino acids, from a few to thousands, and the amino acids can be in any order.

Examples of polypeptides include enzymes (such as amylase, which digests starch – see page 187), hormones (such as insulin – see page 218) and proteins found in the cell membrane (page 226). Histone proteins are involved with DNA packing in eukaryotic chromosomes (page 652).

Top tip!

The terms 'polypeptide' and 'protein' can be used interchangeably but, when a polypeptide is about 20 amino acid residues long, it is generally agreed to have become a protein. Polypeptides that are bigger than 20–30 amino acids are small proteins.

3 The possible number of different polypeptides, P, that can be assembled is given by

$P = A^n$

where

A = the number of different types of amino acids available

n = the number of amino acid residues in the polypeptide molecule.

Given the naturally occurring pool of 20 different amino acids, **calculate** how many different polypeptides are possible if constructed from 5, 25 and 50 amino acid residues, respectively.

Effects of pH and temperature on protein structure

Concept: Form

Once the polypeptide chain is constructed, a protein takes up a specific shape. Shape matters with proteins – their shape is closely related to their function. This is especially the case with proteins that are enzymes.

Link
The topic of enzymes as catalysts is explored in Chapter C1.1, page 380.

♦ **Denaturation**: a structural change in a protein that results in a loss (usually permanent) of its biological properties.

4 **Define** the term *denaturation*.

The rate of an enzyme-catalysed reaction is sensitive to environmental conditions – many factors within cells affect enzymes and therefore alter the rate of the reaction being catalysed. In extreme cases, proteins, including enzymes, may become **denatured**. Denaturation is a structural change in a protein that alters its three-dimensional shape. Many of the properties of proteins depend on the three-dimensional shape of the molecule. This is true of enzymes, which are large, globular proteins, where a small part of the surface is an active site. Here the precise chemical structure and physical configuration of the protein are critical, but provided the active site is unchanged, substrate molecules can bind and reactions can be catalysed.

● Top tip!

Denaturation occurs when the ionic interactions, hydrogen bonds and other weak intermolecular forces within the globular protein, formed between different amino acid residues, break, changing the shape of the protein, including the active site in enzymes.

■ How denaturation is brought about

Rises in temperature or a small deviation in pH from the optimum can denature proteins (Figure B1.2.5). When this happens to small proteins, it is found that they generally revert back to their former correct shape once the conditions that triggered denaturation are removed. This observation suggested that it is simply the amino acid sequence of a protein that decides its three-dimensional structure. This may also be true for many polypeptides and small proteins.

Exposure to heat causes atoms to vibrate violently and this disrupts weak intermolecular forces (i.e. non-covalent) within proteins. Protein molecules change physical and chemical characteristics. We see this most dramatically when a hen's egg is cooked. The translucent egg 'white' is a protein called albumen, which becomes irreversibly opaque and insoluble. The high temperature from cooking has triggered irreversible denaturation of this protein.

Small changes in pH of the medium similarly alter the shape of proteins. The structure of an enzyme may spontaneously reform when the optimum pH is restored, but exposure to strong acids or alkalis usually denatures enzymes irreversibly.

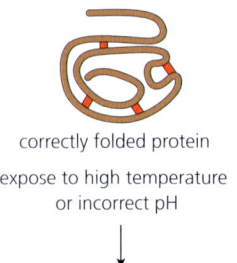
correctly folded protein
expose to high temperature or incorrect pH

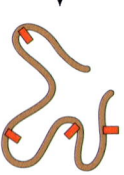
denatured protein
remove denaturant

protein refolds into its original conformation

■ Figure B1.2.5 Folding of a protein from its denatured state

Inquiry 1: Exploring and designing

Designing

Ascorbic acid oxidase is an enzyme released from fruits when their tissue is damaged. The enzyme is activated when exposed to air and causes tissues to lose vitamin C (ascorbic acid). Fruit juice is boiled soon after extraction to denature this enzyme and ensure vitamin C content is retained.

Design an investigation to find out whether all fruits react in the same way, and how effective the process is. What control would you use?

Chemical diversity in the R-groups of amino acids

The reactive or functional groups of the amino acid molecule are the basic amino group ($-NH_2$) and the acidic carboxyl group ($-COOH$), both bonded to the same carbon atom (see Figure B1.2.1, page 206). It is these groups that participate in the condensation reactions that form peptide bonds, linking amino acid residues in polypeptides and proteins. The remainder of the amino acid molecule, the side chain or –R-group, may be very variable between different amino acids.

The variety of amino acids is shown in Figure B1.2.6. While amino acids have the same basic structure, they are all rather different in character because of the different R-groups they carry. The categories of amino acids found in cell proteins are:

- **acidic** amino acids: having additional carboxyl groups (e.g. aspartic acid)
- **basic** amino acids: having additional amino groups (e.g. lysine)
- **amino acids with hydrophilic properties** (water soluble): have polar or charged R-groups (e.g. serine)
- **amino acids with hydrophobic properties** (insoluble): have non-polar R-groups (e.g. alanine).

The joining together of different amino acids in contrasting combinations produces proteins with very different properties. The properties of proteins depend on the different amino side chains (Figure B1.2.6). Some proteins are hydrophobic or non-polar (the electrical charge of the molecule is evenly distributed across the molecule); some are hydrophilic or polar (when positive and negative poles are formed in a molecule – remember, water is a polar molecule). Some amino acids are basic (with an amino group in the side chain) and some are acidic (with a carboxyl group in the side chain).

> **Top tip!**
> You are not required to give specific examples of R-groups, which form the variable side chain in amino acids. However, you should understand that R-groups determine the properties of assembled polypeptides. R-groups can be hydrophobic or hydrophilic. Hydrophilic R-groups are polar or charged, acidic or basic.

> **Top tip!**
> A polar molecule (such as water) has an uneven distribution of electrons, and so has regions of positive and negative charge. The overall charge of these molecules is neutral. In contrast, 'charged' means that the number of positive and negative charges are not the same, i.e. protons and electrons are not balanced.

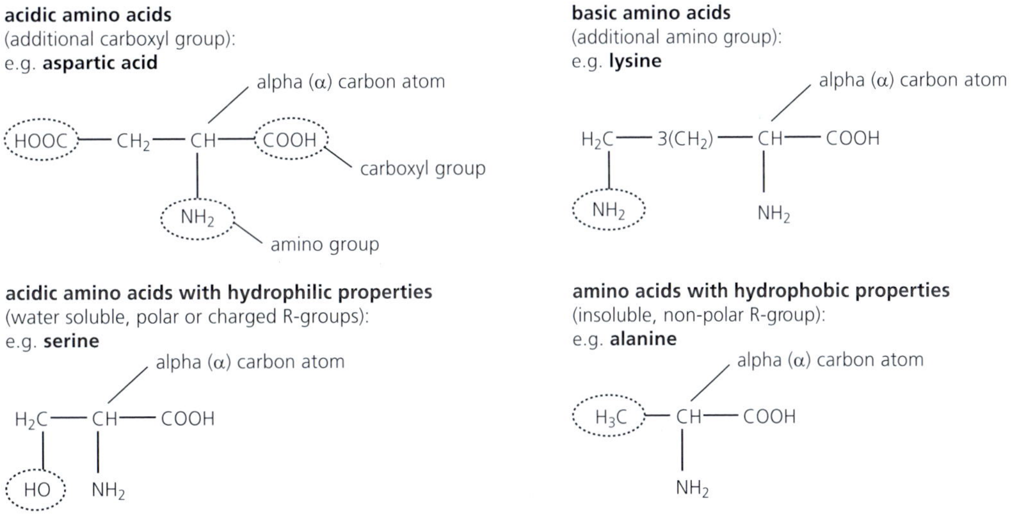

■ **Figure B1.2.6** The range of amino acids used in protein synthesis

Water molecules force hydrophobic R-groups together in order to minimize their disruptive effects on the water molecules. Hydrophobic amino acids have little or no polarity in their side chains: this lack of polarity means they have no way to interact with highly polar water molecules, where the liquid state is formed by the hydrogen bonds between water molecules. Hydrophobic groups held together in this way are sometimes said to be held together by 'hydrophobic forces', but the attraction is actually caused by the repulsion effect from water molecules.

The repulsion of hydrophobic groups from water molecules is also important for the assembly of lipid vesicles and membranes, and protein folding. Polar amino acid side chains tend to be displayed on the outside of the folded protein where they can interact with water; the non-polar amino acid side chains are buried on the inside.

Protein structure

The sequence of amino acids determines the precise position of each amino acid within a structure and the three-dimensional shape of the protein. Proteins therefore have precise, predictable and repeatable structures, despite their complexity.

There are four levels of protein structure, each of significance in biology (Figure B1.2.7).
- The **primary structure** of a protein is the linear sequence of the amino acids in the molecule. This determines the shape of the protein.
- The **secondary structure** occurs when the protein chain interacts with itself through hydrogen bonds to form regions that are helix-shaped or folded.
- The **tertiary structure** refers to the formation or overall three-dimensional shape of a protein, which is formed by hydrogen bonding and other intermolecular forces.
- Some proteins have a **quaternary structure** where two or more polypeptides are combined (via intermolecular forces) to form a larger protein. An example is haemoglobin, present in mammalian blood, which consists of four polypeptide chains held together to form a single larger protein.

> **Common mistake**
>
> Primary structure is not just 'a string of amino acids'. The idea of sequence or order is needed.

> **Concept: Function**
>
> The primary structure of a protein affects the conformation and function of the protein.

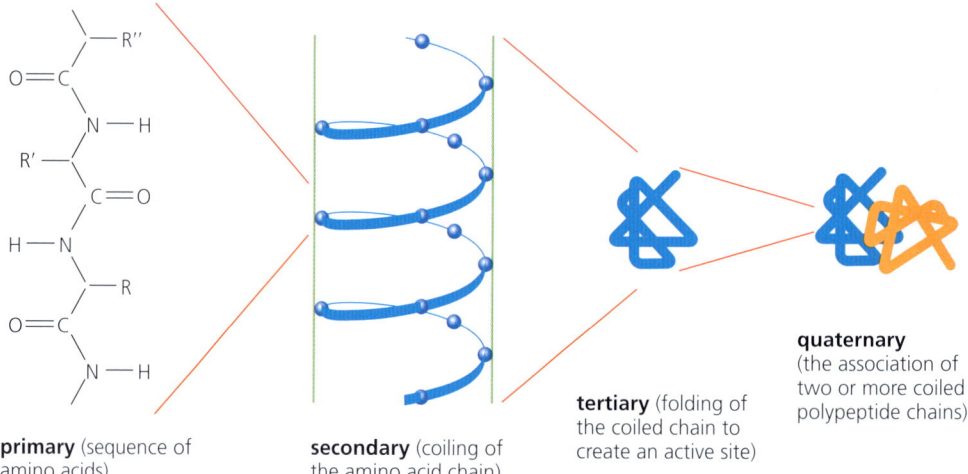

■ Figure B1.2.7 The different levels in the hierarchy of protein structure

The tertiary and secondary structures of proteins are held together by weak intermolecular forces. These are broken at high pH (very alkaline) or low pH (very acidic), or at high or low temperatures. Under these conditions the protein loses its shape (conformation) and is denatured (see page 210).

Proteins vary greatly in sequence, structure and function and are therefore challenging to study. Proteins behave differently in different conditions: some proteins or protein complexes do not fold or assemble without other assembly factors; some proteins cannot function without co-factors present. Modification of proteins by the cell, e.g., phosphorylation, can greatly affect their structures and functions. We will now look at the four levels of protein structure in more detail.

■ Impact of the primary structure on the conformation of proteins

The primary structure of a protein is the sequence of amino acid residues joined by peptide linkages (see Figure B1.2.7 above). Proteins differ in the variety, number and order of their constituent amino acids. We have seen how in the living cell the sequence of amino acids in the polypeptide chain is controlled by the coded instructions stored in the DNA, mediated via mRNA. Changing just one amino acid in the sequence of a protein alters its properties, often quite drastically. This sort of mistake arises by mutation (page 623).

Pleating and coiling of the secondary structure of proteins

The secondary structure of a protein develops when parts of the polypeptide chain take up a particular shape immediately after formation at the ribosome. Parts of the chain become folded or twisted, or both, in various ways.

The most common shapes are formed either by coiling to produce an **α-helix** or folding into **β-sheets**. These shapes are permanent, held in place by hydrogen bonds (Figure B1.2.8).

> **Top tip!**
> Hydrogen bonding in regular positions helps to stabilize α-helices and β-pleated sheets.

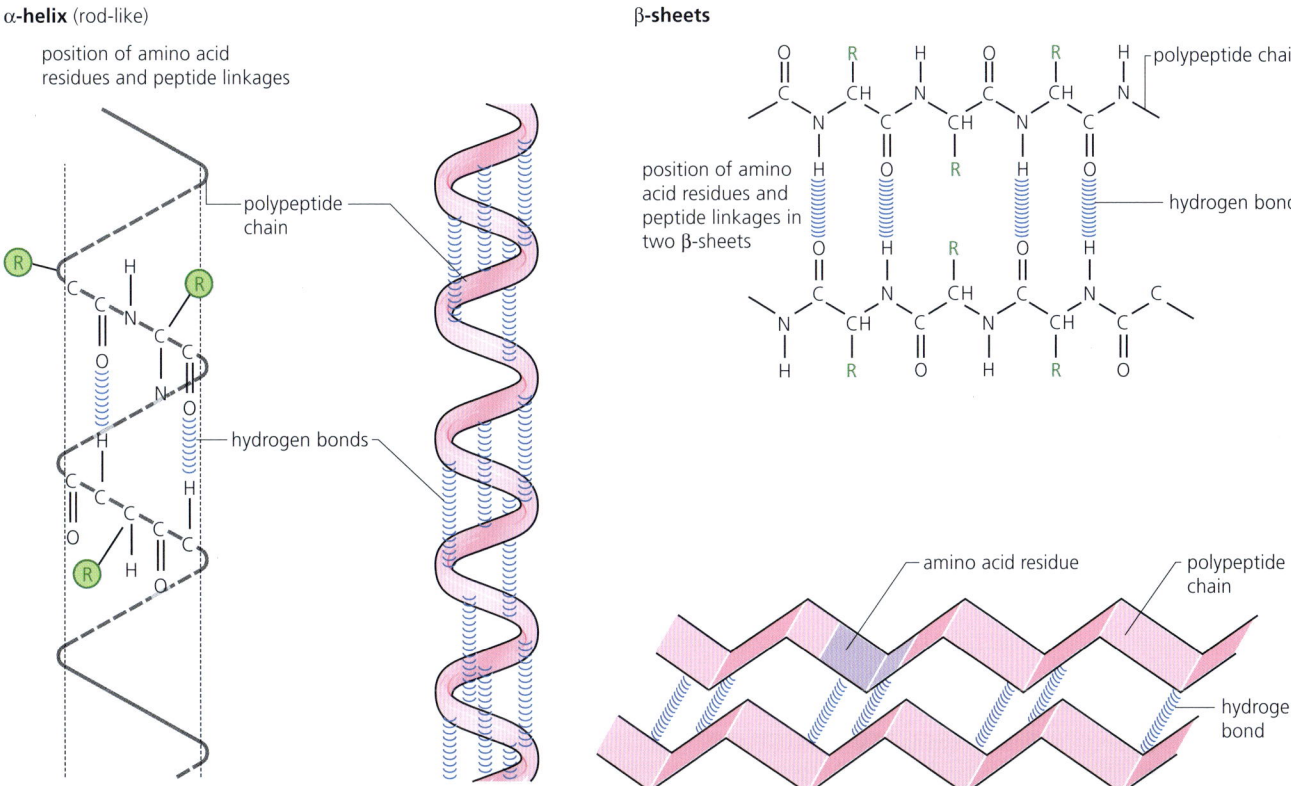

■ **Figure B1.2.8** The secondary structure of protein

Some proteins show extensive regions of α-helical structure such as BipD, an invasion protein from the bacterium *Burkholderia pseudomallei* (see Figure B1.2.9), which causes the disease melioidosis.

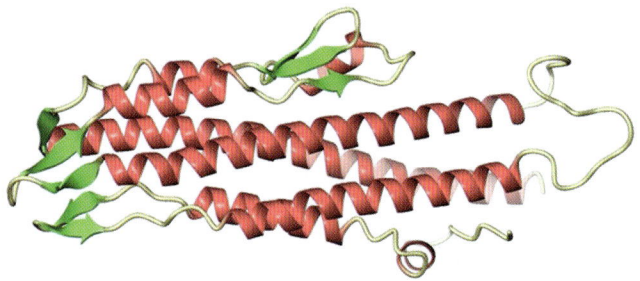

■ **Figure B1.2.9** BipD, an invasion protein from the bacterium *Burkholderia pseudomallei*; α-helices are shown in red and β-sheets in pale green, with the intervening loops coloured cream

> **Going further**
>
> Amino acids are chiral (i.e. exist as two stereoisomers that are mirror images of each other). Enantiomers of amino acids (see page 206) explain why it is that the α-helix is right handed and not left handed: cells use only the L forms to synthesize proteins and this leads to highly specific three-dimensional shapes.

Super-secondary structure

Super-secondary structure describes the different patterns that α-helices or β-sheets commonly adopt in proteins. For example, four α-helices often form a coiled-coil arrangement known as a four-helix bundle (Figure B1.2.10a). Pairs of adjacent helices are often additionally stabilized by salt bridges (ionic bonding) between charged amino acids. Two β-sheets may stack on top of each other in a 'beta sandwich' such as in the immunoglobulin fold (Figure B1.2.10b).

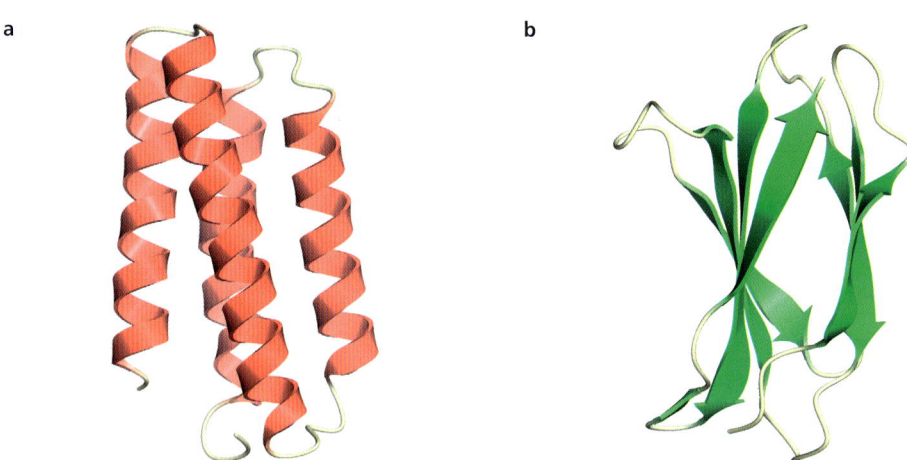

■ **Figure B1.2.10** Super-secondary structure in a protein: a) a four-helix bundle, b) a 'beta sandwich'

Super-secondary structure often corresponds to a particular domain, which are independent structures found in proteins. Protein domains are defined as compact, folded structures within a polypeptide chain and they usually have a specific function, such as DNA-binding capability or the ability to induce dimerisation between two proteins containing the domain. A protein dimer is a macromolecular complex formed by two protein monomers.

■ Tertiary structure

The tertiary structure of a protein is the precise, compact structure, unique to that protein, which arises when the molecule is further folded and held in a particular complex shape. This shape is stabilized by four different types of bonding, which are established by interactions between R-groups and other adjacent parts of the chain (Figure B1.2.11).

Amine (–NH$_2$) and carboxyl (–COOH) groups in R-groups can become positively or negatively charged by binding or dissociation of hydrogen ions and can then participate in ionic bonding.

> ● **Top tip!**
>
> Apart from the example of pairs of cysteines forming disulfide bonds (see Figure B1.2.11), you are not required to name examples of amino acids that participate in types of bonding that maintain tertiary structure.

> ● **Top tip!**
>
> Two of the 20 amino acids in proteins contain sulfur, so disulfide bridges can form between cysteine and methionine residues in the tertiary structure of proteins.

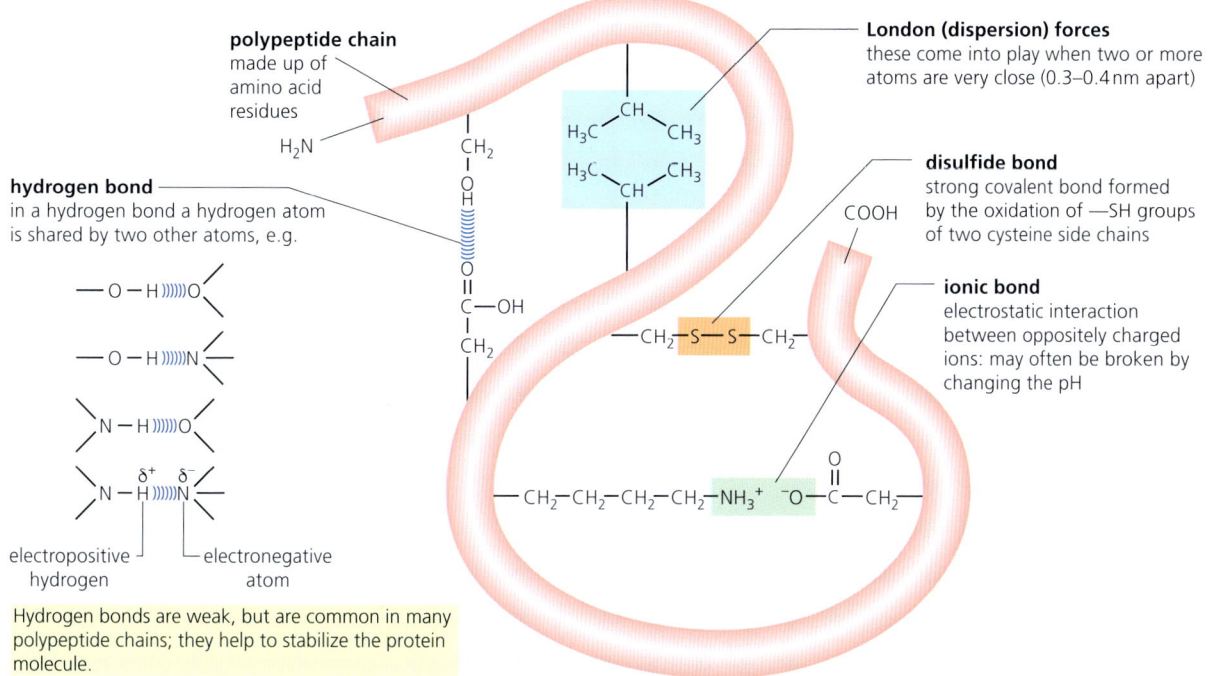

■ Figure B1.2.11 Cross-linking within a polypeptide

The primary, secondary and tertiary structure of the protein lysozyme is shown in Figure B1.2.12. Look carefully at the different levels of organization within the protein.

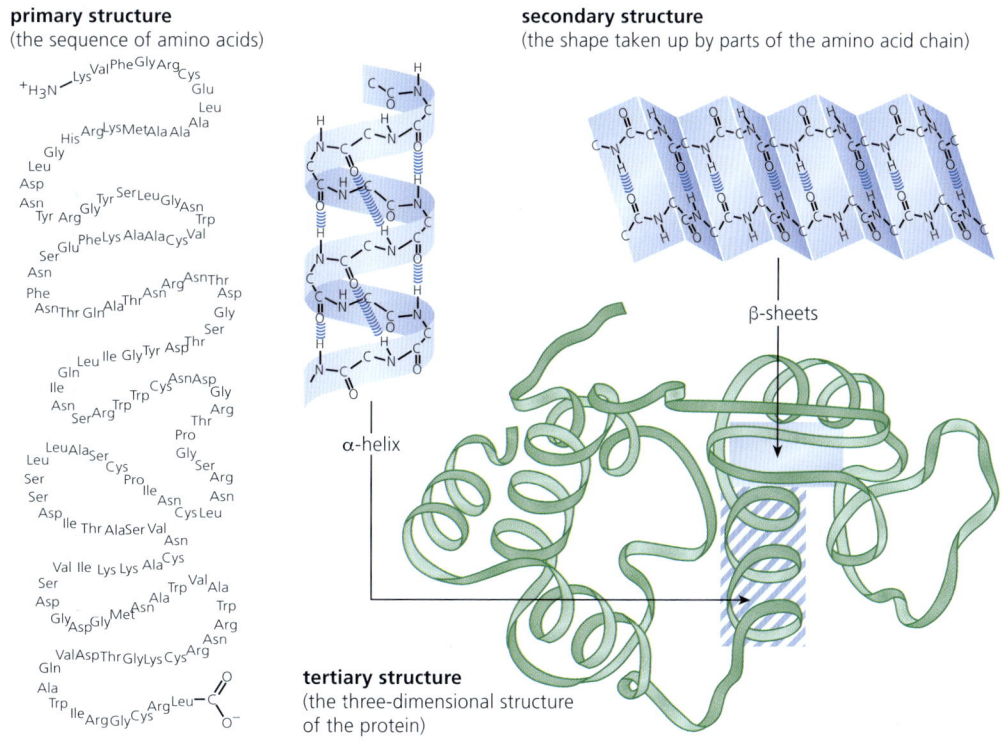

■ Figure B1.2.12 The protein lysozyme – primary, secondary and tertiary structure

> **Top tip!**
> Amine and carboxyl groups in R-groups can become positively or negatively charged by gaining or losing hydrogen ions so that they can then participate in ionic bonding.

ATL B1.2A

Using the internet, research another protein of choice and determine the different levels of structure it shows. How does the form of the protein relate to its function?

B1.2 Proteins

HL ONLY

5 **Predict** what type of amino acid composition proteins that cross from one side of the cell membrane to the other (transmembrane proteins) are expected to have.

Common mistake

Although hydrogen bonds are important in the tertiary structure of proteins, it is the interactions between the side groups of amino acids that determines the tertiary structure. Hydrogen bonds stabilize the α-helix.

Effect of polar and non-polar amino acids on tertiary structure of proteins

Amino acids with polar R-groups have hydrophilic properties. When these amino acids are built into protein in prominent positions, they may influence the properties and functioning of the proteins in cells. Similarly, amino acids with non-polar R-groups have hydrophobic properties.

Hydrophobic amino acids are clustered in the core of globular proteins that are soluble in water. **Integral proteins** have regions with hydrophobic amino acids, helping them to embed in membranes. Integral membrane proteins are permanently embedded within the plasma membrane. The portions of the proteins found inside the membrane are hydrophobic, while those exposed to the cytoplasm or extracellular fluid tend to be hydrophilic. The fatty acid tails that form the interior of the membrane are non-polar and do not repel the hydrophobic (non-polar) parts of the integral proteins. Examples of these outcomes are illustrated in Figures B1.2.13 (for cell membrane proteins) and B1.2.14 (for an enzyme that occurs in the cytoplasm).

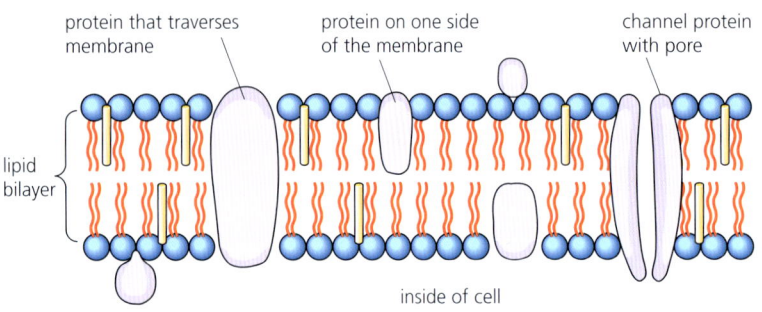

■ Figure B1.2.13 Polar and non-polar amino acids in membrane proteins

◆ **Integral proteins**: proteins that traverse from one side of a membrane to the other side.

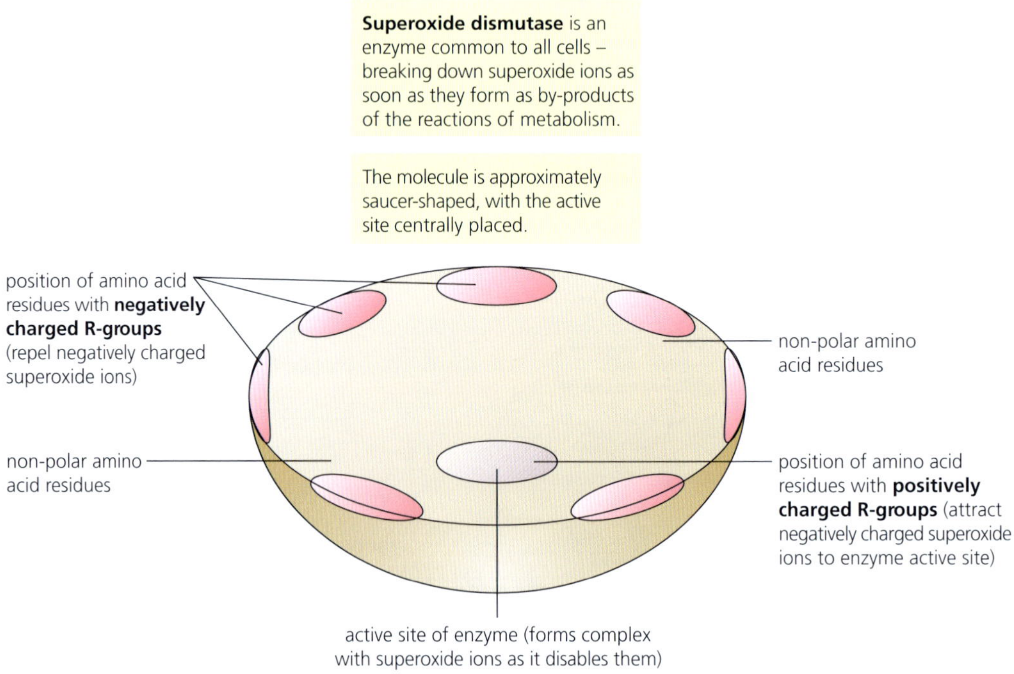

Superoxide dismutase is an enzyme common to all cells – breaking down superoxide ions as soon as they form as by-products of the reactions of metabolism.

The molecule is approximately saucer-shaped, with the active site centrally placed.

6 **Describe** three types of bond that contribute to a protein's tertiary structure.

■ Figure B1.2.14 Polar and non-polar amino acids in the enzyme superoxide dismutase

Theme B: Form and function – Molecules

Quaternary structure of non-conjugated and conjugated proteins

The quaternary structure of proteins arises when two or more polypeptide chains or proteins are held together forming a complex, biologically active molecule.

Conjugated proteins

◆ **Conjugated protein**: a combination of protein and non-protein prosthetic group.

Haemoglobin is known as a **conjugated protein**, which is a combination of protein and non-protein (the prosthetic group). Conjugated proteins have other chemical groups attached, including carbohydrates (such as in the glycocalyx of cell membranes – see page 195), lipids, bound metal ions and other organic groups.

Haemoglobin consists of four polypeptide chains (two α-chains and two β-chains). Each polypeptide chain in the haemoglobin molecule is held around a non-protein haem group (the prosthetic group), in which an atom of iron occurs (Figure B1.2.15).

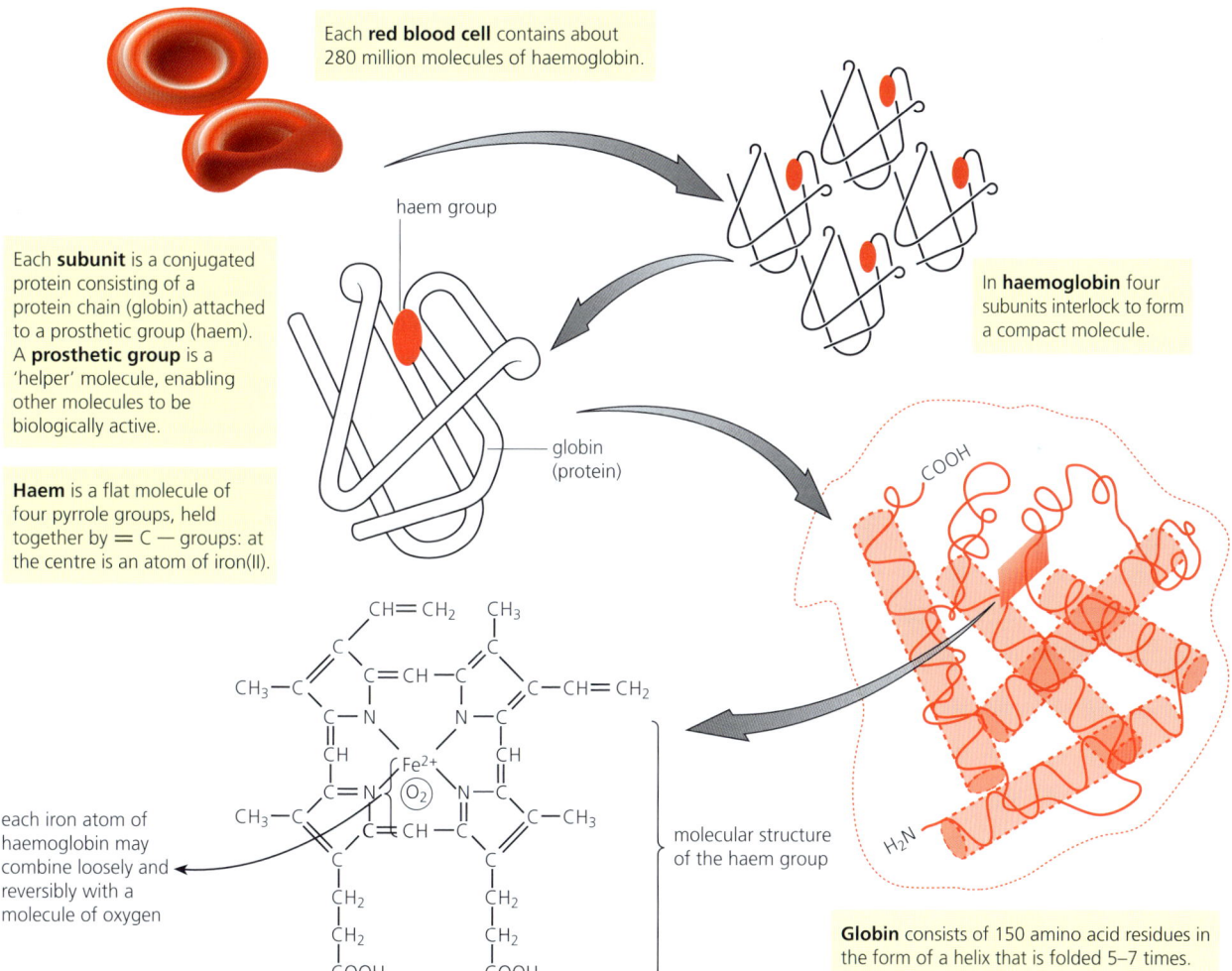

■ **Figure B1.2.15** Haemoglobin – a quaternary protein of red blood cells

Figure B1.2.16 on the next page shows how α-chains and β-chains combine with the haem group to form the haemoglobin molecule.

B1.2 Proteins

■ Figure B1.2.16 Structure of haemoglobin

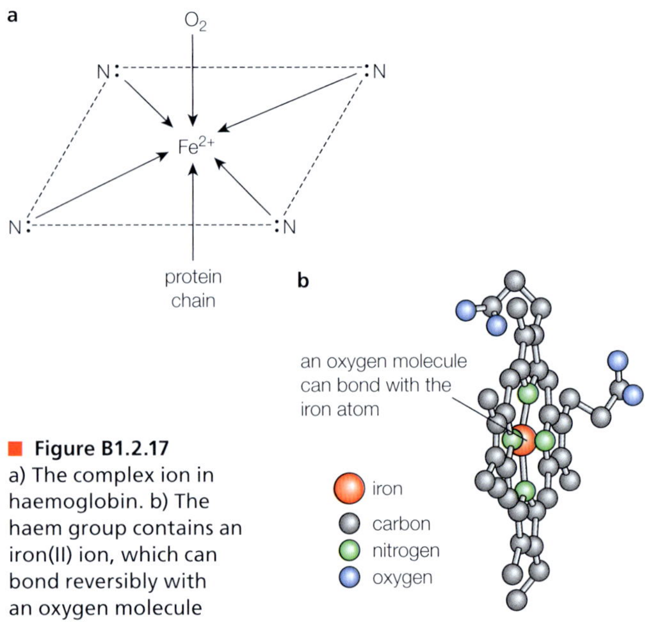

■ Figure B1.2.17
a) The complex ion in haemoglobin. b) The haem group contains an iron(II) ion, which can bond reversibly with an oxygen molecule

7 Explain what is meant by primary, secondary, tertiary and quaternary structure in haemoglobin.

Non-conjugated proteins

Proteins that are not associated with prosthetic groups are known as **non-conjugated proteins**. Insulin and collagen are examples of non-conjugated proteins.

Insulin

In 1951, English biochemist Frederick Sanger made a major discovery when he determined the first amino acid sequence of a protein, **insulin**. This was important because it showed that a protein has a precisely defined amino acid sequence. Before this finding, it was not known that one amino acid sequence characterises one protein.

Insulin is a hormone involved in glucose regulation (Chapter D3.3, page 761). It is an example of a globular protein (see page 221, below). Insulin is composed of two chains, an A chain and a B chain (see Figure B1.2.18). The A and B chains are linked together by two disulfide bonds, and an additional disulfide bond is formed within the A chain. Figure B1.2.19 shows the formation of a disulfide bridge.

> **Top tip!**
> Haemoglobin is not an enzyme. Unlike enzymes, haemoglobin does not catalyse chemical transformations of the molecules to which it binds.

◆ **Non-conjugated protein**: a protein not associated with a non-protein prosthetic group.

◆ **Insulin**: hormone made in the pancreas that promotes the synthesis and storage of glycogen in the liver and muscle cells.

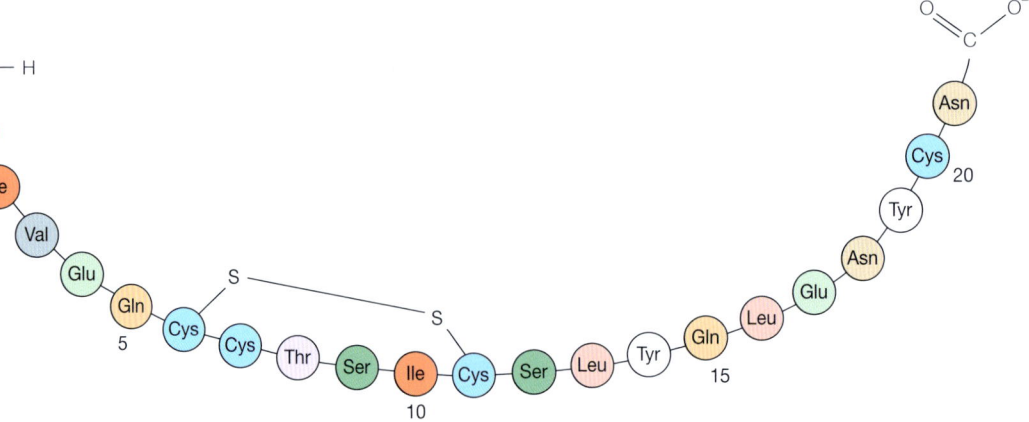

■ Figure B1.2.18 The primary sequence of the insulin A chain (a short polypeptide of 21 amino acids). Note the cysteine residues at positions 6, 7, 11 and 20 in the chain. They form disulfide bridges (the one between residues 6 and 11 is shown here) that stabilize the three-dimensional structure of the insulin molecule

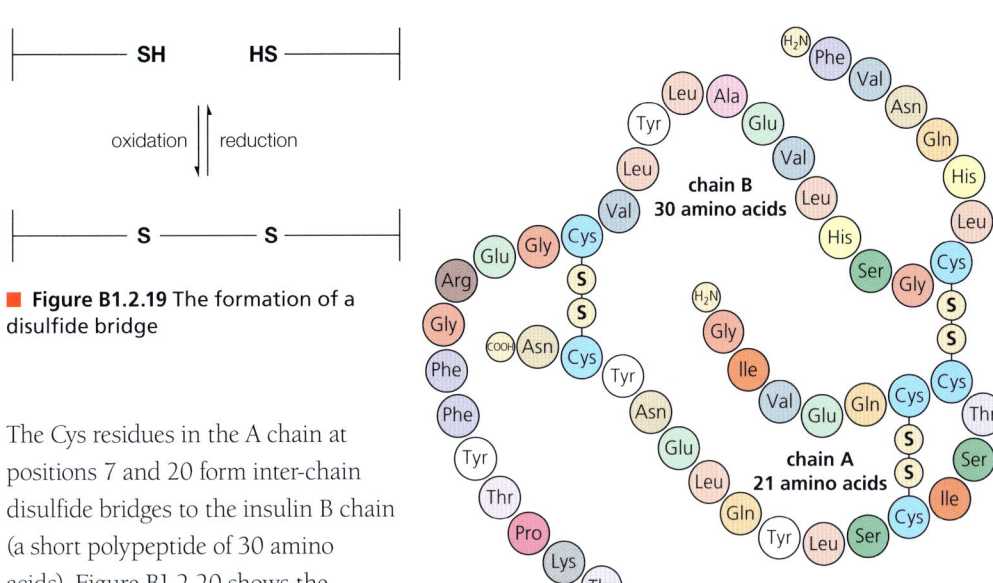

Figure B1.2.19 The formation of a disulfide bridge

Figure B1.2.20 The quaternary structure of human insulin

> **Top tip!**
> SS-bridges (disulfide bridges) are post-translational modifications (PTMs), not secondary structures.

The Cys residues in the A chain at positions 7 and 20 form inter-chain disulfide bridges to the insulin B chain (a short polypeptide of 30 amino acids). Figure B1.2.20 shows the combined quaternary structure of an insulin protein.

ATL B1.2B

PDB-101 is a website (https://pdb101.rcsb.org) that explores the three-dimensional shapes of proteins and nucleic acids. Learning about the variety of shapes and functions of proteins helps us to understand the role of proteins in disciplines such as biomedicine and agriculture.

Research the structure and function of insulin using this site, by going to 'Molecule of the month' and selecting 'insulin'. This link should take you directly to the correct part of the site: **https://pdb101.rcsb.org/motm/14**

Tool 1: Experimental techniques

Physical and digital molecular modelling

Constructing physical models of molecules is a good way of understanding how their structure relates to their function, and how different components of proteins combine together to form the overall structure.

Go to **https://pdb101.rcsb.org** and select 'Learn' from the menu at the top of the page. Select 'Paper models' from the dropdown menu (**https://pdb101.rcsb.org/learn/paper-models**). Select 'Insulin' and follow the instructions. You can download and print the template of the model. Instructions for cutting and assembling are included in a video on the same page.

Collagen

Collagen is the most abundant protein in animals. It is the substance that gives structure and essentially holds the body together. It is found in bones, muscles, skin and tendons, and is an example of a fibrous protein (see page 221).

The quaternary structure of collagen consists of three left-handed helices twisted into a right-handed coil (Figure B1.2.21).

8 **Define** what is meant by the *quaternary structure* of a protein.

9 **Outline** how the structure of haemoglobin is related to its specific function.

10 **Explain** the four different levels of protein organization.

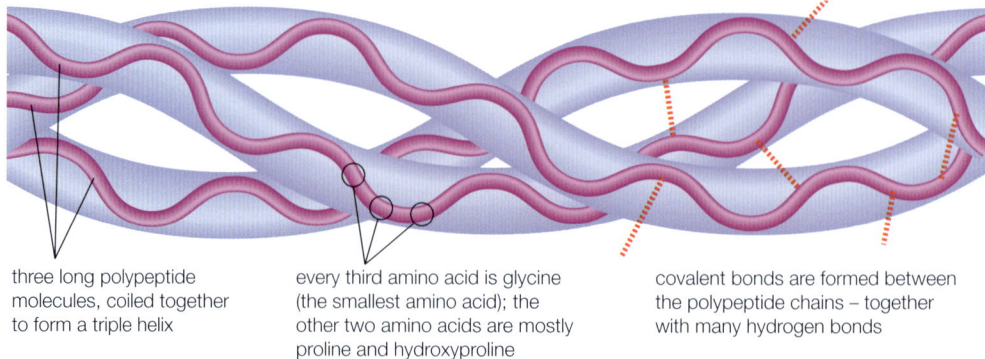

three long polypeptide molecules, coiled together to form a triple helix

every third amino acid is glycine (the smallest amino acid); the other two amino acids are mostly proline and hydroxyproline

covalent bonds are formed between the polypeptide chains – together with many hydrogen bonds

■ **Figure B1.2.21** Collagen – an example of a non-conjugated protein

Each polypeptide chain has a repeated triplet sequence of Gly-X-Y, where X and Y can be any amino acid but are frequently proline (in the X position) and hydroxyproline (in the Y position). Every third amino acid is a glycine (residue), which allows each helical chain to make a turn every three amino acids and intertwine around two other chains to form a compact triple helix, as only glycine is small enough to fit into the centre. The three helical polypeptide chains are held together by interchain hydrogen bonds forming tropocollagen. Many triple helices lie parallel in a staggered pattern to form fibrils with covalent bonds between neighbouring triple helix chains. The fibrils unite to form fibres.

Nature of science: Observations

Use of technology to observe protein molecules

X-ray crystallography (Figure B1.2.22a) has traditionally been used to obtain high-resolution images of the three-dimensional structures and shapes of crystalline proteins and other molecules in the solid state. Crystals are a highly ordered arrangement of individual protein molecules.

a
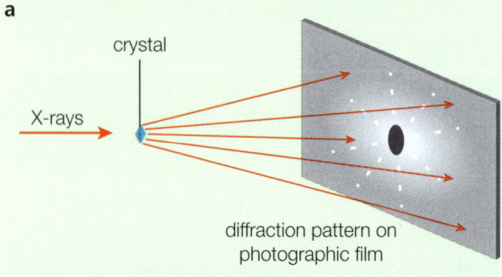

b
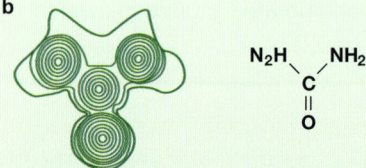

■ **Figure B1.2.22** a) Single-crystal X-ray crystallography b) Electron density contour map of urea and its structural formula derived from X-ray crystallography

When an X-ray beam is focused on a protein crystal the X-rays are diffracted (scattered) by the atoms. The diffracted X-rays generate spots (also called reflections) on a detector (digital camera) that have an intensity and position that can be computer processed to show the structure and shape of the protein or molecule via an electron density contour map (Figure B1.2.22b).

An alternative method called cryogenic electron microscopy uses the same principle as transmission electron microscopy, but cools the samples to cryogenic temperatures (e.g. via liquid nitrogen) and embeds them in an environment of vitreous (non-crystalline) ice, allowing proteins and protein complexes to be studied. Cryo-electron microscopy has allowed imaging of single-protein molecules and their interactions with other molecules.

Randomly orientated proteins are struck by the electron beam, producing a faint image on the detector. Thousands of similar images are averaged (and cleaned-up) by the computer to generate a high-resolution three-dimensional image or density map of the protein.

The specimen can be statistically analysed, allowing for the reconstruction of the structural information, and different conformations (shapes) can be determined in the same sample to understand how proteins, such as membrane-based pumps, function. In this way, the technology allows imaging of structures that would be impossible to observe with our unaided senses.

Another major advantage of cryo-electron microscopy is that large, intact complexes can be studied, allowing the three-dimensional structure of ribosomes, proteins and viruses to be seen, almost to the atomic scale. However, X-ray crystallography is much quicker and simpler, and the large majority of images of structures still come from X-ray.

> **TOK**
>
> **How important are material tools in the production or acquisition of knowledge?**
>
> New technology has enabled biologists to explore in ever more detail the molecular world from which all living organisms are made. The development of cryogenic electron microscopy, for example, has allowed the imaging of single-protein molecules and their interactions with other molecules – something not possible before. However, X-ray is quicker and simpler to use, and less expensive, and so still tends to be used in preference to cryogenic electron microscopy, indicating that it is not just technology that determines how knowledge is acquired, but that other factors are involved. New theories do not necessarily come from new technology – some of the greatest theories in science, such as Darwin's theory of evolution by natural selection and Einstein's theory of relativity, did not result from new technology but rather the imagination and creativity of two remarkable minds.

Link
Cryogenic electron microscopy is also covered in Chapter A2.2, page 61.

Relationship of form and function in globular and fibrous proteins

■ Figure B1.2.23 Photomicrograph of collagen, a fibrous protein, showing many triple helices bound together

■ Fibrous proteins

Some proteins take up a tertiary structure that is a long, much-coiled chain; these are called **fibrous proteins**. They have long, narrow shapes. Examples of fibrous proteins are collagen, a component of bone and tendons (Figure B1.2.23), and keratin, found in hair, horn and nails. Fibrous proteins are often insoluble.

The collagen molecule consists of three polypeptide chains, each in the shape of a helix (see Figure B1.2.21, previous page). The chains are wound together as a triple helix, forming a stiff cable strengthened by numerous hydrogen bonds. The ends of individual collagen molecules are staggered so there are no weak points in collagen fibres, giving the whole structure high tensile strength. This makes the protein well suited to provide structural support in skin, tendons and cartilage.

11 Describe the main structural features of collagen.

■ Globular proteins

Other proteins take up a more spherical shape and are known as **globular proteins**. They are mostly highly soluble in water. Examples include enzymes, such as lysozyme and catalase, and hormones, such as insulin (Figure B1.2.24 and page 218).

Insulin is a very small protein, allowing it to move quickly through the blood. Its shape is recognized by specific receptors on its target cell surfaces. It is difficult to make a small protein that will fold into a stable structure. This problem is solved by synthesizing a longer protein chain, which folds into the proper structure. The extra pieces are removed (Figure B1.2.24), leaving two small chains in the mature form. The structure is further stabilized by three disulfide bridges.

> ● **Top tip!**
>
> Make sure you know the differences in shape between globular and fibrous proteins, and understand how their shapes make them suitable for specific functions. Look closely at the examples of insulin and collagen explained here.

HL ONLY

Insulin is a hormone produced in the β cells of the islets of Langerhans in the pancreas by ribosomes of the rough endoplasmic reticulum (RER) as a polypeptide of 102 amino acid residues (preproinsulin).

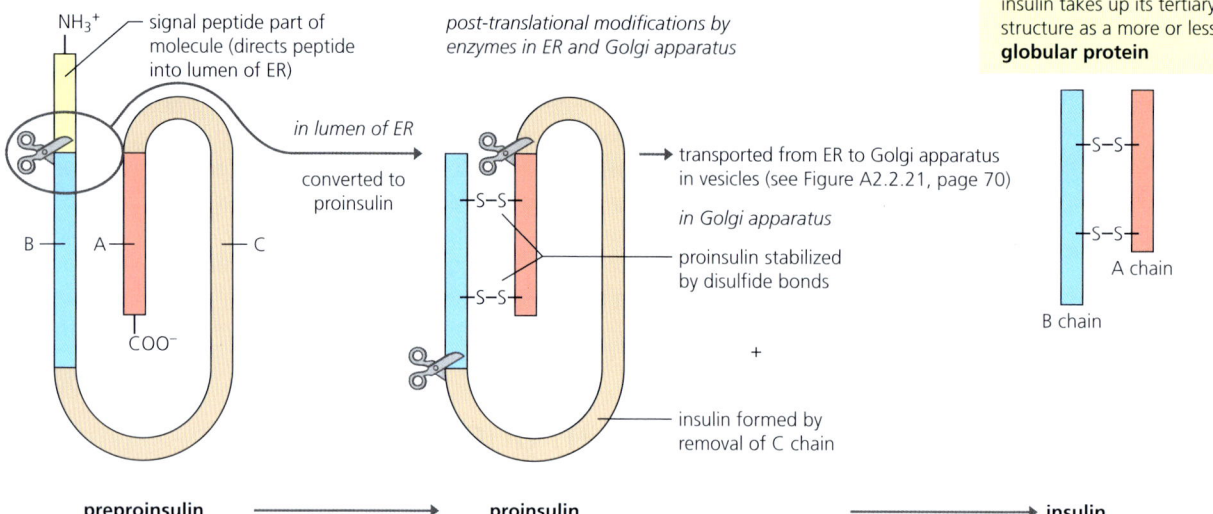

■ Figure B1.2.24 Insulin, a globular protein

12 Explain the difference between fibrous and globular proteins using collagen and insulin as examples.

ATL B1.2C

https://lab.concord.org/embeddable.html#interactives/samples/5-amino-acids.json

Use this website to explore protein folding. Hydrophobic interactions have an important role in determining the shape (conformation) of a protein. Therefore, an important factor controlling the folding of any protein is the distribution of its polar and non-polar amino acids.

1. Run the simulation of protein folding within water (polar solvent), oil (non-polar solvent) and a vacuum. Summarize the differences you observe.
2. Repeat the simulations in the three environments using a protein that is entirely hydrophobic and a protein that is entirely hydrophilic. Summarize the differences you observe.

LINKING QUESTIONS

1. How do abiotic factors influence the form of molecules?
2. What is the relationship between the genome and the proteome of an organism?

B2.1 Membranes and membrane transport

Guiding questions
- How do molecules of lipid and protein assemble into biological membranes?
- What determines whether a substance can pass through a biological membrane?

SYLLABUS CONTENT

This chapter covers the following syllabus content:
- B2.1.1 Lipid bilayers as the basis of cell membranes
- B2.1.2 Lipid bilayers as barriers
- B2.1.3 Simple diffusion across membranes
- B2.1.4 Integral and peripheral proteins in membranes
- B2.1.5 Movement of water molecules across membranes by osmosis and the role of aquaporins
- B2.1.6 Channel proteins for facilitated diffusion
- B2.1.7 Pump proteins for active transport
- B2.1.8 Selectivity in membrane permeability
- B2.1.9 Structure and function of glycoproteins and glycolipids
- B2.1.10 Fluid mosaic model of membrane structure
- B2.1.11 Relationships between fatty acid composition of lipid bilayers and their fluidity (HL only)
- B2.1.12 Cholesterol and membrane fluidity in animal cells (HL only)
- B2.1.13 Membrane fluidity and the fusion and formation of vesicles (HL only)
- B2.1.14 Gated ion channels in neurons (HL only)
- B2.1.15 Sodium–potassium pumps as an example of exchange transporters (HL only)
- B2.1.16 Sodium-dependent glucose cotransporters as an example of indirect active transport (HL only)
- B2.1.17 Adhesion of cells to form tissues (HL only)

The basis of cell membranes – lipid bilayers

Link
The chemical structure of phospholipid was introduced in Figure B1.1.27 and is covered in detail in Chapter B1.1, page 202.

A plasma membrane is a structure common to all cells. The plasma membrane encompasses and maintains the integrity of the cell (it holds the cell's contents together). It is also a barrier across which all substances entering and leaving the cell must cross. The lipid of membranes is predominantly **phospholipid**, together with other lipid types.

a phospholipid molecule has a **hydrophobic tail** – which repels water – and a **hydrophilic head** – which attracts water

in contact with water forming a monolayer

mixing with water to form a bilayer

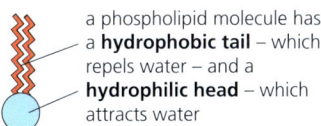
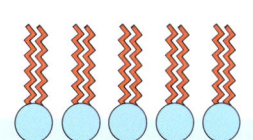
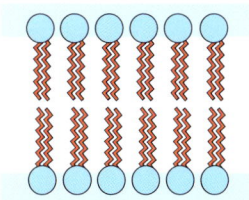

■ Figure B2.1.1 The response of phospholipid molecules when added either to the surface of a water layer (left), forming monolayers, or mixed into water (right), forming bilayers

Phospholipid molecules in contact with water form a **monolayer**, with heads dissolved in the water and the tails sticking outwards.

When mixed with water, phospholipid molecules arrange themselves into a **bilayer**, in which the hydrophobic tails are attracted to each other.

1 **State** the difference between a lipid bilayer and the double membrane of many organelles.

The phosphate heads of phospholipids (which are hydrophilic) interact with the surface of the water whereas the tails are repelled (they are hydrophobic). On the surface of water, phospholipids form monolayers (i.e. a single layer). When submerged in water, phospholipids spontaneously form a double-layered structure, with the hydrophilic heads directed out into the water and the hydrophobic tails pointing away (Figure B2.1.1). Monolayers can be extensive and cover a water surface, whereas bilayers need to be continuous and so form closed structures, called vesicles (see Figure A2.1.5, page 41).

Lipid bilayers as barriers

The lipid bilayer is about 4–10 nm in thickness and this can encompass a prokaryotic cell of 0.1–5 μm in diameter, or a eukaryotic cell of 10–100 μm in diameter. In the lipid bilayer attractions between the hydrophobic hydrocarbon tails on the inside and between the hydrophilic glycerol/phosphate heads and the surrounding water on the outside make a stable, strong barrier.

In this way, the hydrophobic hydrocarbon chains form the core of a membrane. This part of the membrane has low permeability to large and hydrophilic molecules, including ions and polar molecules. Membranes therefore function as effective barriers between aqueous solutions, across which all substances entering and leaving the cell pass, such as metabolites that move between the cytoplasm and the interior of the cell.

■ Crossing the lipid bilayer barrier

The movement of molecules such as water, respiratory gases (oxygen and carbon dioxide), nutrients (e.g. glucose), essential ions and excretory products is continual. These molecules can go into or out of cells.

Cells may secrete substances such as hormones and enzymes, or they may receive growth substances and certain hormones. Plants secrete the chemicals that make up their walls through their cell membranes, and assemble and maintain the wall outside the membrane. Certain mammalian cells secrete structural proteins, such as collagen, in a form that can be assembled outside the cells. All these molecules need to move across the lipid bilayer barrier to get into or out of cells.

Figure B2.1.2 is a summary of the movement across plasma membranes, while Figure B2.1.3 summarizes the mechanisms of transport across membranes.

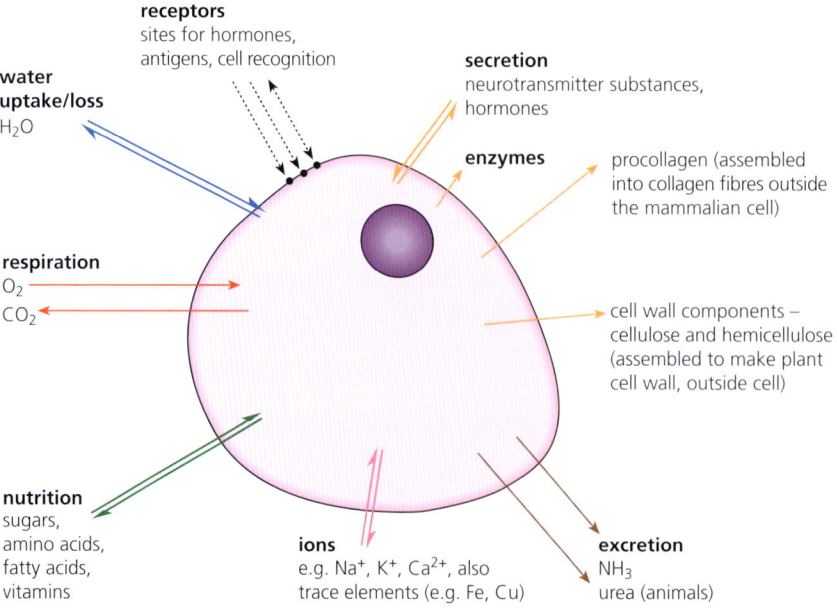

■ Figure B2.1.2 Movements across the plasma membrane

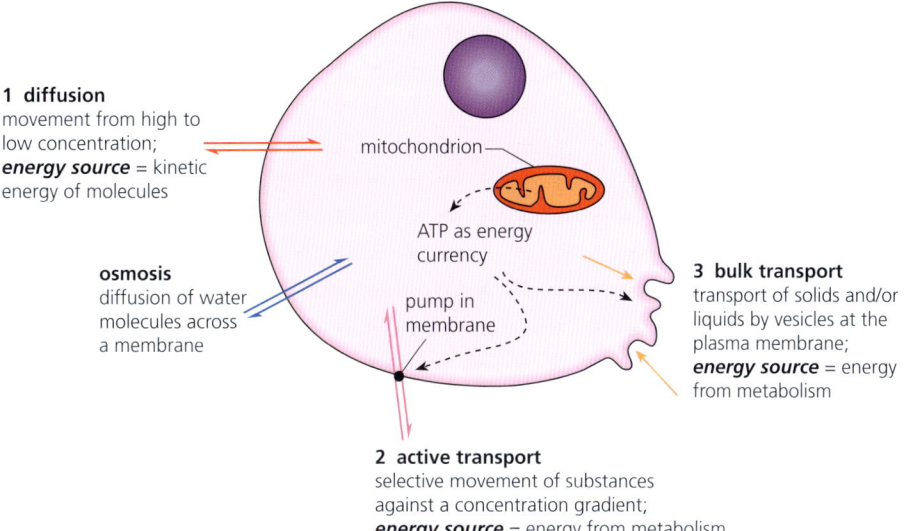

■ Figure B2.1.3 Mechanisms of movement across membranes

We will now look in more detail at the different mechanisms of movement across plasma membranes as summarized in Figure B2.1.3.

Simple diffusion across membranes

Atoms, molecules and ions of liquids and gases undergo continuous random movements. These movements result in the even (homogenous) distribution of the components of a gas mixture and of the atoms, molecules and ions in a solution. So, for example, we are able to take a tiny random sample from a solution and analyse it to find the concentration of dissolved substances in the whole solution – because any sample has the same composition as the whole. Similarly, every breath we take has the same amount of oxygen, nitrogen and carbon dioxide as the atmosphere as a whole.

Continuous, random movements of all molecules ensures complete mixing and even distribution, given time, in solutions and gases.

Diffusion is the passive movement of particles from a higher to lower concentration. The term 'membrane diffusion' is used to describe the movement of molecules through the phospholipid bilayer of membranes. Molecules such as oxygen and carbon dioxide (e.g. in the lung alveoli) are examples of simple diffusion across membranes.

Oxygen and carbon dioxide are very small molecules and can diffuse through the phospholipid bilayer. Oxygen is supplied to cells via the circulatory system – oxygen diffuses into cells from a higher concentration in the blood plasma into tissue fluid, and then on into the cells where it is used for aerobic respiration. Carbon dioxide, produced by respiration, travels in the opposite direction.

Where a difference in concentration (concentration gradient) has arisen in a gas or liquid, random movements carry molecules from a region of high concentration to a region of low concentration. As a result, the particles become evenly dispersed. The energy for diffusion is transferred from the store of kinetic energy of the molecules. 'Kinetic' means that a particle has a store of energy because it is in continuous motion.

Diffusion in a liquid can be illustrated by adding a crystal of a coloured mineral to distilled water. Even without stirring, the ions become evenly distributed throughout the water (Figure B2.1.4). The process takes time, especially as the solid has first to dissolve.

◆ **Diffusion**: the free passage of molecules (and atoms and ions) from a region of their higher concentration to a region of lower concentration.

Link
Concentration gradients are defined and discussed in Chapter B3.1, page 275.

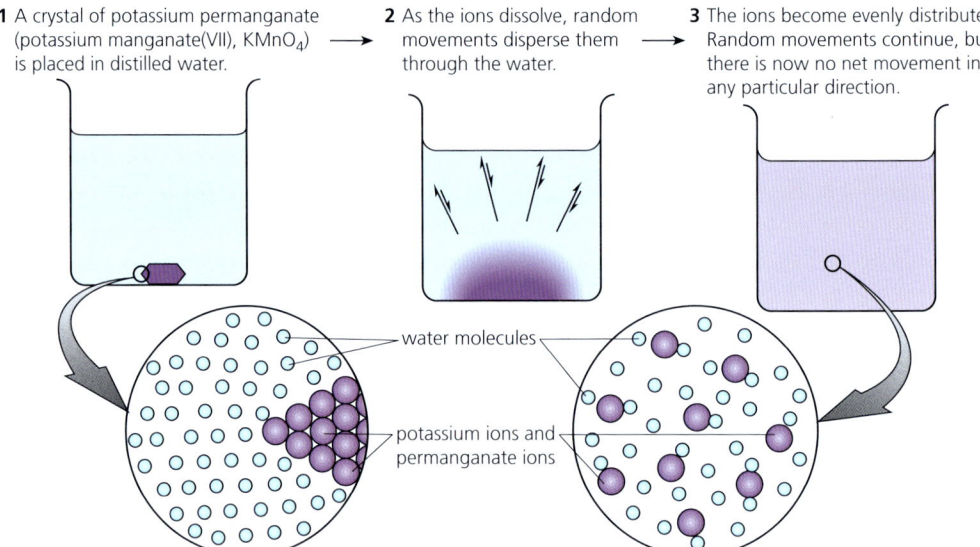

Figure B2.1.4 Diffusion in a liquid

Top tip!

The term 'diffusion', when referring to movement across membranes, refers to movement across the phospholipid part of the membrane.

◆ **Peripheral protein**: a protein that is attached to the surface of the bilayer.

Link

Integral proteins are covered in more detail in Chapter B1.2, page 216.

2 **Compare and contrast** the structure and function of peripheral and integral proteins.

Diffusion across the cell membrane occurs where:
- The plasma membrane is fully permeable to the solute.
- The lipid bilayer of the plasma membrane is permeable to non-polar substances, including steroids, and also to oxygen and carbon dioxide in solution, all of which diffuse quickly via this route.

Inquiry 1: Exploring and designing

Designing

Plan a diffusion experiment to show the effect of concentration gradient, temperature, surface area or distance on the rate of diffusion. Food colouring can be used to see the movement of particles.

Justify the range and/or quantity of the measurements you plan to use in your investigation.

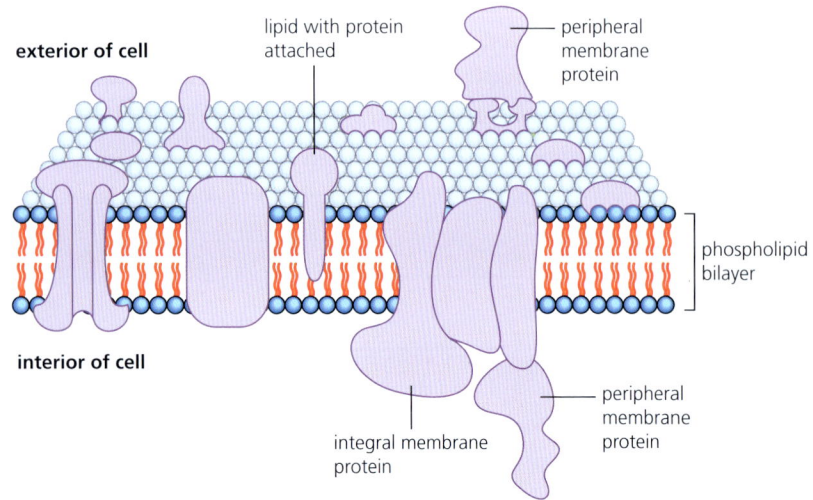

Figure B2.1.5 Peripheral and integral proteins of the plasma membrane

Integral and peripheral proteins in membranes

Membrane proteins have diverse structures, locations and functions. There are two main categories of protein in the cell membrane: integral and peripheral proteins. **Peripheral proteins** are attached to the outer surface of the bilayer (the outside of the cell) or the inner surface (facing the cell cytoplasm) (Figure B2.1.5). Peripheral proteins may shuttle between integral proteins on the surface of the membrane, be scaffold proteins to hold shape, or be receptors for extracellular signals. Proteins that are embedded in one or both of the lipid layers of the membrane are described as **integral proteins**. Integral membrane proteins may act as channels for transport of metabolites, or be enzymes and carriers, and some may be receptors or antigens.

Osmosis

Link

The importance of osmosis in medical applications is covered further in Chapter D2.3, page 690.

♦ **Osmosis**: diffusion of free water molecules from a region where they are more concentrated (low solute concentration) to a region where they are less concentrated (high solute concentration) across a partially permeable membrane.

Osmosis is a special case of diffusion (Figure B2.1.6). It is the diffusion of water molecules across a membrane that is permeable to water (**partially permeable**). Since water makes up 70–90% of living cells and cell membranes are partially permeable membranes, osmosis is very important in biology. For example, too much or too little water in the blood creates osmotic effects, where water will either move into cells in the blood (too much water) causing them to burst, or move out of cells (too little water) by osmosis. Both can be damaging for the body.

■ Why does osmosis happen?

Dissolved substances attract a group of polar water molecules around them (a process known as hydration). The forces that hold the water molecules in this way are weak chemical bonds, including **hydrogen bonds**. Therefore, the tendency for random movement by the dissolved substances and their surrounding water molecules is very much reduced. Organic substances such as sugars, amino acids, polypeptides and proteins, and inorganic ions such as Na^+, K^+, Cl^- and NO_3^- have this effect on the water molecules around them.

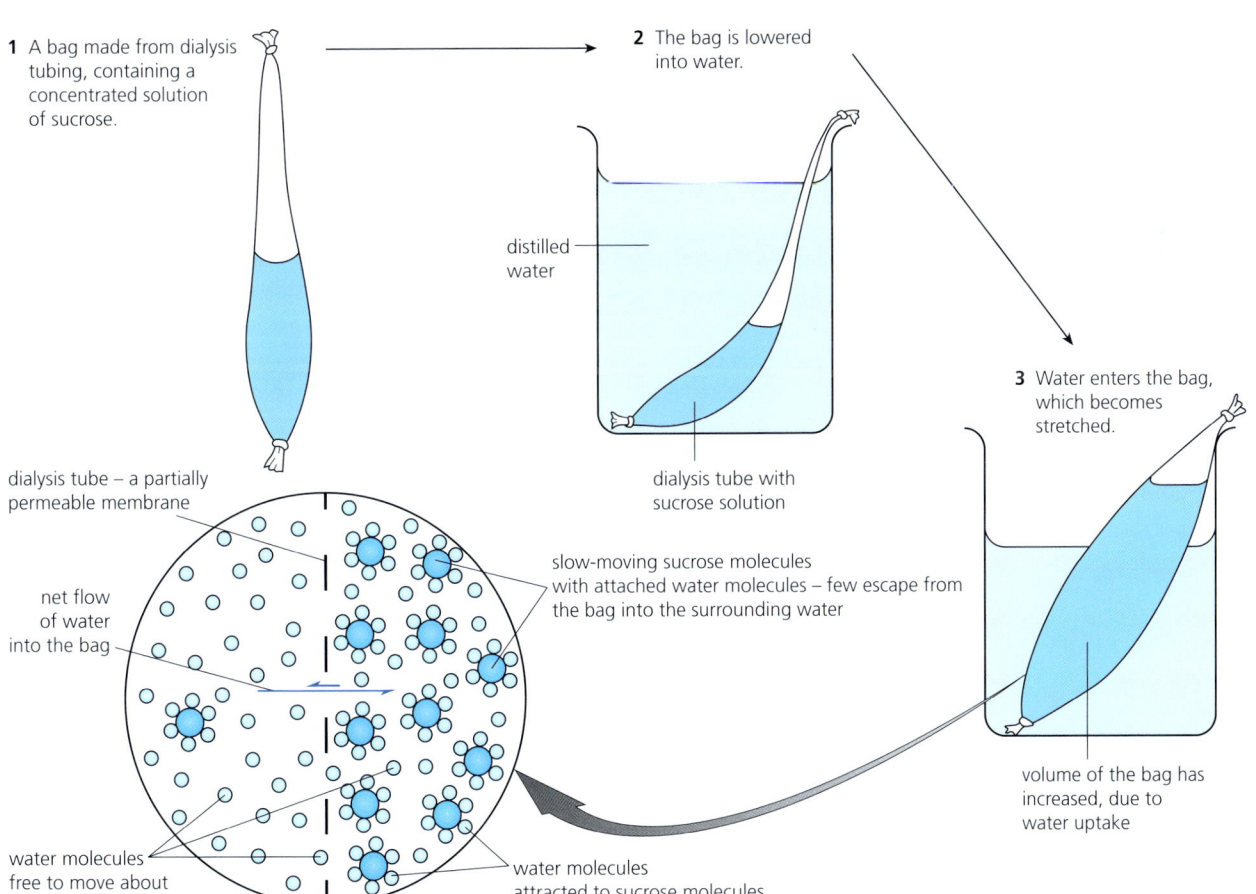

■ Figure B2.1.6 The process of osmosis

B2.1 Membranes and membrane transport

The more concentrated the solution (i.e. the more solute dissolved per volume of water), the greater the number of water molecules that are held almost stationary. In a very concentrated solution, many more of the water molecules have restricted movement than in a dilute solution. In pure water, all of the water molecules are free to move about randomly and do so.

When a solution is separated from water (or a more dilute solution) by a membrane permeable to water molecules (such as the plasma membrane), water molecules that are free to move tend to diffuse, while dissolved molecules and their groups of water molecules hardly move at all. So, there is a net flow (diffusion) of water from a more dilute solution into a more concentrated solution across the membrane. This is why the membrane is described as partially permeable.

So, **osmosis** is defined as the net movement of water molecules (solvent), from a region of high concentration of water molecules to a region of lower concentration of water molecules, across a selectively permeable membrane. Or, we can state that osmosis is the passive movement of water molecules across a partially permeable membrane, from a region of lower solute concentration to a region of higher solute concentration. Because the membrane is impermeable to solutes, differences in solute concentration cause differences in water concentration. If a solution has more solute, then there is less room within the solution for water and so it has a lower water concentration. Water can move through the membrane from a more dilute solution (a higher water concentration) to a solution with higher solute concentration (a lower water concentration).

Link
Osmosis is explored in more detail, including experiments to investigate the changes in plant tissue due to water movement, in Chapter D2.3, page 683.

Common mistake

Osmosis involves the movement of water molecules, not just 'particles', from lower to higher solute concentration across semi permeable membranes.

3 When a concentrated solution of glucose is separated from a dilute solution of glucose by a partially permeable membrane, **determine** which solution will show a net gain of water molecules.

4 **Explain** what happens to a fungal spore that germinates after landing on jam made from fruit and its own weight of sucrose.

Top tip!

If a solution has a higher concentration of solute, it will have a lower water concentration. Think of a limited volume where there is only sufficient space for a fixed number of molecules – if there are more solute molecules then there will be less 'room' for water. The term 'dilute' refers to the concentration of solute molecules: a low concentration of solute and high concentration of water. A 'concentrated' solution is one that has a high concentration of solute (and a corresponding low concentration of water).

The membrane is effectively impermeable to solutes because they are too large or have a charge so cannot travel through the phospholipid bilayer. This means that when a cell gains or loses mass, it is due to the intake or loss of water through osmosis, not through solute movement.

■ The role of aquaporins

♦ **Aquaporin**: a water channel pore (protein) in a membrane.

There are protein-lined pores in the plasma membrane, called **aquaporins**, which allow water to pass through. Water diffusing across the plasma membrane passes via the aquaporins of the membrane and via tiny spaces between the phospholipid molecules. The latter occurs easily where the plasma membrane contains phospholipids with unsaturated hydrocarbon tails, because these hydrocarbon tails are spaced more widely.

Top tip!

'Facilitate' means 'to help'. Facilitated diffusion is therefore diffusion with the help of plasma membrane pores.

♦ **Facilitated diffusion**: the movement of particles from higher to lower concentration through integral proteins (carrier or channel proteins). Movement is passive.

♦ **Channel protein**: a pore in the membrane that allows specific charged particles (e.g. ions) and polar substances (i.e. all hydrophilic substances) to diffuse through the membrane into or out of the cell.

Common mistake

Although facilitated diffusion uses transmembrane proteins, it does not require energy in the form of ATP. Both diffusion and facilitated diffusion are passive processes.

5 Distinguish between diffusion and facilitated diffusion.

Top tip!

If movement down a concentration gradient (i.e. from a higher to lower concentration) occurs via a channel or carrier protein, this is known as facilitated diffusion.

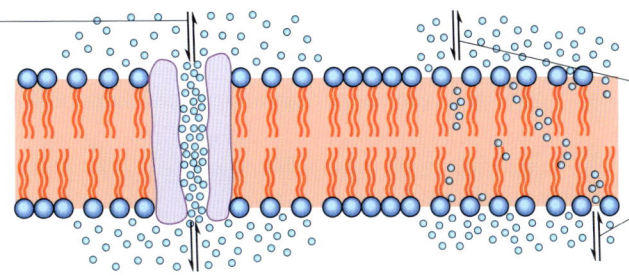

■ **Figure B2.1.7** How polar water molecules cross the lipid bilayer

Aquaporins are involved with the reabsorption of water in the collecting ducts of the kidney (see Chapter D3.3, page 735).

Facilitated diffusion

In **facilitated diffusion**, particles of a substance that cannot diffuse across the plasma membrane through the phospholipid bilayer are helped across the membrane by integral proteins that span the membrane. In the presence of these substances, **channel proteins** (made of globular protein) form pores large enough to allow diffusion (Figure B2.1.8). Channel proteins provide hydrophilic channels for polar and charged molecules to pass through. These pores close again when the substance is no longer present (Figure B2.1.9). Polar and charged molecules, such as carbohydrates, amino acids and ions, cross the plasma membrane by facilitated diffusion.

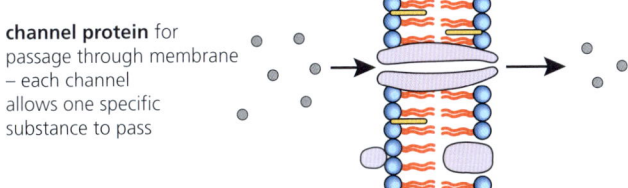

■ **Figure B2.1.8** Channel proteins in the plasma membrane for transport of metabolites or water

The structure of these channel proteins make membranes **selectively permeable**, allowing specific ions to diffuse through when channels are open but not when they are closed. In facilitated diffusion, the energy is transferred from the kinetic energy store of the molecules involved, as is the case in all forms of diffusion because molecules are moving down their concentration gradient. Energy from metabolism is not required. Important examples of facilitated diffusion are the movement of ADP into mitochondria and the exit of ATP from mitochondria (page 248).

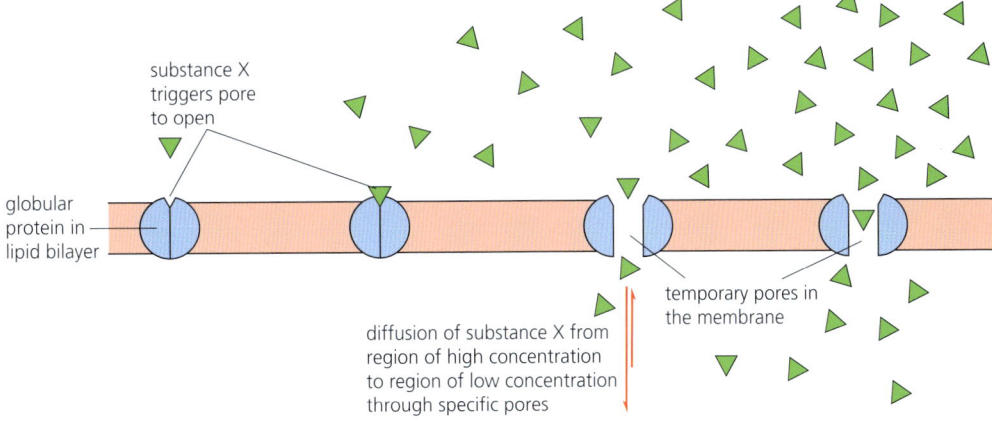

■ **Figure B2.1.9** Facilitated diffusion

B2.1 Membranes and membrane transport

Active transport

♦ **Active transport**: movement of particles from lower to higher concentration, using energy from ATP that has been created during respiration. Movement is through carrier proteins.

♦ **Pump proteins**: proteins in plasma membranes that use energy to actively carry ions and/or solutes against a concentration gradient.

We have seen that diffusion, facilitated diffusion and osmosis are due to random movements of molecules and occur spontaneously from high to low concentrations. However, many of the substances required by cells must be absorbed from a dilute external concentration and taken up into cells that contain a higher concentration of the substance. Uptake against a concentration gradient cannot occur by diffusion. Instead, it requires the transfer of energy to drive it. This type of uptake is known as **active transport**.

In active transport, metabolic energy produced by the cell, stored as ATP (energy currency – see below), is transferred to drive the transport of molecules and ions across cell membranes. **Pump proteins** (and some carrier proteins) use energy transferred from ATP to move specific particles across membranes and, therefore, these particles can be moved against a concentration gradient (Figure B2.1.10).

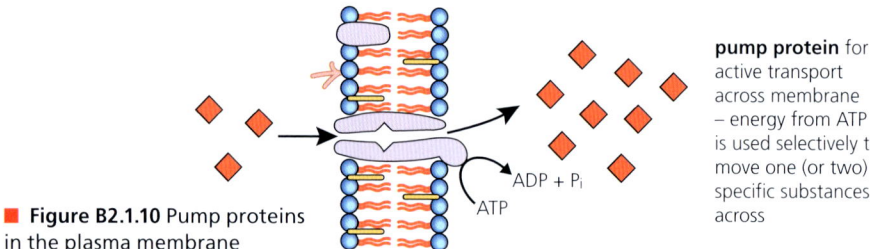

■ **Figure B2.1.10** Pump proteins in the plasma membrane

pump protein for active transport across membrane – energy from ATP is used selectively to move one (or two) specific substances across

Active transport is a feature of most living cells. We meet examples of active transport in the gut where absorption occurs (page 495), in the active uptake of ions by plant roots (page 322), in the kidney tubules where urine is formed (page 775) and in nerve fibres where an impulse or action potential is propagated (page 241).

Active transport has characteristic features distinctly different from those of movement by diffusion.

■ 1 Active transport occurs against a concentration gradient

Active transport occurs from a region of low concentration to a region of higher concentration. The cytoplasm of a cell normally holds reserves of valuable molecules and ions, such as nitrate ions in plant cells or calcium ions in muscle fibres. These useful molecules and ions do not escape; the cell membrane keeps them inside the cell. When more useful molecules or ions become available for uptake, they are actively absorbed into the cells. This happens even though the concentration outside the cell is lower than that inside.

■ 2 Active uptake is highly selective

For example, in a situation where potassium chloride (K^+ and Cl^- ions) is available to an animal cell, K^+ ions are more likely to be absorbed, since they are needed by the cell. Where sodium nitrate (Na^+ and NO_3^- ions) is available to a plant cell, it is likely that more of the NO_3^- ions will be absorbed than the Na^+, since this reflects the metabolic needs of plant cells. Most membrane pumps are specific to particular molecules or ions, bringing about selective transport. If the pump molecule for a particular substance is not present, the substance will not be transported.

■ 3 Active transport involves pump molecules

The pump molecules pick up certain molecules or ions and transport them to the other side of the membrane where they are then released. Pump molecules span the lipid bilayer (Figure B2.1.11). Movements by the pump molecules require reaction with ATP; this reaction transfers metabolic energy to the process.

> ● **Top tip!**
>
> Active transport involves movement against a concentration gradient, using energy transferred from ATP. Protein pumps/carrier proteins are used, but not channel proteins. Channel proteins are used in passive transport to enable solutes to diffuse down concentration gradients.

> ● **Top tip!**
>
> Carrier proteins may or may not carry out active transport; protein pumps always use ATP energy. Carrier proteins, for example, can make use of the concentration gradient of a specific ion built up by pumps to transport other molecules actively against their gradient. See 'Sodium-dependent glucose cotransporters', page 242, for an example of indirect active transport.

The protein pumps of plasma membranes are of different types. Some transport a particular molecule or ion in one direction (Figure B2.1.11), while others transport two substances (like Na⁺ (aq) and K⁺ (aq)) in opposite directions (the sodium–potassium pump, Figure B2.1.22, page 242). Occasionally, two substances are transported in the same direction; for example, Na⁺ and glucose (Figure B2.1.23, page 242).

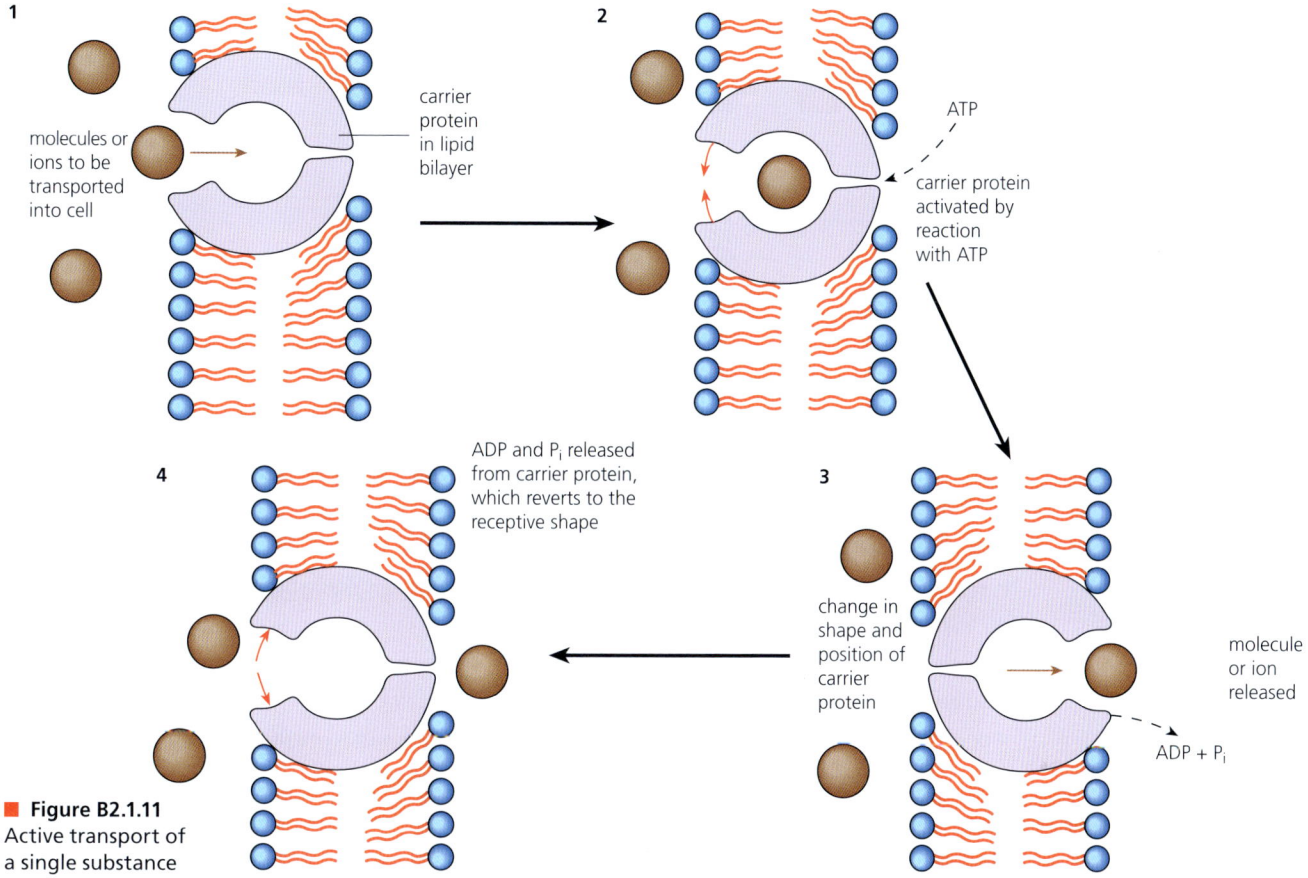

■ Figure B2.1.11 Active transport of a single substance

◆ **Adenosine triphosphate (ATP)**: a nucleotide in every living cell formed in photosynthesis and respiration from ADP and P$_i$; it functions in metabolism as a common intermediate between energy-requiring and energy-yielding reactions.

◆ **Adenosine diphosphate (ADP)**: a nucleotide in every living cell made of adenosine and two phosphate groups linked in series; it is important in energy transfer reactions of metabolism.

Common mistake

When writing about the factors needed for active transport, do not just say that 'energy is needed'. You need to mention ATP and that ATP is produced by respiration.

6 Samples of five plant tissue discs were incubated in dilute sodium chloride solution at different temperatures. The table shows the uptake of ions from the solutions after 24 hours (arbitrary units). **Comment** on how absorption of sodium chloride occurs, giving your reasons.

	Sodium ions	Chloride ions
Tissue at 5 °C	80	40
Tissue at 25 °C	160	80

■ The role of ATP in active transport

The energy released during respiration is used to form **ATP (adenosine triphosphate)** from **ADP (adenosine diphosphate)** and P$_i$ (phosphate). ATP can be regarded as a short-term form of stored chemical energy in cells. ATP is like a 'charged battery' due to the presence of a high energy

B2.1 Membranes and membrane transport

P–O bond, while ADP is a 'flat battery'. Respiration is the 'charging process'. Figure B2.1.12 shows the release of energy from ATP when the third phosphate bond breaks, forming ADP and P_i.

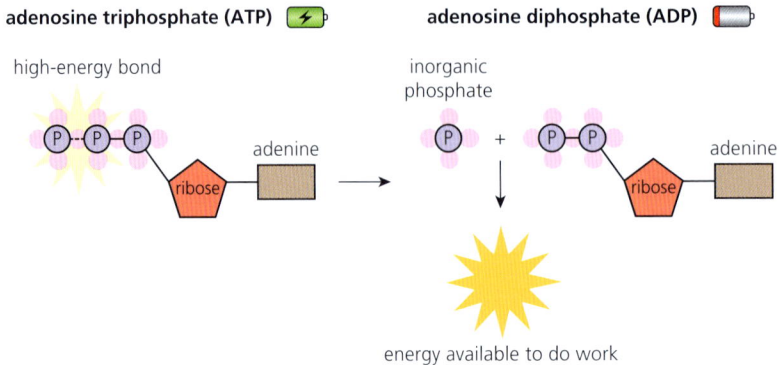

■ **Figure B2.1.12** ATP is the direct source of energy in cells

Selectivity in membrane permeability

Permeability by simple diffusion is not selective and depends only on the size and hydrophilic or hydrophobic properties of particles. On the other hand, facilitated diffusion and active transport allow selective permeability in membranes.

Large polar molecules or charged molecules cannot pass through the hydrophobic interior of the membrane, whereas small non-polar molecules can. This means that, for many substances, movement through the phospholipid bilayer can only be controlled by altering the concentration gradients across the membrane; for example, the movement of oxygen can be controlled by monitoring breathing rate, which in turn increases or decreases movement of oxygen into the lung (see Chapter B3.1). The membrane itself is not selectively permeable to oxygen as it can pass through the phospholipid bilayer.

Substances that are moved across the membrane using transmembrane proteins can be selectively transported, such as ions involved in controlling the generation of electrical impulses in nerve cells. Such selective transport is essential for many processes, for example the movement of minerals into plant roots and the absorption of nutrients in the intestine of animals.

Link
Generation of electrical impulses in nerve cells is covered in Chapter C2.2, page 469.

Top tip!
It is important to zero the colorimeter with a cuvette containing just water before the experiment begins and to select the colour of the light to match the colour absorbed by the compound.

> ## Tool 1: Experimental techniques
>
> ### Investigating the effect of temperature on plant membranes using colorimetry
>
> Colorimetry is the technique that helps to determine the concentration of a solution having colour. Colorimeters pass light of a set colour (frequency or wavelength) through a transparent sample and record the absorbance of the light (Figure B2.1.13). The absorption of light is exponentially related to the number of molecules of the absorbing solute in the solution, i.e., C, solute concentration. The proportion of light passing through the solution is known as transmittance, T, and is calculated as the ratio of the emergent and incident light intensities. It is usually expressed as a percentage. The colorimeter has two scales:
> - an exponential scale from zero to infinity, measuring absorbance
> - a linear scale from 0–100, measuring (per cent) transmittance.

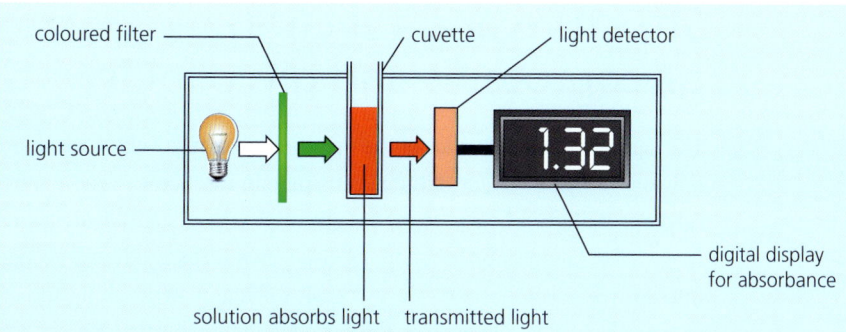

■ **Figure B2.1.13** A colorimeter showing the pathway of visible light (a blue or green filter is used with a red solution); the colorimeter produces a beam of light of a given wavelength (using filters) and directs it at a sample in solution in a cuvette

For most practical purposes you should use absorbance, which is linearly related to the solute concentration [C].

Beetroot can be used to investigate the effect of temperature on the plasma membrane. The beetroot cells contain a pigment, betalain, which gives them their distinctive colour. The pigment is contained within the cell vacuole. Betalain from beetroot vacuoles appears red/purple in aqueous solution as it absorbs green (the complementary colour) and transmits the other wavelengths. To study the concentration of betalain, the colorimeter is set to measure the absorbance of green light via the use of a filter.

● **Top tip!**

A cuvette is a straight-sided clear container for holding liquid samples in a colorimeter. Always use a clean cuvette that has no scratches on it. Insert cuvettes correctly as they often have only two transparent sides for the light to pass through.

Inquiry 1: Exploring and designing

Exploring; designing

What experiment could you carry out to investigate the effect of temperature on plant membranes?

Before you do the investigation, think about the following:

- Which part of the membrane will be affected by temperature? (Hint: what do you know about the effect of heat on one particular type of biological molecule that is found in membranes?)
- The vacuole of beetroot contains a purple pigment called betalain. How can you use this information to measure the effect of temperature?
- What will happen to the membrane as temperature increases? Which organelles are surrounded by membrane in beetroots and which of these will be affected?
- How will you measure the effect? What will be the independent variable and what will be the dependent variable?
- What are the control variables that you must keep the same?

Structure and function of glycoproteins and glycolipids

Link
The role of glycoproteins in cell–cell recognition is covered in Chapter B1.1, page 195.

The carbohydrate molecules of the membrane are relatively short-chain polysaccharides. They occur only on the outer surface of the plasma membrane. Some of these molecules are attached to proteins (**glycoproteins**) and some to lipids (**glycolipids**). Collectively, they are known as the **glycocalyx** (page 195). The various functions of glycocalyx include cell–cell recognition and the binding of cells into tissues (cell adhesion). The glycocalyx is highly hydrophilic and attracts large amounts of water to the cell's surface. This helps the cell's interaction with its watery environment and with the cell's ability to obtain substances dissolved in the water.

The carbohydrate chains of glycoproteins and glycolipids are found on the extracellular side of membranes (see Figure B2.1.14, below).

Concept: Form

Plasma membranes are made from lipids, proteins and some carbohydrates. The structure is known as the fluid mosaic model because the proteins can move about freely within the phospholipid bilayer.

Fluid mosaic model of membrane structure

As we have seen, the plasma membrane is made almost entirely of protein and lipid, together with a small and variable amount of carbohydrate. Figure B2.1.14 shows how these components are assembled into the plasma membrane. This view of the molecular structure of the plasma membrane is known as the **fluid mosaic model**. The plasma membrane is described as *fluid* because the components (lipids and proteins) are on the move, and *mosaic* because the proteins are scattered about in this pattern.

♦ **Fluid mosaic model**: the accepted view of the structure of the plasma membrane, comprising a phospholipid bilayer with proteins embedded but free to move about.

Channel proteins (Figure B2.1.14) are used to move small, charged particles such as ions and polar molecules across the membrane by facilitated diffusion. These proteins form hydrophilic 'tunnels' across the membrane through which the ions can pass. Other channel proteins transport water (these are aquaporins – see page 228).

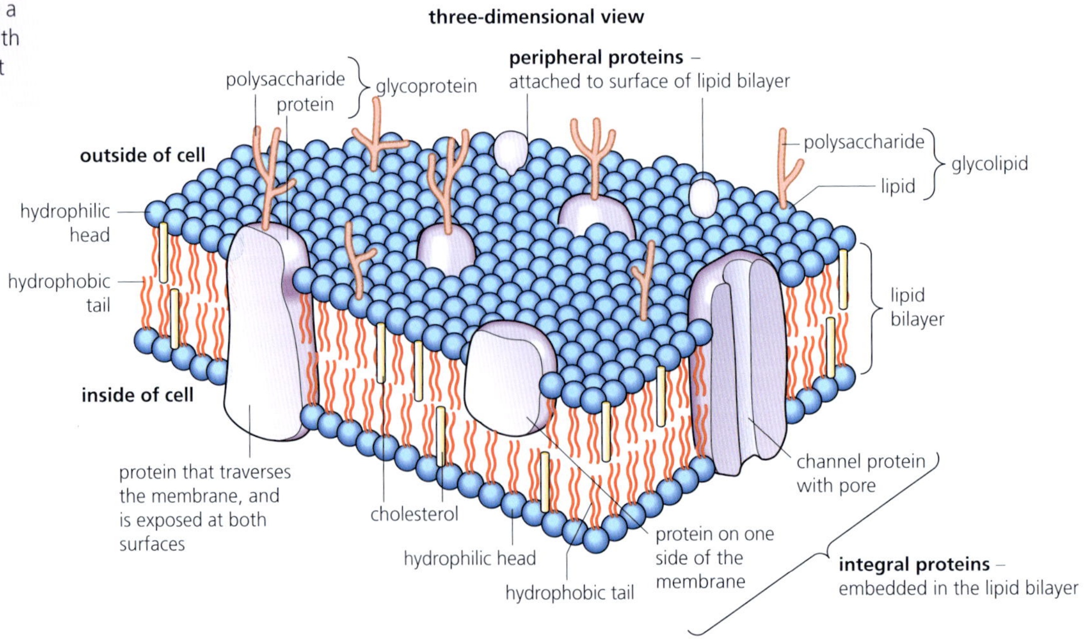

■ Figure B2.1.14 The fluid mosaic model of the plasma membrane

● Top tip!

Practise drawing and labelling a two-dimensional representation of the fluid mosaic model. Make sure to include peripheral and integral proteins, glycoproteins, phospholipids and cholesterol. Hydrophobic and hydrophilic regions should be indicated.

Concept: Function

Each part of the plasma membrane has a specific function and they work together to create a functional membrane around cells. Phospholipids are the main structural component of the cell membrane and act as barriers to the passage of hydrophilic molecules and ions into and out of the cell. Proteins have several roles depending on their location and structure: transport, receptors, anchorage, cell recognition, intercellular joining and enzymatic activity. Carbohydrates on the outer surface are involved in cell recognition and cell adhesion.

Inquiry 3: Concluding and evaluating

Concluding

James Danielli and Hugh Davson were among the first scientists to propose a structure for the plasma membrane.

In response to the evidence, in the 1930s, Danielli and Davson proposed a membrane structure (which was revised in 1954) in which a lipid bilayer was evenly coated with proteins on both surfaces. Pores were thought to be present in places in the membrane. Early electron micrographs of cell membranes seen in section appeared to support this model (Figure B2.1.15).

What evidence from the electron micrograph seemed to support their model?

Evaluating

Using your knowledge of membrane structure, why was the Davson–Danielli model of the plasma membrane falsified? What evidence can you use to refute their model?

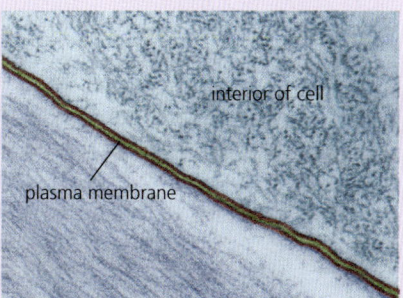

■ Figure B2.1.15 TEM of the cell surface membrane of a red blood cell (×700000)

● TOK

The explanation of the structure of the plasma membrane has changed over the years as new evidence and ways of analysis have come to light. Under what circumstances is it important to learn about theories that were later discredited?

Nature of science: Models

What is known about the composition and structure of the plasma membrane has built up from evidence over time. Studies in cell structure (cytology), cell biochemistry and cell behaviour (cell physiology) all contributed. The first ideas about a 'membrane' were based on the observations that:
- cell contents flow out when the cell surface is ruptured – a membrane barrier is present
- water-soluble compounds enter cells less readily than compounds that dissolve in lipids – this implies that lipids are a major component of the cell membrane
- in the presence of water (the environment of life) phospholipid molecules arrange themselves as a bilayer, with the hydrocarbon tails facing together, forming a stable, strong barrier.

The Davson–Danielli model of membrane structure proposed that membranes were composed of a phospholipid bilayer that lies between two layers of globular proteins. The evidence that seemed to support this model was an electron micrograph that showed two dark lines with a lighter band in between (Figure B2.1.15). Later, additional evidence inspired two cytologists (Singer and Nicolson, in 1972) to propose the fluid mosaic model of membrane structure.
- Attempts to extract the protein from plasma membranes indicated that, while some occurred on the external surfaces and were easily extracted, others were buried within or across the lipid bilayers. These proteins were more difficult to extract.
- Freeze-etching studies of plasma membranes confirmed that when a membrane is by chance split open along its mid-line, some proteins are seen to occur buried within or across the lipid bilayer (Figure B2.1.16).
- Experiments in which specific components of membranes were 'tagged' by reaction with marker chemicals (typically fluorescent dyes) showed component molecules to be continually on the move within membranes – a plasma membrane could be described as strong but 'fluid'.

- Lipid bilayers contain molecules of an unusual lipid, cholesterol, the presence of which disturbs the close-packing of the bulk of the phospholipids of the bilayer; the quantity of cholesterol present may vary with the ambient temperatures that cells experience.
- On the outer surface of the plasma membrane, antenna-like carbohydrate molecules form complexes with some membrane proteins (forming glycoproteins) and lipids (forming glycolipids).

a Cell membrane in cross-section

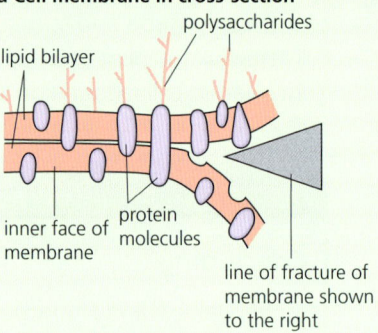

b Electron micrograph of the cell membrane (freeze-etched)

■ Figure B2.1.16 Plasma membrane structure: evidence from freeze-etching

Common mistake

When drawing a plasma membrane, take care: on diagrams showing structure, the most common errors are to place particular types of proteins or cholesterol in the wrong position. Cholesterol should be smaller than the hydrophilic tails and should be embedded in the bilayer; peripheral proteins should be on the membrane surface, not partially or fully embedded.

7 **Draw** a diagrammatic cross-section of the fluid mosaic membrane, using Figure B2.1.14 to help you. **Label** it correctly using these terms: phospholipid bilayer, hydrophobic region, hydrophilic region, cholesterol, glycoprotein, integral protein, peripheral protein.

8 **Describe** the role of membranes in cells.

Relationships between fatty acid composition of lipid bilayers and their fluidity

There is a mixture of phospholipids in the membrane. Some phospholipids have saturated fatty acid tails (no double bonds between a pair of carbon atoms) and some have unsaturated fatty acid tails (double bonds between carbon atoms) (see Figure B1.1.22, page 199). Unsaturated fatty acids in lipid bilayers have lower melting points, so membranes are fluid and therefore flexible at temperatures experienced by a cell. Unsaturated fatty acids melt at a lower temperature because their unsaturated hydrocarbon tails do not pack so closely together in the way those of saturated fatty acids do (see page 200). An excess of unsaturated fatty acid tails makes the membrane more fluid. Saturated fatty acids have higher melting points and make membranes stronger at higher temperatures.

Membrane fluidity is an important factor in membrane composition. Membranes must be sufficiently fluid for many of the proteins present to move about and function correctly. If the temperature of a membrane falls, it becomes less fluid. A point may be reached when a membrane will solidify. Some organisms have been found to vary the balance between saturated and unsaturated fatty acids (and the amount of cholesterol in their membranes – see more on cholesterol below) as ambient temperatures change. In this way, organisms maintain a properly functioning membrane, even at very low temperatures. For example, investigations into the lake sturgeon (*Acipenser fulvescens*), a North American temperate freshwater fish (also known as the rock sturgeon) have shown that when the temperature decreases from 16 °C to 1 °C, both mono- and polyunsaturated fatty acids of phospholipids significantly increase, and saturated fatty acids decrease. The lake sturgeon occupies some of the most northerly distributions of any sturgeon species and experiences extended overwintering periods. Changes to the fatty acid composition of its cell membranes allows this fish species to survive in these cold conditions.

Top tip!

Ensure you are familiar with an example of adaptations in membrane composition in relation to habitat. Can you find any other examples like the lake sturgeon?

A **homeoviscous** adaptation refers to the adaptation of the cell membrane lipid composition to maintain adequate membrane fluidity. When temperature decreases, the composition of phospholipid fatty acids is expected to become more unsaturated to be able to maintain homeoviscosity. Membrane lipids became more unsaturated during acclimatization to cold conditions, with a reverse occurring due to warmer conditions.

Cholesterol and membrane fluidity in animal cells

◆ **Cholesterol**: a lipid of animal plasma membranes; a precursor of the steroid hormones in humans, formed in the liver and transported in the blood as lipoprotein.

As seen in Figure B2.1.14, lipid bilayers contain another type of lipid molecule, known as **cholesterol**, in addition to phospholipids. It is an essential component of the plasma membrane of all cells. The position of cholesterol molecules in membranes has the effect of disturbing the close packing of the phospholipids, increasing the flexibility of the membrane. Cholesterol acts as a modulator (adjustor) of membrane fluidity, stabilizing membranes at higher temperatures and preventing stiffening at lower temperatures.

In mammalian membranes, cholesterol reduces fluidity and reduces the permeability of the membrane to some solutes. It is a steroid, with a hydrophilic hydroxyl (–OH) group and hydrophobic hydrocarbon chain on either side of the carbon ring structure (Figure B2.1.17).

Cholesterol interacts with the phosphate head and fatty acid tails of the phospholipids (Figure B2.1.18).

B2.1 Membranes and membrane transport

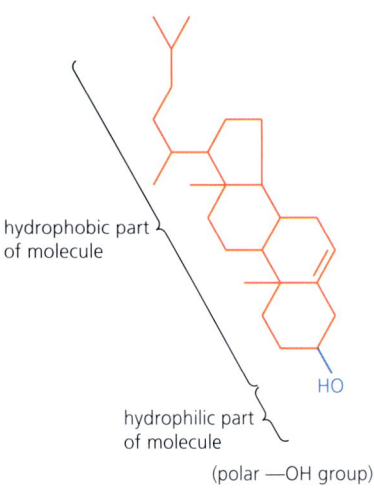

■ Figure B2.1.17 Molecular structure of cholesterol

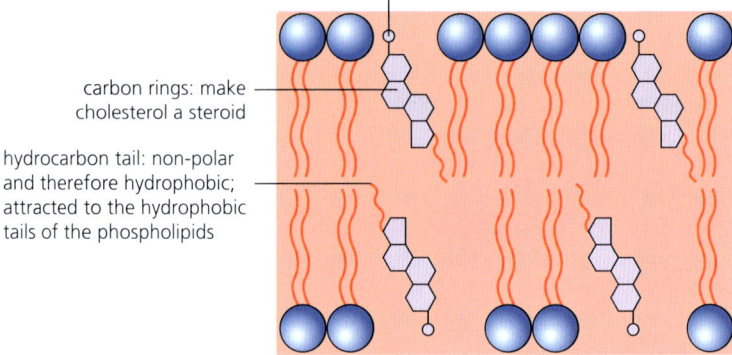

■ Figure B2.1.18 The interaction between cholesterol and the phospholipid bilayer

> ● **Common mistake**
>
> Do not confuse membrane fluidity with membrane permeability.

Concept: Form

Cholesterol has a polar hydroxyl group, whereas the rest of the molecule is hydrophobic. These characteristics determine where cholesterol is located within the membrane, with the hydrophilic –OH group attracted to the phosphate of the phospholipid and the hydrophobic part to the fatty acid tails in the interior of the bilayer.

The quantity of cholesterol present varies with the ambient temperatures that cells experience.
- In low temperatures, the cholesterol maintains the fluidity of the membrane by forcing apart the phospholipids and maintaining distance between them, thereby sustaining movement between the components of the membrane.
- In higher temperatures, bonds between the cholesterol and phospholipids maintain the structural integrity of the membrane and prevent them from becoming too fluid, and potentially disintegrating.

ATL B2.1A

Cholesterol is essential for normal, healthy metabolism. An adult human on a low-cholesterol diet forms about 800 mg of cholesterol in the liver each day. However, diets high in saturated fatty acids often lead to abnormally high blood-cholesterol levels.

Find out the effects of cholesterol on health. Produce a poster using the information you find. Make sure you communicate what you have found clearly. Do the effects of cholesterol depend on the individual and, if so, why?

◆ **Endocytosis**: formation of vesicles as the plasma membrane pinches inward, taking material into the cell.

◆ **Exocytosis**: vesicles fuse with the membrane and material is exported out of the cell.

Membrane fluidity and the fusion and formation of vesicles

Another mechanism of transport across the plasma membrane is known as **bulk transport**. It occurs by movements of **vesicles** of matter (solids or liquids) across the membrane by processes known generally as **cytosis**. Uptake into a cell is called **endocytosis** and export out of the cell is **exocytosis** (see Figure B2.1.19).

> **Top tip!**
>
> The word 'vesicle' should be used for the structure formed by the membrane in endocytosis. Similarly, in exocytosis it is vesicles that fuse with the membrane.

> **Top tip!**
>
> To help remember the differences between exocytosis and endocytosis, think of exocytosis as like 'exit' and endo as like 'into'.

9 **State** an example of endocytosis and exocytosis in the human body.

10 **Distinguish** between endocytosis and exocytosis.

> **Concept: Function**
>
> The fluidity of the membrane allows vesicles to readily form and fuse with organelles within the cell. It also allows the import and export of substances via endo-and exocytosis.

The strength and flexibility of the fluid mosaic membrane make this activity possible. Energy from metabolism (ATP) is also required. For example, when solid matter is being taken in (**phagocytosis**), part of the plasma membrane at the point where the vesicle forms is pulled inwards, and the surrounding plasma membrane and cytoplasm bulge out. The solid matter then becomes enclosed in a small vesicle. The bulk transport of fluids is referred to as **pinocytosis** (Figure B2.1.19).

Vesicles are used to transport materials within cells, for example, between the rough endoplasmic reticulum (RER) and the Golgi apparatus, and on to the plasma membrane (Figures A2.2.22 and A2.2.23).

Endocytosis

In the human body, there is a huge number of **phagocytic** cells (phagocytosis means 'cell eating'). One type of phagocytic cell in humans are called **macrophages**. They engulf the debris of damaged or dying cells by endocytosis and dispose of it. For example, we break down about 2×10^{11} red blood cells each day. These are ingested and disposed of by macrophages, every 24 hours.

Exocytosis

White blood cells that produce antibody molecules secrete them by exocytosis (page 524). Another example of exocytosis is the release of neurotransmitter chemicals, such as acetylcholine into the synaptic cleft between neurons (page 332).

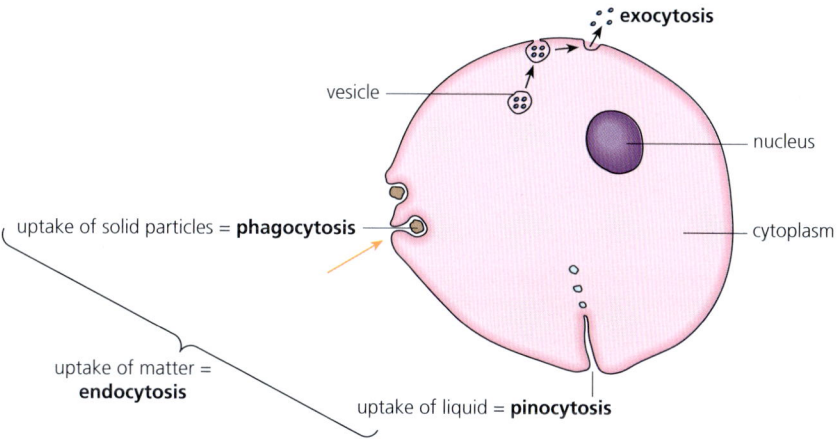

■ **Figure B2.1.19** Transport by endocytosis (taking matter into cells) and exocytosis (matter moving out from cells via vesicles)

> **Common mistake**
>
> Do not confuse endocytosis and exocytosis.
> - Endocytosis: formation of vesicles as the plasma membrane pinches inwards, taking material **into** the cell.
> - Exocytosis: vesicles fuse with the membrane and material is exported **out** of the cell.

Gated ion channels in neurons

Link

Neurons and neural definitions are covered in Chapter C2.2, page 467.

We will be investigating how nerves work later; you can see the structure of a nerve cell and its axon in Figure C2.2.2 (page 469). For the moment, we can note that a nerve impulse is transmitted along the axon of a nerve cell by a momentary reversal in electrical potential difference in the axon membrane, brought about by rapid movements of sodium and potassium ions. The ions pass by facilitated diffusion via pores in the membrane called gated ion channels. These pores are channel proteins that span the membrane (Figure B2.1.14). They have a central channel that can open and close. The channels are exclusively permeable for one type of ion. Because they are activated by reversal in electrical potential, they are called **voltage-gated channels**. Other channels are activated when a neurotransmitter binds to receptors on their surface: these are known as **neurotransmitter-gated ion channels** (see below).

Neurotransmitter-gated ion channels

Synapses are small gaps between neurons. The action potential carried by the neuron cannot cross the synapse, and so the message must be transduced (changed) from an electrical signal into a chemical signal, which can diffuse across the synapse. Neurotransmitters, such as **acetylcholine (ACh)**, are the group of chemicals that cross the synapse in this way.

The nicotinic acetylcholine receptor protein contains binding sites for acetylcholine. These receptor proteins are found in the central and peripheral nervous system, muscle, and many other tissues of many organisms.

Nicotinic receptors are **ionotropic**: when acetylcholine binds to the receptor, ions can flow through the transmembrane protein channel. The receptor acts as a channel for positively charged ions, mainly sodium. These ion channels can be closed, or 'gated', and opened by the neurotransmitter ACh, so they are known as **neurotransmitter-gated ion channels**.

◆ **Neurotransmitter-gated ion channel**: channel protein that temporarily opens when a specific neurotransmitter bonds with it.

ATL B2.1B

Nicotinic receptors also respond to drugs such as nicotine, which is found in cigarettes and vapes. What effect does nicotine have on the body, and why do people become addicted to nicotine?

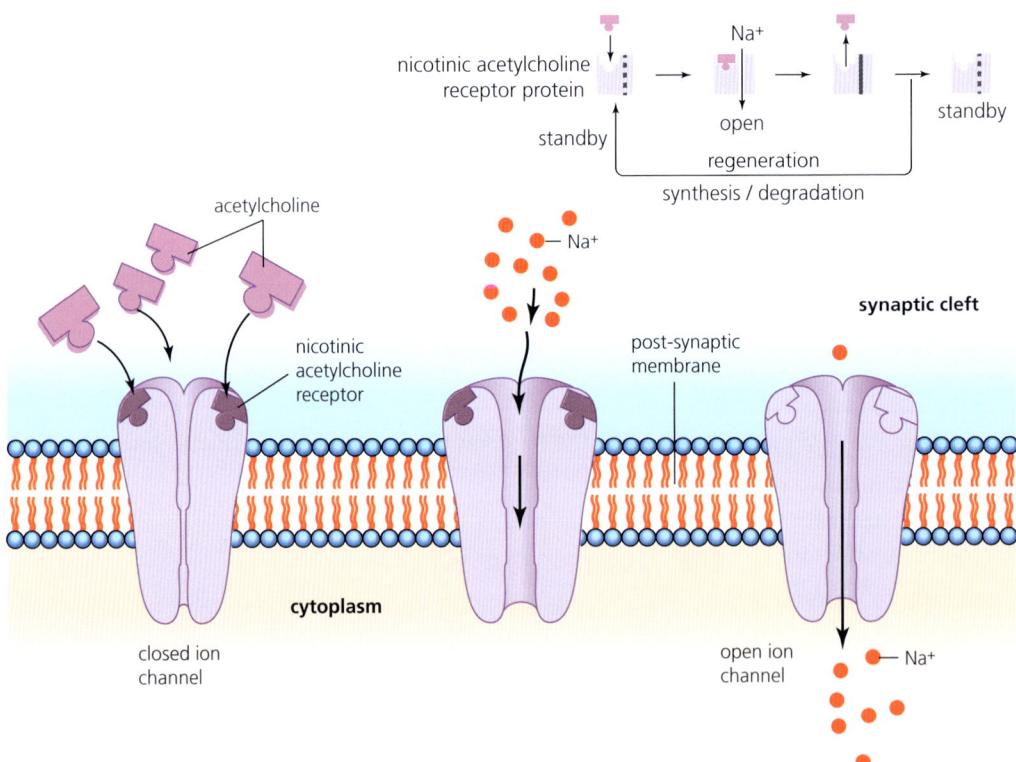

■ Figure B2.1.20 Nicotinic acetylcholine receptors are an example of a neurotransmitter-gated ion channel

◆ **Voltage-gated channels**: a type of transmembrane protein that forms ion channels that are activated by changes in the electrical membrane potential of a neuron.

Voltage-gated channels

Potassium and sodium channels in cell membranes are **voltage-gated channels**. This means that they open or close depending on a certain threshold membrane potential being reached.

Theme B: Form and function – Cells

When a stimulus is received by a receptor, the sodium channels open and let sodium ions into the inside of the axon. This depolarizes the interior of the axon, so the membrane potential goes from −70 mV to +40 mV. This initiates the nerve impulse, or 'action potential'.

Almost immediately after an impulse has passed, there are relatively more positive charges inside the axon and the potassium channels open. The sodium channels close. Now potassium ions can exit the axon down an electrochemical gradient into the tissue fluid outside the nerve cell.

Then, as the interior of the axon starts to become less positive again, the potassium channel is closed, first by action of a 'ball and chain' device (Figure B2.1.21). The 'chain' is believed to be a flexible strand of amino acid residues and the 'ball' is globular protein. Finally, when the axon has more positive charge outside than inside, the potassium channel itself returns to the fully closed condition.

Voltage-gated K⁺ channels have positively charged voltage-sensing paddles which are normally attracted to the negatively charged interior surface of the resting axon. In this position the channel is closed mechanically – no K⁺ ions pass.

Once the membrane has depolarized, the paddles are attracted to the outside of the axon membrane and repelled by the positively charged interior. In this position the selective channel gates are opened, and potassium ions diffuse down an electro-chemical gradient.

A 'ball and chain' attached to the interior of the channel protein fits inside the open channel (the flexible chain allows this) and stops diffusion of K⁺ ions while the exterior of the axon is still negatively charged. The ball remains in place until the interior of the axon becomes negatively charged again and the gate itself is closed.

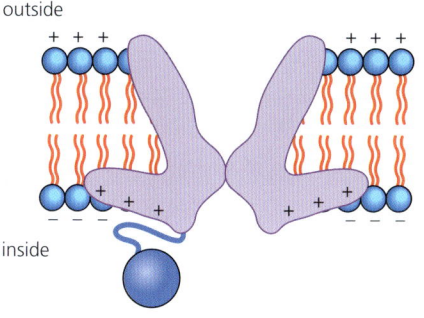

net negative charge inside the axon and a net positive charge outside the axon

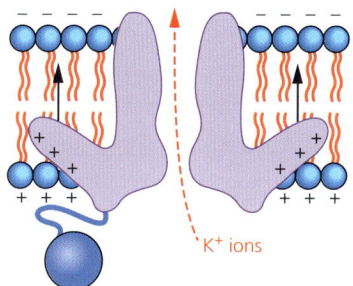

net negative charge outside the axon and a net positive charge inside the axon

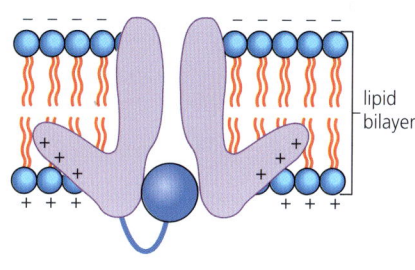

■ Figure B2.1.21 Voltage-gated K⁺ channel – facilitated diffusion in an axon

Exchange transporters

Sodium–potassium pumps are globular proteins that span the axon membrane of nerve cells. They are an example of **exchange transporters**. When preparing the axon for the passage of the next nerve impulse, there is **active transport** of potassium (K⁺) ions in across the axon membrane and sodium (Na⁺) ions out across the membrane. This activity of the Na⁺/K⁺ pump involves a transfer of energy from ATP. The outcome is that potassium and sodium ions gradually concentrate on opposite sides of the membrane. These pumps therefore play an important role in generating membrane potentials.

The steps to the cyclic action of these pumps are:
- with the interior surface of the pump open to the interior of the axon, three sodium ions are loaded by attaching to specific binding sites
- reaction of the globular protein with ATP now occurs, resulting in the attachment of a phosphate group to the pump protein; this triggers the pump protein to close to the interior of the axon and open to the exterior
- the three sodium ions are now released and, simultaneously, two potassium ions are loaded by attaching to specific binding sites
- with the potassium ions loaded, the phosphate group detaches; this triggers a reversal of the shape of the pump protein – it now opens to the interior, again, and the potassium ions are released
- the cycle is then repeated.

11 **Define** the terms *voltage-gated channel* and *neurotransmitter-gated ion channel*.

12 **Compare and contrast** voltage-gated channels and neurotransmitter-gated ion channels.

HL ONLY

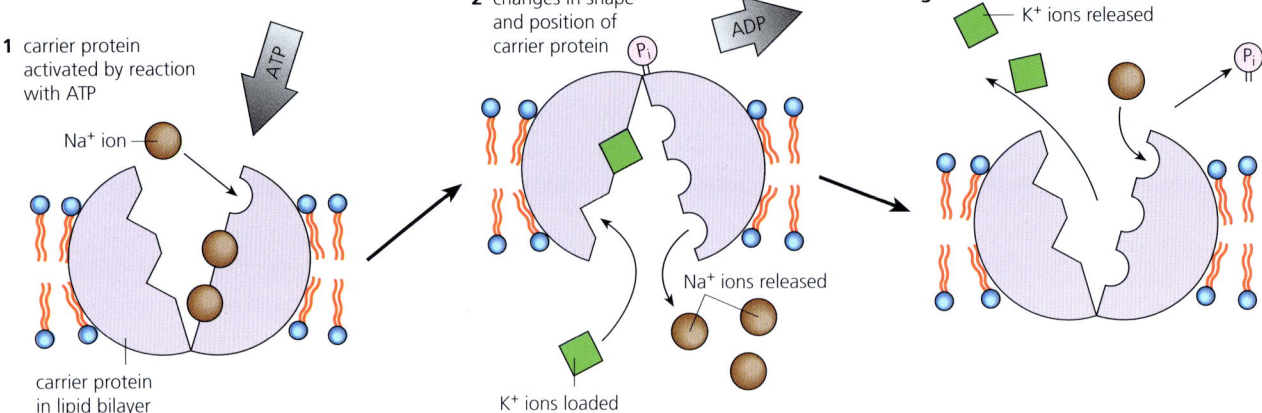

■ **Figure B2.1.22** The sodium–potassium ion pump

Indirect active transport

Sodium-dependent glucose cotransporters are a family of glucose transporter found in the small intestine and the proximal tubule of nephrons in the kidney. They play an important role in glucose absorption.

These cotransporters move glucose in one direction (into the cell) across the membrane and move sodium in the same direction (Figure B2.1.23). Two sodium ions pass through the cotransporter protein for each glucose molecule. Active transport of sodium at the base of the cell, using sodium–potassium pumps (see Figure B2.1.22) builds up a concentration gradient of sodium by removing sodium from the cell, so sodium can enter through the cotransporter protein down its concentration gradient. At the same time, glucose is drawn into the cell through the same cotransporter as the sodium (Figure B2.1.23). In this way, sodium-dependent glucose cotransporters allow glucose to be transported into cells against its concentration gradient. Because glucose is not directly moved into the cell by active transport, but in conjunction with the active transport of sodium ions out from the opposite side of the cell, this process is known as **indirect active transport**. Glucose leaves the cell via facilitated diffusion at the base of the cell, via a glucose transporter (Figure B2.1.23).

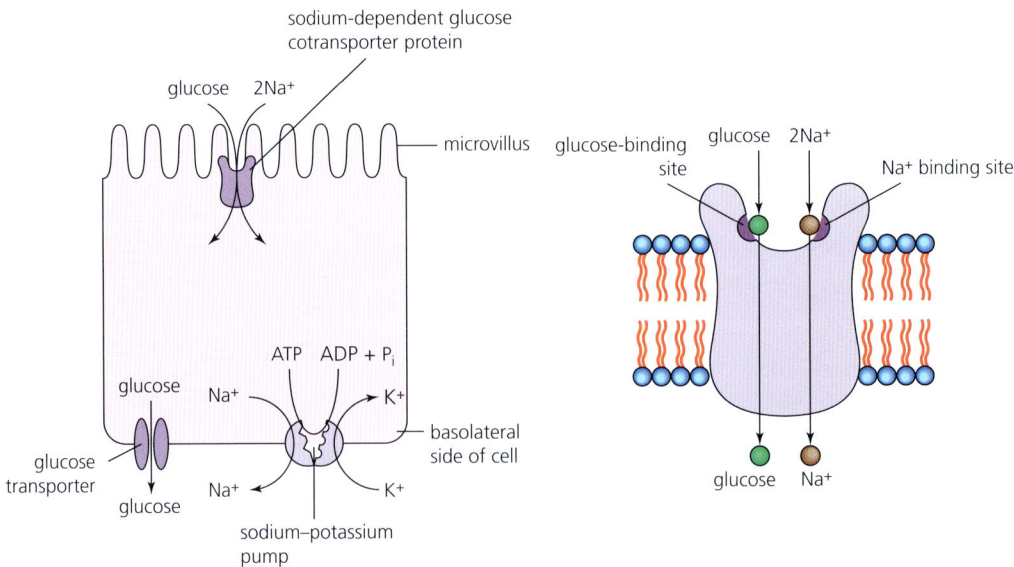

■ **Figure B2.1.23** Epithelial cell with sodium-dependent glucose transporter

Theme B: Form and function – Cells

Adhesion of cells to form tissues

♦ **Cell-adhesion molecule (CAM)**: proteins located on the cell surface that are involved in binding cells with other cells or the extracellular matrix.

♦ **Extracellular matrix (ECM)**: a large network of proteins and other molecules that surround, support and give structure to cells and tissues in the body.

The cells in animal tissues are 'glued' together by **cell-adhesion molecules (CAMs)** – proteins embedded in the cells' surface membranes. Cell adhesion is essential for maintaining tissue structure and function. In multicellular organisms, the binding between CAMs causes cells to adhere (stick) to each other and form a structure called a **cell junction**. Some CAMs bind cells to one another or to other CAMs; other types bind cells to the **extracellular matrix (ECM)** to form a cohesive unit (Figure B2.1.24). The ECM is the non-cellular component present within all tissues and organs that provides essential physical scaffolding for the cellular constituents; for example, collagen fibres and bone mineral comprise the ECM of bone tissue. Different forms of CAMs are used for different types of cell–cell junction.

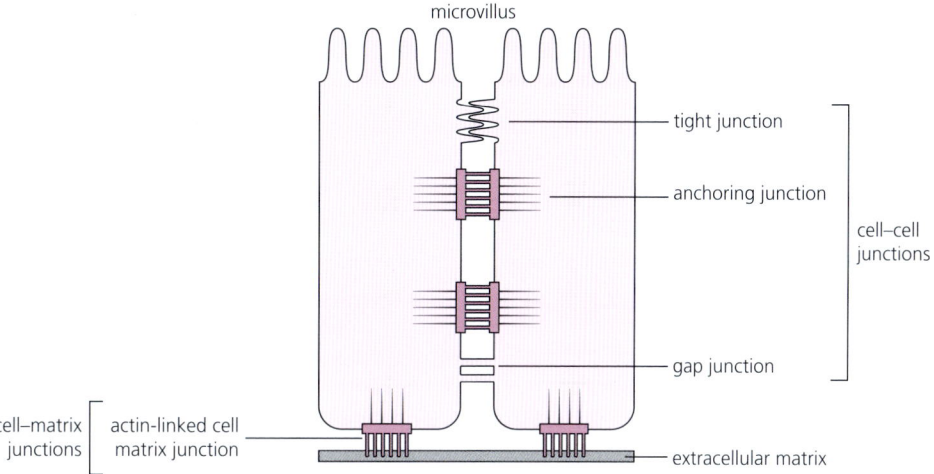

■ **Figure B2.1.24** Different types of CAMs and cell junctions present in epithelial cells, including cell–cell junctions and cell–matrix junctions

● **Top tip!**
You are not required to have detailed knowledge of the different types of CAMs or junctions.

● **Top tip!**
Both cadherins and integrins are transmembrane proteins.

Cell junctions can be divided into two overall types:
- cell–cell junctions (usually involving cadherin proteins)
- cell–extracellular matrix junctions (usually involving integrin proteins).

Cell junctions can be classified according to their functions:
- **anchoring** junctions: strengthen contact between cells
- **tight** junctions: seal gaps between cells through cell–cell contact
- **gap** junctions: link the cytoplasm of adjacent cells, allowing transport of molecules between them
- **signal-relaying** junctions: e.g. synapses.

Tight junctions act as an impermeable or semipermeable barrier between adjacent epithelial cells. They are barriers to the transportation of material and control the movement of membrane transport proteins between the apical and basal layers of epithelia.

The cells of higher plants, such as the flowering plants, contain relatively few CAMs. Instead, plant cells are held together rigidly by extensive interlocking of the cell walls of adjacent cells.

The cytoplasm of adjacent animal or plant cells often are connected by functionally similar but structurally different 'bridges' called **gap junctions** in animals (Figure B2.1.24) and **plasmodesmata** in plants (see Chapter B3.2, page 306). These structures allow cells to exchange small molecules including nutrients and chemical signals, coordinating the function of the cells in a tissue.

LINKING QUESTIONS

1. What processes depend on active transport in biological systems?
2. What are the roles of cell membranes in the interaction of a cell with its environment?

B2.1 Membranes and membrane transport

B2.2 Organelles and compartmentalization

Guiding questions

- How are organelles in cells adapted to their functions?
- What are the advantages of compartmentalization in cells?

SYLLABUS CONTENT

This chapter covers the following syllabus content:
- ▶ B2.2.1   Organelles as discrete subunits of cells that are adapted to perform specific functions
- ▶ B2.2.2   Advantage of the separation of the nucleus and cytoplasm into separate compartments
- ▶ B2.2.3   Advantages of compartmentalization in the cytoplasm of cells
- ▶ B2.2.4   Adaptations of the mitochondrion for production of ATP by aerobic cell respiration (HL only)
- ▶ B2.2.5   Adaptations of the chloroplast for photosynthesis (HL only)
- ▶ B2.2.6   Functional benefits of the double membrane of the nucleus (HL only)
- ▶ B2.2.7   Structure and function of free ribosomes and of the rough endoplasmic reticulum (HL only)
- ▶ B2.2.8   Structure and function of the Golgi apparatus (HL only)
- ▶ B2.2.9   Structure and function of vesicles in cells (HL only)

Organelles

Link
Organelles found in eukaryotic cells were introduced in Chapter A2.2, page 64.

Organelles are discrete (separate) subunits in cells, and include the nuclei, chloroplasts, mitochondria, vesicles, ribosomes and the plasma membrane. The only organelles common to both eukaryotic and prokaryotic cells are the cell membrane and ribosomes. The cell wall, cytoskeleton and cytoplasm are not considered organelles.

 Top tip!

Most organelles are surrounded by membrane and are known as 'membrane-bound' organelles, such as the rough endoplasmic reticulum, smooth endoplasmic reticulum, Golgi apparatus and the double membrane that surrounds the nucleus. Eukaryotic cells contain membrane-bound organelles, but prokaryotic cells do not.

Nature of science: Experiments

New experimental techniques often bring breakthroughs in knowledge in science. For example, studying the function of individual organelles in cells became possible after the **ultracentrifuge** was invented and methods of using it for **cell fractionation** were developed.

Centrifugation is used for two basic purposes in cell biology: as a preparative technique to separate one type of material or substance from others, and as an analytical technique to measure physical properties (e.g., molar mass, density, shape and binding ability) of macromolecules.

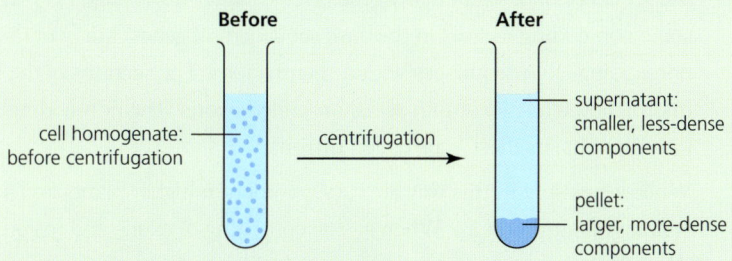

■ **Figure B2.2.1** Principle of centrifugation applied to a cell homogenate (cells with broken plasma membranes but intact organelles)

Centrifuges spin fluids at very high speeds, separating particles depending on their density. More-dense particles settle at the bottom of the tube, while less-dense particles rise to the top. A centrifuge increases the rate of sedimentation by subjecting particles in suspension to centrifugal forces as great as 1 000 000 times the force of gravity, g. This can sediment particles with molar masses as small as $10 000 \text{ g mol}^{-1}$. Modern ultracentrifuges achieve these centrifugal forces by reaching speeds of 150 000 revolutions per minute (rpm) or higher.

Theodor Svedberg, a Swedish chemist and Nobel laureate who designed the first ultracentrifuge in 1925, studied haemoglobin and found that the centrifuged sample revealed a single, sharp band with a molar mass of weight of $68 000 \text{ g mol}^{-1}$. His results strongly supported the theory that proteins were true macromolecules and not mixtures.

Cell fractionation and separation of animal cell homogenates (a suspension of cell fragments) into the various organelles was first achieved during the 1940s. At low speeds, nuclei fall to the bottom of the container. At a slightly higher speed, mitochondria can be collected and, at even higher speeds and longer times, ribosomes can be collected.

These techniques have allowed biologists to study biological processes free from all the complex side reactions that occur in a living cell, by using purified cell-free systems. For example, the mechanism of protein synthesis (translation) was discovered in experiments that used a cell homogenate that could translate mRNA molecules to produce proteins. Ultracentrifugation was used in the Meselson and Stahl experiment (page 603).

Common mistake

Centrifugation does not separate organelles based on their size, but on their relative **density**.

TOK

How important are material tools in the production or acquisition of knowledge?

Tool 3: Mathematics

Applying and using non-SI metric units

As well as his contribution of designing the first ultracentrifuge, Theodor Svedberg also gave his name to the non-SI unit, the Svedberg unit (represented as S or sometimes Sv). This non-SI unit is used to measure the time it takes a particle of a certain size to settle at the bottom of a solution. It is measured at exactly 10^{-13} seconds. The particle's mass, density and shape all affect its S value. This is how ribosomes are differentiated – based on their sedimentation rate. The 70S ribosome in chloroplasts, mitochondria and prokaryotes has a sedimentation rate of 70×10^{-13} seconds.

B2.2 Organelles and compartmentalization

Advantages of separating the nucleus and cytoplasm

The nuclear membrane keeps DNA separate from the other parts of the cell, which protects it from the many cellular reactions that occur in the cytoplasm. Additionally, separating the nucleus and cytoplasm allows gene transcription and translation to be kept separate.

The Human Genome Project (and similar studies of the genomes of other organisms) established the sequence of bases in many genes (see Chapter A3.1, page 117). This led to the discovery of some non-coding regions in the base sequences of genes. Many of the genes of eukaryotes have non-coding DNA sequences within their regions. The sections of the gene that carry meaningful information (code for amino acids) are called **exons**. The non-coding sequences that intervene – interruptions, in effect – are called **introns**.

While genes split in this way are very common in higher plants and animals, some eukaryotic genes contain no introns at all. When a gene consisting of exons and introns is transcribed into mRNA, the mRNA formed contains the sequence of introns and exons exactly as they occur in the DNA. If this unmodified mRNA was to be 'read' and transcribed in a ribosome, it would undoubtedly present problems in the protein-synthesis step (translation). In fact, an enzyme-catalysed reaction, known as **post-transcriptional modification**, removes the introns as soon as the mRNA has been formed. The production of this enzyme is also under the control of a gene. As a result, the short lengths of 'nonsense' transcribed into the RNA sequence of bases are removed. This is known as **RNA splicing**, and the resulting shortened lengths of mRNA are described as 'mature'. It is this form of mRNA that passes out of the nucleus into the cytoplasm, to go to the ribosomes where it is involved in protein synthesis (page 253). Post-transcriptional modification of mRNA can happen in the nucleus before the mRNA meets ribosomes in the cytoplasm. This is an advantage for eukaryotic cells as variants of a protein can be produced from a single gene, due to gene splicing.

In prokaryotes, such modification of mRNA is not possible as there is no nuclear membrane separating DNA from the cytoplasm, so mRNA may immediately meet ribosomes. The genes of prokaryotes therefore do not have introns, and the prokaryotic cell does not have the enzyme machinery to carry out splicing, either.

Like other cell membranes, the nuclear membrane is a phospholipid bilayer, which is permeable only to small, non-polar molecules. Other molecules are unable to diffuse through the phospholipid bilayer. Lamins (which are intermediate protein filaments and are a major component of the cytoskeleton – see page 72) are important in nuclear membrane function (see page 252). The **nuclear pores** in the nuclear envelope control the exit of mRNA from the nucleus, and the entry and exit of proteins (Figure B2.2.2). The pores are made of specific proteins forming a specific structure. The pores are tiny, about 100 nm in diameter, but so numerous that they make up about one-third of the nuclear membrane's surface area. This suggests that communication between the nucleus and cytoplasm is important. The selective movement of proteins and RNA through the nuclear pores not only establishes the internal composition of the nucleus, but also plays an essential role in regulating eukaryotic gene expression.

> **Link**
> Introns and exons are covered in more detail in Chapter D1.2, page 626.

> ◆ **Nuclear pore**: a protein-lined channel in the nuclear envelope; regulates the transport of molecules between the nucleus and the cytoplasm.

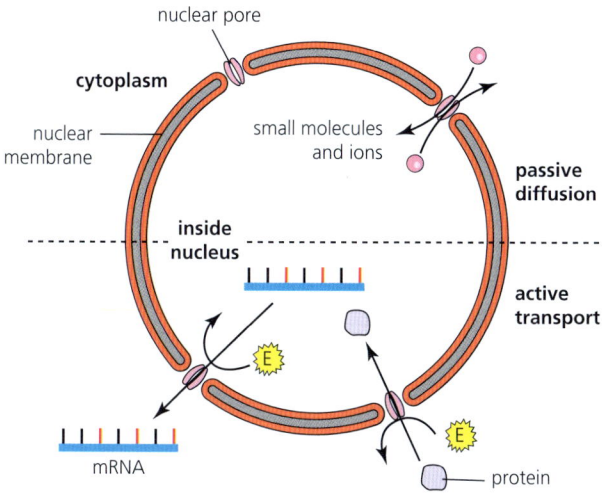

■ **Figure B2.2.2** Transport across the nuclear membrane

Nuclear pores transport proteins in their folded conformation and ribosomal components as assembled particles. This distinguishes the nuclear transport mechanism from the mechanisms that transport proteins into most other organelles. Proteins must unfold to cross the membranes of mitochondria and chloroplasts.

RNA molecules that are synthesized in the nucleus must be exported to the cytoplasm efficiently, where they function in protein synthesis.

Proteins are synthesized in the cytoplasm, so the transport of mRNA, rRNA and tRNA from the nucleus to the cytoplasm is a critical step in gene expression in eukaryotic cells. Export of RNA through nuclear pore complexes is an active process that requires energy (Figure B2.2.2). Ribosomal RNA (rRNA) is assembled with ribosomal proteins in the **nucleolus**, and ribosomal subunits are then transported into the cytoplasm.

1 Explain how the structure of the nuclear envelope facilitates translation.

Link
Organelles were discussed in detail in eukaryotic cell structure in Chapter A2.2, page 68.

Advantages of compartmentalization in the cytoplasm of cells

Each organelle has a different function, carrying out a specific biological process and a different set of chemical reactions. At the most basic level, biochemical reactions within the organelles are what sustain life. The organelles separate the various biochemical reactions that are needed to support the cell's function. Each fulfils a different role and contains different chemical reactions, but all cooperate and work together.

Compartmentalization (the division of the interior cell into separate, discrete areas) allows the correct concentration of **metabolites** to be present for specific metabolic processes and the appropriate enzymes to be present. In this way, incompatible biochemical processes can be separated.

● Top tip!
Separating the reactions of life by compartmentalization means that they can be controlled and do not interfere with each other. Think about the way that different operations in a factory are divided into different areas; for example, in a car factory one area will be used to assemble the engine, another to put together the car framework, and so on.

■ Lysosomes

Lysosomes are tiny spherical vesicles bound by a single membrane (Chapter A2.2, page 71). These organelles contain digestive enzymes in an acidic environment. The hydrolytic enzymes are active only in the lysosome's acidic interior. The pH of the cell is neutral to slightly alkaline, so the enzymes' acid-dependent activity protects the cell from self-digestion in case of lysosomal leakage or rupture. This is an example of how the compartmentalization provided by lysosomes allows functions to occur that would not be possible or controllable in the wider interior of the cell.

■ Phagocytic vacuoles

◆ **Phagocytosis**: the process by which a phagocyte engulfs and destroys foreign substances, such as bacteria.

Phagocytosis is a process used by specific cells to capture and ingest foreign particles (see page 522). Phagocytic white blood cells engulf 'foreign' material, such as bacteria, then digest and destroy it.

B2.2 Organelles and compartmentalization

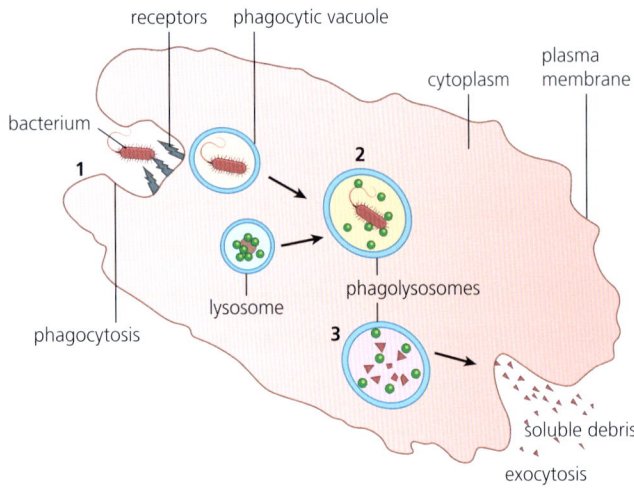

■ Figure B2.2.3 The process of phagocytosis

Phagocytosis involves a series of steps (Figure B2.2.3):

1. recognition of the foreign particle by receptors, ingestion of the foreign particle in a **phagocytic vacuole**
2. fusion of the phagocytic vacuole with a lysosome, creating a **phagolysosome**
3. final destruction of the ingested particle.

The phagocytic vacuole is a temporary organelle formed by the process of phagocytosis, from the outer membrane of the phagocyte. Fusion with the lysosome provides the digestive enzymes needed to break down and destroy the foreign particle. The phagocytic vacuole, followed by the phagolysosome, provides a contained space within the cell where antimicrobial activity can take place at low pH without disrupting other metabolic processes.

◆ **Phagocytic vacuole**: a vesicle formed around a particle engulfed by a phagocyte via phagocytosis.

Link
For full coverage on cell respiration and all definitions, see Chapter C1.2, page 405.

Link
Pump proteins are discussed in Chapter B2.1, page 230.

Adaptations of the mitochondrion

Respiration is one of the key life processes and it occurs in all cells of all organisms. It is the breakdown of glucose to release the energy contained in the bonds that hold the molecule together, converting this energy to ATP (see page 405) and releasing carbon dioxide and water in the process.

Respiration begins in the cytoplasm, in a process called glycolysis. This process is anaerobic, and results in the production of **pyruvate**. This molecule diffuses into the mitochondria, where aerobic respiration takes place.

The mitochondria are formed from two membranes, an inner and outer membrane (Figure B2.2.4). The **outer membrane** contains transport proteins, which move pyruvate into the mitochondrion. In the mitochondrion, carbon is removed from pyruvate (in the form of carbon dioxide – this process is known as **decarboxylation**) and hydrogen is also removed (an **oxidation** reaction). Hydrogen is split into protons (H^+) and electrons.

The electrons pass down a sequence of proteins in the **inner membrane** of the mitochondrion (known as the **electron transport chain**, or **ETC**), releasing energy that is used to pump the protons into the space between the two mitochondrial membranes (called the **intermembrane space**). Protons build up in this small space, accumulating quickly, which creates an electrochemical gradient. As well as the ETC, the inner membrane contains a transmembrane multi-protein assembly (made from several protein components to form a complex structure), **ATP synthase**, through which the accumulated protons pass, generating ATP. The inner membrane is highly folded into structures called **cristae**, which increases the surface area for the ETC and ATP synthase, increasing the production of ATP.

Concept: Function

The structure of the mitochondrion allows it to optimize its function – ultimately to produce large quantities of direct energy for the cell in the form of ATP.

Concept: Form

Mitochondria have adaptations that have evolved from the prokaryotic ancestors that first formed this organelle. The highly folded inner membrane increases the surface area for the production of ATP, while compartmentalization within the structure enables it to separate areas of different proton concentration and pH to create ideal conditions for the processes of aerobic respiration.

Link
The Krebs cycle is fully explored in Chapter C1.2, page 412.

The process where hydrogen and carbon atoms are removed from the derived molecules of glucose in a sequence of enzyme-controlled stages is called the **Krebs cycle**. These processes occur in the matrix of the mitochondria, which has the appropriate enzymes and pH for the Krebs cycle to occur. In this way, mitochondria achieve compartmentalization of enzymes and substrates of the Krebs cycle in the matrix.

Electron micrograph of a mitochondrion
Mitochondria are the site of the aerobic stage of respiration.

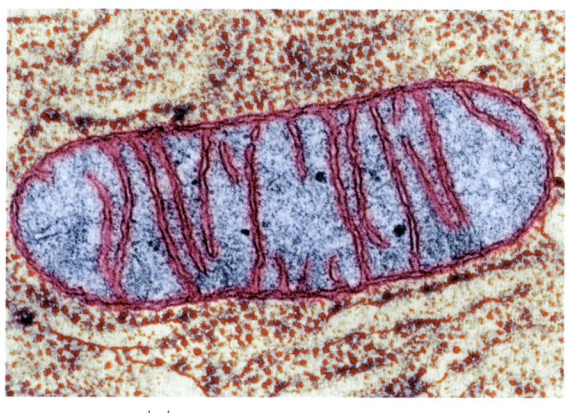

scale bar
1 μm

interpretive drawing

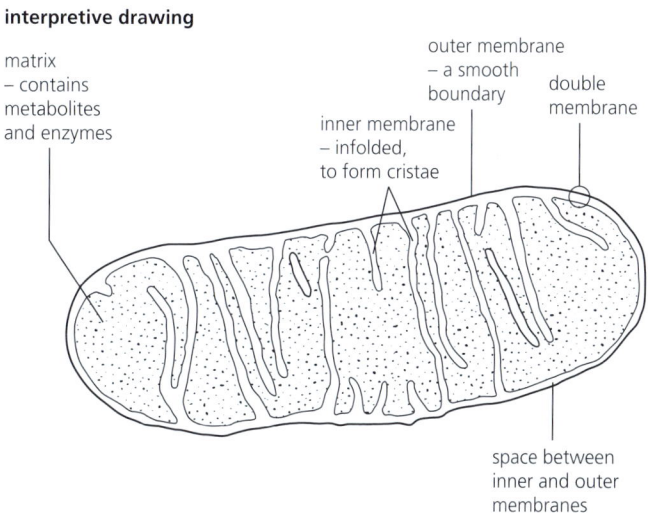

matrix – contains metabolites and enzymes

outer membrane – a smooth boundary

double membrane

inner membrane – infolded, to form cristae

space between inner and outer membranes

■ **Figure B2.2.4** Electron micrograph of a mitochondrion, with an interpretive drawing

2 Using the scale bar, **calculate** the length and width of the mitochondrion in Figure B2.2.4.

3 **Explain** how the structure of a mitochondrion relates to its function.

Adaptations of the chloroplast

Just as the mitochondria are the site of reactions of the Krebs cycle and respiratory ATP formation, so the chloroplasts are the organelles where the reactions of photosynthesis occur. This section contains an outline of photosynthesis and relates the processes to adaptations of chloroplasts.

Chloroplasts are found in green plants. They contain the photosynthetic pigments, along with the enzymes and electron-transport proteins for the reduction of carbon dioxide to glucose and for ATP formation, using light energy. An electron micrograph of a thin section of a chloroplast shows the arrangement of the membranes within this large organelle (Figures A2.2.29, page 76, and B2.2.5 on the next page).

There is a double membrane around the chloroplast, and a third inner membrane that folds extensively at various points to form a system of branching membranes. Here, these membranes are called **thylakoids**. Thylakoid membranes are organized into flat, compact, circular piles called **grana** (singular, granum), almost like stacks of coins. Between the grana are loosely arranged tubular membranes suspended in a watery **stroma**. Separate grana are connected by lamella (Figure B2.2.5).

Link
The process of photosynthesis and all definitions are in covered in detail in Chapter C1.3, page 425.

● Common mistake

It is inaccurate to simply say that the mitochondria carry out 'respiration', because anaerobic respiration occurs in the cytoplasm. It **is** correct to say that mitochondria carry out **aerobic** respiration.

B2.2 Organelles and compartmentalization

■ **Figure B2.2.5** The ultrastructure of a chloroplast

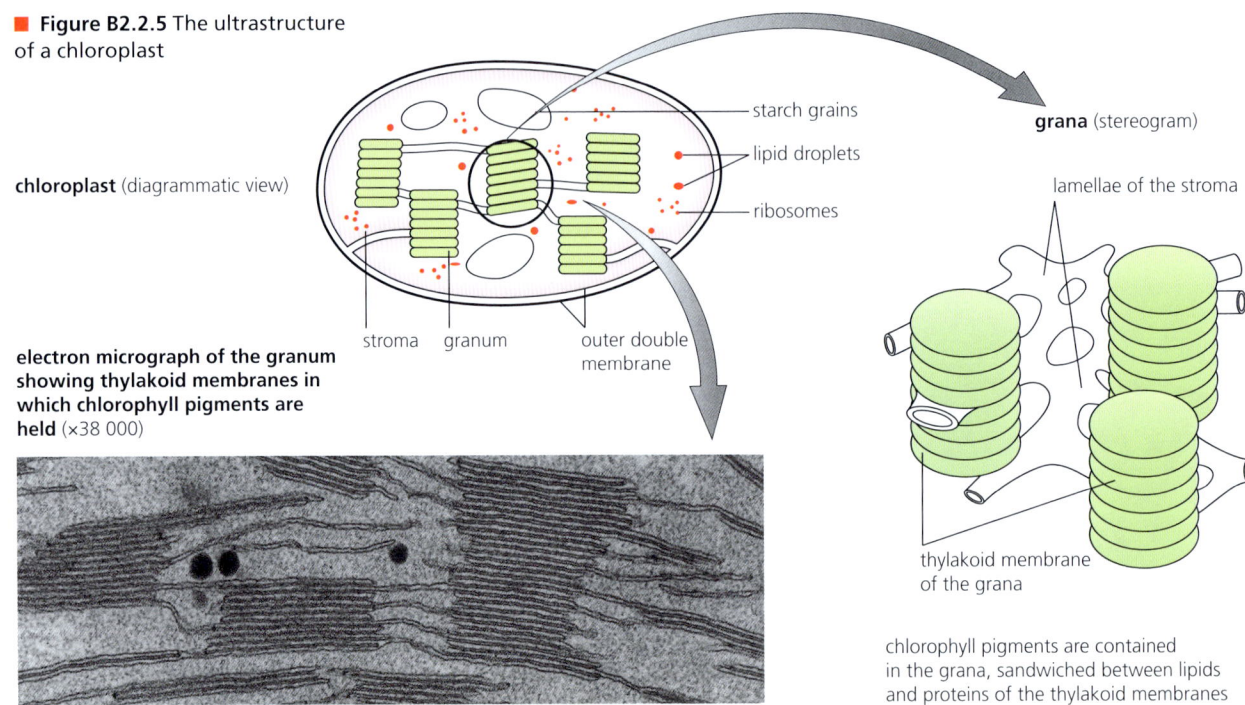

> **Top tip!**
>
> As with aerobic respiration, photosynthesis depends on the build-up of protons, which then move down their electrochemical gradient to generate ATP.

> **Concept: Form**
>
> The structure, composition and arrangement of membranes are central to the biochemistry of photosynthesis, just as the mitochondrial membranes are the sites of many of the reactions of aerobic cell respiration.

Chlorophyll, the photosynthetic pigment that absorbs light energy, exists in the grana. Chlorophyll molecules are grouped together in structures called **photosystems**, held in the thylakoid membranes of the grana by proteins (Figure B2.2.5). The large surface area provided by the thylakoid membranes optimizes the ability of chloroplasts to harvest light energy. The light energy excites electrons, which pass down an electron transport chain, which causes protons (H$^+$) to be pumped into the thylakoids, causing a build-up of protons there. The small volumes of fluid contained inside thylakoids means that protons quickly accumulate there. Protons flow down their electrochemical gradient through ATP synthase proteins, from the inside of the thylakoids into the stroma. ATP is generated in the process.

This ATP is used in a biochemical process that occurs in the stroma, known as the **Calvin cycle** (which involves reduction reactions), which ultimately leads to the production of glucose. The stroma has the appropriate enzymes and the optimum pH for the Calvin cycle to take place. Compartmentalization within chloroplasts allows the separation of the light-harvesting activity of the chloroplasts in the thylakoids (known as the light-dependent reactions) from the enzymes and substrates of the Calvin cycle (known as light-independent reactions) in the stroma.

> **Concept: Function**
>
> The roles of the chloroplast, both in terms of light harvesting and the creation of molecules such as ATP required by the Calvin cycle, are enhanced by adaptations in its structure that enable it to optimize its function.

4 **Construct** a table to show the relationship between the structure and function of a chloroplast.

> **ATL B2.2A**
>
> Create a poster that compares and contrasts in summary form the structures and functions of mitochondria and chloroplasts. Draw diagrams of both organelles on an A3 sheet – one organelle on the left and one on the right. Put similarities in the middle between the two diagrams, and differences on the left- and right-hand sides of the sheet.

> **Inquiry 1: Exploring and designing**
>
> **Designing**
>
> Design a technique for extracting chloroplasts from cells, so that they can be examined under a light microscope. If you do not have access to a centrifuge, think about alternative ways of obtaining concentrated chloroplasts samples from cells.
>
> *Hint: a food blender can be used to break open cells, and muslin cloth used as a filter. How would you ensure that the chloroplasts are not damaged by movement of water in or out of them by osmosis?*

Functional benefits of the nucleus double membrane

The nucleus is the largest organelle in the eukaryotic cell, typically 10–20 μm in diameter. It is surrounded by two membranes, known as the **nuclear envelope**. The outer membrane is continuous with the endoplasmic reticulum (ER), which makes it easy for the nucleus to obtain proteins made in the ER, so endocytosis (which requires energy, see page 239) is not needed.

The nuclear envelope contains many pores (see page 68, and Figure B2.2.6 below). The pores are very small, about 100 nm in diameter. However, they are so numerous that they make up about one-third of the nuclear membrane's surface area. As discussed on page 246, the function of the pores is to make possible rapid movement of molecules between the nucleus and cytoplasm (such as mRNA) and between the cytoplasm and the nucleus (such as proteins, ATP and some hormones).

The double membrane of the nucleus is comprised of one membrane that is folded to form a double structure: the membrane is continuous across the inner and outer surfaces of the nuclear envelope (Figure B2.2.6). This allows the nuclear envelope to connect with the ER.

● **Common mistake**

Students sometimes draw the nuclear membrane as a single line – this is incorrect. The double membrane needs to be clearly drawn, with the nuclear pores within it.

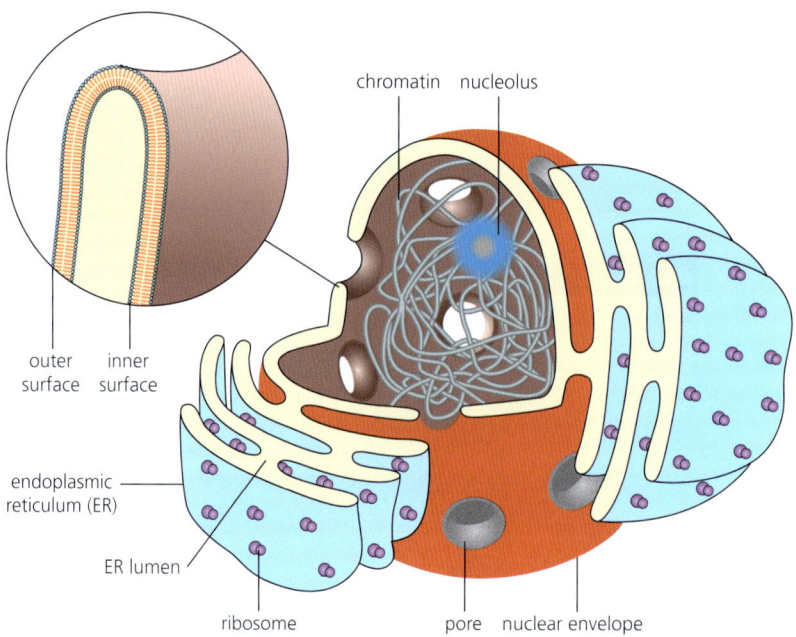

■ **Figure B2.2.6** Diagram showing the nuclear membrane and its connection to the ER. The proteins that constitute the pore complexes are not included and the size of the pores are exaggerated for clarity

The inner membrane contains the DNA and nucleolus. The inner nuclear membrane is lined by the **nuclear lamina**, which is a meshwork of filaments that extend into the interior of the nucleus and provide structural support. The lamina is present in animal cells but not plant or fungal cells, and is comprised of lamins, a type of fibrous protein. The double membrane therefore allows compartmentalization of the functions of the nucleus.

Cells divide by the processes of **mitosis** and **meiosis**. Mitosis results in the production of two daughter cells, genetically identical to the parent cell, while meiosis results in the production of sex cells (gametes) that are genetically different from the original cell. In the first part of both types of cell division, the nuclear envelope dissociates into small vesicles (Figure B2.2.7). This enables the chromosomes to encounter the cytoplasm in the cell and the spindle apparatus, which is central to manipulating the chromosomes during mitosis and meiosis.

Link
Mitosis and meiosis are covered in detail in Chapter D2.1, page 650.

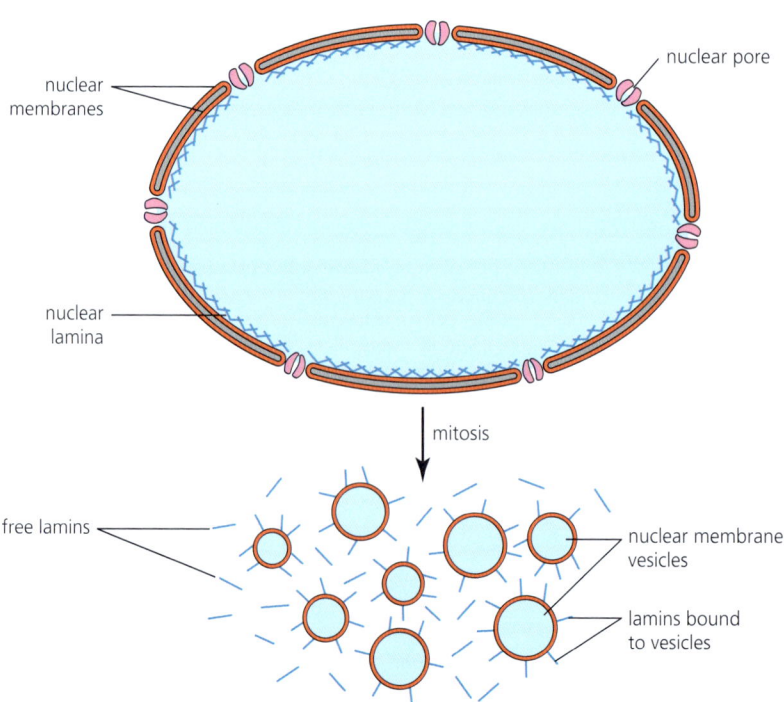

■ **Figure B2.2.7** Breakdown of the nuclear membrane during mitosis in an animal cell

Going further

Origin of the nuclear membrane

The origin of the eukaryotic cell was discussed in Chapter A2.2 (page 82). One hypothesis is that the double membrane of the nucleus is the result of endosymbiosis (page 83). The two membranes reflect those of the organism that was engulfed and the engulfing cell.

An alternative hypothesis is that nuclear membranes and the endoplasmic reticulum may have evolved through infolding of the plasma membrane. In a very old ancestral prokaryotic cell, the plasma membrane could have infolded and later formed a two-layered envelope of membrane completely surrounding the DNA (Figure B2.2.8).

This envelope is assumed to have eventually pinched off completely from the plasma membrane.

A third hypothesis is that the nuclear membrane, ER and other membrane-bound organelles (the 'endomembrane system') in eukaryotic cells formed from prokaryotic vesicles. A wide range of bacteria secrete outer membrane vesicles (OMVs). The vesicles play a critical role in the interactions between bacteria and between bacteria and their hosts. The hypothesis suggests that ancestral mitochondria (prokaryotic cells) released OMVs that, during evolution, formed the ER, which in turn formed the nuclear envelope.

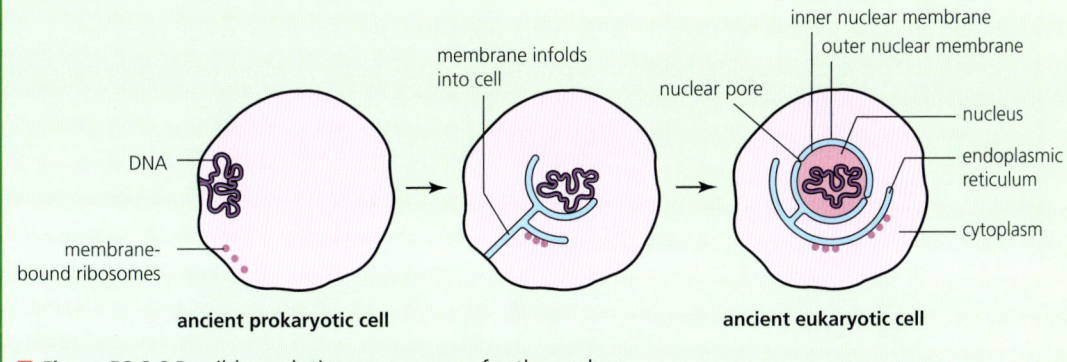

■ Figure B2.2.8 Possible evolutionary sequence for the nucleus

Structure and function of free ribosomes and of the RER

◆ **Polysome**: groups of ribosomes that are translating the same mRNA, indicating the cell needs multiple copies of a particular polypeptide.

Many ribosomes occur freely in the cytoplasm. They are the sites of the synthesis of proteins that are to be retained in the cell and fulfil roles there. It is common for several ribosomes to move along the messenger RNA at one time. The structure (mRNA, ribosomes and their growing protein chains) is called a **polysome** (Figure B2.2.9). Polysomes allow many copies of a particular polypeptide to be produced simultaneously.

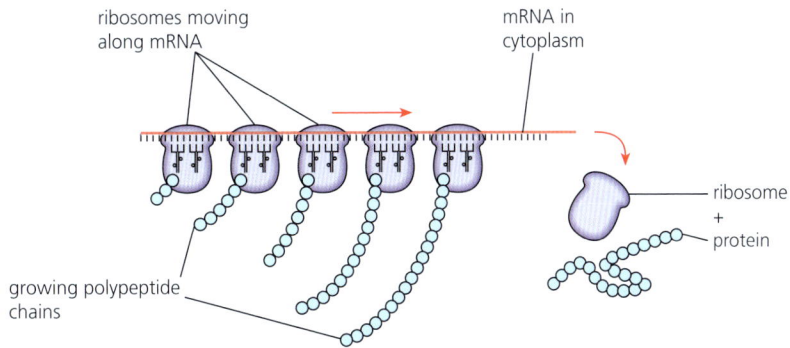

■ Figure B2.2.9 A polysome

B2.2 Organelles and compartmentalization

> **Top tip!**
>
> Free ribosomes produce proteins for use within the cell, while those attached to the RER produce proteins for export or movement to the Golgi for modification.

Other ribosomes are bound to the membranes of the endoplasmic reticulum (the rough endoplasmic reticulum, RER – see page 70). These ribosomes are the site of the synthesis of proteins that go into the lumen of the RER for vesicular transport within the cell and subsequent secretion from cells or for packaging in lysosomes.

> **ATL B2.2B**
>
> Watch an animation of polysomes in action using this site:
> www.sumanasinc.com/webcontent/animations/content/polyribosomes2.html
> What are the advantages of having several ribosomes working concurrently on the same mRNA?

Structure and function of the Golgi apparatus

The Golgi apparatus (see page 71) is involved in the processing and secretion of protein. It consists of a stack-like collection of flattened membranous sacs called **cisternae**. Cisternae vary in number, shape and organization in different cell types. One side of the stack of membranes is formed by the fusion of vesicle membranes from the RER. At the opposite side of the stack, vesicles are formed from swellings at the margins that become pinched off. These vesicles contain proteins for transport within the cell or for secretion.

Membranes flow from the RER to the Golgi, and on through vesicles to the cell membrane (Figure B2.2.10).

> **Concept: Form**
>
> The Golgi apparatus is comprised of membranes arranged in a series of stacks. Each part of the Golgi contains enzymes, which modify proteins in different ways. The membranous nature of the Golgi allows it to fuse with vesicles that are also made of membrane.

> **Concept: Function**
>
> By compartmentalizing protein synthesis at ribosomes in the RER, and protein modification in the Golgi, the production of a wide variety of proteins and modified structures can be carefully controlled.

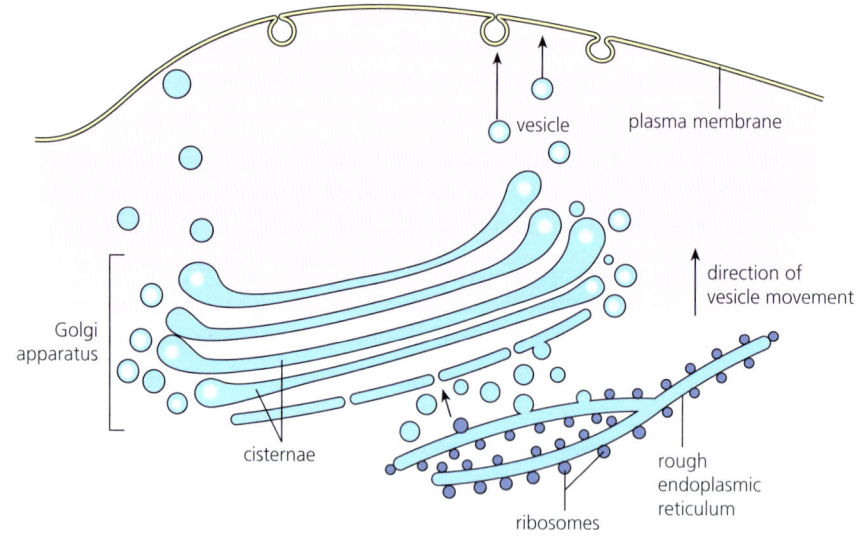

■ **Figure B2.2.10** Flow of membrane and proteins from the endoplasmic reticulum through the Golgi to the plasma membrane

Proteins enter the Golgi apparatus on the side facing the RER and exit on the opposite side of the stack, facing the plasma membrane of the cell. Proteins are transferred from one cisterna to the next. Each cisterna contains different enzymes, which catalyse different types of protein modification. These enzymes catalyse the addition or removal of sugars from proteins (glycosylation), the addition of sulfate groups (sulfation) and the addition of phosphate groups (phosphorylation). Some Golgi-mediated modifications act as signals to direct the proteins to their final destinations within cells, including the lysosome and the plasma membrane.

5 **Describe** the roles of organelles involved in the synthesis and transport of proteins destined to be embedded on the cell surface membrane.

◆ **Invagination**: the intucking of a surface or membrane.

Structure and function of cell vesicles

Membranes must be able to separate or fuse together so that cells and organelles can bring in, transport and release molecules. Many cells take up molecules through receptor-mediated **endocytosis**. A protein (or complex of proteins) initially binds to a receptor on the cell surface. A specialized protein acts to cause the membrane in this region to **invaginate**. One of these specialized proteins is **clathrin**, which polymerizes into a lattice network around the growing membrane bud to form a clathrin-coated pit (Figure B2.2.11). The invaginated membrane eventually breaks off and fuses to form a vesicle from either the Golgi body or plasma membrane.

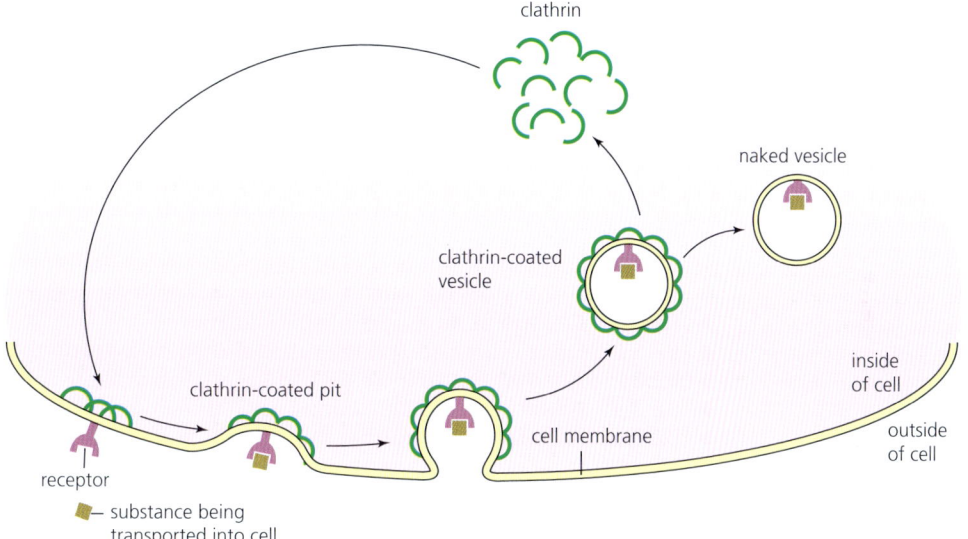

■ **Figure B2.2.11** Endocytosis mediated by clathrin

Some hormones, transport proteins and antibodies use receptor-mediated endocytosis to get inside a cell. This pathway is also used by viruses (see Chapter A2.3, page 88) and toxins to enter cells. The reverse process, the fusion of a vesicle to a membrane, is a critical step in the release of neurotransmitters from a neuron into the synaptic cleft.

ATL B2.2C

Review your knowledge of organelle structure and function using this site:
www.sumanasinc.com/webcontent/animations/content/eukaryoticcells.html
Make summary notes on each organelle so you can use them when you revise organelles.

LINKING QUESTIONS

1. What are examples of structure–function correlations at each level of biological organization?
2. What separation techniques are used by biologists?

B2.2 Organelles and compartmentalization

B2.3 Cell specialization

Guiding questions
- What are the roles of stem cells in multicellular organizations?
- How are differentiated cells adapted to their specialized functions?

SYLLABUS CONTENT

This chapter covers the following syllabus content:
- B2.3.1 Production of unspecialized cells following fertilization and their development into specialized cells by differentiation
- B2.3.2 Properties of stem cells
- B2.3.3 Location and function of stem cell niches in adult humans
- B2.3.4 Differences between totipotent, pluripotent and multipotent stem cells
- B2.3.5 Cell size as an aspect of specialization
- B2.3.6 Surface area-to-volume ratios and constraints on cell size
- B2.3.7 Adaptations to increase surface area-to-volume ratios of cells (HL only)
- B2.3.8 Adaptations of type I and type II pneumocytes in alveoli (HL only)
- B2.3.9 Adaptations of cardiac muscle cells and striated muscle fibres (HL only)
- B2.3.10 Adaptations of sperm and egg cells (HL only)

◆ **Zygote**: cell produced by the fusion of gametes.

◆ **Cell cycle**: repeating process of an orderly sequence of events where cells arise by the division of existing cells, grow, and then divide.

◆ **Embryo**: an organism in the earliest stages of growth, when its basic structures are being formed.

◆ **Morphogen**: a molecule present in a concentration gradient that specifies the fate of each cell along the gradient.

Production of unspecialized cells following fertilization

Multicellular organisms begin life as a single cell called a **zygote**, which grows and divides, at first producing unspecialized cells and then forming many specialized cells. These specialized cells eventually form the adult organism (Figure B2.3.1). So, cells arise by division of existing cells. The time between one cell division and the next is known as the **cell cycle**.

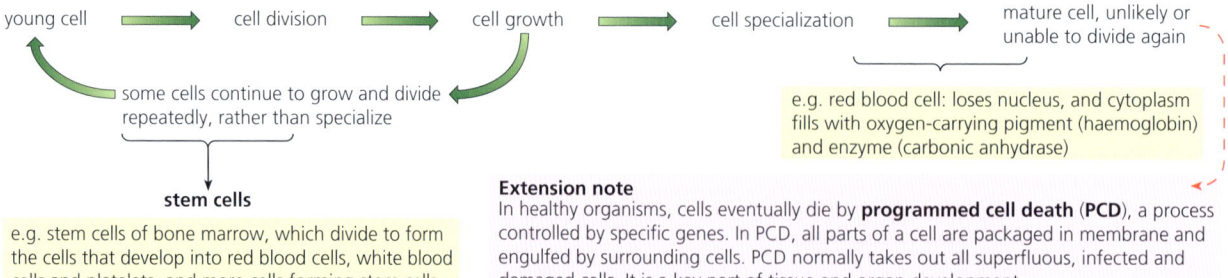

■ Figure B2.3.1 The life history of a cell and the role of stem cells

During early-stage **embryo** development, complex mechanisms of gene expression determine the ways cells differentiate and take on specific roles. A small number of genes determine body patterns during the embryo's development. The specific signalling molecules that are involved in gene expression are called **morphogens**.

Morphogens are **extracellular** (i.e. they exist outside the cells of a tissue) and occur across a gradient of concentrations. The gradient of the morphogen drives the process of differentiation of unspecialized stem cells into different cell types. Where there is a high concentration of morphogens,

1 Outline the importance of morphogens in the development of organisms.

◆ **Blastocyst**: embryo as fluid-filled ball of cells, at the stage of implantation.

◆ **Embryonic stem cell**: undifferentiated cell in early-stage embryo, capable of continual cell division and of developing into all the cell types of an adult organism.

◆ **Stem cell**: undifferentiated cell in embryo or adult that can undergo unlimited division and can give rise to one of many different cell types.

these cells will change differently to cells where there is a lower concentration of morphogens. The initiation or inhibition of gene expression is a result of different concentrations of morphogens, which control the way cells differentiate and develop into specific tissues.

The concentration of the morphogen in each cell is important as it determines a series of subsequent signals (cascades). Responses to these signals determine the direction and extent of cell growth and development, ultimately forming all the tissues and organs of the body. The expression of such gradients also controls the length of body structures such as toes and fingers, the location of the nose and other body patterns.

Properties of stem cells

In the development of a new organism, the first step after the formation of the zygote is one of continual cell division to produce a tiny ball of cells, called the **blastocyst**. All these cells are capable of further divisions. The cells of the inner cell mass (see Figure B2.3.3, page 260) are known as **embryonic stem cells**.

A **stem cell** is a cell that has the ability for endless repeated cell division. Stem cells maintain an undifferentiated state (this is called 'self-renewal'), while also having the capacity to differentiate into mature cell types (potency) along different pathways. Stem cells are the 'building blocks' of life; they divide and form cells that develop into the range of mature cells of the organism. Stem cells are found in all multicellular organisms.

At the next stage of embryological development, most cells lose the ability to divide and differentiate as they develop into the tissues and organs that make up the organism, such as blood, nerves, liver, brain and many others. However, a very few cells within these tissues keep many of the properties of embryonic stem cells. These are called **adult stem cells**. Table B2.3.1 compares embryonic and adult stem cells.

■ Table B2.3.1 Differences between embryonic and adult stem cells

Embryonic stem cells	Adult stem cells
undifferentiated cells capable of continual cell division and of developing into all the cell types of an adult organism	undifferentiated cells capable of cell divisions; these give rise to a limited range of cells within a tissue, for example blood stem cells give rise to red and white blood cells and platelets only
make up the bulk of the embryo as it begins development	occurring in the growing and adult body, within most organs; they replace dead or damaged cells, such as in bone marrow, brain and liver

Concept: Function

Although all cells in a multicellular organism contain the same genetic code, because they formed from one cell that then divided by mitosis, cells can differentiate into different functions. This is because in different cell types, different genes are activated, which produce specific structures for the specialized cell.

B2.3 Cell specialization

> **TOK**
>
> **Should some knowledge not be sought on ethical grounds?**
>
> There are clear advantages to using embryonic stem cells instead of adult stem cells in various medical treatments. The ethical issues surrounding the harvesting of embryonic stem cells, however, means that acting on this knowledge must be considered in a wider social, political and ethical context. That they can be used is true, but is it also the role of the biologist to ask *should* they be used?

> **ATL B2.3A**
>
> Research involving stem cells is increasing and raises ethical issues. What are the ethical issues concerning the therapeutic use of stem cells? Evaluate the use of stem cells from specially created embryos, from the umbilical cord blood of a newborn baby and from an adult's own tissues, and discuss your thoughts in groups.

2 **List** the differences between embryonic and adult stem cells.

◆ **Stem cell niche**: specific microenvironments in the body that either maintain the stem cells or promote their proliferation and differentiation.

Stem cell niches in adult humans

A **stem cell niche** is an area of a tissue that provides a specific environment where stem cells exist in an undifferentiated and self-renewable state, and where they receive stimuli to determine their behaviour. Stimuli includes cell-to-cell and molecular signals: these either activate or repress genes and the subsequent transcription of proteins. Resulting from this interaction, stem cells are maintained in a dormant state, caused to self-renew, or commit to a more differentiated state.

Stem cell niches have defined locations in the body. Two examples of stem cell niches will be explored.

■ Bone marrow

Bone marrow within the skeleton is the site of blood cell formation (also known as haematopoiesis) for all vertebrates except for fish and some amphibians. In human adults, **osteoblasts** (cells that regulate the creation of new bone) and **haematopoietic stem cells** (**HSCs**) (these produce the cellular components of blood and blood plasma) are closely connected in the bone marrow. HSCs produce red blood cells, white blood cells and platelets.

Research has shown that a subgroup of osteoblasts function as a key component of the HSC niche (the osteoblastic niche), controlling HSC numbers. This suggests that these cells rely on each other to determine their functions. Cells that give rise to mature skeletal cells are therefore critical in the regulation of HSCs. The vascular niche (so called because it is the site of blood vessels) in the adult bone marrow is a place for stem cell mobilization or proliferation and differentiation.

■ Hair follicles

Hair follicles are structures of mammalian skin that can regenerate in order to continuously produce new hair.

The stem cell niche of hair follicles is located between the opening of the sebaceous gland and the attachment site of the hair erector muscle. They are known as **bulge hair follicle stem cells** (see Figure B2.3.2). These stem cells are multipotent (see below) and have the potential to increase rapidly. During the growth phase, hair follicle stem cells become activated to regenerate the hair follicle and hair, and hairs grow longer each day. During the resting phase, the stem cells are dormant and hairs can shed more easily.

Complex physiological changes take place in the hair follicle with age. Hair follicles have pigment cells that make melanin. Certain stem cells act as pigment-producing cells: when hair regenerates, some of the stem cells convert into pigment-producing cells that colour the hair. With age, fewer of these cells are active and the hair loses its ability to produce melanin, so the hair greys or turns white. An imbalance in stem cell differentiation and altered stem cell activity can also lead to hair loss with age.

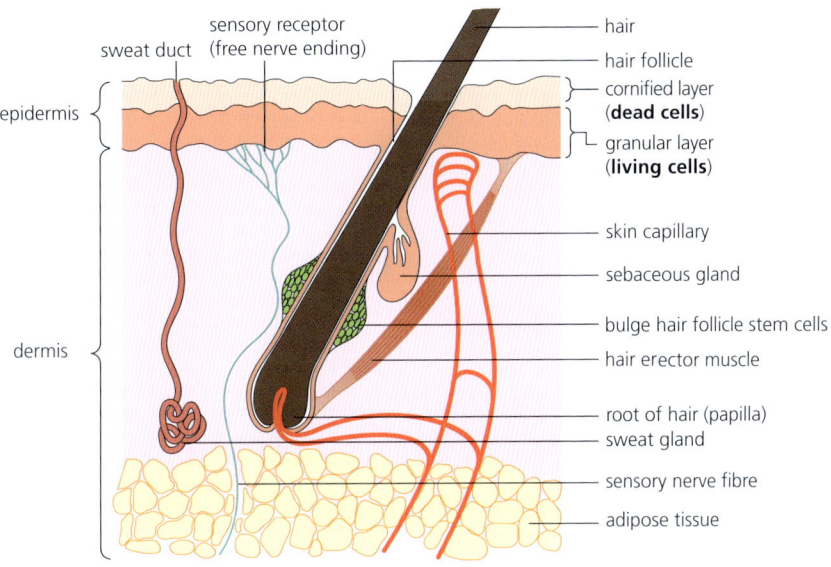

■ **Figure B2.3.2** Structure of the skin, showing location of hair follicle and stem cells

● **Top tip!**

As well as hair follicles, there are also follicles of a different kind in the ovaries – these are structures that surround the developing egg and release hormones that control the menstrual cycle (Chapter D3.1, page 699).

◆ **Totipotent**: capable of giving rise to any cell type in an organism, including making more totipotent stem cells and making the cells that become the placenta.

◆ **Pluripotent**: able to develop into many different types of cells or tissues in the body, except for becoming placental cells or totipotent stem cells.

◆ **Multipotent**: can create, maintain and repair the cells of one particular organ or tissue.

Totipotent, pluripotent and multipotent stem cells

The first formed stem cells of a zygote are **totipotent** and are the most versatile type of stem cell. They can become any type of body cell, including making more totipotent stem cells and the cells that become the placenta. The cells within the first couple of cell divisions after fertilization are the only cells that are totipotent.

Cell division continues, and the cells organize themselves into a fluid-filled ball called the blastocyst (Figure B2.3.3). Stem cells in these subsequent divisions of the embryo are **pluripotent**. This type of stem cell is a little more limited in potential: they can give rise to all of the cell types that make up the body, except that they cannot become placental cells or totipotent stem cells.

By the time differentiation has occurred, stem cells are able only to develop and form the nervous system. This restricted potency is referred to as **multipotent**. Multipotent stem cells can develop into more than one cell type but are more limited than pluripotent cells: they have a limited capacity for self-renewal. Multipotent stem cells can create, maintain and repair the cells of one particular organ or tissue. Stem cells in adult tissue, such as bone marrow, are multipotent (Figure B2.3.3).

● **Common mistake**

Make sure you know the difference between totipotent, pluripotent and multipotent. Simply referring to 'stem cells' is ambiguous so specific types of stem cells should be referred to, especially when discussing a particular stage of development.

> **Concept: Function**
>
> As stem cells mature, their function becomes more limited. The early stem cells can differentiate into any cell-type in the body, including placental cells, but later stem cells can only become a limited range of cells in a particular tissue or organ.

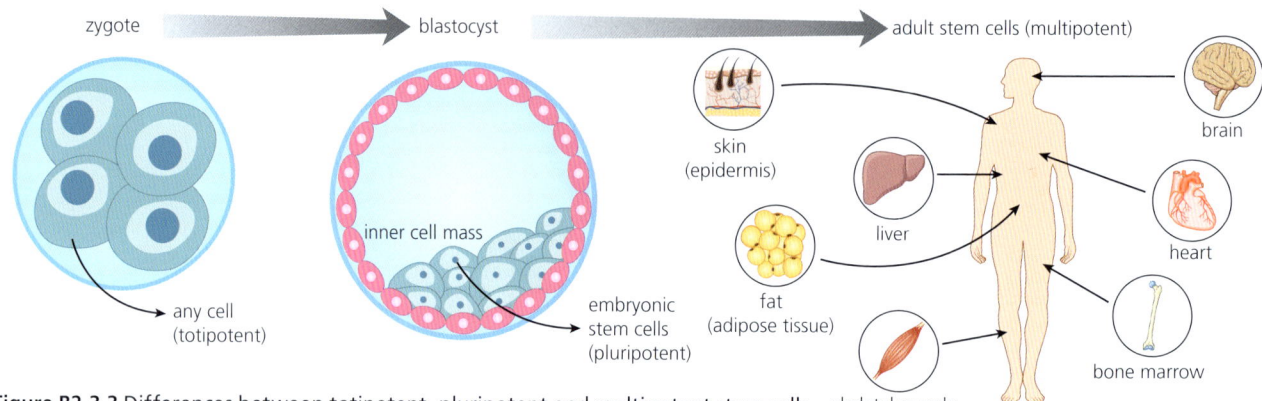

■ **Figure B2.3.3** Differences between totipotent, pluripotent and multipotent stem cells

3 **Define** the term *totipotent stem cells*.

Cell size as an aspect of specialization

Cells exist at a wide variety of different scales, with each type of cell adapted to its function in part by its size (Figure B2.3.4).

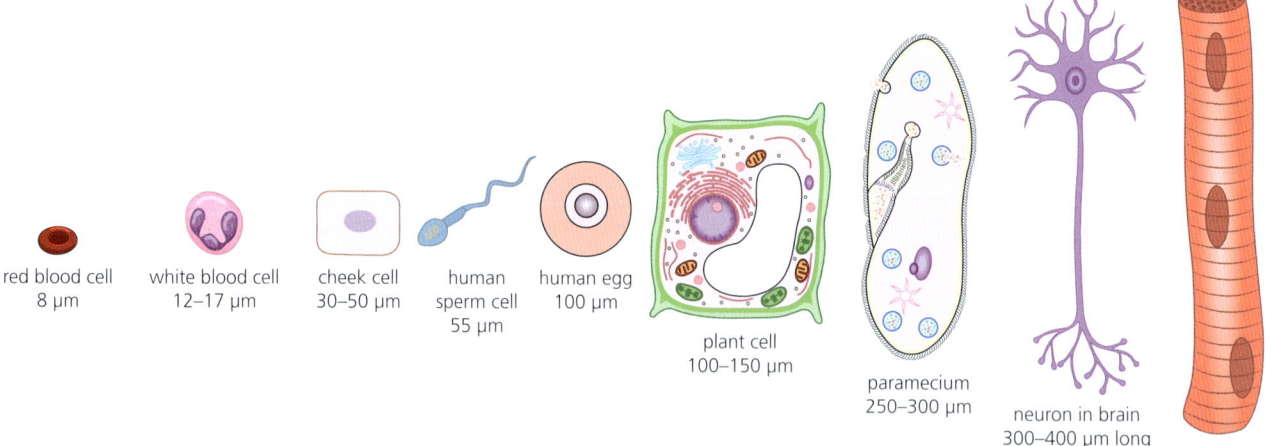

■ **Figure B2.3.4** Different size of cells (not to scale)

> 🌱 **Top tip!**
>
> Make sure you take note of the range of cell sizes in humans including male and female gametes, red and white blood cells, neurons and striated muscle fibres.

Male and female gametes

Male and female gametes in humans are very different in size. The **egg** cell is around 100μm, whereas the main body of the **sperm** is 5μm with a mid-piece (containing mitochondria) and flagellum of 50μm. The larger cell body of the egg allows it to store nutrients for the early development of the fertilized egg, whereas the cell body of the sperm only needs to hold the nucleus, for delivery to the egg, and so can be much smaller. The long flagellum of the sperm provides propulsion to reach the egg.

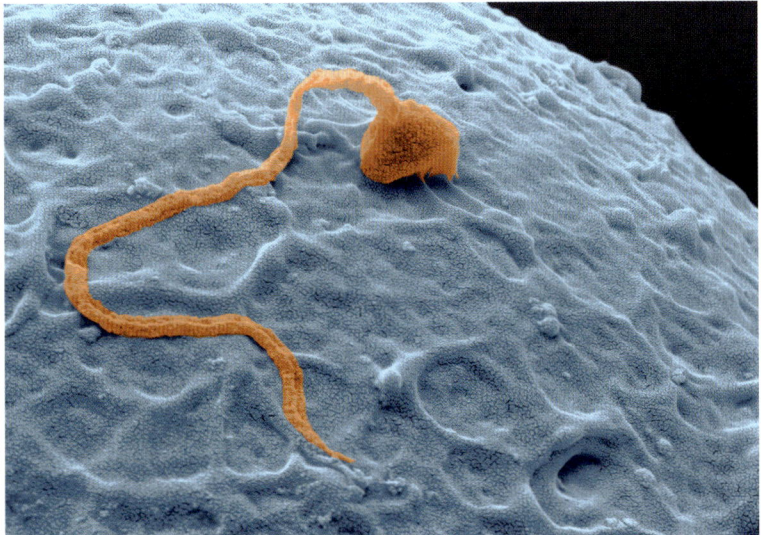

■ Figure B2.3.5 Scanning electron micrograph of a sperm cell fertilizing an egg cell; note the difference in size between the two cells (magnification ×3600)

4 **Outline** why the female gamete in humans is much larger than the male gamete.

Blood cells

Red blood cells (erythrocytes) are around 8μm, much smaller than white blood cells (which are around 12–17μm). Red blood cells need to be small so that they can fit through the small lumen of capillaries. Also, by being small, the red blood cell has a larger surface area compared to its size (see section below), allowing oxygen to diffuse in and out from the cell at a faster rate.

When foreign matter enters the body, such as a bacterium, virus or other organism, white blood cells attack and destroy it. White blood cells are bigger than red blood cells but fewer in number (although they increase in number on infection). The larger size of white blood cells allows them either to engulf and digest pathogens (see page 522) or to produce antibodies that bind to the body's foreign invaders and destroy them (page 523).

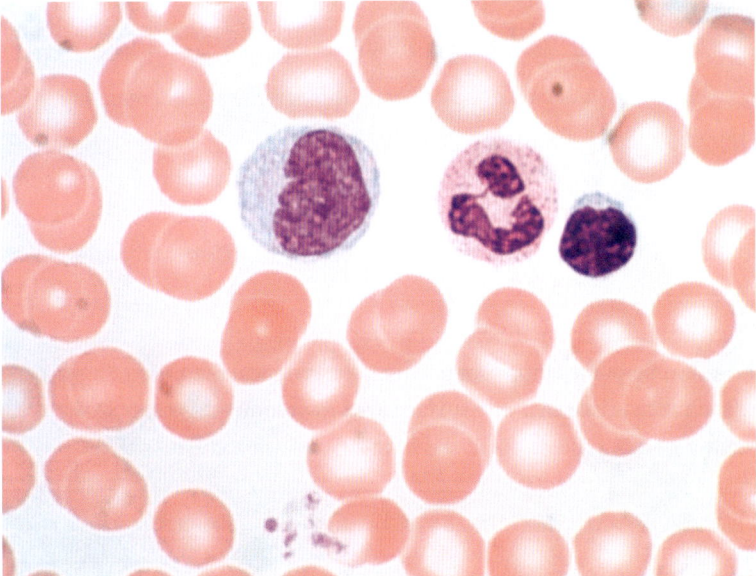

■ Figure B2.3.6 A light micrograph of a blood smear showing erythrocytes (red blood cells) and white blood cells

◆ **Egg/ovum (plural, ova)**: a female gamete.
◆ **Sperm**: motile male gamete.

Top tip!

Red blood cells are clearly discernible in a microscope image because of their colour and lack of nucleus. White blood cells usually have a purple-stained nucleus (Figure B2.3.6) that is either large and rounded (lymphocytes) or 'lobed' with several different sections joined together (phagocytes). The differences between the different types of white blood cell are discussed in Chapter C3.2, page 522.

B2.3 Cell specialization

Link

Neurons are covered in Chapter C2.2, page 467.

Neurons

Neurons (nerve cells) transmit electrochemical impulses through the body, allowing coordination and response to stimuli to occur. They differ in length. The longest neuron in the human body reaches from the base of the spine to the foot, a distance of up to one metre. The long length enables the electrical impulse to be sent without interruption over a long distance. Response to stimuli, especially those concerned with pain, need to happen quickly, and so fast transmission of the impulse is necessary – something achieved by the long length of the myelinated neuron.

Top tip!

Nerve cells can either be spelt neuron or neurone. In the IB Guide, the first spelling is used.

Striated muscle fibres

Striated muscle fibres are multinucleated cells that attach to muscles to allow movement. Instead of being made from many individual cells, muscle fibres are extended cellular structures that can be several centimetres long. This extended length allows them to coordinate contraction and has a significant effect on muscle force generation.

Further adaptions of these cells, other than size, are discussed later in this chapter (page 268).

Surface area-to-volume ratios and the constraints on cell size

The materials required for growth and maintenance of a cell enter through the plasma membrane. Similarly, metabolic waste products must leave the cell through the plasma membrane.

The rate at which materials can enter and leave a cell depends on the surface area of that cell, but the rate at which materials are used and metabolic waste products are produced depends on the amount of cytoplasm present within the cell. Similarly, heat transfer between the cytoplasm and environment of the cell is determined by surface area.

As cells grow and increase in size, an important difference develops between the surface area available for exchange and the volume of the cytoplasm in which the chemical reactions of life occur. The volume increases faster than the surface area; the **surface area-to-volume** ratio falls (**SA:V**, Figure B2.3.7). So, with increasing size of cell, less and less of the cytoplasm has access to the cell surface for exchange of gases, supply of nutrients and loss of metabolic waste products.

Concept: Form

The size of cells or organisms is an essential feature that determines how they interact with their environment. It determines the rate of exchange of materials across a cell surface, and whether organisms need specialized structures such as lungs to solve problems of size.

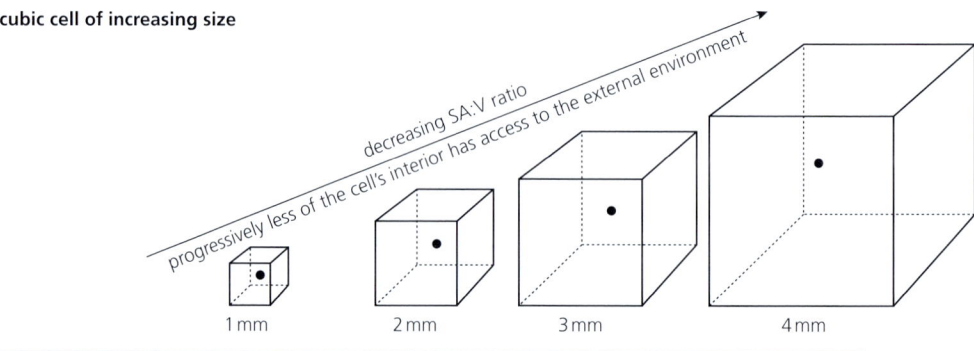

dimensions/mm	1 × 1 × 1	2 × 2 × 2	3 × 3 × 3	4 × 4 × 4
surface area/mm²	6	24	54	96
volume/mm³	1	8	27	64
surface area : volume ratio	6:1 = ⁶/₁ = 6	24:8 = ²⁴/₈ = 3	54:27 = ⁵⁴/₂₇ = 2	96:64 = ⁹⁶/₆₄ = 1.5

■ **Figure B2.3.7** The effect of increasing cell size on the surface area-to-volume ratio

> **Top tip!**
> Volume gets larger at a faster rate than surface area, so the SA:V ratio decreases. This limits cell size as the smaller surface area compared with cell size in larger cells means that oxygen and food cannot be transported into the cell and wastes removed at sufficient rate to maintain metabolic activities.

The exchange of materials across a cell surface depends on its surface area, whereas the need for exchange depends on cell volume. The smaller the cell is, the more quickly and easily materials can be exchanged between its cytoplasm and environment. One consequence of this is that cells cannot grow larger indefinitely. When a maximum size is reached, cell growth stops. The cell may then divide.

Tool 3: Mathematics

Ratios

A ratio is a way to compare amounts. Ratios are similar to fractions and both can be simplified by finding common factors, by dividing by the highest common factor.

Ratios can be used in biology to analyse surface area-to-volume relationships. For example, in Figure B2.3.7, the surface area-to-volume ratios are displayed for cubic cells.

Inquiry 2: Collecting and processing data

Processing data

An experiment is carried out to explore the impact of surface area-to-volume ratio on the rate of diffusion into cells. In the experiment, agar jelly blocks represent cells and acid represents essential materials needed by cells for metabolic reactions. The acid diffuses into each of the blocks through the outside surfaces. The faster the substances reach all parts of a cell, the more likely the cell is to survive. The red dye in the agar jelly blocks becomes pink and eventually colourless when mixed and reacted with the hydrogen ions in the acid. Movement of the hydrogen ions, $H^+(aq)$, is by simple diffusion.

The results from the experiment are recorded in Table B2.3.2.

■ Table B2.3.2 Jelly block experiment table of results

Block	Time to turn colourless/s	Surface area of block/mm²	Volume of block/mm³	Surface area-to-volume ratio
A (20 × 20 × 10)	964	(20 × 20) × 2 + (20 × 10) × 4 =	20 × 20 × 10 =	
B (10 × 10 × 20)	405	(10 × 10) × 2 + (20 × 10) × 4 =	20 × 10 × 10 =	
C (10 × 10 × 10)	247	(10 × 10) × 6 =	10 × 10 × 10 =	
D (5 × 10 × 10)	186	(10 × 10) × 2 + (5 × 10) × 4 =	5 × 10 × 10 =	
E (5 × 5 × 10)	81	(5 × 5) × 2 + (5 × 10) × 4 =	5 × 5 × 10 =	

B2.3 Cell specialization

Tool 3: Mathematics

Calculating surface area-to-volume ratios

Work out the surface area-to-volume ratio for each jelly block. Instead of representing this as a ratio, one number can be given because this is easier to plot on a graph.

Graphing

Plot a graph of your results from the jelly block experiment, with surface area-to-volume ratio on the *x*-axis and time taken for the jelly to completely lose its colour on the *y*-axis.

Inquiry 3: Concluding and evaluating

Concluding

Analyse your results. What do they tell you about the effect of size on the rate of diffusion? What do they tell you about the optimal size for cells?

ATL B2.3B

The relative differences in the changes of surface area and volume as objects get bigger or smaller is known as the **square-cube law**. This law states that that the volume of an object grows faster than its surface area. This is because the surface area changes as a factor of the squared value of its length, whereas the volume changes as a factor of the cubed value. For example, if the size (measured by edge length) of a cube is doubled, its surface area will be four times its original value, whereas its volume will be eight times its original volume.

The relationship between surface area and volume as objects increase or decrease in size can be explored using online calculators, for example:
https://goodcalculators.com/square-and-cube-calculator/

Calculate a range of different dimensions for an object and plot your data. What trends do you see, and what implications do they have for biological objects?

● Nature of science: Models

Using models to explore the surface area-to-volume relationship

Models are simplified versions of complex systems. The surface area-to-volume relationship can be modelled using cubes with different side lengths (see Figure B2.3.8). Although the cubes have a simpler shape than real organisms, scale factors operate in the same way.

The surface area of each cube can be imagined if each cube is spread out flat and the surface of each square added together to work out the total surface area.

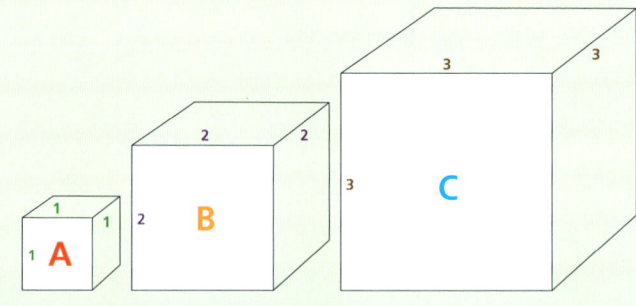

■ Figure B2.3.8 Three cubes representing cells (or organisms) of different sizes

5 Consider imaginary cubic 'cells' with sides of 1, 2, 4 and 6 mm.
   a **Calculate** the volume, surface area and surface area-to-volume ratio for each cell.
   b **State** the effect on the SA:V ratio of a cell as it increases in size.
   c **Explain** the effect of increasing cell size on the efficiency of diffusion in the removal of metabolic waste products from cell cytoplasm.

> **TOK**
>
> It would be impossible to develop useful knowledge about the world without the use of models. Without them, our knowledge would be limited to that of specific events and objects. This book is full of examples of models – every diagram representing a cell or a protein in this book is a model.
>
> Models are only useful when the information they provide is an answer to the question being asked. If asked, *what shape are lipid-anchored proteins?* then Figure B2.1.5 (page 226) is not useful. However, if you wonder what sort of proteins are present in biological membranes, then B2.1.8 to B2.1.10 would be very helpful.
>
> As you use the models in this book, it is worth considering for each what the model is being used for, what features of the model are therefore relevant, and which elements of the model are not meant to convey knowledge.

Adaptations to increase surface area-to-volume ratios of cells

Many cells have adaptations to increase their surface area-to-volume ratio. Two examples are discussed here.

■ Red blood cells (erythrocytes)

Red blood cells (erythrocytes) transport respiratory gases (primarily oxygen) in the bloodstream. Red blood cells are approximately 8 μm in diameter and have the shape of a flat disk or doughnut (Figure B2.3.9), which is round with an indentation in the centre (the surface is said to be biconcave). The flattening of the cell increases its surface area relative to its size, increasing the diffusion of oxygen across its surface. The flat shape ensures a short distance for diffusion so that oxygen can quickly move throughout the cell. The biconcave shape, which can be achieved because the erythrocyte does not have a nucleus, also increases the surface area. In addition, the shape of the cell increases its flexibility, which helps it move through the narrow lumen of capillaries.

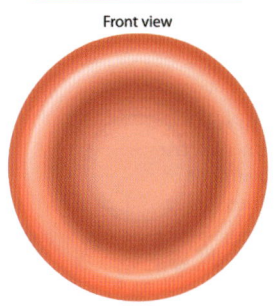
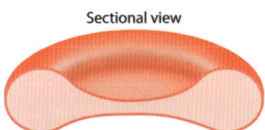

■ Figure B2.3.9 An erythrocyte

♦ **Microvillus (plural, microvilli)**: one of many tiny infoldings of the plasma membrane, making up a brush border.

♦ **Brush border**: tiny, finger-like projections (microvilli) on the surface of epithelial cells, for example in the small intestine.

♦ **Invagination**: the intucking of a surface or membrane.

■ Proximal convoluted tubule cells in the nephron

The role of the kidney is to maintain the homeostatic balance of water in the body, and to filter out metabolic wastes (e.g. urea). The functional units of the kidney are the **nephrons**. These have different sections which play specific roles. The **proximal convoluted tubule (PCT)** is the longest section of the nephron, and it is here that a large part of the filtrate is reabsorbed into the capillary network. Epithelial cells in the PCT reabsorb components of the filtrate that have nutritional significance (such as glucose, amino acids and ions). The walls of the tubule are one cell thick, and these cells are packed with mitochondria. The surface facing the lumen (the inside space) of the tubule is called the apical surface. This is folded into **microvilli**, forming a **brush border** (Figure B2.3.10), which greatly increases the surface area for absorption. On the other side of the cell – the membrane of tubule cells facing the blood at the base of the cell – **invaginations** (infoldings) of the plasma membrane, like brush borders, increase the surface area here too. Invaginations of the basolateral region of the plasma membrane are commonly found in cells engaged in active transport of molecules and ions, where the infoldings increase the surface area available for transport.

B2.3 Cell specialization

Link

The processes of the kidney are covered in detail in Chapter D3.3, page 768.

> #### Concept: Form
>
> Some cells have adaptations to increase surface area-to-volume ratios of cells. Red blood cells are flattened and biconcave. Microvilli are highly folded to increase surface area, and found in the proximal convoluted tubule of the nephron of kidneys and in the wall of the small intestine.

6 The magnification of Figure B2.3.11 is ×3600. **Calculate** the actual size of one of the cells.

7 **Sketch** a diagram of one of the cells in Figure B2.3.11 and label it appropriately.

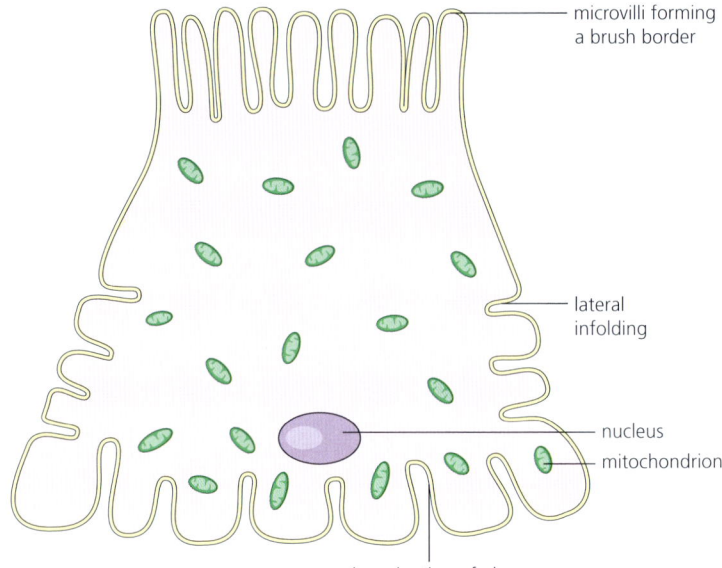

■ **Figure B2.3.10** Adaptations of the proximal convoluted tubule include microvilli on the upper surface and invaginations on the base of the cell, which increase surface area and, therefore, movement of molecules and ions through the cell and into blood capillaries

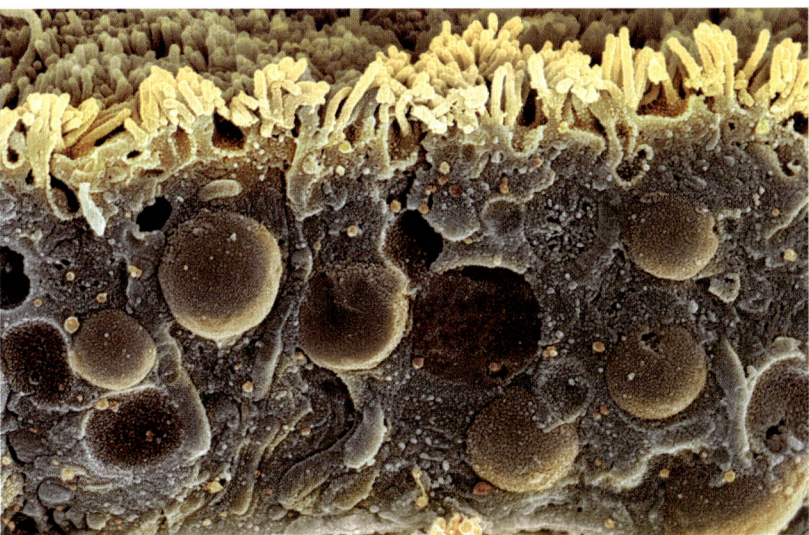

■ **Figure B2.3.11** A coloured SEM of the surface of a proximal convoluted kidney tubule. What features can you pick out?

◆ **Alveolus (plural, alveoli)**: air sac in the lung.

◆ **Pneumocyte**: specialized cells that occur in the alveoli of the lungs.

◆ **Epithelium**: sheet of cells, bound strongly together, covering internal or external surfaces of multicellular organisms.

Adaptations of type I and type II pneumocytes in alveoli

There are some 700 million **alveoli** in our lungs (Figure B2.3.12), providing a surface area of about 70 m^2 in total. This is an area 30–40 times greater than that of the body's external skin! The alveoli are the major sites of gas exchange in the lungs and their enormous surface area helps to maximize gas exchange. Two types of epithelial cells called **pneumocytes** line the alveoli, and they have different adaptions to help with their roles in the alveoli. Alveolar **epithelium** is an example of a tissue where more than one cell type is present, because different adaptations are required for the overall function of the tissue.

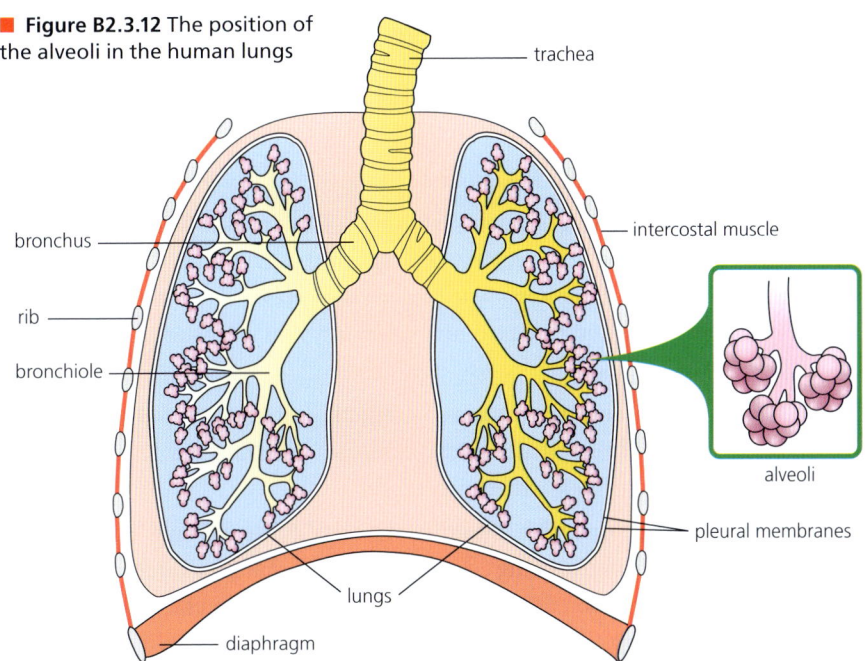

Figure B2.3.12 The position of the alveoli in the human lungs

Common mistake

It is incorrect to refer to the 'cell wall' of the alveolus, as this is a term used for plant and bacterial cell walls. Alveolar walls and capillary walls should be used when referring to the structure of the alveoli.

◆ **Surfactant**: a detergent secreted by cells in the wall of alveoli which breaks surface tension and stops alveoli walls from sticking together, keeping the alveoli open.

Type I pneumocytes

The wall of an alveolus is one cell thick. A capillary lies very close to the wall of the alveolus. The combined thickness of the alveolar and capillary walls separating air and blood is therefore typically only 2–4 µm thick. The capillary is extremely narrow, just wide enough for the red blood cells to squeeze through, so the red blood cells are close to, or in contact with, the capillary walls.

Type I pneumocytes are involved in the process of gas exchange between the alveoli and the capillaries. Type I cells have a flattened shape, which means they are extremely thin to minimize diffusion distance. The cells are tightly connected to prevent leakage of tissue fluid into alveolar air space.

Type II pneumocytes

Type II pneumocytes produce a detergent-like mixture of lipoproteins and phospholipid-rich secretion called **surfactant** that lines the inner surface of the alveoli. The type II cells are cuboidal in shape and bigger than type I so they can store surfactant components. Type II pneumocytes contain many secretory vesicles (lamellar bodies) in their cytoplasm that discharge surfactant to the alveolar lumen. Surface tension is a force acting to minimize the surface area of a liquid. Because of the tiny diameter of the alveoli (about 0.25 mm) they would tend to collapse under surface tension during expiration, with the walls sticking together. The lung surfactant secreted by type II pneumocytes lowers the surface tension, permitting the alveoli to flex easily as the pressure within the thorax falls and rises.

Macrophages

The extremely delicate structure of the alveoli is protected by macrophages (dust cells), abundant in the surface film of moisture. These are the main detritus-collecting cells of the body. They originate from bone marrow stem cells and are dispersed about the body in the blood circulation. These cells migrate into the alveoli from the capillaries (Figure B2.3.13). Here, these phagocytic white cells ingest any debris, fine dust particles, bacteria and fungal spores present. They also line the surfaces of the airways leading to the alveoli.

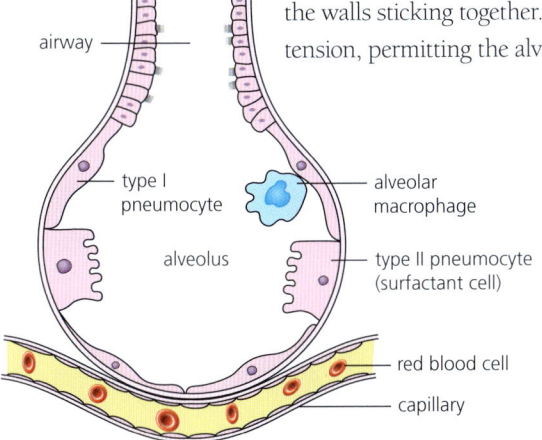

Figure B2.3.13 Cells of the alveolar epithelium

B2.3 Cell specialization

8 Distinguish between type I and type II pneumocytes.

Common mistake

Do not confuse type I and type II pneumocytes. They both have different structure and functions in the alveoli.

◆ **Myogenic**: originating in heart muscle cells themselves, as in the generation of the heartbeat.

◆ **Myofibril**: contractile protein filament from which muscle is composed.

◆ **Striated**: muscle tissue that is marked by alternating dark and light bands.

◆ **Sarcolemma**: membranous sheath around a muscle fibre.

Adaptations of cardiac muscle cells and striated muscle fibres

■ Cardiac muscle cells

Cardiac muscle is unique to the heart. Heart muscle fibres contract rhythmically from formation until they die. Muscles typically contract when stimulated to do so by an external nerve supply; however, in the heart the impulse to contract is generated within the heart muscle itself – it is said to have a **myogenic** origin. The central nervous system is not needed to stimulate cardiac muscles to contract.

Individual cardiac muscle cells have a tubular structure composed of chains of fibres. Each muscle fibre is composed of a mass of **myofibrils**, but we need an electron microscope to see this important detail. The myofibrils consist of repeating sections of sarcomeres, which are the fundamental contractile units of the muscle cells.

A photomicrograph of cardiac muscle is shown in Figure B2.3.14. Cardiac muscle consists of cylindrical branching columns of fibres, forming a unique, three-dimensional network. This allows for contraction in three dimensions and for the contraction signal to spread more quickly. Each fibre has a single nucleus, and the repeating sarcomeres give it a striped or **striated** appearance under the microscope. Fibres are surrounded by a special plasma membrane, called the **sarcolemma**, and all are very well supplied by mitochondria and capillaries.

Top tip!

The myogenic control of heart muscle contraction plays a role in autoregulation. A constant flow of blood must be balanced with blood pressure. The myogenic origin of the impulse to contract means that the heart, and therefore blood circulation, can adapt quickly to changes in the body following, for example, exercise.

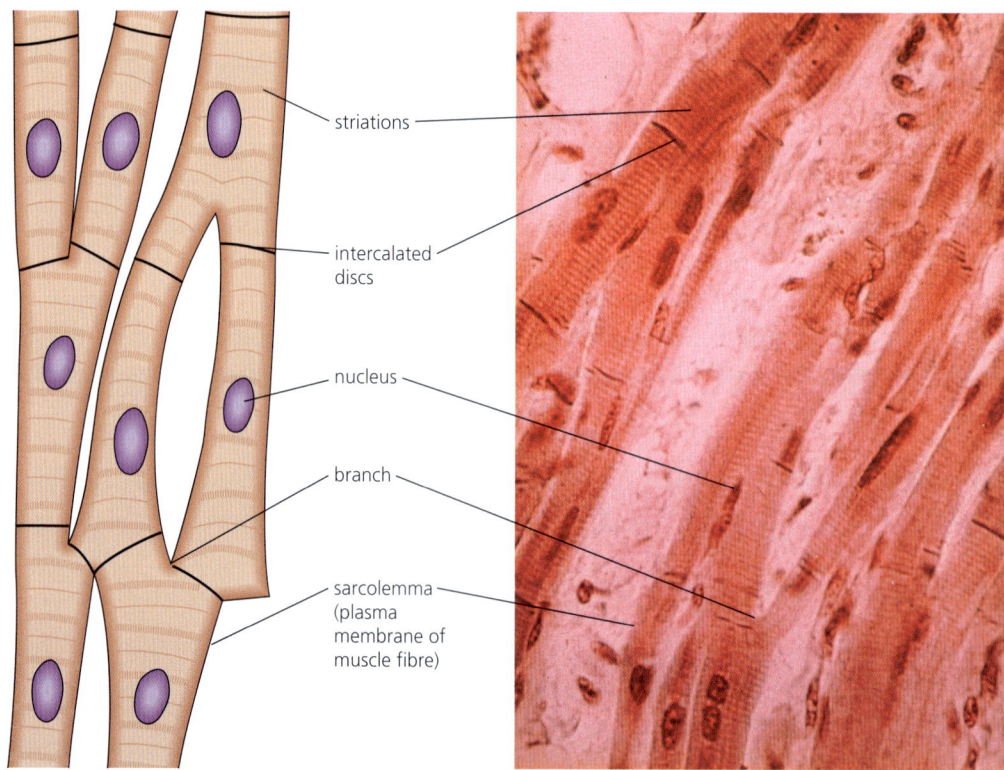

■ Figure B2.3.14 Cardiac muscle

> **Concept: Form and function**
>
> Cardiac muscles have intercalated discs to allow easy transfer of electrical impulses between cells. Intercalated discs also hold cells together so they cannot separate.

Propagation of stimuli throughout the heart wall

The structure of cardiac muscle cells allows propagation of stimuli through the heart wall. **Intercalated discs** are present at the junctions between cardiac muscle cells. A disc consists of a double membrane with gap junctions, through which are cytoplasmic connections between adjacent cardiac cells. Gap junctions form channels that allow continuous flow of cytoplasm between cells. This direct electrochemical coupling between cells allows waves of depolarization to pass through the entire network, synchronizing contraction of the muscle, as if in a single cell.

The branching structure of the cardiac muscle allows connection to multiple cells and provides a larger surface area of contact between cells, allowing groups of cells to work together and synchronize their activity.

Finally, although cardiac muscle fibres form an interconnected network, the network system of the walls of the atria is entirely separate from that of the ventricles. This ensures a transmission delay of the electrical signal between atria and ventricles.

■ Striated muscle fibres

Skeletal muscles are attached to the moveable parts of skeletons, and the contraction of these muscles brings about movement. The muscles are attached by **tendons** and work in **antagonistic pairs**.

Skeletal muscle consists of bundles of muscle fibres (Figure B2.3.15). The remarkable feature of a muscle fibre is its ability to shorten to half or even a third of the relaxed or resting length. Fibres appear striped under the light microscope, so skeletal muscle is also known as striated muscle (like cardiac muscle). Each muscle fibre is composed of a mass of myofibrils.

Link
For further details on heart structure (atria and ventricles), see Chapter B3.2, page 315.

◆ **Intercalated discs**: link cardiac cells together and define their borders, and facilitate cell-to-cell communication, which is needed for coordinated muscle contraction.

◆ **Tendon**: fibrous connective tissue connecting a muscle to bone.

Link
The sliding filament model of muscle contraction is covered in detail in Chapter B3.3, page 327.

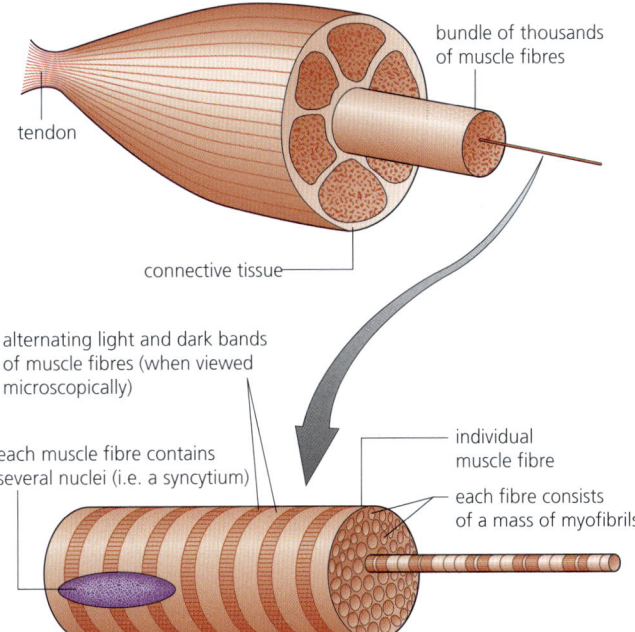

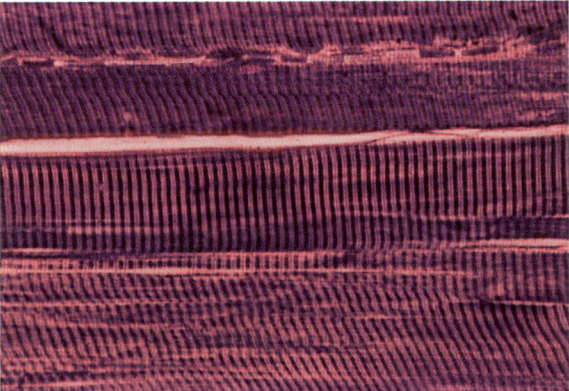

■ **Figure B2.3.15** The structure of skeletal muscle

B2.3 Cell specialization

The ultrastructure of skeletal muscle

Using electron microscopy, we can see that each muscle fibre consists of many parallel myofibrils, within a sarcolemma (a plasma membrane), together with cytoplasm. The cytoplasm contains mitochondria packed between the myofibrils. Skeletal muscle fibres are **multinucleate** (meaning they contain multiple nuclei per cell) and contain specialized endoplasmic reticulum. The sarcolemma is folded to form a system of transverse tubular endoplasmic reticulum, known as sarcoplasmic reticulum. This is arranged as a network around individual myofibrils. The arrangement of myofibrils, sarcolemma and mitochondria, surrounded by the sarcoplasmic membrane, is shown in Figure B2.3.16.

electron micrograph of TS through part of a muscle fibre, HP (x36 000)

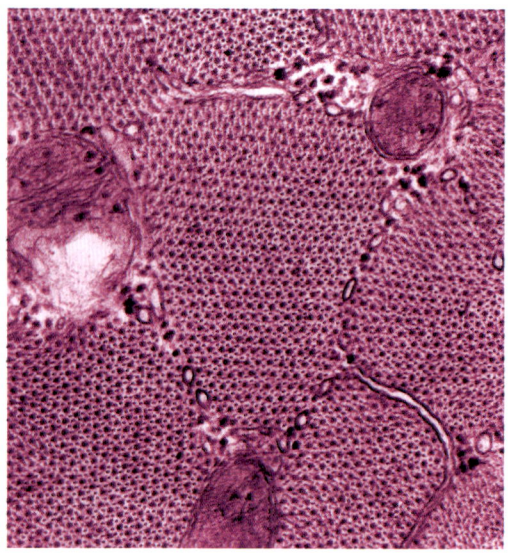

stereogram of part of a single muscle fibre

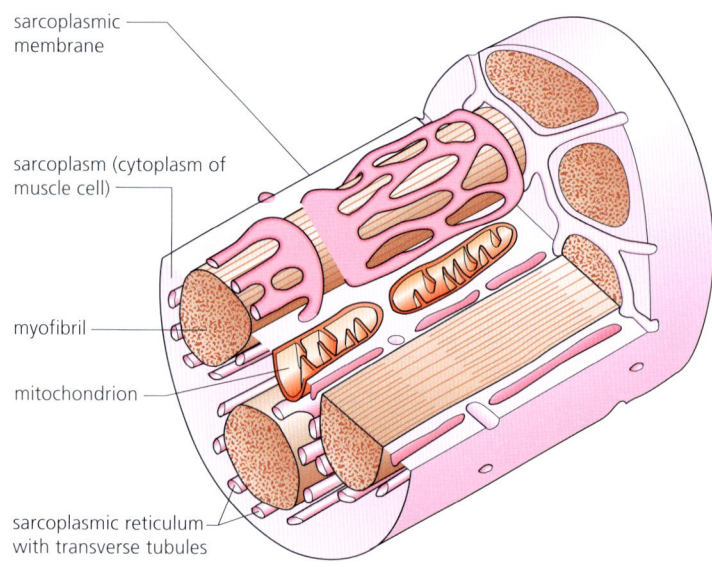

■ Figure B2.3.16 The ultrastructure of a muscle fibre

Is striated muscle fibre a cell?

9 **State** the differences between cardiac muscle and striated muscle.

The striated muscle fibres that make up the skeletal muscles of mammals have an atypical cell structure. The fusion of cells to create one elongated fibre, resulting in a multinucleated structure, is very different from the typical cell structure. This leads to the question: *is a striated muscle fibre a cell?* Myofibrils give the impression of a structure that goes beyond that of a simple cell, and muscle fibres are long and can measure 300 mm or more – much larger than regular cells.

However, the fibres contain the usual organelles of a cell, such as mitochondria and endoplasmic reticulum, and are surrounded by one cell membrane, thus indicating they are cells. Striated muscle fibres do, however, challenge the view that cells act as autonomous (i.e. self-sufficient) units.

Link
Muscles are covered in more detail in Chapter B3.3, page 327.

Striated muscle has similarities with cardiac muscle. Like cardiac muscle fibres, skeletal muscle fibres are surrounded and enclosed by a membrane, the **sarcolemma**, from which **transverse tubules** (T tubules) tunnel in and around the sarcomeres. Also present, here, is the fluid-filled system of branching membranous sacs, the **sarcoplasmic reticulum** – a modified form of endoplasmic reticulum. Like cardiac muscles, skeletal muscle fibres are striated in appearance, and have a similar arrangement of **actin** and **myosin** filaments. Finally, all muscle tissue consists of fibres that can shorten by a half to a third of their length. Both types of muscle have many mitochondria to increase the supply of ATP for muscle contraction.

Adaptations of sperm and egg cells

Sexual reproduction is the production of offspring from two parents using gametes – **egg** and **sperm cells** in animals. The cells of the offspring have two sets of chromosomes (one from each parent). Sexual reproduction involves two stages:
- **meiosis** – the special cell division that makes gametes with half the number of chromosomes (haploid, that is, one set of chromosomes)
- **fertilization** – the fusion of two gametes to form a zygote (back to two sets of chromosomes).

Human gametes have specific adaptations that allow them to carry out their function.

Sperm cell

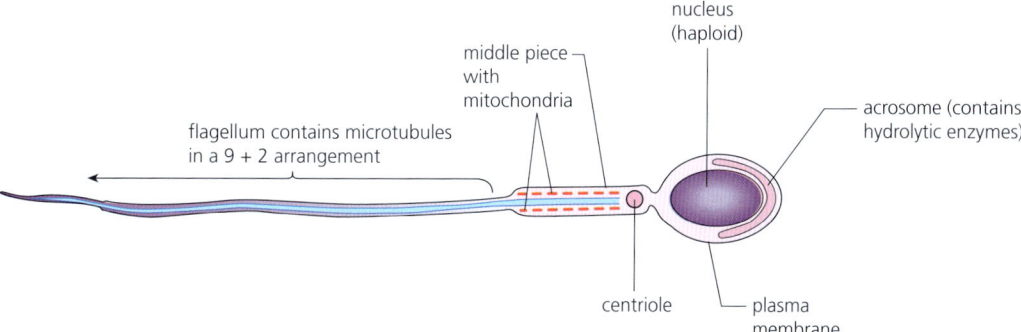

■ **Figure B2.3.17** A human sperm cell

Structures and functions of a mature sperm cell:
- flagellum – motility
- middle piece – contains many mitochondria to provide the energy required for the movement of the flagellum
- head – surrounded by a plasma membrane, containing a haploid nucleus; the head is smooth and oval in shape to make the cell streamlined for fast movement
- acrosome (at the top of the head) – contains enzymes for digesting the zona pellucida (jelly coat, containing glycoproteins) in the **oocyte** (egg).

◆ **Oocyte**: a female sex cell in the process of a meiotic division to become an ovum.

> **Concept: Function**
>
> The role of a sperm is to deliver a nucleus from a male into the egg of a female, so that fertilization can take place and a new zygote formed. Only the nucleus enters the egg cell – the rest of the cell (the neck and the flagellum) remains outside the egg. This means that all mitochondria are from the mother via the egg cell.

Egg cell

Egg cells, or ova, go through several stages of development. Primary oocytes (immature egg cells) are formed after the first stage of meiosis. During the menstrual cycle, once a month, one primary oocyte then begins the second stage of meiosis to produce a secondary oocyte and a polar body (containing a second haploid nucleus, but which contains unequal division of cytoplasm and fails to develop further). Meiosis is only completed at fertilization (for further details see Chapter D3.1, page 694). By delaying the two divisions of meiosis (meiosis I and meiosis II), the oocyte is provided with more genetic material for longer, for mRNA production and subsequent protein synthesis – an important factor as the cell grows and matures.

Concept: Form

Eggs cells go through several stages of development, only becoming the final egg cell on fertilization. Eggs are much larger than sperm, as they contain all the nutrients needed for the growth and development of the zygote.

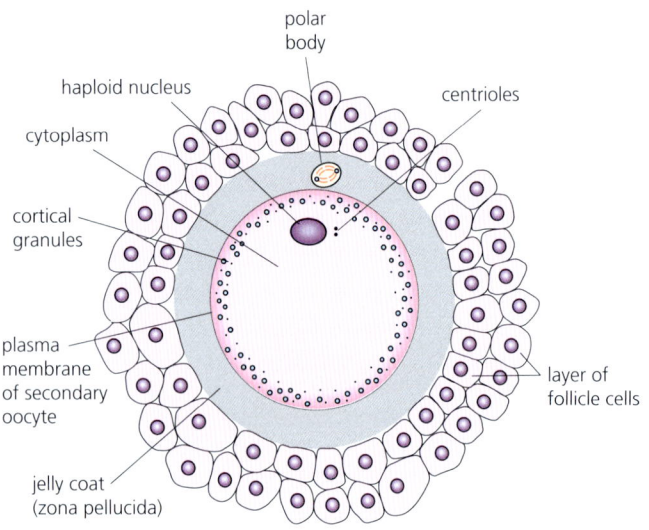

■ Figure B2.3.18 The structure of a mature secondary oocyte

Structures and functions of a mature secondary oocyte:
- follicle cells nourish and protect the oocyte
- zona pellucida (jelly coat) allows binding of sperm cells, prevents polyspermy (several sperm haploid nuclei entering) and premature implantation of the embryo (ectopic pregnancy); this is also the place where the polar body is found
- plasma membrane surrounds the haploid nucleus and contains microvilli to absorb nutrients from follicular cells
- cortical granules prevent polyspermy during fertilization
- cytoplasm – the place for metabolic reactions and where all organelles are located
- nucleus – a haploid nucleus (half the chromosome number).

LINKING QUESTIONS

1. What are the advantages of small size and large size in biological systems?
2. How do cells become differentiated?

B3.1 Gas exchange

Guiding questions
- How are multicellular organisms adapted to carry out gas exchange?
- What are the similarities and differences in gas exchange between a flowering plant and a mammal?

SYLLABUS CONTENT

This chapter covers the following syllabus content:
- B3.1.1 Gas exchange as a vital function in all organisms
- B3.1.2 Properties of gas-exchange surfaces
- B3.1.3 Maintenance of concentration gradients at exchange surfaces in animals
- B3.1.4 Adaptations of mammalian lungs for gas exchange
- B3.1.5 Ventilation of the lungs
- B3.1.6 Measurement of lung volumes
- B3.1.7 Adaptations for gas exchange in leaves
- B3.1.8 Distribution of tissues in a leaf
- B3.1.9 Transpiration as a consequence of gas exchange in a leaf
- B3.1.10 Stomatal density
- B3.1.11 Adaptations of foetal and adult haemoglobin for the transport of oxygen (HL only)
- B3.1.12 Bohr shift (HL only)
- B3.1.13 Oxygen dissociation curves as a means of representing the affinity of haemoglobin for oxygen at different oxygen concentrations (HL only)

◆ **Gas exchange**: exchange of respiratory gases (oxygen, carbon dioxide) between cells/organism and the environment.

Gas exchange as a vital function in all organisms

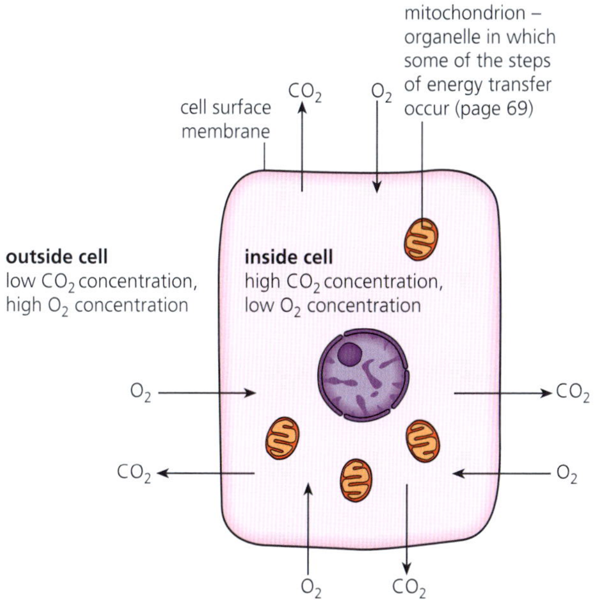

■ Gas exchange – meeting the needs of respiration

Cellular respiration is the controlled release of energy, in the form of ATP, from organic compounds in cells. Respiration occurs continuously in living cells, largely in the mitochondria of eukaryotic organisms and the cytoplasm of prokaryotic cells. To support aerobic cellular respiration, cells take in oxygen from their environment and release carbon dioxide by a process called gaseous exchange (Figure B3.1.1).

Gas exchange is the exchange of gases between an organism and its surroundings, including the uptake of oxygen and the release of carbon dioxide in animals. Plants uptake oxygen and release carbon dioxide at night and take in carbon dioxide and release oxygen during the day.

■ Figure B3.1.1 Gas exchange in an animal cell

B3.1 Gas exchange

The exchange of gases between the individual cell and its environment takes place by **diffusion**. For example, in cells that are respiring aerobically there is a higher concentration of oxygen outside the cells than inside, and so there will be a continuous net inward diffusion of oxygen.

Concept: Form

In living organisms, factors that determine the rate of diffusion include:

- The **size of the surface area** available for gas exchange (the respiratory surface) – the greater this surface area, the greater the rate of diffusion. In a single cell, the respiratory surface is the whole plasma membrane.
- The **difference in concentration** (concentration gradient) – a rapidly respiring organism has a very much lower concentration of oxygen in the cells and a higher-than-normal concentration of carbon dioxide. The greater the gradient in concentration across the respiratory surface, the greater the rate of diffusion.
- The **length of the diffusion path** – the shorter the diffusion path, the greater the rate of diffusion, so the respiratory surface must be as thin as possible.

Top tip!

Energy for diffusion comes from the kinetic energy of molecules. The higher the temperature, the faster the rate of diffusion due to the increased kinetic energy of molecules. In warm-blooded animals, temperature is kept constant through homeostatic mechanisms and so the effects of temperature on diffusion do not vary.

Challenges of gas exchange

The challenges of gas exchange become greater as organisms increase in size because their **surface area-to-volume ratio** decreases with increasing size, and the distance from the centre of an organism to its exterior increases. The surface area of a single-celled organism is large in relation to the amount of cytoplasm it contains, so the surface of the cell is sufficient for efficient gaseous exchange. On the other hand, most cells in large multicellular organisms are too far from the surface of the body to receive enough oxygen by diffusion alone. In addition, animals often develop an external surface of tough or hardened skin that, while it provides protection to the body, is not suitable for gaseous exchange. These organisms require an alternative respiratory surface.

Active organisms have an increased metabolic rate, and the demand for oxygen in their cells is higher than in sluggish and inactive organisms. Therefore, large, active organisms have specialized organs for gaseous exchange.

Link

The surface area-to-volume ratio is covered in detail in Chapter B2.3, page 262.

Properties of gas-exchange surfaces

All gas-exchange surfaces need properties that increase the rate of diffusion of respiratory gases across them. These properties include:

- **permeability** – to allow the gases across
- **thin tissue layer** – to make the shortest distance for diffusion as possible
- **moisture** – gases dissolve in the moisture, helping them to pass across the gas-exchange surface
- **large surface area** – so that large quantities of the respiratory gases can cross at the same time.

Maintenance of concentration gradients at exchange surfaces in animals

◆ **Concentration gradient**: the difference in concentration of a substance between one area and another.

Diffusion requires a **concentration gradient** to be maintained – the steeper the gradient, the faster the diffusion (Figure B3.1.2). A concentration gradient needs to be maintained at the gas-exchange surface as this allows oxygen to diffuse into the body and carbon dioxide to diffuse out.

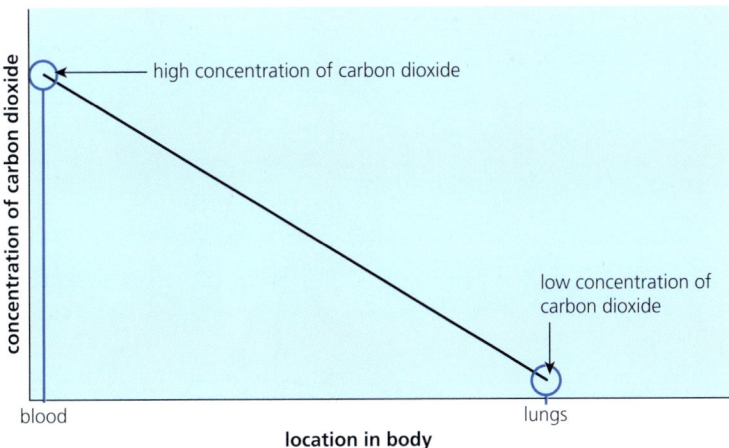

■ **Figure B3.1.2** The concentration gradient in alveoli allows carbon dioxide to diffuse from the blood

How are concentration gradients maintained?

There are three ways in which a concentration gradient is maintained at exchange surfaces:
- a **dense networks of blood vessels**: capillaries provide a large surface area for the diffusion of respiratory gases; blood carries the gases either in red blood cells (mainly oxygen) and plasma (carbon dioxide)
- a **continuous blood flow**: this maintains the difference in concentration of molecules between the air and blood by carrying oxygen away from the gas-exchange surfaces in the capillaries and carbon dioxide to them
- **ventilation**: with air for lungs and with water for gills, ventilation brings oxygen to the gas-exchange surface and removes carbon dioxide.

Gas exchange in mammals

The lungs

In mammals, the lungs are the organs where gas exchange occurs. Lungs provide a large, thin, moist surface area that is suitable for gaseous exchange. However, the lungs are in a protected position inside the **thorax** (chest), so air must be brought to the respiratory surface there: the lungs must be ventilated.

A ventilation system is a pumping mechanism that moves air into and out of the lungs efficiently, thereby maintaining the concentration gradients of oxygen and carbon dioxide for diffusion.

In addition, in mammals the conditions for diffusion at the respiratory surface are improved by:
- a **blood circulation system**, which rapidly moves oxygen to the body cells as soon as it has crossed the respiratory surface, thereby maintaining the concentration gradient in the lungs
- a **respiratory pigment**, which increases the oxygen-carrying ability of the blood. This is the haemoglobin of the red blood cells, which are by far the most numerous of the cells in our blood circulation.

1 **List** three characteristics of an efficient gas-exchange surface. **Explain** how each influences diffusion.

◆ **Thorax**: in mammals, the upper part of the body separated from the abdomen; in insects, the region between head and abdomen.

◆ **Intercostal muscles**: muscles between the ribs involved in ventilation.

◆ **Diaphragm**: sheet of tissues, largely muscle, separating thorax from abdomen in mammals.

◆ **Pleural membrane**: lines lungs and thorax cavity; it secretes the pleural fluid.

◆ **Trachea**: windpipe

◆ **Bronchus (plural, bronchi)**: a tube connecting the trachea with the lungs.

◆ **Bronchiole**: small terminal branch of a bronchus.

◆ **Alveolus (plural, alveoli)**: air sac in the lung.

The structure of the lung of mammals

The structure of the human thorax is shown in Figure B3.1.3. The lungs are in the **thorax**, an airtight chamber formed by the **ribcage** and its muscles (**intercostal muscles**), with a domed floor, the **diaphragm**. The diaphragm is a sheet of muscle attached to the body wall at the base of the ribcage, separating the thorax from the abdomen. The internal surfaces of the thorax are lined by the **pleural membrane**, which secretes and maintains pleural fluid. Pleural fluid is a lubricating liquid derived from blood plasma; it protects the lungs from friction during breathing movements.

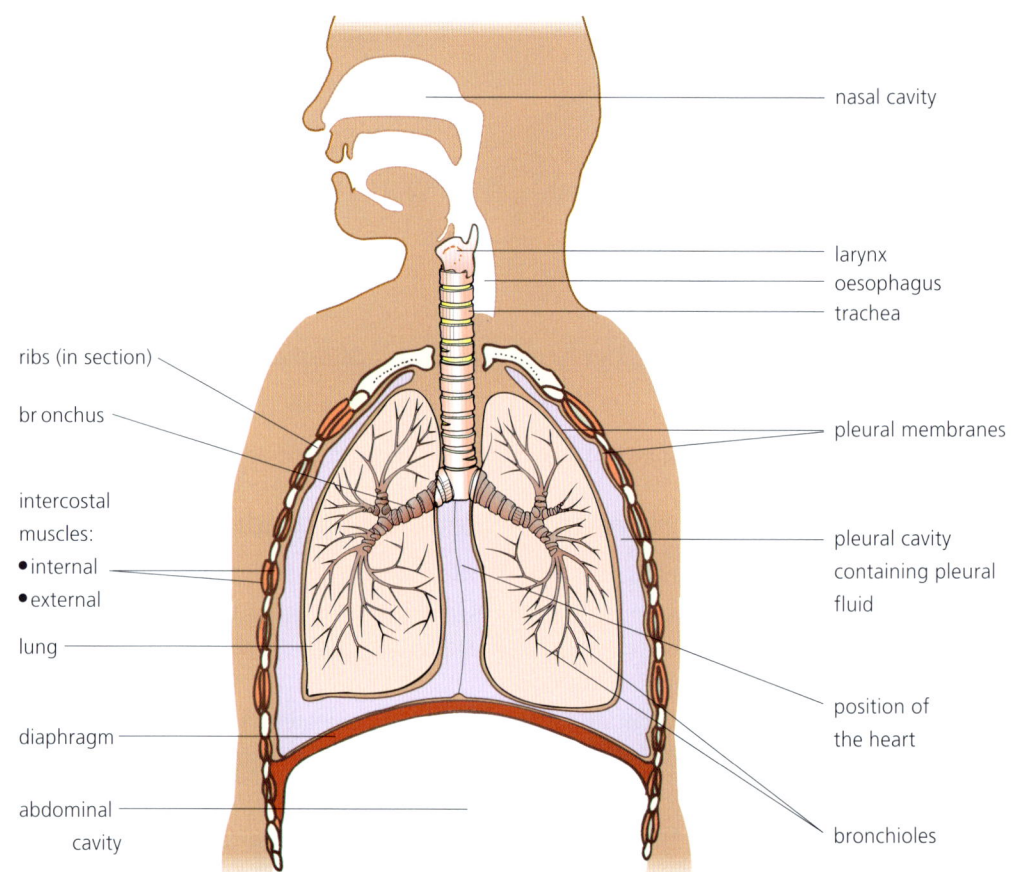

■ Figure B3.1.3 The structure of the human thorax

From trachea to alveoli

The lungs connect with the rear of the mouth by the **trachea** (Figure B3.1.3). The trachea then divides into two **bronchi**, one to each lung. Within the lungs the bronchi divide into smaller **bronchioles**. The finest bronchioles end in the **alveoli** (air sacs). The walls of bronchi and larger bronchioles contain smooth muscle and are also supported by rings or tiny plates of cartilage, preventing collapse that might be triggered by a sudden reduction in pressure that occurs with powerful inspirations of air.

Lungs are extremely efficient, but they cannot prevent some water loss during breathing – an issue for most terrestrial organisms.

Link
Adaptations of the cells in the alveoli are discussed in detail in Chapter B2.3, page 266.

Adaptations of mammalian lungs for gas exchange

Alveolar structure and gaseous exchange

The lung tissue consists of the alveoli, arranged in clusters, each served by a tiny bronchiole. Bronchioles exist in a branched network, spreading out within the lung to provide an even distribution of alveoli. Alveoli have elastic connective tissue as an integral part of their walls (Figure B3.1.4). There are very many small alveoli in each lung – the small size and large number of alveoli provide a large surface area for gas exchange.

A capillary system wraps around the clusters of alveoli (Figure B3.1.4). The extensive capillary beds are an adaptation of the lungs for gas exchange, providing a large surface area for the diffusion of oxygen from the alveoli into the blood and carbon dioxide out of the blood into the alveoli. Each capillary is connected to a branch of the pulmonary artery and is drained by a branch of the pulmonary vein. The pulmonary circulation is supplied with deoxygenated blood from the right side of the heart and returns oxygenated blood to the left side of the heart to be pumped to the rest of the body.

Common mistake

A common misunderstanding is that the spherical shape of alveoli gives the lungs a large surface area for gas exchange. In fact, a sphere has less surface area for a given volume than any other shape. It is the small size and large number of alveoli that gives the large surface area.

Surfactant lines the inner surface of the alveoli. It is secreted by cells in the wall of alveoli and reduces surface tension. Because of the tiny diameter of the alveoli (about 0.25 mm) they would tend to collapse under surface tension during expiration, with their walls sticking together. The lung surfactant lowers the surface tension, permitting the alveoli to flex easily as the pressure of the thorax falls and rises.

Top tip!

Make sure you know how the lung is adapted for gas exchange. The adaptions include: the presence of surfactant, a branched network of bronchioles, extensive capillary beds and a large surface area.

Common mistake

Do not refer to an 'alveolar membrane' because this leads to confusion with cell plasma membranes. The term 'wall' is preferable. One of the adaptations of alveoli is that alveolar walls are one cell thick. Do not confuse 'alveolar walls' with 'cell walls'.

Blood arriving in the lungs is low in oxygen but high in carbon dioxide. As blood flows past the alveoli, gaseous exchange occurs by diffusion. Oxygen dissolves in the alveolar surface film of water, diffuses across into the blood plasma and into the red blood cells, where it chemically combines with haemoglobin to form oxyhaemoglobin. At the same time, carbon dioxide diffuses from the blood into the alveoli.

■ Table B3.1.1 The composition of air in the lungs

Component	Inspired air/%	Alveolar air/%	Expired air/%
oxygen	20.9	14	16
carbon dioxide	0.04	5.5	4.0
nitrogen	79	81	79
water vapour	variable	saturated	saturated

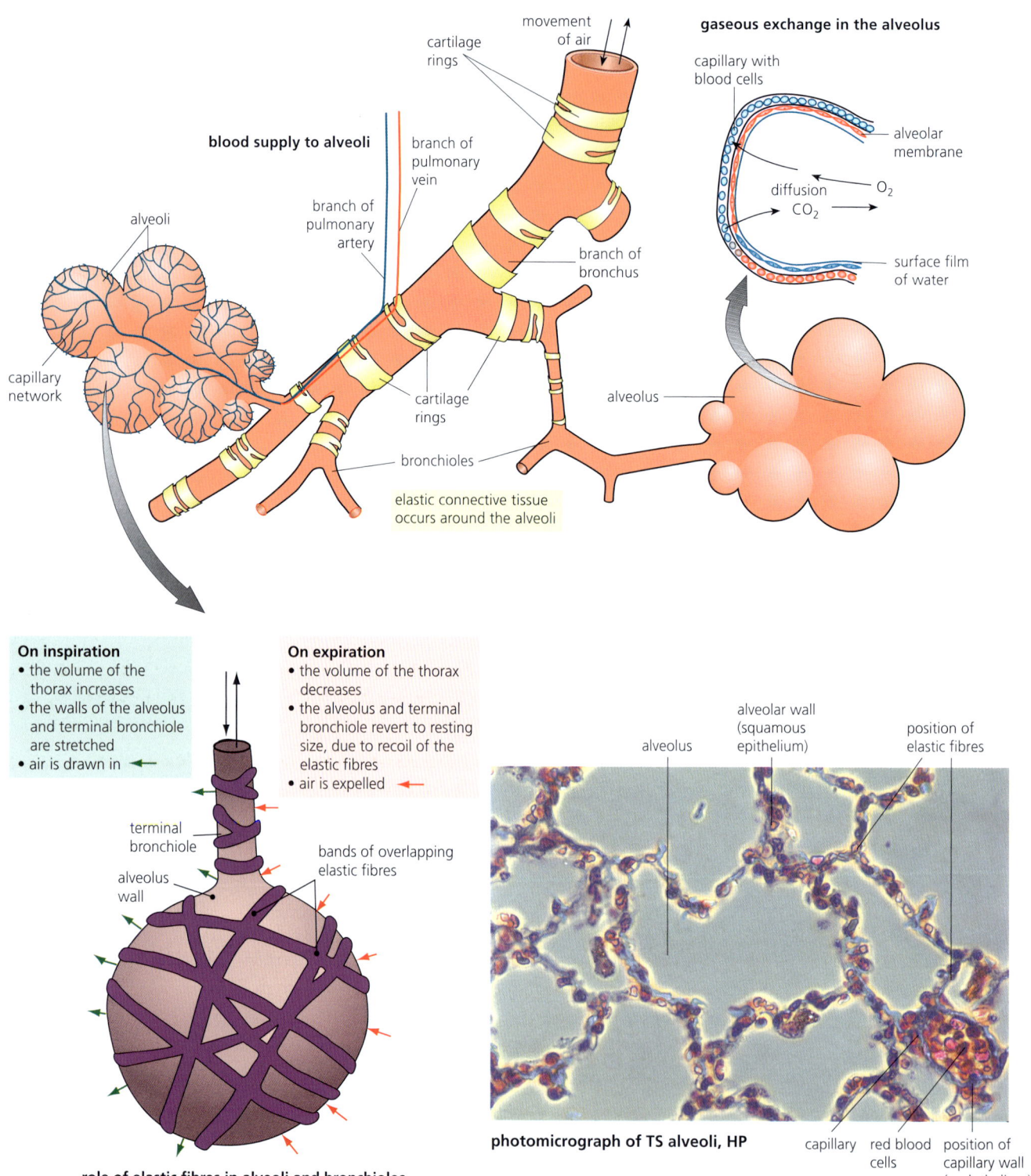

■ Figure B3.1.4 Gaseous exchange in the alveoli

Common mistake

A common misconception is that the gas breathed in is oxygen and the gas breathed out is carbon dioxide. The air breathed in and out contains both oxygen and carbon dioxide – exhaled air has a higher concentration of carbon dioxide than inhaled air, and inhaled air has a higher concentration of oxygen.

How effective are mammalian lungs?

Air flow in the lungs of mammals is tidal: air enters and leaves by the same route. Consequently, there is a residual volume of air that cannot be expelled. Incoming air mixes with and dilutes the residual air, rather than replacing it. The effect of this is that air in the alveoli contains significantly less oxygen than the atmosphere outside (see Table B3.1.1).

Nevertheless, the lungs are efficient organs. Their success is due to numerous features of the alveoli that adapt them to gaseous exchange. These are listed in Table B3.1.2.

■ Table B3.1.2 Features of alveoli that adapt them to efficient gaseous exchange

Feature	Effects and consequences
surface area of alveoli	a huge surface area for gaseous exchange (50 m² = area of doubles tennis court)
alveolar wall	very thin, flattened (squamous) epithelium (5 μm) means the diffusion pathway is short
capillary supply to alveoli	network of capillaries around each alveolus (supplied with deoxygenated blood from the pulmonary artery and draining into pulmonary veins) maintains the concentration gradient of O_2 and CO_2
surface film of moisture	O_2 dissolves in water lining the alveoli; O_2 diffuses into the blood in solution
surfactant producing	a detergent secreted by cells in the walls of alveoli; it reduces surface tension and stops alveoli walls from sticking together, keeping alveoli open

Common mistake

Be careful with word choice. It is not the alveolus that is one cell thick, but the alveolar *wall*.

ATL B3.1A

The production of adequate amounts of surfactant in the foetus, from about five months of pregnancy in humans, marks the beginning of the possibility of independent life. Premature babies born before this stage are unable to inflate their lungs and breathe; those born after it can do so and, with intensive care, can survive.

Find out about the effects of premature birth on babies. How are these effects treated? Use your biological knowledge to produce an information leaflet (of the type found in hospitals and doctors' surgeries), which explains to the general public the problems caused by premature birth for babies and how these can be treated.

2 **Outline** how the structure of alveoli adapts them for efficient gas exchange.

Ventilation of the lungs

Air is drawn into the alveoli when the air pressure in the lungs is lower than atmospheric pressure. Air is then forced out when pressure is higher in the lungs than atmospheric pressure. Since the thorax is an airtight chamber, pressure changes in the lungs occur when the volume of the thorax changes. How the volume of the thorax is changed during breathing is illustrated in Figure B3.1.5 and summarized in Table B3.1.3.

Top tip!

Make sure you understand the role of the diaphragm, intercostal muscles, abdominal muscles and ribs in ventilation. Figure B3.1.5 and Table B3.1.3 describe the roles in detail.

inspiration:
- external intercostal muscles contract
- internal intercostal muscles relax
- diaphragm muscles contract

} ribs moved upwards and outwards, and the diaphragm down

expiration:
- external intercostal muscles relax
- internal intercostal muscles contract
- diaphragm muscles relax

} ribs moved downwards and inwards, and the diaphragm up

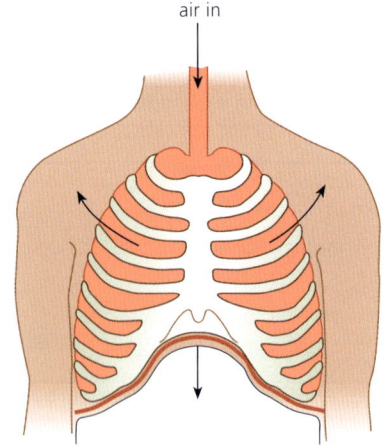
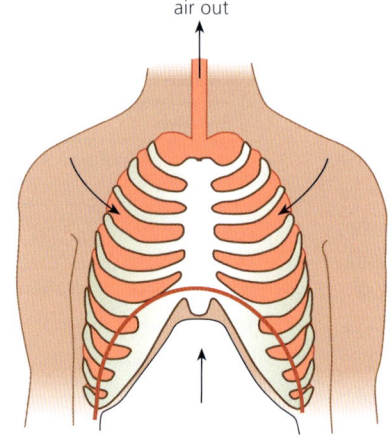
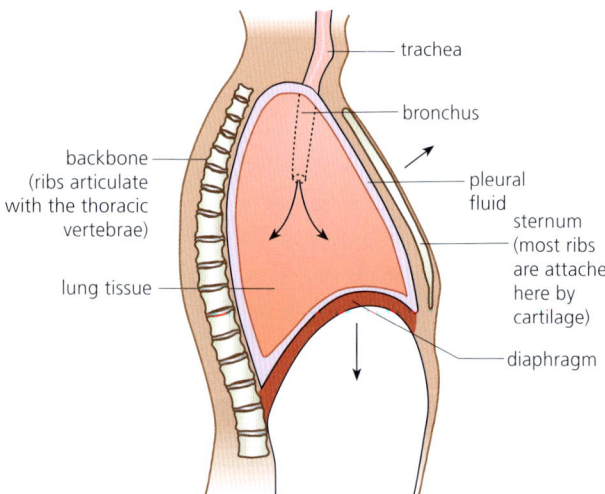
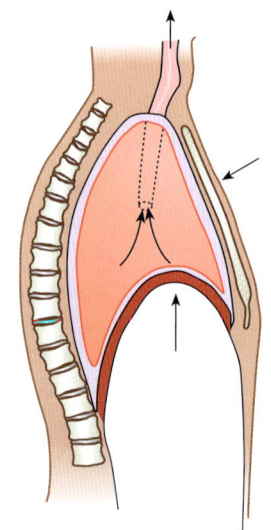

volume of the thorax (and therefore of the lungs) increases; pressure is reduced below atmospheric pressure and air flows in

volume of the thorax (and therefore of the lungs) decreases; pressure is increased above atmospheric pressure and air flows out

■ **Figure B3.1.5** The ventilation mechanism of the lungs

Top tip!

Be clear about which intercostal muscles you are referring to, i.e. internal or external. They are antagonistic muscles, so one is relaxing as the other contracts, and vice versa.

Common mistake

When describing ventilation, make sure you refer to changes of *thoracic* volume rather than changes in the lung volume.

Theme B: Form and function – Organisms

■ Table B3.1.3 The mechanism of lung ventilation – a summary

Inspiration (inhalation)	Structure/outcome	Expiration (exhalation)
muscles contract, flattening the diaphragm and pushing down on contents of abdomen	*diaphragm	muscles relax
relax	*abdominal muscles	contract – pressure from abdominal contents pushes diaphragm into a dome shape
contract, moving ribcage up and out	*external intercostal muscles	relax
relax	*internal intercostal muscles	contract, moving ribcage down and in
increases	**volume of thoracic cavity**	decreases
falls below atmospheric pressure	**air pressure of thorax**	rises above atmospheric pressure
in	**air flow**	out

different muscles are required for inspiration and expiration because muscles only work when they contract – this is referred to as antagonistic muscle action.

Common mistake

Be careful with cause and effect when discussing ventilation of the lungs. For example, it is not the movement of air into the lungs that causes the diaphragm to move down, but rather the diaphragm contracting and moving down, which causes reduced pressure in the thoracic cavity and thus air to be drawn into the lungs.

3 **Compare** the roles of the following muscles during ventilation of the lungs:
   a the internal and external intercostal muscles
   b the diaphragm and abdominal muscles.

Measurement of lung volumes

■ Monitoring of ventilation in humans at rest and after exercise

One way of monitoring ventilation (the movement of air into and out of the lungs) is by data logging using an apparatus called a **spirometer** (Figure B3.1.6). In the diagram, you can see this consists of a plastic lid that moves up and down in the spirometer chamber. The plastic lid is hinged over a tank of water: when the person taking part in the experiment breathes into a mouthpiece, connected to the spirometer chamber by flexible tubing, the lid rises and falls as the volume of air in the chamber changes.

With the spirometer chamber filled with air, the capacity of the lungs when breathing at different rates can be investigated. Note that if the spirometer chamber is filled with oxygen, and a carbon dioxide-absorbing chemical such as soda lime is added to a compartment on the air return circuit, this apparatus can also be used to measure oxygen consumption by the body.

B3.1 Gas exchange

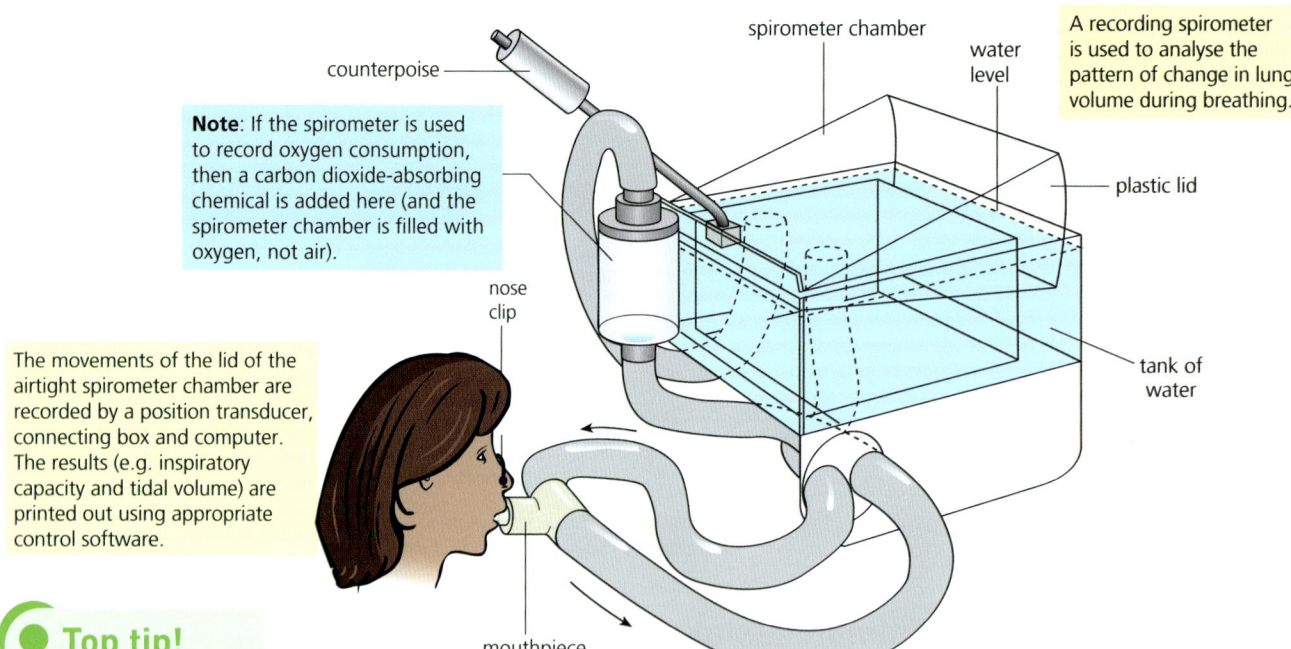

Figure B3.1.6 Investigating breathing with a spirometer. The nose clip is needed to ensure breathing is through the mouth and not through the nose. The arrows indicate the movement of air

The movements of the lid may be recorded by a position transducer, connecting the box to a computer. From 'traces' printed out from investigations of human breathing under different conditions (including at rest, and after mild and vigorous activity), ventilation rate and tidal volume may be measured. Figure B3.1.7 shows an example trace – examine it closely.

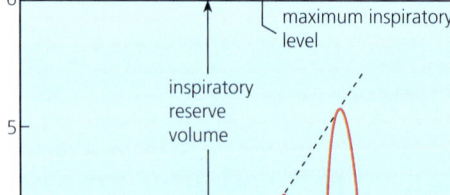

Tidal volume and inspiratory and expiratory reserves

The volume of air breathed in and out during normal, relaxed, rhythmical breathing, the **tidal volume**, is typically 400–500 cm³ (0.4–0.5 dm³). Tidal volume is the volume of air taken in and out with each inhalation or exhalation. However, we have the potential for an extra-large intake (**maximum inspiratory capacity**) and an extra-large expiration of air (**expiratory reserve volume**), when required. The difference between the maximum inspiratory level and the tidal volume is called the **inspiratory reserve**, and the difference between the maximum expiratory level and the tidal volume is called the **expiratory reserve** (Figure B3.1.7). These, together, make up about 4.5 dm³.

> ● **Top tip!**
> Make sure you know which measurements determine tidal volume, vital capacity, and inspiratory and expiratory reserves.

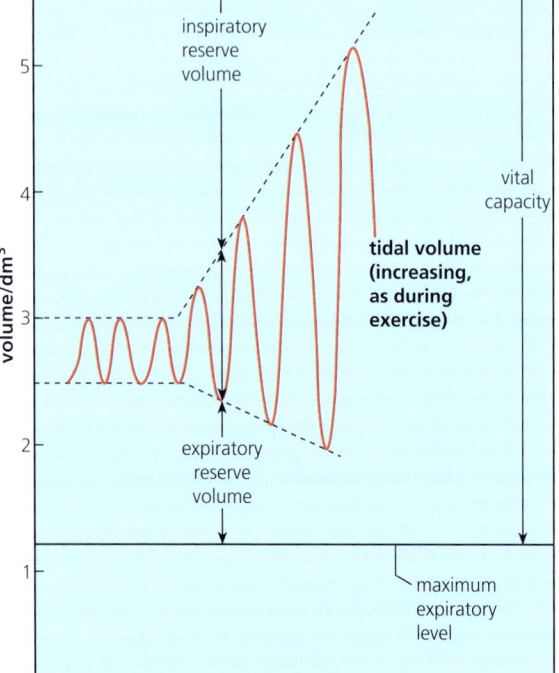

■ **Figure B3.1.7** A spirometer trace

- ◆ **Tidal volume**: volume of air normally exchanged in breathing.
- ◆ **Inspiratory reserve**: difference between the maximum inspiratory level and the tidal volume.
- ◆ **Expiratory reserve**: difference between the maximum expiratory level and the tidal volume.
- ◆ **Vital capacity (VC)**: the total amount of air exhaled after maximal inhalation.
- ◆ **Ventilation rate**: number of inhalations or exhalations per minute.

■ Vital capacity

Vital capacity (VC) is the total amount of air exhaled after maximal inhalation. The value varies according to age, body size and level of regular physical activity. It is calculated by adding the tidal volume (TV), inspiratory reserve volume (IRV) and expiratory reserve volume (ERV). That is:

$$VC = TV + IRV + ERV$$

■ Ventilation rate

The spirometer can also be used to investigate steady breathing over a short period of time and, in these cases, the y-axis of the spirometer trace is 'time'. In this way, the rate at which the lungs are ventilated can be investigated. The **ventilation rate** is the number of inhalations or exhalations per minute.

Alternative experimental approaches are possible, including the use of a chest belt and pressure monitor linked to a data logger.

> ### Inquiry 3: Concluding and evaluating
>
> #### Concluding
>
> Use data from an investigation on the effects of exercise on breathing rate. Describe the differences between the different levels of exercise. Explain the differences between each experiment, using your biological knowledge. Why did breathing rate increase under more intense exercise? (What metabolic process is increasing in cells, and how do the products formed increase breathing rate/depth? Why must breathing rate/depth increase?) Use as many different synoptic ideas from the course (i.e. subject knowledge from different parts of the syllabus) as you can think of to explain your results.
>
> #### Evaluating
>
> What were the limitations of the experiment? How would they have affected the accuracy of the result? How could you have improved the experiment to reduce these limitations and improve accuracy?
>
> Why did you repeat the experiment? (*Hint*: this enables you to comment on reliability and identify any anomalous results.) What is standard deviation, and what does this tell you about your results? (*Hint*: i.e. variation around the mean.)
>
> If you were to carry out another experiment investigating the effect of exercise on ventilation, what improvements could you make? Describe an experiment and explain the improvements you have made.

- ◆ **Epidermis**: outer layer(s) of cells.
- ◆ **Mesophyll tissue**: parenchyma cells containing chloroplasts.
- ◆ **Vascular tissue**: xylem and phloem of plants.
- ◆ **Vascular bundle**: strands of xylem and phloem (often with fibres) separated by cambium; the site of water and food movements up and down the stem.
- ◆ **Palisade mesophyll**: columnar (oblong-shaped) cells containing many chloroplasts found beneath the upper epidermis in leaves.
- ◆ **Spongy mesophyll**: rounded cells in the leaf that are loosely packed, creating air spaces where air circulates, providing a large surface area for gas exchange; contains chloroplasts but fewer than palisade mesophyll.

Gas exchange in plants

■ Adaptations for gas exchange in leaves

The leaf is the plant organ where gas exchange occurs. The top and bottom of the leaf are covered by a single layer of cells, the **epidermis**, and the inner part of the leaf contains **mesophyll tissue** and **vascular tissue** in a system of **vascular bundles** (Figure B3.1.8). The vascular bundles in leaves are referred to as veins. The bulk of the leaf is taken up by mesophyll tissue and the cells here are supported by veins arranged in a branching network. There are two types of mesophyll cells: the **palisade mesophyll**, which form a single layer of cells towards the top of the leaf and are full of chloroplasts to maximize photosynthesis, and the **spongy mesophyll**, which contain air spaces where gas exchange occurs.

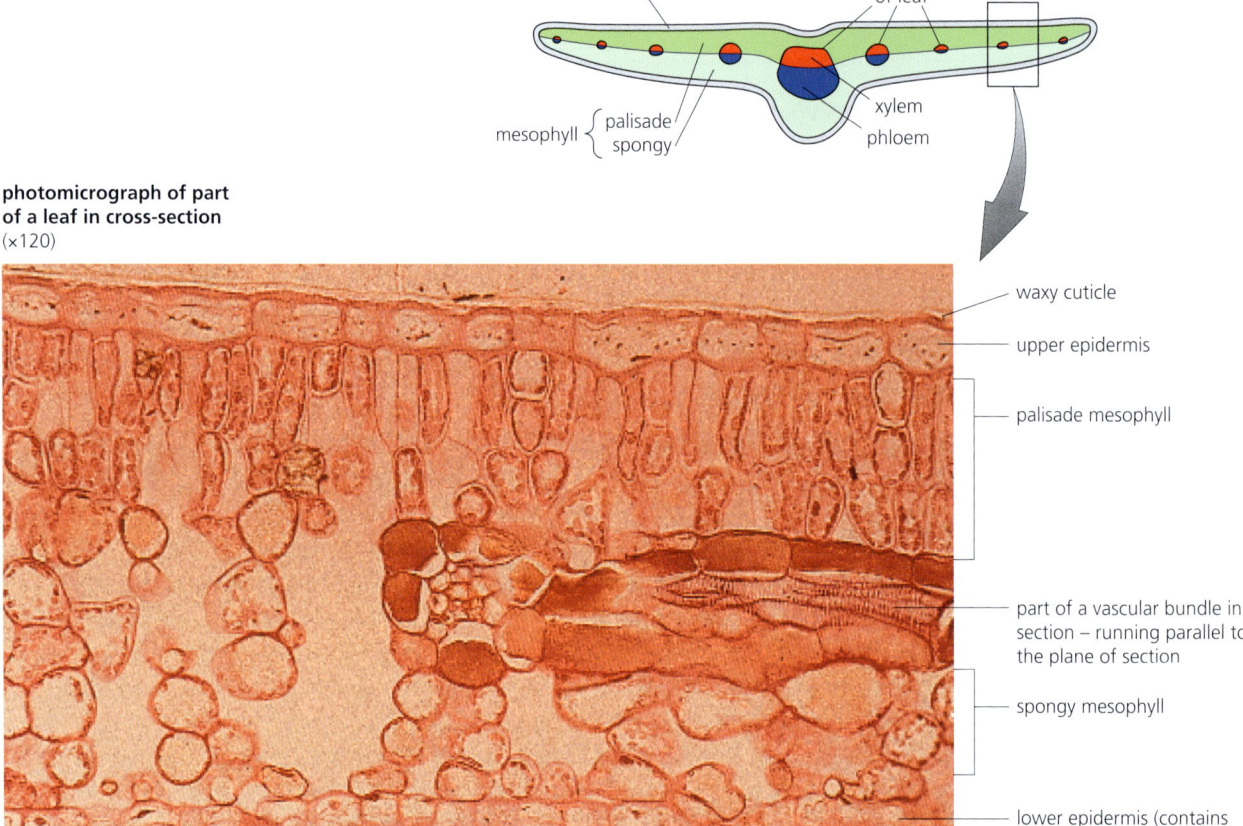

■ Figure B3.1.8 The distribution of tissues in the leaf

◆ **Stoma (plural, stomata)**: pore in the epidermis of a leaf, surrounded by two guard cells.

◆ **Turgid**: condition where the vacuole of a plant cell is full of water, pushing the cell membrane against the cell wall.

◆ **Flaccid**: plant cell that has become soft and less rigid than normal because the cytoplasm within its cells has shrunk and contracted away from the cell walls through loss of water.

The tiny pores of the epidermis of leaves, through which gas exchange can occur, are known as **stomata** (Figure B3.1.9). Most stomata occur in the epidermis of leaves, but some do occur in stems. Each stoma is surrounded by two elongated guard cells. Guard cells change shape to open and close the stomata.

Common mistake

Do not confuse 'guard cell' with 'stoma'. Guard cells create the stoma when they open to form a pore (or hole) between them (the stoma), through which gas exchange occurs. 'Stoma' should be used when referring to singular structures, and 'stomata' when referring to several.

Stomata open and close due to the change in turgor pressure of the guard cells. They open when water is absorbed by the guard cells from the surrounding epidermal cells. The guard cells then become fully **turgid** and they each push into the epidermal cell beside them (because of the way cellulose is laid down in the walls, Figure B3.1.9). A pore develops between the guard cells. When water is lost and the guard cells become **flaccid**, the pore closes again.

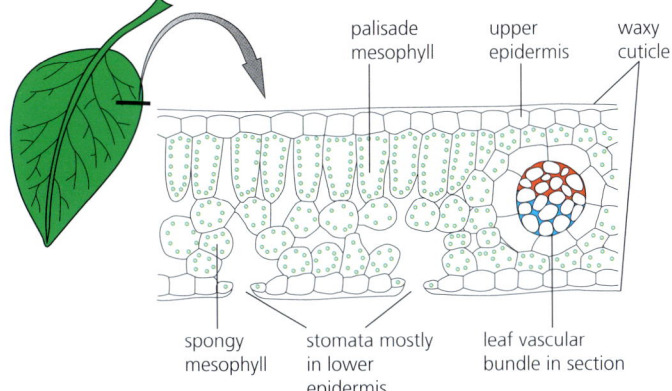

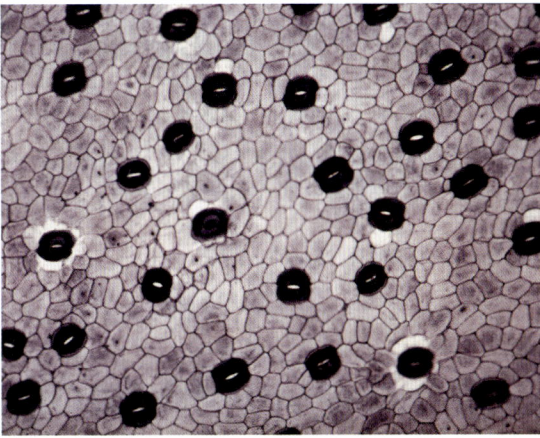

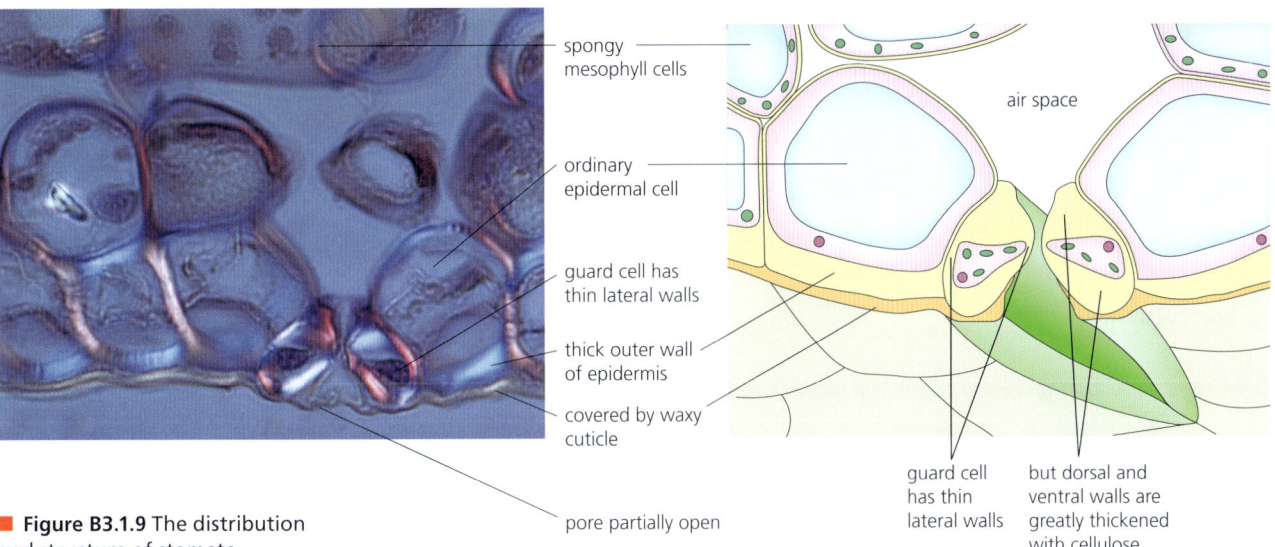

■ **Figure B3.1.9** The distribution and structure of stomata

The adaptations of the leaf for gas exchange include:
- **waxy cuticle**: forms a largely impermeable barrier, so gases and water vapour must leave via the stomata on the underside of the leaf, allowing gas exchange and water loss to be controlled.
- **epidermis**: the lower epidermis contains many pores (stomata) that facilitate gas exchange. Most of these are in the lower epidermis, away from the brightest sunlight, so that humidity and temperature is lower than they are above the leaf, to reduce water loss.
- **air spaces**: allow gases to circulate around the loosely packed spongy mesophyll cells, maintaining a concentration gradient between air and cells
- **spongy mesophyll**: provides a large surface area for gas exchange – carbon dioxide entering the cells and oxygen leaving
- **stomatal guard cells**: open and close to control gas exchange and water loss
- **veins** (vascular bundle): carry water to the leaves (in xylem vessels), which is then lost through stomata by the process of transpiration. Water is required for the process of photosynthesis – a small proportion of the water entering the leaf is used in photosynthesis – carbon dioxide is also needed, and so ongoing photosynthesis in the leaves, driven by water supply via veins and carbon dioxide through stomata, maintains concentration gradients of both molecules.

B3.1 Gas exchange

> **Top tip!**
> Make sure you can draw and label a plan diagram to show the distribution of tissues in a transverse section of a dicotyledonous leaf, as shown in Figure B3.1.10, but not details of individual cells (as described later in this chapter).

> **Concept: Form**
>
> Further adaptations to increase gas exchange include:
> - leaves provide a **large surface area** so that many stomata can be incorporated. Stomata allow movement of gases in and out of the air spaces inside the leaf to maintain a steep concentration gradient of carbon dioxide from the outside of the leaf to the inside
> - leaves are **flat**, which provides a large surface area-to-volume ratio
> - leaves are **thin** to facilitate fast diffusion of carbon dioxide to all cells.

Distribution of tissues in a leaf

Figures B3.1.8 and B3.1.9 show the distribution of tissues in a leaf in relation to the adaptations to gas exchange. Figure B3.1.10 shows a summary of the different tissues within a dicotyledonous leaf.

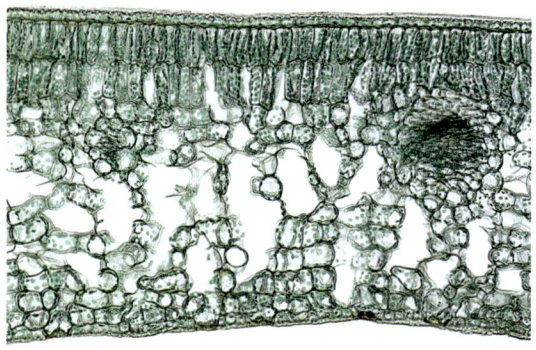
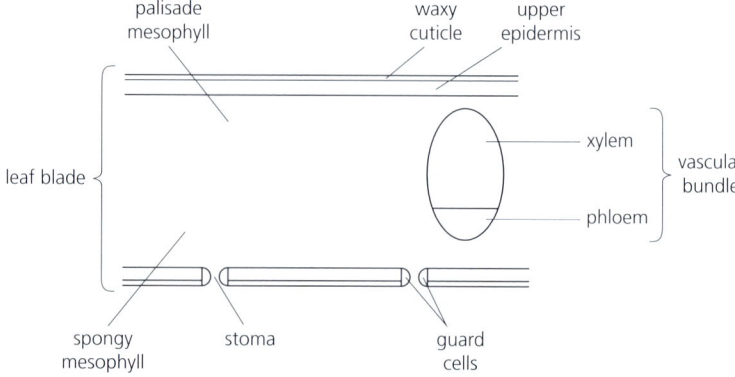

■ Figure B3.1.10 Section through a typical leaf showing different tissues and cell structure

> 4 **Explain** how a leaf is adapted for efficient gas exchange.
> 5 **Compare and contrast** adaptations for gas exchange in mammalian lungs and in the leaves of plants.

Transpiration

♦ **Transpiration**: the evaporation of water from the spongy mesophyll tissue and its subsequent diffusion through the stomata.

When stomata open for gas exchange, water vapour is lost from the leaf. Water vapour diffuses from the interior of the leaf to the air outside down a concentration gradient. The process of the loss of water vapour from leaves is called **transpiration** (Figure B3.1.11).

Transpiration is found to be dramatically affected by environmental conditions around the plant.

> **Common mistake**
>
> It is incorrect to say that 'water' diffuses from the leaf during transpiration: you need to say that water vapour diffuses through stomata.

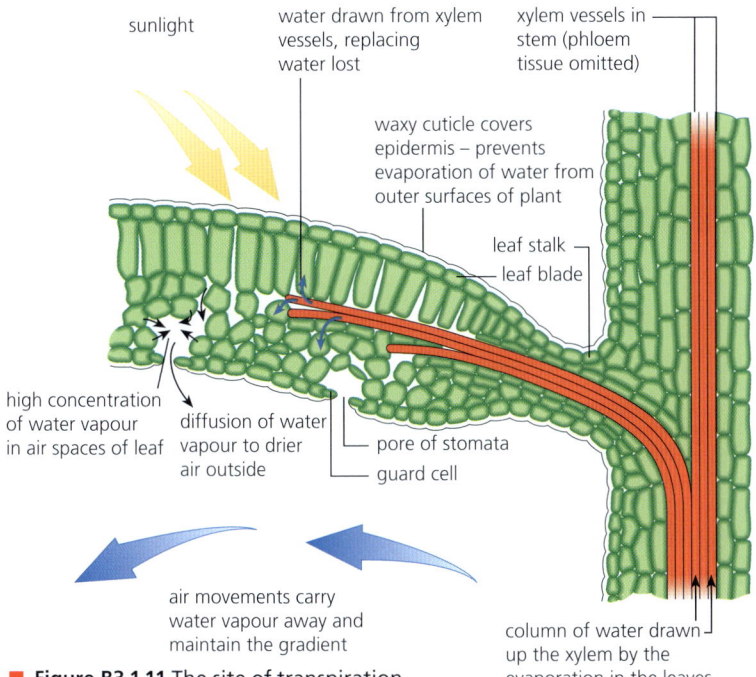

Figure B3.1.11 The site of transpiration

Why is this so?

Transpiration occurs because water molecules evaporate from the cellulose walls of cells in the leaf continuously. This increases the water potential of the air spaces between mesophyll cells. If the air outside the plant has a lower water potential (i.e. it is less humid – as it very often is) *and the stomata are open*, then water vapour will diffuse out into the drier air outside, from a higher to lower water potential.

Common mistake

It is incorrect to refer to the movement of water molecules from a liquid state inside the leaf to the gases of the air outside as *osmosis*, because water is not travelling through a cell membrane (something that defines osmosis).

The following factors can affect the rate of transpiration:
- **Temperature** affects transpiration because it causes the evaporation of water molecules from the surfaces of the cells of the leaf. A rise in the concentration of water vapour within the air spaces increases the difference in concentration in water vapour between the leaf's interior and the air outside, so diffusion is enhanced. In addition, water vapour molecules will move faster, resulting in faster diffusion. An increase in temperature of the leaf therefore raises the transpiration rate.
- High **humidity** slows transpiration. If humid air collects around a leaf, it decreases the difference in concentration of water vapour between the interior and exterior of the leaf, slowing diffusion of water vapour from the leaf.
- **Wind** sweeps away the water vapour molecules accumulating outside the stomata of the epidermis of the leaf surface, enhancing the difference in concentration of water vapour between the leaf interior and the outside. Movements of air around the plant enhance transpiration.
- **Light intensity** affects transpiration because the stomata tend to be open in the light in order for carbon dioxide to diffuse into the leaf for the process of photosynthesis, at the same time resulting in the loss of water vapour from the leaf. Light from the Sun also contains infrared rays that warm the leaf and raise its temperature. Light is an essential factor for transpiration.

Top tip!

Make sure you are aware of the factors affecting the rate of transpiration.

Inquiry 1: Exploring and designing

Designing

Potometers can be used to measure the rate of water uptake in plants (Figure B3.1.12).

The apparatus consists of a leafy shoot inserted into a tube (the seal must be airtight), which is attached to a capillary tube.
- The capillary is attached to a reservoir of water.
- The apparatus must be set up underwater to ensure that the column of water is continuous between the plant and the capillary.
- An air bubble is introduced into the capillary. As water transpires from the leaf, water is pulled up the tube and along the capillary. The rate of movement of the air bubble (distance moved per minute) is an indirect measure of transpiration rate (although not all the water entering the plant leaves it).
- The tap below the reservoir allows the bubble to be reset so that a new measurement can be made.

B3.1 Gas exchange

Design an investigation into the effect of one **abiotic variable** (e.g. temperature, wind, humidity, light) on the rate of transpiration in plants. Consider these points:

- Which plant species will you use? (It must be one that fits into the potometer apparatus.)
- What will be your independent variable (the abiotic factor you will be changing)?
- What are your controlled variables (factors you will be keeping the same)?
- How many repeats will you carry out to ensure that your results are reliable?
- A potometer does not measure transpiration directly; it measures the rate of water uptake by the cut stem. How would this affect the conclusions of a study using a potometer to investigate the effects of environmental conditions on the loss of water from plants?

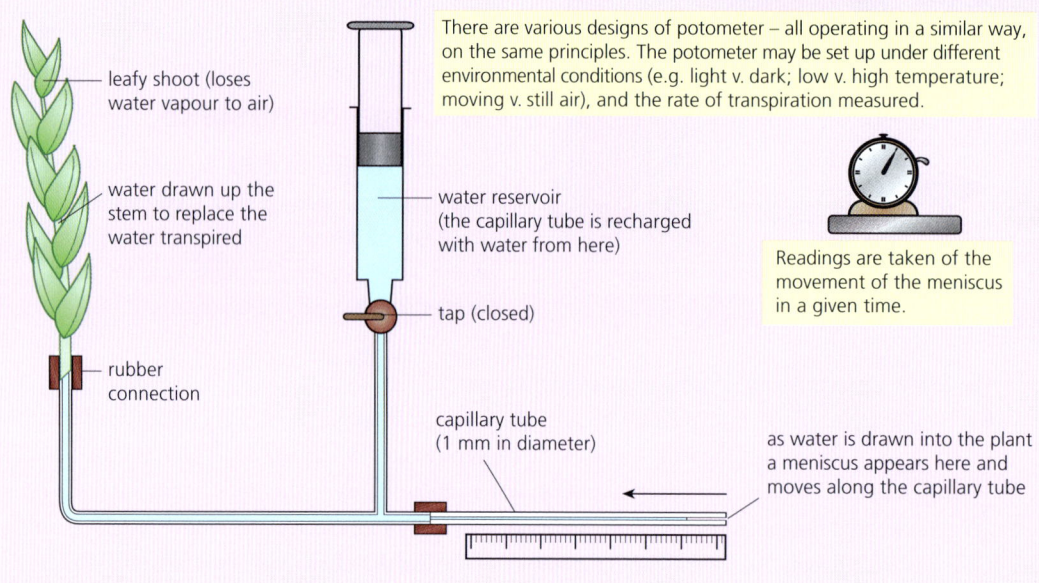

■ Figure B3.1.12 Investigating transpiration and the factors that influence it

TOK

Consider how behaving like a 'biologist' means being able to perform certain skills. This type of knowledge is 'ability knowledge' and something you can analyse in your TOK course.

Top tip!

A potometer measures water uptake rather than water loss. The rate of uptake of water will not be the same as the rate of transpiration; some of the water (around 5%) remains in the plant for photosynthesis and to keep the cells turgid. The rest (95%) is lost from the plant through transpiration. The difference between water uptake and water loss can be important for a large tree, but for a small shoot in a potometer the difference is usually trivial and can be ignored.

Top tip!

When explaining the effect of environmental factors on transpiration rate, make sure you refer to **concentration gradients**. The air spaces inside the leaf are at or close to saturation with water vapour. High humidity in the air outside stomata will reduce the concentration gradient between the air spaces in the leaf and the air outside, because the air will be as saturated with water as the inside of the leaf. Wind, in contrast, leads to a steeper concentration gradient between the air spaces and the air outside, increasing the transpiration rate.

Tool 1: Experimental techniques

Accurately measuring time

The SI unit for time is the second (s). Other units are the hour (h) and the minute (min). Time is usually measured in the laboratory with a digital stop watch for accuracy.

Tool 1: Experimental techniques

Physical modelling

Models of water transport in the xylem, and demonstrations of the power and inevitability of evaporation from moist surfaces, can be designed and tested using familiar laboratory equipment.

Design a model that represents transpiration in plants. You could use the model to carry out the investigation you designed, using a potometer, to measure the effects of an abiotic variable on transpiration rate.

Here are some ideas for equipment you could use.

- Porous pot
  - models evaporation from leaves
  - water fills the clay of the pot, demonstrating adhesive forces
  - as water is drawn into the porous pot, cohesive forces are demonstrated as water moves up the capillary tube attached to the pot.
- Capillary tubing
  - demonstrates cohesive and adhesive forces in xylem
  - water binds to the side of the tubing, demonstrating adhesive forces
  - water is drawn up the capillary, demonstrating cohesive forces.
- Filter or blotting paper
  - models evaporation from leaves.

ATL B3.1B

Stomata tend to open in daylight and be closed in the dark. This diurnal pattern is overridden, however, if the plant becomes short of water and starts to wilt. For example, in very dry conditions when there is an inadequate water supply, stomata inevitably close relatively early in the day (turgor cannot be maintained). This limits water vapour loss by transpiration and halts further wilting. Adequate water reserves from the soil may be taken up subsequently, thereby allowing the opening of stomata again the following day. The effect of this mechanism is that stomata regulate transpiration by preventing excessive water loss (Figure B3.1.13).

Examine Figure B3.1.13. Suggest why the stomatal apertures of the plant in very dry conditions differed in both maximum size and duration of opening from those of the plant with adequate moisture.

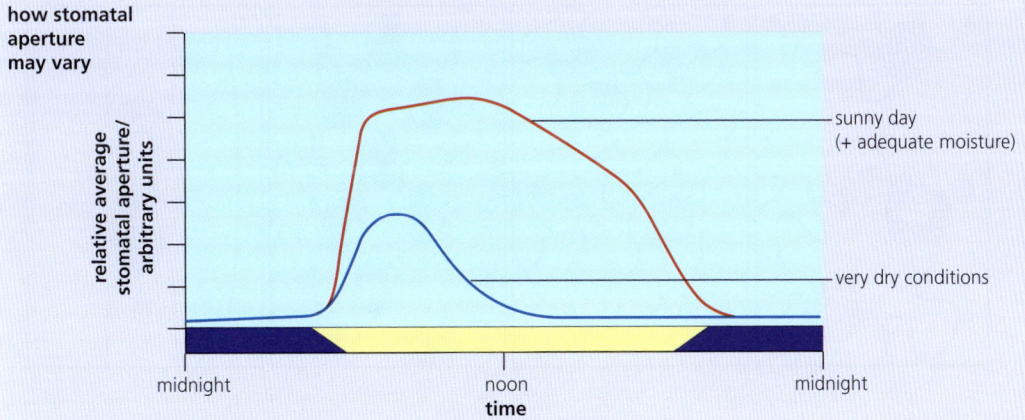

■ Figure B3.1.13 Stomatal opening and environmental conditions

Stomatal density

The more stomata per unit area (stomatal density), the more carbon dioxide can be taken into the spongy mesophyll and the more water can be released by transpiration. Therefore, higher stomatal density can greatly increase both water loss and carbon dioxide uptake.

Stomatal density can be measured using micrographs of leaves or by using leaf casts, as shown on the next page.

B3.1 Gas exchange

⚙ Performing a leaf cast to determine stomatal density

1. Take a bottle of nail varnish and a leaf.
2. Spread nail varnish thinly on underside of a leaf (Figure B3.1.14).
3. Leave varnish to dry – this can be accelerated by creating air currents over the leaf (by waving in the air or blowing on the leaf).
4. Take a piece of transparent sticky tape and place it over the dried varnish.
5. Slowly and carefully pull up the sticky tape from the leaf surface – this will pull a layer of varnish with it.
6. Put tape sticky-side down on a slide.
7. View the slide under a microscope – start with low power and then increase to see details of stomata.
8. Count the number of stomata in the field of view. Work out the size of the field of view (put a ruler under the objective lens and use this to estimate the diameter of the area under view – the area can be calculated by using the formula $\pi \times r^2$, where r is the radius of the field of view).
9. Measure the surface area of the leaf by tracing the outline on graph paper. Extrapolate the number of stomata in the field of view to the full surface area of the leaf.

■ **Figure B3.1.14** Making a leaf cast using nail varnish

■ Variation in stomatal density between different species of plant

Plants are adapted to the environment they live in. One adaptation is the size and shape of stomata and the stomatal density (Figure B3.1.15).

■ **Figure B3.1.15** Clear differences in stomatal size and stomatal density can be observed between the four species of plant. Stomatal traits also vary between different species. **(A)** *Arabidopsis thaliana* and **(B)** *Phaseolus vulgaris* display 'kidney-shaped' guard cells (coloured in green). The grasses **(C)** *Oryza sativa* and **(D)** *Triticum aestivum* show dumb-bell-shaped guard cells (solid green) and specialized subsidiary cells (light green gradient). Subsidiary cells may support guard cell function by offering a mechanical advantage that helps guard cell movements, and/or by acting as a reservoir for water and ions.

Theme B: Form and function – Organisms

Nature of science: Measurements

Creating knowledge that can be relied upon is increased by repeating measurements

The level of trust we can place in **quantitative data** (numerical measurements with units) is increased by repeating measurements. All measurements are limited in precision and accuracy. Repeating measurements under identical conditions increases the reliability of quantitative data. In the case of measuring stomatal density, repeating the counts of the number of stomata visible in the field of view at high power illustrates the variability of biological material. Being able to compare data is crucial in the production of scientific knowledge. The results from a single counting may contain an error. However, if repeated measurements tend to return the same results, we say that the data are 'reliable'; this is not the same thing as saying the data are 'true' but it does mean that we can rely on that data more when more measurements give us similar values.

TOK

In Theory of Knowledge, you will often see questions (IA Prompts or Essay Prescribed Titles) that refer to the 'pursuit of knowledge'. This generally refers to the 'Methods and Tools' element of the Knowledge Framework. So, for example, you could explore this notion of 'reliability' as a feature of the scientific method: repeating experiments and finding that the results are similar each time suggests that the hypothesis being tested is something we can rely on.

Haemoglobin and the transport of oxygen

 Partial pressure of oxygen

As terrestrial animals living at or near sea level, we live at the bottom of a 'column' of air. The atmosphere that surrounds us exerts a significant pressure. The air we breathe is a mixture of gases and, in a gas mixture, each gas exerts a pressure. In fact, the pressure of a mixture of gases is the sum of the partial pressure of the component gases. Consequently, the pressure of a specific gas in a mixture of gases is called its **partial pressure**. The symbol for partial pressure is p, and the partial pressure for a gas (X) is pX. So, for example, pO_2 denotes the partial pressure of oxygen.

◆ **Partial pressure**: the pressure of a specific gas in a mixture of gases.

Tool 3: Mathematics

Applying and using SI prefixes and units

Partial pressure is the mole fraction of the total gas pressure that is exerted by a particular gas. The SI unit of pressure is the **pascal (Pa)** and its multiple is the **kilopascal (kPa)**.

What is the partial pressure of the oxygen in the air around us?

At sea level, the atmospheric pressure is typically about 101.3 kPa and air is 20.9% oxygen. So, the partial pressure of oxygen is given by:

$$\frac{101.3 \times 20.9}{100} = 21.2 \, \text{kPa}$$

The significance of this figure is reflected in the properties of the respiratory pigment haemoglobin, which transports oxygen in the blood.

B3.1 Gas exchange

HL ONLY

Oxygen dissociation curves

Haemoglobin occurs in the red blood cells. Each red blood cell contains about 280 million molecules of haemoglobin. The haemoglobin molecule is built of four interlocking polypeptide subunits (Figure B3.1.16; see also Chapter B1.2, page 217). These subunits are composed of a large globular protein with a non-protein haem group attached, containing iron. One molecule of oxygen combines with each haem group, at the concentration of oxygen that occurs in our lungs. This means each haemoglobin molecule can transport four molecules of oxygen:

haemoglobin + oxygen → oxyhaemoglobin

$$Hb + 4O_2 \rightarrow HbO_8$$

The affinity of haemoglobin for oxygen is measured experimentally by finding the percentage saturation with oxygen of blood exposed to air mixtures that contain different partial pressures of oxygen. The result is called an **oxygen dissociation curve** (Figure B3.1.16).

◆ **Oxygen dissociation curve**: graph of percentage saturation (with oxygen) of haemoglobin against concentration of available oxygen.

◆ **Cooperative binding**: process where the addition of a substance to the subunit of a macromolecule increases the affinity of a neighbouring subunit for the same substance.

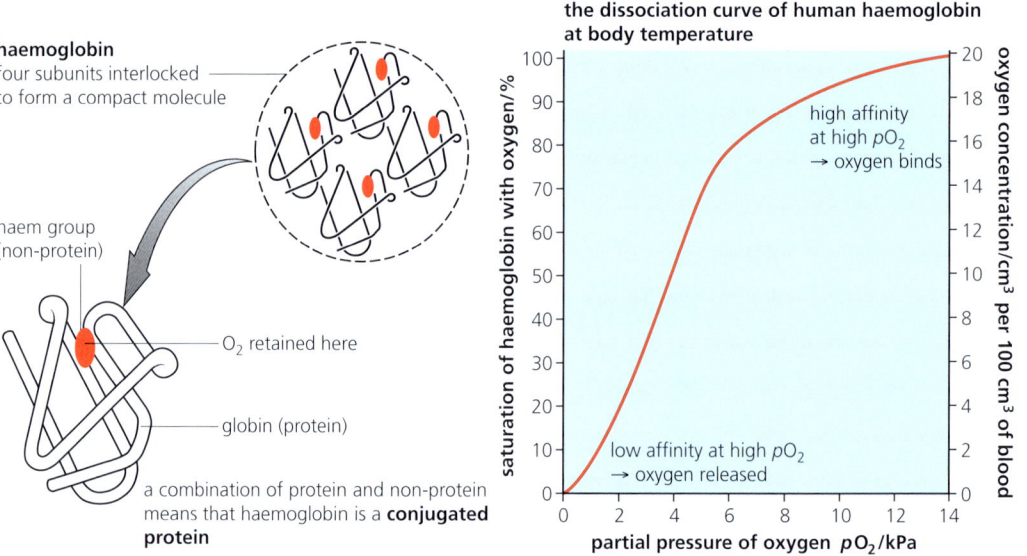

■ Figure B3.1.16 The structure of haemoglobin and its affinity for oxygen

The shape of haemoglobin is altered as each oxygen molecule binds, making each successive binding of oxygen easier. This is known as **cooperative binding**. As a result, haemoglobin has a higher affinity for oxygen in oxygen-rich areas (such as the air spaces in lung alveoli), promoting oxygen loading, and has a lower affinity for oxygen in oxygen-depleted areas (such as the muscles), promoting oxygen unloading (see Figure B3.1.16).

Notice that the oxygen dissociation curve is **S-shaped** (sigmoid curve). This tells us that, in the complex haemoglobin molecule, the first oxygen molecule attaches with difficulty but, once it has, the second combines more easily, and so on until all four are attached and the molecule is saturated. In other words, the amount of oxygen held by haemoglobin depends on the partial pressure of oxygen.

Concept: Function

The function of haemoglobin depends on whether it is exposed to oxygen-rich or oxygen-depleted areas. In the former it picks up oxygen and, in the latter, it unloads it.

Top tip!

When haemoglobin saturation with oxygen is plotted as a function of the partial pressure of oxygen, a sigmoidal (or 'S-shaped') curve is seen. This indicates that the more oxygen is bound to haemoglobin, the easier it is for more oxygen to bind, until all binding sites are saturated.

Theme B: Form and function – Organisms

What is the significance of the oxygen dissociation curve in the working body?

In the body the amount of oxygen held by haemoglobin depends on the partial pressure. In the lungs, air is saturated with water vapour, so the partial pressure of the component gases is different from that outside, in dry air (Table B3.1.4).

■ Table B3.1.4 Partial pressures of the components of air in the alveoli

Component gases	Composition/%	Partial pressure/kPa
nitrogen	75.5	76.4
oxygen	13.1	13.3
carbon dioxide	5.2	5.3
water vapour	6.2	6.3

From the oxygen dissociation curve, we can see that the haemoglobin in red blood cells in the capillaries around the alveoli in the lungs will be about 95% saturated. However, in respiring tissues, the oxygen partial pressure is much lower due to aerobic respiration there. In fact, the oxygen partial pressure in actively respiring tissues may be 0.0–4.0 kPa. At these partial pressures, oxyhaemoglobin breaks down, releasing oxygen in solution that rapidly diffuses into the surrounding tissues. Clearly, the chemistry of haemoglobin makes it an efficient way of transporting oxygen, given the partial pressure of oxygen in respiring tissue compared with that in the lungs.

Adaptations of foetal and adult haemoglobin for the transport of oxygen

The foetus obtains oxygen from its mother's blood through the placenta, where the maternal and foetal circulations come very close together but do not mix. The haemoglobin of the adult mammal and the haemoglobin in the foetal circulation differ slightly in their molecular composition. Foetal haemoglobin has a higher affinity for oxygen (Figure B3.1.17). This means that the haemoglobin present in the circulation of the foetus combines with oxygen more readily than maternal haemoglobin does at the same partial pressure. Because foetal haemoglobin has a higher affinity for oxygen, the dissociation curve is shifted to the left. It is obvious why it is advantageous for foetal haemoglobin to have this property, given that the only access to an oxygen supply is via the placenta. Foetal haemoglobin will load oxygen when adult haemoglobin is unloading it (i.e. in the placenta). If foetal haemoglobin had a lower affinity than maternal haemoglobin, oxygen would pass from the foetus to the mother.

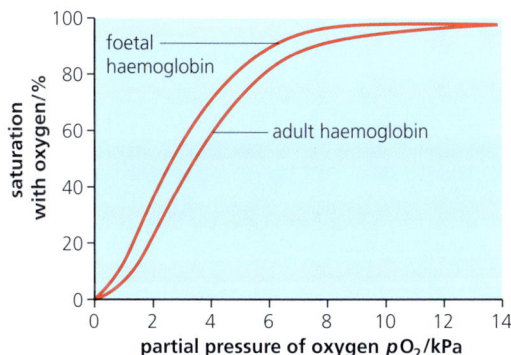

■ Figure B3.1.17 The oxygen dissociation curve of foetal haemoglobin compared to adult haemoglobin

Following birth, foetal haemoglobin is almost completely replaced by adult haemoglobin.

We have already seen that the binding of oxygen to one of the subunits affects the interaction of oxygen with the other subunits (cooperative binding). This is because haemoglobin is an **allosteric protein**. The binding of oxygen to one haemoglobin subunit causes conformational changes that are relayed to the other subunits, making them more able to bind oxygen by raising their affinity for this molecule.

Carbon dioxide is an allosteric effector of haemoglobin. It attaches to haemoglobin, making it more difficult for oxygen to bind, lowering the affinity of haemoglobin for oxygen. In foetal red blood cells, carbon dioxide has a lower allosteric effect, which leads to a higher affinity of oxygen for foetal haemoglobin.

◆ **Allosteric protein**: a protein that can exist in multiple conformations (shapes) depending on the binding of a molecule (at a site other than the catalytic site).

HL ONLY

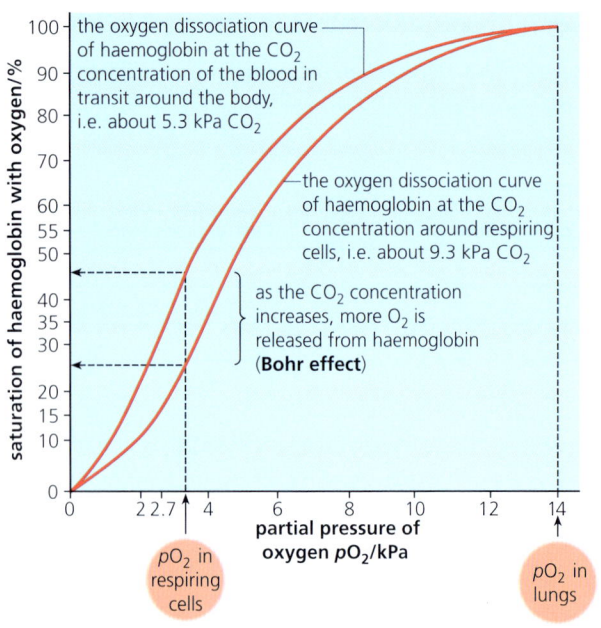

■ **Figure B3.1.18** How carbon dioxide favours release of oxygen in respiring tissues

Bohr shift

The blood circulation also transports carbon dioxide from respiring tissues, where it is at relatively high partial pressures, to the lungs. In respiring cells, the concentration of carbon dioxide is approximately 9.3 kPa, whereas in the lungs, as we have seen in Table B3.1.4, it is 5.3 kPa. The effects of these partial pressures of carbon dioxide on the oxygen dissociation curve of haemoglobin is noticeable (Figure B3.1.18).

An increase in carbon dioxide concentration shifts the oxygen dissociation curve to the right (see Figure B3.1.18). Where the carbon dioxide concentration is high (in the actively respiring cells), oxygen is released from oxyhaemoglobin even more readily. Carbon dioxide lowers the pH, caused by protons from the dissociation of carbonic acid, which causes haemoglobin to release its oxygen. This very useful outcome for living tissues is known as the **Bohr effect**.

An increase in carbon dioxide causes increased dissociation of oxygen. This benefits cells with increased metabolism, such as respiring tissues, in the following ways:
- greater amounts of carbon dioxide (a product of cell respiration) are released
- haemoglobin releases its oxygen (required for aerobic cell respiration) at regions of greatest respiratory need.

◆ **Bohr effect**: the decrease in the oxygen affinity of haemoglobin in response to decreased blood pH, resulting from increased carbon dioxide concentration in the blood.

ATL B3.1C

Marine mammals, such as whales and seals, must hold their breath for long periods as they dive beneath the surface of water to hunt for food. What would you predict about the concentrations of haemoglobin in their blood compared to land mammals?

Research the levels of haemoglobin in a marine mammal of your choice. How do they compare to those in land mammals? This article discusses some of the issues:
www.nature.com/articles/news.2007.385

6 **Deduce** the change in percentage saturation of haemoglobin if the oxygen partial pressure drops from 4.0 kPa to 2.7 kPa when the partial pressure of CO_2 is 5.3 kPa (Figure B3.1.18).

LINKING QUESTIONS

1 How do multicellular organisms solve the problem of access to materials for all their cells?
2 What is the relationship between gas exchange and metabolic processes in cells?

Theme B: Form and function – Organisms

B3.2 Transport

> **Guiding questions**
> - What adaptations facilitate transport of fluids in animals and plants?
> - What are the differences and similarities between transport in animals and plants?

SYLLABUS CONTENT

This chapter covers the following syllabus content:
- B3.2.1 Adaptations of capillaries for exchange of materials between blood and the internal or external environment
- B3.2.2 Structure of arteries and veins
- B3.2.3 Adaptations of arteries for the transport of blood away from the heart
- B3.2.4 Measurement of pulse rates
- B3.2.5 Adaptations of veins for the return of blood to the heart
- B3.2.6 Causes and consequences of occlusion of the coronary arteries
- B3.2.7 Transport of water from roots to leaves during transpiration
- B3.2.8 Adaptations of xylem vessels for transport of water
- B3.2.9 Distribution of tissues in a transverse section of the stem of a dicotyledonous plant
- B3.2.10 Distribution of tissues in a transverse section of the root of a dicotyledonous plant
- B3.2.11 Release and reuptake of tissue fluid in capillaries (HL only)
- B3.2.12 Exchange of substances between tissue fluid and cells in tissues (HL only)
- B3.2.13 Drainage of excess tissue fluid into lymph ducts (HL only)
- B3.2.14 Differences between the single circulation of bony fish and the double circulation of mammals (HL only)
- B3.2.15 Adaptations of the mammalian heart for delivering pressurized blood to the arteries (HL only)
- B3.2.16 Stages in the cardiac cycle (HL only)
- B3.2.17 Generation of root pressure in xylem vessels by active transport of mineral ions (HL only)
- B3.2.18 Adaptations of phloem sieve tubes and companion cells for translocation of sap (HL only)

An introduction to the blood system

Living cells require a supply of water and nutrients, such as glucose and amino acids, and they need oxygen. The waste products of cellular metabolism must be removed. In single-celled organisms and very small organisms, internal distances are small, so movements of nutrients can occur efficiently by diffusion. In larger organisms, an internal transport system is required to meet the needs of the cells. Internal transport systems at work are examples of **mass flow**.

◆ **Mass flow**: the movement of fluids down a pressure gradient.

There are three types of vessel in the circulation system:
- **arteries** – carry blood away from the heart
- **veins** – carry blood back to the heart
- **capillaries** – fine networks of tiny tubes linking arteries and veins.

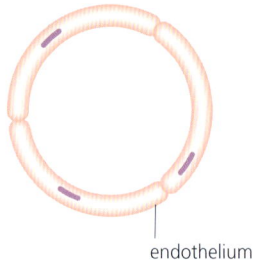

a cross section of a capillary
endothelium

b three-dimensional view of a capillary wall (enlarged)

substances leave (and enter) capillaries through the walls

basement membrane   endothelial cells

■ Figure B3.2.1 Structure of a capillary

Adaptations of capillaries

Capillaries are the smallest blood vessel (Figure B3.2.1).

Capillaries serve the tissues and cells of the body. Blood is under lower pressure (about 35 mm Hg) to prevent the walls of the capillaries bursting. They are narrow tubes, about the diameter of a single red blood cell (approximately 8 μm). Because of the low blood pressure, the flow rate of blood is reduced, which increases the rate of exchange of molecules between blood and tissue.

Adaptations of capillaries for effective exchange of materials between the blood plasma and tissue fluid include:
- a large surface area due to branching (to increase diffusion)
- narrow diameters (so that red blood cells are in close contact with the capillary wall to reduce diffusion distance)
- thin walls (single layer of endothelial cells) for fast diffusion
- fenestrations (gaps in the wall) in some capillaries where exchange needs to be particularly rapid; these gaps are sufficient to allow some components of the blood to escape and contribute to tissue fluid (page 310).

The walls of the capillaries consist of **endothelium** only (endothelium is the innermost lining layer of arteries and veins – see below). Capillaries branch abundantly and bring the blood circulation close to cells – no cell is far from a capillary.

◆ **Endothelium**: the innermost lining layer of arteries and veins, and the layer of cells that comprise the capillary; also lines the inside of the heart.

◆ **Lumen**: the hollow interior of a blood vessel, through which the blood passes.

Structure of arteries and veins

Both arteries and veins have strong, elastic walls, but the walls of the arteries are very much thicker and stronger than those of the veins (Figure B3.2.2). The strength of the walls comes from the thickness of the wall and collagen fibres present, and the elasticity is due to the elastic and involuntary (smooth) muscle fibres. Compared to the thickness of the wall, the **lumen** of an artery is relatively small compared to the lumen of a vein – this enables to the pressure inside the artery to be maintained. The large lumen of a vein reduces friction between blood cells and the wall of the vessel, enabling the blood to flow freely under low pressure.

TS artery and vein, LP (×20) – in sectioned material (as here), veins are more likely to appear squashed, whereas arteries are circular in section

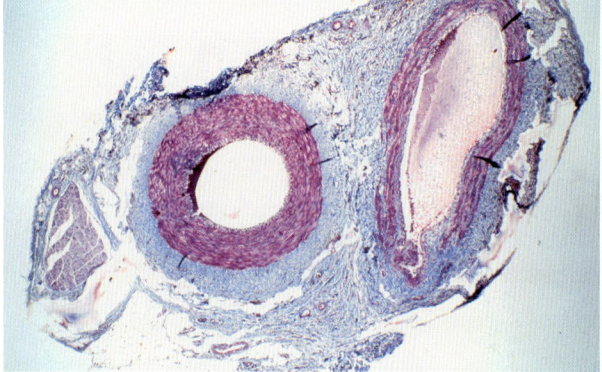

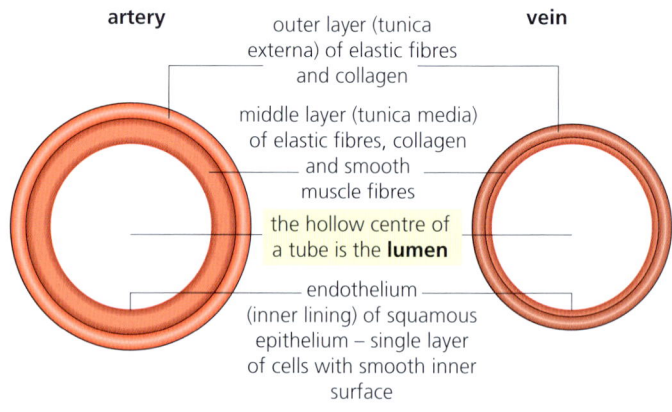

■ Figure B3.2.2 The structure of the wall of arteries and veins

Look at Figure B3.2.2. The micrograph indicates distinguishing features between arteries and veins, for example, the thickness of the vessel wall relative to the diameter of the lumen. Arteries have a much thicker wall compared to the vein.

1 **Construct** a table showing the similarities and differences between arteries and veins.

> ### Concept: Form and function
>
> The form of a blood vessel is related to its function – arteries have thick walls to withstand high pressure whereas veins have thinner walls because blood is at a lower pressure; veins also contain valves to stop blood backflow. Capillary walls are one cell thick to ensure fast diffusion.

> ### ATL B3.2A
>
> Endothelins are peptide-based hormones with receptors and effects in many organs in the human body. Endothelins constrict blood vessels and raise blood pressure. The endothelins are normally kept in balance by other mechanisms but, when overexpressed, can result in health issues. Carry out your own research on endothelins. How do they contribute to high blood pressure (hypertension), heart diseases and cancer? Produce a poster or PowerPoint to present to your class. In your talk, explain how different types of biological knowledge are required to understand how endothelins function and what happens when they fail to work properly.

Blood leaving the heart is under high pressure and travels in waves or pulses, following each heartbeat. By the time the blood has reached the capillaries, it is under very much lower pressure, without a pulse. This difference in blood pressure accounts for the differences in the walls of arteries and veins (see below). Figure B3.2.2 shows an artery and vein in section and details the wall structure of these two vessels. In sectioned material, veins are more likely to appear squashed, whereas arteries are circular in section. Examine Figure B3.2.2 carefully so that you can identify these vessels in cross-section by their wall structure.

> **Top tip!**
>
> Thick walls in arteries are needed to withstand high blood pressures; elasticity is to help maintain and even out blood pressure.

Adaptations of arteries for the transport of blood away from the heart

◆ **Aorta**: main artery that carries blood away from the heart to the rest of the body.

The main artery that leaves the heart is called the **aorta**, which takes blood from the heart to the head and rest of the body. Blood pressure in the aorta is very high – about 120 mm Hg.

> ### Tool 3: Mathematics
>
> #### Applying and using SI or non-SI metric units
>
> The SI unit **pascal (Pa)** and its multiple, the **kilopascal (kPa)**, are generally used by scientists to measure pressure. However, in medicine, the older, non-SI unit of **millimetres of mercury (mm Hg)** is still used (1 mm Hg = 0.13 kPa). It refers to the height of a column of mercury that can be supported by the pressure being measured.

Arteries that divide from the aorta distribute blood under high pressure to regions of the body and to the main organs. The layers of muscle and elastic tissue in the walls of arteries help them to withstand and maintain high blood pressures.

B3.2 Transport

The adaptations of the arteries include:
- overall wall thickness/muscle layer thickness to withstand blood pressure and prevent rupture of artery wall
- thick layer of elastic tissue to even out and maintain blood pressure
- walls stretch to accommodate the huge surge of blood when ventricles contract
- elastic tissue and collagen fibres of the tunica externa (outer layer – Figure B3.2.2) prevent rupture as blood surges from the heart
- the high proportion of elastic fibres are first stretched and then recoil, keeping the blood flowing and propelling it forwards after each pulse passes
- with increasing distance from the heart, the tunica media (middle layer) progressively contains more smooth muscle fibres and less elastic tissue as less stretching and recoiling occurs due to smaller differences in blood pressure
  - by varying constriction and dilation of arteries, blood flow is maintained
  - muscle fibres stretch and recoil, tending to even out the pressure, but a 'pulse' can still be detected.

◆ **Arteriole**: a very small artery.

Arteries divide into smaller blood vessels called **arterioles**, which deliver blood to the tissues. Arterioles have a high proportion of smooth muscle fibres, so are able to regulate blood flow from arteries into capillaries.

Measurement of pulse rates

Ventricular contractions of the heart force a wave of blood through the arteries. The expansion of the arteries can be felt as a pulse, particularly where the artery is near the skin surface and passes over a bone. The pulse is traditionally taken above the wrist (Figure B3.2.4).

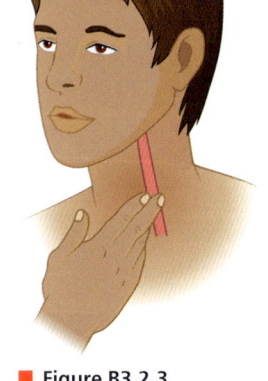

■ Figure B3.2.3 Taking a pulse from a carotid artery

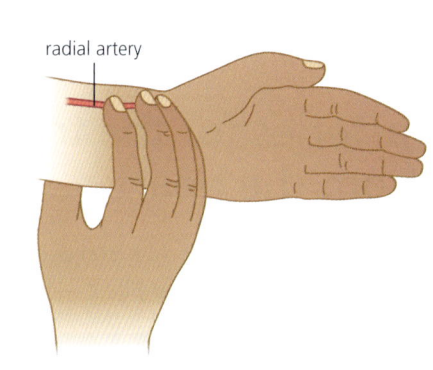

■ Figure B3.2.4 Taking a radial pulse

To take a pulse you should use two fingers to compress the artery. After you have found the pulse, you need to count to determine the heart rate in beats per minute. You can either count for a full minute (this is the most accurate method) or count for 30 seconds and multiply by 2. The pulse can either be taken from a radial artery above the wrist (Figure B3.2.4) or from a carotid artery on the neck (Figure B3.2.3).

The carotid artery is located below the ear, on the side of the neck directly below the jaw. You should feel the artery as you exert pressure on the neck (Figure B3.2.3).

ATL B3.2B

Practise determining a person's heart rate by feeling the carotid or radial pulse with fingertips. Take the pulse of a volunteer in your class. Which method is easier – measuring the pulse using the radial or carotid artery?

There are also digital methods of measuring your pulse. For example, you could use a data logger, or some smartwatches and fitness bands measure heart rate by scanning blood flow near your wrist, by illuminating it with LEDs. Compare the traditional methods with digital ones: what are the advantages and disadvantages of both methods?

● Top tip!

Do not use your thumb to take a pulse, because it has its own pulse that you may feel.

Inquiry 2: Collecting and processing data

Collecting data; processing data

Carry out an investigation of the effect of exercise on heart rate. Select a range of different activities, from gentle to more active. Do not forget to measure resting heart rate as well.

Select your method for measuring heart rate (either by taking the pulse rate directly – see ATL box above – or using electronic methods).

How many times will you repeat the investigation, and why? How will you control other variables?

Collect your data and plot your results using an appropriate graph.

Interpreting results

Identify, describe and explain any patterns, trends or relationships you have recorded. Were any of your results anomalous? How do you identify and justify the removal or inclusion of outliers in data?

Now assess the accuracy, precision, reliability and validity of your data.

Adaptations of veins for the return of blood to the heart

Because of the low pressure in veins there is a possibility of backflow. Veins have **valves** at intervals that prevent this (Figure B3.2.5). Pressure from the blood opens the valves, which are made from a strong protein called collagen and other material. At the end of a heart contraction, as the blood tries to flow backwards, blood pressure pushes against the valves causing them to close (Figure B3.2.5). This ensures that blood only moves in one direction, back towards the heart.

Valves are also found in the heart (see page 316).

Valves are especially common in the veins of the limbs.

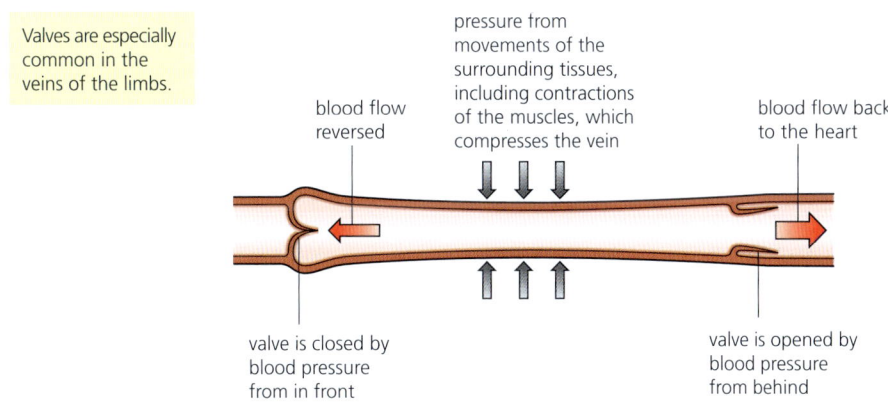

■ **Figure B3.2.5** Valves in veins prevent backflow of blood

Veins have thin walls, with the external layer (tunica externa, with elastic and collagen fibres) the thickest layer. The middle layer (tunica media) contains a few elastic fibres and muscle fibres; because of the low blood pressure, these do not need to be extensive. The wall of the vein is flexible, which allows surrounding muscles to compress the vein (Figure B3.2.5). This helps to keep the blood moving.

Blood flows into veins from smaller vessels called **venules**, which collect blood from the tissues. Venules are formed by a union of several capillaries and have a blood pressure of about 15 mm Hg. Veins have the lowest blood pressure of any blood vessel – approximately 5 mm Hg.

◆ **Venule**: branch of a vein.

Causes and consequences of occlusion of the coronary arteries

◆ **Atherosclerosis**: deposition of plaque (cholesterol derivative) in the inner wall of blood vessels.

Occlusion means the blockage or closing of a blood vessel. Diseases of the heart and blood vessels are primarily due to a condition called **atherosclerosis** (Figure B3.2.6), which causes the occlusion of the arteries that supply blood to the heart muscle (the coronary arteries). This results in the progressive degeneration of the artery walls. The structure of a healthy artery wall is shown in Figure B3.2.2 (page 296).

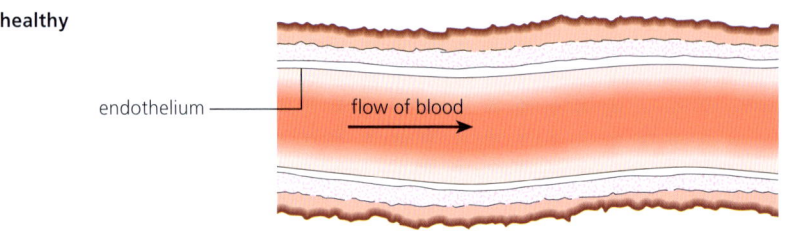

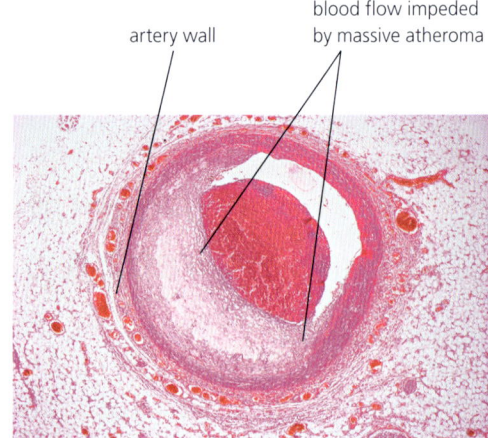

photomicrograph of diseased human artery in TS

■ Figure B3.2.6 Atherosclerosis, leading to a thrombus

Common mistake

The plaque does not deposit on the wall of the artery but within it. The plaque forms beneath the damaged endothelium.

Healthy arteries have pale, smooth linings, and the walls are elastic and flexible, as seen at the top of Figure B3.2.6.

The steps to atherosclerosis in arteries are:

- **Damage to the artery walls** as fatty tissue is deposited under the endothelium. These atheromas (fatty deposits) significantly reduce the diameter of the lumen.
- **Raised blood pressure** follows when fat deposits and fibrous tissue start to impede blood flow, leading to damage to the arterial wall.
- The damaged region is repaired with fibrous tissue, which significantly reduces the elasticity of the vessel wall.
- **Lesion formation** occurs when the smooth lining of the arteries breaks down. These lesions are known as **atherosclerotic plaques**.
- If the plaque ruptures, blood clotting is triggered. A blood clot formed within a blood vessel is known as a **thrombus**. When a thrombus breaks free and circulates in the bloodstream, it is then called an **embolus**.

The consequences of atherosclerosis

One likely outcome of atherosclerotic damage to major arteries is that an embolus may be swept into a small artery or arteriole, which is narrower than the diameter of the clot, causing a blockage. Immediately, the blood supply to the tissue downstream of the block is deprived of oxygen, and without oxygen tissues die. The coronary arteries (the arteries that supply blood to the heart) are especially vulnerable – particularly those going to the left ventricle. When sufficient heart muscle dies in this way, the heart may cease to be an effective pump. We say a 'heart attack' has occurred (known as a **myocardial infarction**).

Theme B: Form and function – Organisms

Coronary arteries that have been damaged can be surgically bypassed – known as a **heart bypass operation**. Typically, a blood vessel taken from the patient's leg is used. In some cases, multiple bypasses are required. This operation is increasingly common and survival rates are high.

An adult human on a low-cholesterol diet forms about 800 mg of cholesterol in the liver each day. However, diets high in saturated fatty acids (i.e. animal fats) often lead to abnormally high blood-cholesterol levels. Excess blood cholesterol, present as LDL (page 402), may cause atherosclerosis – the progressive degeneration of the artery walls. A direct consequence of this is an increased likelihood of the formation and circulation of blood clots, leading to strokes and heart attacks (Figure B3.2.6, opposite).

Claims about links between health problems and causes are based on:
- **epidemiological studies** – these provide circumstantial evidence of health risks; they suggest connections, but they do not establish a cause or biochemical connection (Figure B3.2.7)
- clinical studies of individual patients with health problems attempt to show causal relations between diseases and diets (however, it is not possible to carry out 'controlled' experiments for ethical reasons).

2 **Evaluate** whether data in Figure B3.2.7 show a correlation between saturated fat intake and the incidence of coronary heart disease.

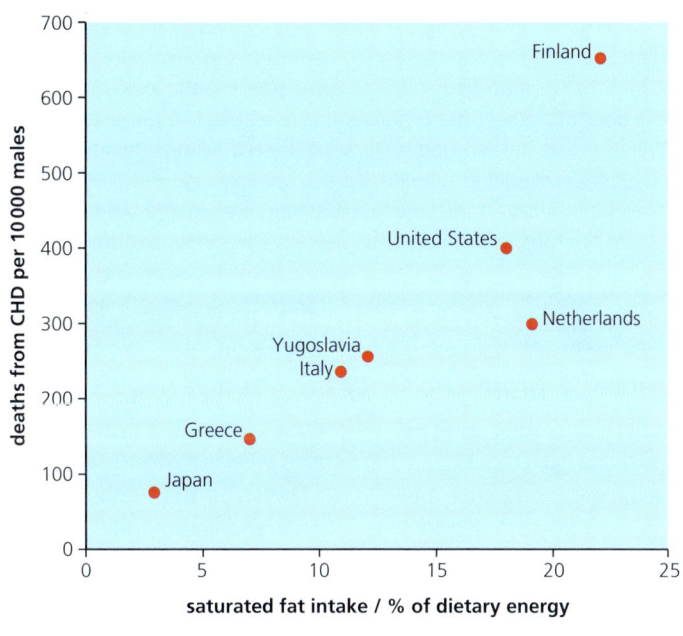

(Since these data were collated, the former Yugoslavia has become Bosnia and Herzegovina, Croatia, Montenegro, Republic of Macedonia, Serbia and Slovenia.)

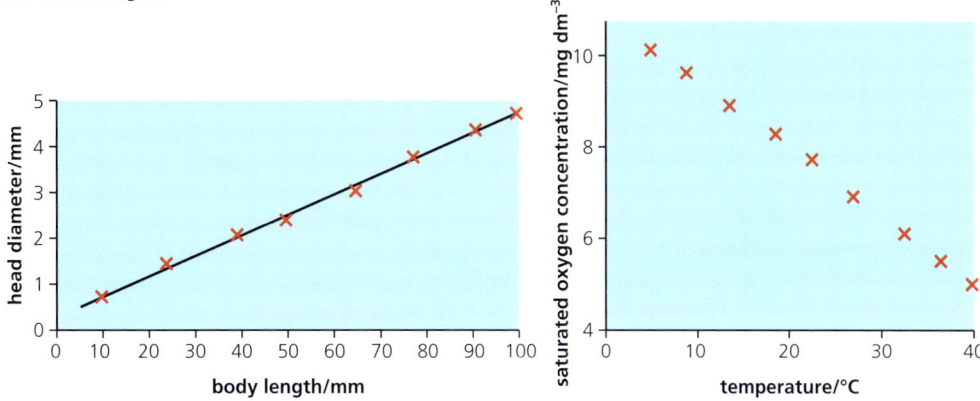

■ **Figure B3.2.7** Correlation studies – diet and coronary heart disease (CHD)

A study on the incidence of coronary heart disease (CHD) related to saturated fat intake as a percentage of dietary energy (Figure B3.2.7) showed a **positive correlation**.

There are marked variations in the prevalence of different health problems in societies around the world from claims based on epidemiological studies and clinical studies of individual patients with health problems. Health claims about the incidence of coronary heart disease need to be assessed in context. The issues continue to present complex challenges for dieticians and politicians in all societies. Even strong correlations, such as that between saturated fat intake and coronary heart disease, do not prove a causal link.

Nature of science: Patterns and trends

Correlation coefficients quantify correlations between variables and allow the strength of the relationship to be assessed. Low correlation coefficients or lack of any correlation can provide evidence against a hypothesis, but even strong correlations do not prove a causal link, as discussed above in relation to the incidence of coronary heart disease.

Top tip!

When evaluating (that is, making an appraisal by weighing up the strengths and limitations) the evidence and the methods used to obtain the evidence for the incidence of coronary heart disease, you should use the following approach:

Strengths: whether there is a statistically significant correlation, for example between intake of saturated fat and incidence of CHD; there is comparison of mean values and analysis of how different they are; statistical assessment of any difference is carried out; analysis of variation within the data is done (i.e. how widely spread the data are, which can be assessed by the spread of data points or the relative size of error bars, where the more widely spread the data the smaller the significance that can be placed on the correlation and conclusion).

Limitations: whether the measure of health was a valid one, e.g. cholesterol levels in blood are more informative than body mass index; a smaller sample is less reliable than a larger one; the sample does not reflect the population as a whole, but rather a particular sex, age, state of health, lifestyle or ethnic background; data are gathered from animal trials rather than humans trials, and so may be less applicable to humans; certain variables are not controlled, e.g. other aspects of the diet; levels and frequency of the lipid intake being investigated are not realistic; methods used to gather data were rigorous, e.g. if only a survey was used, how truthful were the respondents?

Tool 3: Mathematics

Understanding direct and inverse proportionality between variables

A causal relationship is suggested by statistical studies of deaths from coronary heart disease (CHD) per 1000 of the population each year, plotted against the levels of cholesterol measured in blood serum (Figure B3.2.8). The establishment of the role of cholesterol in triggering CHD was provided by experimental laboratory and clinical evidence that destructive plaques (see page 300) are created as a result of these raised levels of blood serum cholesterol.

1 **Describe** the trends shown in the graph, for both the bar and line graph.
2 **Explain** to what extent the data in Figure B3.2.8 supports the hypothesis that high blood cholesterol is a causal factor in CHD.

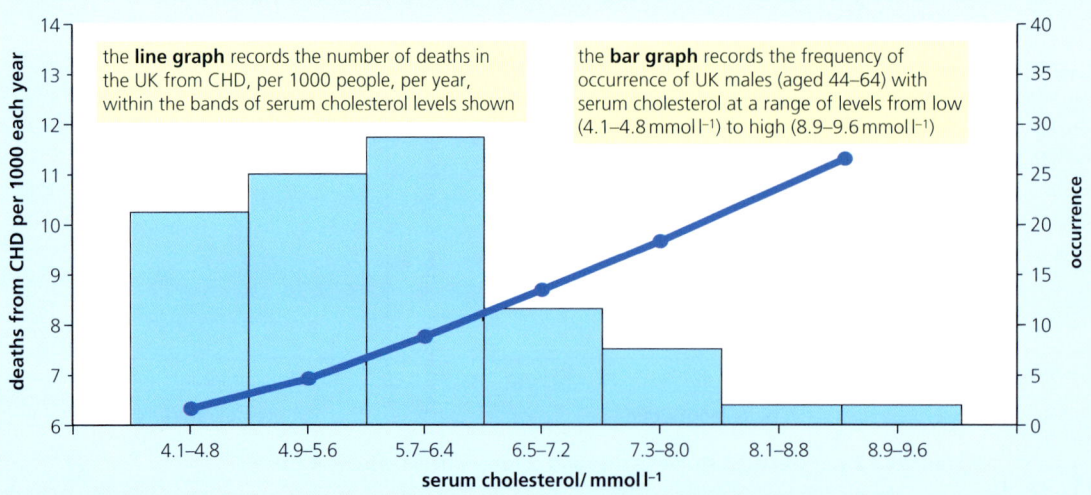

■ Figure B3.2.8 The relationship between deaths caused by coronary heart disease (CHD) and blood serum cholesterol levels.

Link
Cohesion between water molecules, adhesion of water molecules to materials, and capillary action are covered fully in Chapter A1.1, see page 5.

 Top tip!
The xylem vessels are rigid and do not change in volume: this means that as water molecules are lost the pressure is reduced. This causes the flow of water down a pressure gradient – from higher to lower pressure.

◆ **Tension**: the force that is transmitted through a substance when it is pulled tight by forces acting from opposite ends.

◆ **Transpiration stream**: the flow of water through a plant, from the roots to the leaves, via the xylem vessels.

Transport of water from roots to leaves during transpiration

Transpiration is the evaporation of water from the spongy mesophyll tissue and its subsequent diffusion through the stomata, mainly from the leaves of the green plant (Chapter B3.1, page 286). This loss of water vapour is the inevitable consequence of gas exchange in the leaf. The plant must transport water from the roots to the leaves to replace losses from transpiration.

Each part of a plant has a specific role:
- The stem supports the leaves in the sunlight and transports organic materials (such as sucrose and amino acids), ions and water between the roots and leaves.
- The root anchors the plant and is the site of absorption of water and ions from the soil.
- A leaf is an organ specialized for photosynthesis and consists of a leaf blade connected to the stem by a leaf stalk.

The transpiration stream

The water that evaporates from the walls of the mesophyll cells of the leaf is continuously replaced. It comes, in part, from the cell cytoplasm, but mostly it comes from the water in the spaces in walls of nearby cells and then from the xylem vessels in the network of vascular bundles nearby. The loss of water by transpiration from cell walls in leaf cells causes water to be drawn out of xylem vessels and through walls by **capillary action**.

Xylem vessels are usually full of water. As water leaves the xylem vessels in the leaf, a **tension** (negative pressure potential) is set up in the entire water column in the xylem tissue of the plant. This tension draws water up in the xylem. The tension is transmitted down the stem to the roots because of the **cohesion** of water molecules. Cohesion and adhesion ensures a continuous column of water. Consequently, under tension, the water column does not break or pull away from the sides of the xylem vessels. This explanation of how water is drawn up the stem is known as the **cohesion–tension theory**. The result is that water is drawn (literally 'pulled') up the stem. So, water flow in the xylem is always upwards. The flow of water is known as the **transpiration stream**.

The tension on the water column in xylem vessels is demonstrated experimentally when a xylem vessel is pierced by a fine needle. Immediately, a bubble of air enters the column (and will interrupt water flow). If the contents of the xylem vessel had been under pressure, a jet of water would be released from broken vessels.

ATL B3.2C

Evidence of the cohesion–tension theory comes from measurement of the diameter of a tree trunk over a 24-hour period. In a large tree, there is an easily detectable shrinkage in the diameter of the trunk during the day. Under tension, xylem vessels get narrower in diameter (although their collapse is prevented by the lignified thickening of their walls).

The diameter of the tree trunk recovers during the night when transpiration virtually stops, and water uptake makes good the earlier losses of water vapour from the aerial parts of the plant.

Explain to what extent the graph in Figure B3.2.9 supports the cohesion–tension theory of water movement in stems.

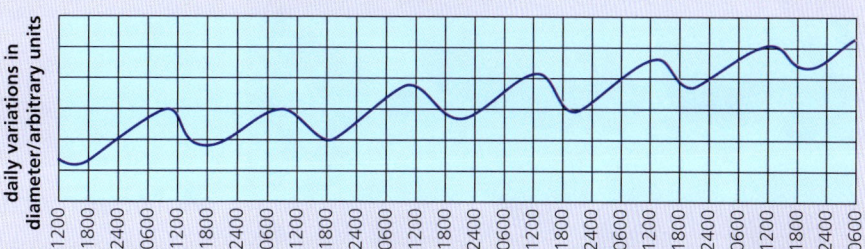

The data were obtained in early May. The tree trunk is undergoing secondary growth in girth during the experimental period. The maximum daily shrinkage amounted to nearly 5 mm.

■ **Figure B3.2.9** Diurnal changes in the circumference of a tree over a 7-day period

B3.2 Transport

What are the advantages of transpiration?

It is evident that transpiration is a direct result of plant structure, plant nutrition and the mechanism of gas exchange in leaves, rather than being a valuable process itself. In effect, the living plant is a 'wick' that steadily dries the soil around it. However, transpiration confers advantages, too:
- Evaporation of water from the cells of the leaf in the light has a strong cooling effect (energy is needed to break hydrogen bonds between water molecules as water vapour forms, which cools the plant).
- The stream of water travelling up from the roots in the xylem passively carries the dissolved ions that have been absorbed from the soil solution in the root hairs. These are required in the leaves and growing points of the plant.
- All the cells of a plant receive water by lateral movements of water from xylem vessels, via pits in their walls. This allows living cells to be fully hydrated. It is the turgor pressure of these cells that provides support to the whole leaf, enabling the leaf blade to receive maximum exposure to light. In fact, the entire aerial system of non-woody plants is supported by this turgor pressure.

So, transpiration does have significant roles in the life of the plant. The loss of water from plants also plays a role in the water cycle, where water evaporating from leaves condenses to form clouds and then falls as rain – without plants this cycle would be disrupted.

> **Link**
> Phytohormones as signalling chemicals controlling growth, development and response to stimuli in plants is covered in Chapter C3.1, page 512.

◆ **Lignin**: complex polymer in which cellulose microfibrils are embedded, giving great strength and rigidity to xylem.

ATL B3.2D

Long-distance transport has been demonstrated for many plant hormones, including auxins, abscisic acid, cytokinins and gibberellins.

How are these hormones transported around the plant? What effects do they have on growth, development and response of plants to stimuli (changes in the environment)?

> **Top tip!**
> The walls of the xylem are strengthened by lignin, which enables the vessels to resist negative pressure caused by cohesion–tension.

Movement of water in the xylem

Transport of water through the plant occurs in the xylem tissue. As we have seen, the cohesive property of water and the structure of xylem vessels allow transport under tension.

Xylem begins as cells with cellulose walls and living contents, connected end to end. During development, the end walls are dissolved away so that mature xylem vessels become long, hollow tubes. The living contents of a developing xylem vessel are used up in the process of depositing cellulose thickening to the inside of the lateral walls of the vessel. This is hardened by the deposition of a chemical substance, **lignin**. Consequently, xylem is extremely tough tissue. Furthermore, it is strengthened internally, which means it can resist negative pressure (suction) without collapsing in on itself.

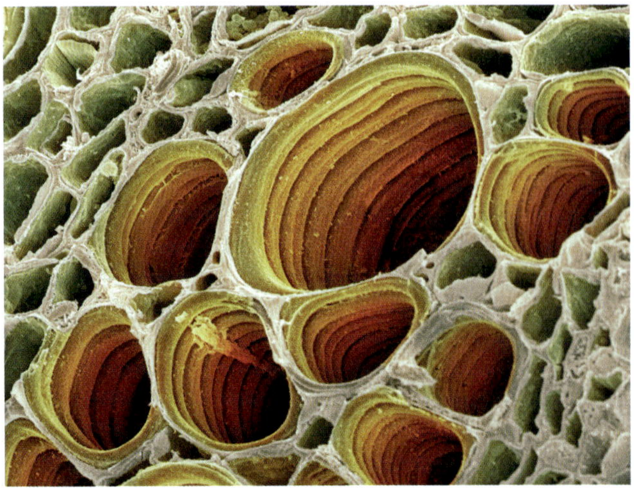

Figure B3.2.10 is a scanning electron micrograph of spirally thickened xylem vessels. When you examine xylem vessels in longitudinal sections by light microscopy, you will see that some xylem vessels may have differently deposited thickening – many have rings of thickening, for example.

Movement into and through the root

The uptake of water occurs in the root system but, as we have seen, the movement of water up the stem is driven by conditions in the leaves.

■ Figure B3.2.10 Scanning electron micrograph of spiral xylem vessels

The root system provides a huge surface area in contact with soil, because plants have a system of branching roots that continually grow at each root tip, pushing through the soil. This is important because it is the dilute solution that occurs around soil particles that the plant accesses to obtain the large volume of water and essential ions it requires. Contact with the soil is vastly increased by the region of root hairs that occurs just behind the growing tip of each root (Figure B3.2.11). Root hairs are extensions of individual epidermal cells and are relatively short-lived. As root growth continues, fresh resources of soil solution are exploited.

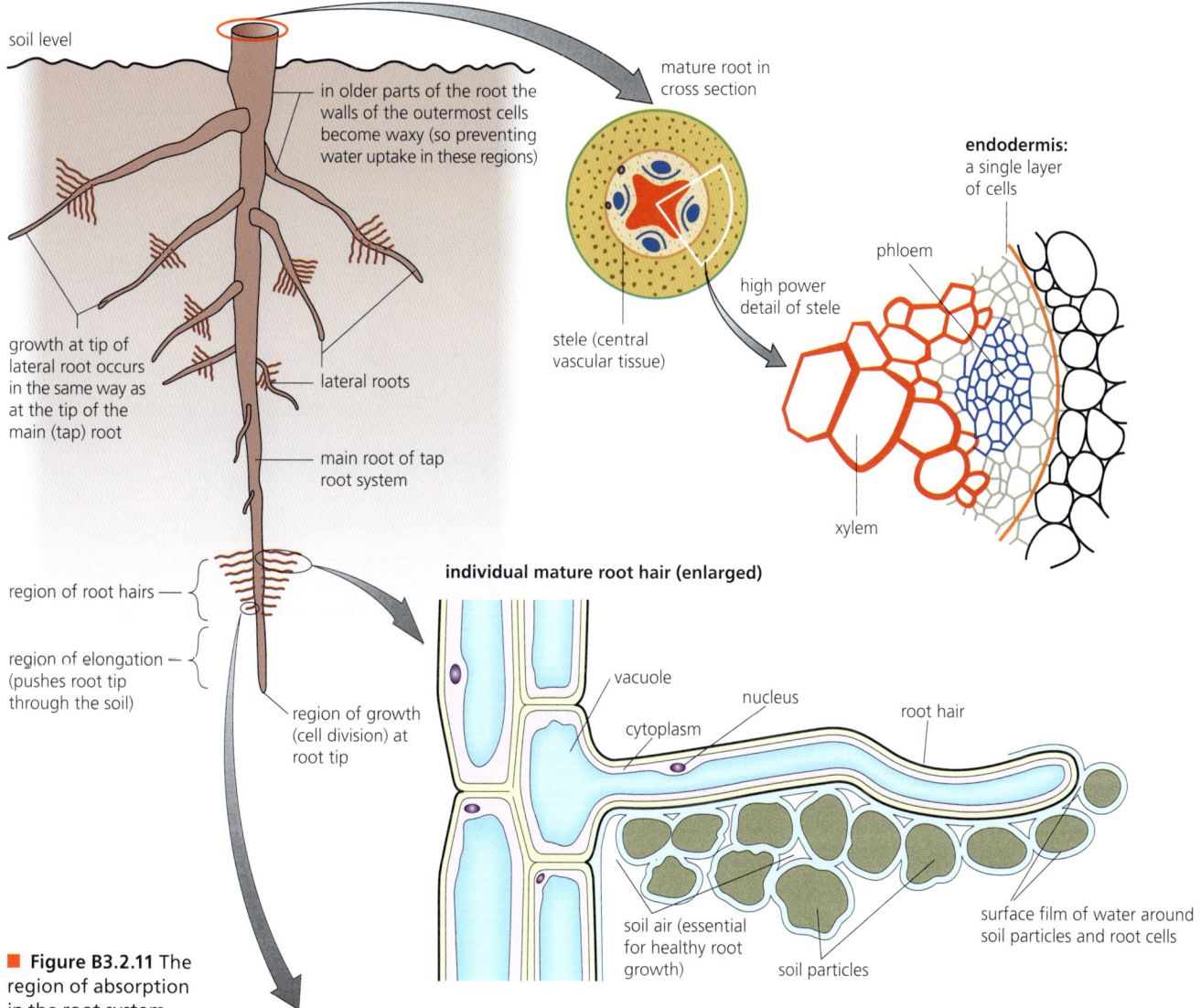

■ Figure B3.2.11 The region of absorption in the root system

Uptake of water is largely by mass flow through the interconnecting 'free spaces' in the cellulose cell walls, but there are three possible routes of water movement through plant cells and tissues (Figure B3.2.12).

- **Mass flow** of water occurs through the interconnecting free spaces between the cellulose fibres of the plant cell walls. This free space in cellulose makes up about 50% of the wall volume. This route of water movement through the plant is a highly significant one. This pathway entirely avoids the living contents of cells and is called the **apoplast pathway**. The apoplast also includes the water-filled spaces of dead cells (and the hollow xylem vessels – as we shall shortly see).

◆ **Apoplast pathway**: the pathway (e.g. of water) through the non-living part of a cell, e.g. cell walls and spaces between cells.

◆ **Symplast pathway**: the pathway (e.g. of water) through the cell membrane and plasmodesmata (the living contents of the cell).

3 List the features of root hairs that facilitate absorption of water from the soil.

- **Diffusion** occurs through the cytoplasm of cells and via the cytoplasmic connections between cells (called **plasmodesmata**). Osmosis occurs through the cell membrane. This route is called the **symplast pathway**. The plant cells are packed with many organelles, which resist the flow of water, slowing movement of water through this pathway relative to the apoplast pathway.
- **Osmosis** occurs from vacuole to vacuole of cells, driven by a gradient in osmotic pressure. This is the **vacuolar pathway**. Active uptake of mineral ions in the roots causes absorption of water by osmosis. This is not a significant pathway of water transport across the plant, but it is how individual cells absorb water.

While all three routes are open, the bulk of water crosses from the epidermis of the root tissue to the xylem via the apoplast (Figures B3.2.12 and B3.2.13).

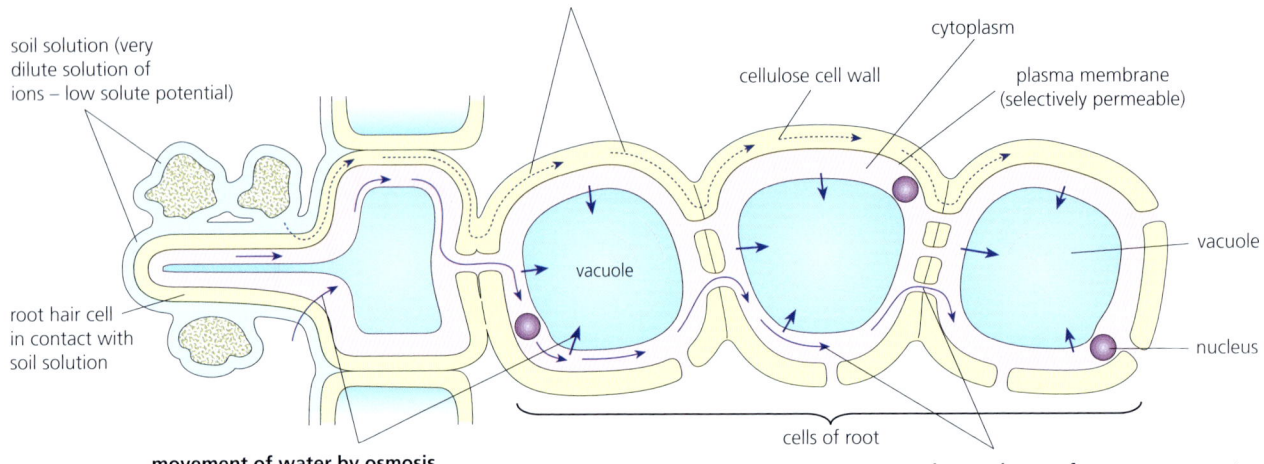

■ Figure B3.2.12 How water moves across plant cells

◆ **Vacuolar pathway**: water movement through the plasma membrane, cytoplasm and then through the vacuole.

◆ **Endodermis**: single layer of cells that surrounds the vascular tissue (xylem and phloem) in the root of a plant.

◆ **Casparian strip**: a band of cells containing suberin, a waxy substance impermeable to water found in the endodermal cell walls of plant roots.

A transverse section of a root (Figure B3.2.12 and Figure B3.2.13, page 307) shows that the centrally placed vascular tissue is contained by the **endodermis**. The endodermis is a layer of cells that is unique to the root. At the endodermis, a waxy strip in the radial walls blocks the passage of water by the apoplastic route (Figure B3.2.13). This waxy strip is called the **Casparian strip**. Water passes through the endodermis symplastically (via the symplast pathway) by osmosis. The significance of this diversion is that the cytoplasm of endodermal cells actively transports ions from the cortex to the endodermis. The result is a higher concentration of ions in the cells at the centre of the root. The raised osmotic potential there thus causes passive water uptake to follow, by osmosis. Once again, it is active uptake and transfer of ions within the root that causes water uptake by osmosis.

Top tip!

All transport must pass through the symplast, which allows control over substances reaching the xylem. However, there are places where the endodermal restriction can be bypassed – this is known as 'bypass flow'. Sites of bypass occur in the root tip where endodermis is not formed and where lateral roots emerge, although this differs between species.

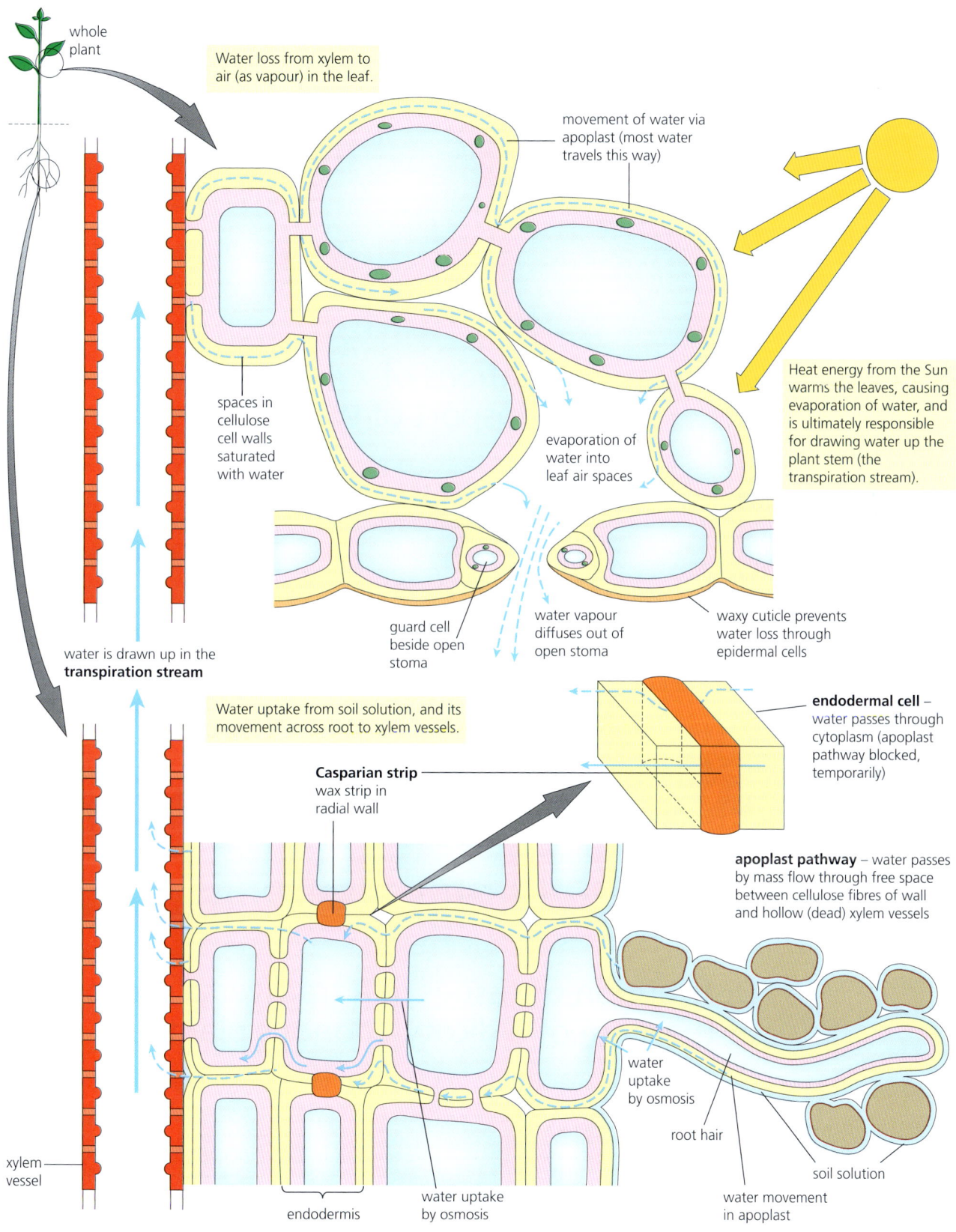

■ **Figure B3.2.13** Water uptake and loss by a green plant – a summary

4 **Explain** the consequence of the Casparian strip for the apoplast pathway of water movement.

ATL B3.2E

Scientists are often asked to present complex ideas succinctly and in an easily understandable way. The transport of water involves several different stages, all parts of the plant and many different processes.

In groups, read the following article and summarize the key points in a poster to present to the rest of your class:

www.nature.com/scitable/knowledge/library/water-uptake-and-transport-in-vascular-plants-103016037

Did you learn anything extra about how water is transported in plants? Were you able to explain these ideas clearly in your poster?

Adaptations of xylem vessels for transport of water

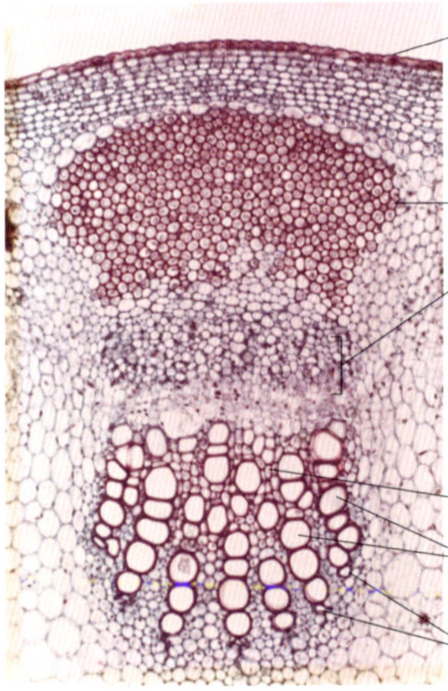

The structure of xylem vessels is seen in transverse (TS) and longitudinal sections (LS) of plant stems. In prepared slides of sectioned and stained plant organs, the cell walls that have been strengthened with lignin stand out particularly clearly.

Xylem have thickened walls that are strengthened with lignin. Lignin is waterproof, which stops water escaping from the vessels. Pits in the walls allow for entry of water and lateral movement between vessels. After xylem form, the cell contents (cytoplasm, nucleus and other organelles) are removed and the cells join to form dead, hollow tubes.

> **Concept: Function**
>
> The adaptations of xylem allow the following:
> - unimpeded flow of water and minerals (lack of cell contents and incomplete or absent end walls)
> - ability to withstand tension (lignified walls)
> - areas for entry and exit of water (pits).

■ **Figure B3.2.14** Photomicrograph of vascular bundle of sunflower (*Helianthus annuus*) seen in TS (×150)

Distribution of tissues in a transverse section of the stem of a dicotyledonous plant

The internal structure of plants is seen by examination of transverse sections. A plan diagram (sometimes called a low-power diagram) is a drawing that records the relative positions of structures within an organ or organism, seen in section. It does not show individual cells.

From the plan diagram in Figure B3.2.15, we can see that the stem is surrounded or contained by a layer of cells called the epidermis. The stem contains vascular tissue (xylem for water transport and phloem for transport of organic solutes) in a discrete system of veins or vascular bundles. In the stem, the vascular bundles are arranged in a ring, positioned towards the outside of the stem (in a region that we call the cortex).

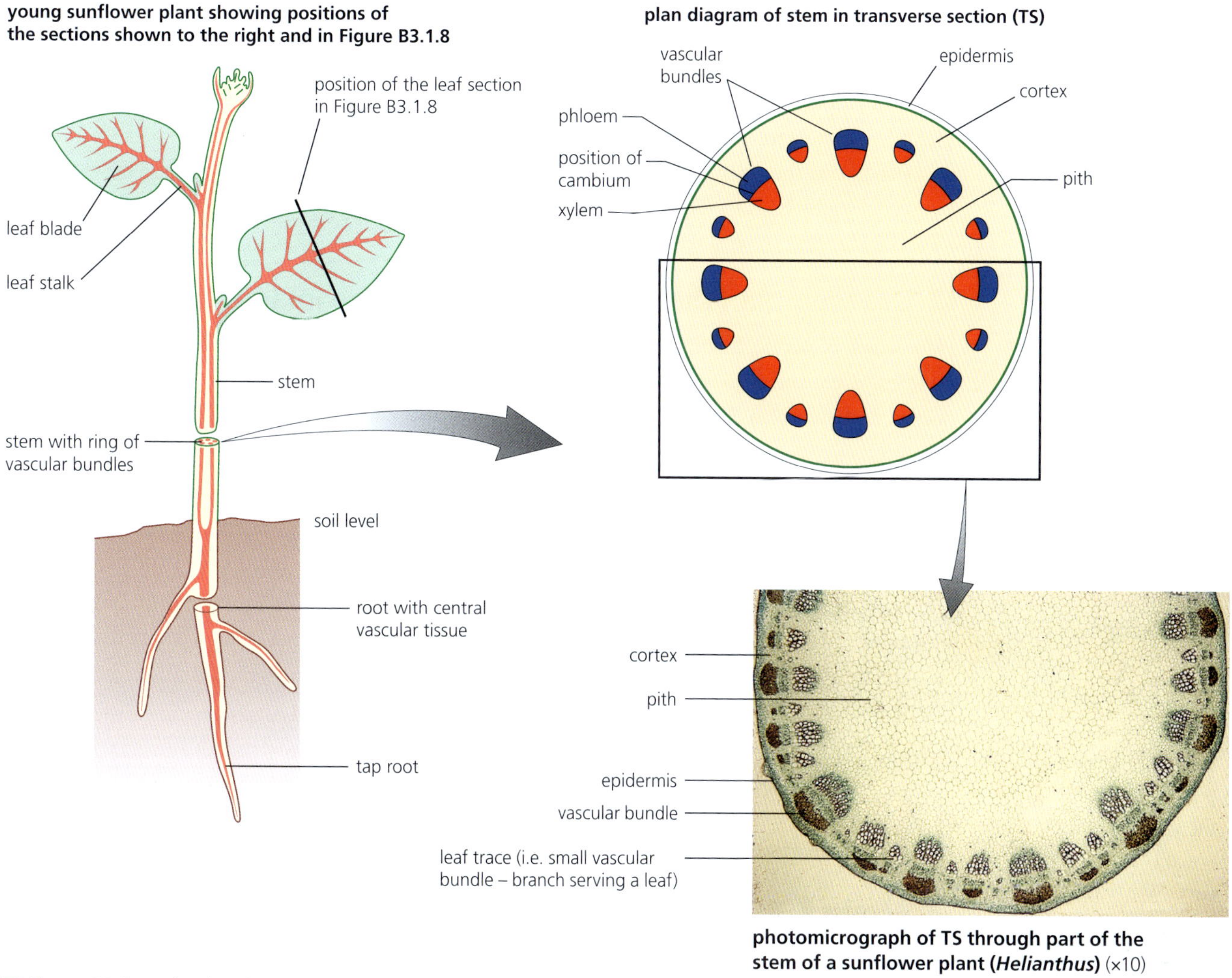

■ Figure B3.2.15 The distribution of tissues in the stem

Drawing plan diagrams from micrographs

Plan diagrams can be drawn from micrographs to identify the relative positions of vascular bundles, xylem, phloem, cortex and epidermis in a stem. You should annotate the diagram with the main functions of these structures discussed in this chapter.

When drawing a plan diagram: show only outlines and not individual cells; use clear, continuous lines; use a sharp pencil; do not use shading; label every boundary. Draw the size of boundaries and regions in the proportions indicated in the original micrograph. Add a scale to your drawing: provide a bar of defined length, e.g. 100 μm, or give an estimated size of the object (see page 57).

Top tip!

A plan diagram can be drawn from the stem in the micrograph (just one half), respecting shapes, sizes and proportions as described in the text.

Distribution of tissues in a transverse section of a dicotyledonous root

The xylem of root and stem are connected, but the arrangement of vascular tissues varies. In a root (see below) the xylem is centrally placed but, in the stem, xylem occurs in the ring of vascular bundles – Figure B3.2.15. The location of the phloem are also different in the root compared to the stem (Figures B3.2.15 and B3.2.16).

■ Drawing plan diagrams from microscope images

Plan diagrams can be drawn from a root micrograph to identify the relative positions of vascular bundles, xylem and phloem, cortex and epidermis. High power, to see details of structures, is not needed to draw a plan diagram.

When drawing a plan diagram, make it large enough so that it fills the paper – when looking down a microscope it is easy to make the image too small. Keep the eye you are not using to look down the microscope open – this helps you to move your gaze away from the eyepiece and refocus on the drawing paper. If you make a mistake, use a good quality eraser to rub out the lines completely.

Specimens should be studied carefully before any drawing is undertaken, noting particularly where the outlines of structures are going to be identified in the final drawing. Use high power if necessary: accurate details of structures can often only be revealed at high power.

● Common mistake

Do not confuse the cells of the cortex with xylem and phloem tissue. The cortex is found around the outside of the root; the vascular tissues are in the central part. Xylem form a cross-shaped structure with phloem located between the arms of the cross. Xylem can be distinguished from phloem by their thicker cell walls.

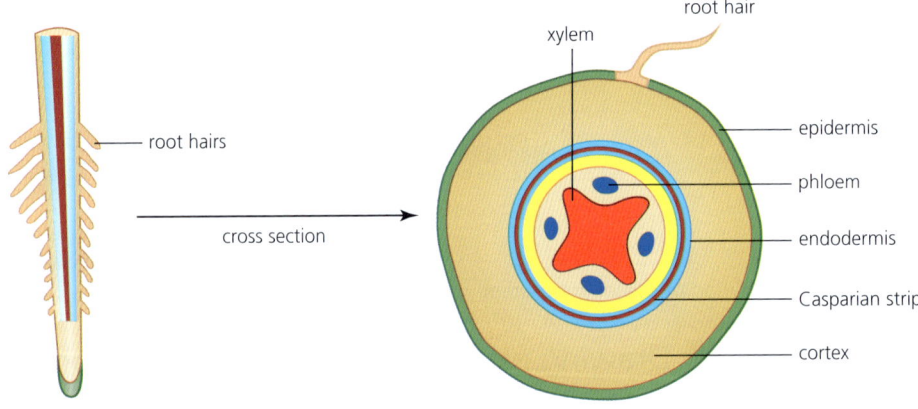

■ Figure B3.2.16 Transverse section of a root

Tissue fluid

■ Release and reuptake of tissue fluid in capillaries

The blood circulation delivers essential resources (nutrients and oxygen, for example) to the tissues of the body. This occurs as the blood flows through the capillaries between the cells of the body (Figure B3.2.17). There are tiny gaps in the capillary walls, found to vary in size in different parts of the body. Through these gaps passes a watery liquid, very similar in composition to plasma. This is **tissue fluid**. Red blood cells, platelets and most blood proteins are not present in tissue fluid, however; these are retained in the capillaries. Tissue fluid surrounds (bathes) the living cells of the body. Nutrients are supplied from this fluid, and carbon dioxide and waste products of metabolism are carried away in it.

◆ **Tissue fluid**: a mixture of water and solutes, forced out of the blood by ultrafiltration, which surrounds body cells.

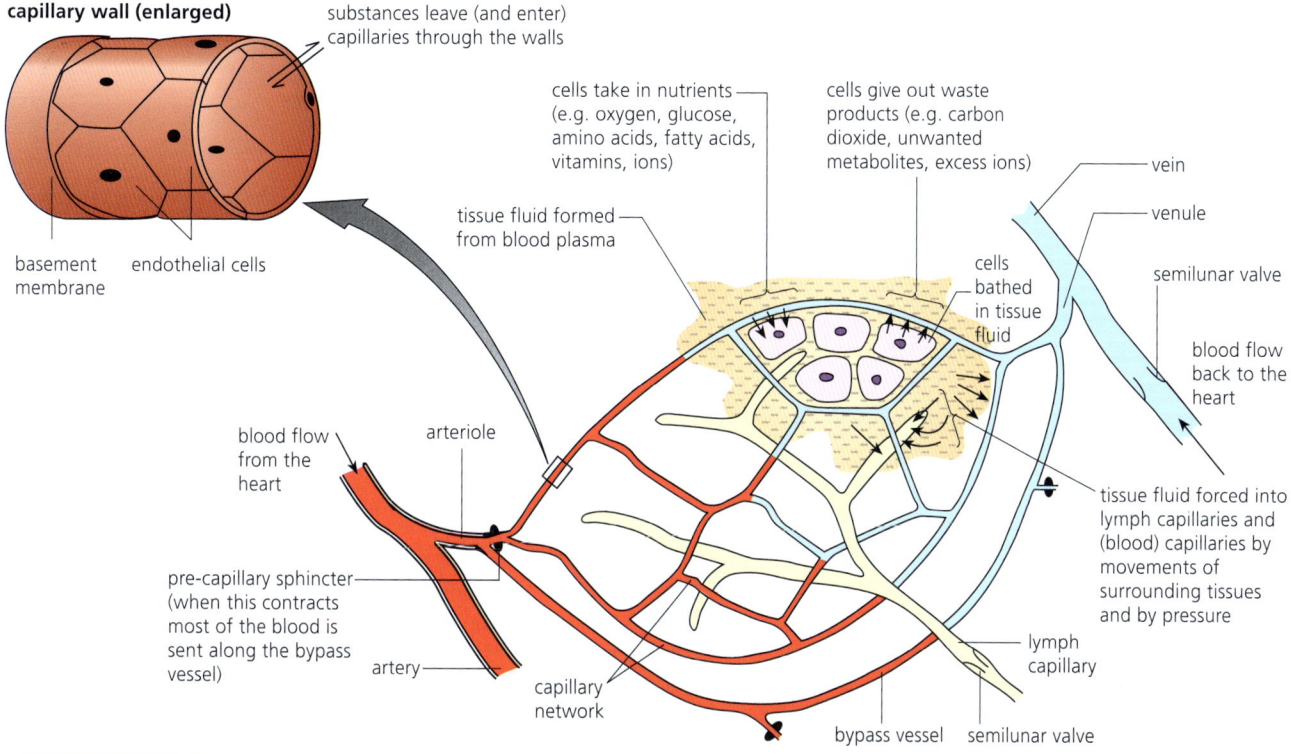

The role of the bypass vessel
Which tissues of the body are being served by the blood circulation is regulated. When tiny pre-capillary sphincter muscles are contracted, blood flow to that capillary bed is restricted to a minimum. Most blood is then diverted through bypass vessels – it is shunted to other tissues and organs in the body.

■ Figure B3.2.17 Exchange between blood and cells via tissue fluid

Link
Water potential is covered fully in Chapter D2.3, page 690.

◆ **Hydrostatic pressure**: mechanical pressure exerted on or by liquid (e.g. water), also known as pressure potential.

Given the quantity of dissolved solutes in the plasma (including all the blood plasma proteins), we would expect the water potential of the blood to limit the loss of water by osmosis. In fact, we might expect water to be passing back into the capillaries from the tissue fluid due to osmosis. However, the force applied to the blood by the heart creates enough **hydrostatic pressure** to overcome osmotic water uptake, at least at the arteriole end of the capillary bed. Here the blood pressure is significantly higher than at the venule ends. Then, as the blood flows through the capillary bed, there is progressive loss of water and hydrostatic pressure. As a result, much of the tissue fluid can eventually return to the plasma – about 90% returns by this route (Figure B3.2.18).

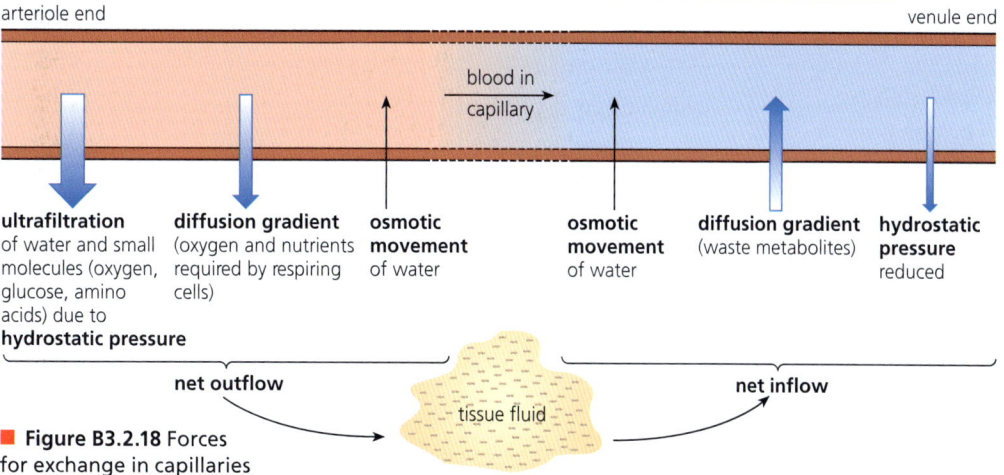

■ Figure B3.2.18 Forces for exchange in capillaries

B3.2 Transport

> **Top tip!**
> Tissue fluid is formed by pressure filtration (ultrafiltration) of plasma in capillaries. This is promoted by the higher pressure of blood from arterioles. Lower pressure in the venules allows tissue fluid to drain back into capillaries.

This means that further along the capillary bed there is a net inflow of tissue fluid to the capillary. However, not all tissue fluid is returned to the blood circulation by this route – some enters the lymph capillaries (see below and Figure B3.2.17).

Exchange of substances between tissue fluid and cells in tissues

Blood plasma is a straw-coloured liquid that forms around 55% of the blood. Plasma is largely composed of water (95%). Water is a good solvent, and so many substances dissolve in it, allowing them to be transported around the body in the plasma (Chapter A1.1, page 8). Both plasma and tissue fluid contain water and dissolved solutes (such as salts).

There are many differences between the composition of plasma and tissue fluid. Tissue fluid contains far fewer proteins, for example. Many plasma proteins are too large to fit through gaps in the capillary walls and so remain in the blood.

Table B3.2.1 shows a comparison of the composition of blood plasma and tissue fluid.

■ Table B3.2.1 Comparing blood plasma and tissue fluid

Content	Blood plasma	Tissue fluid
Cells	red blood cells (erythrocytes), white blood cells (phagocytes and lymphocytes) and platelets	phagocytes
Proteins	more proteins, including large plasma proteins and hormones	fewer proteins (smaller)
Glucose	80–120 mg per 100 ml	less (absorbed by cells for respiration)
Fats	lipoproteins (for fat transport in plasma)	none
Amino acids	more	fewer (absorbed by cells)
Oxygen	more	less (absorbed by cells for respiration)
Carbon dioxide	less	more (released by cells as a product of respiration)

Drainage of excess tissue fluid into lymph ducts

♦ **Lymphatic system**: network of fine capillaries throughout the body of vertebrates, which drain lymph and return it to the blood circulation.
♦ **Lymph**: fluid that flows through the lymphatic system.
♦ **Lymph node**: a kidney-shaped organ in the lymphatic system, part of the body's defences against disease.

Molecules too large to enter capillaries are removed from the tissues via the **lymphatic system** (Figure B3.2.19). There are tiny valves in the walls of the lymph capillaries that permit this. The network of lymph capillaries drains into larger **lymph ducts** (known as lymphatics). Then, **lymph** is moved along the lymphatics by contractions of smooth muscles in their walls and by compression from body movements. Backflow is prevented by valves, as it is in the veins.

Lymph finally drains back into the blood circulation in veins close to the heart. At intervals along the way there are **lymph nodes** present in the lymphatics. Lymph passes through these nodes before it is returned to the blood circulation. In the nodes, phagocytic macrophages engulf bacteria and any cell detritus present. The nodes are also a site where certain cells of the immune system are found (see Chapter C3.2, page 523).

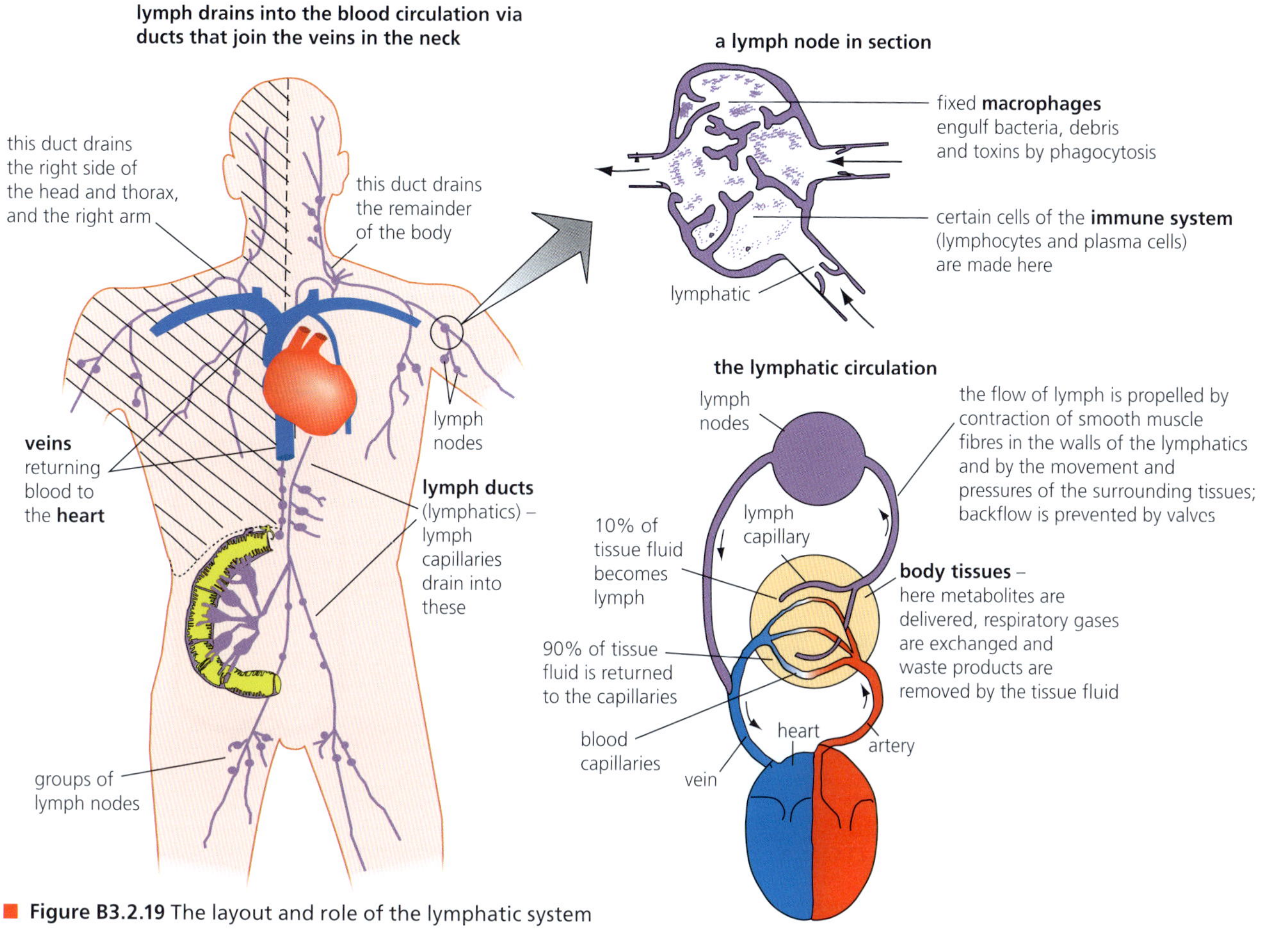

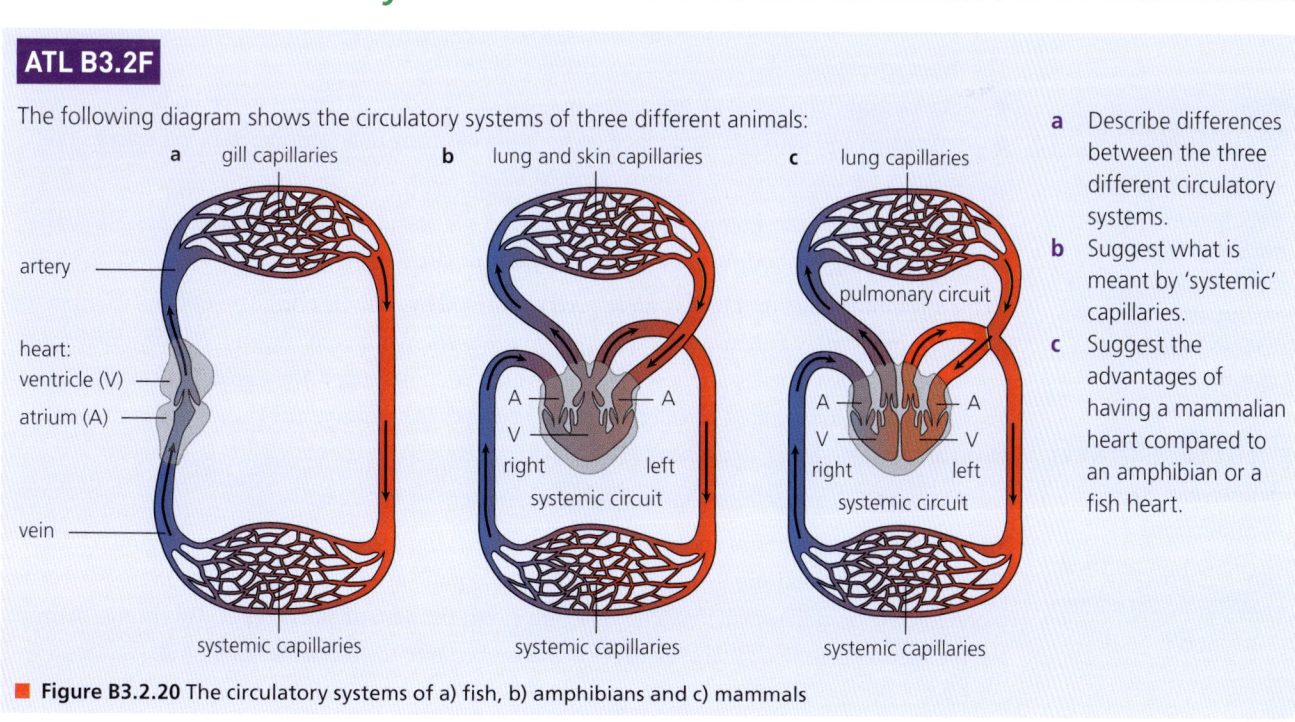

Figure B3.2.19 The layout and role of the lymphatic system

Differences between the single circulation of bony fish and the double circulation of mammals

ATL B3.2F

The following diagram shows the circulatory systems of three different animals:

a Describe differences between the three different circulatory systems.
b Suggest what is meant by 'systemic' capillaries.
c Suggest the advantages of having a mammalian heart compared to an amphibian or a fish heart.

Figure B3.2.20 The circulatory systems of a) fish, b) amphibians and c) mammals

B3.2 Transport

◆ **Closed circulation**: blood is contained inside blood vessels, circulating in one direction from the heart through the circulatory system before returning to the heart again.

◆ **Single circulation**: blood passes through the heart once in each complete circuit of the body; blood is pumped by the heart to the gills, after which the blood flows to the rest of the body and back to the heart.

◆ **Double circulation**: in which the blood passes twice through the heart (pulmonary circulation, then systemic circulation) in any one complete circuit of the body.

◆ **Pulmonary circulation**: the blood circulation to the lungs in vertebrates having a double circulation.

◆ **Systemic circulation**: the blood circulation to the body (not the pulmonary circulation).

Mammals and fish have a **closed circulation** in which blood is pumped by a powerful, muscular heart and circulated in a continuous system of tubes – the arteries, veins and capillaries – under pressure.

Bony fish have a **single circulation** (Figure B3.2.21). This means that blood enters and leaves the heart once in each complete circuit. Blood enters from the body and then leaves to go to the gills, where blood is oxygenated, and then it is moved to the rest of the body where oxygen is delivered to tissues.

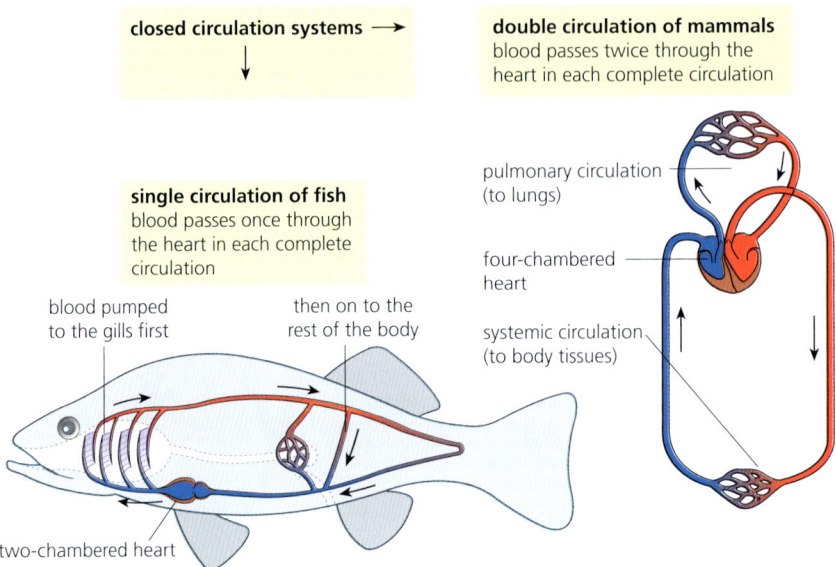

■ **Figure B3.2.21** Closed circulatory systems of bony fish and mammals

Mammals, however, have a **double circulation** (Figures B3.2.21 and B3.2.22). The heart has four chambers and is divided into right and left sides. Blood flows from the right side of the heart to the lungs, then back to the left side of the heart. The role of the right side of the heart is to pump deoxygenated blood to the lungs, where it is oxygenated. The left side of the heart pumps the oxygenated blood around the rest of the body and back to the right side of the heart. The arteries, veins and capillaries serving the lungs are known as the **pulmonary circulation**.

The major advantages of the mammalian circulation are:
- simultaneous high-pressure delivery of oxygenated blood to all regions of the body
- oxygenated and deoxygenated blood do not mix, ensuring that oxygenated blood reaches the respiring tissues undiluted by deoxygenated blood.

The left side of the heart pumps oxygenated blood to the rest of the body. The arteries, veins and capillaries serving the body are known as the **systemic circulation**.

In the systemic circulation, organs are supplied with blood by many arteries. These all branch from the **aorta**, the main artery that takes blood away from the heart. Within each organ the artery divides into numerous arterioles (smaller arteries), and the smallest arterioles supply the capillary networks. Capillaries drain into venules (smaller veins) and venules join to form veins. The veins join the **vena cava** carrying blood back to the heart. The branching sequence in the circulation is, therefore:

aorta → artery → arteriole → capillary → venule → vein → vena cava

You can see that the arteries and veins are often named after the organs they serve (Figure B3.2.22). For example, the blood supply to the liver is via the **hepatic artery**. Notice, too, the liver also receives blood directly from the small intestine via a vein called the **hepatic portal vein**. This brings many of the products of digestion, after they have been absorbed into the capillaries of the villi.

● **Top tip!**

In mammals, as the blood passes twice through the heart in every single circulation of the body, the system is called a double circulation.

5 **List** the differences between the pulmonary circulation and the systemic circulation of blood.

Top tip!

Blood flows from the intestine to the liver through the hepatic portal vein. 'Hepatic' means of or relating to the liver, and 'portal' refers to blood moving from the capillary bed of one structure (in this case, the small intestine) to supply the capillary bed of another structure (the liver).

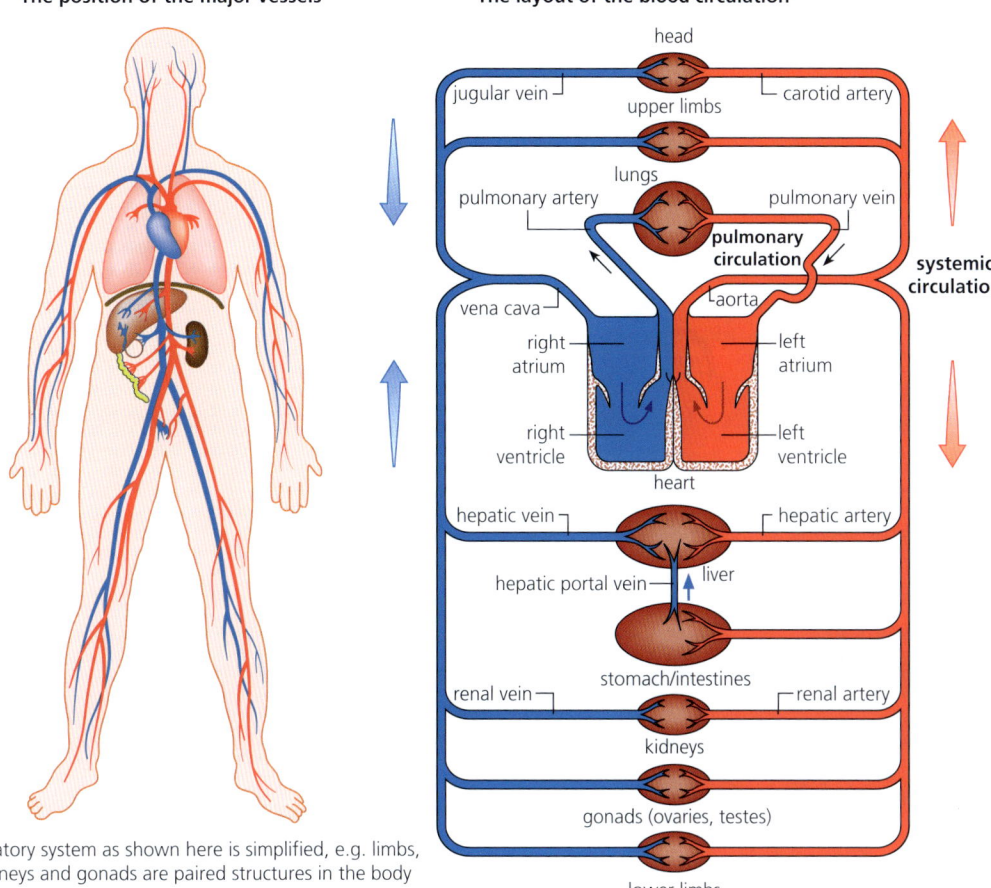

■ Figure B3.2.22 The human blood circulation

the circulatory system as shown here is simplified, e.g. limbs, lungs, kidneys and gonads are paired structures in the body

Adaptations of the mammalian heart for delivering blood to the arteries

■ The heart as a pump

An adult human's heart is about the size of a clenched fist. It lies in the thorax between the lungs and beneath the breastbone (sternum). The heart is a hollow organ with a muscular wall that is contained in a tightly fitting membrane called the **pericardium** – a strong, non-elastic sac that anchors the heart within the thorax. Most importantly, the wall of the heart is well supplied with oxygenated blood from the aorta, via coronary arteries.

All muscle tissue consists of special cells called fibres that can shorten by a half to a third of their length. Cardiac muscle itself consists of cylindrical branching columns of fibres, forming a unique three-dimensional network. This allows contraction in three dimensions. Each fibre has a single nucleus and it appears striped, or striated, under the microscope. Fibres are surrounded by a special plasma membrane called the sarcolemma, and all are very well supplied by mitochondria and capillaries.

■ The chambers of the heart

The cavity of the heart is divided into four chambers, with the chambers on the right side of the heart completely separate from those on the left. The two upper chambers are thin-walled **atria** (singular, **atrium**). These receive blood into the heart. The two lower chambers are thick-walled **ventricles**, with the muscular wall of the left ventricle much thicker than that of the right ventricle. However, the volumes of the right and left sides (the quantities of blood they contain) are identical. The ventricles pump blood out of the heart.

Link
Cardiac muscle is covered in Chapters B2.3, page 268, and B3.3, page 327.

◆ **Pericardium**: a tough membrane surrounding and containing the heart.
◆ **Atrium (plural, atria)**: upper chamber of the heart from which blood is passed to the ventricles.
◆ **Ventricle**: lower chamber of the heart which receives blood from a corresponding atrium and from which blood moves into the arteries.

The heartbeat originates in a tiny part of the muscle of the wall of the right atrium, called the pacemaker. From here, a wave of excitation (electrical impulses) spreads out across both atria. In response, the muscle of both atrial walls contracts simultaneously. Subsequently, the electrical impulse passes to the base of the ventricles, from which it spreads upwards causing the ventricles to contract.

TOK

Our current understanding is that emotions are the product of activity in the brain, rather than in the heart. Is knowledge based on science more valid than knowledge based on intuition?

Concept: Form

The structure of the heart includes cardiac muscle, a pacemaker, atria, ventricles, atrioventricular and semilunar valves, the septum and coronary vessels.

Top tip!

Make sure you can identify the features of the mammalian heart on a diagram of the heart shown from the front, including cardiac muscle, pacemaker, atria, ventricles, atrioventricular and semilunar valves, septum and coronary vessels. Practise tracing the unidirectional flow of blood from named veins to arteries.

6 The edges of the atrioventricular valves have non-elastic strands attached (the chordae tendineae (tendinous cords) or 'heart strings'), which are anchored to the ventricle walls (Figure B3.2.23). **Explain** exactly what the strands do.

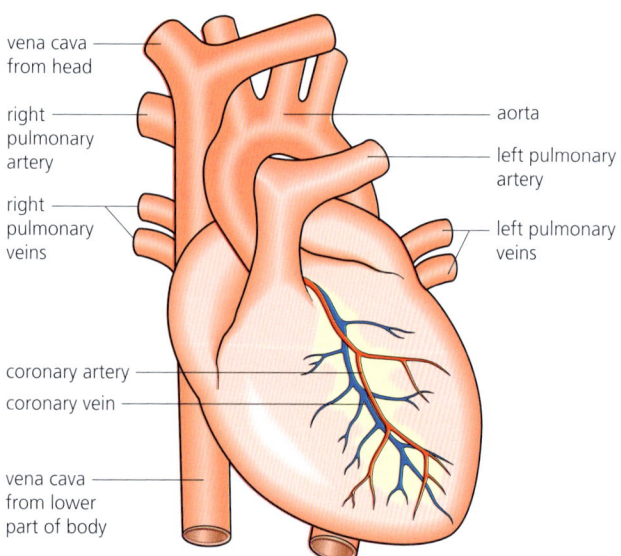
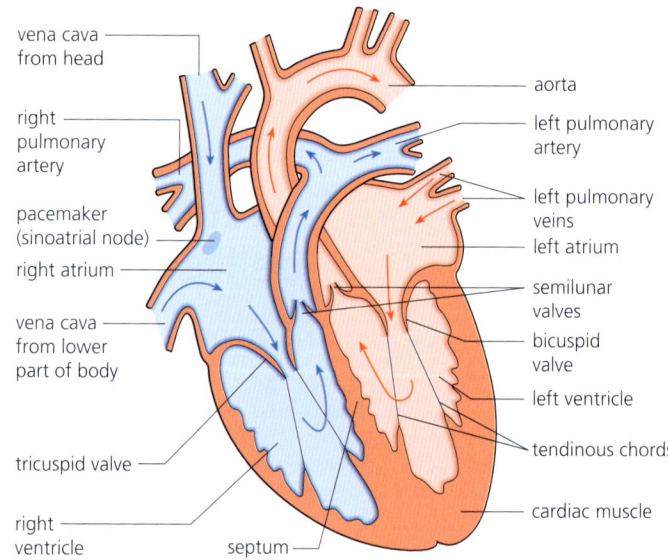

■ Figure B3.2.23 The structure of the heart

◆ **Atrioventricular valve**: heart valve that opens to allow the passage of blood into a ventricle; it closes to prevent backflow of blood into the atrium.

◆ **Tricuspid valve**: atrioventricular valve on the right side of the heart.

◆ **Bicuspid valve**: atrioventricular valve on the left side of the heart.

■ The valves in the heart

The valves of the heart prevent backflow of the blood, maintaining the direction of blood flow through the heart. You can see these valves in action in the diagrams in Figure B3.2.24. Notice that the **atrioventricular valves** are large valves, in a position to prevent backflow of blood from ventricles back up into the atria. The edges of these valves are supported by tendons, anchored to the muscle walls of the ventricles below. These tendons prevent the valves from folding back due to the huge pressure that develops with each heartbeat. The atrioventricular valves are individually named: on the right side is the **tricuspid valve**; on the left is the **bicuspid valve**.

◆ **Semilunar valve**: half-moon shaped valve, preventing backflow in a tube (e.g. a vein)

A different type of valve separates the ventricles from the pulmonary artery (right side) and aorta (left side). These are pocket-like structures called **semilunar valves**, rather like the valves seen in veins. These stop backflow from the aorta and pulmonary artery into the ventricles as the ventricles relax between heartbeats.

In Figure B3.2.23, we can see the coronary arteries. These arteries, and the capillaries they serve, deliver oxygen and nutrients essential for the pumping action to the muscle fibres of the heart and they remove the waste products.

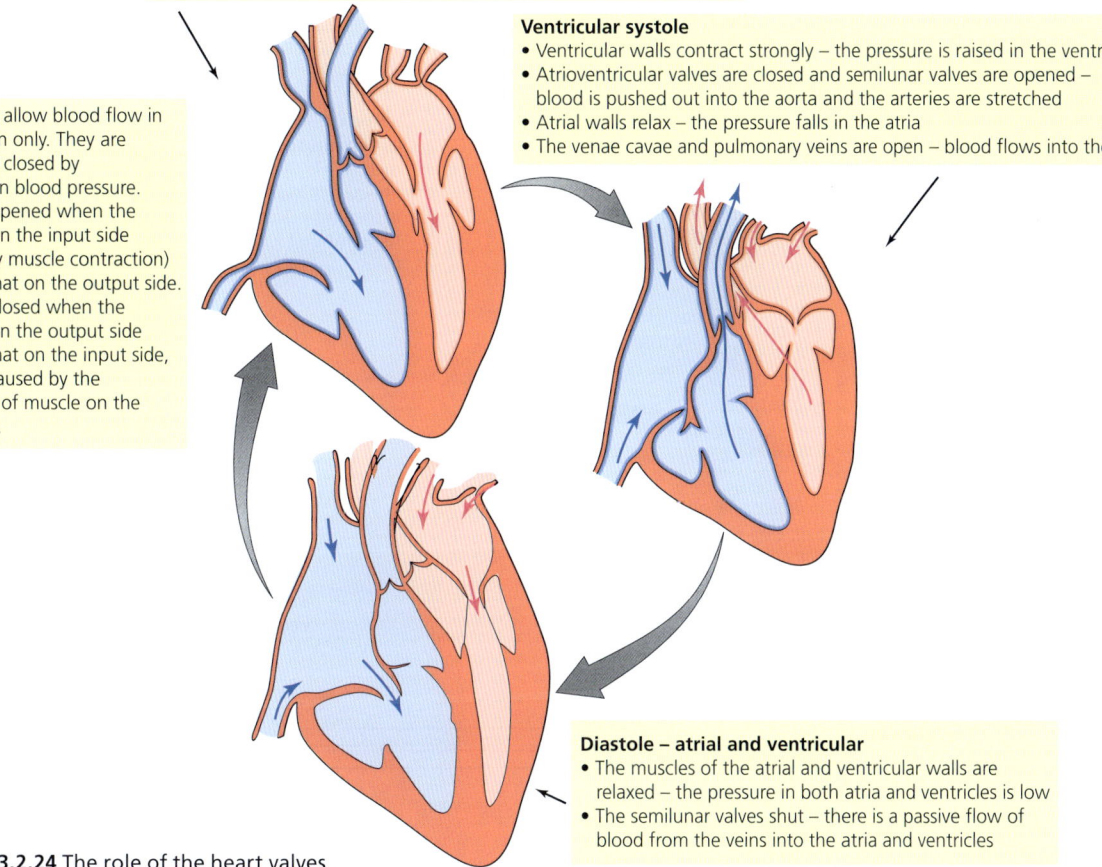

■ Figure B3.2.24 The role of the heart valves

Concept: Function

The structures of the mammalian heart enable:
- the separation of oxygenated and deoxygenated blood (septum separating left and right atria, left and right ventricles)
- continuing contraction of muscle throughout life, with muscle contracting in multiple directions (cardiac muscle)
- the flow of blood in one direction, from atria to ventricles and then from ventricles to arteries leaving the heart (atrioventricular valves and semilunar valves)
- supply of blood to the heart to ensure sufficient supply of nutrients and oxygen (coronary arteries)
- control of heart contraction (pacemaker) : cardiac muscle is myogenic, which means that it is self-excitable and does not require electrical impulses from the central nervous system.

B3.2 Transport

◆ **Cardiac cycle**: the sequence of events of a heartbeat, by which blood is pumped all over the body.

◆ **Systole**: contraction of heart muscle.

◆ **Diastole**: relaxation of heart muscle.

Stages in the cardiac cycle

The **cardiac cycle** is the sequence of events of a heartbeat, by which blood is pumped all over the body. The heart beats at an approximate rate of about 60–100 times per minute, so each cardiac cycle is about 0.8 seconds long. This period of 'heartbeat' is divided into two stages, which are called **systole** and **diastole**. In the systole stage, heart muscle contracts; during the diastole stage, heart muscle relaxes. When the muscular walls of the chambers of the heart contract, the volume of the chambers is decreased. This increases the pressure on the blood contained there, forcing the blood to a region where pressure is lower. Valves prevent blood from flowing backwards to a region of low pressure, so blood always flows in one direction through the heart.

Look at the steps in Figure B3.2.24 and B3.2.25. You will see that Figure B3.2.25 illustrates the cycle on the left side of the heart only, but both sides function together, in the same way, as Figure B3.2.24 makes clear. Here is the sequence of events in the left side of the heart that follow the initiation of the heartbeat by the sinoatrial node (pacemaker).

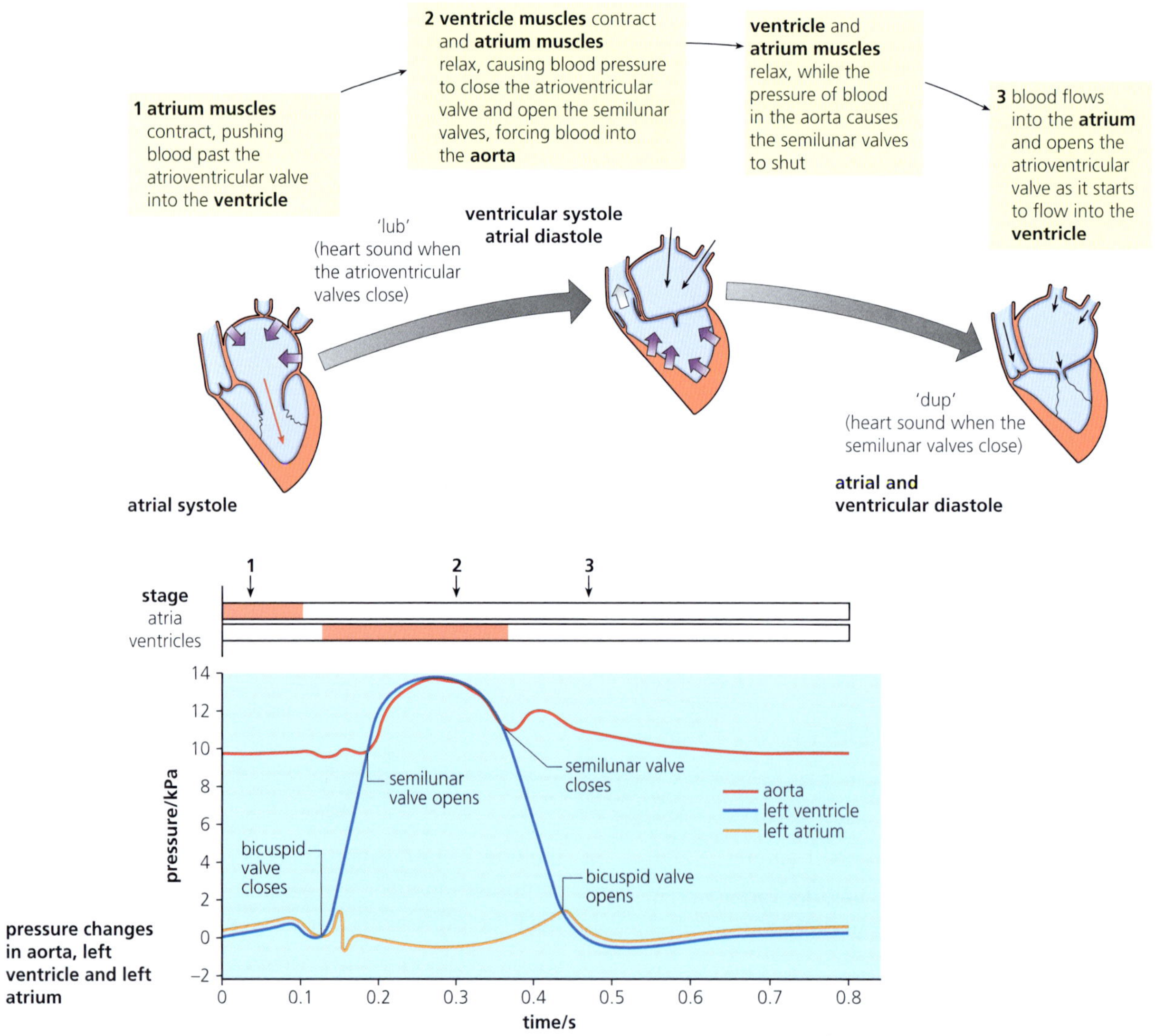

■ Figure B3.2.25 The cardiac cycle

We start with contraction of the atrium (**atrial systole**, about 0.1 s). As the walls of the atrium contract, blood pushes past the atrioventricular valve into the ventricles, where the contents are under low pressure. At this time, any backflow of blood from the aorta (back into the ventricle chamber) is prevented by the semilunar valves. Notice that backflow from the atria into the vena cava and the pulmonary veins is prevented because contraction of the atrial walls seals off these veins. Veins also contain semilunar valves, which prevent backflow here too.

Next, the ventricle contracts (**ventricular systole**, about 0.5 s). The high pressure this contraction generates slams shut the atrioventricular valve and opens the semilunar valves, forcing blood into the aorta. A 'pulse', detectable in arteries all over the body, is generated (see below). Following contraction of the ventricle muscles, the atrium relaxes (**atrial diastole**, about 0.7 s).

This is followed by relaxation of the ventricles (**ventricular diastole**).

Each contraction of cardiac muscle is followed by relaxation. The changing pressure of blood in the atria, ventricles, pulmonary artery and aorta (shown in the graph in Figure B3.2.25) automatically opens and closes the valves.

> **Top tip!**
> Make sure you can interpret systolic and diastolic blood pressure measurements from data and graphs. Use question 8 to help practice.

7 Examine the data on pressure change during the cardiac cycle in the graph in Figure B3.2.25. **Deduce** why:
   a pressure in the aorta is always significantly higher than that in the atria
   b pressure falls *most* abruptly in the atrium once ventricular systole is underway
   c the semilunar valve in the aorta does not open immediately that ventricular systole commences
   d when ventricular diastole commences, there is a significant delay before the bicuspid valve opens, despite rising pressure in the atrium
   e it is significant that about 50% of the cardiac cycle consists of diastole.

8 Examine the graph shown in Figure B3.2.26, then answer the questions below.
   a **Comment** on the velocity and pressure of the blood as it enters the aorta, compared with when it is about to re-enter the heart.
   b **Deduce** what aspects of the structure of the walls of arteries may be said to be a response to the condition of blood flow through them.
   c **Explain** what factors may cause the characteristic velocity of blood flow in the capillaries and why is this advantageous.

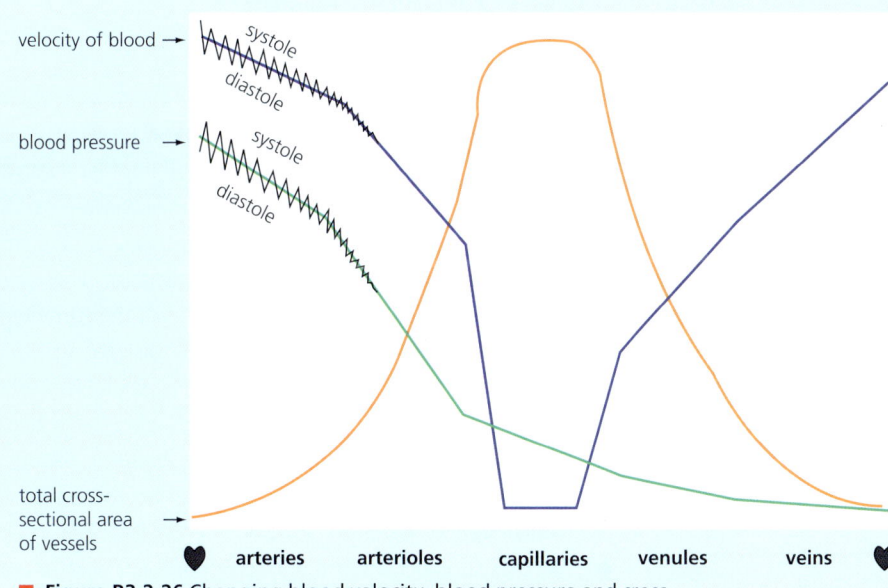

■ **Figure B3.2.26** Changing blood velocity, blood pressure and cross-sectional area of vessels as blood circulates in the body

■ Origin and control of the heartbeat

The heart beats rhythmically throughout life, without rest, apart from the momentary relaxation that occurs between each beat. While heartbeats occur naturally, without the cardiac muscle needing to be stimulated by an external nerve, the alternating contractions of cardiac muscle of the atria and ventricles are controlled and coordinated precisely. Only in this way can the heart act as an efficient pump. The positions of the structures within the heart that bring this about are shown in Figure B3.2.27.

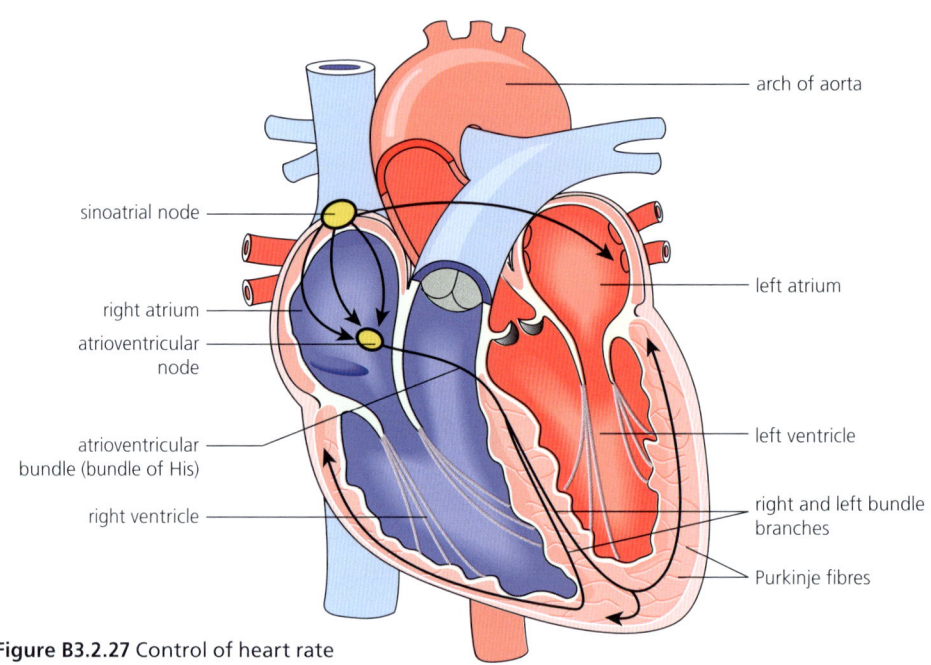

■ Figure B3.2.27 Control of heart rate

◆ **Sinoatrial node (pacemaker)**: structure found on right atrium of heart that is the origin of the myogenic heartbeat.

◆ **Atrioventricular node**: mass of specialized cardiac muscle cells in the wall of the right atrium.

> **Top tip!**
> Although cardiac muscle fibres form an interconnected network, the network system of the atrial walls is entirely separate from that of the ventricles. The signals from the sinoatrial node cannot pass directly to the ventricles. This ensures a transmission delay between atria and ventricles.

The steps to control of the cardiac cycle are as follows:
- The heartbeat originates in a tiny part of the muscle of the wall of the right atrium, called the **sinoatrial node (SA node)** or **pacemaker**.
- From here, a wave of excitation (electrical impulses) spreads out across both atria.
- In response, the muscle of both atrial walls contracts simultaneously (**atrial systole**).
- This stimulus does *not* spread to the ventricles immediately, due to the presence of a narrow band of non-conducting fibres at the base of the atria. These block the excitation wave, preventing its conduction across to the ventricles. Instead, the stimulus is picked up by the **atrioventricular node (AV node)**, situated at the base of the right atrium.
- After a delay of 0.1–0.2 s, the excitation is passed from the AV node via the **bundles of His** (heart muscle that takes part in electrical conduction in the heart) to the base of both ventricles by tiny bundles of conducting fibres, known as **Purkinje tissue**.
- When stimulated, the ventricle muscles start to contract from the *base* of the heart upwards (**ventricular systole**).
- The delay that occurs, before the AV node acts as a 'relay station' for the impulse, permits the emptying of the atria into the ventricles to completely finish, and prevents the atria and ventricles from contracting simultaneously.
- After every contraction, cardiac muscle has a period of insensitivity to stimulation, the **refractory period** (a period of enforced non-contraction – **diastole**). In this phase, the heart begins, passively, to refill with blood. This period is a relatively long one in heart muscle, and enables the heart to beat throughout life.

The heart's own rhythm, set by the SA node, is about 60–100 beats per minute. Conditions in the body can and do override this basic rate to increase heart performance. The action of the pacemaker is modified according to the needs of the body. For example, pacemaker activity may be increased during physical activity. This occurs because increased muscle contraction causes an increased volume of blood to pass back to the heart. The response may be more powerful contractions without an increase in heart rate. Alternatively, the rate of 50–60 beats per minute of the heart 'at rest' may be increased to up to 200 beats a minute in very strenuous exercise.

There is a delay between the arrival and passing on of a stimulus at the AV node. This delay is the product of:
- the cells of the AV node taking longer to become excited than those of the SA node
- the smaller diameter of the AV cells, compared with those of the SA node, slowing conduction
- fewer sodium ion channels in the membranes of AV cells, and a more negative resting potential (page 469) than in the cells of the SA node
- fewer gap junctions in the intercalated discs of the cells of the AV node compared to the SA node.

The result of this delay in transmission (of about 100 ms) gives time for the atria to contract fully, adding to the volume of blood delivered to the ventricles. The delay also prevents the atria and ventricles from contracting simultaneously. The cardiac cycle relies on the atria contracting, pushing blood down into the ventricles, before the ventricles then contract to force blood out through the aorta or pulmonary artery. If the atria and ventricles contracted together, the chambers would not fill properly, leading to reduced cardiac output and low blood pressure.

TOK

The original discovery of the circulation of mammalian blood was made in Europe by experimental scientists William Harvey and Marcello Malpighi in the seventeenth century. Before this, much so-called medical knowledge was derived from the theories expressed in the writings of Galen, a Roman physician (AD 129–199), and on the ideas of earlier Greek writers. In what ways have influential individuals contributed to the development of the natural sciences as an area of knowledge?

Generation of root pressure in xylem vessels

Endodermal cells of the root contain a waterproof substance (suberin) that blocks the apoplast pathway, directing water through the symplast pathway. This ensures water flows into the xylem and not out, enabling the plant to control water movement (Figure B3.2.28). In addition to the Casparian strip, endodermal cells pump salts into the xylem. This lowers the water potential of the xylem (i.e. increases solute concentration), enabling water to move down a water potential gradient (Figure B3.2.29). This creates a positive pressure potential, moving water and dissolved minerals a short distance up the xylem. This is called **root pressure**.

When high humidity prevents transpiration, or in spring before the leaves on deciduous plants have opened, root pressure is generated to cause water movement in roots and stems when transport in the xylem due to transpiration is insufficient.

♦ **Root pressure**: force generated in the roots that helps to drive water upwards into xylem vessels.

Link

Water potential is covered in detail in Chapter D2.3, page 690.

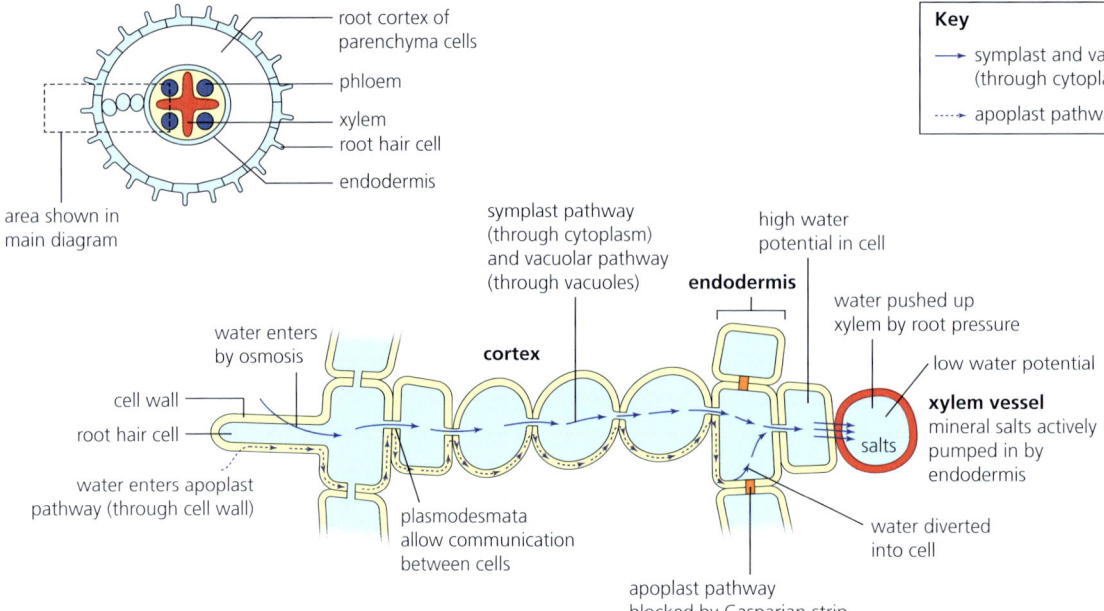

■ Figure B3.2.28 Passage of water through the root

Common mistake

Diffusion is not the main method of mineral absorption – active transport is the means by which minerals are absorbed into plant roots. If plants are able to absorb water by osmosis, they must have higher solute concentrations inside their cells than outside, and this can only be achieved by active transport.

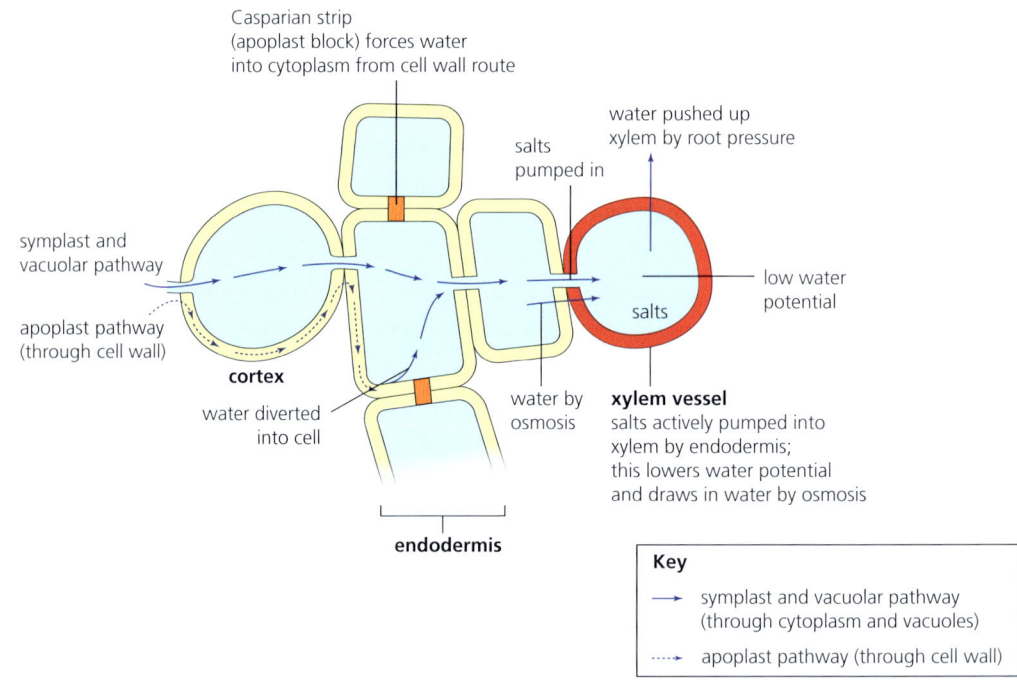

■ Figure B3.2.29 Development of root pressure

Theme B: Form and function – Organisms

Adaptations of phloem sieve tubes and companion cells for translocation of sap

Concept: Form

Phloem tissue show the following adaptations: sieve plates, reduced cytoplasm and organelles, no nucleus for sieve tube elements and presence of many mitochondria for companion cells; plasmodesmata between companion cells and sieve tube elements.

Phloem tissue consists of sieve tubes and companion cells, and is served by transfer cells (Figure B3.2.30, and see also Chapter A2.2, page 79) in the leaves. Sieve tubes are narrow, elongated elements, connected end to end to form tubes. The end walls, known as sieve plates, are perforated by pores. The cytoplasm of a mature sieve tube has no nucleus, nor many of the other organelles of a cell. However, each sieve tube is connected to a companion cell by strands of cytoplasm, called plasmodesmata, which pass through narrow gaps (called pits) in the walls. The companion cells are believed to service and maintain the cytoplasm of the sieve tube, which has lost its nucleus.

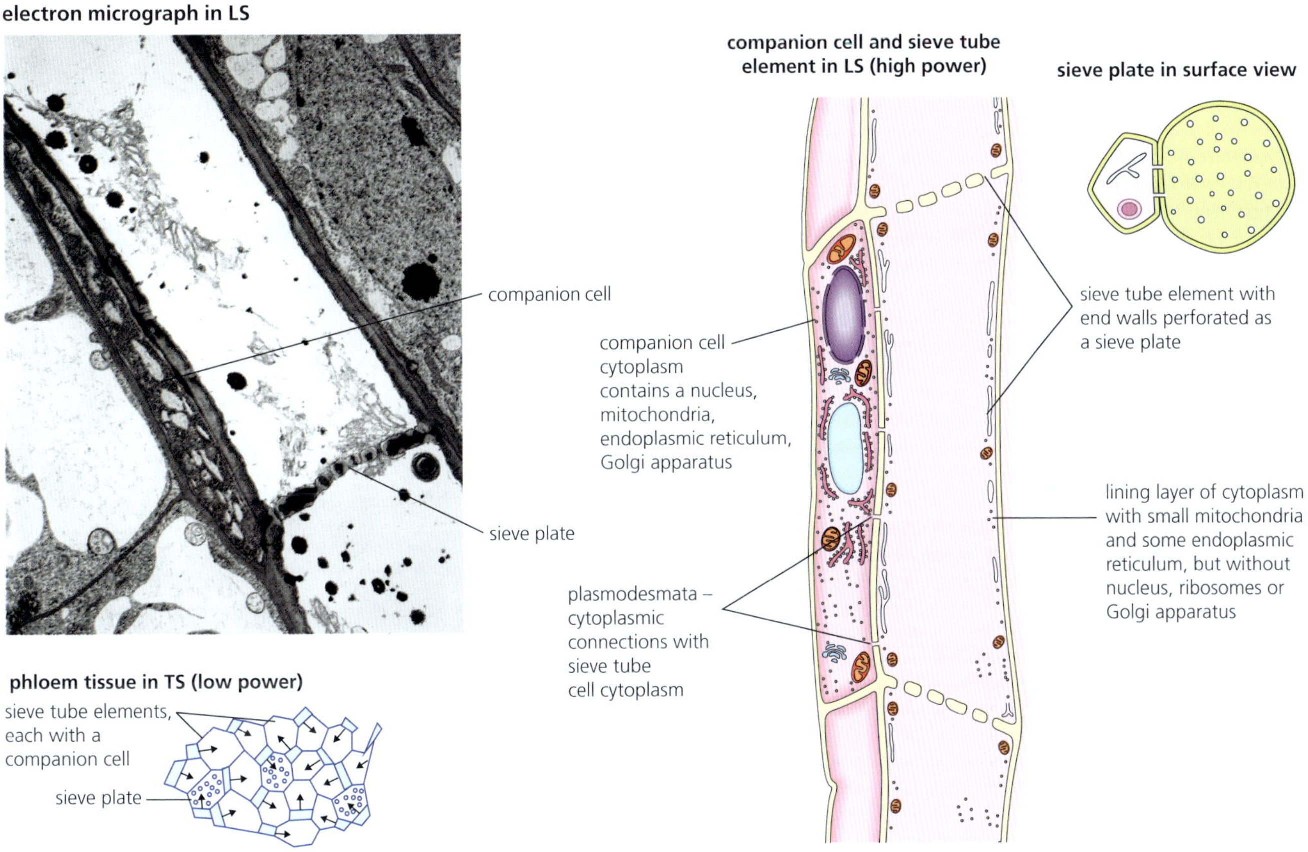

electron micrograph in LS

companion cell and sieve tube element in LS (high power)

sieve plate in surface view

- companion cell
- companion cell cytoplasm contains a nucleus, mitochondria, endoplasmic reticulum, Golgi apparatus
- sieve plate
- plasmodesmata – cytoplasmic connections with sieve tube cell cytoplasm
- sieve tube element with end walls perforated as a sieve plate
- lining layer of cytoplasm with small mitochondria and some endoplasmic reticulum, but without nucleus, ribosomes or Golgi apparatus

phloem tissue in TS (low power)

sieve tube elements, each with a companion cell
sieve plate

■ **Figure B3.2.30** The structure of phloem tissue

> ● **Top tip!**
>
> The cell walls of the sieve tubes are rigid, which enables the hydrostatic pressure needed to achieve mass flow.

Phloem is a living tissue and has a relatively high rate of aerobic respiration during transport. In fact, transport of manufactured sucrose and amino acids in the phloem is an active process, using energy from metabolism.

The process of translocation

Translocation can be illustrated by examining the movement of sugar from the leaves. The story starts at the point where sugars are made and accumulate within the mesophyll in the leaf. This is the **source** area.

Sucrose is loaded into the companion cell, and from there into the phloem sieve tube elements in the leaf by active transport. The accumulation of sugar in the sieve tube elements raises the solute potential and water follows the sucrose by osmosis. This creates a high hydrostatic pressure in the sieve tubes of the source area.

Meanwhile, in living cells elsewhere in the plant – often, but not necessarily, in the roots – sucrose may be converted into insoluble starch deposits. This is a **sink** area. As sucrose flows out of the sieve tubes here, the solute potential is lowered. Water then diffuses out and the hydrostatic pressure is lowered.

Sources and sinks can vary according to the time of year. For example, in annual plants storage organs can be sinks or sources at different times of the year. Following periods of dormancy, such as winter, roots become sources when stores of starch are mobilized and converted into sucrose to be transported to growing regions of the plant, such as the leaves (which are the sinks).

These processes create the difference in hydrostatic pressures in source and sink areas that drive mass flow in the phloem.

> **9 Deduce** what the presence of many mitochondria in the companion cells implies about the role of these cells in the movement of 'sap' in the phloem.

The pressure–flow hypothesis

The principle of the **pressure–flow hypothesis** is that the sugar solution flows down a hydrostatic pressure gradient. There is a high hydrostatic pressure in sieve elements near mesophyll cells in the light (source area), but low hydrostatic pressure in elements near starch storage cells of the stem and root (sink area). This mass flow is illustrated in Figure B3.2.31, and the annotations explain the steps.

In this hypothesis, the role of the companion cells (living cells, with a full range of organelles in the cytoplasm) is to maintain conditions in the sieve tube elements favourable to mass flow of solutes. Companion cells use metabolic energy (ATP) to do this.

◆ **Translocation**: the movement of manufactured food (e.g. sucrose and amino acids) which occurs in the phloem tissue of the vascular bundles.

◆ **Source**: location in a plant where glucose is produced or stored (cotyledons, potato tubers, carrots) and converted into sucrose for transport.

◆ **Sink**: location in a plant where sucrose is transported to, either to be converted to glucose for storage as starch or for use in respiration.

model demonstrating pressure flow
(A = mesophyll cell, B = starch storage cell)

In this model, the pressure flow of solution would continue until the concentration in A and B is the same.

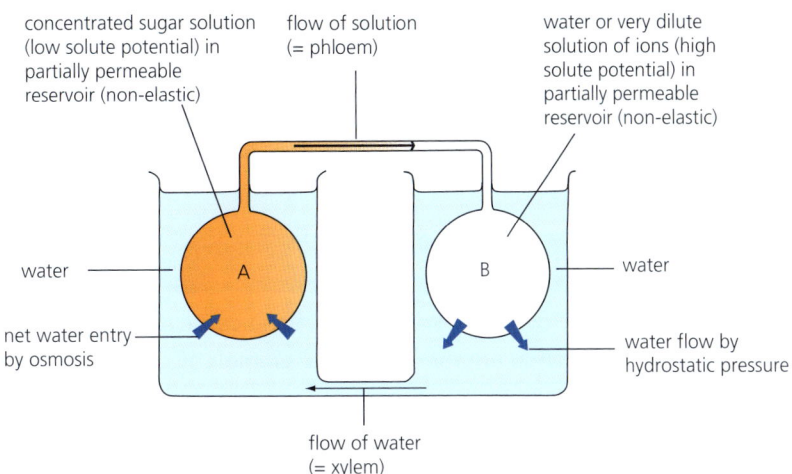

pressure flow in the plant

In the plant, a concentration difference between A and B is maintained by conversion of sugar to starch in cell B, while light causes production of sugar by photosynthesis in A.

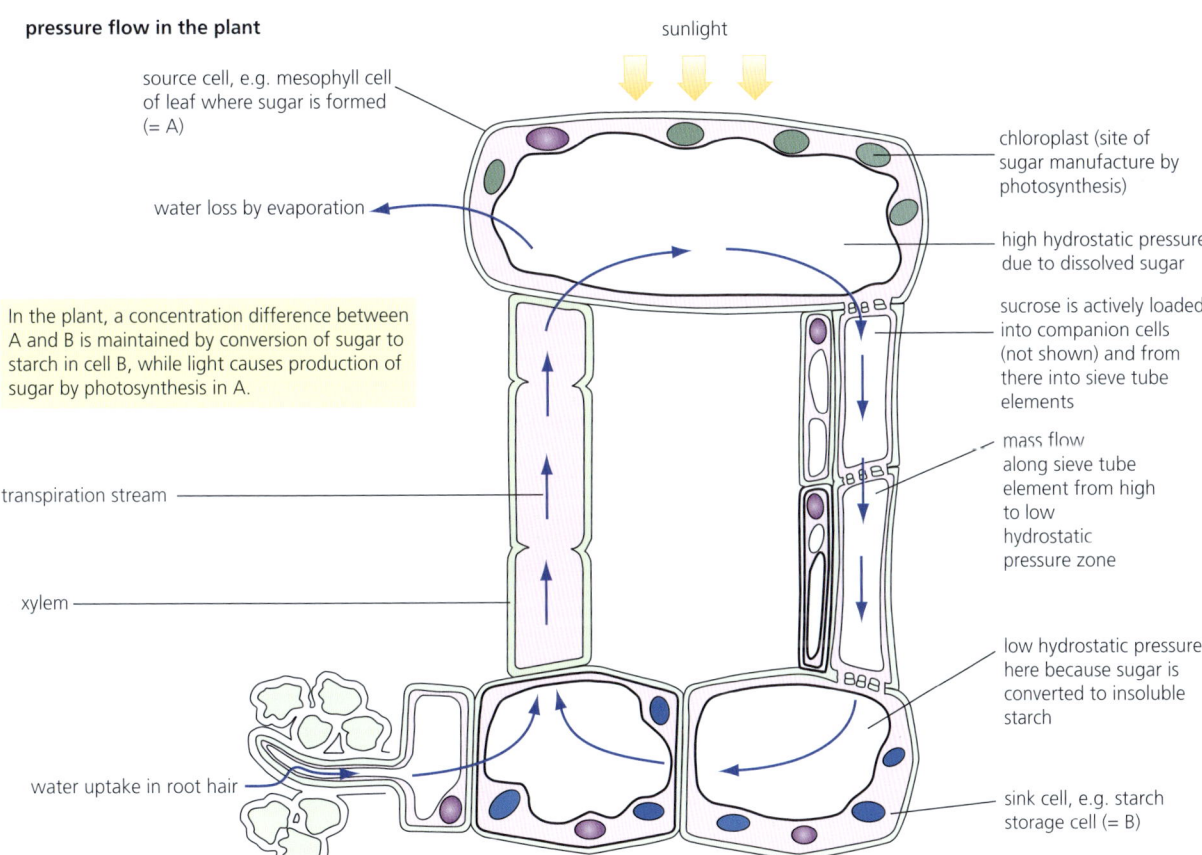

■ Figure B3.2.31 The pressure–flow theory of phloem transport

Concept: Function

The adaptations of phloem tissue ease the flow of sap and enhance loading of carbon compounds into phloem sieve tubes at sources and unloading of them at sinks.

LINKING QUESTIONS

1. How do pressure differences contribute to the movement of materials in an organism?
2. What processes happen in cycles at each level of biological organization?

B3.2 Transport

B3.3 Muscle and motility

Guiding questions
- How do muscles contract and cause movement?
- What are the benefits to animals of having muscle tissue?

SYLLABUS CONTENT

This chapter covers the following syllabus content:
- ▶ B3.3.1 Adaptations for movement as a universal feature of living organisms (HL only)
- ▶ B3.3.2 Sliding filament model of muscle contraction (HL only)
- ▶ B3.3.3 Role of the protein titin and antagonistic muscles in muscle relaxation (HL only)
- ▶ B3.3.4 Structure and function of motor units in skeletal muscle (HL only)
- ▶ B3.3.5 Roles of skeletons as anchorage for muscles and as levers (HL only)
- ▶ B3.3.6 Movement at a synovial joint (HL only)
- ▶ B3.3.7 Range of motion of a joint (HL only)
- ▶ B3.3.8 Internal and external intercostal muscles as an example of antagonistic muscle action to facilitate internal body movements (HL only)
- ▶ B3.3.9 Reasons for locomotion (HL only)
- ▶ B3.3.10 Adaptations for swimming in marine mammals (HL only)

◆ **Motile**: capable of locomotion (whole organism movement).

◆ **Sessile**: fixed in one place; immobile.

> **Concept: Form**
>
> Although all organisms carry out movements, some are capable of locomotion whereas others are immobile. Each group is adapted to its form.

Adaptations for movement in living organisms

All organisms carry out movement: it is one of the key processes of life. **Motile** (or mobile) organisms are those that can move from place to place in search of food or reproductive mates, to migrate to new areas to escape extreme cold, and so on. **Sessile** organisms are ones that are fixed to one place once they have established a place to live. Some species have a mobile and sessile life stage, for example coral. Sessile organisms are fixed in place, but they still show movement of parts of the organism, such as cilia.

The cheetah (*Acinonyx jubatus*) is the fastest animal on land and can run at up to 120 km per hour (Figure B3.3.1). The cheetah's backbone is very flexible and is adapted to act like a spring, absorbing energy as the animal flexes during running, and converting this energy back into movement. The long tail acts as a balance to ensure smooth locomotion.

Sessile organisms, such as coral, carry out movements to feed. The polyp secretes a skeleton of calcium carbonate, which renders them immobile. The tentacles of the coral polyp (a small animal) contain stinging cells (nematocysts) that kill small prey animals, then the tentacles move the prey into the central mouth of the polyp. Interestingly, the larvae of coral are fully mobile – they are formed when eggs and sperm released from sessile coral mix in seawater and fuse together in the process of fertilization.

■ Figure B3.3.1 A cheetah (*Acinonyx jubatus*) in pursuit of prey

Another sessile animal is the barnacle. These animals are found on rocky shores and feed by filtering seawater through fine hairs on feather-like structures called cirri (Figure B3.3.2): when the cirri are drawn back, the food is scraped off into the mouth. Sessile organisms can move via external forces but are usually attached to something, which provides stability and security.

Plants are largely sessile but also carry out movement. Plants can grow towards stimuli, such as light, to maximize photosynthesis (this is known as phototropism). The Venus fly trap (*Dionaea muscipula*), a carnivorous plant, feeds on insects. When an insect lands on the 'trap' and touches a hair on the surface a couple of times, it triggers a response where the trap closes (Figure B3.3.3). Digestive enzymes are then secreted, which digest the animal.

■ Figure B3.3.2 Barnacles are sessile, here seen filtering plankton from the seawater

Link
Phototropism is covered in Chapter C3.1, page 510.

Top tip!
Motile organisms have many adaptations that sessile species do not have, such as wings, legs, flippers, fins, webbed feet and so on. Motile animals often have a streamlined body to help minimize friction with either air or water as they move – this is something that sessile species, in general, do not show.

ATL B3.3A
Research one terrestrial and one aquatic motile organism. Find out about their adaptations for movement. Compare and contrast their respective adaptations.

■ Figure B3.3.3 A Venus fly trap (*Dionaea muscipula*) with captured fly

◆ **Myosin**: a thick protein that forms (together with actin) the contractile filaments of muscle cells; it can convert chemical energy in the form of ATP to mechanical energy.

◆ **Actin**: a thin protein that forms (together with myosin) the contractile filaments of muscle cells; contains binding sites for the myosin heads.

◆ **Z line**: defines the lateral boundaries of the sarcomere.

◆ **Sarcomere**: unit of a skeletal (voluntary) muscle fibre, between two Z lines.

Sliding filament model of muscle contraction

In the previous section, we saw how movement is a feature of all organisms. Some movements are passive and are caused by changes in the flow of air and water surrounding the organism. Other movements are active, involving muscles. The mechanism of muscle contraction is complex and will now be explored.

In Chapter B2.3 (page 269), we examined the structure of striated muscle. In this section, we will investigate how muscles contract.

Skeletal muscle fibres are multinucleate and contain specialized endoplasmic reticulum. The striped appearance is due to an interlocking arrangement of two types of protein filaments. The thick filaments are made of a protein called **myosin**. They are about 15 nm in diameter. The longer thin filaments are made of another protein, **actin**. Thin filaments are about 7 nm in diameter and are held together by transverse bands known as **Z lines** (Figure B3.3.4). Each repeating unit of the myofibril is, for convenience of description, referred to as a **sarcomere**. So, we can think of a myofibril as consisting of a series of sarcomeres, attached end to end.

Top tip!
In skeletal muscle, fibres appear striped under a microscope, so this muscle is also known as striated muscle.

Common mistake

Many students think that actin and myosin are produced in the sarcoplasmic reticulum (SR). However, SR is smooth endoplasmic reticulum, with no ribosomes associated with it. Nor is it produced in rough endoplasmic reticulum, because those proteins are used in the muscle cells rather than exported from them. Actin and myosin are produced by free-floating ribosomes in the cytoplasm of muscle cells.

◆ **Tropomyosin**: a protein that wraps around actin filaments, blocking binding sites for myosin heads.

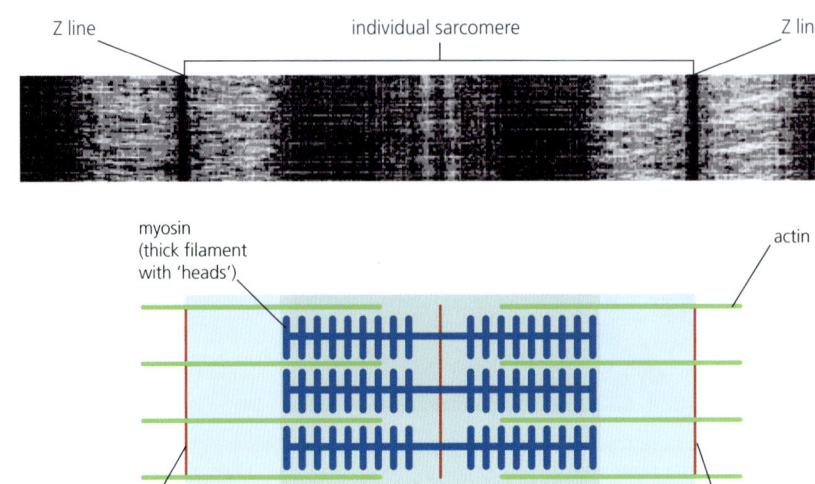

■ Figure B3.3.4 The ultrastructure of a myofibril

When skeletal muscle contracts in response to nervous stimulation, actin and myosin filaments slide past each other, causing shortening of the sarcomeres (Figure B3.3.5). This occurs in a series of steps, sometimes described as a ratchet mechanism. ATP is necessary for the contraction process.

The shortening of the sarcomeres is possible because the thick filaments are composed of many myosin molecules, each with a bulbous head that protrudes from the length of the myosin filament. Along the actin filament is a complementary series of binding sites into which the bulbous heads fit. When the muscle fibres are at rest, the binding sites carry blocking molecules (a protein called **tropomyosin**); so, binding and contraction are not possible at rest.

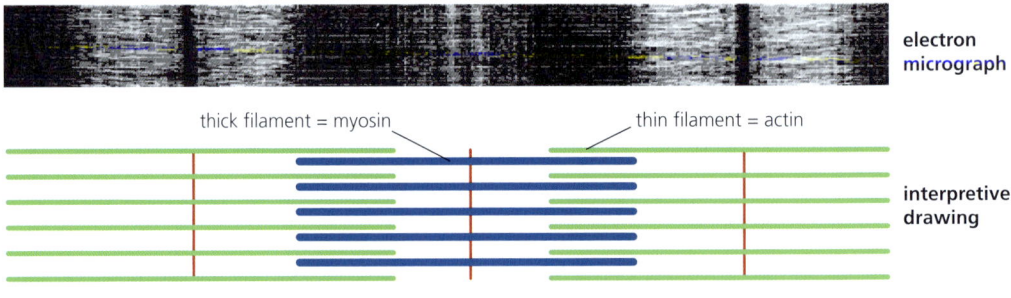

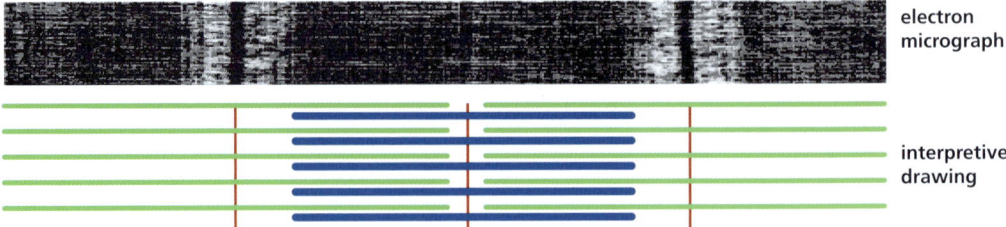

■ Figure B3.3.5 Muscle contraction of a single sarcomere

♦ **Troponin**: a protein to which calcium binds to regulate muscle contraction; when calcium binds, it triggers the removal of the blocking molecule, tropomyosin.

♦ **ATPase**: a complex of integral proteins located in the mitochondrial inner membrane where it catalyses the synthesis of ATP from ADP and phosphate, driven by a flow of protons.

● **Common mistake**

Do not confuse actin and myosin. Myosin is a thick filament with a head, whereas actin is a thinner filament without a head.

Calcium ions play a critical part in the muscle fibre contraction mechanism, together with the proteins tropomyosin and **troponin**. The contraction of a sarcomere is best described in the following four steps.

1. The myofibril is stimulated to contract by the arrival of an **action potential** (Figure B3.3.6). This triggers the release of calcium ions from the sarcoplasmic reticulum, to surround the actin molecules. Calcium ions now react with the protein troponin, which when activated triggers the removal of the blocking molecule, tropomyosin. The binding sites are now exposed.
2. Each bulbous head of myosin, to which ADP and P_i are attached (called a charged bulbous head), reacts with a binding site on the actin molecule beside it. The phosphate group (P_i) is dislodged (removed) from the head at this moment.
3. The ADP molecule is then released from the bulbous head; this is the trigger for the rowing movement of the head, which tilts by an angle of about 45°, pushing the actin filament along. At this step, the power stroke, the myofibril has been shortened (contraction).
4. Finally, a new molecule of ATP binds to the bulbous head. The protein of the bulbous head includes the enzyme **ATPase**, which catalyses the hydrolysis of ATP. When this reaction occurs, the ADP and inorganic phosphate (P_i) formed remain attached, and the bulbous head is now 'charged' again. The charged head detaches from the binding site and straightens.

● **Top tip!**

Muscles need ATP to release from a contracted state. ATP causes separation of the cross-bridges during muscle relaxation. The development of rigor mortis after death is an indicator that energy from ATP is directly used to break rather than make cross-bridges – muscles remain contracted after death because ATP is no longer available to break cross-bridges.

● **Common mistake**

Students often write incorrect explanations for the role of calcium ions in muscle contraction. Calcium ions do not directly form cross-bridges between actin and myosin filaments during muscle contraction, but are involved with moving the molecules blocking the myosin binding sites on actin, allowing cross-bridges to form. Some YouTube clips on this topic are misleading – all material taken from the Internet should be treated with caution!

Muscles are involved in maintaining body posture and in subtle and delicate movements, as well as in vigorous or even violent actions. Consequently, nervous control of muscle contraction may cause relaxed muscle to contract slightly, moderately or fully, depending on the occasion. In these differing states of contraction, the overall lengths of the sarcomeres are changed accordingly.

The relative changes in sarcomere length are illustrated diagrammatically in a single sarcomere in Figure B3.3.7.

Arrival of action potential at myofibril releases Ca^{2+} ions from sarcoplasmic reticulum.

Ca^{2+} ions react with a protein (troponin), activating it. Activated troponin reacts with tropomyosin at the binding sites on the actin molecules, thereby exposing the binding sites.

Each myosin molecule has a 'head' that reacts with ATP → ADP + P_i which remain bound.

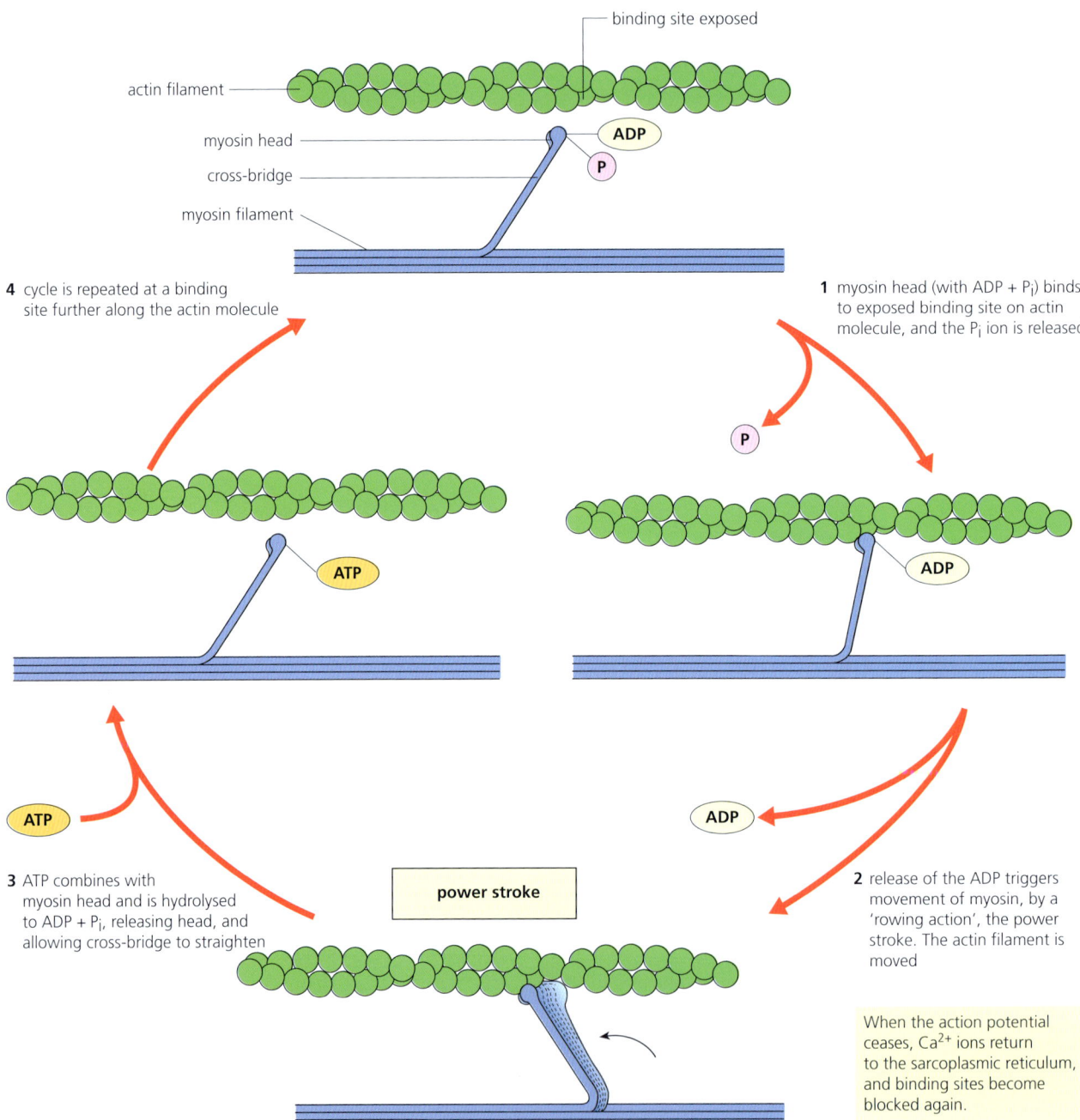

Figure B3.3.6 The sliding-filament hypothesis of muscle contraction

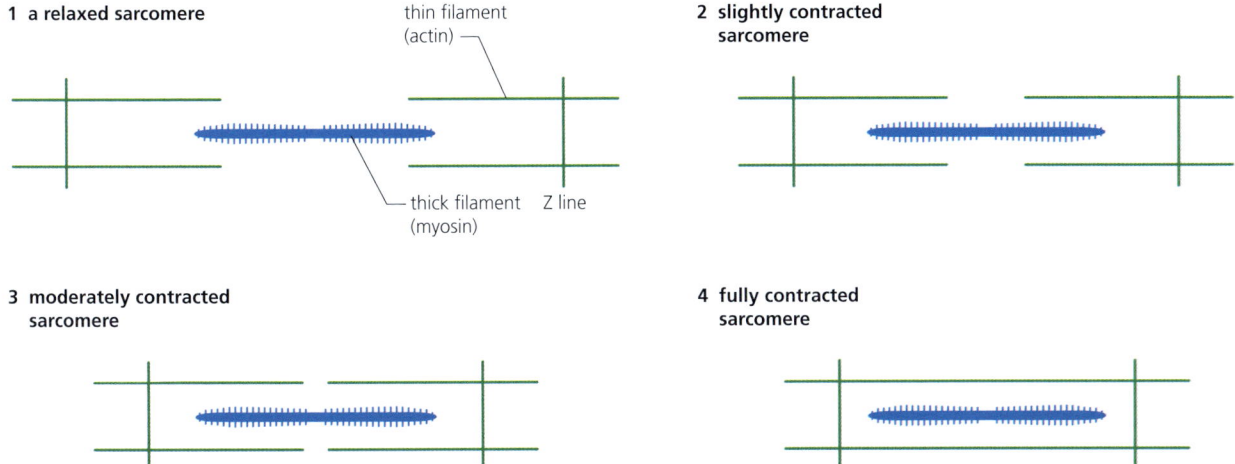

■ **Figure B3.3.7** Analysing states of contraction in striated muscle fibre

Inquiry 2: Collecting and processing data

Processing data

Figure B3.3.8 shows a representation of part of a myofibril, seen at a particular stage of contraction. This representation is based on an interpretation of an electron micrograph.

Interpret the diagram to identify the approximate state of contraction illustrated in the sketch of the electron micrograph of a myofibril. Use Figure B3.3.5 to help you explain your answer.

■ **Figure B3.3.8** Part of a myofibril at a particular stage of contraction

Now **sketch** a similar myofibril, fully contracted. Label your drawing (showing sarcomere, Z lines, light band and dark band).

Concept: Form and function

Striated muscle tissue takes the form of a repeating series of functional units (sarcomeres). By fusing cells together to form a multinuclear structure, the function of muscle fibres can be achieved, i.e. to generate force and contract in order to support breathing, locomotion and posture (skeletal muscle), and to pump blood throughout the body (cardiac muscle).

◆ **Titin**: a large mechanical protein in muscle cells; its main function is to act as a molecular spring in the sarcomeres.

◆ **Antagonistic muscle**: a muscle that works as one of a pair: as one muscle contracts, the other muscle relaxes/lengthens. The muscle that is contracting is called the agonist and the muscle that is relaxing/lengthening is called the antagonist.

Role of the protein titin and antagonistic muscles in muscle relaxation

Titin is an immense protein found within the sarcomere of striated (and cardiac) muscle (Figure B3.3.9). It is the largest protein in the body, composed of some 27 000 amino acids. The titin protein is extended when the sarcomeres of skeletal muscle are 'passively' stretched by **antagonistic muscle** contraction (i.e. contraction of the opposite muscle). For example, titin in external intercostal muscles in the chest is extended and stretched when the internal intercostal muscles contract.

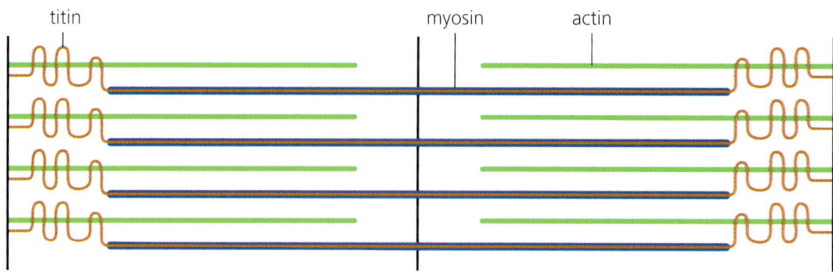

■ **Figure B3.3.9** Diagram showing the position of titin in the sarcomere, before stretching

Antagonistic muscles are needed because muscle tissue can only exert force when it contracts. After being stretched, the titin recoils like a spring, shortening the sarcomere after it is stretched. Titin is also used to stabilize the myosin (the thick filaments of the sarcomere – see page 327) and centre it between the thin actin filaments. It also prevents overstretching of the sarcomere. Titin binds at several sites along its length to the actin filaments in the sarcomere, mainly at the edges of the Z line, which serves an anchoring function.

Research has shown than in addition to acting as a 'passive' spring, the titin protein can create 'active' contraction by folding itself to generate force. The titin molecule undergoes unfolding–refolding changes as the muscle stretches and recoils.

Structure and function of motor units in skeletal muscle

Stimuli (a change in the environment) received by receptors travel through the **nervous system** via a series of **neurons**. Receptors create an electrical impulse – the **action potential** (see Chapter C2.2, page 471), which passes along sensory neurons via relay neurons (within the central nervous system – the brain and spinal cord) and on to **motor neurons** (Chapter C3.1, page 499). Motor neurons connect to **effectors** (muscle or glands), which carry out the response to the stimulus.

Motor neurons have the cell body at one end of the neuron, with an axon forming the structure through which the action potential passes (Figure B3.3.10).

A neuron is surrounded by many supporting cells, one type of which, **Schwann cells**, become wrapped around the axons of motor neurons, forming a structure called a **myelin sheath**. Myelin consists largely of lipid and has high electrical resistance. Frequent gaps occur along a myelin sheath, between the individual Schwann cells. The gaps are called **nodes of Ranvier**.

Striated muscle fibres are supplied with nerves by a motor neuron nerve ending known as a **motor end plate** or **neuromuscular junction**, see Figure B3.3.11. This is where the motor neuron connects to a muscle cell and is a special type of synapse (see page 476). The chemical transmitter substance that diffuses across the synaptic cleft is **acetylcholine**.

When an action potential arrives at the neuromuscular junction, calcium ions (Ca^{2+}) enter the motor neuron and cause synaptic vesicle exocytosis: vesicles of acetylcholine (ACh) move to the presynaptic membrane and release ACh into the neuromuscular junction. ACh binds to receptors on the sarcolemma (this is the plasma membrane of the muscle fibre). This triggers the release of calcium ions from the sarcoplasmic reticulum into the cytoplasm around the myofibrils (Figure B3.3.11). These calcium ions then remove the blocking molecules on the binding sites of actin filaments (see above). This starts the sarcomere contraction process. When action potentials stop arriving at the muscle fibres, calcium ions return to the sarcoplasmic reticulum and the binding sites are again covered by blocking molecules.

Top tip!

When a muscle contracts, its antagonistic muscle stretches. The immense protein titin helps sarcomeres to recoil after stretching and also prevents overstretching.

Link

Neurons are covered in more detail in Chapter C2.2, page 467.

◆ **Nervous system**: a complex network of neurons that carry messages to and from the brain and spinal cord to various parts of the body.

◆ **Neuron**: cell within the nervous system that carries electrical impulses.

◆ **Motor neuron**: nerve cell that carries impulses away from the central nervous system to an effector (e.g. muscle, gland).

◆ **Effector**: a muscle or gland that acts in response to a stimulus.

◆ **Neuromuscular junction**: a specialized synapse between a motor neuron nerve terminal and its muscle fibre, responsible for converting electrical impulses generated by the motor neuron into electrical activity in the muscle fibres.

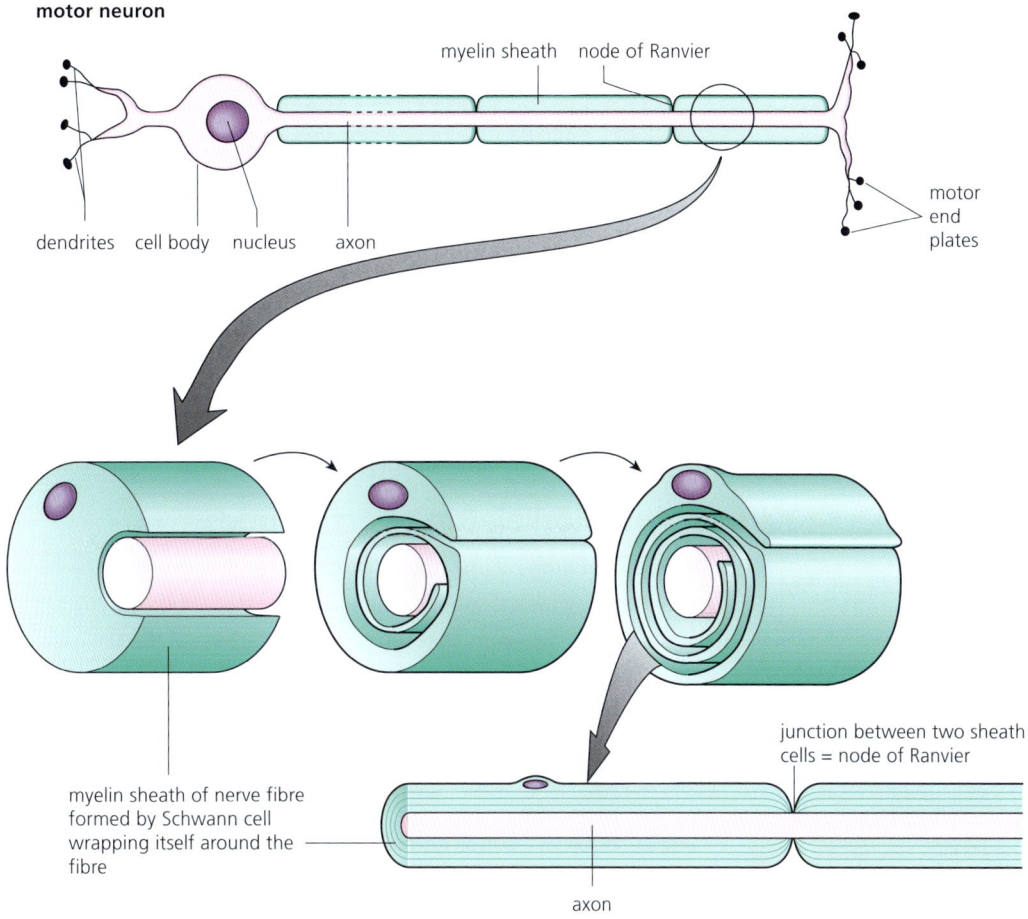

■ **Figure B3.3.10** A motor neuron

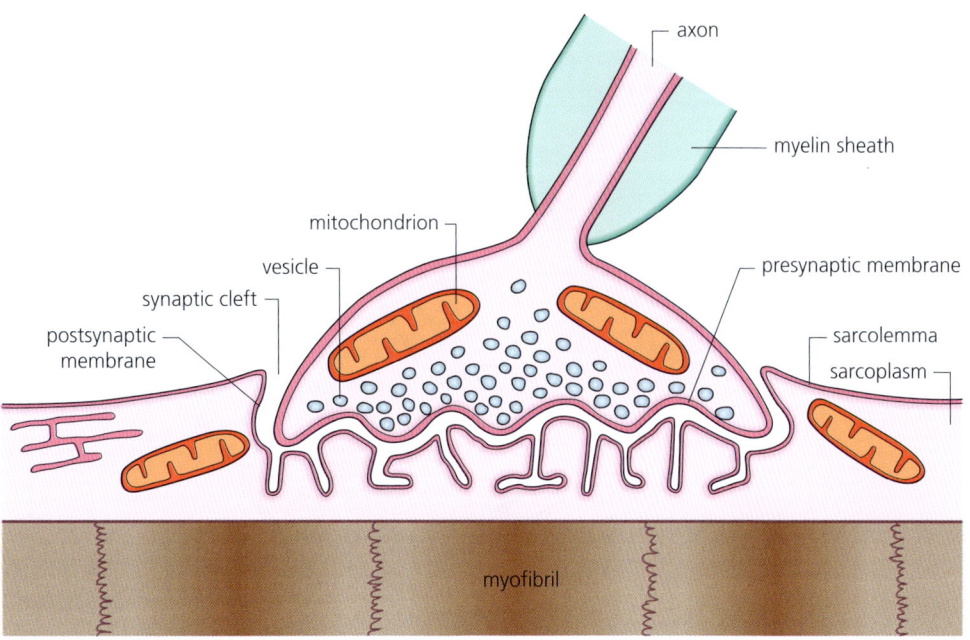

■ **Figure B3.3.11** The neuromuscular junction

B3.3 Muscle and motility

Roles of skeletons as anchorage for muscles and as levers

Locomotion – the movement of whole organisms – is the result of the interactions of nervous, muscular and skeletal systems.

Bones support and partially protect the body parts. They also articulate with other bones at moveable joints and provide anchorage for the muscles. In mammals, the skeleton consists of the axial skeleton (skull and vertebral column) and the appendicular skeleton (limb girdles and limbs).

■ The endoskeleton – vertebrates

Vertebrates have an **endoskeleton**. This is an internal support structure to which muscles are attached via tendons. 'Endo' means 'inside', because the structure lies within the body. Most vertebrates have a skeleton made from bone, although in some fish (e.g. sharks) it is made from cartilage.

■ The exoskeleton of arthropods

The body and limbs of arthropods (for example, insects) are covered by a tough external skeleton, or **exoskeleton**, with a flexible membrane between the body segments and in the joints of the limbs. The muscles for movement are attached to the inside of the skeleton. The arthropod's legs are a series of hollow cylinders made from **chitin** (a modified polysaccharide that contains nitrogen), held together by joints. Across the joints are attached muscles in antagonistic pairs.

Figure B3.3.12 compares how exoskeletons and bones provide anchorage for muscles and act as levers to bring about movement in insects (an example of an arthropod group) and in mammals.

◆ **Endoskeleton**: an internal skeleton, such as the bony or cartilaginous skeleton of vertebrates.

◆ **Exoskeleton**: skeleton secreted external to the epidermis of the body.

◆ **Chitin**: a modified structural polysaccharide that contains nitrogen, used in the exoskeleton of insects.

1 **Distinguish** between an endoskeleton and an exoskeleton.

Concept: Form

Some animals have exoskeletons, which surround the body and form an impervious outer shell, whereas others have an endoskeleton within the body. Both systems allow muscle attachment, contraction and movement.

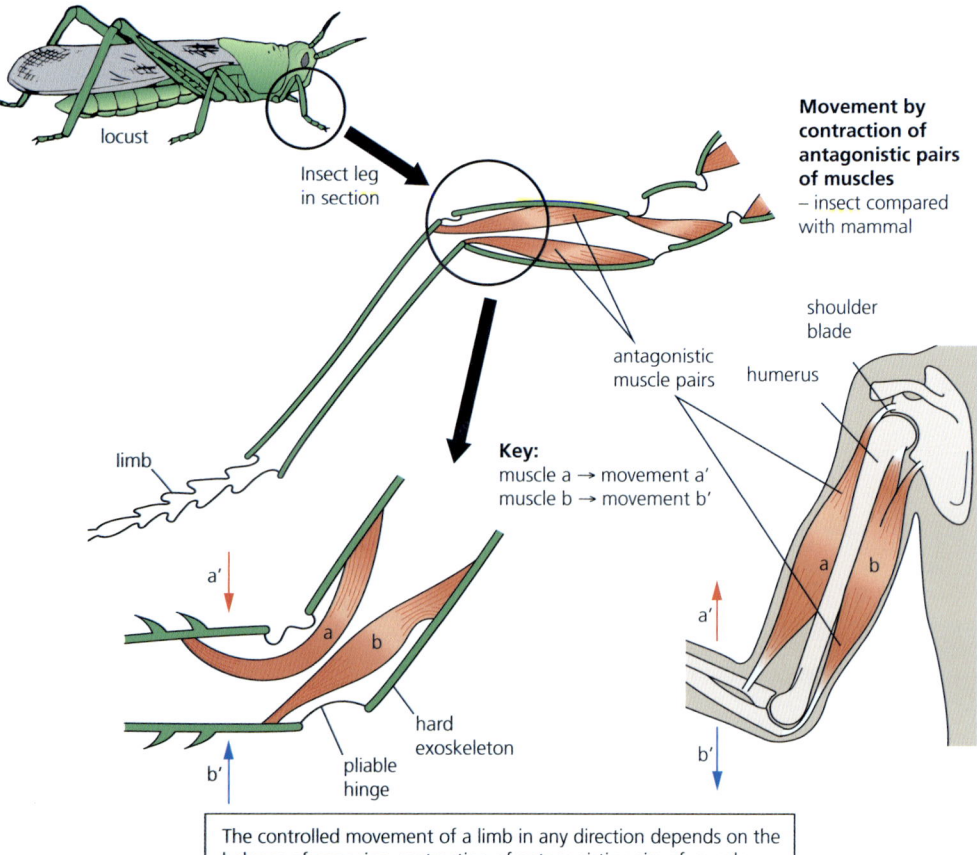

■ Figure B3.3.12 Limb movement – in insects and mammals

Common mistake

Students are often unaware that muscles only do work when they contract. Muscles can only contract and relax, so work in so-called antagonistic muscle pairs. They cannot *actively* extend or lengthen – extension happens as the result of contraction of the antagonistic muscle.

◆ **Synovial joint**: joint where a very thin layer of viscous synovial fluid separates and lubricates the two cartilage-covered bone surfaces.

◆ **Synovial fluid**: lubricates the joint, nourishes the cartilage and removes any (harmful) detritus from worn bone and cartilage surfaces.

Link
Collagen is covered in detail in Chapter B1.2, page 219.

2 Distinguish between:
  a bone and cartilage
  b ligament and tendon.

Top tip!

You are not required to name the muscles and ligaments of the hip joint, but you should be able to name the femur and pelvis. Use Figure B3.3.14 to help you.

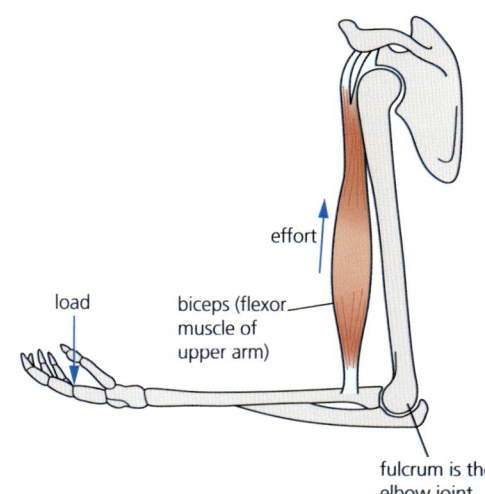

Figure B3.3.13 A lever system in the body

Bones, joints and muscles as levers

We use machines to apply forces to objects in the world about us. A simple, commonly used machine is the lever. The skeleton of an animal, together with the muscles attached across the joints, functions as a system of levers. Each joint acts as a pivot point or **fulcrum**. The force applied (when the muscle contracts) is called the effort. The load, or force to be overcome, is known as the resistance. The further away from the fulcrum the effort is applied, the greater the leverage; that is, the smaller the force that is required to raise the load. Figure B3.3.13 shows the arm as an example of a lever system in the body.

Movement at a synovial joint

Moveable joints in the body are of different types, but they all permit controlled movements. They are known as **synovial joints** because a thick, viscous fluid – the **synovial fluid** – is secreted and retained in the joint for lubrication. We will consider the hip joint as a synovial joint example.

At the hip is a ball-and-socket joint. In this type of joint, the ball-like surface of one bone (the head of the femur of the upper leg in this example) fits into a cup-like depression on another bone (here the pelvis) – see Figure B3.3.14. Muscles between the pelvis and the femur allow movement of the leg in three planes (see Figure B3.3.15 for more details). **Tendons** (fibrous connective tissue) attach the muscles to the bones, while **cartilage**, a flexible connective tissue, cushions the ends of bones at the joints. It has a smooth, shiny surface to reduce friction and is made of collagen – therefore it has very high tensile strength. **Ligaments** (also a fibrous connective tissue) connect bones to other bones, and also help to support internal organs.

Range of motion of a joint

There are contrasting degrees of movement at knee and hip joints, and at the shoulder and elbow. This is because of a fundamental difference in the types of joints involved (Figure B3.3.15 and Table B3.3.1).

We have already seen that the hip is a ball-and-socket joint. The shoulder is a similar type of joint. This type of joint permits the widest range of movement, in all three planes. This type of movement is described as **circumduction**.

The knee and elbow are **hinge joints**, which restrict movement to one plane. This is because of the shape of the surface of the joint at which the ends of the bones meet and the position of the ligaments that hold the bones together. Movements at the elbow and knee are described as **flexions and extensions**.

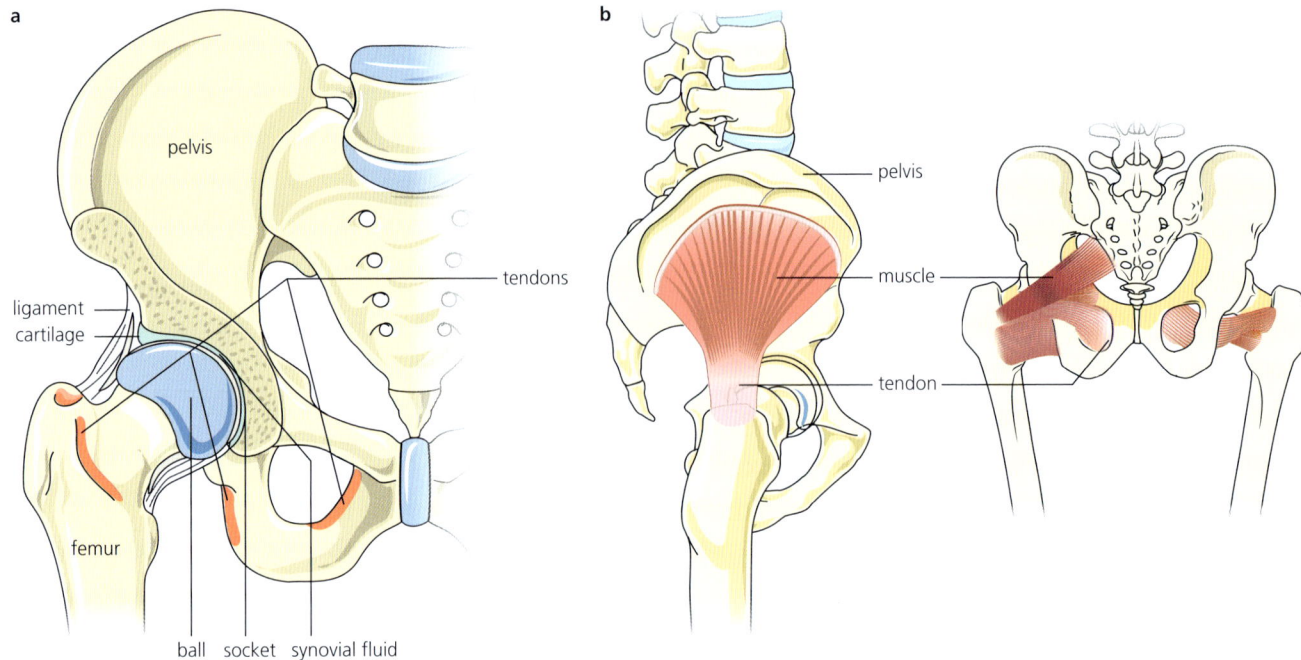

■ Figure B3.3.14 Hip joint: a) front view, b) muscle attachments

3 **Sketch** a diagram of a ball-and-socket joint and compare it to a hinge joint. **State** examples of both type of joint.

Concept: Form and function

Different types of joints have different functions (Table B3.3.1). The form of the joint determines its function, e.g. range of movement.

Here, movements occur at joints where a cavity filled with fluid (synovial fluid) separates the articulating surfaces.

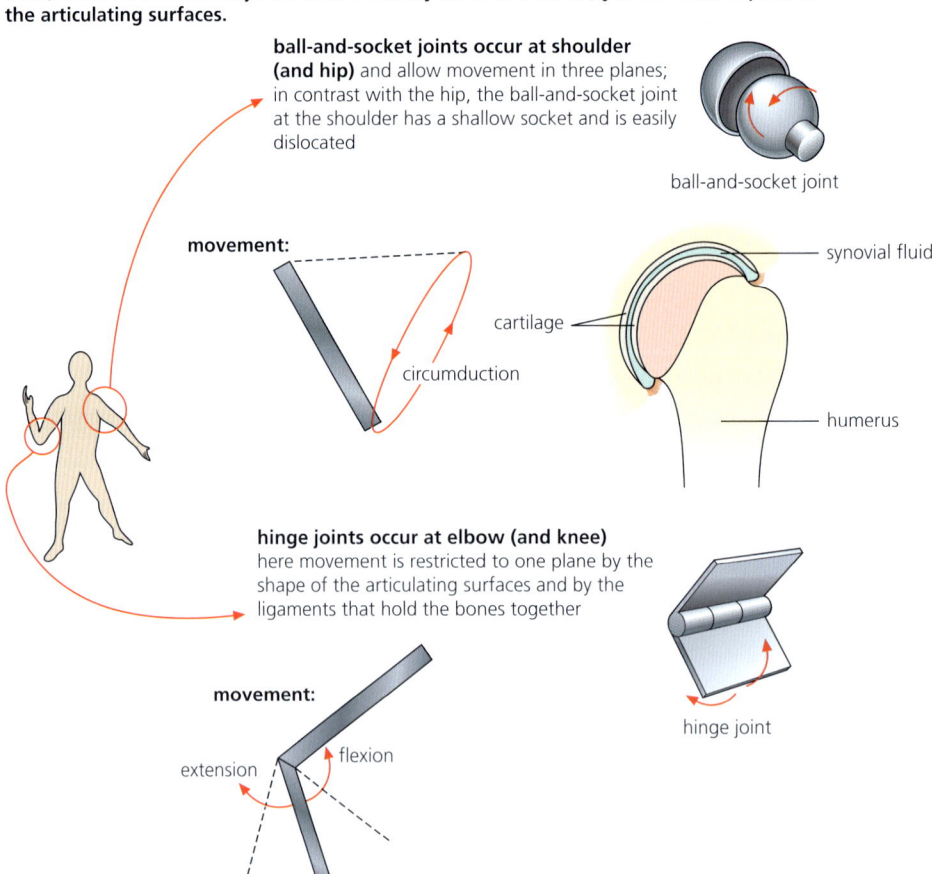

■ Figure B3.3.15 Movement at shoulder and elbow joints

The range in motion of a joint can be compared in a number of dimensions. Look at Table B3.3.1, which compares the range of motion of different joints.

■ Table B3.3.1 Movement of shoulder and elbow joints compared

Comparison	Shoulder joint	Elbow joint
Type of joint	synovial – ball and socket	synovial – hinge
Articulating bones	shoulder blade and humerus (head)	humerus, radius and ulna
Articulating surface(s)	between shoulder blade and humerus	between humerus and radius and ulna
Range of motion	circumduction	flexion and extension

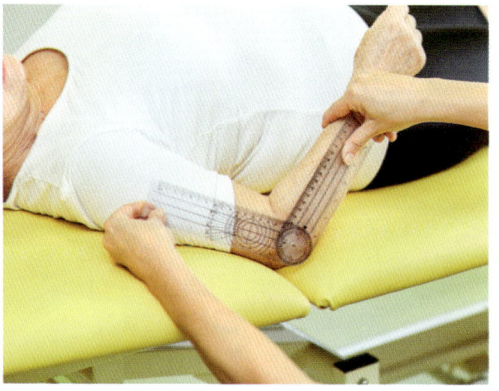

■ Figure B3.3.16 A goniometer being used to measure a joint's range of motion

A goniometer has two 'arms' that are hinged together – one arm is stationary and the other movable. Each is positioned at specific points on the body with the centre of the goniometer aligned at the joint of interest. Marks on the hinge allow the range of movement to be recorded accurately (Figure B3.3.16).

Apps are available for mobile phones that act as goniometers. These use the accelerometer and gyroscopic technology of the smartphone to measure changes in the position of the phone. Once the app is opened, the phone can be put in the position of the part of body being measured, and moved through its available range of motion for the joint in question.

Photographs can also be used to measure changes of angle of limbs around a joint. Computer analysis of the images can be undertaken.

Tool 2: Technology

Carrying out image analysis

Download a goniometer app on to your smartphone (e.g. https://apps.apple.com/us/app/goniometer-plus/id1081939665)

First, select the joint you want to investigate. Place your device against the limb that will be moving and select 'Set 0°' to establish a relative 0°. Rotate the limb to the natural maximum extent (be careful, don't overextend) and then select 'get reading'. The smartphone will record the degree of motion around the joint.

Compare the movement of different joints, for example the hip, shoulder, knee and wrist. Can you relate these differences to the structure, form and function of each joint? Create a table to compare the different joints (using Table B3.3.1 as a template).

Internal and external intercostal muscles as an example of antagonistic muscle action

Intercostal muscles are involved with the ventilation of the lungs (page 276). There are two groups of intercostal muscles, internal and external, which work antagonistically (see Figures B3.1.5, page 280 and B3.3.17). The internal intercostal muscles attach to the internal surface of the ribs and the external intercostal muscles attach to the outer surface of the ribs. The muscles work as follows:
- external intercostal muscles contract to pull the ribs up and out
  - this increases the volume of the thorax and reduces the pressure, forcing air into the lungs
- internal intercostal muscles contract to pull the ribs down and in
  - this decreases the volume of the thorax and increases the pressure, forcing air out of the lungs.

The different orientations of muscle fibres in the internal and external layers of intercostal muscles (look carefully at the cut out in Figure B3.3.17) mean that they move the ribcage in opposite directions. When one layer contracts it stretches the other, storing potential energy in the sarcomere protein **titin** (see Figure B3.3.9).

Intercostal muscles are generally not used in shallow breathing – the diaphragm contracts and relaxes to move air into and out from the lungs in these instances. The intercostal muscles are used during more vigorous exercise, when greater volumes of air (and therefore oxygen) are drawn into the lungs for increased levels of respiration.

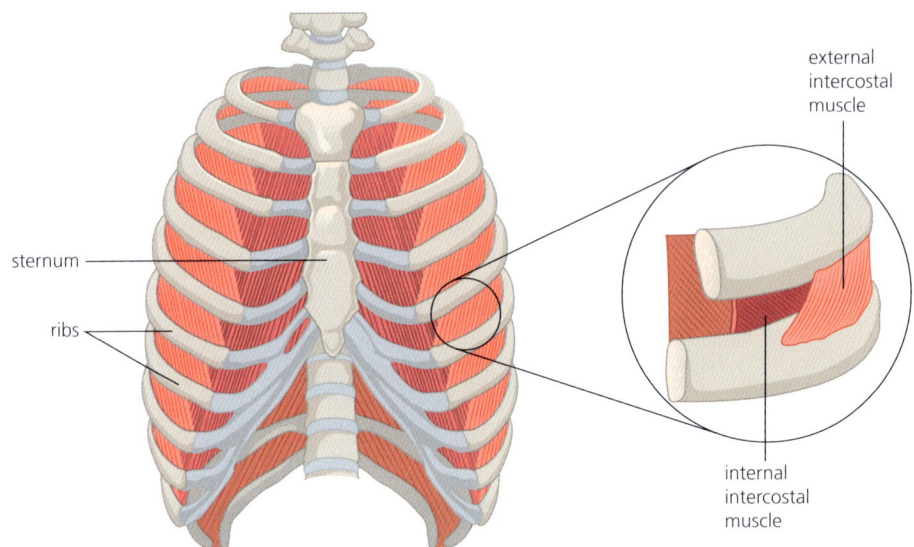

■ Figure B3.3.17 External and internal intercostal muscles work antagonistically to ventilate the lungs

Reasons for locomotion

There are a variety of reasons for locomotion. These include:
- **foraging for food** – all animals must forage for food. Figure B3.3.18a shows a blackbird (*Turdus merula*) feeding on an earthworm. Insects and other invertebrates dominate their diet in spring and summer, while seeds and berries are more important in the colder autumn and winter months when invertebrates are less abundant.
- **escaping from danger** – such as prey animals escaping from predators, for example a snowshoe hare running from a lynx as in Figure B3.3.18b.
- **searching for a mate** – most great apes, such as gorillas and chimpanzees, live in groups. Orangutans (Figure 3.3.18c), in contrast, are solitary animals and spend most of their life on their own when adults. This is because their diet is primarily fruit, which is not found in sufficient quantities to support big groups. Orangutans only meet up with others of the species to reproduce, often travelling large distances through the forest to find a mate.
- **migration** – many animals leave an area where they spend spring and summer to escape the colder winter months. For example, many species of whale undertake some of the longest migrations on Earth, often swimming many thousands of miles, over several months, so they can breed in the warmer waters of the tropics (Figure B3.3.18d).

■ Figure B3.3.18 Locomotion in animals: a) a blackbird (*Turdus merula*) feeding on an earthworm, b) a lynx chasing a snowshoe hare, c) an orangutan in a tree, d) a humpback whale mother with her calf

Adaptations for swimming in marine mammals

Marine mammals include seals, sea lions, dugongs, manatees and cetaceans (dolphins, porpoises and whales – see Figure B3.3.18d). The aquatic medium they live in is much denser than air, so they need specific adaptations to move efficiently through water. Body forms that are suitable for life on land offer too much resistance in water (such as hands and feet) or are an obstruction to movement (such as flat parts of the body, e.g. the front of the thorax and pelvis). The streamlined, hydrodynamic shape of marine mammals reduces the friction (or drag) of water over the body, enabling them to move more easily.

Some marine mammals are fully aquatic (they live in water all the time, e.g. whales) whereas others are semi-aquatic, living in both water and on land (e.g. seals and sea lions). This influences their adaptations. For example, sea lions and seals have different types of tails to dugongs, manatees and cetaceans.

Front limbs have modified pentadactyl limbs to form flippers (pectoral fins), which have a large surface area but a streamlined shape to direct body movement. The pelvic bone, which in land mammals has attachments for legs, is much reduced in marine mammals and rear limbs are lacking. Rear limbs would hinder movement through the water. The tail is modified into a fluke (caudal fin) which moves up and down through muscular movements to propel the animal through the water (Figure B3.3.18d).

> **Concept: Form**
>
> Aquatic marine mammals are adapted for swimming. Their streamlined shape ensures they can move swiftly through the aquatic environment, and limbs are modified to propel the animal through water.

Common mistake

Because they are of similar size and shape, marine mammals such as whales are sometimes confused with large fish such as sharks. One anatomical difference is that shark tails move side to side, a different movement from marine mammal flukes, which go up and down.

Other features that would hinder movement through water are also generally lacking, such as external ears (although semi-aquatic mammals such as sea lions do have external ears to help locate prey on land) and external reproductive organs. The necks of these animals are short and relatively immobile to enable fast, muscular swimming.

■ **Figure B3.3.19** An orca off the coast of Alaska, exhaling through its blowhole

Marine mammals have specialized skin. The inner-most layer of this skin is a layer of blubber (see Chapter A1.1, page 10). The blubber covers bone that would otherwise stick up from the body and ensures that the skin's surface is streamlined. The outer skin layer also improves hydrodynamics by continually exuding oil droplets and shedding epithelial cells. The continual shedding of cells ensures that dead skin does not build up on the surface of the body and impede movement by increasing drag. Oil helps to lubricate the skin, allowing the water to pass over it more smoothly.

Cetaceans (whales and dolphins) have adaptations to allow periodic breathing between dives. Their nostrils (blowholes) are located at the top of the head, which allows them to breathe quickly and easily when they surface (Figure B3.3.19).

Being mammals, they have lungs and so (unlike fish) cannot carry out ventilation while under water. This means that they must take a breath before diving and then close the airway to ensure that water does not enter. When the animal surfaces, it opens its blowhole and expels air from its lungs, then inhaling quickly before closing the blowhole once more before diving. The blowhole remains closed while the animal is underwater. Other marine mammals, such as seals and sea lions, do not have these adaptations as they can come up on land so do not need blowholes.

> **ATL B3.3B**
>
> Research the evolution of marine animals, focusing in particular on cetaceans that evolved from land animals to marine mammals. What were the form and function of the land-based ancestors of marine species?
>
> Marine mammals have a residual pelvic bone. What do residual structures indicate about the evolutionary origin of species?

> **LINKING QUESTIONS**
>
> 1 What are the advantages and disadvantages of dispersal of offspring from their parents?
> 2 In what ways does locomotion contribute to evolution within living organisms?

B4.1 Adaptation to environment

> **Guiding questions**
> - How are adaptations and habitats of species related?
> - What causes the similarities between ecosystems within a terrestrial biome?

SYLLABUS CONTENT

This chapter covers the following syllabus content:
▶ B4.1.1 Habitat as the place in which a community, species, population or organism lives
▶ B4.1.2 Adaptations of organisms to the abiotic environment of their habitat
▶ B4.1.3 Abiotic variables affecting species distribution
▶ B4.1.4 Range of tolerance of a limiting factor
▶ B4.1.5 Conditions required for coral reef formation
▶ B4.1.6 Abiotic factors as the determinants of terrestrial biome distribution
▶ B4.1.7 Biomes as groups of ecosystems with similar communities due to similar abiotic conditions and convergent evolution
▶ B4.1.8 Adaptations to life in hot deserts and tropical rainforest

Note: There is no higher-level only content in B4.1.

Habitats

◆ **Habitat**: the place in which a community, species, population or organism lives.

The smallest biological unit that an ecologist tends to study is the **species**. A species is a group of organisms that can potentially interbreed to produce fertile offspring. Where a species lives is called its **habitat**, and a complete description of a species' ecology (where, when and how it lives) is called its **niche**.

As already discussed, a **population** is defined as a group of individuals of the same species, and a **community** is defined as all the populations of different species living together and interacting with each other.

Communities form the **biotic** (living) part of an ecosystem, while the **abiotic** components (such as rocks, water, light and air) comprise the non-living part, with an **ecosystem** being formed by the interaction between communities and their abiotic environment. The **environment** can be defined as the external surroundings that act on an organism, population or community, influencing survival and development. Life can be seen as organized within a hierarchy, from species comprised of many individuals that form populations, populations that interact with other species' populations to form communities, and communities that interact with the abiotic environment to form ecosystems.

Link

For more information on species and populations, see Chapter A3.1, page 107; for communities see Chapter C4.1, page 562; for niches see Chapter B4.2, page 363.

● Common mistake

Make sure you carefully learn the definitions of the terms 'species', 'population' and 'community'. Students often incorrectly use these words interchangeably. Each word has a precise ecological meaning that you need to know and use correctly.

◆ **Microhabitat**: a small-scale habitat that differs in abiotic and biotic factors from the surrounding, more extensive habitat.

Microhabitat

Within any ecosystem, organisms are normally found in a particular part or habitat. The habitat is the locality in which an organism occurs. So, for example, within a woodland, the tree canopy is the habitat of some species of insects and birds, while other organisms occur in the soil. Within a lake, habitats might include a reed swamp and open water. If the occupied area is extremely small, we call it a **microhabitat**. The insects that inhabit the crevices in the bark of a tree are in their own microhabitat. Conditions in a microhabitat are likely to be very different from conditions in the surrounding habitat.

Top tip!

A description of the habitat of a species can include both geographical and physical locations, and the type of ecosystem. For example, the habitat of an orangutan is the rainforests of South East Asia, on the islands of Borneo and Sumatra, with a habitat that is largely arboreal (i.e. they live above ground and forage in tree canopies).

1 **Define** the term *habitat*.

Adaptations of organisms to the abiotic environment of their habitat

Grass species adapted to sand dunes

Marram grass (*Ammophila arenaria*, Figure B4.1.1) is a species found in coastal sand dune areas in Europe and western Asia.

◆ **Xerophyte**: a group of plants that survive in dry areas by having features that prevent water loss.

Marram grass is a **xerophyte** (*xero* meaning dry, *phyte* meaning plant), which is a plant adapted and able to survive in an environment with little available water or moisture. Plants with xerophytic adaptations are not confined to hot, dry deserts but also:
- cold regions due to frozen water (i.e. water is not available to plants)
- windy, exposed areas (such as sand dunes).

In windy or hot environments, plants are susceptible to increased water loss, either due to increased evaporation (due to warmer temperatures) or differences in water concentration between the leaf tissue and air (dry air has a reduced water concentration, as do windy areas). Species that live in such areas therefore need to be adapted to these abiotic conditions.

Marram grass has a rolled leaf (Figure B4.1.2) rather than the usual flat leaves that grasses have. By rolling the leaf, the stomata are on the inside of the rolled structure. The inside air of the leaf is humid because water vapour that has diffused through stomata cannot easily escape the rolled leaf, so their air becomes saturated with water. The stomata are protected from wind and other air movements. The humid air means that the water concentration gradient between the mesophyll tissue and the air space is reduced, meaning that the rate of evaporation of water is lowered. The inner epidermis is folded and hairy to trap water vapour. The hairs limit air movement, again limiting evaporative loss of water from the stomata.

■ **Figure B4.1.1** Marram grass growing on sand dunes in a coastal area

Theme B: Form and function – Ecosystems

The outer epidermis (outer circle) of the marram grass leaf consists of a layer of thick cuticle and layers of cells with thick cell walls, to prevent water loss from the outside of the leaf.

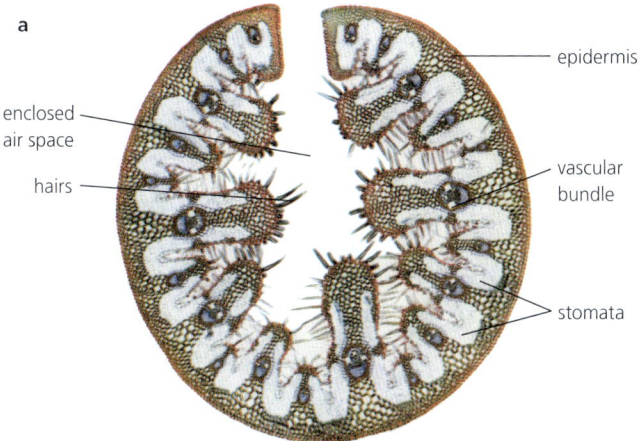
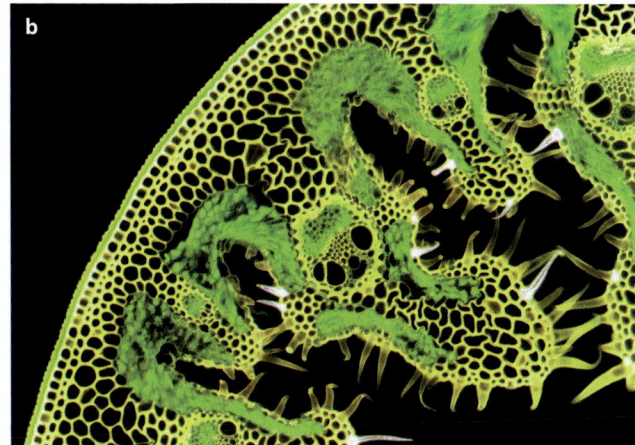

■ Figure B4.1.2 a) Light micrograph of a cross-section of a blade of marram grass (*Ammophila arenaria*); b) magnified section of a marram grass blade to show details of xerophytic adaptations – a layer of thick cuticle and layers of cells with thick cell walls in the epidermis, and the inner epidermis is folded and hairy to trap water vapour

2 Create a table to **outline** the structural features of marram grass to prevent water loss and the effects of those features.

■ Tree species adapted to mangrove swamps

Mangrove swamps (Figures B4.1.3 and B4.1.4) are found in tropical coastal areas, where the average rainfall is generally over 1500 mm yr^{-1} and insolation (the amount of solar radiation received on a given surface in a given time period) is constant throughout year. Mangrove trees grow in a saline and oxygen-deficient environment. This harsh environment limits productivity compared with other tropical ecosystems, although it is still high. Mangroves provide food, habitats and nursery sites for many aquatic species.

Trees that grow in mangrove swamps are exposed to saline (salty) conditions. The roots can be submerged under water and so have less access to oxygen than trees that grow in well-aerated soils.

Aerial roots allow mangroves to absorb oxygen directly from the air and to survive when the forest floods with salt water. The roots of mangroves also allow the trees to grow in areas where the ground is unstable – the stilt-like roots give the trees stability (Figure B4.1.3).

Some genera of mangrove tree, including *Avicennia*, *Laguncularia* and *Sonneratia*, grow specialized roots called pneumatophores: these roots emerge vertically from the ground (Figure B4.1.4). Pneumatophores act like snorkels when the forest is flooded and have pores (lenticels) that cover their surface where oxygen absorption occurs. The lenticels contain substances that are **hydrophobic** and so, when submerged, water cannot flood the root.

Concept: Form

The form of an organism adapts it to abiotic variables in its environment, for example, the rolled leaf of a marram grass helps reduce water loss in an arid environment, and aerial roots help mangrove trees grow where the ground is unstable and oxygen is limited.

■ Figure B4.1.3 Aerial roots of mangrove in Thailand

■ Figure B4.1.4 Mangrove pneumatophores in the Sundarbans, India

B4.1 Adaptation to environment

Mangroves trees live in a variety of salty environments, with soils that range from 60–65 parts per thousand (ppt) of NaCl, where red mangroves (*Rhizophora stylosa*) occur, to 90 ppt salinities, where black (*Avicennia germinans*) and white (*Laguncularia racemosa*) mangroves are found. Mangrove trees have adaptations to limit access of salt and to remove it once present in the plant. Other plants do not have these adaptations, which gives mangrove trees a competitive advantage in these extreme habitats. Root membranes prevent salt from entering while allowing the water to pass through. The red mangrove is an example of a salt-excluding species. Other species excrete salt through glands on their leaves. Black and white mangroves are both salt excreters.

3 Oleander (*Nerium oleander*), an ornamental flowering plant, is a common xerophyte seen in Singapore (Figure B4.1.5a). This plant originates from the Mediterranean; it is found around stream beds where it can endure long seasons of drought followed by winter rain.

Use your knowledge of plant structure and function, and the information in Figure B4.1.5b, to **predict** and **explain** four likely characteristics (adaptations) of an oleander plant.

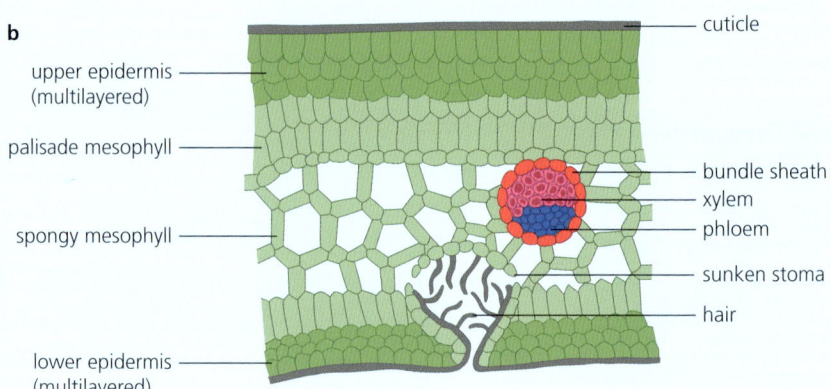

■ **Figure B4.1.5** a) Oleander plants *(Nerium oleander)*; b) leaf section through an oleander plant

Abiotic variables affecting species distribution

◆ **Limiting factor**: any variable that slows life processes or stops a population from growing.

◆ **Biotic factor**: a living part of an ecosystem (i.e. part of the community) that can influence an organism or ecosystem.

◆ **Abiotic factor**: a non-living, physical factor that can influence an organism or ecosystem, e.g. temperature, sunlight, pH, salinity or precipitation.

Limiting factors are the components of an ecosystem, either biotic or abiotic, that limit the distribution or numbers of a population. Limiting **biotic factors** include interactions between organisms, such as competition or predation, while limiting **abiotic factors** include physical components of the environment, such as temperature, salinity, pH, oxygen, carbon dioxide, light, hydrostatic pressure, water current, wind velocity, substratum type (layer of rock or soil beneath the surface of the ground), rainfall amount and humidity.

Animals and plants are affected by similar limiting factors (usually biotic ones such as competition – see page 557) and ones they do not have in common (usually abiotic ones). Limiting factors in plants include light, water, nutrients, carbon dioxide and temperature. In animals, limiting factors include food, space, mates and water. 'Limiting factor' does not only mean a factor that limits growth, development, reproduction or activity of a population by its deficiency – it may also be a limiting factor if it is in excess.

The concept of **tolerance** suggests that there are levels of environmental factors beyond which a population cannot survive. It also suggests that there is an optimum range of these environmental factors within which species can exist and thrive. Adaptations of a species give it a range of tolerance.

Zones of stress and limits to tolerance

Nature of science: Models

The law of tolerance was developed by American zoologist Victor Ernest Shelford in 1911. The law states that an organism or population has certain minimum, maximum and optimum environmental factors that determine success (meaning their survival). The model allowed complex systems to be simplified as a diagram (for example, Figure B4.1.6). The distribution of a species can be plotted on a graph, showing zones of stress and limits of tolerance. They are **models** that attempt to reflect the real world. However, without complete data for a given species, such models remain generalized representations of reality. They are usually shown as bell-shaped curves even though, in reality, the distribution of many species may be skewed towards one preferred area of tolerance.

◆ **Limits of tolerance**: the upper (i.e. critical maximum) and lower (i.e. critical minimum) limits to the range of particular environmental factors (e.g. light, temperature, availability of water) within which an organism can survive.

◆ **Range of tolerance**: range between critical minimal and critical maximum limits of environmental factors affecting an organism.

The **critical minimal** and **critical maximum** limits are a species' or population's **limits of tolerance**. The **range of tolerance** is the range between critical minimal and critical maximum limits of environmental factors affecting an organism. The distribution of a species can be plotted on a graph as shown in Figure B4.1.6 below, which represents the frequency at which individuals of the species are found under a range of environmental factors.

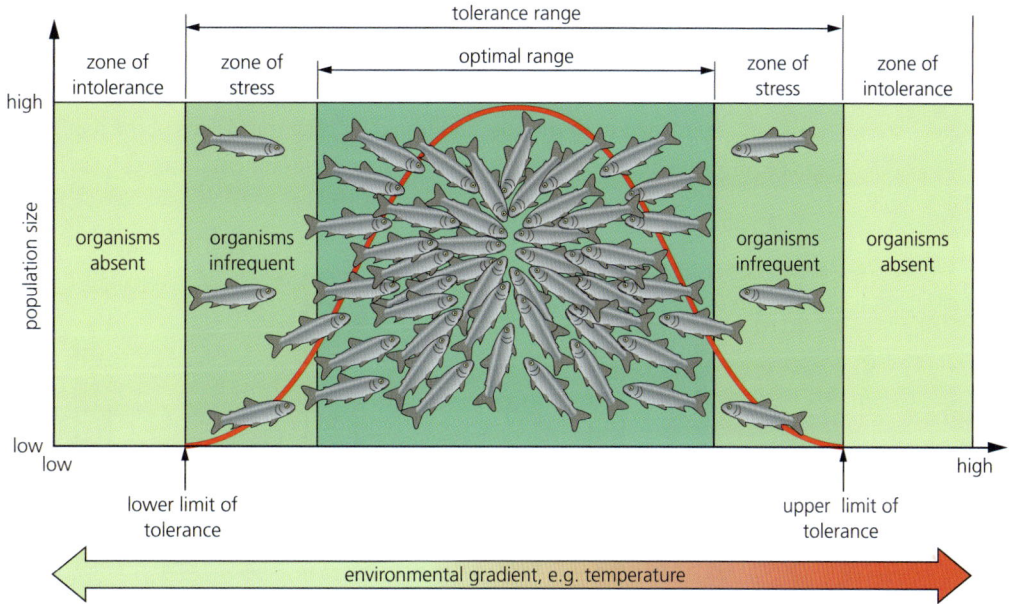

■ **Figure B4.1.6** Graph showing zones of stress and limits of tolerance model

The **optimum zone of tolerance** – the central portion of the curve in Figure B4.1.6 – has conditions that favour maximum **fitness** (that is, **reproductive success**, in terms of numbers of offspring that also have reproductive success), growth, abundance and survival. At either side of the optimal zone are the **zones of stress**, where fewer individuals occur and survival is lower. Organisms are unable to reproduce in zones of stress. Beyond the critical minimum and critical maximum limits of environmental factors, the organisms cannot occur – these are known as the **zones of intolerance**. Tolerance limits exist for all important environmental factors. For some species, one factor may be most important in regulating a species' distribution and abundance, but, in general, many factors interact to affect species distribution.

Top tip!

Tolerance ranges are not necessarily fixed. They can change as seasons change, as environmental conditions change or as the life stage of the organism changes.

4 **List** three abiotic variables that affect species distribution.

B4.1 Adaptation to environment

Thermal limits of the Pompeii worm

The Pompeii worm (*Alvinella pompejana*) lives in one of the hottest environments on Earth – deep sea hydrothermal vents (Figure B4.1.7). They colonize black smoker walls (vent chimneys expelling super-heated fluids that have a high concentration of metal ions). They are believed to be one of the most heat-tolerant animals on Earth. These worms can reach up to 13 cm in length and are pale grey with red, tentacle-like gills on their heads. They live in tubes that they make to protect their bodies from excessive heat. Because they live at a deep distance below the surface of the water (2500 m) and at very high pressures (25 MPa), live specimens have been difficult to study and so their heat tolerance limits have remained contentious.

A research study removed worms from deep sea vents and used a special pressure chamber (isobaric sampling device) to transport them to surface aquaria that were also kept under high pressure. The animals were subjected to three different thermal regimes: a constant mild 20 °C-exposure, and two heat exposures at 42 °C and 55 °C, followed by a 3-hour recovery period at 20 °C. The heat exposures lasted about 2 hours. Cells were removed from the animals following heat treatment to establish the amount of cell death.

■ **Figure B4.1.7** The Pompeii worm (*Alvinella pompejana*)

The survival of all the animals was established by observing movements, just before sampling their body fluid for cells. As well as the observation of vital life signs and cell survivorship, mRNA was extracted from the gills and tested for gene transcripts of the hsp70 stress gene; hsp70 proteins are known to be activated in response to environmental stresses, such as temperature. The results of the experiment are shown in Figure B4.1.8.

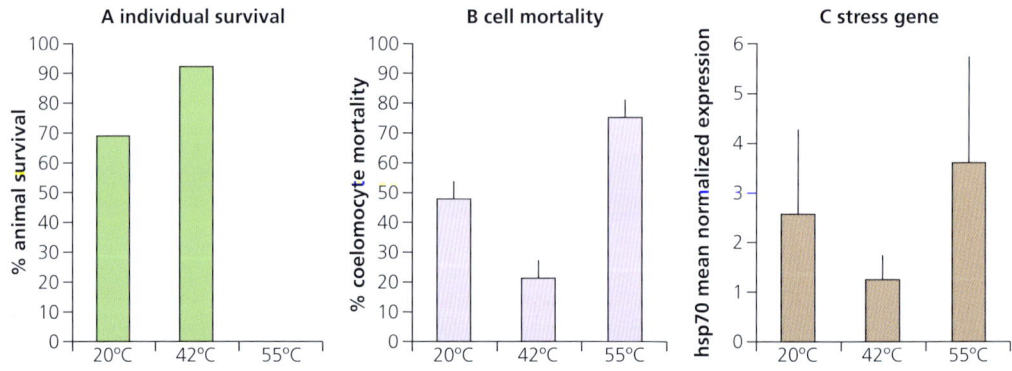

■ **Figure B4.1.8** a) Survival of Pompeii worms under different temperatures, b) percent death of cells (coelomocytes – phagocytic white blood cells that appear in the bodies of animals that have a coelom) that circulate within the body fluid of Pompeii worms, and c) activation of hsp70 gene compared to normal untreated animals

Look at the data carefully. What do the results tell you about the optimal range, lower limit of tolerance and upper limit of tolerance of the Pompeii worm? How does the genetic analysis of mRNA support the overall conclusions of the experiment?

In the 55 °C experiment, the worms initially showed normal behaviour, when they ventilated their tube by moving up and down. After 10 minutes, the worms left their tubes and crawled on the surface of the colony. *A. pompejana* is usually extremely inactive and rarely leaves its tube, so this unnatural behaviour showed that the worms were being disturbed. At the end of the experiment at 55 °C, all worms were dead, and all showed very serious damage of their tissues and cells.

The worms at 55 °C also showed high mortality of circulating cells. The mRNA extracted from the tissues of these animals contained a significantly higher quantity of transcripts for the hsp70 gene compared to normal (not heat-treated) animals. The activation of the hsp70 gene confirms that thermal exposure up to 55 °C is harmful for *A. pompejana*.

In contrast, the 42 °C experiment displayed the highest survival rates at both organism and cellular levels, with no observable structural damage in the tissues. The animals did trigger a mild heat stress response, but with a level of hsp70 gene expression significantly lower than for the 55 °C specimens.

Interestingly, in the 20 °C experiment, cell mortalities were significantly higher in animals compared to the ones kept at 42 °C. Survivorship of individuals was also lower in the 20 °C experiment compared to the 42 °C one. In addition, the stress gene expression showed values significantly higher in animals subjected to the 20 °C compared to the 42 °C treatment. Specimen survival, cell mortality and stress gene expression data, therefore, show that *A. pompejana* endured more damage from the 20 °C exposure compared to the 42 °C experiment.

These findings confirm that *A. pompejana* is a **thermophilic species (extremophile)**: such organisms thrive at relatively high temperatures of between 45 °C and 122 °C. The results of this study provide the first direct experimental evidence that *A. pompejana* cannot withstand prolonged exposure to temperatures in the 50–55 °C range, and that its thermal optimum lies below 55 °C and above 20 °C.

Recent work has shown that the worms have a layer of bacteria on their back that helps insulate them from extreme heat. The worm on its own can only tolerate up to 55 °C but can tolerate higher temperatures with the help of bacteria.

Limits of tolerance of crop plants to salt

Plants that can grow in soil or water of high salinity (high NaCl levels) are called **halophytes**. These plants come into contact with saline water through their roots or by salt spray, such as occurs in saline semi-deserts (dry areas that have some of the characteristics of a desert but with greater annual precipitation), mangrove swamps (see page 343), salt marshes and seashores. Plants that are not salt tolerant are called **glycophytes**; these are damaged fairly easily by high salinity. High concentrations of salts in the soil make it harder for roots to extract water, due to osmotic effects, and high concentrations of salts can be toxic within the plant. Relatively few plant species are halophytes – perhaps under 0.5% of all plant species. Most plant species are glycophytes.

> ● **TOK**
>
> Do researchers have different ethical responsibilities when they are working with human subjects compared to when they are working with animals?

> ### Inquiry 3: Concluding and evaluating
>
> #### Concluding; evaluating
>
> Data were compiled for a variety of different plant species, including both wild and crop species. Salt tolerance was calculated by comparing shoot dry matter of plants exposed to NaCl for at least three weeks to the growth of plants grown in the absence of NaCl. Species studied were as follows.
>
> - Cereal crop plants: rice (*Oryza sativa*), durum wheat (*Triticum turgidum* ssp. *durum*), and barley (*Hordeum vulgare*).
> - Wild-growing plants: tall wheatgrass (*Thinopyrum ponticum*), which is used as forage and for hay in many places; arabidopsis (*Arabidopsis thaliana*); alfalfa (*Medicago sativa*), a plant in the pea family that is cultivated to feed livestock; and saltbush (*Atriplex amnicola*), a plant that is endemic to Western Australia and is native to the floodplains of the Murchison and Gascoyne Rivers.
>
> *Look at Figure B4.1.9, which shows the response of the plant species to salt levels.* →

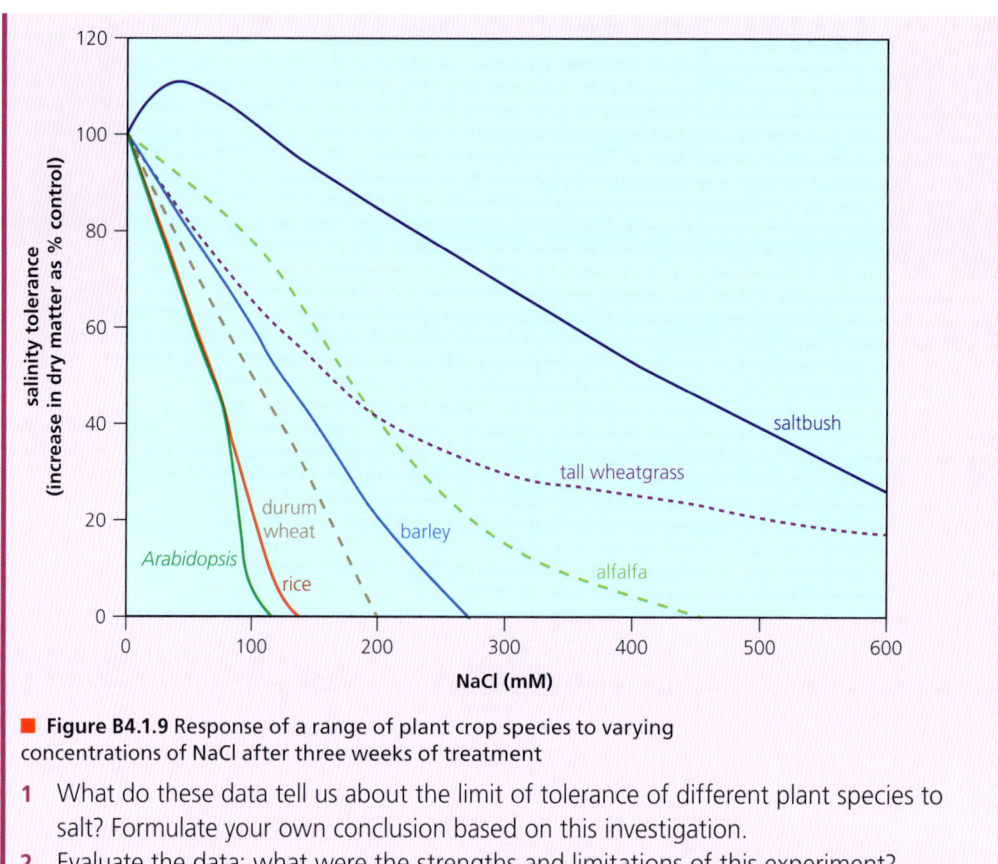

■ **Figure B4.1.9** Response of a range of plant crop species to varying concentrations of NaCl after three weeks of treatment

1. What do these data tell us about the limit of tolerance of different plant species to salt? Formulate your own conclusion based on this investigation.
2. Evaluate the data: what were the strengths and limitations of this experiment?

ATL B4.1A

More than 800 million hectares of land throughout the world are affected by salt, more than 6% of the Earth's total land area. Forty-five million hectares (20%) of the current 230 million hectares of irrigated land are affected by salt. Irrigated land has twice the productivity of land supported by rainfall and produces one-third of the world's food. Understanding the response of wild and crop plants to salinity may enable us to reduce the impact of salinity stress on plants, improving the performance of species that are important in agriculture. For example, crossing tall wheatgrass with its domesticated relative, wheat, can give wheat traits such as stress tolerance to saline conditions.

Arabidopsis (*Arabidopsis thaliana*) is a small wild-growing cress plant that was the first plant to have its genome sequenced, making it a popular tool for understanding the molecular biology of many plant traits.

How can knowledge of the genome of *Arabidopsis* (a salt-sensitive species) be used to compare it to more salt-tolerant species, such as its close relative *Eutrema halophilum* (saltwater cress)? How could this information give a more detailed understanding of the molecular basis of saline tolerance?

Range of tolerance of a limiting factor

■ Correlating the distribution of animal species with an abiotic variable

◆ **Transect**: arbitrary line through a habitat, selected to systematically sample the community.

◆ **Line transect**: a tape is laid out in the direction of an environmental gradient and all organisms touching the tape are recorded.

◆ **Belt transect**: all organisms within a band, usually between 0.5 m and 1 m, are sampled along an environmental gradient.

Rocky shores are ideal environments to investigate optimum zones of tolerance and critical limits because abiotic factors vary dramatically across a small spatial scale between the high and low tide lines. **Transects** can be used to study the distribution of animals and plants along environmental gradients. **Quadrats** (see more in Chapter C4.1, page 550) are used to sample plants, algae and other immobile organisms along a transect. There are different types of transect, used depending on need:

- A **line transect** is made by placing a tape measure in the direction of the gradient. For example, on a beach this would be at 90° to the sea (Figure B4.1.10). All organisms touching the tape are recorded. Many line transects need to be taken to obtain valid quantitative data.
- Larger samples can be taken by using a **belt transect**; this is a band of chosen width, usually between 0.5 m and 1 m, placed along the gradient.
- If the whole transect is sampled, this is called a **continuous transect**. If samples are taken at points of equal distance along the gradient, it is called an **interrupted transect**.
- Horizontal distances are used if there is no visible vertical change in an interrupted transect, such as along a shingle ridge succession (a shingle ridge is a beach covered with small rocks, where different communities exist at different distances from the sea). If there is a climb or descent in an interrupted transect, then vertical distances are normally used, such as on a rocky shore.

■ **Figure B4.1.10** Line transects being used to study an environmental gradient along a rocky shore

Figure B4.1.11 shows the results of a belt transect study of a seashore community. Data are plotted as kite diagrams, where the relative width of each 'kite' represents the abundance of an organism at any one point along a transect.

B4.1 Adaptation to environment

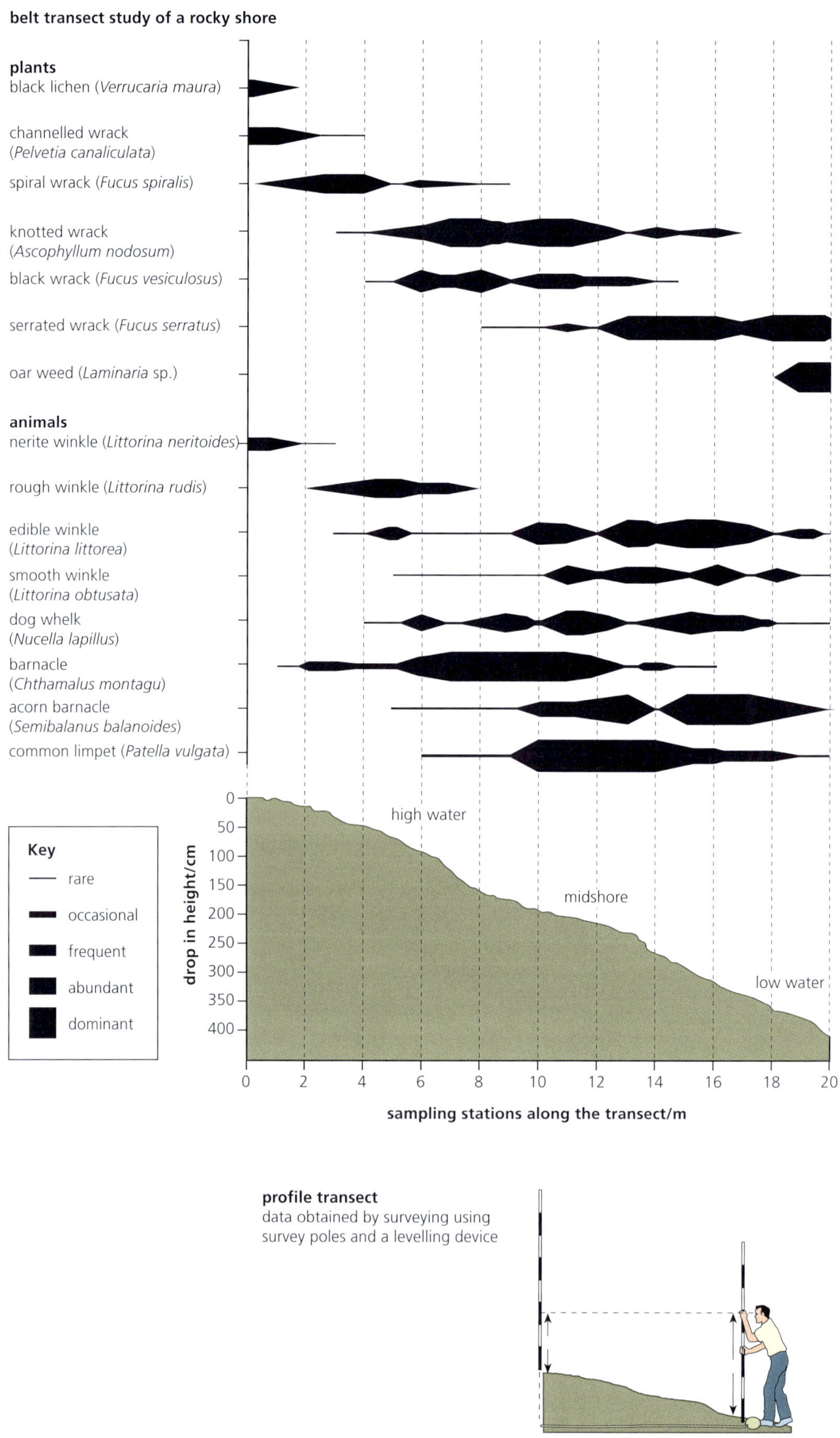

belt transect study of a rocky shore

plants
- black lichen (*Verrucaria maura*)
- channelled wrack (*Pelvetia canaliculata*)
- spiral wrack (*Fucus spiralis*)
- knotted wrack (*Ascophyllum nodosum*)
- black wrack (*Fucus vesiculosus*)
- serrated wrack (*Fucus serratus*)
- oar weed (*Laminaria* sp.)

animals
- nerite winkle (*Littorina neritoides*)
- rough winkle (*Littorina rudis*)
- edible winkle (*Littorina littorea*)
- smooth winkle (*Littorina obtusata*)
- dog whelk (*Nucella lapillus*)
- barnacle (*Chthamalus montagu*)
- acorn barnacle (*Semibalanus balanoides*)
- common limpet (*Patella vulgata*)

Key
- — rare
- ▬ occasional
- ▬ frequent
- ▬ abundant
- ▬ dominant

drop in height/cm

high water • midshore • low water

sampling stations along the transect/m

profile transect
data obtained by surveying using survey poles and a levelling device

5 From the data in Figure B4.1.11, **suggest** one plant and one animal species that appear to be well adapted to the degree of exposure experienced at:
  a a high-water location of the shore
  b a low-water location of the shore.

6 **Define** the term *transect*.

■ Figure B4.1.11 Profile and belt transect analysis of a rocky shore community

Theme B: Form and function – Ecosystems

Most parts of the seashore are periodically submerged below seawater, and changes in the tides affect the highest point up the rocky shore that is submerged at any one point in a month. Tides are controlled by the gravitational pull of the Moon on Earth's oceans and seas. The shore is an area of great diversity and almost all species are of marine origin. The higher an organism occurs on the shore, the longer their daily exposure to the air and the less frequent their submersion in water.

Look at the data in Figure B4.1.11. Which abiotic factor do you think could be correlated with the distribution of animal and plants along the transect?

Exposure brings the threat of desiccation (drying out) and wider extremes of temperature than those experienced during submersion. **Exposure** is an abiotic factor that influences the distribution of organisms on the seashore.

> **Concept: Form and function**
>
> The form and function of a species determines how it interacts with its abiotic environment. This is turn determines its distribution along an environmental gradient.

> **● TOK**
>
> If the sampling method causes some members of the population to be less likely to be included than others, due to a systematic error, then a sampling bias results.
>
> To what extent is random sampling a useful tool for scientists, despite the potential for sampling bias?

> **Tool 1: Experimental techniques**
>
> **Making appropriate qualitative observations**
>
> Qualitative observations are used to record the conditions in which data are recorded. They are descriptive information that reinforce quantitative data. Qualitative observations for a fieldwork investigation should include a site description: this could be in the form of maps, sketches or photographs with annotations.

> **Inquiry 2: Collecting and processing data**
>
>
>
> **Collecting data**
>
> Set up a transect in a local ecosystem (as demonstrated in Figure B4.1.10 for a rocky shore) and collect data to correlate the distribution of plant or animal species with an abiotic variable. Environmental gradients can be found on rocky shores, in the transition from a forested area to open land, and in areas that are in succession (see Chapter D4.2). Abiotic factors you could use as the independent variable include wind speed, soil moisture, light intensity, temperature and humidity. You could collect this data yourself from a natural or semi-natural habitat. Semi-natural habitats have been influenced by humans but are dominated by wild rather than cultivated species. Semi-natural habitats can include hedges, fields left fallow, woods where there is controlled logging, and secondary tropical rainforest.
>
> Use sensors to measure abiotic variables such as temperature, light intensity and soil pH.
>
> **Interpreting results**
>
> What correlation did you find in your transect data? What type of graph should you draw to show correlation? How can you test to see whether the correlation is significant or not?
>
> Did you repeat the transect? If not, what effect would this have had on your data and conclusions?

B4.1 Adaptation to environment

Tool 1: Experimental techniques

Addressing safety and environmental issues in an investigation

When planning your transect investigation, what relevant safety, ethical or environmental issues must you address? Hazards and risks associated with fieldwork include:

- Terrain (how the land lies). There might be uneven surfaces, flat areas, hills and steep gradients.
- Weather conditions can change very quickly in the field. A weather forecast should be consulted before setting out, and appropriate clothing, footwear and supplies selected. In extreme weather, fieldwork might have to be postponed or abandoned.
- Areas where fieldwork is carried out can be isolated. The school and parents who are not going into the field must be aware of the route and the expected time of return.
- Tides can change very quickly. Tide tables should be consulted before setting out.

Consider the impact of any investigation on the organisms you are studying and the environment they live in, if relevant to your study. Bear in mind the IB ethical policy.

Top tip!

When assessing safety, ethics and environmental issues, consider the following:
- evidence of a risk assessment
- the application of the IB animal experimentation policy
- a reasonable consumption of materials
- the use of consent forms in human physiology experimentation
- the correct disposal of waste
- minimizing the impact of the investigation on field sites.

Top tip!

If you use your smartphone to record data, you need to ensure that the measurements taken are accurate enough to be used in quantitative experiments. Ask your teacher if you are unsure.

Common mistake

Some apps display graphs that do not have proper axes showing independent and dependent variables and exclude important features such as units.

Inquiry 3: Concluding and evaluating

Concluding

Compare the outcomes of your investigation in Inquiry 2 to the accepted scientific context. Have other scientists carried out similar investigations to yours? If so, were their conclusions the same? How can you use this information to evaluate your investigation?

Nature of science: Measurements

Using sensors when making observations helps with data collection. Data loggers can make continual measurements of abiotic variables over long periods of time and are more accurate than other forms of measurement. They provide more evidence on which to base claims about the scientific knowledge under investigation. Inexpensive data loggers can take measurements that can be repeated rapidly or automatically over long time periods, even with limited technical expertise. There are many different sensors, each measuring different abiotic variables (see Table B4.1.1).

■ Table B4.1.1 The uses of different sensors in biological investigations

Sensor	Possible investigation
temperature sensor	air or soil temperature in ecological investigations
balance	loss in mass due to respiration
light sensor	light intensity in ecological investigations
pH	catalytic and other biochemical reactions involving a change in pH
chest belts and pressure meters	breathing rate
dissolved oxygen	percentage of oxygen in solution
carbon dioxide sensor	respiration experiments

Tool 2: Technology

Using sensors

Data logging is an electronic method of recording physical measurements. Electrical sensors provide signals that are calibrated and recorded by a computer system, but the main advantage of data logging to practical work is in the process of analysing and interpreting the raw data.

There are many different sensors, each measuring different abiotic variables (see Nature of science box on previous page), such as carbon dioxide concentration, dissolved oxygen concentration, pH and temperature.

Decide which sensors you will use when you collect data from your transects in Inquiry 2 on page 351.

Sensors are built into smartphones for specific purposes, but specific apps can use the sensors as external measuring instruments for investigations. Others record geographical data, such as iNaturalist, which provides a means of recording the geolocation of a species in ecological field studies. Apps are available on smartphones that allow biological principles to be investigated.

Conditions required for coral reef formation

Coral reefs are an example of a marine ecosystem. Tropical coral reefs are found between the tropics of Cancer and Capricorn (23.5° N and S of the equator), where seas are warm and there is strong sunlight throughout the year. Many reef-building corals have a narrow temperature range in which they can thrive and prefer water temperatures that range between 23°C and 29°C.

■ Factors affecting coral reef formation

> **Link**
> Zooxanthellae and their symbiotic relationship with coral is covered in Chapter C4.1, page 567.
>
> ♦ **Symbiotic**: a close and long-term biological interaction between two different biological organisms.

Reef-building corals have a **symbiotic** relationship with a microscopic unicellular algae called **zooxanthellae**. 'Symbiosis' means a long-term biological interaction between different species. Because zooxanthellae need light to photosynthesize, coral can only grow at relatively shallow depth. Most reef-building corals occur in less than 25 m of seawater. As well as light level, the temperature of the water is a limiting factor that affects algal survivorship. Corals live at the uppermost boundary of their temperature tolerance, and even a 1 °C increase in sea surface temperature can stress zooxanthellae, causing the algae to leave the coral. A large-scale loss of zooxanthellae makes coral appear white because coral tissue itself is mostly transparent and their calcium carbonate skeletons are white. When this 'bleaching' occurs, the corals begin to starve as they do not have algae to produce glucose; most corals struggle to survive without their zooxanthellae. If conditions return to normal relatively quickly, corals can regain their zooxanthellae and survive.

> **Top tip!**
> Carbon dioxide is not acidic itself. When it reacts with water it forms carbonic acid, which dissociates to release ions, specifically H^+ (aq), which cause acidity.

As well as temperature effects, increasing carbon dioxide concentration in seawater lowers the pH, making it more acidic – a process referred to as ocean acidification. Increased carbon dioxide in the atmosphere because of accelerated burning of fossil fuels leads to more dissolved carbon dioxide in the oceans. The dissolved carbon dioxide reacts with water to form hydrogen carbonate and hydrogen ions, which make the water more acidic. Because the coral reefs are made of calcium carbonate ($CaCO_3$ (s)), and are therefore basic, a decrease in pH can lead to reduced calcification rates of corals and destruction of existing coral reefs. Calcium carbonate reacts with acid (i.e. hydrogen ions) to form Ca^{2+} (aq). Sea level rises caused by global warming can also result in coral reef reduction, as greater depth means that less light reaches the coral, leading to lower rates of photosynthesis in the zooxanthellae.

> **Top tip!**
> The factors that affect coral reef formation include: water depth, pH, salinity, clarity and temperature. All the factors are in a delicate balance to ensure coral reef growth.

Abiotic factors as the determinants of terrestrial biome distribution

Insolation (amount of sunlight), temperature and precipitation affect the type of ecosystem that develops in any given area. Each element of the climate plays a role in structuring communities within the abiotic environment they inhabit:

- **Temperature** affects the rate of enzyme reactions in the cells of all organisms, affecting the rate of primary productivity through photosynthesis and decomposition. Temperature also affects the rate of transpiration.
- **Precipitation** affects the rate of photosynthesis and primary productivity.
- **Insolation** also affects the rate of photosynthesis and primary productivity.

Other elements of the climate that can affect ecosystems are air pressure, humidity, cloudiness and wind.

Information about the relative contributions of two climatic factors, precipitation and temperature, can be used to predict the type of stable ecosystem that can be expected in an area. These two climatic variables can be plotted on a graph and the different ecosystems plotted within it (Figure B4.1.12). **Biomes** are groups of ecosystems that share similar abiotic conditions, and so develop similar communities through **convergent evolution**.

◆ **Biome**: Groups of ecosystems with similar abiotic conditions and communities, defined by their climate and dominant plant species.

Link

Convergent evolution is discussed in Chapter A3.2, page 136 and 145.

◆ **Climograph**: a graphical model that shows the relationship between temperature, precipitation and ecosystem type.

> ### Concept: Form
>
> The form of a biome is determined by the climate in the area it is found. A rainforest, for example, is maintained by year-round insolation, plentiful rainfall and warm temperatures that maximize growth and productivity.

A **climograph** is a graphical model that shows the relationship between temperature, precipitation and ecosystem type. It was first developed by the plant ecologist RH Whittaker. It shows the likely stable ecosystems that are found under specific climatic conditions. Vegetation underpins communities found in all different geographical areas, and so factors that affect plant growth strongly influence the distribution of different ecosystems. Temperature and rainfall are two of the main limiting factors that affect plant growth, so these abiotic factors can be used to model and predict the geographical distribution of different ecosystems around the planet. Figure B4.1.12 shows a climograph that illustrates the distribution of major terrestrial ecosystems with respect to mean annual precipitation and temperature.

Common mistake

The climograph suggests that each ecosystem has a distinct 'edge' – this may, in fact, not be the case, with steady gradation from one ecosystem to another rather than distinct boundaries. Tropical ecosystems, for example, graduate from highly productive rainforest to low-productive desert.

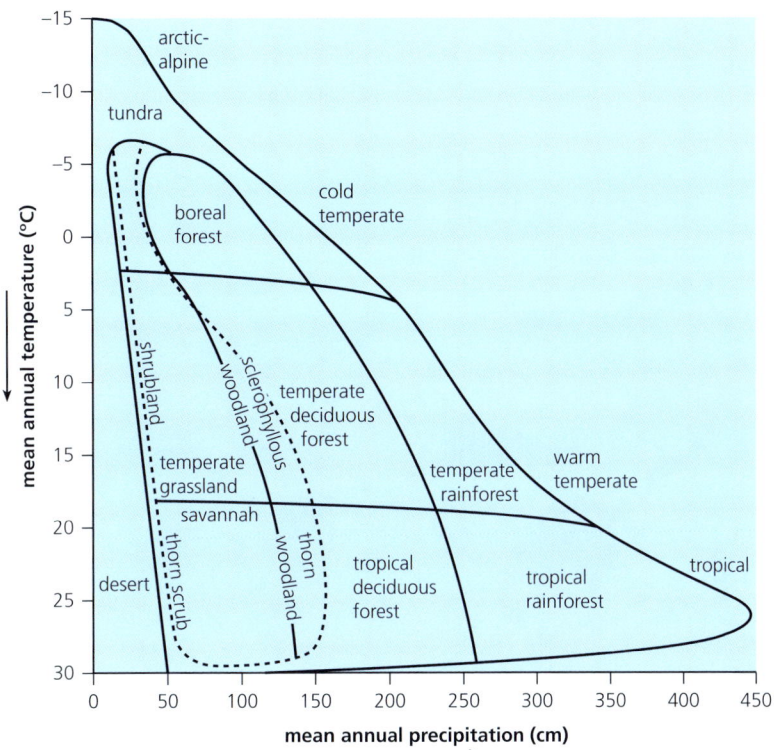

■ **Figure B4.1.12** Whittaker's climograph

Top tip!

To find the biome that is found under specific climatic conditions, locate the mean annual precipitation in cm per year on the *x*-axis of the climograph. Now locate the mean annual temperature on the *y*-axis. Draw two lines in from these points: where the lines meet is the biome found under those conditions. For example, if the mean annual precipitation is 300 cm and mean annual temperature 25 °C, the biome found under these conditions is tropical rainforest.

There are different forms of this graph: sometimes the axes are reversed, or the temperature is plotted from low to high. In some regions, distribution is determined by soils rather than climate: in savannah areas (grassy plains in tropical and subtropical regions with few trees), for example, grasslands are found on sandy soil (which is well drained) and forests are found on clay soils (where water is retained). The dashed line in Figure B4.1.12 defines ecosystems where factors other than rainfall and temperature strongly influence ecosystem structure, such as soil type, the occurrence of fire, animal grazing and seasonal drought.

7 Using Figure B4.1.12:
   a **State** the biome found where mean annual precipitation is 25 cm and mean annual temperature is 25 °C.
   b **Explain** which factor is most important in determining whether an area will be a broadleaf forest or a coniferous forest.
   c **Explain** which factor is most important in determining whether an area will be a grassland or a forest.

Biomes

Biomes are groups of ecosystems with similar communities due to similar abiotic conditions and convergent evolution.

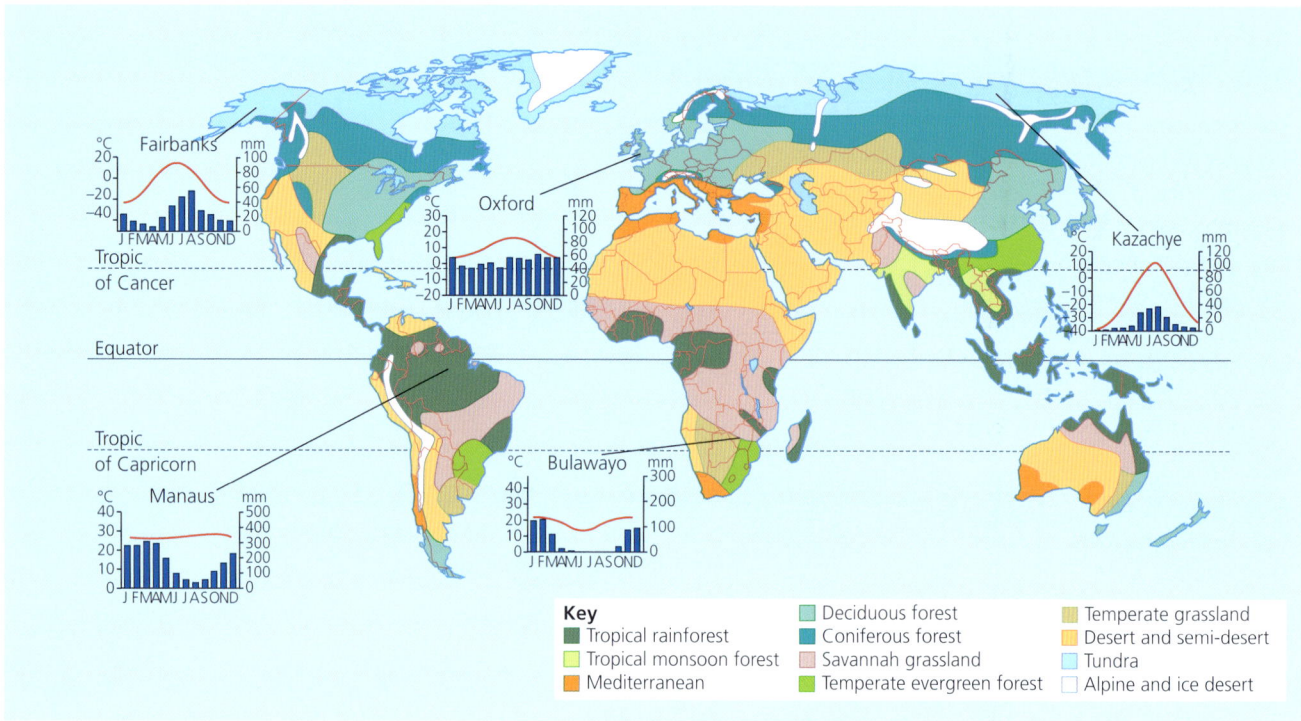

■ Figure B4.1.13 The global biome map

Link
Productivity is covered in more detail in Chapter C4.2, page 587.

Rainfall, the amount of sunlight (insolation) throughout the year and temperature all affect the rate of photosynthesis in plants in different biomes. The rate of storage of energy in plant biomass through photosynthesis is known as **productivity**. Higher temperatures speed up enzyme reactions that drive photosynthesis, although temperatures that are too hot can lead to denaturation of enzymes.

Tropical rainforests

Tropical rainforests (Figure B4.1.14) have the highest productivity of any biome because they are found between the tropics of Cancer and Capricorn where rainfall is high (over 2500 mm yr^{-1}), insolation is constant throughout the year and temperatures are warm (typically 26 °C). The high productivity means that rainforests have a complex structure with a number of layers from ground level to canopy, with emergent trees up to 50 metres high and lower layers of shrubs and vines.

■ Figure B4.1.14 Tropical rainforest in Thailand

> **Top tip!**
>
> Because tropical rainforests, as their name implies, lie in a band around the equator within the tropics of Cancer and Capricorn (23.5° N and S), they experience high light levels throughout the year. There is little seasonal variation in sunlight and temperature (although the monsoon period can reduce levels of insolation), providing an all-year growing season.

Temperate forests

Temperate ecosystems vary from deciduous and evergreen forests to grasslands. Temperate forests (Figure B4.1.15) are largely found between 40° and 60° N and S of the equator. Although temperate forests are highly productive for part of the year, seasonal variation in the amount of sunlight limits their overall productivity. They are found in seasonal areas where winters are cold and summers are warm (unlike tropical rainforests, which have similar conditions all year round). Rainfall and temperature are also seasonal, which further reduces the overall productivity compared to tropical rainforests. Lower levels of productivity mean less stored chemical energy by plants and other autotrophs. At these mid-latitudes, the amount of rainfall determines whether an area develops forest. Rainfall is sufficient in temperate forest areas to establish forest (500–1500 mm yr^{-1}) rather than grassland.

■ Figure B4.1.15 A temperate forest in the UK

Taiga

Taiga (northern coniferous forest) is a biome characterized by coniferous forests consisting mostly of pines (Figure B4.1.16). It is found between northern latitudes of 50° and 70°, near to the Arctic Circle. The climate is extremely cold (winds blow cold Arctic air into the biome), harsh, with a low rate of precipitation (snow and rain) and a short growing season. Long, severe winters can last up to 6 months. Mean annual temperatures range from a few degrees Celsius above freezing down to −10 °C. Summers are short, lasting around 50 to 100 days without frost. Snow cover affects the climate because the snow reflects incoming solar radiation and increases cooling. Taiga is located south of the tundra biome, which is characterized by a land frozen by ice and constant snow (see below).

■ Figure B4.1.16 Taiga forest in Canada

The taiga is the world's largest land biome (covering around 27% of Earth's land surface), ranging from North America (where it covers most of Canada, Alaska and parts of the northern United States), Eurasia (where it covers most of Sweden, Finland, much of Russia – including two-thirds of Siberia – and large areas of Norway and Estonia). It is also found in Iceland, and areas of northern Kazakhstan, northern Mongolia and northern Japan (on the island of Hokkaidō).

Grasslands

The grassland biome is found on every continent except Antarctica and covers about 16% of Earth's surface (Figure B4.1.17). Grasslands develop where there is not enough precipitation to support forests, but there is enough to prevent deserts forming. There are several types of grassland: the Great Plains and the Russian Steppes are temperate grasslands; the savannahs of east Africa are tropical grasslands.

Grasses have a wide diversity but low levels of productivity. Away from the sea, grasslands have wildly fluctuating temperatures, which can limit the survival of animals and plants. The mixing of cold polar air with warmer southerly winds (in the northern hemisphere) causes increased precipitation compared to polar and desert regions. Rainfall is approximately in balance with levels of evaporation. Grasses grow beneath the surface and, during cold periods (northern grasslands suffer a harsh winter), can remain dormant until the ground warms.

■ Figure B4.1.17 The Serengeti grasslands of Tanzania

Tundra

In colder ecosystems found in the northern hemisphere, such as tundra, water is locked up as ice and is not available to plants, reducing productivity. It is a highly stressful environment, with very low temperatures and low rainfall, so only mosses and lichens may be able to survive (Figure B4.1.18). Most of the world's tundra is found in the northern polar region, and so is known as Arctic tundra. There is a small amount of tundra in parts of Antarctica that are not covered with ice, however, and in lower latitudes on high altitude mountains (alpine tundra).

■ Figure B4.1.18 Alaskan tundra

Tundra is found at high latitudes where insolation is low. Short daytime lengths also limit the levels of sunlight. Temperatures are very low for most of the year, which is a limiting factor because it affects the rate of photosynthesis, respiration and decomposition (these enzyme-driven chemical reactions are slower in colder conditions). Due to the low temperatures, water may be locked up in ice for months at a time and this, combined with little rainfall, means that water is also a limiting factor. The low light intensity and rainfall mean that rates of photosynthesis and productivity are low. Soil may be permanently frozen (permafrost) and nutrients are also limited. The vegetation consists of low scrubs and grasses.

During winter months temperatures can fall to −50 °C. All life activity is low in these harsh conditions. In the summer the tundra changes: the Sun is out almost 24 hours a day, so levels of insolation and temperature both increase, leading to plant growth. Only small plants are found in this biome because there is not enough soil for trees to grow and, even in the summer, there is permafrost only a few centimetres below the surface.

Common mistake

Deserts are areas where water is generally inaccessible and vegetation is sparse. Antarctica is therefore classified as a desert (the water needed by organisms is largely locked up as ice) but it is a cold desert, as opposed to a hot desert such as the Sahara. It is therefore incorrect to say all deserts are hot.

8 **Define** the term *biome*.

9 **Compare and contrast** the distribution of two terrestrial biomes.

Hot deserts

A 'hot' desert is a part of the world that has high average temperatures and very low rainfall. Deserts are found in bands at latitudes of approximately 30° N and S (Figure B4.1.19). They cover 20–30% of the land surface. Here, dry air descends having lost its water vapour over the tropics.

The Sahara Desert in northern Africa is the world's largest hot desert. Covering approximately 9.2 million square kilometres (3.6 million square miles), it is only slightly smaller than the USA (9.8 million square kilometres, 3.8 million square miles)!

Hot deserts are characterized by high temperatures at the warmest time of day (typically 45–9 °C) in the early afternoon together with low levels of precipitation (typically under $250\,mm\,yr^{-1}$), which may be unevenly distributed. The lack of water limits rates of photosynthesis and so rates of productivity are very low. Organisms also must overcome fluctuations in temperature (the temperatures at night when the sky is clear can drop to 10 °C, sometimes as low as 0 °C), which makes survival difficult.

■ **Figure B4.1.19** Location of hot deserts around the world

Adaptations to life in hot deserts and tropical rainforest

Concept: Form

Organisms in hot deserts and tropical rainforests are adapted to the specific conditions they live in. Abiotic conditions determine the form of these species.

Hot desert

There are three methods plants and animals use to live in a hot desert:
- **Expire**: when the conditions are too extreme, parent organisms die but leave behind tough seeds or eggs.
- **Evade**: avoid extreme temperatures by changing activity from day to night or from above ground to below. Kangaroo rats (see below) sleep in their burrows during the hot day.
- **Endure**: adaptations enable plants and animals to survive in this extreme environment, such as the cactus (see below). Fatty deposits stored by animals in their tails and other tissues can be used as a respiratory substrate, releasing water. Water can be stored in the roots, stems and/or leaves of plants (plants that do this are called succulents, e.g. cacti).

Camels

Camels live in deserts that are hot and dry during the day and cold at night. They need to cope with wind-blown sand as well as a lack of water. They are well adapted for survival in the desert. Camels can tolerate a body temperature of up to 42 °C. They have thick fur on the top of their body for shade and thin fur elsewhere to allow easy heat loss. Their long neck and legs increase their surface area for heat loss.

Camels also have large, flat feet to spread their mass on the sand to reduce the pressure. They lose very little water through urination and sweating. The fat in the hump on their back can be converted to metabolic water via respiration. Slit-like nostrils, which can close, and two rows of eyelashes help to keep the sand of out their nose and eyes.

Cacti

■ Figure B4.1.20 Cacti growing in the Andes in South America

Cacti (Figure B4.1.20) have no leaves, so photosynthesis occurs in their stem to reduce surface area. Their rounded and spherical shapes also reduce their surface area-to-volume ratio to limit water loss through transpiration. A thick, waxy cuticle further prevents water loss through their surface. They can store large quantities of water in collapsible water-storage cells found in the stem (although some cacti also store water in their roots) and are covered with needles (modified leaves) to deter mammals from eating the succulent flesh of the cactus and getting access to water stores. Cacti have shallow, wide-spread root systems to ensure that any precipitation is rapidly absorbed before it evaporates. They also have deep roots to reach sources of water beneath the surface.

Scorpions

■ Figure B4.1.21 Granulated thick-tailed scorpion (*Parabuthus granulatus*) in the Kalahari desert, South Africa

Scorpions (Figure B4.1.21) are nocturnal arthropods, so they are active at night and live underground to avoid hot daytime temperatures. They are opportunistic predators, having a wide range of prey, and use a venomous sting to kill prey. They wait for prey to come near to them and carry out ambush attacks rather than chasing down prey, which saves energy. Their exoskeletons give them protection and prevents water loss, and their low food and water needs allow them to survive in desert environments. Scorpions use their eight legs to detect movement. The movements of prey animals cause the sand's surface to vibrate slightly, which the scorpion can detect and estimate the direction and distance to its prey.

Kangaroo rats

■ Figure B4.1.22 Kangaroo rat

Animals of arid or desert regions clearly survive with little or no liquid water in their diets. This group of animals includes the kangaroo rat (*Dipodomys* species, Figure B4.1.22), which lives in hot dry deserts, hiding in a burrow during daylight. It can survive without access to drinking water.

Physiologists have investigated the metabolism, diet, and breathing and excretory losses of water in the kangaroo rat. They excrete extremely concentrated urine and produce no sweat, which explains why survival is possible for a well-adapted organism. Other desert species show similar adaptations.

■ Tropical rainforests

The complex structure of tropical rainforest has a number of layers from ground level to canopy. Emergent trees grow above the canopy, and there are lower layers of shrubs and vines. The range of different conditions and the complexity of the ecosystem means that organisms display a wide range of adaptations to survive in this highly competitive environment.

Pitcher plants

Pitcher plants are carnivorous plants, with some found in the canopy of tropical rainforest and in areas with nutrient-poor soils (*Nepenthes* species). Pitcher plants can photosynthesize to provide glucose for respiration and growth, but lack other nutrients from the immediate environment needed for metabolic processes in cells (such as nitrogen to make amino acids). They can survive in areas of low nutrients, such as the canopy of rainforest where plants grow epiphytically (directly on the host tree rather than in soil), because they catch and digest insects. The pitcher is grown from modified leaves, which fold into a cup-shaped trap. Digestive enzymes are secreted by the pitcher, and downward-pointing spines prevent trapped insects from escaping. Animals provide the nutrients, such as a source of nitrogen, otherwise lacking from the environment of the pitcher. Some larger pitcher plants, such as *Nepenthes rajah*, use the faeces of small mammals such as tree shrews as a source of nourishment – these animals are attracted to the pitcher by a sweet secretion and use the pitcher as a toilet; both organisms benefit from this **mutualistic** relationship.

■ Figure B4.1.23 Pitcher plants growing in the canopy of tropical rainforest

Flying lizards

The extended vertical dimension provided by tropical rainforest provides an alternative form of locomotion for some animal species. Rather than walk along the forest floor, they have developed adaptations to glide from tree to tree. The forest floor is a dangerous place, home to many potential predators; living above the forest floor enables prey to evade predation (and saves energy by not having to regularly run away from danger). One such animal is the flying lizard (*Draco* species), see Figure B4.1.24. These tiny lizards (usually reaching lengths of about 20 cm) have folds of skin that rest flat against the body when not in use but act as wings when stretched tight between extended ribs, allowing them to glide from one tree to another.

■ Figure B4.1.24 *Draco* lizard gliding in rainforest in Thailand

Gibbons

Gibbons (*Hylobates* species) are lesser apes found in the rainforests of South East Asia (Figure B4.1.25). They have elongated forearms to help them swing from tree to tree (a form of locomotion known as brachiation). Gibbons also have hook-like fingers and high mobility in their shoulder joints to help with the swinging motion. Gibbons very rarely, if ever, descend to the forest floor – their body is adapted to their arboreal (tree-living) habitat.

Orchid mantis

Mimicry, where one species copies the appearance of another, is one strategy to survive in rainforest. The orchid mantis (*Hymenopus coronatus*) is so called because it mimics the look of an orchid flower. Its four legs resemble flower petals (Figure B4.1.26a). The insect attracts pollinators to its flower-like structure and uses its forelimbs to attack and kill its prey.

The young of the orchid mantis (the first instar) are dark orange with black legs and a black head (Figure B4.1.26b). These small insects do not resemble orchid flowers, but rather the unopened buds of a flowering tree species.

■ Figure B4.1.25 A Lar gibbon (*Hylobates lar*) swinging from tree to tree

B4.1 Adaptation to environment

■ **Figure B4.1.26** a) An orchid mantis (*Hymenopus coronatus*) sitting on an orchid, from the tropical forests of South East Asia; b) young orchid mantis

10 **Suggest** why the adult orchid mantis looks different to the young (first instar) orchid mantis.

Mimicry is a very effective adaptation, and it is crucial to the survival of many species. Other animals that use this type of camouflage in tropical rainforests include beetles, caterpillars, moths, lizards, snakes and frogs.

LINKING QUESTIONS

1. What are the properties of the components of biological systems?
2. Is light essential for life?

Ecological niches

Guiding questions

- What are the advantages of specialized modes of nutrition to living organisms?
- How are the adaptations of a species related to its niche in an ecosystem?

SYLLABUS CONTENT

This chapter covers the following syllabus content:
- ▶ B4.2.1 Ecological niche as the role of a species in an ecosystem
- ▶ B4.2.2 Differences between organisms that are obligate anaerobes, facultative anaerobes and obligate aerobes
- ▶ B4.2.3 Photosynthesis as the mode of nutrition in plants, algae and several groups of photosynthetic prokaryotes
- ▶ B4.2.4 Holozoic nutrition in animals
- ▶ B4.2.5 Mixotrophic nutrition in some protists
- ▶ B4.2.6 Saprotrophic nutrition in some fungi and bacteria
- ▶ B4.2.7 Diversity of nutrition in archaea
- ▶ B4.2.8 Relationship between dentition and the diet of omnivorous and herbivorous representative members of the family Hominidae
- ▶ B4.2.9 Adaptations of herbivores for feeding on plants and of plants for resisting herbivory
- ▶ B4.2.10 Adaptations of predators for finding, catching and killing prey, and of prey animals for resisting predation
- ▶ B4.2.11 Adaptations of plant form for harvesting light
- ▶ B4.2.12 Fundamental and realized niches
- ▶ B4.2.13 Competitive exclusion and the uniqueness of ecological niches

Note: There is no higher-level only content in B4.2.

Ecological niches

Communities are made up of many interacting species. Each species plays a unique role within a community because of the unique combination of its spatial habitat and interactions with other species. A complete description of a species' place within a community is known as its **niche**. A species' niche depends not only on where it lives (its habitat), but also on what it does. For example, the niche of a lion includes all the information that defines this species: its habitat, courtship displays, grooming, alertness to prey, when it is active, interactions with other species, and so on. The niche of a species includes the biotic and abiotic interactions that influence the growth, survival and reproduction of the species, including how it obtains food.

♦ **Niche**: the role played by a species in its community, which includes its abiotic requirements and tolerances, and its interactions with other organisms.

No two species can have the same niche, because the niche completely defines a species and the role that species has in an ecosystem. Two organisms could temporarily occupy the same niche, leading to evolutionary processes (see page 378) or competitive exclusion (see page 378 this section). The principle of distinct niches can be illustrated by two common and rather similar seabirds, the cormorant and the shag (Figure B4.2.1).

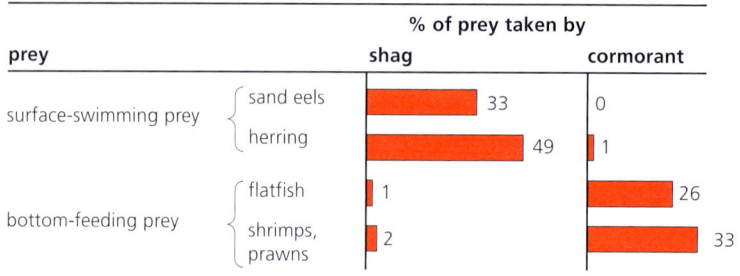

Figure B4.2.1 Different niches of two very similar species of bird: the cormorant (*Phalacrocorax carbo*) and the shag (*Gulosus aristotelis*)

1 Using Figure B4.2.1, **state** one difference between the niche of a cormorant (*Phalacrocorax carbo*) and the shag (*Gulosus aristotelis*).

Concept: Function

The role of a species in its environment is defined by its niche. Each species has a unique niche that differentiates it from other species.

Link

Respiration (aerobic and anaerobic) is covered in detail in Chapter C1.2, page 408.

The cormorant and the shag live and feed along the coastline and they rear their young on similar cliffs and rock systems. It looks like they share the same habitat. However, their diet and behaviour are different. The cormorant feeds close to the shore on seabed fish, such as flatfish. The shag builds its nest on much narrower cliff ledges. It also feeds further out to sea, capturing fish and eels from the upper layers of the waters. Since these birds feed differently and have different behavioural patterns, although they occur in close proximity, they avoid competition with each other. They therefore occupy different niches.

Obligate anaerobes, facultative anaerobes and obligate aerobes

The earliest life on Earth evolved in an environment where oxygen was absent from the atmosphere. The first life on Earth therefore carried out anaerobic respiration (i.e. respiration that does not require oxygen), and all niches would have been anaerobic. Following the evolution of photosynthetic organisms that produced oxygen, the concentration of oxygen in the atmosphere increased, enabling the evolution of prokaryotes that made use of this new resource, and the development of niches that were aerobic. By evolving into different niches, they avoided competition with the anaerobic species. As oxygen levels increased, anaerobic species retreated into environments with little or no oxygen, for example the intestines and stomachs of mammals.

◆ **Obligate aerobes**: organisms that can only respire aerobically.

◆ **Obligate anaerobes**: organisms that only respire in the absence of oxygen.

◆ **Facultative anaerobe**: organism that normally respires aerobically but has the facility to switch to anaerobic respiration in the absence of oxygen.

2 **Define** the term *obligate aerobe*.

3 **Outline** the differences between obligate and facultative anaerobes.

● **Top tip!**

The tolerance of obligate anaerobes, facultative anaerobes and obligate aerobes is determined by the presence or absence of oxygen gas in their environment.

Respiration takes place in every living cell. For respiration, many micro-organisms require oxygen; they can only respire aerobically (these are known as **obligate aerobes**). An example of an obligate aerobe is *Mycobacterium tuberculosis* (the bacteria that causes the disease TB – see page 537). Other organisms only respire in the absence of oxygen (**obligate anaerobes**): these microbes are poisoned by the presence of oxygen because they lack defence mechanisms to protect enzymes from oxidants. An example of obligate anaerobes are methane-producing archaea (see Diversity of nutrition in archaea, later in the chapter). A third group of organisms normally respire aerobically but have the facility to switch to anaerobic respiration in the absence of oxygen (**facultative anaerobes**), such as *Escherichia coli* (*E. coli*) which normally lives in the intestines of animals but can also exist in water, food, soil or on surfaces that have been contaminated with animal or human faeces. The biochemical pathways involved in respiration are summarized in Figure B4.2.2.

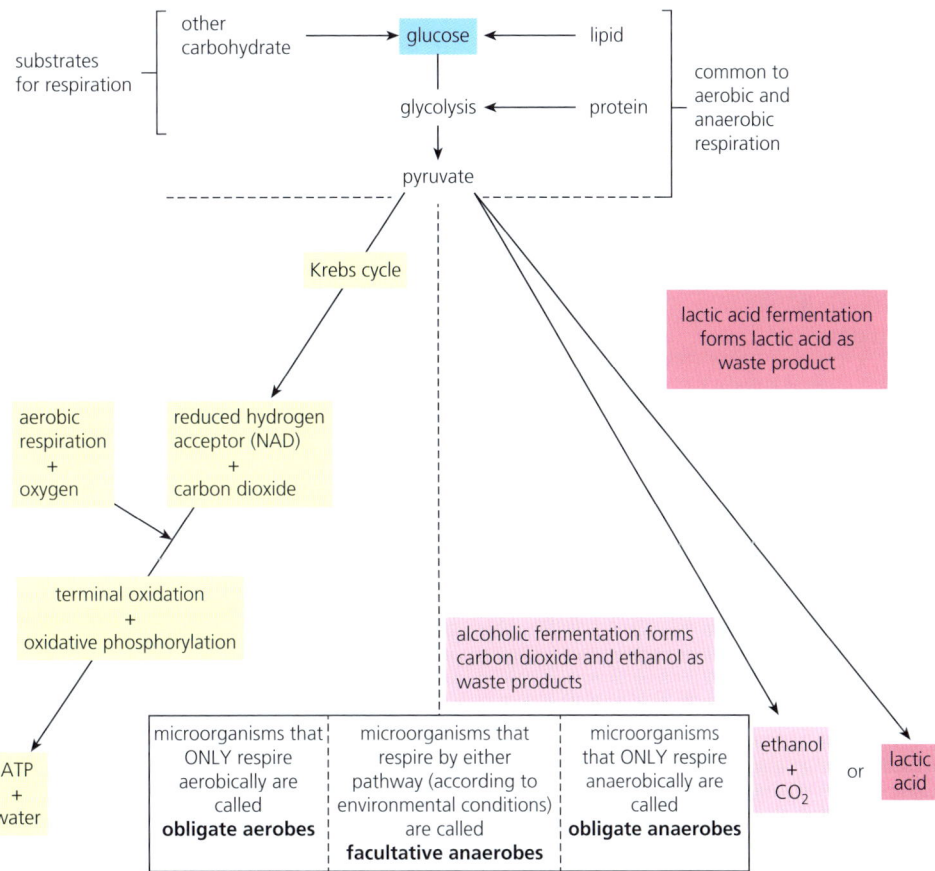

■ **Figure B4.2.2** Classification of micro-organisms by respiration

Different modes of nutrition

As well as type of respiration, another factor that determines the niche of a species is its mode of nutrition. Species living in the same habitat can be distinguished by feeding in different ways. There are a variety of modes of nutrition.

♦ **Autotrophic**: using external energy sources to synthesize glucose from simple inorganic substances.

♦ **Heterotrophic**: using carbon compounds obtained from other organisms to synthesize required carbon compounds.

● Top tip!

Autotrophs use inorganic molecules while heterotrophs use organic molecules in their nutrition.

Link

The process of photosynthesis is covered in detail in Chapter C1.3, page 425.

♦ **Photoautotroph**: an organism that uses light energy to generate ATP and to produce glucose from inorganic substances.

♦ **Holozoic**: nutrition in consumers where food is ingested, digested internally, absorbed and assimilated.

♦ **Herbivore**: an animal that feeds (holozoically) exclusively on plants.

♦ **Carnivore**: flesh-eating organism.

Link

Herbivores, carnivores and consumers are covered in detail in Chapter C4.2, page 585.

■ The nutrition of micro-organisms

Microbes can also be grouped according to their nutrition. Micro-organisms are either **autotrophic**, that is, they make their own organic molecules using an external source of energy, or **heterotrophic**, relying on a supply of ready-made complex food substances. The different types of autotrophic and heterotrophic nutrition are explored next.

● Common mistake

Energy cannot be created or destroyed, so it is incorrect to say that autotrophs make their own energy. Autotrophs convert energy from one form (sunlight) into another (chemical energy in the form of food). Autotrophs produce their own food, not their own energy.

■ Photosynthesis in plants, algae and photosynthetic prokaryotes

Plants contain pigments, such as chlorophyll, so they can absorb solar energy (sunlight) and convert this into chemical energy – the process of photosynthesis. As well as the pigments, they contain the necessary metabolic processes to reduce carbon dioxide and produce glucose (and oxygen), see page 425. The great majority of green plants are entirely **autotrophic** in their nutrition. As they use sunlight in their autotrophic nutrition, they are called **photoautotrophs**. Because of this, they play a key part at the beginning of food chains (page 581). Other organisms also contain pigments, and so are also autotrophic. These include algae (a group of protists) and photosynthetic bacteria (see below). Cyanobacteria (a group of aquatic photosynthetic bacteria) were the first photosynthetic organisms on Earth, evolving some 3.7 billion years ago, leading to a dramatic change in the atmosphere (see Chapter A2.1, page 34).

■ Holozoic nutrition in animals

In contrast to green plants, animals and most other types of organism use only existing nutrients, which they obtain by digestion and then absorption into their cells and tissues for assimilation (use in the body). Consequently, animal nutrition is dependent on plant nutrition, either directly or indirectly. In ecology, animals are known as **consumers** and animal nutrition is described as **heterotrophic** (meaning 'other nutrition').

● Top tip!

A heterotroph is an organism that obtains organic molecules from other organisms. A consumer is an organism that ingests other organic matter that is living or recently killed.

All animals are heterotrophic and most – but not all – are consumers (holozoic). In **holozoic** nutrition, food is ingested, digested internally, absorbed and assimilated.

Note that some of the consumers, known as **herbivores**, feed directly and exclusively on plants. Herbivores are **primary consumers**. Animals that feed exclusively on other animals are **carnivores**. Carnivores that feed on primary consumers are known as **secondary consumers**. Carnivores that feed on secondary consumers are called tertiary consumers, and so on.

> ### ● Top tip!
> An animal takes in food (complex organic matter) and digests it in the alimentary canal or gut, producing molecules that can be taken up into the body's cells via the blood circulation system. This is known as holozoic nutrition (meaning 'feeding like an animal'). Holozoic nutrition is just one of the forms of heterotrophic nutrition (meaning 'feeding on complex, ready-made foods') to obtain the required nutrients. All heterotrophs are dependent, directly or indirectly, on organisms that manufacture their own food (autotrophs).

■ Mixotrophic nutrition in some protists

◆ **Mixotrophic nutrition**: nutrition that is both autotrophic and heterotrophic.

Some organisms carry out both autotrophic and heterotrophic nutrition. These organisms are known as **mixotrophs**, and their form of feeding **mixotrophic nutrition**. We have already seen how pitcher plants are carnivorous, feeding on captured insects (page 360), while also carrying out photosynthesis.

Some animals contain algae within their tissues. Many marine flatworms have algae living symbiotically within their bodies (a form of endosymbiosis – see also Chapter A2.2, page 82) such as many species of *Convoluta*. Many species of coral can live in symbiosis with algae (see page 567). *Euglena* is a well-known freshwater example of a protist that is both autotrophic and heterotrophic (Figure B4.2.3). Many other mixotrophic species are part of oceanic plankton (see below).

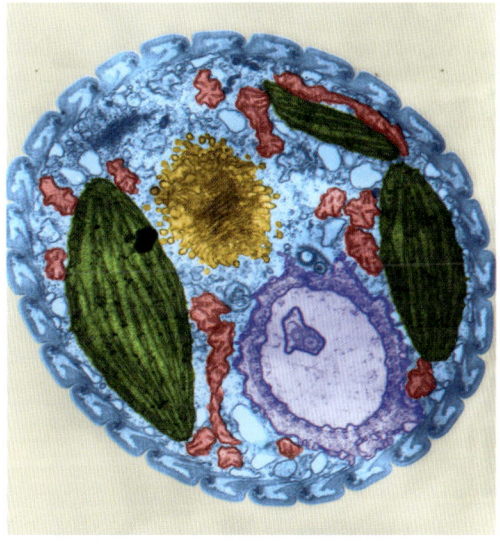

In **heterotrophic nutrition**, bacteria are taken into food vacuoles by phagocytosis and the contents digested by hydrolytic enzymes from lysosomes.

In **autotrophic nutrition**, photosynthesis occurs in the chloroplasts. There is a light-sensitive 'eyespot' present which enables *Euglena* to detect the light source.

Notice the plasma membrane has a ridged appearance here – this arrangement is supported by a system of microtubules below.

■ **Figure B4.2.3** False-colour micrograph of *Euglena*, a species that is both autotrophic and heterotrophic (magnification ×40 000)

4 Calculate the actual size of the organism shown in Figure B4.2.3.

Link
Decomposers and detritivores are covered in more detail in Chapter C4.2, page 590.

Some mixotrophs are **obligate** (meaning they always carry out both forms of nutrition, such as *Euglena*) and others are **facultative** (meaning they sometimes carry out both but can be either autotrophic or heterotrophic). Many phytoplankton (a type of alga), such as *Cryptomonas* sp., are facultative mixotrophs because they take up dissolved organic carbon or, under inorganic nutrient stress (when they lack access to key nutrients), use dissolved amino acids or other organic sources of nitrogen. The uptake of dissolved organic material is known as **osmotrophy**.

■ Saprotrophic nutrition in some fungi and bacteria

◆ **Saprotroph**: an organism that lives on or in dead organic matter, secreting digestive enzymes into it and absorbing the products of digestion.

Eventually, all producers and consumers die and decay. Organisms that feed on dead plants and animals, and on the waste matter of animals, are described as **saprotrophs** (meaning 'putrid feeding'). Fungi and bacteria with this mode of heterotrophic nutrition can be referred to as **decomposers**.

5 Define the term *saprotroph*.

Feeding by saprotrophs releases inorganic nutrients from the dead organic matter, including carbon dioxide, water, ammonia, amines, and ions such as nitrates and phosphates. These inorganic nutrients are then absorbed by green plants and reused.

Top tip!

Make sure you can distinguish key terms such as autotroph, heterotroph, detritivore and saprotroph. Learn this comparative language carefully and then apply it accurately.

Top tip!

A detritivore is an organism that ingests dead organic matter, demonstrating holozoic nutrition. Decomposers secrete enzymes and digest food outside the body, absorbing the products of digestion. Decomposers demonstrate saprotrophic nutrition.

Common mistake

If asked to describe the role of a saprotroph, it is incorrect to simply say that it is an organism that feeds on dead organic matter, because detritivores (which are not saprotrophs) ingest dead matter. You must make it clear that saprotrophs feed on dead organic matter by external digestion.

Common mistake

Saprotrophs obtain energy from external digestion – it is incorrect to say that this is the 'recycling of energy' as they will release the energy by respiration and then lose it as heat.

Diversity of nutrition in archaea

Microbes occur in enormous numbers. They occur everywhere in the biosphere, including some of Earth's most hostile environments. These latter micro-organisms are the **extremophiles** and are in the domain archaea. There are many more species of micro-organism than there are other forms of life. However, micro-organisms fall into just one of four groups (Table B4.2.1).

■ Table B4.2.1 The range of micro-organisms

	Archaea – the extremophiles	Bacteria – the true bacteria	Unicellular protoctista – protozoa and some algae	Unicellular fungi
Prokaryotic/ eukaryotic	prokaryotes	prokaryotes	eukaryotes	eukaryotes
Domain	archaea	eubacteria	eukarya	eukarya

Link
The classification of living things in domains was introduced in Chapter A3.2, page 137.

Link
Ion pumps are discussed in Chapter B2.1, page 230.

◆ **Chemosynthesis**: inorganic molecules are oxidized to release energy; this energy is used to synthesize glucose.

◆ **Chemoautotrophs**: organisms that are chemosynthetic, i.e. use energy from chemical reactions involving the oxidation of inorganic compounds to make glucose.

The archaea include a variety of photosynthetic, **chemosynthetic** and heterotrophic organisms:
- Some archaea use sunlight as a source of energy; however, oxygen-generating photosynthesis does not occur. Instead, archaea such as halobacteria use light-activated ion pumps to generate ion gradients by pumping ions out of the cell across the plasma membrane. The potential energy in these electrochemical gradients is then transferred to ATP by ATP synthase.
- Some organisms are autotrophic but do not use sunlight to produce glucose. They use energy generated from chemical reactions. Because these archaea can make their own glucose, they are autotrophic ('self-feeders') and, because they do this by utilizing the energy from chemical reactions, they are known as **chemoautotrophs**. Some archaeans live at great depths in the sea, where there is no light for photosynthesis; they are hydrogen-dependent, using molecular hydrogen as an energy source rather than sunlight.

Chemosynthetic archaea are found in the stomachs and intestines of some mammals. Methane-producing archaea, or methanogens, release energy for ATP synthesis by producing methane gas. They use energy from the electrons found in hydrogen gas to produce methane and other organic compounds. Methanogens can also be found at the bottom of the ocean, where they can create huge methane bubbles beneath the ocean floor. It is not correct to say, therefore, that all energy in food comes from the Sun – some of it is generated by chemoautotrophs.

> ### Top tip!
> Both chemosynthesis and photosynthesis are processes where energy is used to synthesize glucose. The way that this energy is obtained varies in each process: photosynthesis (e.g. in plants) uses energy from sunlight to excite electrons, which then leads to the production of glucose; chemosynthesis (e.g. in some extremophile bacteria) uses oxidative chemical reactions involving the oxidation of inorganic compounds to make glucose.

Heterotrophic archaeans include marine species that feed on the lignin from woody plants washed out to sea. Lignin is a complex molecule found in the cell walls of vascular plants, serving to protect plants from pathogens as well as strengthening the cell wall. It is naturally resistant to degradation from many types of microbes. Archaea, by digesting the lignin, help to recycle plant remains, thereby contributing to the carbon cycle.

Relationship between dentition and diet

Humans are part of the family Hominidae, one of the two families of the ape superfamily Hominoidea, the other being the Hylobatidae (gibbons – see page 361). Hominidae includes the great apes: the orangutans (genus *Pongo*), the gorillas (genus *Gorilla*), the chimpanzees and bonobos (genus *Pan*), and humans (genus *Homo*). The evolutionary tree of the Hominidae is shown in Figure B4.2.4.

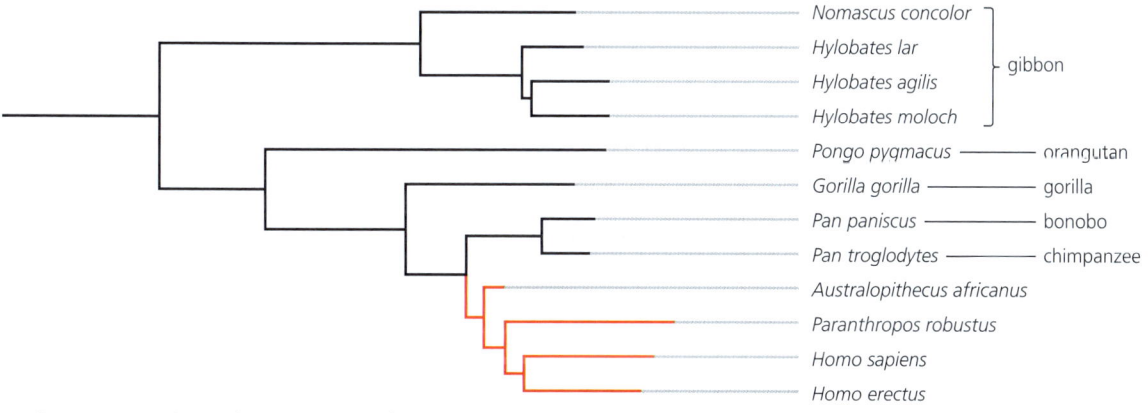

■ **Figure B4.2.4** The evolutionary tree of the Hominidae

Species within the Hominidae have a variety of different diets. Gorillas are mainly herbivores, feeding on large amounts of vegetation each day. Their skull and jaws are adapted for this plant-based diet. Large **masseter muscles** connect the skull with the jaw, enabling the animal to grind plant material between its teeth (masseter muscles create a side-to-side motion of the jaw). **Temporal muscles** pull up the jaw – this allows the animals to bite food. The temporal area is the part of the skull where these muscles attach: in gorillas (and in the other great apes) these areas are much larger than in humans. Gorillas also have a sagittal crest (a ridge of bone running along the centre-line of the top of the skull), to allow attachment for the large temporal muscle. This corresponds to the need to bite fibrous plant material forcibly.

Chimpanzees are primarily frugivores (feed on fruit) and only eat vegetation if other forms of food are limiting. Occasionally, chimpanzees eat meat (e.g. from monkeys they have killed), so can have an omnivorous diet. As a result, they have less-developed masseter and temporal muscles than gorillas.

> **Top tip!**
>
> Make sure you understand the relationship between dentition (teeth) and the diet of omnivorous and herbivorous representative members of the family Hominidae.

There are several different types of teeth: incisors, canines, premolars and molars. **Incisors** are used for slicing food, **canines** for tearing, and **premolars** and **molars** for grinding food. In terms of dentition, gorillas have more developed canine and incisors than humans (Figure B4.2.5), similar to chimpanzees, and larger premolars and molars for grinding the tough vegetation.

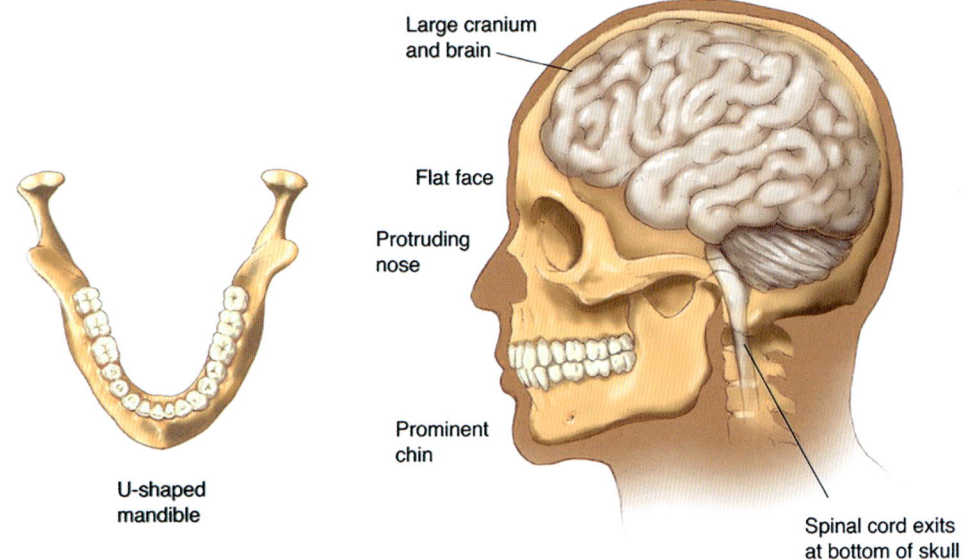

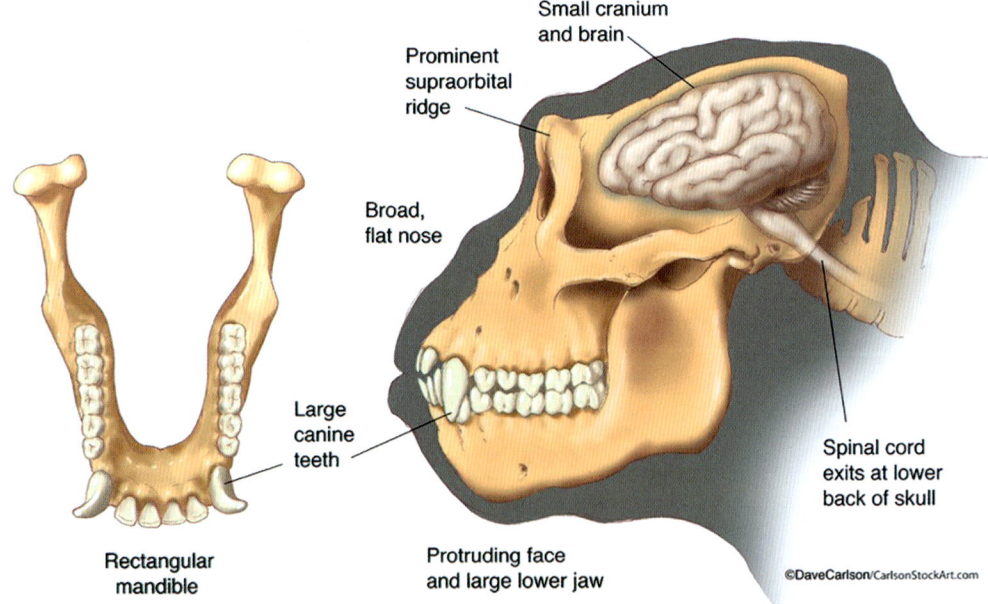

■ **Figure B4.2.5** Comparing the dentition of gorillas and humans

Humans evolved from a common ancestor with chimpanzees some 4 million years ago. One branch of evolution led to modern humans (see Figure B4.2.4). A side branch of evolution led to another group called the australopithecines, which are believed eventually to have gone extinct. One of the species from that group, *Paranthropus robustus*, had a skull that resembled that of a gorilla including a sagittal crest, due to the diet of tough vegetation that sustained them. The species name 'robustus' refers to the robust skull that they display, i.e. large teeth and strongly built jaws; in contrast, humans have a 'gracile' skull – a lighter, more slender structure. *P. robustus* had megadont (meaning 'large teeth') cheek teeth, with thick enamel to cope with the tough vegetation. Large molars and premolars helped in the powerful chewing motion.

Top tip!

Current evidence indicates a plant-based diet dominated in *Homo floresiensis*, but there is also some evidence of meat eating. The species had large premolars and a robust mandible, similar to the australopithecines, although they also had reduced molar size and a cranium similar to those of later species of Homo, suggesting a reduction in the frequency of forceful biting behaviours.

Link

Look at Figure A3.1.5 in Chapter A3.1, page 105, for a photograph of skulls from the genus *Homo*.

Seven million years ago, human ancestors' jaws and teeth were similar to those of modern chimpanzees. The incisors (the four front teeth on the top and the bottom of the skull and jaw) were relatively large, and the upper incisors were broad and projected outward. The canines were very long and pointed, and much larger in males than in females. Larger canines in apes have been linked with more fighting between males for access to females. Developed incisors and canines also allow food to be grasped and bitten. In humans, incisors are relatively small, narrow and vertical; canines are short (almost level with the other teeth) and relatively blunt. Molars (back teeth) in humans are small, and 'wisdom teeth' may be partially hidden or impacted due to the shortening of the jaw. Premolars and molars are relatively flat with low, rounded cusps (bumps) on the grinding surface. Evolution led to a larger cranial capacity in humans with increased brain size, along with smaller teeth and more V-shaped jaws (reflecting changes in dietary requirements, with less emphasis on tough vegetation and a more omnivorous diet).

Homo floresiensis was a very small hominid, measuring just over a metre tall and having a very small brain. It is thought that it was a descendant of *Homo erectus* (an ancestor of modern humans) that underwent island dwarfism (a process where isolated species that lack predators and have limited resources evolve to become smaller). They coexisted with modern humans but are thought to have become extinct around 12 000 years ago. Remains of a skull of *Homo floresiensis* were found in Liang Bua cave, Flores, Indonesia, in 2003. Tooth wear from the skull suggests that their diet was tough and fibrous (i.e. plant based), requiring powerful chewing action (mastication).

Examine models (if you have access to them) or digital collections of skulls to infer diet from the anatomical features of members of the Hominidae family. Start with *Homo sapiens* (humans), *Homo floresiensis* and *Paranthropus robustus*. Look at the examples above and work through the ATL activity to practise doing this.

ATL B4.2A

Use the following digital collections to examine models of *Homo sapiens* (humans), *Homo floresiensis* and *Paranthropus robustus*. While looking at the collections, think about these questions:
1. What is the advantage of having digital models of these species?
2. What are the limitations of the digital models?

Smithsonian Institution National Museum of Natural History:
https://humanorigins.si.edu/evidence/human-fossils

Smithsonian's 3D Digitization website: **https://3d.si.edu/collections/hominin-fossils**

Homo floresiensis cranium:
https://3d.si.edu/object/3d/homo-floresiensis-cranium:425a517b-8308-4ac9-8f3b-08f74e4ff9e5

Paranthropus robustus cranium:
https://3d.si.edu/object/3d/paranthropus-robustus-cranium:8bc77140-b75c-4f96-9899-bf275b5d43dd

Paranthropus robustus jaw:
https://3d.si.edu/object/3d/paranthropus-robustus-mandible:54907853-f218-466b-ab54-41a8aaaa59ea

Homo sapiens cranium (Cro-Magnon 1):
https://3d.si.edu/object/3d/homo-sapiens-cranium:09d681b2-5ae9-44a8-b444-8e31bb40305e

ATL B4.2B

The following site has digital models of non-human animals (both vertebrate and invertebrate):

University of Michigan Online Repository of Fossils: **http://umorf.ummp.lsa.umich.edu/wp**

Examine several herbivores and several carnivores. How does the shape of the skull and jaw relate to their mode of feeding? How do these relate to what you know about the hominid skulls you have learnt about?

> ### Nature of science: Theories
>
> It is not possible to observe extinct hominid forms today, so we must use modern-day examples and extrapolate from these to form theories about how and what extinct species ate. In this way, deductions can be made from theories. Observation of living mammals led to theories relating dentition to herbivorous or carnivorous diets. Tooth shape, tooth enamel wear-and-tear, and the shape of skull and jaws are all evidence for the animal's diet. These theories allowed the diet of extinct organisms to be deduced. Scientific knowledge must be supported by evidence and so new evidence can amend/change theories.

■ **Figure B4.2.6** SEM image of a black aphid feeding on leaf sap by inserting its stylet into the plant

■ **Figure B4.2.7** Proboscis monkey (*Nasalis larvatus*) feeding on leaves

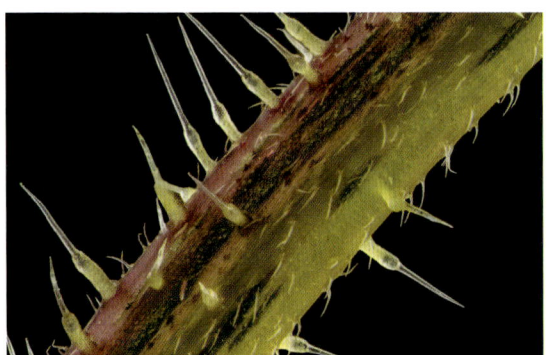

■ **Figure B4.2.8** Stinging hairs on the stem of a stinging nettle (*Urtica dioica*)

◆ **Herbivory**: feeding on plants.

Adaptations of herbivores for feeding on plants and of plants for resisting herbivory

■ Adaptations of herbivores

Aphids have modified piercing mouth parts called **stylets** (Figure B4.2.6). These secrete the enzyme pectinase. Pectin, a polysaccharide that sticks the cell walls of plants together, is digested by pectinase, so the stylet can slide between cell walls into the plant's phloem vessels and access the sucrose contained there (page 324).

Other insects eat plant leaves by using chewing mouthparts to bite, remove and masticate sections of leaf. Groups of insects that have chewing mouthparts include the grasshoppers, locusts, cockroaches, most wasps, beetles and termites. Caterpillars (the larval form of moths and butterflies) also have chewing mouthparts.

Some animals have adaptations for detoxifying plant toxins. For example, the proboscis monkey (*Nasalis larvatus*, Figure B4.2.7) that lives in the forests of northern Borneo, Malaysia, has bacteria in its extended intestines that help it to neutralize toxins from certain leaves, as well as to digest the cellulose cell walls.

■ Plant adaptations to resist herbivory

Plants have ways of resisting **herbivory**. For example, cacti have spines to stop predators accessing their stores of water (page 360). Stinging nettles (*Urtica dioica*) deter herbivores by having long, thin, hollow hairs over most of the stem and the underside of the leaves (Figure B4.2.8). Nettle stings contain methanoic (formic) acid and histamine, which cause a painful stinging and burning sensation.

Other plants produce toxic secondary compounds in seeds and leaves. Deadly nightshade (*Atropa belladonna*) contains the toxins atropine and scopolamine in its leaves, stems, berries and roots, which cause paralysis in the involuntary muscles of the body (including the heart) by competitively blocking the binding of acetylcholine (a neurotransmitter, see page 382) to receptors in the nervous system.

■ **Figure B4.2.9** Deadly nightshade (*Atropa belladonna*), which contains toxins in its leaves, stems, berries and roots

Oleander (*Nerium oleander*) (see page 344) contains toxic cardiac glycosides (oleandrin and nerine), which when eaten can cause diarrhoea, vomiting, an erratic pulse, seizures, coma and eventually death. Many plants, such as the roots of cassava (*Manihot esculenta*), produce precursors of hydrogen cyanide. When an animal eats the plant, the precursors are converted into cyanide, which kills the animal by blocking the respiratory metabolic pathway. Humans eat cassava in many parts of the world (it is native to South America but is also grown and eaten in Nigeria, Thailand and Indonesia), but before it is consumed it needs to be detoxified by being cut into small pieces, soaked and then boiled in water, before further preparation takes place.

Adaptations of predators and prey

> **Concept: Form**
>
> Predators have adaptations for finding, catching and killing prey. Prey have adaptations for resisting **predation**.

> **Top tip!**
>
> Make sure you can suggest or identify examples of chemical, physical and behavioural adaptations in predators or prey.

■ Physical adaptations

A physical adaptation involves a structural modification to the body. There are many different physical adaptations, some of which are outlined below.

Eye positions

Predators have eyes on the front of their head to enable better depth perception, meaning they use binocular vision to estimate the distance to prey organisms (Figure B4.2.10a). **Prey**, in contrast, tend to have eyes on the side of their head so they can have a good view of their surroundings and detect the approach of predators from both sides, front and back (Figure B4.2.10b).

♦ **Predation**: a biological interaction where one organism, the predator, kills and eats another organism, its prey.

♦ **Predator**: an organism that catches and kills other animals to eat.

♦ **Prey**: an organism hunted and eaten by a predator.

■ **Figure B4.2.10** a) A barn owl (*Tyto alba*) in flight – notice that its eyes are at the front of its head; b) a red deer (*Cervus elaphus*) – by having eyes on the side of the head, these prey animals have a larger field of vision to see approaching predators

Specialized sense organs

Predators also have at least one sense organ that is very efficient at detecting prey. For example, predatory snakes (such as rattlesnakes) use their tongue to pick up chemicals emitted from prey and a specialized organ in the mouth (the Jacobson's organ) to detect these chemicals. Birds of prey (such as the peregrine falcon) have very acute vision, while nocturnal bats (such as pipistrelle bats) use echolocation to locate prey, using their large, sensitive ears to detect sound.

■ Figure B4.2.11 A peregrine falcon (*Falco peregrinus*) in flight

Speed

Fast speed is an advantage for both predators and prey as it enables predators to catch their prey or for prey to escape their predators. Animals can be very fast on land, for example the cheetah, see page 326, or through the air, such as birds of prey diving at great speed. The peregrine falcon is the fastest diving bird in the world and has been recorded travelling at more than $380\,km\,h^{-1}$, reaching its prey before the prey can respond (Figure B4.2.11).

Mechanical defences

Hard shells on tortoises and turtles, or the spines of a hedgehog (a small mammal found throughout parts of Europe, Asia and Africa), physically prevent the predator from being able to eat the prey or causing pain to the predator (Figure B4.2.12). This discourages predation.

Camouflage

Some species evolve camouflage as an adaptation: the organism evolves to resemble its background. This means that it is much less visible to predators. Chameleons have some of the best-known camouflage of any animal. They are lizards in the family Chamaeleonidae and are found in Africa, Madagascar, southern Europe and southern Asia. Fischer's chameleon, for example, is found in Tanzania (Figure B4.2.13).

■ Figure B4.2.12 A European hedgehog (*Erinaceus europaeus*)

■ Figure B4.2.13 Fischer's chameleon (*Kinyongia fischeri*): a species of chameleon that is endemic to Tanzania

Mimicry

Coral snakes are one of the most poisonous snakes on Earth (Figure B4.2.14a). They have the second-strongest venom of any snake (the black mamba has the deadliest venom). Their black, red and white striped colouration is a warning to predators that they are venomous. Other, non-venomous snakes, such as king snakes (Figure B4.2.14b), mimic the colour, size and shape of coral snakes. Predators are tricked into thinking that that king snakes are poisonous and so avoid them. King snakes do not have to invest in making poison but have all the advantages of coral snakes in avoiding predation.

■ Figure B4.2.14 An example of mimicry: a) coral snake (*Micrurus altirostris*), b) California mountain king snake (*Lampropeltis zonata*)

ATL B4.2C

Research other examples of mimicry – what behavioural and physical adaptations do these animals demonstrate? What are the advantages for the mimics in copying the appearance or behaviour of another animal, for example the king snake copying the coral snake?

6 **Outline** the role of physical adaptations in helping predator and prey animals survive.

> **Tool 1: Experimental techniques**
>
> **Drawing annotated diagrams from observation**
>
> Select an animal or plant that shows physical adaptations to its environment. Draw the organism and then label its adaptations. Add annotations that explain how these adaptations help the organism to survive.

■ Figure B4.2.15 Caterpillars of the cinnabar moth (*Tyria jacobaeae*) on the yellow flowers of ragwort (*Jacobaea vulgaris*)

Chemical adaptations

Many animals contain chemicals that are harmful to predators. These chemicals are synthesized and stored by the animal.

Toxicity

Warning colours are used to tell predators that the prey species may be toxic and not good to eat. For example, the caterpillar of the cinnabar moth has distinctive black and yellow stripes: this combination of colours is often used in the animal kingdom to indicate that an animal is poisonous. This caterpillar absorbs toxins from the ragwort on which it feeds (Figure B4.2.15). Predators that eat the caterpillar will experience an unpleasant taste and the presence of toxic chemicals, so they learn not to eat them. This type of defensive mechanism is called aposematic colouration.

Poison dart frogs live in the rainforests of Central and South America. The frog's skin secretes a poison that can paralyze and kill predators. Poison dart frogs may get their toxicity from the insects they eat, such as ants and termites. They have bright colouration (Figure B4.2.16), which is a warning to potential predators of their toxicity.

Chemical defences

Other species use chemicals for defence. Bombardier beetles (*Brachinus* sp.), Figure B4.2.17 eject a hot, noxious chemical spray from the tip of their abdomen. The spray is produced by a reaction between two chemicals, hydroquinone and hydrogen peroxide, which are stored in two chambers in the beetle's abdomen.

■ Figure B4.2.16 The red striped poison dart frog (*Ranitomeya reticulata*)

The skunk is a well-known animal that has chemical defences. These creatures spray a pungent liquid containing volatile sulfurous chemicals when threatened. The liquid sticks to the fur and skin of predators and deters them from approaching skunks in the future.

> **Top tip!**
>
> Behavioural, chemical and physical adaptations often interact and reinforce each other. Before spraying a predator, a skunk will face away from the threat and arch its back, raise its tail and stomp its feet while making a hissing noise. Their colouration – bold white and black stripes – is a warning to potential predators.

■ Figure B4.2.17 A bombardier beetle (*Brachinus alternans*)

B4.2 Ecological niches

■ **Figure B4.2.18** A Sumatran pit viper (*Trimeresurus sumatranus*) – a stealth predator

■ **Figure B4.2.19** A pill millipede rolled up into a defensive ball

7 **Describe** the chemical and behavioural adaptations of predators for finding, catching and killing prey, and of prey animals for resisting predation.

■ Behavioural adaptations

As well as physical and chemical adaptations, predators and prey can behave in ways that increase their chance of survival.

Not all predators rely on speed to catch their prey. Stealth predators, such as the Sumatran pit viper (Figure B4.2.18), wait for their prey to get close enough before they strike – to do this they must remain still until the prey is within reach. These animals are often camouflaged so they can remain undetected.

Many animals roll into a ball when threatened by a predator, such as woodlice and millipedes. This is an example of a behavioural adaptation as well as a physical adaptation (Figure B4.2.19).

Top tip!

Not all adaptations fit neatly into one category. They often overlap or influence each other. For example, a 'rolling into a ball' behavioural adaptation only works if the prey has a tough exoskeleton to protect itself.

Top tip!

Not all predators are animals: some plants are carnivorous and hunt insects (e.g. the Venus fly trap (*Dionaea muscipula*), page 327) and pitcher plants (e.g. *Nepenthes* species, page 361).

Adaptations of plants for harvesting light

Link
Competition and its definition are covered in more detail in Chapter C4.1, page 557.

Concept: Form
Plants in forests have a range of different forms to enable them to harvest light, to avoid competition.

In all ecosystems there is competition for often limited resources. The large number of species in tropical rainforests (page 360) means that competition in this biome is especially severe.

Plant species are adapted in many ways to give them an advantage in the competition to survive. One of the main limiting factors in forest ecosystems is light, and plants use different strategies to reach light sources.

The forest can be divided into different layers, or strata. **Emergent trees** are the tallest trees that reach over the canopy, then the **canopy trees** form a layer of vegetation high above the forest floor. Several metres below the canopy is the understory of smaller plants and, at the bottom, the ground-layer plants grow.

Canopy trees have a competitive advantage over smaller plants as they have first access to sunlight and can acquire all frequencies of light. This maximizes photosynthesis and enables them to grow to a large size. The understory has less access to light, while the ground layer only receives around 1% of the light available to the canopy layer.

Lianas are woody vines: they have leaves and flowers in the canopy and roots on the ground, but instead of having a strong trunk they use the support of trees to reach the sunlight. They begin life on the forest floor and send shoots upwards towards the sunlight, and depend on the support of other plants for their growth and survival.

Epiphytes are plants that grow on the branches of trees. For example, the bird's nest fern (South East Asia) and bromeliads (South America, see Figure B4.2.20) are both found high above the forest floor. Seeds of these plants are deposited by birds or mammals who rest on the branches of the tree.

Epiphyte species have the advantages of canopy trees but with none of the expense of having to grow long stems to reach this height – epiphytes use the support of trees to find their position at the highest levels in the forest. They have access to strong sunlight and warmth – both of which enable them to maximize growth. Epiphytes form microhabitats for many species of animals: they trap water, which enables aquatic animals, such as frogs, to live in an otherwise hostile environment.

Some epiphytes send roots down to the forest floor, where they can draw up nutrients and water. Some of these epiphytes form a thick network of stems and roots around the host tree, eventually killing the host. These are known as strangler epiphytes and include the strangler fig (Figure B4.2.21). By killing the host, the strangler epiphyte has greater access to resources and does not have to share these with the host.

■ Figure B4.2.20 A bromeliad in the rainforest canopy

■ Figure B4.2.21 Strangler fig (*Ficus* species) in the rainforest, Cape Tribulation National Park, Queensland, Australia

8 **Explain** how plants at different levels (strata) in a forest are adapted for harvesting light.

Shade-tolerant shrubs and herbs growing on the forest floor only have access to a small proportion of the light available in the canopy. Canopy plants filter out some frequencies of light (red and blue), leaving a more limited range of light wavelengths at the forest floor (e.g. green light). This means that shrubs and herbs on the forest floor contain different photosynthetic pigments from canopy plants, giving them a different colour (for example, red leaves rather than the mostly green leaves of the canopy). Another adaptation to these low light levels is that understorey plants have often evolved much larger leaves. There are few flowering plants in the understorey – the lack of light makes flowers difficult to see by pollinators. The plants that do flower often produce very brightly coloured flowers so they can be seen easily in such surroundings (Figure B4.2.22). They are also strongly scented so they can attract pollinators with their smell.

Fundamental and realized niches

Early work by ecologists Joseph Grinnell and Charles Elton defined niche as simply the 'job' carried out by a particular species in the environment. This concept of niche allows for 'vacant' or empty niches: roles in an ecosystem that are not yet, but could be, filled. In the 1950s, zoologist G. Evelyn Hutchinson put a different perspective on the concept of niche by including the interactions between organism and environment. In the Hutchinson concept of niche, where 'niche' is a property of an organism, there can be no 'empty niches' because the definition of niche is species-centred, defined by the interaction of the species and its abiotic and biotic environment.

■ Figure B4.2.22 Spiral ginger flower (*Costus pulverulentus*), a shade-tolerant shrub, in the rainforest of Costa Rica

◆ **Fundamental niche**: the potential extent of a species based on adaptations and tolerance limits.

◆ **Realized niche**: the actual extent of a species when in competition with other species.

Hutchinson realized that there was a difference between the area that a species could, *in theory*, occupy and one that it *actually* occupied. The abiotic environment forms physical limitations for the species, and the distribution of a species is affected by its range of tolerance to a variety of different environmental factors (as discussed on page 349). In reality, however, factors affecting how a species disperses itself, and how it interacts with other species, restrict the actual niche. Species live within communities and participate in complex food webs based on feeding and other interactions (see page 581), and so it is not sufficient to define a species' niche simply by its range of tolerance to different abiotic variables. Abiotic factors, such as climate, are dominant across all geographical scales, whereas biotic factors, such as competition, predation and mutualism, tend to be dominant at smaller scales.

From these ideas the concepts of **fundamental niche** and **realized niche** were developed.

In summary, the fundamental niche is the potential location of a species along an environmental gradient, and the realized niche is its actual location along gradients, determined by interactions between abiotic and biotic factors. The fundamental niche can, therefore, be simply defined as where an organism *could* live, and the realized niche as where an organism *does* live (Figure B4.2.23).

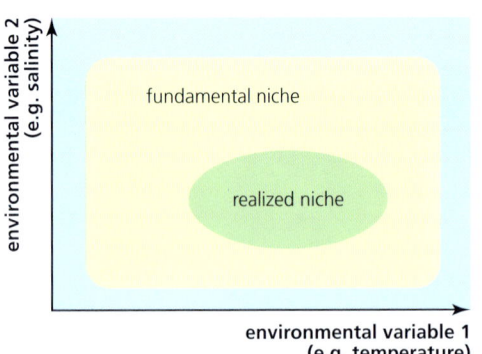

■ Figure B4.2.23 The fundamental and realized niche

9 Distinguish between fundamental and realized niches.

Competitive exclusion and the uniqueness of ecological niches

Competitive exclusion

When resources are limiting, populations will compete to survive. Competition can be either within a species (**intraspecific** competition) or between different species (**interspecific** competition). Interspecific competition exists when the niches of different species overlap (Figure B4.2.24). No two species can occupy the same niche, so the degree to which niches overlap determines the degree of interspecific competition.

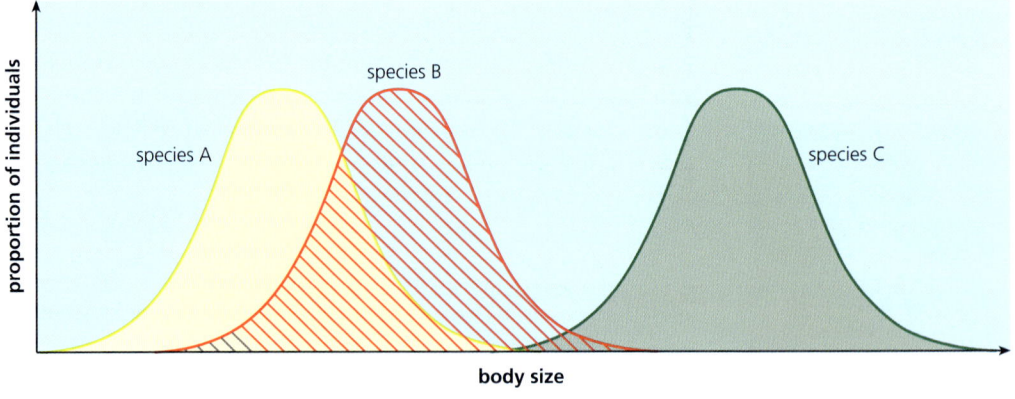

■ Figure B4.2.24 The niches of species A and species B, based on body size, overlap with each other to a greater extent than with species C; strong interspecific competition will exist between species A and B, but not with species C

If two species share the same resource at the same place and the same time, then the dominant species will outcompete the other. The inferior competitor will either die out or move away to avoid the competition. This is **competitive exclusion**.

The competitive exclusion principle was first suggested by a Russian biologist GF Gause in 1934, based on his experiments of culturing different species of *Paramecium* in the laboratory (Figure B4.2.25).

Theme B: Form and function – Ecosystems

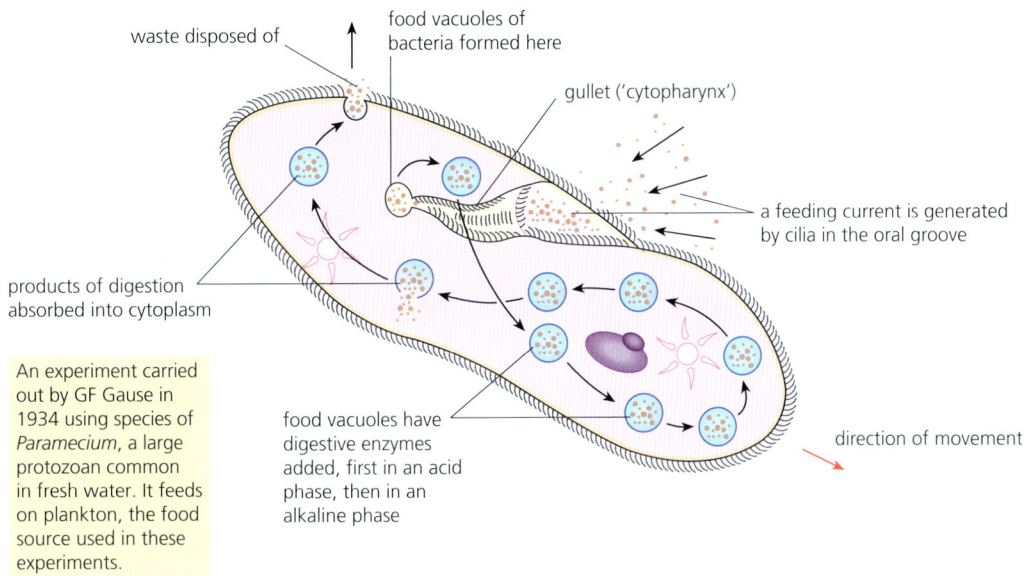

An experiment carried out by GF Gause in 1934 using species of *Paramecium*, a large protozoan common in fresh water. It feeds on plankton, the food source used in these experiments.

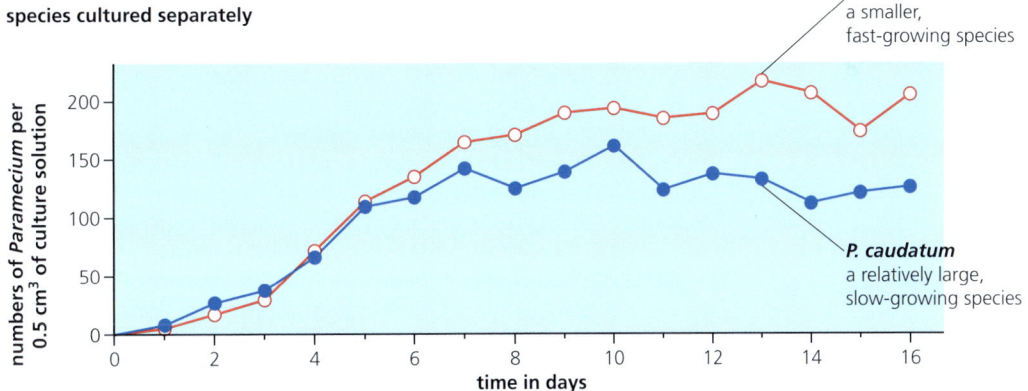

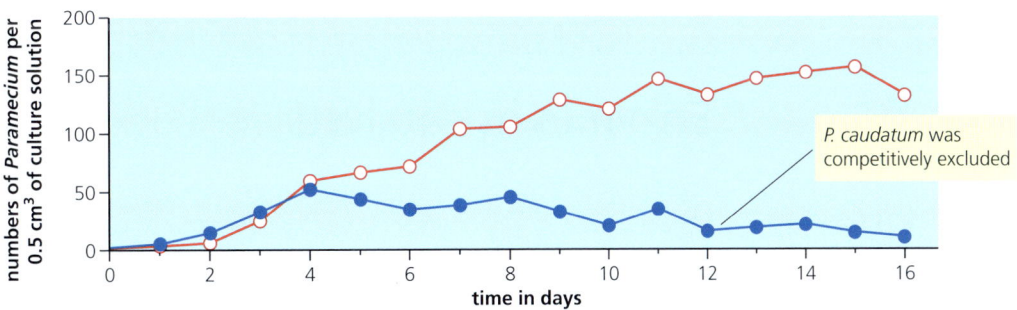

■ **Figure B4.2.25** Population growth curves for two species of *Paramecium*. If two species with very similar resource needs (that is, similar niches) are grown separately, both can survive and flourish (top graph); if the two species are grown in a mixed culture, the superior competitor – in this case *P. aurelia* – will eliminate the other

Link
A test for interspecific competition is covered in more detail in Chapter C4.1, page 569.

10 Define the term *competitive exclusion*.

LINKING QUESTIONS
1. What are the relative advantages of specificity and versatility?
2. For each form of nutrition, what are the unique inputs, processes and outputs?

Competitive exclusion could be the pressure that causes closely related species, living together in the same habitat, to have evolved clearly defined but separate niches. This would occur over an extended period of time. For example, the competitive exclusion principle would account for the difference in niches between the cormorant and the shag that we noted earlier on page 364.

B4.2 Ecological niches

C1.1 Enzymes and metabolism

Concept: Interaction and interdependence

Systems are based on interactions, interdependence and integration of components. Systems result in the emergence of new properties at each level of biological organization.

Guiding questions

- In what ways do enzymes interact with other molecules?
- What are the interdependent components of metabolism?

SYLLABUS CONTENT

This chapter covers the following syllabus content:
- ▶ C1.1.1 Enzymes as catalysts
- ▶ C1.1.2 Role of enzymes in metabolism
- ▶ C1.1.3 Anabolic and catabolic reactions
- ▶ C1.1.4 Enzymes as globular proteins with an active site for catalysis
- ▶ C1.1.5 Interactions between substrate and active site to allow induced-fit binding
- ▶ C1.1.6 Role of molecular motion and substrate–active site collisions in enzyme catalysis
- ▶ C1.1.7 Relationships between the structure of the active site, enzyme–substrate specificity and denaturation
- ▶ C1.1.8 Effects of temperature, pH and substrate concentration on the rate of enzyme activity
- ▶ C1.1.9 Measurements in enzyme-catalysed reactions
- ▶ C1.1.10 Effect of enzymes on activation energy
- ▶ C1.1.11 Intracellular and extracellular enzyme-catalysed reactions (HL only)
- ▶ C1.1.12 Generation of heat energy by the reactions of metabolism (HL only)
- ▶ C1.1.13 Cyclical and linear pathways in metabolism (HL only)
- ▶ C1.1.14 Allosteric sites and non-competitive inhibition (HL only)
- ▶ C1.1.15 Competitive inhibition as a consequence of an inhibitor binding reversibly to an active site (HL only)
- ▶ C1.1.16 Regulation of metabolic pathways by feedback inhibition (HL only)
- ▶ C1.1.17 Mechanism-based inhibition as a consequence of chemical changes to the active site caused by the irreversible binding of an inhibitor (HL only)

Enzymes as catalysts

Many chemical reactions do not occur spontaneously, or they may happen very slowly (Figure C1.1.1). In a laboratory or in an industrial process, chemical reactions may be made to occur by applying high temperatures, high pressures, extremes of pH, by maintaining high concentrations of the reacting molecules, or by using inorganic catalysts. If these drastic conditions were not applied, very little of the chemical product would be formed. On the other hand, in cells and organisms, many chemical reactions occur simultaneously, at extremely low concentrations, at normal temperatures and under the very mild, almost neutral, aqueous conditions we find in cells.

It is the presence of **enzymes** in cells and organisms that enables these reactions to occur at relatively high rates, in an orderly manner, yielding products that the organism requires, when they are needed. Sometimes, reactions happen even though the reacting molecules are present in very low concentrations. Enzymes are biological **catalysts** made of protein. They are truly remarkable molecules. In general, catalysts:
- are effective in small amounts
- remain unchanged at the end of the reaction.

◆ **Enzyme**: mainly proteins (some are RNA) that function as biological catalysts.

◆ **Catalyst**: a substance that speeds up the rate of a chemical reaction. Catalysts are effective in small amounts and remain unchanged at the end of the reaction.

Concept: Interaction

Interaction: The effect or effects that two or more systems, bodies, substances or organisms have on one another, so that the overall result is not simply the sum of the separate effects.

Concept: Interdependence

Interdependence: Biological systems are not self-sufficient. Molecules, cells, organisms and ecosystems interact with each other within and across levels of organization. The greater the level of interaction, the greater the degree of interdependence.

Concept: Interaction

Metabolism depends on the interaction of many different enzymes. Enzymes for specific reactions are located within compartments (organelles) within eukaryotic cells, for example the enzymes of oxidative respiration are found within mitochondria.

● Common mistake

It is incorrect to use the term 'amount' when discussing variables in practical procedures. Use more precise terms available, such as 'concentration' or 'volume'.

Figure C1.1.1 shows the benefit of increasing rates of reaction in cells. Many enzymes are always present in cells and organisms, but some enzymes are produced only under particular conditions, at certain stages or when a particular substrate molecule is present. By making some enzymes and not others, cells can control what chemical reactions happen in the cytoplasm.

Role of enzymes in metabolism

There are many thousands of chemical reactions taking place within cells and organisms. Metabolism is the name we give to these chemical reactions. These reactions can only occur in the presence of specific enzymes. If an enzyme is not present, the reaction it catalyses only occurs at a very slow rate. The molecules involved are collectively called **metabolites**. Many metabolites are made in organisms, but others are imported from the environment, such as from food substances, water and the gases carbon dioxide and oxygen.

Random collision possibilities:
when sucrose and water molecules collide at the wrong angle

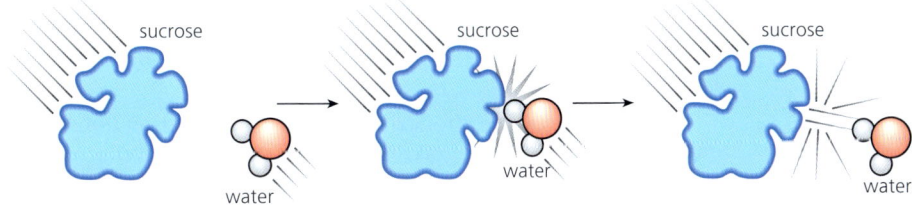

when sucrose and water collide at the wrong speed

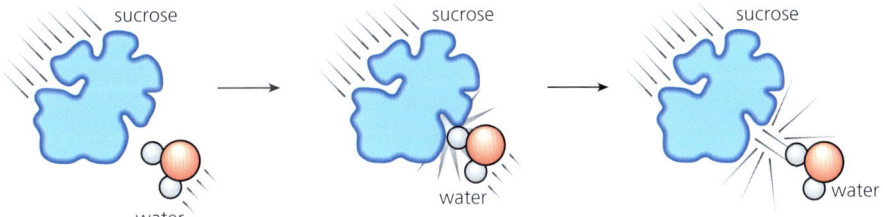

for the reaction to occur, sucrose and water must collide in just the right orientation – glucose and fructose are formed

These events are what happens at most random collisions.

Under normal conditions this happens so very infrequently it is an insignificant event.

In the presence of one molecule of the enzyme sucrase (invertase), approximately 3.0×10^4 molecules of sucrose are hydrolysed each minute!

■ Figure C1.1.1 Can a reaction occur without an enzyme?

C1.1 Enzymes and metabolism

■ What is the meaning of enzyme specificity?

Many different enzymes are required by living organisms, and control over metabolism can be exerted through these enzymes. Each enzyme is specific to one substrate, so that metabolic processes can be closely controlled. Enzymes are usually classified according to the type of reaction they catalyse and they can be named according to their substrates (i.e. the reactants that enzymes act on).

Metabolism consists of chains (linear sequences) and cycles of enzyme-catalysed reactions, such as we see in respiration (page 405), photosynthesis (page 425), protein synthesis (page 615), and in very many other pathways. All these reactions may be classified as one of just two types, according to whether they involve the build-up or breakdown of organic molecules. These are called anabolism and catabolism, respectively (see below).

1 Define the term *metabolism*.

Anabolic and catabolic reactions

In anabolic reactions, larger molecules are built up from smaller molecules. **Anabolism** involves the formation of macromolecules from monomers by **condensation reactions** (page 186).

Examples of anabolism include:
- the synthesis of proteins from amino acids (Chapter B1.2, page 207)
- the synthesis of polysaccharides from simple sugars, e.g. glycogen formation (Chapter B1.1, page 187)
- photosynthesis (Chapter C1.3, page 425).

Catabolism is the breakdown of larger molecules into simple molecules. The catabolic reactions involved include **hydrolysis reactions** (page 186).

Examples of catabolism include:
- hydrolysis of macromolecules into monomers in digestion (Chapter B1.1, page 187)
- oxidation of substrates in respiration (Chapter C1.2, page 412).

Overall: metabolism = anabolism + catabolism

◆ **Anabolism**: the synthesis of complex molecules from simpler molecules including the formation of macromolecules from monomers by condensation reactions.

◆ **Catabolism**: the breakdown of complex molecules into simpler molecules including the hydrolysis of macromolecules into monomers.

2 **Define** the following terms and give one example of each:
  a *anabolic reaction*
  b *catabolic reaction*.

3 **Distinguish** between anabolism, catabolism and metabolism.

■ Figure C1.1.2 Metabolism = anabolism + catabolism

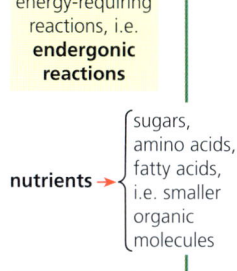

■ Figure C1.1.3 Summary of anabolic and catabolic reactions

(Diagram labels: synthesis of complex molecules used in growth and development, and in metabolic processes, e.g. proteins, polysaccharides, lipids, hormones, growth factors, haemoglobin, chlorophyll; **anabolism**: energy-requiring reactions, i.e. **endergonic reactions**; nutrients → sugars, amino acids, fatty acids, i.e. smaller organic molecules; **catabolism** energy-releasing reactions, i.e. **exergonic reactions**; release of simple substances, e.g. small inorganic molecules, CO_2, H_2O, mineral ions)

Link
Globular proteins are discussed in Chapter B1.2, page 221.

◆ **Substrate**: the starting substance (reactant) in a reaction catalysed by an enzyme. It is the molecule that the enzyme reacts with.

◆ **Product**: what the substrate is converted to in a reaction catalysed by an enzyme.

◆ **Active site**: region of an enzyme molecule where the substrate molecule binds and catalysis occurs.

◆ **Enzyme–substrate (ES) complex**: a temporary structure formed when a substrate binds to the active site of an enzyme.

Enzymes as globular proteins with an active site for catalysis

Enzymes are typically large, globular protein molecules, where the tertiary structure has given the molecule a generally rounded shape. In a reaction catalysed by an enzyme, the starting substance is called the **substrate**, and what it is converted to is the **product**.

An enzyme works by binding to the substrate molecule(s) at a specially formed pocket in the enzyme. This binding point is called the **active site**. Here, as the enzyme and substrate form an **enzyme–substrate (ES) complex**, the substrate is raised to a high-energy transition state. This complex has the briefest of existences before the substrate molecule is formed into another molecule or broken down into others by the catalytic properties of the active site. Then the product(s) are released, together with the unchanged enzyme (Figure C1.1.4). The enzyme is available for reuse.

The sequence of steps to an enzyme-catalysed reaction:
enzyme + substrate ⟶ E–S complex ⟶ product + enzyme available for reuse
                                  |
                          (substrate raised to
                           transition state)

E + S ⟶ E–S ⟶ Pr + E

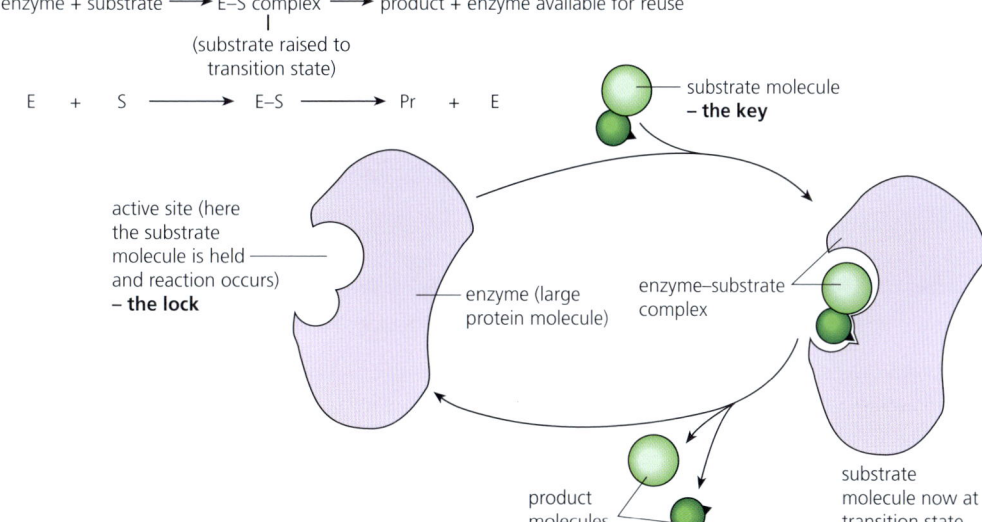

■ Figure C1.1.4 The enzyme–substrate complex and the active site

🔴 Common mistake

When describing enzyme-catalysed reactions, crucial details are sometimes missed out. Do not forget to mention the active site, substrate and ES complex.

The active site is a pocket or crevice in the protein where the substrate molecule binds and catalysis occurs (see Figure C1.1.4). The active site of an enzyme may only be made from a few amino acids, but the interactions between the amino acids within the overall three-dimensional structure of the enzyme ensure that the active site has the necessary properties for catalysis. These properties include binding to the substrate molecule, holding on to it while the chemical reaction takes place and lowering the energy of the transition state (see below).

Most substrate molecules are quite small compared to the enzyme. Even when the substrate molecules are very large, for example certain macromolecules such as polysaccharides, only one bond in the substrate is in contact with the enzyme active site. The active site takes up a relatively small part of the total volume of the enzyme.

C1.1 Enzymes and metabolism

4 **Explain** why the shape of enzymes is important in enzyme action.

Induced-fit binding

Enzymes are highly specific in their action, which makes them different from most inorganic catalysts. Enzymes are specific because of the way they bind with their substrate at the active site, which is a pocket or crevice in the protein (Figure C1.1.5). At the active site, the arrangement of a few amino acid molecules in the protein (enzyme) matches certain groupings on the substrate molecule, enabling the enzyme–substrate complex to form. As it forms, a slight change of shape is induced in the enzyme and substrate molecules (the **induced-fit** model, Figure C1.1.6). It is this change in shape that is important in raising the substrate molecule to the transition state in which it can react.

◆ **Induced-fit**: the binding of the substrate to the enzyme causes a change in the shape of the enzyme and the substrate, resulting in the proper alignment of the catalytic groups on its surface, which enables catalysis to take place.

Specificity:
- Some amino acid molecules allow a particular substrate molecule to 'fit'.
- Some amino acid molecules bring about particular chemical changes.

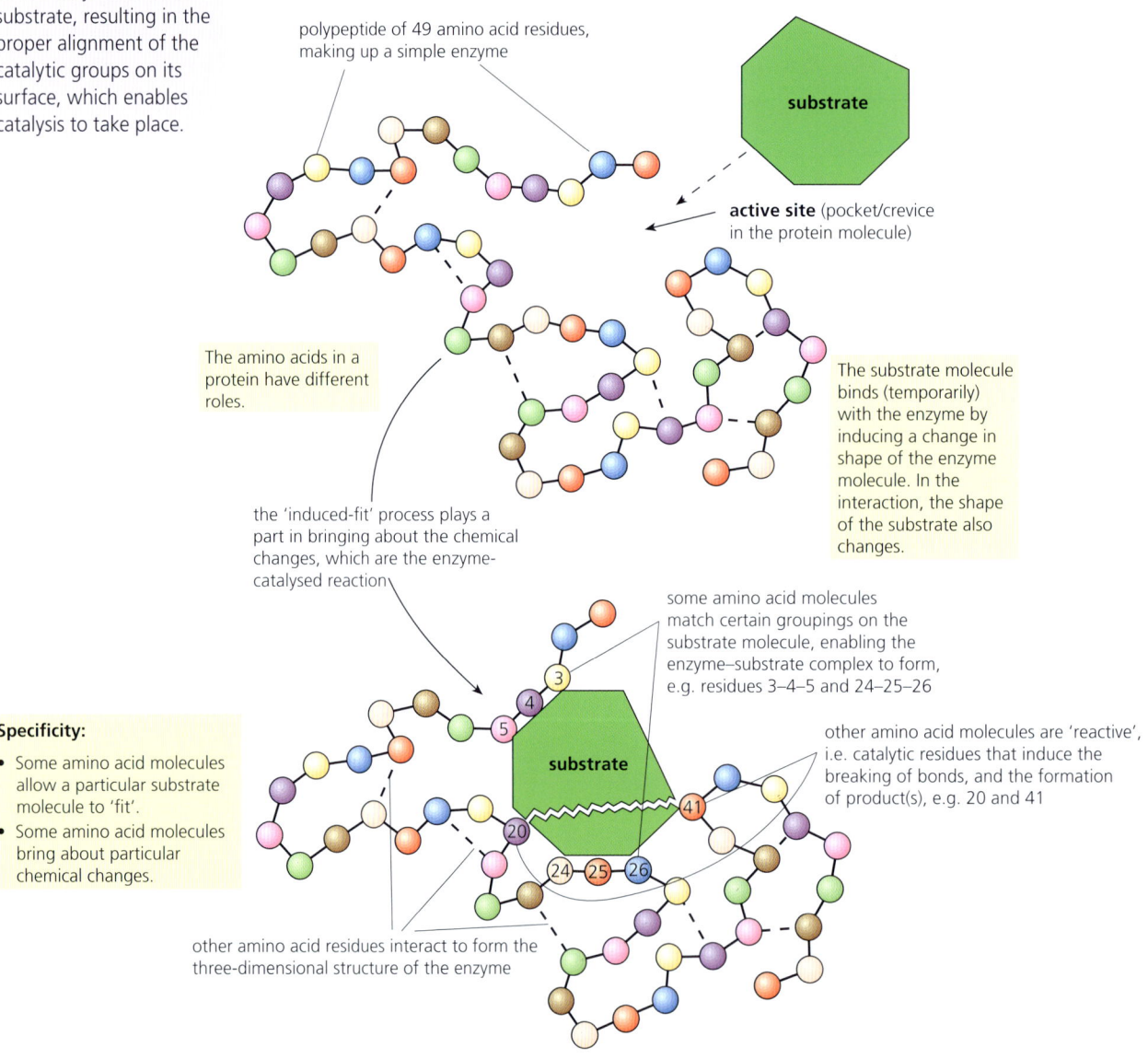

■ **Figure C1.1.5** Enzyme specificity and the active site

Meanwhile, other amino acids of the active site bring about the specific catalytic reaction mechanism, perhaps breaking particular bonds in the substrate molecule and forming others. Different enzymes have different arrangements of amino acids in their active sites. Consequently, each enzyme catalyses either a single chemical reaction or a group of closely related reactions.

Theme C: Interaction and interdependence – Molecules

The enzyme hexokinase catalyses the reaction:

glucose + ATP ⟶ glucose-6-phosphate + ADP

Computer-generated image of the induced-fit hypothesis in action:

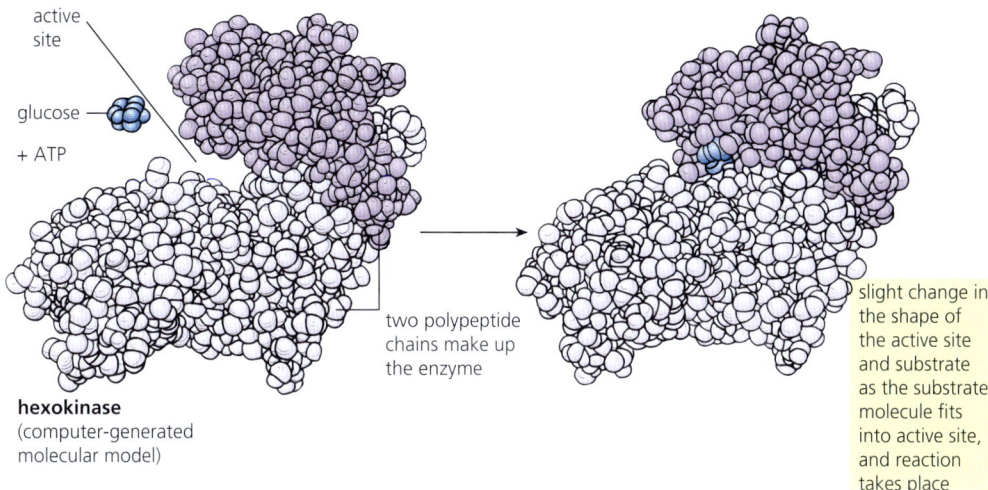

■ Figure C1.1.6 Induced-fit catalysis by hexokinase

> **Top tip!**
> Both substrate and enzyme change shape when binding occurs.

In the original model of enzyme-substrate interactions, the **lock-and-key model**, the active site of the enzyme is the complementary shape of the substrate and so fits to the substrate precisely. The enzyme acts as a 'lock' and the substrate is the 'key' – the shape of the key matches the lock and one key opens one lock. In the **induced-fit model**, the active site of the enzyme undergoes a conformational change to improve binding with the substrate – it has been compared to a hand entering a glove, where both hand and glove change shape to ensure optimal fit (the 'hand-in-glove' model). Most enzymes seem to follow the latter model.

ATL C1.1A

Access this website: www.abpischools.org.uk/topics/enzymes-16/enzymes-biological-catalysts-that-control-the-reactions-of-life

Explore the structure and function of enzymes using the information and animations provided. Produce a poster explaining the role that enzymes play in organisms. When were enzymes first used by humans? How are they used today in industrial processes?

5 **Define** the term *induced-fit binding*.

Tool 1: Experimental techniques

Physical modelling

This video discusses the lock-and-key and induced-fit models:

www.youtube.com/watch?v=E-_r3omrnxw

Make and present models (using modelling clay, for example), emphasizing the differences between the lock-and-key and induced-fit ('hand-in-glove') models. You could also consider the limitations of representing the enzyme as the 'lock' and the substrate as the 'key'.

Discuss in pairs, or a small group, why making a physical model is useful.

C1.1 Enzymes and metabolism

The role of molecular motion and collisions in enzyme catalysis

Movement is needed for a substrate molecule and an active site to come together. The greater the kinetic energy of enzyme and substrate, the greater the chance of collisions between molecules and the formation of enzyme–substrate complexes.

Sometimes large substrate molecules are immobilized, and sometimes enzymes can be immobilized by being embedded in membranes. **Immobilized enzymes** are often more stable than free enzymes in a solution and may provide a better environment for enzyme activity.

Enzymes immobilized on (or in) a membrane (EIM) is a widely used method in food processing, pharmaceuticals and wastewater treatment. The EIM system facilitates recycling of the enzymes and also, in many cases, enhanced enzyme properties (e.g. stability and viability). The membrane is a porous structure on to which specific enzymes are attached.

◆ **Immobilized enzyme**: an enzyme attached to an inert, insoluble material, enabling recovery, reuse and improved enzyme stability.

■ Immobilization of enzymes for industrial use

Enzymes may be used in industrial processes as **cell-free preparations** that are added to a reaction mixture, or they may be **immobilized** and the reactants passed over them. Enzyme immobilization involves the attachment of enzymes to insoluble materials which then provide support. For example, the enzyme may be entrapped between inert fibres or it may be covalently bonded to a matrix (Figure C1.1.7). In both cases, the enzyme molecules are prevented from being leached (washed) away.

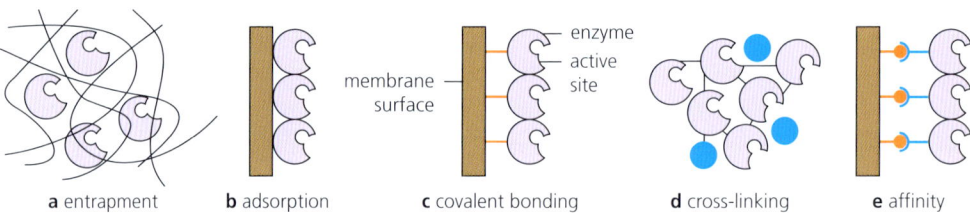

■ **Figure C1.1.7** Enzyme immobilization techniques

The advantages of using an immobilized enzyme in industrial processes are:
- it permits reuse of the enzyme preparation
- the product is enzyme free
- the enzyme may be much more stable and long lasting, due to protection by the inert matrix.

⏺ TOK

People who are unable to produce lactase fail to digest lactose, a carbohydrate found in milk. As a result, bacteria in the large intestine feed on the lactose, producing fatty acids and methane, causing diarrhoea and flatulence. Such people are said to be lactose intolerant and need to consume lactose-free milk.

The knowledge and technology required to make lactose-free milk is common in more economically developed countries. However, sharing that knowledge would undoubtedly benefit people from all over the world. If a company from a more economically developed country has this knowledge, are they ethically obligated to share this knowledge? One famous example is Volvo sharing its knowledge of the three-point seat-belt system without charge, while Moderna recently promised not to enforce its COVID-19 vaccine patents in certain countries. In both cases, the knowledge was important enough distribute freely.

If you are going to explore this question in a TOK context, remember that TOK is about knowledge, not (as this case might lead you into discussing) about solutions for distribution of a product. Make sure your analysis is about the knowledge required to develop these products.

Relationships between the structure of the active site, enzyme–substrate specificity and denaturation

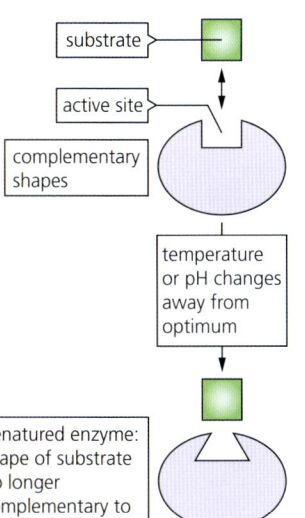

■ Figure C1.1.8 The effect of pH and temperature on enzymes

Denaturation occurs when the weak intramolecular interactions within the globular protein, formed between different amino acid residues, break. This changes the three-dimensional shape of the active site. Peptide bonds are not hydrolyzed and broken during denaturation. In denaturation, the structure of the active site is changed, so the enzyme and substrate can no longer bind together (Figure C1.1.8).

Link
Denaturing of proteins and the effects of pH and temperature on protein structure are detailed in Chapter B1.2, page 210. Peptide bonds are covered in B1.2, page 207.

Effects of temperature, pH and substrate concentration on rate of enzyme activity

Nature of science: Patterns and trends
You need to be able to describe the relationship between variables as shown in graphs. You should recognize that generalized sketches of relationships are examples of models in biology. Models in the form of sketch graphs can be evaluated using results from enzyme experiments. Enzyme experiments include investigating the effects of temperature, pH and substrate concentration on the rate of enzyme activity. In each case, the shape of the graph will be specific to the reaction being investigated.

Top tip!
Accurate, quantitative measurements in enzyme experiments require replicates to ensure reliability.

■ Effect of temperature

Examine the investigation of the effects of temperature on the hydrolysis of starch by the enzyme amylase, shown in Figure C1.1.9.

When starch is hydrolyzed by the enzyme amylase, the product is maltose, a disaccharide (page 186). Starch gives a blue–black colour when mixed with iodine solution (iodine in potassium iodide solution), but maltose gives a yellow or brown colour (depending on concentration of iodine), indicating no reaction has occurred. The first step in this experiment is to bring samples of the enzyme and the substrate (the starch solution) to the temperature of the water bath before being mixed – a step called pre-incubation.

The progress of the hydrolysis reaction is followed by taking samples of a drop of the mixture on the end of the glass rod at half-minute intervals. These are tested with iodine solution on a white tile. Initially, a strong blue–black colour is seen, confirming the presence of starch. Later, as maltose accumulates, an orange colour predominates. The endpoint of the reaction is indicated when all the starch colour has disappeared from the test spot. Using fresh reaction mixture each time, the investigation is repeated at a series of different temperatures, say at 10, 20, 30, 40, 50 and 60 °C. The time taken for complete hydrolysis at each temperature is recorded and the rate of hydrolysis in unit time is plotted on a graph.

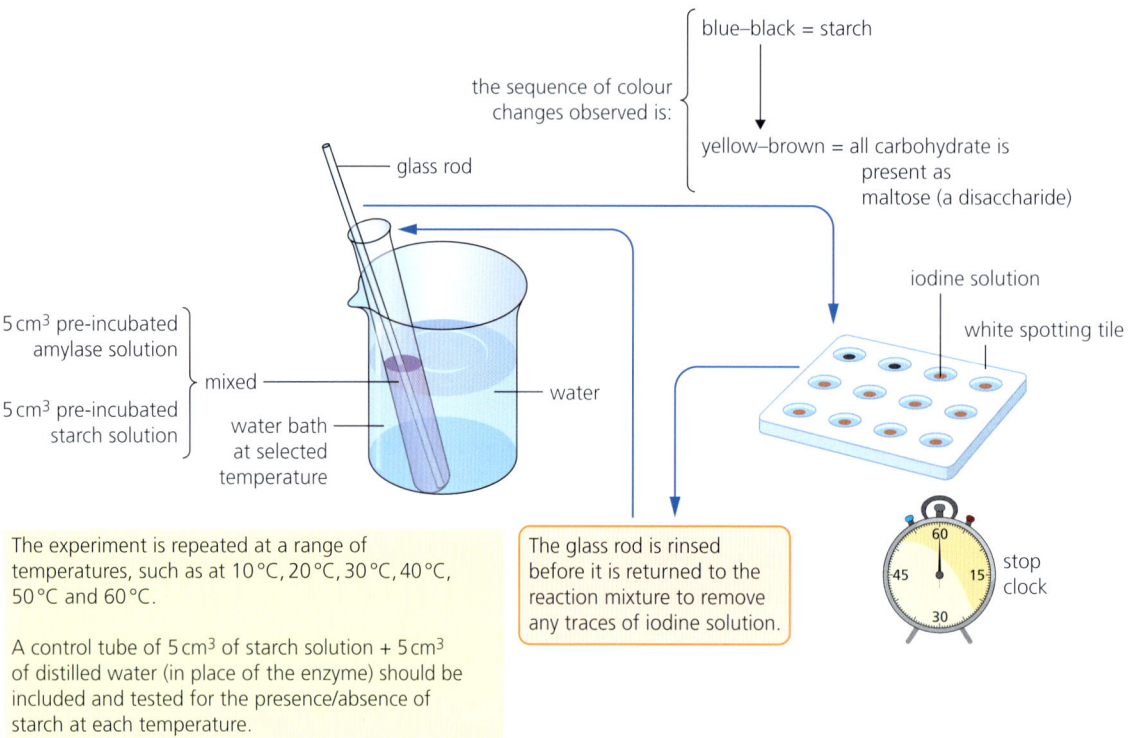

■ **Figure C1.1.9** The effects of temperature on hydrolysis of starch by amylase

 Interpreting graphs: the effect of temperature on the rate of enzyme activity

A characteristic curve is the result (Figure C1.1.10) – although the optimum temperature varies with the reaction and with different enzymes.
- As temperature is increased, molecules have increased kinetic energy and reactions between them go faster.
- The molecules are moving more rapidly and are more likely to collide and react. Q_{10}, the **temperature coefficient**, is a measure of the rate of change of a reaction when the temperature is increased by 10 °C (see Tools box, page 394). Many enzymes have a Q_{10} of about 2, which means that the rate of the reaction approximately doubles for every 10 °C rise in temperature.
- However, in enzyme-catalysed reactions the effect of temperature is more complex, as proteins are denatured by heat. The rate of denaturation increases at temperatures higher than the optimum. So, as the temperature rises, the amount of active enzyme progressively decreases, and the rate is slowed. As a result of these two effects of temperature on enzyme-catalysed reactions, there is an apparent optimum temperature for an enzyme. Humans have enzymes with an optimum temperature at or about normal body temperature.

● Common mistake

When explaining the increase in the reaction rate of an enzyme, it is insufficient to say 'because there are more collisions with the enzyme'. A better answer is: 'because there are more *frequent* collisions between the substrate and the active site'.

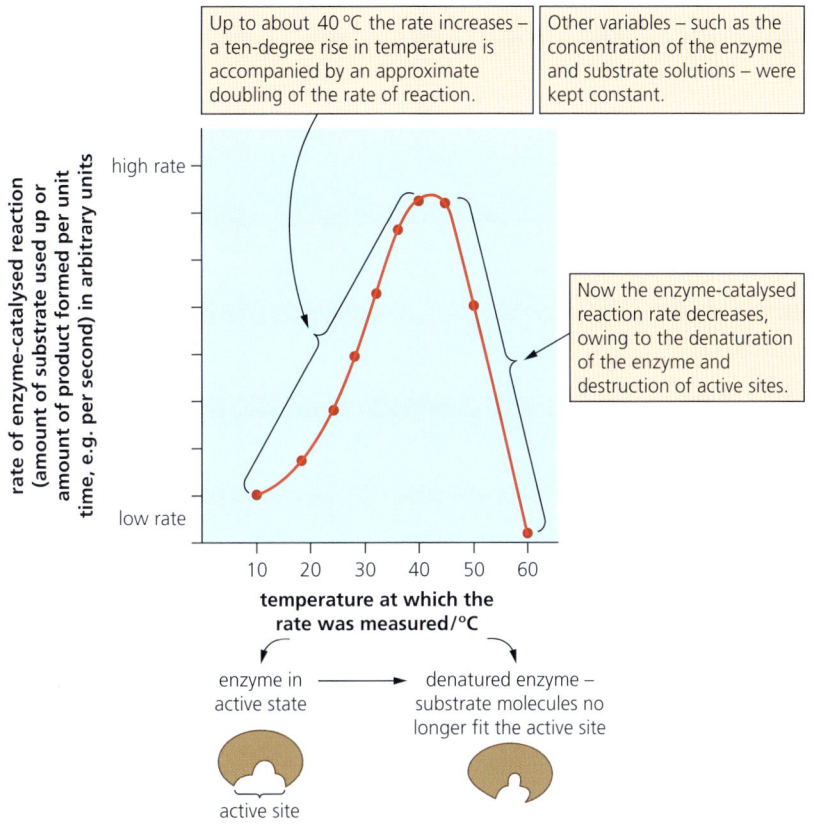

■ **Figure C1.1.10** Temperature and the rate of an enzyme-catalysed reaction

Not all enzymes have the same optimum temperature. For example, the bacteria in hot thermal springs have enzymes with optima between 80 °C and 100 °C or higher (see page 47), whereas seaweeds of northern seas and the plants of the tundra have enzymes with optimum temperatures closer to 0 °C. This feature of enzymes is often exploited in the commercial and industrial uses of enzymes.

6 Figure C1.1.10 shows the effect of temperature on the rate of an enzyme-catalysed reaction. **State** the optimum temperature of the enzyme.

7 In studies of the effect of temperature on enzyme-catalysed reactions, **suggest** why the enzyme and substrate solutions are pre-incubated to a particular temperature before they are mixed.

● **Top tip!**

When plotting data, ensure that the axes are correctly orientated, i.e. independent variable on the *x*-axis and the dependent variable on the *y*-axis. Ensure that the data points fill the area for the graph, and that an appropriate linear scale is chosen.

Tool 3: Mathematics

Processing uncertainties

Whenever a measurement is recorded, there will always be some doubt about the result that has been obtained. An uncertainty in a measurement is an interval that indicates a range within which we are confident that the true value lies. Scientific apparatus has a level of uncertainty. For example in Table C1.1.1, the thermometer has an uncertainty of ±0.1 °C, i.e. measurements may be + or − 0.1 above or below the recorded measurement. ➔

C1.1 Enzymes and metabolism

Plotting linear graphs showing the relationship between two variables

The following results were obtained in an investigation of the effects of pre-incubation of starch and amylase solutions at different temperatures on the subsequent hydrolysis of the starch to maltose.

1 **Draw** a graph of these results, including a curve of best fit, and **analyse** the trend shown.
2 **Suggest** the shape of the graph if the reciprocal of time (1/*t*) was plotted on the *y*-axis. This represents the relative rate of reaction.

■ Table C1.1.1 Time taken to hydrolyse starch to maltose at different temperatures

Temperature of solution/°C ± 0.1 °C	10.0	20.0	30.0	35.0	40.0	45.0	50.0	55.0	60.0
Time taken to hydrolyse starch to maltose/s ± 0.1 s	100.0	58.0	30.0	21.0	15.0	11.0	19.0	46.0	100.0

8 **Describe** the effect of temperature on enzyme-catalysed reactions.

9 A buffer is a solution that is used to control the pH of a process occurring in an experimental aqueous medium. **Suggest** why such solutions are often used in enzyme experiments.

ATL C1.1B

Read the following article: www.nature.com/news/1998/980917/full/news980917-7.html

Many organisms possess enzymes that work in extremes of temperature or pH. This *Nature* article discusses lactate dehydrogenase (LDH) in Antarctic fish. Can you find an example where an enzyme works in extremes of pH?

Effect of pH

Change in pH can have a dramatic effect on the rate of an enzyme-catalysed reaction (Figure C1.1.11). Each enzyme has a limited range of pH at which it functions efficiently. This is often at or close to the neutrality point (pH 7.0). This effect of pH occurs because the structure of a protein (and, therefore, the shape of the active site) is maintained by various bonds and weaker intermolecular forces within the three-dimensional structure of the protein. A change in pH from the optimum value alters the bonding patterns, progressively changing the three-dimensional shape of the molecule. The active site may be quickly rendered inactive. However, unlike the effects of temperature changes, the effects of pH on the active site are normally reversible – provided, that is, the change in surrounding acidity or alkalinity is not too extreme. As the pH reverts to the optimum for the enzyme, the active site may reform.

Interpreting graphs on the effect of pH on the rate of enzyme activity

Rates of enzyme reaction can be calculated for an enzyme at different pH (Figure C1.1.11). In these graphs, the maximum values are of interest because they represent the optimum pH for the enzyme. Rates either side of these values are non-optimum.

Some of the digestive enzymes of the gut have different optimum pH values from most other enzymes. For example, those adapted to operate in the stomach, where there is a high concentration of acid during digestion, have an optimum pH that is close to pH 2.0 (Figure C1.1.11).

10 Figure C1.1.11 shows the effect of pH on the rate of reaction of two enzymes. **Estimate** the optimum pH for trypsin.

11 **Sketch** a graph showing the effect of pH on an enzyme that has an optimal pH of 7.

Tool 3: Mathematics

Interpolating graphs

'Interpolate' means to make assumptions about the graph in-between the data points that have been collected. For example, in the graph for trypsin (Figure C1.1.11), the plotted data do not include the optimum; this is predicted to lie between two recorded values between pH 7 and 8, with the interpolate line showing it to be at approximately pH 7.8.

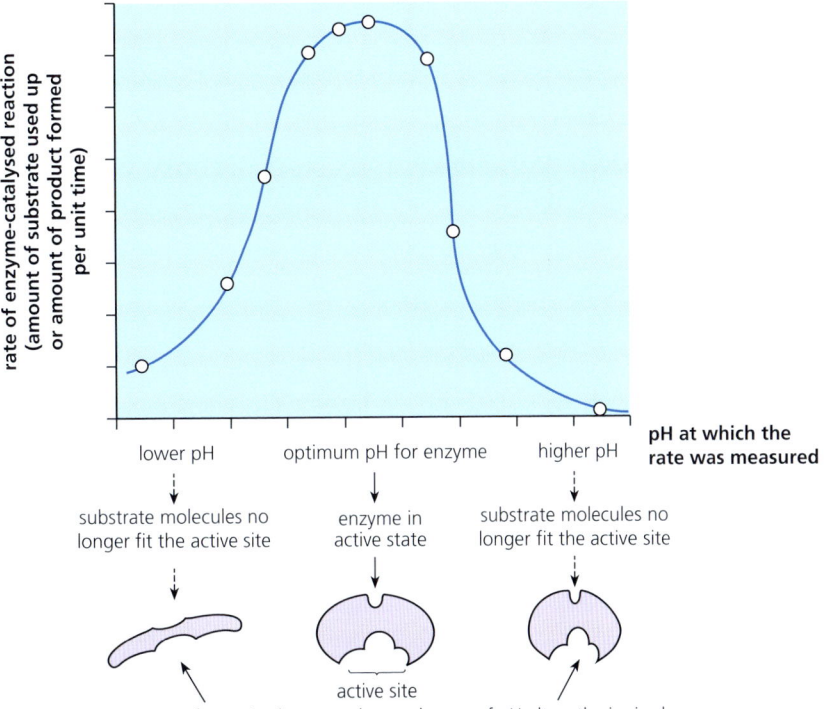

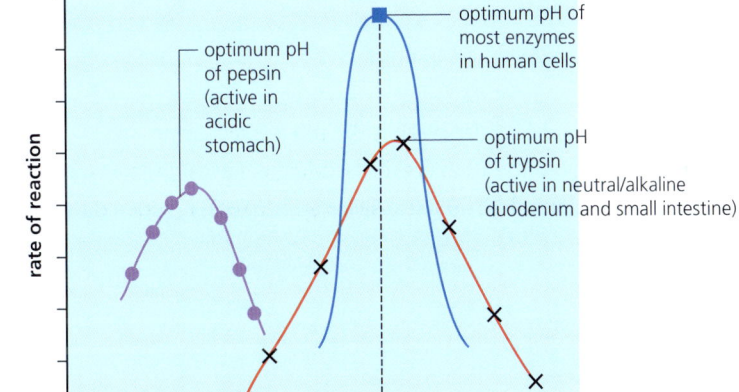

■ Figure C1.1.11 Effect of pH on enzyme shape and activity

Effect of substrate concentration

The effect of different concentrations of substrate on the rate of an enzyme-catalysed reaction can be shown using an enzyme called catalase. This enzyme catalyses the breakdown of hydrogen peroxide:

$$2H_2O_2 \xrightarrow{\text{catalase}} 2H_2O + O_2$$

Catalase occurs very widely in living things; it functions as a protective mechanism for the delicate biochemical machinery of cells. This is because hydrogen peroxide is a common by-product of reactions of metabolism, but it is also a very toxic substance since it is a powerful oxidizing agent (see page 412). Catalase inactivates hydrogen peroxide as it forms before any damage can occur.

C1.1 Enzymes and metabolism

When measuring the rate of enzyme-catalysed reactions, we measure the amount of substrate that has disappeared from a reaction mixture or the amount of product that has accumulated in a unit of time. For example, in Figure C1.1.10 (page 389) it is the rate at which the substrate starch disappears from a reaction mixture that is measured.

Interpreting graphs on the effect of substrate concentration on the rate of enzyme activity

Working with catalase, it is convenient to measure the rate at which the product (oxygen) accumulates – the volume of oxygen that has accumulated at 30-second intervals is recorded (Figure C1.1.12).

Over time, the initial rate of reaction is not maintained, but falls off quite sharply. This is typical of enzyme actions studied outside their location in the cell. The fall-off can be due to a number of reasons, but most commonly it is because the concentration of the substrate in the reaction mixture has fallen. Consequently, it is the initial rate of reaction that is measured. This is the slope of the tangent to the curve in the initial stage of reaction.

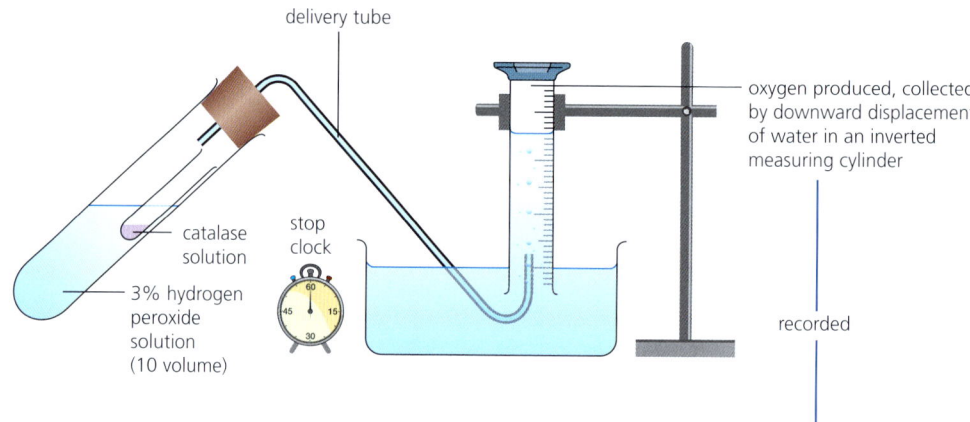

Tool 1: Experimental techniques

Accurately measuring volume

Figure C1.1.12 shows the volume of gas being measured using an inverted measuring cylinder. This is only applicable for gases that have low solubility in water (e.g. oxygen) and not applicable for gases that have high solubility in water. Volume of gas can also be measured using a gas syringe (see Figure C1.3.11, page 435).

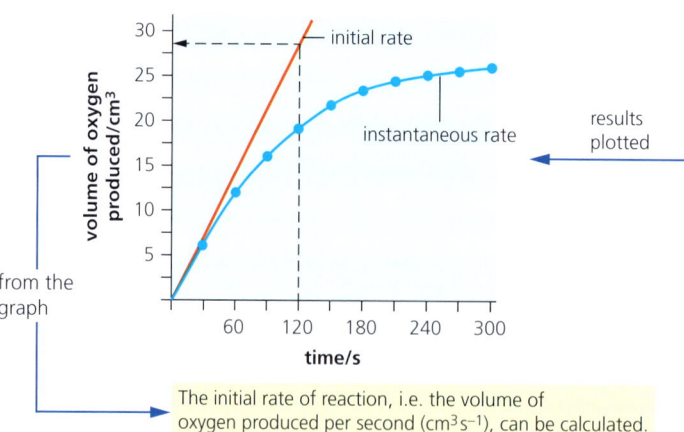

■ **Figure C1.1.12** Measuring the rate of reaction using catalase

Inquiry 1: Exploring and designing

Designing

In designing the experiment illustrated in Figure C1.1.12, **identify** and **justify** the choice of the dependent variable, the independent variable and the controlled variables.

Inquiry 3: Concluding and evaluating

Evaluating

Look at the experiment illustrated in Figure C1.1.12. Identify and discuss one likely potential source of error when the experiment is carried out. **Explain** how such an error is detected and what relevant improvements could be made to this experiment.

To investigate the effects of substrate concentration on the rate of an enzyme-catalysed reaction, the experiment shown in Figure C1.1.12 is repeated at different concentrations of substrate, and the initial rate of reaction is plotted in each case. Other variables, such as temperature and enzyme concentration, are kept constant.

When the initial rates of reaction are plotted against the substrate concentration, the curve shows two phases. At lower concentrations, the rate increases in direct proportion to the substrate concentration. At higher substrate concentrations, the rate of reaction becomes constant and shows no increase.

Now we can see that the enzyme catalase works by forming a short-lived enzyme–substrate complex. At a low concentration of substrate, all molecules can find an active site without delay. Effectively, there is an excess of enzyme present. The rate of reaction is set by how much substrate is present – as more substrate is made available, the rate of reaction increases. Increased concentrations of substrate mean that there are more collisions between substrate and the enzyme, with a higher likelihood of the substrate binding with the active site of the enzyme and enzyme–substrate complexes forming, causing the reaction to take place.

However, at higher substrate concentrations there comes a point when there is more substrate than enzyme. In effect, substrate molecules must 'queue up' for access to an active site. Adding more substrate increases the number of molecules awaiting contact with an enzyme molecule, so there is now no increase in the rate of reaction (Figure C1.1.13).

12 When there is an excess of substrate present in an enzyme-catalysed reaction, **explain** the effect on the rate of reaction of increasing the concentration of:
   a the substrate
   b the enzyme.

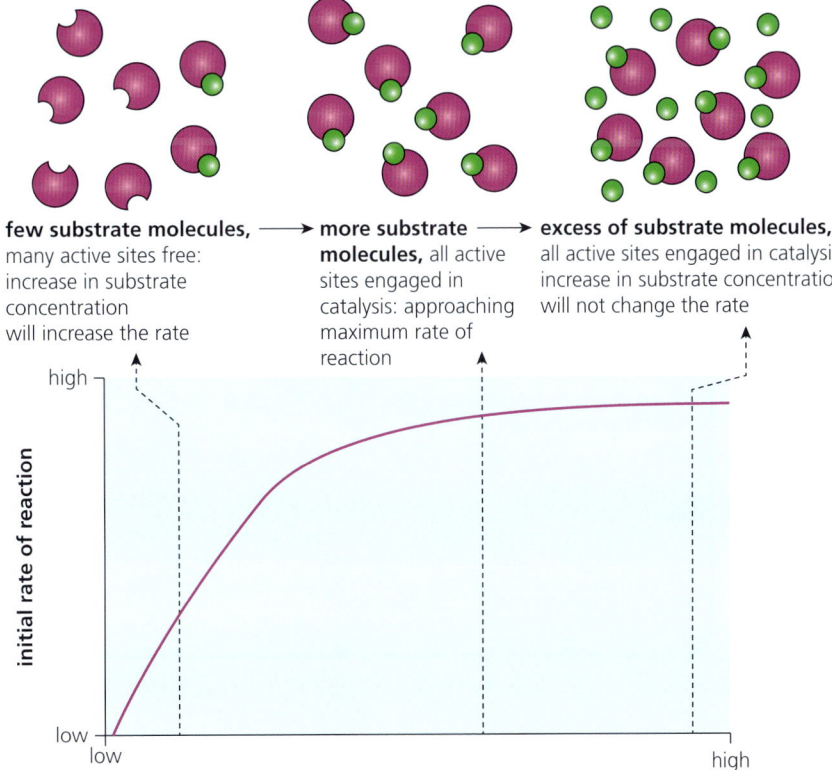

■ **Figure C1.1.13** The effect of substrate concentration

C1.1 Enzymes and metabolism

> ### Inquiry 2: Collecting and processing data
>
> **Collecting data**
>
> Access this site: **https://exchange.iseesystems.com/public/jondarkow/lactase-with-variable-sample-sizes/index.html#page1**
>
> This is a simulation of experiments with the enzyme lactase, which hydrolyzes lactose to glucose and galactose. Click on the 'Simulate' tab to access the simulation. Keep the pH and temperature at the default settings of pH 7 and 25 °C. Leave the lactase enzyme concentration at 5 mM (millimoles) for this and all your experiments. Set the sample size to 10.
>
> Run the simulation at initial lactose (the substrate) concentrations of 10, 20, 40, 60, 80, 100 and 120 mM. After each simulation at each substrate concentration, click on the 'mean of samples' tab. Here you will find a graph of the average amount of glucose produced over time. On the graph at the 10-minute mark, determine the value of the product (glucose) concentration (mM). Since the simulation begins with a glucose concentration of zero, the value of glucose concentration at 10 minutes divided by 10 is also the slope or initial velocity of the reaction (mM min^{-1}). Carry out further experiments to determine the effect of pH and temperature on lactase rate.

Measurements in enzyme-catalysed reactions

> ### Tool 3: Mathematics
>
> **Calculating rates of change from graphical and tabulated data**
>
> The rate of reaction is measured by dividing a value of the dependent variable by time. In Figure C1.1.12, the initial rate is taken by extrapolating a straight line from the first section of the curve. The rate of reaction is calculated by dividing a value of the dependent variable (volume of oxygen produced) taken from along this line by the corresponding time, as shown by the dashed lines in the graph. Relative rates of reaction can be calculated using the reciprocal of time ($1/t$).
>
> For example, in the experiment on the effect of temperature on starch digestion by amylase (page 390), relative rates can be calculated:
>
> | Temperature of solution /°C ± 0.1 °C | 10.0 | 20.0 | 30.0 | 35.0 | 40.0 | 45.0 | 50.0 | 55.0 | 60.0 |
> |---|---|---|---|---|---|---|---|---|---|
> | Time taken to hydrolyse starch to maltose/s ± 0.1 s | 100.0 | 58.0 | 30.0 | 21.0 | 15.0 | 11.0 | 19.0 | 46.0 | 100.0 |
> | Relative rate of reaction ($1/t$) s^{-1} | 0.010 | 0.017 | 0.033 | 0.048 | 0.067 | 0.091 | 0.053 | 0.022 | 0.010 |
>
> These data show the rate of reaction increasing from 10 to 45 °C, and then decreasing to the original rate at 60 °C.

13 Calculate the initial rate of reaction in Figure C1.1.12.

14 Explain the factors that affect the rate of enzyme-controlled reactions in cells.

15 List three factors that can affect the rate of enzyme reactions.

Effect of enzymes on activation energy

We visualize an enzyme (**E**) as a large molecule that works by reacting with another compound or compounds, the substrate (**S**). Initially, a short-lived enzyme–substrate complex (**ES**) is formed at the active site. This complex exists at a local energy minimum and is quite stable. The transition state (TS) is the point where there is a maximum value of energy. The transition state exists at the **top of the energy profile and is transient**. Almost instantly, the product (**P**) is formed and the enzyme is released unchanged. The enzyme immediately takes part in another reaction. We represent this reaction as follows:

$$E + S \rightarrow [ES] \rightarrow ES^\ddagger \rightarrow P + E$$

where $ES^\ddagger$ is the transition state.

◆ **Activation energy**: energy required by a substrate molecule before it can undergo a chemical change.

Energy is released when the 'substrate' becomes the 'product'. However, to bring about the reaction, a small amount of energy is needed initially to break or weaken bonds in the substrate, to form the transition state. This energy input is called the **activation energy** (Figure C1.1.14). It is a small but significant energy barrier that must be overcome before the reaction can happen. Enzymes work by lowering the amount of energy required to activate the reacting molecules by providing a new, alternative reaction pathway.

Another model of enzyme catalysis includes a boulder (substrate) perched on a slope, prevented from rolling down by a small hump (representing activation energy). The boulder can be pushed over the hump, or the hump can be dug away to lower it (= lowering the activation energy), allowing the boulder to roll down and shatter at a lower level (giving products).

Energy is needed to break the bonds within the substrate. When bonds are made from the products of an enzyme-catalysed reaction, there is an energy yield. You should be able to interpret graphs showing this effect, for example Figure C1.1.14.

Note: virtually all enzyme-catalysed reactions still have activation energy barriers, they are just smaller. The enzyme lowers the 'hump' of activation energy, it does not remove it.

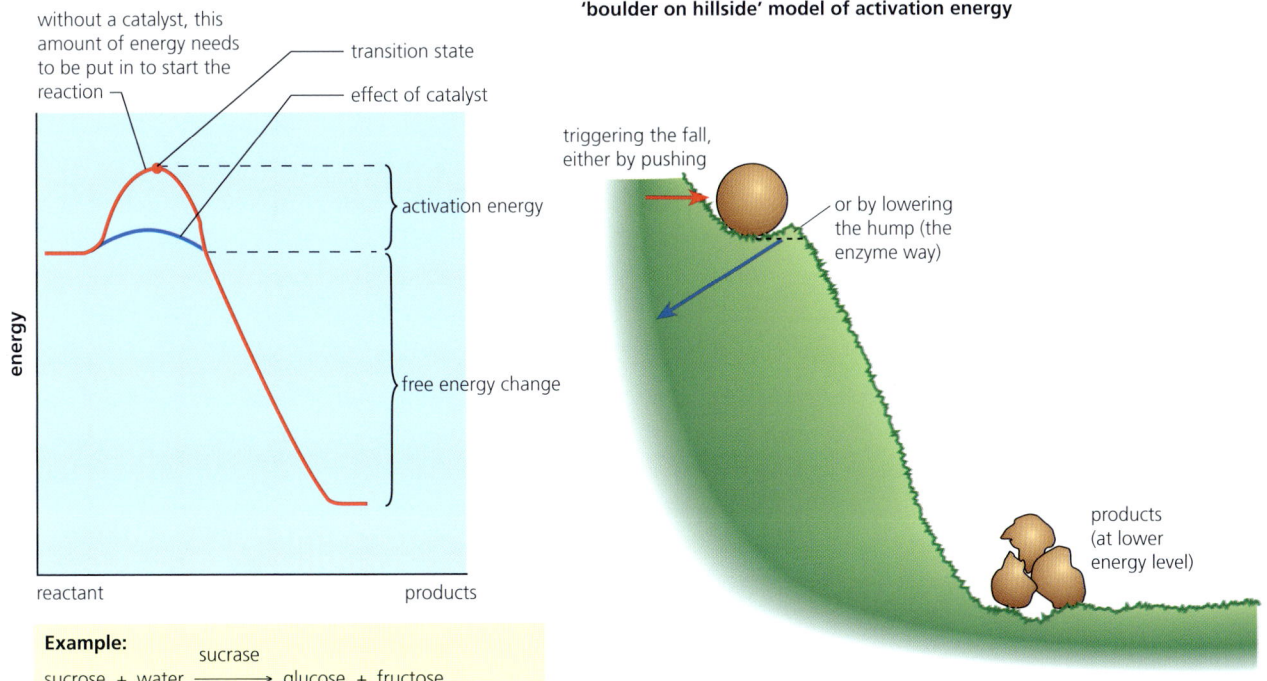

■ **Figure C1.1.14** Activation energy

C1.1 Enzymes and metabolism

16 **Sketch** a graph showing the effect of an enzyme on the activation energy of a metabolic reaction.

17 **Define** the term *activation energy*.

ATL C1.1C

Find out more about activation energy here:
https://ed.ted.com/lessons/activation-energy-kickstarting-chemical-reactions-vance-kite

Produce a cartoon or poster, using ideas from this book and from the animation, to summarize the role of enzymes in reducing activation energy in metabolic reactions. Producing your own visual summaries of important biological concepts can help you understand and remember them.

Top tip!

Molecules collide more frequently at higher temperatures; however, the main reason why the reaction speeds up is that more molecules have enough energy to get over the activation energy.

Intracellular and extracellular enzyme-catalysed reactions

Some enzymes are exported from cells, such as the digestive enzymes. Enzymes like these are put into vesicles, secreted by **endocytosis** and work externally. They are called **extracellular enzymes**. Chemical digestion in the gut is an example of an extracellular reaction.

However, most enzymes remain within cells and work there. These are the **intracellular enzymes**. Many are found inside organelles and in the membranes of organelles, in the fluid medium around the organelles and in the plasma membrane. Many are also in the cytoplasm (e.g. glycolysis, page 414). Two of the main metabolic processes in respiration are glycolysis and the Krebs cycle. These are intracellular enzyme-catalysed reactions.

♦ **Extracellular enzyme**: an enzyme secreted by a cell that functions outside the cell.

♦ **Intracellular enzyme**: an enzyme that functions within the cell in which it was produced.

Link
Respiration and the Krebs cycle are covered in detail in Chapter C1.2, page 405.

18 **Distinguish** between intracellular and extracellular enzymes.

Generation of heat energy by the reactions of metabolism

When glucose is oxidized to carbon dioxide and water in aerobic cell respiration, energy is transferred from the store of chemical potential energy to heat energy (i.e. the kinetic energy of molecular motion). This energy is no longer in store but is on the move; it is active energy. Only part of the stored energy in a molecule is available. This is known as free energy and can be used to do work. Reactions that release free energy are known as **exergonic reactions** (Figure C1.1.15). The oxidation of glucose is an example of an exergonic reaction.

Heat generation is inevitable in metabolic reactions. Exergonic reactions involve the release of heat because metabolic reactions are not 100% efficient in energy transfer. Mammals, birds and some other animals depend on this heat production for maintenance of constant body temperature and are said to be endotherms (page 766) or 'warm-blooded'. Other animals cannot control their cell metabolism in this way and are described as ectotherms ('cold-blooded').

Link
Energy transfer is covered in Chapter C4.2, page 578. The maintenance of constant body temperature is covered in Chapter D3.3, page 766.

On the other hand, reactions in which energy is absorbed, and there is more energy in the system at the end of the reaction than at the beginning of it, are called **endergonic reactions**. The synthesis of a protein from amino acids is an example of an endergonic reaction.

Theme C: Interaction and interdependence – Molecules

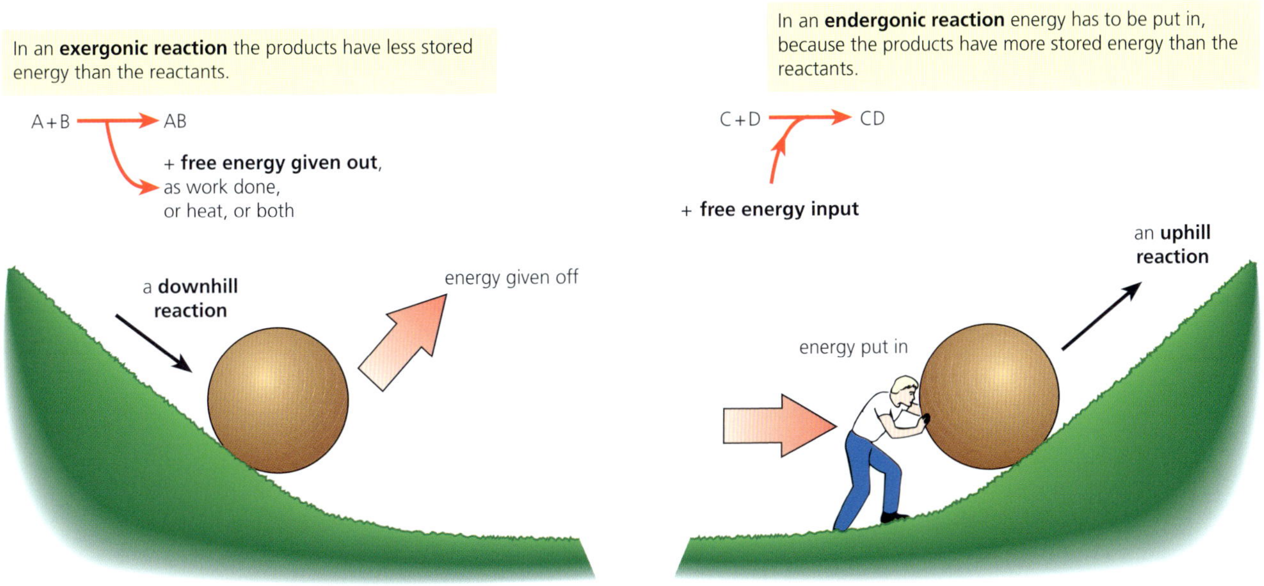

In an **exergonic reaction** the products have less stored energy than the reactants.

A + B ⟶ AB
+ **free energy given out**, as work done, or heat, or both

a **downhill reaction**

energy given off

In an **endergonic reaction** energy has to be put in, because the products have more stored energy than the reactants.

C + D ⟶ CD
+ **free energy input**

an **uphill reaction**

energy put in

■ Figure C1.1.15 Exergonic and endergonic reactions

Link
The definition and structure of ATP are detailed in Chapter B2.1, Figure B2.1.12, page 232.

The many endergonic reactions that occur in metabolism are made possible by being coupled to exergonic reactions. Coupling occurs through **adenosine triphosphate (ATP)** (the energy currency molecule in biology). Molecules of ATP work in metabolism by acting as common intermediates, linking energy-absorbing and energy-yielding reactions. Metabolic processes mostly involve ATP, directly or indirectly.

Cyclical and linear pathways in metabolism

Metabolism consists of series of reactions in which the product of one reaction is an intermediate of the next, and so on. Many pathways consist of **linear metabolic pathways** (that is straight chains) of reactions, while others are **cyclical metabolic pathways** (Figure C1.1.16). We shall see examples of both types of **metabolic pathway** within the reactions of respiration, a catabolic process (page 405), and photosynthesis, an anabolic process (page 425).

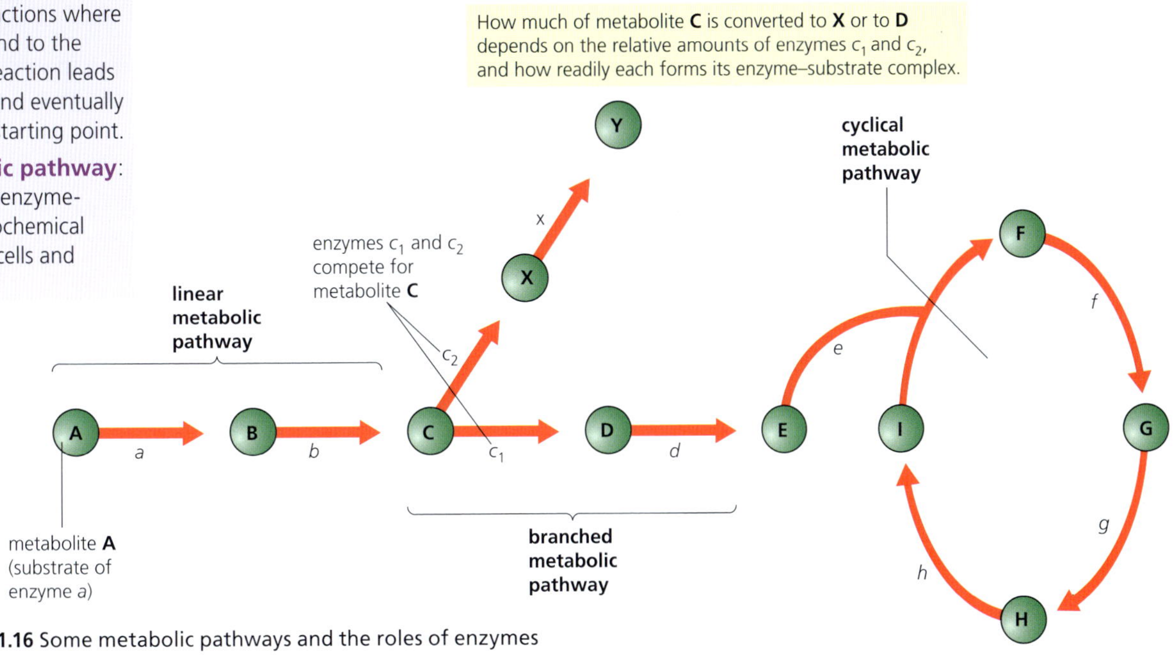

◆ **Linear metabolic pathway**: a series of enzyme-catalysed reactions that run in one direction, from reactant to product.

◆ **Cyclical metabolic pathway**: a circular series of enzyme-catalysed reactions where there is no end to the series; one reaction leads to the next and eventually back to the starting point.

◆ **Metabolic pathway**: sequence of enzyme-catalysed biochemical reactions in cells and tissues.

How much of metabolite **C** is converted to **X** or to **D** depends on the relative amounts of enzymes c_1 and c_2, and how readily each forms its enzyme–substrate complex.

■ Figure C1.1.16 Some metabolic pathways and the roles of enzymes

C1.1 Enzymes and metabolism

In respiration, glycolysis is a linear metabolic pathway (page 414) and the Krebs cycle a cyclical one (page 418). Photosynthesis has two sets of reactions – the light-dependent reactions and the light-independent reactions (page 439). The light-independent reactions are driven by the Calvin cycle (page 445) – another example of a cyclical pathway.

TOK

You will have noticed that the study of biology requires some understanding of chemistry. Chemical theory is used to explain the inner workings of biological phenomena. Do you think this means that chemistry is a body of knowledge that is somehow more foundational than biology? Can all the facts of biology be translated into facts about chemistry?

You might also ask similar questions in relation to other areas of knowledge (AOKs): does physics describe the world better than chemistry? Can art describe facts about human experience in a preferable way to history or biology?

Thinking about the scope of the AOKs is a helpful tool in TOK. The scope element of the TOK Knowledge Framework encourages you to think about the specific questions that one AOK will ask rather than another. Biology, for instance, might be more interested in identifying and explaining facts about whole organisms or systems, while chemistry focuses more on chemical and atomic interactions. That these different types of knowledge overlap and inform one another is crucial to understanding any one of them, but being able to explore them as distinct knowledge communities is a helpful TOK strategy.

19 Compare and contrast cyclical and linear pathways in metabolism.

Enzyme inhibitors

The actions of enzymes may be inhibited by other molecules, some formed in the cell and others absorbed from the external environment. These substances are known as **enzyme inhibitors** since their effect is generally to lower the rate of reaction.

◆ **Enzyme inhibitor**: a substance that slows or blocks enzyme action.

◆ **Competitive inhibitor**: a substance that binds to the active site of an enzyme, slowing or blocking enzyme action.

◆ **Non-competitive inhibitor**: a substance that does not bind to the active site but to another part of the enzyme, slowing or blocking enzyme action.

Top tip!

Studying the effects of inhibitors has helped our understanding of:
- the chemistry of the active site of enzymes
- the natural regulation of metabolism and which pathways operate
- the ways certain commercial pesticides and drugs work, namely by inhibiting specific enzymes and preventing particular reactions.

For example, molecules that sufficiently resemble the substrate in shape may compete to occupy the active site. These are known as **competitive inhibitors**. The enzyme that catalyses the reaction between carbon dioxide and the acceptor molecule in photosynthesis is known as ribulose bisphosphate carboxylase (rubisco, page 445) and is competitively inhibited by oxygen in the chloroplasts.

Alternatively, an inhibitor may be unlike the substrate molecule, yet still combine with the enzyme. In these cases, the attachment occurs at some other part of the enzyme, probably quite close to the active site. Here, the inhibitor either partly blocks access to the active site by substrate molecules, or it causes the active site to change shape and so be unable to accept the substrate. These are called **non-competitive inhibitors** since they do not compete for the active site. Adding excess substrate does not overcome their inhibiting effects (Figure C1.1.17).

Link

Rubisco is covered in more detail in Chapter C1.3, page 445.

Examine the graph in Figure C1.1.17 carefully.

You can see that when the initial rates of reaction of an enzyme are plotted against substrate concentration, the effects of competitive and non-competitive inhibitors are clearly different.

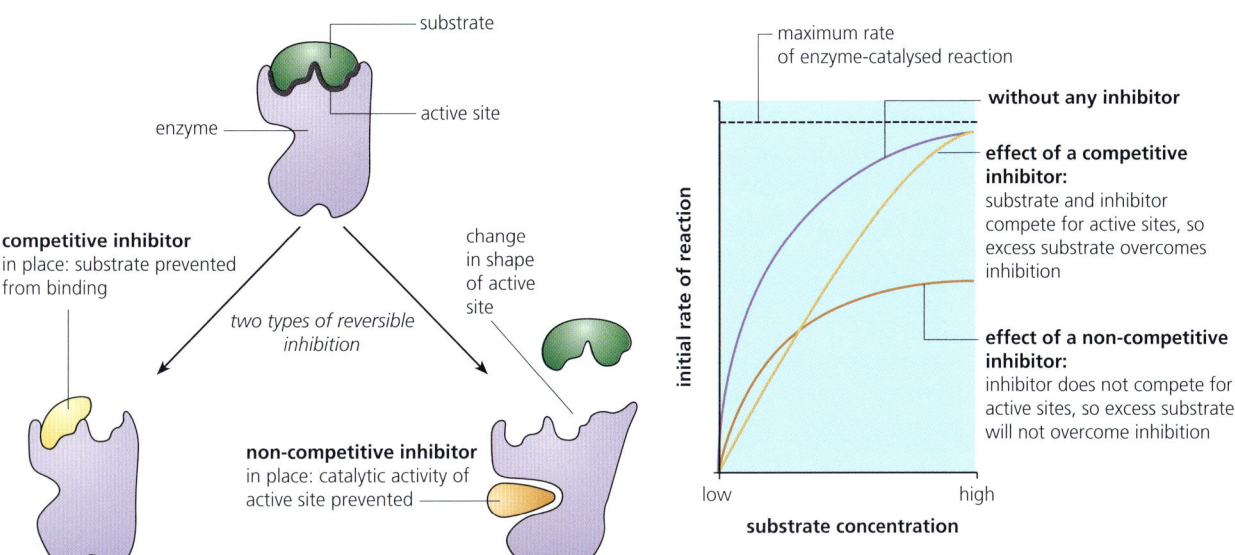

■ **Figure C1.1.17** Competitive and non-competitive inhibitors

The difference between competitive and non-competitive inhibition is in the interactions between substrate and inhibitor and, therefore, in the effect of substrate concentration. In non-competitive inhibition, the inhibitor does not compete for active sites and so excess substrate will not overcome inhibition. In competitive inhibition, substrate and inhibitor compete for active sites and so excess substrate will overcome inhibition.

■ Allosteric sites and non-competitive inhibition

Allosteric regulators are molecules that change the shape and activity of an enzyme by reversibly binding at a site on the enzyme, typically some distance from the active site. Binding of an allosteric activator temporarily stabilizes the enzyme shape as an active and effective catalyst. Binding of an allosteric inhibitor changes the enzyme shape into an inactive form.

♦ **Allosteric regulators**: molecules that change the shape and activity of an enzyme by reversibly binding at a site on the enzyme.

We can see allosteric regulation as a form of reversible non-competitive inhibition or activation of an enzyme. Subtle fluctuations in the concentrations of activators and inhibitors may fine-tune the activity of a critical pathway as constantly changing conditions or requirements of cell metabolism demands.

Only specific substances known as effectors can bind to an allosteric site. Binding causes interactions within an enzyme that lead to conformational changes, which alter the active site enough to prevent catalysis. Binding is reversible.

An investigation of the initial rate of reaction of the enzyme catalase in the presence and absence of heavy metal ions (Cu^{2+})

The enzyme catalase catalyses the breakdown of hydrogen peroxide (see page 391 earlier in the chapter):

$$2H_2O_2 \xrightarrow{\text{catalase}} 2H_2O + O_2$$

C1.1 Enzymes and metabolism

The **rate of an enzyme-catalysed reaction** is the amount of substrate that has disappeared from a reaction mixture, or the amount of product that has accumulated, in a period of time. For example, with catalase it is convenient to measure the rate at which the product (oxygen) accumulates. In this experiment, an indication of the volume of oxygen that was released at half-minute intervals was recorded at each substrate concentration.

When these 'raw' results for each concentration of hydrogen peroxide were plotted on a graph, it was found that, over time, the **initial rate** of reaction was not maintained; in fact, it decreased quite quickly. This is typical of enzyme actions studied outside their location in the cell. The reduction in rate can be due to a number of reasons, but most commonly it is because the concentration of the substrate in the reaction mixture has fallen. Consequently, it is the initial rate of reaction that is determined. This is the slope of the tangent to the curve in the initial stage of reaction. How this is calculated is shown in Figure C1.1.12 (page 392).

Look at this account of how an initial reaction rate is calculated now.

The apparatus used is illustrated in Figure C1.1.18 and the reaction mixtures in this experiment are tabulated below. Note that:

1 A yeast suspension was used as the source of the enzyme. The number of bubbles released at half-minute intervals was counted and recorded for six minutes with each reaction mixture. (The initial rush of bubbles when the yeast was injected was discounted. The depth of the bubble nozzle was the same in each experiment.)

2 The concentration of a hydrogen peroxide solution is given as the volume of oxygen that can be released. For example, a 20-volume solution will, when completely decomposed, give 20 times its own volume of oxygen.

3 A duplicate investigation was carried out in the presence of a dilute solution of copper(II) (Cu^{2+} (aq)) ions.

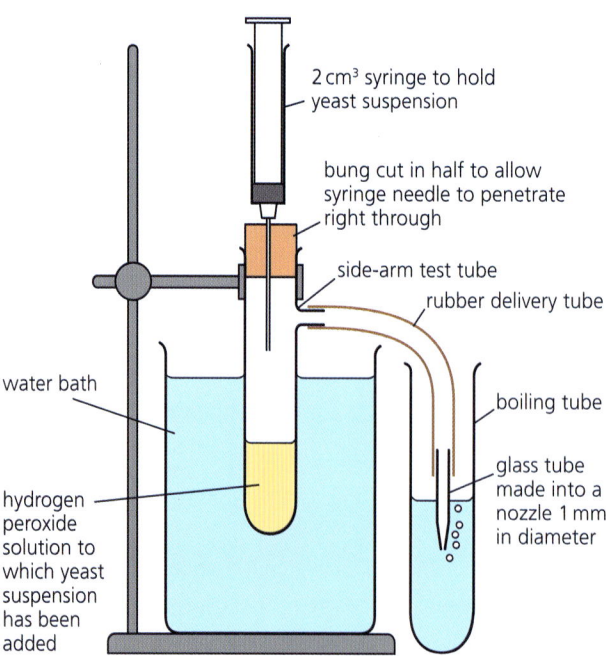

■ **Figure C1.1.18** Apparatus for monitoring the effects of substrate concentration on the action of catalase

20 **Construct** a graph showing the initial rates of reaction of the enzyme catalase over the substrate concentration range of 4–14 vol hydrogen peroxide in the presence and absence of heavy metal ions.

21 **Explain** to what extent these data support the hypothesis that copper(II) ions are a non-competitive inhibitor of the enzyme that decomposes hydrogen peroxide.

The reaction mixtures used, and the results obtained, are shown in Tables C1.1.2 and C1.1.3. *Examine them and then answer the questions.*

■ **Table C1.1.2** Effect of substrate concentration on enzyme-catalysed reaction

Experiment	1	2	3	4	5	6
volume distilled water (cm³)	4.0	3.5	3.0	2.5	2.0	1.5
volume 20-vol H_2O_2 (cm³)	1.0	1.5	2.0	2.5	3.0	3.5
concentration of H_2O_2 (vol)	4.0	6.0	8.0	10.0	12.0	14.0
volume yeast suspension (cm³)	1.0	1.0	1.0	1.0	1.0	1.0
initial rate of reaction (bubbles/min)	0.5	17.0	24.0	27.0	30.0	32.0

■ **Table C1.1.3** Effect of substrate concentration on enzyme-catalysed reaction in presence of heavy metal ions

Experiment	1	2	3	4	5	6
volume distilled water (cm³)	3.9	3.4	2.9	2.4	1.9	1.4
volume 0.1 M copper(II) (Cu^{2+}) solution (cm³)	0.1	0.1	0.1	0.1	0.1	0.1
volume 20-vol H_2O_2 (cm³)	1.0	1.5	2.0	2.5	3.0	3.5
concentration of H_2O_2 (vol)	4.0	6.0	8.0	10.0	12.0	14.0
volume yeast suspension (cm³)	1.0	1.0	1.0	1.0	1.0	1.0
initial rate of reaction (bubbles/min)	0.2	8.0	14.0	15.0	15.4	15.6

Competitive inhibition

Competitive inhibition is a consequence of an inhibitor binding reversibly to an active site (see above).

Statins are an example of competitive inhibitors. They are competitive inhibitors of an enzyme crucial for the synthesis of cholesterol: 3-hydroxy-3-methylglutaryl coenzyme A (HMG-CoA) reductase (HMGR) (see Figure C1.1.19). Statins are similar in structure to the substrate of HMGR, HMG-CoA. Statins can therefore occupy a portion of the binding site of HMG-CoA, blocking access of this substrate to the active site of the enzyme (Figure C1.1.20).

◆ **Statin**: any of a class of drugs that lower the level of low-density lipoproteins in the blood by inhibiting the activity of an enzyme involved in the production of cholesterol in the liver.

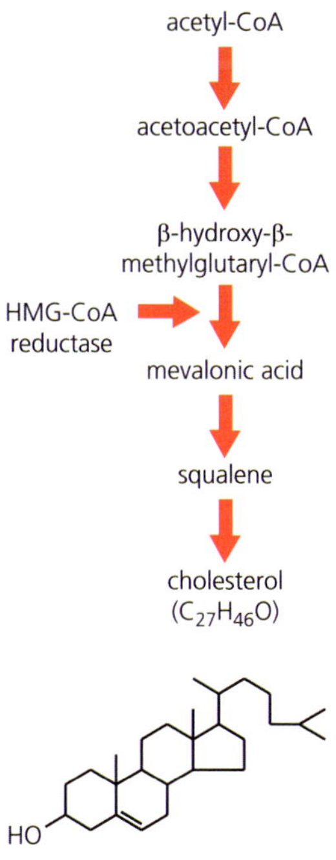

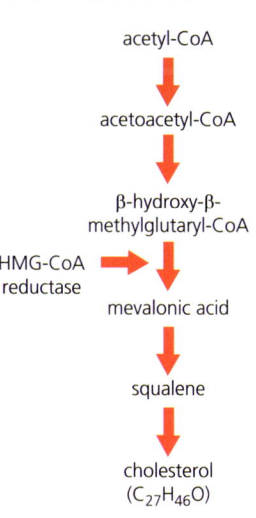

■ **Figure C1.1.19** The metabolic pathway of cholesterol synthesis

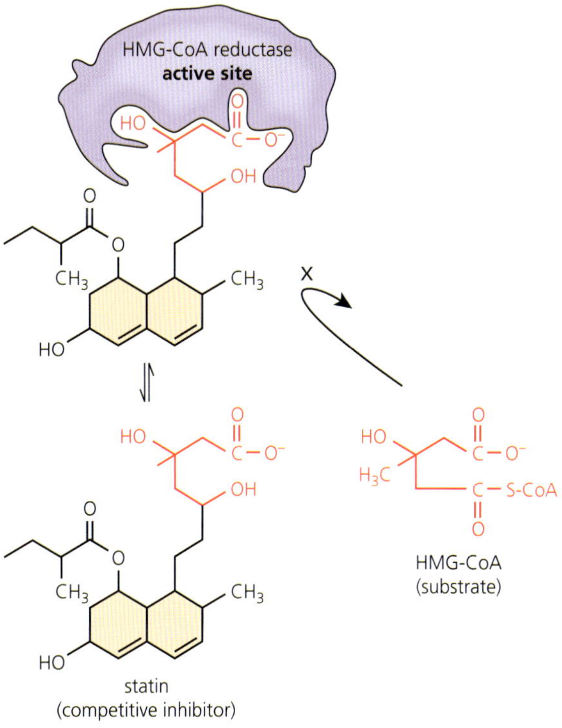

● **Top tip!**

You will not be expected to recall the names of individual chemicals in these reactions. They are used here to explain the mechanism of enzyme inhibition.

■ **Figure C1.1.20** Statins compete with HMG-CoA for the active site of HMG-CoA reductase (X indicates that site is blocked to substrate)

C1.1 Enzymes and metabolism

Health consequences of the lipid content of diets

The lipid cholesterol is a component of the diet, particularly when animal fats are present. This lipid is a steroid (see Chapter B2.1, page 237).

There are two forms of cholesterol carried in the blood:
- low-density lipoproteins (LDLs), complexes of thousands of cholesterol molecules bound to proteins in particles ('bad cholesterol')
- high-density lipoproteins (HDLs), or 'good cholesterol'.

Excess blood cholesterol, present as LDLs, may cause atherosclerosis – the progressive degeneration of the artery walls (page 300). A direct consequence of this is an increased likelihood of the formation and circulation of blood clots.

Statins are taken to reduce LDLs in the blood and therefore reduce the risk of coronary heart disease and other effects of high levels of cholesterol in the blood.

22 Distinguish between competitive and non-competitive enzyme inhibition.

ATL C1.1D

Many drugs are enzyme inhibitors, for example angiotensin-converting enzyme (ACE) inhibitors, which are used to treat hypertension. Research this example or generate a list of other medicines that work as enzyme inhibitors, such as the antibacterial effects of penicillin or sulfonamide.

◆ **End-product inhibition**: when the product of the last reaction in a metabolic pathway inhibits the enzyme that catalyses the first reaction of the pathway.

■ Regulation of metabolic pathways by feedback inhibition

Individual pathways in metabolism may be switched off by the final product acting as a reversible inhibitor of the enzyme that catalyses the first step in the pathway.

In **end-product inhibition**, as the product molecules accumulate, the steps in their production are switched off (Figure C1.1.21). However, these product molecules may now become the substrate in subsequent metabolic reactions. If so, the accumulated product molecules will be removed and production of new product molecules will recommence.

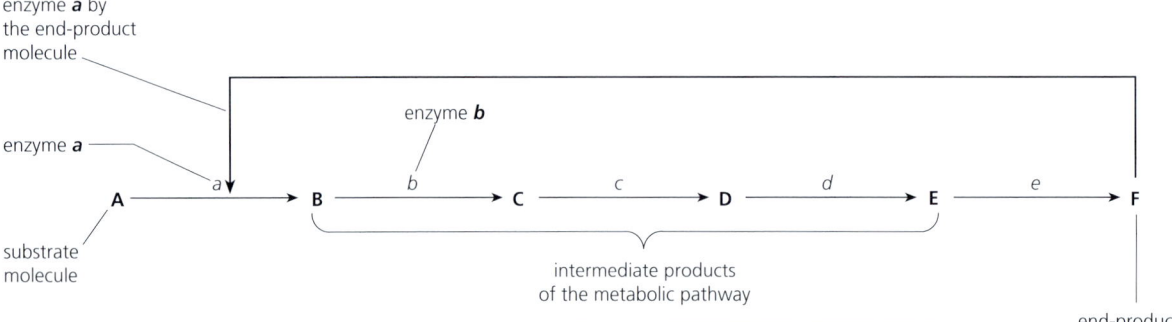

This is an example of the regulation of a metabolic pathway by **negative feedback**.

■ **Figure C1.1.21** Regulation of a metabolic pathway by end-product inhibition

● Common mistake

A common misconception is that a competitive inhibitor of an intermediate enzyme in a pathway would prevent any final product from being formed. This is unlikely because the substrate of the inhibited enzyme would still sometimes manage to bind to the active site and there would be some final product.

End-product inhibition of the pathway that converts threonine to isoleucine

Bacteria can synthesize isoleucine from threonine. Figure C1.1.22 shows the metabolic pathway for the synthesis of isoleucine.

- Isoleucine acts as a non-competitive inhibitor by binding to the allosteric site of the enzyme threonine deaminase.
- Threonine deaminase is an essential enzyme in the first stage of the metabolic pathway – its inhibition turns off isoleucine production. This regulates the production of isoleucine.
- Initially, when isoleucine concentration is still low, the metabolic pathway can proceed as non-competitive inhibition is low.
- As isoleucine concentration increases, non-competitive inhibition takes place and the metabolic pathway is regulated.
- As isoleucine is used in the cell for protein synthesis, its concentration falls and the allosteric sites of threonine deaminase are no longer occupied, so the enzyme can once again act in the conversion of threonine to isoleucine.

> **Top tip!**
> Isoleucine is an essential amino acid: it cannot be made by the human body and so must be consumed from food.

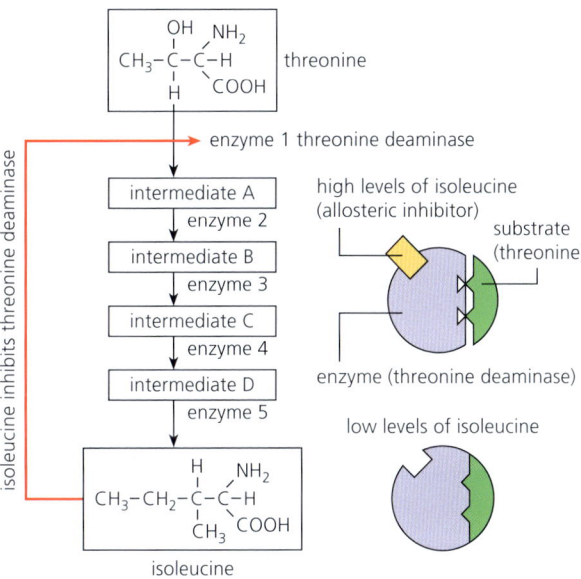

■ **Figure C1.1.22** Isoleucine inhibits threonine deaminase, acting as a non-competitive inhibitor

Mechanism-based inhibition

A substrate analogue is a molecule that has a similar structure to the substrate, allowing it to bind covalently to the active site of the enzyme. When it binds to the active site, the substrate analogue can be modified by the enzyme to produce a reactive group, which reacts irreversibly to form a stable inhibitor–enzyme complex. This is known as **mechanism-based inhibition**.

Penicillin was the first group of antibiotics developed to fight bacterial disease (see page 532).

But what is the mechanism by which penicillin kills bacteria?

The cell walls of bacteria are made from **peptidoglycan** (also known as murein), which is formed from polysaccharide chains linked by short peptides (Figure C1.1.23). The bacterial enzyme **DD-transpeptidase** forms cross-links between these strands, strengthening the structure. Penicillin inhibits DD-transpeptidase. By inhibiting transpeptidase, the cell walls are not cross-linked, so they are weak. Without a strong cell wall, a bacterial cell is vulnerable to the movement of water into the cell by osmosis, which kills the bacteria by lysis (the increased pressure of water inside the cell causes the membrane to burst).

◆ **Mechanism-based inhibition**: process that occurs when unreactive molecules are transformed into an active form through catalytic reactions; these active forms inhibit the enzyme, typically through covalent modification of the active site. It is an irreversible form of enzyme inhibition.

Link
Antibiotics are covered in more detail in Chapter C3.2, page 532.

How does penicillin inhibit DD-transpeptidase?

The structure of penicillin resembles parts of the growing peptide chain of the peptidoglycan cell wall. Instead of binding to a peptide chain, the transpeptidase binds to penicillin; the structure of the penicillin is modified by the enzyme, forming an inactive enzyme–penicillin complex, which blocks the formation of further cross-linkages (Figure C1.1.23). The penicillin substrate is modified by its interaction with the transpeptidase, and an irreversible form of enzyme inhibition occurs. This is an example of mechanism-based inhibition.

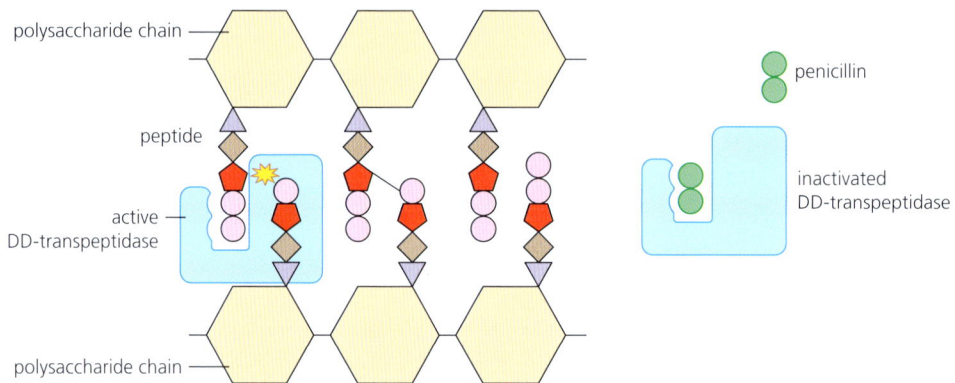

■ **Figure C1.1.23** Penicillin blocks the formation of peptide cross-links between polysaccharide chains of peptidoglycan

Link

Bacterial resistance to antibiotics is covered in more detail in Chapter C3.2, page 534.

Bacteria can become resistant to antibiotics such as penicillin. Mutations in the bacterial genome can lead to changes in the active site of some transpeptidases, causing the penicillin to reduce or lose its affinity with the enzyme, promoting resistance to the antibiotic (Figure C1.1.24). When mutations occur in plasmid DNA, resistance can quickly spread through bacteria through the process of conjugation (see page 119).

■ **Figure C1.1.24** Genetic mutations can result in structural changes in the active site of transpeptidases, leading to antibiotic resistance

ATL C1.1E

Many metabolic poisons are enzyme inhibitors. Research the example of α-amantin as an RNA polymerase inhibitor, or the use of acetylcholinesterase inhibitors as insecticides or nerve gas. Concepts you will encounter will link to Chapters D1.2 and C2.1, respectively.

LINKING QUESTIONS

1. What are examples of structure–function relationships in biological macromolecules?
2. What biological processes depend on differences or changes in concentration?

C1.2 Cell respiration

Guiding questions
- What are the roles of hydrogen and oxygen in the release of energy in cells?
- How is energy distributed and used inside cells?

SYLLABUS CONTENT

This chapter covers the following syllabus content:
- C1.2.1 ATP as the molecule that distributes energy within cells
- C1.2.2 Life processes within cells that ATP supplies with energy
- C1.2.3 Energy transfers during interconversions between ATP and ADP
- C1.2.4 Cell respiration as a system for producing ATP within the cell using energy released from carbon compounds
- C1.2.5 Differences between anaerobic and aerobic cell respiration in humans
- C1.2.6 Variables affecting the rate of cell respiration
- C1.2.7 Role of NAD as a carrier of hydrogen and oxidation by removal of hydrogen during cell respiration (HL only)
- C1.2.8 Conversion of glucose to pyruvate by stepwise reactions in glycolysis with a net yield of ATP and reduced NAD (HL only)
- C1.2.9 Conversion of pyruvate to lactate as a means of regenerating NAD in anaerobic cell respiration (HL only)
- C1.2.10 Anaerobic cell respiration in yeast and its use in brewing and baking (HL only)
- C1.2.11 Oxidation and decarboxylation of pyruvate as a link reaction in aerobic cell respiration (HL only)
- C1.2.12 Oxidation and decarboxylation of acetyl groups in the Krebs cycle with a yield of ATP and reduced NAD (HL only)
- C1.2.13 Transfer of energy by reduced NAD to the electron transport chain in the mitochondrion (HL only)
- C1.2.14 Generation of a proton gradient by flow of electrons along the electron transport chain (HL only)
- C1.2.15 Chemiosmosis and the synthesis of ATP in the mitochondrion (HL only)
- C1.2.16 Role of oxygen as terminal electron acceptor in aerobic cell respiration (HL only)
- C1.2.17 Differences between lipids and carbohydrates as respiratory substrates (HL only)

ATP as the molecule that distributes energy within cells

Cellular respiration: introduction

Living organisms require energy to build and repair body structures and to maintain all the activities of life, such as movement, reproduction, nutrition, excretion and sensitivity. Energy is also required for life processes within cells (see below). Energy is transferred in cells by the breakdown of nutrients, principally of carbohydrates such as glucose. The process by which energy is made available from nutrients in cells is called **cell respiration**. Cell respiration is the controlled release of energy from organic compounds in cells.

◆ **Cell respiration**: the enzyme-controlled release of energy from organic compounds to produce ATP.

◉ Common mistake

People sometimes talk about 'burning up food' in respiration, but comparing respiration to combustion is unhelpful. In combustion, energy in matter is released in a one-step reaction, as heat and light (Figure C1.2.1). Such a rapid change would be disastrous for body tissues. In cellular respiration, many small steps occur, each catalysed by a specific enzyme.

◆ **Adenosine triphosphate (ATP)**: a nucleotide, present in every living cell, formed in photosynthesis and respiration from ADP and P_i, and functioning in metabolism as a common intermediate between energy-requiring and energy-yielding reactions.

Link

The structure of ATP is detailed in Chapter B2.1, Figure B2.1.12, page 232.

Energy in respiration is transferred in small quantities; much of the energy is made available to the cells and may be temporarily trapped in the energy currency molecule adenosine triphosphate (ATP). However, some energy is still lost as heat in each step – we notice how warm we become with strenuous physical activity.

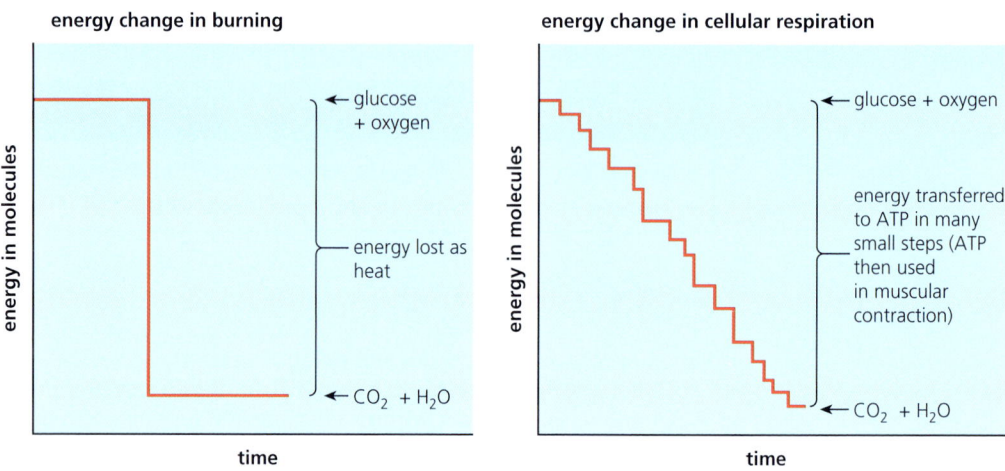

■ Figure C1.2.1 Combustion and respiration compared

■ ATP – the universal energy currency

Energy that is made available within the cytoplasm is transferred from glucose to a molecule called **adenosine triphosphate (ATP)**. ATP is referred to as **energy currency** because, like money, it can be used in different contexts, and it is constantly recycled. Respiration changes energy from one form of 'currency' (glucose) into a different currency (ATP). Glucose is relatively stable, whereas ATP breaks down, releasing the energy it stores when coupled to another reaction.

ATP is a **nucleotide** (page 15) with an unusual feature. It carries three phosphate groups linked together in a linear sequence (Figure C1.2.2). ATP may lose both of the outer phosphate groups, but usually only loses one at a time. ATP is a relatively small, soluble organic molecule. It occurs in cells at a concentration of 0.5–2.5 mg cm^{-3}, which remains fairly constant (except in fatigued muscle cells).

Like many organic molecules of its size, ATP contains a good deal of chemical energy locked up in its structure. What makes ATP special as a reservoir of chemical energy is its role as a common intermediate between energy-yielding reactions and energy-requiring reactions and processes. Energy-yielding reactions include many of the individual steps in respiration. Energy-requiring reactions include the synthesis of cellulose from glucose, the synthesis of proteins from amino acids and the contraction of muscle fibres.

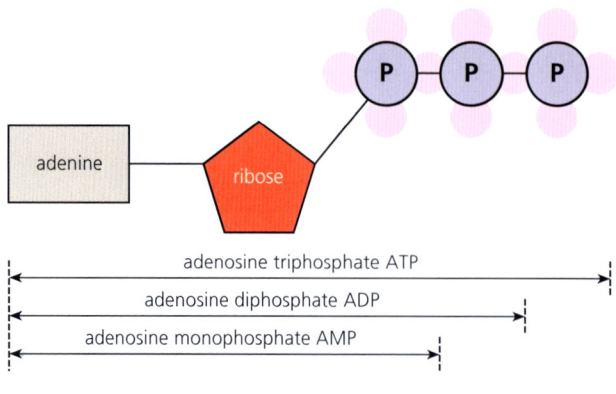

■ Figure C1.2.2 ATP, ADP and AMP

In summary, ATP is a molecule universal to all living things. It is the source of energy for chemical change in cells, tissues and organisms. In addition, ATP is:
- a substance that moves easily within cells and organisms – by facilitated diffusion
- a very reactive molecule, able to take part in many steps of cellular respiration and in many reactions of metabolism
- an immediate source of energy, able to deliver energy in relatively small amounts, sufficient to drive individual reactions and physical processes.

Life processes within cells that ATP supplies with energy

ATP provides a direct source of energy for many life processes within cells, such as:
- active transport of molecules and ions across membranes by membrane pumps (Chapter B2.1)
- synthesis of macromolecules (anabolism, page 382, and Chapters B1.1 and B1.2)
- movement of the whole cell
- movement of cell components, such as chromosomes (Chapter D2.1).

ATP is universal to all living organisms. Because it is reactive, ATP is never stored. The conversion of ATP to **adenosine diphosphate (ADP)** and phosphate ion (P_i) is thermodynamically favourable. The products ADP and P_i are thermodynamically more stable (lower in free energy) than the reactant ATP. Glucose and fatty acids are short-term energy stores, while glycogen, starch and triglycerides are long-term stores.

 **Adenosine diphosphate (ADP)**: a nucleotide, present in every living cell, made of adenosine and two phosphate groups bonded in a linear sequence; it is important in energy-transfer reactions of metabolism.

Energy transfers during interconversions between ATP and ADP

In cells, ATP is formed from ADP and P_i using energy from respiration. Then, in the presence of enzymes, ATP participates in energy-requiring reactions. Energy is released by hydrolysis of ATP to ADP and phosphate. The free energy available in ATP is approximately $30–4\,kJ\,mol^{-1}$. Some of this energy is lost as heat in the reaction, but much free energy is made available to do useful work, more than sufficient to drive a typical energy-requiring reaction of metabolism (Figure C1.1.16, page 397).

Sometimes ATP reacts with water (a **hydrolysis** reaction) and is converted to ADP and P_i. For example, direct hydrolysis of the terminal phosphate group happens in muscle contraction.

Mostly, ATP reacts with other metabolites and forms phosphorylated intermediates, making them more reactive in the process. The phosphate groups are released later, so both ADP and P_i become available for reuse as metabolism continues.

> **Top tip!**
> You are not required to know the quantity of energy in kilojoules during interconversions between ATP and ADP, but you do need to know that it is sufficient for many tasks in the cell, i.e. chemical reactions or physical processes.

1 **Outline** why ATP is an efficient energy currency molecule.

Cell respiration as a system for producing ATP within the cell

Carbohydrates such as glucose are the preferred respiratory substrate even though the energy yield is lower compared to that of lipids (see page 424): this is because oxidation is easier compared to oxidation of other respiratory substrates, because the molecules can directly enter glycolysis.

> **Link**
> Fatty acids and their structure are discussed in Chapter B1.1, page 197.

> **Common mistake**
>
> Do not confuse cell respiration with gas exchange. Gas exchange is the process where oxygen and carbon dioxide are exchanged at a surface, for example the alveoli in humans. Cell respiration is a chemical process in cells, where glucose is oxidized in a step-by-step process, yielding ATP.

◆ **Aerobic respiration**: respiration requiring oxygen, involving the oxidation of glucose to carbon dioxide and water.

◆ **Anaerobic respiration**: respiration in the absence of oxygen, producing either lactic acid (humans) or ethanol and carbon dioxide (plants and yeast).

◆ **Lactic acid**: organic acid produced by the body when glucose is broken down to generate ATP in the absence of oxygen; lactic acid is formed by lactate in solution.

2 Identify two products of anaerobic respiration in muscle.

Fats need to be converted to carbohydrates through a process called gluconeogenesis before the glycolytic process can take place. Proteins have to be hydrolyzed into amino acids, then deaminated (NH_2 group removed) before they can enter the respiratory pathway. This requires ATP, so the net production of ATP from protein is reduced. Proteins are structural molecules, and so are only used in respiration when other respiratory substrates, such as glucose and fats, are limited, i.e. under starvation conditions.

Differences between anaerobic and aerobic cell respiration in humans

While no oxygen is required in the early steps of cellular respiration, most animals and plants, and very many micro-organisms, do require oxygen for cell respiration. We say that they respire aerobically.

In **aerobic respiration**, sugar is completely oxidized to carbon dioxide and water, and much energy is made available. The steps of aerobic respiration can be summarized by a single equation:

glucose + oxygen → carbon dioxide + water + ATP

$C_6H_{12}O_6$ + $6O_2$ → $6CO_2$ + $6H_2O$ + ATP

This equation is a balance sheet of the inputs (raw materials) and the outputs (products). It tells us nothing about the separate steps, each catalysed by a specific enzyme, by which cellular respiration occurs. Aerobic respiration occurs in mitochondria and produces large amounts of ATP.

In the absence of oxygen, many organisms (and sometimes tissues in organisms, when these have become deprived of oxygen) will continue to respire glucose by different pathways, known as fermentation or **anaerobic respiration**, at least for a short time. Anaerobic respiration occurs in the cytoplasm of cells and produces only a small quantity of ATP.

Vertebrate muscle tissue can respire anaerobically and results in the formation of lactic acid rather than ethanol. Under conditions in the cytoplasm, **lactic acid** is weakly ionized and, therefore, exists as the **lactate** ion.

Lactic acid fermentation occurs in muscle fibres, but only when the demand for energy for contractions is very great and cannot be met fully by aerobic respiration. In lactic-acid fermentation the sole waste product is lactate.

The steps of anaerobic respiration can be summarized by a single equation:

glucose → lactate + ATP

> **Top tip!**
>
> You need to know simple word equations for aerobic and anaerobic respiration, with glucose as the substrate.

> **Common mistake**
>
> It is not correct to say that 'respiration occurs in the mitochondria'. Mitochondria are required for aerobic, but not anaerobic, respiration. Anaerobic respiration occurs in the cytoplasm of cells. Respiration begins in the cytoplasm and ends in mitochondria.

Variables affecting the rate of cell respiration

There are different variables that can affect the rate of cellular respiration:
- **Metabolic rate of the cell**: e.g. muscle cells will require more energy and therefore have higher rates of respiration.
- **The size of the organism**: smaller organisms have a larger surface area compared to their size (page 262) and have a correspondingly higher respiratory rate to allow for heat loss.
- **Supply of oxygen**: cells need a constant supply of oxygen to release the maximum amount of ATP; inadequately supplied cells will respire anaerobically (see above).
- **Supply of substrates for respiration**, e.g. glucose. Other substrates can also be used in the respiratory pathway (see page 424); the rate of respiration and the quantity of products produced (carbon dioxide and water) will depend on the respiratory substrate.
- **Temperature**: because respiration is controlled by enzymes, temperature affects the rate of respiration by increasing it up to an optimum temperature.
- **pH**: the release of carbon dioxide during the process of respiration decreases the pH (i.e. increases acidity) of cell content and body tissues, which affects the functioning of enzymes involved in respiration.

Measurements can be made to determine the rate of cell respiration. Determination of the rate of cell respiration can be made using a respirometer (Figure C1.2.3). Respiration rate is measured by the uptake of oxygen per unit time. The manometer in this apparatus detects change in the pressure or volume of a gas. Respiration by tiny organisms (germinating seeds or fly maggots are ideal) that are trapped in the chamber of the respirometer alters the composition of the gas there, once the screw clip has been closed.

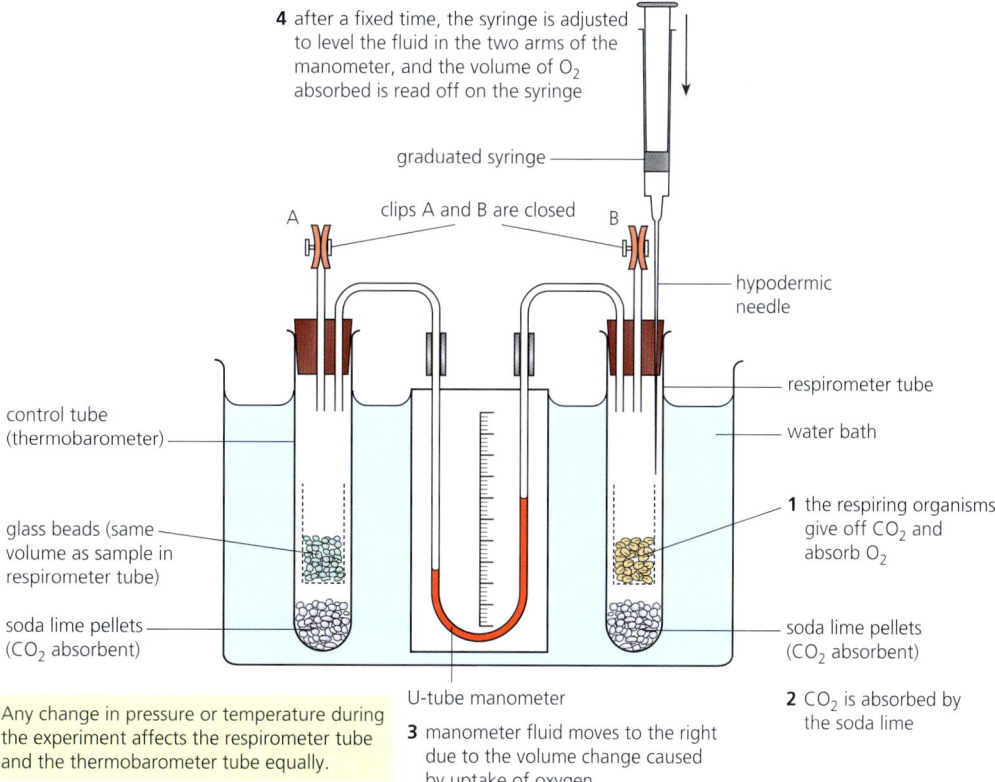

■ **Figure C1.2.3** A respirometer to measure respiration rate

If soda lime is present in the chambers, the carbon dioxide gas released by the respiring organism is removed. In this case, only oxygen uptake by the respiring organisms causes a change in volume. As a result, the coloured liquid in the attached capillary tube will move towards the respirometer tube. The resulting reduction in the volume of air in the respirometer tube in a given time period can now be estimated. It is the volume of air from the syringe that must be injected back into the respirometer tube to make the manometric fluid level in the two arms equal again. That volume is equivalent to the volume of oxygen taken up by the respiring organisms.

To estimate the quantity of carbon dioxide produced by the organisms:
- Remove the soda lime from the apparatus (Figure C1.2.3).
- The difference between the distance the water moved with and without the soda lime indicates the volume of carbon dioxide produced.
- Without soda lime, carbon dioxide is not removed and the water might not move at all: if the same volume of oxygen is taken in as carbon dioxide is produced, there will be no change in the volume of gas and the water will not move.
- If less carbon dioxide is produced than oxygen is taken in, the fall in pressure will move the water a little towards the organisms.

3 In the respirometer (Figure C1.2.3), **explain** how changes in temperature or pressure in the external environment are prevented from interfering with the measurement of oxygen uptake by respiring organisms in the apparatus.

4 The experiment shown in Figure C1.2.3 was repeated with maggot fly larvae in tube B, first with soda lime present and subsequently with water in place of soda lime. The volume change with soda lime was $30\,mm^3\,h^{-1}$, but without soda lime it was $3\,mm^3\,h^{-1}$. **Analyse** these results, explaining the significance of each value.

● Top tip!

The use of invertebrates in respirometer experiments has ethical implications. IB guidelines state that: 'Any experimentation should not result in any pain or undue stress on any animal (vertebrate or invertebrate) or compromise its health in any way.' Animals must be handled with care and must not be exposed directly to soda lime. They should only be used for brief periods of time, and returned to a safe environment once the experiment is completed.

■ Calculating the rate of respiration

The respirometer can be used to measure the rate of respiration. The rate of oxygen consumption ($mm^3\,min^{-1}$) can be used as the rate of respiration for organisms.

1 Use the respirometer to measure the distance travelled by the fluid in the manometer in 10 minutes.

2 The volume of oxygen absorbed and respired in 10 minutes can be calculated using the radius of the capillary tube, r (mm) and the distance moved by the manometer fluid, d (mm), using the formula: $\pi r^2 d$ (this is the formula to work out the volume of a cylinder).

3 The average rate of oxygen consumption can be calculated by dividing the volume of oxygen consumed by ten (the number of minutes).

4 The experiment should be repeated at least five times to obtain reliable data.

Depending on how long you have to carry out this experiment, the investigation could be carried out for longer at each repeat. This would make the results more accurate.

Inquiry 1: Exploring and designing

Designing

Design an experiment using a respirometer. Select an organism to use and an independent variable to change. You must think about the ethical issues of using animals and what variables you can safely change without harming the animal. If you are planning to alter temperature, use a plant, not an animal.

Justify the number of repeats you will use and controlled variables you will keep the same.

Formulate a hypothesis for your experiment. What effect do you expect the independent variable to have on the dependent variable, and why?

Inquiry 2: Collecting and processing data

Collecting data; processing data

An investigation can be carried out on the rate of respiration of germinating peas at different temperatures.

1. Calculate the average rate of respiration at room temperature, using the method outlined above in Inquiry 1.
2. Reset the apparatus by allowing air to re-enter the tubes by releasing the clips, and reset the manometer fluid using the syringe.
3. Set up a series of water baths at 30, 40, 50 and 60 °C.
4. Put the test tube containing the peas in the 30 °C water bath. Open the clips to allow the tubes to acclimatize (the warmth will cause the air in the respirometer to expand), then close the clips to begin the experiment.
5. Record the rate of respiration.
6. Repeat the procedure at the other temperatures.
7. Plot a graph showing the effect of temperature on the rate of respiration. Temperature should be on the x-axis and rate of respiration on the y-axis.

Inquiry 3: Concluding and evaluating

Concluding; evaluating

Analyse and evaluate your results from your respirometer experiment: either the investigation you designed yourself in Inquiry 1, or the one outlined in Inquiry 2.

What do you conclude about the effect of the independent variable on the rate of respiration of your organism? Do your results confirm your hypothesis? How do your result compare with other research on this subject?

What were the limitations of the experiment and what would you do differently next time?

ATL C1.2A

Access the following site: http://amrita.olabs.edu.in/?sub=79&brch=17&sim=204&cnt=1

Investigate the effect of type of seed, number of seeds or temperature on the rate of respiration using a virtual respirometer. Analyse your results and write a conclusion, using your knowledge of cellular respiration to explain your data.

The role of NAD as a carrier of hydrogen

■ Cell respiration involves the oxidation and reduction of compounds

Glucose is a relatively large molecule containing six carbon atoms. During aerobic cellular respiration, glucose undergoes a series of enzyme-catalysed oxidation reactions and decarboxylation reactions (Figure C1.2.4). These reactions are grouped into three major phases and a link reaction:

- **glycolysis**, in which glucose is converted to pyruvate
- a **link reaction**, in which pyruvate is converted to acetyl coenzyme A (acetyl-CoA). Carbon dioxide is released
- the **Krebs cycle**, in which acetyl-CoA is converted to carbon dioxide
- the **electron transport chain**, in which hydrogen that is removed in the oxidation reactions of glycolysis and the Krebs cycle is converted to water. The electron transport chain is a series of proteins that transfer electrons received from reduced coenzymes, generating a gradient of protons that drives the synthesis of adenosine triphosphate (ATP). The bulk of the ATP is synthesized here.

> **Top tip!**
> Coenzymes, for example acetyl coenzyme A (acetyl-CoA), are substances that work with an enzyme to initiate or aid the function of the enzyme.

> **Link**
> The roles of oxidation and reduction in biological systems are explored in the linking question in Chapter B1.1, page 204.

◆ **Reduction**: the gain of electrons in a chemical reaction; this can be by removing oxygen atoms or by adding hydrogen atoms.

◆ **Oxidation**: the loss of electrons in a chemical reaction; this can be by adding oxygen atoms or by removing hydrogen atoms.

◆ **Redox reaction**: linked reaction where one substance loses an electron while the other gains an electron: the substance that gains the electron is said to be **reduced** while the one that lost the electron is said to be **oxidized**.

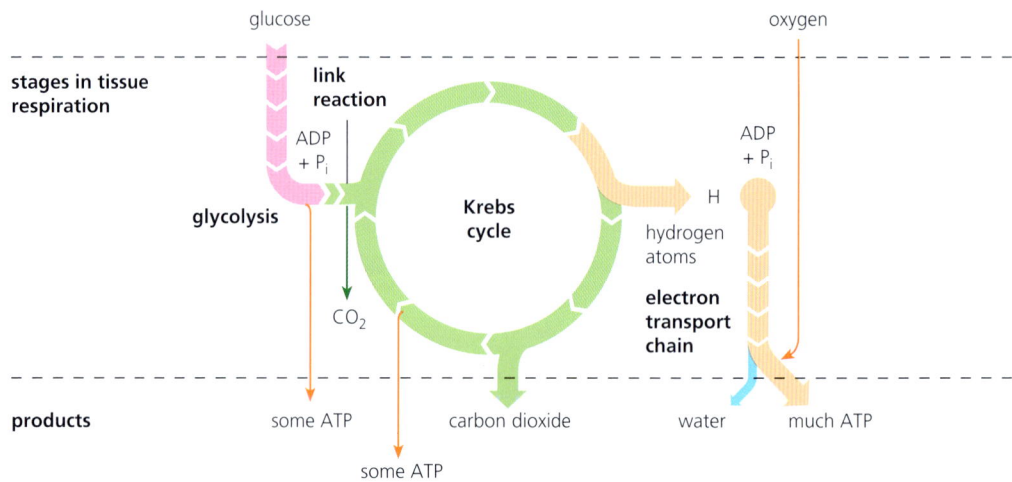

■ Figure C1.2.4 The stages of aerobic cellular respiration

■ Respiration as a series of redox reactions

The terms **reduction** and **oxidation** recur frequently in respiration.

In cellular respiration, glucose is oxidized to carbon dioxide but, at the same time, oxygen is reduced to water (Figure C1.2.5). In fact, tissue respiration is a series of oxidation–reduction reactions, so described because when one substance in a reaction is oxidized, another is automatically reduced. The short-hand name for **red**uction–**ox**idation reactions is **redox reactions**. Redox reactions involve both oxidation and reduction.

In biological oxidation, oxygen atoms may be added to a compound but, alternatively, hydrogen atoms may be removed (dehydrogenation). In respiration, all the hydrogen atoms are gradually removed from glucose. When hydrogen is removed from a substrate, the substrate has been oxidized. The hydrogen atoms are added to hydrogen acceptors, which are thus reduced.

Since a hydrogen atom consists of an electron and a proton, gaining hydrogen atom(s) (a case of reduction) involves gaining one or more electrons. In fact, the best definition of **oxidation** is **the loss of electrons**, and **reduction** is **the gain of electrons**.

Theme C: Interaction and interdependence – Molecules

> **Top tip!**
>
> Remembering these definitions has given countless people problems, so this mnemonic has been devised:
>
> **OIL RIG** = **O**xidation **I**s **L**oss of electrons; **R**eduction **I**s **G**ain of electrons

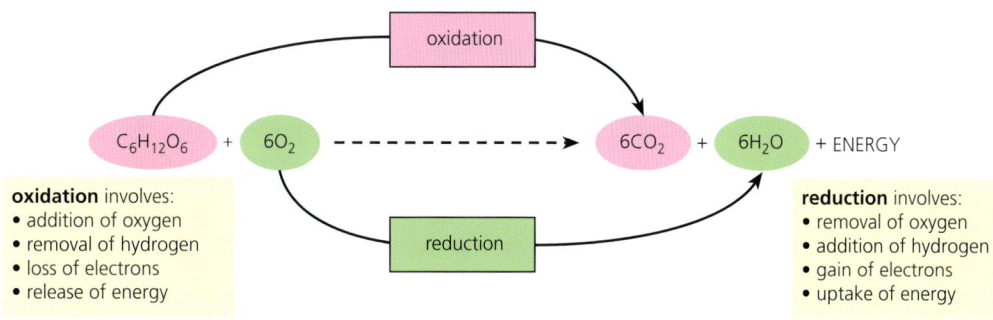

oxidation involves:
- addition of oxygen
- removal of hydrogen
- loss of electrons
- release of energy

reduction involves:
- removal of oxygen
- addition of hydrogen
- gain of electrons
- uptake of energy

■ **Figure C1.2.5** Respiration as a redox reaction

Redox reactions take place in biological systems due to the presence of a compound with a strong tendency to take electrons from another compound (an **oxidizing agent**) or the presence of a compound with a strong tendency to donate electrons to another compound (a **reducing agent**).

Another feature of oxidation and reduction is **energy change**. When reduction occurs, energy is absorbed (an **endergonic reaction**). When oxidation occurs, energy is released (an **exergonic reaction**). An example of energy release in oxidation is the burning of a fuel in air. Here, energy is given out as heat.

Link
Endergonic and exergonic reactions are covered in Chapter C1.1, Figure C1.1.15, page 397.

> **Top tip!**
>
> The amount of energy in a molecule depends on its degree of oxidation. An oxidized substance has less stored energy than a reduced substance. For example, the fuel molecule methane (CH_4), has more stored chemical energy than carbon dioxide (CO_2).

◆ **Coenzyme**: substance that works with an enzyme to initiate or aid the function of the enzyme.

Coenzymes

In respiration, all the hydrogen atoms are gradually removed from glucose. They are added to hydrogen acceptors (electron carriers), which are thus reduced. These hydrogen acceptors are all **coenzymes**. Coenzymes link oxidation and reduction in cells. The electron carriers in respiration are:

 NAD (nicotinamide adenine dinucleotide): the main electron carrier used in respiration.
 FAD (flavin adenine dinucleotide): a second electron carrier used in respiration.

◆ **Nicotinamide adenine dinucleotide (NAD)**: a coenzyme that is a hydrogen carrier in cellular respiration. NAD^+ is the oxidized form that accepts hydrogen atoms.

Electron carriers can receive hydrogen atoms (i.e. are reduced), oxidizing the substance they remove them from. For example, NAD is reduced to NADH and H^+ (reduced NAD):

$$NAD^+ + 2H^+ + 2e^- \rightarrow NADH + H^+ \text{ (NADH can also be represented as } NADH_2)$$

◆ **Flavin adenine dinucleotide (FAD)**: a coenzyme that is a hydrogen carrier in cellular respiration. FAD is the oxidized form that accepts two hydrogen atoms, forming $FADH_2$.

Reduced NAD can pass hydrogen ions and electrons on to other acceptor molecules and, when it does so, becomes oxidized back to NAD.

> **Concept: Interaction**
>
> Coenzymes such as NAD and FAD interact with substrates involved in the respiratory pathway, passing hydrogen ions to other acceptor molecules, which ultimately results in the generation of ATP.

C1.2 Cell respiration

Glycolysis

Conversion of glucose to pyruvate in glycolysis

In the first steps of cellular respiration, the glucose molecule (a 6-carbon sugar) is split into two 3-carbon molecules. The products are converted to an organic acid called pyruvic acid (also a 3-carbon compound). Under conditions in the cytoplasm, organic acids are weakly ionized and, therefore, pyruvic acid exists as the **pyruvate** ion.

Two molecules of pyruvate are formed from each molecule of glucose. In addition, there is a small amount of ATP formed, using a little of the energy that had been locked up in the glucose molecule. No molecular oxygen is required for these first steps of cellular respiration.

Because glucose has been split into smaller molecules, these steps are known as **glycolysis**. The enzymes that catalyse these reactions are found in the cell cytoplasm generally, not inside an organelle. Throughout cellular respiration, a series of oxidation–reduction reactions occur. (Oxidation and reduction are explained on page 412.)

In summary:

$$\text{glucose} \xrightarrow{\text{enzymes in the cytoplasm}} \text{pyruvate} + \text{small amount of ATP}$$

Details of the process are outlined below.

> ◆ **Pyruvate**: the end product of glycolysis, which is converted into acetyl-CoA that enters the Krebs cycle when there is sufficient oxygen available.
> ◆ **Glycolysis**: the first stage of cell respiration in which glucose is converted to pyruvate by stepwise enzyme-controlled reactions, without use of oxygen; ATP is generated.
> ◆ **Phosphorylation**: the addition of a phosphoryl group (PO_3^{2-}) to an organic compound.

Details of glycolysis

Glycolysis occurs in four stages:

1 **Phosphorylation** by reaction with ATP is the way glucose is first activated, forming glucose phosphate. Phosphorylation of molecules makes them more energetically unstable and higher in energy, which means that they are more reactive. Glucose phosphate is converted to fructose 6-phosphate, and a further phosphate group is then added at the expense of another ATP molecule to form fructose 1,6-bisphosphate. So, two molecules of ATP are *consumed* per molecule of glucose respired at this stage of glycolysis.

2 **Lysis** (splitting) of the fructose 1,6-bisphosphate is the next step, forming two molecules of a 3-carbon sugar called **triose phosphate**.

3 **Oxidation** of the triose phosphate molecules occurs by removal of hydrogen. The enzyme for this reaction (a dehydrogenase) works with a coenzyme, NAD (see above).

4 **ATP formation** occurs twice in the reactions by which each triose phosphate molecule is converted to pyruvate. This form of ATP synthesis is described as being **at substrate level**, in order to differentiate it from the bulk of ATP synthesis, which occurs later in cell respiration – during operation of the electron transport chain (see below). As two molecules of triose phosphate are converted to pyruvate, four molecules of ATP are synthesized *at this stage of glycolysis*. So, in total, there is a **net gain of two ATPs in glycolysis**.

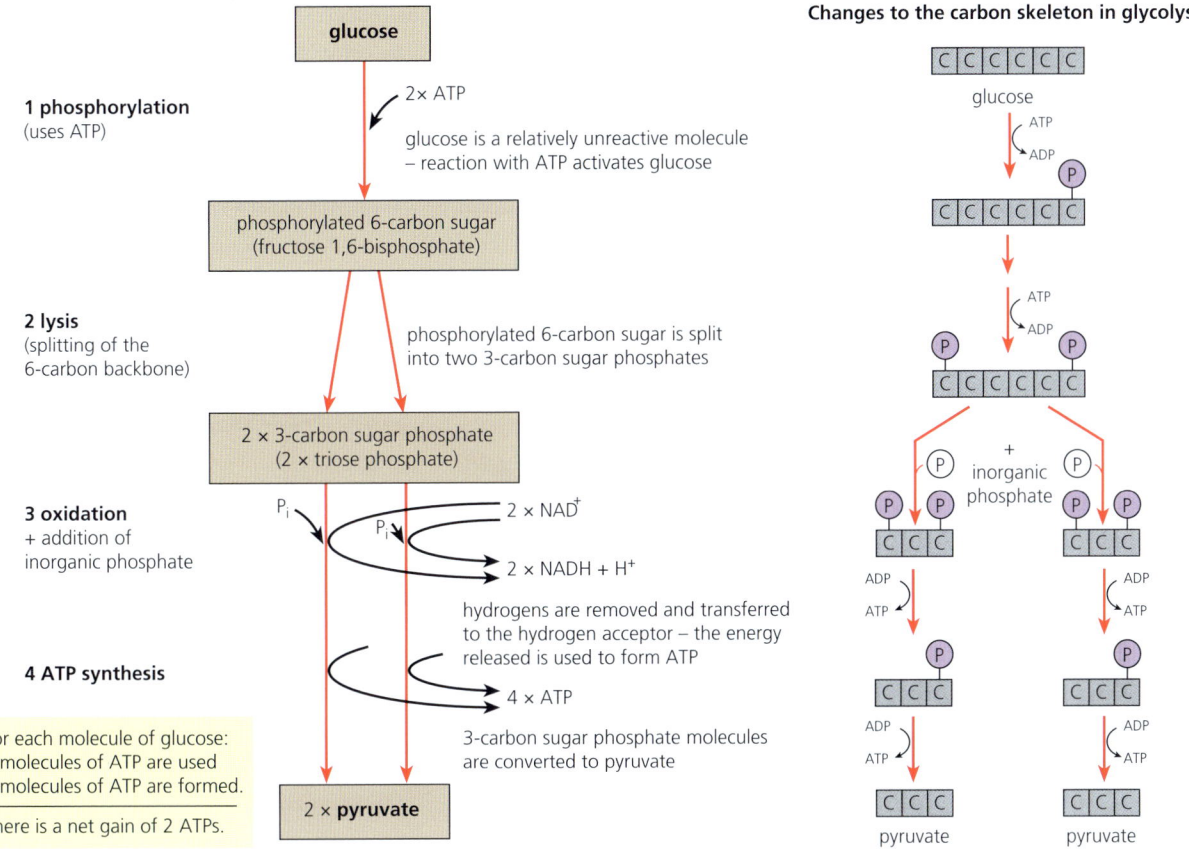

■ Figure C1.2.6 Glycolysis: a summary

> **Top tip!**
>
> You are not required to know the names of the intermediates in glycolysis, but you do need to know that each step in the pathway is catalysed by a different enzyme.

Once pyruvate has been formed from glucose in the cytoplasm, the remainder of the pathway of aerobic cell respiration is located in the **mitochondria**. This is where the enzymes concerned with the **link reaction**, **Krebs cycle** and **electron transport chain** are all located.

5 **State** which of the following are produced during glycolysis:

carbon dioxide   NADH   ATP   pyruvate   lactate   glycogen NAD^+ glucose

■ Subsequent steps of respiration

If oxygen is available to cells and tissues, the pyruvate is completely oxidized to carbon dioxide, water and a large quantity of ATP. Before these reactions take place, the pyruvate first passes into mitochondria by facilitated diffusion because the required enzymes are only found in the mitochondria (Figure C1.2.7).

In summary:

$$\text{pyruvate} \xrightarrow{\text{enzymes in the mitochondria}} \text{carbon dioxide + water + large amount of ATP}$$

In this phase of cellular respiration, the pyruvate is oxidized by:
- the removal of hydrogen atoms by hydrogen acceptors (oxidizing agents)
- the addition of oxygen to the carbon atoms to form carbon dioxide.

These reactions occur one at a time, each catalysed by a different enzyme.

C1.2 Cell respiration

In the mitochondria, the hydrogen (proton) carried by the reduced hydrogen-acceptor molecules reacts with oxygen to form water. The reduced hydrogen acceptor is reoxidized (loses its H) and is available for reuse in the production of more pyruvate. The majority of ATP molecules (the key product of respiration for the cell) are generated in this step.

The enzymes of cellular respiration occur partly in the cytoplasm (the enzymes of glycolysis) and partly in the mitochondria (the enzymes of pyruvate oxidation and most ATP formation). After formation, ATP passes to all parts of the cell. Both ADP and ATP pass through the mitochondrial membranes by facilitated diffusion. The locations of the different stages of aerobic respiration are shown in Figure C1.2.7.

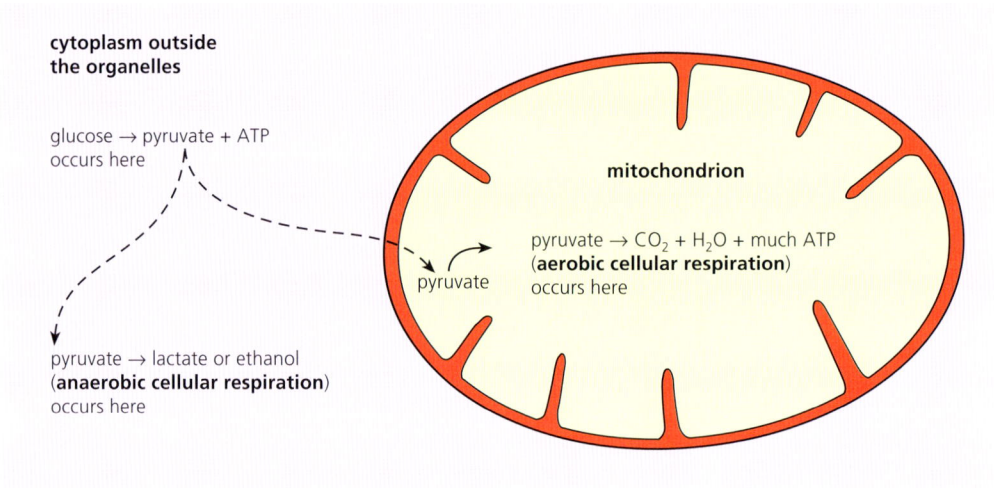

6 **State** which steps in cellular respiration occur whether or not oxygen is available to cells.

■ **Figure C1.2.7** The sites of cellular respiration in cells

Conversion of pyruvate to lactate

When oxygen is not available in human skeletal muscle tissue, the pyruvate remains in the cytoplasm and is converted to lactate. In yeast, whether or not oxygen is available, the pyruvate is converted to the alcohol called ethanol.

How is the supply of pyruvate maintained in cells in the absence of oxygen?

This is a potential problem in the breakdown of pyruvate in the absence of oxygen. Remember, in pyruvate formation, hydrogen-acceptor molecules (NAD) are reduced (take up hydrogen atoms). Without using oxygen, these must be reoxidized (lose their H) if production of pyruvate is to continue.

The answer is, in anaerobic cellular respiration the reduced hydrogen-acceptor molecules (NADH) donate their hydrogen to form lactate or ethanol from pyruvate. They are thus reoxidized in the absence of oxygen and are available for further pyruvate synthesis. Pyruvate formation is able to continue.

This can be summarized:

$$\text{pyruvate} \xrightarrow[\text{oxidized H acceptor (lost H)}]{\text{reduced H acceptor (carrying H)}} \text{lactate} \quad \text{pyruvate} \xrightarrow[\text{oxidized H acceptor}]{\text{reduced H acceptor}} \text{ethanol} + CO_2$$

■ Anaerobic respiration is 'wasteful'

Anaerobic respiration is 'wasteful' of respiratory substrate because the total energy yield per molecule of glucose respired, in terms of ATP generated, is limited to the net two molecules of ATP produced in pyruvate formation (in both alcoholic and lactic-acid fermentation). No additional energy is transferred in the latter steps and made available in cells.

So, we can think of both lactate and ethanol as energy-rich molecules. (Ethanol is sometimes used as a fuel in cars, for example.) The energy locked up in these molecules may be used later, however. In humans, lactate is transported to the liver and later metabolized aerobically. Energy yields of cellular respiration are compared in Table C1.2.1.

■ **Table C1.2.1** Energy yield of aerobic and anaerobic cellular respiration compared

Yield from each molecule of glucose respired		
Aerobic respiration		Anaerobic respiration
2 ATPs	glycolysis	2 ATPs
up to 36 ATPs	fates of pyruvate	nil
38 ATPs	total	2 ATPs

> **Concept: Interdependence**
>
> The anaerobic and aerobic pathways of cellular respiration can be seen as interdependent. Pyruvate is produced by anaerobic respiration, which is the substrate that forms the basis of aerobic respiration. Anaerobic respiration in humans forms lactic acid if insufficient oxygen is available. This must ultimately be converted back to pyruvate in the liver, under aerobic conditions.

Anaerobic cell respiration in yeast and its use in brewing and baking

Many species of yeast (*Saccharomyces*) respire anaerobically, even in the presence of oxygen. The products are ethanol and carbon dioxide. **Alcoholic fermentation** of yeast has been exploited by humans for many thousands of years:

- in bread making – the carbon dioxide causes the bread to rise
- in wine and beer production.

glucose → ethanol + carbon dioxide + ATP

In anaerobic cell respiration, reduced NAD is oxidized through other reactions than it is for aerobic respiration. The pathways of anaerobic respiration are the same in humans and yeasts, except for the way in which NAD is regenerated and, therefore, the final products (Figure C1.2.8). In effect, oxygen is replaced as the hydrogen acceptor. In alcoholic fermentation, ethanal is the hydrogen acceptor, and in lactic acid fermentation pyruvate is the hydrogen acceptor.

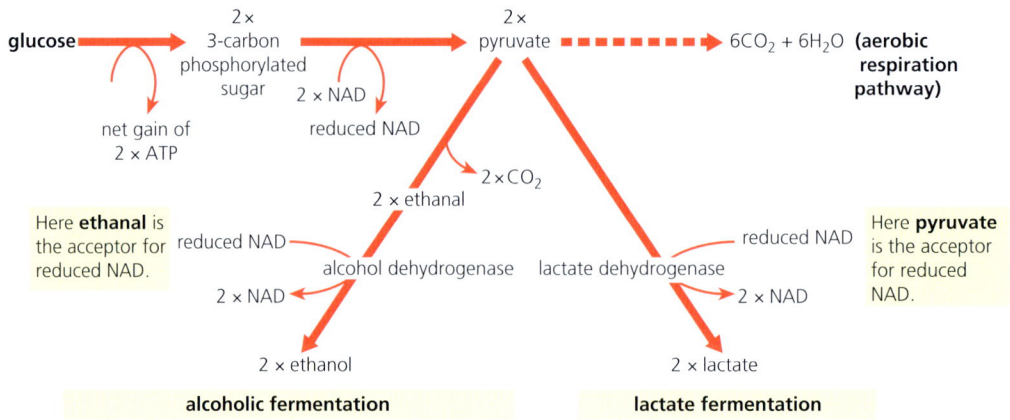

Figure C1.2.8 The respiratory pathways of anaerobic respiration

The link reaction in aerobic cell respiration

Pyruvate diffuses into the matrix of the mitochondrion as it forms and is metabolized there.

First, the 3-carbon pyruvate is **decarboxylated** by removal of carbon dioxide and, at the same time, oxidized by removal of hydrogen. Reduced NAD is formed. The product of this **oxidative decarboxylation** reaction is an acetyl group – a 2-carbon fragment. This acetyl group is then combined with coenzyme A, forming **acetyl coenzyme A (acetyl-CoA)**. The production of acetyl coenzyme A from pyruvate is known as the **link reaction** because it connects glycolysis to reactions of the **Krebs cycle**, which follow.

◆ **Decarboxylation**: a chemical reaction that releases carbon dioxide (usually with hydrogen replacing it).

◆ **Link reaction**: the reactions that connect glycolysis to the reactions of the Krebs cycle by producing acetyl coenzyme A from pyruvate. Carbon dioxide and NADH are produced by this reaction.

◆ **Krebs cycle**: cyclical metabolic process in aerobic cell respiration involving the oxidation and decarboxylation of acetyl groups, with a yield of ATP and reduced NAD.

The Krebs cycle

The **Krebs cycle** is named after Hans Krebs, who discovered it, but it is also sometimes referred to as the **citric acid cycle**, after the first intermediate acid formed.

The acetyl coenzyme A enters the Krebs cycle by reacting with a **4-carbon organic acid** (oxaloacetate, OAA). The products of this reaction are a **6-carbon acid** (citrate) and coenzyme A, which is released and reused in the link reaction.

The citrate is then converted back to the 4-carbon acid (an acceptor molecule, in effect) by the **reactions of the Krebs cycle**. These involve the following changes:

- two molecules of carbon dioxide are released, in separate decarboxylation reactions
- a molecule of ATP is formed, as part one of the reactions of the cycle; as in glycolysis, this ATP synthesis is at substrate level
- three molecules of reduced NAD are formed
- one molecule of another hydrogen acceptor, the coenzyme **flavin adenine dinucleotide (FAD)** is reduced (NAD is the chief hydrogen-carrying coenzyme of respiration but FAD has this role in the Krebs cycle).

Top tip!

In the Krebs cycle, you are required to name only the intermediates **citrate** (6C) and **oxaloacetate** (4C).

ATL C1.2B

Steps in aerobic respiration involve decarboxylation and oxidation. Make a copy of the pathway of the link reaction and Krebs cycle, and highlight where these types of reaction occur. Look again at Figure C1.2.6 (glycolysis). Are both of these types of reaction observed there, too?

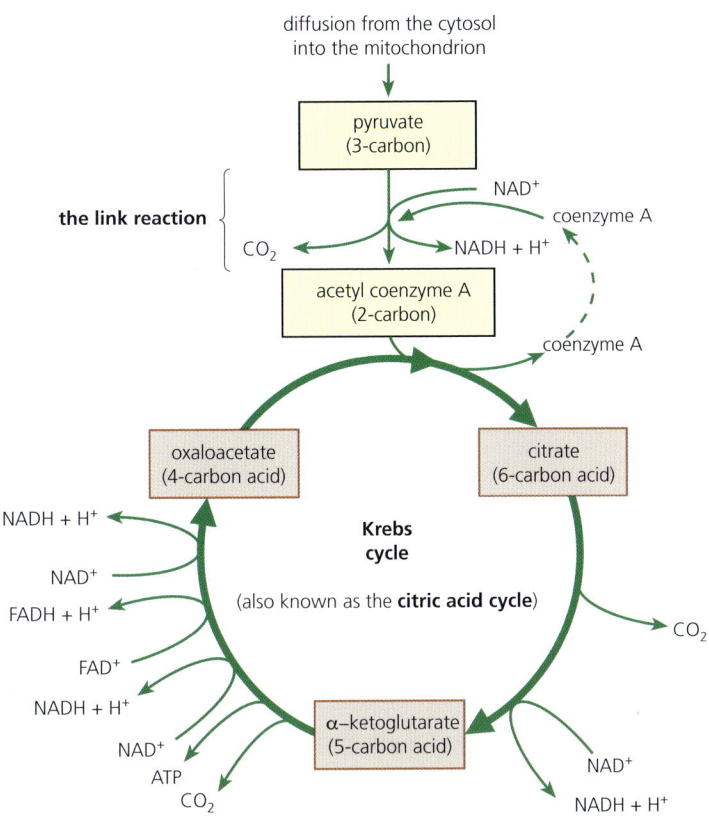

■ Figure C1.2.9 Link reaction and Krebs cycle: a summary

Concept: Interaction

Interaction between intermediates in the Krebs cycle and coenzymes NAD and FAD allow decarboxylation and oxidation reactions to occur, which results in the production of carbon dioxide and reduced coenzymes.

Because glucose is converted to two molecules of pyruvate in glycolysis, the whole Krebs cycle sequence of reactions 'turns' twice for every molecule of glucose that is metabolized by aerobic cellular respiration (Figure C1.2.9).

In both glycolysis and the Krebs cycle, pairs of hydrogen atoms are removed from various intermediates of the respiratory pathway. Either oxidized NAD is converted to reduced NAD or (on one occasion) an alternative hydrogen-acceptor coenzyme, FAD, is reduced.

Now, we are in a position to summarize the changes to the molecule of glucose that occur in the reactions of glycolysis and the Krebs cycle. A 'budget' of the products of glycolysis and two turns of the Krebs cycle is shown in Table C1.2.2.

■ Table C1.2.2 Net products of aerobic respiration of glucose at the end of the Krebs cycle

	Product			
Step	CO_2	ATP	reduced NAD	reduced FAD
glycolysis	0	2	2	0
link reaction (pyruvate → acetyl-CoA)	2	0	2	0
Krebs cycle	4	2	6	2
Total:	6 CO_2	4 ATP	10 reduced NAD	2 reduced FAD

7 Outline the types of reaction catalysed by:
  a dehydrogenases
  b decarboxylases.

Top tip!

Key features of the Krebs cycle are:
- citrate is produced by transfer of an acetyl group to oxaloacetate, and that oxaloacetate is regenerated by the reactions of the Krebs cycle, including four oxidations and two decarboxylations
- the oxidations are dehydrogenation reactions.

C1.2 Cell respiration

Fats can be respired

As discussed above (page 407), in addition to glucose, **fats** (lipids) are also commonly used as respiratory substrates. They are first broken down to **fatty acids** (and glycerol). A fatty acid is then 'cut up' into 2-carbon fragments and fed into the Krebs cycle via **coenzyme A**. Vertebrate muscle is well adapted to the respiration of fatty acids in this way (as is our heart muscle), and they are just as likely as glucose to be the respiratory substrate.

The electron transport chain

◆ **Electron transport chain**: a series of proteins that transfer electrons received from reduced coenzymes, generating a gradient of protons that drives the synthesis of adenosine triphosphate (ATP).

Reduced coenzymes pass to the **electron transport chain** – a series of proteins in the inner mitochondrial membrane. Energy is transferred when a pair of electrons is passed to the first carrier in the chain, converting reduced NAD (NADH) back to NAD. The reduced NAD comes from glycolysis, the link reaction and the Krebs cycle. The removal of electrons from hydrogen atoms creates H^+ (protons), which then play a role in the generation of ATP (see below).

The removal of pairs of hydrogen atoms from various intermediates of the respiratory pathway is a feature of several of the steps in glycolysis and the Krebs cycle. On most occasions, oxidized NAD is converted to reduced NAD but, once in the Krebs cycle, an alternative hydrogen-acceptor coenzyme, FAD, is reduced.

Therefore, in this final stage of aerobic respiration, the electrons received from the reduced coenzymes, i.e. from the reduced NAD (or FAD), are transported along a **series of carriers** to be combined finally with oxygen to form water (Figure C1.2.10).

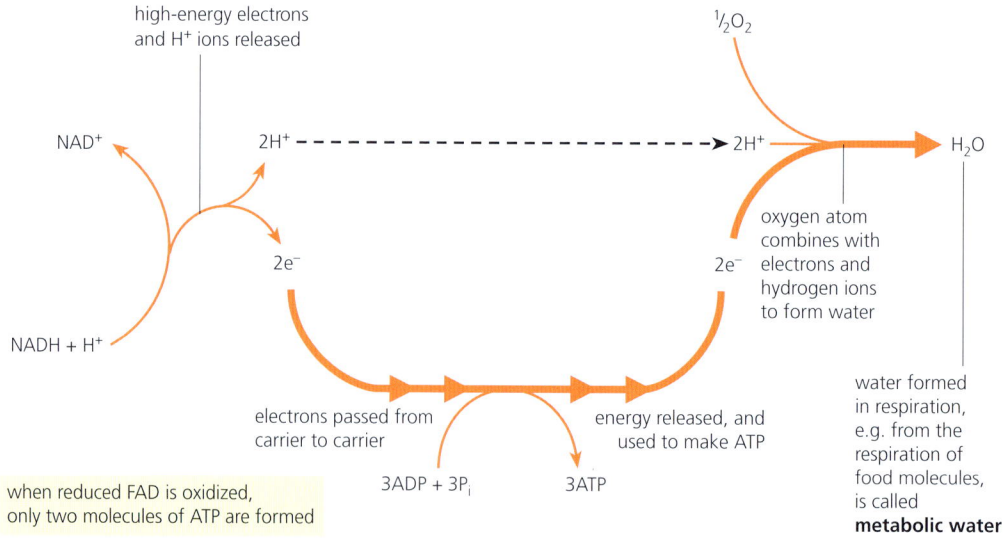

■ Figure C1.2.10 Transfer of energy by reduced NAD to the electron transport chain in the mitochondrion

8 **Suggest** how the absence of oxygen in respiring tissue might 'switch off' both the Krebs cycle and terminal oxidation.

As electrons are passed between the carriers in the series, **energy is released**. Release of energy in this manner is controlled and can be used by the cell. The energy is transferred to ADP and P_i, to form ATP. Normally, for every molecule of reduced NAD that is oxidized (that is, for every pair of hydrogens) approximately three molecules of ATP are produced (but less when FAD is oxidized).

The process is summarized in Figure C1.2.10.

In total, the yield from aerobic respiration is 32 ATPs per molecule of glucose respired (Table C1.2.3).

■ Table C1.2.3 Yield from each molecule of glucose respired aerobically

	Reduced NAD (or FAD)	ATP
glycolysis	(substrate level)	(net) = 2
	2	2 × 2.5 = 5
link reaction	2	2 × 2.5 = 5
Krebs cycle	6	6 × 2.5 = 15
	2	2 × 1.5 = 3
	(substrate level)	2
Total		32

Generation of a proton gradient by flow of electrons along the electron transport chain

> **Top tip!**
>
> You do not need to know the names of protein complexes in the electron transport chain.

The electron-carrier proteins are arranged in the inner mitochondrial membrane in a highly ordered way. These carrier proteins oxidize the reduced coenzymes, and energy from the oxidation process is used to pump hydrogen ions (protons) from the matrix of the mitochondrion into the space between the inner and outer mitochondrial membranes. The hydrogen ions accumulate here. Because the inner membrane is largely impermeable to ions, a significant **gradient in hydrogen ion concentration builds up** across the inner membrane, generating a potential difference across the membrane. This represents a store of potential energy.

Chemiosmosis and the synthesis of ATP in the mitochondrion

■ Phosphorylation by chemiosmosis

Chemiosmosis is a process by which the synthesis of ATP is coupled to electron transport via the movement of protons (Figure C1.2.11).

The protons concentrated in the space between the inner and outer mitochondrial membranes flow back into the matrix, via channels in the **ATP synthase** enzyme (ATPase), located in the inner mitochondrial membrane. As the protons flow down their concentration gradient, through the enzyme, the energy is transferred as ATP synthesis occurs. ATP synthase therefore couples release of energy from the proton gradient with phosphorylation of ADP. The ATPase has a rotational mechanism – energy generated by the rotation of the enzyme leads to the production of ATP.

◆ **Chemiosmosis**: process by which the synthesis of ATP is coupled to electron transport via the movement of protons, using ATP synthase.

◆ **ATP synthase (ATPase)**: a complex of integral proteins located in the mitochondrial inner membrane where it catalyses the synthesis of ATP from ADP and phosphate, driven by a flow of protons.

> **TOK**
>
> *How could a mitochondrion use the energy that is made available in the flow of electrons between carrier molecules to drive the synthesis of ATP?*
>
> It was a biochemist, Peter Mitchell, who in 1961 first suggested the chemiosmotic theory as the answer to this question. He was at an independently funded research institute in Cornwall, UK, studying the metabolism of bacteria. His hypothesis was not generally accepted for about 10 years – his ideas were, in some ways, regarded as too novel. Today, we describe the revolution that Mitchell's ideas started as a **paradigm shift** in the field of bioenergetics. Nearly two decades later, he was awarded the Nobel Prize in Chemistry for his discovery. Discuss what roles paradigm shifts have in the progression of scientific knowledge.

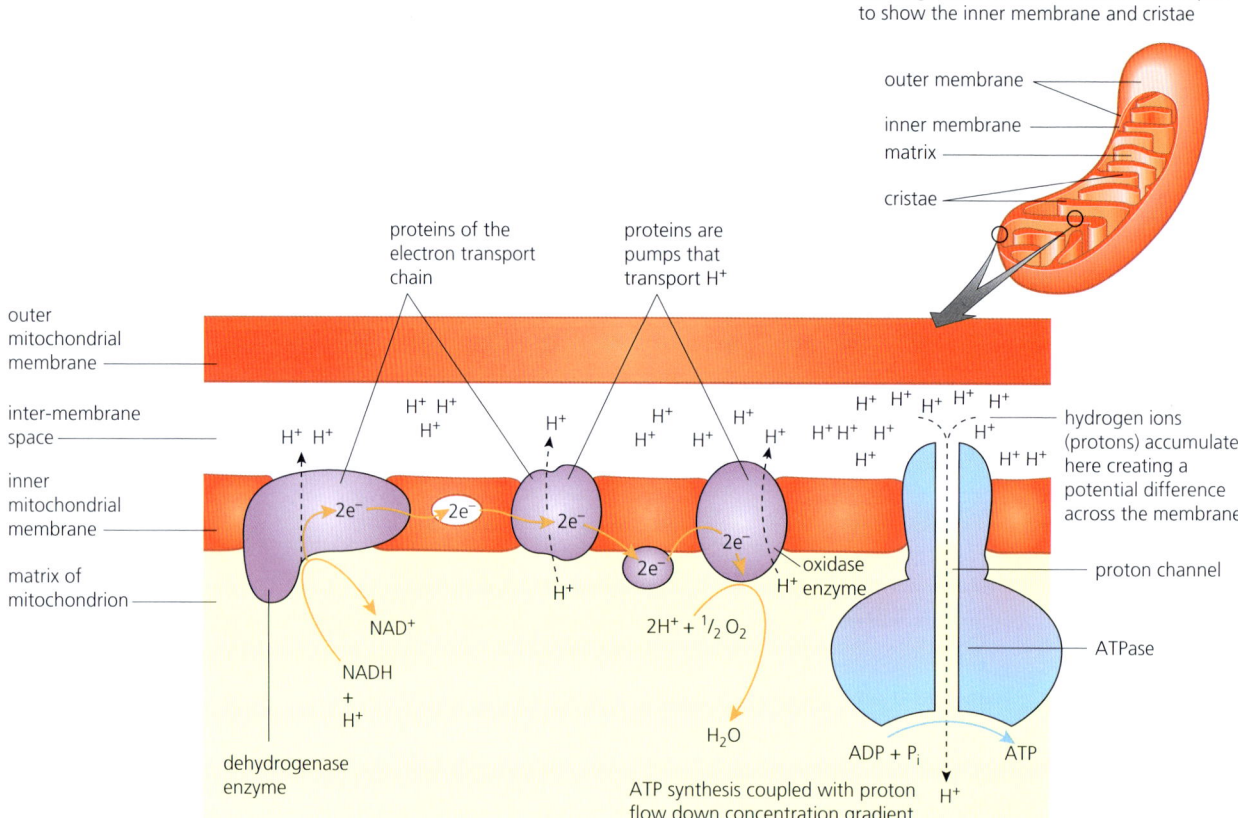

■ **Figure C1.2.11** Chemiosmosis

Nature of science: Evidence

Evidence for the chemiosmosis hypothesis

Peter Mitchell's chemiosmotic hypothesis proposed that the energy of a proton gradient formed through the transfer of electrons via the electron transport chain could be used to drive ATP synthesis. To support the hypothesis, evidence was needed.

If the gradient drives ATP synthesis then disrupting the inner mitochondrial membrane or eliminating the proton gradient across it should inhibit ATP production. Both predictions were found to be supported by experimental evidence.

Physical disruption of the inner mitochondrial membrane stops ATP synthesis. The loss of the proton gradient by a chemical 'uncoupling' agent, such as 2,4-dinitrophenol, also inhibits mitochondrial ATP production. The 2,4-dinitrophenol molecule carries protons (H+) across the inner mitochondrial membrane, bypassing the ATP synthase complex.

This uncoupling process takes place naturally in brown fat cells (found in hibernating mammals, such as bears) and results in most of the energy from the oxidation of fatty acids being released as heat, rather than converted to ATP.

Chemiosmotic coupling mechanisms are of ancient origin. Modern micro-organisms, including archaea, that live in environments similar to those thought to have been present on the early Earth, also use chemiosmotic coupling to produce ATP.

ATL C1.2C

Find out and present the research work by Efraim Racker and Walther Stoeckenius in 1974, who used bacteriorhodopsin to provide further strong support for Mitchell's chemiosmotic hypothesis. Collaborate with a classmate and use diagrams in your presentation.

Tool 1: Experimental techniques

Digital molecular modelling

The following site has animations of molecular models illustrating the mechanism of ATPase:
www.mrc-mbu.cam.ac.uk/research-groups/walker-group/molecular-animations-atp-synthase

How is the rotational movement of the enzyme generated, and how is this coupled with ATP synthesis?

9 When ATP is synthesized in mitochondria, **explain** where the electrochemical gradient is set up, and in which direction protons move.

ATL C1.2D

Explore the structure and function of ATPase further using the following sites:
www.sigmaaldrich.com/GB/en/technical-documents/technical-article/research-and-disease-areas/metabolism-research/atp-synthase
www.youtube.com/watch?v=k_DQ1FjFuYM
Draw an annotated diagram to show the structure and function of ATPase.

Role of oxygen as terminal electron acceptor in aerobic cell respiration

At the end of the electron transport chain, electrons need a final destination (i.e. a terminal acceptor) where they can be removed from the system. This enables the continued flow of electrons along the chain. Similarly, protons used to create the electrochemical gradient also need to be removed. The terminal acceptor of the electrons is oxygen, and when these are combined with protons from the matrix of the mitochondrion, metabolic water is formed (Figure C1.2.12).

ATL C1.2E

Review your knowledge of respiration using this site. Produce a mind map showing the metabolic pathways of aerobic and anaerobic respiration.
www.sumanasinc.com/webcontent/animations/content/cellularrespiration.html

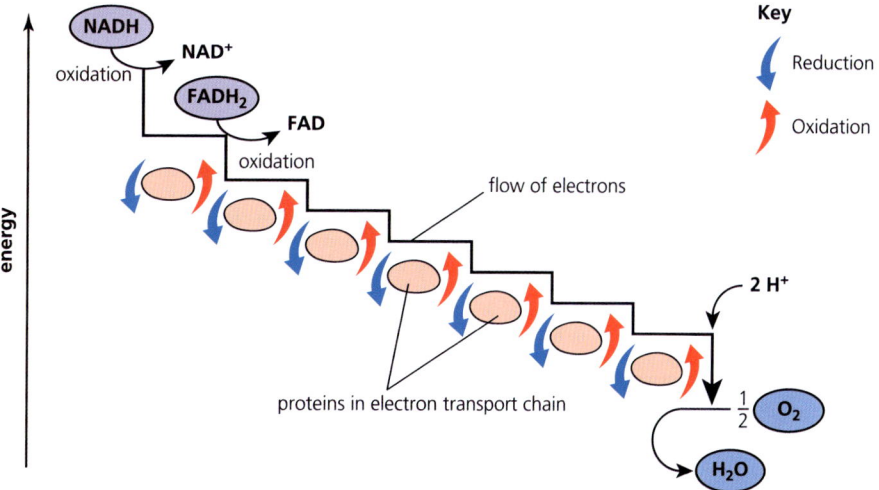

■ **Figure C1.2.12** Oxygen is the terminal acceptor for the electron transport chain in mitochondria

C1.2 Cell respiration

Differences between lipids and carbohydrates as respiratory substrates

When triglycerides are oxidized in respiration, a lot of energy is transferred (and used to make ATP for example, see page 406). Mass for mass, fats and oils transfer more than twice as much energy as carbohydrates when they are respired (Table C1.2.4). This is because fats are comparatively low in oxygen atoms (the carbon of lipids is more reduced than that of carbohydrates), so more of the oxygen in the respiration of fats comes from the atmosphere. In the oxidation of carbohydrates, more oxygen is present in the carbohydrate molecule itself. Fat and oils, therefore, form a concentrated energy store.

Because they are insoluble, the presence of fat or oil in cells does not cause osmotic water uptake. A fat store is especially typical of animals that endure long unfavourable spells in which they survive on reserves of food stored in the body. Oils are often a major energy store in plants, their seeds and fruits, and it is common for fruits and seeds – including maize, olives and sunflower seeds – to be used commercially as a source of edible oils for humans.

Complete oxidation of fats and oils produces a large amount of water, far more than when the same mass of carbohydrate is respired. Desert animals such as the camel and desert rat retain much of this metabolic water within the body, helping them survive when there is no liquid water for drinking. The development of bird and reptile embryos, while in their shells, also benefits from metabolic water formed by the oxidation of the stored fat in their egg's yolk.

■ Table C1.2.4 Lipids and carbohydrates as energy stores – a comparison

Lipids	Role	Carbohydrates
more energy per gram than carbohydrates	energy store	less energy per gram than lipids
much metabolic water is produced on oxidation	metabolic water source	less metabolic water is produced on oxidation
insoluble, so osmotic water uptake is not caused	solubility	sugars are highly soluble in water, causing osmotic water uptake
not quickly 'digested'	ease of breakdown	more easily hydrolyzed – energy is transferred quickly

LINKING QUESTIONS

1. In what forms is energy stored in living organisms?
2. What are the consequences of respiration for ecosystems?

C1.3 Photosynthesis

> **Guiding questions**
> - How is energy from sunlight absorbed and used in photosynthesis?
> - How do abiotic factors interact with photosynthesis?

SYLLABUS CONTENT

This chapter covers the following syllabus content:
- ▶ C1.3.1 Transformation of light energy to chemical energy when carbon compounds are produced in photosynthesis
- ▶ C1.3.2 Conversion of carbon dioxide to glucose in photosynthesis using hydrogen obtained by splitting water
- ▶ C1.3.3 Oxygen as a by-product of photosynthesis in plants, algae and cyanobacteria
- ▶ C1.3.4 Separation and identification of photosynthetic pigments by chromatography
- ▶ C1.3.5 Absorption of specific wavelengths of light by photosynthetic pigments
- ▶ C1.3.6 Similarities and differences of absorption and action spectra
- ▶ C1.3.7 Techniques for varying concentrations of carbon dioxide, light intensity or temperature experimentally to investigate the effects of limiting factors on the rate of photosynthesis
- ▶ C1.3.8 Carbon dioxide enrichment experiments as a means of predicting future rates of photosynthesis and plant growth
- ▶ C1.3.9 Photosystems as arrays of pigment molecules that can generate and emit excited electrons (HL only)
- ▶ C1.3.10 Advantages of the structured array of different types of pigment molecules in a photosystem (HL only)
- ▶ C1.3.11 Generation of oxygen by the photolysis of water in photosystem II (HL only)
- ▶ C1.3.12 ATP production by chemiosmosis in thylakoids (HL only)
- ▶ C1.3.13 Reduction of NADP by photosystem I (HL only)
- ▶ C1.3.14 Thylakoids as systems for performing the light-dependent reactions of photosynthesis (HL only)
- ▶ C1.3.15 Carbon fixation by rubisco (HL only)
- ▶ C1.3.16 Synthesis of triose phosphate using reduced NADP and ATP (HL only)
- ▶ C1.3.17 Regeneration of RuBP in the Calvin cycle using ATP (HL only)
- ▶ C1.3.18 Synthesis of carbohydrates, amino acids and other carbon compounds using the products of the Calvin cycle and mineral nutrients (HL only)
- ▶ C1.3.19 Interdependence of the light-dependent and light-independent reactions (HL only)

◆ **Photosynthesis**: the production of glucose from carbon dioxide and water, in the presence of chlorophyll and enzymes, using light energy, producing oxygen as a waste product.

◆ **Chlorophyll**: the main photosynthetic pigments of green plants, occurring in the grana membranes (thylakoid membranes) of the chloroplasts; also found in cyanobacteria and in the chloroplasts of algae.

The transformation of light energy to chemical energy in photosynthesis

Green plants and other photosynthetic organisms use the energy of sunlight to produce glucose from the inorganic raw materials carbon dioxide and water, by a process called **photosynthesis**. The waste product is oxygen. Photosynthesis occurs in plant cells that contain the organelles called **chloroplasts** (page 76), including many of the cells of the leaves of green plants. Here, energy of light is trapped by the green pigment **chlorophyll** and becomes the chemical energy in molecules such as glucose and ATP. This supplies most of the chemical energy needed for life processes in ecosystems.

1 **Compare** the source of glucose for cellular respiration in mammals and flowering plants.

Glucose formed in photosynthesis may temporarily be stored as starch, but sooner or later much is used in metabolism. For example, plants synthesize other carbohydrates, together with the lipids, proteins, growth factors and all other metabolites they require. For this, they additionally need certain mineral ions, which are absorbed from the soil solution. Figure C1.3.1 is a summary of photosynthesis and its place in plant metabolism.

photosynthesis: a summary

The process in the chloroplast can be summarized by the equation:

carbon dioxide + water + LIGHT ENERGY $\xrightarrow{\text{chlorophyll in chloroplast}}$ organic compounds, e.g. sugars + oxygen

| raw materials | energy source | products | waste product |

$6CO_2 + 6H_2O + $ LIGHT ENERGY $\xrightarrow{\text{chlorophyll in chloroplast}}$ $C_6H_{12}O_6 + 6O_2$

plant nutrition: a summary

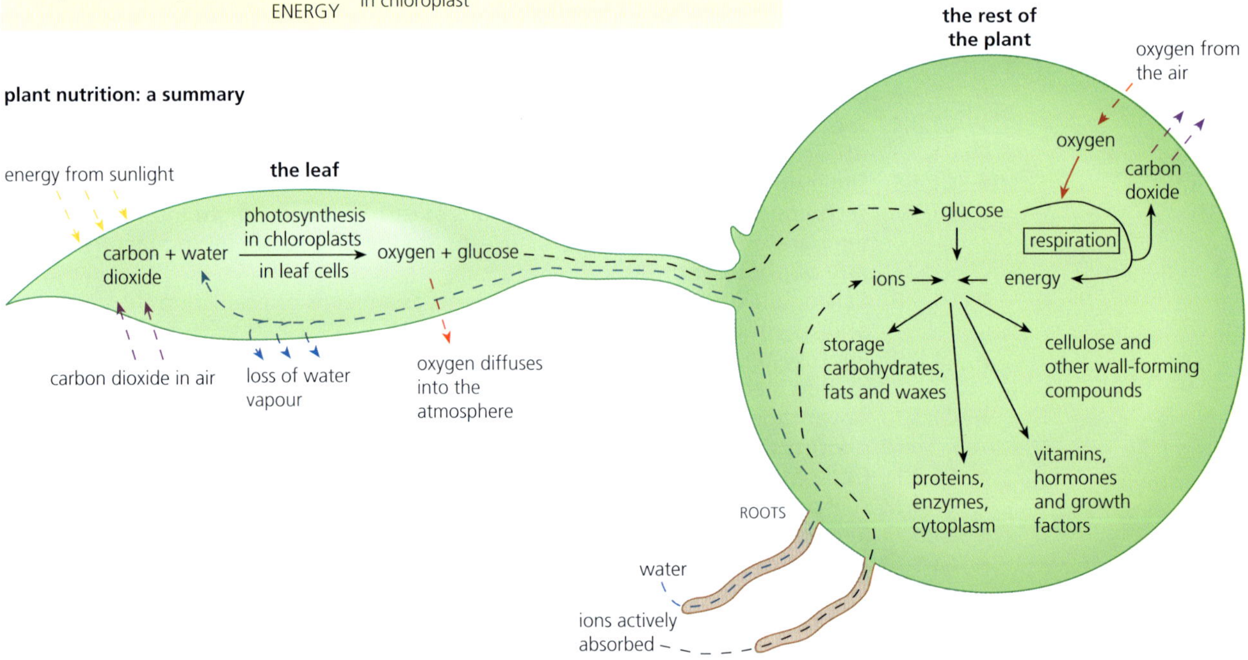

■ Figure C1.3.1 Photosynthesis and its place in plant nutrition

⊙ Common mistake

Do not forget to say that light is the original source of energy for photosynthesis, not simply 'the Sun'.

■ The wider significance of photosynthesis

The importance of photosynthesis to the green plant is that it provides the energy-rich sugar molecules from which the plant builds its other organic molecules. However, the advent of oxygen generation by **photolysis** (the splitting of water molecules using light energy – see page 427) had immense consequences for living organisms and geological processes on Earth:

1 Photosynthesis maintains the **composition of the atmosphere**. For example, the quantity of carbon dioxide removed by plants and other photosynthetic organisms each day is almost equal to that added to the air from respiration and from the burning of fossil fuels. This is illustrated by

Link

The carbon cycle is covered in Chapter C4.2, page 594. The early atmosphere of the Earth is detailed in Chapter A2.1, page 34.

Theme C: Interaction and interdependence – Molecules

> **Common mistake**
>
> It is incorrect to say that plants only undergo photosynthesis and only animals undergo cellular respiration. Plant cells also contain mitochondria and are continually respiring.

◆ **Pigment**: coloured compounds produced by metabolism.

◆ **Accessory pigments**: light-absorbing compounds that trap light energy and channel it to chlorophyll *a*, the primary pigment, which initiates the reactions of photosynthesis.

> **Link**
>
> Other pigments encountered in the course are haemoglobin, see Chapter B1.2, page 217, and melanin, see Chapter D3.2, page 744.

2 State what colour of light the pigment chlorophyll chiefly reflects or transmits (rather than absorbs).

◆ **Photolysis**: the splitting of water molecules using light energy.

the **carbon cycle**. Photosynthesis is also the only natural process that releases oxygen into the atmosphere. All the oxygen present in the air (about 21%) is a waste product of photosynthesis. This is the source of oxygen for aerobic respiration.

2 Some of the oxygen originating from green plants is converted into **ozone** in the upper atmosphere due to the action of ultraviolet (UV) light from the Sun. As a result, ozone occurs naturally in Earth's atmosphere as a layer in the stratosphere. The ozone layer protects terrestrial life from UV light by significantly reducing the quantity of UV that reaches the Earth's surface. UV light is very harmful to living things because it is absorbed by the organic bases of nucleic acids (DNA and RNA), causing them to be modified. Consequently, the conversion of oxygen to ozone and the maintenance of the high-level ozone layer have been important in the evolution of terrestrial life (UV light does not penetrate water and so cannot reach aquatic organisms). The ozone in our upper atmosphere is important to the survival of life today.

■ What is light?

Light is a form of electromagnetic radiation produced by the Sun. Visible light makes up only a part of the total electromagnetic radiation reaching the Earth. When this visible 'white' light is dispersed through a prism, we see a continuous spectrum of light – a rainbow of colours from red to violet. Different colours have different wavelengths and different energies.

The significance of the spectrum of light in photosynthesis is that not all the colours of the spectrum present in white light are absorbed equally by chlorophyll (and other **pigments** used in photosynthesis – known as **accessory pigments**). Some are even transmitted (or reflected), rather than being absorbed.

Biological pigments are coloured compounds produced by metabolism (enzyme-controlled reactions inside cells). Chlorophyll is a biological pigment: it is responsible for the green colour of photosynthetic organisms, which have this pigment as their primary molecule to trap sunlight energy. The colour of pigments results from the absorption of certain wavelengths of visible light. All pigment molecules have intense absorption bands in the visible region of the spectrum. The colour seen is the light that is not absorbed but instead is reflected.

Examples of naturally occurring coloured photosynthetic compounds include the anthocyanins, carotenoids (e.g. carotene) and the porphyrins (e.g. chlorophyll).

Conversion of carbon dioxide to glucose in photosynthesis using hydrogen obtained by splitting water

Photosynthesis is a set of many reactions occurring in chloroplasts in the light. However, these can be conveniently divided into two main steps.

1 Light energy is used to split water (**photolysis**)

This releases the waste product of photosynthesis, oxygen, and allows the hydrogen atoms to be retained on hydrogen-acceptor molecules. The hydrogen is one requirement of Step 2. At the same time, ATP is generated from ADP and phosphate, also using energy from light.

2 Glucose is built up from carbon dioxide

We say that carbon dioxide is 'fixed' to make organic molecules. To do this, both the energy of ATP and hydrogen atoms from the reduced hydrogen-acceptor molecules are required. The hydrogen acceptors carry the hydrogens to a series of reactions that result in the formation of glucose.

C1.3 Photosynthesis

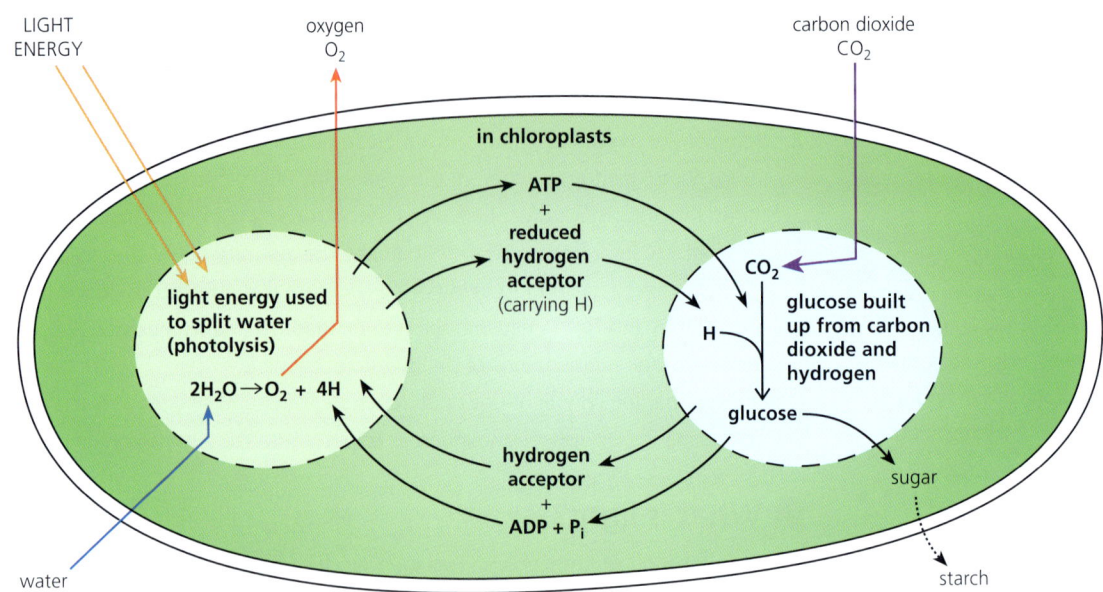

■ Figure C1.3.2 Two steps of photosynthesis

> ● **Top tip!**
>
> This is the simple word equation for photosynthesis, with glucose as the product. Make sure you know it.
>
> $$\text{carbon dioxide} + \text{water} \xrightarrow{\text{light}} \text{glucose} + \text{oxygen}$$

Oxygen as a by-product of photosynthesis in plants, algae and cyanobacteria

As we have seen above, the oxygen produced by photosynthesis comes from the splitting of water. The early Earth had no free oxygen and so the evolution of organisms that could release oxygen was critical in the development of life on Earth.

It is not only plants that carry out photosynthesis. Algae and cyanobacteria also contain photosynthetic pigments and produce glucose and oxygen. Cyanobacteria were the first organisms on Earth to produce oxygen, some 2.8 billion years ago (see page 34). Cyanobacteria have accessory pigments called phycobilins, and transfer energy from these to chlorophyll a. Around 1.5 billion years ago, red and then green algae evolved. Red algae have the same pigments as cyanobacteria: chlorophyll a and phycobilins. In particular, red algae derive most of their colouring from a red phycobilin called phycoerythrin. Green algae (Figure C1.3.3) have chloroplasts that contain chlorophyll a and b, giving them a bright green colour, as well as the accessory pigments β-carotene (red-orange) and xanthophylls (yellow). Unlike cyanobacteria and red and brown algae, they do not contain phycobilin pigments.

Link

The early Earth's atmosphere composition is covered in Chapter A2.1, page 34.

3 Define the term *photolysis*.

■ Figure C1.3.3 Photomicrograph of *Volvox globator*, a green algae. Dark green spots are daughter colonies forming inside of the parent colony

Separation and identification of photosynthetic pigments by chromatography

◆ **Chromatography**: technique used to separate components of a mixture. It involves letting soluble substances spread across filter paper (or through a powder).

Chromatography is the technique we use to separate, identify and quantify the component pigments (or dyes) in a mixture. It is an ideal technique for separating biologically active molecules, since biochemists often can only obtain very small amounts of those molecules. Plant chlorophyll consists of a mixture of pigments. Some plant pigments are soluble in water, but chlorophyll is not. Chlorophyll can be extracted in an organic solvent such as propanone (acetone) (Figure C1.3.4). We can then identify the photosynthetic pigments in plant leaves using chromatography.

> ### Tool 1: Experimental techniques
>
> #### Carrying out paper chromatography
> Look carefully at Figures C1.3.4 and C1.3.5, which show how to carry out chromatography. Details of how to interpret chromatograms are given below.

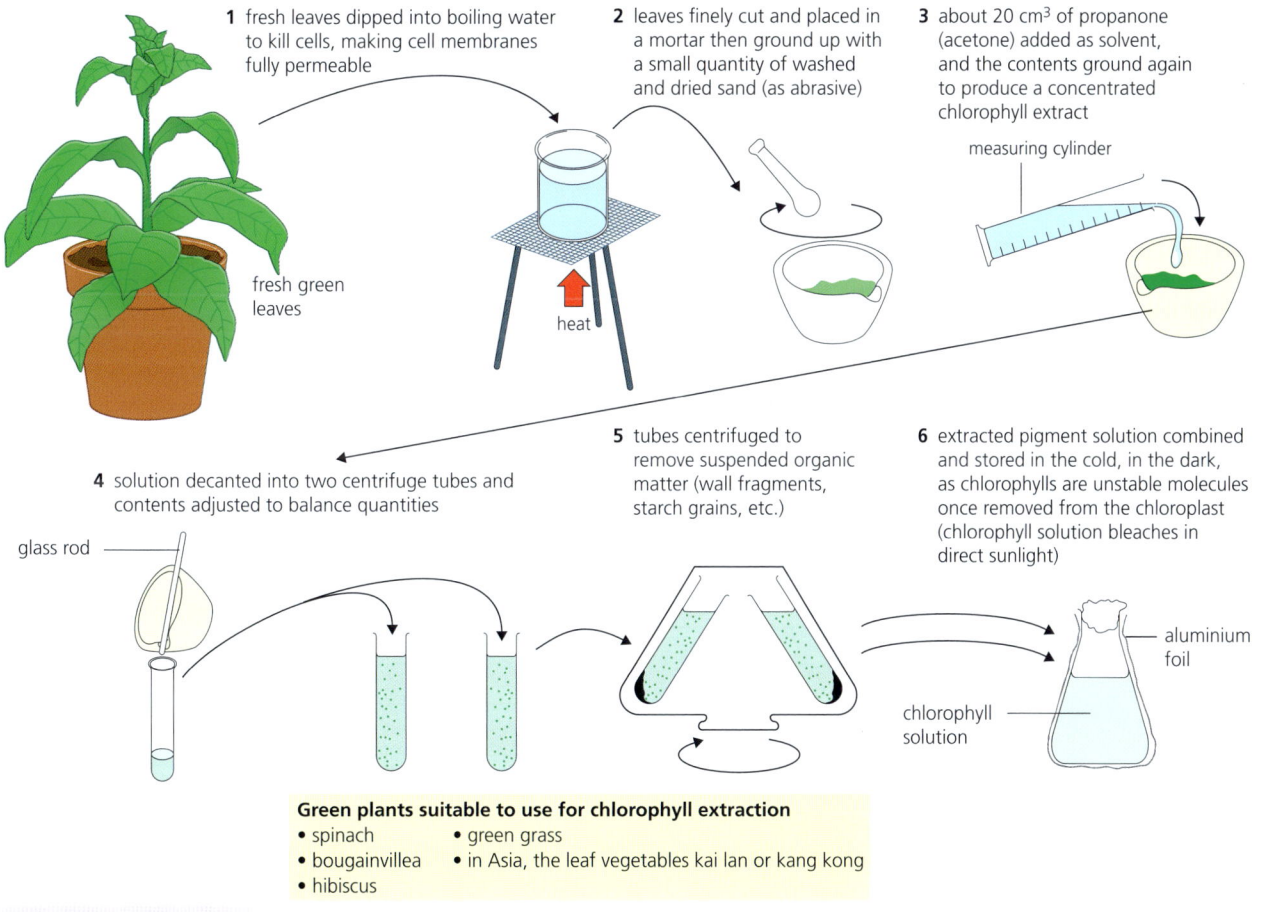

■ **Figure C1.3.4** Steps in the extraction of plant pigments

◆ **Chromatogram**: the pattern formed on an adsorbent medium showing the result of separating the components of a mixture by chromatography.

Chromatograms are typically run on adsorbent paper (paper chromatography), powdered solid (column chromatography), or on a thin film of dried solid (thin-layer chromatography). The process is illustrated in Figure C1.3.5.

C1.3 Photosynthesis

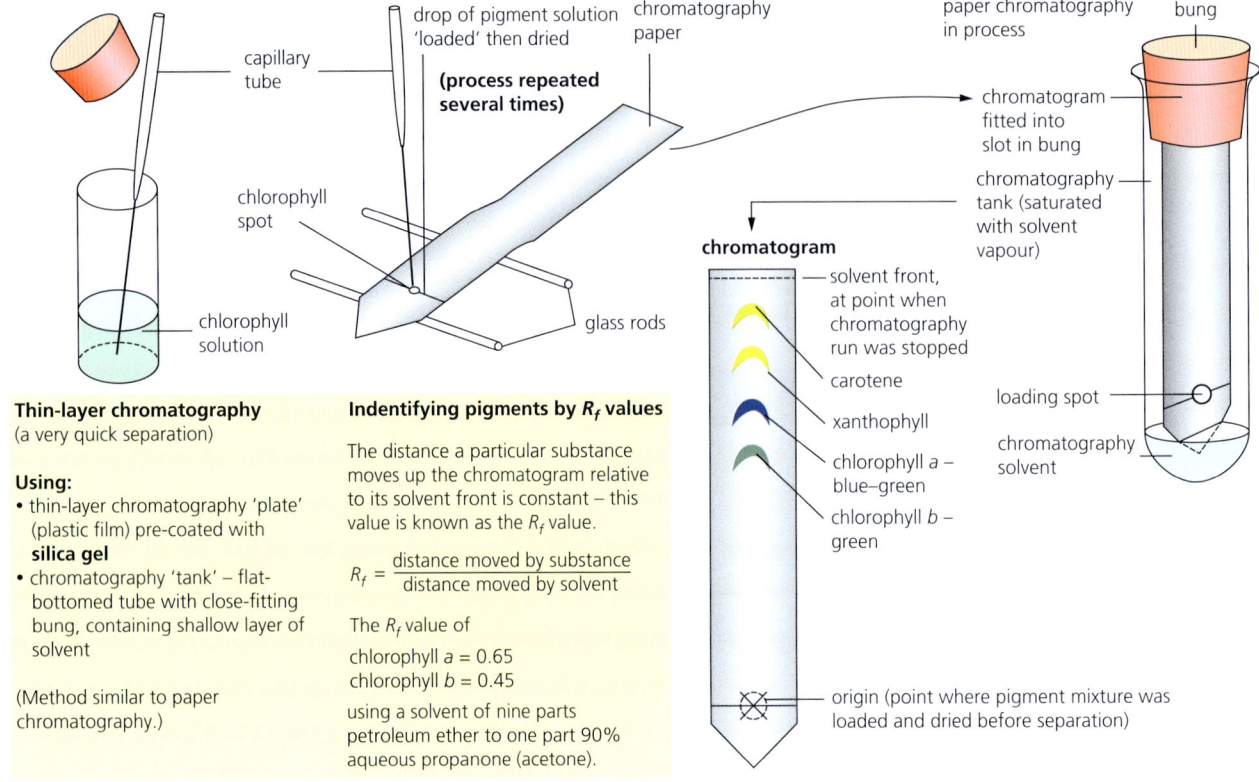

Thin-layer chromatography
(a very quick separation)

Using:
- thin-layer chromatography 'plate' (plastic film) pre-coated with **silica gel**
- chromatography 'tank' – flat-bottomed tube with close-fitting bung, containing shallow layer of solvent

(Method similar to paper chromatography.)

Identifying pigments by R_f values

The distance a particular substance moves up the chromatogram relative to its solvent front is constant – this value is known as the R_f value.

$$R_f = \frac{\text{distance moved by substance}}{\text{distance moved by solvent}}$$

The R_f value of
chlorophyll a = 0.65
chlorophyll b = 0.45
using a solvent of nine parts petroleum ether to one part 90% aqueous propanone (acetone).

■ **Figure C1.3.5** Preparing and running a chromatogram

Top tip!

Chromatography is based upon the differential retention of compounds in a mobile phase as they pass through or across a stationary phase.

In paper chromatography, the stationary phase is a liquid adsorbed on to the surface of the paper. The paper has many pores that can adsorb and form hydrogen bonds with water molecules to form the stationary phase. The water can be displaced by other liquids to give different stationary phases. Pigments that are more soluble in the solvent than they are in the water molecules of the stationary phase move rapidly up the paper, while those that are more soluble in the water are not carried as far up the paper (Figure C1.3.6).

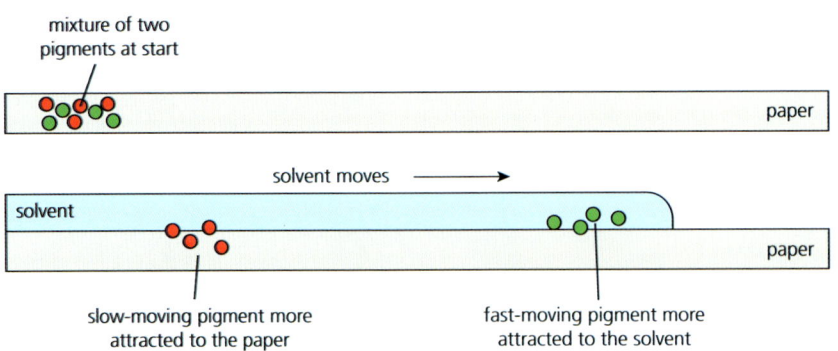

■ **Figure C1.3.6** The principles of paper chromatography

Thin-layer chromatography (TLC) uses a stationary phase of silica or alumina particles bonded to a thin layer of glass or plastic. TLC separates a mixture of pigments based on how strongly they are adsorbed on the stationary phase and dissolved in the mobile phase (a liquid or mixture of liquids). This equilibrium is known as partitioning. The greater the affinity of the pigment for the stationary phase, the more slowly it moves along the surface of the TLC plate.

The distance the spot moves from its starting location divided by the distance moved by the solvent is known as the **R_f value** (retardation factor). An R_f value is characteristic of a particular solute in a particular solvent. It can be used to identify components of a mixture by comparing the measured value with tables of known R_f values. The R_f value is calculated using the distance travelled by the solvent front and the distance from the origin to the centre of each spot (see Figure C1.3.5). The R_f value depends on the compound, solvent, temperature and other factors, such as nature of the TLC plate. You need to be able to calculate R_f values from the results of chromatographic separation of photosynthetic pigments and identify them by colour and by values. Thin-layer chromatography or paper chromatography can be used. Figure C1.3.5 shows how to calculate the R_f value.

> **Top tip!**
>
> Paper chromatography can be used to separate photosynthetic pigments but thin-layer chromatography (TLC) gives better results. TLC is faster, more sensitive and has better resolution than paper chromatography.

4 a **Calculate** the R_f value for the pigment in the experiment shown in Figure C1.3.7.
  b **Identify** the pigment present using Table C1.3.1.

■ Table C1.3.1 R_f values for different pigments

Name	Colour	R_f value
carotene	yellow	0.95
phaeophytin	yellow–grey	0.83
xanthophylls	yellow–brown	0.71
chlorophyll a	blue–green	0.65
chlorophyll b	green	0.45

◆ **R_f value**: a constant distance that a particular substance moves up a chromatogram relative to its solvent front. The R_f value of a compound is equal to the distance travelled by the compound divided by the distance travelled by the solvent front (both distances measured from the origin).

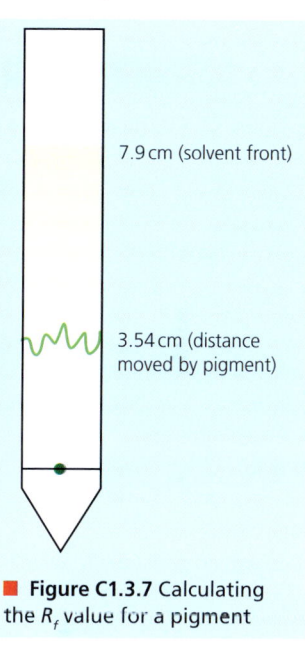

7.9 cm (solvent front)

3.54 cm (distance moved by pigment)

■ Figure C1.3.7 Calculating the R_f value for a pigment

> **ATL C1.3A**
>
> Access this site: http://amrita.olabs.edu.in/?sub=79&brch=17&sim=124&cnt=1
> Use the simulation to test different types of plant pigment (spinach leaf extract). Move the hand to collect a sample of the extract and place it on the filter paper. Put the filter paper into the isopropyl alcohol and water, and run the simulation. Calculate the R_f values from the chromatography experiment.

Absorption of specific wavelengths of light by photosynthetic pigments

◆ **Absorption spectrum**: a graph showing the relative absorbance of different wavelengths of light by a pigment.

We have seen that light consists of a roughly equal mixture of all the visible wavelengths, namely violet, indigo, blue, green, yellow, orange and red – although the names of the colours are arbitrary and they merge into each other. How much of each wavelength does chlorophyll absorb? We call this information an **absorption spectrum**.

C1.3 Photosynthesis

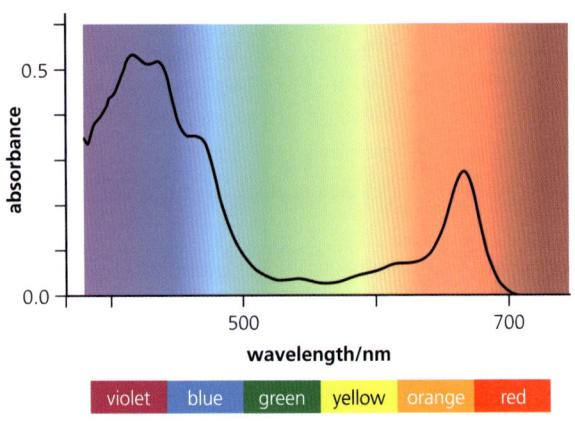

■ **Figure C1.3.8** The absorbance spectrum of chlorophyll

The **absorption spectra** (singular, spectrum) of chlorophyll pigments are obtained by measuring their absorption of violet, indigo, blue, green, yellow, orange and red light in turn. The results are plotted as a graph showing the amount of light absorbed over the wavelength range of visible light, as shown in Figure C1.3.8. You can see that chlorophyll absorbs blue and red light most strongly. Other wavelengths are absorbed less or not at all. It is the chemical structure of the chlorophyll molecule that causes absorption of the energy of blue and red light.

Within a pigment molecule, absorption of light excites electrons. Accessory pigments transfer excited electrons to chlorophyll *a*, which leads to a series of reactions that ultimately lead to the formation of glucose. In this way, light energy is transformed to chemical energy.

5 Absorption spectra for two coloured substances, A and B, are shown in Figure C1.3.9. **Identify** the colour of each substance. **Explain** your answer.

> ● **Top tip!**
>
> Unless colours are shown in an absorption spectrum (Figure C1.3.9), both wavelengths and colours of light should be included on the horizontal axis of absorption spectra. This enables the graph to be easily interpreted with each pigment related to the colour of light it absorbs.

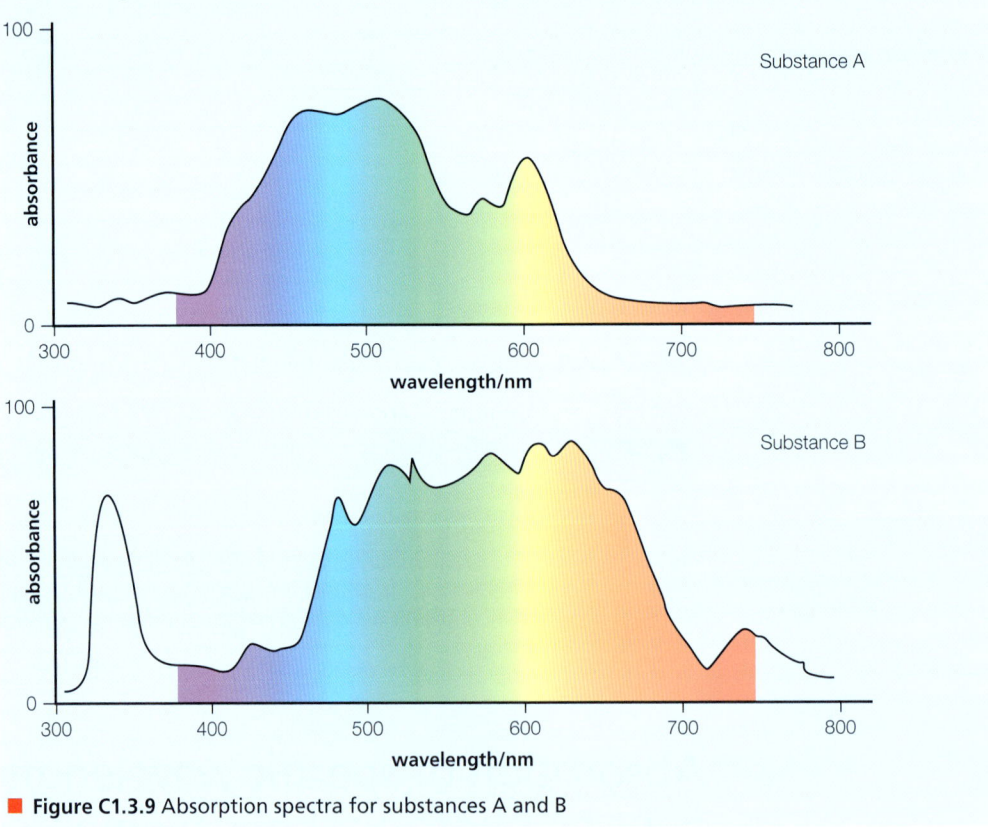

■ **Figure C1.3.9** Absorption spectra for substances A and B

Similarities and differences of absorption and action spectra

◆ **Action spectrum**: range of wavelengths of light within which a process like photosynthesis takes place.

The **action spectrum** of chlorophyll is the wavelengths of light that are used in photosynthesis. This may be discovered by projecting light of different wavelengths, in turn and for a unit of time, on aquatic green pond weed. This is carried out in an experimental apparatus in which the rate of photosynthesis can be measured. The gas evolved by a green plant in the light is largely oxygen, and the volume given off in a unit of time is a measure of the rate of photosynthesis. Suitable apparatus is shown in Figure C1.3.11 (page 435).

The rate of photosynthesis at different wavelengths may then be plotted on a graph on the same scale as the absorption spectrum (Figure C1.3.10). We have seen that blue and red light wavelengths are most strongly absorbed by chlorophyll. From the action spectrum, we see that it is these wavelengths that give rise to the highest rates of photosynthesis.

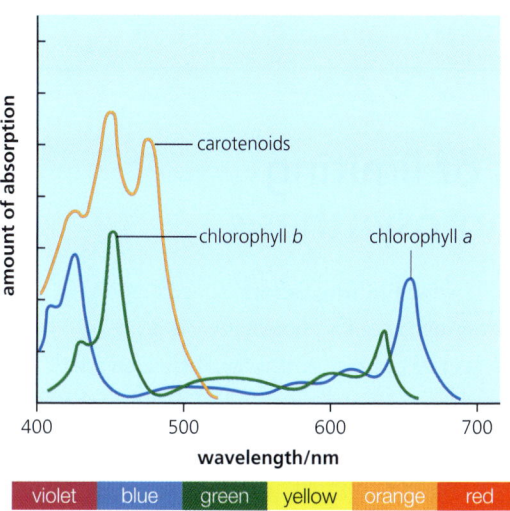
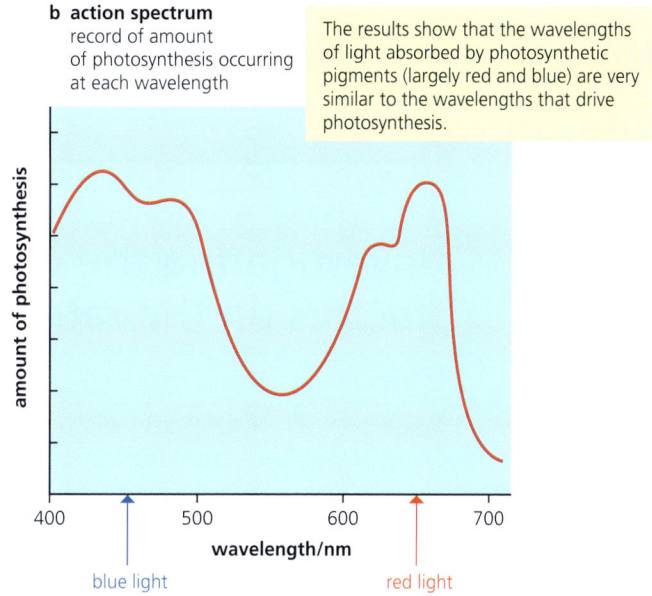

■ **Figure C1.3.10** Absorption and action spectra of chlorophyll pigments

Inquiry 2: Collecting and processing data

Collecting and processing data; interpreting results

Rates of photosynthesis can be determined from data for oxygen production and carbon dioxide consumption for varying wavelengths. These data can be plotted to make an action spectrum.

Set up an experiment to measure the effects of varying wavelengths on the rate of photosynthesis. Pondweed or other aquatic plants can be used to measure the rate of photosynthesis (see Figure C1.3.11 below). Use different colour filters in front of the light source so that the plant is exposed to a range of different colours of light. Coloured plastic film can be used to make a filter. You should aim to cover the full spectrum.

Carry out a series of experiments, each using a different coloured filter, and measure the rate of photosynthesis for each colour by counting the number of bubbles of oxygen produced by the aquatic plant per minute. Each colour represents a different wavelength of light. You should also carry out an experiment with no filter, to measure the effects of white light. You should repeat each experiment (ideally five times) to ensure your results are reliable.

Which variables will you need to keep the same and how will you do this?

Use your data to plot an action spectrum, with wavelength of light on the *x*-axis and rate of photosynthesis (measured by number of bubbles of oxygen produced per minute) on the *y*-axis. Identify, describe and explain the patterns you see. Which colours of light generated the greatest rate of photosynthesis, and which the lowest? Use your knowledge of photosynthesis and photosynthetic pigments to explain your results.

C1.3 Photosynthesis

> ### Going further
>
> **Chlorophyll in solution versus chlorophyll in the chloroplast**
> The chlorophyll that has been extracted from leaves and dissolved in an organic solvent still absorbs light. However, chlorophyll in solution cannot use light energy to make sugar (in the form of glucose). This is because, in the extraction process, chlorophyll has been separated from the membrane systems and enzymes that surround it in chloroplasts. These are also essential for carrying out the biochemical steps of photosynthesis.

Investigating the effects of limiting factors on the rate of photosynthesis

Hypotheses can be suggested for the effects of the following **limiting factors**: concentrations of carbon dioxide, light intensity and temperature. A limiting factor for photosynthesis is an environmental factor that, when in short supply, affects rate of photosynthesis. There are various protocols for investigating these limiting factors, based upon an understanding of photosynthesis. In the following sections, you will be asked to develop a hypothesis about the effect of one limiting factor on the rate of photosynthesis. You will then be told about the expected results and reasons for these.

Link
Further details and the definition of limiting factors are covered in Chapter B4.1, page 344.

● Nature of science: Hypotheses

Hypotheses are provisional explanations that require repeated testing. Scientists make provisional explanations for the patterns that they have observed in natural phenomena. During scientific research, hypotheses can either be based on theories then tested in an experiment, or be based on evidence from an experiment already carried out. Evidence is required to obtain support for a hypothesis or show that it is false. In the sections below, you will be asked to suggest hypotheses for the effects of limiting factors on photosynthesis (concentration of carbon dioxide, light intensity and temperature). Note: you should be able to identify the dependent and independent variables in experiments investigating the effects of limiting factors on the rate of photosynthesis.

6 **Define** the term *limiting factor*.

Measuring the rate of photosynthesis

A protocol for investigating limiting factors on the rate of photosynthesis is discussed in this section.

An illuminated, freshly cut shoot of a pondweed, when inverted, produces a vigorous stream of gas bubbles from the base. The bubbles tell us the pondweed is actively photosynthesizing. At the same time, dissolved carbon dioxide is being removed from the water. Suitable plants include *Elodea*, *Myriophyllum* and *Cabomba*. The rate of photosynthesis can be estimated using one of the following:

- A microburette to measure the volume of oxygen given out in the light (Figure C1.3.11). The pondweed is placed in a very dilute solution of sodium hydrogencarbonate, which supplies the carbon dioxide (as HCO_3^-) required by the plant for photosynthesis. The quantity of gas evolved in a given time, say in 30 minutes, is measured by drawing the gas bubble that collects into the capillary tube, and measuring its length. This length is then converted to a volume.
- An oxygen sensor probe connected to a data-logging device.
- A pH meter connected to a data-logging monitor. The uptake of carbon dioxide from the water will cause the pH to rise.

7 A thermometer is not shown in the apparatus in Figure C1.3.11. **Predict** why one is required, and state where it should be positioned.

8 **Explain** why the cut stem of pondweed was inverted here.

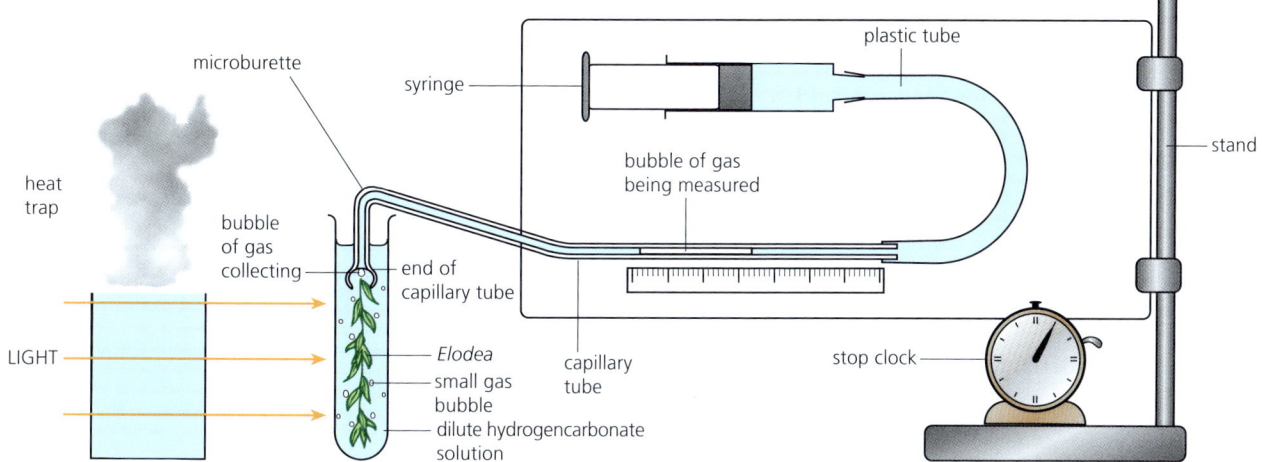

■ **Figure C1.3.11** Measuring the rate of photosynthesis with a microburette

 You can use one or more of these techniques to investigate the effects of external conditions on the rate of photosynthesis (Table C1.3.2). Explore protocols investigating these limiting factors, based upon your understanding of photosynthesis, and test these by experimentation.

■ **Table C1.3.2** Issues in the design of experiments to investigate the effect of external factors on the rate of photosynthesis

Carbon dioxide concentration		Light intensity
O_2 output (bubbles or volume) in unit time	**dependent variable**	O_2 output (bubbles or volume) or pH change in unit time
external CO_2 (i) absence of CO_2 by boiling and cooling water (ii) subsequent stepwise addition of $NaHCO_3$ solution to raise CO_2 by $0.01\,mol\,dm^{-3}$ until no further change in O_2 output	**independent variable**	light intensity – systematically positioning the light source (photoflood lamp or 150 W bulb) at 5, 10, 15, 20 and 25 cm from the experimental chamber
temperature, light intensity	**controlled variables**	concentration of $NaHCO_3$, temperature
possibly errors in $NaHCO_3$ solution additions	**sources of error**	possibly the heating effect of the light source

9 Look at Table C1.3.2. **Discuss** further controlled variables and sources of error not included in this table.

ATL C1.3B

Access the following site: www.biologysimulations.com/cell-energy-sim

Alter the number of plants, fish and temperature to explore how limiting factors affect the balance between respiration and photosynthesis, and correspondingly the concentration of oxygen in the environment. Write a summary of the investigation, and how inputs and outputs from respiration and photosynthesis make both processes interdependent.

■ External factors and the rate of photosynthesis

 Using your knowledge of photosynthesis, develop hypotheses to predict the effect on the rate of photosynthesis of separately changing three different independent variables: carbon dioxide concentration, light intensity and temperature. Once you have done this, read on.

C1.3 Photosynthesis

> ## Tool 3: Mathematics
>
> ### Sketch graphs
>
> Sketch graphs have labelled but unscaled axes to qualitatively describe trends, whereas graphs have units in both axes and use numerical data. Figures C1.3.12, C1.3.13 and C1.3.14 show examples of sketch graphs and indicate how these should be drawn.

The effect of carbon dioxide concentration

The effect of the **concentration of carbon dioxide** on the rate of photosynthesis is shown in Figure C1.3.12. *Look at this sketch graph and note the shape of the curve.*

In this experiment (Figure C1.3.12):
- when the concentration of carbon dioxide is at zero, there is no photosynthesis
- as the concentration is steadily increased, the rate of photosynthesis rises, and the rate of that rise shows positive correlation with the increasing carbon dioxide concentration (carbon dioxide is a limiting factor)
- at much higher concentrations of carbon dioxide, the rate of photosynthesis reaches a plateau. Some other factor is limiting the rate, e.g. light intensity or temperature.

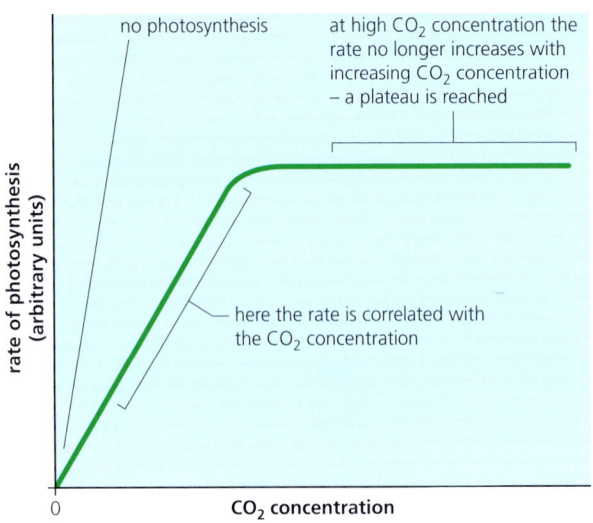

■ **Figure C1.3.12** The effect of carbon dioxide concentration on photosynthesis

● Common mistake

Refer to the concentration of carbon dioxide rather than 'amount', 'level' or 'quantity'.

The effect of light intensity

The effect of **light intensity** on the rate is shown in Figure C1.3.13. *Look at this sketch graph – the shape of the curve is familiar. Then answer question 10.*

Figure C1.3.13 shows the following:
- As the light intensity increases, the rate of photosynthesis rises – the rate of that rise is positively correlated with the increasing light intensity. Light intensity is the limiting factor.
- At much higher light intensities, the rate of photosynthesis reaches a plateau – now there is no increase in rate with rising light intensity. Some other factor is limiting the rate, e.g. carbon dioxide concentration or temperature.

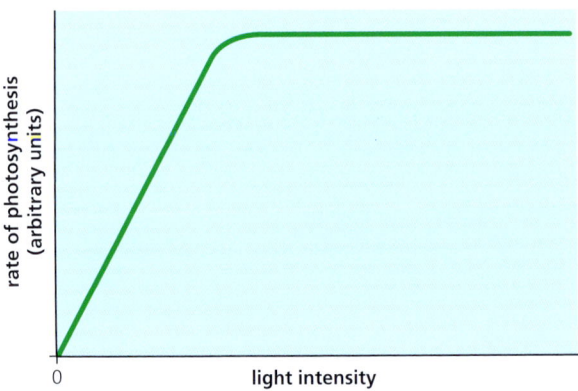

■ **Figure C1.3.13** The effect of light intensity on photosynthesis

10 Explain what the sketch graph implies about the relationship between light intensity and the rate of photosynthesis.

The effect of temperature

The effect of **temperature** on the rate of photosynthesis is shown in Figure C1.3.14. Here the curve of the sketch graph is an entirely different shape. At relatively low temperatures, as the temperature increases, the rate of photosynthesis increases more and more steeply. However, at higher temperatures, the rate of photosynthesis abruptly stops rising and actually falls steeply. The result is a clear optimum temperature for photosynthesis.

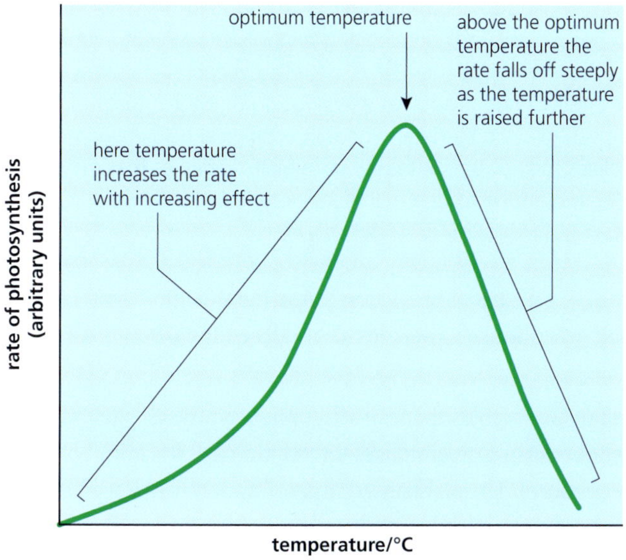

The shape of the sketch graph in Figure C1.3.14 can be explained by enzyme collision theory (Chapter C1.1).

- As temperature increases, the enzymes involved in photosynthesis have increased kinetic energy, as do their substrates.
- There are increased collisions between enzymes and substrates, resulting in more enzyme–substrate complexes and increased rate of reaction.
- These reactions reach an optimum rate at a specific temperature.
- Above the optimum rate, as the temperature rises further the enzymes involved with the fixation of carbon dioxide cannot work effectively.

■ **Figure C1.3.14** The effect of temperature on photosynthesis

● Common mistake

It is not correct to say that the rate of photosynthesis reduces at higher temperatures because of enzyme denaturation. The rate reduction occurs at much lower temperatures than those at which this denaturation would occur. The problem at higher temperatures is due to an enzyme (RuBisCo – see page 445 if you are studying HL) failing to fix carbon dioxide effectively.

Tool 1: Experimental techniques

Accurately measuring temperature

The SI unit for the temperature is the Kelvin (k). However, most thermometers are calibrated with a temperature scale in degree Celsius (°C) a non-SI unit. Electronic (digital) thermometers are now commonly used to measure temperature, as they are more accurate than alcohol-in-glass type thermometers. Accuracy reflects how close to the true value measurements can be taken. Uncertainties are given for lab equipment, which indicate a range within which we are confident that the true value lies. The uncertainty for digital thermometers is ± 0.1 °C.

Tool 2: Technology

Generating data from simulations

Access this site: **https://iwant2study.org/lookangejss/biology/ejss_model_photosynthesis/photosynthesis_Simulation.xhtml**

Alter the concentration of light or the concentration of carbon dioxide by selecting experiment 1 or experiment 2, respectively, to investigate the effect of these limiting factors on the rate of photosynthesis.

Start with a light intensity of '1'. Press 'play' to count the number of bubbles in one minute. The data point will automatically be plotted. Now move the light intensity to '2' using the ↕ button, and press play. Repeat this until you have completed all light intensities.

The simulation automatically plots a graph of your data. Explain the shape of each graph using your knowledge of photosynthesis and limiting factors.

C1.3 Photosynthesis

Carbon dioxide enrichment experiments

● Nature of science: Experiments

Finding methods for careful control of variables is part of experimental design. This may be easier in the laboratory but some experiments can only be done in the field. Field experiments include those performed in natural ecosystems. You need to be able to identify a controlled variable in an experiment. For example, in photosynthesis experiments one limiting factor is changed at a time to determine its effect on the rate of photosynthesis (page 434). Controlled variables enable you to keep other variables that may affect the dependent variable the same, so that the impact of the independent variable on the dependent variable can be determined. In ecological field experiments, where other variables cannot be controlled (such as light intensity or rainfall), variables that may affect the dependent variable need to be monitored to determine any potential impact (see Chapter C4.1, page 572).

Carbon dioxide enrichment experiments are a means of predicting future rates of photosynthesis and plant growth. They include **enclosed greenhouse experiments** and **free-air carbon dioxide enrichment experiments (FACE)**.

Enclosed greenhouse experiments manipulate variables that can affect photosynthesis, such as sunlight, wavelength of light and temperature, using greenhouses (or polytunnels) to control other variables (hence the term 'enclosed' greenhouse). The plants used in these experiments need to be small enough to fit into greenhouses, and data gathered do not directly relate to changes that would been seen in natural ecosystems.

■ **Figure C1.3.15** One of the FACE plots set up by Duke University to provide elevated atmospheric carbon dioxide concentration

FACE experiments pump carbon dioxide into the air to raise the local concentration of the gas in forests, grasslands and agricultural fields (Figure C1.3.15). Compared to enclosed greenhouse experiments, FACE experiments are in natural ecosystems, and can investigate the effects of carbon dioxide enrichment on large producers such as trees. Unlike greenhouses, other variables cannot be controlled (such as rainfall and sunlight) and so need to be monitored.

Duke University's FACE project in North Carolina, USA, began operating in 1996 and is the oldest forest FACE project. It has since been completed and was used to study the response of a temperate coniferous forest to high levels of atmospheric carbon dioxide. Tree growth and productivity under conditions of at least 550 parts per million of carbon dioxide – over 1.25 times higher than present levels – were monitored.

With global carbon dioxide levels in the atmosphere continuing to rise (see Chapter C4.2), predicting the effects that this increase will have on biological systems is urgent.

ATL C1.3C

Read more about the Duke Forest FACE carbon dioxide enrichment experiment in this scientific paper: https://link.springer.com/chapter/10.1007/3-540-31237-4_11

Find out about carbon dioxide enrichment experiments in your country or region. How are they being used to predict future rates of photosynthesis and plant growth?

How could you set up a small carbon dioxide enrichment experiment in your school or college? What equipment would you need and what data would you collect?

◆ **Light-dependent reactions**: part of photosynthesis occurring in the grana of chloroplasts and requiring light; water is split and ATP and NADPH are generated.

◆ **Photophosphorylation**: the formation of ATP using light energy (in the light-dependent step of photosynthesis).

◆ **Light-independent reactions**: part of photosynthesis occurring in the stroma of chloroplasts; it uses the products of the light-dependent step to reduce (or fix) carbon dioxide to glucose.

The light-dependent and light-independent reactions of photosynthesis: introduction

Photosynthesis is a complex set of many reactions that takes place in illuminated chloroplasts. Biochemical investigations of photosynthesis by several teams of scientists have established that the many reactions by which light energy brings about the production of glucose, using the raw materials water and carbon dioxide, fall naturally into two interconnected stages (Figure C1.3.16).

- The **light-dependent reactions** use light energy to split water (**photolysis**). Hydrogen is then removed and retained by the photosynthesis-specific hydrogen acceptor, known as $NADP^+$. ($NADP^+$ is very similar to the respiration coenzyme NAD^+, but it carries an additional phosphate group, hence NAD**P**.) At the same time, ATP is generated from ADP and phosphate, also using light energy. This is known as **photophosphorylation**. Oxygen is given off as a waste product of the light-dependent reactions. This stage occurs in the **grana** of the chloroplast.
- The **light-independent reactions** synthesize glucose using carbon dioxide. The products of the light-dependent reactions (ATP and reduced hydrogen acceptor NADPH + H^+) are used in glucose production. This stage occurs in the **stroma** of the chloroplast. It requires a continuous supply of the products of the light-dependent reactions but does not directly involve light energy (hence the name). Names can be misleading because this stage is an integral part of photosynthesis, and photosynthesis is a process that is powered by light energy.

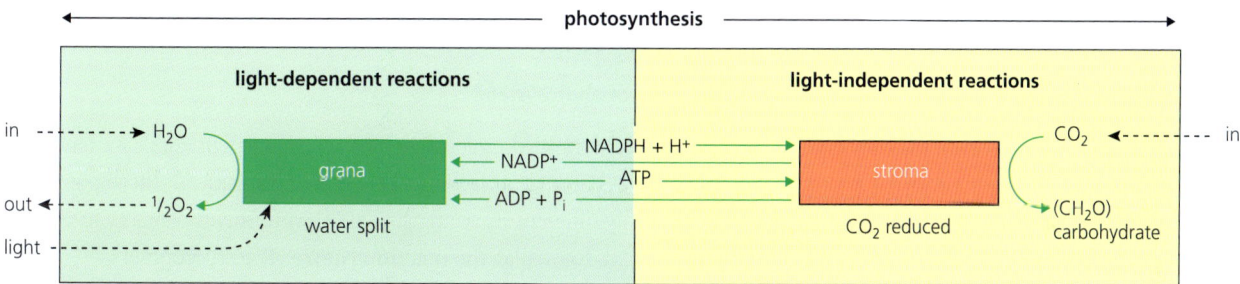

■ **Figure C1.3.16** The two sets of reactions of photosynthesis, inputs and outputs

We shall now look at both sets of reactions in turn to understand more about how these complex changes are brought about.

The light-dependent reactions

■ Photosystems

◆ **Photosystems**: molecular arrays of chlorophyll, accessory pigments and proteins (light-harvesting complexes) with a special chlorophyll as the reaction centre from which an excited electron is emitted.

Photosystems are arrays of pigment molecules (with associated proteins) that can generate and emit excited electrons. Photosystems are always located in membranes and they occur in cyanobacteria and in the chloroplasts of photosynthetic eukaryotes.

In the light-dependent stage, light energy is trapped by the photosynthetic pigment, chlorophyll. Chlorophyll molecules do not occur haphazardly in the grana. Rather, they are grouped together in structures called **photosystems**, held in the **thylakoid membranes** of the **grana** (Figure C1.3.17).

Link
Grana and thylakoid membranes are covered in Chapter A2.2, page 76.

C1.3 Photosynthesis

- ◆ **Light-harvesting complex (LHC)**: an array of protein and chlorophyll molecules embedded in the thylakoid membrane of plants, algae and cyanobacteria, which transfer light energy to one chlorophyll a molecule at the reaction centre of a photosystem.
- ◆ **Photosystem I**: a chlorophyll–protein complex that uses light energy to release excited electrons, replacing each lost electron by one in the ground state (electrons are received from photosystem II).
- ◆ **Photosystem II**: a membrane super-complex of proteins and several hundred chlorophyll molecules, plus accessory pigments, that carries out the initial reaction of photosynthesis; light from the Sun excites electrons, which pass down an electron transport chain: these electrons are replaced by splitting water to release protons and electrons.
- ◆ **Ground-state electrons**: the energy level normally occupied by an electron; the state of lowest energy for an electron.
- ◆ **High-energy (excited) electrons**: chlorophyll in photosystems I and II absorbs light energy (photons), which increases the energy level of electrons.

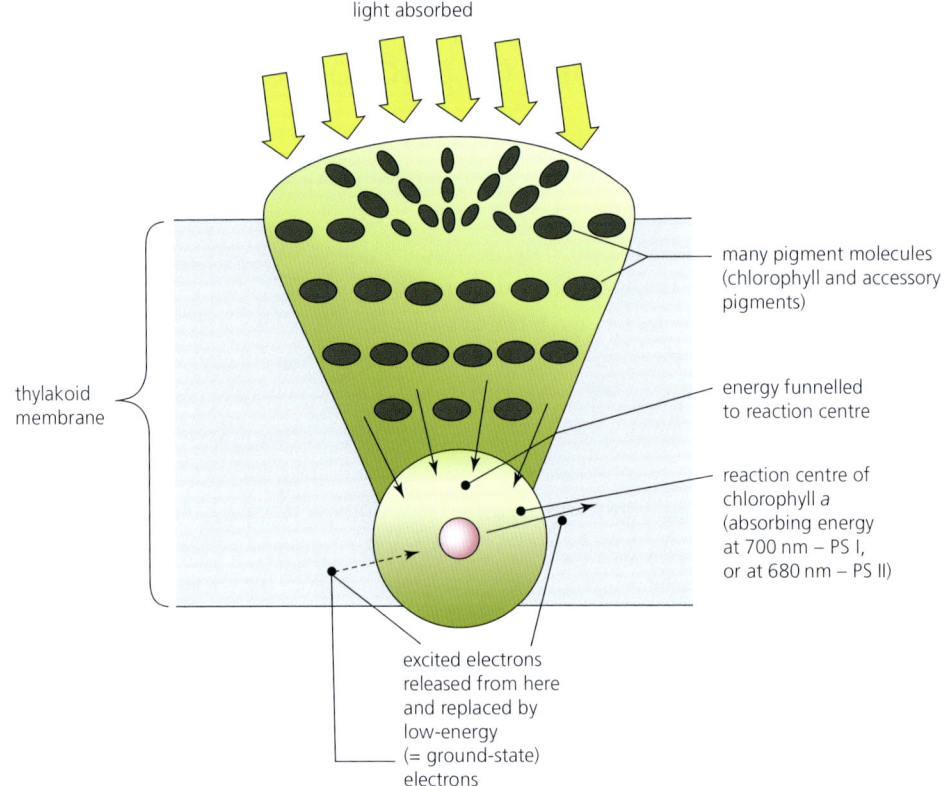

■ Figure C1.3.17 Generalized structure of photosystems

In each photosystem, several hundred chlorophyll molecules, plus accessory pigments (carotene and xanthophylls) and proteins (**light-harvesting complexes** or **LHCs**) are arranged. All these pigment molecules harvest light energy of slightly different wavelengths, and they funnel the energy to a single chlorophyll molecule of the photosystem, known as the **reaction centre**. The chlorophyll is then **photoactivated**.

There are **two types of photosystem** present in the thylakoid membranes of the grana, identified by the wavelength of light that the chlorophyll of the reaction centre absorbs:

- **Photosystem I** has a reaction centre that is activated by light of wavelength 700 nm. This reaction centre is referred to as P700.
- **Photosystem II** has a reaction centre that is activated by light of wavelength 680 nm. This reaction centre is referred to as P680.

Generating and emitting excited electrons – the transfer of light energy

When light energy reaches a reaction centre, **ground-state electrons** in the reaction centre of chlorophyll molecules are raised to an 'excited state' by the light energy received. As a result, high-energy electrons are released, and these electrons bring about the biochemical changes of the light-dependent reactions. The spaces vacated by the **high-energy (excited) electrons** in the reaction centres are continuously refilled by ground-state electrons.

We will examine this sequence of reactions in the two photosystems next.

Advantages of the structured array of different types of pigment molecules in a photosystem

Photosystems I and II have specific and differing roles. However, they occur grouped together in the thylakoid membranes of the grana, along with LHC proteins that function quite specifically.

These LHC proteins consist of:
- enzymes
  - catalysing formation of ATP from ADP and phosphate (P_i)
  - catalysing conversion of the oxidized hydrogen carrier $NADP^+$ to the reduced carrier $NADPH + H^+$
- electron carrier molecules.

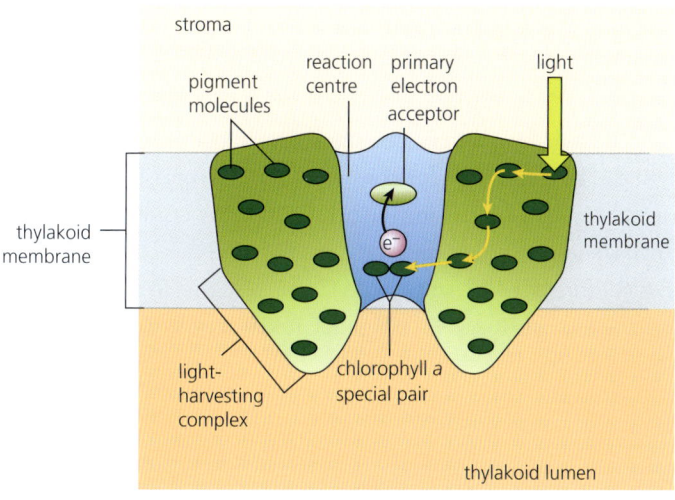

■ Figure C1.3.18 Transfer of electrons in a photosystem

The structured array of different types of pigments and accessory pigments enables light energy to excite electrons in a controlled way and direct these electrons along electron transport chains (Figure C1.3.18). The variety of different pigments allow plants to harvest light from a range of different wavelengths, increasing the flow of electrons to the reaction centre. The two photosystems work in tandem – receiving light energy, releasing excited electrons and replacing each lost electron by one in the ground state (photosystem II electrons come from the splitting of water molecules; photosystem I electrons from photosystem II – see below). A single molecule of chlorophyll or any other pigment would not be able to perform any part of photosynthesis; a series of different molecules is needed to harvest a range of wavelengths of light and to pass the excited electrons resulting from this on to the reaction centre.

11 Define the term *photosystem*.

12 Construct a table that identifies the components of the two photosystems and the role of each one.

ATL C1.3D

Research the structure and function of each photosystem. How does their structure relate to their function? How are the photosystems of cyanobacteria different to those of higher plants? How and why did these changes come about?

Generation of oxygen by the photolysis of water in photosystem II

After receiving light energy, excited electrons from photosystem II are accepted and passed along a chain of electron carriers. As a result of these energy transfers, the excitation level of the electrons falls back to the ground state and they come to fill the vacancies in the reaction centre of photosystem I. Electrons have been transferred from photosystem II to photosystem I (see below).

Meanwhile the vacancies in the reaction centres of photosystem II are filled by electrons (in their ground state) from water molecules. Photosystem II is a catalyst that splits the water molecule (**photolysis**). This event triggers the release of hydrogen ions and oxygen atoms, as well as ground-state electrons.

Top tip!

Protons is an alternative term to hydrogen ions, and the terms can be used interchangeably.

Top tip!

Photosystem II is called 'II' as it was discovered after photosystem I. The names of the two photosystems do not represent the electron flow as this begins in photosystem II.

13 **Suggest** the likely fate of starch stored in a green leaf during periods of darkness.

From the split water molecule, the oxygen atoms combine to form molecular oxygen, the waste product of photosynthesis. The hydrogen ions are used in the reduction of NADP$^+$ as shown in the annotated equation here:

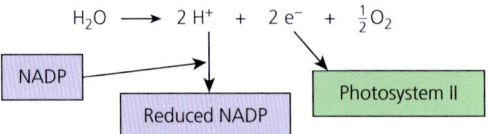

ATP production by chemiosmosis in thylakoids

In the grana of the chloroplasts, the synthesis of ATP is coupled to electron transport via the movement of protons by chemiosmosis, as it was in mitochondria. Here, it is the hydrogen ions trapped within the thylakoid space that activate the rotatory mechanism of ATP synthase enzymes (see page 250), moving down their electrochemical gradient. At the same time, ATP is synthesized from ADP and P$_i$.

As excited electrons from photosystem II pass along the electron transport chain or electron carriers, some of the energy causes the pumping of hydrogen ions (protons) from the chloroplast's matrix into the thylakoid spaces. Here, protons accumulate, causing the pH to drop. The result is a proton gradient that is created across the thylakoid membrane, which sustains the synthesis of ATP. Protons pass down their electrochemical gradient through ATP synthase (ATPase), resulting in the synthesis of ATP. This is another example of **chemiosmosis** (page 421). Figure C1.3.19 shows the process of chemiosmosis in chloroplasts and mitochondria.

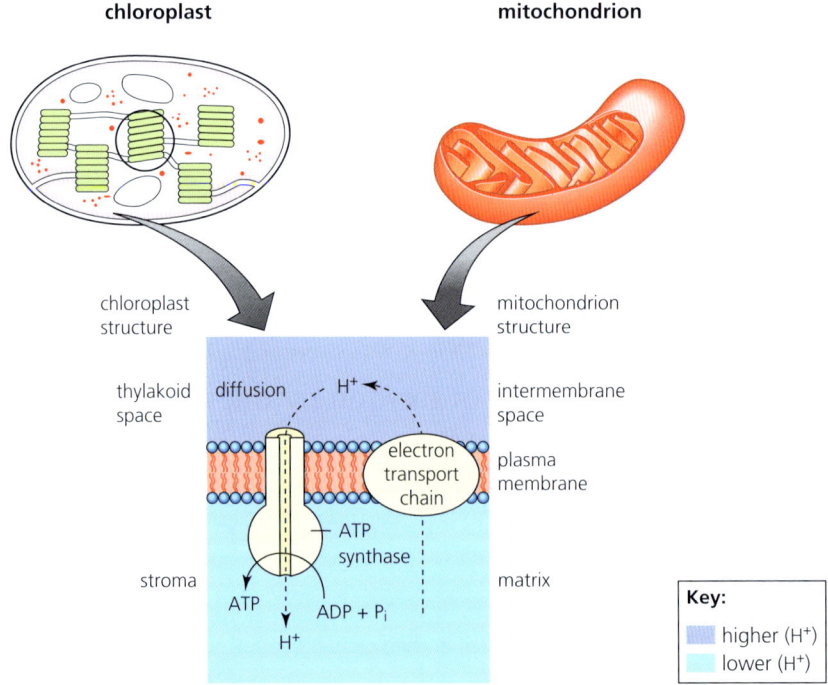

■ **Figure C1.3.19** Chemiosmosis occurs in both mitochondria and chloroplasts

We have seen that the excited electrons that provided the energy for ATP synthesis originate from water and move on to fill the vacancies in the reaction centre of photosystem II. They are subsequently moved on to the reaction centre in photosystem I, and finally are used to reduce NADP$^+$ (see below).

● **Top tip!**

You can either use the terms 'NADP and reduced NADP' or 'NADP$^+$ and NADPH', but whichever you choose should be paired consistently.

14 Construct a table comparing chemiosmosis in mitochondria and chloroplasts. Include the site of proton accumulation, the origin of the protons, the energy source and fate of the ATP that is formed.

■ Reduction of NADP by photosystem I

The excited electrons from photosystem I are passed two at a time to the electron acceptor $NADP^+$, which, with the addition of a hydrogen ion from the stroma, is reduced to form $NADPH + H^+$.

Because the pathway of the electrons from photosystem II and then on to photosystem I is linear, with electrons being accepted by $NADP^+$, the photophosphorylation reaction in which they are involved is described as **non-cyclic photophosphorylation** (Figure C1.3.20).

The electrons released by photosystem I are transferred back to P700 (photosystem I) instead of moving into the $NADP^+$ from $NADPH + H^+$. This movement of electrons from an acceptor to P700 results in the formation of ATP molecules. Electrons cycle repeatedly through photosystem I and the first portion of the electron transport chain but do not pass through photosystem II, and so NADPH is not formed, and water is not needed to replenish the electron supply. Because the electrons move in a cyclic manner, this production of ATP is known as **cyclic photophosphorylation** (Figure C1.3.20).

◆ **Non-cyclic photophosphorylation**: process used to produce NADPH in addition to ATP; requires the presence of water.

◆ **Cyclic photophosphorylation**: process that produces a steady supply of ATP in the presence of sunlight.

Non-cyclic photophosphorylation
- The normal flow of electrons is from water, via PS II and electron carriers to both PS I and NADP.
- The excited electrons from PS II, as they lose the energy of the excited state, cause hydrogen ions (protons) to be pumped across into the thylakoid space.
- The gradient in protons between thylakoid space and the stroma causes protons to flow through ATP synthase and generate ATP.

Cyclic photophosphorylation
- Occurs when reduced NADP accumulates in the stroma – e.g. when CO_2 concentration is low and the light-independent stage is blocked.
- Cyclic photophosphorylation results – excited electrons from PS I are captured by electron acceptors serving PS II, and then returned to a ground state (pumping protons as they go), to reoccupy their original space in PS I.
- The proton gradient between the thylakoid space and the stroma is maintained and ATP formation continues.

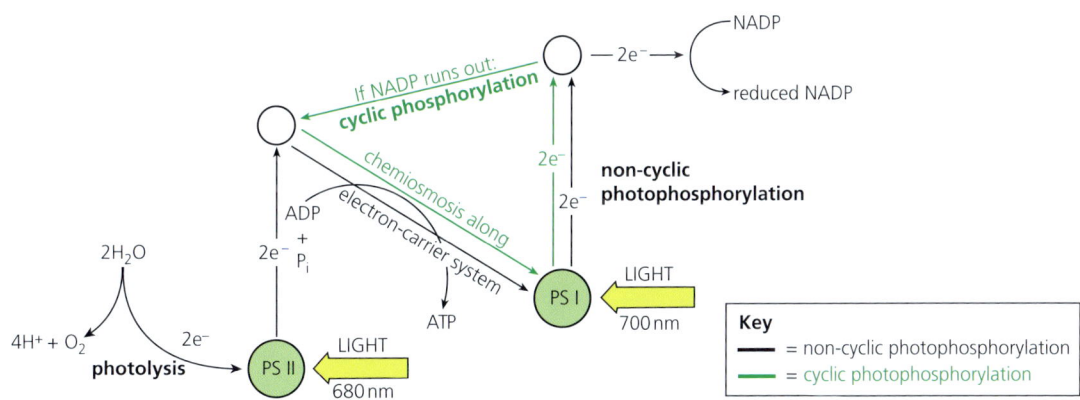

■ Figure C1.3.20 Cyclic and non-cyclic photophosphorylation

15 In non-cyclic photophosphorylation, **outline** the ultimate fate of electrons displaced from the reaction centre of photosystem II.

16 Both reduced NADPH and ATP, products of the light stage of photosynthesis, are formed on the side of thylakoid membranes that face the stroma. **Suggest** why this fact is significant.

17 Explain how the gradient in protons between the thylakoid space and the stroma is generated.

18 Using a table, **compare and contrast** cyclic and non-cyclic photophosphorylation.

C1.3 Photosynthesis

ATP and reduced NADP do not normally accumulate, however, as they are immediately used in the fixation of carbon dioxide in the surrounding stroma (light-independent reactions, page 439). Then the ADP and NADP$^+$ diffuse back into the grana for reuse in the light-dependent reactions.

By this sequence of reactions, repeated again and again at very high speed throughout every second of daylight, the products of the light-dependent reactions (ATP and NADPH + H$^+$) are formed (Figure C1.3.21).

■ Thylakoids as systems for performing the light-dependent reactions of photosynthesis

Link
The adaptations of chloroplasts are covered in detail in Chapter A2.2, page 76.

> **Top tip!**
> The grana are the site of the light-dependent reactions, while the stroma is the site of the light-independent reactions of photosynthesis.

Figure C1.3.21 shows where the photolysis of water, synthesis of ATP by chemiosmosis and reduction of NADP occur in a thylakoid as part of the light-dependent reactions of photosynthesis.

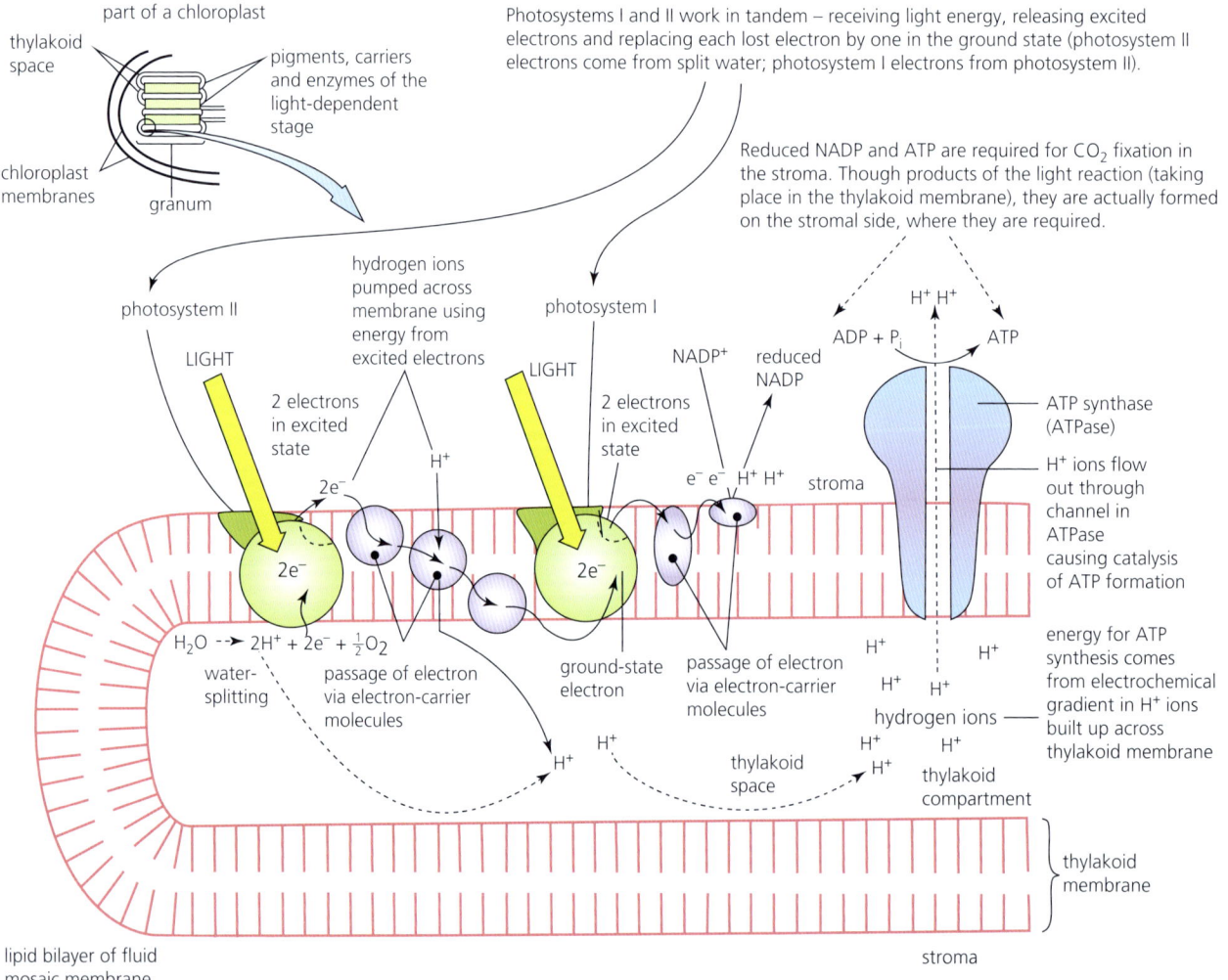

■ **Figure C1.3.21** The light-dependent reactions – the diagram shows the role of the electron transport chain and not the specific detail of each protein, which is not needed for IB Biology

> **Common mistake**
>
> Make it clear whether you are referring to the light-independent reactions of photosynthesis or the light-dependent reactions. These processes are different and clarity is needed.

19 Compare and contrast photophosphorylation with oxidative phosphorylation.

> **ATL C1.3E**
>
> The following site uses computer animations to show the light-dependent reactions of photosynthesis:
>
> **https://vcell.science/project/photosynthesis**
>
> Use the animations and the information in this book to summarize in your own words the light-dependent reactions of photosynthesis. You may want to add annotated illustrations to help you explain the reactions. Note you do not need to know the names of the individual electron-carrier proteins.

◆ **Calvin cycle**: a cycle of reactions in the stroma of the chloroplast, also known as the light-independent reactions, in which carbon dioxide is converted to carbohydrate.

◆ **Rubisco (RuBisCo)**: ribulose-1,5-bisphosphate carboxylase is an enzyme involved in the first major step of carbon fixation; it is the central enzyme of photosynthesis.

Light-independent reactions

Carbon fixation by rubisco

In the light-independent reactions of photosynthesis (Figure C1.3.22), carbon dioxide is converted to carbohydrate. The light-independent reactions are also known as the **Calvin cycle** (see Figure C1.3.22). The Calvin cycle requires the products of the light-dependent reactions. Carbon dioxide is combined with an acceptor molecule in the presence of a special enzyme, ribulose bisphosphate carboxylase (**RuBisCo**, or rubisco for short) in the stroma. The enzyme rubisco is probably the most abundant enzyme present on Earth and makes up the bulk of all the protein in a green plant. High concentrations of it are needed in the stroma of chloroplasts because it works relatively slowly and is not effective at low carbon dioxide concentrations.

Ribulose bisphosphate (RuBP) is a 5-carbon sugar and carbon dioxide is added to it in a process known as fixation. After RuBP and carbon dioxide have combined, the 6-carbon product immediately splits into two 3-carbon molecules: **glycerate 3-phosphate**. Glycerate 3-phosphate is the first product of the fixation of carbon dioxide and so this is known as the **fixation step** (Figure C1.3.22).

> **Top tip!**
>
> You are expected to know the following details of the Calvin cycle: the names of the substrates RuBP and CO_2, the enzyme rubisco and the product glycerate 3-phosphate.

> **Common mistake**
>
> Glycerate 3-phosphate is sometimes abbreviated to GP. This is discouraged as it is ambiguous in accounts of the Calvin cycle. It is recommended that you use the chemical's full name.

Synthesis of triose phosphate

Glycerate 3-phosphate is immediately reduced to the 3-carbon sugar phosphate, **triose phosphate**, using NADPH and ATP. Reduced NADP supplies the hydrogen to reduce glycerate 3-phosphate, and ATP supplies the necessary energy. This is the **reduction step** (see Figure C1.3.22).

> **Common mistake**
>
> Do not confuse triose phosphate with ATP; they are very different chemicals! The role of triose phosphate is often poorly understood. Make sure you learn the Calvin cycle carefully and know what each chemical does and how each chemical links with the next.

C1.3 Photosynthesis

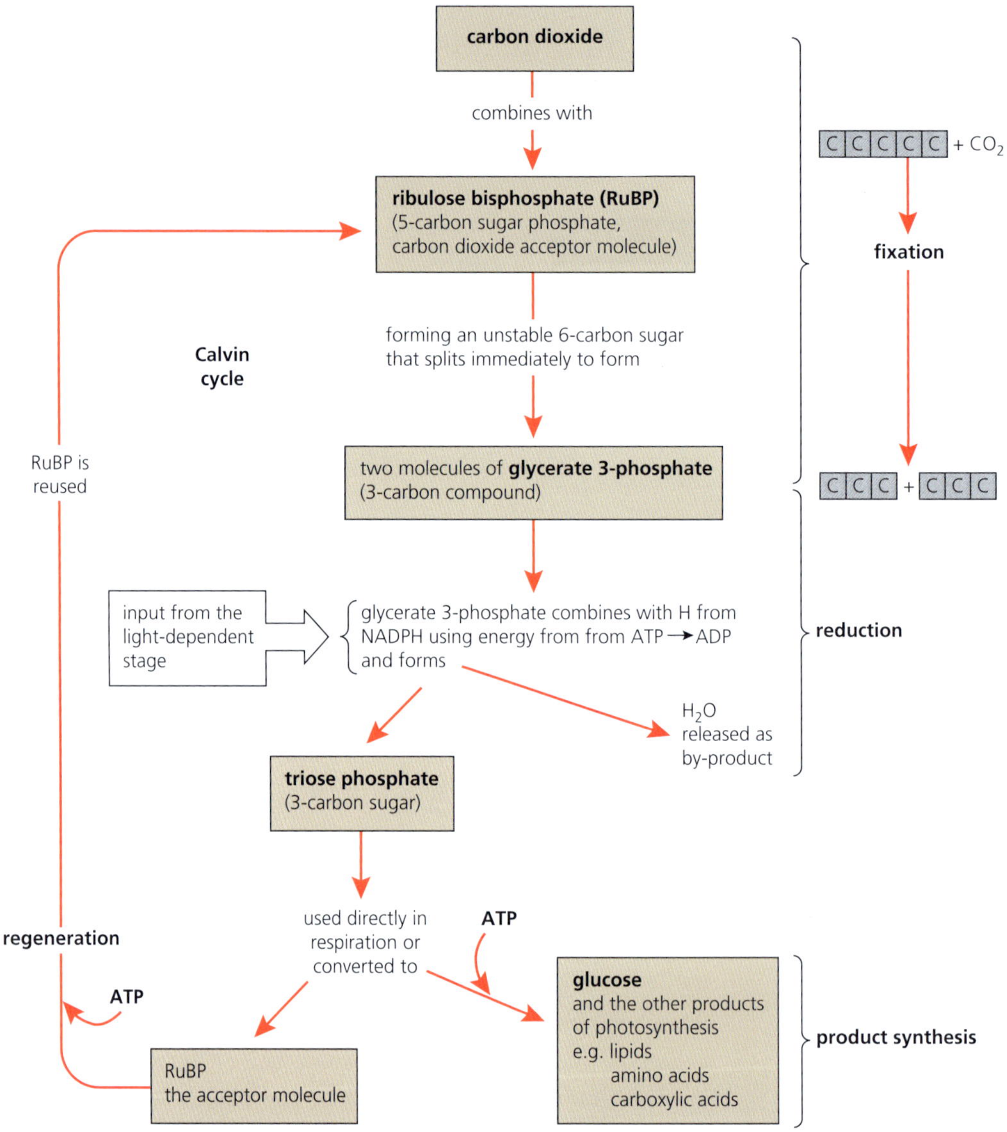

HL ONLY

■ Figure C1.3.22 The path of carbon in photosynthesis – Calvin cycle

Regeneration of RuBP

Some of the triose phosphate is metabolized to produce the molecule that first reacts with carbon dioxide (the acceptor molecule RuBP). This is the **regeneration-of-acceptor step** (see Figure C1.3.23). Five molecules of triose phosphate are converted to three molecules of RuBP, allowing the Calvin cycle to continue. The remaining triose phosphate can be used to synthesize other products, including glucose (see below). Because each turn of the cycle generates two molecules of triose phosphate, three turns are needed to produce six molecules of triose phosphate, allowing five to be converted to RuBP and one to be used in the formation of other molecules such as glucose.

> ### Concept: Interdependence
> Because the light-independent reactions of photosynthesis are a cycle, each intermediate depends on every other part of the metabolic pathway. Ultimately, molecules of triose phosphate must be converted back to RuBP so the cycle can continue.

> ### Top tip!
> You do not need to know details of the individual reactions of the Calvin cycle, but you should understand that five molecules of triose phosphate are converted to three molecules of RuBP, allowing the Calvin cycle to continue.

> ### Top tip!
> If glucose is the product of photosynthesis, five-sixths of all the triose phosphate produced must be converted back to RuBP so that the cycle can continue.

Theme C: Interaction and interdependence – Molecules

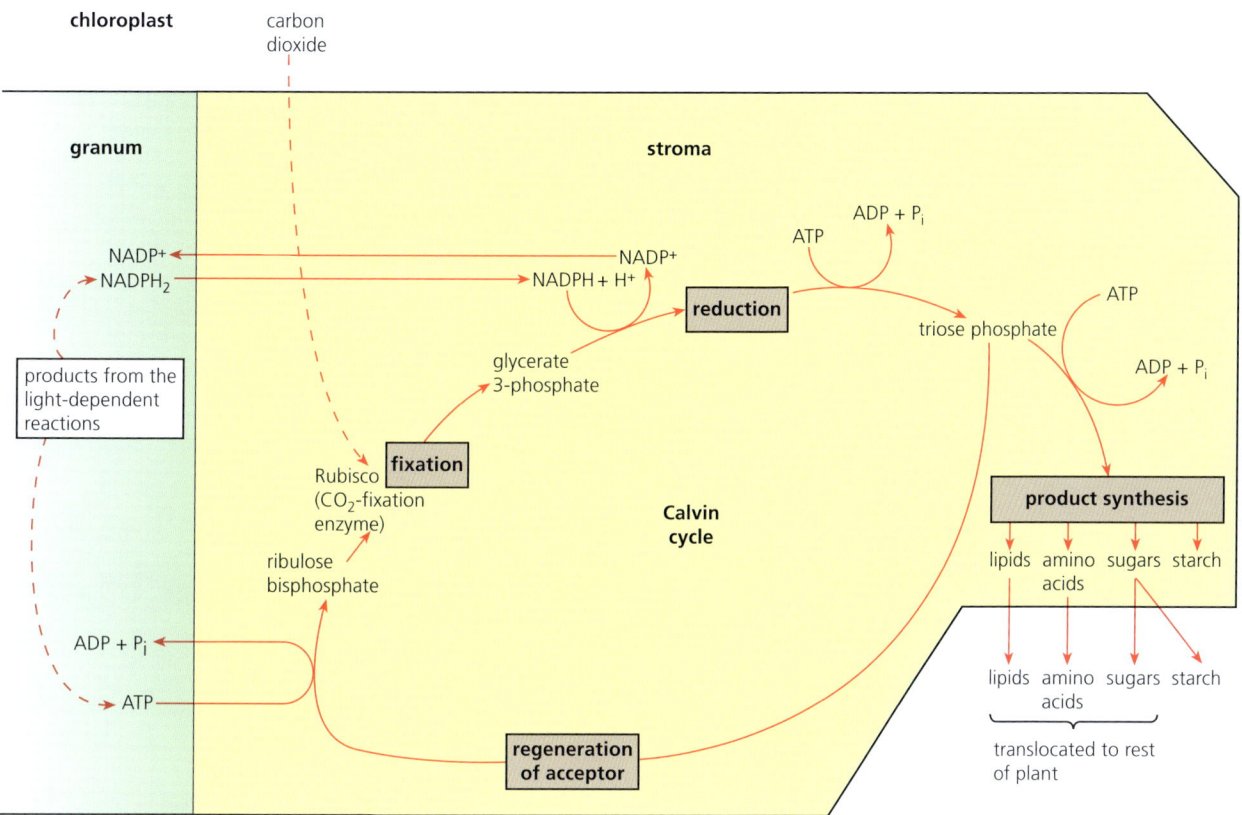

■ Figure C1.3.23 Summary of the light-independent reactions

20 List the carbohydrate intermediates of the Calvin cycle.

> **Concept: Interaction**
>
> The reduced coenzyme NADPH interacts with glycerate 3-phosphate, reducing it to triose phosphate. In the process, NADPH is oxidized to $NADP^+$.

■ Synthesis of other carbon compounds

The importance of photosynthesis to the green plant is that it provides the energy-rich glucose molecules from which the plant builds its other organic molecules. The metabolism of the green plant is sustained by the products of photosynthesis. Synthesis of carbohydrates, amino acids and other carbon compounds uses the products of the Calvin cycle and mineral nutrients (Figure C1.3.24). For example, triose phosphate can be further metabolized to produce carbohydrates such as sugars, sugar phosphates and starch, and later lipids, amino acids such as alanine, and organic acids such as malate. This is the **product-synthesis step** (see Figure C1.3.24).

C1.3 Photosynthesis

HL ONLY

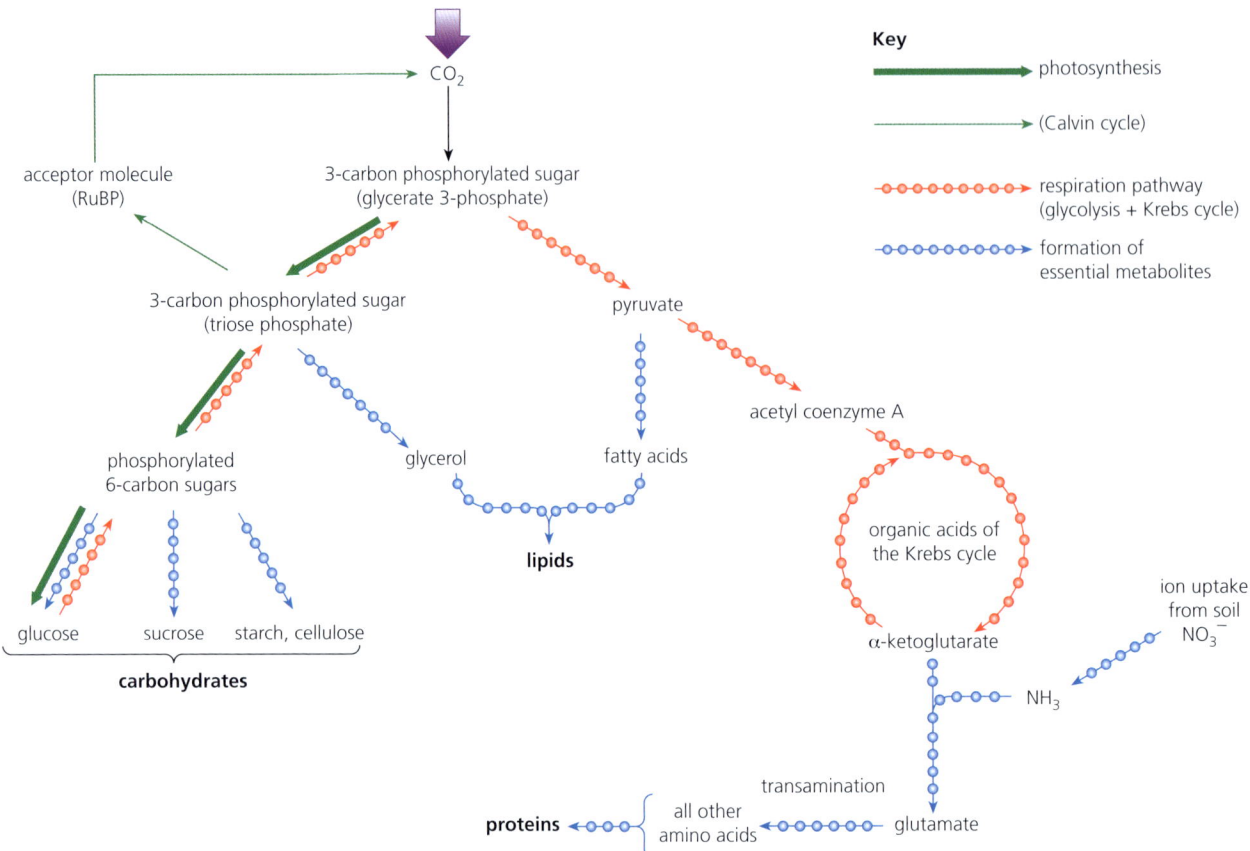

■ **Figure C1.3.24** The product-synthesis steps of photosynthesis

🌱 Top tip!

Details of the metabolic pathways in Figure C1.3.24 are given to show the complexity of the reactions involved. They can be summarized by the understanding that all of the carbon in compounds in photosynthesizing organisms is fixed in the Calvin cycle, and that carbon compounds other than glucose are made by metabolic pathways that can be traced back to an intermediate in the cycle.

🌱 Nature of science: Experiments

In the light-independent reactions, carbon dioxide is converted to carbohydrate in the stroma of the chloroplasts, surrounding the grana. The pathway by which carbon dioxide is reduced to glucose was first investigated by a team at the University of California, led by Melvin Calvin. He was awarded a Nobel Prize in 1961. The technique used radioactively labelled carbon dioxide ($^{14}CO_2$), which is taken up by the cells in exactly the same way as non-labelled carbon dioxide ($^{12}CO_2$) and then fixed into the same products of photosynthesis. Photosynthesizing cells from a culture of *Chlorella* (a unicellular alga) were used in the experiment, as they are easier to sample than, for example, mesophyll cells of a multicellular plant.

A brief pulse of labelled $^{14}CO_2$ was introduced into the otherwise continuous supply of $^{12}CO_2$ to photosynthesizing cells in the light and its progress monitored. Samples of the photosynthesizing cells, taken at frequent intervals after the $^{14}CO_2$ had been fed, contained a sequence of radioactively labelled intermediates and (later) products of the photosynthetic pathway. These compounds were isolated by **chromatography** from the sampled cells and identified using autoradiography (a detection method that uses X-ray film to show molecules that are radioactively labelled).

The chromatography and autoradiography techniques that the team exploited were relatively recent inventions then, and radioactive isotopes were only just becoming available for biochemical investigations.

Find out more about Calvin's experiments, and how sources of carbon-14 and autoradiography enabled Calvin and his team to elucidate the pathways of carbon fixation.

> **ATL C1.3F**
>
> Review the Calvin cycle using this site:
> **www.youtube.com/watch?v=c2ZTumtpHrs**
> Review both light-dependent and light-independent reactions using this site:
> **https://media.hhmi.org/biointeractive/click/photosynthesis**

> **21 Distinguish** between the following:
>   **a** light-dependent reactions and light-independent reactions
>   **b** photolysis and photophosphorylation.
>
> **22 Deduce** the significant difference between the starting materials and the end products of photosynthesis.

Interdependence of the light-dependent and light-independent reactions

> **Concept: Interdependence**
>
> The light-dependent and light-independent reactions of photosynthesis are interdependent – one cannot function without the other. These reactions happen at the same time.

The light-dependent reactions of photosynthesis provide the NADPH and ATP, which in turn provide the electrons and energy needed to produce carbohydrates in the light-independent reactions. A lack of light therefore stops light-independent reactions as well because NADPH and ATP are not produced in the light-dependent reactions. These are needed for these light-independent reactions.

Cyanobacteria, algae and higher plants require carbon dioxide in solution, in the form of hydrogen carbonate ions (HCO_3^-), for the optimal function of photosystem II. Carbon dioxide, in its ionic form hydrogen carbonate, has a regulating function in the splitting of water in photosynthesis. This means that carbon dioxide has an additional role to being reduced to glucose. Hydrogen carbonate acts as an acceptor for the protons that are produced when water is split in photosystem II. Carbon dioxide therefore acts both as a terminal electron acceptor at the very end of photosynthetic reactions and, at the same time, as a proton acceptor (HCO_3^-) at the beginning of the reactions.

Removal of hydrogen carbonate also slows electron transfer through the electron acceptors associated with photosystem II. This is because hydrogen carbonate is bound to proteins in photosystem II, which are needed for the activity of the electron transport chain.

Depletion of carbon dioxide (and, therefore, hydrogen carbonate) not only causes cessation of carbon dioxide fixation, but also a strong decrease in the activity of photosystem II.

> **23 Explain** how the light-independent reactions of photosynthesis rely on the light-dependent reactions.

> **LINKING QUESTIONS**
>
> **1** What are the consequences of photosynthesis for ecosystems?
> **2** What are the functions of pigments in living organisms?

C2.1 Chemical signalling

Guiding questions
- How do cells distinguish between the many signals that they receive?
- What interactions occur inside animal cells in response to chemical signals?

SYLLABUS CONTENT

This chapter covers the following syllabus content:
- ▶ C2.1.1 Receptors as proteins with binding sites for specific signalling chemicals (HL only)
- ▶ C2.1.2 Cell signalling by bacteria in quorum sensing (HL only)
- ▶ C2.1.3 Hormones, neurotransmitters, cytokines and calcium ions as examples of functional categories of signalling chemicals in animals (HL only)
- ▶ C2.1.4 Chemical diversity of hormones and neurotransmitters (HL only)
- ▶ C2.1.5 Localized and distant effects of signalling molecules (HL only)
- ▶ C2.1.6 Differences between transmembrane receptors in a plasma membrane and intracellular receptors in the cytoplasm or nucleus (HL only)
- ▶ C2.1.7 Initiation of signal transduction pathways by receptors (HL only)
- ▶ C2.1.8 Transmembrane receptors for neurotransmitters and changes to membrane potential (HL only)
- ▶ C2.1.9 Transmembrane proteins that activate G protein (HL only)
- ▶ C2.1.10 Mechanism of action of epinephrine (adrenaline) receptors (HL only)
- ▶ C2.1.11 Transmembrane receptors with tyrosine kinase activity (HL only)
- ▶ C2.1.12 Intracellular receptors that affect gene expression (HL only)
- ▶ C2.1.13 Effects of the hormones oestradiol and progesterone on target cells (HL only)
- ▶ C2.1.14 Regulation of cell signalling pathways by positive and negative feedback (HL only)

Receptors

Cells in a multicellular organism (and even bacteria) must respond to environmental stimuli and communicate with each other to coordinate their activities. This is done by specific chemical signalling molecules known as **ligands**. These ligands include proteins, small peptides, amino acids, nucleotides, steroids, amines and even dissolved gases such as nitrogen monoxide (nitric oxide, NO).

The cell **chemical signalling** process ensures that important activities occur in the correct cells, at the right time and in coordination with other cells in the organism. This usually involves a change in gene transcription, which often affects several genes simultaneously, but not necessarily all to the same extent.

In most cases the ligand cannot cross the cell membrane and must bind with a **receptor protein** (Figure C2.1.1). Receptor proteins are unique to specific ligands. The specificity of the ligand–receptor interaction allows a ligand to produce responses in specific target cells. Ligands can activate many different target cells simultaneously, which allows for the regulation and control of the cellular response. The binding process transmits a signal across the membrane and generates a second protein or messenger inside the cell. This can cause further changes within the cell, such as activation of an enzyme, which often occurs via phosphorylation (addition of phosphate groups).

◆ **Ligand**: general term for a molecule that binds to a specific site on a protein.

◆ **Chemical signalling**: the release of chemicals (ligands) that bind to a specific molecule which delivers a signal within the cell or to another cell.

◆ **Receptor protein**: protein that recognizes and binds with a specific chemical signal molecule on the outside of the plasma membrane.

 **Top tip!**

You should use the term 'ligand' for chemical signalling molecules.

1 **Define** the term *ligand*.

Link

Transcription is covered in detail in Chapter D1.2, page 615.

◆ **Signal transduction**: the conversion of an impulse or stimulus from one physical or chemical form to another. In cell biology, the process by which a cell responds to an extracellular signal.

◆ **Second messenger**: small intracellular ligand (signalling molecule) generated or released inside a cell in response to an extracellular signal.

◆ **Effector protein**: proteins that cause a change inside a cell during signalling.

◆ **Scaffold protein**: protein that binds two or more other proteins, and organizes binding partners into a functional unit to enhance signalling efficiency.

2 **Suggest** why chemical signalling in multicellular organisms is more complicated than in single-celled organisms, such as yeast.

In some cases, the messenger may enter the nucleus (via the nuclear pores) and change transcription (copying of the genetic code on to messenger RNA). These processes are examples of **signal transduction**.

● Common mistake

Do not confuse 'receptor protein' with the term 'receptor'. Receptors are specialized cells that receive stimuli, and change (transduce) this signal into an electrical impulse (see Chapter C2.2). Receptor proteins are molecules that bind with ligands to convey messages into a cell, stimulating specific metabolic reactions.

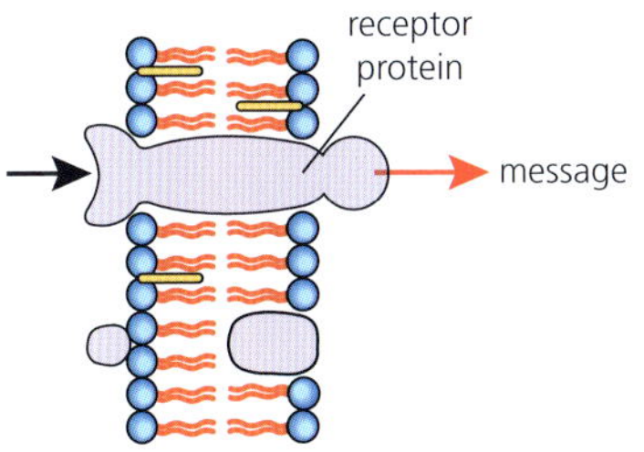

■ Figure C2.1.1 Outline of cellular chemical signalling (signal transduction)

A chemical signalling pathway has one or more critical functions:
- Relay ('pass on') the signal onward and help spread it through the cell.
- Amplify the signal received (via a **second messenger**), making it stronger, so that a small number of ligands are enough to produce a large intracellular response. One ligand can activate many signal transduction pathways to trigger many cell reactions simultaneously.
- Detect signals from more than one intracellular chemical signalling pathway and integrate them (two or more signals become one signal) before relaying a signal onward.
- Distribute the signal to more than one **effector protein**, creating branches and resulting in a complex response (Figure C2.1.2).

Primary transduction is when the 'message' converts from being extracellular (outside the cell) to intracellular (inside the cell). The **scaffold protein** holds some proteins close together, so the reaction is faster.

The same signal molecule can trigger different cellular responses in two different cells in the body. Because different types of differentiated cells activate different sets of genes, different kinds of cells have different collections of proteins.

Concept: Interdependence

Chemical signalling relies on many different molecules to pass along the signal, all needing to work together to control cell metabolism and keep the organism working properly.

ATL C2.1A

Before learning more about cell signalling in this chapter, find out about this topic yourself by watching the animation at this website: **https://dnalc.cshl.edu/resources/3d/cellsignals.html**

The animation shows how a fibroblast cell responds to external signals after an injury.

What knowledge of biology already covered in the course does this topic draw on? What aspects of cell structure, cell membranes and proteins do you need to know about to understand how cells respond to external chemical signals?

◆ **Quorum sensing**: the ability of some bacteria to monitor cell density and to adjust their gene expression.

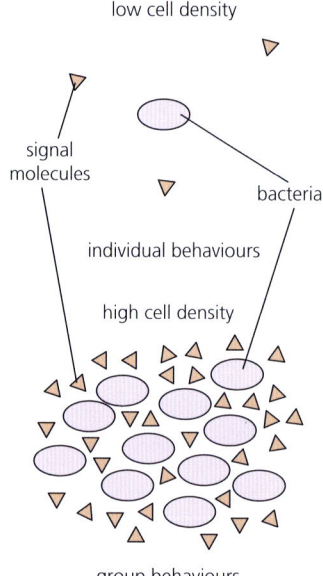

■ Figure C2.1.3 Bacterial quorum sensing

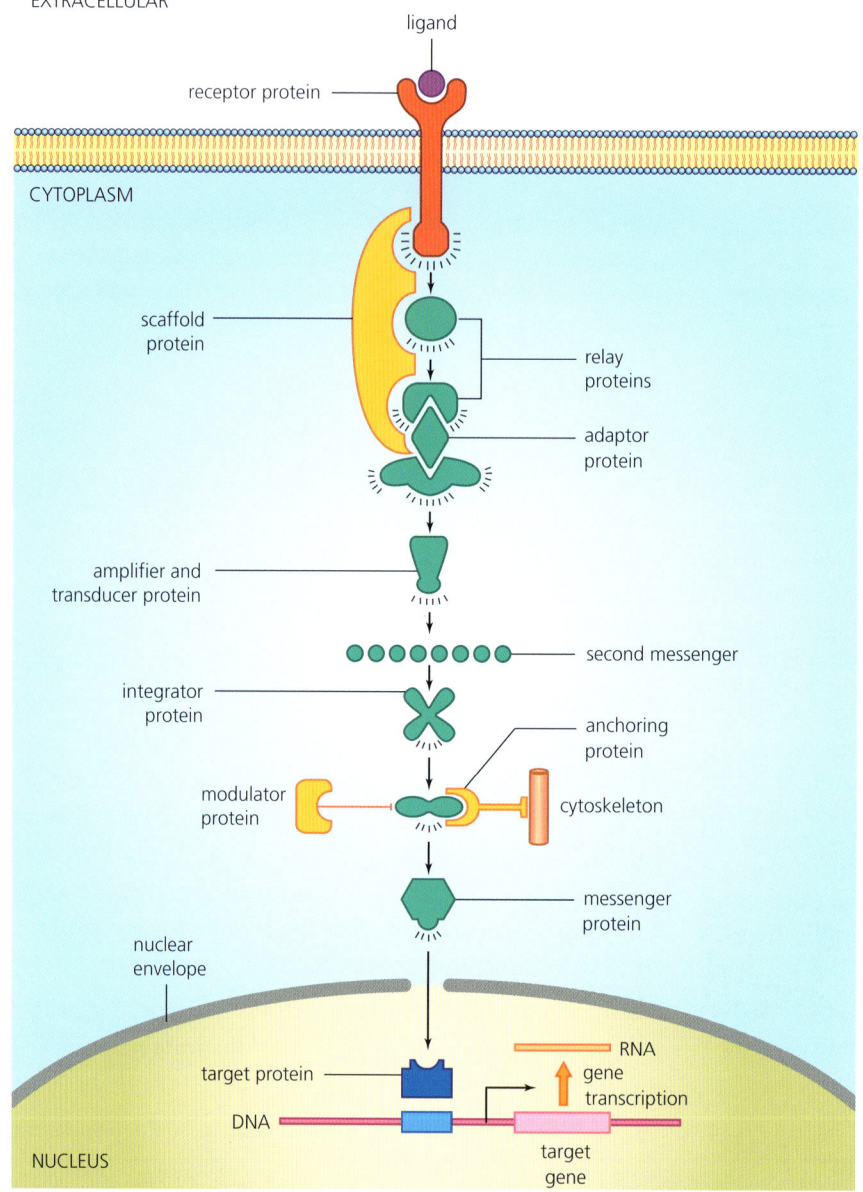

■ Figure C2.1.2 Intracellular signalling proteins can relay, amplify, integrate and distribute the incoming signal

Cell signalling by bacteria in quorum sensing

In the early days of microbiology, bacteria were considered non-social organisms that led individual lifestyles with little interaction. However, recent research has shown that many strains of bacteria communicate with each other by emitting, detecting and responding to ligands. One important, well-studied method of intercellular communication, known as **quorum sensing** (Figure C2.1.3), allows bacteria to turn on genes together in response to increases in the density of cells in the population.

Quorum sensing relies on the bacteria producing and releasing signal molecules that diffuse outwards from the bacterium. This allows the bacteria population to communicate with each other and coordinate their behaviour when a certain population size is reached.

● **Top tip!**

The size of the 'quorum' is not fixed but depends on the rate of production and loss of the signal molecules.

◆ **Autoinducers**: a signalling molecule produced and used by bacteria participating in quorum sensing.

Quorum sensing signalling molecules are often known as **autoinducers**. These are detected by specific proteins inside the cell or in the bacterial cell membrane. When the autoinducer binds to the receptor, it activates or represses transcription of target genes, which often include those for autoinducer synthesis.

When the bacterial population is low, diffusion reduces the concentration of the autoinducer in the surroundings to a very low value. As the population increases, the concentration of the autoinducer reaches a threshold, gene expression occurs and more autoinducer is synthesized. This forms a **positive feedback loop** for autoinducer production.

Link
Positive feedback loops are covered later in this chapter, page 465, in fruit ripening in Chapter C3.1, page 516, and in hormonal control of pregnancy in Chapter D3.1, page 722.

A well-known example of quorum sensing is shown by the bioluminescent marine bacterium *Vibrio fischeri*, which produces the light-emitting enzyme luciferase when it reaches a critical cell population density in the light organ of its host, the squid (Figure C2.1.4). It is of little benefit to *V. fischeri* to produce luciferase when it is on its own as a single cell. The bacteria do not emit light when they are outside of the host. The bioluminescence is very energy intensive and so the bacteria benefits from being located within the animal where it can obtain nutrients. The squid also benefits from having the bacteria: the light the bacteria emit attracts prey or can be used as camouflage. This interaction is known as **mutualism**.

Link
Mutualism as an interspecific relationship is covered in Chapter C4.1, page 565.

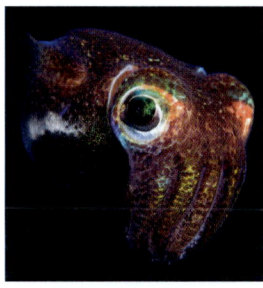

■ **Figure C2.1.4** Image of the Hawaiian bobtail squid showing bioluminescence produced by *Vibrio fischeri*

Going further

Cystic fibrosis is a genetic disorder. The most common symptoms are difficulty in breathing and excessive mucus production. This is due to frequent lung infections often involving large colonies of the bacterium *Pseudomonas aeruginosa*, which are resistant to antibiotics because they produce enzymes and a biofilm when the cell density exceeds a critical value detected by quorum sensing. A biofilm is a complex aggregation of micro-organisms marked by the excretion of a protective and adhesive extracellular matrix. The production of the biofilm provides the organism with protection against antibiotics and the enzymes damage the lung epithelium.

Signalling chemicals in animals

In animals, many ligands may be classified as hormones, neurotransmitters, cytokines or calcium ions.

Hormones

Hormones are secreted by endocrine glands and are carried through the circulatory system to act on distant target cells. Hormones act over a longer time, work in cell metabolism and other functions e.g. sexual reproduction. Hormones are discussed further later in this chapter on page 454.

Neurotransmitters

Neurotransmitters carry signals between neurons or from neurons to other types of target cells (such as muscle cells). The release of neurotransmitters is caused by the arrival of an action potential at the end of a neuron. Neurotransmitters are discussed further later in this chapter on page 455.

◆ **Hormone**: extracellular signal molecule that is secreted and transported via the bloodstream (in animals) or the sap (in plants) to target tissues where it causes a specific effect.

◆ **Neurotransmitter**: chemical released at the presynaptic membrane of an axon when an action potential arrives, which transmits the action potential across the synapse.

3 Outline the role of quorum sensing in bacteria.

4 Define the term *hormone*.

Link
Neurotransmitters are also covered in Chapter C2.2, page 476; human hormones are covered in Chapter D3.1, page 693 and D3.3, page 760.

♦ **Cytokines:** small signalling molecules (usually a protein or glycoprotein) made and secreted by cells that act on neighbouring cells to alter their behaviour.

Top tip!

Note that some neurotransmitters can also act as hormones. For example, epinephrine (adrenaline) functions both as a neurotransmitter and as a hormone produced by the adrenal gland to signal glycogen breakdown in muscle cells.

Cytokines

The **cytokines** form a family of relatively small, secreted proteins that control many aspects of growth and differentiation of specific types of cells. Some cytokines, such as α-interferon, are produced and secreted by many types of cells after infection with viruses.

Calcium ions

Most ligands are organic, but some metal ions are involved in cells signalling, especially calcium ions (Ca^{2+} (aq)). Many ligands in animals, including neurotransmitters, growth factors and some hormones, induce responses in their target cells via signal transduction pathways that increase the concentration of Ca^{2+} (aq) in the cytoplasm.

Increasing the cytoplasmic concentration of Ca^{2+} (aq) causes a number of responses in animal cells, including muscle cell contraction, secretion of substances (for example, the neurotransmitter acetylcholine) and cell division.

Although cells always contain calcium ions, this ion can function as a second messenger because its concentration in the cytoplasm is normally much lower than the concentration outside the cell.

Chemical diversity of hormones and neurotransmitters

A wide range of extracellular molecules and ions can act as ligands for receptor proteins in cell signalling pathways. They include metal ions, low molar mass compounds (for example, amino acids and molecules derived from them), steroids, peptides and proteins.

Hormones

There are three main classes of hormone (Table C2.1.1): proteins (such as insulin), amines (also known as amino acid derivatives, such as epinephrine) and steroids (such as cholesterol, which is needed to synthesize oestradiol and testosterone). Hormones can be small, non-polar, hydrophobic molecules that diffuse through the cell membrane to reach receptors in the nucleus or cytoplasm. Examples are progesterone and testosterone, as well as thyroid hormones.

■ Table C2.1.1 Three main chemical classes of hormones

Hormone class	Protein/peptide hormones	Steroid hormones	Amine hormones
Example	Insulin	Oestradiol	Epinephrine (adrenaline)

Hormones can also be water-soluble molecules that bind to receptors on the plasma membrane. They are either proteins like insulin and glucagon, or small, charged molecules like histamine and epinephrine.

Nature of science: Experiments

Evidence for the role of nitric oxide (NO) in causing the relaxation of smooth muscle came from a set of experiments in which the neurotransmitter acetylcholine was added to experimental preparations of the smooth muscle cells that surround blood vessels. The direct application of acetylcholine to these cells caused them to contract, the expected effect of acetylcholine on these muscle cells. However, addition of acetylcholine to the lumen of small, isolated blood vessels in culture medium caused the underlying smooth muscles to relax, not contract. Later studies showed that, in response to acetylcholine, the endothelial cells that line the lumen of blood vessels were releasing a substance (later shown to be NO) that triggered muscle cell relaxation.

◆ **Synaptic signalling**: a type of cell–cell communication that occurs across chemical synapses in the nervous system.

Link
Neurons and synapses are covered in detail in Chapter C2.2, page 467.

Neurotransmitters

Neurotransmitters are a chemically diverse group of chemicals, including simple amines (for example, dopamine), amino acids (for example, gamma-aminobutyric acid or GABA), polypeptides (for example, endorphins) and acetylcholine (an amino acid derivative).

Although some neurons produce and release only one kind of neurotransmitter, most make two or more and may release one or more at any given time. The coexistence of more than one neurotransmitter in the synapse makes it possible for the cell to exert several influences at the same time.

In the brain and other parts of the central nervous system, the gas nitric oxide (NO) functions as a neurotransmitter, or as an agent that influences neurotransmitters.

> ### Concept: Interaction
> Neurotransmitters interact with the synapse to convey messages through the body. Several neurotransmitters in the same synapse allow a number of messages to be passed on at the same time, and so allow varied responses to stimuli.

> ### Going further
>
> #### Nitric oxide and cell signalling
> Nitric oxide (nitrogen monoxide, NO) gas molecules can rapidly diffuse across the plasma membrane into the cytoplasm of target cells and directly control the activity of specific intracellular proteins. Nitric oxide is synthesized from the amino acid arginine and diffuses from its site of synthesis into nearby cells. The gas only acts locally because it is quickly converted to a range of products by reacting with oxygen and water outside cells.
>
> Nitric oxide causes the smooth muscle in blood vessels to relax, causing the blood vessel to dilate (widen), increasing blood flow. Many nerve cells also use NO to signal neighbouring cells.

Localized and distant effects of signalling molecules

Cells in multicellular organisms usually communicate via ligands targeted for cells that may be either adjacent (local chemical signalling) or not adjacent (long-distance chemical signalling).

Local chemical signalling

Local signalling involves direct contact between cells. **Neurons** are separated by **synapses**. **Synaptic signalling** occurs across chemical synapses and involves neurotransmitters such as acetylcholine and norepinephrine. The gap between the presynaptic membrane and the postsynaptic membrane is very small, between 20 and 40 nanometres (nm). Neurotransmitters therefore only have to transfer the signal over a small distance. While the immediate effect of the neurotransmitter is localized, the overall effect is widespread because the signal transmitted across the synapse enables an electrical impulse to be established in the postsynaptic neuron and the message is transferred on through the body.

HL ONLY

◆ **Ligand-gated channel**: an ion channel that is stimulated to open by the binding of a small molecule such as a neurotransmitter.

◆ **Ion channel**: transmembrane protein that forms a pore across the bilayer through which specific ions can diffuse down their concentration gradients.

Top tip!

Neurons releasing acetylcholine are described as cholinergic neurons, while those releasing norepinephrine are described as adrenergic neurons.

Link

The endocrine system is covered in Chapter C3.1, page 493.

Top tip!

The same ligand can bind to different receptor proteins causing different responses (for example, acetylcholine). On the other hand, different ligands binding to different receptor proteins can produce the same cellular response (for example, glucagon and epinephrine).

■ **Figure C2.1.5** Lipid-insoluble ligands utilize receptor proteins in the plasma membrane, whereas lipid-soluble ligands can diffuse through the membrane and access intracellular protein receptors

---

One of the main neurotransmitters in the human body is acetylcholine. On reaching the postsynaptic membrane, acetylcholine binds with the neurotransmitter receptors (**ligand-gated** sodium **channels**) found on the postsynaptic membrane, causing the opening of **ion channels**. This results in the influx of sodium ions, Na$^+$ (aq), which generates a new impulse in the postsynaptic neuron.

Other examples of local chemical signalling include gap junctions in animals and plasmodesmata in plants. In animals, cells may communicate between membrane-bound molecules on the cell surface membrane (cell–cell recognition).

Link

Gap junctions are discussed in Chapter B2.1, page 243. Plasmodesmata are discussed in Chapter B3.2, page 306.

■ Long-distance chemical signalling

For long-distance signalling, hormones are secreted into the transport system (for example, the blood plasma of an animal or the sap of a plant) to act on distant target cells. In animals with a **closed circulation** system (i.e. heart, arteries, veins and capillaries – see Figure B3.2.22, page 315), hormones are transported in the blood, and so can travel throughout the body to receptors that are very far away from the source. The hormones are specific to receptor proteins either in the plasma membrane or within the cell (see below). The hormones trigger a cascade of reactions, which leads to the target cell altering its metabolism to respond to the signal.

Specialized animal cells release hormone molecules (for example, insulin and glucagon) into blood vessels of the circulatory system to other parts of the body, where they reach target cells that can recognize and respond to the hormones. This is known as **endocrine** signalling.

Plant hormones (for example, auxin, a growth hormone) sometimes travel in the vascular tissue but often reach their targets by moving through cells or by diffusion through the air as a gas.

Differences between transmembrane and intracellular receptors

Receptor proteins can either be in the plasma membrane (**transmembrane**) or within the cytoplasm (**intracellular**) (see Figure C2.1.5). The properties of a ligand determine the type of receptor that is used. Some ligands are non-polar and therefore lipid-soluble, whereas others are charged and water-soluble. Non-polar ligands can diffuse through the phospholipid bilayer and access receptor proteins within the cell. Polar ligands, such as peptide hormones (e.g. insulin and glucagon) are hydrophilic and cannot pass through the phospholipid bilayer. These ligands must bind with receptor proteins on the surface of the cell.

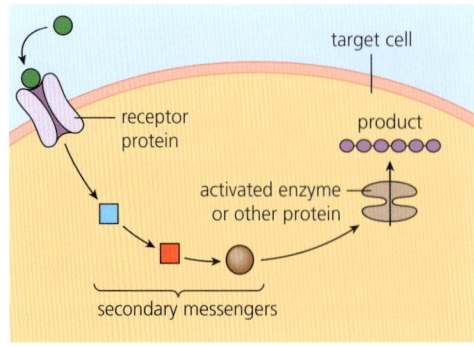
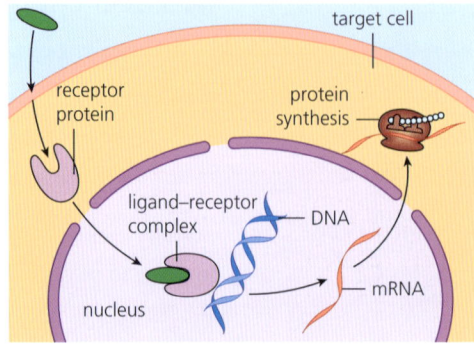

Theme C: Interaction and interdependence – Cells

◆ **Transmembrane receptor protein**: cell signalling receptor protein that is in the membrane.

◆ **Intracellular receptor protein**: cell signalling receptor protein located inside the cell.

● **Common mistake**

It is incorrect to say that all receptor proteins are in the plasma membrane. The receptors on the surface of the cell respond to charged, polar ligands. Ligands that are lipid soluble, such as steroid hormones, can pass through the phospholipid bilayer and so their receptors are inside the cell.

Corresponding to the two classes of signal molecules there are, therefore, two classes of receptor proteins: **transmembrane receptor proteins** and **intracellular receptor proteins**.

Transmembrane receptor proteins

In Chapter B1.2, we saw how R-groups determine the properties of assembled polypeptides. R-groups are hydrophobic or hydrophilic, with hydrophilic R-groups being polar or charged, acidic or basic. Transmembrane receptor proteins have three different regions: a ligand-binding region, a hydrophobic region extending through the membrane, and an intracellular region that transmits the signal to the inside of the cell. Integral proteins, including receptor proteins, have regions with hydrophobic amino acids, helping them to embed in membranes. Hydrophilic regions of the receptor proteins are towards the outside of the membrane where the phosphate heads of the phospholipids are located. Hydrophilic ligands bind to the hydrophilic region of the receptor protein.

In most cases, ligands bind to a receptor in the plasma membrane of the responding cell. This interaction produces a change in the shape of the receptor that causes the signal to be relayed across the membrane to the receptor's cytoplasm. This is the process of cell transduction.

There are many different receptors in the plasma membrane of cells and they can be classified into a small number of structural classes. They are covered later in this chapter.

● **Top tip!**

The number of cell-surface receptors may be regulated by endocytosis to reduce their number. They may also be inactivated by phosphorylation.

● **Top tip!**

Ion channels are necessary to transport ions into and out of a cell because ions are charged particles (surrounded by polar water molecules) and therefore cannot diffuse across the hydrophobic interior of the plasma membrane.

● **Nature of science: Experiments**

Identification of cell-surface receptors

Hormone receptors are difficult to identify and purify (usually with chromatography) because they are present in very low numbers, and they have to be solubilized with non-ionic detergents. Because of their high specificity and high affinity for binding with their ligands, the presence of a certain receptor in a cell can be detected and quantified by measuring their binding to radioactively labelled hormones. The binding of the hormone to a cell suspension increases with hormone concentration until it reaches receptor saturation. Specific binding is obtained by measuring both the total and the non-specific binding (which is obtained by using a large excess of unlabelled hormone).

Intracellular receptor proteins

Many lipid-soluble ligands diffuse across the plasma membrane, diffuse into the nucleus, bind to the DNA and initiate specific protein synthesis, which then initiates reactions (Figure C2.1.6). Steroid hormones such as oestradiol, progesterone and testosterone are examples of this type of ligand.

Steroid receptors

Steroid receptors are not present in the membrane but are normally present as soluble proteins in the cytoplasm. Since steroid hormones are lipid soluble, they diffuse from the bloodstream across the hydrophobic plasma membrane into the cytoplasm of cells. If the cell is a target cell, the hormone binds to a receptor molecule, which may be present in the cytoplasm or may be within the nucleus. If the receptor is in the cytoplasm, it moves to the nucleus by simple diffusion. In either case, if the receptor molecule is activated, it acts as transcription factor by binding to specific regions of DNA in the chromosomes (see Chapters D1.1 and D1.2).

Link
Transcription is covered in Chapter D1.2, page 615 and genes are covered in Chapter D2.2, page 670.

C2.1 Chemical signalling

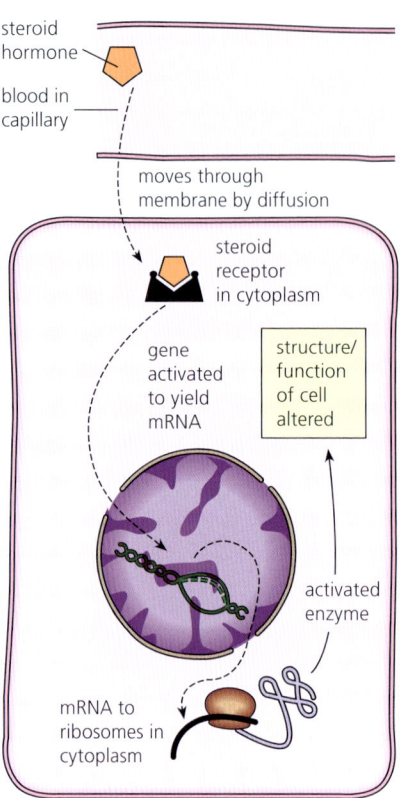

Figure C2.1.6 The mechanism of action of a lipid-soluble hormone

Depending on a hormone's mode of action, transcription of a gene may be switched on or switched off. If a gene has been activated, new RNA is formed, leaves the nucleus and then directs the formation of new proteins (most likely an enzyme) at ribosomes. The new protein or enzyme will bring about a structural or functional change in the cell. Of course, if a gene is switched off by hormone action, some cell process will be interrupted or terminated.

Initiation of signal transduction pathways by receptors

For a cell to respond when it encounters a signal, the ligand must first be recognized by a specific receptor molecule on the plasma membrane and then transmitted to the cell's interior before a cellular response can occur.

Cell signalling can be divided into three stages:

1. Ligand–receptor interaction
2. Signal transduction
3. Cellular response.

1 Ligand–receptor interaction

The ligand is complementary in shape to a binding site on the receptor and binds to it. There is high specificity in the ligand–receptor binding. The binding of the ligand to the receptor activates the receptor.

2 Signal transduction

The signal transduction pathway often requires a sequence of changes in a series of different molecules in a multistep pathway. These molecules in the pathway are often called relay molecules.

Signal transduction occurs via two main ways: protein phosphorylation in a phosphorylation **signal cascade** and the release of second messengers, for example cyclic AMP (see Figures C2.1.2 and C2.1.13). Such pathways also allow for signal amplification.

The activated receptor activates a relay molecule, which activates another relay molecule, and so on, until the molecule (usually a protein) that produces the final cellular response is activated. Many of the relay molecules in signal transduction pathways are protein kinases and they often act on other protein kinases in the pathway. Each activated protein kinase will initiate a sequential phosphorylation and activation of other kinases, resulting in a **phosphorylation cascade**.

Relay molecules are usually activated when they are phosphorylated, and deactivated when they are dephosphorylated. The processes of phosphorylation and dephosphorylation act as a molecular switch in the cell, turning metabolic activities on and off as required.

3 Cellular response

The signal transduction pathway leads to a specific cellular response, which is the regulation of one or more cellular activities. The response may occur in the nucleus or the cytoplasm of a cell.

5 **Compare and contrast** transmembrane and intracellular receptor proteins.

◆ **Signal cascade**: series of linked reactions, often including phosphorylation and dephosphorylation, that carries information within a cell, often amplifying an initial signal.

◆ **Phosphorylation cascade**: a sequence of events where one protein kinase enzyme phosphorylates another, causing a sequence of events that leads to the phosphorylation of many protein kinases. As the signal is carried onwards it is amplified and sometimes can spread to other signalling pathways.

◆ **Apoptosis**: a form of programmed cell death that allows cells that are unneeded or unwanted to be eliminated from an adult or developing organism.

Depending on the type of cell and ligand, some cellular responses may include:
- regulation of activity of protein (for example, the opening or closing of an ion channel in the plasma membrane changes the membrane permeability)
- regulation of protein synthesis by activating or deactivating specific gene expression by turning genes on or off in the nucleus
- regulation of the activity of an enzyme
- rearrangement of the cytoskeleton of the cell
- death of the cell (for example, **apoptosis**).

After the specific cellular response has occurred, the signal is terminated. The ability of a cell to receive new ligands depends on the reversibility of the changes produced by previous signals.

Link
The role of sodium ion channels is discussed in Chapter B2.1, page 239 and their role in changing membrane potential is covered in C2.2, page 479.

Transmembrane receptors for neurotransmitters and changes to membrane potential

The ligand-gated sodium ion channel (Figure C2.1.7) is composed of five transmembrane protein subunits. The protein subunits combine to form an aqueous pore across the lipid bilayer, which is lined by five transmembrane α-helices, one from each subunit. There are two acetylcholine binding sites.

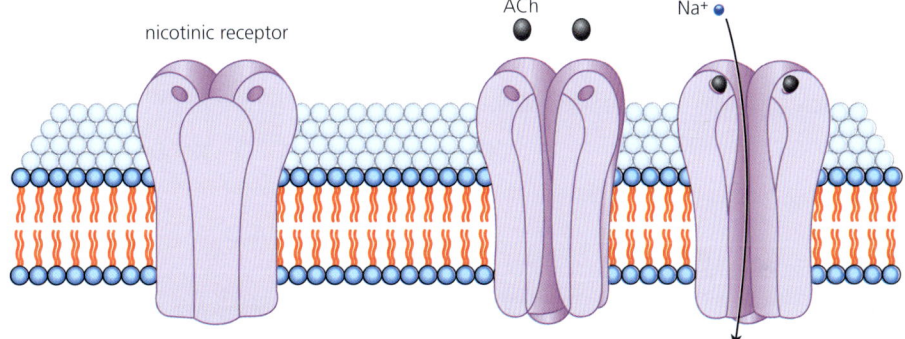

■ **Figure C2.1.7** The open and closed conformations of the acetylcholine receptor (ligand-gated sodium channels)

When acetylcholine released by a motor neuron binds to both acetylcholine binding sites, the channel undergoes a change in conformation (shape): the hydrophobic side chains move apart and the gate opens, allowing sodium ions to flow across the membrane down their electrochemical and concentration gradient. This depolarizes the plasma membrane (Figure C2.2.4, page 471) and changes the membrane potential (the voltage across the membrane), which leads to other changes (see Chapter C2.2, page 467).

Going further

Curare
Curare (a plant alkaloid) causes muscle paralysis by blocking excitatory acetylcholine receptors at the neuromuscular junction. It is used by surgeons to relax muscles during an operation.

◆ **GTP**: guanosine triphosphate with a role in cell signalling.

◆ **G protein**: a membrane-bound GTP binding protein, usually activated by the binding of a hormone or other ligand to a transmembrane receptor.

Transmembrane receptors that activate G protein

GTP is an energy-rich nucleotide like ATP, composed of ribose and three phosphate groups but with guanine rather than adenine as the base. **G proteins** are membrane-bound GTP binding proteins, usually activated by the binding of a hormone or other ligand to a transmembrane receptor. They respond to extracellular signals from hormones and neurotransmitters, and trigger intracellular signalling cascades that regulate bodily functions.

G proteins are important in vision, as well as smell and taste. Their role is to couple the primary stimuli (the light waves, odorants (molecules that cause smell) or flavourants (molecules that cause

◆ **G protein-coupled receptor (GPCR)**: cell-surface receptor that associates with a G protein.

flavour) to ion channels. Changes in ion flow through these channels lead to the production of a nerve impulse that flows from the sensory organ (such as the eye or nose) to the brain.

The **G protein-coupled receptor (GPCR)** is closely associated with a G protein, a protein that binds to **guanosine triphosphate** or **guanosine diphosphate (GDP)**.

G protein-coupled receptors have a common structure consisting of seven hydrophobic α-helices that span the plasma membrane (Figure C2.1.8). By forming hydrophobic interactions with the hydrophobic core of the membrane lipid bilayer, the α-helices allow the receptor membrane to be embedded within the plasma membrane. Signal transduction from GPCRs is controlled by G proteins that bind mainly to the third and largest cytoplasmic loop in the polypeptide chain of the receptor.

Top tip!

GPCRs are only found in eukaryotes, not prokaryotes. There are nearly 1000 GPCRs and many different signal molecules are specific for different types of GPCRs. These receptors vary in their binding sites for recognizing signal molecules and different G proteins inside the cell.

Top tip!

Proteins fold in water (a polar solvent) so that hydrophobic groups are buried and polar groups are exposed to the water molecules. However, proteins arrange themselves differently in the bilayer of the plasma membrane as it is a highly non-polar environment; they expose their hydrophobic groups and hide polar groups in their core.

◆ **GTPase**: an enzyme that hydrolyzes GTP into GDP and P_i.

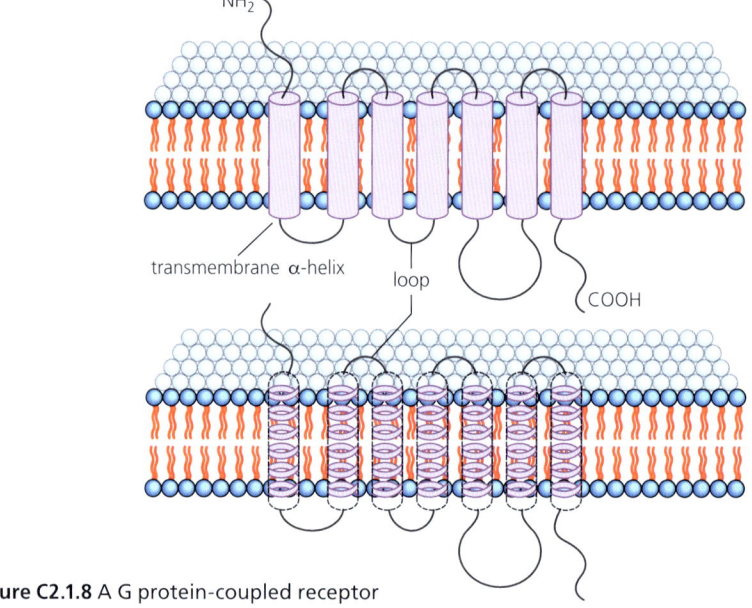

■ **Figure C2.1.8** A G protein-coupled receptor

The GPCR is inactive when not bound to a ligand. The G protein is inactive when bound to GDP. When the ligand binds to the extracellular side of GPCR, the receptor is activated, causing it to change its conformation. The change in conformation (shape) of the receptor when the ligand binds activates a G protein, which in turn activates an effector protein that generates a second messenger, causing the G protein to exchange its bound GDP for GTP.

The G protein is activated and dissociates from the receptor, then binds to an enzyme or other protein, activating it. Once activated, the enzyme triggers signal transduction leading to cellular response. Once the signal molecule is absent, GTP is hydrolyzed back into GDP by the **GTPase** enzyme found in the G protein subunit. The G protein therefore dissociates from the enzyme and returns to its inactive form. The signal is switched off (Figure C2.1.9).

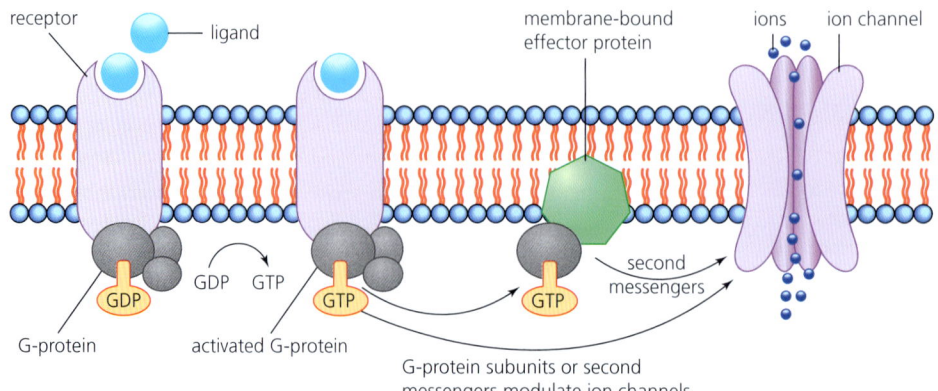

■ **Figure C2.1.9** Mechanism of action of G protein-coupled receptors; GDP is dislodged when GTP interacts with G protein

> **Tool 1: Experimental techniques**
>
> **Physical molecular modelling**
>
> Make a paper model of a G protein-coupled receptor using this site:
> https://pdb101.rcsb.org/learn/paper-models/g-protein-coupled-receptor-gpcr
>
> The site provides a template PDF to download and print. *How does the structure of the protein enable it to carry out its function?* Use your model to explain to someone else in your class how the G protein-coupled receptor functions.

Mechanism of action of epinephrine (adrenaline) receptors

Link
Epinephrine secretion is covered further in Chapter C3.1, page 503.

The adrenal glands release the hormone epinephrine (adrenaline), which circulates in the bloodstream. It binds to a class of GPCRs (see page 460) known as adrenergic receptors, which are present on many types of cells. Epinephrine prepares the body for vigorous activity. It increases heart and breathing rate, enabling increased levels of oxygen and glucose to be delivered to muscle cells for respiration. The action of an anaesthetic can be lengthened considerably if it is administered as a drug along with epinephrine. Because epinephrine is a vasoconstrictor, it reduces the blood supply, allowing the drug to remain at its targeted site for a longer period.

Epinephrine acts as a peptide hormone. Peptide hormone receptors activate a cascade, mediated by a second messenger inside the cell. Second messengers are small, non-protein, water-soluble molecules or ions that relay signals received at receptor proteins on the cell surface to target molecules in the cytoplasm or nucleus. Being small and water-soluble, they can readily diffuse throughout the cell.

◆ **Cyclic AMP (cAMP):** small intracellular signalling molecule generated from ATP in response to hormonal stimulation of cell-surface receptors.

Second messengers serve to greatly amplify the strength of the signal. The most common second messengers are **cyclic AMP** (cyclic adenosine monophosphate) and calcium ions, Ca^{2+} (aq) (Figure C2.1.10). Second messengers participate in pathways initiated by G protein-coupled receptors and receptor tyrosine kinases (see page 463).

6 **Suggest** how the same second messenger (for example, cAMP) is used in many different cells, but the response to the same second messenger is different in each cell.

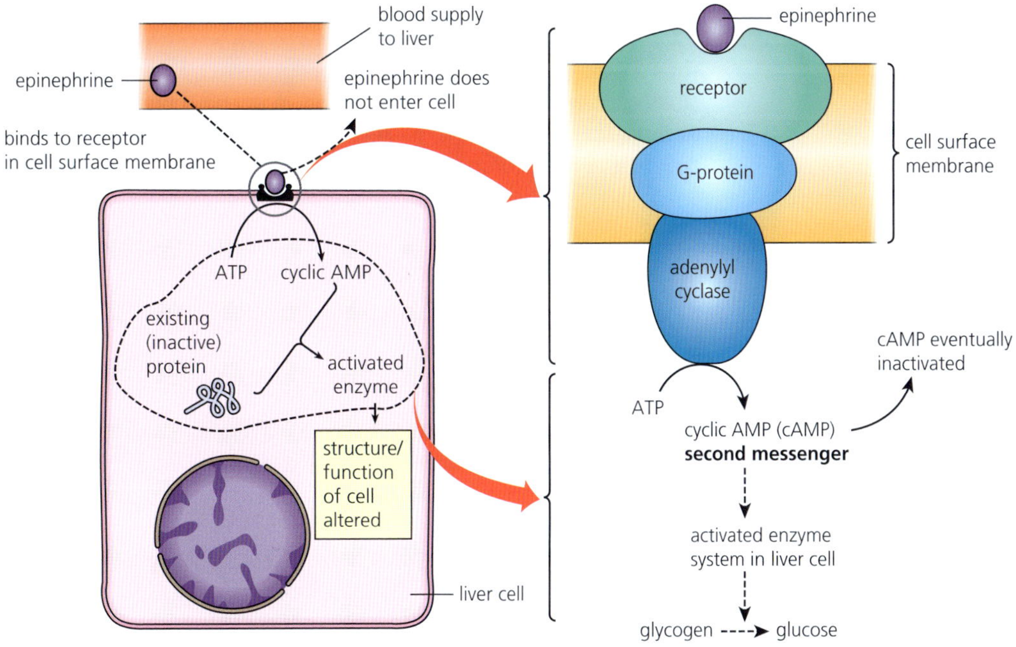

■ **Figure C2.1.10** The cell signalling pathway, involving epinephrine, stimulating glycogen breakdown in skeletal muscle cells

C2.1 Chemical signalling

- **Adenylyl cyclase/ adenylate cyclase**: enzyme that catalyses the formation of cAMP from ATP.
- **Phosphodiesterase**: enzyme that catalyses the breaking of a phosphodiester bond in an oligonucleotide.
- **Kinase**: enzyme that catalyses the transfer of a phosphate group from ATP to a specific amino acid side chain on a target protein.
- **Phosphorylase**: enzyme that breaks down glucose-based polysaccharides (e.g. glycogen) to glucose 1-phosphate.

An enzyme embedded in the cell surface membrane, **adenylyl cyclase/adenylate cyclase**, when activated by the G protein, can catalyse the conversion of ATP to many cyclic AMP (cAMP) molecules. The concentration in the cytoplasm of the cAMP is increased very rapidly, amplifying the signal in the cytoplasm. It does not last for long because another enzyme, called **phosphodiesterase**, converts the cAMP to AMP, resulting in signal termination.

The hormone activates a GPCR, which turns on a G protein (G_s) that activates adenylyl cyclase to boost the production of cAMP. The increase in cAMP activates PKA (cAMP-dependent protein kinase), which phosphorylates and activates an enzyme called phosphorylase kinase, which activates glycogen phosphorylase, the enzyme that breaks down glycogen.

In skeletal muscle, epinephrine increases the concentration of intracellular cyclic AMP, causing the breakdown of glycogen by activating PKA, which leads to both the activation of an enzyme that promotes glycogen breakdown and the inhibition of the enzyme that catalyses glycogen synthesis (Figure C2.1.11).

By stimulating glycogen breakdown and inhibiting its synthesis, the increase in cyclic AMP maximizes the amount of glucose available as a respiratory substrate for muscular activity. Epinephrine also acts on adipose cells, stimulating the breakdown of fat to fatty acids, which can also be used to generate ATP by respiration.

The relay molecules are activated kinases and phosphorylases. These proteins catalyse specific types of reactions:

- **Kinases** are enzymes that transfer a phosphate group from ATP to an acceptor.
- **Phosphorylases** are enzymes that break down glucose-based polysaccharides (e.g. glycogen) to glucose 1-phosphate.

The critical role of the second and third stages of signal transduction is the amplification of the hormone signal (see page 458). So, from one activated receptor molecule, 10 000 (10^4) molecules of cAMP are formed. As a result of the presence of cAMP in the cytoplasm, approximately 10^6 molecules of active phosphorylase enzyme will be formed, and these will then trigger the formation of perhaps 10^8 molecules of glucose 1-phosphate.

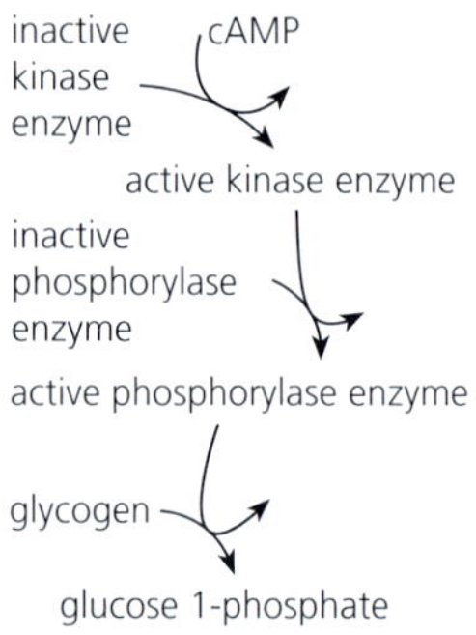

inactive kinase enzyme → cAMP → active kinase enzyme
inactive phosphorylase enzyme → active phosphorylase enzyme
glycogen → glucose 1-phosphate

■ Figure C2.1.11 Cascade of reaction triggered by the binding of epinephrine to a target cell

Nature of science: Science as a shared endeavour

The IUPAC chemical name of epinephrine is (R)-1-(3,4-dihydroxyphenyl)-2-methylaminoethanol. In Japan and the USA, the substance is known as epinephrine, but in most other countries it is known as adrenaline. Adrenaline is marketed in Britain as 'Epipen' for intramuscular injection and as Eppy or Simplene for eyedrops.

George Oliver, a medical doctor, and Edward Schäfer, a professor of physiology, both in the UK, showed that the adrenal glands contained a substance with strong pharmacological effects. It was named 'epinephrine' by John Abel in the USA in 1897.

In 1901, after visiting Abel, Jokichi Takamine, a Japanese chemist, prepared a pure extract of the active principle from the adrenal gland and patented it. A company marketed his extract and, because they used the proprietary name adrenaline, epinephrine became the generic name in America and Japan.

There are arguments to prefer adrenaline over epinephrine. The gland is the adrenal gland, not the epinephric gland. However, the gland is above ('epi') the kidneys (which contain structures called nephrons) hence 'epinephrine'. Unusually, these two terms persist in common use in different parts of the world. Naming conventions show international cooperation in science for mutual benefit. In the case of adrenaline/epinephrine, conventions developed by one set of scientists have not been adopted by others, although overall the two names are widely understood, as is their interchangeability. The IUPAC chemical name is internationally understood, although it would be unrealistic to use this in everyday discussions of this hormone!

Top tip!

Because these reactions do not involve changes in gene transcription or new protein synthesis, they occur rapidly.

> **TOK**
>
> The story of (R)-1-(3,4-dihydroxyphenyl)-2-methylaminoethanol is interesting in that it shows how reliant the development of biological knowledge is on the community of biologists. The interactions and sharing of knowledge between Oliver, Schäfer, Abel and Takamine all resulted in new knowledge and development of the chemical, although under various names. This emphasizes how new knowledge, and the language used to express the knowledge, sometimes takes time to formalize. One important function of the IUPAC is to create a single, authoritative source of knowledge in chemistry for the various scientific communities that use it.

Transmembrane receptors with tyrosine kinase activity

Tyrosine kinases

Receptor **tyrosine kinases** (RTKs) belong to a major class of receptors that have enzymatic activity. One RTK complex (Figure C2.1.12) may activate ten or more different signal transduction pathways and cellular responses, helping the cell to regulate and coordinate several cell activities simultaneously.

◆ **Tyrosine kinase**: an enzyme that catalyses the transfer of phosphate groups from ATP to the tyrosine (amino acid) residues on a substrate protein, activating it.

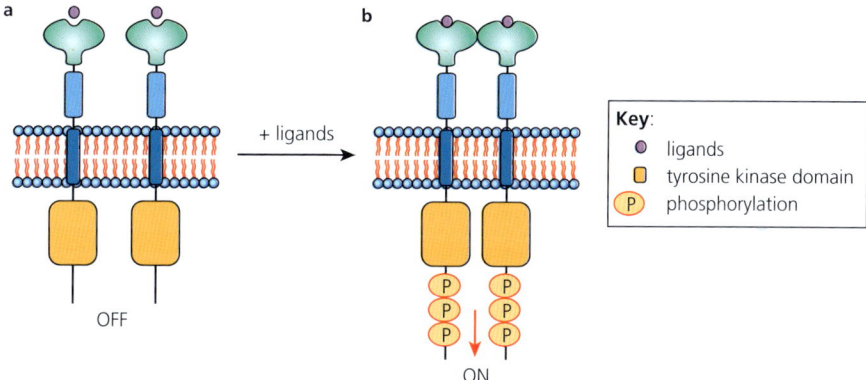

■ Figure C2.1.12 Activation of an RTK stimulates the assembly of an intracellular signalling complex

◆ **Dimer**: a molecule composed of two structurally similar subunits.

◆ **Dimerisation**: the formation of a dimer (from two monomers).

◆ **Autophosphorylation**: the phosphorylation of a kinase by itself.

◆ **Relay protein**: a protein that passes the signal to the next member of the signalling pathway.

◆ **Gluconeogenesis**: set of enzyme-catalysed reactions by which glucose is synthesized from small organic molecules such as pyruvate, lactate or amino acids.

An RTK is a single polypeptide chain with a single transmembrane α-helix, an extracellular ligand binding site and an intracellular tail that functions as tyrosine kinase and contains many tyrosine amino acid residues.

Before a ligand binds, the receptors exist as individual RTK monomers. The binding of a ligand to the extracellular binding sites of RTKs causes two RTK proteins to associate together in the membrane, forming a **dimer** (Figure C2.1.12). **Dimerisation** brings a suitable substrate up to the kinase active site in the intracellular tails of RTK. Tyrosine kinase adds a phosphate group from an ATP molecule to the tyrosine residues on the tail of the other RTK protein by **autophosphorylation**.

The activated RTK will trigger the assembly of specific **relay proteins** on the receptor tails, activating them. Each activated protein triggers a signal transduction pathway, leading to a cellular response.

Insulin signalling

Insulin functions as an extracellular messenger molecule (hormone), informing cells that glucose levels are high. Cells that express insulin receptors on their surface, such as cells in the liver, respond to this message by increasing glucose uptake, increasing glycogen and triglyceride synthesis, and/or decreasing **gluconeogenesis**. The ligand (insulin) binds to an insulin receptor, an RTK.

C2.1 Chemical signalling

◆ **Glycogenesis**: the enzymatic formation or synthesis of glycogen.

Insulin uses RTK signalling (Figure C2.1.13) to allow cells to uptake glucose. Activated relay proteins cause vesicles embedded with glucose transporter proteins to move to the cell surface membrane. The vesicles fuse with the plasma membrane, inserting the transporter proteins into the cell surface membrane, which results in the increase in uptake of glucose into muscle cells.

Many glycogen synthase molecules are activated, which will catalyse the synthesis of glycogen from glucose (**glycogenesis**). Hence the binding of a single insulin molecule to a receptor will lead to the synthesis of large amounts of glycogen.

The signal is terminated when insulin is released from receptors, the tyrosine residues are dephosphorylated by phosphatases and the two halves of the dimer separate. Protein phosphatases inactivate protein kinases by dephosphorylation.

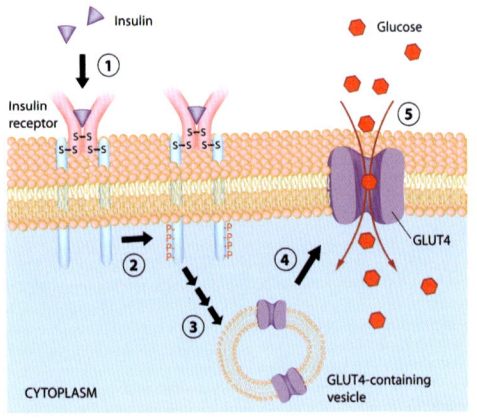

Figure C2.1.13 shows the effect of insulin on cells. The numbers in the figure correspond to the following events:

1. Insulin activates RTK receptors.
2. Relay proteins are activated, which travel to vesicles within the cell.
3. These vesicles contain glucose transporter proteins (channel proteins known as GLUT4 (glucose transporter type 4).
4. The GLUT4 vesicles fuse with the plasma membrane.
5. The GLUT4 proteins become incorporated in the membrane, where they transport glucose into the cell.

■ **Figure C2.1.13** Insulin–RTK signalling pathway

Intracellular receptors that affect gene expression

Steroid (and thyroid) hormones are extracellular chemical signal molecules that can also pass through the plasma membrane. Testosterone, a steroid hormone secreted by the cells of the testes, travels through the blood and enters cells all over the body. However, only cells that contain receptor molecules for testosterone will respond.

◆ **Nuclear receptor**: eukaryotic protein that, on binding with a signal molecule, enters the nucleus and regulates transcription.

In the cytoplasm of target cells, the hormone binds to the receptor, activating it. With the hormone attached, the activated ligand–receptor complex then enters the nucleus.

The activated ligand–receptor complex (or **nuclear receptor**) acts as a transcription factor and binds to a regulatory sequence in the DNA promoting transcription of specific genes, giving rise to male secondary sexual characteristics. Not all genes are transcribed in the genome – specific genes can be activated (a process called gene expression) to produce proteins required by the cell or organism (see Chapter D1.2).

Link
Transcription and gene expression are covered in Chapter D1.2, page 615 and Chapter D2.2, page 670.

 **Top tip!**

Intracelullar receptors and a very similar signalling pathway operate for the female steroid hormones, oestradiol and progesterone. This is covered in more detail on page 465.

◆ **Ovarian follicle**: a fluid-filled spherical sac that contains and nourishes an immature egg, or oocyte.

◆ **Oestradiol**: a steroid hormone; a sex hormone of female mammals.

◆ **Progesterone**: hormone released by the corpus luteum that stimulates the uterus to prepare for pregnancy.

Link
Control of the developmental changes at puberty is covered in Chapter D3.1, page 709.

Link
The menstrual cycle, ovulation and changes during the ovarian and uterine cycles and their hormonal regulation are covered in detail in Chapter D3.1, page 698.

◆ **Feedback**: when part of the output from a system returns as an input, so as to affect subsequent outputs.

◆ **Positive feedback**: feedback that increases change; it promotes deviation away from an equilibrium.

◆ **Negative feedback**: feedback that tends to counteract any deviation from equilibrium and promotes stability.

Effects of the hormones oestradiol and progesterone on target cells

In the ovary, each ovum (egg) is surrounded by a protective structure called the **ovarian follicle**, which nourishes the ovum. The follicle also secretes the steroid hormones **oestradiol** and **progesterone**. Oestradiol is synthesized mainly in the ovary, but also in the placenta, testis and possibly the adrenal cortex. Oestradiol controls the development and maintenance of female sex characteristics.

The production of oestradiol in women's ovaries is controlled by hormones released from both the hypothalamus in the base of the brain and the pituitary. The hypothalamus releases a hormone called **gonadotropin-releasing hormone**. This hormone then acts on the pituitary gland to cause the release of luteinizing hormone (LH) and follicle-stimulating hormone (FSH), which are involved in the **menstrual cycle**. The target cells, as well as being located in the pituitary, are also located in the lining of the uterus, breasts and bone marrow. The interaction between the hormone and these tissues leads to the development of female secondary sexual characteristics in puberty.

Once a month, one egg is released from a follicle in a process known as **ovulation**. After ovulation, the follicle becomes a structure known as the corpus luteum. **Progesterone** is produced primarily by the corpus luteum but also by the placenta and prepares the inner lining of the uterus (endometrium) for implantation of a fertilized ovum. If implantation fails, the corpus luteum degenerates and progesterone production stops. If implantation occurs, the corpus luteum continues to secrete progesterone, under the influence of LH and prolactin, for several months of pregnancy, after which the placenta takes over this function. During pregnancy, progesterone maintains the endometrium of the uterus and prevents further release of ova from the ovary.

7 **State** the type of chemical signalling shown by the secretion of hormones from the pituitary gland.

Regulation of cell signalling pathways by positive and negative feedback

The steps in a chemical signalling pathway may be controlled by **feedback** regulation. In **positive feedback**, changes increase and amplify conditions further away from the starting point (i.e. away from equilibrium). In **negative feedback**, the feedback loop leads to a return to the initial conditions (Figure C2.1.14).

ATL C2.1B

Positive and negative feedback loops are covered throughout this course and occur in every level of biological organization. Work through the book, making a list of all the examples of negative and positive feedback. Create diagrams that show the process of feedback in each case. These notes will help you when you come to revise for your exams.

■ **Figure C2.1.14** Positive and negative feedback within an intracellular chemical signalling pathway

> ## Top tip!
> Positive feedback can generate all-or-none, switch-like responses, but negative feedback can generate responses that oscillate on and off. More typically they act after some time delay to shut down the pathway once it has produced its effects.

Link
Positive feedback in fruit ripening is also covered in Chapter C3.1, page 516.

■ Fruit ripening as an example of positive feedback

Plants make extensive use of transmembrane receptors, especially enzyme-coupled receptors. Plant receptors are thought to play an important part in a large variety of cell signalling processes, including those controlling plant growth, development and disease resistance. Plant cells do not use RTKs, cAMP or steroid hormone-type nuclear receptors.

One of the best-studied signalling systems in plants controls the response of cells to ethylene (ethene, C_2H_4), a gaseous molecular hormone that controls seed germination and fruit ripening. Ethylene receptors function as enzyme-coupled receptors. It is the empty receptor that is active: in the absence of the ethylene molecule, the empty receptor activates a protein kinase that switches off the ethylene-responsive genes in the nucleus. When ethylene is present, the receptor and kinase are inactive, and the ethylene-responsive genes are transcribed.

Link
The regulation of blood glucose is covered in Chapter D3.3, page 761.

■ Control of blood sugar as an example of negative feedback

The use of insulin in the control of blood sugar is an example of a negative feedback loop. Insulin causes the absorption of glucose by cells where it is stored as glycogen (page 761). This lowers the blood sugar concentration, which means that insulin is no longer produced by the pancreas as blood sugar levels have returned to normal.

> **LINKING QUESTIONS**
> 1 What patterns exist in communication in biological systems?
> 2 In what ways is negative feedback evident at all levels of organization?

C2.2 Neural signalling

> **Guiding questions**
> - How are electrical signals generated and moved within neurons?
> - How can neurons interact with other cells?

> **SYLLABUS CONTENT**
>
> This chapter covers the following syllabus content:
> ▶ C2.2.1 Neurons as cells within the nervous system that carry electrical impulses
> ▶ C2.2.2 Generation of the resting potential by pumping to establish and maintain concentration gradients of sodium and potassium ions
> ▶ C2.2.3 Nerve impulses as action potentials that are propagated along nerve fibres
> ▶ C2.2.4 Variation in the speed of nerve impulses
> ▶ C2.2.5 Synapses as junctions between neurons and between neurons and effector cells
> ▶ C2.2.6 Release of neurotransmitters from a presynaptic membrane
> ▶ C2.2.7 Generation of an excitatory postsynaptic potential
> ▶ C2.2.8 Depolarization and repolarization during action potentials (HL only)
> ▶ C2.2.9 Propagation of an action potential along a nerve fibre/axon as a result of local currents (HL only)
> ▶ C2.2.10 Oscilloscope traces showing resting potentials and action potentials (HL only)
> ▶ C2.2.11 Saltatory conduction in myelinated fibres to achieve faster impulses (HL only)
> ▶ C2.2.12 Effects of exogenous chemicals on synaptic transmission (HL only)
> ▶ C2.2.13 Inhibitory neurotransmitters and generation of inhibitory postsynaptic potentials (HL only)
> ▶ C2.2.14 Summation of the effects of excitatory and inhibitory neurotransmitters in a postsynaptic neuron (HL only)
> ▶ C2.2.15 Perception of pain by neurons with free nerve endings in the skin (HL only)
> ▶ C2.2.16 Consciousness as a property that emerges from the interaction of individual neurons in the brain (HL only)

◆ **Central nervous system (CNS)**: in vertebrates, the brain and spinal cord.

◆ **Nerve**: nerve bundle of many nerve fibres connecting the central nervous system with parts of the body.

◆ **Peripheral nervous system (PNS)**: in vertebrates, nerves that convey sensory information to the CNS, and nerves that convey impulses to muscles and glands (effector organs).

◆ **Neuron**: cell within the nervous system that carries electrical impulses.

◆ **Dendrite**: a fine fibrous structure on a neuron that receives impulses from other neurons.

Link
The CNS and PNS are also covered in Chapter C3.1, page 496.

Neurons

Introducing the nervous system

The nervous system is composed of the **central nervous system (CNS)**, which consists of the brain and spinal cord. To and from the central nervous system run **nerves** of the **peripheral nervous system (PNS)**, which enable communication between the central nervous system and all parts of the body (Figure C2.2.1). Nerve cells are called **neurons**. A neuron has a cell body containing the nucleus and most of the cytoplasm. Multiple, short, elongated cytoplasmic nerve fibres called **dendrites** (see Figure C2.2.2) project from the cell body, forming connections to other neurons.

Neurons are specialized for the transmission of information in the form of impulses. An impulse is a momentary reversal in the electrical potential difference across the membrane of a neuron. The transmission of an impulse along a fibre occurs at speeds of between 30 and 120 metres per second in mammals, so nervous coordination is extremely fast and the responses are virtually immediate. Impulses travel to particular points in the body, connected by the fibres. Consequently, the effects of impulses are localized rather than diffuse.

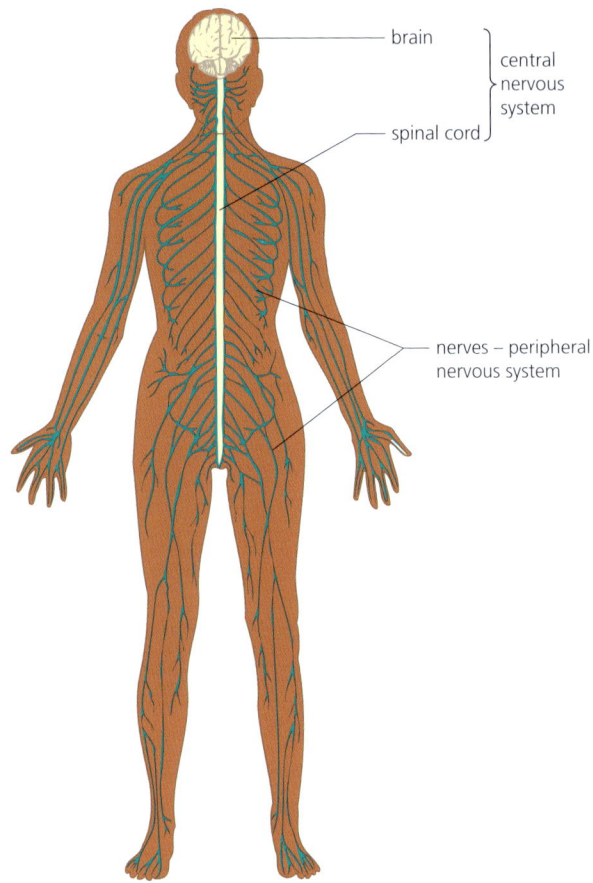

Figure C2.2.1 The organization of the mammalian nervous system

Specialized nerve cells (**neurons**) are organized into the **central nervous system** and **peripheral nerves** linking **sense organs**, **muscles** and **glands** with the **brain** or **spinal cord**.

Neuron structure

Three types of neuron make up the nervous system (see Figure C2.2.2):

- **Motor neurons** have many fine dendrites, which bring impulses towards the cell body, and a single long **axon** that carries impulses away from the cell body. The function of the motor neuron is to carry impulses from the central nervous system to a muscle or gland (known as an effector).
- **Sensory neurons** have a cytoplasmic fibre running to the cell body (in sensory neurons this starts from the **receptor**), called the **dendron**. In contrast to the motor neuron, sensory neurons have the cell body part way along the neuron, with a dendron running to the cell body and an axon away from it.
- **Relay neurons** (also known as intermediate or interneurons) have numerous short fibres. Each fibre is a thread-like extension of a nerve cell. Relay neurons occur in the central nervous system. They connect sensory neurons and motor neurons.

♦ **Motor neuron**: nerve cell that carries impulses away from the central nervous system to an effector (e.g. muscle, gland).

♦ **Axon**: fibre carrying impulses away from the cell body of a neuron.

♦ **Sensory neuron**: nerve cell carrying impulses from a sense organ or receptor to the central nervous system.

♦ **Receptor**: a cell specialized to respond to stimulation by the production of an action potential (impulse).

♦ **Dendron**: fibre carrying impulses towards the cell body of a neuron.

♦ **Relay neuron**: a specialized cell in the central nervous system which relays an electrical impulse from a sensory neuron to a motor neuron.

♦ **Schwann cells**: cells of the peripheral nervous system that wrap around the axons of motor and sensory neurons to form the myelin sheath.

♦ **Myelin sheath**: an insulating sheath around the axons of nerve fibres formed by Schwann cells.

♦ **Nodes of Ranvier**: gap in the myelin sheath around a myelinated nerve fibre.

A neuron is surrounded by many supporting cells. One type of supporting cell are **Schwann cells** (Figure C2.2.2), which wrap around the axons of motor neurons, forming a structure called a **myelin sheath**. Myelin consists largely of lipid and has high electrical resistance. The myelin sheath also contains protein (between 15% and 25% of dry mass). Frequent gaps occur along a myelin sheath, between the individual Schwann cells. The gaps are called **nodes of Ranvier**.

● Common mistake

Do not confuse effectors and receptors. Receptors detect stimuli and effectors respond to them. Sensory neurons pass impulses from receptors to the CNS. Motor neurons stimulate effectors (muscles or glands) and affect movement.

1 **Distinguish** between a sensory neuron and a motor neuron.

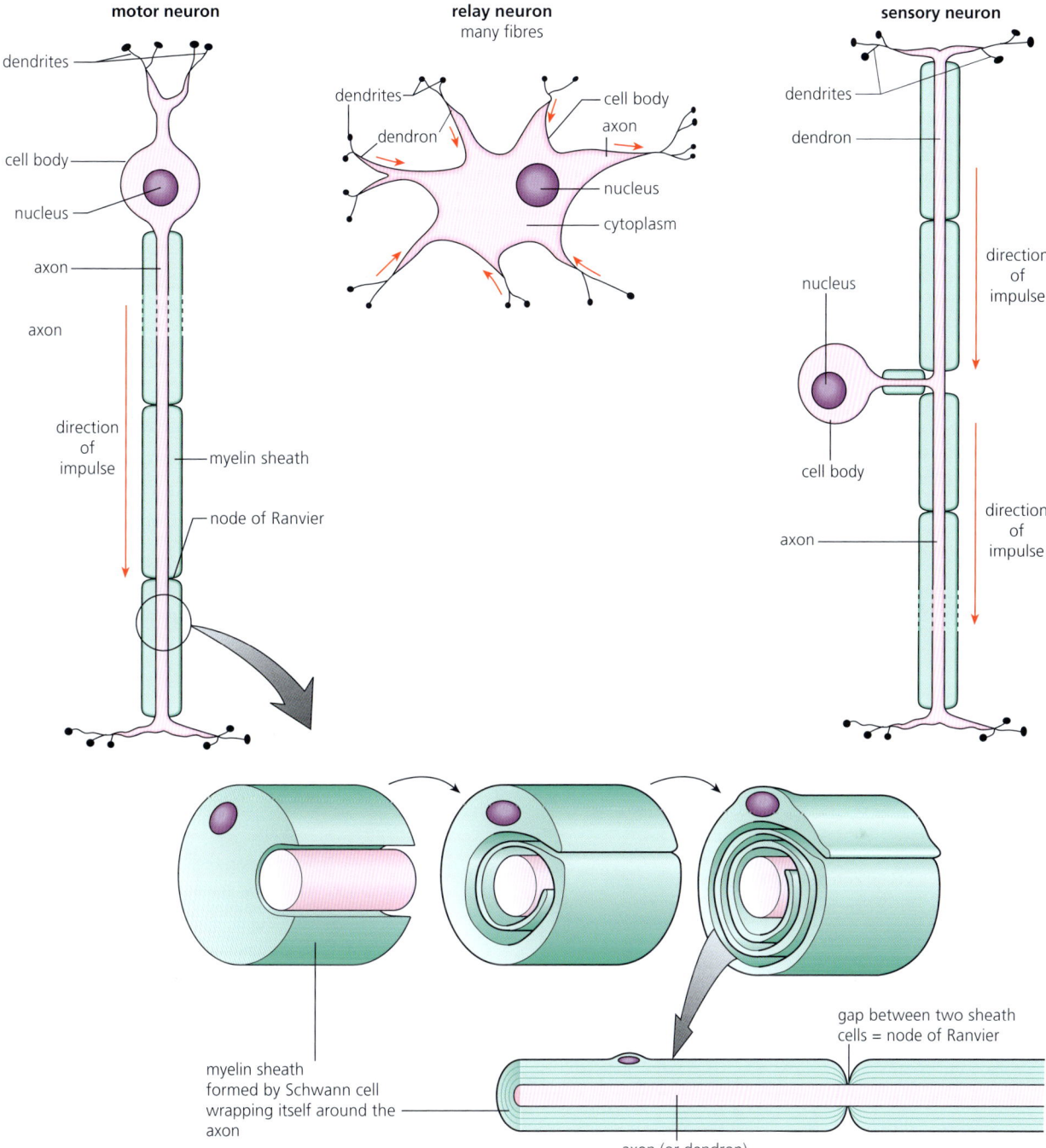

■ **Figure C2.2.2** Structure of the neurons of the nervous system

The resting potential

Neurons transmit information in the form of impulses. An impulse is transmitted along nerve fibres, but it is not an electrical current that flows along the 'wires' of the nerves. An impulse is a momentary reversal in electrical potential difference in the membrane – a change in the position of charged ions between the inside and outside of the membrane of the nerve fibres. This reversal flows from one end of the neuron to the other in a fraction of a second.

◆ **Resting potential**: the potential difference across a nerve cell membrane when it is not being stimulated. It is normally about −70 millivolts (mV).

◆ **Membrane polarization**: a lipid membrane that has a positive electrical charge on one side and a negative charge on the other side.

◆ **Membrane potential**: the difference in charge between the inside and outside of a neuron, which is created due to the unequal distribution of ions on both sides of the cell membrane.

Between the conduction of one impulse and the next, the neuron is sometimes said to be resting, but this is not the case. The 'resting' neuron membrane is actively setting up the electrical potential difference between the inside and the outside of the fibre, known as the **resting potential**.

The resting potential is the potential difference across a nerve cell membrane when it is not being stimulated. It is normally about −70 millivolts (mV) (Figure B2.1.21, page 241). The resting potential difference is re-established across the neuron membrane after a nerve impulse has been transmitted. We say the nerve fibre has been **repolarized**.

Links

Active transport is covered in Chapter B2.1, page 230; sodium–potassium pumps as an example of exchange transporters (HL only) is also covered in Chapter B2.1 on page 241.

The resting potential is the product of the **active transport** of potassium ions (K^+) in across the membrane and sodium ions (Na^+) out across the membrane. This occurs by a K^+/Na^+ pump, using energy from ATP. Three Na^+ are pumped out for every two K^+ pumped in (B2.1.22, page 242) across the plasma membrane of the neuron. The tissue fluid outside the neuron therefore contains many more positive ions than are present in the tiny amount of cytoplasm inside. As a result, a negative charge is developed inside, compared to outside, and the resting neuron is said to be **polarized**. The **membrane potential** is the difference in charge between the inside and outside of a neuron, which is created due to the unequal distribution of ions on both sides of the cell membrane. The next event, sooner or later, is the passage of an impulse, known as an **action potential** (see below).

 Top tip!

Energy from ATP drives the pumping of sodium and potassium ions in opposite directions across the plasma membrane of neurons. ATP is derived from aerobic respiration. Even when 'doing nothing', the nervous system needs to use ATP to maintain the resting potential.

2 **Outline** how a resting potential is generated.

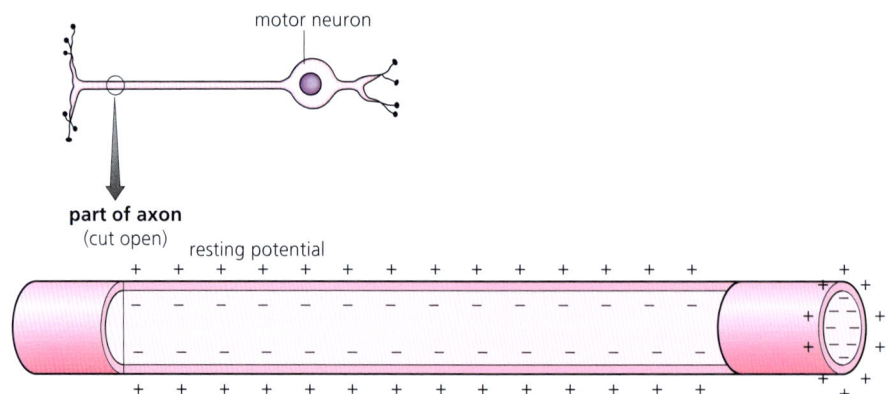

■ **Figure C2.2.3** The resting potential of an axon

Nerve impulses as action potential

◆ **Action potential**: the potential difference produced across the plasma membrane of the nerve cell when stimulated, reversing the resting potential from about −70 mV to about +40 mV.

A stimulus causes sodium ions, which are positively charged, to flow into the cytoplasm of the axon, reversing the polarity of the axon (Figure C2.2.4). Further details of the process are covered at HL (page 479). The **action potential** is the potential difference produced across the plasma membrane of the nerve cell when stimulated, reversing the resting potential from about −70 mV to about +40 mV. The nerve impulse is electrical because it involves the movement of positively charged ions, leading to a potential difference (voltage) across the membrane. Nerve impulses are action potentials that are propagated (spread) along nerve fibres.

The action potential then runs the length of the neuron fibre. At any one point, it exists for only two-thousandths of a second (2 milliseconds) before the resting potential starts to be re-established, so action potential transmission is exceedingly quick.

Note in Figure C2.2.4 that some neurons have Schwann cells wrapped around the axon forming the myelin sheath while some axons do not have this structure (they are unmyelinated). This will be explored further in the next section. The junctions in the sheath, the nodes of Ranvier, occur at 1–2 μm intervals; it is only at these junctions that the axon membrane is exposed so the action potential 'jumps' from node to node. Unmyelinated dendrons and axons are common in non-vertebrate animals. Here, the action potential must travel step-by-step along the entire surface of the fibres.

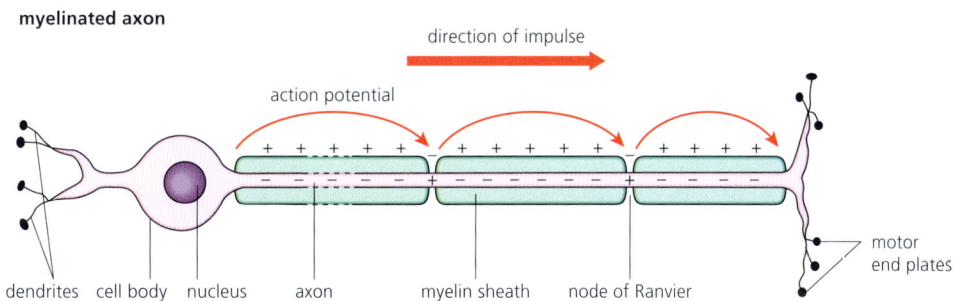

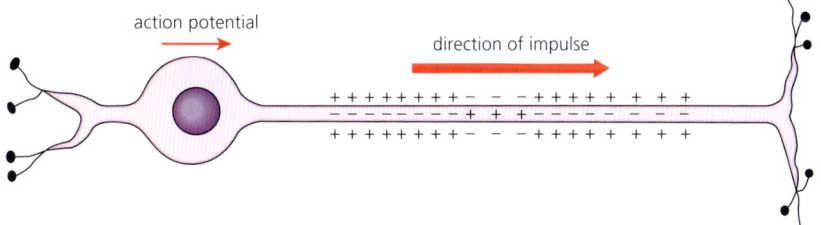

■ **Figure C2.2.4** The action potential

3 **Define** the term *action potential*.

4 **Deduce** the source of energy used to:
   a establish the resting potential
   b power an action potential.

ATL C2.2A

What effect does stimulus intensity and duration have on the generation of an action potential? Carry out your own investigations using these sites:

Stimulus intensity:
https://ilearn.med.monash.edu.au/physiology/action-potentials/action-potential#simulation

Stimulus duration:
https://ilearn.med.monash.edu.au/physiology/action-potentials/adaptation#simulation

In each case, read the theory and then carry out the simulation. What conclusions can you reach?

C2.2 Neural signalling

ATL C2.2B

The neural signalling system topic brings together many ideas you will have encountered throughout IB Biology. These include membrane structure; membrane proteins such as protein pumps/carrier proteins, channel proteins and gated channel proteins such as protein receptors; complementary three-dimensional shape; facilitated diffusion; active transport; surface area-to-volume ratio; factors that affect diffusion; vesicles; exocytosis; enzymes; mitochondria; ATP; concentration gradient and voltage gradient.

Write a brief summary about each of the above topics. This process will help you to link existing biological knowledge to new understandings you will encounter in this topic.

Variation in the speed of nerve impulses

The following activities explore the effects of myelination and axon size on the speed of conduction. Ensure you carry out these activities before reading on in this section.

Tool 2: Technology

Generating data from simulations

Access this website, which investigates the effect of axon diameter and myelination on conduction velocity:
https://ilearn.med.monash.edu.au/physiology/action-potentials/axon-diameter#simulation

Carry out the simulation with unmyelinated fibres. What is the effect of axon diameter on conduction speed? Now repeat the experiment with myelinated fibres.

Using spreadsheets to manipulate data

For both experiments gather your data and plot a graph. You can export data to a spreadsheet program such as Excel and use it to find the line of best fit for the unmyelinated and myelinated results.

What are the effects of myelination on conduction speed?

Suggest why, for very small axon diameters, there is little benefit of myelination for conduction velocity.

Inquiry 1: Exploring and designing

Exploring

Research the conduction speed of nerve impulses for a variety of different animals. Ensure that you include both invertebrates and vertebrates, with some having myelinated axons (i.e. covered by a myelin sheath) and others unmyelinated (no myelin sheath). You also need to find the diameter of the axon in each case. Consult a variety of sources and select sufficient and relevant sources of information.

Suggestions for different animals you could research are included in Table C2.2.1. Redraw the table and add in your findings.

■ **Table C2.2.1** Researching the effect of axon diameter and degree of myelination on nerve impulse conduction speed in a range of different animals

Organism/nerve	Axon diameter (μm)	Myelination	Speed of propagation (m s⁻¹)
squid/giant axon			
Lumbricus (earthworm)			
lobster			
frog			
rat			
domestic cat			

Theme C: Interaction and interdependence – Cells

> **Inquiry 3: Concluding and evaluating**
>
> Concluding
>
> Analyse the data you found for Table C2.2.1.
>
> Compare the speed of transmission in animals that have myelination and those that do not. Which have the faster conduction speeds overall? Why is this?
>
> Now compare the animals that have unmyelinated axons, such as the giant axons of squid and those animals with smaller, unmyelinated nerve fibres.
>
> What conclusions can you reach about the effect of axon diameter size and myelination on conduction velocity? What explanation can you develop for these observations?

Top tip!

Some invertebrates, such as the squid and the earthworm, have giant fibres, which allow fast transmission of action potentials (although not as fast as in myelinated fibres).

You should have carried out the activities on this page and the previous page and drawn your own conclusions. You can now read on!

In the Tools and Inquiry activities, you should have found that myelinated fibres have a faster conduction speed compared to unmyelinated fibres. This is discussed further below. In addition, an unmyelinated fibre with a large diameter transmits an action potential much faster than a narrow fibre does. This is because the speed of transmission depends on resistance offered by the axoplasm (cytoplasm inside the axon). The narrower the fibre, the greater its resistance, and the lower the speed of conduction of the action potential. Incidentally, the original investigations of the nature of the action potential by physiologists were carried out on giant fibres.

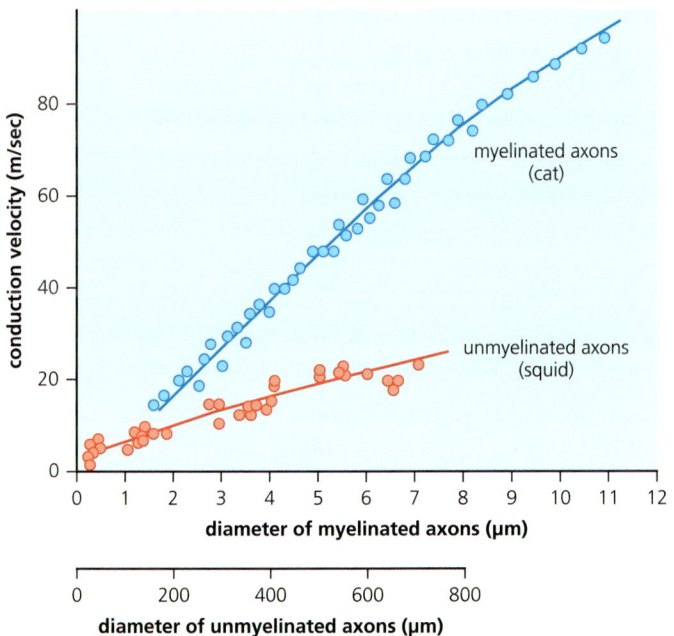

■ **Figure C2.2.5** Conduction velocities of myelinated and unmyelinated axons as functions of axon diameter. Myelinated axons have faster conduction velocities than unmyelinated axons that are 100 times greater in diameter

Nervous systems have therefore evolved two contrasting mechanisms for increasing the conduction speed of the electrical impulse. One is through developing giant axons: using axons several times larger in diameter than is normal for other large axons. Giant axons are found in squid (*Loligo* sp., a type of aquatic mollusc). The other mechanism is by using the **myelin sheath**, where axons are wrapped by cells that have high amount of lipid in their plasma membrane (**Schwann cells**). Each mechanism, on its own or in combination with the other, is used in the nervous systems of many different groups of animals, both vertebrate and invertebrate.

The presence of a myelin sheath speeds up transmission of the action potential because it is only at the nodes of Ranvier that the axon membrane is exposed and the action potential can 'jump' from node to node. The electrical resistance of the myelin sheath prevents reversal of the axon polarity at the nodes of Ranvier. By contrast, the step-by-step travel of the action potential along the entire surface of the fibres in unmyelinated dendrons and axons is a relatively slow process (Figure C2.2.4).

Action potential propagation (or conduction) velocity is directly correlated with the axon diameter (Figure C2.2.5 and Table C2.2.2). The larger the axon diameter, the higher the action potential propagation velocity will be, which means signals can travel more quickly. This is because there is less resistance facing the ion flow. In addition, myelination, which leads to the action potential jumping between nodes along axons, greatly increases the action potential propagation velocity.

■ **Table C2.2.2** Effect of nerve fibre diameter and myelination on conduction velocity of nerve impulses

Diameter (mm)	Conduction velocity (m s^{-1})	Myelination
12–20	70–120	yes
5–12	30–70	yes
2–5	12–30	yes
3–6	15–30	yes
<3	3–14	some
0.4–1.2	0.5–2	no
0.3–1.3	0.7–2.3	no

Despite having much smaller diameters, myelinated axons achieve much higher action potential propagation velocities than unmyelinated axons. Figure C2.2.5 shows the correlation between axon diameter and conduction speed in both myelinated and unmyelinated axons.

Nature of science: Observations

Investigating the effect of axon diameter on conduction velocity

Unmyelinated fibres with a large diameter transmit an action potential much more quickly than a narrow fibre because the speed of transmission depends on the resistance of the axoplasm, as mentioned above. The narrower the fibre, the greater its resistance to ion movement, and the lower the speed of conduction of the action potential. In addition, a larger diameter means there can be more ion channels around the axon, which means that there are more ion channels per length of axon.

Read more about the early work on the action potential here:
www.ncbi.nlm.nih.gov/pmc/articles/PMC3424716
https://neuroscientificallychallenged.com/posts/history-of-neuroscience-hodgkin-huxley

 ■ **Applying correlation coefficients to determine the strength of a correlation**

Axon diameter and degree of myelination can be correlated with speed of conduction (as explored above). The strength of the correlation can be tested using the coefficient of determination, which measures how close the data are to the **fitted regression line**. The regression line is a line that best fits the data, i.e. has the smallest overall distance from the line to the points. The formula for the regression line is:

$y = mx + b$

where m is the slope of the line and b is the value on the y-axis where the line crosses; y is the dependent variable and x is the independent variable.

Top tip!

Correlations can either be negative or positive. Correlation coefficients can be applied as a mathematical tool to determine the strength of these correlations. The coefficient of determination (R^2) is one such tool that can be used to evaluate the degree to which variation in the independent variable may explain the variation in the dependent variable.

The coefficient of determination, or R squared (R^2) method, is the proportion of the variance in the dependent variable that is predicted from the independent variable. The value of R^2 shows whether the line of best fit is a good fit for the data. The calculation determines the extent to which two variables are correlated. The coefficient of determination ranges from 0 to 1 (or 0% to 100%):

- If R^2 is equal to 0, then the dependent variable cannot be predicted from the independent variable.
- If R^2 is equal to 1, then the dependent variable can be predicted from the independent variable without any error.
- Generally, a coefficient of determination of about 70% is considered a good correlation.

R^2 indicates the extent that the dependent variable can be predicted. If R^2 is 0.20, it means that 20% of the variance in the y variable is predicted from the x variable. If the value is 0.50, it means that 50% of the variance in the y variable is predicted from the x variable, and so on.

Tool 3: Mathematics

Applying the coefficient of determination (R^2) to evaluate the fit of a trend line

Steps to find the coefficient of determination:
1 Find R, the Pearson correlation coefficient.
2 Square R.
3 Change the above value to a percentage.

The formula for the Pearson correlation coefficient is:

$$R = n \frac{n(\sum xy) - (\sum x)(\sum y)}{\sqrt{[n\sum x^2 - (\sum x)^2][n\sum y^2 - (\sum y)^2]}}$$

Where:

n = total number of observations

$\sum x$ = total of the first variable value

$\sum y$ = total of the second variable value

$\sum xy$ = sum of the product of the first and second value

$\sum x^2$ = sum of the squares of the first value

$\sum y^2$ = sum of the squares of the second value

$$\text{The coefficient of determination} = \left(\text{Pearson correlation coefficient}\right)^2 = R^2$$

Enter the data in Table C2.2.3 into an Excel spreadsheet.

■ **Table C2.2.3** Effect of diameter on speed of conduction in myelinated nerve fibres

Diameter (mm)	Conduction (m s⁻¹)
16.0	95.0
14.0	75.0
12.0	70.0
8.5	50.0
6.0	30.0
4.5	22.5
3.5	21.0

1 Highlight the data and select 'Insert – Chart – Scatter'.
2 Select chart area and select '+' to add axes labels – add labels for each axis.
3 Click on any data point. Right click and select 'Add trendline'. Select 'linear' trendline.
4 Under 'Trendline options', select 'Display R-squared value on chart'. The R^2 will appear above the trendline.

The graph should appear as follows:

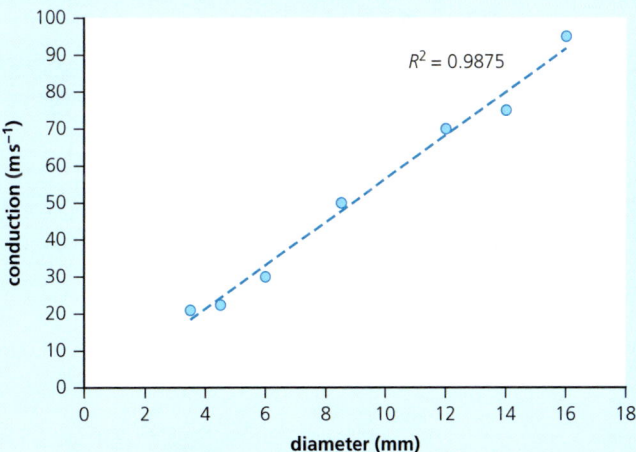

■ **Figure C2.2.6** Correlating axon diameter with conduction speed

Both the line of best fit and R^2 value show a **strong positive correlation** between axon diameter and conduction speed.

Properties of the coefficient of determination (R^2):
- It helps to find the ratio of how a dependent variable varies from the independent variable.
- It also tests the strength of the (linear) association between the variables.
- If the value of R^2 is closer to 1 (or 100%), the values of y are close to the regression line. If the value of R^2 is closer to 0, the values of y are not close to the regression line.

Synapses

◆ **Synapse**: the connection between the end of a nerve cell and another cell; functionally a tiny gap, the synaptic cleft, traversed by transmitter substances.

◆ **Presynaptic neuron**: neuron 'upstream' of a synapse.

◆ **Postsynaptic neuron**: neuron 'downstream' of a synapse.

The **synapse** is the link point between neurons or between neurons and effector cells. A synapse consists of the swollen tip (synaptic knob) of the axon of one neuron (**presynaptic neuron**) and the dendrite or cell body of another neuron (**postsynaptic neuron**). At the synapse, the neurons are extremely close, but they have no direct contact. Instead, there is a tiny gap, called a synaptic cleft, about 20 nm wide (Figure C2.2.7).

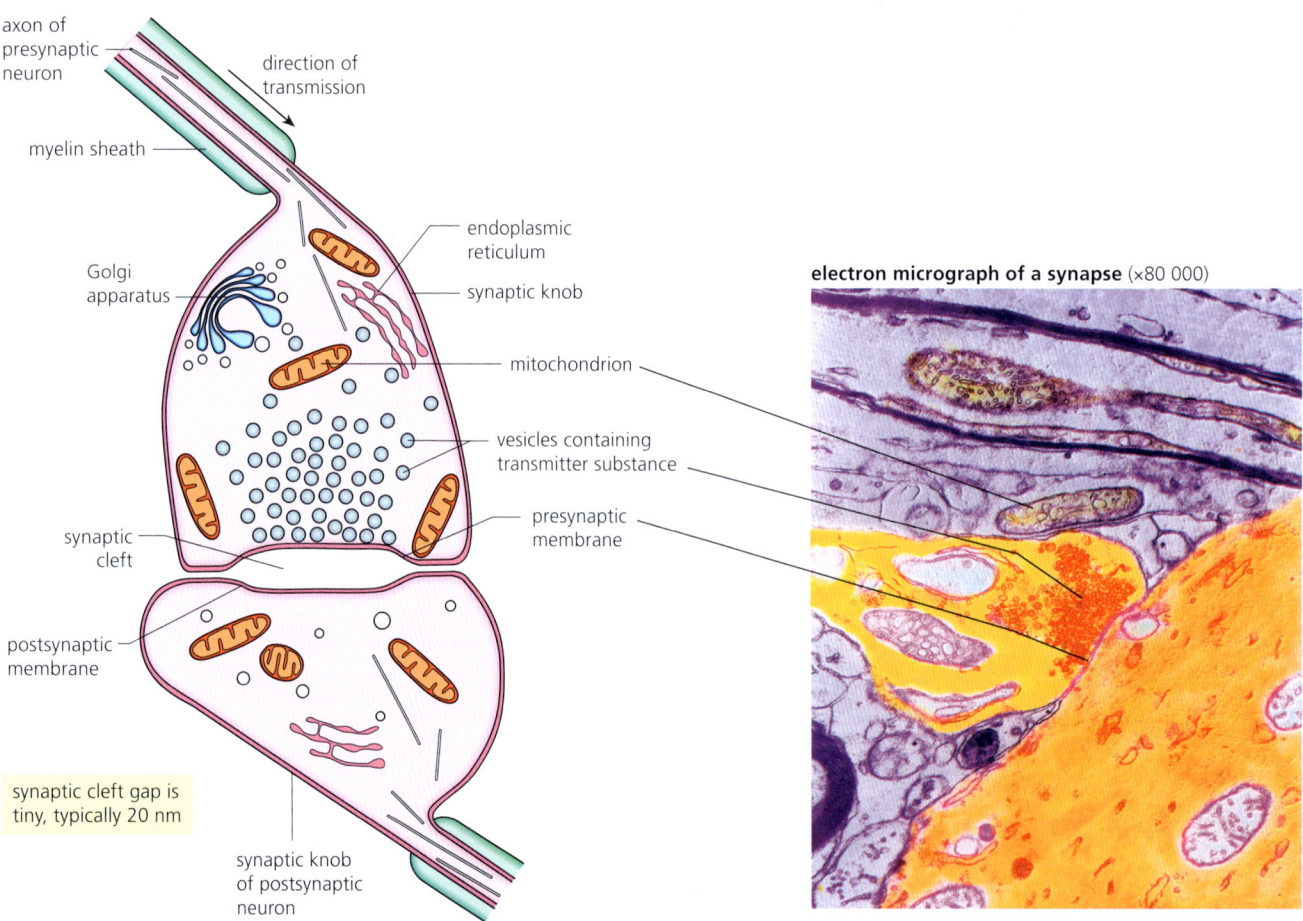

■ Figure C2.2.7 A synapse in section

◆ **Neurotransmitter**: chemical released at the presynaptic membrane of an axon on arrival of an action potential, which transmits the action potential across the synapse.

The practical effect of the synaptic cleft is that an action potential can only cross it via specific chemicals, known as **neurotransmitter** substances. Neurotransmitter substances are all relatively small molecules that diffuse quickly. They are produced in the Golgi apparatus in the synaptic knob and are held in tiny vesicles before release. The way that neurotransmitters cross the synapse, from presynaptic membrane to postsynaptic membrane, ensures that a signal can only pass in one direction across a typical synapse.

◆ **Acetylcholine (ACh)**: a neurotransmitter that functions in both the central and peripheral nervous systems.

Acetylcholine (ACh) is a commonly occurring neurotransmitter substance (the neurons that release acetylcholine are known as cholinergic neurons – see page 456). Another common neurotransmitter substance is norepinephrine (found in adrenergic neurons). In the brain, the commonly occurring transmitters are glutamic acid and dopamine.

Top tip!

Acetylcholine exists in many types of synapse, including neuromuscular junctions.

Link

The structure and function of motor units in skeletal muscle are covered in Chapter B3.3, page 332.

Neuromuscular junctions are a specialized synapse between a motor neuron nerve ending and its muscle fibre. They are responsible for converting electrical impulses generated by the motor neuron into electrical activity in the muscle fibres. When an action potential arrives at the neuromuscular junction, vesicles of acetylcholine are released and the transmitter molecules bind to receptors on the sarcoplasm (this is the plasma membrane of the muscle fibre). This triggers the release of calcium ions from the sarcoplasmic reticulum, which leads to muscle contraction (see page 327).

5 **Define** the term *neurotransmitter*.

Steps of synapse transmission

▨ Release of neurotransmitters from a presynaptic membrane

1 The arrival of an action potential at the synaptic knob leads to a change in polarity of the presynaptic membrane, which opens calcium ion channels in the membrane, and calcium ions flow in from the synaptic cleft.

2 The calcium ions cause vesicles of transmitter substance to fuse with the presynaptic membrane, releasing transmitter substance into the synaptic cleft. The calcium therefore acts as a signalling chemical inside a neuron.

▨ Generation of an excitatory postsynaptic potential

3 The transmitter substance diffuses across the synaptic cleft and binds with a specific receptor protein.

In the postsynaptic membrane, there are specific receptor sites for each transmitter substance. Each of these receptors also acts as a channel in the membrane, which allows a specific ion (e.g. Na^+, Cl^- or some other ion) to pass. The attachment of a transmitter molecule to its receptor instantly opens the ion channel.

6 **Draw** and **annotate** a diagram of a synapse to indicate its structure and function.

7 **Identify** the role of the:
 a Golgi apparatus
 b mitochondria in the synaptic knob.

When a molecule of ACh attaches to its receptor site, an Na^+ channel opens. As the Na^+ ions rush into the cytoplasm of the postsynaptic neuron, a reversal of polarity of the postsynaptic membrane occurs. As more and more molecules of ACh bind, it becomes increasingly likely that the reversal of polarity will reach the threshold level (see more on page 481). When it does, an action potential is generated in the postsynaptic neuron. This process of build up to an action potential in the postsynaptic membranes is called facilitation.

4 The transmitter substance on the receptors is immediately inactivated by enzyme action. For example, the enzyme cholinesterase hydrolyses ACh to choline and ethanoic acid, which are inactive as neurotransmitters. This causes the ion channel of the receptor protein to close, and so allows the resting potential in the postsynaptic neuron to be re-established.

5 The inactivated products from the transmitter re-enter the presynaptic knob, are resynthesized into neurotransmitter substance and packaged for reuse (Figure C2.2.8).

● Common mistake

When discussing how the neurotransmitter crosses the synapse, it is not enough to say that it 'moves' across the synaptic cleft – you must say that the neurotransmitter reaches the postsynaptic membrane by diffusion, from high concentration to low concentration.

● Common mistake

Do not forget to refer to to the removal of the neurotransmitter by an enzyme such as cholinesterase when discussing how the synapse functions.

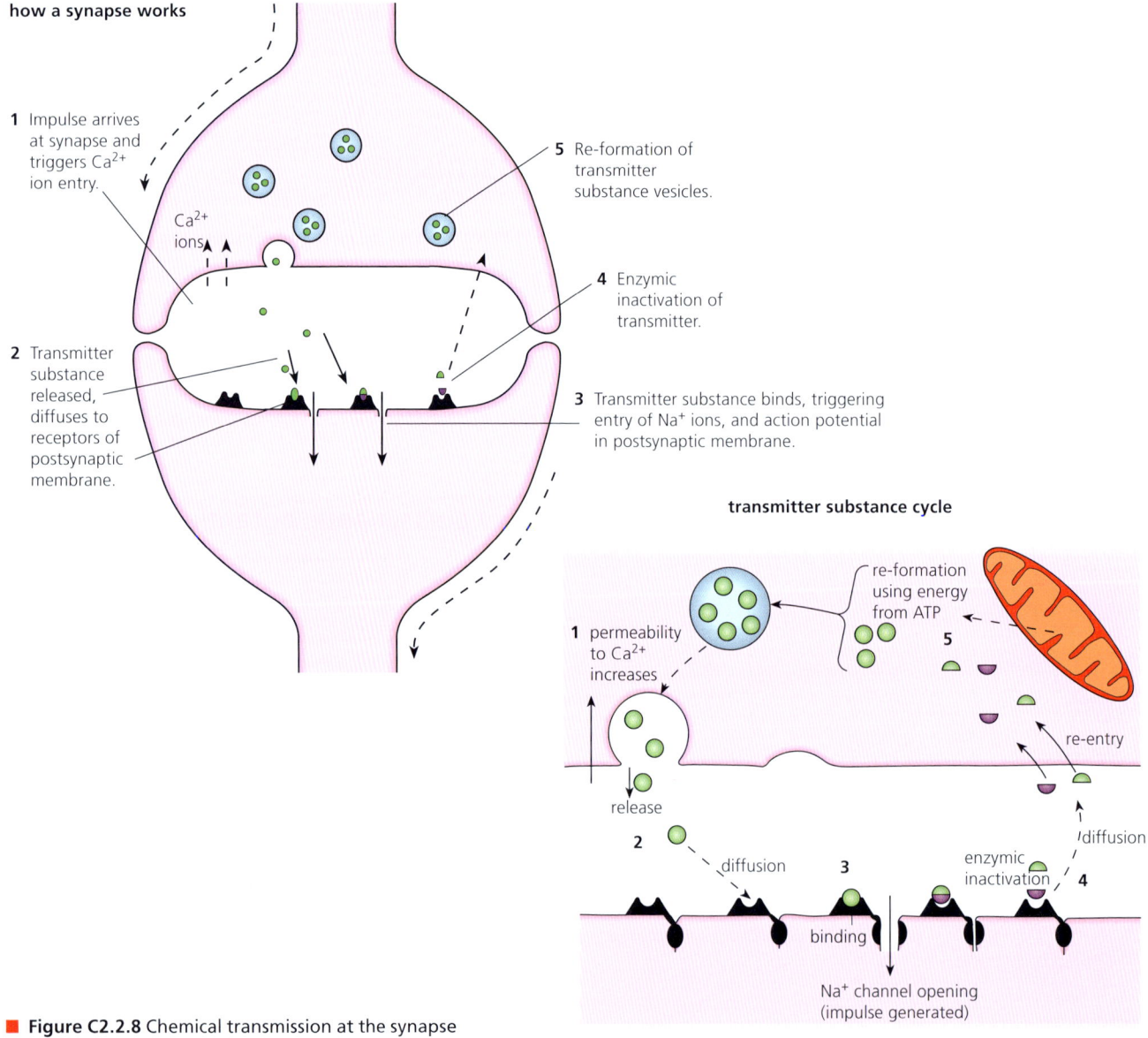

■ Figure C2.2.8 Chemical transmission at the synapse

◆ **Depolarization**: a temporary and local reversal of the resting potential difference of the membrane that occurs when an impulse is transmitted along the axon.

◆ **Repolarization**: return of polarity towards the resting potential following depolarization.

◆ **Voltage-gated channels**: a type of transmembrane protein that forms ion channels activated by changes in the electrical membrane potential of a neuron.

Link
Voltage-gated channels are also covered in Chapter B2.1, page 240.

Depolarization and repolarization during action potentials

An action potential is triggered by a stimulus that is received at a receptor cell or sensory nerve ending. The energy of the stimulus causes a temporary and local reversal of the resting potential. The result is that the membrane is briefly **depolarized**. Following the action potential there is a return of polarity towards the resting potential – this is known as **repolarization**. When the resting potential has been restored, the axon is said to be **repolarized**.

This change in potential across the membrane occurs because of ion channels in the membrane. These special channels are globular proteins that span the membrane. They have a central pore with a gate that can open and close. One type of channel is permeable to sodium ions and another to potassium ions. During a resting potential these channels are all closed.

The transfer of energy of the stimulus first opens the gates of the sodium channels in the plasma membrane and sodium ions diffuse in, down their electrochemical gradient. Because the channels open due to changes to the potential difference (voltage) across the membrane, they are called **voltage-gated channels**.

Top tip!
Sodium ions are predominantly found outside the membrane and, when they enter the neuron, their positive charges increase in that part of the membrane inside the cell. Positively charged potassium ions are predominantly found inside the cell and, when they flood out, the inner side of the membrane becomes more negatively charged.

What causes an electrochemical gradient?

The electrochemical gradient of an ion is due to its electrical and chemical properties.
- Electrical properties are due to the charge on the ion (an ion is attracted to an opposite charge).
- Chemical properties are due to concentration in solution (an ion tends to move from a high to a low concentration).

With sodium channels opened, the cytoplasm of the neuron fibre (the interior) quickly becomes progressively more positive with respect to the outside. When the charge has been reversed from $-70\,\text{mV}$ to $+40\,\text{mV}$ (due to the electrochemical gradient), an action potential has been created in the neuron fibre (Figure C2.2.9).

The change in potential difference (voltage) causes further sodium channels to open. These channels open or close depending on a certain threshold membrane potential being reached. Almost immediately an action potential has passed, the sodium channels close and the potassium channels open, again caused by a change in the potential difference of the membrane (and are therefore called voltage-gated potassium channels). Now, potassium ions can exit the cell, again down an electrochemical gradient, into the tissue fluid outside. The interior of the neuron fibre starts to become less positive again. **Repolarization** occurs. Then, the potassium channels also close. Finally, the resting potential is re-established by the action of the sodium–potassium pump and the process of facilitated diffusion.

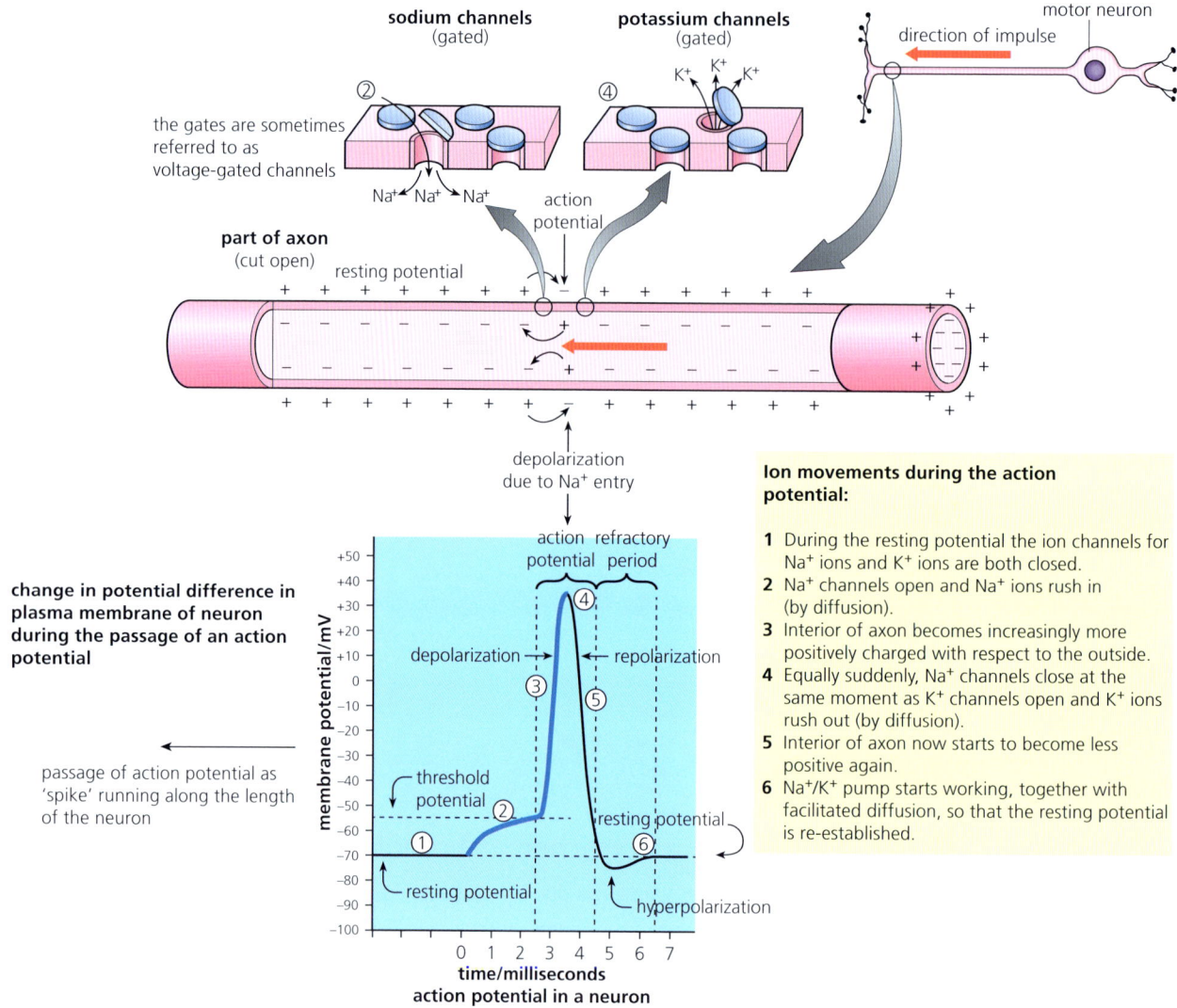

■ Figure C2.2.9 The action potential

ATL C2.2C

Explanations in biology are often complex, such as how the action potential works. By making links to existing biological knowledge, and subdividing complicated processes into smaller 'chunks', difficult biological ideas can be better understood and remembered.

The following video shows you how the action potential can be broken down into a series of processes that draw on your existing knowledge of biology to explain key ideas in a coherent and easy-to-follow way: www.youtube.com/watch?v=7EyhsOewnH4

Create a poster showing the events of the action potential, summarizing key processes in a simple and visually memorable way. Keep this for future reference – it will be useful when you come to revise this topic!

The refractory period

During the resting potential, sodium ions are predominantly outside the neuron and potassium ions mainly inside. Following **depolarization** (sodium influx) and **repolarization** (potassium efflux), this ionic distribution is largely reversed. Before a neuron can fire again, the resting potential must be restored via the sodium–potassium pump. For a brief period, therefore, following the passage of an action potential, the neuron fibre is no longer excitable. This is the **refractory period**, and it lasts only 5–10 milliseconds in total. The neuron fibre is not excitable during the refractory period because there is a large excess of sodium ions inside the fibre and further influx is impossible. Subsequently, as the resting potential is progressively restored, it becomes increasingly possible for an action potential to be generated again. Because of the refractory period, the maximum frequency of impulses is between 500 and 1000 per second.

Because areas of the membrane that have recently depolarized will not depolarize again due to the refractory period, the action potential only travels in one direction.

◆ **Refractory period**: the period immediately following stimulation when a nerve or muscle is unresponsive to further stimulation. During this period the voltage-gated sodium channels are closed and will not respond to changes in voltage.

8 **Sketch** and **annotate** a graph showing an action potential.

9 **Describe** how an action potential is generated.

◆ **Threshold potential**: the critical level to which a membrane potential must be depolarized in order to start an action potential. If the threshold is not reached, the neuron remains at rest and an action potential is not triggered.

Common mistake

It is incorrect to say that the sodium–potassium pump causes the falling phase of the action potential by pumping Na+ ions back out of the neuron. It is the opening of potassium channels, causing potassium to leave the interior of the axon, that returns it to a negative potential difference.

The all-or-nothing principle

Stimuli have widely different strengths: for example, contrast a light touch and the pain of a finger hit by a hammer. A stimulus must be at or above a minimum intensity, known as the **threshold potential**, to initiate an action potential (see Figure C2.2.9). Either the depolarization is sufficient to fully reverse the potential difference in the cytoplasm (from –70 mV to +40 mV) or it is not. If not, no action potential arises. There is a need for a threshold potential to be reached for sodium channels to start to open. With all sub-threshold stimuli, the influx of sodium ions is quickly reversed and the full resting potential is re-established.

However, as the intensity of the stimulus increases, the frequency at which the action potentials pass along the fibre increases (the individual action potentials are all of standard strength). For example, with a very persistent stimulus, action potentials pass along a fibre at an accelerated rate, up to the maximum possible permitted by the refractory period. This means the effector (or the brain) can recognize the intensity of a stimulus from the frequency of action potentials (Figure C2.2.10).

| stimuli below the threshold value: not sufficient to reverse polarity of the membrane to +40 mV | brief stimulus just above threshold value: needed to cause depolarization of the membrane of the sensory cell, and thus trigger an impulse | stronger, more persistent stimulus | much stronger stimulus: has stimulated almost the maximum frequency of impulses |

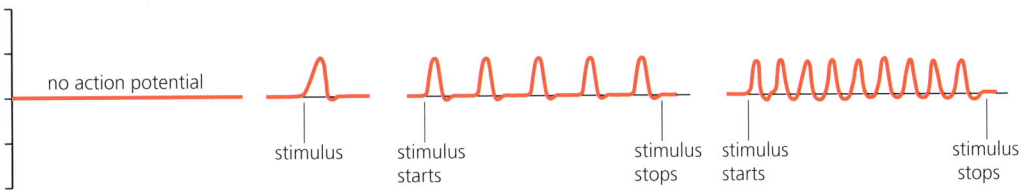

■ **Figure C2.2.10** Weak and strong stimuli and the threshold value

Propagation of an action potential as a result of local currents

The area of the axon that is depolarized during the action potential sets up local currents: positive charges flow towards adjacent negative areas and depolarize the adjacent membrane. Consequently, the action potential is conducted along the axon.

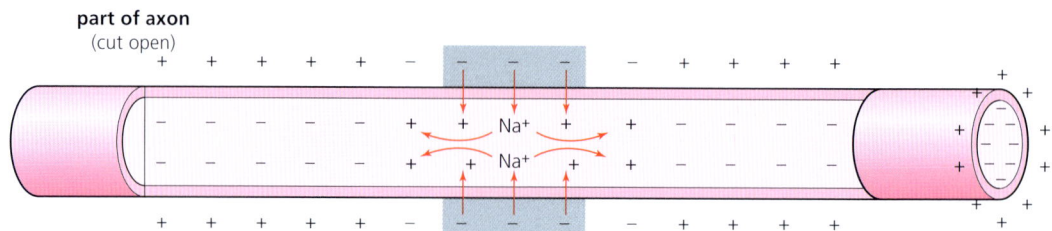

■ **Figure C2.2.11** Diffusion of sodium ions both inside and from outside an axon can cause the threshold potential to be reached

As sodium ions enter the axon, causing depolarization, they flow to adjacent areas, which causes these areas to become less negative. This leads to further depolarizing and the opening of voltage-gated sodium channels. This causes the action potential to move along the axon. The areas of the membrane that have recently depolarized will not depolarize again because of the refractory period, which means that the action potential will only travel in one direction.

Action potentials are therefore propagated along the axons of neurons via local currents. Local currents induce depolarization of the adjacent axonal membrane and, where this reaches a threshold, further action potentials are generated.

Oscilloscope traces showing resting potentials and action potentials

Oscilloscopes are scientific instruments that can be used to measure the membrane potential across an axon membrane. Data are displayed as a graph, with time (in milliseconds) on the *x*-axis and membrane potential (in millivolts) on the *y*-axis. The graph shows the events of an action potential (see graph in Figure C2.2.9, page 480).

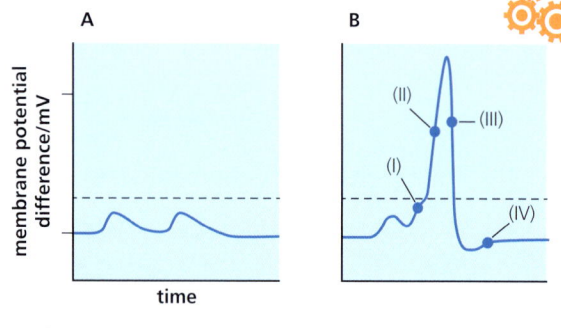

■ **Figure C2.2.12** Oscilloscope traces obtained from postsynaptic neurons

You need to be able to interpret an oscilloscope trace in relation to cellular events relating to generation of an action potential. The oscilloscope trace allows the number of impulses per second to be measured.

Figure C2.2.12 shows two oscilloscope traces of specific events in an axon of a postsynaptic neuron.

10 Examine trace A and **explain** what has happened.

11 In trace B, **outline** what specific events are occurring at the points labelled (I), (II), (III) and (IV).

Saltatory conduction

◆ **Saltatory conduction**: how an electrical impulse jumps from node to node along a myelinated axon, speeding the arrival of the impulse (in comparison with the slower continuous progression of depolarization spreading down an unmyelinated axon).

The presence of a **myelin sheath** affects the speed of transmission of the action potential. Only at the junctions in the sheath, at the **nodes of Ranvier**, is the axon membrane exposed. Ion pumps and channels are grouped at nodes of Ranvier. Elsewhere along the fibre, the electrical resistance of the myelin sheath prevents depolarization of the membrane. The action potentials actually 'jump' from node to node. This is called **saltatory conduction**, meaning 'to leap' (Figure C2.2.13). This greatly speeds up the rate of transmission.

By contrast, unmyelinated dendrons and axons are common in invertebrate animals. Here, step-by-step depolarization occurs as the action potential flows along the entire surface of the fibres. This is a relatively slow process compared with saltatory conduction.

12 **Explain** why myelinated fibres conduct impulses faster than unmyelinated fibres of the same size.

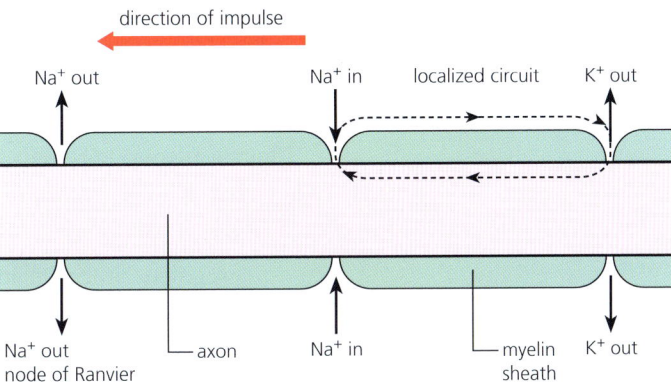

■ **Figure C2.2.13** Saltatory conduction between small gaps in the myelin sheath

Effects of exogenous chemicals on synaptic transmission

◆ **Exogenous chemical**: chemical that originates outside the body, e.g. drugs.

Drugs that interfere with the action of neurotransmitters have been discovered. For example, some drugs amplify the processes and increase postsynaptic transmission. Nicotine has this effect (see page 240). Drugs are **exogenous chemicals** because they originate outside the body.

Other drugs inhibit the processes of synaptic transmission, in effect decreasing synaptic transmission. Beta blocker drugs have these effects and are used in the treatment of high blood pressure. Other drugs are stimulants, speeding up the transmission of electrical impulses, such as amphetamines.

An understanding of the workings of neurotransmitters and synapses has led to the development of numerous pharmaceuticals for the treatment of mental and neurological disorders.

◼ Neonicotinoids

Neonicotinoids are a type of pesticide that completely block synaptic transmission at the cholinergic synapses (synapses that have receptors that are activated when they bind to acetylcholine) of insects. They are similar in structure to nicotine.
- Neonicotinoids bind to acetylcholine receptors in synapses in the CNS of insects, blocking the binding of acetylcholine, inhibiting synaptic transmission.
- They cannot be broken down by acetylcholinesterase and so their effects are irreversible.
- There are issues about their impact on the wider insect community – concerns have been raised about their effects on honeybees.

C2.2 Neural signalling

Cocaine

Cocaine is an example of a drug that blocks reuptake of neurotransmitters. Cocaine primarily affects dopamine – a neurotransmitter that plays an important role in the brain's 'reward and reinforcement' systems. It also affects the neurotransmitters serotonin and norepinephrine. The drug prevents neurotransmitter molecules from being taken back up into the cell. Cocaine works by binding to the transporters that normally remove the excess of these neurotransmitters from the synaptic gap. It prevents them from being reabsorbed by neurons and therefore increases their concentration in the synapses. As a result, the natural effect of the neurotransmitter on the postsynaptic neurons is amplified. The group of modified neurons produce much more dependency (from dopamine), feelings of confidence (from serotonin) and energy (from norepinephrine), typically experienced by people who take cocaine.

In chronic cocaine consumers, the brain comes to rely on this exogenous drug to maintain the high degree of pleasure associated with the artificially elevated levels of neurotransmitters. The brain responds to increased dopamine levels by synthesizing new dopamine receptors: if cocaine consumption ceases and dopamine levels return to normal, this increased level of sensitivity produces cravings and depression.

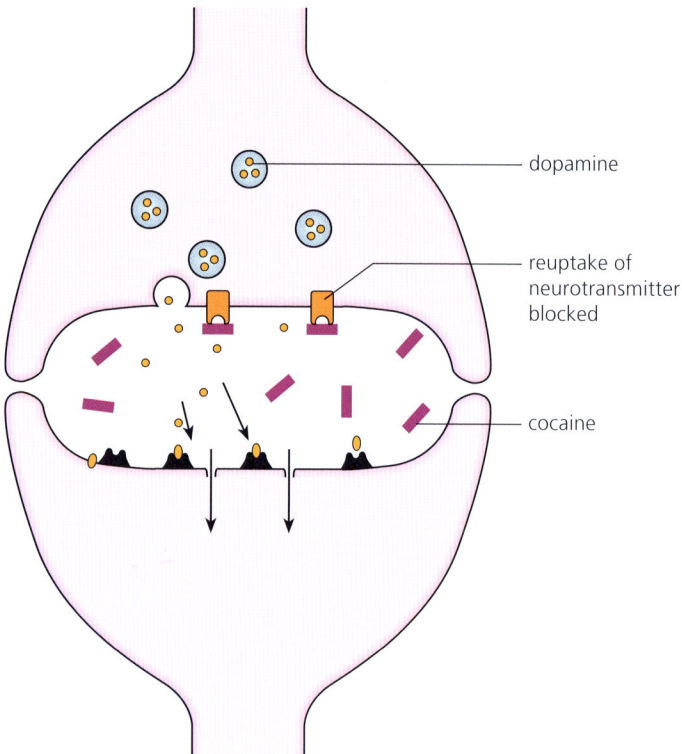

■ **Figure C2.2.14** Cocaine blocks the reuptake of neurotransmitters such as dopamine

13 Outline how exogenous chemicals affect synaptic transmission.

Inhibitory neurotransmitters and generation of inhibitory postsynaptic potentials

In the introduction to the working of the chemical synapse, an excitatory synapse was described (page 477). The incoming action potential excited the postsynaptic membrane and generated an action potential that was transmitted along the postsynaptic neuron.

Some synapses have the opposite effect. These are known as **inhibitory synapses**. Here, release of the neurotransmitter into the synaptic cleft triggers the opening of ion channels in the postsynaptic membrane through which chloride ions enter, or channels through which potassium ions leave. In either case, the interior of the postsynaptic neuron becomes more negative (we say it is **hyperpolarized**). This makes it more difficult for the postsynaptic cell to generate a nerve impulse in response to excitation by other (excitatory) synapses present on the same postsynaptic cell.

◆ **Inhibitory synapse**: synapse at which arrival of an impulse blocks forward transmissions of impulses in the postsynaptic membrane.

◆ **Hyperpolarization**: change of a neuron's membrane potential to a more negative value; when a neuron is hyperpolarized, it is less likely to generate an action potential.

Concept: Interaction

Some neurotransmitters excite nerve impulses in postsynaptic neurons; others inhibit them. There are complex interactions between the activities of excitatory and inhibitory presynaptic neurons at the synapses, operating with a very great number of connections between the very large numbers of neurons present.

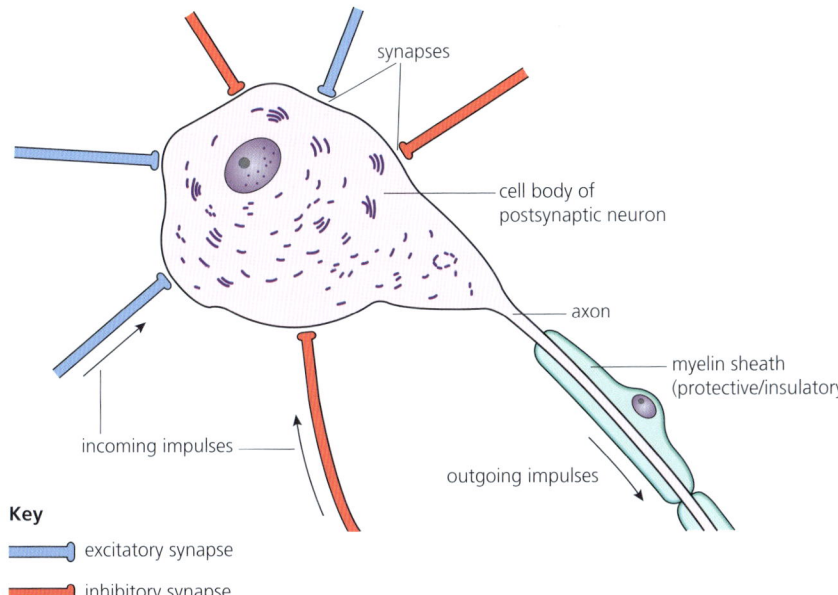
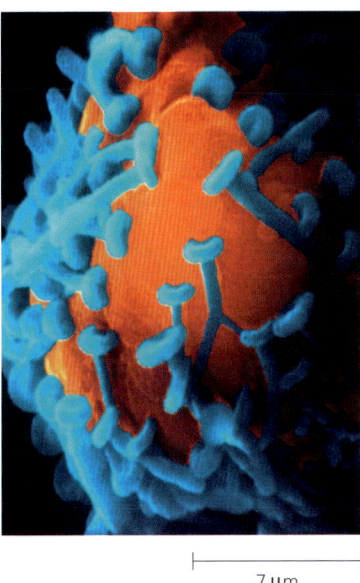

Figure C2.2.15 Integration of multiple synaptic inputs

◆ **Summation**: the process that determines whether or not an action potential will be generated by the combined effects of excitatory and inhibitory signals, both from multiple simultaneous inputs (spatial summation) and from repeated inputs (temporal summation).

Summation

In general, at a synapse an action potential will only be generated in the postsynaptic neuron if the combined effects of the excitatory action potentials and inhibitory action potentials exceed the threshold level. As we have already seen, the action potential is an all-or-nothing event: if the threshold potential is not reached through the additive effect of postsynaptic potentials – **summation** – then the action potential will not be generated. In **excitatory synapses**, the incoming action potential excites the postsynaptic membrane and generates an action potential that is transmitted along the postsynaptic neuron. **Inhibitory synapses** make it more difficult for the postsynaptic cell to generate a nerve impulse in response to excitation by other (excitatory) synapses present on the same postsynaptic cell. Several impulses may arrive at the synapse in quick succession from a single axon, causing an action potential in the postsynaptic neuron (temporal summation).

Alternatively, impulses from several different axons may contribute to the total (spatial summation). Summation contributes to the decision-making processes of the brain, for example. In each case, the additive effect of the different postsynaptic potentials needs to be sufficient to reach the threshold potential of the axon in order to stimulate an action potential.

There are complex interactions between the activities of excitatory and inhibitory presynaptic neurons at the synapses, operating with unimaginably numerous connections between the vast numbers of neurons present (Figure C2.2.15).

Perception of pain by neurons with free nerve endings in the skin

There are a variety of different receptors that respond to different stimuli. Signals from the skin may be conveyed by physical change (mechanoreceptors), temperature (thermoreceptors) or pain (nociceptors). Many neurons end in encapsulated receptors (i.e. covered). Free nerve endings, as their name implies, are unencapsulated dendrites of a sensory neuron. They detect stimuli that may lead to damaged tissue. They are commonly found in skin, the cornea of the eye, pulp of teeth and mucous membranes. Free nerve endings are the most common nerve endings in the skin, and they extend into the middle of the epidermis. Free nerve endings respond to a variety of different stimuli.

Stimuli cause channel proteins to open and allow positively charged ions to diffuse into the axon. These channel proteins are known as **transient receptor potential (TRP) channels** (Figure C2.2.16). There are many different types of TRP, each composed of unique proteins. TRPs act as sensors in the epidermal layer of the skin: in response to a specific stimulus, they open an ion channel that lets positive ions flow inside the neuron. The positively charged ions enter the axon and cause the threshold level to be reached, generating an action potential. The action potential passes along the sensory neuron axon, which connects to other neurons in the central nervous system. The impulse reaches the brain, where it is perceived as pain. One member of this group of channel proteins, TRPA1, is a sensor for a wide variety of noxious external stimuli such as intense cold, pungent compounds, reactive chemicals and endogenous signals associated with cell damage.

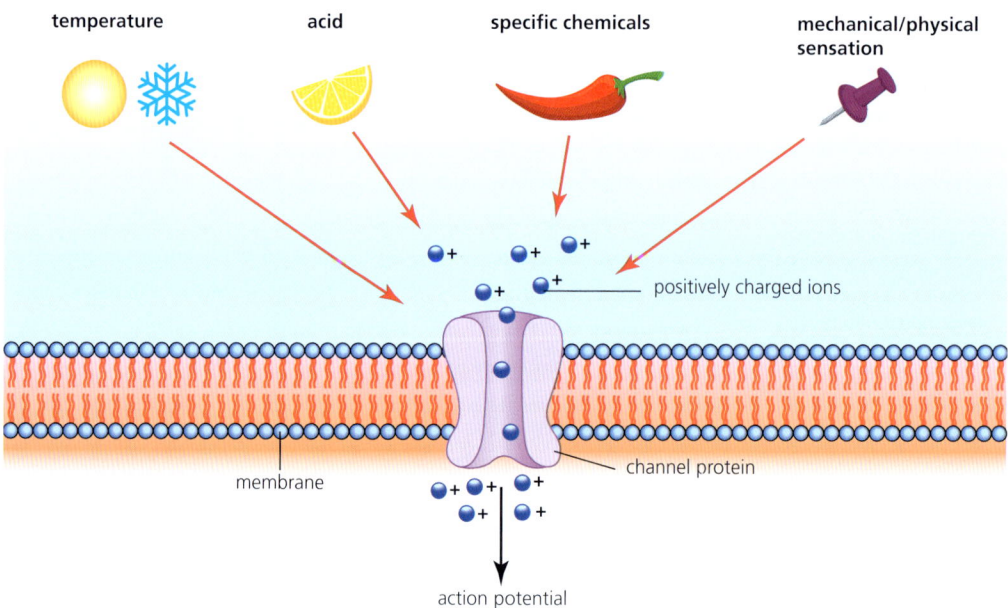

■ Figure C2.2.16 Skin nerve endings have positively charged ion channels that open in response to a stimulus such as high temperature, acid, certain chemicals (such as capsaicin in chilli peppers) and mechanical/physical sensation

Capsaicin is a chemical compound found in chilli peppers. Another in this group of channel proteins, TRPV1, is a single protein molecule that can respond to both heat and capsaicin. How did a thermosensor like TRPV1 become sensitive to a plant product like capsaicin? One answer is that TRPV1 evolved in some animals as temperature sensors and that certain plants then developed compounds that would activate these sensors in order to deter consumption of the plants by predators. Plants that produced capsaicin would therefore have a survival and reproductive advantage, and become more widespread in the population of that species. Plant evolution therefore drove the dual function of the sensors.

Going further

When mammals eat chilli peppers, they tend to destroy the seeds with their molars. Unlike mammals, birds have beaks and so pass most of the seeds through their digestive system undamaged. When they defecate, they spread viable chilli pepper seeds to new locations. Mammals respond to the capsaicin in chillies and detect the chemical as pain – this deters them from eating chillies. *Why do birds continue to eat chillies?*

Birdwatchers often add chilli pepper to seeds in their feeders to deter squirrels, raccoons and other mammals while leaving the birds unaffected. This is because, while mammals have the standard form of TRPV1, activated by both capsaicin and heat, birds do not respond to capsaicin. When the TRPV1 gene is extracted from birds, a variant form of TRPV1 is seen, which responds to heat but not capsaicin. Examination of the sequence of bird DNA indicates that exact location on the inner surface of the cell's outer membrane that is necessary for capsaicin binding. Chilli peppers use birds as a dispersal agent – coevolution between chillies and birds enabled this mutually beneficial system to develop.

TRP function can have effects on the perception of pain in other ways. Skin that has become sunburnt causes a series of inflammatory processes in the skin, including the production of compounds called prostanoids and bradykinin. These chemicals have the property of reducing the temperature threshold of TRPV1 activation from 43 °C to 29 °C. This results in a typical water temperature for a bath or shower then feeling too hot, resulting in a painful burning sensation.

Consciousness

Link
The cerebral hemispheres of the brain are also covered in Chapter C3.1, page 498.

The cerebral hemispheres make up a larger proportion of the brain and are more highly developed in humans than in other animals. Each hemisphere is divided into four lobes, each lobe being named after the bone of the cranium that covers it. The human cerebral cortex has become enlarged, principally by an increase in total area, with extensive folding to accommodate it within the confines of the skull. Consequently, each hemisphere has a vastly extended surface, achieved by folding with deep grooves.

The cerebral hemispheres are responsible for higher order functions. The division of duties of the two hemispheres is listed in Table C2.2.4.

■ Table C2.2.4 Inputs and control duties in the cerebral hemispheres

Left cerebral hemisphere	Right cerebral hemisphere
receives sensory inputs from sensory receptors in the right side of the body and the right side of the visual field of both eyes	receives sensory inputs from sensory receptors in the left side of the body and the left side of the visual field of both eyes
controls muscle contraction in the right side of the body	controls muscle contraction in the left side of the body

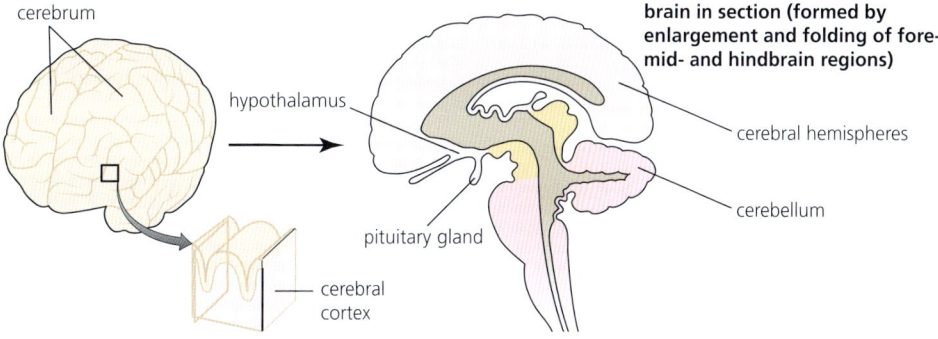

■ Figure C2.2.17 The human brain

The surface of the cerebral hemispheres, known as the cerebral cortex (Figure C2.2.17), is covered by grey matter about 3 mm deep and is densely packed with unmyelinated neurons (known as pyramidal cells). These cells have a mass of dendrites and a very large number of synaptic connections. Below this, the bulk of the cerebral hemispheres consists of white matter. This is made up of myelinated neurons that connect the cerebral cortex with the midbrain, hindbrain and spinal cord. Beneath, the right and left hemispheres are connected by tracts of fibres.

Consciousness is often described as the 'mind's subjective experience'. A robot can unconsciously detect conditions such as colour, temperature or sound, whereas consciousness describes the qualitative feeling that is associated with those perceptions, together with the deeper processes of reflection, communication and thought. The development of brain-scanning technologies such as electroencephalography (EEG), magnetic resonance imaging (MRI) and functional magnetic resonance imaging (fMRI), have enabled scientists to research the mechanisms in the brain that are associated with the conscious processing of information (Figure C2.2.18).

◆ **Consciousness**: the qualitative feeling that is associated with perceptions such as colour, temperature or sound, together with the deeper processes of reflection, communication and thought.

Link
Emergent properties are discussed in detail in Chapter C3.1, page 491.

Concept: Interaction

Research has shown that consciousness is a property that emerges from the interaction of individual neurons in the brain. Emergent properties such as consciousness are another example of the consequences of interaction.

In EEG, electrodes consisting of small metal discs with thin wires are attached to the scalp. The electrodes detect tiny electrical charges that result from the activity of brain cells. The charges are amplified and appear as a graph on a computer screen, or as a recording that may be printed out on paper.

Using fMRI, the precise parts of a living, healthy, functioning brain that are activated when a particular activity occurs can be mapped accurately. This technique is entirely non-invasive. Furthermore, it can detect activity anywhere in the brain, and with high resolution. Results of a scan are generated quickly.

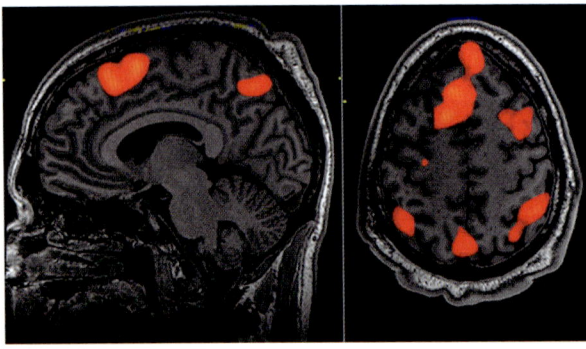

■ Figure C2.2.18 Image showing use of fMRI to provide early detection of consciousness in patients with severe traumatic brain injury

MRI uses a powerful magnetic field to produce detailed images. It works by measuring the way the vast number of hydrogen atoms present in the body's water molecules absorb and then emit electromagnetic energy. The nucleus of each hydrogen atom (a single proton) is, in effect, a tiny magnet; in a strong magnetic field, these line up – as compass needles do in a magnetic field. In the process of a scan, a pulse of radio waves is applied, sufficient to cause the hydrogen nuclei to change orientation. When the pulse is switched off, the nuclei revert to their original orientation and each nucleus emits a signal (at radio frequencies). From this signal, the scanner can work out the three-dimensional location of each nucleus.

fMRI is an advanced form of MRI that detects the parts of the brain that are active when the body performs particular tasks. Brain cells always require energy and a good supply of oxygen, but during periods of intense activity the demand for these resources increases locally. The scanner can detect an increase in red blood cell oxygenation at the site of special neural activity; the technique for this is known as blood oxygen level dependent (BOLD) contrast. The increase in blood flow to the most active areas is detected due to the difference between signals arising from hydrogen nuclei in water molecules in the neighbourhoods of (a) oxyhaemoglobin and (b) deoxyhaemoglobin. When the concentration of oxyhaemoglobin increases, the fMRI signal rises.

The data obtained from fMRI scans are transformed into three-dimensional images that record the regions of the brain that are most active. This new brain-mapping technique permits us to work out the spatial organization of human brain function, down to a sub-millimetre level.

Scientists have begun to determine which regions and circuits in the brain are most important for consciousness. The cerebral cortex has been known to be important for consciousness since the nineteenth century. However, more recent research has shown that consciousness is not confined to only one region of the brain, and that various cells and pathways are engaged, depending on the stimulus. Data collected by fMRI showed that the brains of healthy individuals had more complex patterns of coordinated signalling that also changed constantly, compared with people in minimally conscious states and those under anaesthesia. Studies of brain function using people at varying levels of consciousness provide data that enable us to determine the mechanisms involved.

ATL C2.2D

What is consciousness and how can research on the brain help to answer this question?

Read this article and summarize the main points in a mindmap or poster:

www.nature.com/articles/d41586-018-05097-x

Points you may want to consider:
- What are qualia and can they be meaningfully studied by science?
- What happens to consciousness after an operation on the cerebellum?
- Can a theory of consciousness be developed?

TOK

The attempt to develop a 'theory of consciousness' represents an interesting intersection between biology and philosophy. The philosophy of mind is a field of philosophy that attempts to conceptualise the relationship between facts related to mentality (consciousness) and facts related to the physical world as described by neuroscience. Since the scientific method requires observation of events in the world, the status of an individual's mental state, which can only be observed by that individual, is seen by some to be problematic and represents the limits of scientific knowledge. If science requires shared observations and if consciousness cannot be shared (no-one else can experience your feelings, sensations or thoughts), then consciousness seems to be beyond the scope of the natural sciences.

Physicalists deny this, claiming that any knowledge about consciousness will really only be knowledge about the brain or brain processes; there is nothing other than physical processes to be known. Dualists, however, believe that knowledge of consciousness represents knowledge about distinct events, properties or even substances, meaning that there is a whole world of other facts that science cannot fully explain.

These are philosophical theories (as opposed to scientific ones) because they move beyond what we can observe and make claims about what is real; physicalists say only physical facts are real, and dualists say that reality is comprised of other, non-scientific facts related to consciousness.

Today, the brain is known to be the seat of consciousness, but the nature of consciousness (whether it is best thought of as physical process, like running, or whether it is somehow beyond the descriptive power of science) is still hotly debated.

ATL C2.2E

How does the anatomy of the human nervous system compare to that in other animals with a very different evolutionary origin? Another intelligent animal is the octopus. Read the following article about the nervous system of an octopus: **www.scientificamerican.com/article/the-mind-of-an-octopus**

Carry out your own research about this intelligent animal. What is it like to be an octopus? Produce a short article or poster that summarizes your findings.

LINKING QUESTIONS

1. In what ways are biological systems regulated?
2. How is the structure of specialized cells related to function?

C3.1 Integration of body systems

> ### Guiding questions
> - What are the roles of nerves and hormones in the integration of body systems?
> - What are the roles of feedback mechanisms in the regulation of body systems?

SYLLABUS CONTENT

This chapter covers the following syllabus content:
- C3.1.1 System integration
- C3.1.2 Cells, tissues, organs and body systems as a hierarchy of subsystems that are integrated in a multicellular living organism
- C3.1.3 Integration of organs in animal bodies by hormonal and nervous signalling and by transport of materials and energy
- C3.1.4 The brain as a central information integration organ
- C3.1.5 The spinal cord as an integrating centre for unconscious processes
- C3.1.6 Input to the spinal cord and cerebral hemispheres of the brain through sensory neurons
- C3.1.7 Output from the cerebral hemispheres of the brain to muscles through motor neurons
- C3.1.8 Nerves as bundles of nerve fibres of both sensory and motor neurons
- C3.1.9 Pain reflex arcs as an example of involuntary responses with skeletal muscle as the effector
- C3.1.10 Role of the cerebellum in coordinating skeletal muscle contraction and balance
- C3.1.11 Modulation of sleep patterns by melatonin secretion as a part of circadian rhythms
- C3.1.12 Epinephrine secretion by the adrenal glands to prepare the body for vigorous activity
- C3.1.13 Control of the endocrine system by the hypothalamus and pituitary gland
- C3.1.14 Feedback control of heart rate following sensory input from baroreceptors and chemoreceptors
- C3.1.15 Feedback control of ventilation rate following sensory input from chemoreceptors
- C3.1.16 Control of peristalsis in the digestive system by the central nervous system and enteric nervous system
- C3.1.17 Observations of tropic responses in seedlings (HL only)
- C3.1.18 Positive phototropism as a directional growth response to lateral light in plant shoots (HL only)
- C3.1.19 Phytohormones as signalling chemicals controlling growth, development and response to stimuli in plants (HL only)
- C3.1.20 Auxin efflux carriers as an example of maintaining concentration gradients of phytohormones (HL only)
- C3.1.21 Promotion of cell growth by auxin (HL only)
- C3.1.22 Interactions between auxin and cytokinin as a means of regulating root and shoot growth (HL only)
- C3.1.23 Positive feedback in fruit ripening and ethylene production (HL only)

System integration

◆ **Integration**: the ability of interacting parts of a biological system to perform together and maintain a functioning organism or other biological system.

Biological systems are very complex. Organisms are made of many interacting parts, and an increased number or specialization of those parts leads to greater complexity and their many parts interacting in numerous ways. This is true for every level of organization, from molecules to ecosystems (Figure C3.1.1). Interactions between systems or parts of systems means there is **integration** between the systems.

> ### Concept: Interaction
> System interaction and integration is a necessary process in living systems. Coordination is needed for component parts of a system to collectively perform an overall function.

System integration is necessary for living systems. To collectively perform an overall function, coordination is needed between component parts of a system.

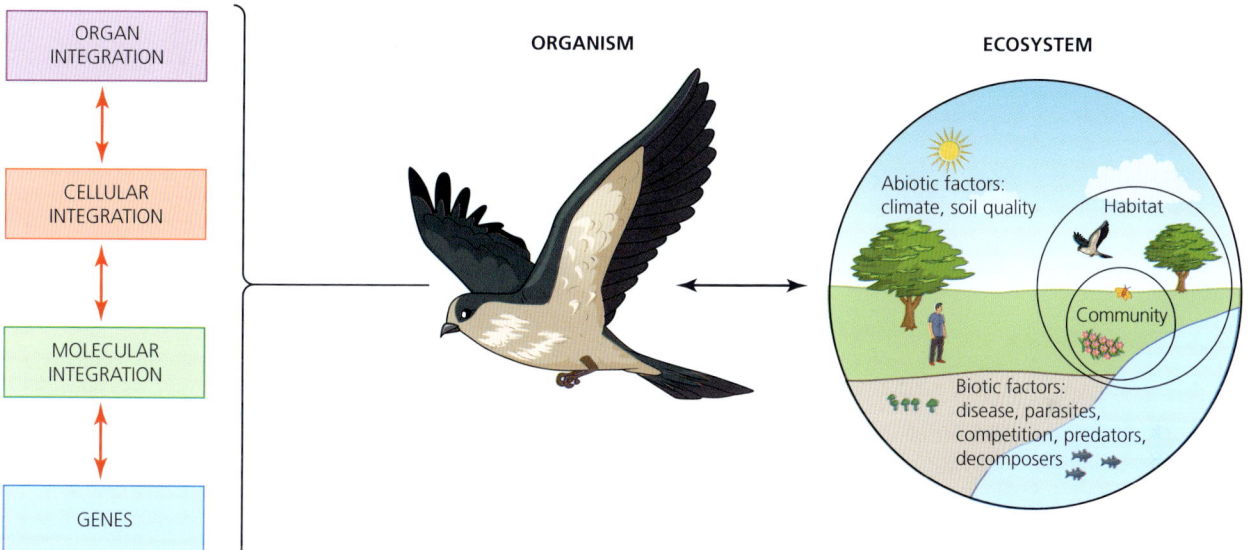

■ **Figure C3.1.1** System integration exists at every level of biological organization

Integration in multicellular organisms

◆ **Emergent property**: a property gained by a complex system when the individual parts work together.

It can be said that 'complex systems are more than the sum of their parts'. But what does this mean? The function of a cell, for example, depends on the collective behaviour of many molecular parts all acting together. These collective properties – known as **emergent properties** – are an essential property of biological systems. You cannot understand or predict the overall system's behaviour based on the individual parts alone; a 'holistic' approach is needed. Emergent properties arise from the interactions of the parts of the larger system.

> ### Concept: Interaction
> Emergent properties arise from the interactions of the parts of the larger system. A hierarchy of subsystems work together to provide functions that the individual parts do not have. Emergent properties help living organisms better adapt to their environments.

C3.1 Integration of body systems

Cells, tissues, organs and organ systems have their own properties and represent a hierarchy of subsystems that are integrated in a multicellular living organism. Multicellular organisms, such as humans, have properties that emerge from the interaction of their cellular components. This integration is responsible for emergent properties. The cells and tissues of the small intestine (Figure C3.1.2) have their own properties and functions, but when they work together, they allow the whole organ to carry out the emergent properties of peristalsis, digestion, food absorption and transport.

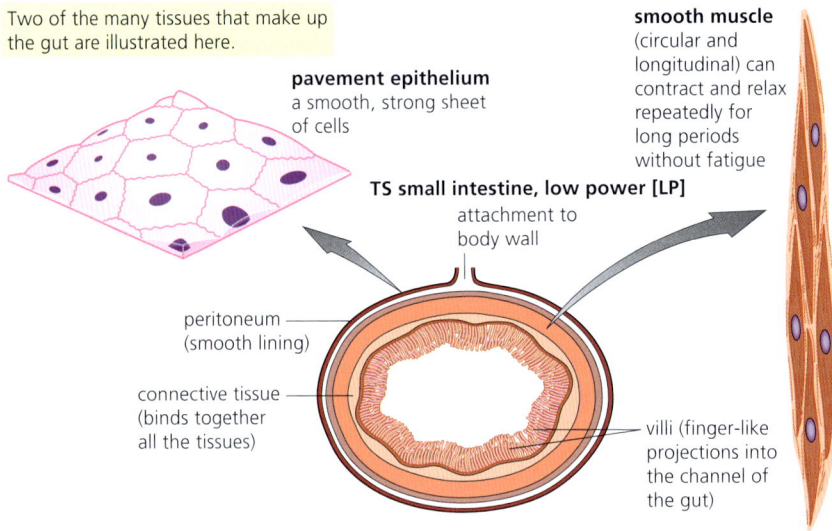

■ **ATL C3.1A**

Research one organ found in the body. How do the component cells and tissues work together to form the emergent properties of the organ?

■ **Figure C3.1.2** Tissues of part of the mammalian gut

■ The hierarchy of subsystems in multicellular organisms

Cellular level

The basic unit of life is the cell, and cells are formed by macromolecules. Each macromolecule in a cell determines the structure and function of the cell. Nucleotides form nucleic acids containing genes that determine an organism's characteristics. Interactions between genes and amino acids determine which proteins are formed and the ultimate function of the cell.

Tissue level

Multicellular organisms arrange cells into tissues, which are groups of similar cells that work together to perform a particular function. For example, a single epithelial cell cannot form a protective layer whereas multiple epithelial cells can form layers and other structures that can carry out functions such as protection, secretion, absorption, filtration and sensory reception.

The five types of tissues found in animals (including humans) perform different functions:
- epithelial tissue forms the linings of organs
- connective tissue connects or separates, and supports all the other types of tissues in the body
- blood is considered a tissue because it is a collection of similar specialized cells that serve particular functions
- nervous tissue is comprised of neurons and allows the body to receive stimuli and coordinate responses
- muscle tissue is made up of cells capable of generating motion, such as pumping blood or moving the body.

Organs

Different types of tissues combine to form organs that perform specialized functions (see Figure C3.1.2). The properties of an organ depend on the properties that emerge as a result of interactions between tissues and other previous levels of organization. For example, DNA determines how cells form tissues and how tissues form organs.

Organ systems

Organs form organ systems. For example, the digestive system is formed from interactions between the mouth, oesophagus, stomach, liver, pancreas, gall bladder, small intestine and large intestine. The stomach cannot digest food without the enzymes and digestive fluids that the gallbladder, liver and pancreas make. The interactions between the organs in a system are controlled at the molecular level.

Organism level

Organ systems together create an organism. The organism is the combined interaction at every level of organization – the whole body cannot function without closely controlled system integration at every level. Ultimately, genes determine the structure and function of the components of an organism, and how they interact.

An example of integration of body systems: the cheetah

A cheetah (see Figure B3.3.1, page 326) is an effective predator thanks to the integration of its body systems as shown in Table C3.1.1.

■ **Table C3.1.1** The integration of body systems in the cheetah

Hierarchy of subsystems	Description
Cellular level	Cheetah DNA provides the blueprints for its body shape and pigmentation of the fur to create spots for camouflage. These adaptations give the animal an advantage in hunting and hiding.
Tissue level	The cheetah has loose hips and shoulder joints, and a flexible spine for running. The spine is very flexible, flexing and storing potential energy, which is released when the spine springs back as the animal moves forwards. Teeth are smaller than other big cats to allow for a larger nasal passage to enable quick air intake.
Organ level	The cheetah has an enlarged heart to ensure effective delivery of glucose and oxygen to muscles for rapid physical response. Eyes are positioned for maximum binocular vision to enable distances to prey to be assessed accurately.
Organ system level	The breathing system enables rapid delivery of oxygen to muscles; the circulatory system delivers blood to muscles and other parts of the body, which enables it to run at very fast speeds.
Organism level	Built perfectly for chasing prey, with narrower paws than other big cats to have minimal contact with the ground, blunt claws to increase traction, and a long tail to act as a counterbalance, enabling the animal to run at 69–70 mph with an acceleration of 0–45 mph in 2.5 seconds.

1 **Define** the term *emergent property*.

◆ **Hormone**: chemical messenger that is produced and secreted from the cells of the ductless or endocrine glands.

◆ **Endocrine glands**: the hormone-producing glands that release secretions directly into blood plasma.

◆ **Endocrine system**: the network of glands that produce hormones; they are released directly into the blood, where they travel to target tissues and organs throughout the body.

Integration of organs in animal bodies

The integration of organs in animal bodies is controlled by hormonal and nervous signalling, and by transport of materials and energy. Nervous signalling is examined in Chapter C2.2 (page 467), and the hormonal system will be outlined below.

■ Introducing the endocrine system

Hormones are chemical substances that are produced and secreted from the cells of the ductless or **endocrine glands** (from 'endo' meaning 'within', and 'crine' = meaning 'secreting'). The network of glands throughout the body is known as the **endocrine system**. In effect, hormones carry messages about the body – but in a completely different way from the nervous system. Hormones are known as 'chemical messengers'.

Hormones are transported in the bloodstream but they act only at specific sites called target organs. Although present in small quantities, hormones are extremely effective messengers, helping to control and coordinate body activities. Once released, hormones may cause changes to specific metabolic reactions of their target organs. However, they circulate in the bloodstream briefly.

C3.1 Integration of body systems

In the liver, hormones are broken down and the breakdown products are excreted in the kidneys. So, long-acting hormones must be secreted continuously to be effective.

An example of a hormone is insulin, released from endocrine cells in the pancreas. Insulin regulates blood glucose. The positions of endocrine glands in the body are shown in Figure C3.1.3.

Link

How insulin regulates blood glucose is covered in more detail in Chapter D3.3, page 761.

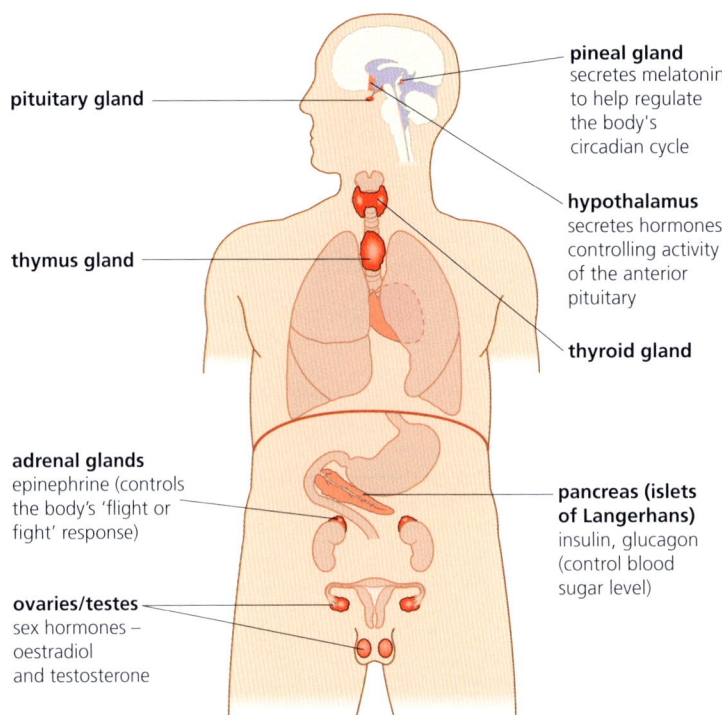

■ Figure C3.1.3 The human endocrine system

The endocrine glands are ductless glands (Table C3.1.2).

■ Table C3.1.2 Endocrine glands

Endocrine glands	
Description	these glands secrete hormones directly into the bloodstream
	at target organs, hormones trigger changes to specific metabolic reactions
Examples	islets of Langerhans: secrete insulin (Figure D3.3.3, page 762), target organs: muscle and liver tissue
	posterior pituitary gland: secretes anti-diuretic hormone (ADH, Figure D3.3.13, page 776), target organs: collecting ducts of kidney tubules
	gonads: secrete sex hormones (page 709), target organs: the gonads and other body tissues
	pineal gland: secretes melatonin, target organs: tissues and organs that respond to our 'body clock' (Figure C3.1.10)

ATL C3.1B

Another set of glands in the body are the exocrine glands. These glands deliver their secretions via ducts, typically into the lumen of the gut or on to the body surface. Research exocrine glands and how their role differs from the role of endocrine glands.

You will encounter some exocrine glands in Chapter D3.3, for example sweat glands, which secrete sweat on to the skin's surface.

Hormonal control of body function is quite different from nervous control; the latter is concerned with quick, precise communication, whereas hormones mostly work by causing specific changes in metabolism and development, often over an extended period of time. However, these contrasting systems are both coordinated by the brain – the nervous system and hormones work together. Table C3.1.3 distinguishes between the roles of the nervous system and endocrine system in sending messages.

■ Table C3.1.3 Comparison of the nervous and endocrine systems

	Nervous system	Endocrine system
Type of message	electrical impulses; neurotransmitters	chemical messengers
Method of transmission	neurons and nerves	bloodstream
Parts of the system	central nervous system and peripheral nervous system	glands
Speed of action	very rapid	can be slow
Length of effect	short duration	long duration
Effectors	muscles or glands at specific locations	target cells in specific tissues
Distance of action	can act over very short distances as well as long distances (all parts of the body)	can act over long distances
Examples of processes controlled	reflexes, e.g. blinking and pain reflex	growth, development of reproductive system, homeostatic mechanisms

Transport of materials in the blood system

◆ **Assimilation**: uptake of nutrients into cells and the utilization of this material to provide energy and to synthesize new biological molecules.

Products of digestion are absorbed from the blood into body cells in a process called **assimilation**. The blood system plays a role in transporting materials between organs. In the first stage of assimilation, absorbed nutrients are transported from the intestine. Blood transports vitamins and inorganic ions to all cells, and the products of digestion are absorbed into the bloodstream where they are transported to all cells in the body.

- All the products of carbohydrate digestion pass from epithelial cells in the lining of the small intestine to the bloodstream (capillary network) by facilitated diffusion. Monosaccharide sugars (mainly glucose) dissolve and are transported in blood plasma. From here, they are transported to the liver and then to other parts of the body. The liver maintains a constant level of blood sugar.
- Amino acids are actively transported into the epithelial cells of the small intestine by the action of membrane protein pumps, and pass into the capillary network. They dissolve in the plasma and the bloodstream transports them to the liver. Here, they contribute to the pool or reserves of amino acids from which new proteins are made in cells and tissues all over the body.
- The short-chain fatty acids (and glycerol) are absorbed by simple diffusion into the epithelial cells in the small intestine. From inside the epithelial cells, the short-chain fatty acids diffuse out into the capillaries and are transported in the bloodstream, just like the monosaccharides and amino acids.
- Long-chain fatty acids are combined with glycerol to reform triglycerides. Triglycerides are hydrophobic and so cannot travel in blood plasma without first being modified. Triglycerides are coated with proteins in the smooth endoplasmic reticulum of epithelial cells. These complex molecules enter the lymph system (a body system that collects excess fluid from the blood), which transports them into the blood circulation at a point close to the heart.
- Waste products are also transported in the blood plasma. Urea is produced by the breakdown of amino acids in the liver. It is a toxic substance that is filtered from the blood in the kidneys. Carbon dioxide is a waste product of cellular respiration. Carbon dioxide is carried by both blood plasma and red blood cells, and it is removed from the body in the lungs.

The hepatic portal vein carries the products of digestion to the liver, where glucose may enter the liver cells and be converted to glycogen (page 763). Glucose remaining in the blood may be converted to glycogen in muscle cells around the body. Brain cells depend on a continuous supply of glucose from the blood circulation – they cannot store glycogen.

Oxygen combines with haemoglobin in red blood cells (Chapter B3.1, page 291) and is transported to cells for use in aerobic respiration.

Integration by the central nervous system

Link
The central nervous system and peripheral nervous system are introduced in Chapter C2.2, page 467.

The **central nervous system (CNS)** consists of the brain and spinal cord (Figure C3.1.4). Nerves of the **peripheral nervous system (PNS)** run to and from the central nervous system. Communication between the CNS and all parts of the body occurs via the PNS. The rather complex layout of the PNS is summarized in Figure C3.1.4.

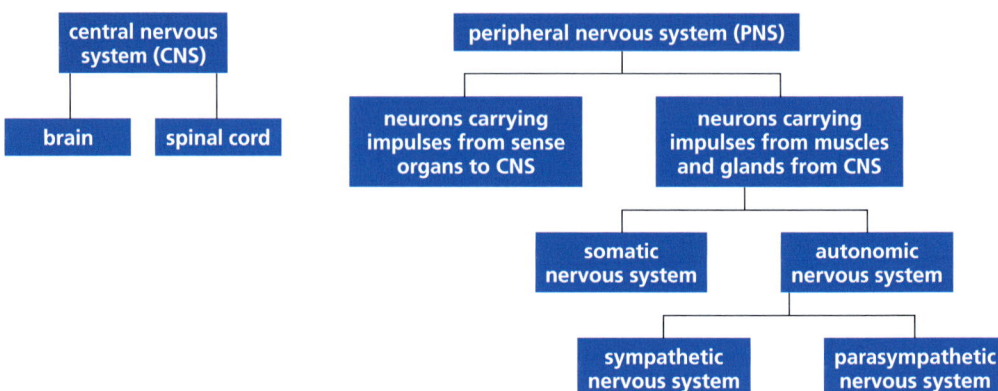

■ Figure C3.1.4 The organization of the central nervous system and peripheral nervous system

The brain as a central information integration organ

The human brain controls body functions (apart from those functions under the control of simple spinal reflexes) by:
- receiving impulses from sensory receptors
- integrating and correlating incoming information in association centres
- sending impulses to effector organs (muscles and glands), causing bodily responses
- storing information and building up an accessible memory bank
- initiating impulses from its own self-contained activities – the brain is also the seat of personality and emotions, and enables us to imagine, create, plan, calculate, predict and reason abstractly.

The cerebral hemispheres make up a large proportion of the brain (Figure C3.1.12, page 504) and are more highly developed in humans than in other animals.

Nature of science: Science as a shared endeavour

Research into memory and learning

Brain function, and in particular its higher functions such as memory and learning, are still only poorly understood by scientists. Research was initially undertaken by psychologists, but increasingly molecular and biochemical techniques are being used. Cooperation and collaboration between groups of scientists has been essential in developing an understanding of brain function.

Scientists have investigated the brain of Henry Molaison (HM), a patient famous for having severe amnesia following surgery that removed most of his hippocampus (the part of the brain associated with memory, emotions and motivation).

Throughout his life, MRI scans and thousands of psychological experiments were carried out to investigate the anatomy of HM's brain and how it related to his lack of memory. After his death, 2000 slices from his brain were taken for onward research. These are being used at The Brain Observatory at UC San Diego, USA, where researchers are looking for physical traces of life events in brain microstructure using high-resolution images at the neuron level.

The work of one group of scientists can add significantly to existing knowledge.

In 1997, Suzanne Corkin and her co-workers published a scientific paper in the *Journal of Neuroscience*, summarizing their work in which they had performed an MRI scan on the brain of HM.

Many researchers were already studying his cognitive impairment and were able to gain insight from her description of the actual damage seen in the scan, and how this leads to amnesia.

This illustrates how cooperation and collaboration between groups of scientists is essential in progressing scientific knowledge and understanding.

Theme C: Interaction and interdependence – Organisms

TOK

Should some knowledge not be sought on ethical grounds?

Henry Molaison's initial surgery was experimental; his doctor hoped it would relieve his severe epilepsy, but the effect of the removal of parts of his hippocampus were unknown prior to the surgery. The surgery was successful in the treatment of his epilepsy, but it was devastating to Molaison, who could no longer form long-term memories. He lived in a care institute for the rest of his 55 years, under constant supervision and as the subject of innumerable interviews relating to his inability to create new memories. The knowledge gathered from these interviews and from the analysis of his brain after his death was hugely beneficial to our understanding of human memory and instrumental in the formation of an entirely new field of science: cognitive neuropsychology. This knowledge, however, was developed in the context of deep human tragedy. Was this knowledge worth the damage inflicted on Henry Molaison?

Researchers at Stanford University, USA, have been investigating how the suprachiasmatic nucleus (SCN), a region in the front part of the hypothalamus, as well as controlling circadian rhythm (the waking–sleep cycle; see page 502), has an important role in learning and memory. They found that when the SCN is not functioning properly in hamsters, memory is impaired. When they surgically removed the SCN, memory abilities returned. Other researchers have shown there is a link between altered circadian rhythms and diminished memory in people suffering from Alzheimer's disease. It is possible that work on the role of the SCN relating to memory and learning will play an important role in understanding Alzheimer's and other diseases.

Other work has shown how molecular biology and genetics are giving insights into how key proteins and other molecules influence memory. Recent animal studies have shown that manipulating these molecules can modify memories, with the potential of weakening traumatic memories that may underlie post-traumatic stress disorder (PTSD). Such studies may also lead to new treatments for memory loss.

ATL C3.1C

This TED talk discusses how the mysteries of the mind could be solved, including mental illness, memory and perception: a supercomputer has been developed that models all the brain's 100 000 000 000 000 synapses.

www.youtube.com/watch?v=LS3wMC2BpxU

The TED talk is over 12 years old. Is its message still up to date or has more recent research superseded it? Find out what current research shows.

The brain has many interconnected areas, containing on average 86 billion neurons, and can simultaneously perceive, interpret, store, analyse and distribute information. To what extent can the brain itself be thought of as a supercomputer?

Find out more about the brain using this New Scientist site:
www.newscientist.com/article-topic/brains

Tool 2: Technology

Using computer modelling

Download the 3D Brain app for iPhones and iPads:
https://apps.apple.com/app/3d-brain/id331399332

Use rotate and zoom functions to discover how each brain region functions, and what happens when it is injured. How are regions of the brain involved in mental illness?

> ### Common mistake
>
> Do not confuse the parts of the nervous system involved with conscious processes, such as voluntary muscle movements, thought and speech, with unconscious processes, such as heart rate, regulation of pupil size and movement of food through the digestive system.

> ### Link
> Motor, sensory and relay neurons are covered in more detail in Chapter C2.2, page 467.

◆ **Cerebral hemisphere (cerebrum)**: the bulk of the human brain, formed during development by the outgrowth of part of the forebrain, consisting of densely packed neurons and myelinated nerve fibres.

2 **Compare and contrast** motor, sensory and relay neurons by means of a concise table.

■ The spinal cord as an integrating centre for unconscious processes and autonomic functions

The spinal cord carries nerve impulses between the brain and the rest of the body. It also controls reflexes without input from the brain, where electrical impulses pass from receptors to the CNS along sensory neurons and synapse with interneurons that send electrical impulses along motor neurons directly to effectors. Reflexes are unconscious responses to stimuli that protect the body from harm. An example of a reflex is the pain reflex (see page 501). These unconscious processes are automatic and involuntary, whereas conscious processes are deliberate and voluntary. Autonomic functions (i.e. involuntary and not under direct control) include the sweating of the lower half of the body (which is transmitted through and processed by the spinal cord), defecation and urination.

■ Input to the spinal cord and cerebral hemispheres of the brain through sensory neurons

The nervous system is built up from specialized cells called **neurons**. Neurons have a cell body and a number of extensions called nerve fibres. The three types of neuron are **sensory neurons**, **motor neurons** and **relay neurons** (Figure C2.2.2, page 469).

The **cerebral hemispheres** (see Figure C3.1.12 on page 504), an extension of the forebrain, form the bulk of the human brain. They consist of densely packed nerve cells (sometimes referred to as 'grey matter') and myelinated nerve fibres (known as 'white matter'). The ratio of the size of the cerebral hemispheres to the size of the whole nervous system is larger in humans than in any other mammal. The human brain contains about 10^{11}–10^{12} relay neurons and the same number again of supporting cells (known as neuroglia cells). Most of these neurons occur in the cerebral hemispheres, where each relay neuron is in synaptic contact with about a thousand other neurons. Mammals are the most intelligent of all animals, and their memory, complex behaviour and subtle body controls are linked to this development.

Many of the body's voluntary (conscious) activities are coordinated in the cerebral hemispheres, together with many involuntary (unconscious) ones. They are an integrating centre of memory, learning, emotions and other complex functions.

The pathway of nerve impulses from receptor to CNS can be traced using the example of writing with a pen or pencil (see Figure C3.1.5):

1 As the pen is touched, a sensory receptor in the skin of the fingers is stimulated.
2 The sensory receptor changes energy from touch to an electrical impulse (action potential) in the axon of the sensory neuron.
   - Sensory neurons convey messages from receptor cells to the central nervous system.
   - The electrical signal travels along the axon to the spinal cord and the brain.
3 The electrical signal causes the release of a neurotransmitter at a synapse between the sensory neuron and an interneuron.
   - The neurotransmitter stimulates the interneuron to form an action potential in its dendrites and cell body.
   - The action potential travels along the axon of the interneuron, which results in neurotransmitter release at the next synapse with another interneuron.

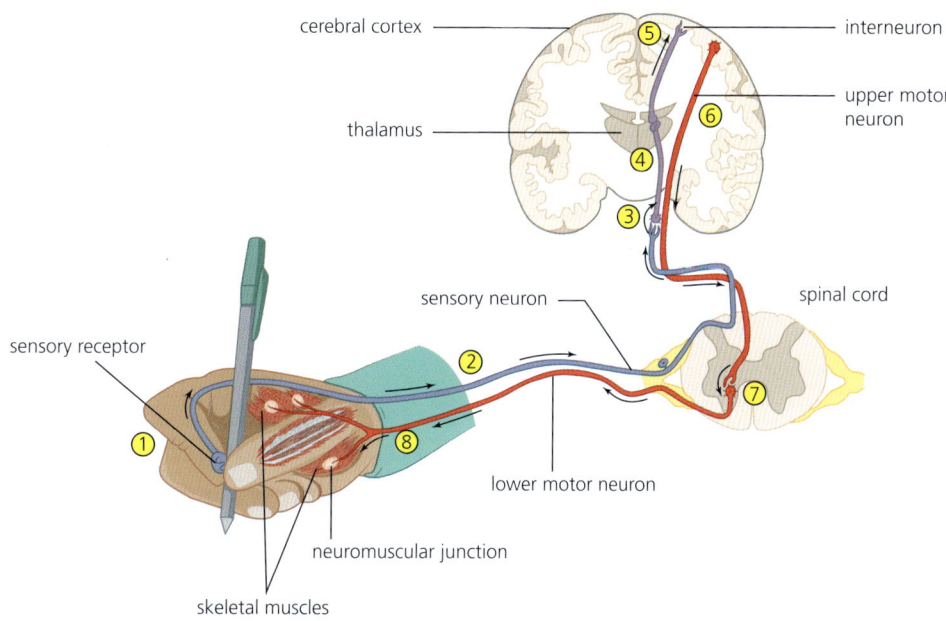

■ **Figure C3.1.5** Pathway of nerve impulses from receptor to effector

4 This process of neurotransmitter release at a synapse followed by an action potential occurs many times as interneurons in higher parts of the brain (such as the cerebral cortex) are activated.

5 Once interneurons in the cerebral cortex (the outer part of the brain – see page 487) are activated, perception occurs, and the outer surface of the pen is sensed as the fingers touch it. Conscious awareness of a sensation is primarily a function of the cerebral cortex.

■ Output from the cerebral hemispheres of the brain to muscles through motor neurons

The pathway of nerve impulses from CNS to effector can be traced continuing the previous example of writing with a pen or pencil.

6 A stimulus in the brain causes an action potential to form in the dendrites and cell body of an upper motor neuron (a type of motor neuron that synapses with a lower motor neuron farther down in the CNS – Figure C3.1.5).

- The action potential travels down the axon of the upper motor neuron.

7 The action potential causes neurotransmitter release in the synapse between the upper and lower motor neurons.

- The neurotransmitter generates an action potential in a lower motor neuron (a type of motor neuron that directly supplies skeletal muscle fibres).

8 The action potential in the lower motor neuron causes neurotransmitter to be released at neuromuscular junctions formed with skeletal muscle fibres (which control the movement of fingers).

- The neurotransmitter stimulates the muscle fibres to form muscle action potentials.
- The muscle action potentials cause these muscle fibres to contract, which allows writing with the pen.

● **Top tip!**

Remember – motor neurons transmit electrical impulses from the central nervous system to muscles and glands (effectors). Motor neuron pathways lead to muscles contracting.

ATL C3.1D

Like most brain functions, the right side of the brain controls the left side of the body, and the left side of the brain controls the right side. The axons of the neurons from each side of the brain must therefore split in two (bifurcate) somewhere during their descent to the spinal cord so that they can change sides. This crossover occurs just before the junction between the medulla and the spinal cord.

Carry out research to find out the evolutionary reasons for the right side of the brain controlling the left half of your body, and vice versa. Use this video as a starting point:

www.youtube.com/watch?v=rB5IsOqYxFY

What impact does the crossover have following traumatic events such as brain injuries and strokes?

Nerves as bundles of nerve fibres of both sensory and motor neurons

A **nerve** consists of many axons of neurons collected into bundles. Large nerves, such as the sciatic nerves that go from the lower back to the feet, for example, contain tens of thousands of axons. Each bundle has several coverings of protective sheath, made of connective tissue (Figure C3.1.6):

- endoneurium: tissue surrounding each individual fibre
- perineurium: smooth connective tissue surrounding each bundle of fibres
- epineurium: fibrous tissue surrounding and enclosing a number of bundles of nerve fibres; most large nerves are covered by epineurium.

● Common mistake

Do not confuse the term 'neuron' and 'nerve'. A neuron is a single specialized cell in the nervous system that conducts electrical impulses, whereas a nerve is a group of neurons.

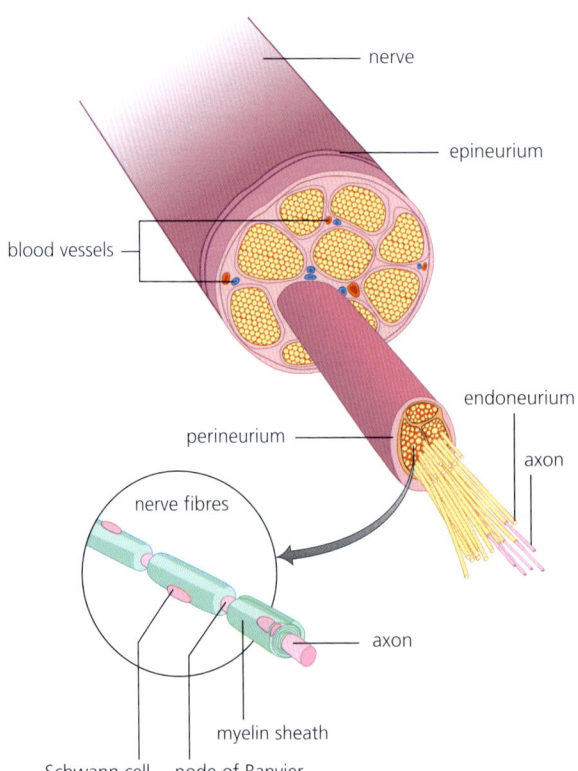

■ **Figure C3.1.6** Transverse section of a peripheral nerve showing the protective connective tissue coverings

Pain reflex arcs

Reflex arcs comprise the neurons that mediate the reflexes. The layout of a reflex arc is shown in Figure C3.1.7.

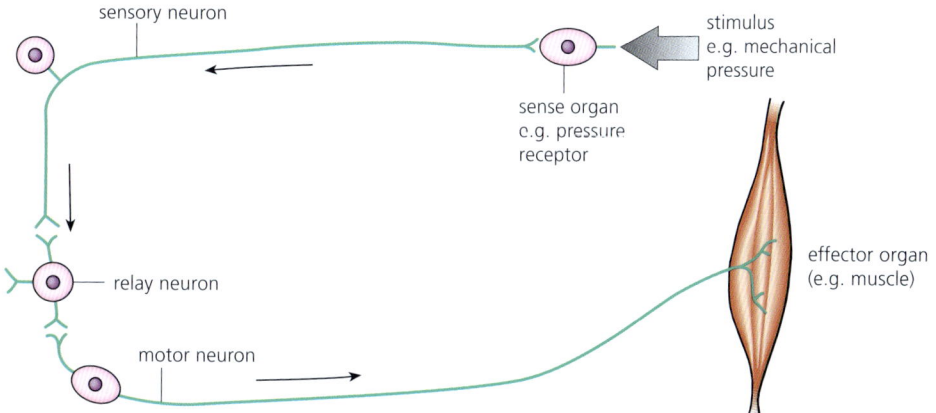

■ **Figure C3.1.7** The layout of a reflex arc

In animals, by means of a specific reflex arc, a particular stimulus produces the same immediate, quick and involuntary response – a reflex action – every time. In humans, an example of a reflex action is the jerking away of our hand from scalding hot water, or the withdrawal of a limb from pain (Figure C3.1.8). The pain reflex arc has a single relay neuron (interneuron) in the grey matter of the spinal cord and a free sensory nerve ending in a sensory neuron as a pain receptor (Figure C3.1.8).

Within the nervous system are a vast number of reflex arcs. Many neurons connect reflex arcs with the control centre, the brain. The brain contains a highly organized mass of relay neurons, connected with the rest of the nervous system by motor and sensory neurons. With a nervous system of this type, complex patterns of behaviour are common in addition to many reflex actions. This is because:
- impulses that originate in a reflex arc also travel to the brain
- impulses may originate in the brain and be conducted to effector organs.

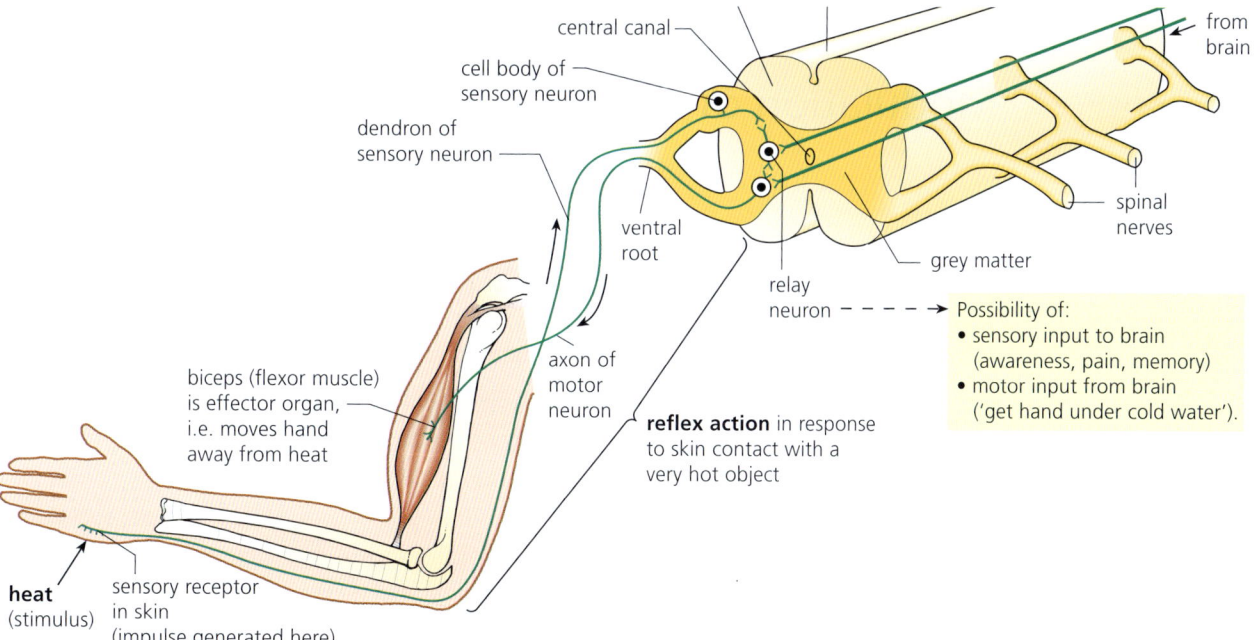

■ **Figure C3.1.8** Pain reflex arcs as an example of involuntary responses with skeletal muscle as the effector

C3.1 Integration of body systems

Consequently, much activity is initiated by the brain, rather than being merely a response to external stimuli. Also, reflex actions such as those responding to pain may be overruled by the brain and the response modified (for instance, we may decide not to drop an extremely hot object that is causing pain but that is also very valuable).

So, we can see that the nervous system of an animal, such as a mammal, has roles in:
- quick and precise communication between the sense organs that detect stimuli and the muscles or glands that cause changes
- complex behaviour patterns that animals display.

3 **Describe** how the body responds to a pain stimulus that can cause harm to the body.

ATL C3.1E

Review reflex arcs using this animation:
www.sumanasinc.com/webcontent/animations/content/reflexarcs.html
Produce a summary of the key points about the components of a reflex arc and how reflexes are coordinated.

Role of the cerebellum

The **cerebellum**, part of the hindbrain (Figure C3.1.9), also has an external surface layer of grey matter. It is here that involuntary muscle movements of posture and balance are controlled. It is also where all precise, voluntary manipulations involved in hand movements, speech and writing are coordinated (rather than initiated). The cerebellum is in overall control of movements of the body, including skeletal muscle contraction and balance.

The cerebellum receives input from receptors and adjusts commands to motor neurons to move the body to maintain balance. It is responsible for fine-tuning movement but does not initiate motor controls. While the cerebellum makes up 10% of the human brain's volume, it contains 50% of the total number of neurons in our brains, indicating its importance.

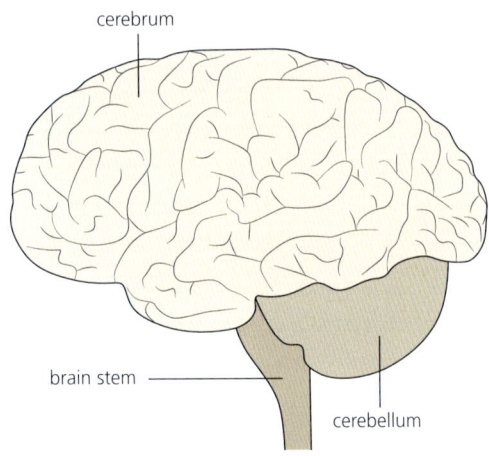

■ **Figure C3.1.9** Location of the cerebellum in the brain

Modulation of sleep patterns by melatonin

◆ **Cerebellum**: part of hindbrain, concerned with muscle tone, posture and movement.

◆ **Circadian rhythms**: physical, mental and behavioural changes that follow a 24-hour cycle.

◆ **Melatonin**: a hormone produced by the pineal gland in response to darkness; it coordinates the timing of circadian rhythms and sleep.

Daily rhythms in physiological processes and behaviour are common throughout the plant and animal world. For example, animals are active for only part of the 24-hour cycle. Some function at dusk or dawn (crepuscular), some in the night (nocturnal), and many during the day (diurnal), as humans do. In fact, much of our behaviour (physical activity, sleep, body temperature, secretion of hormones and other features) follows regular rhythms or cycles (Figure C3.1.10). These cycles operate over an approximately 24-hour cycle and are called **circadian rhythms** (meaning 'about a day').

Circadian rhythms are controlled by a 'biological clock' within the brain. However, cycles are coordinated with the cycle of light and dark – day and night.

Melatonin is a hormone that is found naturally in the body. It is produced in the brain, in the **pineal gland**. This hormone contributes to setting our biological clock. More melatonin is released in darkness and more is released in seasonal longer nights, such as in winter time. Light decreases melatonin production, and this lowering of the melatonin level leads to the body's preparation for being awake. In darkness, melatonin production resumes and sleepiness returns. The condition known as 'winter depression' (seasonal affective disorder, SAD), which affects the mental state of some people, is associated with higher-than-normal melatonin levels in the body.

Theme C: Interaction and interdependence – Organisms

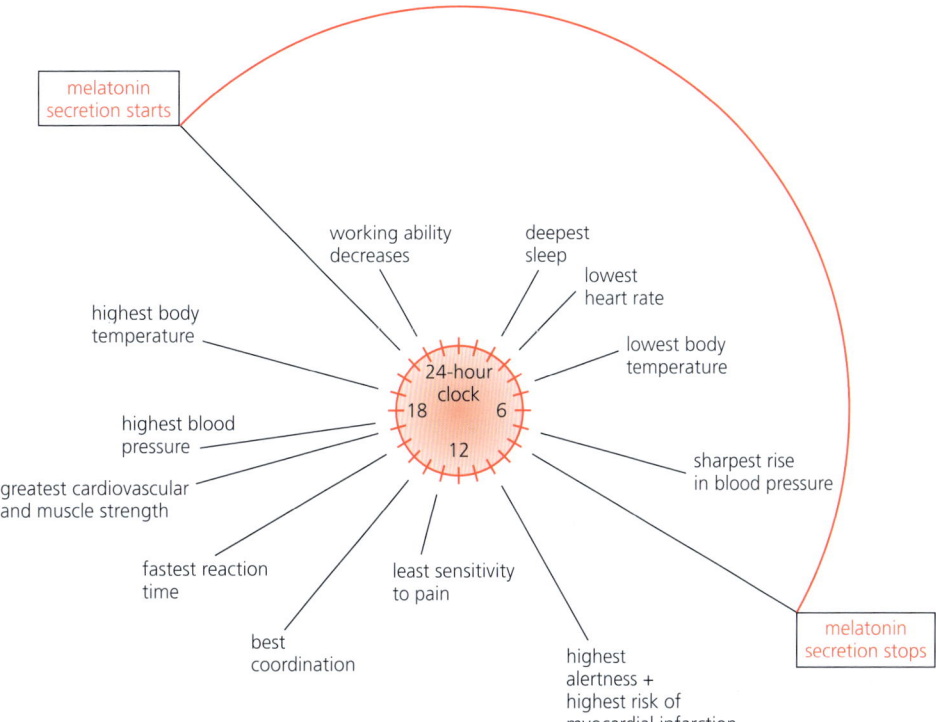

■ **Figure C3.1.10** Our physiology and behaviour show circadian rhythm

We see that an external stimulus, such as light, has an impact on the biological clock, but the effects are not immediate. When normal patterns of light and dark are changed abruptly – for example, as a result of aircraft travel across different time zones – it may take several days before patterns of 'sleep' and 'activeness' return to normal. We say the traveller is suffering from 'jet lag'. People who work day-and-night shift patterns that change frequently may have an even greater problem.

4 **Define** the term *circadian rhythm*.

ATL C3.1F

What impact has modern technology had on our biological clocks, for example using phones and electronic devices at night? What impact does exposure to the bright screens of electronic devices have on us? What does it do to melatonin levels?

Engage in a class discussion on the use of technology at night and how it affects circadian rhythms, and the potential impacts on our health.

◆ **Epinephrine**: a hormone secreted by the adrenal medulla (and a neurotransmitter secreted by nerve endings of the sympathetic nervous system); it has many effects including the increase of heart rate and the breakdown of glycogen to glucose in muscle and liver.

Epinephrine (adrenaline) secretion by the adrenal glands

Epinephrine is a hormone secreted by the adrenal glands (see Figure C3.1.3) to prepare the body for vigorous activity. It is released when the body is under stress. Epinephrine is carried in the blood throughout the body and causes the pacemaker of the heart to increase the heart rate. An increased heart rate delivers increased oxygen and glucose to tissues of the body, in particular muscles, enabling increased rates of aerobic respiration and, therefore, increased production of ATP for activity, ready for intense muscle contraction to happen.

5 **Suggest** likely conditions or situations in which the body is likely to secrete epinephrine.

Epinephrine is a peptide hormone, which, while it circulates in the bloodstream, also stimulates liver cells to convert stored glycogen to glucose. The glucose formed then passes out from the liver cells and contributes to blood glucose levels. Increased blood glucose levels contribute to the elevated rates of respiration in muscle tissues and elsewhere in the body.

C3.1 Integration of body systems

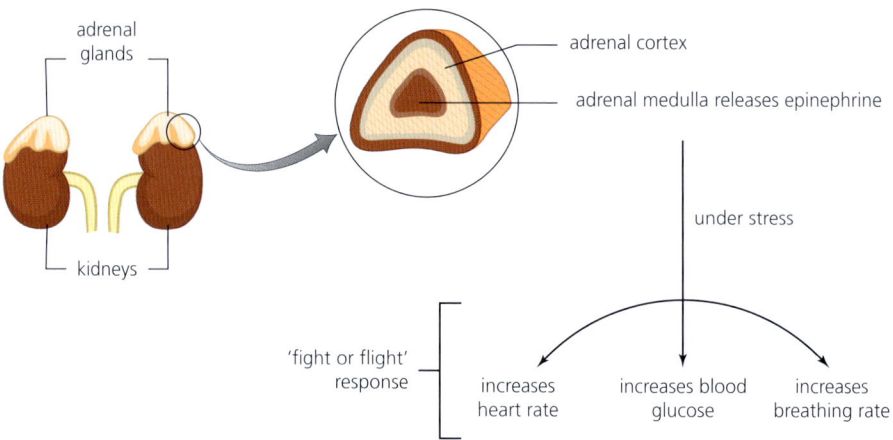

■ Figure C3.1.11 The effects of epinephrine on the body

Control of the endocrine system by the hypothalamus and pituitary gland

◆ **Hypothalamus**: a structure at the base of the brain; a control centre for the autonomic nervous system and source of hormones for the pituitary gland.

Link

Homeostasis is covered in more detail in Chapter D3.3, page 760.

The **hypothalamus** is a part of the brain that has a major endocrine function (see Figure C3.1.3 and C3.1.12). Most importantly, it is exceptionally well supplied with blood vessels and is the site of special neurons.

The hypothalamus has a key role in the control of many aspects of body function and acts as a connection between the endocrine and nervous systems. Partly, this is achieved by the constant monitoring of blood composition as it circulates through the capillary networks of the hypothalamus. These data, and data from sensory receptors located in key organs in the body via sensory neurons and the spinal cord, enable the hypothalamus to regulate many body activities concerned with the maintenance of a constant internal environment – **homeostasis**. The hypothalamus, for example, secretes a hormone involved with the reabsorption of water from the kidneys. It also regulates the function of the pineal gland in relation to wake–sleep cycles (see page 502).

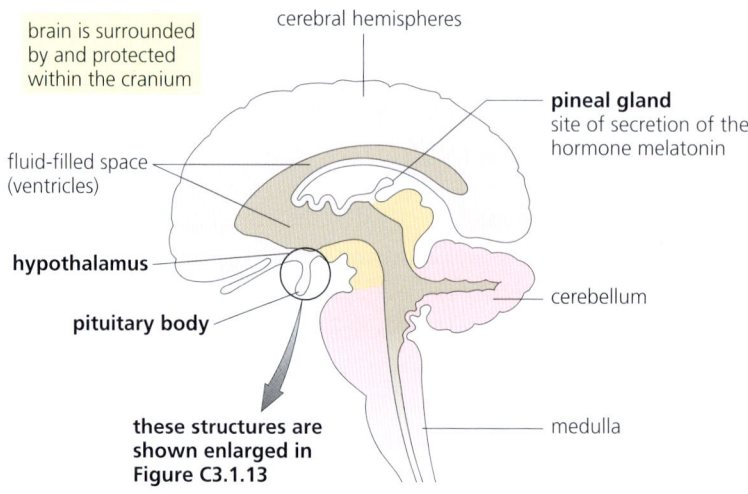

■ Figure C3.1.12 The formation of the human brain

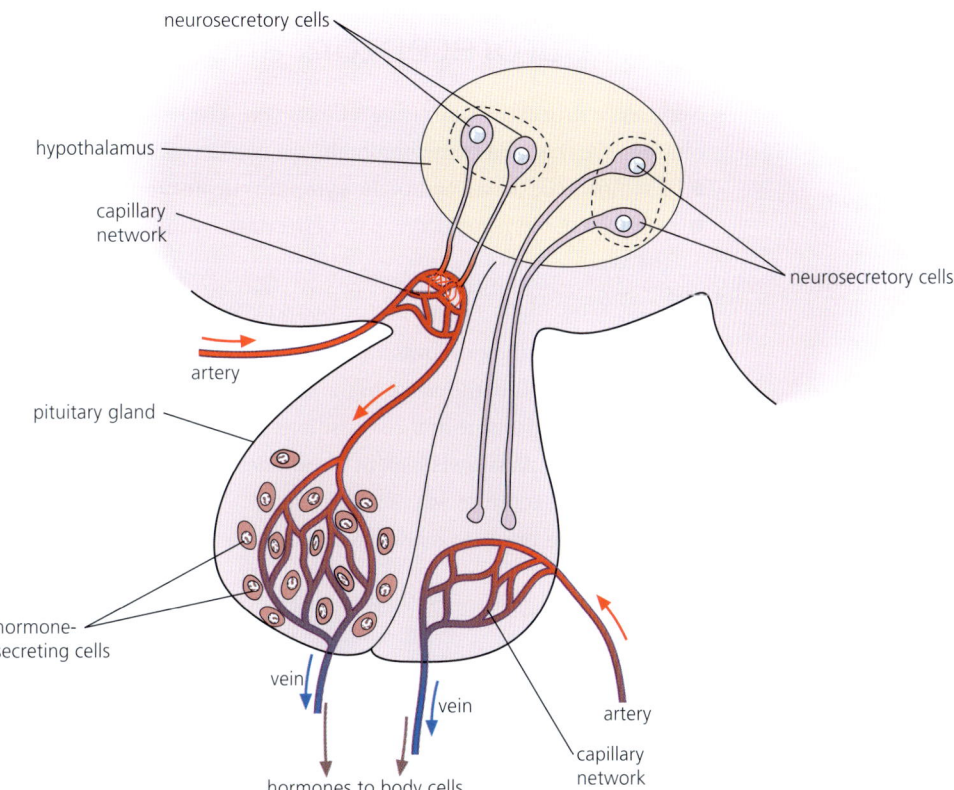

■ Figure C3.1.13 The hypothalamus and pituitary gland

Top tip!
Make sure you understand the pituitary gland's function and purpose.

◆ **Pituitary gland**: the 'master' endocrine gland, attached to the underside of the brain.

6 In the nervous system, impulses reach a specific muscle or gland. **Explain** how the effects of hormones are restricted to particular cells or tissues.

The **pituitary gland** is situated below the hypothalamus and is connected to it. The pituitary gland is part of the endocrine system and it has been called the 'master' hormone gland. However, it is the hypothalamus that largely controls the endocrine activity of the pituitary gland. The hypothalamus does this by releasing different hormones from its special neurosecretory cells into the portal vein running between the hypothalamus and pituitary gland (Figure C3.1.13), as well as by nerve impulses via other neurons that connect with the pituitary.

The hypothalamus is the site of production of several hormones. These largely control the secretion of other hormones by the pituitary gland, either by stimulating or inhibiting the release of specific hormones.

Hormones secreted by the pituitary gland control:
- growth
- developmental changes in the body's tissues and organs
- reproduction
- homeostasis.

One part of the pituitary gland does not synthesize hormones, but it does store and release two hormones produced in the hypothalamus: oxytocin and antidiuretic hormone (ADH).

C3.1 Integration of body systems

◆ **Medulla (medulla oblongata)**: structure located at the base of the brain, where the brain stem connects the brain to the spinal cord.

◆ **Sympathetic nervous system**: part of the involuntary nervous system, antagonistic in effect to the parasympathetic nervous system.

◆ **Parasympathetic nervous system**: part of the involuntary nervous system, antagonistic in effect to the sympathetic nervous system.

◆ **Baroreceptor**: a sensory receptor responding to stretch in the walls of blood vessels to monitor blood pressure.

Control of heart rate using sensory input from baroreceptors and chemoreceptors

Nervous control of the heart is by reflex action. The heart receives impulses from the cardiovascular centre in the **medulla** of the hindbrain, via two nerves (Figure C3.1.14):

- a sympathetic nerve, part of the **sympathetic nervous system**, which speeds up the heart
- a branch of the vagus nerve, part of the **parasympathetic nervous system**, which slows down the heart.

Since the sympathetic nerve and the vagus nerve have opposite effects, we say they are antagonistic.

Nerves supplying the cardiovascular centre bring impulses from **baroreceptors** (stretch receptors) located in the walls of the aorta, in the carotid arteries, and in the wall of the right atrium, when change in blood pressure at these positions is detected.

- When blood pressure is high in the arteries, the heart rate is lowered by impulses from the cardiovascular centre, via the vagus nerve.
- When blood pressure is low, the heart rate is increased.

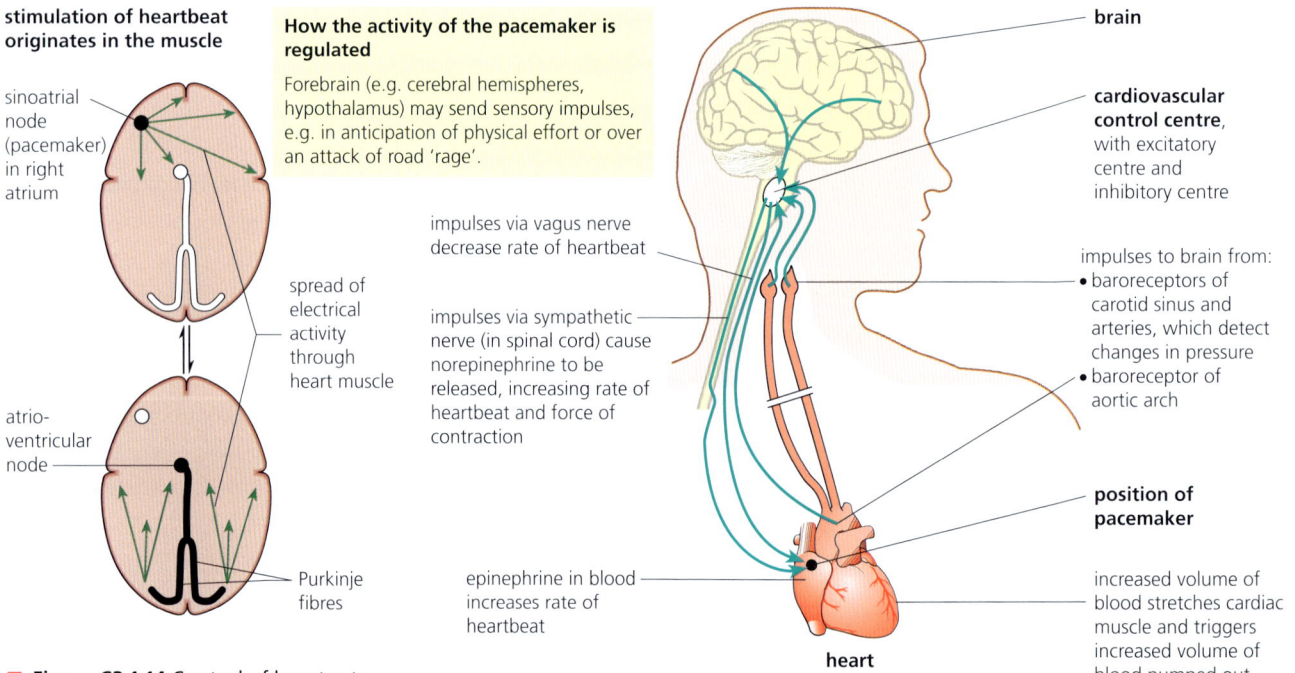

■ Figure C3.1.14 Control of heart rate

◆ **Chemoreceptor**: receptors that monitor blood pH and concentrations of oxygen and carbon dioxide in the blood.

Chemoreceptors in carotid arteries, the aorta (Figure C3.1.15) and the brain (see Figure C3.1.16) monitor the level of carbon dioxide in the blood and pH level. Increased carbon dioxide or decreased pH levels causes the chemoreceptors to signal the heart to beat faster. By increasing blood flow, carbon dioxide can be moved more quickly to the lungs where it is removed by diffusion into the alveolar spaces. This lowers carbon dioxide levels and returns blood pH to safe limits.

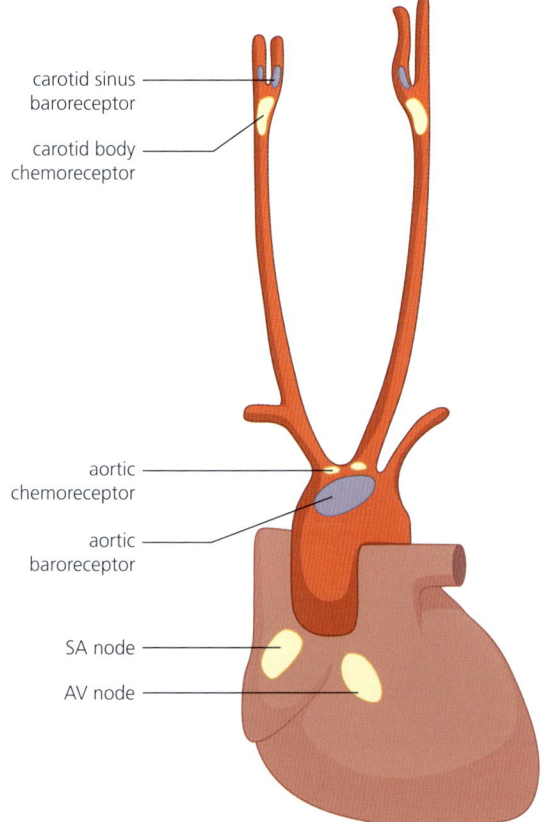

■ **Figure C3.1.15** Location of chemoreceptors and baroreceptors responsible for controlling heart rate

Using chemoreceptors and baroreceptors, the medulla coordinates responses and sends nerve impulses to the heart to change the heart's stroke volume and heart rate. Electrical impulses are sent from the chemoreceptors and baroreceptors to the medulla and, depending on the message, the medulla responds by stimulating either the sympathetic or parasympathetic nerves.

- If blood pressure is low or carbon dioxide concentration high, the cardiovascular control centre of the medulla increases the rate of sinoatrial (SA) node firing through activation of the sympathetic nervous system.
- If there is high blood pressure or low carbon dioxide concentration, the cardiovascular control centre reduces the rate of SA node firing by activating the parasympathetic nervous system.

The nerve impulses along the sympathetic and parasympathetic neurons release different neurotransmitters in the SA node. The sympathetic nerve releases the neurotransmitter norepinephrine to increase heart rate. The parasympathetic nerve (vagus nerve) releases the neurotransmitter acetylcholine to decrease heart. Depending on which signals are sent, the SA node modifies its rate to either slow down or speed up the heart rate.

Feedback control of ventilation rate following sensory input from chemoreceptors

◆ **Respiratory centre**: region of the medulla of the brain concerned with the involuntary control of breathing.

The **respiratory centre** is situated in the medulla of the hindbrain and controls the rate at which we breathe (Figure C3.1.16). Here, two adjacent and interacting groups of nerve cells, known as the **inspiratory centre** and the **expiratory centre** respectively, bring about ventilation movements by reflex action. Breathing occurs automatically (it is involuntary).

- The inspiratory centre sends impulses to increase rate and depth of breathing.
- The expiratory centre sends impulses to inhibit the inspiratory centre and stimulate expiration.
- Alternating impulses from these two centres cause rhythmic breathing.

The breathing rate is also continually adjusted. On average, our normal rate of breathing is about 15 breaths per minute. Since the tidal volume is typically 400 cm³, the volume of air taken into the lungs in one minute (ventilation rate) is about 6 litres. We can consciously override this breathing rate with messages sent from the **cerebral hemispheres**, for example, when we prepare to shout, sing or play woodwind or brass instruments.

Link
The ventilation mechanism by which the air is moved in and out of the lungs is illustrated in Chapter B3.1, page 280.

Breathing rates may also be adjusted without conscious thought. This occurs during increased physical activity when voluntary muscles use much more oxygen and more carbon dioxide is produced and transported in the blood. The main stimulus that affects breathing is the concentration of carbon dioxide in the blood. Blood carbon dioxide level is detected by the chemoreceptors present in the carotid arteries and aorta (Figure C3.1.15).

During increased respiration, as takes place during strenuous physical activity, levels of carbon dioxide in the blood increase. When this happens, the chemoreceptors that are hydrogen ion detectors (carbon dioxide is an acidic gas in solution) send impulses to the inspiratory centre in the brain stem as the blood pH level has changed. In response, this centre sends additional impulses to the intercostal muscles and diaphragm, causing an increase in their contraction rates. (To a lesser extent, lowered oxygen concentration is also detected.) In this way, increased levels of carbon dioxide in the blood result in a more rapid breathing rate, which causes more carbon dioxide to be exchanged at the alveoli and then exhaled.

After strenuous exercise stops, the concentration of carbon dioxide in the blood falls (and the concentration of oxygen rises). These changes are detected, and the ventilation rate is regulated accordingly.

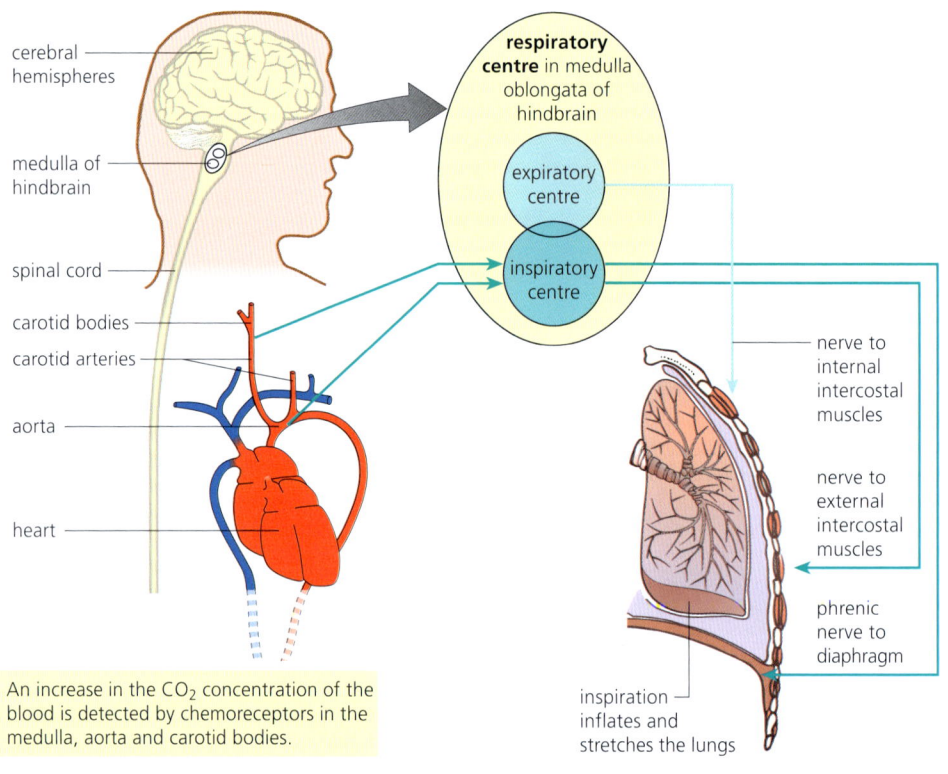

7 **Outline** the role of baroreceptors and chemoreceptors in the feedback control of heart rate.

An increase in the CO_2 concentration of the blood is detected by chemoreceptors in the medulla, aorta and carotid bodies.

■ **Figure C3.1.16** The control of ventilation rate

Control of peristalsis in the digestive system by the central nervous system and enteric nervous system

♦ **Autonomic nervous system (ANS)**: the involuntary nervous system.

♦ **Enteric nervous system (ENS)**: network of sensory neurons, motor neurons and interneurons embedded in the wall of the gastrointestinal system.

As we have already seen, the nervous system consists of the CNS (brain and spinal cord) and the PNS (Figure C3.1.4, page 496). The PNS contains the **autonomic nervous system (ANS)** – autonomic means 'self-governing'. The ANS is special because it controls activities and structures inside the body that are mostly under unconscious (involuntary) control. In effect, the ANS regulates the working of the interior of the body, such as heart rate (page 320) and ventilation rate (page 283), more or less without our knowledge. The **enteric nervous system (ENS)**, part of the ANS, is a network of sensory neurons, motor neurons and interneurons embedded in the wall of the gastrointestinal system. We will look at the action of the ENS in control of peristalsis.

Movement of food through the alimentary canal

In the nutrition of mammals, food is taken into the alimentary canal (gut). This is a long tube with a muscular wall, beginning at the mouth and ending at the anus. The whole structure is specialized for the movement and digestion of food, and for the absorption of the useful products of digestion. The ENS extends from the lower third of the oesophagus (the muscular tube that connects the mouth to the stomach) through to the rectum. Digestion is controlled through the integration of multiple signals from the ENS and CNS. Throughout the gut, waves of contraction and relaxation of the circular and longitudinal muscles of the wall propel food along. This process is known as **peristalsis**. Gut muscles are described as involuntary muscles because they are not under conscious control. The swallowing of food and removal of faeces from the body are under voluntary control by the CNS, but peristalsis between these two points in the digestive system is under involuntary control by the ENS.

◆ **Peristalsis**: waves of muscular contractions passing down the gut wall.

The action of the ENS ensures passage of material through the gut is coordinated. Neural signals pass between distinct gut regions to coordinate digestive activity.

Common mistake

Do not confuse the terms 'enteric' and 'autonomic'. Both refer to the involuntary nervous system, but the enteric nervous system is the part which specifically serves the gastrointestinal system.

Observations of tropic responses in seedlings

Link
Sensitivity is a function of life in organisms and is discussed in Chapter A2.2, page 73.

The reactions of plants and animals to changes within their environments (their **sensitivity**) differ. Most plants are **sessile**, growing in one place and staying there. Therefore, plant responses are often less evident than those of animals. However, plants are highly sensitive to environmental disturbances. Unlike animals, plants have no nervous system and no muscle tissue, so their responses may be less dramatic than those of animals. Fast movements by plants are extremely rare so most responses by plants are changes in growth. An example of this is the growth of young stems towards light. Plant organs respond to external stimuli. The response to the stimulus is determined by the direction of the stimulus and is called a tropic movement (**tropism**) or tropic response.

You can gather different types of data when observing tropic responses in seedlings. For **qualitative data**, you could draw diagrams to record your observations of seedlings, illustrating tropic responses. For **quantitative data** you could measure the angle of curvature of the seedlings as they respond to external stimuli.

◆ **Tropism**: a growth response of plants in which the direction of growth is determined by the direction of the stimulus.

ATL C3.1G

Discuss with a partner what qualitative and quantitative data you could collect in an experiment investigating the effect of light on plant growth. Write down as many as you can think of together.

Link

The definitions for qualitative and quantitative observations are given in Chapter A2.2, page 59.

Nature of science: Observations

You need to be able to distinguish between qualitative and quantitative observations. Quantitative data are measurable, for example, measuring the angles of the seedlings' curvatures. Qualitative data are descriptive and based on interpretation, such as your drawings of the seedlings' responses. When collecting data, understanding what factors will limit the accuracy and precision of your measurements is important.

Accuracy is how close to the true value a result is, whereas **precision** describes the reproducibility of repeated measurements of the same quantity and how close they are to each other. It is the degree to which repeated measurements, under the same experimental conditions, give the same result. Remember, measurements can be precise but not accurate. Replicates (repeat data) can improve the reliability of an investigation and enable anomalies to be identified.

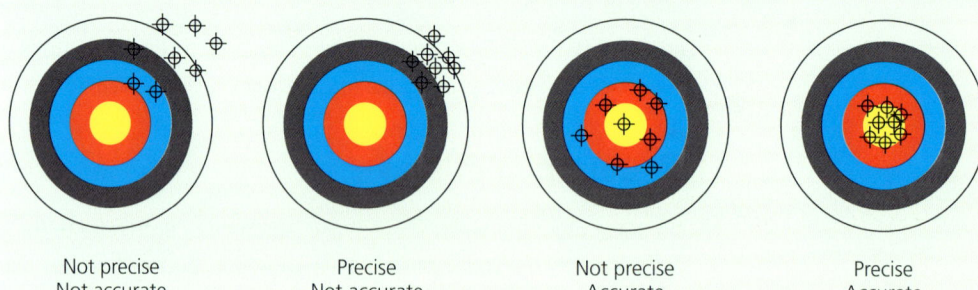

■ **Figure C3.1.17** Accuracy versus precision

Precision in tropism experiments can be improved by carrying out the same methodology in exactly the same way for each repeat. One way to analyse the precision of the measurements is to determine the range, or difference, between the lowest and the highest measured values, or the standard deviation. Small ranges or standard deviation indicates high precision.

In order to improve a measurement's accuracy, all other variables must be controlled (kept the same). In the tropism experiment, other factors that can affect the growth of plants must be kept the same, such as volume of water given to seeds, the number of seeds in each sample, the distance between each seed, and the temperature they are grown at. If the experiment is carried out in the same way each time, the repeats show similar results so there is low variability in data, and other variables except the independent variable are controlled, the accuracy of the experiment will be increased.

◆ **Phototropism**: response of plants to the stimulus of light.

◆ **Positive phototropism**: the directional growth response in plant shoots towards lateral light sources.

Positive phototropism

The response of plants to the stimulus of light is called **phototropism**. Growing towards a stimulus is said to be a 'positive' response, and so the directional growth response to lateral light in plant shoots is known as **positive phototropism** (Figure C3.1.18). By growing towards light, leaves are orientated towards the sunlight and so maximize the opportunity for **photosynthesis**.

Top tip!

You may encounter other plant tropisms. This course studies the response by plants to light (phototropism). Other responses do not need to be known.

■ **Figure C3.1.18** Phototropism in plants: seedlings grow towards a light source

Inquiry 3: Concluding and evaluating

Concluding

Look carefully at the three experiments shown in Figure C3.1.19.

Analyse the information in each figure and **deduce** what each experiment indicates about the response of plants to light. Which part of the plant responds to the stimulus? How is the response to light coordinated?

Write a conclusion for each experiment and **justify** your answers (give valid reasons or evidence to support your conclusions).

Hint: agar (Figures b and c) is a jelly-like substance which substances can diffuse into. The impermeable block (Figure b) is resistant to absorbance: substances cannot diffuse into it.

You can check whether you have developed the correct conclusions using this site:
www.youtube.com/watch?v=4-2DZo2ppAY

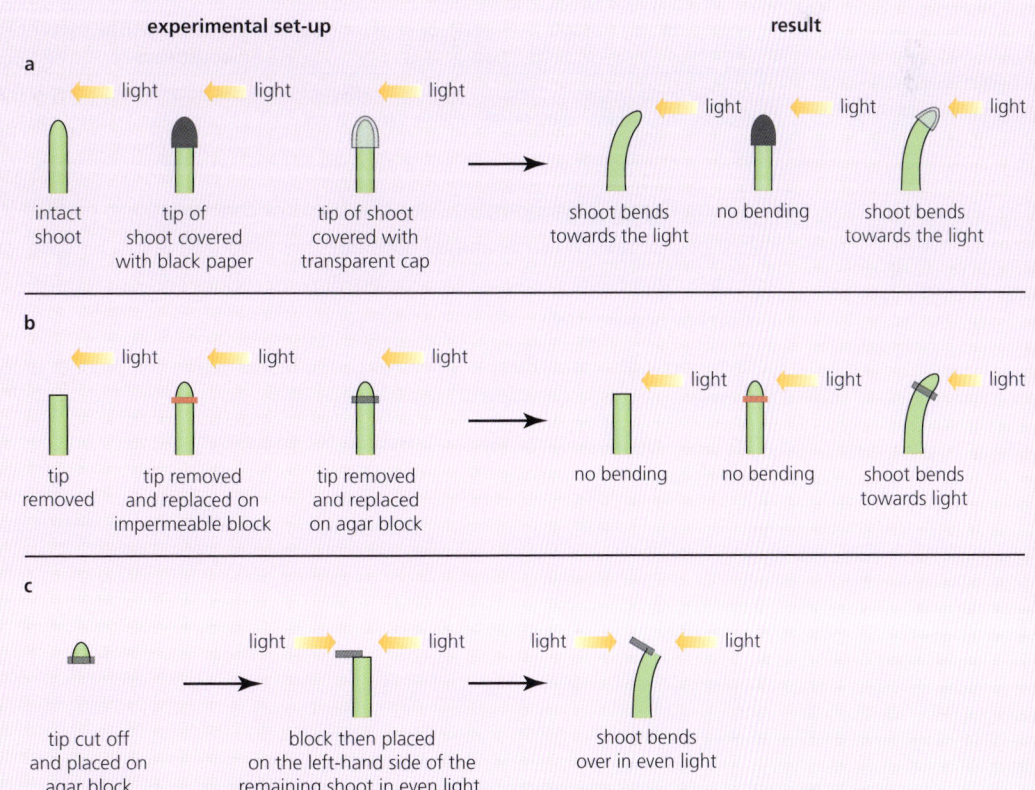

■ **Figure C3.1.19** Three experiments investigating the response of plants to light

C3.1 Integration of body systems

If a plant is exposed to light equally from all directions, it grows directly upwards, whereas when exposed to unidirectional light, it bends towards the light source. Figure C3.1.20 shows a possible explanation of why this happens. The exact mechanism will be explored on page 513.

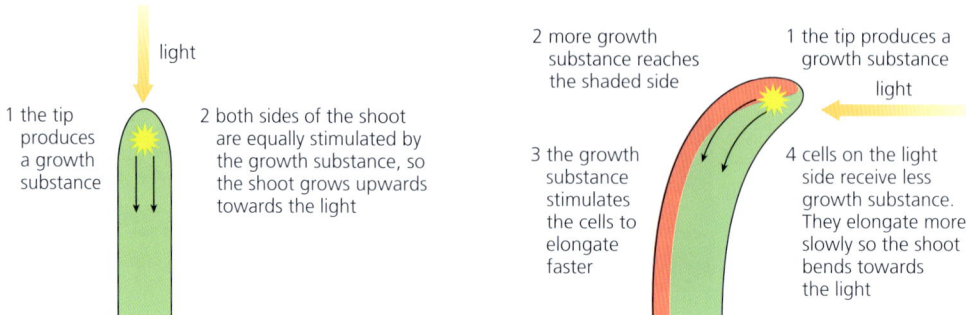

■ Figure C3.1.20 Possible explanation of phototropism in a shoot

Phytohormones

♦ **Phytohormones**: plant hormones that regulate plant growth, development, reproductive processes, longevity and death; responsible for the adaptation of plants to environmental stimuli.

Among the internal factors that play a part in plant sensitivity, the most important are the substances called **phytohormones** (plant growth regulators) and their effects. There are five major types of phytohormones naturally occurring in plants: auxin, gibberellin (GA), cytokinin, ethylene and abscisic acid (ABA). The effects of these chemicals are rather different from those of animal hormones (see Table C3.1.4).

■ Table C3.1.4 Differences between animal hormones and phytohormones

Phytohormones (plant growth regulators)	Animal hormones
produced in a region of plant structure, e.g. stem or root tips, in unspecialized cells	produced in specific glands in specialized cells, e.g. islets of Langerhans in the pancreas
not necessarily transported widely, or at all, and some are active at sites of production	transported to all parts of the body by the bloodstream
not particularly specific, tend to influence different tissues and organs, sometimes in contrasting ways	effects are mostly highly specific to a particular tissue or organ, and without effects in other parts or on different processes

In plant tissues, phytohormones occur in very low concentrations, making it difficult to determine their precise role. Whereas animal hormones are produced in discrete endocrine glands, phytohormones are produced by a variety of plant tissues and can reinforce another phytohormone's effects or oppose them.

Phytohormones can diffuse from cell to cell or be carried in the phloem or xylem, although not all are transported away from their source. Depending on their concentration (low or high), they may have profoundly different effects, based on the tissue they are in or on the stage of development of that tissue.

■ Table C3.1.5 Phytohormones: roles, synthesis and structural formulae

Auxins (indoleacetic acid, IAA)	
Roles promotion of extension growth of stems and roots (at different concentrations) dominance of terminal buds promotion of fruit growth inhibition of leaf fall	indoleacetic acid (the principal auxin)
Synthesis at stem and root tips, and in young leaves (from the amino acid tryptophan)	
Gibberellins (GAs)	
Roles promotion of extension growth of stems delay of leaf senescence and leaf fall inhibition of lateral root initiation switching on of genes to promote germination	gibberellins e.g. gibberellic acid
Synthesis in the embryos of seeds and in young leaves (except in genetically dwarf varieties)	
Abscisic acid (ABA)	
Roles a stress hormone triggering of stomatal closure when leaf cells are short of water induction of bud and seed dormancy	abscisic acid
Synthesis in most organs of mature plants, in very small amounts	

The structural formulae of phytohormones show how chemically diverse these substances are (they do not need to be memorized).

The role of auxin

◆ **Auxin**: plant growth substance, indoleacetic acid, which generally stimulates cell elongation.

Auxin was initially discovered by Darwin as he investigated the curvature of the protective sheath covering emergent plant shoots towards a light source. It is manufactured by cells undergoing repeated cell division, such as those found at the stem and root tips.

Consequently, the concentration of auxin is highest there. Auxin is then transported to the region of growth behind the tip, where it causes cells to elongate. A coleoptile is the sheath that covers the primary bud of a seedling, which goes on to form the stem and leaves.

Top tip!

Unlike many hormones, auxin is not a protein but is the chemical indoleacetic acid in its naturally occurring form.

Auxin has a major role in the growth of the shoot apex, where it promotes the elongation of cells (see below). It also inhibits the growth and development of lateral buds that occur immediately below the terminal growing point. This leads to a quality known as apical dominance. However, a high concentration of auxin actually inhibits growth in length of the stem. Other hormones, the gibberellins, interact with auxin to enhance stem elongation. Cytokinins, from the root apex, pass back up to the stem and promote lateral bud growth – it is antagonistic to the effect of auxin in this respect. The full picture of plant hormone interactions is a complex one.

Nature of science: Experiments

Developments in analytical techniques allow the detection of trace amounts of hormones

A DNA microarray consists of a collection of DNA probes (containing sequences of DNA) attached to a solid surface. The 'surface' of the microarray can be a glass or silicon chip, to which the DNA is covalently bonded. DNA from a sample is broken up into fragments that can react with the DNA probes: if they contain complementary sequences, the probes will bind to the sample DNA. One use for such microarrays is the detection and measurement of the expression of particular genes. Genes being expressed may be caused to fluoresce so they can be detected. In plants, the hormone auxin has been shown to influence gene expression and so regulates growth and development. Data on this have been obtained from studies on cells of a small flowering plant, *Arabidopsis thaliana*, when grown under the influence of unilateral environmental stimuli, such as light or gravity. A combination of several genes is typically involved.

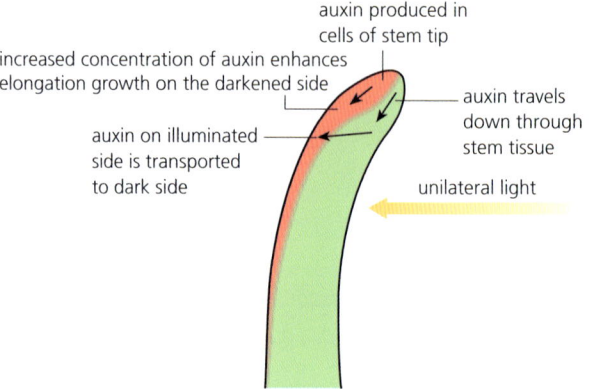

■ Figure C3.1.21 The role of auxin in phototropism

Figure C3.1.21 shows the effect of auxin in coleoptiles.

The positively phototropic response of plants occurs as follows.
- The light stimulus causes auxin to be released from the growing tip of the coleoptile.
- Light from the illuminated side of the plant causes auxin to accumulate on the darkened side of the stem, increasing its concentration.
- Auxin on the darkened side causes cells to elongate there, bending the stem towards the light (Figure C3.1.21).

■ Auxin efflux carriers

Auxin effects on growth and development are by direct action on the components of growing cells, including the walls, and on the gene-expression mechanisms operating in the nucleus. Auxin transport across cells is polar: auxin entry into the cell is passive (by diffusion) and its efflux is active (ATP-driven). Auxin can diffuse freely into plant cells but not out of them. This mechanism of auxin movement and the ways in which auxin may influence the growth and development of cells are outlined in Figure C3.1.22. Note particularly the mechanisms of movement of auxin into and out of the cell; auxin efflux pumps set up concentration gradients in plant tissues. Auxin efflux carriers can be positioned in a cell membrane on one side of the cell. If all cells coordinate to concentrate these carriers on the same side, auxin is actively transported from cell to cell through the plant tissue and becomes concentrated in part of the plant.

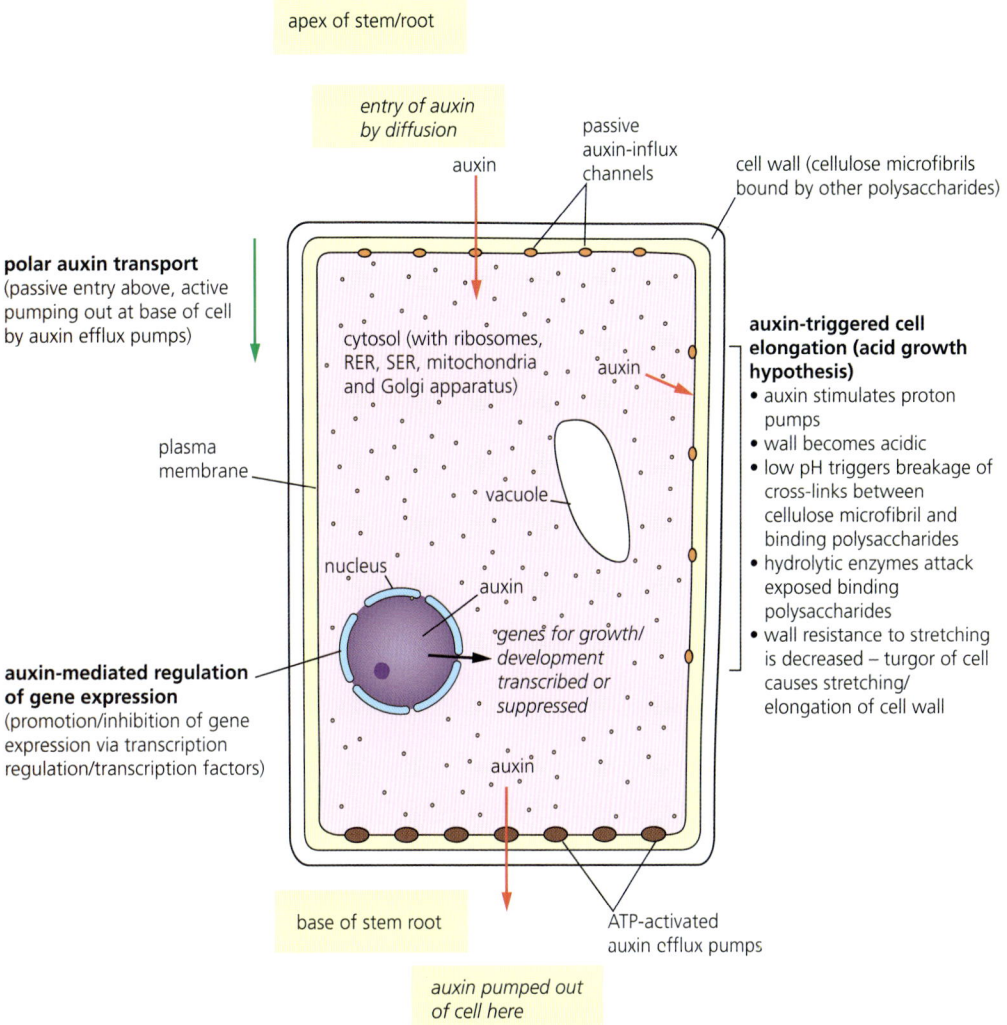

■ Figure C3.1.22 Auxin – mechanisms of movement and control

Auxin efflux carriers are an example of maintaining concentration gradients of phytohormones.

Promotion of cell growth by auxin

Auxin promotes hydrogen ion secretion into the **apoplast** (the pathway through the non-living part of a cell, e.g. cell walls and spaces between cells), acidifying the cell wall. Proteins called expansins, activated by H^+ ions, alter the pattern of hydrogen bonding between the polysaccharides in the cell wall, allowing these macromolecules to slip past each other, loosening cross-links between cellulose molecules and facilitating the stretching of the cell wall and, therefore, cell elongation. Concentration gradients of auxin cause the differences in growth rate needed for phototropism.

In summary, auxin causes changes to the cell wall:
- an increase of protons (H^+) in the cell wall causes the wall to become acidic
- low pH activates expansins, which trigger breakage of cross-links within cellulose
- cell wall resistance to stretching is decreased
- cell turgor causes stretching/elongation of cell wall.

8 **Describe** the role of auxin in a phototropic response.

Interactions between auxin and cytokinin

The phytohormone **cytokinin** is produced in the root tips. It is then transported to shoots, and shoot tips produce auxin, which is transported to roots. Interactions between these phytohormones help to ensure that root and shoot growth are integrated.

Both auxins and cytokinins have been described as essential regulators of many different plant processes with complex regulations at various levels, ranging from the synthesis of biological molecules to transport and signalling. Their activities can best be termed 'antagonistic', where one hormone influences the other and vice versa.

> **Concept: Interaction**
>
> The interactions between the phytohormones auxin and cytokinin ensure the regulation of root and shoot growth in plants.

> **Concept: Interdependence**
>
> The activities of auxin and cytokinin are interdependent. Auxin positively affects cytokinin signalling, and cytokinin positively affects auxin production, thereby positively affecting auxin signalling.

◆ **Cytokinin**: phytohormone that influences plant growth, development and physiology, including cell division.

◆ **Ethylene**: a hormone found in plants that causes fruit ripening.

In addition to growth and developmental processes, plants need to regulate responses to, and interactions with, their environment. Biotic and abiotic factors in the environment can cause stress for plants. Cytokinins and auxins have been established as hormones that regulate or adjust these responses in crop plants.

Positive feedback in fruit ripening and ethylene production

Ethylene (IUPAC name: ethene) is a hormone in plants that stimulates the changes in fruits that occur during ripening. In turn, ripening stimulates increased production of ethylene. Increased ethylene increases fruit ripening, and so on (Figure C3.1.23). This process, where one change leads to an increased change away from the original condition, is called **positive feedback**. Positive feedback tends to amplify change away from equilibrium (a state of physical balance). The benefit of this positive feedback mechanism is that it ensures that fruit ripening is rapid and synchronized.

9 **Suggest** how a farmer could prevent fruit from ripening, so that they can get their produce from the farm to the supermarket in the best condition.

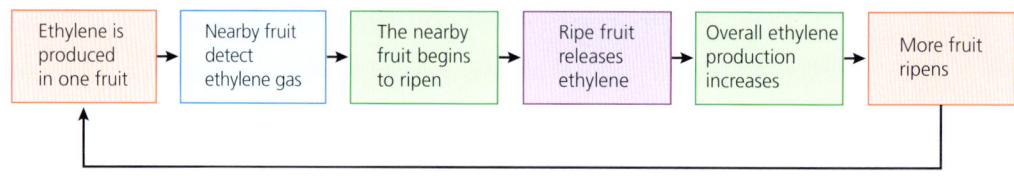

■ Figure C3.1.23 Positive feedback in fruit ripening

LINKING QUESTIONS

1 What are examples of branching (dendritic) and net-like (reticulate) patterns of organization?
2 What are the consequences of positive feedback in biological systems?

ATL C3.1H

Review your knowledge of phytohormones and how they coordinate plant growth, development and response to the environment. Use this site to help you:
www.sumanasinc.com/webcontent/animations/content/plantgrowth.html

Work with a partner to create a mind map summarizing all your knowledge on this topic.

C3.2 Defence against disease

Guiding questions

- How do body systems recognize pathogens and fight infections?
- What factors influence the incidence of disease in populations?

SYLLABUS CONTENT

This chapter covers the following syllabus content:
▶ C3.2.1 Pathogens as the cause of infectious diseases
▶ C3.2.2 Skin and mucous membranes as a primary defence
▶ C3.2.3 Sealing of cuts in skin by blood clotting
▶ C3.2.4 Differences between the innate immune system and the adaptive immune system
▶ C3.2.5 Infection control by phagocytes
▶ C3.2.6 Lymphocytes as cells in the adaptive immune system that cooperate to produce antibodies
▶ C3.2.7 Antigens as recognition molecules that trigger antibody production
▶ C3.2.8 Activation of B-lymphocytes by helper T-lymphocytes
▶ C3.2.9 Multiplication of activated B-lymphocytes to form clones of antibody-secreting plasma cells
▶ C3.2.10 Immunity as a consequence of retaining memory cells
▶ C3.2.11 Transmission of HIV in body fluids
▶ C3.2.12 Infection of lymphocytes by HIV with AIDS as a consequence
▶ C3.2.13 Antibiotics as chemicals that block processes occurring in bacteria but not in eukaryotic cells
▶ C3.2.14 Evolution of resistance to several antibiotics in strains of pathogenic bacteria
▶ C3.2.15 Zoonoses as infectious diseases that can transfer from other species to humans
▶ C3.2.16 Vaccines and immunization
▶ C3.2.17 Herd immunity and the prevention of epidemics
▶ C3.2.18 Evaluation of data related to the COVID-19 pandemic

Note: There is no higher-level only content in C3.2.

Pathogens as the cause of infectious diseases

◆ **Pathogen**: disease-causing organism.

◆ **Infection**: the process of infecting or the state of being infected by a pathogen.

◆ **Disease**: any harmful deviation from the normal structural or functional state of an organism, causing illness.

There are many disease-causing organisms that can infect humans. A disease-causing organism is known as a **pathogen**, although typically the term is reserved for viruses, bacteria (eubacteria), fungi and protists. Archaea (one of the two domains of prokaryotic organisms – see page 127) are not known to cause any diseases in humans. Pathogens may bring disease to healthy organisms through **infection**, leading to **disease**. Accordingly, there are cells in our blood circulation that have roles in defence against communicable (i.e. infectious) diseases.

A disease is caused when another organism or virus particle invades the body and lives there parasitically. The pathogen is the invader and the infected organism – a human in this case – is known as the host.

> **ATL C3.2A**
>
> Why are there no known archaean pathogens? Is this because they have a special biochemistry that excludes them from causing disease, or is it simply that they are a relatively recently known group (they were only discovered to be a separate domain about 40 years ago) and that currently there is not enough evidence for pathogenic species?
>
> Carry out your own research and write a short article titled 'Why do archaea not cause disease?', perhaps for a school or college publication.

◆ **Infectious or communicable disease**: disease capable of being transmitted from one organism to another.

Pathogens may pass from diseased host to healthy organisms, so these diseases are known as **infectious or communicable diseases**. Generally, disease can be defined as an 'unhealthy condition of the body', and this rather broad definition includes some distinctly different forms of ill-health. For example, ill-health may be caused by unfavourable environmental conditions. Diseases of this type are non-infectious or non-communicable diseases, and they include conditions such as cardiovascular disease, malnutrition and cancer. Other diseases are genetic in origin, such as phenylketonuria (page 733) and Down syndrome (page 658).

World Health Organization (WHO)

Good health is more than the absence of harmful effects of a disease. This point is emphasized by the World Health Organization (WHO), which identifies health as a 'state of complete physical, mental and social well-being, and not merely the absence of disease or infirmity'.

The WHO was founded in 1948 and is the United Nations agency responsible for international public health. One role of the WHO is to monitor and communicate information on the spread and containment of infectious diseases, such as COVID-19.

Pathogens and disease

The range of disease-causing organisms that infect humans includes not only micro-organisms such as certain bacteria and fungi, but also some protozoa (single-celled animals), certain invertebrate animals in the phyla of flatworms (and roundworms), and many viruses.

Not all bacteria or fungi are parasitic and pathogenic; in fact, only relatively few species are. However, no virus can function outside a host organism, so we can say that all viruses are parasitic. Viruses, once introduced into a host cell, may take over the machinery of protein and nucleic acid synthesis, and force their host cells to manufacture more virus components from which viruses can self-assemble.

1 **State** the differences in structure between a bacterial cell and a virus.

Link
The structure of viruses and how they invade host cells is described in Chapter A2.3, page 88.

Nature of science: Observations

Careful observation can lead to important progress in treating diseases. For example, careful observations during nineteenth century epidemics of cholera in London and childbed fever (due to an infection after childbirth) in Vienna led to breakthroughs in the control of infectious disease.

In August 1854, an outbreak of cholera in Soho, London, killed 616 people within a month. At that time, cholera was a major risk to human health in Britain. It was thought that 'bad air' from rotting organic matter caused the outbreak and two preceding ones. John Snow, a doctor, mapped the deaths from cholera and observed that the people who had died had mainly lived near to a water pump on Broad Street, a road in Soho. His observations convinced the local council to remove the handle from the water pump to prevent its use. As a result, the number of deaths from cholera reduced significantly. Later, it was discovered that the water from that pump had been polluted by sewage from a nearby pit used to dispose of human waste, including faeces, which had been contaminated with the cholera bacterium.

In the nineteenth century, many women died after childbirth from puerperal septicaemia, more commonly known as childbed fever. Ignaz Semmelweis, a Hungarian physician, was a medical student at the University of Vienna's School of Medicine and at its teaching hospital, the General Hospital. In 1840, the hospital's general maternity ward had two sections: the first ward where male medical students delivered the babies and the second ward where female midwives performed the deliveries. Semmelweis, while working as an assistant obstetrician (a doctor specializing in pregnancy and childbirth) in the maternity ward, observed more deaths from puerperal septicaemia occurred in the first ward compared to the second. He also observed that the medical students spent their mornings in the autopsy rooms, dissecting and examining the bodies of women who had died of childbed fever. They then moved into the maternity ward, often without washing their hands. The midwives did not carry out autopsies. He found a clear correlation between deaths and autopsies. He believed that 'particles' were being transferred from the dead bodies to the mothers in the maternity wards, causing infection. In 1847, Semmelweis proposed that anyone delivering a baby should wash their hands beforehand, to prevent transmission. In many respects, his idea was close to the truth, and was accurate enough for him to devise a method of prevention. Today, it is known that childbed fever is caused by a bacterium, and so washing hands helps to prevent transmission of this microbe.

TOK

How is current knowledge shaped by its historical development?

Germ theory is the theory that certain diseases are caused by infection of the body by micro-organisms. Prior to the development of this theory, scientists thought that diseases such as cholera were caused by exposure to 'bad air'. The basis for germ theory was developed by Ignaz Semmelweis. Despite his conclusion that childbed fever could be prevented by hand washing, his observations conflicted with the established scientific and medical opinions of the time and so his ideas were rejected by the medical community, leading to further deaths. It was only when French microbiologist Louis Pasteur confirmed germ theory by proving that infection was caused by microbes, and Joseph Lister, acting on Pasteur's research, developed antiseptic methods for surgery, that hygienic methods were introduced into medical procedures.

Skin and mucous membranes as a primary defence

The skin and the **mucous membranes** (the internal linings of the lungs, trachea and gut) are the **primary defence** against pathogens that cause infectious diseases. Not surprisingly, protective measures have evolved at these surfaces.

■ The skin

The skin acts as both a physical and chemical barrier to pathogens (Figure C3.2.1). The external skin is covered by keratinized protein of the dead cells of the epidermis. This is a tough and impervious layer, and an effective barrier to most organisms unless the surface is broken, cut or deeply scratched. Periodic shedding of the epidermis removes microbes. However, folds or creases in the skin that are permanently moist may become the home of micro-organisms that degrade this barrier and cause infection, such as athlete's foot.

◆ **Mucous membranes (mucosa)**: the inner lining of some organs and body cavities (e.g. the nose, mouth, lungs and intestines). Glands in the mucous membrane make mucus.

◆ **Primary defence**: the initial barriers to prevent pathogens from causing infection.

Top tip!

While the structure and function of the skin need to be appreciated, you do not have to be able to draw or label diagrams of the skin.

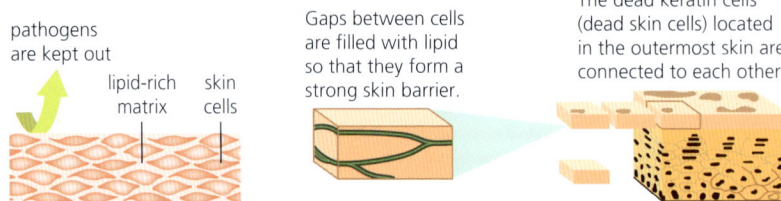

■ Figure C3.2.1 The skin is a protective barrier

Lysozyme is an enzyme secreted by the skin during perspiration (sweating), which acts as a chemical defence breaking down the cell walls of bacteria, thus acting as an antibiotic.

■ Mucous membranes

The internal surfaces of our breathing apparatus (the trachea, bronchi and bronchioles) and the gut are all lined by moist epithelial cells (mucous membranes, also known as mucosa). These vulnerable internal barriers are protected by the secretion of large quantities of **mucus**.

Cilia are organelles that project from the surface of certain cells (Chapter A2.2, page 76). Cilia occur in large numbers on the lining (epithelium) of the air tubes (bronchi) serving the lungs (Figure C3.2.2). Here, they sweep the fluid mucus across the epithelial surface, away from the delicate air sacs of the lungs.

◆ **Mucus**: a watery solution of glycoprotein with protective and lubrication functions.

◆ **Cilium (plural, cilia)**: motile, hair-like outgrowth from the surface of certain eukaryotic cells, which move rhythmically to propel objects such as mucus in the trachea and eggs in oviducts.

2 **Suggest** how mucus secreted by the airways may protect lung tissue.

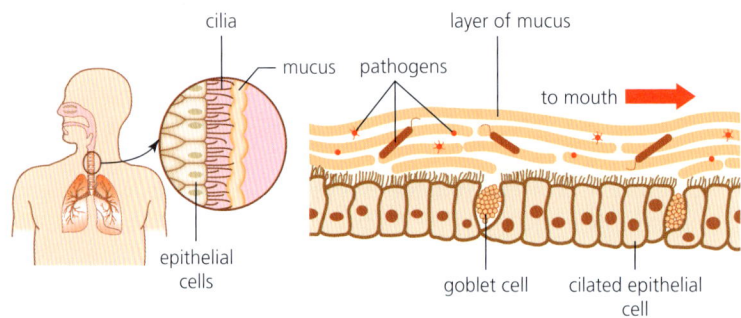

■ Figure C3.2.2 Mucous membrane in the trachea

Common mistake

Do not use poor terminology to describe the reasons for clotting, such as 'stops diseases getting in'. The stronger answer is to say that 'clotting prevents the entrance of pathogens'.

◆ **Platelets**: tiny cell fragments found in the blood that lack a nucleus; they are involved in the blood-clotting mechanism.

Sealing of cuts in skin by blood clotting

When a blood vessel is ruptured, the **blood-clotting mechanism** is activated. This leads to localized clotting of blood and further blood loss is prevented. A significant fall in blood pressure is also prevented, whether at small haemorrhages or at larger breakages or other wounds. The clot also reduces the chances of invasion by disease-causing organisms. After that, repair of the damaged tissues can take place.

The formation of a blood clot is triggered by a 'cascade' of events at the site of a broken blood vessel (shown in Figure C3.2.3). The steps are as follows.

- Firstly, **platelets** collect at the site. These components of the blood are formed in the bone marrow (along with the red and white blood cells) and they are circulated throughout the body, suspended in the plasma. Platelets are actually cell fragments, disc-shaped and very small (only 2 μm in diameter) – too small to contain a nucleus. Each platelet consists of a sack of cytoplasm that is rich in vesicles containing enzymes, and is surrounded by a plasma membrane. Platelets stick to the damaged tissues and clump together there – at this point, they change shape from sacks to flattened discs with tiny projections that interlock. This action alone seals off the smallest breaks.

- Next the collecting platelets release a **clotting factor** (a protein called thromboplastin that is also released by damaged tissues at the site). This clotting factor, along with vitamin K and calcium ions (always present in the plasma), causes a soluble plasma protein called **prothrombin** to be converted to an active, proteolytic enzyme, **thrombin**.
- The action of the enzyme thrombin is to convert another soluble blood protein, **fibrinogen**, into insoluble **fibrin** fibres at the site of the cut. Within this mass of fibres, red blood cells (erythrocytes) are trapped, and the blood clot has formed.

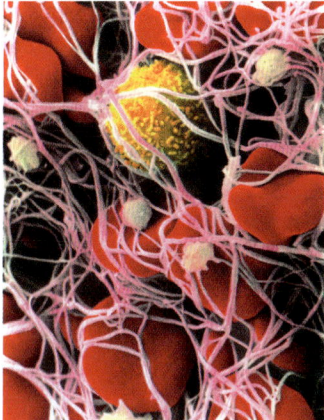

scanning electron micrograph of blood clot showing meshwork of fibrin fibres and trapped blood cells

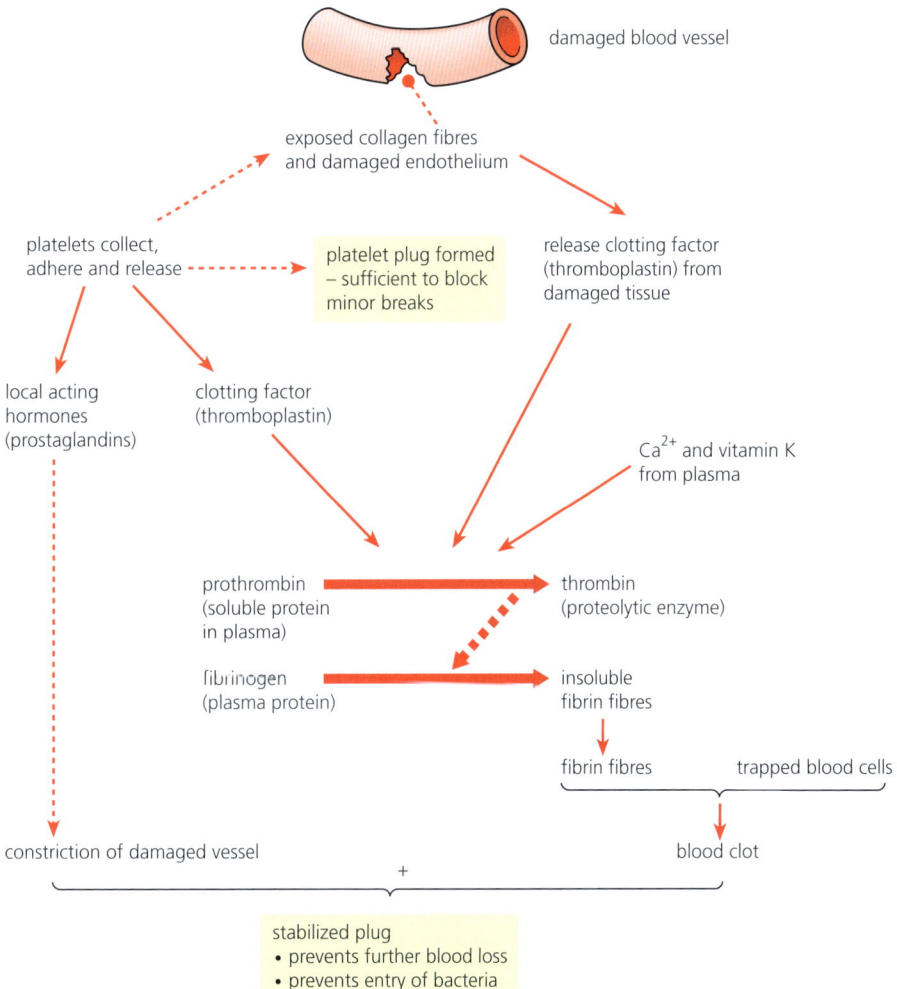

■ Figure C3.2.3 The blood-clotting mechanism

Common mistake

Do not mix up fibrin and fibrinogen. Make sure you know the function and properties of both. Learn the cascade events of clotting carefully.

Concept: Interaction

The blood-clotting mechanism depends on the interaction of several different components: release of clotting factors from platelets, the subsequent cascade pathway that results in rapid conversion of fibrinogen to fibrin by thrombin, the trapping of erythrocytes to form a clot.

3 **Identify** the correct sequence of the following events during blood clotting: fibrin formation; clotting factor release; thrombin formation.

C3.2 Defence against disease

◆ **Innate immune system**: responds to broad categories of pathogens and does not change during an organism's life.

◆ **Phagocytes**: white blood cells that engulf, absorb and digest pathogens, foreign particles and cell debris.

◆ **Adaptive immune system**: responds in a specific way to particular pathogens and builds up a memory of pathogens encountered, so the immune response becomes more effective.

◆ **Lymphocytes**: white blood cells in the adaptive immune system that cooperate to produce antibodies.

Differences between the innate immune system and the adaptive immune system

The immune response is our main defence once invasion of the body by harmful micro-organisms or 'foreign' materials has occurred. The immune response begins with a first response from the **innate immune system**, carried out by **phagocytes** (a type of white blood cell) that can engulf and destroy (by phagocytosis) many different foreign organisms.

At the same time, the primary phase of the **adaptive immune system** response begins, involving specialized white blood cells called **lymphocytes**. This is followed by the secondary phase of the adaptive immune system response, where specific white blood cells retain a memory of a specific pathogen so that a rapid immune response can be quickly coordinated if reinfection occurs.

The innate immune system is different to the adaptive immune system: it responds to a broad range of pathogens and does not change over an organism's life. On the other hand, the adaptive system responds in a certain way to individual types of pathogen. It builds up a memory of the pathogens that are encountered. This means the immune response becomes more effective.

● Top tip!

White blood cells are called leucocytes. These cells are divided into two groups: **phagocytes** that engulf foreign organisms, and **lymphocytes** that are involved in the adaptive immune system. Both circulate in the blood and are contained in lymph nodes.

Tool 1: Experimental techniques

Drawing annotated diagrams from observation

Examine the appearance of the types of white blood cells observed in a blood smear preparation.

Draw and fully annotate some of the blood cells you observe (e.g. phagocytes, lymphocytes and red blood cells), making clear the differences between them.

◆ **Plasma**: the clear, yellowish, liquid part of the blood that carries blood cells.

Infection control by phagocytes

Certain white blood cells have the role of engulfing foreign material, including invading bacterial cells. Some of these white blood cells are short-lived cells of the **plasma**. Others are the long-lived, rubbish-collecting cells found throughout the body tissues. Both types of cell take up material into their cytoplasm, much as the protozoan *Amoeba* is observed to feed, by phagocytosis (Figure C3.2.4). Amoeboid movement (crawling-like movement using protrusions of cytoplasm) allows phagocytes to move from blood to sites of infection, where they recognize pathogens, engulf them by endocytosis and digest them in a controlled way using enzymes from lysosomes (page 71).

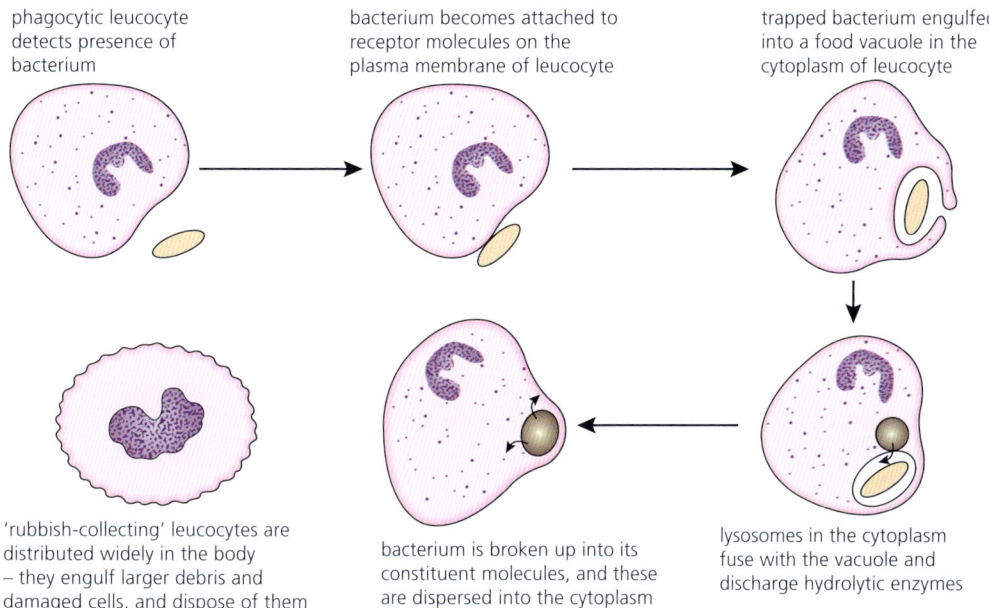

■ Figure C3.2.4 Phagocytosis of a bacterium

◆ **B-lymphocytes (B-cells)**: lymphocytes that produce antibodies and are responsible for presenting antigens to T-cells. Once activated, they can mature into plasma cells or memory B-lymphocytes.

◆ **T-lymphocytes (T-cells)**: white blood cells processed by the thymus that are responsible for cell-mediated immunity.

Link
The lymphatic system is discussed in more detail in Chapter B3.2, page 312.

Lymphocytes – the adaptive immune system

Lymphocytes are responsible for our specific immune response. These cells make up 20% of the white blood cells circulating in the blood plasma (or in the tissue fluid – phagocytes can move freely through the walls of blood vessels). Lymphocytes circulate in the blood and are contained in **lymph nodes**, which are part of the **lymphatic system**.

Lymphocytes can detect any foreign matter that enters from outside the body (macromolecules, as well as micro-organisms) as different from our own cells and proteins (see below).

There are two types of lymphocyte at work in our immune system: the **B-lymphocytes** (B-cells) and **T-lymphocytes** (T-cells). Both cell types originate in the bone marrow, where they are formed from stem cells (page 257). As they mature, these cells undergo different development processes in preparation for their distinctive roles (Figure C3.2.5).

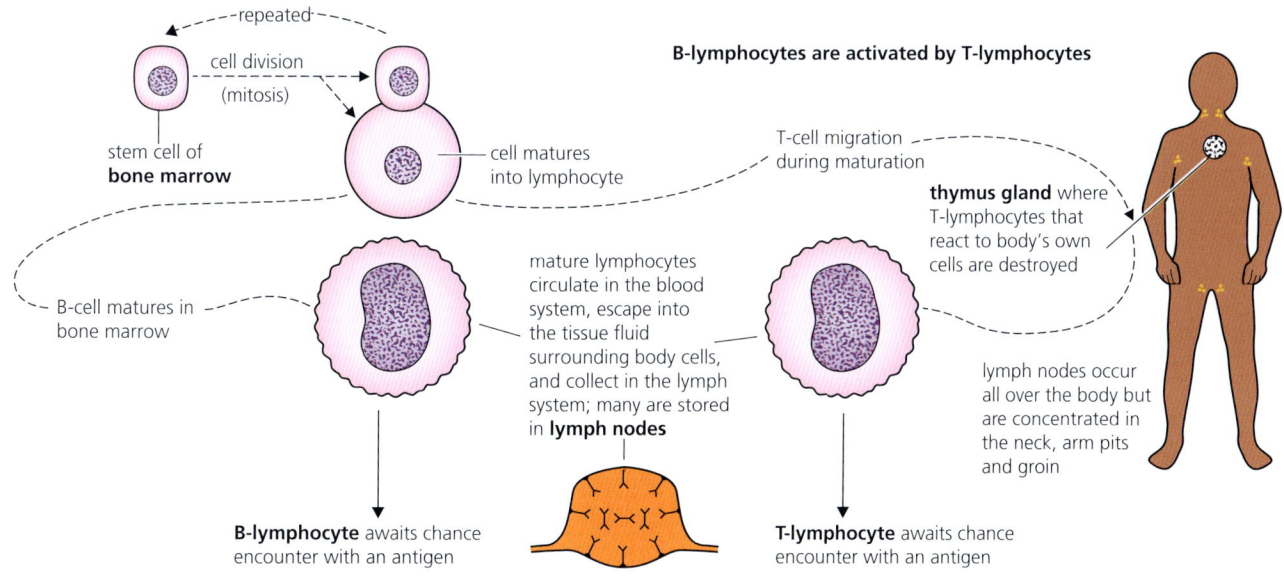

■ Figure C3.2.5 T- and B-lymphocytes

C3.2 Defence against disease

◆ **Thymus gland**: a small organ that lies in the upper chest under the breastbone in which T-lymphocytes grow, multiply and mature; it is part of the lymphatic system.

◆ **Plasma cells**: type of immune cell that makes large amounts of a specific antibody. Plasma cells develop from B-cells that have been activated.

◆ **Memory cells**: a long-lived lymphocyte capable of responding to a particular antigen on its reintroduction, long after the exposure that stimulated its initial production.

■ T-lymphocytes

T-lymphocytes leave the bone marrow soon after they have been formed and migrate to the **thymus gland**. The thymus gland is found in the chest, just below the breastbone (sternum). It is active and enlarged during the early stages of our growth and development. While the T-cells are present in the thymus gland, all of those that would react to the body's own cells are removed and destroyed. The surviving T-cells are released and circulate in the blood plasma. Many are stored in lymph nodes. The thymus gland shrinks in size by the time puberty is reached, its task completed.

● Top tip!

Remember that antibodies are proteins that are produced by B-lymphocytes to attach to foreign, non-self material and allow the immune system to remove it. Antigens are substances (usually glycoproteins or other protein) capable of binding specifically to an antibody.

■ B-lymphocytes

B-lymphocytes complete their maturation in the bone marrow, prior to circulating in the blood. Many of these lymphocytes are also stored in lymph nodes. The role of the majority of B-cells, after activation by T-cells, is to form clones of **plasma cells** that then secrete antibodies into the blood system. In addition, **memory cells** are formed – see page 529.

An individual has a very large number of B-lymphocytes that each make a specific type of antibody.

4 **Explain** the significance of the role of the thymus gland in destroying T-cells that would otherwise react to 'self' body proteins.

◆ **Immunoglobulin**: a protein made by B-lymphocytes and plasma cells that helps the body fight infection.

■ Antibodies

An **antibody** is a globular protein called an **immunoglobulin**. It is made of four polypeptide chains held together by disulfide bridges (-S–S-) and forming a Y-shaped molecule (see Figure C3.2.6). The top of each 'arm' contains an antigen binding site.

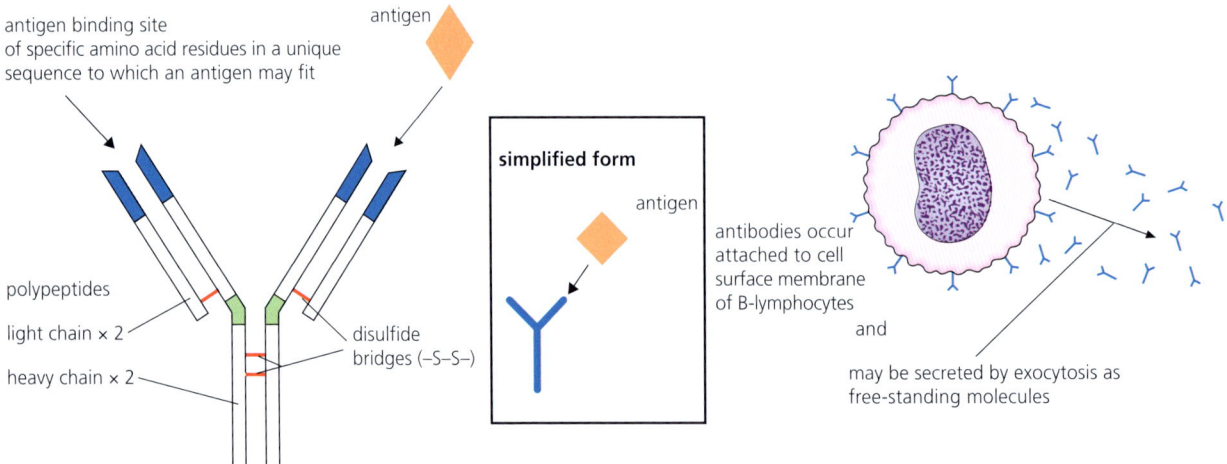

■ Figure C3.2.6 The structure of an antibody

Millions of different types of antibodies may be produced by our bodies, each by a different type of lymphocyte – there are as many antibodies as there are types of foreign matter (antigens) invading the body. The amino acid sequence of the antigen binding site in the 'fork' region differs according to the chemistry of the antibody it binds with and is unique to that antibody. It is the antigen binding site that gives each antibody its **specificity**.

Antibodies initially occur attached to the cell surface membrane of B-cells, but later are mass-produced and secreted by cells derived from the B-cell. This occurs after that B-cell has undergone an activation step (see page 526). A huge range of different antibody-secreting lymphocytes exists, each type recognizing one specific antigen. The more antigens we encounter, the more antibodies we are able to form, should they be required.

Antibodies destroy antigens in different ways (Table C3.2.1). Toxins may be inactivated by reaction with the antibody, and bacterial cells may be clumped together so that they 'precipitate' and can be engulfed by phagocytic cells. Antibodies also attach to foreign matter, ensuring its recognition by phagocytic cells. Alternatively, antibodies can act by destroying bacterial cell walls, causing lysis of the bacterium.

■ Table C3.2.1 How antibodies aid the destruction of pathogens

Agglutination	antibodies attach to pathogens, causing them to stick together – clumped in this way, they are more easily ingested by phagocytic cells
Complement activation	complement proteins in the plasma cause cell lysis by destroying the plasma membrane of pathogen cells, after antibodies have identified them by binding
Toxin neutralization	antibodies bind to toxins in the plasma, preventing them from affecting susceptible cells
Opsonization	antibodies make pathogens instantly recognizable by binding to them, and then linking them to phagocytic cells

Link
The definition of antigen is discussed in cell–cell recognition in Chapter B1.1, page 195.

Link
Blood groups and immune reactions are examined in Chapter B1.1, page 195.

Link
Glycoproteins and glycolipids in the plasma membrane are covered in Chapter B2.1, page 234.

◆ **Major histocompatibility complex (MHC)**: group of genes that code for proteins found on the surfaces of cells that help the immune system recognize foreign substances.

5 **Identify** where antigens and antibodies may be found in the body.

6 **Deduce** which synthetic machinery in the lymphocytes is active in the production of the components of their antibodies.

Antigens

The immune system can recognize 'self' – our body's cells and proteins – and tell them apart from foreign or 'non-self' substances, such as those on or from an invading organism. Any molecule that the body recognizes as foreign or 'non-self' is known as an **antigen**. Most antigens are **glycoproteins** or other proteins, and they are usually located on the outer surfaces of pathogens. It is the lymphocytes that can recognize antigens and take steps to overcome them. The specificity of antigens allows for responses that are customized to specific pathogens.

Each type of lymphocyte in our body recognizes only one specific antigen. In the presence of that antigen (and only that antigen), the lymphocyte divides rapidly, producing many cells known as clones. These cloned lymphocytes then secrete an antibody specific to that antigen (see below).

Antigens on the surface of red blood cells (erythrocytes) may stimulate antibody production if transfused into a person with a different blood group.

■ How are 'self' and 'non-self' recognized?

Every organism has unique molecules on the surface of their cells. Cells are identified by these specific molecules (markers) that are lodged in the outer surface of the plasma membrane. These molecules include the highly variable glycoproteins on the cell surface membrane. Remember, carbohydrates occur attached to proteins here.

The glycoproteins that identify cells are known as the **major histocompatibility complex (MHC)** antigens. In humans, the genes for MHC antigens are on chromosome 6. The MHC antigens of individuals are genetically determined – they are a feature we inherit. In inherited characteristics that are products of sexual reproduction, variations occur, so each of us has distinctive MHC antigens present on our cell surface membranes. Unless you have an identical twin, your MHC antigens are unique.

Lymphocytes of our immune system have antigen receptors that recognize our own MHC antigens and can tell them apart from any foreign antigens detected in the body. It is critically important that our own cells are not attacked by our immune system. This is the basis of the 'self' and 'non-self' recognition mechanism.

C3.2 Defence against disease

Steps to the immune response

When an infection occurs, the white blood cell population immediately increases and many of these cells collect at the site of the invasion. The complex response to infection has begun.

The roles of T- and B-cells in this response are as listed here (see also Figure C3.2.7).

■ Activation of B-lymphocytes by helper T-lymphocytes

◆ **Antigen-specific B-cell**: lymphocytes that form either antibody-secreting cells or memory B-cells after infection or vaccination.

◆ **Macrophage**: type of phagocytic white blood cell that surrounds and kills pathogens, removes dead cells and stimulates the action of other immune system cells.

◆ **Antigen presentation**: process where an antigen is taken into a white blood cell by receptor-mediated endocytosis, digested, complexed with MHC II molecules and then presented on the cell surface membrane for interaction with T-cells.

◆ **Helper T-cell**: lymphocyte that recognizes a foreign antigen and then activates T-cell and B-cell production.

◆ **Activated B-cell**: B-cells absorb an antigen and present pieces of it on their surface via a major histocompatibility complex (MHC); helper T-cells then recognize the antigens via the MHC and activate the B-cells, resulting in B-cell differentiation into memory B-cells or plasma cells.

> **Top tip!**
>
> T- and B-cells have molecules on the outer surface of their cell surface membrane that enable them to recognize antigens, but each B- and T-lymphocyte has only one type of surface receptor. Consequently, each lymphocyte can recognize only one type of antigen.

> **Concept: Interdependence**
>
> There are coordinated interactions between helper T-cells and B-cells that occur during the initiation and refinement of antibody production.

1. When a specific antigen enters the body, **B-cells** with surface receptors (antibodies) that recognize the antigen bind to it.
2. On binding to the B-cell, the antigen is taken into the cytoplasm by endocytosis. Then it is expressed and displayed on the cell surface membrane of the B-cell (this is known as an **antigen-specific B-cell**).
3. Meanwhile phagocytic cells, the **macrophages**, engulf any antigens they encounter. (Macrophages occur in the plasma, lymph or tissue fluid, liver, spleen and alveoli.) Macrophages (and T-cells) secrete cytokines, small signalling molecules (usually a protein or glycoprotein) that act on neighbouring cells to alter their behaviour. This assists with the immune response (Chapter C2.1, page 454). Once antigens have been taken up, they are presented externally, attached to the HMC antigens, on the surface of the macrophages. **T-cells** respond to antigens that are presented on the surface of other cells, as on the macrophages. This is called **antigen presentation** by a macrophage.
4. As T-cells encounter these macrophages and briefly bind to them, they are immediately activated. They are now called 'armed' or activated **helper T-cells**.
5. Activated helper T-cells bind to antigen-specific B-cells with the same antigen expressed on their cell surface membrane. As a result, the B-cell is activated. It is now an 'armed' or **activated B-cell**.

> **Common mistake**
>
> Do not confuse the role of T- and B-cells. The role of B-cells is to secrete antibodies. The role of helper T-cells is *not* to secrete antibodies but to further activate B-cells after 'activation' by contact with antigens of a particular pathogen or other foreign matter.

> **Top tip!**
>
> B-cells become memory cells only when they have been activated. Activation requires both direct interaction with the specific antigen and contact with a helper T-cell that has also become activated by the same type of antigen.

Concept: Interaction

Interactions between B- and T-cells coordinate the immune response and facilitate antibody production for a specific pathogen.

Top tip!

Phagocytes, such as macrophages, have the capacity to leave blood vessels.

Common mistake

Do not confuse the functions of macrophages, B-cells, T-cells and memory cells. Use Figure C3.2.7 to help you learn the role of each different type of cell and the processes involved in the immune response.

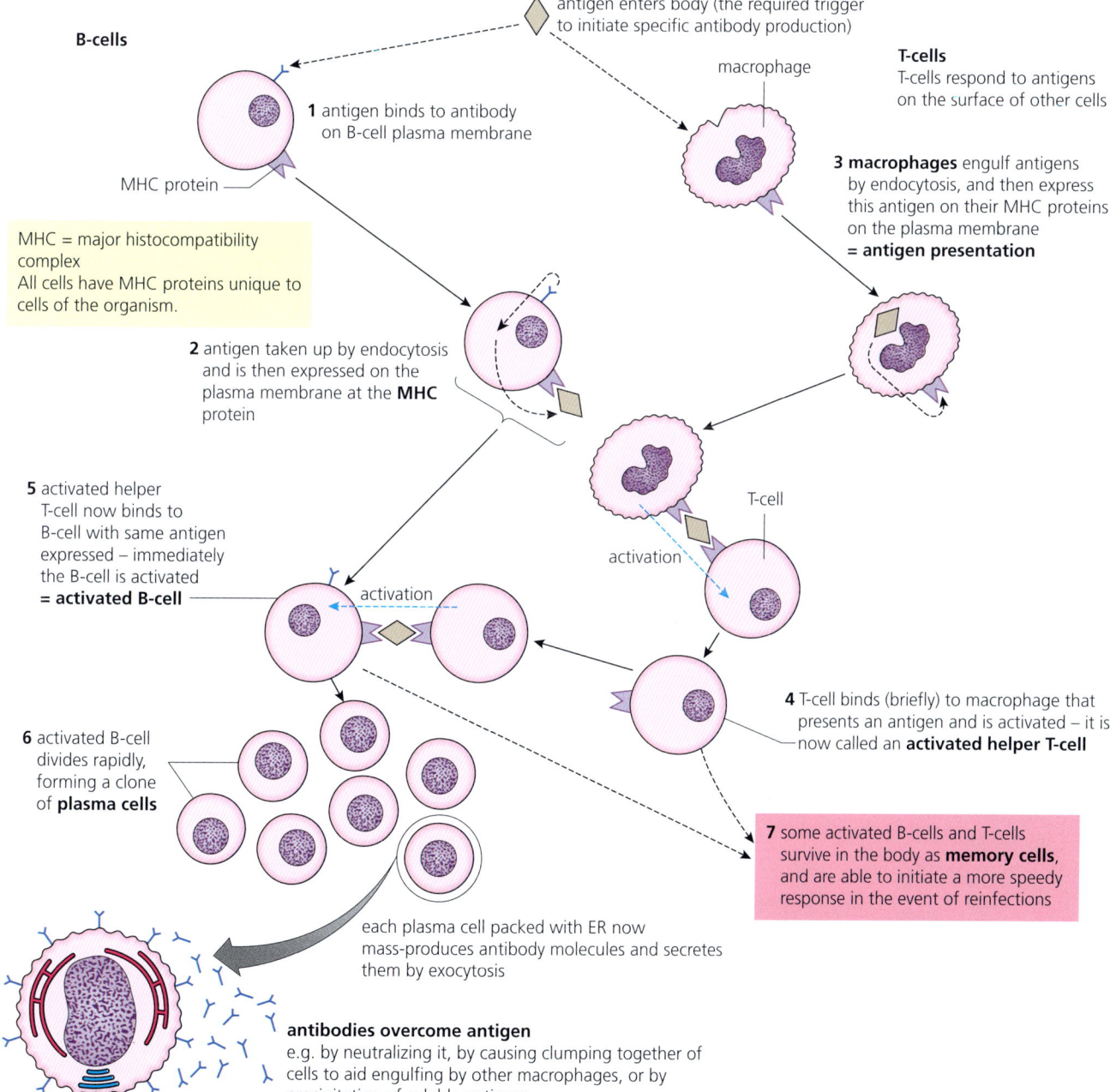

■ Figure C3.2.7 Stages in antibody production

C3.2 Defence against disease

> **Top tip!**
>
> There are relatively small numbers of B-cells that respond to a specific antigen. To produce sufficient quantities of antibody, activated B-cells first divide by mitosis to produce large numbers of plasma B-cells that are capable of producing the same type of antibody.

■ Multiplication of activated B-lymphocytes to form clones of antibody-secreting plasma cells

6 After B-cells are activated (step 5), they divide rapidly by mitosis, forming clones of **plasma cells**. An electron micrograph of plasma cells shows each one is packed with rough endoplasmic reticulum (RER, Figure C3.2.8). It is in the RER that the antibody is mass-produced and exported from the B-cell by exocytosis. The generation of a large number of plasma cells that produce one specific antibody type is known as clonal selection.

The antibodies are normally produced in such numbers that the antigen is overcome. The action of antibodies is to bind to antigens, neutralizing them or making them clear targets for phagocytic cells.

We have noted that the T- and B-cells have molecules on the outer surface of their cell surface membrane that enable them to recognize antigens, but each B- and T-lymphocyte has only one type of surface receptor. Consequently, each lymphocyte can recognize only one type of antigen.

7 After antibodies have attacked the foreign matter and the disease threat is overcome, the special proteins disappear from the blood and tissue fluid. So, too, do the bulk of the specific B-cells and T-cells that were responsible for their formation.

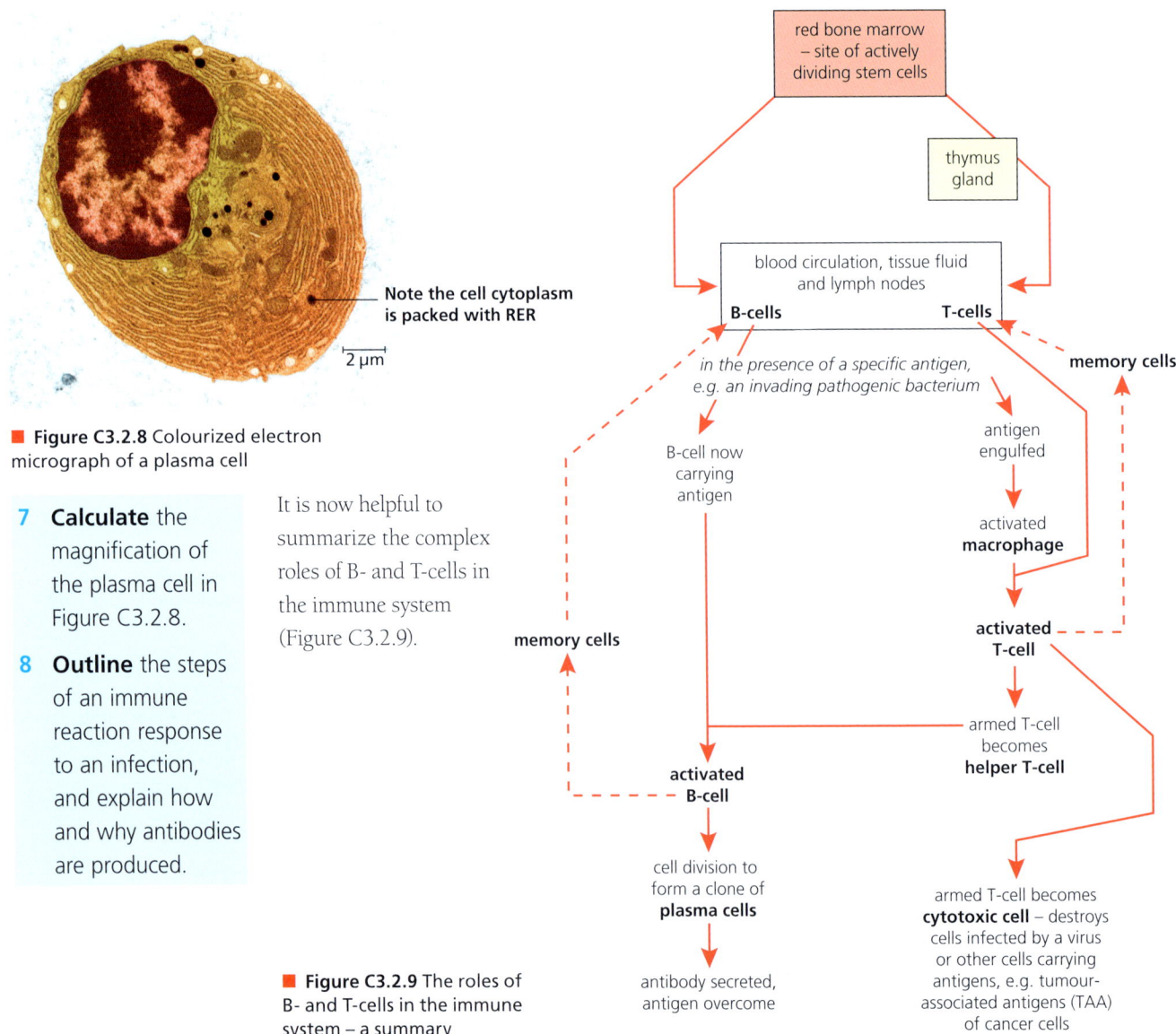

■ **Figure C3.2.8** Colourized electron micrograph of a plasma cell

7 **Calculate** the magnification of the plasma cell in Figure C3.2.8.

8 **Outline** the steps of an immune reaction response to an infection, and explain how and why antibodies are produced.

It is now helpful to summarize the complex roles of B- and T-cells in the immune system (Figure C3.2.9).

■ **Figure C3.2.9** The roles of B- and T-cells in the immune system – a summary

528   Theme C: Interaction and interdependence – Organisms

◆ **Immunity**: resistance to the onset of a disease after infection by the causative agent.

9 For Figure C3.2.10, **compare and contrast** the primary response following initial infection with the secondary response.

10 **Identify** the steps of plasma cell formation that are avoided in cases of reinfection, due to the existence of memory cells.

11 **State** what is meant by the term *immunity*.

◆ **HIV (human immunodeficiency virus)**: virus that attacks the body's immune system. If HIV is not treated, it can lead to AIDS.

◆ **AIDS (acquired immune deficiency syndrome)**: the name used to describe a number of potentially life-threatening infections and illnesses that happen when the immune system has been severely damaged by the HIV virus.

■ Immunity as a consequence of retaining memory cells

Immunity is the ability to eliminate an infectious disease from the body and to resist an infection by a pathogen. Long-lived and specific immunity is the result of the long-term survival of lymphocytes that are capable of making the specific antibodies needed to fight the infection. These are **memory cells**, retained after a previous infection by that pathogen. Memory cells are specifically activated B-cells. They are long-lived cells, in contrast to plasma cells and other activated B-cells. Memory cells make possible an early and effective response in the event of a reinfection of the body by the same antigen (Figure C3.2.10). This is the basis of natural immunity (see below).

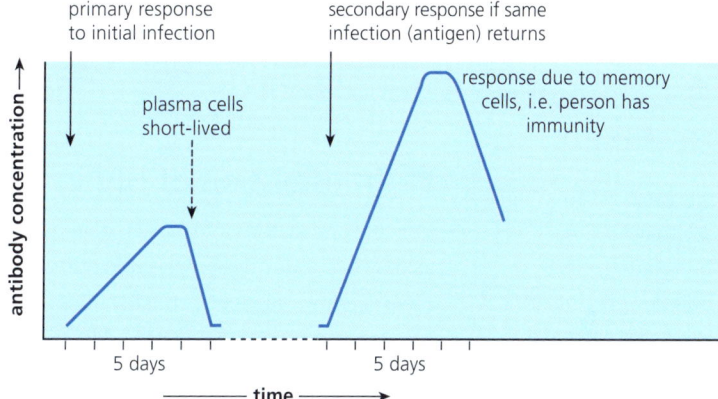

■ Figure C3.2.10 Profile of antibody production in infection and reinfection

⬤ Common mistake

Explanations of what happens upon second exposure to an antigen are often lacking detail. It should be noted that antibodies are produced more rapidly and to a higher level.

HIV

Human immunodeficiency virus (HIV) was first identified in 1983 as the cause of a disease of the human immune system known as **acquired immune deficiency syndrome (AIDS)**.

HIV is a tiny virus, less than 0.1 μm in diameter (Figure C3.2.11). It consists of two single strands of RNA which, together with enzymes, are enclosed by a protein capsid. A membrane, derived from the human host cell in which the virus was formed, encapsulates each new virus particle (virion) leaving the host cell.

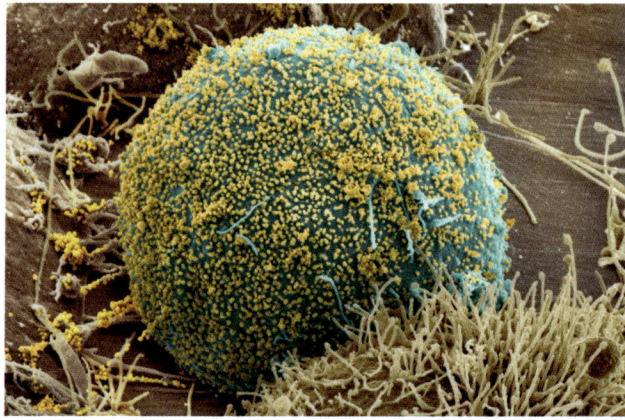

■ Figure C3.2.11 Electron micrograph of white blood cell (blue), from which HIV (yellow) is budding off

C3.2 Defence against disease

♦ **Retrovirus**: a virus that has RNA as its nucleic acid and uses the enzyme reverse transcriptase to copy its genome into the DNA of the host cell's chromosomes.

Link
Retroviruses are covered in Chapter A2.3, page 99.

Common mistake
It is incorrect to say that AIDS is transmitted. HIV is transmitted, not AIDS. HIV is the virus and AIDS is the disease it causes.

HIV is a **retrovirus**. A retrovirus reverses the normal flow of genetic information from the DNA of genes to messenger RNA in the cytoplasm (page 99). The idea that information always flows in this direction in cells was called the central dogma of cell biology (implying it was always the case). However, in retroviruses, the information in RNA in the cytoplasm is translated into DNA within a host cell, which then becomes integrated within the DNA of a chromosome in the host's nucleus.

ATL C3.2B

Explore the life cycle of HIV using this animation:
www.sumanasinc.com/webcontent/animations/content/lifecyclehiv.html

How does application of your knowledge of cell structure (Chapter A2.2 Cell structure) help in understanding how HIV particles are assembled? Make a list of the relevant parts of the syllabus that you need to know about to understand the life cycle of HIV and its effects on the body.

■ Transmission of HIV in body fluids

Infection with HIV is possible through contact with the blood or body fluids of infected people, such as may occur during sexual intercourse, sharing of hypodermic needles by intravenous drug users or breastfeeding of a newborn baby. Blood transfusions and organ transplants can also transmit HIV, but donors are now screened for HIV infection in most countries.

HIV is not transferred by contact with saliva on a drinking glass, or by sharing a towel, for example. Female mosquitoes also do not transmit HIV when feeding on human blood.

■ Infection of lymphocytes by HIV with AIDS as a consequence

HIV is an enveloped virus with embedded glycoproteins that act as antigens when the virus is present in the human bloodstream. These glycoproteins attach the virus to protein receptors of the surface of T-lymphocytes and the core of the virus penetrates to the cytoplasm (Figure C3.2.12) of these important regulatory cells of the immune system. Macrophages, another type of white blood cell, may also be infected by HIV.

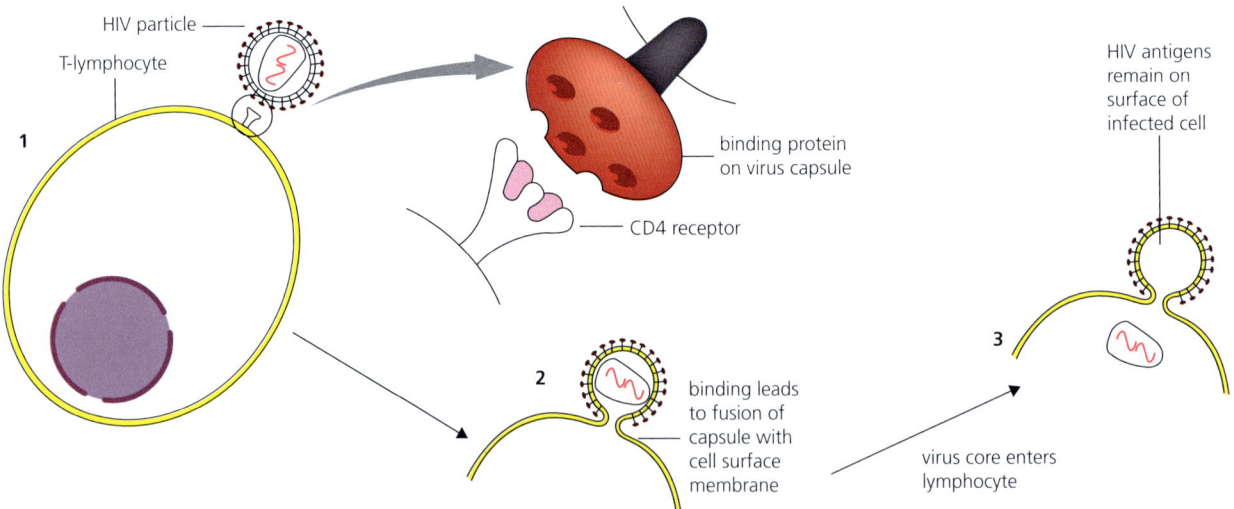

■ **Figure C3.2.12** HIV infection of host cell: CD4 T-helper cells

Newly synthesized HIV virions leave the host cell by lysis and will then invade and kill other lymphocytes (Figure C3.2.13).

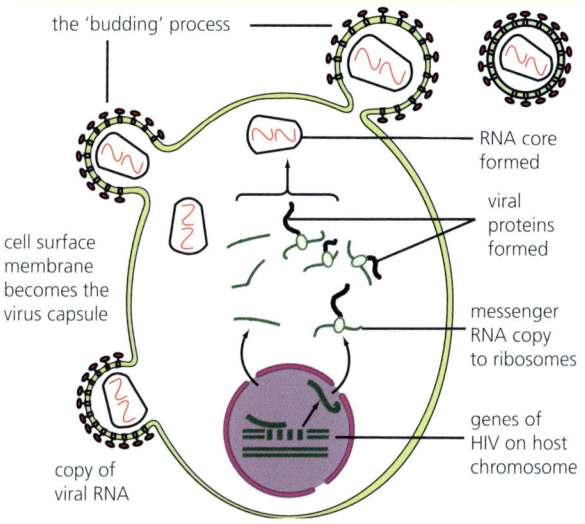

■ Figure C3.2.13 Activation of the HIV genome and production of newly replicated HIV virions

Without treatment, this process causes the body's reserve of lymphocytes to decrease very quickly. Macrophages also form an important reservoir of HIV.

The reduction in the number of active lymphocytes means the body loses the ability to produce antibodies. Eventually, no infection, however trivial, can be resisted; death follows. This is why AIDS is known as an 'immunodeficiency syndrome'. Only certain types of lymphocyte are infected and killed, but a reduction in these lymphocytes limits the body's ability to produce antibodies and fight opportunistic infections, leading to AIDS.

Ideally, a vaccine against HIV would be the best solution – one designed to remove both infected lymphocytes and HIV particles in the patient's bloodstream. The work of several laboratories is dedicated to this solution. The problem is that, in the latent state of the infection, the infected lymphocyte cells frequently change their membrane marker proteins because of the presence of the HIV genome within the cell. Effectively, HIV can hide from the body's immune response by changing its identity.

12 Figure C3.2.14 shows changes in the number of helper T-cells and HIV RNA copies in blood samples of a person infected by HIV. Analyse the graph shown in Figure C3.2.14.
   a **Describe** the trend seen in the graph.
   b **Explain** the trend seen in the graph.

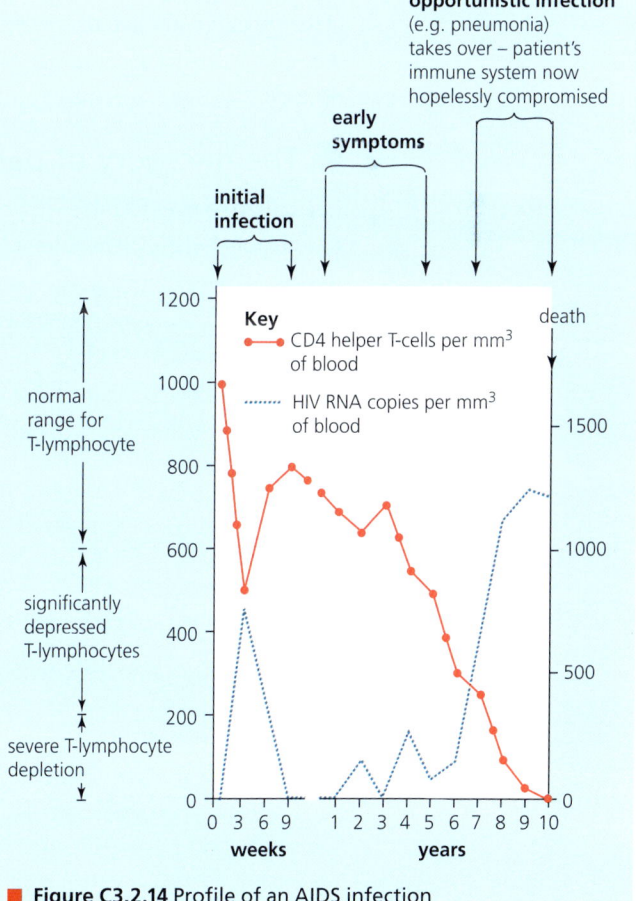

■ Figure C3.2.14 Profile of an AIDS infection

C3.2 Defence against disease

> **Tool 1: Experimental techniques**
>
> **Physical modelling**
>
> https://pdb101.rcsb.org/learn/paper-models/hiv-capsid
>
> https://cdn.rcsb.org/pdb101/learn/resources/structural-biology-of-hiv/index.html#
>
> The first URL provides a template and video for making an HIV capsid paper model and the second URL allows you to explore the structural biology of HIV (viral enzymes, accessory proteins and structural proteins).

Antibiotics

◆ **Antibiotics**: chemicals that block processes occurring in bacteria but not in eukaryotic cells.

Antibiotics are substances that slow down or kill micro-organisms. They are obtained from fungi or bacteria and are substances that these organisms manufacture in their natural habitats. An antibiotic, when present in low concentrations, inhibits the growth of other micro-organisms. Many bacterial diseases of humans and other animals can be successfully treated with them.

Before antibiotics became available to treat bacterial infections, typical hospital wards were filled with patients with pneumonia, typhoid fever, tuberculosis, meningitis, syphilis and rheumatic fever. These bacterial diseases claimed many lives, sometimes very quickly. Patients of all ages were affected.

Today, these infections are not the 'killers' they once were here. For example, in the 1930s about 40% of the patients with bacterial pneumonia died of the disease. Today, about 5–10% may die. Antibiotic drugs have brought about this improvement in survival rates. The viral forms of pneumonia and meningitis are not overcome by antibiotics, however, since antibiotics do not affect viruses.

■ The discovery of penicillin – the first antibiotic

In 1929, Alexander Fleming (1881–1955), a Scottish bacteriologist working at St Mary's Hospital, Paddington, London, was studying the bacterium *Staphylococcus*, which causes boils and sore throats. When examining some older bacteriological plates, he came across one in which a fungal colony had also become established (Figure C3.2.15). He noticed that the bacteria were killed in areas surrounding the mould. This he identified as *Penicillium notatum*. He cultured this mould in broth and discovered that a substance from it – which he named **penicillin** – was bactericidal. He also showed that penicillin did not harm human blood cells. Fleming published these results in a scientific paper.

> ## ● Nature of science: Global impact of science
>
> ### Early testing of the safety of penicillin as a drug
>
> The leading figures in the development of penicillin were Australian pathologist Harold Florey (1898–1968) and German biochemist Ernst Chain (1906–79). This team isolated penicillin in a stable form for therapeutic uses.
>
> Florey and Chain used mice to test penicillin on bacterial infections in an experiment that today would not be compliant with drug-testing protocols (Figure C3.2.15). *Can you see why?* Today, scientists have an obligation to assess the risks associated with their work, and these same standards are recognized internationally.

Since their original discovery, over 4000 different antibiotics have been isolated, but only about 50 have proved to be safe to use as drugs. The antibiotics that are effective over a wide range of pathogenic organisms are called **broad-spectrum antibiotics**. Others are effective with just a few pathogens. Many antibiotics in use today are semi-synthetic (i.e. made by converting starting materials from natural sources into final products via chemical reactions), as they can act against bacteria that are resistant to the original compound, have a greater spectrum of activity or cause fewer side effects.

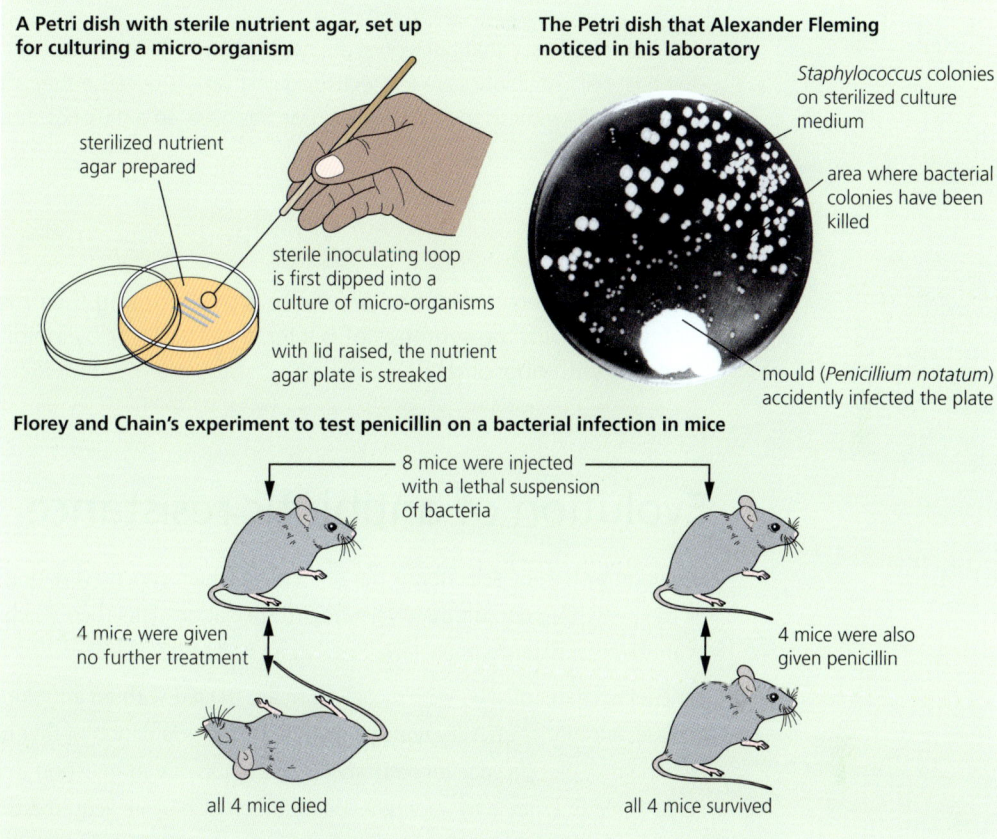

ATL C3.2C

Find out about how modern drugs are tested.

Suggest possible reasons why Florey and Chain's tests on the safety of penicillin would not be compliant with current protocols on drug testing.

■ **Figure C3.2.15** Early steps in the discovery and use of penicillin as an antibiotic

How antibiotics work

Most antibiotics disrupt the metabolism of prokaryotic cells – whole populations of bacteria may be quickly suppressed. The division and growth phases of bacteria are vulnerable to antibiotic action (Table C3.2.2). At the same time, the cells of the human host organism (a eukaryote) are not affected.

■ **Table C3.2.2** The biochemical mechanisms of antibiotic action

Mechanism targeted	Effects
Cell wall synthesis	The antibiotic interferes with the synthesis of bacterial cell walls. Once the cell wall is destroyed, the delicate plasma membrane of the bacterium is exposed to the destructive force generated by excessive uptake of water by osmosis so that lysis (bursting) of the cell occurs. Several antibiotics, including penicillin, ampicillin and bacitracin, bind to and inactivate specific wall-building enzymes – the bacterium's walls fall apart. (This is the most effective mechanism.)
Protein synthesis	The antibiotic inhibits protein synthesis by binding with ribosomal RNA. The ribosomes of prokaryotes are made of particular RNA subunits. The ribosomes of eukaryotic cells are larger and are built with different types of RNA molecules. Antibiotics such as streptomycin, chloramphenicol, the tetracyclines and erythromycin all bind to the prokaryotic ribosomal RNA subunits that are unique to bacteria, terminating their protein synthesis.

C3.2 Defence against disease

Viruses, on the other hand, are non-living particles and have no metabolism of their own, so have no function that can be inhibited by antibiotics. Viruses reproduce using metabolic pathways in their host cell. Antibiotics cannot be used to prevent viral diseases.

13 Explain why antibiotics are effective against bacteria but not viruses.

● Common mistake

It is incorrect to say that antibiotics are not effective against viruses because these can 'hide inside the host cell'. Antibiotics are ineffective against viruses because they affect bacterial enzymes and cell wall synthesis, neither of which viruses have, so antibiotics should not be taken to treat viral infections.

Concept: Interaction

Antibiotics interact with prokaryotic cells by disrupting their metabolism, either by interfering with the synthesis of bacterial cell walls or by inhibiting protein synthesis by binding with ribosomal RNA.

Evolution of antibiotic resistance

Evolution by natural selection is the process by which genetic variation in organisms is selected for by their environment, ultimately leading to speciation. This theory helps to explain the development of antibiotic resistance in bacteria.

Patients who are infected with a bacterium are treated with an antibiotic to help them overcome the disease; antibiotics are very widely used. In a large population of that bacterium, some individual bacteria may carry a gene for resistance to the antibiotic in question, typically on a plasmid. Such genes sometimes arise by spontaneous mutation or may be acquired through conjugation between bacteria of different populations (see page 119).

Most of a bacteria population is not adapted to an environment that has changed (with the addition of an antibiotic), but an unusual or mutant variety of the population is suited and, therefore, has a selective advantage. The resistant bacteria in the population have no selective advantage in the absence of the antibiotic and must compete for resources with non-resistant bacteria. But when the antibiotic is present, most bacteria of the population will be killed. The resistant bacteria are very likely to survive and will be the basis of the future population. In the new population, all individuals now carry the gene for resistance to the antibiotic. The genome has changed abruptly (Figure C3.2.16). The evolution of bacteria that are resistant to antibiotics means that careful use of antibiotics is necessary to slow the emergence of multi-resistant bacteria.

Link
Evolution by natural selection is covered in Chapter A4.1, page 140 and Chapter D4.1, page 779.

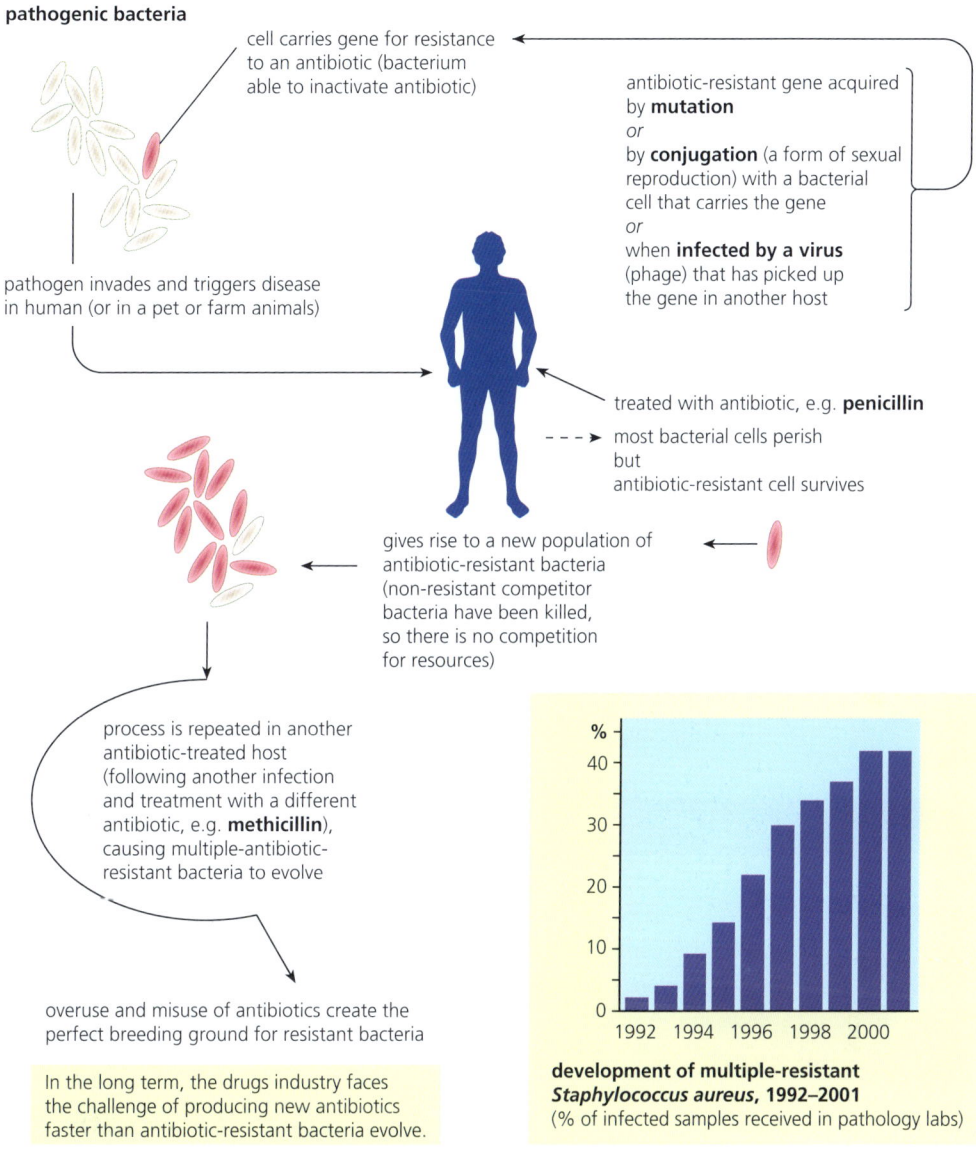

■ **Figure C3.2.16** Multiple antibiotic resistance in bacteria, where there is evolution of resistance to several antibiotics in strains of pathogenic bacteria

14 **Explain**:
  a why doctors ask patients to complete the full course of antibiotics, even if they start to feel better.
  b why the medical profession tries to combat resistance by alternating the types of antibiotic used against an infection.

For example, a strain of *Staphylococcus aureus* has acquired resistance to a range of antibiotics including **methicillin**, forming the so-called methicillin-resistant *Staphylococcus aureus* (**MRSA**). MRSA presents the greatest threat to patients who have undergone surgery. With cases of MRSA, the intravenous antibiotic called **vancomycin** is prescribed, but recently there have been cases of partial resistance to this drug, too.

Similarly, a strain of the bacterium *Clostridium difficile* is now resistant to all but two antibiotics. This bacterium is a natural component of our gut 'microflora'. It is only when *C. difficile*'s activities are no longer suppressed by the surrounding, hugely beneficial ('friendly') gut flora that it may multiply to life-threatening numbers, triggering toxic damage to the colon. Suppression of beneficial gut bacteria is a typical consequence of heavy doses of broad-spectrum antibiotics, administered to overcome infections of other superbugs.

C3.2 Defence against disease

Nature of science: Science as a shared endeavour

The development of new techniques can lead to new avenues of research. For example, the discovery of penicillin and its subsequent development as an antibiotic used to treat bacterial disease, has led to the development of a wide range of antibiotics that target specific bacteria. Modern techniques use computer programs (machine-learning algorithms) to survey more than a hundred million chemical compounds to pick out potential antibiotics that kill bacteria using different mechanisms to those of existing drugs. International efforts have led to the development of chemical libraries that can be surveyed in the quest to find new antibiotics. In the face of increasing antibiotic resistance, this is also a global effort. Researchers hope to use computer models not only to design new antibiotics but also to optimize existing molecules by adding features that would make a particular antibiotic target only certain bacteria, preventing it from killing beneficial bacteria in a patient's digestive tract.

ATL C3.2D

Use this site to explore further the range of antibiotics available, and the mechanisms through which bacteria can gain resistance: **https://cdn.rcsb.org/pdb101/learn/resources/superbugs/index.html**

Create a poster that summarizes the range of antibiotics available, and how changes to bacterial populations can lead to antibiotics becoming ineffective.

If bacteria become resistant to all antibiotics, what alternatives are there to stop the development of disease?

Inquiry 3: Concluding and evaluating

Concluding

Figure C3.2.17 shows the percentage of samples of *Staphylococcus aureus* that were methicillin resistant. Analyse the data shown in the graph.

Interpret the trend seen in the data between 1992 and 2001; justify your answer with valid reasons.

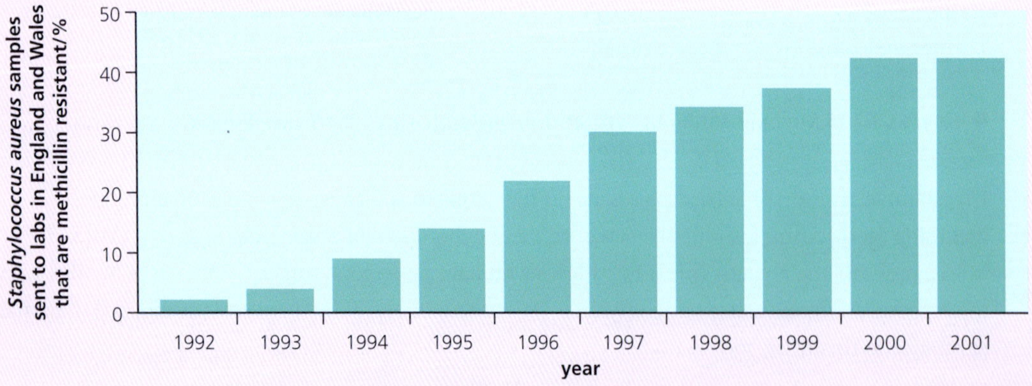

■ Figure C3.2.17 The increasing incidence of MRSA

15 Antibiotics are widely used in animal farming, partly to control animal disease, especially in intensive agriculture. **Suggest** what this means, why it happens, and what possible dangers could arise?

One result of bacterial resistance to antibiotics is that the pharmaceutical industry faces the challenge of producing **new antibiotics** faster than bacteria develop resistance to them. However, this is proving increasingly difficult – the number of new antibiotics being developed each year has fallen dramatically (Table C3.2.3).

■ **Table C3.2.3** The number of new antibiotics developed per 5-year period

	1980 – 1984	1985 – 1989	1990 – 1994	1995 – 1999	2000 – 2004	2005 – 2009	2010 – 2012
Number of new antibiotics approved (in USA)	19	11	11	11	4	3	1

16 Deduce the challenges in developing new antibiotics today.

Zoonoses

Pathogens are often specific in their choice of host. For example, humans are the only host for the pathogens that cause the diseases of syphilis, gonorrhoea, measles and poliomyelitis.

◆ **Zoonosis (plural, zoonoses)**: infectious disease that can transfer from other species to humans.

On the other hand, other pathogens can cross species barriers, infecting a range of hosts. **Zoonoses** are diseases of other animals that can be transmitted to humans. Generally, infectious diseases are most often 'shared' between species that are closely related and inhabit the same geographic area.

Existing viruses can spread from one host to another. The SARS virus (75% genetic similarity to SARS-CoV-2, COVID 19) originated in South East Asia in 2002, where people were infected with a coronavirus that originated in bats and spread to humans either directly or indirectly through civet cats. The incidence of zoonoses increases when humans exist in close contact with animals and when humans encounter animals in new geographical regions.

The prevalence of zoonoses as infectious diseases in humans, and their varied modes of infection, are illustrated in the following examples.

■ Tuberculosis

Tuberculosis (TB) is caused by a rod-shaped bacterium, *Mycobacterium tuberculosis* (see Figure C3.2.18).

In Illinois, USA, three elephants died due to *M. tuberculosis* from 1994 and 1996. In late 1996, a fourth living elephant tested positive for *M. tuberculosis*. Animal handlers on the farm were tested for TB and 11 out of 22 had positive tests, with one having active TB. A comparison of the DNA fingerprint from the four elephants and the handler with active TB showed that the TB was the same strain. This showed that transmission of *M. tuberculosis* between humans and elephants was possible.

Although most cases of human TB are caused by *M. tuberculosis*, zoonotic TB in people is predominantly caused by a closely related species, *Mycobacterium bovis* which causes bovine tuberculosis, (bTB). The disease can be transmitted directly by contact with infected domestic and wild animals or indirectly by ingestion of contaminated material. The usual route of infection within cattle herds is by inhalation of infected aerosol, which are expelled from the lungs by coughing. Humans can become infected by ingesting raw milk from infected cows, or through contact with infected tissues at abattoirs or butchers.

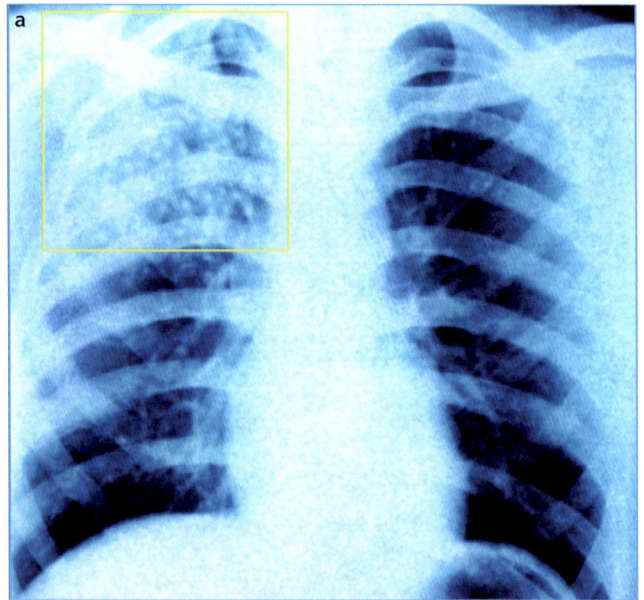

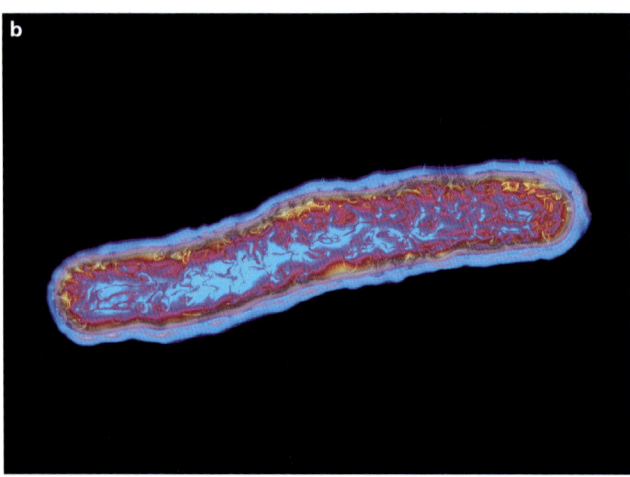

■ **Figure C3.2.18** a) A chest X-ray showing TB in the lungs – the white patches visible in the yellow square contain live *Mycobacterium tuberculosis* where lung structure and function are permanently destroyed; b) an electron micrograph of the bacterium that causes TB, *Mycobacterium tuberculosis* (×17 750)

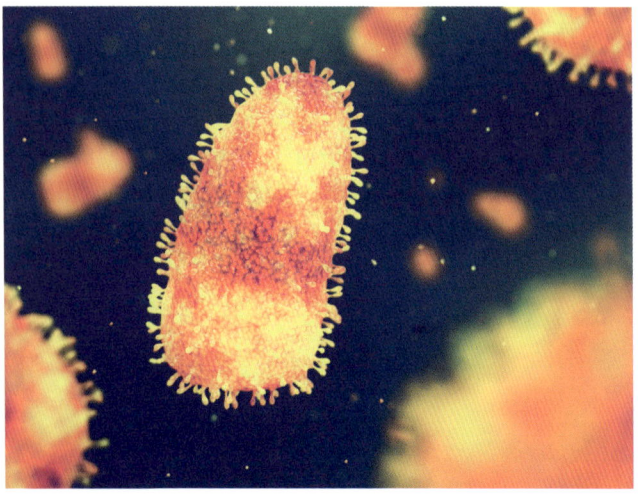

■ **Figure C3.2.19** The rabies virus

In Europe, badgers are affected by bTB and can be infected by the same strains found in local cattle. The incidence of bTB in cattle has increased, and the epidemic has spread geographically since the 1980s. bTB can spread from badgers to cattle, and from cattle to badgers, although nearly ten times as much infection is from badgers to cattle than the converse. There is even more transmission of infection within each species.

M. tuberculosis and *M. bovis* are included in the *Mycobacterium tuberculosis complex* (MTBC), which also includes other species and variants. The MTBC species are so closely related that they are now considered a single species, *M. tuberculosis*, with variants.

Zoonotic transmission of *M. bovis* from cattle to humans was recognized more than a century ago, but transmission of MTBC from humans to cattle is less often recognized. Within the last decade, however, there have been many published reports from around the world that describe human-to-cattle transmission of MTBC.

■ Rabies

Rabies is a viral disease that affects many carnivorous animals, including dogs, cats, foxes, skunks, jackals and wolves, affecting both wild and domestic animals. The rabies virus is transmitted through direct contact (such as through broken skin or mucous membranes in the eyes, nose or mouth) with saliva or nervous system tissue from an infected animal. Humans are usually infected with rabies from the bite of a rabid animal.

Rabies causes progressive and fatal inflammation of the brain and spinal cord. Once symptoms of the disease appear, rabies is fatal. However, the disease is avoidable due to vaccines, medicines and other intervention techniques that have long been available. Despite these precautions, rabies still kills approximately 59 000 humans each year in over 150 countries, according to the Centers for Disease Control and Prevention (USA), with 95% of cases occurring in Africa and Asia. The WHO states that the bite of an infected dog causes approximately 99% of rabies transmission to humans.

Japanese encephalitis

Japanese encephalitis (JE) is an infectious disease of the central nervous system caused by the Japanese encephalitis virus (JEV), a zoonotic virus. JEV is prevalent in much of Asia and the Western Pacific. It is transmitted to humans through the bite of infected mosquitoes (Figure C3.2.20), particularly *Culex tritaeniorhynchus*.

'Encephalitis' is inflammation of the tissues of the brain caused by an infection. The inflammation causes the brain to swell, which can lead to headache, stiff neck, sensitivity to light, mental confusion and seizures.

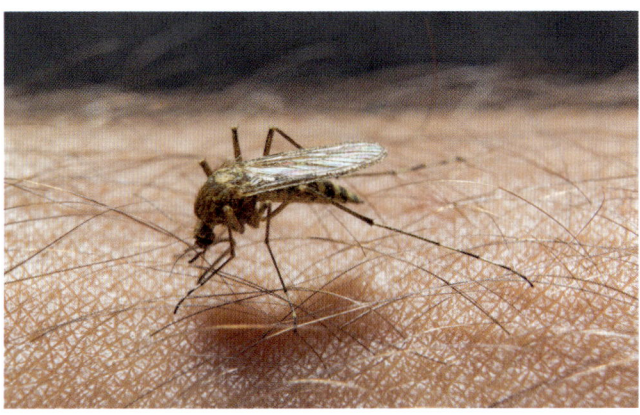

■ Figure C3.2.20 Japanese encephalitis is carried by mosquitoes

JEV causes disease in horses, donkeys and pigs. Pigs and wading birds are the main carriers of the JEV. Other animals can be infected but do not usually show signs of illness, including wild mammals, reptiles, amphibians and birds, as well as domesticated animals such as cattle, sheep, goats and pets (dogs and cats).

There is no cure for the disease. Treatment is focused on relieving severe clinical signs and supporting the patient to overcome the infection. Safe and effective vaccines are available to prevent JE. Despite its name, Japanese encephalitis is now relatively rare in Japan as a result of mass immunization programmes.

AIDS

◆ **Virulent**: the ability of a virus (or bacterium) to cause rapid and severe disease.

AIDS is also caused by two closely related zoonotic viruses: HIV-1 and HIV-2. HIV-1 is the strain with more **virulence**: it is more easily transmitted and is the cause of most HIV infections. Both HIVs are the result of cross-species transmission of simian immunodeficiency viruses (SIVs) naturally infecting African primates.

AIDS caused by HIV went unidentified (until 1983) and virtually unnoticed for decades before it began to spread around the world in the 1980s. This was due to technological and social factors, such as affordable international air travel, blood transfusions (with no screening for HIV), sexual promiscuity, and the abuse of intravenous drugs.

COVID-19

◆ **Novel virus**: a virus that has not previously been recorded.

COVID-19 is a disease that recently transferred from another species, with profound consequences for humans. It is one of three **novel** and infectious respiratory coronaviruses that have emerged: SARS (severe acute respiratory syndrome), MERS (Middle East respiratory syndrome) and COVID-19.

SARS-CoV-2 is the coronavirus responsible for COVID-19, named because it is a **CO**rona **VI**rus **D**isease, and was first reported in Wuhan, China, in 2019. It is related to the SARS virus and research suggests it came from a bat coronavirus, although the origin that caused the pandemic has not been identified.

> **ATL C3.2E**
>
> How does COVID-19 affect the nervous system? Revise what you know about the brain and nervous system (Chapters C2.2 and C3.1). Now watch this presentation:
> https://neuroscientificallychallenged.com/posts/2-minute-neuroscience-covid-19-brain
>
> How has your understanding of neural signalling and coordination in the body helped you to understand the complex biological concepts presented here?

C3.2 Defence against disease

The COVID-19 global pandemic changed most people's way of life, resulting in many countries closing their borders, restrictions on international travel, lockdowns in many countries, and schools closing. COVID-19 has had a massive impact on life and continues to have effects around the globe.

Nature of science: Science as a shared endeavour

Scientific collaboration in vaccine creation

Vaccines traditionally take many years to develop and bring into general use, typically 10–15 years. This is because of the complexity of vaccine development. Following development of a vaccine, which itself takes time and financial support, clinical testing must take place to see whether there are any serious side-effects or complications that may arise from receiving the vaccine.

During the COVID-19 pandemic, many scientists came together from around the world to create COVID-19 vaccines in a very short space of time. The first COVID-19 vaccine was given in the UK on 8 December 2020 – within the same year that the pandemic was declared. This was achieved through international collaboration: information about the disease was shared with different groups working on specific types of vaccine, governments provided extensive financial support and several different stages of clinical trials were carried out simultaneously rather than consecutively. The combination of the global collaboration of scientists and the development of mRNA vaccines has been compared to the breakthrough that happened when humans first landed on the Moon. The techniques and collaborations developed during the pandemic have shown that, when a global medical crisis emerges, science has the capability to adapt and come up with solutions that are successful and can be implemented quickly. The pandemic created a blueprint for future vaccine development.

ATL C3.2F

How is knowledge of biological macromolecules driving research and discovery related to SARS-CoV-2? Use this site to find out more:

https://pdb101.rcsb.org/learn/resources-to-fight-the-covid-19-pandemic

Write an article about how knowledge of the atomic structures of viral proteins and nucleic acids can lead to the discovery of new therapies and the development of vaccines for SARS-CoV-2.

Vaccines and immunization

♦ **Vaccination**: conferring immunity from a disease by injecting an antigen of attenuated (weakened) micro-organisms, an inactivated component or a nucleic acid (mRNA), so that the body acquires antibodies prior to potential infection.

Vaccines contain antigens, or nucleic acids (DNA or messenger RNA) with sequences that code for antigens. These antigens trigger the development of immunity to a specific pathogen without causing the disease. **Vaccination** is the deliberate administration of antigens that have been made harmless (after they are obtained from disease-causing organisms) or nucleic acids, in order to give future immunity. The practice of vaccination has made important contributions to public health, for example, in the case of measles (page 543), smallpox (page 543) and many other diseases.

Vaccines are administered either by injection, nasal spray or by mouth. Briefly and without causing infection, they cause the body's immune system to make antibodies against the disease, and then retain the appropriate memory cells. Active artificial immunity (see page 529) is established in this way. The profile of the body's response in terms of antibody production, if it is re-exposed to the antigen, is normally exactly the same as if the immunity was acquired by overcoming an earlier infection.

Vaccines are made from:
- dead or attenuated (weakened) bacteria
- purified polysaccharides from bacterial walls
- inactivated viruses
- toxoids (inactivated toxins)
- recombinant DNA produced by genetic engineering
- messenger RNA (mRNA).

Vaccines are often produced using the immune system responses of other animals, too.

In communities and countries where vaccines are widely available and have been taken up by 85–90% of the relevant population, vaccination has reduced some previously common and dangerous diseases to rare occurrences. As a result, in these places, the public sometimes becomes casual in its regard for the threat such diseases still represent.

> ### Common mistake
>
> It is incorrect to say that all vaccines are given by injection. Some vaccines can be given orally (such as the polio vaccine, which is easier to administer in this way and more economical) or by nasal spray, such as nasal spray flu vaccines which are given to children.

■ Types of immunity

The immune system provides protection to the body from the worst effects of many of the pathogens that may invade. **Immunity** may be acquired actively or passively, naturally or artificially:
- actively – *naturally*, as when our body responds to invasion by a pathogen, or *artificially*, after injection of killed or weakened antigens in a vaccine or nucleic acids that code for antigens, which stimulate the development of immunity to a specific pathogen – causing memory cells to be made.
- passively – *naturally*, as when maternal antibodies enter the foetus through the placenta, or *artificially*, such as when ready-made antibodies are injected into the body.

■ Viral vaccines

Viral vaccines were traditionally whole-virus vaccines that were of two types. Live attenuated vaccines use a weakened form of the virus, which can still replicate, but does not cause illness. Inactivated vaccines contain viruses whose genome has been destroyed so they cannot infect cells and replicate, but can still generate an immune response.

Vaccines cause the body's immune system to briefly make antibodies against the virus and memory cells. Active artificial immunity is established in this way. New vaccines may have to be developed if the virus is not stable and undergoes rapid genetic change.

Some vaccines used against COVID-19 are mRNA based and injected into the bloodstream. The mRNA is carried by lipid micro-vesicles (liposomes). The target antigen coded by the mRNA is either intact spike proteins, or fragments of spike protein. A spike protein is a **glycoprotein** that sticks out from the envelope of some viruses (such as a coronavirus) and facilitates entry of the virion into the host cell. These proteins can be identified as foreign by the immune system.

Once mRNA is inside the cell, it starts producing spike protein antigens that are then displayed on its surface. The antigens can be detected by the immune system, triggering a response involving killer T-cells, which destroy infected cells, as well as antibody-producing B-cells and helper T-cells that support antibody production. Memory cells are made, which retain the ability to produce the specific antibodies to respond quickly to an infection by the pathogen.

■ Antivirals

Most antiviral drugs resemble nucleosides and interfere with viral nucleic acid synthesis. Some antivirals have been developed against SARS-CoV-2 to target the viral replication process, which is less tolerant to mutations than the spike protein.

Azidothymidine (AZT) inhibits HIV reproduction by interfering with the synthesis of DNA by reverse transcriptase. Currently, multi-drug treatments, known as 'drug cocktails', have been found to be most effective against HIV. Such an approach commonly includes a combination of two nucleoside mimics and a protease inhibitor, which interferes with an enzyme required for assembly of virus particles. It is also less likely that any new mutant HIV strain will overcome all of the inhibitory effects.

■ The effectiveness of vaccination programmes – epidemiological data

Epidemiology is the study of the occurrence, distribution and control of diseases. A study of community disease patterns of this sort may be a reliable source of evidence. However, when the community-wide administration of a vaccine is closely followed by a decrease in the targeted disease, a causal relationship has not been proven. We cannot be certain it was the vaccine that prevented the infections because there are so many uncontrolled variables affecting a community and its health.

TB is a major worldwide public health problem of long standing. There is evidence that TB was present in some of the earliest human communities and it has persisted as a major threat to health where people live in crowded conditions. This disease is relatively rare in most economically developed countries, but globally almost 14 million people have active TB. In economically developing countries with a high incidence of this disease, it primarily affects young adults. In economically developed countries where TB, until very recently, ceased to be a major health threat, the rising incidence is largely among people with HIV/AIDS who are immunocompromised (page 529), and also among people arriving from other countries.

A vaccine against tuberculosis, the Bacillus Calmette-Guérin (BCG) vaccine, is prepared from a strain of attenuated (weakened) live bovine tuberculosis bacillus. It is most often used to prevent the spread of TB among children.

---

17 Distinguish between antibiotics and antivirals. Suggest a disease for which each could be used.

◆ **Epidemiology**: the study of the occurrence, distribution and control of infectious diseases.

Inquiry 3: Concluding and evaluating

Concluding

Look at the pattern of deaths from TB in a developed country, 1900–2000, shown in Figure C3.2.21a. Examine the graph carefully and then answer question 1.

In the prevention and control of measles, the first measles vaccines became available in 1963. It was then replaced by a superior version in 1968. Epidemiological evidence of the effect of the widespread use of these vaccines in one developed country is shown in Figure C3.2.21b. Examine these data carefully and then answer question 2.

1 Using the epidemiological evidence in Figure C3.2.21a, **suggest** the actions needed to combat the spread of TB in a community in order of priority, paying particular attention to the importance to be placed on a vaccination programme.
2 **Analyse** the extent to which the epidemiological data in Figure C3.2.21b support the idea of vaccination against measles as an effective procedure for maintaining community health.

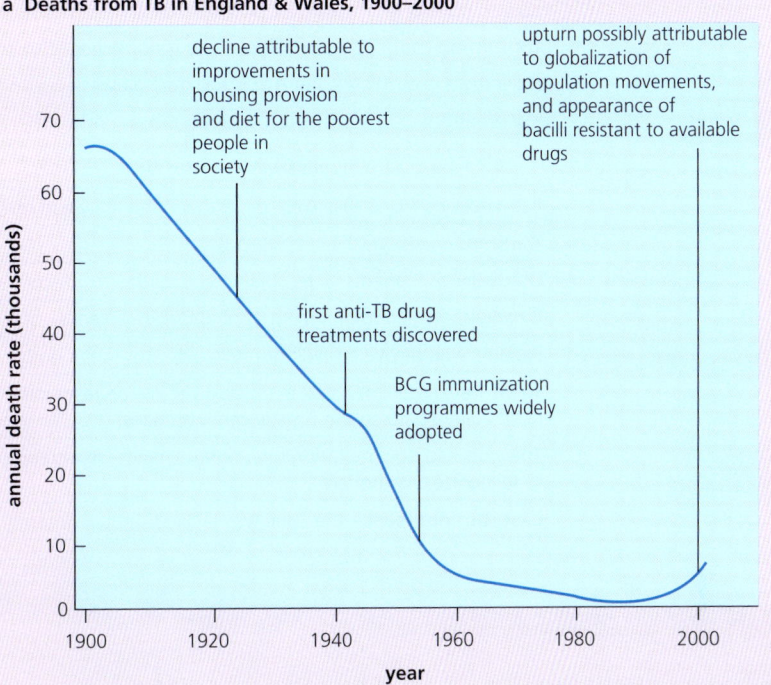

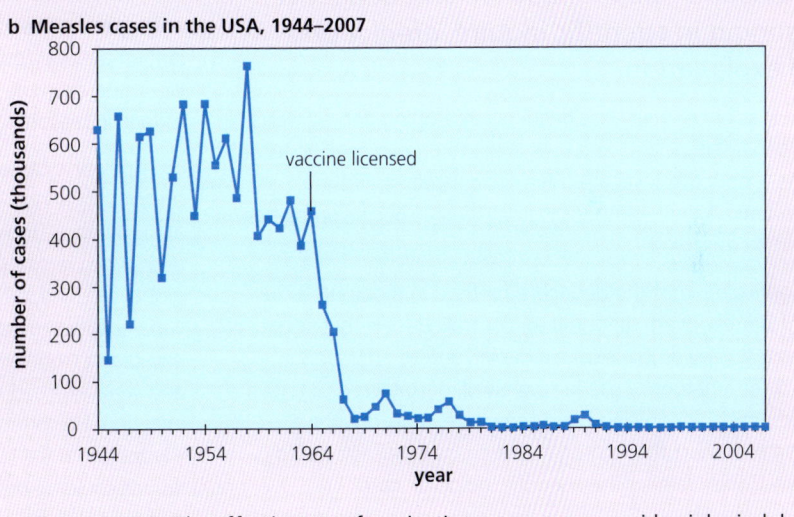

■ Figure C3.2.21 The effectiveness of vaccination programmes – epidemiological data

Link

Smallpox and the development of the first vaccines is discussed in Chapter A2.3, page 89.

 ■ Smallpox – an infectious disease eradicated by vaccination

Smallpox, a highly contagious disease, was once regularly found throughout the world. It killed or disfigured all those who contracted it. Smallpox is caused by a DNA variola virus.

Eventually, a suitable vaccine was identified by Edward Jenner (1749–1823), made from a harmless but related virus, vaccinia. This was used in a 'live' state and could be freeze-dried for transport and storage. Consequently, the vaccine was relatively easy to handle and stable for long periods in tropical climates.

How eradication came about

Smallpox has been eradicated (the last case occurred in Somalia in 1977). The development of a vaccine played an important part in this achievement, which was the outcome of a determined WHO programme that began in 1956. This involved careful surveillance of cases in isolated communities and within countries sometimes scarred by wars – altogether a remarkable achievement. The reasons why smallpox was eradicated when so many other diseases continue are listed in Table C3.2.4.

■ Table C3.2.4 Why smallpox was eradicated

Patients with the disease were easily identified; they had obvious clinical features.
Transmission was by direct contact only.
On diagnosis, patients were isolated and all their contacts traced. All were vaccinated.
It had a short period of infectivity of about 3–4 weeks.
Patients who recovered did not retain any virus in the body, so 'carriers' did not exist.
There were no animals that acted as a vectors or 'reservoirs' of the infection (which, otherwise, would have been able to pass on the virus to other humans).
The virus had slow mutation rate. Minor changes in antigens (**antigenic shift**) may cause memory cells to fail to recognize a pathogen.
Last, but not least – the issue was tackled through international cooperation by the WHO.

TOK

What is the role of inductive and deductive reasoning in scientific inquiry, prediction and explanation?

Medical science often uses deductive reasoning, which starts with general principles and uses them to predict specific future results. Deductive tests are often formulated as, 'if then' statements: if the hypothesis is true, then a specific prediction should be observed.

A good example of the deductive method in medicine is the use of double-blind, randomized, controlled trials that scientists use to test new drugs, e.g., remdesivir, a drug approved in the US for emergency use in severely ill, hospitalized COVID-19 patients. Randomized clinical trials are implied when companies assert that a drug has been 'scientifically tested', and such trials remain the scientific 'gold standard'.

Medical science also uses inductive reasoning, which starts with specific observations that lead to a hypothesis that is then generalized. The discovery of the smallpox vaccine is a good example of a medical discovery made through inductive reasoning.

Smallpox was caused by one of two virus variants, Variola major and Variola minor. The risk of death was 30%, with higher rates among babies; those who survived had extensive skin scarring and some were left blind. It was eradicated by 1980 with a global vaccination programme. The inventor of the smallpox vaccine, British medical doctor Edward Jenner, observed at the end of the eighteenth century that individuals who milked cows and were exposed to cowpox were immune to human smallpox infection. He correctly reasoned from this observation that a vaccine (from the Latin *vacca*, for 'cow') made from cowpox lesions could be protective.

Herd immunity and the prevention of epidemics

◆ **Herd immunity**: indirect protection from an infectious disease that happens when the majority of a population is immune either through vaccination or immunity developed through previous infection.

To control a contagious disease, it is not essential for the whole population to be immunized against it. The **herd immunity** theory proposes that the more people in a population who are immune to a disease, the lower the risk of someone who is susceptible becoming infected, as the transmission of the disease is greatly impeded. For herd immunity to succeed, members of a population are dependent on each other. To take an extreme example, if all but one person in a country is immune to a contagious disease, the chance of the one susceptible person being exposed to the pathogen within that country is almost zero. Some people cannot have vaccines so, with the majority of the population vaccinated, those who cannot have vaccines are still protected.

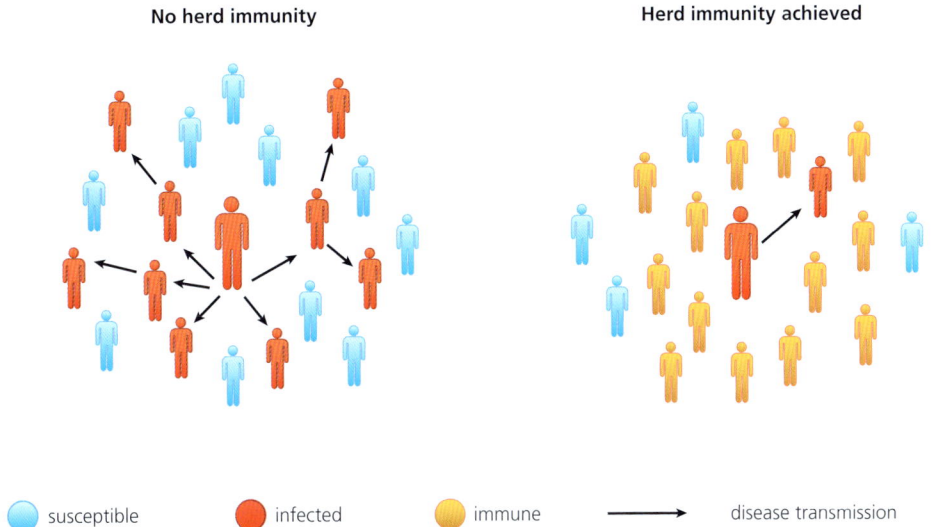

■ Figure C3.2.22 Herd immunity

What proportion of a population could be not vaccinated but still be protected by the majority who have been vaccinated? This depends on the nature of the disease-causing organism and its method of transmission. Table C3.2.5 provides information about some contagious diseases.

■ Table C3.2.5 The threshold percentage of the population needed to achieve herd immunity

Disease	Transmission route	Percentage threshold to achieve herd immunity
diphtheria	via saliva	85%
measles	airborne	83–95%
mumps	droplet	75–86%
rubella	droplet	83–85%

Based on the data in Table C3.2.5, some might think it is safe to avoid immunization. However, this is a mathematical model concerned with reducing the likelihood of infection. It is best to restrict failure to immunize or vaccinate to those who might be harmed by it, such as people with weakened or compromised immune systems.

Given the success of vaccination programmes in eradicating contagious diseases, the public has sometimes become casual about the threat that such diseases still pose. During your middle school science courses, you may have studied the effects of a newspaper article from 1998 that suggested a link between a vaccine to protect against measles, mumps and rubella (the combined MMR vaccine) and autism. Based on this article, so many parents declined the invitation to have their children vaccinated that the proportion of vaccinated children fell below the herd immunity threshold in some communities. The doctor who proposed this link to autism was subsequently discredited.

In 2016, the WHO declared the UK free of measles, but the UK lost this status three years later. There were 991 confirmed cases in England and Wales in 2018. The majority of these cases were among older teenagers and people in their early 20s who did not receive the MMR vaccination when they were younger.

Nature of science: Science as a shared endeavour

Scientists publish their research so that other scientists can evaluate it. This 'peer review' of scientists reviewing and commenting on the work of others ensures that research published in journals is accurate and valid. The media often report on the research while evaluation is still happening, and the public need to be aware of this. Vaccines are tested rigorously and the risks of side effects are minimal, but not nil. The distinction between pragmatic truths and certainty is poorly understood. In science, if something has utility then it is useful. The term 'pragmatism' refers to a way of dealing with problems or situations that focuses on practical approaches and solutions. Pragmatism can be seen as truth established through a majority consensus within a scientific community. The COVID-19 vaccine had been fully tested and, although side effects were known about, it was a pragmatic solution to the pandemic. Vaccine hesitancy was, in part, the result of some members of the public not appreciating the difference between pragmatism and certainty.

Science is not about absolute certainty. It aims to generate a theory that matches observations and to produce repeatable experiments, and is therefore pragmatic. However, theories can be overturned by data and replaced with new theories.

TOK

How might the context in which knowledge is presented influence whether it is accepted or rejected?

The COVID-19 pandemic and public attitudes towards vaccination programmes highlight interesting TOK dynamics between the scientific community and non-scientists. Many cases of vaccination reluctance depended on a poor understanding either of how vaccinations work or the statistical data showing how effective vaccinations actually are. Some people held political, social, cultural or religious beliefs about the role of government in encouraging vaccinations. An interesting TOK analysis might also explore the role that social media plays in the dissemination of flawed information about vaccinations.

◆ **Percentage difference**: the difference between two values divided by the average of the two values, shown as a percentage.

◆ **Percentage change**: the change in a variable over time; i.e. the percent change in a value compared to its initial value.

Evaluation of data related to the COVID-19 pandemic

Tool 3: Mathematics

Calculating and interpreting percentage change and percentage difference

Percentage change measures the change in a variable over time, whereas percentage difference calculates the difference between two averages. To calculate the **percentage change**:

1 Calculate the difference (increase or decrease) between the two numbers you are comparing.

   increase = new number − original number

2 Divide the increase by the original number and multiply the answer by 100.

$$\% \text{ change} = \frac{\text{change}}{\text{original number}} \times 100$$

If your answer is a negative number, then this is a percentage decrease.

To calculate percentage difference:

1 Calculate **the difference** between the two values by subtracting one value from another.

2 The average is the value halfway between the two numbers:

$$\text{average} = \frac{\text{first value} + \text{second value}}{2}$$

3 Express the difference as a percentage of the average:

$$\left(\frac{\text{difference}}{\text{average}}\right) \times 100$$

> **Top tip!**
>
> Make sure you can calculate both percentage difference and percentage change. Here we will practise using data relating to the COVID-19 pandemic.

■ Vaccine efficacy

Vaccine efficacy refers to the effectiveness of a vaccine in reducing the incidence of a disease. It is a measure of the percentage change in numbers of people with the disease following vaccination.

A study was carried out to investigate the efficacy of an mRNA vaccine, mRNA-1273, on the incidence of COVID-19 disease. The trial enrolled 30 420 volunteers who were randomly assigned to receive either the vaccine or a placebo (an injection that did not contain the vaccine). There were 15 210 participants in each group. Symptomatic COVID-19 illness was later confirmed in 185 participants in the placebo group and in 11 participants in the mRNA-1273 group.

The percentage change in the incidence of COVID-19 was:

$$(\text{change} \div \text{original number}) \times 100 = \frac{(185 - 11)}{185} \times 100$$

Vaccine efficacy was therefore 94.1%

■ Comparing the incidence of COVID-19 in different countries

The daily incidence of COVID-19 in one country was 40 and in another country it was 20. The percentage difference between the two countries is:

$$\text{difference between the two values} = 40 - 20 = 20$$

$$\text{average value} = \frac{(40 + 20)}{2} = 30$$

$$\text{percentage difference} = \left(\frac{\text{difference}}{\text{average}}\right) \times 100 = \left(\frac{20}{30}\right) \times 100 = 66.7\%$$

ATL C3.2G

Full data for the COVID-19 pandemic for every country can be found on the WHO website: **https://covid19.who.int**

Locate your own country on this website.
1. Select 'Table view'.
2. Click on your country: graphs showing changes in the incidence of cases will be shown.
3. Move the cursor over the graph: figures for the number of cases for a specific day will be shown. Slide from left to right to show changes in incidence.
4. Select a one-month period and take values for the number of cases at the start and end of this period.
5. Calculate the percentage change in coronavirus infections over this period.

Now select a second country. Find the value for coronavirus infections at the end of the same period you used in the first exercise. Calculate the percentage difference between the two values. How much higher or lower were infections in the second country? Why do you think this was?

Further data about the pandemic can be found on this site:

https://worldhealthorg.shinyapps.io/covid

LINKING QUESTIONS

1. How do animals protect themselves from threats?
2. How can false-positive and false-negative results be avoided in diagnostic tests?

C4.1 Populations and communities

Guiding questions

- How do interactions between organisms regulate sizes of component populations in a community?
- What interactions within a community make its populations interdependent?

SYLLABUS CONTENT

This chapter covers the following syllabus content:
- ▶ C4.1.1 Populations as interacting groups of organisms of the same species living in an area
- ▶ C4.1.2 Estimation of population size by random sampling
- ▶ C4.1.3 Random quadrat sampling to estimate population size for sessile organisms
- ▶ C4.1.4 Capture–mark–release–recapture and the Lincoln index to estimate population size for motile organisms
- ▶ C4.1.5 Carrying capacity and competition for limited resources
- ▶ C4.1.6 Negative feedback control of population size by density-dependent factors
- ▶ C4.1.7 Population growth curves
- ▶ C4.1.8 Modelling of the sigmoid population growth curve
- ▶ C4.1.9 A community as all of the interacting organisms in an ecosystem
- ▶ C4.1.10 Competition versus cooperation in intraspecific relationships
- ▶ C4.1.11 Herbivory, predation, interspecific competition, mutualism, parasitism and pathogenicity as categories of interspecific relationship within communities
- ▶ C4.1.12 Mutualism as an interspecific relationship that benefits both species
- ▶ C4.1.13 Resource competition between endemic and invasive species
- ▶ C4.1.14 Tests for interspecific competition
- ▶ C4.1.15 Use of the chi-squared test for association between two species
- ▶ C4.1.16 Predator–prey relationships as an example of density-dependent control of animal populations
- ▶ C4.1.17 Top-down and bottom-up control of populations in communities
- ▶ C4.1.18 Allelopathy and secretion of antibiotics

Note: There is no higher-level only content in C4.1.

Link
Reproductive isolation is covered in Chapter A4.1, page 149.

Populations

Members of a species may be reproductively isolated in separate **populations**. A population consists of all the individuals of the same species in a habitat at any one time. The members of a population have the chance to interbreed, assuming the species concerned reproduces sexually. Reproductive isolation is used to distinguish one population of a species from another.

◆ **Population**: interacting groups of organisms of the same species living in an area.

■ Figure C4.1.1 A population of zebra

Theme C: Interaction and interdependence – Ecosystems

Estimation of population size by random sampling

To study populations, an ecologist must be able to sample them. Specific methods are used to estimate population size in non-mobile species, for example plants and sessile animals such as barnacles (see page 327), and mobile animal species in the wild. Population growth is affected by abiotic and biotic factors, and the increase in a population follows a pattern that can be modelled.

In ecological investigations, the size and complexity of study sites means that not every organism can be examined, nor every part of the site looked at. **Samples** are taken that form a representation of the full site. Populations are estimated by taking samples from the study site. Randomness is needed when sampling to avoid bias (see below). **Random sampling** is needed to avoid areas being selected that do not reflect the whole study site – there would be a temptation, for example, to choose areas that have a high biodiversity but which are not necessarily representative of the whole area.

The distribution of two or more species in a habitat may be entirely random. Alternatively, factors such as specific abiotic conditions may bring about close association of some species – plant A may tend to grow close to plant B. For example, soils rich in calcium ions typically support distinctively different populations from those found on dry acidic soils. If we want to discover whether there is a particular association between two species in a habitat, we need reliable data on their distribution; such data are obtained by random sampling.

◆ **Sample**: a subset of a whole population or habitat used to estimate the values that might have been obtained if every individual or response was measured.

◆ **Random sampling**: a method of choosing a sample from a population without any bias.

◆ **Sampling error**: statistical errors that arise when a sample does not represent the whole population, i.e. it is the difference between the real values of the population and the values derived by using samples from the population.

Top tip!

Random sampling ensures that every individual in the community has an equal chance of being selected, ensuring a representative sample. It is appropriate if the study area is homogeneous (the same throughout), but not if sampling along a transect (page 349) where there is an environmental gradient and change in species composition.

Top tip!

Systematic errors lead to inaccurate measurements of the true value. The best way to check for systematic errors is to use different methods to perform the same measurement. Random uncertainty is always present and cannot be corrected; it is connected with the precision of repeated measurements.

Nature of science: Measurement

Random sampling, instead of measuring an entire population, inevitably results in **sampling errors**. In this case, the difference between the estimate of population size and the true size of the whole population is the sampling error. If the sample does not adequately represent the entire population, any analysis of the sample might not apply to the entire population.

Random errors in measurement, due to unknown or unpredictable differences, lead to imprecision and uncertainty, whereas systematic errors lead to inaccuracy. Repeating and then averaging measurements can reduce random error. Systematic errors result from faults or flaws in the investigation design or procedure that shift all measurements in a systematic way so that, in the course of repeated samples, the measurement value is constantly displaced in the same way. Systematic errors can be reduced or eliminated with careful experimental design and techniques. Scientists can also share their experimental data and design with other scientists, giving others the chance to evaluate the design in a process called peer review, which further guards against systematic errors.

◆ **Stratified sampling**: used in areas that contain two or more different habitat types; the technique takes into account the proportional area of different habitat types and samples each one accordingly.

◆ **Systematic sampling**: used where the study area includes an environmental gradient. A transect is used to sample systematically along the environmental gradient.

Inquiry 1: Exploring and designing

Controlling variables

It is important to record a representative sample from the area being studied when sampling populations in the wild. **Random sampling** is used if the same habitat is found throughout the area. **Stratified sampling** is used in two areas of different habitat quality.

When sampling along an environmental gradient (see Chapter B4.1, page 349), the positions of transects are located at random, but sampling along the transect is done in a systematic (non-random) way to ensure that the whole gradient is sampled, called **systematic sampling**. To do this, sample points are taken at fixed distances along the transect (for example, every 5 m).

1 Design an experiment to sample a local habitat using the quadrat method. You can either select a sessile species to quantify (the organism should be sufficiently abundant to measure in repeated samples) or the whole plant community.

Samples must be located at random so that accurate representative data can be acquired. A random number generator, found on most modern calculators, helps to ensure population sampling is free from bias. Random number generators can also be found online (for example www.random.org) and as smartphone apps (sometimes advertised as 'integer generators'). Alternatively, putting numbers from, say, 1 to 100 into a bag and pulling numbers out one at a time to generate sample locations will do the same job!

Estimating population size for sessile organisms

■ Random quadrat sampling

Quadrats (Figure C4.1.2) are commonly used to study populations and communities. They are used for estimating the abundance of plants and sessile animals.

A quadrat is a square frame which outlines a known area for the purpose of sampling. The choice of size of quadrat varies depending on the size of the individuals of the population being analysed. For example, a 10 cm^2 quadrat is ideal for assessing epiphytic *Pleurococcus*, a single-celled alga commonly found growing on damp walls and tree trunks. Alternatively, a 1 m^2 quadrat is far more useful for analysing the size of two herbaceous plant populations observed in grassland, or of the earthworms and slugs that can be extracted from between the plants or from the soil below. Quadrats are placed according to random numbers, after the area has been divided into a grid of numbered sampling squares (Figure C4.1.3). The presence or absence in each quadrat of the species under investigation is then recorded.

■ **Figure C4.1.2** A quadrat being used to estimate the species richness of plants in a water meadow at Slapton Ley, Devon

1 **Outline** how quadrats can be used to sample populations of sessile organisms.

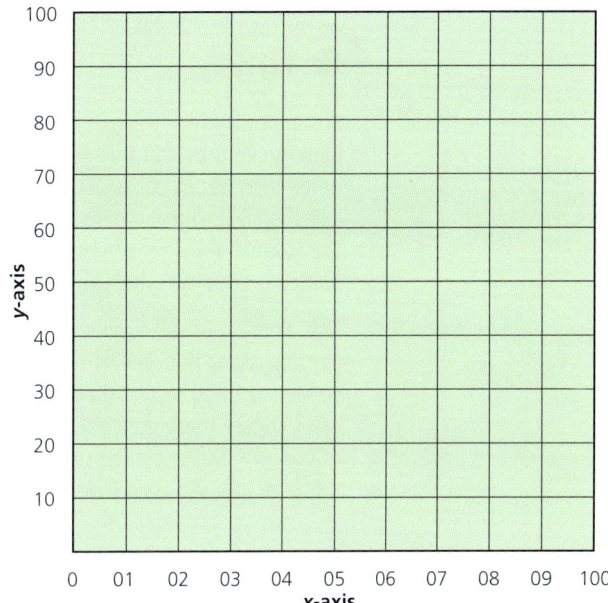

1 A map of the habitat (e.g. meadowland) is marked out with gridlines along two edges of the area to be analysed.

2 Coordinates for placing quadrats are obtained as sequences of random numbers, using computer software, a calculator or published tables.

3 Within each quadrat, the individual species are identified, and then the density, frequency, cover or abundance of each species is estimated.

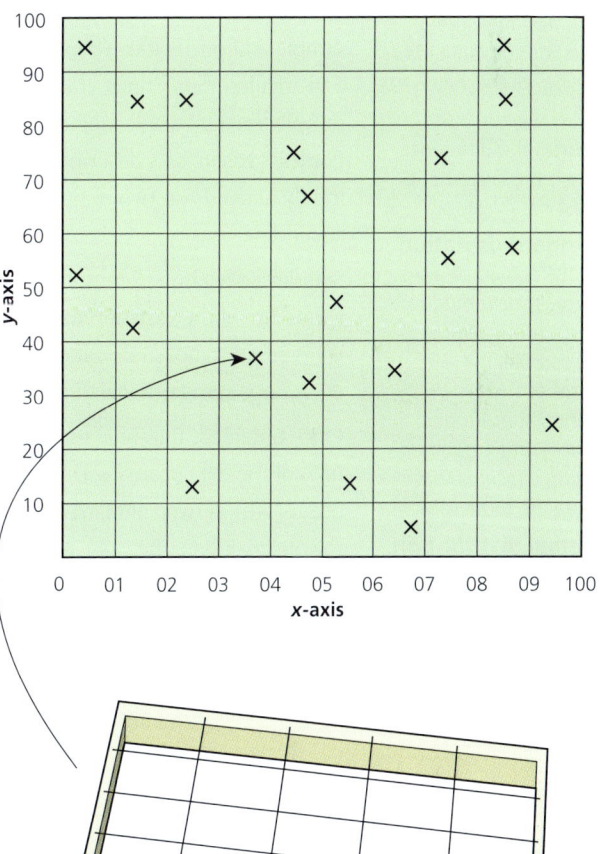

4 Density, frequency, cover or abundance estimates are then quantified by measuring the total area of the habitat (the area occupied by the population) in square metres. The mean density, frequency, cover or abundance can be calculated using the equation:

$$\text{population size} = \frac{\text{mean density (etc.) per quadrat} \times \text{total area}}{\text{area of each quadrat}}$$

■ **Figure C4.1.3** Random locating of quadrats

C4.1 Populations and communities

TOK

It is impossible for a scientist to see every event or every object in the world, yet it is their job to identify features and truths of the world that apply to the world as a whole. This raises a challenge when it comes to constructing general knowledge – how do we select the observations that we can then use to make general claims about the world? Therefore effective sampling is so important; the scientist must make choices and decisions regarding the sample to ensure that the knowledge gained is accurate when applied to the wider world.

This element of selection and analysis introduces the possibility of bias, where the scientist, either consciously or not, makes decisions that shape the data in ways that may make it less accurate. Following the scientific method and thinking carefully about how best to create samples, however, can mitigate these concerns. A full TOK analysis of bias in the sciences would both identify the root of concern, but also then explore the conventions that the scientific community follows to mitigate these factors. An even fuller TOK analysis might evaluate the success of these conventions.

Calculating standard deviation

◆ **Variation**: a quantitative measure of the distribution (spread or clustering) of the values in a data set.

◆ **Range**: the difference between the largest and smallest data values.

◆ **Standard deviation (SD)**: the spread of a set of normally distributed data from the mean of the sample; it is a measure of the variability of a population from a sample. A small standard deviation indicates that the data are more reliable.

◆ **Normal distribution**: a data-set distribution that is symmetrical about the mean, forming a bell-shaped curve.

Variation is a quantitative measure of the distribution (spread or clustering) of the values in a data set. **Range** and **standard deviation** are two statistics used when summarizing variability within a sample, but one is much more reliable than the other. The range is the difference between the largest and smallest data values. It is measured in the same units as the data. If there are outliers in the sample, the range will suggest a high degree of spread, though most data may be highly clustered. For this reason, as a descriptive statistic describing variability, range is an unreliable measure. Since it only depends on two of the measurements, range is most useful in representing the dispersion of small data sets.

Standard deviation from the mean measures how spread out data are from the central tendency. It is a measure of the variation from the mean of a set of **normally distributed** values.

- A small standard deviation indicates that the data are clustered closely around the mean value.
- A large standard deviation indicates a wider spread around the mean.

Once obtained, the value may be applied to the normal distribution curve (Figure C4.1.4). Note that 68% of the data occurs within one standard deviation of the mean, and more than 95% of the data occurs within two standard deviations of the mean. So, a small standard deviation indicates that the observations (the values) differ very little from the mean. Standard deviation is also a measure of reliability: a small standard deviation indicates reliability.

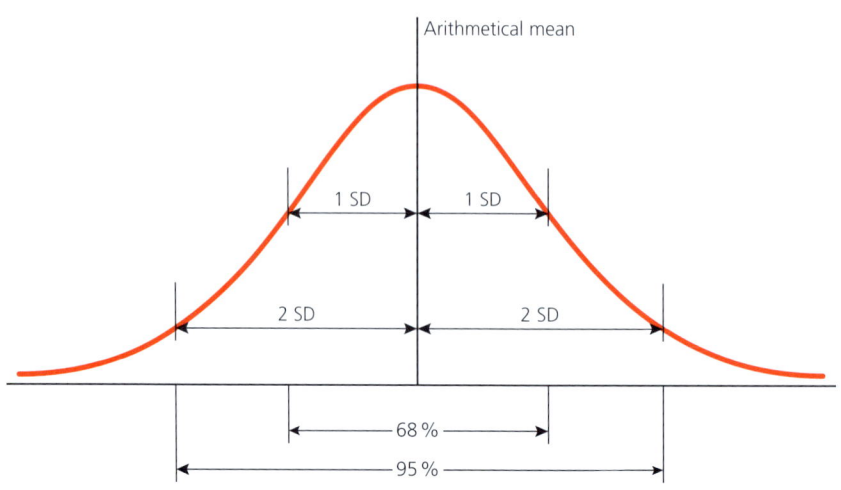

■ **Figure C4.1.4** The normal distribution and its standard deviation (SD)

The standard deviation can be used to help to decide whether the difference between two related means is significant or not. If the standard deviations are much larger than the difference between the means, this difference in the means is highly unlikely to be significant. On the other hand, when the standard deviations are much smaller than the difference between the means, then this difference is almost certainly significant.

> **Top tip!**
>
> Standard deviation is a calculation based on the mean value and the variation around it. It is more reliable than range. Standard deviation gives a value that represents the average distance of each point of data from the mean. You do not need to memorize the formula for standard deviation.

Link

Calculating the mean (and mode and median) is covered in Chapter D3.2, page 745.

> **Top tip!**
>
> The mean is calculated by adding together all the values and dividing by the total number of values. The purpose of averaging data (provided it follows a normal distribution) is to reduce the effect of random errors.

Tool 3: Mathematics

Applying measures of dispersion: standard deviation

The standard deviation (SD) is calculated in five steps.
1. First, calculate the mean, $\bar{x}$
2. Calculate the deviation of each value from the mean: $x - \bar{x}$
3. Square each deviation: $(x - \bar{x})^2$
4. Add the squared deviations: $\sum(x - \bar{x})^2$
5. Finally, divide by the number of values (n).

You could use a calculator or practise applying technology to process data by using an Excel spreadsheet to calculate the standard deviation (Tool 2: Technology, page 561).

The formula for SD given here is for when the data set represents an entire population. This is known as the 'population standard deviation'. If the SD is calculated from a sample, the formula needs to be modified by dividing by $n - 1$ rather than n. This is called the 'sample standard deviation'. This is because when the sample standard deviation is calculated, the true variability in the population tends to be underestimated, and so the estimate of the true population standard deviation is biased: to correct this bias, the numerator is divided by $n - 1$. This has been shown to make the sample standard deviation an unbiased estimate of the population standard deviation.

 Common mistake

A common misconception is that standard deviation decreases with increasing sample size. Standard deviation can either increase or decrease as sample size increases; it depends on the measurements in the sample. If there is a lot of variation in a population, the standard deviation will be large.

> **Top tip!**
>
> When the value of the standard deviation is low, most scores in the sample will be clustered and close to the mean. When the value of the standard deviation is high, the sample scores are more widely spread from the mean.

C4.1 Populations and communities

2. An ecologist investigated the reproductive capacity of two species of a common grassland flower, *Ranunculus acris* and *Ranunculus repens*. Using comparable sized plants growing under similar conditions in the same soil, the numbers of fruits formed in 100 flowers of each species were counted and recorded. The results are given in Table C4.1.1.

■ **Table C4.1.1** Data on fruit production in two species of *Ranunculus*

Number of fruits	Frequency	
	R. acris	*R. repens*
15	0	0
16	0	1
17	0	1
18	1	1
19	2	2
20	1	4
21	4	4
22	8	8
23	7	7
24	9	9
25	10	10
26	16	16
27	9	9
28	10	10
29	4	4
30	5	5
31	3	3
32	1	1
33	1	1
34	2	2
35	1	1
36	1	1
37	0	0
38	0	0
39	2	0

a **Calculate** the mean and standard deviation for both species.
b **Deduce** whether there is a difference in fruit production between the two species.

TOK

Mathematics is inherently formal and abstract – its scope is related to universal truths unrelated to whatever happens to be the case in the observable world. Science, on the other hand, aims to develop formal and abstract truths about the observable world, but must do so building on certain observations. Scientists will translate observations of the world into mathematical entities (for example by observing species in a sample and translating that into a number). Using these quantitative data, scientists can conduct mathematical analyses of the types considered here to identify numerical trends and further data. These data then are applied back to the natural world. A good example of this process is shown in Figure C4.1.3 (page 551). In step 3, the observations are quantified and, after the mathematical manipulations in step 4, the numbers are applied to the actual world in the form of a claim about the overall population size of the species.

Try to identify where the choices of the investigating scientist have been made and why they made them, and evaluate their effectiveness. This overlap of the Areas of Knowledge of Natural Sciences and Mathematics helps make the sciences even more reliable.

Estimating population size for motile organisms

■ Capture–mark–release–recapture method

It is not possible to collect every individual of the population. A sample is taken which is representative of the population, from which an estimate of population size can be made. To do this, animals are captured, marked in some way, released, and then resampled. A calculation is made that uses the relationship between two ratios: the number of marked animals compared to the size of the resampled population, and the ratio of sampled population (captured and marked) compared to the total population. This is called the 'capture–mark–release–recapture method'. Marking varies according to the type of organism, for example, wing cases of insects can be marked with pen, snails with enamel paint, and fur clippings used for mammals. The markings must be difficult to see as high visibility increases the predation risk to the organism. The number of individuals of a species are recorded at each stage. The steps of the capture–mark–release–recapture technique and how to estimate the population size are outlined in Figure C4.1.5.

■ Calculating the Lincoln index

The Lincoln index is a statistical measure used to estimate the population size for motile organisms. You need to be able to use it to estimate population size and understand the assumptions made when using this method.

> ### Tool 3: Mathematics
>
> #### Applying the Lincoln index
>
> The total population size is estimated using the following equation:
>
> $$\text{Population size} = \frac{(M \times N)}{R}$$
>
> M = number of animals captured (marked and released)
>
> N = number of animals recaptured
>
> R = number of marked animals recaptured
>
> Prior to experimenting on natural populations, the Lincoln index can be learned and understood by application to non-living models. This can be done using a paper activity.
>
> Cut out 100 identical small paper squares and place on a tray. Mark a known number, e.g. 20. Put the marked squares back in the tray and mix well. (This is the 'habitat'.) Pull out a random number of squares from the container and count the sample size and the number in the sample which are marked. Calculate an estimated population using the Lincoln index formula.
>
> Repeat the experiment at least five times to obtain a range of estimates. Calculate a mean estimate and the range of the values obtained.

ATL C4.1A

Before reading on, what are the assumptions made when using the Lincoln index? Find out about these using this site: **www.countrysideinfo.co.uk/lincoln.htm**. Make a list of the assumptions and comment on how these would affect the way you collect data for the Lincoln index.

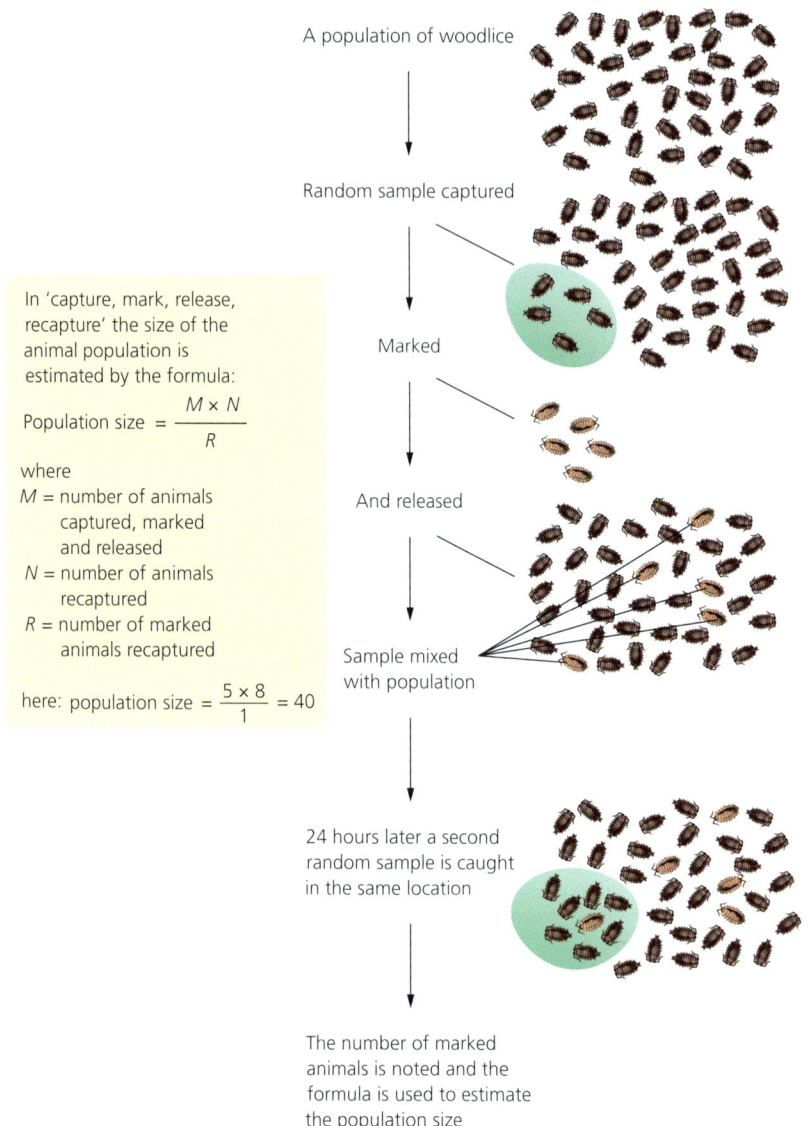

In 'capture, mark, release, recapture' the size of the animal population is estimated by the formula:

$$\text{Population size} = \frac{M \times N}{R}$$

where
M = number of animals captured, marked and released
N = number of animals recaptured
R = number of marked animals recaptured

here: population size = $\frac{5 \times 8}{1}$ = 40

■ **Figure C4.1.5** Estimating animal populations using capture, mark, release and recapture

Inquiry 3: Evaluating and concluding

Evaluating

Evaluate the Lincoln index by discussing the advantages and pitfalls of the method.

Lincoln index assumptions

Assumptions made when using the Lincoln index on wild animal populations are that:
- mixing is complete, so that the marked individuals have spread evenly throughout the population between capture and recapture
- marks are not removed and do not disappear between capture and recapture
- marks are not harmful to the animal nor do they increase predation by making marked individuals more visible
- it is equally easy to catch each individual in the population
- there is no immigration or emigration (i.e. it is a closed population rather than one open to other areas), or births or deaths in the population between the times of sampling.

3 Discuss the use of capture–mark–release–recapture methods to estimate the population size of motile organisms.

> ### Inquiry 2: Collecting and processing data
>
> #### Collecting data
> Carry out a capture–mark–release–recapture experiment to investigate and estimate the population size for a motile organism using the Lincoln index. You will need to:
> - collect quantitative data (the number of species counted)
> - identify and record any relevant qualitative observations
> - identify and address issues that arise during your data collection.
>
> Repeat the investigation after a few weeks. Are your estimates similar? If not, what explanations can you think of for any discrepancies?

Carrying capacity and competition for limited resources

Link
Limiting factors are covered in Chapter B4.1, page 344.

In most ecosystems and for many species, abiotic and biotic factors limit the size of the population. Exponential growth cannot continue forever. Eventually resources run low, **competition** occurs and other limiting factors take effect. **Limiting factors** restrict the growth of a population or prevent it from increasing further.

> ### ● Top tip!
> Limiting factors in plants include light, nutrients in the soil, water, carbon dioxide and temperature.
> Limiting factors in animals include space, food, mates, nesting sites and water.

♦ **Carrying capacity**: the maximum number of individuals of a species that can be supported by a given environment.

For example, if rabbits were introduced into a new meadow, after initial exponential growth, the vegetation would eventually be used up faster than it could grow, because of the large numbers of rabbits. Further increases in population would stop: in this situation the food has become a limiting factor in the growth of the rabbit population. Eventually, the rabbits would reach the **carrying capacity** of their environment. This is the maximum number of a species, or the 'load', which can be sustainably supported by a given environment.

Negative feedback control of population size by density-dependent factors

♦ **Negative feedback**: feedback that tends to counteract any deviation from equilibrium and promotes stability.

Negative feedback refers to the return of a system to its original starting state. In terms of a population, this represents the return of a population to its **carrying capacity**. The interaction between predators and their prey is an example of negative feedback (Figure C4.1.6).

Negative feedback control of populations is determined by **density-dependent factors**. Density-dependent factors are limiting factors that are related to population density.

♦ **Density-dependent**: factors that lower the birth rate or raise the death rate as a population grows.

They are biotic factors (e.g. competition for resources) that limit population growth:
- **internal density-dependent factors**: might include fertility or size of breeding territory
- **external density-dependent factors**: might include the increased risk of predation and the transfer of pathogens or pests in dense populations.

C4.1 Populations and communities

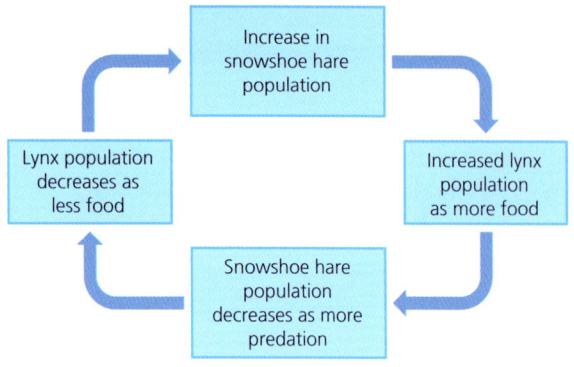

■ Figure C4.1.6 An example of negative feedback: the predator–prey relationship between snowshoe hare and lynx in the boreal forests of North America

Density-independent factors will affect the populations of all species in the same ecosystem, regardless of population size (for example, weather, climate and natural disasters), whereas density-dependent factors will have varying effects on the populations of different species. Because density-independent factors are not reliant on the numbers of individuals in a population they do not affect carrying capacity, whereas density-dependent factors, that do limit population size, define the upper limit to population growth (the carrying capacity). The number of individuals in a population may fluctuate due to density-independent factors, but density-dependent factors tend to push the population back towards the carrying capacity.

Link

The density-dependent relationship between the snowshoe hare and lynx in Figure C4.1.6 is an example of population control by negative feedback.

4 **Define** the term *carrying capacity*.

> **Concept: Interaction**
>
> Density-dependent factors interact with organisms to cause population numbers to fluctuate, ultimately pushing populations back to carrying capacity.

5 **List** factors that are:
   a affected by population density
   b unrelated to population density.

Population growth curves

■ Exponential growth

Imagine a situation in which a small number of rabbits are allowed to enter a meadow from which other rabbits have been excluded. The rabbits initially get used to the new territory and establish burrows. There are few reproducing individuals, so initially population growth is small. Once established, the rabbits benefit from unrestricted access to rich food supplies and an absence of limiting factors – they begin to reproduce. Once numbers start to build up, there is a rapid increase in the population. Such rapid increase is called exponential growth.

Exponential growth is an increasing or accelerating rate of growth. The exponential growth pattern occurs in an ideal and unlimited environment.

Exponential growth occurs when:
- limiting factors are not restricting the growth of the population
- there are plentiful resources, such as light, space, food, and a lack of competition with other species
- there are favourable abiotic components, such as temperature and rainfall, and a lack of predators or disease.

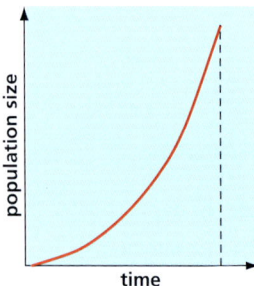

■ Figure C4.1.7 An exponential growth curve

Growth is initially slow but becomes increasingly rapid and does not slow down as the population increases (see Figure C4.1.7).

Sigmoid growth curve

The carrying capacity (see above) represents the population size at which **environmental resistance** limits further population growth. Exponential growth eventually succumbs to environmental resistance, which can cause a decrease in natality, an increase in mortality, or both. When a graph of population growth for such species is plotted against time, an S-shaped population curve is produced. This is also known as a **sigmoid growth curve**. An S-shaped population curve shows an initial rapid growth (exponential growth) and then slows down as the carrying capacity is reached.

Figure C4.1.8 shows an **S-population curve**. The curve shows the establishment of a population following introduction into a new environment. There are three stages.

1. **Exponential growth** stage, where limiting factors are not restricting the growth of a population. With low or reduced limiting factors, the population expands **exponentially** into the habitat. Both the number of individuals and the rate of growth increases rapidly.

2. **Transition phase**, where the increase in number begins to slow, as does the rate of growth. The slowdown occurs because limiting factors begin to affect the population and restrict its growth. In a larger population there will be increased competition between the individuals of that population for the same limiting factors, such as resources. An increase in predators, attracted by the large population, and an increase in the rate of disease and mortality due to increased numbers of individuals living in a small area, also cause a slowdown in growth.

3. **Plateau phase**, where limiting factors restrict the population to its carrying capacity (see above). Changes in limiting factors, predation, disease and abiotic factors cause populations to fluctuate (increase and decrease) around the carrying capacity (K).

Sigmoid growth starts with the exponential growth phase, where the population is not controlled by limiting factors. Population growth slows as a population reaches the carrying capacity of the environment.

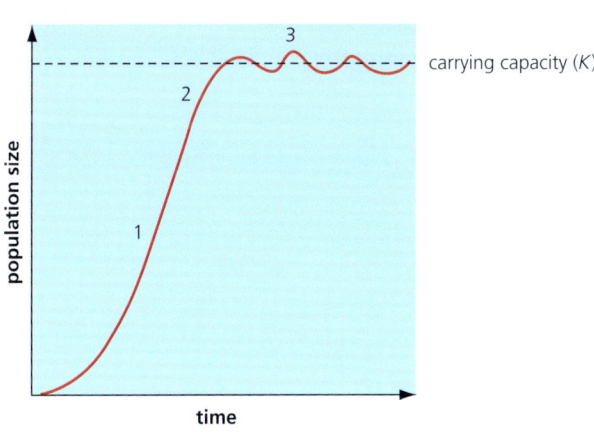

■ **Figure C4.1.8** An S-shaped or sigmoid growth curve

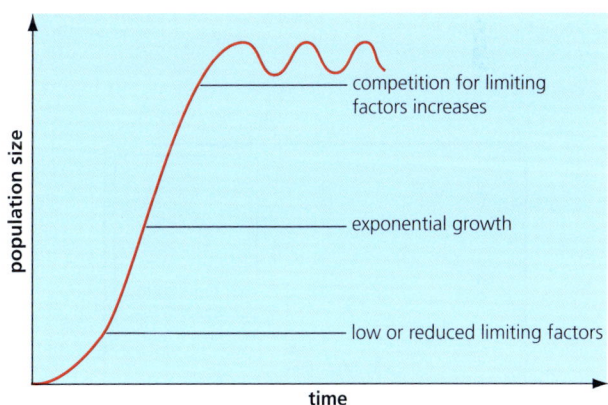

■ **Figure C4.1.9** Changing conditions along an S-shaped population growth curve

The phases of the S-shaped curve can be summarized as equations using natality (N), immigration (I), emigration (E) and mortality (M), where:
- **natality** – the rate at which individuals are born (the birth rate)
- **mortality** – the rate at which individuals die (the death rate)
- **immigration** – movement of new individuals into a population from outside a given area
- **emigration** – the departure of individuals from a population in a given area.

■ **Table C4.1.2** The phases shown in the sigmoid curve can be explained by relative rates of natality, mortality, immigration and emigration

Phase	Equation	Relative effects	What is happening?
exponential	$I + N > E + M$	initially, natality exceeds mortality; as population rapidly increases, mortality rates begin to increase	no limiting factors; abundant resources; low competition; high reproduction
transition	$I + N > E + M$	natality rate approaches mortality rate	resources are more scarce; competition increases
plateau	$I + N = E + M$	fluctuation around a set point; where the curve slopes downwards, $M > N$, where it slopes upwards, $N > M$	population stabilizes around the carrying capacity

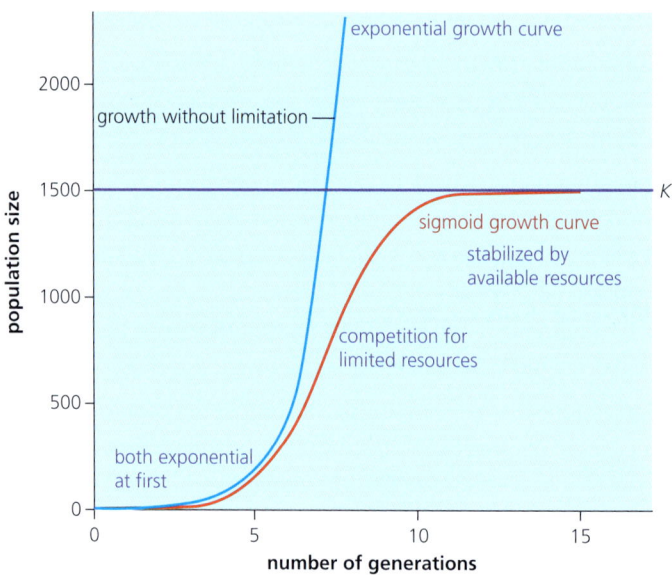

■ **Figure C4.1.10** Comparing exponential and sigmoid growth curves

■ Sigmoid population growth – case study

A study carried out on Lady Musgrave Island, off the east coast of Australia, investigated the growth of corals following disturbance from tropical cyclones. The location of the island, and the data recorded, are shown in Figure C4.1.11.

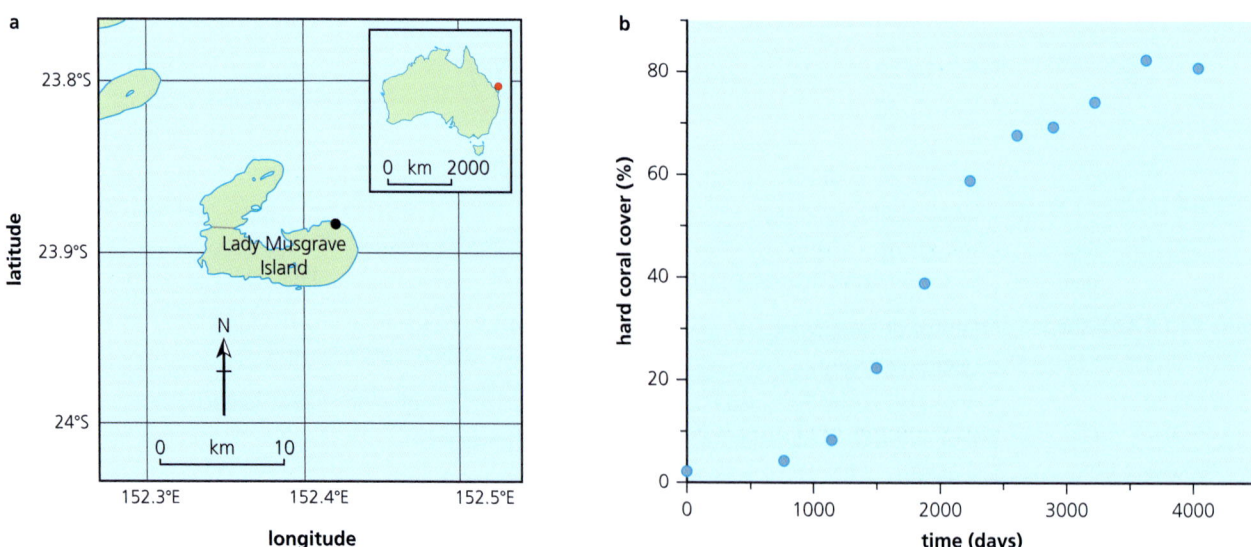

■ **Figure C4.1.11** Field data showing the percentage area covered by hard corals as a function of time after external disturbance on Lady Musgrave Island

The data show a typical sigmoid growth response after some disturbance (e.g. a tropical cyclone): exponential growth for low coral cover is followed by a gradual reduction in net growth rate as the coral cover approaches its carrying capacity.

Nature of science: Models

The population growth curve represents an idealized graphical model. It is unlikely that the population growth of an actual population will show such smooth, idealized curves. There will be fluctuations due to density-independent and density-dependent factors. Models are often simplifications of complex systems, and indicate overall patterns rather than the specific detail of individual examples. Models allow predictions of patterns that should occur given certain parameters.

Tool 2: Technology

Using spreadsheets to manipulate data

A sigmoid function is a mathematical function that has an S-shaped curve when plotted. The logistic sigmoid function, which is the most common example of a sigmoid function, is calculated as:

$$F(x) = \frac{1}{(1 + e^{-x})}$$

To calculate the value of a sigmoid function for a given x value in Excel, use the following formula: =1/(1+EXP(-A1))

Logarithmic graphs

Sometimes during a biology investigation, the data collected are not easy to plot on a graph because of the very wide range of values measured. An example arises when measuring the population growth rate of yeast using a counting chamber known as a haemocytometer. Rather than plotting the averaged raw data, it is preferable to calculate the base ten logarithm of the measurements.

Some measurements, such as pH and absorbances, have no units. The pH scale, for example, is logarithmic. A logarithmic scale compresses the range of values and gives more space to smaller values while compressing the space available for the larger values. Each cycle on the scale increases by a power of 10: for example, in the first cycle values would be 1, 2, 3, 4, etc., whereas in the second cycle they would be 10, 20, 30, 40 and so on, and in the third cycle 100, 200, 300, 400 and so on.

Tool 3: Mathematics

Constructing and interpreting logarithmic graphs

You need to be able to test the growth of a population against the model of exponential growth using a graph with a logarithmic scale for size of population on the vertical y-axis and a non-logarithmic scale for time on the horizontal x-axis.

A logarithmic scale can be calculated using a calculator or Excel: the formula is '=LOG10(cell reference)'. The logarithm of the number only increases relatively slowly and a straight line plot is generated. Alternatively, log-linear graph paper (Figure C4.1.12) can be used.

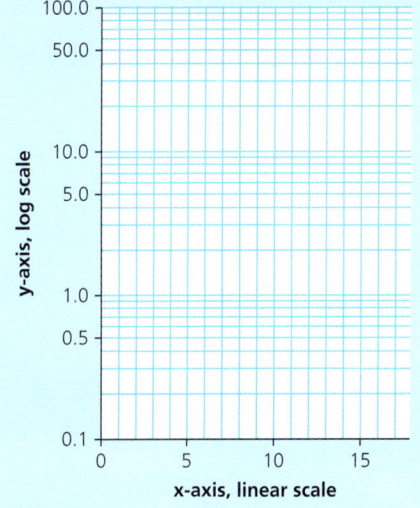

■ **Figure C4.1.12** A logarithmic scale on the y-axis with three cycles: 0.1 to 1.0, 1.0 to 10.0 and 10.0 to 100.0. The x-axis has a linear scale. Numbers can be plotted directly on to graph paper without calculating their $\log_{10}$ values

Top tip!

Quantities that represent ratios of two values, such as absorbance, do not have units. Both pH and absorbance are unit-less since they are a logarithmic function, and logarithms are always pure numbers that have no units.

Modelling of the sigmoid population growth curve

You need to be able to collect data regarding population growth. Yeast and duckweed are recommended, but there are other organisms that could be used that reproduce under experimental conditions.

Modelling the growth curve using duckweed

Pond surfaces are often covered by free-floating aquatic plants. One such group is the duckweeds, in the genus *Lemna*. They are known as duckweed because they provide a staple diet for ducks and other aquatic birds. These rapidly growing plants are used as a model system for studies in population ecology. This simple organism follows the classic growth curve and can, therefore, be used effectively to model the changes in population size.

Inquiry 2: Collecting and processing data

Collecting data

An experiment on population growth in *Lemna* can be carried out in the lab:

1. A small number (e.g. 15) of duckweed thalli is put in a cup or beaker containing pond water (200–300 cm^3).
2. Replicates (at least five) of the experiment are set up.
3. The number of thalli is counted twice a week, for at least six weeks.
4. The mean number of thalli is calculated and plotted against time.
5. The graph should show the S-shaped/sigmoid population growth curve. Carrying capacity is reached when there is no room for further thalli, and competition for space limits further growth.

The effect of different environmental conditions on growth rate can be studied, such as salinity, pH, concentration of nitrates and phosphate level. Of course, one environmental condition should be changed at a time.

■ **Figure C4.1.13** Duckweed of the genus *Lemna* covering the surface of a pond, and a diagram of duckweed thalli (a leaf unit that is over 1.5 mm in diameter)

◆ **Community**: a group of different species living in an area.

■ **Figure C4.1.14** A community of wild animals congregate around a waterhole in Etosha National Park, northern Namibia, Africa

Communities in ecosystems

A **community** consists of all the living things in an ecosystem – the total of all the populations of all species. So, for example, the community of a well-stocked pond would include the populations of rooted, floating and submerged plants, the populations of bottom-living animals, the populations of fish and invertebrates of the open water, and the populations of surface-living organisms – a very large number of organisms in total. The savannah grasslands and lakeland ecosystems of Africa, for example, contain wildebeest, lions, hyenas, giraffes and elephants, as well as zebras (Figure C4.1.14). Communities include all the **biotic** parts of the ecosystem, including plants, animals and fungi.

> ### Common mistake
>
> Do not confuse the terms 'population' and 'community'. In ecology, these terms have different meanings to how they are used in geography, for example. In ecology, a community is many species living together, whereas the term population refers to just one species.

♦ **Intraspecific competition**: competition between individuals of the same species.

♦ **Interspecific competition**: competition between individuals of different species.

♦ **Cooperation**: the action or process of working together.

Link
Interspecific competition is also discussed in Chapter B4.2, page 378.

> ### Concept: Interaction
>
> A community is made up of many interacting populations of different species within an ecosystem. These interactions maintain stability in the ecosystem.

Competition versus cooperation in intraspecific relationships

Interactions within a species may be competitive or cooperative. **Competition** can be within the same species (**intraspecific competition**) or between different species (**interspecific competition**). Intraspecific competition can be the strongest type of competition because organisms all compete for the same resources.

Many animals have developed complex behaviours of **cooperation** to minimize the potential impact of direct competition. For example, groups of animals maintain dominance hierarchies, which reduces fighting and the risk of injury.

■ **Figure C4.1.15** Intraspecific competition in blue wildebeest (*Connochaetes taurinus*): two males competing for territory, Kalahari Desert, South Africa

Because individuals within the same species share the same niche, intraspecific competition can be more intense than interspecific competition, where competing species have different niches and therefore less overlap in shared resources. Within a species, there is competition for limited resources such as food, reproductive partners, territory (see Figure C4.1.15), water, space and any other resource required for survival or reproduction.

Interactions within a population do not need to be competitive. Other interactions involve cooperation between members of the population. Although all organisms compete for resources, cooperation is widespread. Genes cooperate in genomes; cells cooperate in tissues; individuals cooperate in societies. For example, hunting animals such as hyenas work in cooperative packs (Figure C4.1.16). Hyena groups are called clans – a dominant female (or 'queen') controls the clan, but all individuals benefit from the ability to hunt more effectively and kill prey as a group rather than as an individual. Such dominance hierarchies and collective action emerge from cooperation among individuals, and represent a high order of social complexity. Such societies are found in insects, mammals and birds, and in unicellular organisms.

■ **Figure C4.1.16** Hyena clan showing intraspecific cooperation on a hunt

C4.1 Populations and communities

■ **Figure C4.1.17** The queen (marked with dot) of a bee colony (*Apis mellifera*) surrounded by sterile worker bees

■ **Figure C4.1.18** Leafcutter ants (*Atta* sp.) carrying leaves back to a nest

Insects that work as a group are called social insects, and include bees, wasps, ants and termites. In these colonies, a single reproductive female is supported by large numbers of non-breeding workers (Figure C4.1.17). Because the offspring are produced by one female, all offspring are genetically similar, and so the whole colony can be seen as one 'superorganism'. Seemingly **altruistic** behaviour, like protecting the colony at risk to self, or raising the offspring of others instead of trying to reproduce, can be explained by the shared genetic inheritance shown by interacting individuals.

Leafcutter ants are found in the rainforests of Central and South America. They build nests that can contain thousands of compartments (cells) and cover up to $0.5\,km^2$, and a mature colony can contain more than 8 million individuals. The ants work as a collective, cutting the leaves from plants and taking them back to the colony (Figure C4.1.18). They are extremely strong insects and can carry 20–50 times their body weight. The leaves are fed to the fungus *Leucoagaricus gongylophorus*, which they grow in their nests, cultivating the fungus as a farmer would their crop. The ants cannot digest the cellulose in the cell walls of plants, but the fungus can. The fungus acts as food for the ant larvae. By cooperation, a large number of leaf segments can be collected (up to 17% of the vegetation in a forest can be harvested by leafcutter ants) and sufficient fungus grown to support the colony.

◆ **Altruism**: behaviour of an animal that benefits another at its own expense. Biologists call a behaviour pattern altruistic if it increases the number of offspring produced by the recipient and decreases that of the altruist.

Link

Intraspecific competition is the driving force for evolution by natural selection and is discussed in Chapter D4.1, page 783.

Link

Competitive exclusion is discussed in Chapter B4.2, page 378.

Interspecific relationships within communities

When resources are limiting in the environment, organisms must compete for them. Interspecific competition may be so intense as to cause the exclusion of one species by another. Examples of interspecific competition are species of *Paramecium* competing for plankton or species of barnacle competing for space on the rocky shore.

Herbivory, predation, mutualism, parasitism and pathogenicity are all categories of interspecific relationships within communities.

■ Herbivory

Herbivory is when an organism feeds on a plant. The carrying capacity of herbivores is affected by the quantity of plants they feed on. An area abundant in plant resources will have a higher carrying capacity for herbivores than an area that has less plant material.

Link

The definition of predation and more details are covered in Chapter B4.2, page 373.

■ Predation

Predation is when one animal (or sometimes a plant) eats another animal. The number of prey is reduced by the predator, lowering the prey's carrying capacity. The carrying capacity of the predator is affected by the prey because the number of predators is reduced when prey become fewer (see predator–prey relationships, page 574).

■ **Figure C4.1.19** Mutualism – mushroom of fly agaric fungus (*Amanita muscaria*) takes sugars and amino acids from a tree's roots in return for essential ions, via its hyphae attached below ground

■ **Figure C4.1.20** An evergreen robber frog in the Atlanta Botanical Garden's amphibian conservation laboratory, USA

Mutualism

Mutualism (symbiosis) is an interaction in which both species derive benefit. Mutualism can increase the carrying capacity of both species in the relationship. An example of a mutualistic relationship is shown in Figure C4.1.19.

Parasitism

Parasitism is a relationship in which the parasite organism benefits at the expense of the host organism from which it derives its food. The parasite lowers the carrying capacity of the host.

Pathogenicity

Diseases are caused by pathogens, which include bacteria, viruses, fungi and single-celled organisms (protistans). **Pathogenicity** is the capacity of a microbe to cause damage to a host resulting in disease. A pathogen may reduce the carrying capacity of the population it is infecting. Changes in the incidence of the pathogen, and therefore the disease, can also cause populations to increase and decrease around the carrying capacity. The evergreen robber frog (*Craugastor gollmeri*), see Figure C4.1.20, has suffered population declines across its habitat range in Costa Rica and Panama. The declines have been caused by the amphibian disease chytridiomycosis, caused by the fungus *Batrachochytrium dendrobatidis*. Over the last 30 years, the disease has caused the populations of over 200 amphibian species to decline in numbers significantly, and in some cases the species have become extinct.

Link

The definition of herbivory and how plants resist it are covered in Chapter B4.2, page 372.

- ◆ **Mutualism**: an interaction in which both species derive benefit. This is a specific type of symbiotic relationship.
- ◆ **Parasitism**: a relationship where one organism (the parasite) benefits at the expense of another (the host) from which it derives its food.
- ◆ **Pathogenicity**: the capacity of a microbe to cause damage in a host resulting in disease.

ATL C4.1B

Consider your own environment. Think of local examples to illustrate the range of ways in which species can interact within a community.

For example, if you live in Malaysia or Indonesia, an example of a parasite could be the *Rafflesia* plant. *Rafflesia* has the largest flowers in the world, but no leaves. Without leaves, the plant cannot photosynthesize, so it grows close by South East Asian vines (*Tetrastigma* sp.) from which it draws the sugars it needs for growth. An example of mutualism would be coral reefs (see page 567). The clouded leopard (*Neofelis nebulosi*) is a local predator that hunts small mammals. The Asian elephant feeds on vegetation within the rainforest and provides an example of herbivory.

Find local examples of each type of ecological interaction, using at least one example for each type of interspecific competition. Explain to other students in your class why you have placed them in that category.

Mutualism as an interspecific relationship that benefits both species

◆ **Root nodule**: small swelling on the root of plants that contain symbiotic nitrogen-fixing bacteria.

◆ **Symbiotic**: a close and long-term biological interaction between two different species.

Concept: Interdependence

In a mutualistic relationship, there is an interdependence between two species from which both benefit.

Root nodules in Fabaceae (legume family)

Rhizobium is a bacterium that lives in the **root nodules** of plants in the legume family (Fabaceae) such as clover plants (Figure C4.1.21). *Rhizobium* associates with roots in a **symbiotic** relationship. These bacteria work in the following way:

- *Rhizobium* fixes nitrogen gas to form ammonium ions, which are then converted by other bacteria to nitrates. The plant benefits from the nitrates. The plant supplies carbohydrates to the bacteria in the root nodule. This is very nutritionally expensive for the plant (they lose a source of glucose) but it does enable them to live in nitrate-deficient soils. Other plants without this symbiotic relationship are not able to survive in such poor soils, so the legume plants are able to exploit a niche that other plants cannot. The bacteria need a ready supply of glucose for respiration, which supplies the energy to split the triple-bonded nitrogen molecule. The relationship is, therefore, a mutualistic and symbiotic one.
- *Rhizobium* is aerobic. Nitrogen fixation needs lots of energy, which requires large amounts of ATP that could not be produced in sufficient quantity through anaerobic pathways.
- *Rhizobium* contains an enzyme, nitrogenase, which catalyses nitrogen fixation. This reaction is blocked by exposure to oxygen (by competitive inhibition) because nitrogen and oxygen are roughly the same size and shape. This means that oxygen exclusion is necessary for nitrogen fixation. This problem is solved in *Rhizobium* in two ways:
  - *Rhizobium* has an exceptionally fast and efficient aerobic metabolism, meaning very little oxygen builds up in the root nodules.
  - The root nodule has an oxygen-scavenging chemical called leghaemoglobin which, like haemoglobin in blood, has a high affinity for oxygen. Leghaemoglobin removes oxygen from the nodule, allowing nitrogen fixation to occur catalysed by nitrogenase.

■ **Figure C4.1.21** Root nodules on the roots of clover. The nodules are created by the nitrogen-fixing bacteria *Rhizobium trifolii*. The bacteria convert atmospheric nitrogen into ammonium ions, which are then further converted by nitrifying bacteria into nitrate ions, something the clover cannot do itself. Nitrate can then be absorbed by plants and converted into a usable organic form (amino acids)

● Top tip!

Farmers plant clover to increase the nitrate content of the soil and plough the crop into the ground rather than harvest it. Regular ploughing and good drainage keep conditions in the soil aerobic for the *Rhizobium*.

Mycorrhizae in Orchidaceae (orchid family)

Mycorrhiza is a type of fungus that grows in mutualism with plant species. The hyphae of the fungus (Figure C4.1.22a) grow through the soil and can enter plant tissue. Orchids are a group of flowering plants in the family Orchidaceae. Mycorrhizae play a key role during orchid germination. An orchid seed is very small, with limited food reserves, and so the growing orchid obtains nutrients and water from the fungal symbiont. In turn, the mycorrhizae gains photosynthetically-derived glucose from the orchid.

◆ **Mycorrhiza (plural, Mycorrhizae)**: a fungus that grows in association with the roots of a plant in a symbiotic relationship.

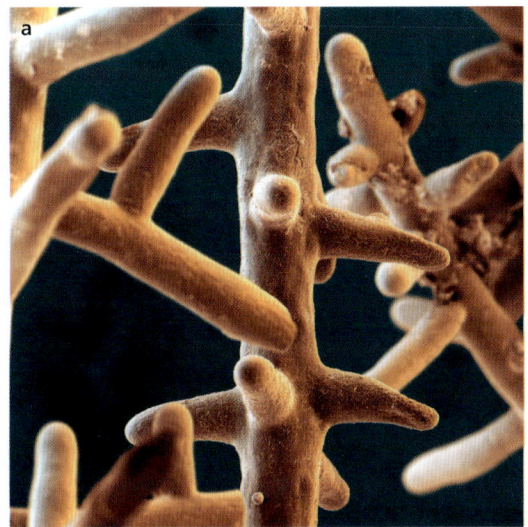 

■ Figure C4.1.22 a) SEM of mycorrhizae fungal hyphae, b) lady-slipper orchid (*Cypripedium calceolus*) is a terrestrial orchid that lives in temperate and northern areas and has a symbiotic relationship with mycorrhizae

One species of orchid, the lady-slipper orchid (*Cypripedium calceolus*), is a very rare species that is under threat (Figure C4.1.22b). To help conserve the species, it has been artificially grown using tissue culture techniques. However, it is very difficult to introduce new seedlings from laboratory conditions back to the wild due to lack of specific symbiotic mycorrhizae for the slipper orchid.

ATL C4.1C

What other examples of mutualistic relationships are there between plants and mycorrhizae? Find an example from your local area or region. What is the interdependence between the plant and fungus? Write a summary of the information you find and keep this as a useful case study for answering questions.

Corals and zooxanthellae

Link
The conditions required for coral reef formation are covered in Chapter B4.1, page 353.

Coral reefs are created by the symbiotic relationships between two organisms – an animal and an algae. The interaction in coral reefs is mutually beneficial; both organisms gain from the relationship. The animal that makes the coral reef is a **cnidarian** – a group that also includes sea anemones and jellyfish. Corals are colonial organisms (i.e. made of many individuals of the same species physically connected, interdependent on one another) made up of individual **polyps**, each 1–3 mm in diameter, with stinging tentacles arranged around a central mouth (Figure C4.1.23). The polyps are connected to one another via a thin layer of tissue, which enables nutrients to be shared between organisms. The polyps secrete a protective skeleton of calcium carbonate, which forms the foundation of the coral reef ecosystem. The algae (**zooxanthellae**) live within the cells of the coral's **endodermis** (the innermost lining of the animal).

C4.1 Populations and communities

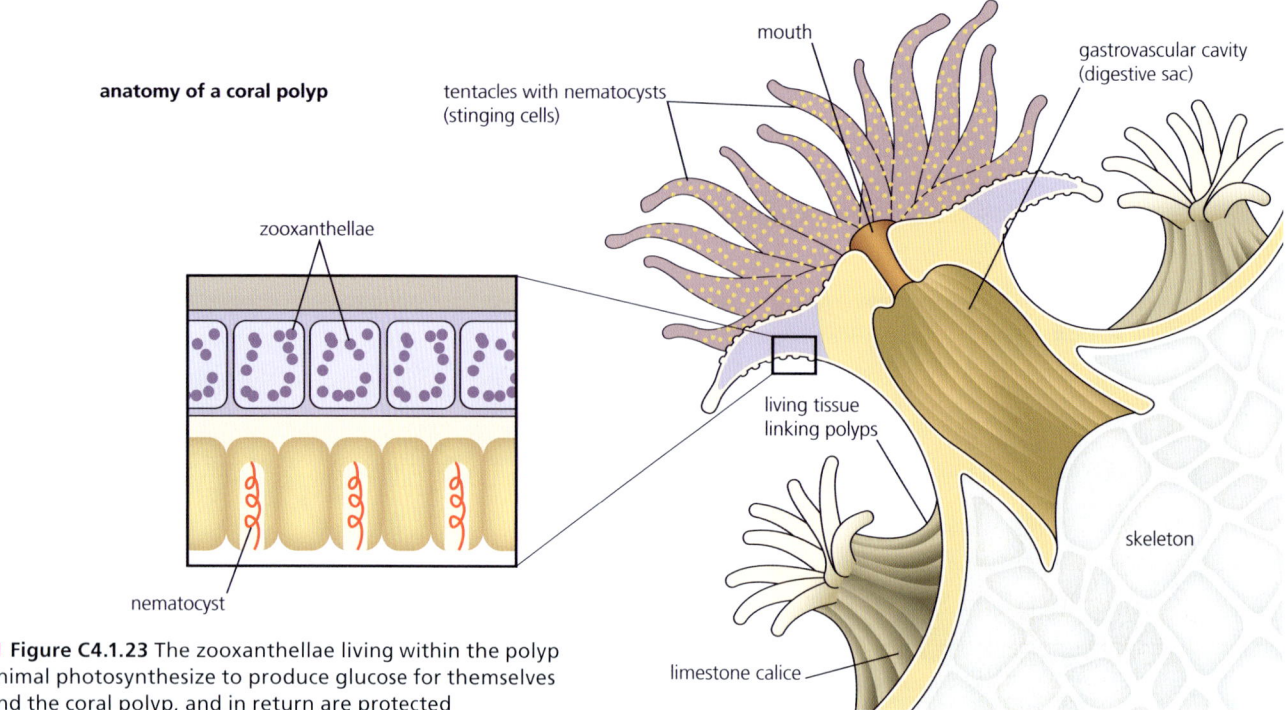

■ Figure C4.1.23 The zooxanthellae living within the polyp animal photosynthesize to produce glucose for themselves and the coral polyp, and in return are protected

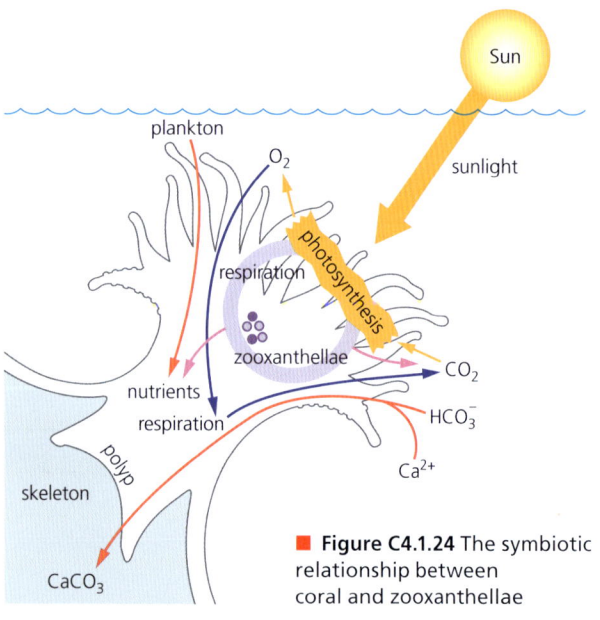

■ Figure C4.1.24 The symbiotic relationship between coral and zooxanthellae

The coral provides the algae with a protected environment and the carbon dioxide needed for photosynthesis (a metabolic waste product of respiration in the coral). In return, the algae produce oxygen, utilize the coral's waste carbon dioxide, and provide the coral with a food source (the glucose produced by photosynthesis). As much as 90% of the organic material manufactured by the algae through photosynthesis is transferred to the host coral tissue (Figure C4.1.24).

The zooxanthellae give coral its colour. Several million zooxanthellae live and produce pigments in just a few square centimetres of coral. The pigments are visible through the clear body of the polyp. Zooxanthellae have a range of different photosynthetic pigments, which range in colour from golden-yellow to brown, giving their hosts the same shade.

Resource competition between endemic and invasive species

Link

Further examples of alien and invasive species, and their impacts on the current biodiversity crisis, are discussed in Chapter A4.2, page 172.

Alien species are species that are introduced into an area by human activity. **Invasive species** are alien species that have increased rapidly in number, having a negative effect on the environment and on native species.

The success of the grey squirrel (*Sciurus carolinensis*) at the expense of the red squirrel (*Sciurus vulgaris*) (Figure C4.1.25) is a very good example of an alien species becoming an invasive one. The grey squirrel is native to eastern North America and was introduced to the UK in the 1800s. Red squirrels are now restricted to just two locations in the UK, having once been widespread over much of the country. The grey squirrel has spread rapidly, chiefly by being able to consume a wide range of locally growing nuts (including the very common woodland oak tree's acorn – this contains polyphenolic substances that the new arrival can digest, but which the red squirrel cannot). Although they have similar niches, with its wider diet, the grey has outbred the red and multiplied much faster. This enlarged population has successfully competed for hazelnuts and pine cones too (the limited diet of the red squirrel). The grey squirrel is an example of an alien species that has a similar niche to an endemic species: because the niches of the grey and red squirrel overlap, **competitive exclusion** can take place with the result that the less-effective competitor declines in number or is wiped out.

> **Link**
> Niches and competitive exclusion are discussed in Chapter B4.2, page 363.

■ **Figure C4.1.25** The red squirrel (*Sciurus vulgaris*) and grey squirrel (*Sciurus carolinensis*)

> **Link**
> Fundamental and realized niches are discussed in Chapter B4.2, page 377.

A consequence of introducing an alien species may be that both the alien and native species end up occupying smaller **realized niches** due to interspecific competition. If the alien species lacks natural predators, it could outcompete the native species and become invasive. Understanding the **fundamental niche** of invasive species can help predict their dispersal patterns and invasion success, providing the basis for better-informed conservation and management policies.

6 Compare and contrast, using examples, species that are endemic and those that are invasive.

> **ATL C4.1D**
> Select an invasive species from your local area. Create a flow chart showing how the introduction of this species has affected populations of endemic species and the management issues that have resulted. You need to bring in the following concepts: competition, conservation, fundamental and realized niches, and ecosystem management.

Tests for interspecific competition

Interspecific competition is indicated but not proven if one species is more successful in the absence of another. There are a range of possible approaches to research of interspecific competition: laboratory experiments, field observations by random sampling and field manipulation by removal of one species.

■ Investigating competitive exclusion

American ecologist Joseph Connell examined two species of barnacle – a common animal on rocky shores – to investigate the difference between fundamental and realized niche. His study in Scotland, UK, used two experiments. Connell had observed that one of the species, *Semibalanus (Balanus) balanoides*, was most abundant on the lower intertidal area and that the other species,

Chthamalus stellatus, was most common on the upper intertidal area of the shore. He knew that the free-swimming larvae of each species could settle anywhere on the rocky shoreline and grow into adult barnacles. He asked the question: 'Why do *Semibalanus* and *Chthamalus* not grow together?'

Inquiry 1: Exploring and designing

Designing

How would you address Connell's question regarding why *Semibalanus* and *Chthamalus* do not grow together? What experiments could you do to explain Connell's observations? Think about this now and write down your ideas. What would your variables be (including controlled variables)? How would you ensure reliability of your data?

In his first experiment (Figure C4.1.26), Connell removed *Semibalanus* from the upper area of shore. He found that, over time, no *Chthamalus* replaced it. His explanation was that *Semibalanus* could not survive in an area that regularly dried out and experienced desiccation due to low tides. He concluded that the realized niche of *Semibalanus* was the same as its fundamental niche.

In his second experiment (also shown in Figure C4.1.26), Connell removed *Semibalanus* from the middle area. He found that over time *Chthamalus* replaced it. His explanation was that *Semibalanus* was a more successful competitor in the middle intertidal zone and usually excluded *Chthamalus*.

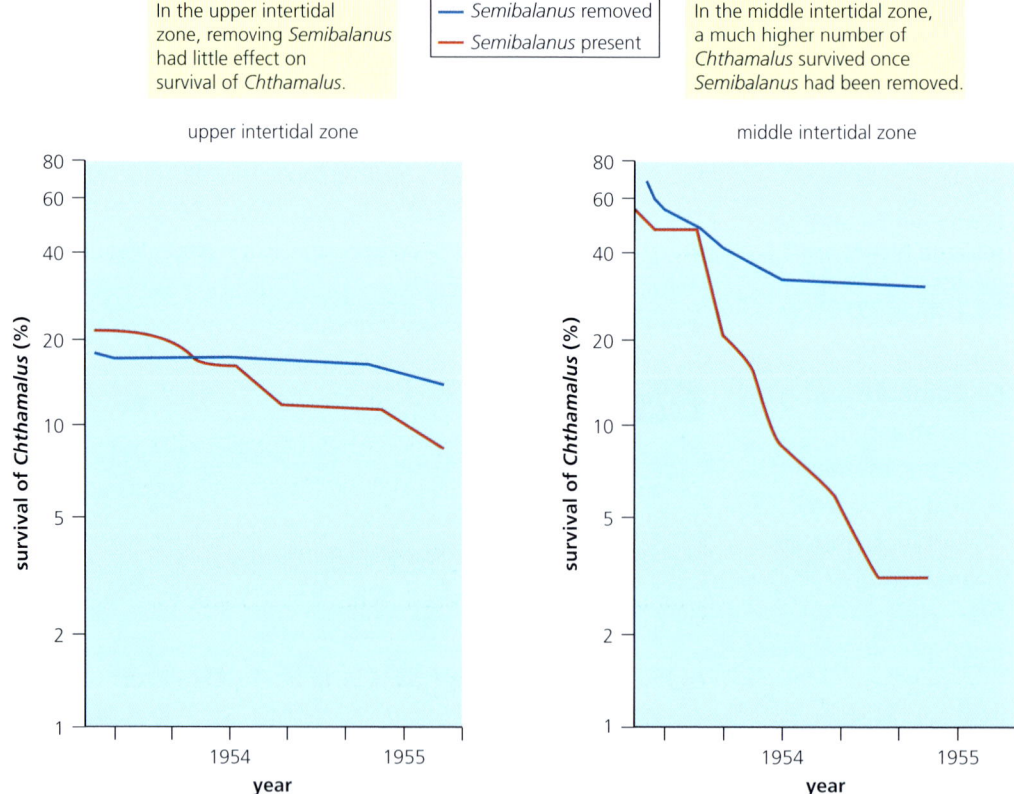

■ Figure C4.1.26 Data from Joseph Connell's experiment showing the effect of *Semibalanus* removal from the upper shore (left) and middle/lower shore (right). The left-hand figure shows that in the upper intertidal zone, removing *Semibalanus* had little effect on the survival of *Chthamalus*; the right-hand figure shows that, in the middle/lower intertidal zone, a much higher percentage of *Chthamalus* survived once *Semibalanus* had been removed

He concluded that the fundamental niche and realized niche for *Chthamalus* were not the same, and that the species' realized niche was smaller due to interspecific competition (that is, competition between the two species) leading to **competitive exclusion** (when one species outcompetes and excludes another, when their niches overlap).

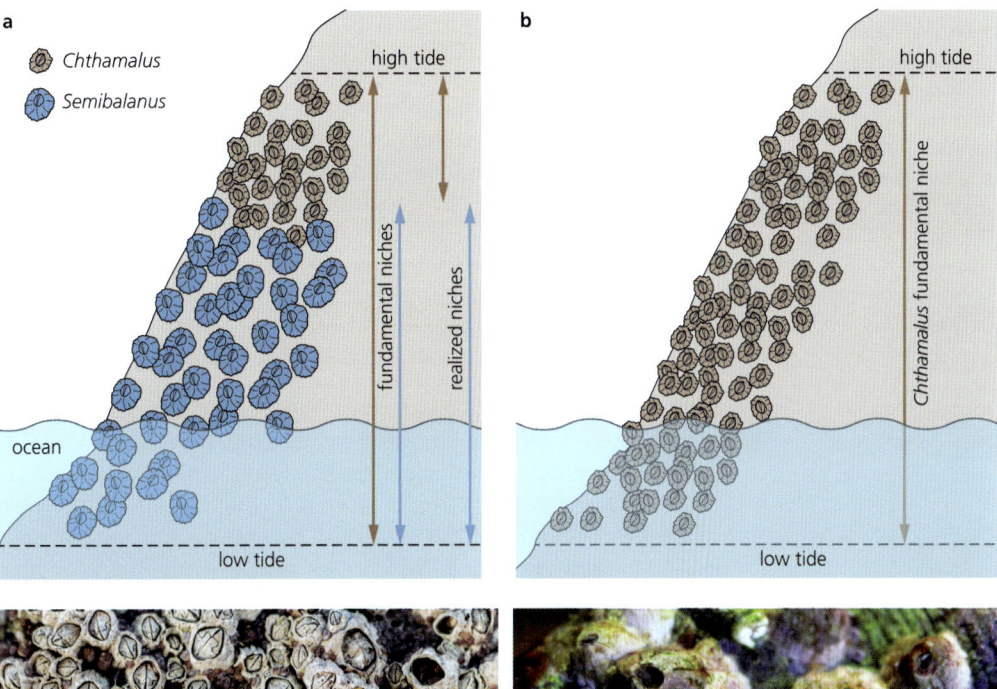

Semibalanus sp. (left-hand photo), easily crowded out from more exposed positions on the shore, and *Chthamalus* sp. (right-hand photo), able to withstand prolonged exposure and so can survive higher on rocky shorelines.

■ **Figure C4.1.27** a) Two species of barnacle in the intertidal region along the Scottish coast show stratified distribution. Competition between the two species explains the distinction between the fundamental niche and realized niche for each organism. b) The results of Connell's second experiment. When *Semibalanus* were removed from the middle zone of the rocky shore, the *Chthamalus* population moved into that area, indicating that the fundamental niche of *Chthamalus* covered both the upper and middle zones of shore. Competition between the two species usually excluded the *Chthamalus* and restricted it to its observed realized niche

Tool 2: Technology

Using computer modelling

Read more about Connell's experiment using this site: **https://virtualbiologylab.org/ModelsHTML5/BarnacleCompetition/BarnacleCompetitionModel.html**

Recreate the experiment using the simulation feature. Gather your own data and compare them with Connell's. Do you reach the same conclusion about the fundamental and realized niches of both species?

Top tip!

Elimination of one of the competing species or the restriction of both to a part of their fundamental niche are both possible outcomes of competition between two species.

Restriction of two species due to interspecific competition

Connell's experiments with barnacles show the elimination of a species that has a similar fundamental niche to another species in the same area. In other cases, competing species can coexist but with restrictions with both species. For example, in Africa, leopards and lions display interspecific competition, since both species feed on the same prey. When they coexist in the same area, they each negatively impact the presence of the other because both will have less food. The result is the restriction of both species as an outcome of competition.

C4.1 Populations and communities

◆ **Hypothesis**: a provisional explanation of an observed phenomenon or event that can be investigated using the scientific method; when using a statistical test, the hypothesis states that there is a statistically significant difference between two variables.

Nature of science: Hypotheses

A **hypothesis** is a provisional explanation of an observed phenomenon or event that can be investigated using the scientific method. When making a hypothesis, the investigator proposes how an independent variable may affect a dependent variable. A hypothesis is a testable prediction (ideally quantitative) of how you think an independent variable and a dependent variable are causally related. A hypothesis includes a scientific explanation (often involving a reference to a scientific model) of how the two variables are related. For example, you might be planning to investigate the relationship between temperature and the rate of fermentation of brewer's yeast. Your hypothesis might be that, as temperature increases, the rate of fermentation is expected to increase exponentially up to an optimum temperature and then decrease rapidly (in a curved manner) to zero.

Hypotheses can be tested by both experiments and observations. An experiment changes one variable (the independent variable) and measures the effect on the dependent variable. All other variables are kept the same (they are controlled), unless this is not possible, for example in ecological investigations. In these cases, other variables that may affect the dependent variable are monitored. Observations are qualitative rather than quantitative, and also provide evidence for the hypothesis. An experiment with controlled variables is better support for a hypothesis than simple observations. For example, in Joseph Connell's investigation of the fundamental and realized niches of barnacles, he observed that one of the species was most abundant on the lower intertidal area and that the other was most common on the upper intertidal area of the shore. However, it was only by removing each species of barnacle from their respective zones on the shore that he was able to demonstrate the fundamental niches of both species and how they compare to their realized niches.

Using the chi-squared test for association between two species

Quadrats can be used to record the presence or absence in each quadrat of two species under investigation. The data are then subjected to a statistical test called the chi-squared (χ^2) test. It is used to examine data that fall into discrete categories (as in this case). It tests the significance of the deviations between numbers observed (**O**) in an investigation and the number expected (**E**). The measure of deviation, known as chi-squared, is converted into a probability value using a chi-squared table. In this way, we can decide whether the differences observed between our sets of data are likely to be real or, alternatively, obtained just by chance.

You need to be able to apply chi-squared tests on the presence or absence of two species in several sampling sites, exploring the differences or similarities in distribution. This may provide evidence for interspecific competition.

■ Applying and interpreting appropriate tests of statistical significance

All ecological statistical techniques involve hypothesis testing. They test a statement called the **null hypothesis**. Statistical analyses test whether data match the null hypothesis or significantly vary from it. Where data are being compared, the null hypothesis states that 'there is no difference between the sets of data' and, when an association is being investigated, it states that 'there is no association'. The alternative hypothesis is the 'opposite' of the null hypothesis, i.e. that there is a difference or association shown by the data.

◆ **Null hypothesis**: there is no statistically significant difference between two variables.

Tool 3: Mathematics

Applying the chi-squared test to test for an association between two moorland species

This example examines whether the moorland species bell heather (*Erica cinerea*) and common heather, also known as ling (*Calluna vulgaris*), tend to occur together. Moorlands are upland areas with acidic and low-nutrient soils, where heather plants dominate. Heathers have long woody stems and grow in dense clumps. They have colourful, bright flowers. Is there a statistically significant association between ling and bell heather on an area of moorland? As scientists, we would carry out a statistical test to work out the probability of getting results that indicate there is *no* association between the two species – indicating the **null hypothesis** is true. The null hypothesis in this example would be that there is no statistically significant association between bell heather and ling in an area of moorland; that is, their distributions are independent of each other. If our results do not support the null hypothesis, then there is an association.

1 The measurements and results

To sample the two species, the presence or absence of each species was recorded in each of 200 quadrats. The quadrats were located at random on a 100 m by 100 m area of moorland (Table C4.1.3).

■ **Table C4.1.3** Observed results – the distribution of ling and bell heather

	Bell heather present	Bell heather absent	Total
Ling present	89	45	134
Ling absent	31	35	66
	120	80	200

2 The calculations

a) Expected results: assuming that the two species are randomly distributed with respect to each other, the probability of ling being present in a quadrat is:

$$\frac{\text{column total}}{\text{total number of quadrats}} = \frac{134}{200} = 0.67$$

Similarly, the probability of bell heather being present in a quadrat is:

$$\frac{120}{200} = 0.60$$

The probability of both species occurring together, assuming random distribution between each species, is: $0.60 \times 0.67 = 0.40$. The number of quadrats in which both species can be expected is therefore $0.40 \times 200 = \mathbf{80}$.

Having calculated the number of expected quadrats where the species are found together, other expected values can be calculated by subtracting from the totals. For example, the expected number of quadrats with bell heather but no ling is $120 - 80 = 40$. Expected values follow the assumption that totals for each row and column do not change, because the relationship shown by the data is assumed to represent the true relative frequency of each species (Table C4.1.4).

■ **Table C4.1.4** The full expected results

	Bell heather present	Bell heather absent	Total
Ling present	80	54	134
Ling absent	40	26	66
	120	80	200

■ **Figure C4.1.28** a) A moorland ecosystem and two common plants found there: b) ling (*Calluna vulgaris*) and c) bell heather (*Erica cinerea*)

Calculated values can be checked by using the ratios represented in the table of observed results (Table C4.1.3).

For example, the expected number of quadrats where there is no ling and no bell heather can be calculated as follows:

Probability of no ling in a quadrat = 66 ÷ 200 = 0.33

Probability of no bell heather in a quadrat = 80 ÷ 200 = 0.40

Probability of neither species in a quadrat = 0.33 × 0.40 = 0.13

Number of expected quadrats with neither species present = 0.13 × 200 = 26

(Note that this figure agrees with the estimated value in Table C4.1.4).

b) Statistical test: the observed and expected results are recorded in Table C4.1.5.

■ **Table C4.1.5** Observed (O) and expected (E) distribution of ling and bell heather

	Bell heather present	Bell heather absent	Total
Ling present			
O	89	45	134
E	80	54	
Ling absent			
O	31	35	66
E	40	26	
	120	80	200

Then, chi-squared is calculated from the formula:

$$\chi^2 = \sum \frac{(O - E)^2}{E}$$

So, chi-squared in this example:

$$= \frac{(89 - 80)^2}{80} + \frac{(45 - 54)^2}{54} + \frac{(31 - 40)^2}{40} + \frac{(35 - 26)^2}{26}$$

$$= 1.01 + 1.50 + 2.03 + 3.11$$

$$= 7.65$$

To find whether this result is statistically significant or not, the value must be compared to a **critical value** (Table C4.1.6). To locate the critical value, the appropriate degrees of freedom need to be calculated.

Degrees of freedom = (number of columns − 1) × (number of rows − 1)

In this case, degrees of freedom = (2 − 1) × (2 − 1) = 1

■ **Table C4.1.6** Critical values for the χ^2 test

Degrees of freedom	0.05 level of significance
1	3.84
2	5.99
3	7.81
4	9.49

The chi-squared value of 7.65 is larger than the critical value of 3.84 for 1 degree of freedom at the probability level of $p = 0.05$ (the 5% probability level). The null hypothesis is therefore rejected; there is a statistically significant association between bell heather and ling in this area of moorland. So, the distributions of the two species are not independent of each other – the distribution of the two species is associated.

c) The value of chi-squared may also be obtained using a programmed calculator or a computer program such as: www.socscistatistics.com/tests/chisquare/Default2.aspx

Predator–prey relationships

◆ **Predator–prey relationship**: the inter-relationship of population sizes due to predation of one species (the predator) on another (the prey).

In Chapter B3.3, different reasons for locomotion were discussed, including escaping from danger. Figure B3.3.18b, page 338, shows a snowshoe hare running from a lynx. In this interaction, the snowshoe hare is the **prey** and the lynx the **predator**. Such an interaction is known as a **predator–prey relationship**.

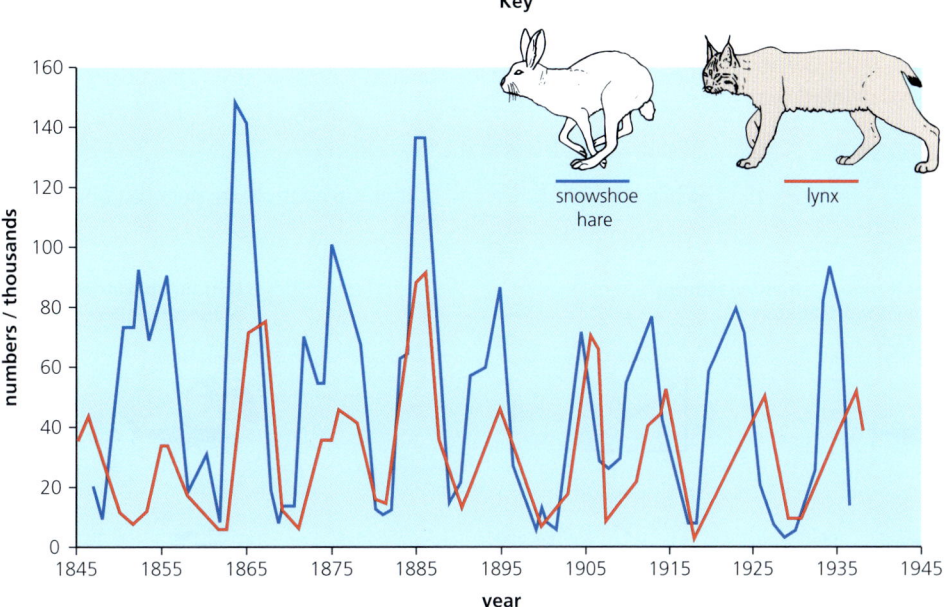

■ **Figure C4.1.29** The snowshoe hare and lynx: an example of a predator–prey relationship

Predator–prey relationships are an example of **density-dependent** (see page 557) control of animal populations.

These predator–prey interactions are often controlled by **negative feedback** mechanisms (see page 557) that control population densities (Figure C4.1.29). In the relative absence of the predatory lynx (due to a limited prey population), the population of snowshoe hare begins to increase in size. As the availability of prey increases, there is an increase in predator numbers, after a time-lag. As the number of predators increases, the population size of the prey begins to decrease, again after a time-lag. With fewer prey, the number of predators decreases again. With fewer predators, the number of prey may begin to increase again, and the cycle continues.

Predation may actually be good for the prey: it removes old and sick individuals first as these organisms are easier to catch. Those remaining are healthier and form a superior breeding pool.

7 Describe the role of negative feedback mechanisms in a named example of predator–prey relationships.

● Common mistake

The term 'negative' does not mean that the feedback loop is detrimental to the environment. Quite the opposite – it usually counteracts deviation away from steady-state equilibrium. The term 'positive' does not mean that the feedback loop has a constructive effect on the environment. Positive feedback increases change in a system, leading to it moving further away from steady-state equilibrium.

Link
Trophic levels and food chains are covered in Chapter C4.2, page 578.

◆ **Top-down control**: changes to the food chain occur at the top trophic level and then impact on the trophic levels lower in the food chain.

Top-down and bottom-up control of populations in communities

A **trophic level** is a feeding level within a food chain (see Chapter C4.2, page 582). In **top-down control**, a higher trophic level influences the community structure of a lower trophic level through predation. This approach is also called a predator-controlled food web. Figure C4.1.30a shows how reduction in the top predator leads to an increase in the number of prey that it feeds on, which in turn leads to a reduction in zooplankton (small floating or weakly swimming organisms that live in the ocean). Reduction in zooplankton has a knock-on effect on phytoplankton (algae and photosynthetic bacteria), leading to population growth due to decreased predation.

C4.1 Populations and communities

◆ **Bottom-up control**: changes to the food chain occur at the lowest trophic level (producers) and then impact on the trophic levels higher in the food chain.

Bottom-up control means that a lower trophic level in an ecosystem affects the community structure of higher trophic levels through resource restriction. Limitation is placed by resources that allow growth, such as food source, habitat or space. Figure C4.1.30b shows bottom-up control in a marine ecosystem. In this system, reduction in the producers (phytoplankton) leads to a reduction in all subsequent levels of the food chain.

While both top-down and bottom-up controls are possible, one or the other is likely to be dominant in a community.

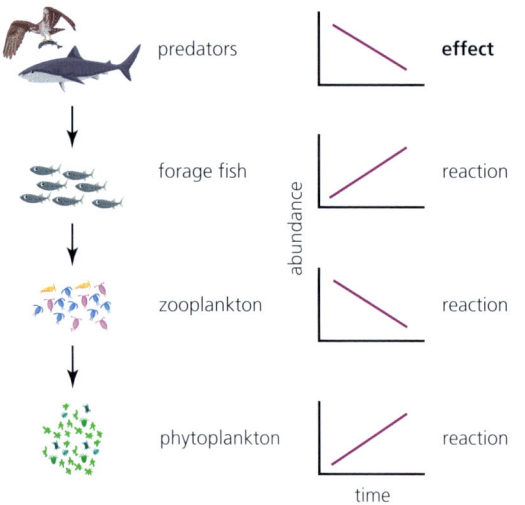

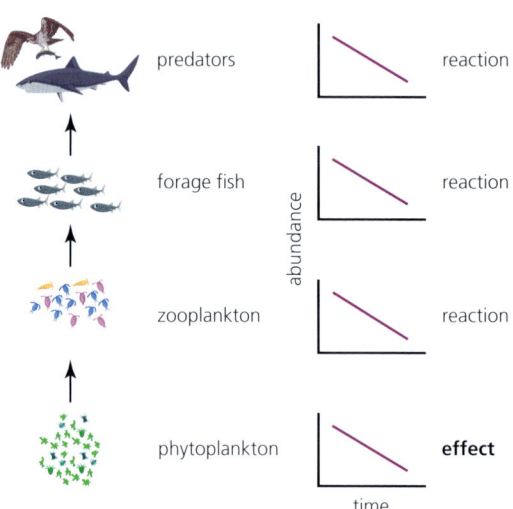

■ Figure C4.1.30 a) Top-down control in the food chain for four trophic levels in a marine ecosystem, b) bottom-up control, or control through primary production, in a simplified four-level food chain in a marine ecosystem

8 Distinguish between top-down and bottom-up control of populations in communities.

Allelopathy and secretion of antibiotics

Allelopathy

Competition in plants can be severe and lead to the reduction in carrying capacity of both species if the competition is interspecific. Mechanisms that give one plant an advantage over another have evolved, such as **allelopathy**. The word is derived from the Greek words 'allelon', which means 'each other', and 'pathos', which means 'to suffer' (note – the term 'pathogen', meaning organisms that cause disease in another organisms, also has this origin). Allelopathy in plants is where a chemical compound (allelochemical) is used by one species to suppress the other and take advantage of that suppression. Allelopathic plants create adverse conditions for other neighbouring, competitor plants by reducing their seed germination, seedling growth or growth of the plant. Some allelochemicals change the amount of chlorophyll production in a plant and therefore slow down or stop photosynthesis, which leads to the suppression or death of the plant.

◆ **Allelopathy**: chemical inhibition of one plant (or other organism) by another, due to the release of chemicals (allelochemicals) that act as germination or growth inhibitors.

Allelopathy can be carried out by plants by the following processes:
- releasing chemical compounds (allelochemicals) from their roots into the soil: these chemicals suppress or kill neighbouring plants when they are absorbed
- releasing allelochemicals in a gaseous form from the stomata in the leaves: when the neighbouring plants absorb these gases, they are suppressed or killed
- leaves dropping from allelopathic plants: these contain toxic chemicals which, when the leaves decompose, are released and inhibit the growth of competitors.

■ Figure C4.1.31 Broccoli – an example of an allelopathic plant

■ Figure C4.1.32 The common bracken fern (*Pteridium aquilinum*), which excludes other plants from areas that it grows in with its allelopathic properties

ATL C4.1E

Find out about allelopathic plants in your area. How do they inhibit competitors? Produce a fact sheet on at least three different named species of plant.

All brassica plants, such as cabbage, mustard, kale and radish, have allelopathic properties. Mustard, for example, can suppress fungal pathogens. Extracts of black mustard can limit the germination of some legumes, such as lentils, as well as oats. Broccoli (Figure C4.1.31) can be allelopathic to later-planted broccoli or any other crops in the brassica family. Farmers growing broccoli therefore rotate to other crops (i.e. alternate the crop so that broccoli is not grown every year) and do not plant other brassicas where broccoli have recently been grown.

Ferns are also allelopathic. Bracken (*Pteridium aquilinum*, Figure C4.1.32), for example, competitively excludes other plants by allelopathy, making it the dominant species in the areas where it is found. This dominance reduces the biodiversity of the ecosystem because other plants cannot grow in the presence of the bracken. Associated plants are often severely inhibited or even excluded from dense stands of the fern.

■ Secretion of antibiotics

The first mass-produced antibiotic for the treatment of human bacterial disease was penicillin, derived from the *Penicillium* fungi. Penicillin was accidently discovered by Alexander Fleming in 1929 (see Figure C3.2.15, page 533) while studying *Staphylococcus*, the bacterium that causes boils and sore throats. Fungi naturally produce antibiotics to kill or inhibit the growth of bacteria, which limits the competition with the fungi. The fungi is allelopathic, producing chemicals that scientists call antibiotics to kill and inhibit the growth of bacteria: these properties have been taken advantage of to produce antibiotics as medicines.

9 Using examples, **outline** how allelopathy is used by plants to give a competitive advantage over other species in their habitat.

10 **Define** the term *allelopathy*.

Link

Antibiotics and penicillin are covered in more detail in Chapter C3.2, page 532.

● Top tip!

Allelopathy and the secretion of antibiotics are similar processes in that a chemical substance is released into the environment to deter potential competitors.

LINKING QUESTIONS

1 What are the benefits of models in studying biology?
2 What factors can limit capacity in biological systems?

C4.2 Transfers of energy and matter

> **Guiding questions**
> - What is the reason matter can be recycled in ecosystems but energy cannot?
> - How is the energy that is lost by each group of organisms in an ecosystem replaced?

SYLLABUS CONTENT

This chapter covers the following syllabus content:
- ▶ C4.2.1 Ecosystems as open systems in which both energy and matter can enter and exit
- ▶ C4.2.2 Sunlight as the principal source of energy that sustains most ecosystems
- ▶ C4.2.3 Flow of chemical energy through food chains
- ▶ C4.2.4 Construction of food chains and food webs to represent feeding relationships in a community
- ▶ C4.2.5 Supply of energy to decomposers as carbon compounds in organic matter coming from dead organisms
- ▶ C4.2.6 Autotrophs as organisms that use external energy sources to synthesize carbon compounds from simple inorganic substances
- ▶ C4.2.7 Use of light as the external energy source in photoautotrophs and oxidation reactions as the energy source in chemoautotrophs
- ▶ C4.2.8 Heterotrophs as organisms that use carbon compounds obtained from other organisms to synthesize the carbon compounds that they require
- ▶ C4.2.9 Release of energy in both autotrophs and heterotrophs by oxidation of carbon compounds in cell respiration
- ▶ C4.2.10 Classification of organisms into trophic levels
- ▶ C4.2.11 Construction of energy pyramids
- ▶ C4.2.12 Reductions in energy availability at each successive stage in food chains due to large energy losses between trophic levels
- ▶ C4.2.13 Heat loss to the environment in both autotrophs and heterotrophs due to conversion of chemical energy to heat in cell respiration
- ▶ C4.2.14 Restrictions on the number of trophic levels in ecosystems due to energy losses
- ▶ C4.2.15 Primary production as accumulation of carbon compounds in biomass by autotrophs
- ▶ C4.2.16 Secondary production as accumulation of carbon compounds in biomass by heterotrophs
- ▶ C4.2.17 Constructing carbon cycle diagrams
- ▶ C4.2.18 Ecosystems as carbon sinks and carbon sources
- ▶ C4.2.19 Release of carbon dioxide into the atmosphere during combustion of biomass, peat, coal, oil and natural gas
- ▶ C4.2.20 Analysis of the Keeling curve in terms of photosynthesis, respiration and combustion
- ▶ C4.2.21 Dependence of aerobic respiration on atmospheric oxygen produced by photosynthesis, and of photosynthesis on atmospheric carbon dioxide produced by respiration
- ▶ C4.2.22 Recycling of all chemical elements required by living organisms in ecosystems

Note: There is no higher-level only content in C4.2.

> **Concept: Interaction**
>
> A community forms an ecosystem by its interactions with the non-living (abiotic) environment.

> **Link**
>
> Habitats are covered in Chapter B4.1, page 341.

◆ **Open system**: system where both energy and matter can enter and exit.
◆ **Closed system**: system where only energy can pass in and out, not matter.

Ecosystems as open systems

An ecosystem is defined as a community of organisms and their surroundings, the environment in which they live and with which they interact. It is a basic functional unit of ecology since the organisms that make up a community cannot realistically be considered apart from their physical environment.

Examples such as a woodland or a lake illustrate two important features of an ecosystem, namely that it is:
- a largely **self-contained** unit, since most organisms of the ecosystem spend their entire lives there and their essential nutrients are endlessly recycled around and through it
- an **interactive** system, in that the kinds of organism that live there are largely decided by the physical environment, and the physical environment is constantly altered by the organisms.

The organisms of an ecosystem are called the **biotic** component, and the physical environment is known as the **abiotic** component. Within any ecosystem, organisms are normally found in a particular part or **habitat**.

Ecosystems are **open systems** because both energy and matter can enter and exit (Figure C4.2.1). It is this constant input of energy that sustains the ecosystem, and although matter is recycled (see page 601) organisms can bring matter into and out from these natural systems. In contrast, a **closed system** exchanges only energy across its boundary. Closed systems only exist experimentally, although the global geochemical cycles (such as the carbon cycle – see page 594) approximate to closed systems.

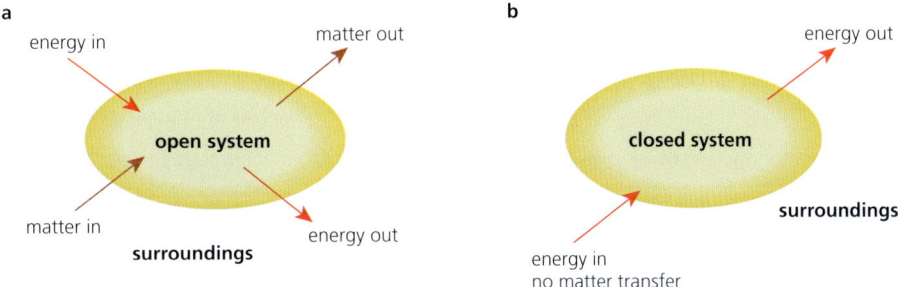

■ **Figure C4.2.1** The exchange of matter and energy across the boundaries of different systems. An open system (a) exchanges both energy and matter, whereas a closed system (b) exchanges only energy

1 Distinguish between an open and closed system.

Sunlight as the principal source of energy that sustains most ecosystems

Think of an ecosystem with which you are very familiar. Perhaps it is one near your home, school or college. It might be savannah, a forest, a lake, woodland or meadow. Whatever you have in mind, it will certainly contain a community of plants, animals and micro-organisms, all engaged in their characteristic activities. Some of these organisms will be much easier to observe than others, possibly because of their size or the times of day (or night) at which they feed, for example.

The essence of survival of organisms is their activity. To carry out their activities, organisms need energy. We have already seen that the immediate source of energy in cells is the molecule ATP (page 405), which is produced by respiration. The energy of ATP has been transferred from sugar and other organic molecules – the respiratory substrates. These organic molecules are obtained from nutrients because of the organism's mode of nutrition.

◆ **Producer**: an autotrophic organism that can synthesize glucose; a photosynthetic green plant or chemosynthetic bacterium, forming the first trophic level in a food chain.

The ultimate source of energy for most food chains is sunlight: this is the principal source of energy that sustains most ecosystems. Energy from sunlight is trapped by chlorophyll or other photosynthetic pigments and used ultimately to synthesize glucose. In this way, light energy is transformed into a store of chemical energy. Organisms that can trap energy in glucose are called **producers**. This chemical energy is then used to synthesize ATP, which is used in all life processes.

Not all food chains begin with the input of sunlight energy. Exceptions include ecosystems in caves and those in the ocean below the level light can reach, such as hydrothermal vents (see Chapter A2.1, page 47). Producers in these ecosystems synthesize glucose using energy from chemical reactions (see page 48).

ATL C4.2A

EO Wilson (1929–2021), an American biologist and naturalist, proposed two fundamental laws of biology:
1. All the phenomena of biology are ultimately obedient to the laws of chemistry and physics.
2. All the phenomena of biology have arisen by evolution through natural selection.

To what extent do you agree or disagree with Wilson's proposal? Discuss with a partner.

Nature of science: Theories

Laws in science are generalized principles, or rules of thumb, created to describe patterns observed in living organisms. Unlike theories, they do not offer explanations, but describe phenomena. The term 'law' is sometimes used for statements that allow predictions to be made about natural phenomena without explaining them. Like theories, laws can be used to make predictions. You need to be able to outline the features of useful generalizations. For example, it is a generalization to say that 'sunlight is the principal source of energy that sustains ecosystems' because some food chains use other sources of energy. It is a useful generalization because most food chains do obtain energy in this way, although it would be more accurate to say that 'sunlight is the principal source of energy that sustains *most* ecosystems'.

Flow of chemical energy through food chains

Unlike matter, which cycles through ecosystems, energy enters an ecosystem, usually as sunlight, and leaves it as heat (Figure C4.2.2). Energy does not cycle but is conserved; it is not available to be reused by living organisms as it is lost from ecosystems.

Top tip!

You need to be clear about the distinction between *one-way flow* of energy in ecosystems and the *cycling* of inorganic nutrients.

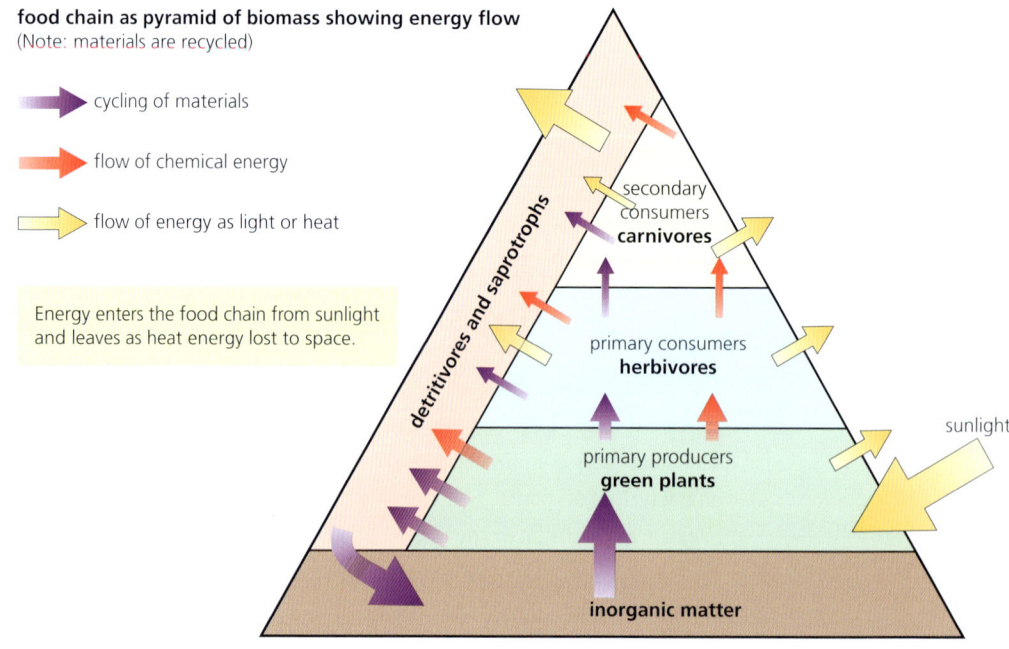

■ Figure C4.2.2 Cycling of nutrients and the flow of energy within an ecosystem

Tool 3: Mathematics

Representing energy flow in the form of food chains

A feeding relationship in which a carnivore eats a herbivore, which itself has eaten plant matter, is called a **food chain**. In Figure C4.2.3, light is the initial energy source. Note that, in a food chain, the arrows point to the **consumers** and so indicate the direction of energy flow.

Food chains are simple models that represent the flow of energy through ecosystems. Chemical energy passes to a consumer as it feeds on an organism that is the previous stage in the food chain, for example from the oak beauty caterpillar to the caterpillar-hunting beetle.

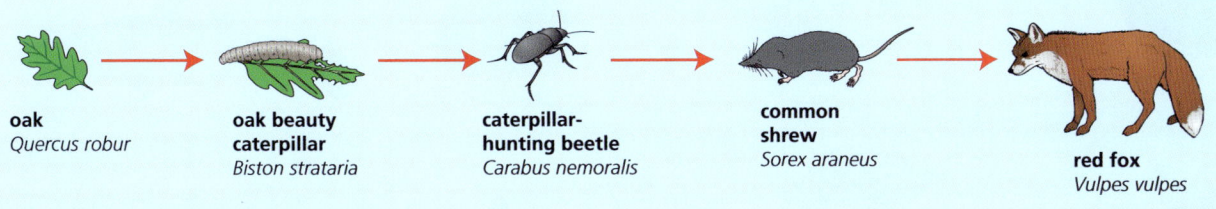

oak
Quercus robur

oak beauty caterpillar
Biston strataria

caterpillar-hunting beetle
Carabus nemoralis

common shrew
Sorex araneus

red fox
Vulpes vulpes

■ Figure C4.2.3 A food chain

Common mistake

Arrows in food chains indicate energy flow and not which organism is eating which. Arrows in food chains go from producer → primary consumer → secondary consumer, and so on.

◆ **Food chain**: sequence of organisms within a community in which each is the food (and, hence, energy source) of the next, starting with a producer.

◆ **Consumer**: organisms that are unable to synthesize glucose and so eat other organisms or organic matter to obtain it and other nutrients.

◆ **Food web**: interconnected food chains in an ecological community.

ATL C4.2B

Construct a food chain for two named ecosystems – tropical rainforest and savannah. Make sure you include both the common name and scientific name for each species (as shown in Figure C4.2.3). Don't forget that arrows indicate energy flow in your diagrams. You do not need to include pictures of the organisms, just their names.

Food chains and food webs represent feeding relationships in a community

In an ecosystem, food chains are not isolated. Rather, they interconnect with other chains. This is because most prey species have to escape the attentions of more than one predator. Predators, as well as having preferences, need to exploit alternative food sources when one source becomes limited. They also take full advantage of gluts of food as particular prey populations become temporarily abundant.

Consequently, individual food chains may be temporary and are interconnected so that they may form a **food web**. An example of a marine food web is shown in Figure C4.2.4. As before with food chains, the arrows indicate the direction of the transfer of energy and biomass.

Food chains tell us about the feeding relationships of organisms in an ecosystem, but they are entirely qualitative relationships (we know which organisms are present as prey and as predators), rather than providing quantitative data (we do not know the numbers of organisms at each level).

Concept: Interdependence

Changes in the population of one species in a food chain or web can affect the populations of other organisms. Organisms higher in a food chain depend on those lower down as sources of food, for example. If one essential organism is removed from the chain, perhaps resulting from human impacts such as habitat loss, then the food web could collapse, endangering the entire ecosystem.

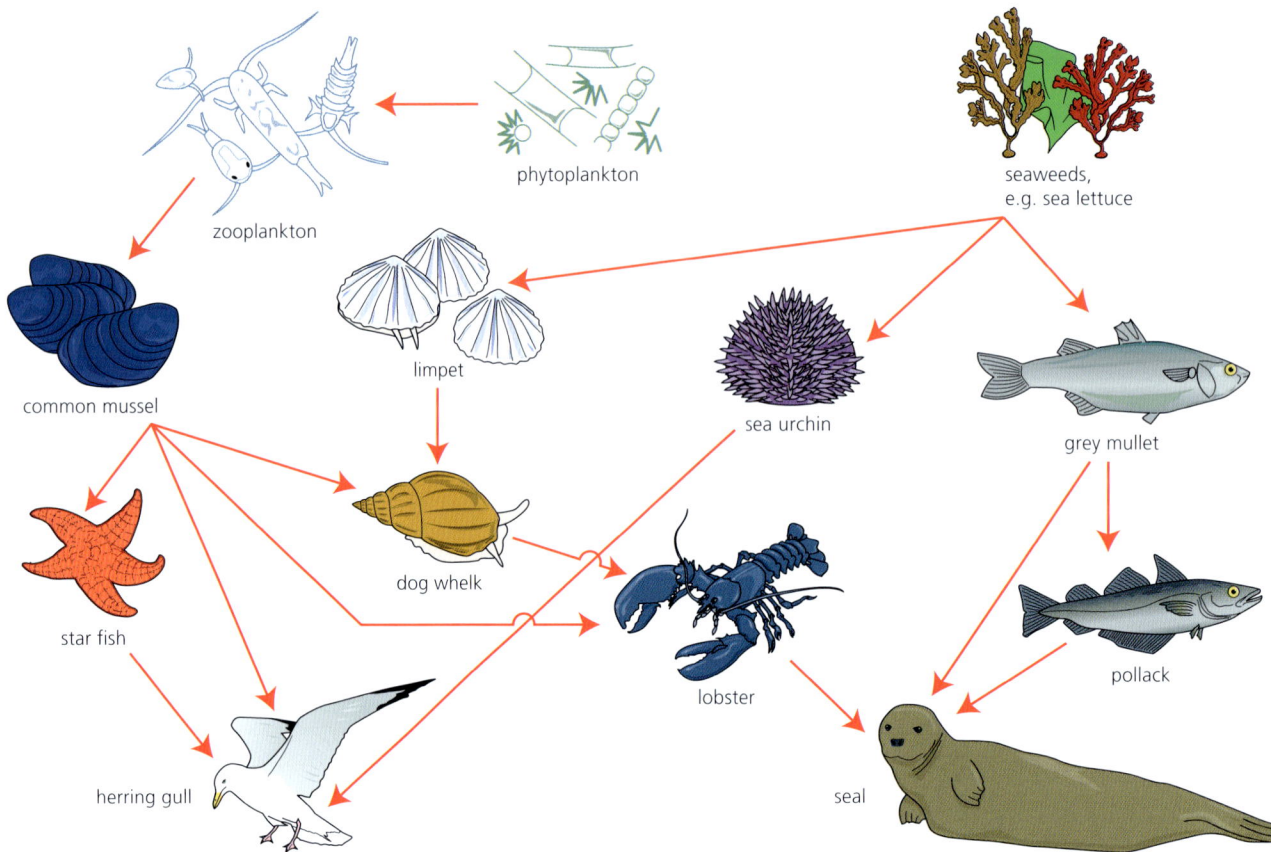

■ Figure C4.2.4 Example of a marine food web

2 Using the information in the food web in Figure C4.2.4, **construct** two individual food chains from the marine ecosystem, each with at least three linkages (four organisms).

3 **Identify** the initial energy source for almost all communities.

♦ **Trophic level**: feeding level within a food chain.

♦ **Decomposer**: organisms (typically micro-organisms) that feed on dead plant and animal material, causing matter to be recycled by other living things.

Trophic levels

Food chains show a sequence of organisms in which each is the food of the next organism in the chain, so each organism represents a feeding or **trophic level**. Chains typically start with a producer and end with a consumer – perhaps a secondary or tertiary consumer.

There is not a fixed number of trophic levels in a food chain, but typically there are three, four or five levels only. There is an important reason why stable food chains remain quite short. We will come back to this point shortly.

Supply of energy to decomposers as carbon compounds in dead organic matter

Organisms that feed on dead plants and animals, and on the waste matter of animals, are **decomposers**.

Producers support all ecosystems through constant input of energy and new biomass. Decomposers feed at each trophic level of a food chain and are essential for recycling matter, including elements such as carbon and nitrogen, in ecosystems (page 601, this section). Feeding by decomposers releases inorganic nutrients from the dead organic matter, including carbon dioxide, water and ammonia, and ions such as nitrates and phosphates. Sooner or later, these inorganic nutrients are absorbed by producers and reused (Figure C4.2.5). We will look in more detail at the cycling of nutrients in the biosphere below and later in this chapter.

Common mistake

The terms 'decomposer' and 'detritivore' are often used interchangeably. Although 'decomposer' can be used to include all organisms that feed on dead biomass and waste from organisms, the term can also be used to refer specifically to micro-organisms that carry out saprotrophic nutrition (known as saprotrophs). Detritivores, in contrast, eat the detritus (dead or waste plant and animal material) and process it in their intestines. Decomposers include bacteria and fungi, and detritivores include earthworms and woodlice. Matter consumed by decomposers and detritivores includes faeces, dead parts of organisms and dead whole organisms.

Link
Saprotrophic nutrition is described in Chapter B4.2, page 367.

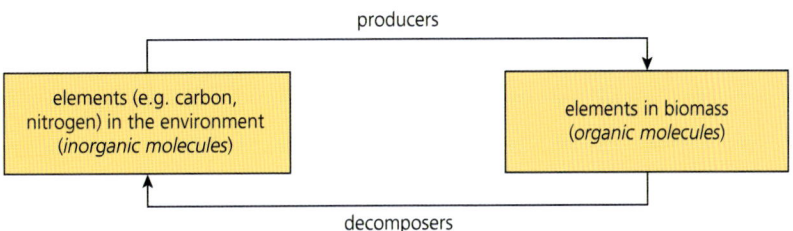

■ Figure C4.2.5 The role of producers and decomposers in ecosystems

Cycling of nutrients

Recycling of nutrients is essential for the survival of living things because the available resources of many elements are limited. When organisms die, their bodies are broken down into simpler substances (for example, carbon dioxide, water, ammonia and various ions), as illustrated in Figure C4.2.6. Nutrients are released.

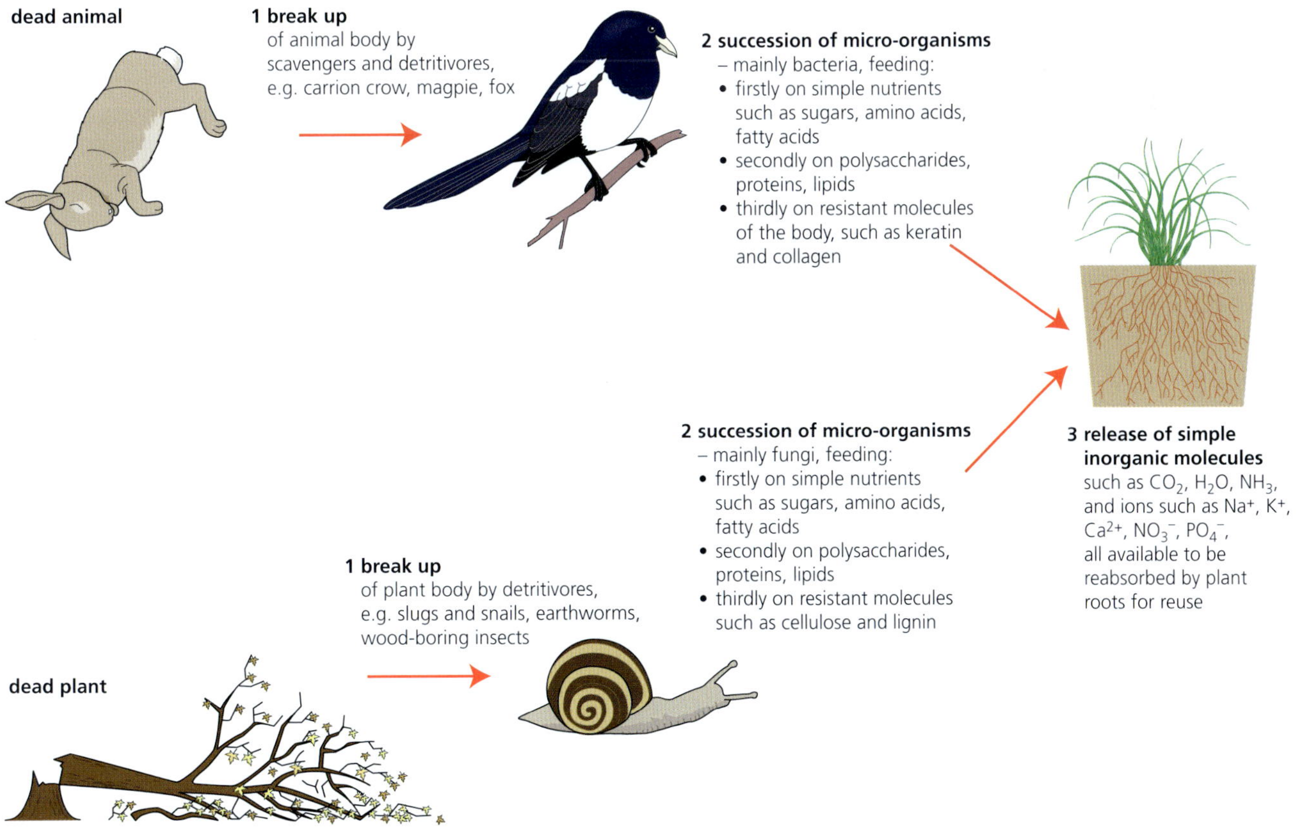

dead animal

1 break up of animal body by scavengers and detritivores, e.g. carrion crow, magpie, fox

2 succession of micro-organisms – mainly bacteria, feeding:
- firstly on simple nutrients such as sugars, amino acids, fatty acids
- secondly on polysaccharides, proteins, lipids
- thirdly on resistant molecules of the body, such as keratin and collagen

2 succession of micro-organisms – mainly fungi, feeding:
- firstly on simple nutrients such as sugars, amino acids, fatty acids
- secondly on polysaccharides, proteins, lipids
- thirdly on resistant molecules such as cellulose and lignin

3 release of simple inorganic molecules such as CO_2, H_2O, NH_3, and ions such as Na^+, K^+, Ca^{2+}, NO_3^-, PO_4^-, all available to be reabsorbed by plant roots for reuse

1 break up of plant body by detritivores, e.g. slugs and snails, earthworms, wood-boring insects

dead plant

■ Figure C4.2.6 The sequence of organisms involved in decay

C4.2 Transfers of energy and matter

Detritivores begin the process of breakdown and decay, but saprotrophic bacteria and fungi (decomposers) always complete the breakdown. Elements that are released may become part of the soil solution, and some may react with chemicals of soil or rock particles, before becoming part of living things again through reabsorption and assimilation by plants. Ultimately, both plants and animals depend on the activities of saprotrophic micro-organisms to release matter from dead organisms for reuse.

The complete range of recycling processes by which essential elements are released and reused involve both living things (the biota) and the non-living (abiotic) environment. The latter consists of the atmosphere, hydrosphere (oceans, rivers and lakes) and lithosphere (rocks and soil). All the essential elements take part in such cycles. One example is the carbon cycle (page 594).

The supply of inorganic nutrients in an ecosystem is finite and limited as usually there are no new inputs from outside the ecosystem, but recycling does ensure their continuous availability. By contrast, there is a continuous, but variable, supply of energy in the form of sunlight. We focus on this next.

Link
How a species feeds is an important part of its niche and is covered in Chapter B4.2, page 363.

Link
The definition of autotroph and more details are given in Chapter B4.2, page 366.

Autotrophs

Energy flow through ecosystems depends on the interrelationship between organisms that can trap and store energy in the form of glucose, and those that release energy through respiration.

Energy is required for carbon fixation (photosynthesis) and for the anabolic reactions that synthesize macromolecules, such as condensation reactions that produce polysaccharides, polypeptides and nucleic acids. We know that green plants make their own organic nutrients from an external supply of inorganic molecules, in general using energy from sunlight in photosynthesis. The nutrition of a typical green plant is described as **autotrophic**, meaning 'self-feeding'. There are a very few exceptions to this, for example carnivorous plants and the species discussed in the Nature of science box below (Figure C4.2.7). The great majority of green plants are entirely autotrophic in their nutrition and in this way play a key part in food chains (see below).

● Nature of science: Patterns and trends

The statement 'all plants are autotrophic' is a generalization. Scientists, while looking for patterns or trends and trying to draw general conclusions by inductive reasoning, also look for discrepancies. Broomrape (*Orobanche* sp.) is a 'root parasite', attaching to the root systems of its various host plants, below ground. Above ground, the shoots are virtually colourless (chlorophyll-free), and the leaves reduced to small bracts. This plant is heterotrophic rather than autotrophic. *Why?*

Once established, the plant is seen to concentrate on reproduction, seed production and seed dispersal. This suggests that the task of reaching fresh hosts is a major challenge in the life cycle of a parasite. Which of these features are evident in the plant shown here?

■ **Figure C4.2.7** Not quite all green plants are autotrophic

Link
Photoautotrophs are also covered in Chapter B4.2, page 366. Redox reactions are covered in Chapter C1.2, page 412.

4 Compare and contrast photoautotrophs and chemoautotrophs.

Photoautotrophs and chemoautotrophs

Producers make their own glucose and convert (fix) inorganic molecules into organic molecules. Plants, algae and some bacteria are producers. There are two different types of producer.
- **Photoautotrophs** (e.g. nearly all plants) transfer solar energy into chemical energy.
- **Chemoautotrophs** (e.g. nitrifying bacteria) use chemical energy from oxidation reactions to create glucose.

Redox reactions release energy, and so are useful in living organisms. Both respiration and photosynthesis use redox reactions. Light is used as the external energy source in photoautotrophs, while oxidation reactions are the energy source in chemoautotrophs.

Iron-oxidizing bacteria are an example of a chemoautotroph. Common iron-oxidizing bacteria, including *Gallionella*, *Sphaerotilus*, *Crenothrix* and *Leptothrix* species, oxidize iron(II) ions (Fe^{2+}) to iron(III) ions (Fe^{3+}) to obtain energy for the synthesis of glucose. Iron-oxidizing bacteria take iron atoms out of water and remove electrons to release energy for metabolic processes.

Heterotrophs

Link
Heterotrophs are also discussed and defined in Chapter B4.2, page 366.

Top tip!

A heterotroph is an organism that obtains organic molecules from other organisms. A consumer is an organism that ingests other organic matter that is living or recently killed. Saprotrophs are heterotrophs that do not 'ingest' organic matter but secrete enzymes for external digestion.

In contrast to green plants, animals and most other types of organism are **consumers**, using only existing organic matter. Consumers do not contain photosynthetic pigments (e.g. chlorophyll) and so cannot make their own food. They must obtain the energy, minerals and nutrients they need by eating other organisms. They are also known as **heterotrophs**. The complex carbon compounds, such as proteins and starch, are digested to monomers, which are later used by the organism to synthesize their own polymers. Monomers, such as amino acids and glucose, are obtained by digestion and absorbed into heterotrophs' cells and are assimilated by constructing the carbon compounds required. Digestion can be internal or external, however absorption is always into cells. Energy and biomass are passed through a food chain from the producers through to the top carnivores. Consequently, animal nutrition is dependent on plant nutrition, either directly or indirectly.

Note that some of the consumers, known as **herbivores**, feed directly and exclusively on plants (page 372). **Carnivores** (and omnivores) feed on other consumers.

Release of energy in both autotrophs and heterotrophs

Autotrophs have stores of chemical energy in the form of glucose, whereas heterotrophs need to consume other organisms to acquire stored chemical energy. What autotrophs and heterotrophs do have in common, however, is the ability to release this stored chemical energy to produce ATP in order to carry out life processes. The release of energy in both autotrophs and heterotrophs is from the oxidation of carbon compounds in cell respiration. Autotrophs are the base of all food chains, in that they provide the energy to maintain ecosystems, and heterotrophs pass on this energy from one trophic level to the next. All energy is ultimately released from food chains as heat, through the breakdown of organic molecules via respiration (see page 405).

Common mistake

People often think that only animals respire. All organisms respire, including plants, bacteria, fungi and protists. Respiration provides all organisms with energy for growth, synthesis of biological molecules, and other living processes.

C4.2 Transfers of energy and matter

Link

Food chains are covered earlier in this chapter on page 580.

◆ **Primary consumer**: second trophic level of a food chain; organisms that consume/feed on producers.

◆ **Secondary consumer**: third trophic level of a food chain; an animal (carnivore) that feeds on plant-eating animals (herbivores) in a food chain.

◆ **Tertiary consumer**: fourth trophic level of a food chain; an animal that feeds on secondary consumers in a food chain.

Classification of organisms into trophic levels

Food chains are formed from different trophic (or feeding) levels. Producers are at the bottom of food chains, the first trophic level. Herbivores, which consume producers, are called **primary consumers**.

Animals that feed exclusively on other animals are carnivores. Carnivores that feed on primary consumers are known as **secondary consumers**, and carnivores that feed on secondary consumers are called **tertiary consumers**, and so on.

The trophic levels of organisms in food chains and food webs from different ecosystems are classified in Table C4.2.1.

■ Table C4.2.1 A summary of trophic levels in different ecosystems

Trophic level	Woodland	Rainforest	Savannah
producer	oak	vines and creepers on rainforest trees	grass
primary or first consumer	caterpillar	silver-striped hawk-moth	wildebeest
secondary consumer	beetle	praying mantis	lion
tertiary consumer	shrew	chameleon lizard	
quaternary consumer	fox	hook-billed vanga-shrike	

Sometimes it can be difficult to decide at which trophic level to place an organism. For example, in the marine food chain of Figure C4.2.4, page 582, the seal feeds on grey mullet (a primary consumer), pollack (a secondary consumer) and lobster (a tertiary consumer), making the seal itself a secondary, tertiary or quaternary consumer depending on which food chain is referred to.

● Top tip!

Many organisms have a varied diet and therefore occupy different trophic levels in different food chains.

5 **Define** the term *trophic level*.

6 **Suggest** what trophic levels humans occupy. Give examples of different food chains that humans can be in.

7 Refer back to Figure C4.2.4 on page 582, showing a marine food web.
   a **State** the trophic level of each species.
   b **Identify** organisms that occupy more than one trophic level.

Construction of energy pyramids

Feeding relationships of a food chain may be structured like a pyramid. At the start of the chain is a very large amount of energy in green plants. This supports a smaller quantity of energy in primary consumers, which, in turn, supports an even smaller quantity of energy in secondary consumers. A generalized ecosystem pyramid diagram, representing the structure of an ecosystem in terms of the energy content of the organisms at each trophic level, is shown in Figure C4.2.8. Percentages represent the relative quantities of energy available at each trophic level.

● Top tip!

An energy pyramid is a graphical representation showing the flow of energy at each trophic level in an ecosystem.

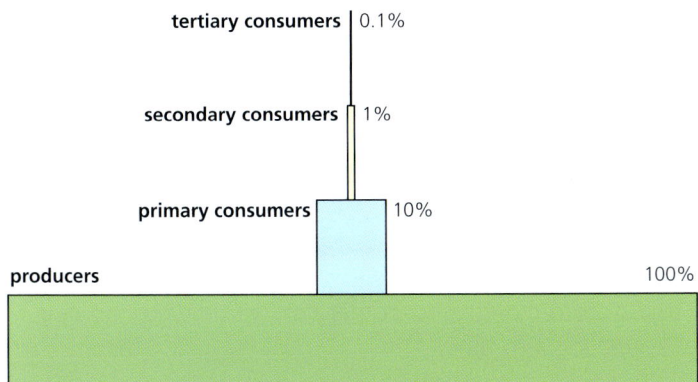

Only energy taken in at one trophic level and then built in as chemical energy in the molecules making up the cells and tissues is available to the next trophic level. This is about 10% of the energy.

The reasons are as follows:
- Much energy is used for cell respiration to provide energy for growth, movement, feeding and all other essential life processes.
- Not all food eaten can be digested. Some passes out with the faeces. Indigestible matter includes bones, hair, feathers and lignified fibres in plants.
- Not all organisms at each trophic level are eaten. Some escape predation.

■ Figure C4.2.8 A generalized pyramid of energy

Nature of science: Models

Pyramids are graphical models. They show quantitative differences between the trophic levels in an ecosystem. Pyramids of energy model the energy flow through ecosystems and represent the flow of energy through trophic levels. As energy is lost at each trophic level and gradually diminishes throughout the food chain, these diagrams are always pyramid shaped (unlike pyramids of numbers, which plot the number of organisms at each trophic level, and pyramids of biomass, which plot the total biomass at each trophic level).

Pyramids of energy represent the productivity of each trophic level, that is to say, how much new biomass is being made (and therefore energy stored) in a particular period of time in a specific area. They are therefore measured in units of flow (e.g. $kg\,m^{-2}\,yr^{-1}$ or $kJ\,m^{-2}\,yr^{-1}$). The advantages of using pyramids of energy to represent ecosystems as graphical models, compared to other methods, can be summarized as follows:
- Pyramids of numbers and biomass show the storage in the food chain at a given point in time (they are a 'snapshot' of the ecosystem), whereas pyramids of energy show the rate at which those stores are being generated.
- Pyramids of energy take into account seasonal fluctuations, as data are recorded over a period of time, rather than at one point in time. Pyramids of numbers and biomass can, therefore, be inverted (whereby the producer trophic level is narrower than the primary consumer trophic level), if the amount and/or biomass of the producers is lower than that of the next trophic level at the time when the samples were recorded. For example, grass in a field may have a lower biomass than the herbivores that feed on it because the herbivores have been eating the grass over time. This results in an inverted pyramid of biomass; over time, as the grass grows and gets longer, the productivity of the grass will be higher than the productivity of the herbivores that feed on it. Because productivity, recorded over a period of time, is always higher in producers than subsequent trophic levels, pyramids of energy are never inverted.

Inquiry 1: Exploring and designing

Exploring

You need to be able to use research data from specific ecosystems to represent energy transfer and energy losses between trophic levels in food chains.

Research data from a specific ecosystem in your country or region. Make sure you consult a variety of sources and select the relevant information. How is the energy in the ecosystem measured? How could this be used to represent energy transfer and energy losses between trophic levels in food chains?

C4.2 Transfers of energy and matter

Tool 3: Mathematics

Representing energy flow in the form of pyramids of energy

On page 586, the use of pyramids of energy as graphical models of energy flow through an ecosystem was discussed.

Pyramids of energy are created by plotting each horizontal bar (representing each trophic level) with the same height and different lengths to represent the different energy content. The vertical axis is centred on the horizontal axis, with data distributed equally to each side of the axis, creating a pyramid.

Pyramids of energy can be drawn for different biomes (see Figure C4.2.9). Biomes vary in their capacity to accumulate biomass, as noted above. Biomass accumulates when autotrophs and heterotrophs grow or reproduce. Trophic levels can be constructed by using the '10% rule' of energy conversion (that is, usually around 10% of energy is passed on from one trophic level to the next).

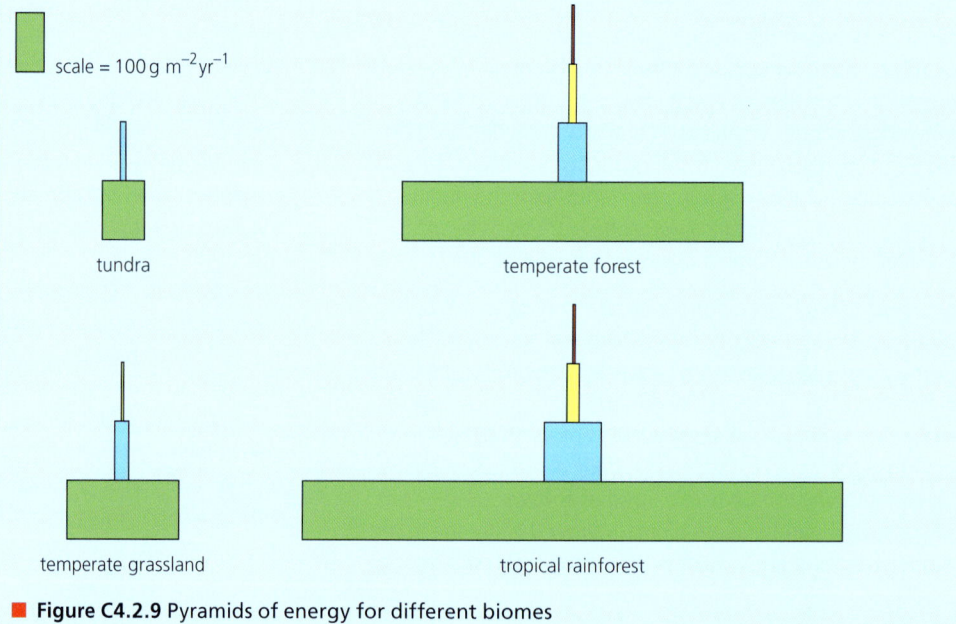

■ **Figure C4.2.9** Pyramids of energy for different biomes

Inquiry 2: Collecting and processing data

Processing data; interpreting results

Construct a pyramid of energy for an ecosystem in your country. Use the primary production data (measured in grams of carbon per m² per year) you collected in Inquiry 1 on page 587 to plot the pyramid, using the instructions in the Tool 3 above.

How many trophic levels does your pyramid of energy have? Why is this? How does your graph compare to those in Figure C4.2.9?

Diagrams of pyramids of energy often show inaccurate proportions for the different trophic levels. The width of each level should be 10% of the one below it (see Figure C4.2.9).

Reductions in energy availability in food chains

Between each trophic level there is an energy transfer. At the base of the food chain, green plants (the producers) transfer light energy into the chemical energy of sugars in photosynthesis. However, much of the light energy reaching the green leaf is not retained in the green leaf. Some is reflected away, some transmitted, and some lost as heat energy. Meanwhile, sugars are converted into lipids, amino acids and other metabolites within the cells and tissues of the plant. Some of these metabolites are used in the growth and development of the plant and, through these reactions, energy is stored in the organic molecules of the plant body.

The reactions of respiration and of the rest of the plant's metabolism produce heat energy, so some energy is lost from the plant through these reactions. Chemical energy is transferred every time the tissues of a green plant are eaten by herbivores. Finally, on the death of the plant, the remaining energy passes to detritivores and saprotrophs when dead plant matter is broken down and decayed. The diverse routes of energy flow through a primary producer are summarized in Figure C4.2.10.

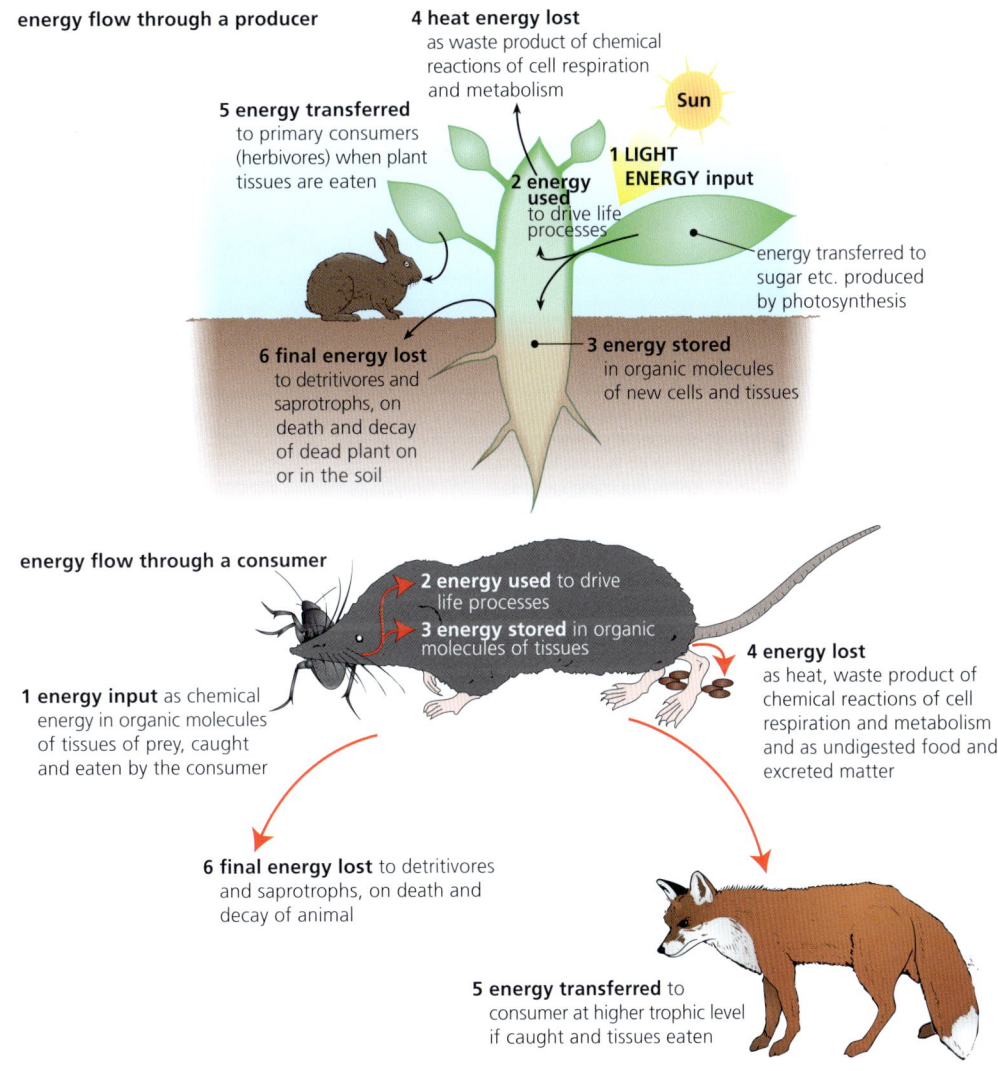

■ **Figure C4.2.10** Energy flow through producers and consumers

C4.2 Transfers of energy and matter

The reduction in energy availability at each successive stage in food chains is due to large energy losses between trophic levels. Producers are eaten by primary consumers, or herbivores. In this way, energy is transferred to the consumer. The consumer, in turn, transfers energy in the muscular movements by which it hunts and feeds, and as it seeks to escape from predators. Some of the food it has eaten remains undigested, passing through the consumer unchanged, and is lost as waste products in faeces and as urea. Heat energy, a waste product of the reactions of respiration and of the animal's metabolism, is continuously lost as the consumer grows, develops and forms body tissues. If the consumer is itself caught and consumed by another, larger consumer, energy is again transferred. Finally, on the death of the consumer, the remaining energy passes to detritivores and saprotrophs when dead matter is broken down and decayed. Ultimately, the energy in matter that is decayed is lost as heat. Only a relatively small proportion of this energy is passed on to the next trophic level (see Figure C4.2.11). Around 90% of energy is lost between trophic levels due to:

- biomass not being ingested (eaten)
- food not being digested (and then absorbed into consumers for assimilation)
- excretion
- loss as heat from respiration
- loss as inedible parts, such as bones, teeth and fur.

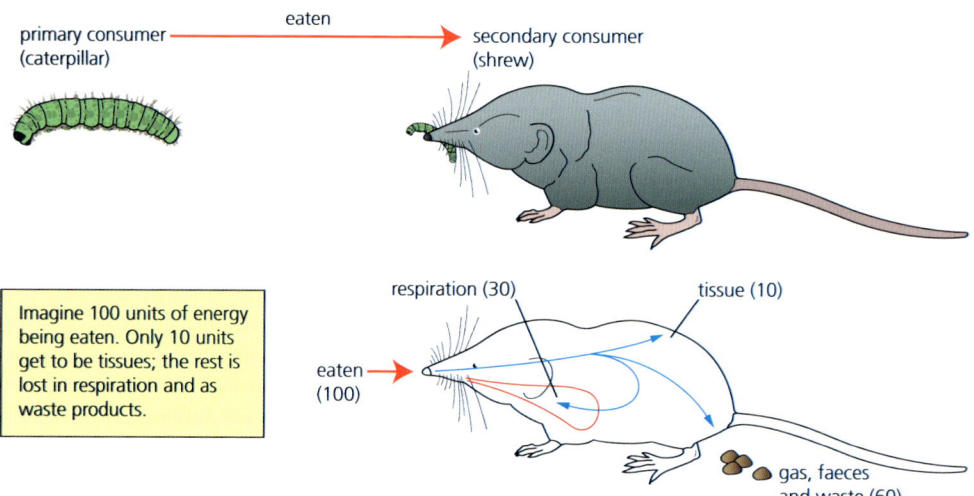

■ Figure C4.2.11 Loss of energy between trophic levels

The percentage of ingested energy converted to biomass is dependent on the respiration rate. Animals with greater respiration rates will, on average, pass on less energy than ones with a lower metabolic rate, although much also depends on the diet of the animal (herbivores are less efficient at digesting plant material, with more faecal matter produced, than carnivores are at digesting meat). Ectothermic (cold-blooded) animals, such as snakes and reptiles, lose less heat from respiration than endothermic animals, such as mammals, that maintain their body temperatures at constant levels.

Decomposers and detritus feeders are not usually considered to be part of food chains. However, these organisms play a role in energy transformations in food chains. Biomass that is not eaten by consumers is recycled by decomposers and detritivores, and they ultimately return the energy to the environment as heat via respiration.

8 **List** the factors that lead to energy loss from food chains.

■ Heat loss in ecosystems

Heat loss to the environment in both autotrophs and heterotrophs is due to conversion of chemical energy to heat in cell respiration (Figure C4.2.12). Cell respiration and metabolism produce heat as a waste product.

A linear food chain – energy flow

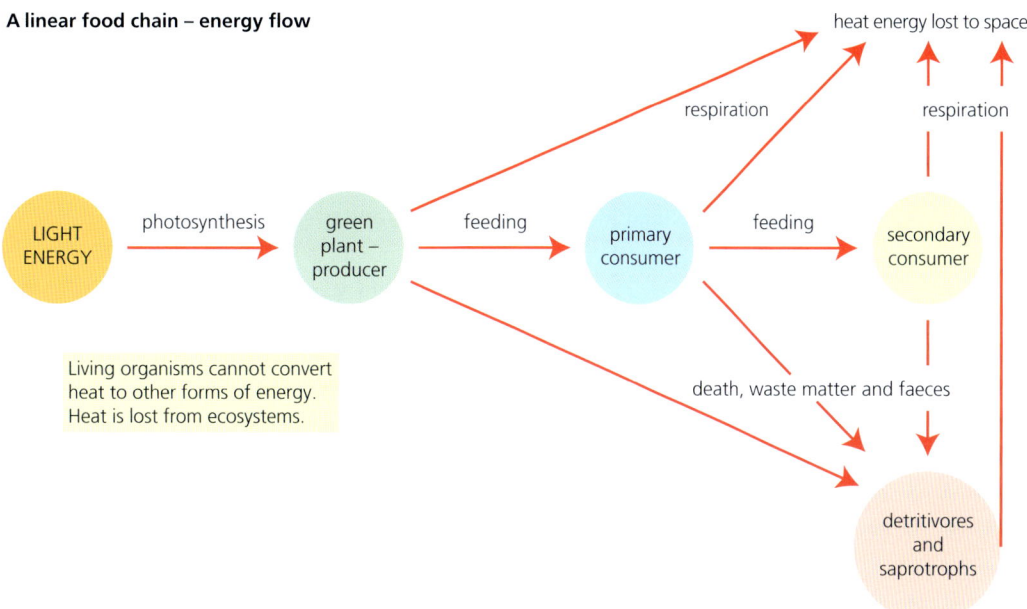

Figure C4.2.12 Heat is lost from food chains

> **Top tip!**
> Energy transfers are not 100% efficient, so heat is produced both when ATP is produced in cell respiration and when it is used in cells.

The transfer of energy in food chains is inefficient. The first law of thermodynamics states that energy cannot be created or destroyed. The total amount of energy in an **isolated system** does not change, although this energy is gradually dissipated from useable energy (such as ATP) to less useful, simpler forms (such as heat) as work is done. That is the second law of thermodynamics.

Restrictions on the number of trophic levels

Only about 10% of what is eaten by a consumer is built into that organism's body, and so is potentially available to be transferred on through predation or grazing on plants. There are two consequences of this:
- At each successive stage in food chains there are fewer organisms or smaller organisms. (*Note: there is therefore less biomass, but the energy content per unit mass is not reduced.*)
- The number of trophic levels in ecosystems is therefore restricted due to energy losses.

Energy is lost from the food chain as heat during respiration, due to incomplete digestion, and through excretion of the waste products of metabolism. Therefore, energy decreases through a food chain, with a general rule of 10% of the energy being passed on to the next level. The energy loss at transfer between trophic levels means that food chains are short. Few transfers can be sustained when so little of what is eaten by one consumer is potentially available to those organisms in the next step in the food chain. Consequently, it is very uncommon for food chains to have more than four or five links between producer (green plant) and top carnivore. Food chains that contain species that expend a lot of energy, such a warm-blooded predator that loses energy through maintaining body temperature and in hunting prey, will be shorter than food chains that contain organisms that expend less energy, such as a spider, which is both cold-blooded and catches its prey using a web, and so loses less energy through movement and loss of body heat.

9 **Explain** why, in a food chain, a large amount of plant material supports a smaller mass of herbivores and an even smaller mass of carnivores.

 ## Food chains and world hunger

Today, the number of humans in the world population is huge. The human population is predicted to reach 8 billion on 15 November 2022 and the United Nations estimates it will reach 9.7 billion in 2050. This frequently places excessive demands on the food supply and on the resources that are used in its production – so much so that, around the world, there are local populations of people with too little to eat. World hunger is a major problem of which we can't fail to be aware.

TOK

Do you think that learning about the loss of energy between trophic levels, along with your knowledge about climate change and world hunger, necessarily raises ethical issues surrounding people's eating habits? Why or why not?

◆ **Primary production**: the accumulation of carbon compounds in biomass by autotrophs; accumulation of biomass occurs when autotrophs grow and reproduce.

◆ **Gross primary productivity (GPP)**: the total gain in biomass per unit area per unit time fixed by producers. (The units are g per unit time.)

◆ **Net primary productivity (NPP)**: the gain by producers in biomass per unit area per unit time remaining after allowing for respiratory losses (R). This is potentially available to consumers in an ecosystem.

There is a potential ethical challenge for well-nourished people living alongside other humans who may be starving and who may die prematurely due to malnourishment. In the light of this, some humans opt for a vegetarian diet. Vegetarians do not eat meat and most do not eat fish, but the majority consume animal products such as milk, cheese and eggs.

The rationale for this type of vegetarianism is that, when we eat meat rather than food of plant origin, we **extend food chains by a least one trophic level**. In effect, vegetarians observe that we waste energy by choosing to eat animal products rather than matter of plant origin, thereby leaving less food for others. This is an important issue when considering eating animal products.

Primary production

Primary production is the accumulation of carbon compounds in biomass by autotrophs. Units are mass (of carbon) per unit area per unit time (usually $g\,m^{-2}\,yr^{-1}$).

- The amount of glucose produced by the plant in a specific area in a specific period of time is known as **gross primary productivity (GPP)**. This equates to the rate of photosynthesis. Much of this energy is used in respiration and ultimately lost as heat from the producers.
- The remaining energy, taking into account losses through respiration, is stored in new biomass: this is known as **net primary productivity (NPP)**.

NPP can be calculated as follows:

$$NPP = GPP - R$$

where R = respiratory loss.

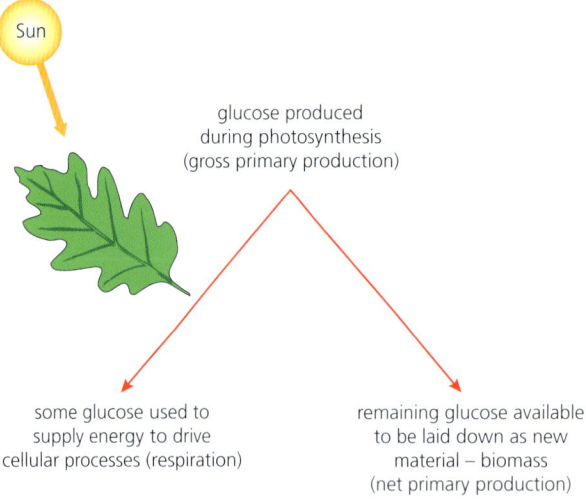

■ Figure C4.2.13 Primary production

NPP is the rate at which plants accumulate new biomass. It represents the actual store of energy contained in potential food for consumers. NPP is easier to calculate than GPP as biomass is simpler to measure than the amount of energy fixed into glucose.

NPP values vary greatly for different biomes. NPP is affected by factors that influence the rate of photosynthesis: the amount of insolation (sunlight), the amount of rainfall (precipitation) and the temperature. Warmer temperatures speed up enzyme reactions that drive photosynthesis, although temperatures that are too hot can lead to denaturation. Rainforests have the highest NPP because of year-round insolation, warm temperatures and high levels of precipitation. The high NPP means that rainforest has a complex structure, with layers from ground level to canopy.

As so much energy is stored at the producer level, rainforests can support a number of trophic levels, as at the tertiary (and even at the quaternary) consumer level there is sufficient energy to support a wide diversity of animals.

Temperate ecosystems vary from deciduous and evergreen forests to grasslands. Rainfall, temperature and sunlight are seasonal, which reduces overall NPP compared to tropical rainforests. Lower levels of NPP mean less stored chemical energy at the producer level and, therefore, less biomass at each trophic level, potentially leading to shorter food chains.

In the colder ecosystems found in the Northern Hemisphere, such as tundra, there are slower nutrient cycles, a lack of soil fauna and acidic leaf litter. Tree and shrub growth can be stunted. Evergreen species maximize productivity as photosynthesis can take place as soon as spring sunlight becomes available. The low productivity of tundra means that large areas are needed to support grazing animals such as reindeer, and these animals need to migrate long distances to support their herds. In highly stressful environments, with very low temperatures or rainfall, only mosses and lichens may be able to survive. Lower productivity in such ecosystems and the harsher conditions means that food chains are usually shorter, with limited energy available in biomass to support possibly only the primary consumer trophic level.

Secondary production

Secondary production is the accumulation of carbon compounds in biomass by heterotrophs. Secondary production may also be categorized according to gross (total) and net (usable) amounts of biomass.

Much of the biomass eaten by consumers is absorbed (e.g. through the guts of animals) and converted into new biomass within cells – this is **gross secondary productivity (GSP)**.
- Consumers do not use all the biomass they eat.
- Some energy passes out in faeces and excretion.
- Only the biomass remaining can be used by the consumer (GSP), where GSP = food eaten − faecal loss

◆ **Secondary production**: the accumulation of carbon compounds in biomass by heterotrophs; accumulation of biomass occurs when heterotrophs grow and reproduce

◆ **Gross secondary productivity (GSP)**: the total gain by consumers in energy or biomass per unit area per unit time through absorption.

◆ **Net secondary productivity (NSP)**: the gain by consumers in energy or biomass per unit area per unit time remaining after allowing for respiratory losses (R).

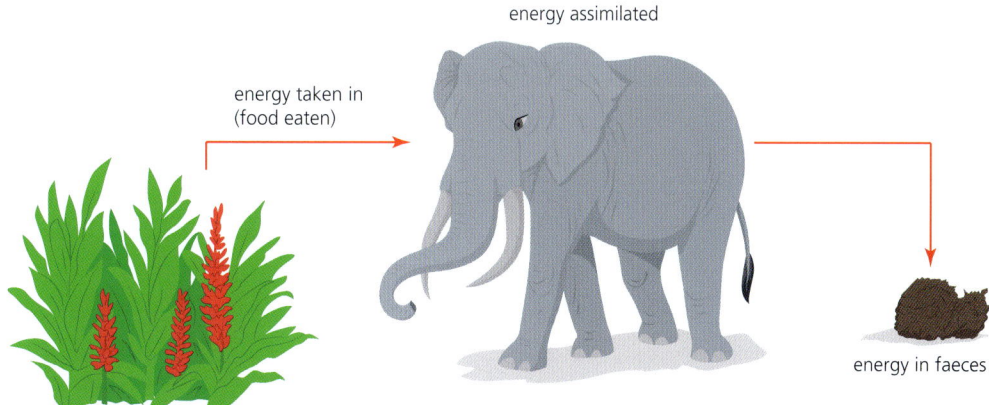

■ **Figure C4.2.14** Animals do not use all the biomass they consume. Some of it passes out in faeces and excretion (GSP)

Some of the biomass absorbed by animals is used in respiration:
- The energy released is used to support life processes.
- The remaining energy is available to form new biomass: **net secondary productivity (NSP)**
- So, NSP = GSP − R (where R = respiratory loss).

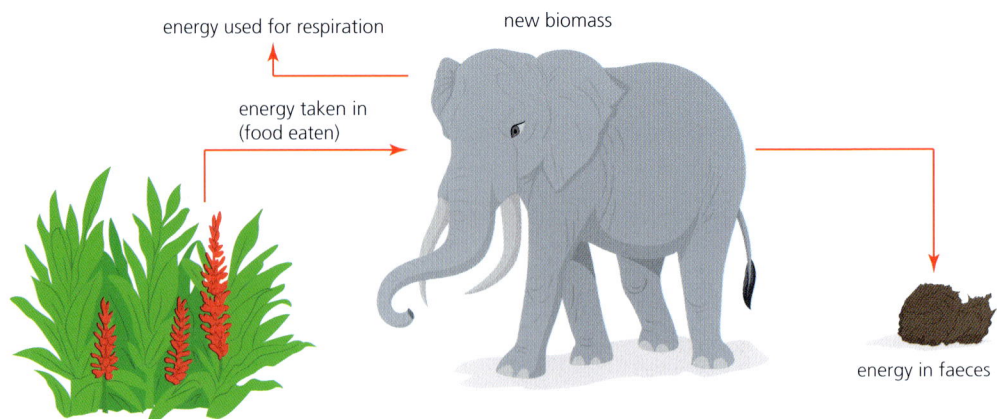

■ **Figure C4.2.15** Some of the energy assimilated by animals is used in respiration, to support life processes, and the remainder is available to form new biomass (NSP)

It is this new biomass (NSP) that is then available to the next trophic level.

Due to loss of biomass when carbon compounds are converted to carbon dioxide and water in cell respiration, secondary production is lower than primary production in an ecosystem.

10 Distinguish between primary production and secondary production.

The carbon cycle

Constructing carbon cycle diagrams

The nutrients in an ecosystem are recycled and reused. All the chemical elements of nutrients circulate between living things and the environment. The process of exchange or 'flux' (back and forth movements) of materials is a continuous one. These movements take more or less circular paths and have a biological component and a geochemical component. Consequently, these movements are known as **biogeochemical cycles**.

Photosynthesis converts atmospheric carbon, in the form of carbon dioxide, into carbon stored in biomass, initially in the form of glucose and then other organic compounds. **Feeding**, initially by herbivores and omnivores on producers, and then by carnivores consuming other organisms, passes carbon along food chains. Carbon is returned to the atmosphere by **respiration**, where glucose and other organic compounds are broken down to release ATP, producing carbon dioxide as a waste product. Respiration is carried out by all organisms in a food chain – producers and consumers – as well as decomposers, which feed on the organic molecules from dead organisms and waste products.

We are going to examine the **carbon cycle** in more detail (Figure C4.2.16).

11 Look at Figure C4.2.16. **State** an example of when the element carbon is an abiotic component and when it is a biotic component.

> ● **Top tip!**
>
> Make sure you can illustrate with a diagram how carbon is recycled in ecosystems by photosynthesis, feeding and respiration.

◆ **Carbon cycle**: the series of processes by which carbon compounds are cycled between the abiotic and biotic environments, involving photosynthesis, feeding, decomposition and respiration.

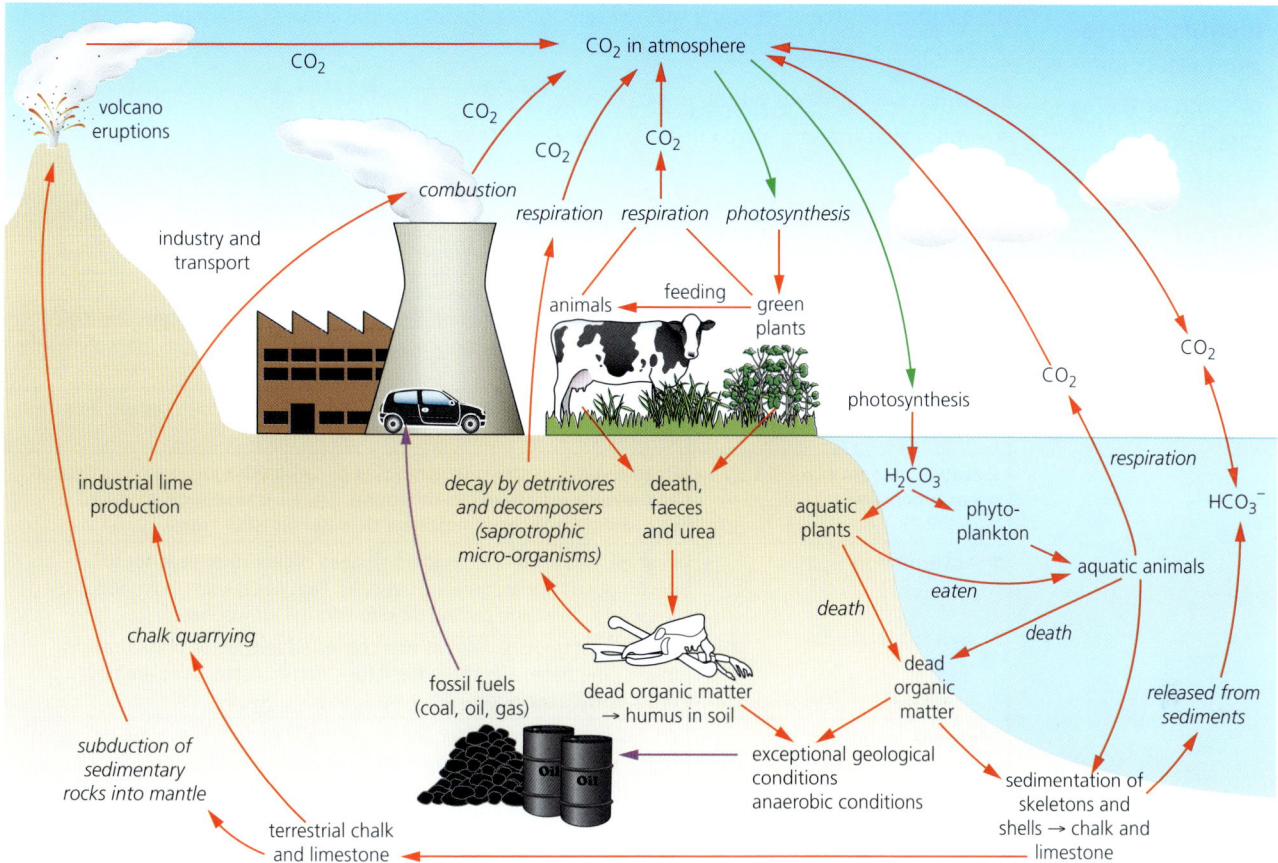

■ Figure C4.2.16 The carbon cycle

12 Use the representation of the carbon cycle in Figure C4.2.16 to **construct** a specific carbon cycle for the terrestrial ecosystem in which you live. Include the roles of photosynthesis, feeding and respiration.

In diagrams of the carbon cycle, the stores of carbon, such as carbon dioxide in the atmosphere, can be represented as boxes, and the fluxes, such as photosynthesis, as arrows between the boxes. Arrows have arrow heads to show direction. The forms in which carbon exists are stated, for example carbon compounds in plants, together with the processes that convert carbon from one form to another. Figure C4.2.16 includes all the main biological processes in the cycling of carbon.

● Common mistake

The carbon cycle should be drawn using boxes and arrows. It is not necessary to draw organisms in the carbon cycle using pictures, or a factory smokestack to represent the combustion of fossil fuels, for example. Storages (also called reservoirs) should be represented as boxes, named and the linking arrows (fluxes) labelled with the processes labelled.

Make sure that carbon dioxide is added to the diagram of the carbon cycle. This is the ultimate source of all carbon in living things, so it is essential that it is included.

■ Ecosystems as carbon sinks and carbon sources

The carbon cycle illustrates the fact that biogeochemical cycles have sources and sinks.
- A carbon **sink** is anything that absorbs more carbon from the atmosphere than it releases – for example, plants, the ocean and soil. These are reservoirs that retain carbon and keep it from entering Earth's atmosphere. Forests are typically carbon sinks, as well as oceans. When plants produce new biomass through photosynthesis, carbon dioxide is removed from the atmosphere. Deforestation is a source of carbon emission into the atmosphere, but forest regrowth is a form of carbon **sequestration**, with the forests themselves serving as carbon sinks.

C4.2 Transfers of energy and matter

13 Identify sources and sinks in the diagram of the carbon cycle (Figure C4.2.16).

14 Draw a simplified diagram of the carbon cycle, including the roles of animals, plants and decomposers.

- A carbon **source** is anything that releases more carbon into the atmosphere than it absorbs – for example, the burning of fossil fuels.

The processes by which carbon moves between these sinks, and between living things and the environment, are summarized in Table C4.2.2. Read through the table and locate each process in the carbon cycle.

■ **Table C4.2.2** How carbon is circulated between living organisms and the environment

Diffusion	Carbon dioxide diffuses from the atmosphere or water into autotrophs.
Photosynthesis in terrestrial and aquatic plants	Autotrophs convert carbon dioxide from the atmosphere into carbohydrates and other organic compounds. Aquatic plants use dissolved carbon dioxide and hydrogencarbonate ions (HCO_3^-) from the water in the same way.
Respiration in all living things	Carbon dioxide is produced as a waste product and diffuses out into the atmosphere or water.
Decay by saprotrophic micro-organisms	Dead organic matter is decomposed to carbon dioxide, water, ammonia and mineral ions by micro-organisms. The carbon dioxide produced diffuses out into the atmosphere or water (as HCO_3^- ions).
Peat formation	In acidic and anaerobic conditions, dead organic matter is not fully decomposed but accumulates as peat. Peat decays slowly when exposed to oxygen, releasing carbon dioxide into the atmosphere.
	Peat in past geological areas was converted to coal, oil or natural gas (mainly methane) and these fossil fuels accumulated. They are now progressively exploited.
Methane formation from organic matter under anaerobic conditions	Organic matter held under anaerobic conditions (such as in waterlogged soil or in the mud of deep ponds) is decayed by methane-producing bacteria. Methane accumulates in the ground in porous rocks or under water, but may progressively escape into the atmosphere. In air and UV light, methane is slowly oxidized to carbon dioxide and water.
Combustion of fossil fuels	Releases carbon dioxide into the atmosphere. Since the start of the Industrial Revolution in Europe, carbon dioxide has been released at an increasing rate.
Shell and bone formation by organisms	Many organisms combine HCO_3^- with calcium ions and other minerals to form carbonate shells and bones. These may accumulate as sediments and come to form sedimentary rocks (chalk and limestone) in geological time. Reef-building corals are particular examples of this.
Lime formation for agriculture and building materials	Terrestrial chalk and limestone are quarried and converted to lime in kilns. Carbon dioxide is released into the atmosphere in the process.
Volcanic eruptions	On a geological timescale, sedimentary rocks (including chalk and limestone) become subducted into the Earth's mantle. When volcanoes erupt, molten rocks and carbon dioxide are released into the atmosphere.

Top tip!

If photosynthesis exceeds respiration, there is a net uptake of carbon dioxide; if respiration exceeds photosynthesis, there is a net release of carbon dioxide.

Carbon fluxes

The size of the sources and sinks, and the fluxes of carbon between them, is summarized in Figure C4.2.17. Examine these data carefully. The data on directions of flux will help confirm your identification of the sources and sinks of carbon.

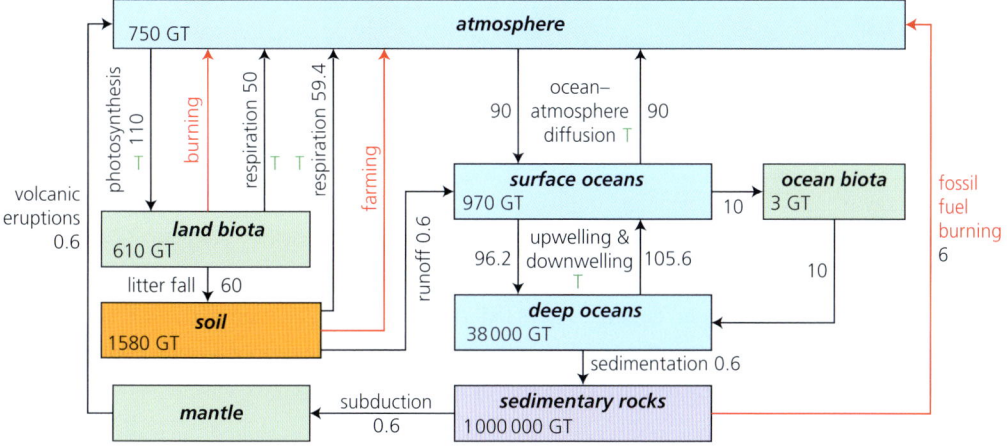

■ Figure C4.2.17 Carbon fluxes due to processes in the carbon cycle

> **15 Outline** the difference in the rates of 'flux' between the atmosphere and land biota, and between the deep ocean and sedimentary rocks.

■ Release of carbon dioxide into the atmosphere during combustion

Biomass, peat, coal, oil and natural gas are carbon sinks. Carbon sinks vary in when they were formed. All these sinks can undergo combustion to release carbon dioxide. Combustion following lightning strikes sometimes happens naturally, but human activities have greatly increased combustion rates.

Peat

Peat is formed over millions of years by plant matter that is not fully decomposed (Table C4.2.2). Large areas of wetland, such as the East Anglian fens in the UK, have been drained for farming, which has degraded the peat. No longer waterlogged, the peat has been eroded by the wind, which has increased the flux of carbon dioxide to the atmosphere. Drier conditions make it more likely for peat to catch fire. Climate-related drying is increasing the emissions of carbon from peat, while increased average temperatures, reduced rainfall and altered soil hydrology increase the risk of peat fires.

Burning of fossil fuels

Most carbon dioxide emissions come from burning fossil fuels. The discovery of coal and other fossil fuels as a source of energy enabled the widespread use of machinery in the nineteenth century to power industrial revolutions. Energy – along with carbon – trapped by plants millions of years ago can be released. Since the start of the Industrial Revolution in Europe, in the early part of the nineteenth century, carbon dioxide has been released at an increasing rate (Table C4.2.3), attributed to the burning of coal and oil. Fossil fuels are so-called because they are the remains of biological life, mostly laid down in the Carboniferous period. As a result, we are now adding to our atmosphere carbon that had been locked away for about 350 million years. This is an entirely new development in geological history.

■ Table C4.2.3 Changing levels of atmospheric carbon dioxide recorded and predicted

	CO_2/parts per million
pre-Industrial Revolution level	280 (±10)
by mid-1970s	330
by 1990	360
by 2007	380
by 2013	400
by 2018	409
by 2050 (if current rate is maintained)	500

Energy security is a key goal for countries, with most relying on fossil fuels. Energy security depends on an adequate, reliable and affordable supply of energy that provides a degree of independence. The most accessible, and lowest-cost, deposits of fossil fuels are invariably developed first. Large oil, coal and gas reserves in certain countries have led to fossil fuel exploitation in those countries. Carbon-emitting energy resources are currently at the core of the global economy, with countries largely dependent on fossil fuels to drive economic development.

As well as modifying the carbon cycle, fossil fuel combustion has altered the way in which energy from the Sun interacts with the atmosphere and the surface of our planet. The concentration of atmospheric carbon, and associated positive feedback mechanisms, leads to global warming. Global warming and devastating natural disasters such as Hurricane Katrina have increased the awareness about the misuse of energy resources, leading to movement away from fossil fuel resources.

Link
Global warming and positive feedback mechanisms are discussed in Chapter D4.3, page 825.

Fires in the Amazon

Parts of the Amazonia region of South America experience a dry season (any month with less than 100 mm of rain) from July to September. However, even the Amazon forest's drought tolerance has its limits. Land-use activities increase forest susceptibility to fire during periods of drought by providing ignition sources, by fragmenting the forest and by thinning the forest through logging. Forests are burned in preparation for crops or pasture and to improve pasture forage. However, these fires frequently burn beyond their intended boundaries into neighbouring forests. During the severe drought of 1998, approximately 39 000 km² of standing forest caught fire, which is twice the area that was clear-cut that year. In 2016, the number of fires increased by 36% compared with the preceding 12 years. The dry season is also lasting longer, thus increasing the risk of fires. These low, slow-moving fires can kill up to 50% of trees above 10 cm in diameter. Forest fires can increase susceptibility to further burning in a positive feedback mechanism as they kill trees, thus opening the canopy and enabling increased solar radiation to reach the forest floor.

One projection for 2030 suggests that 31% of the Amazon closed-canopy forest formation will be deforested, and 24% will be damaged by drought or logging. If 'business as usual' continues, logging will further reduce the Amazonian forest carbon store size by 15%; drought damage will cause another 10% reduction in forest biomass. The effects of more frequent fires could be to release an additional 20% of forest carbon to the atmosphere.

Analysis of the Keeling curve in terms of photosynthesis, respiration and combustion

Link
Climate change and its impacts are covered in Chapter D4.3, page 820.

◆ **Keeling curve**: a daily record of global atmospheric carbon dioxide concentration kept since the 1950s.

16 Mauna Loa is a volcano on the Island of Hawaii in the US State of Hawaii in the Pacific Ocean. It may seem unusual that such a tiny island provides such key data for scientists. **Suggest** why Mauna Loa is used as the site for monitoring atmospheric carbon dioxide concentrations.

Nature of science: Measurements

The concentration of atmospheric carbon dioxide has become of critical concern today, with the climate warming year on year. It is therefore important to obtain reliable data on the concentration of carbon dioxide in the atmosphere. Accurate measuring devices were established at the Mauna Loa Observatory, Hawaii, in 1957 to monitor the global environment. Measurements of atmospheric carbon dioxide concentrations had begun with work by Dr Charles David Keeling, so the graph plotted from carbon dioxide data is known as the **Keeling curve**. Dr Keeling measured carbon dioxide concentrations representative of the 'free atmosphere': concentrations that prevail over a large part of the Northern Hemisphere. Measurement of carbon dioxide is now the responsibility of two scientific institutions: the Scripps Institution of Oceanography, and the National Oceanic and Atmospheric Administration. Monthly data are posted.

In early work at Mauna Loa, Keeling noticed that the carbon dioxide concentration rose by 1 part per million (ppm) in April 1958 to a maximum in May before declining and reaching a minimum in October 1958. After this, the concentration increased again and repeated the same seasonal pattern in 1959. Keeling said: 'We were witnessing for the first time nature's withdrawing carbon dioxide from the air for plant growth during summer and returning it each succeeding winter.'

The most recent data for carbon dioxide levels available from the ongoing investigation are shown in Figure C4.2.18. Notice the annual rhythm shown in the atmospheric carbon dioxide concentration (lower in the summer months and higher in the winter months).

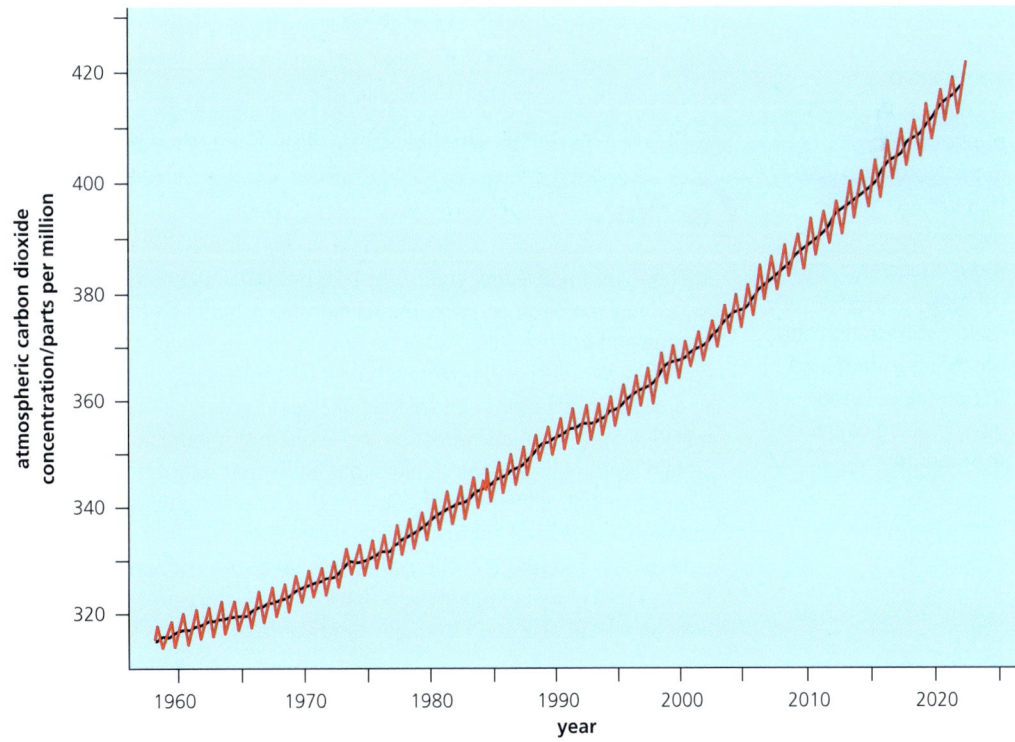

■ **Figure C4.2.18** Atmospheric carbon dioxide measured at the Mauna Loa Observatory, Hawaii

Notice the annual rhythm (Figure C4.2.18 and Figure C4.2.19), first noticed by Dr Keeling, shown in the atmospheric carbon dioxide concentration (in parts per million): lower in the summer months and higher in the winter months. This is due to photosynthesis that occurs on land in the Northern Hemisphere, which impacts on the composition of the global atmosphere.

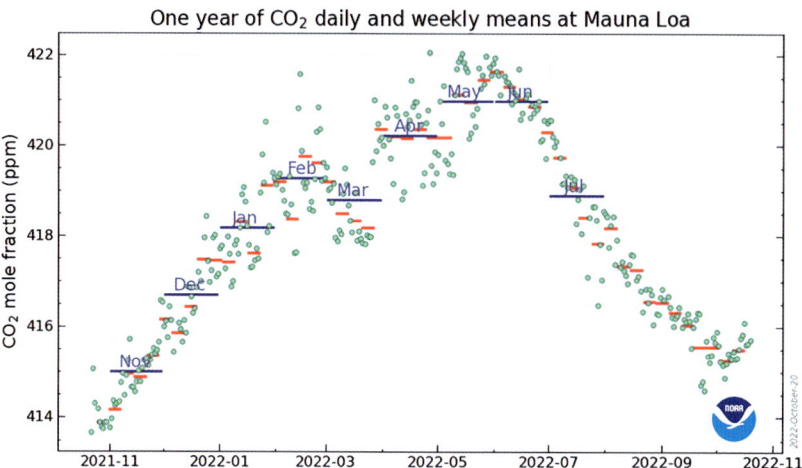

■ **Figure C4.2.19** A record of carbon dioxide daily and weekly means at Mauna Loa, 2021–22; note that the *x*-axis shows data recorded for more than one year

In June 2022, carbon dioxide was present in the atmosphere at a concentration that varied between 413 and 422 ppm (Figure C4.2.19). We might expect the amount of atmospheric carbon dioxide to be maintained by a balance between the fixation of this gas during photosynthesis and release of carbon dioxide into the atmosphere by respiration, combustion and decay by micro-organisms – an interrelationship illustrated in the carbon cycle (Figure C4.2.16). In fact, photosynthesis does withdraw almost as much carbon dioxide during daylight hours as is released into the air by all the other processes, day and night – but not quite as much. Overall, outputs of carbon dioxide from respiration and combustion are higher than inputs into producers through photosynthesis. As a result, the level of atmospheric carbon dioxide continues to rise (Figure C4.2.18).

● **Top tip!**

There are both annual fluctuations and a long-term trend shown by the Keeling curve. Short-term (i.e. yearly) trends are due to changes in the levels of photosynthesis in the Northern Hemisphere. The overall upward trend is due to increasing combustion of fossil fuels, deforestation and other factors that lead to outputs of carbon dioxide being higher than inputs.

● **TOK**

In Theory of Knowledge, you might consider the impact of knowledge from one Area of Knowledge on another. Here, the relationship is between *biological* knowledge on climate change and *political* knowledge on how to manage the limited resources of a country or between countries. Whether and how to regulate a country's use of fossil fuels will be decided in the context of a wide range of other knowledge, including economics and biology, and this information might not all point in the same direction. An economist might claim that using more fossil fuels will have certain types of important benefits, but a biologist might claim that using more fossil fuels will have important detrimental effects and that increasing the use of green technology to generate renewable sources of energy, such as solar power, is ultimately better for the economy in the long term and for the future of the planet. It is up to the political figures to weigh these facts in relation to one another when determining public or regulatory policy. You might argue that our politicians have an ethical obligation to understand the relevant facts from the competing areas of knowledge.

> **ATL C4.2C**
>
> Access the latest updates of the Keeling curve here: https://scrippsco2.ucsd.edu
> The weekly average carbon dioxide concentration for the past year can be found here: https://gml.noaa.gov/ccgg/trends/weekly.html
> How has the carbon dioxide concentration changed since June 2022 to the present day? Print the Keeling curve and yearly pattern, and produce a poster that explains both annual fluctuations and the long-term trend.

Atmospheric oxygen and carbon dioxide

> **Concept: Interdependence**
>
> There is dependence of aerobic respiration on atmospheric oxygen produced by photosynthesis, and of photosynthesis on atmospheric carbon dioxide produced by respiration.

The oxygen produced by autotrophs as the product of photosynthesis is used by organisms in aerobic respiration, and autotrophs use the carbon dioxide released by aerobic respiration in the process of photosynthesis. The fluxes between the release of carbon dioxide by all organisms through aerobic respiration and the intake of carbon dioxide through photosynthesis are very large (Figure C4.2.17). This is a major interaction between autotrophs and heterotrophs.

Recycling of chemical elements

As we saw on page 583, ecosystems recycle chemical elements. All elements used by living organisms, not just carbon, are recycled (Figure C4.2.20).

> **Top tip!**
>
> You need to know about the carbon cycle but are not required to know details of other nutrient cycles, such as the nitrogen cycle.

> **LINKING QUESTIONS**
>
> 1. What are the direct and indirect consequences of rising carbon dioxide levels in the atmosphere?
> 2. How does the transformation of energy from one form to another make biological processes possible?

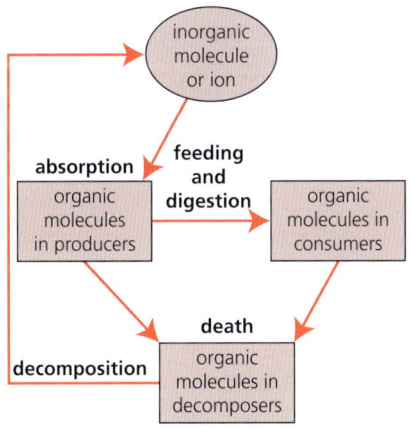

■ **Figure C4.2.20** Nutrient cycles

Nutrients provide the chemical elements that make up the molecules of cells and organisms. All organisms are made of carbon, hydrogen and oxygen, together with mineral elements nitrogen, calcium, phosphorus, sulfur and potassium, and several others, in increasingly small amounts. Plants obtain their essential nutrients as carbon dioxide and water, from which they manufacture glucose. With the addition of mineral elements, absorbed as ions from the soil solution, they build up the complex organic molecules they require. Animals, on the other hand, obtain nutrients as complex organic molecules of food which they digest, absorb and assimilate into their own cells and tissues.

Decomposers play a key role in this recycling process (Figure C4.2.20) by breaking down detritus and other waste products into simple inorganic chemicals that are released into the environment. These inorganic molecules are absorbed by organisms and used to synthesize complex biological molecules such as proteins, fats, polysaccharides and DNA, before the cycle repeats itself.

DNA replication

Concept: Continuity and change

Living things have mechanisms for maintaining equilibrium and for bringing about transformation. Environmental change is a driver of evolution by natural selection.

◆ **DNA replication**: production of exact copies of DNA with identical base sequences.

Link

DNA, base pairing and chromosomes are covered in detail in Chapter A1.2, page 14. Daughter cells are covered in Chapter D2.1, page 648.

Link

DNA replication takes place in the interphase nucleus, well before the events of nuclear division, see Chapter D2.1, page 648.

◆ **Semi-conservative replication**: each strand of an existing DNA double helix acts as the template for the synthesis of a new strand from free nucleotides.

Guiding questions

- How is new DNA produced?
- How has knowledge of DNA replication enabled applications in biotechnology?

SYLLABUS CONTENT

This chapter covers the following syllabus content:
▶ D1.1.1 DNA replication as production of exact copies of DNA with identical base sequences
▶ D1.1.2 Semi-conservative nature of DNA replication and role of complementary base pairing
▶ D1.1.3 Role of helicase and DNA polymerase in DNA replication
▶ D1.1.4 Polymerase chain reaction and gel electrophoresis as tools for amplifying and separating DNA
▶ D1.1.5 Applications of polymerase chain reaction and gel electrophoresis
▶ D1.1.6 Directionality of DNA polymerases (HL only)
▶ D1.1.7 Differences between replication on the leading strand and the lagging strand (HL only)
▶ D1.1.8 Functions of DNA primase, DNA polymerase I, DNA polymerase III and DNA ligase in replication (HL only)
▶ D1.1.9 DNA proofreading (HL only)

DNA replication

A copy of each **chromosome** must pass into daughter cells formed by cell division, so the chromosomes must first be copied (**replicated**). **DNA replication** produces exact copies of DNA, with identical base sequences, and is required for reproduction and for growth and tissue replacement in multicellular organisms. The base pairing of DNA allows this process to proceed accurately and without (in general) any errors. Errors, known as mutations, do occur sometimes however (see page 637).

Concept: Continuity

DNA replication enables the genetic code of an organism to be copied into new cells. This allows continuity within the genome of the organism and ensures the genetic blueprint is passed on to new cells in a way that maintains their integrity and function.

Semi-conservative nature of DNA replication

At the end of the process of DNA replication, each new pair of DNA strands reforms as a **double helix**. One strand of each double helix has come from the original chromosome, and one is a newly synthesized strand. This arrangement is known as **semi-conservative replication** because half of the original molecule stays the same (Figure D1.1.1). If replication was 'conservative', then one DNA molecule formed would be of two new strands and the other molecule would consist of the two original strands.

Crick and Watson suggested replication of DNA would be 'semi-conservative', and this has since been shown experimentally using DNA of bacteria 'labelled' with a 'heavy' nitrogen isotope.

In **semi-conservative replication** one strand of each new double helix comes from the parent chromosome and one is a newly synthesized strand (i.e. half the original molecule is conserved)

If an entirely new double helix were formed alongside the original, then one DNA double helix molecule would be conserved without unzipping in the next generation (i.e. **conservative replication**).

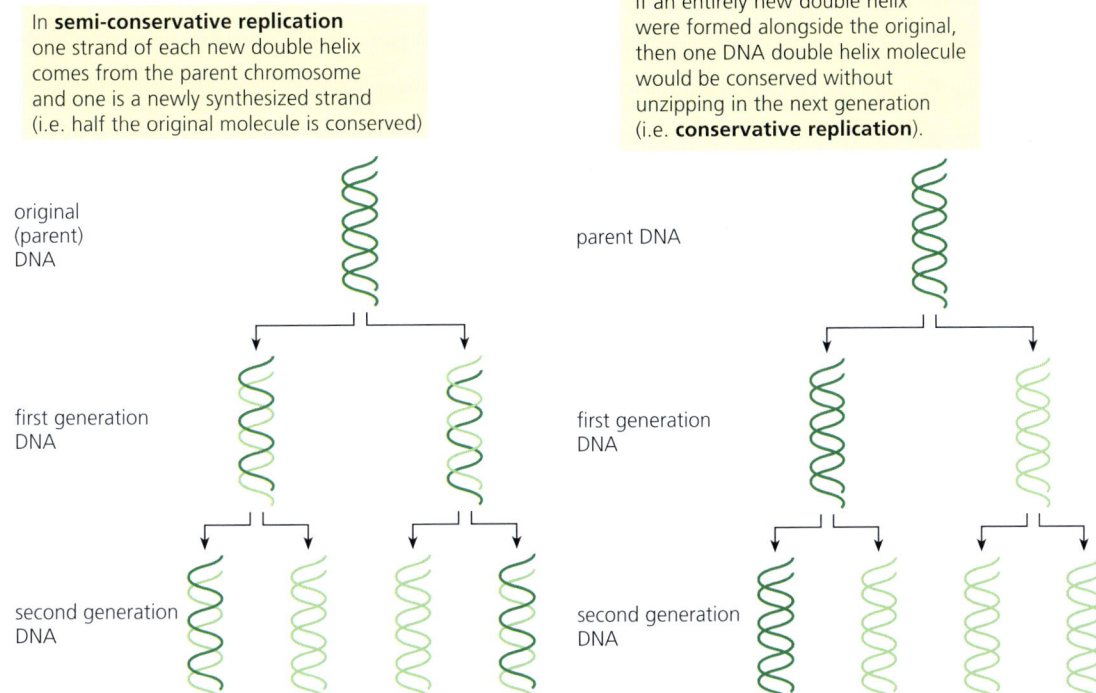

■ **Figure D1.1.1** Semi-conservative versus conservative replication

Link

The role of complementary base pairing in allowing genetic information to be replicated is covered in Chapter A1.2, page 23.

Complementary base pairing and the retention of one strand of the original DNA double helix allows a high degree of accuracy in copying base sequences.

Nature of science: Evidence

The evidence for semi-conservative DNA replication

Watson and Crick proposed semi-conservative replication as how DNA replicates, but they had no evidence to support this claim. Scientific knowledge must be supported by evidence. Meselson and Stahl designed an experiment to test Watson and Crick's model, to evaluate the claim. Experimental evidence that DNA replication is semi-conservative came from an experiment by Meselson and Stahl. They first grew a culture of the bacterium *E. coli* in a medium (food source) where the available nitrogen contained only the heavy nitrogen isotope, ^{15}N. Consequently, the DNA of the bacterium became entirely 'heavy'.

These *E.coli* were then transferred to a medium of the more abundant 'light' isotope, ^{14}N. Both ^{14}N and ^{15}N are stable (non-radioactive) isotopes of nitrogen. New DNA synthesized by the cells was now made of ^{14}N. The change in concentration of ^{15}N and ^{14}N in the DNA of successive generations was measured. Interestingly, the bacterial cell divisions in a culture of *E. coli* are naturally synchronized; they all divide every 60 minutes.

The DNA was extracted from samples of the bacteria from each successive generation and the DNA in each sample was separated. This was done by placing the sample on top of a salt solution of increasing density, in a centrifuge tube. When centrifuged, the different DNA molecules were carried down to the level where the salt solution was of the same density. Thus, DNA with 'heavy' nitrogen ended up nearer the base of the tubes, whereas DNA with 'light' nitrogen stayed near the top of the tubes.

D1.1 DNA replication

1 **Outline** what is meant by semi-conservative replication.

ATL D1.1A

Find out more about the background to the Meselson and Stahl experiment and how the investigation was developed, using this site:
www.sumanasinc.com/webcontent/animations/content/meselson.html

Summarize the stages of the experiment in a poster and discuss what the experiment tells us about how DNA replicates.

Role of helicase and DNA polymerase in DNA replication

◆ **Helicase**: an enzyme that unwinds the DNA double helix at replication forks.

◆ **DNA polymerase**: an enzyme that links nucleotides together to form a new DNA strand, using the pre-existing strand as a template.

Link

DNA proofreading is covered in more detail for higher-level students in this chapter on page 614.

The process of DNA replication covered here is that of the prokaryotic system, which is simpler than the eukaryotic system. The first step in DNA replication is the disruption of the hydrogen bonds that hold the two strands of double-stranded DNA together (i.e. the **unzipping** of the two strands). An enzyme called **helicase** unwinds the DNA double helix at one region, breaking the hydrogen bonds that hold the strands together there and then temporarily keeping the strands of the helix separated. The unpaired nucleotides are now exposed, surrounded by a pool of free-floating nucleotides.

In the next step, both strands of DNA act as **templates** in replication. Complementary nucleotides line up opposite each base of the exposed strands – adenine with thymine, cytosine with guanine. Hydrogen bonds then form between the complementary bases, holding them in place.

Finally, a **condensation reaction** links together the sugar and phosphate groups of adjacent nucleotides, so forming the new strands. This reaction is catalysed by an enzyme called **DNA polymerase**. The enzyme has another role in replication, too – any mistakes that start to happen (such as the wrong bases attempting to pair up) are corrected – this is known as 'proofreading'. The result is that the two strands formed are identical to the original strands. DNA replication is summarized in Figure D1.1.2.

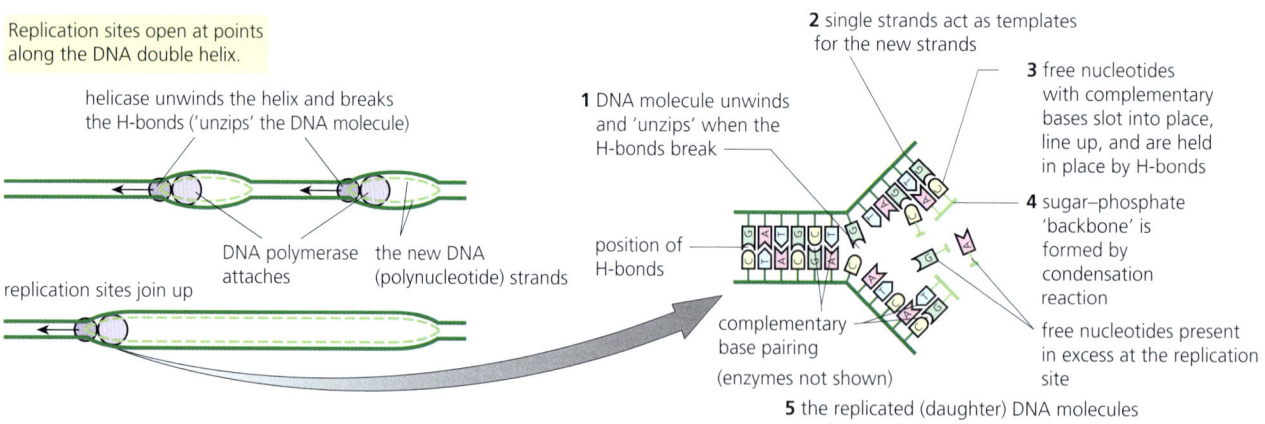

■ Figure D1.1.2 DNA replication

⬤ Common mistake

It is incorrect to say that 'helicase is in charge of elongation of DNA'. Helicase only unwinds DNA for DNA polymerase to act.

TOK

How do the tools that we use shape the knowledge that we produce?

How important are material tools in the production or acquisition of knowledge? Research on human DNA replication has benefited significantly by using the DNA virus SV40 system as a 'tool'. SV40 is a DNA virus that has the potential to cause tumours in animals, but most often persists as a latent infection.

The knowledge gained about DNA replication from studying the SV40 virus depends on how the 'tool' is used. For example, the initial use of the viral SV40 system in vitro identified the polymerases and other proteins, but entirely missed identifying helicase since this activity was supplied by the SV40 large T-antigen. Hence the 'tool' (in vitro replication relying on SV40 large T-antigen) completely missed key proteins providing a vital enzymatic activity, giving only partial knowledge of DNA replication.

A more recently developed 'tool' is cryogenic electron microscopy (CEM, see page 61). CEM gives atomic resolution of proteins, and images can be used to reconstruct the three-dimensional structure of the protein in a computer and generate an atomic model that allows the researcher to understand how the protein works as it changes shape. The proteins involved in DNA replication can be seen in all the three-dimensional shapes they adopt in solution as they perform their biological function.

◆ **Polymerase chain reaction (PCR)**: technique for amplifying selected regions of DNA by multiple cycles of DNA synthesis.

◆ **Taq polymerase**: heat-stable DNA polymerase enzyme used in PCR, extracted from the thermophilic bacteria *Thermus aquaticus*.

◆ **Gel electrophoresis**: a process used to separate fragments of DNA or proteins, on a polymer gel, according to size and overall charge.

PCR and gel electrophoresis as tools for amplifying and separating DNA

DNA can be analysed in order to identify the individual from which the DNA was taken. It is now commonly used in forensic science (for example, to identify someone from a blood or other body fluid sample, or to identify botanical evidence from samples of pollen and plant fragments) and to establish the genetic relatedness of individuals. It is also used to determine whether individuals of endangered species have been bred in captivity or captured in the wild.

Sometimes only a very small sample of DNA is available. The discovery of the **polymerase chain reaction (PCR)** has permitted fragments of DNA to be copied repeatedly, faithfully and quickly, in a process that has been fully automated. A heat-tolerant DNA polymerase enzyme, *Taq* **polymerase**, obtained from an **extremophile** bacterium, is used. Note: *Taq* polymerase does not perform proofreading, thus there may be mistakes in the complementarity of the nucleotides added.

Once the DNA has been amplified, DNA fragments can be separated using a technique known as **gel electrophoresis**, so that the genotype of different individuals can be compared.

Top tip!

Taq polymerase was named after the heat-tolerant bacterium from which it was isolated, *Thermus aquaticus*. *T. aquaticus* lives in hot springs and hydrothermal vents, and so its enzymes are adapted to these hot conditions. This heat-stability makes *Taq* polymerase ideal for PCR, as temperature fluctuations are used in the technique.

■ Polymerase chain reaction

In the polymerase chain reaction (PCR), double-stranded DNA is **amplified**, meaning many copies are made (Figure D1.1.3). Often, it is only possible to produce or recover a very small amount of DNA (such as in genetic engineering or at a crime scene). In PCR, DNA is replicated in an entirely automated process, in vitro, to produce a large amount of the sequence. A single molecule is sufficient as the starting material, should this be all that is available. The products are exact copies.

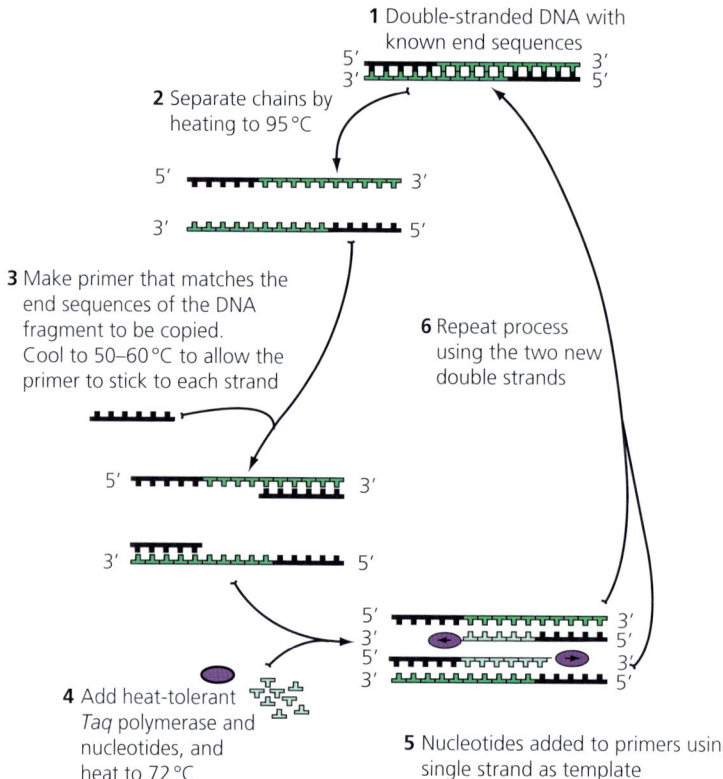

Figure D1.1.3 The polymerase chain reaction

◆ **DNA primer**: short sequence of single-stranded DNA, used in PCR, made synthetically with base sequences complementary to one end of DNA.

ATL D1.1B

Find out how DNA can be extracted from human cells using this site: **https://learn.genetics.utah.edu/content/labs/extraction**

Work through the animation.

Make a list of topics from the IB Biology syllabus that you need to know about to understand the processes involved in DNA extraction. Why is DNA extraction from human cells needed?

The site also contains instructions for how you can extract DNA from a range of different organisms, at home or at your school or college: **https://learn.genetics.utah.edu/content/labs/extraction/howto**

Knowledge of the DNA or amino acid sequence of a desired gene or protein is needed to synthesize flanking nucleotide primers, which is one limitation of the procedure. **DNA primers**, used in PCR, are short sequences of single-stranded DNA, made synthetically with base sequences complementary to one end of DNA.

Another limitation is that a non-target DNA sequence may be amplified instead of the desired sequence, as primers are short nucleotide sequences and may not be specific enough.

Link
The uses of whole genome sequencing are outlined in Chapter A3.1, page 117.

■ Gel electrophoresis

Gel electrophoresis is a process that is used to separate molecules such as proteins and fragments of nucleic acids. It is widely applied in many studies of DNA. For example, it is central to investigation of the sequence of bases in particular lengths of DNA, known as **DNA sequencing**. It is also used in the identification of individual organisms and species, known as **genetic profiling**. We will review these applications on page 608.

Theme D: Continuity and change – Molecules

In electrophoresis, proteins or nucleic acid fragments (either DNA or RNA) are separated on the basis of their overall charge and mass. Separation is due to the following factors:
- Differential migration of these molecules through a supporting medium – typically this is either agarose gel (a very pure form of agar) or polyacrylamide gel (PAG). In these media, the tiny pores in the gel act as a **molecular sieve**; small particles can move quite quickly, whereas larger molecules move much more slowly.
- The **electrical charge** that molecules carry – it is the phosphate groups in DNA fragments that give them a net negative charge. Consequently, when these molecules are placed in an electric field, they migrate towards the positive pole (anode).

> **Top tip!**
>
> Gel electrophoresis can be used to separate fragments of proteins, as well as DNA.

There is thus a double principle of electrophoretic separation: separation occurs on the basis of **size** and **charge**.

DNA fragments are produced by the actions of one or more restriction enzymes. Different restriction enzymes cut at particular base sequences, as and where these occur along the length of the DNA. Consequently, fragments of different lengths are produced.

A series of groves or wells are cut close to one end of the gel, which is then submersed in a salt solution that conducts electricity. Then a small quantity of a mixture to be separated is placed in a well. Several different mixtures can be separated in a single gel at one time (Figure D1.1.4).

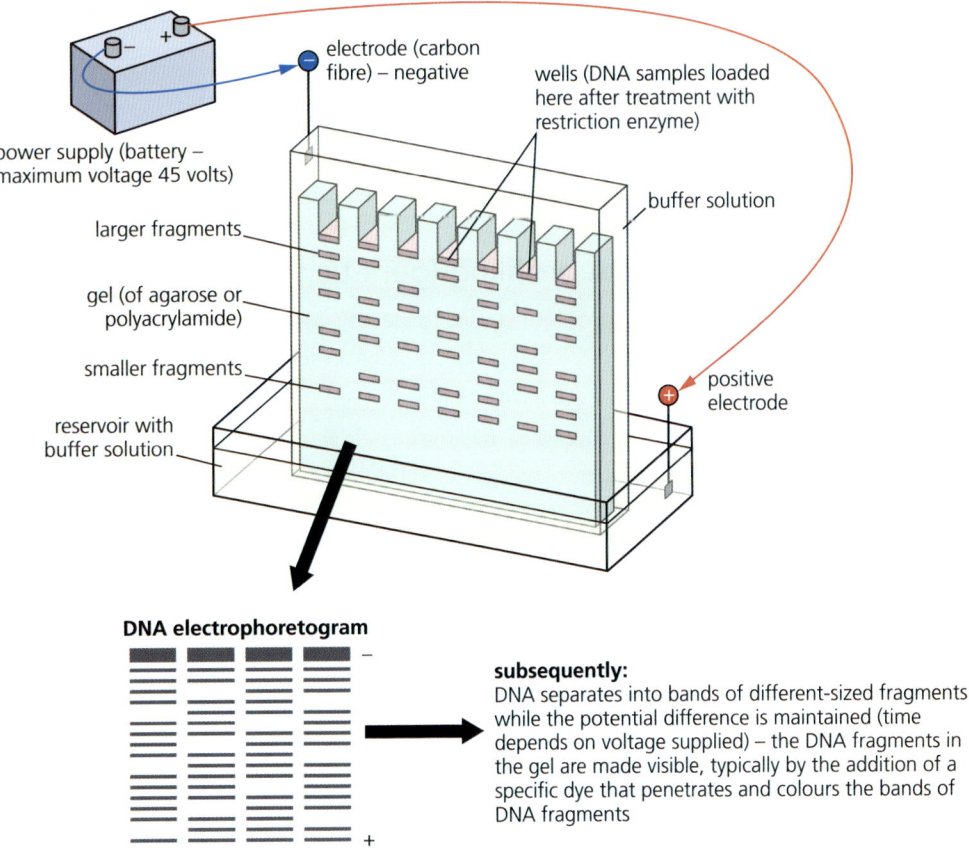

■ Figure D1.1.4 Electrophoretic separation of DNA fragments

D1.1 DNA replication

> **ATL D1.1C**
>
> Use the following sites to produce your own summary of gel electrophoresis, and the principles by which it works:
> www.sumanasinc.com/webcontent/animations/content/gelelectrophoresis.html https://dnalc.cshl.edu/resources/animations/gelelectrophoresis.html
> https://learn.genetics.utah.edu/content/labs/pcr
>
> Explore PCR using this site and draw a flow diagram to explain how PCR works:
> https://learn.genetics.utah.edu/content/labs/pcr

After separation, the fragments are not immediately visible – they are tiny and transparent. They can be identified by gene probes and DNA stains.

- **Gene probes** – these consist of single-stranded DNA with a base sequence that is complementary to that of a particular fragment or gene whose position or presence is sought. The probe must be made radioactive so that, when the treated gel is exposed to X-ray film, the presence of that particular probe and complementary fragment will be disclosed. Alternatively, the probe can have a fluorescent stain attached. It will then fluoresce distinctively in UV light, thus indicating the presence of the particular fragment or gene that is being looked for.
- **Stains** – these immediately locate the position of all DNA fragments once applied. Stains include:
  - ethidium bromide – DNA fragments fluoresce in short-wave UV radiation (a potential carcinogen); ethidium bromide is fluorescent and is visible under UV light. When it binds to DNA, the DNA fluoresces.
  - methylene blue – stains gel and DNA, but is less sensitive and the colour fades quickly.

2. **Explain** how electrophoresis works.
3. **Outline** the role of the buffer and power supply in electrophoresis.

Applications of PCR and gel electrophoresis

DNA profiling (genetic fingerprinting)

Genetic profiling is used to identify organisms, species or individuals using DNA. The technique uses both PCR and gel electrophoresis techniques.

The majority of our DNA is not composed of 'genes' and so does not code for proteins. These extensive 'non-gene' regions include short sequences of bases, repeated many times. While some of these sequences are scattered throughout the length of the DNA molecule, many are joined together in major clusters. It is these lengths of non-coding, 'nonsense' DNA that are used in genetic profiling. They may be referred to as 'satellite' and 'microsatellite' DNA, but are also known as **variable number tandem repeats (VNTRs)**.

We inherit a distinctive combination of these apparently non-functional VNTRs, half from our mother and half from our father. Consequently, each of us has a unique sequence of nucleotides in our DNA (except for identical twins, who share the same pattern).

To produce a genetic 'fingerprint' or profile, a sample of DNA is cut with a restriction enzyme that acts close to the VNTR regions. Electrophoresis is then used to separate fragments according to length and size, and the result is a pattern of bands (Figure D1.1.5).

♦ **Genetic profiling**: the identification of individual organisms or species using DNA.

♦ **Variable number tandem repeats (VNTRs)**: short base sequences that show variation between individuals in terms of number of repeats. These major lengths of non-coding DNA are used in genetic profiling.

4. **Explain** why the composition of the DNA of identical twins challenges an underlying assumption of DNA fingerprinting, but that of non-identical twins does not?

Southern blotting (named after the scientist who devised the routine):
- extracted DNA is cut into fragments with restriction enzyme
- the fragments are separated by electrophoresis
- fragments are made single-stranded by treatment of the gel with alkali.

1 Then a copy of the distributed DNA fragments is produced on nylon membrane:

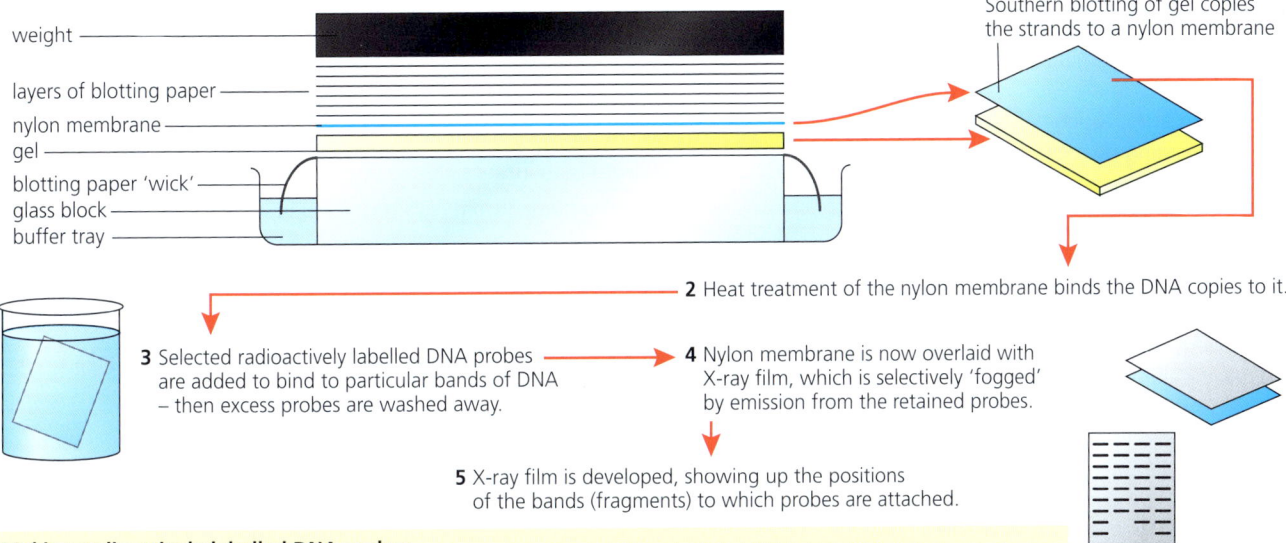

2 Heat treatment of the nylon membrane binds the DNA copies to it.

3 Selected radioactively labelled DNA probes are added to bind to particular bands of DNA – then excess probes are washed away.

4 Nylon membrane is now overlaid with X-ray film, which is selectively 'fogged' by emission from the retained probes.

5 X-ray film is developed, showing up the positions of the bands (fragments) to which probes are attached.

Making radioactively labelled DNA probes
- Single-stranded DNA has the ability to form a stable double strand with another single strand of DNA, provided the bases are complementary (i.e. pair). If one strand is 'labelled', the presence of the paired strands is easily detected.
- Short lengths of single-stranded DNA are made in the laboratory for this purpose, by enzymically combining and then adding selected nucleotides one at a time, in a precise sequence.
- Consequently, the base sequence of probes is predetermined and known.
- All the nucleotides used contain radioactive phosphorus (^{32}P), or carbon (^{14}C) in the ribose of the nucleic acid backbone, so the subsequent positions of the probes (and the location of a complementary strand of DNA, e.g. on a nylon membrane) can be located by autoradiography.

What a probe is and how it works

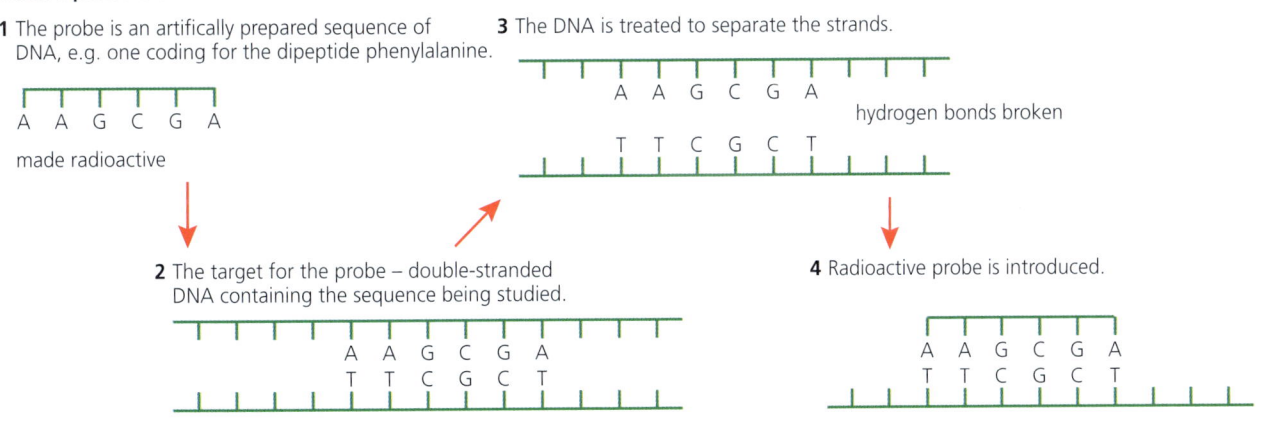

■ **Figure D1.1.5** Steps to genetic profiling

Concept: Change

Although the genetic code is relatively consistent within a species, changes occur that result in individuals having different alleles to others in the population. Modifications to DNA also occur in the number of STRs and VNTRs. These differences in DNA allow genetic profiling techniques to identify differences and similarities between genetic samples from individuals.

D1.1 DNA replication

Top tip!

Short tandem repeats (STRs) can also be used in DNA profiling. An STR is another section of DNA within the genome that is organized as repeating units consisting of 2 to 13 nucleotides. The difference between VNTRs and STRs is the number of nucleotides in a repeating sequence: repeating units of VNTRs consist of 10 to 100 nucleotides, while repeating units of STRs consist of 2 to 13 nucleotides. Both VNTRs and STRs are used widely in forensic studies.

Applications of genetic fingerprinting

Applications in forensic science include the following.

- **Identification of suspects** – samples are taken from the scenes of serious and violent crimes. DNA from both victims and suspects, as well as from others who have certainly not been involved in the crime (used as control samples), is profiled. The greatest care is required; there must be no possibility of cross-contamination if the outcome of testing is to be accurate.
- **Identification of corpses** – bodies that are otherwise too decomposed for recognition may be identified, as might parts of the body remaining after violent incidents, including natural disasters. DNA samples from body tissues can be taken and their profile compared with those of close relatives or with DNA obtained from cells recovered from personal effects, where these are available.
- **Determining paternity** – a range of samples of DNA are analysed side by side, of the people who are possibly related. The banding patterns are then compared (Figure D1.1.6). Because a child inherits half its DNA from its mother and half from its father, the bands in a child's DNA fingerprint that do not match those of its mother must come from the child's father.

Nature of science: Measurement

Reliability is enhanced by increasing the number of measurements in an experiment or test. In DNA profiling, increasing the number of markers used reduces the probability of a false match. Markers are DNA sequences with a known physical location on a chromosome and include single-nucleotide polymorphisms (SNPs, see page 115) and STRs. The markers have uniform band intensities and cover a wide range of DNA sizes. The greater the number of markers (i.e. bands on a DNA profile) the greater the reliability and accuracy of the match, as there are more points of comparison between the DNA profiles.

DNA profiling in forensic investigation

Identification of criminals

At the scene of a crime (such as a murder), hairs – with hair root cells attached – or blood may be recovered. If so, the resulting DNA profiles may be compared with those of DNA obtained from suspects.

Examine the DNA profiles shown below, and suggest which suspect should be interviewed further.

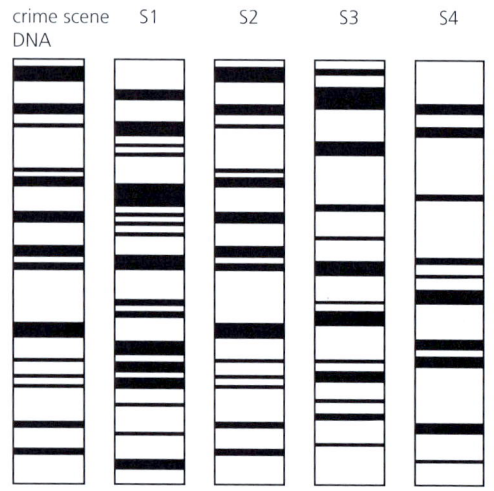

DNA profiles used to establish family relationships

Is the male (M) the parent of both children?

Examine the DNA profiles shown below.

Look at the children's bands (C).

Discount all those bands that correspond to bands in the mother's profile (F).

The remaining bands match those of the biological father.

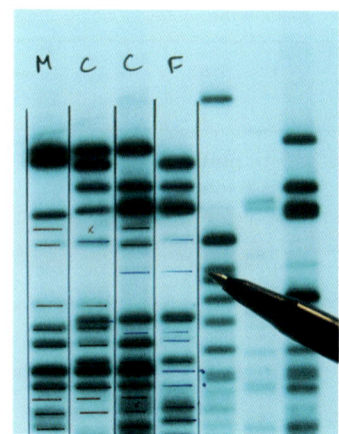

■ **Figure D1.1.6** DNA profiles used to investigate relatedness

TOK

The use of DNA for securing convictions in legal cases is well established, yet even universally accepted theories are overturned in the light of new evidence. What criteria are necessary for assessing the reliability of evidence? What counts as good evidence for a claim? On what grounds might we doubt a claim?

Link

The antiparallel chains of DNA are shown in Chapter A1.2, Figure A1.2.6, page 19. Directionality of RNA and DNA is also mentioned in A1.2, page 25.

Top tip!

DNA replication of the new strand only occurs in a 5′ to 3′ direction. This is because the shape of the active site of DNA polymerase requires the 3′–OH end of the template strand to initiate replication.

◆ **RNA primer**: in DNA replication, a short length of RNA made at the beginning of the synthesis of each DNA fragment; these RNA segments are subsequently removed and filled in with DNA.

◆ **DNA polymerase III**: the primary enzyme involved in DNA replication.

Inquiry 2: Collecting and processing data

Collecting data

Access the following site: **https://learn.genetics.utah.edu/content/labs/gel**

First, remind yourself of the processes involved in gel electrophoresis. Then carry out your own experiment using the simulation and gather your own quantitative data. Sort and measure DNA strands by running your own gel electrophoresis experiment. What equipment do you need? How are the data processed? Can you identify any issues that may have arisen during the data collection and how would you address these?

Directionality of DNA polymerases

We can identify direction in the DNA double helix. One chain runs from 5′ to 3′, while the other runs from 3′ to 5′. (Remember, the carbon atoms of organic molecules can be numbered, see page 17.) The two chains of DNA are antiparallel, which becomes important in replication and when the genetic code is transcribed into mRNA.

DNA polymerase requires a free 3′–OH group to start the synthesis of a new DNA strand, adding the 5′ of a DNA nucleotide to the 3′ end of a strand of nucleotides. This means it can synthesize in only one direction, moving along the template strand in a 3′ to 5′ direction, and synthesizing the new DNA strand in a 5′ to 3′ direction.

The proofreading mechanism of DNA replication (page 614) also explains why DNA polymerases synthesize DNA only in the 5′ to 3′ direction. A hypothetical DNA polymerase that synthesized in the 3′ to 5′ direction would be unable to proofread: if it removed an incorrectly paired nucleotide, the polymerase would form a chain that could no longer be elongated. Thus, for a DNA polymerase to function as a self-correcting enzyme that removes its own polymerization errors as it moves along the DNA, it must proceed only in the 5′ to 3′ direction.

Differences between replication on the leading strand and the lagging strand

Replication is be initiated by an **RNA primer** (Figure D1.1.7). This is probably a 'relic' of the RNA world (see Theme A2.1, page 41). Semi-conservative replication is actually initiated at many points along the DNA double helix. These points are known as **replication forks**. Here, the DNA strands separate (a 'bubble' forms), brought about by the enzyme **helicase**. Another enzyme, **DNA gyrase**, assists in overcoming the strains that come as the double-stranded DNA is unwound. Then, **single-strand binding proteins** attach and prevent the separated strands from rejoining. The unwound sections of both strands are now ready to act as templates for the synthesis of complementary DNA strands.

Both strands are replicated simultaneously. However, since DNA polymerase (known as **DNA polymerase III**) can add nucleotides only to the free 3′ end, the DNA strands can elongate only in the 5′ to 3′ direction. Consequently, the details of the replication process differ in the two strands. Figure D1.1.7 illustrates the steps, and Table D1.1.1 lists the enzymes involved.

D1.1 DNA replication

HL ONLY

◆ **Leading strand**: at a replication fork, the DNA strand that is made by continuous synthesis in the 5' to 3' direction.

◆ **Primase**: an RNA polymerase that uses DNA as a template to produce an RNA fragment that serves as a primer for DNA synthesis.

◆ **Lagging strand**: a single DNA strand that is discontinuously replicated in the opposite direction to the replication fork, forming Okazaki fragments.

◆ **Okazaki fragments**: short lengths of DNA produced on the lagging strand during DNA replication. Adjacent fragments are rapidly joined together by DNA ligase to form a continuous DNA strand.

◆ **DNA ligase**: enzyme that joins two DNA fragments together, via the formation of a 3'–5' phosphodiester bond.

Leading strand

The exposed 5' to 3' strand is referred to as the **leading strand**. Here, **DNA polymerase III** adds nucleotides by complementary base pairing to the free 3' end of the new strand, in the same direction as the replication fork. This process is **continuous**, immediately happening behind the advancing helicase as fresh template is exposed. The initial nucleotide chain formed is actually a short length of RNA called a primer. This primer is synthesized by an enzyme, **primase**. Then the new DNA starts from the 3' end of the RNA primer.

● Top tip!

RNA primer is required for the synthesis of a new strand in DNA replication because DNA polymerase can only join nucleotides on to an existing chain with a free 3'–OH group at one end.

Primase, unlike DNA polymerases, does not require a free 3'–OH end for synthesis. Primase synthesizes a short piece of RNA that is complementary to the template DNA strand (an RNA primer): this gives DNA polymerase the starting point it needs to initiate synthesis.

Lagging strand

In contrast to events in the leading strand, in the **lagging strand** the replication is **discontinuous**. Strands are synthesized in opposite directions. A series of relatively short lengths of DNA, called **Okazaki fragments** are formed (see Figure D1.1.7), each one primed separately. Okazaki fragments are necessary so that replication of both strands can happen at the same time. In contrast to replication on the leading strand, which has to be initiated by an RNA primer only once, replication has to be initiated repeatedly on the lagging strand. There is therefore more than one primer/primase in the lagging strand. The enzyme **DNA ligase** is present in the lagging strand to join together Okazaki fragments.

● Common mistake

An RNA primer begins replication on both the lagging strand and the leading strand, not only on the lagging strand.

● Top tip!

The DNA replication process covered here occurs in prokaryotes. Replication in eukaryotic cells has some differences to the replication covered on these pages.

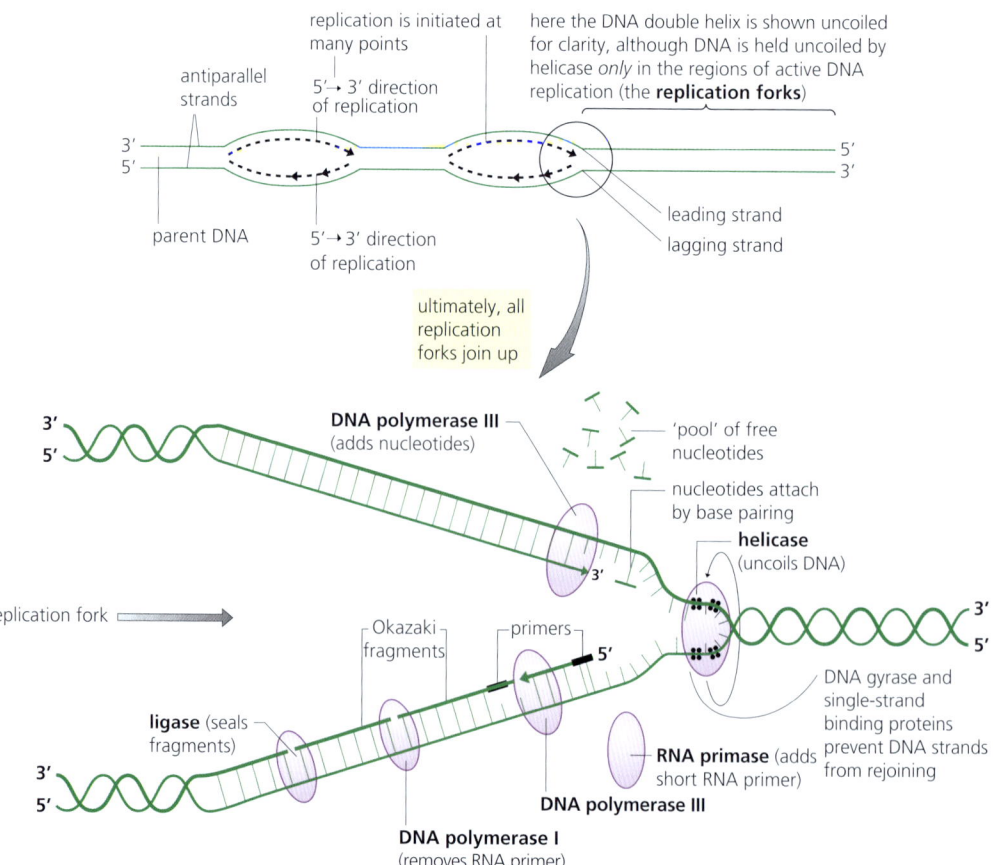

■ Figure D1.1.7 The steps of DNA replication

Theme D: Continuity and change – Molecules

5. **Explain** why Okazaki fragments are formed during DNA replication.

6. **Describe** how enzymes are used in DNA replication.

Nature of science: Experiments

The discovery of Okazaki fragments

The fragments of newly synthesized DNA along the lagging strand are called Okazaki fragments. They are named after their discoverers, Japanese molecular biologists Reiji Okazaki (1930–75) and Tsuneko Okazaki (1933–). They made the discovery by a pulse-chase experiment, which involved exposing replicating DNA to a short 'pulse' of isotope-labelled nucleotides. They then varied the length of time that the cells would be exposed to non-labelled nucleotides. This later period is termed the 'chase'. The labelled nucleotides were incorporated into growing DNA molecules only during the first few seconds of the pulse and non-labelled nucleotides were only incorporated during the chase. The newly synthesized DNA was centrifuged and it was observed that the shorter chases resulted in most of the radioactivity appearing in 'slow' DNA. The sedimentation rate was determined by size: smaller fragments precipitated more slowly than larger fragments because of their lower mass. As they increased the length of the chases, radioactivity in the 'fast' DNA increased with little or no increase of radioactivity in the slow DNA. The researchers interpreted these observations to mean that, with short chases, only very small fragments of DNA were being synthesized along the lagging strand. As the chases increased in length, giving DNA more time to replicate, the lagging strand fragments started integrating into longer, heavier, more rapidly sedimenting DNA strands.

■ Functions of enzymes in replication

The DNA replication covered here is the prokaryotic system. In summary, first an **RNA primer** is formed by **primase** and then **DNA polymerase III** attaches nucleotides to the RNA primer, forming a fragment. Next, **DNA polymerase I** replaces the RNA nucleotides at the start of each fragment with DNA nucleotides. Finally, the enzyme **DNA ligase** joins the **Okazaki fragments** together.

In this way, short lengths of DNA are synthesized and joined together.

7. **Outline** the functions of enzymes in prokaryotic DNA replication.

■ Table D1.1.1 The enzymes that bring about DNA replication

1 Formation of a replication fork	
helicase enzyme	separates the two strands of DNA to expose a replication fork and prevents them rejoining
DNA gyrase enzyme	
single-strand binding proteins	
2 a) DNA replication in the leading strand – a continuous process	
RNA primase	forms a single short length of RNA primer
DNA polymerase III	forms the DNA strand, beginning at the RNA primer
2 b) DNA replication in the lagging strand – a discontinuous process	
RNA primase	forms short lengths of RNA primer at intervals along the DNA strand
DNA polymerase III	forms short DNA strands (Okazaki fragments), starting from each RNA primer
DNA polymerase I	replaces the RNA primer at the start of each Okazaki fragment with a DNA strand
ligase	joins the DNA stands together

D1.1 DNA replication

DNA proofreading

◆ **Proofreading**: the process by which DNA polymerase corrects its own errors as it moves along DNA.

Proofreading of DNA is carried out by prokaryotes as well as eukaryotes. DNA polymerase III is involved in removing any mismatched base from the 3′ terminal, followed by replacement with a correctly matched nucleotide (Figure D1.1.8).

DNA proofreading ensures that any mismatched nucleotides are removed before DNA replication proceeds. This proofreading is carried out by a nuclease that cleaves the phosphodiester bond. Polymerization and proofreading are coordinated, and the two reactions are carried out by different catalytic domains in the same polymerase molecule.

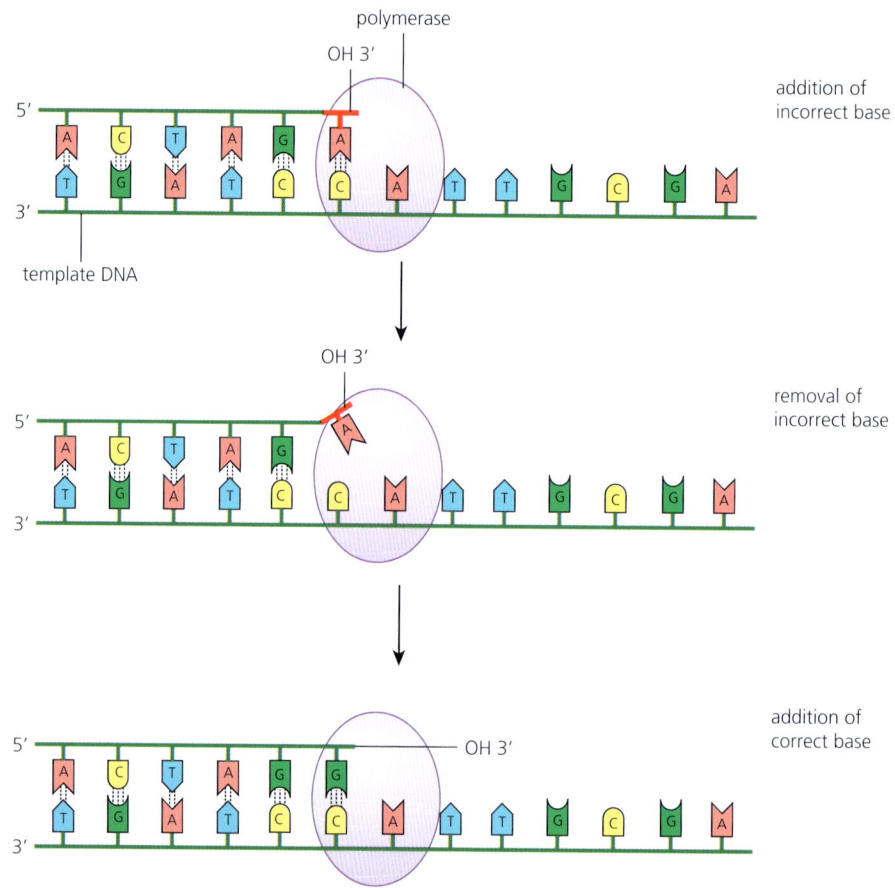

■ Figure D1.1.8 DNA proofreading

ATL D1.1D

Xeroderma pigmentosum is a genetic disorder where a genetic mutation leads to the proofreading and repair mechanism of DNA replication being absent or much reduced. There is a decreased ability to repair DNA damage, such as that caused by UV light.

Find out about this genetic disorder and how it relates to DNA replication. What are the treatments for the disease?

LINKING QUESTIONS

1. How is genetic continuity ensured between generations?
2. What biological mechanisms rely on directionality?

D1.2 Protein synthesis

Guiding questions

- How does a cell produce a sequence of amino acids from a sequence of DNA bases?
- How is the reliability of protein synthesis ensured?

SYLLABUS CONTENT

This chapter covers the following syllabus content:
- D1.2.1 Transcription as the synthesis of RNA using a DNA template
- D1.2.2 Role of hydrogen bonding and complementary base pairing in transcription
- D1.2.3 Stability of DNA templates
- D1.2.4 Transcription as a process required for the expression of genes
- D1.2.5 Translation as the synthesis of polypeptides from mRNA
- D1.2.6 Roles of mRNA, ribosomes and tRNA in translation
- D1.2.7 Complementary base pairing between tRNA and mRNA
- D1.2.8 Features of the genetic code
- D1.2.9 Using the genetic code expressed as a table of mRNA codons
- D1.2.10 Stepwise movement of the ribosome along mRNA and linkage of amino acids by peptide bonding to the growing polypeptide chain
- D1.2.11 Mutations that change protein structure
- D1.2.12 Directionality of transcription and translation (HL only)
- D1.2.13 Initiation of transcription at the promoter (HL only)
- D1.2.14 Non-coding sequences in DNA do not code for polypeptides (HL only)
- D1.2.15 Post-transcriptional modification in eukaryotic cells (HL only)
- D1.2.16 Alternative splicing of exons to produce variants of a protein from a single gene (HL only)
- D1.2.17 Initiation of translation (HL only)
- D1.2.18 Modification of polypeptides into their functional state (HL only)
- D1.2.19 Recycling of amino acids by proteasomes (HL only)

Concept: Continuity

The transcription of the genetic code from DNA to mRNA allows continuity between the genome of an organism and the proteins that it produces. mRNA copies the genetic code and transfers it to ribosomes, allowing proteins needed by the cell and organism to be synthesized.

Transcription

In Chapter B1.2, we saw that proteins are linear sequences of amino acids condensed together. Most proteins contain several hundred amino acid residues, but all are built from only 20 different amino acids.

The unique properties of a protein lie in:
- **which amino acids** are involved in its construction
- **the sequence** in which these amino acids are condensed together. The sequence of bases in DNA dictates the order in which specific amino acids are assembled and combined.

The DNA molecule in each chromosome is very long and it codes for a large number of proteins. Within this molecule, the relatively short lengths of DNA that code for single proteins are called **genes**. Proteins are very variable in size and, therefore, so are genes. A very few genes are as short as 75–100 nucleotides. Most are at least 1000 nucleotides long, and some are more.

◆ **Transcription**: the synthesis of messenger RNA using a DNA template.

◆ **RNA polymerase**: enzyme that catalyses the synthesis of RNA molecules from a DNA template.

All proteins are formed in the cytoplasm: some are formed at free-floating ribosomes and others at the ribosomes embedded within the RER. For this to happen, a mobile copy of the information in the genes must be made and then transported to these sites of protein synthesis. That copy is made of RNA and is called **messenger RNA (mRNA)**. It is formed by a process called **transcription**. Both DNA and RNA have roles in protein synthesis.

An enzyme called **RNA polymerase** catalyses the formation of a complementary copy of the genetic code of a gene in the form of a molecule of mRNA (Figure D1.2.1).

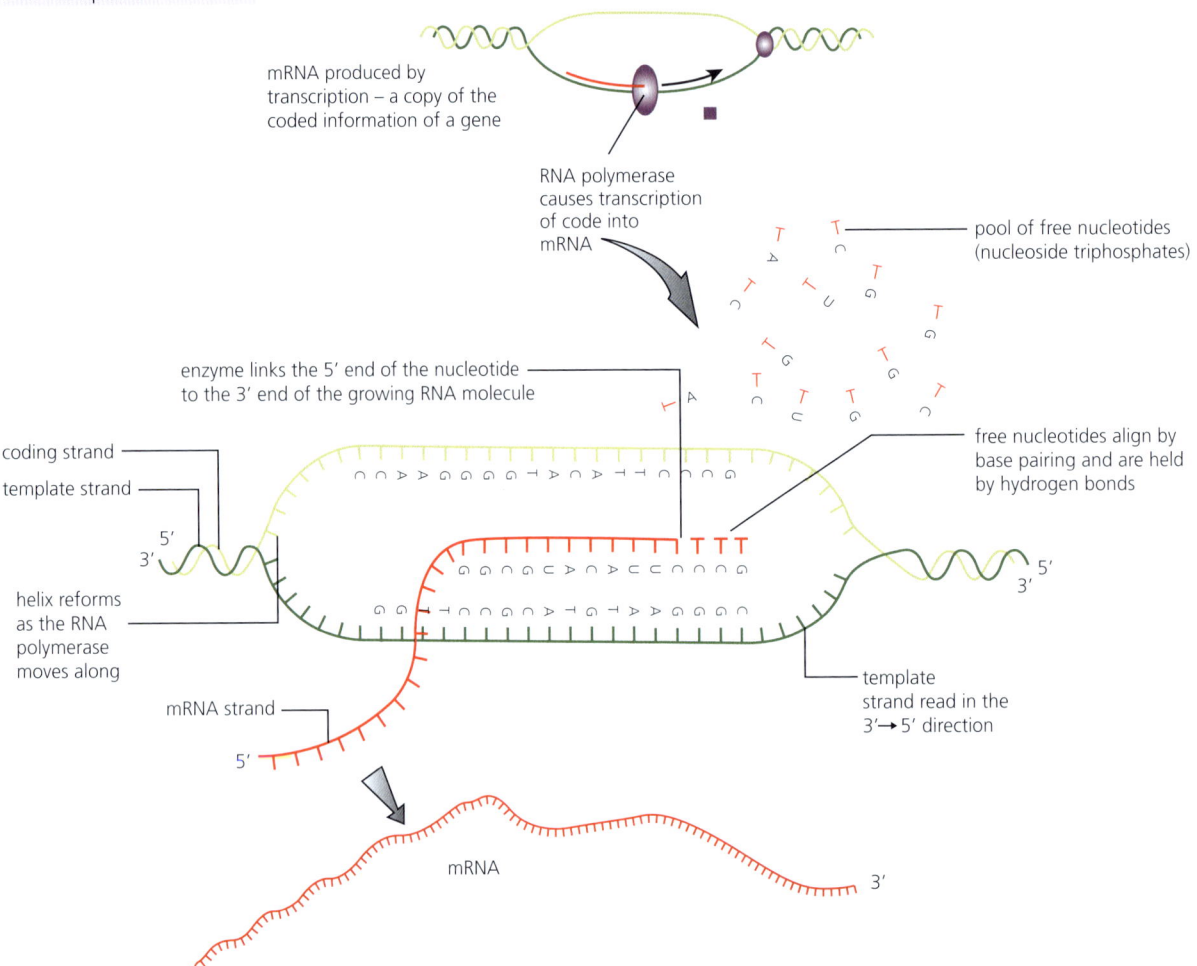

■ Figure D1.2.1 Transcription

◆ **Coding strand**: the strand of a DNA double helix **not** used as a template by DNA or RNA polymerase during DNA replication or RNA transcription.

◆ **Template strand**: the strand of a DNA double helix used during transcription to produce mRNA. It is complementary to the coding strand of DNA.

1 **State** where in a eukaryotic cell you would expect to find the enzyme RNA polymerase.

■ The genetic code

Information in the DNA lies in the sequence of the bases, cytosine (C), guanine (G), adenine (A) and thymine (T). The sequence of bases determines which specific amino acids are assembled and combined. The code lies in the sequence in *one* of the strands, the **coding strand**. The other strand is complementary to it and forms the **template strand** from which the mRNA is formed. This template strand is called the non-coding strand.

> **Top tip!**
>
> DNA is double-stranded, but only one strand serves as a template for transcription at any given time. This template strand is called the **non-coding strand**. The non-template strand is referred to as the coding strand because its sequence will be the same as that of the new RNA molecule.

> **Link**
>
> The definition for codon is in Chapter A1.2, page 17.

> **Common mistake**
>
> The terminology used to describe the two strands of DNA during transcription can cause confusion. The coding strand is **not** the strand used to form mRNA – this is done by the template strand. The coding strand is so-called because it corresponds to the same sequence as the mRNA that will contain the codon sequences necessary to build proteins. The only difference between the coding strand and the new mRNA strand is instead of thymine, uracil takes its place in the mRNA strand. It is the template strand that mRNA is synthesized from – it serves as a template for transcription.

The code is a three-letter or **triplet code**, meaning that each sequence of three of the four bases represents one of the 20 amino acids, and is called a **codon**.

Amino acid	Abbreviation
alanine	Ala
arginine	Arg
asparagine	Asn
aspartic acid	Asp
cysteine	Cyc
glutamine	Gln
glutamic acid	Glu
glycine	Gly
histidine	His
isoleucine	Ile
leucine	Leu
lysine	Lys
methionine	Met
phenylalanine	Phe
proline	Pro
serine	Ser
threonine	Thr
tryptophan	Trp
tyrosine	Tyr
valine	Val

		Second base			
		A	G	T	C
First base	A	AAA Phe AAG Phe AAT Leu AAC Leu	AGA Ser AGG Ser AGT Ser AGC Ser	ATA Tyr ATG Tyr ATT Stop ATC Stop	ACA Cys ACG Cys ACT Stop ACC Trp
	G	GAA Leu GAG Leu GAT Leu GAC Leu	GGA Pro GGG Pro GGT Pro GGC Pro	GTA His GTG His GTT Gln GTC Gln	GCA Arg GCG Arg GCT Arg GCC Arg
	T	TAA Ile TAG Ile TAT Ile TAC Met/Start	TGA Thr TGG Thr TGT Thr TGC Thr	TTA Asn TTG Asn TTT Lys TTC Lys	TCA Ser TCG Ser TCT Arg TCC Arg
	C	CAA Val CAG Val CAT Val CAC Val	CGA Ala CGG Ala CGT Ala CGC Ala	CTA Asp CTG Asp CTT Glu CTC Glu	CCA Gly CCG Gly CCT Gly CCC Gly

■ **Figure D1.2.2** The 20 amino acids found in proteins, and the genetic code from the DNA template strand (you will use this figure to deduce which codon corresponds to which amino acid)

> **2** The sequence of bases in part of a DNA template strand was found to be:
> -A-G-A-C-T-G-T-T-C-A-T-T. Use Figure D1.2.2 to **determine** the sequence of amino acids this codes for, and where along the length of a gene it occurred.

> **Link**
>
> Hydrogen bonding is covered in Chapter A1.1, page 3; hydrogen bonding between complementary base pairing is covered in Chapter A1.2, page 19.

> **3 State** the enzyme responsible for breaking the hydrogen bonds between complementary strands of DNA.

Role of hydrogen bonding and complementary base pairing in transcription

In transcription, the hydrogen bonds between the complementary strands of DNA are broken, which allows the DNA double helix of a particular gene to unwind. Inevitably, there is a pool of free nucleotides present.

One strand of the DNA, the **template strand**, becomes the template for transcription. A single-stranded molecule of RNA is formed by complementary base pairing. Remember, in RNA synthesis it is uracil that pairs with adenine. That mRNA strand leaves the nucleus through pores in the nuclear membrane and passes to ribosomes in the cytoplasm. Here, the information can be 'read' and used in the synthesis of a protein.

Common mistake

Remember that mRNA does not contain thymine (T), and that adenine (A) on the DNA template strand pairs with uracil (U) on the RNA strand during transcription.

Stability of DNA templates

The sugar–phosphate 'backbone' of the DNA molecule (see Figure A1.2.2, page 17) ensures the stability of the base sequence. Hydrogen bonds between the two strands of the double helix also maintain the integrity of the molecule (Figure A1.2.6, page 19). Single DNA strands are used as the template for transcribing the base sequence, without the DNA base sequence changing. It is essential that the same code is replicated exactly from one DNA molecule to the next (page 602) so that cells have the same genome for the organism, from which its functional and behavioural properties derive.

Some body cells (somatic cells) do not divide during the lifetime of the organism, such as neurons, skeletal and cardiac muscle cells, mature bone cells (osteocytes) and retinal receptor cells. In such cells, base sequences must be conserved throughout the life of a cell to ensure the ongoing functioning of the cells through transcription and translation.

Transcription as a process required for the expression of genes

◆ **Gene expression**: the process by which the information encoded in a gene is used to direct the synthesis of a protein molecule.

Not all genes in a cell are expressed at any given time and transcription, being the first stage of **gene expression**, is a key stage at which expression of a gene can be switched on and off. Organisms control which of their genes are **expressed** at any one moment. Remember, the cells in an organism all contain the same genome, whatever the fate of a particular cell. So, in a multicellular organism every nucleus contains the coded information relating to the development and maintenance of *all* mature tissues and organs. Both during development and later, this genetic information is used selectively. Typically, less than 25% of the protein-coding genes in human cells are expressed at any time. The expression of genes is related to when and where the proteins they code for are needed (Table D1.2.1). A mechanism for control of gene expression was first discovered in **prokaryotes**.

■ Table D1.2.1 Gene regulation

When and why genes are expressed	
expressed all the time	genes responsible for routine and continuous metabolic functions, e.g. respiration, which is common to all cells and is continuous throughout life
expressed at a selected stage in cell or tissue development	e.g. as cells derived from stem cells are developing into muscle fibres or neurons
expressed only in the mature cell	e.g. genes responsible for antibody production in a mature plasma cell, after these have been cloned (page 528)
expressed on receipt of an internal or external signal	e.g. when a particular hormone signal, metabolic signal or nerve impulse is received by the cell and activates a gene, such as the gene for insulin production in β cells in the islets of Langerhans (page 763)

Link

Gene expression is covered in more detail in Chapter D2.2, page 670.

Translation as the synthesis of polypeptides from mRNA

◆ **Translation**: the information of mRNA is decoded into protein (amino acid sequence).

The sequence of bases contained within the DNA in the form of genes, and subsequently transcribed to mRNA, is 'expressed' at the ribosomes to form proteins. The genetic code determines the sequence in which amino acids are assembled, which consequently determines the polypeptide produced. Gene expression, which is completed in the process of **translation** at ribosomes, is therefore a function of the complementary base pairing seen in nucleic acids.

Theme D: Continuity and change – Molecules

Roles of mRNA, ribosomes and tRNA in translation

> **Link**
> The definition of transfer RNA (tRNA) is in Chapter A1.2, page 18.

The amino acids from the pool available for protein synthesis are activated by combining with short lengths of a different type of RNA, **transfer RNA (tRNA)**. It is this tRNA that translates a three-base sequence into an amino acid sequence. How does this occur?

All the tRNAs have a clover-leaf shape, but there is a different tRNA for each of the 20 amino acids involved in protein synthesis. At one end of each tRNA molecule is a site where one particular amino acid of the 20 can be bonded covalently. At the other end, there is a sequence of three bases called an **anticodon**. This anticodon is complementary to the codon of mRNA that codes for the specific amino acid (Figure D1.2.3).

◆ **Anticodon**: set of three consecutive nucleotides in a tRNA molecule that recognizes, through base pairing, the three-nucleotide codon on a messenger RNA molecule.

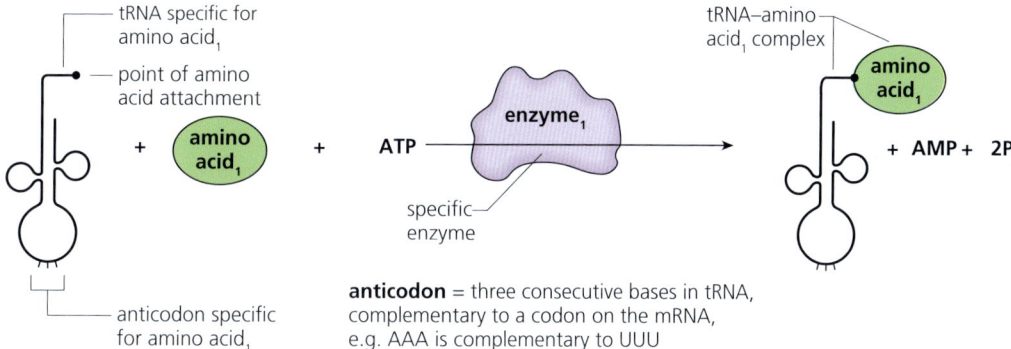

■ **Figure D1.2.3** An amino acid is linked to a specific tRNA

The amino acid becomes attached to its tRNA by an enzyme in a reaction that also requires ATP. These enzymes are specific to the particular amino acids (and types of tRNA) to be used in protein synthesis. The specificity of the enzymes is a way of ensuring the correct amino acids are used in the right sequence.

Ribosomes are made from two subunits (Figure A2.2.20, page 69). In translation, mRNA binds to the small subunit of the ribosome so two tRNAs can bind simultaneously to the large subunit.

Complementary base pairing between tRNA and mRNA

A ribosome moves along the mRNA 'reading' the **codons** from the 'start' codon. The sequence of three bases in the codon is complementary to the sequence of bases in the **anticodon**: this means that in the ribosome, for each mRNA codon, the complementary anticodon (Figure D1.2.3) of the tRNA–amino acid complex slots into place and is temporarily held in position by hydrogen bonds. While held there, the amino acids of neighbouring tRNA–amino acid complexes are joined by a peptide linkage. This frees the first tRNA, which moves back into the cytoplasm for reuse. Once this is done, the ribosome moves on to the next mRNA codon (Figure D1.2.4). The process continues until a ribosome meets a 'stop' codon.

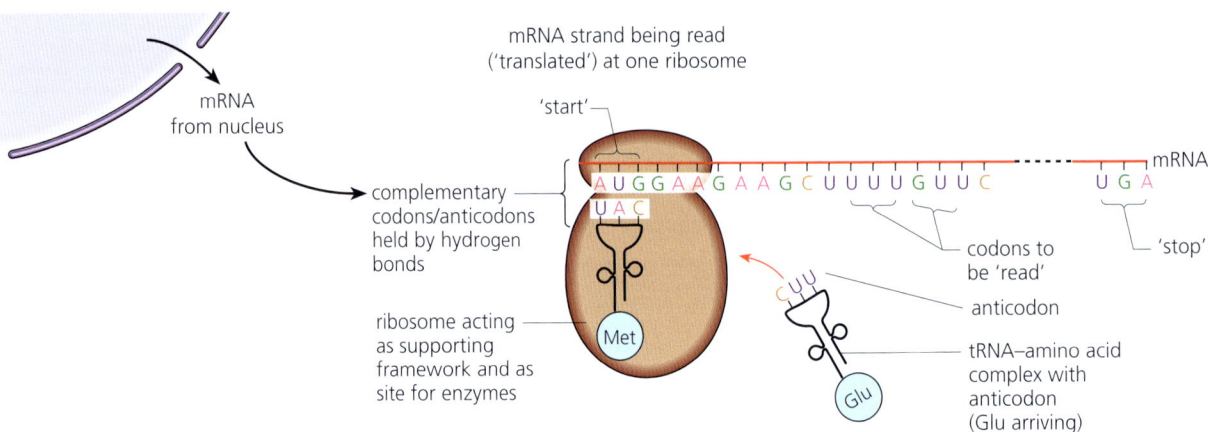

■ **Figure D1.2.4** Translation

Features of the genetic code

All organisms use the same codons to specify the placement of each of the 20 amino acids in protein formation. This supports the idea of a common origin of life on Earth; that the very first DNA has sustained an unbroken chain of life from the first cells on Earth to all cells in organisms alive today. Only the most minor variations in the genetic code have arisen in the evolution and expansion of life since it originated 3500 million years ago. The processes of 'reading' the code and protein synthesis, using RNA and ribosomes, are very similar in prokaryotes and eukaryotes. This is referred to as the **universality** of the genetic code.

◆ **Degenerate code**: the genetic code contains more codons than there are amino acids to be coded, so most amino acids are coded for by more than one codon.

The fact that most amino acids are coded for by more than one codon (see Figure D1.2.2, page 617) is known as the **degeneracy** of the genetic code. The **degenerate code** provides enough different combinations to code for all amino acids and makes allowances for the possibility of mutation. If the third nucleotide in the triplet changes through mutation, it is possible that this will not change the amino acid coded for. For some amino acids, the first two bases are sufficient to provide enough information for which amino acid to use in translation: this means that the third nucleotide can be relatively tolerant to mutation, which reduces the chance that the protein synthesized will be altered when a base does mutate.

Why are there three not two nucleotides used in each codon? A codon size of 2 would give 16 possible combinations – not enough to code for all amino acids, and it has no allowance for further growth. A triplet code gives a total of 64 combinations – more than enough to include all amino acids with the capability to cope with any expansion in the number of amino acids during evolution.

Concept: Continuity

All organisms use the same genetic code. This continuity shows the connectivity between all organisms, and that all life ultimately share a common ancestor (LUCA).

Common mistake

The word 'degenerate' has different meanings. Students often think that, in relation to the genetic code, it means that the code has somehow 'deteriorated'. This is not what 'degenerate code' is referring to. In this context, it means a code in which several codons have the same meaning. For example, when coding for serine (Ser), any code beginning 'UC' followed by any of the four bases will code for this amino acid (Figure D1.2.5). The genetic code is degenerate because there are many instances in which different codons specify the same amino acid.

Using the genetic code expressed as a table of mRNA codons

> **Top tip!**
>
> You need to be able to deduce the sequence of amino acids coded by an mRNA strand.

To read the table of mRNA codons (Figure D1.2.5), select the first base from the column on the left, the second base from the list of bases along the top of the table, and the third base from the cell that you have identified from the first and second base. For example, UAU is Tyr (tyrosine).

		2nd base			
		U	C	A	G
1st base	U	UUU Phe UUC Phe UUA Leu UUG Leu	UCU Ser UCC Ser UCA Ser UCG Ser	UAU Tyr UAC Tyr UAA *Stop* UAG *Stop*	UGU Cys UGC Cys UGA *Stop* UGG Trp
	C	CUU Leu CUC Leu CUA Leu CUG Leu	CCU Pro CCC Pro CCA Pro CCG Pro	CAU His CAC His CAA Gln CAG Gln	CGU Arg CGC Arg CGA Arg CGG Arg
	A	AUU Ile AUC Ile AUA Ile AUG Met/*Start*	ACU Thr ACC Thr ACA Thr ACG Thr	AAU Asn AAC Asn AAA Lys AAG Lys	AGU Ser AGC Ser AGA Arg AGG Arg
	G	GUU Val GUC Val GUA Val GUG Val	GCU Ala GCC Ala GCA Ala GCG Ala	GAU Asp GAC Asp GAA Glu GAG Glu	GGU Gly GGC Gly GGA Gly GGG Gly

■ **Figure D1.2.5** The mRNA genetic dictionary; compare it to Figure D1.2.2

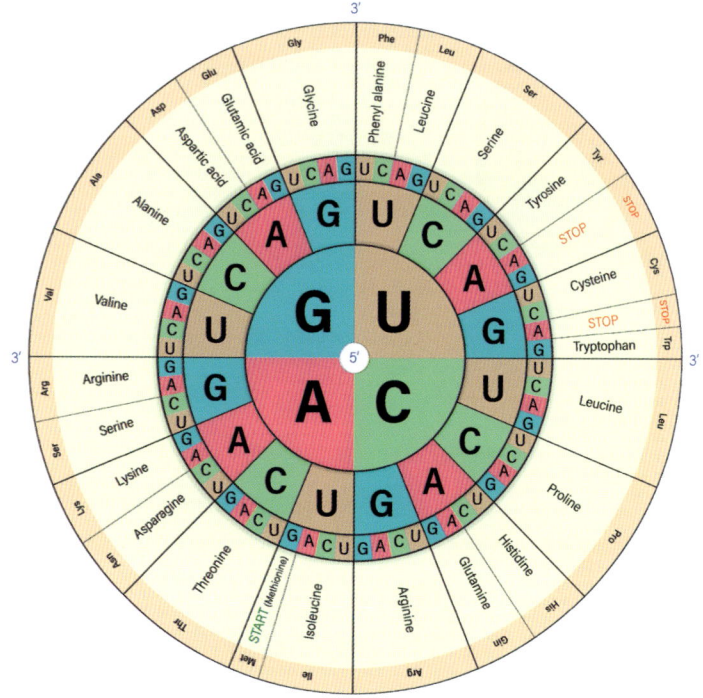

An alternative arrangement for the mRNA codons and the amino acids they code for is shown in Figure D1.2.6.

This 'wheel' diagram is read by taking the first base from the centre of the wheel (containing A, G, U and C), the second from the circle of bases around this central portion, and the final base from the outer ring.

You may be given a list of bases from a sequence of DNA and asked what the sequence of amino acids will be coded for. You first need to transcribe the DNA code into the mRNA code (not forgetting that U replaces T), and then from this read the code in triplets to identify the amino acid sequence. For example, if the DNA code is A-G-A-C-T-G-T-T-C-A-T-T, the mRNA transcribed from this will be U-C-U-G-A-C-A-A-G-U-A-A. The amino acids translated from this sequence will be U-C-U = Ser (serine), G-A-C = Asp (aspartic acid), A-A-G = Lys (lysine) and U-A-A = Stop codon.

■ **Figure D1.2.6** mRNA codon wheel for encoding nucleotide sequences

D1.2 Protein synthesis

> ● **Common mistake**
>
> When determining an amino acid sequence based on a DNA code, make sure that the DNA code is transcribed into mRNA first. A table showing the codons for each amino acid (Figure D1.2.5) or a wheel (Figure D1.2.6) can be used to do this. The clue that a genetic dictionary relates to mRNA rather than DNA is that the table/wheel will contain U and not T.

ATL D1.2A

Create your own DNA sequence of bases that codes for a specific sequence of amino acids. Do this for a template strand of DNA. Use start and stop codons to indicate the beginning and end of translation for the protein you are coding for. Ask a partner to do the same. Swap your codes. Can each of you solve your respective codes and work out the amino acid sequence using the table of codons in Figure D1.2.5 or D1.2.6? Remember that you will first need to transcribe the code from the template strand to mRNA.

4 The sequence of bases in a sample of mRNA was found to be:
   GGU, AAU, CCU, UUU, GUU, ACU, CAU, UGU
   Using Figures D1.2.5 or D1.2.6:
   a **Deduce** the order of amino acids this sequence codes for.
   b **Determine** the sequence of bases in the template strand of DNA from which this mRNA was transcribed.
   c **State** where the triplet codes, codons and anticodons are found within a cell.

The growing polypeptide chain

During translation, the ribosome moves in steps along the mRNA. In each stepwise movement, the ribosome moves three bases along the mRNA. In this way, the ribosome progresses along the mRNA molecule, codon by codon (Figure D1.2.7). Consecutive amino acids are linked together via condensation reactions, forming peptide bonds. By these steps, constantly repeated, a polypeptide is formed and elongated, eventually emerging from the large subunit.

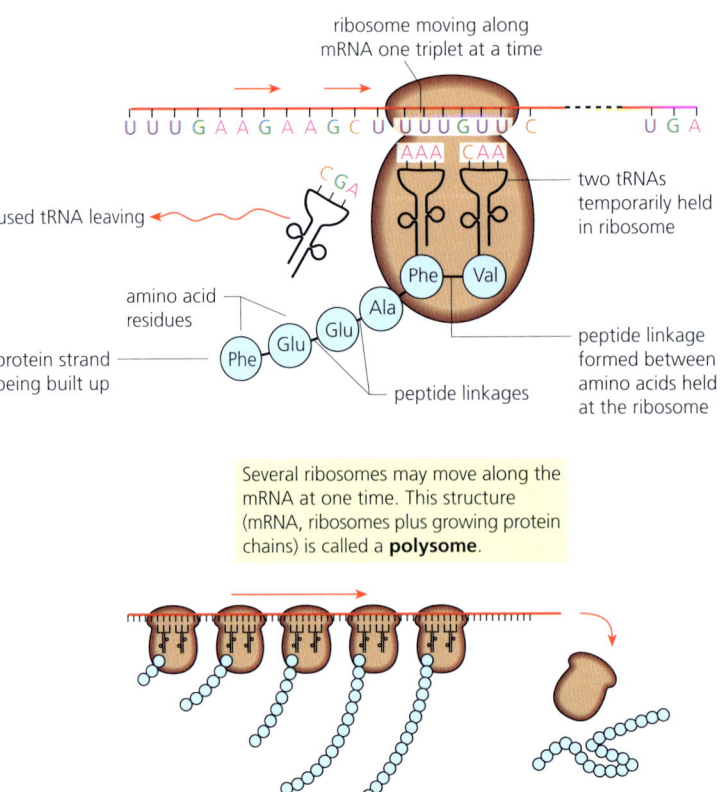

■ **Figure D1.2.7** The stepwise movement of the ribosome along the mRNA

> **Link**
>
> For HL students, polysomes are discussed in the structure and function of free ribosomes and of the RER in Chapter B2.2, page 253.

> ## Tool 1: Experimental techniques
>
> ### Digital molecular modelling
>
> Computer-based modelling is an ideal tool for learning about the structure of biological molecules and biological theory such as natural selection. Computer software gives a much more accurate representation of biological structures (such as DNA and proteins) and assessment of theoretical predictions.
>
> Molecular models can display bond lengths, bond angles, locations of electrons, an accurate representation of the three-dimensional shape of molecules, and reactions between molecules (see Chapter B1.2, page 207).

Mutations that change protein structure

◆ **Point mutation**: occurs in DNA when a single base pair is added, deleted or changed.

DNA can change. Normally, the sequence of nucleotides in DNA is maintained without change, but very occasionally alterations do happen. If a change occurs, we say a **mutation** has occurred. A **point mutation** occurs when a single base pair is added, deleted or changed (i.e. A for T or C for G) within a DNA molecule.

Link
The different types of mutations are covered in detail in Chapter D1.3, page 637.

The sequence of amino acids in the polypeptide chain is controlled by the coded instructions stored in the DNA of the chromosomes in the nucleus, mediated via mRNA. Changing just one amino acid in the sequence of a protein alters its properties, often quite drastically. This sort of mistake arises by mutation.

■ An example of a point mutation: sickle cell anaemia

Link
For HL students, the quaternary structure of haemoglobin is shown in Chapter B1.2, page 217.

An extremely common point mutation occurs in haemoglobin. This oxygen-transporting pigment of red blood cells is made of four polypeptide molecules – two known as α **haemoglobin** and two as β **haemoglobin**. These interlock to form a compact molecule.

The gene that codes for the amino acid sequence of β haemoglobin occurs on chromosome 11 and is prone to a change of the base A to T in a codon for the amino acid glutamic acid – the sixth amino acid in this polypeptide. Due to this base substitution, the amino acid valine appears at that point instead (Figure D1.2.8).

Concept: Change

Mutations result in changes in the base sequence of DNA. If these changes occur within a gene, it can result in altered protein structure, which can affect the physical, functional or behavioural characteristics of an organism.

The presence of a non-polar valine in the β haemoglobin creates a hydrophobic spot in the otherwise hydrophilic outer section of the protein. This tends to attract other haemoglobin molecules, which bind to it. In tissues with low partial pressures of oxygen (such as a tissue with a high rate of aerobic respiration) the sickle-cell haemoglobin molecules in the capillaries readily clump together into long fibres. These fibres distort the red blood cells into sickle shapes. In this condition, the red blood cells cannot transport oxygen. Sickle cells may also get stuck together, blocking smaller capillaries and preventing the circulation of normal red blood cells. The result is that people with sickle cells suffer from anaemia – a condition of inadequate delivery of oxygen to cells. This unusual haemoglobin molecule is known as **haemoglobin S**.

People with a single allele for haemoglobin S have less than 50% haemoglobin S. Such a person is said to have **sickle-cell trait** and they are only mildly anaemic. However, those with both alleles for haemoglobin S are described as having **sickle-cell anaemia** – a serious condition that may trigger heart and kidney problems, too.

Anaemia is a disease typically due to a deficiency in healthy red blood cells.

Haemoglobin occurs in red blood cells – each contains about 280 million molecules of haemoglobin. A molecule consists of two α haemoglobin and two β haemoglobin subunits, interlocked to form a compact molecule.

The **mutation** that produces sickle cell haemoglobin (**Hg^S**) is in the gene for β haemoglobin. It results from the substitution of a single base in the sequence of bases that make up all the codons for β haemoglobin.

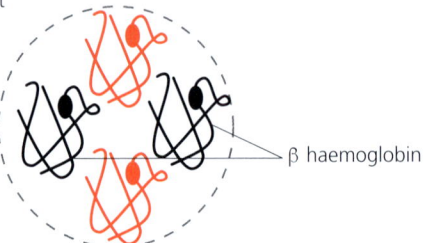

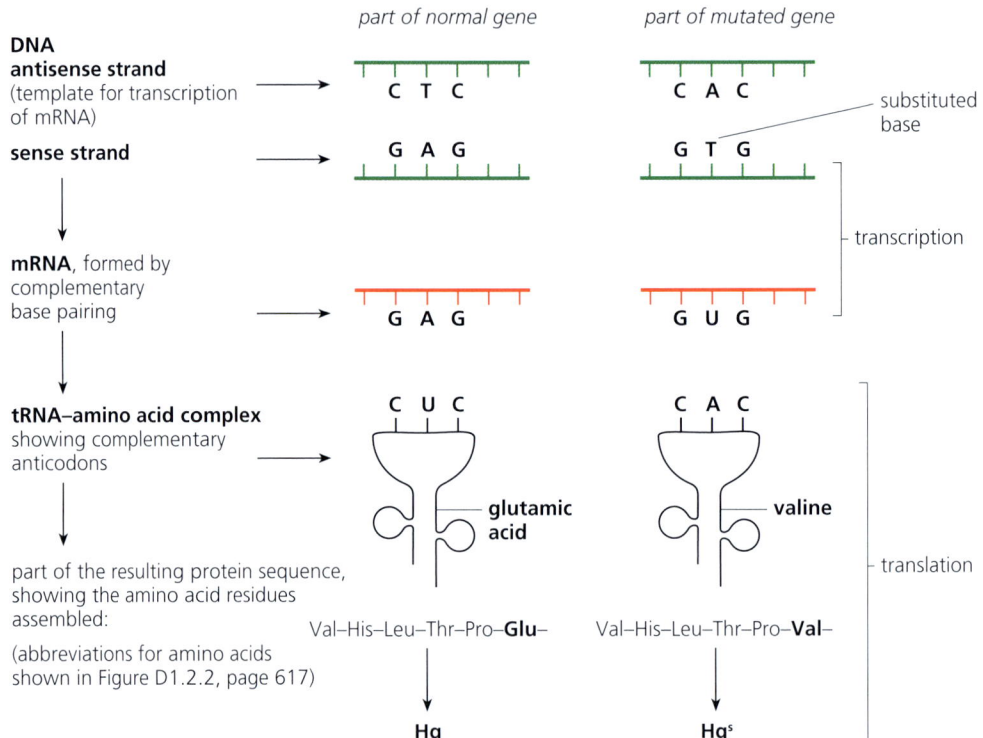

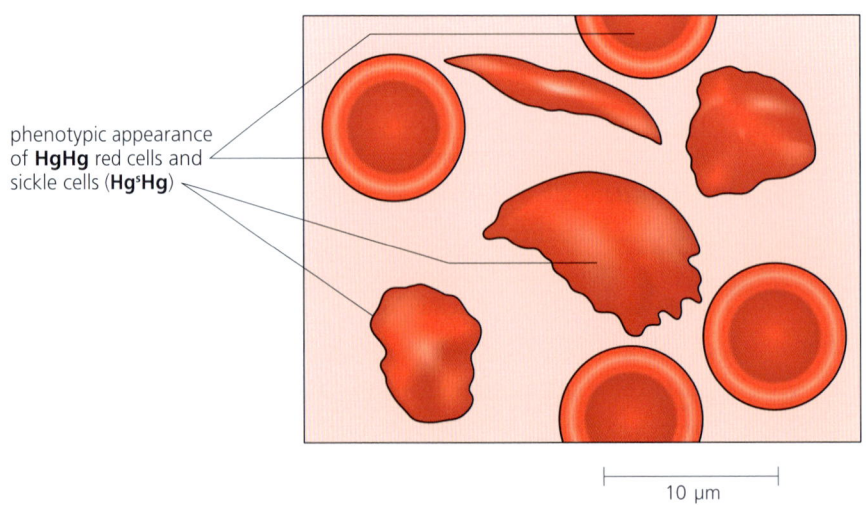

drawing based on a photomicrograph of a blood smear, showing blood of a patient with sickle cells present among healthy red cells

■ Figure D1.2.8 Sickle-cell anaemia, an example of a point mutation

Directionality of transcription and translation

The phosphate groups along each strand of the DNA double helix are bridges between the carbon-3 of one sugar molecule and the carbon-5 of the next. One chain runs from 5′ to 3′ while the other runs from 3′ to 5′ (see Figure A1.2.6, page 19).

As we have seen with DNA polymerase, polymerase enzymes work in the 5′ to 3′ direction. Similarly, in transcription, RNA polymerase synthesizes the RNA strand in the 5′ to 3′ direction, while reading the template DNA strand in the 3′ to 5′ direction. RNA polymerase requires a free 3′–OH group to start the synthesis of a new RNA strand, adding the 5′ of an RNA nucleotide to the 3′ end of a strand of nucleotides. In translation, the codons of mRNA are also read from the 5′ end to the 3′ end by tRNAs.

> **Link**
> Directionality of DNA and RNA is also discussed in Chapter A1.2, page 25.

Top tip!
The start of an amino acid chain is referred to as the N-terminus, and the end is the C-terminus (see Chapter B1.2, page 207). In translation, all mRNAs are read in the 5′ to 3′ direction, with polypeptide chains synthesized from the N-terminus to the C-terminus.

Initiation of transcription at the promoter

In eukaryotes, before mRNA can be transcribed by the enzyme RNA polymerase, it first binds together with a small group of proteins called general transcription factors at a sequence of bases known as the **promoter**. Promoter regions occur on DNA strands just before the start of a gene's sequence of bases. (The promoter is an example of a length of non-coding DNA with a special function.) Transcription of the template strand of the gene can only begin when this transcription complex of proteins (enzyme plus factors) has been assembled. Once transcription has been initiated, the RNA polymerase moves along the DNA, untwisting the helix as it goes and exposing the DNA nucleotides, so that RNA nucleotides can pair and the mRNA strand can be formed and released.

◆ **Promoter**: DNA sequence that initiates gene transcription; includes sequences recognized by RNA polymerase.

The rate of transcription may be increased (or decreased) by the binding of specific transcription factors on the enhancer site for the gene. The position of an enhancer site of a gene is shown in Figure D1.2.9. This is at some distance 'upstream' of the promoter and gene sequence. However, when activator proteins bind to this enhancer site, a new complex is formed and makes contact with the polymerase–transcription factor complex. Then the rate of gene expression is increased.

Top tip!
Regulator transcription factors, activators and RNA polymerase are all proteins, and are coded for by other genes. It is clear that protein–protein interactions play a key part in the initiation of transcription.

D1.2 Protein synthesis

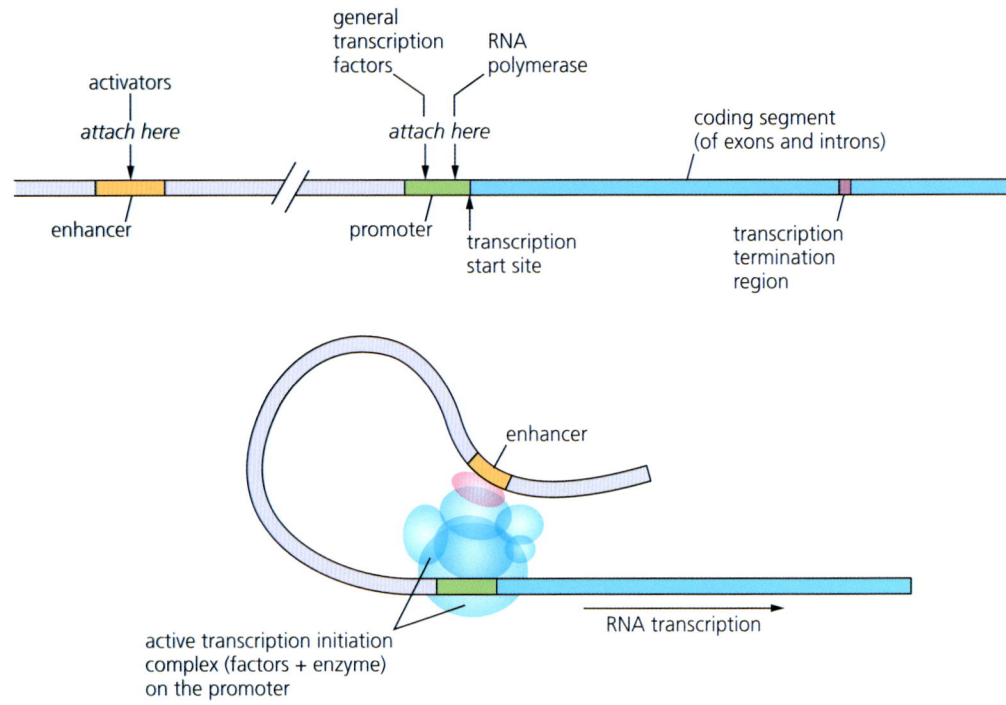

The lower part of the figure shows a portion of the upper DNA molecule with the enhancer (orange) region, promoter (green) region and coding (blue) region. The DNA is looped so that it comes into contact with different parts of the transcription initiation complex. The complex consists of several proteins and other factors, hence the multiple overlapping spheres (blue). The activator (pink) is also part of the complex. These spheres have been drawn transparent so that contact with the DNA molecule can be shown.

■ Figure D1.2.9 Control sites and the initiation of transcription

Non-coding sequences in DNA do not code for polypeptides

Protein-coding sequences of our DNA account for only approximately 1.5% of our DNA. The remainder includes some DNA sequences that regulate the expression of protein-coding genes (**regulatory DNA sequences**), but that leaves 70% of our DNA with other or no roles. (At one time these regions were described as 'junk' or 'nonsense' DNA). In fact, these extensive 'non-gene' regions of eukaryotic chromosomes also consist of:

- **introns**: these are non-coding nucleotide sequences, one or more of which interrupts the coding sequences (exons) of eukaryotic genes (Figure D1.2.10).
- **telomeres**: these are special nucleotide sequences, typically consisting of multiple repetitions of one short nucleotide sequence. They occur near the ends of DNA molecules and 'seal' the ends of the linear DNA. Here, they stop erosion of the genes that would occur with each repeated round of replication (see below).
- **genes for tRNA**: these are parts of the DNA template that code for relatively short lengths of RNA that are formed in the nucleus and pass out into the cytoplasm. Here they transfer amino acids from the pool there, to supply a growing polypeptide in a ribosome (Figure D1.2.7).
- **major lengths of non-coding DNA**: today, these are important to genetic engineers, for they are exploited in DNA profiling. They are short sequences of bases that are repeated very many times. Where they often occur together in major clusters, they are known as **variable number tandem repeats** (**VNTRs**). Their use in genetic fingerprinting (DNA profiling) is described on page 608.

Link

Non-coding sequences are mentioned in the conservation of the genetic code, Chapter A1.2, page 24.

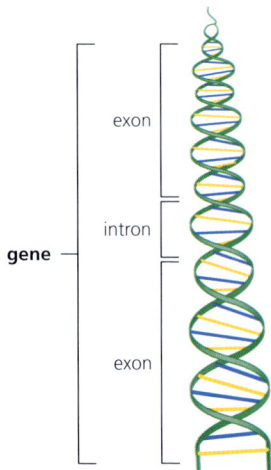

■ Figure D1.2.10 Eukaryotic genes consist of exons and introns

Theme D: Continuity and change – Molecules

◆ **Telomere**: made of repetitive sequences of non-coding DNA, located at the end of chromosomes, that protect the chromosome from damage.

Telomeres

Telomeres protect the organism's genes from being lost with each cycle of DNA replication, due to a gap at the 5' end of each replicated DNA strand. They protect chromosomal ends from degradation by binding proteins to form telomere caps. Telomeres also prevent the ends of chromosomes attaching to each other, which stops the chromosomal ends from activating the cell's system for monitoring DNA damage, therefore avoiding apoptosis (the process of programmed cell death). In addition, recognition sites for the enzyme telomerase enable telomeres to lengthen.

5 Telomeres are parts of chromosomes. **Describe** the function of telomeres.

Post-transcriptional modification in eukaryotic cells

The sections of a gene that carry meaningful information (code for amino acids) are called exons. The non-coding sequences that intervene – interruptions, in effect – are called introns (see Figure D1.2.10). While genes split in this way are very common in higher plants and animals, some eukaryotic genes contain no introns at all.

When a gene consisting of exons and introns is transcribed into mRNA, the mRNA formed contains the sequence of introns and exons exactly as they occur in the DNA. Now, you can see that if this unmodified mRNA was to be 'read' and transcribed in a ribosome, it would actually present problems in the protein-synthesis step.

◆ **Post-transcriptional modification**: changes that occur to a newly transcribed mRNA, after transcription has occurred and prior to its translation into a protein.

◆ **Cleave**: reaction that breaks one of the covalent sugar–phosphate linkages between nucleotides that form the sugar–phosphate backbone of a nucleic acid.

Because introns are non-coding regions within a gene, they must be removed from the mRNA. An enzyme-catalysed reaction, known as **post-transcriptional modification**, removes the introns as soon as the mRNA has been formed (Figure D1.2.11). This is done by a spliceosome (a large RNA–protein complex with enzymatic properties). The spliceosome is also under the control of a gene. This is known as RNA splicing. The process is as follows:
- Pre-mRNA introns are spliced out using spliceosomes.
- Spliceosomes recognise highly conserved regions (splice sites) between the 3' end of the exon and 5' end of the intron, and **cleave** the phosphodiester bond between the nucleotides.
- All exons are joined together to form one polypeptide from a gene.

The resulting shortened lengths of mRNA are described as **mature**. It is this form of mRNA that passes out into the cytoplasm, to the ribosomes, where it is involved in protein synthesis (page 628).

Top tip!

The genes of prokaryotes do not have introns, and the prokaryotic cell does not have the enzyme machinery to carry out splicing either. This means, when a genetic engineer plans to place a copy of a eukaryotic gene in the chromosome of a bacterium, that copy must be intron-free.

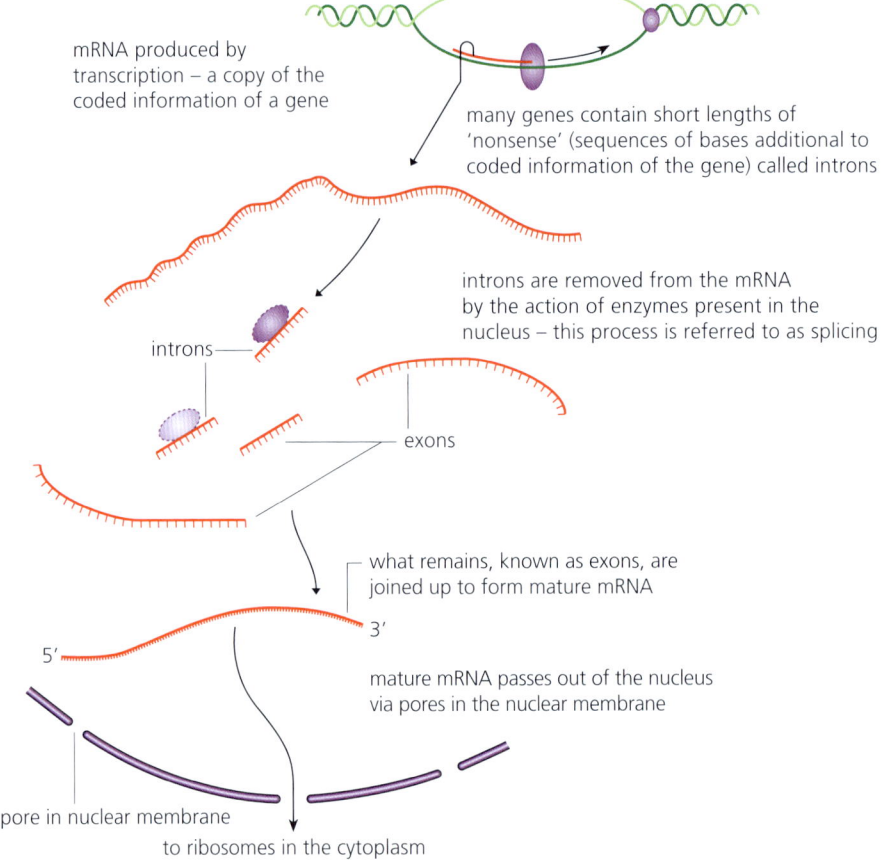

■ **Figure D1.2.11** Post-transcriptional modification of mRNA

A process called alternative splicing can join together different combinations of exons to form different types of polypeptides from a single gene (see page 629).

■ Polyadenylation and 5′ capping

To complete the process of post-transcriptional modification, both ends of the mRNA are modified to increase the stability of the molecule.

The processing of the 3′ end adds a long chain of adenine residues, known as the **poly-A tail**, to the RNA molecule. This section of the mRNA has not been encoded in the genome. To attach the poly-A tail, the 3′ end of the pre-mRNA is **cleaved** to free a 3′ hydroxyl group. Then an enzyme called poly-A polymerase adds a chain of adenine nucleotides to the RNA in a process called **polyadenylation**. The poly-A tail is between 100 and 250 adenine residues long.

The 5′ is capped by the addition of a 5′ modified guanine nucleotide. The result of both capping processes forms a mature mRNA that will be protected from enzymatic degradation and can be exported out of the nucleus into the cytoplasm for translation into a protein by ribosomes.

Alternative splicing

◆ **Alternative splicing**: process of splicing of exons to produce variants of a protein from a single gene.

Change occurs to mRNA immediately after it has been formed by transcription as introns are removed. However, the remaining lengths of mRNA (exons) may be spliced together in different combinations. The consequence is that a single gene can code for more than one type of polypeptide. This process is known as **alternative splicing**. In fact, many genes give rise to two or more different polypeptides, depending on the order in which exons are assembled (Figure D1.2.12). Even more variety in gene products may result if one or more introns are treated as an exon during RNA processing.

This process, also part of the story of how gene action is regulated, is under the control of specific genes, although these are often found on other chromosomes.

As a result of alternative mRNA splicing, the number of proteins produced can be greater than the number of genes present.

Alternative mRNA splicing may explain why the human genome consists of the same (low) number of genes as some small, invertebrate animals have.

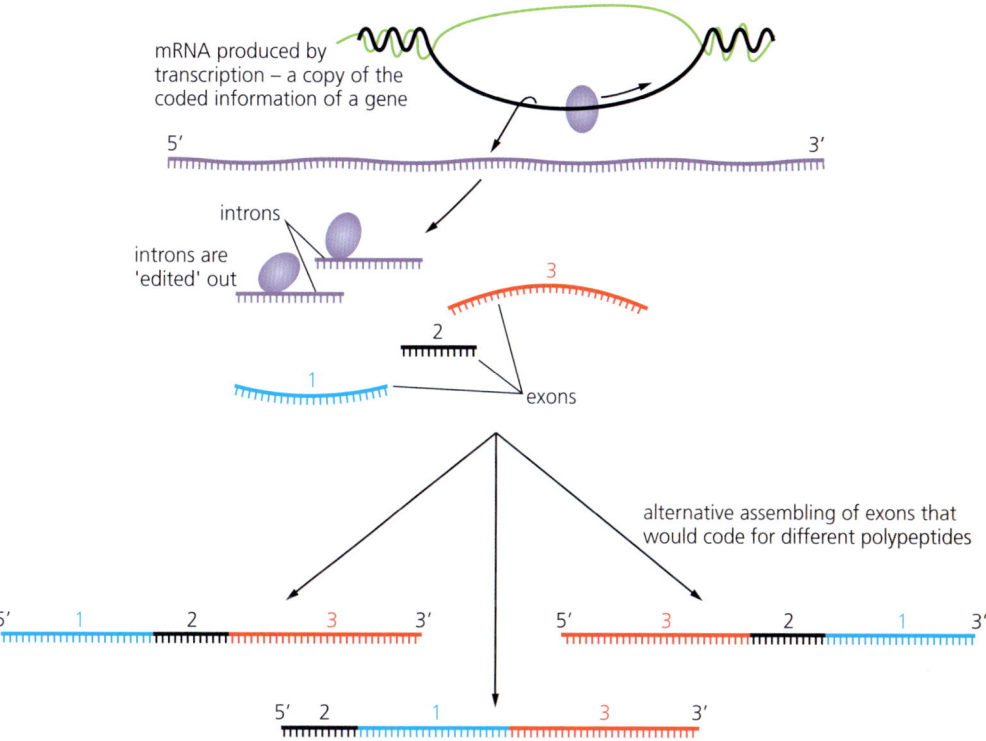

■ **Figure D1.2.12** Alternative RNA splicing

Alternative splicing of transcripts of the troponin T gene in foetal and adult heart muscle is an example of this process (Figure D1.2.13). Troponin plays a central role in the contraction of vertebrate striated muscles (see Chapter B3.3, page 329). The alternative splicing of the troponin T (TnT) gene is developmentally regulated. Foetal heart tissue uses one form of spliced mRNA and, as the heart matures, a different form is used in the adult. The two mRNAs shown in Figure D1.2.13 are translated into different but related muscle proteins – one used in the foetus and the other in the adult. These two different types of protein are known as **isoforms**. The changes prepare the heart for increased demand at birth.

D1.2 Protein synthesis

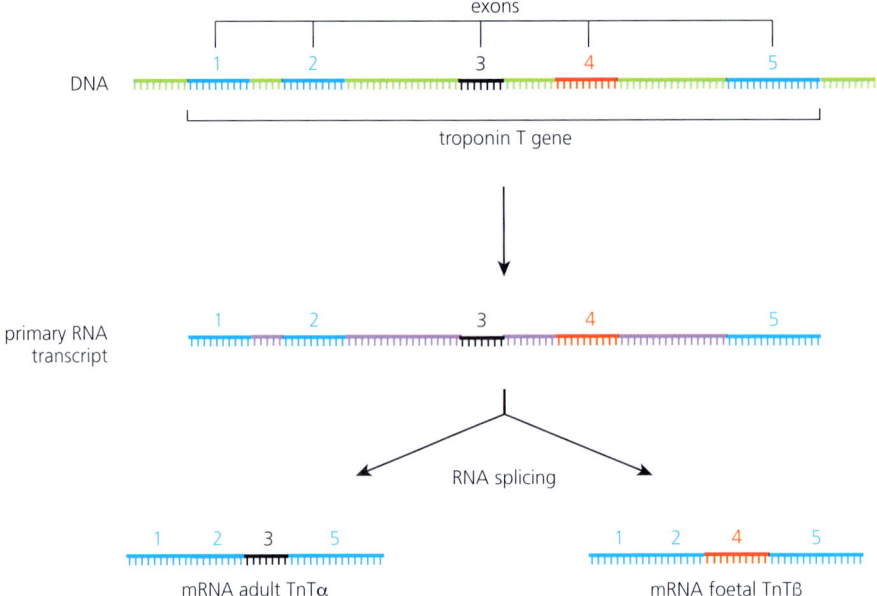

■ **Figure D1.2.13** Alternative RNA splicing of the troponin T gene gives rise to the adult TnTα and foetal TnTβ isoforms

Initiation of translation

Activation of amino acids

Some 20 different amino acids are involved in protein synthesis. Amino acids are activated for protein synthesis by combination with a short length of a different sort of RNA, called **transfer RNA (tRNA)** (Figure D1.2.14). The activation process involves ATP. There are 20 different tRNA molecules, one for each of the amino acids coded for in proteins.

All tRNA molecules have a clover-leaf shape, but they differ in the sequence of bases, known as the anticodon, which is exposed on one of the 'clover leaves'. This anticodon is complementary to a codon of mRNA. The enzyme catalysing the formation of the amino acid–tRNA complex 'recognises' only one type of amino acid and the corresponding tRNA.

1 tRNA structure
– a 'clover-leaf' shape

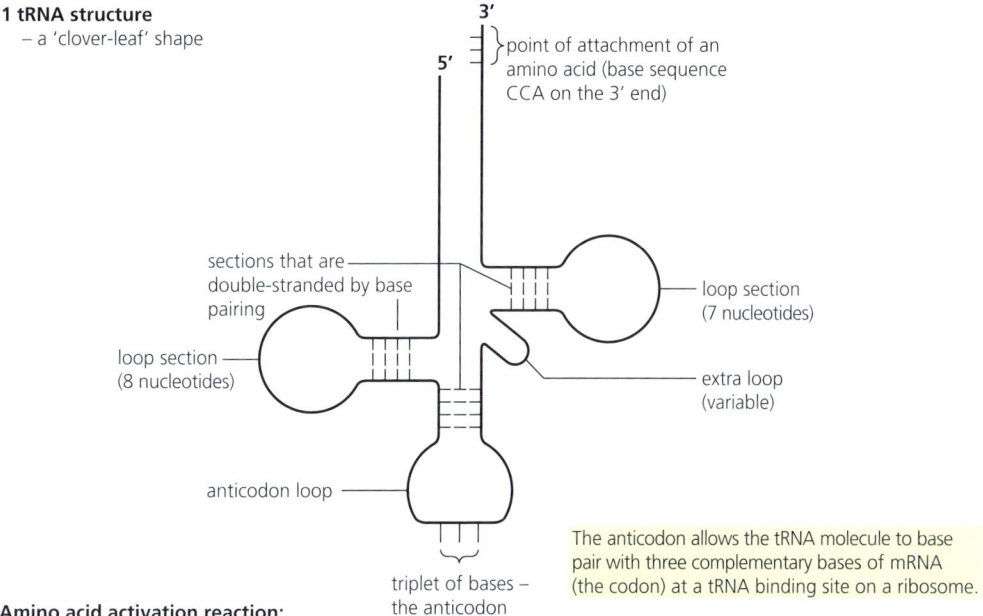

Amino acid activation reaction:

Each amino acid is linked to a specific tRNA before it can be used in protein synthesis by the action of a tRNA-activating enzyme (there are 20 different tRNA-activating enzymes, one for each of the 20 amino acids).

2 role of tRNA-activating enzyme, illustrating enzyme–substrate specificity and the role of phosphorylation

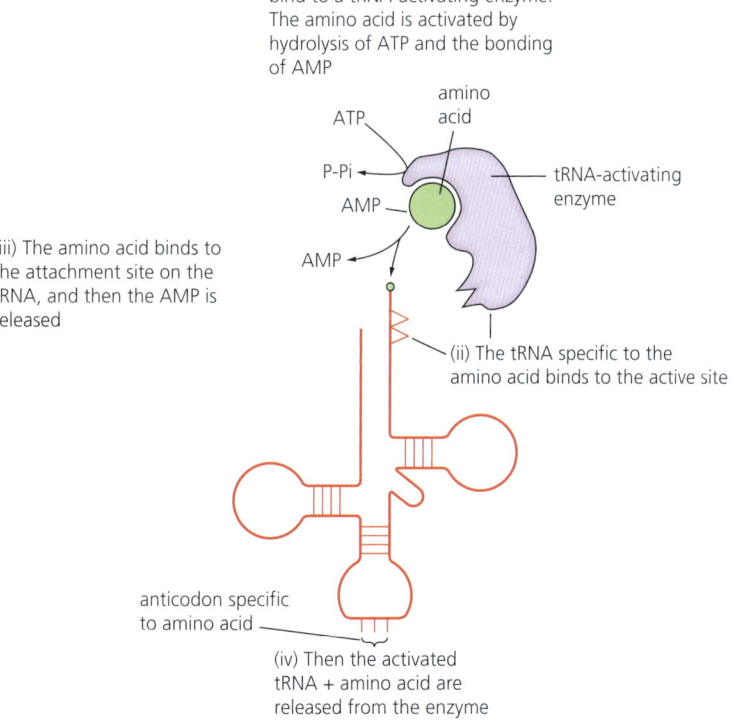

■ Figure D1.2.14 Transfer RNA (tRNA) and amino acid activation

tRNA-activating enzyme, enzyme–substrate specificity and the role of ATP

Each of the amino acids is attached to the 3' terminal of its specific tRNA molecule by a tRNA-activating enzyme. Remember, this enzyme is specific to the particular amino acid, as well as to a particular tRNA molecule. This specificity is a property of the structure of the enzyme's active site. The structure of a tRNA molecule and the steps to amino acid activation are shown in Figure D1.2.14:

- A specific amino acid and a molecule of ATP bind to a tRNA-activating enzyme. The amino acid is activated by hydrolysis of ATP and the bonding of AMP (adenosine monophosphate). P-P$_i$ is released (two phosphate groups linked together).
- The tRNA specific to the amino acid binds to the active site of the enzyme.
- The amino acid binds to the attachment site on the tRNA, and then the AMP is released.
- Then the activated tRNA with attached amino acid is released from the enzyme.

Nature of science: Experiments

Deducing the first codon of the genetic code

Marshall W. Nirenberg and Heinrich J. Matthaei (1962) made their own simple, artificial mRNA and identified the polypeptide product that was encoded by it. They used polynucleotide phosphorylase to do this. This enzyme randomly joins together any RNA nucleotides that it locates. Nirenberg and Matthaei began with the simplest codes possible. They added polynucleotide phosphorylase to a solution of pure uracil (U), so that the enzyme would generate RNA molecules consisting entirely of a sequence of Us; these molecules were known as poly(U) RNAs. Each poly(U) RNA contained a pure series of UUU codons. Molecular biologist Maxine Singer (along with Leon Heppel) provided the poly(U) RNAs for their experiment. These poly(U) RNAs were added to tubes containing components for protein synthesis (e.g. ribosomes, activating enzymes and tRNAs). Each tube contained one of the 20 amino acids, radioactively labelled. Of these 20 tubes, only one tube, containing the labelled amino acid phenylalanine, resulted in a product. Nirenberg and Matthaei had therefore discovered that the UUU codon could be translated into the amino acid phenylalanine. Similar experiments using poly(C) and poly(A) RNAs showed that proline was encoded by the CCC codon, and lysine by the AAA codon.

Ribosomes – the site of protein synthesis

A ribosome consists of a large and a small subunit, both composed of RNA (known as rRNA) and protein (Figure A2.2.20, page 69, and Figure D1.2.15). There are three sites in the ribosome where the tRNAs interact:

- **A site** – the first site. Here, a codon of the incoming mRNA binds to specific tRNA–amino acids through its anticodon (**complementary base pairing**).
- **P site** – the second site. Here, the amino acid attached to its tRNA is condensed with the growing polypeptide chain by **formation of a peptide linkage**.
- **E site** – the third site. Here, the **tRNA leaves the ribosome**, following transfer of its amino acid to the growing protein chain.

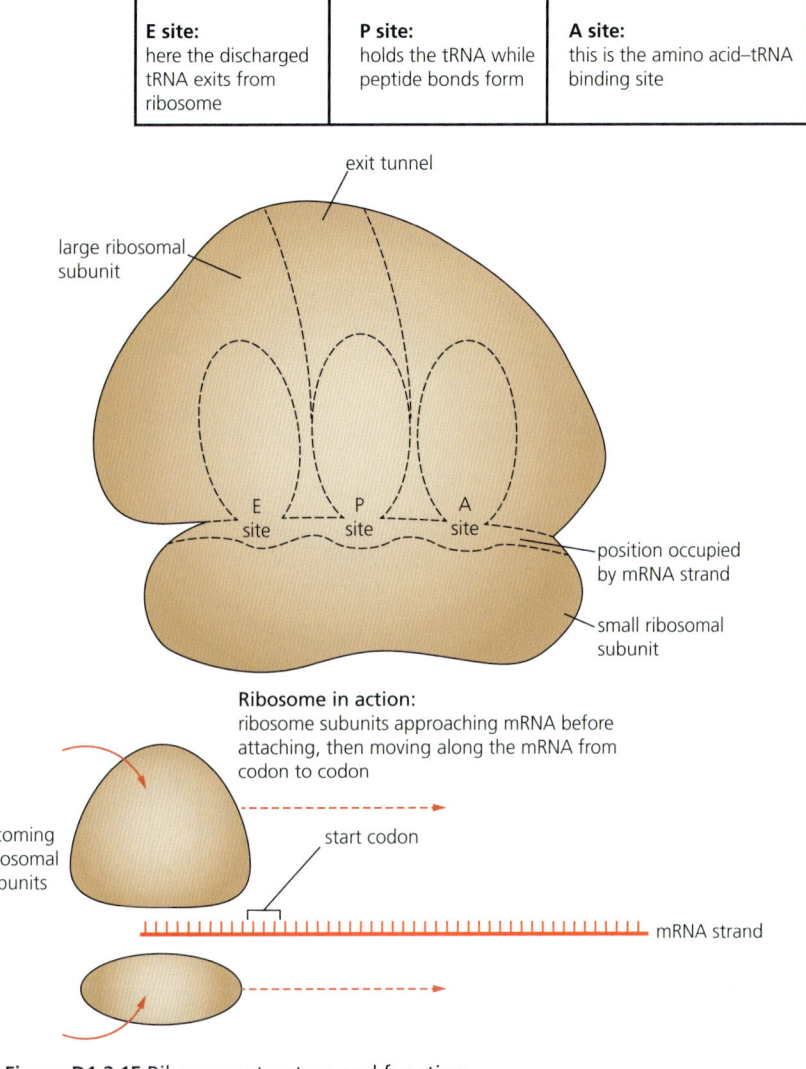

■ **Figure D1.2.15** Ribosome structure and function

■ Initiation of translation

Translation begins when an mRNA molecule binds with the small ribosomal subunit at an mRNA binding site at the 5′ end of the mRNA. This is joined by an initiator tRNA (to which the amino acid methionine is attached) at the start codon 'AUG'. This is followed by the attachment of a large ribosomal unit. The initiator tRNA occupies the P site in the assembled ribosome.

Now, the next codon of the tRNA, present in the A site, is available to a tRNA with the appropriate anticodon. Its arrival brings two activated amino acids (in sites P and A) into position and a peptide bond forms between them by a condensation reaction. The reaction is catalysed by enzymes present in the large subunit. A dipeptide has now been formed.

■ Elongation of the peptide

The ribosome moves three bases along the mRNA. In the process, the tRNA in the P site moves to the E site and is released. This movement of the ribosome also brings the next codon to occupy the now vacant A site – allowing a tRNA with the appropriate anticodon to bind to that codon. This, in turn, brings a further amino acid to lie alongside and, while these amino acids are held close together, another peptide bond is formed. In this way, the ribosome progresses along the mRNA molecule in the 5′ to 3′ direction, codon by codon. By these steps, constantly repeated, a polypeptide is formed and emerges from the large subunit (Figure D1.2.16).

6 **Draw** and **label** the structure of a peptide linkage between two amino acids.

Initiation of translation

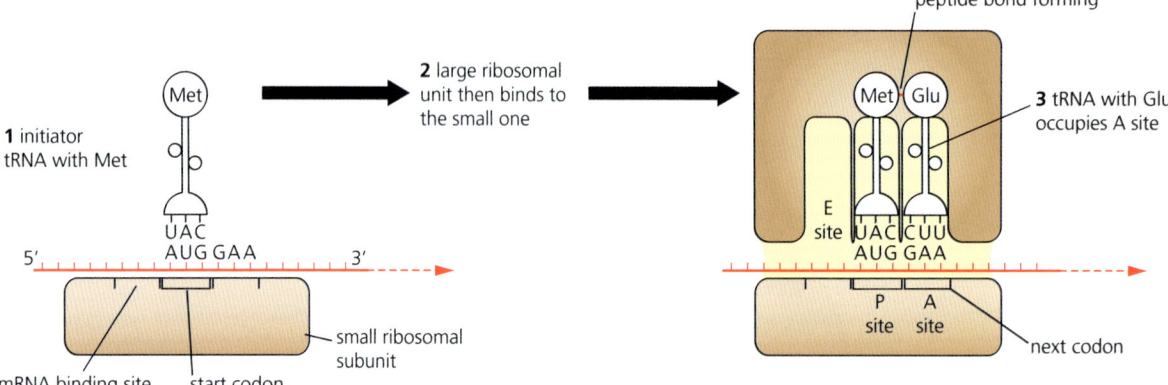

Elongation of the polypeptide

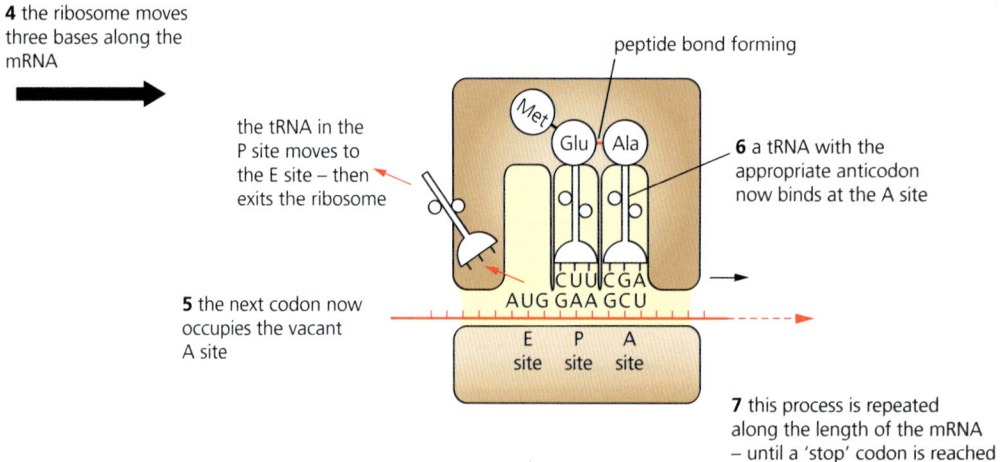

Termination of translation

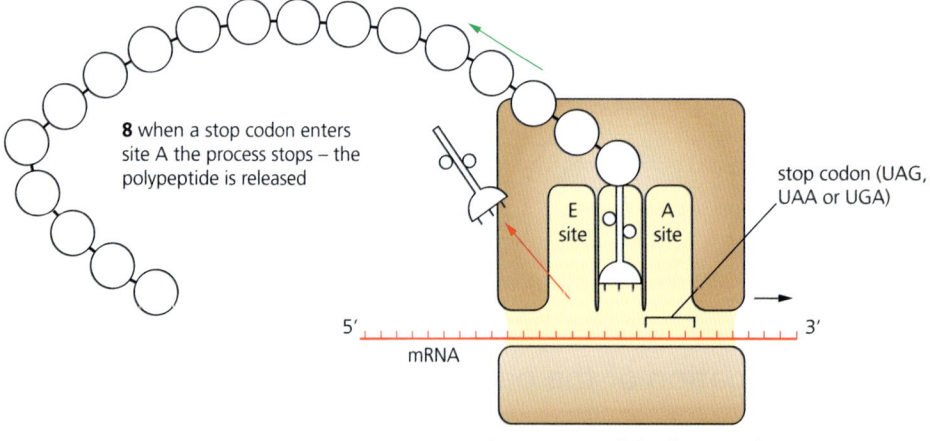

Note: the movement of the ribosome along the mRNA has been from the 5' to the 3' end

■ Figure D1.2.16 Initiation, elongation and termination in translation

Termination of translation

Eventually a 'stop' codon is reached. This takes the form of one of three codons – UAA, UAG or UGA. At this point, the completed polypeptide is released from the ribosome into the cytoplasm. Disassembly of the components of the ribosome follows termination of translation.

Modification of polypeptides into their functional state

◆ **Post-translational modification**: a biochemical modification that occurs to one or more amino acids on a protein after the protein has been translated by a ribosome.

When a protein exits from translation at a ribosome, it may take up its active three-dimensional shape and be functional immediately. For example, it may be active as an enzyme in the cytoplasm in some essential and continuous biochemical pathway. On the other hand, many polypeptides must be modified before they can function. These steps occur after translation and so are known as **post-translational modifications** (Figure D1.2.17). Some are produced in the form of inactive precursors, which require processing steps. In fact, it is often important for proteins to become active only at particular sites. For example, the protein-digesting enzyme trypsin is produced in the pancreas in an inactive form (trypsinogen) that does not digest the proteins of the pancreas cells in which it is formed.

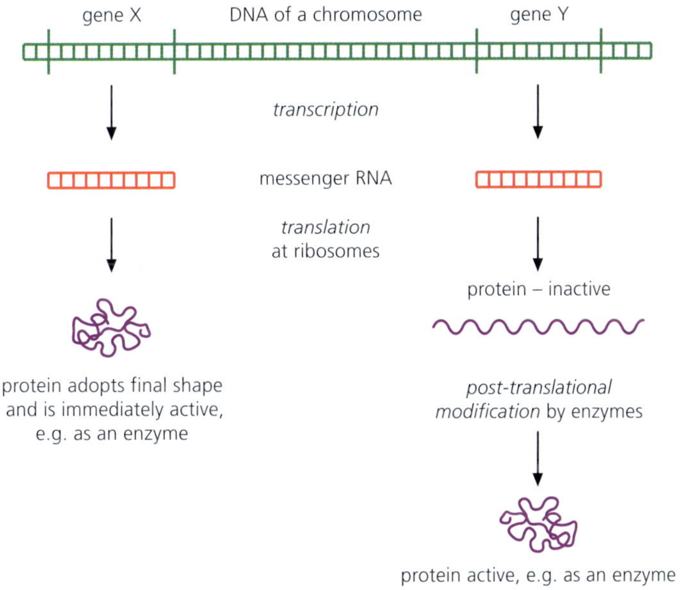

■ **Figure D1.2.17** Protein synthesis and post-translational modification

Insulin is a protein that needs to be modified before it can function. This involves a two-stage modification of pre-proinsulin to proinsulin, and proinsulin to insulin (see Figure B1.2.20, page 219).

Link
The quaternary structure of insulin as a non-conjugated protein is covered in Chapter B1.2, page 218.

> ● **Common mistake**
>
> Do not confuse post-*transcriptional* modifications with post-*translational* modifications. Post-transcriptional modifications are changes that occur to a newly transcribed mRNA after transcription has occurred and before it is translated into a protein. In contrast, post-translational modifications are modifications that occur to a protein after the protein has been translated by a ribosome.

HL ONLY

◆ **Proteolysis**: hydrolysis reaction of peptide bonds in which proteins are broken down into smaller peptides and/or into individual amino acids.

◆ **Proteasome**: protein complexes that degrade unneeded, misfolded or damaged proteins by proteolysis.

Recycling of amino acids by proteasomes

To maintain the proteome of an organism, proteins are continually synthesized by translation and broken down by hydrolysis reactions (a process known as **proteolysis**). In this way, there is a constant turnover of proteins, which ensures the maintenance of optimally functioning proteins. Amino acids are constantly being absorbed by organisms, but recycling of existing amino acids also takes place to ensure sufficient turnover. Breakdown of proteins is carried out by **proteasomes** – complexes made of protein into which the protein substrate (i.e. the protein marked for degradation) is fed (Figure D1.2.18). Proteasomes degrade proteins that are damaged, misfolded or no longer needed by the cell.

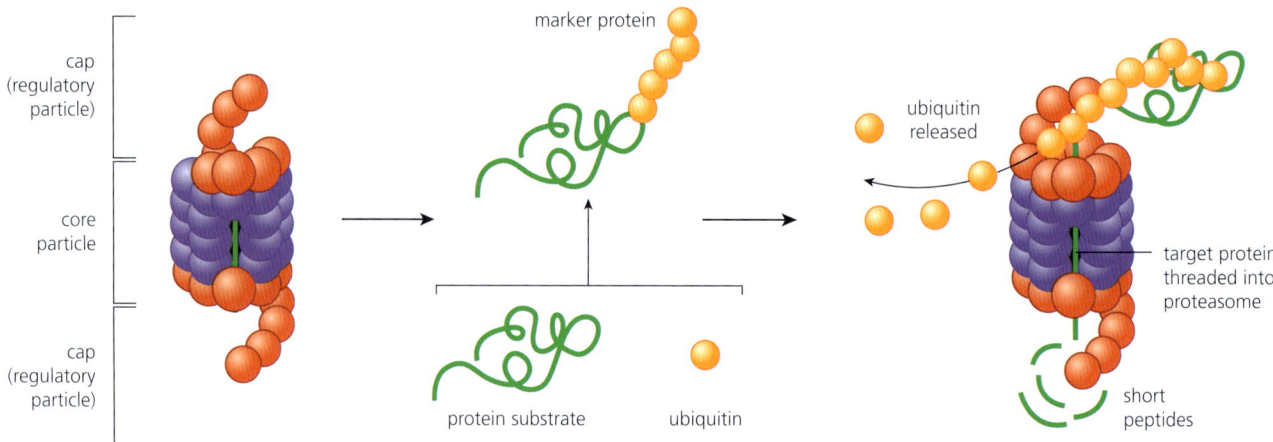

■ **Figure D1.2.18** The mechanism of protein breakdown by a proteasome

7 **Outline** the role of proteasomes in cells.

Proteins are marked for degradation by the attachment of a short chain of regulatory protein (Figure D1.2.18), known as **ubiquitin**. The marker protein binds to the top of the proteasome (in an area called the cap, or regulatory particle) which feeds the protein into the core particle where it is broken down into short peptides. The peptides are further processed to provide a pool of amino acids from which new proteins can be synthesized.

LINKING QUESTIONS

1 How does the diversity of proteins produced contribute to the functioning of a cell?
2 What biological processes depend on hydrogen bonding?

D1.3 Mutation and gene editing

> ### Guiding questions
> - How do gene mutations occur?
> - What are the consequences of gene mutation?

SYLLABUS CONTENT

This chapter covers the following syllabus content:
- ▶ D1.3.1 Gene mutations as structural changes to genes at the molecular level
- ▶ D1.3.2 Consequences of base substitutions
- ▶ D1.3.3 Consequences of insertions and deletions
- ▶ D1.3.4 Causes of gene mutation
- ▶ D1.3.5 Randomness in mutation
- ▶ D1.3.6 Consequences of mutation in germ cells and somatic cells
- ▶ D1.3.7 Mutation as a source of genetic variation
- ▶ D1.3.8 Gene knockout as a technique for investigating the function of a gene by changing it to make it inoperative (HL only)
- ▶ D1.3.9 Use of the CRISPR sequences and the enzyme Cas9 in gene editing (HL only)
- ▶ D1.3.10 Hypotheses to account for conserved or highly conserved sequences in genes (HL only)

♦ **Gene mutation**: change in the sequence of bases of a particular gene.

♦ **Substitution**: a type of mutation in which one nucleotide is replaced by a different nucleotide.

♦ **Insertion**: a type of mutation that involves the addition of one or more nucleotides into a segment of DNA.

♦ **Deletion**: a type of mutation that involves the loss of one or more nucleotides from a segment of DNA.

♦ **Frameshift**: a mutation in a DNA chain that occurs when the number of nucleotides inserted or deleted is not a multiple of three, so that every codon beyond the point of insertion or deletion is read incorrectly during translation.

Gene mutations as structural changes to genes at the molecular level

A **gene mutation** involves a change in the sequence of bases of a particular gene. Mutations are more likely to occur at certain times in the cell cycle than at other times. One such occasion is when the DNA molecule is replicating. We have noted that the enzyme DNA polymerase, which brings about the building of a complementary DNA strand, also 'proofreads' and corrects most errors (page 614). However, although proofreading is highly efficient, it does not correct all errors and gene mutations can and do occur spontaneously during the replication process.

Also, environmental factors or chemicals may cause changes to the DNA sequence of bases. These factors can include ionising radiation (page 640), UV light and various chemicals.

A **point mutation** is a change in one base in the gene sequence. It can be the result of:
- the change of one base to a different base (often caused by the DNA being copied incorrectly) – this is called a **substitution** mutation (Figure D1.2.8)
- the **insertion** of an additional base or bases into the DNA
- the **deletion** of a base or bases from the DNA.

A point mutation may change the amino acid in the protein encoded by the gene.

Because bases work in triplets (threes) to determine which amino acid is needed in a protein, an addition or deletion can have a profound effect. The deletion or insertion of a base results in what is called a **frameshift**.

1. **Define** the term *mutation*.
2. **Outline** how mutations can affect DNA.

Some changes result in a 'nonsense' mutation, so the changed genetic code cannot be read and has no meaning. If the change results in a different amino acid being coded for, then this mutation is called a 'missense' mutation.

> ### Concept: Change
>
> Mutations change the genetic code through insertion, deletion or substitution. As a result of mutations, polypeptides can cease to function, either through frameshift changes or through major insertions or deletions. Single-nucleotide polymorphisms (SNPs) are the result of base substitution mutations.

Consequences of base substitutions

Base substitutions can have an adverse effect. For example, sickle cell anaemia is caused by a change in the codon for the amino acid glutamic acid, resulting in the amino acid valine appearing instead (page 623).

However, because of the **degeneracy** of the genetic code (page 620), base substitutions may or may not always change an amino acid in a polypeptide. Each amino acid has several different codons and, if the base substituted matches one of these alternative codons for an amino acid, then there will be no effect on the proteome of the cell.

If the substitution occurs in a non-coding region of DNA, the mutation may have a neutral effect on the proteome of the cell. However, if the mutation is in a regulatory region that directs expression of a gene, a mutation may prevent the expression of the gene and so have a significant effect on the proteins found in a cell. **Single-nucleotide polymorphisms** (SNPs) are also the result of base substitution mutations.

Link
Single-nucleotide polymorphisms (SNPs) are also covered in Chapter A3.1, page 115.

● Common mistake
Many people associate mutations with 'bad' effects on the cell and organism. This is over-simplistic. Mutations can have a neutral effect, due to the degeneracy of the genetic code. Mutations that occur on non-coding regions of DNA may also have no effect on gene expression. Other mutations can be beneficial, leading to an adaptive advantage. Some mutations are harmful, leading to deleterious effects on the organism (see below).

Consequences of insertions and deletions

Mutations by major insertions, deletions or resulting frameshifts increase the likelihood of polypeptides not functioning. Two examples are outlined below, one of an insertion and another of a deletion, which show the results of these mutations.

■ Insertion: effect on trinucleotide repeats of the *HTT* gene

Insertions can result in repeating base sequences of three nucleotides, expanding **trinucleotide repeat sequences**. These are known as **trinucleotide repeat disorders**. Trinucleotide repeat disorders are caused due to an abnormal number of triplet repeat sequences either in the coding or non-coding regions of the genome. **Huntington's disease** is an example of this type of disorder.

Huntington's disease is caused by an abnormality in the *HTT* gene. The *HTT* mutation that causes Huntington's disease involves a DNA segment called a CAG trinucleotide repeat. This segment is made up of a series of three DNA bases (cytosine, adenine and guanine) that appear multiple times in a row. Normally, the CAG segment is repeated 10 to 35 times within the gene. Repeating segments can have a beneficial effect on brain development; more frequent repeats are associated with higher cognitive function. However, mutation leading to expansion of the trinucleotide repeats beyond 39 leads to Huntington's disease – a neurodegenerative disorder (Figure D1.3.1). The disease takes the form of progressive mental deterioration, accompanied by involuntary muscle movements.

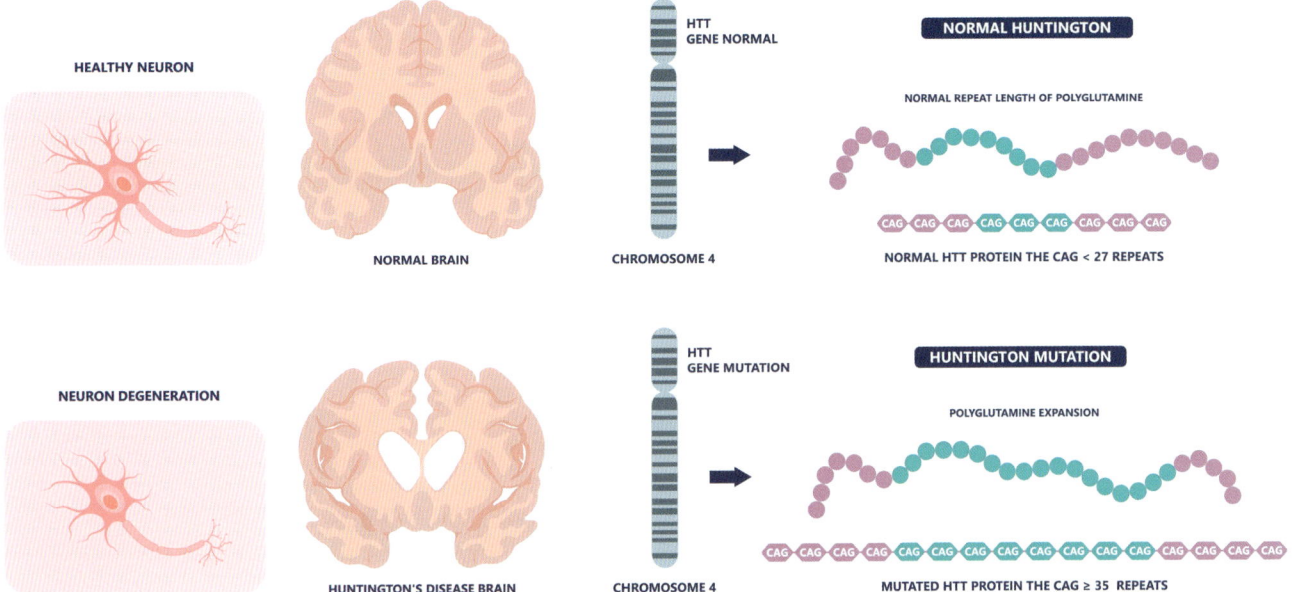

■ **Figure D1.3.1** Huntington's disease is a neurodegenerative disorder caused by multiple insertions of a trinucleotide repeat sequence (CAG)

Huntington's disease is extremely rare (1 case per 20 000 live births). It is caused by a mutated dominant allele (see Chapter D3.2, page 732). The symptoms usually do not appear until the affected person is 40–50 years old, by which time they – unaware of the presence of the disease – may have passed on the dominant allele to their children. In a pedigree chart (see Chapter D3.2, page 741) of a family with Huntington's disease, the condition tends to occur in one or more members of the family in every generation – a characteristic due to a dominant allele (page 728).

TOK

A simple blood test is all that is needed to find out if a person has the mutated dominant allele that will lead to Huntington's disease. Is this knowledge desirable? It would allow a person to prepare for the onset of the disease and may be relevant to their decision of whether to have children. However, the knowledge might be very distressing to someone who discovers that they have an ultimately fatal disease. The development of our understanding of genetics means that this type of knowledge of what will happen to us in the future might become far more common. Is this knowledge good for us?

■ Deletion: delta 32 mutation of the *CCR5* gene

The **delta 32 mutation** of the ***CCR5* gene** is an example of a deletion mutation. People who have two copies of the delta 32 mutation (i.e. a copy of the allele has been received from both parents) are less susceptible to HIV-1 infection.

The deletion mutation prevents the functional expression of the CCR5 protein co-receptor normally used by HIV-1 to enter CD4+ T-cells (page 530). T-cells (T-lymphocytes) are white blood cells processed by the thymus and responsible for cell-mediated immunity (page 523): CD4+ T-cells are the type of lymphocyte that is infected by HIV. The HIV can no longer bind to this co-receptor and so cannot efficiently enter cells (Figure D1.3.2). The *CCR5* mutation does not lead to complete resistance, although it does significantly lessen the likelihood of infection. The virus can use alternative, less-efficient co-factors to enable infection to occur. That has not stopped interest in using this observation as a possible effective therapy or preventative measure.

Link
T-lymphocytes are covered in Chapter C3.2, page 523.

D1.3 Mutation and gene editing

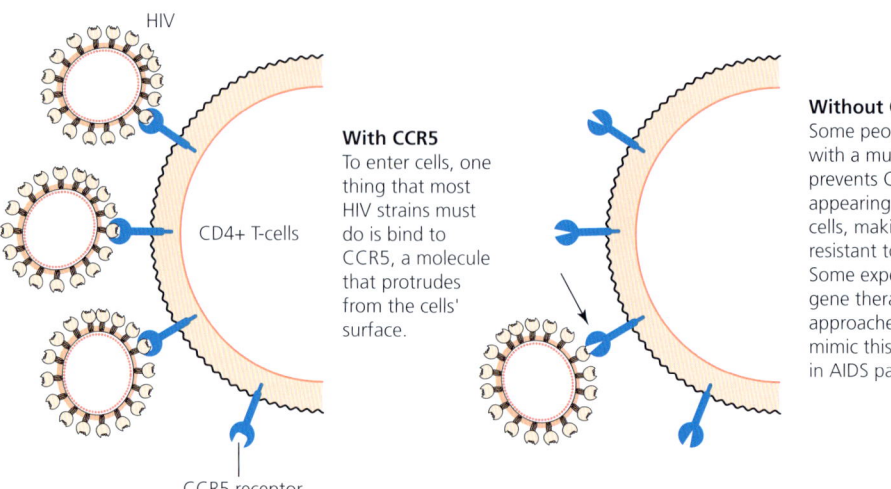

■ Figure D1.3.2 The delta 32 mutation of the *CCR5* gene changes the shape of the CCR5 receptor protein on the surface of the white blood cell (CD4+ T-cells), so HIV can no longer efficiently bind and infect the cell

3 **Explain**, using an example, the consequences of an insertion mutation.

Scientists believe that the *CCR5* delta 32 mutation existed as much as 2500 years ago but occurred only rarely (approximately in 1 out of every 20 000 Europeans). Today, the mutation is much more common, occurring in 1 in 10 people. Scientists think that a viral disease provided the selection pressure needed to increase the frequency of the mutation over time. In humans, certain populations have inherited the delta 32 mutation, resulting in the genetic deletion of a portion of the *CCR5* gene, leading to a conformational change that means it no longer binds to the HIV virus.

Causes of gene mutation

A factor capable of causing a mutation is called a **mutagen**, and its effects are described as 'mutagenic'. These factors are identified and defined in Table D1.3.1.

◆ **Mutagen**: an agent that causes gene mutations; anything that causes a mutation may cause a cancer, such as chemicals in tobacco smoke, X-rays, short-wave UV light and some viruses.

■ Table D1.3.1 Principal factors that can increase mutation rates and the likelihood of cancer

Ionising radiation	Ionising radiation includes X-rays and radiation (gamma rays, α particles, β particles) from various radioactive sources. These may trigger the formation of damaging ions and radicals inside the nucleus, leading to the break-up of the DNA.
Non-ionising radiation	Non-ionising radiation includes UV light. This is less energetic and therefore less penetrating than ionising radiation but, if it is absorbed by the nitrogenous bases of DNA, may modify them – causing adjacent bases on the DNA strand to bind to each other instead of binding to their partner on the opposite strand (thymine dimers – see page 19).
Bacterial infection	Some specific bacterial infections can trigger cancers, for example *Helicobacter pylori* can cause stomach cancer.
Chemicals	Several chemicals that are carcinogens are present in tobacco smoke. Also, prolonged exposure to asbestos fibres may trigger cancer in the lining of the thorax cavity (pleural membranes). The harm usually becomes apparent only many years later.
Virus infection	A specific virus infection, such as with hepatitis B and C viruses, human papillomavirus (HPV) and HIV, may trigger DNA mutations. The HPV mutation is the result of insertion of part of the virus genome into that of the host. The virus genes produce proteins that alter the cell cycle controls and so lead to uncontrolled replication of the cells.

As well as by mutagens, mutation can be caused by errors in DNA replication or repair. DNA polymerase has a role in 'proofreading' during DNA replication, to ensure that the base code is copied correctly. Any mistakes lead to the error being removed. It is estimated that, in eukaryotes, DNA polymerases make errors approximately once every 10^4–10^5 nucleotides copied. This means that each time a diploid mammalian cell replicates, between 100 000 and 1 000 000 polymerase

4 Outline three external causes of DNA mutation.

errors occur. Most of these are corrected, but when this mechanism fails, a mutation occurs. When the error occurs in a coding sequence of DNA, a gene mutation may occur, which can lead to a change in the protein synthesized.

Randomness in mutation

Mutation is a random process. Mutations are more likely to occur at certain times in the cell cycle than at other times, such as during DNA replication. Certain conditions or chemicals may cause change to the DNA sequence of bases (page 640). It is important to note that, although there are artificial ways of increasing the chance of mutation, there is no known natural mechanism for making a deliberate change to a particular base with the purpose of changing a trait. Evolution by natural selection is a process that selects favourable alleles that have been generated by **random** mutation, rather than created by a deliberate mechanism within the nucleus of a cell.

There are two types of DNA substitution mutations (Figure D1.3.3). **Transitions** are swaps between bases of similar shape, so between adenine and guanine, the two-ring purines, or between cytosine and thymine, the one-ring pyrimidines. **Transversions** are swaps between a purine and a pyrimidine base. Transition mutations are more common because they keep the same ring shape and, therefore, overall structure when bases form complementary pairs. In addition, transitions are less likely to result in amino acid substitutions (due to the degeneracy of the genetic code, page 620). They are therefore more likely to persist as 'silent substitutions' with no phenotypic effect (i.e. no changes in the physical traits of the organism) in populations as single-nucleotide polymorphisms (SNPs). Not all SNPs are transitions though.

The most common mutations are at the sequence CG (often written as CpG, i.e. with a phosphate in between). The proofreading process is 'blind' to the location of the bases it is correcting, which raises the question as to why these sequences are often mutated. CpG couplets are more difficult to correct because both C and G involve three hydrogen bonds to pair with the base on the opposite strand, which means that this region of DNA is more strongly held together than if it was two purines side by side or a purine and a pyrimidine. Regions rich in CpG are often highly methylated, making proofreading a bit more difficult.

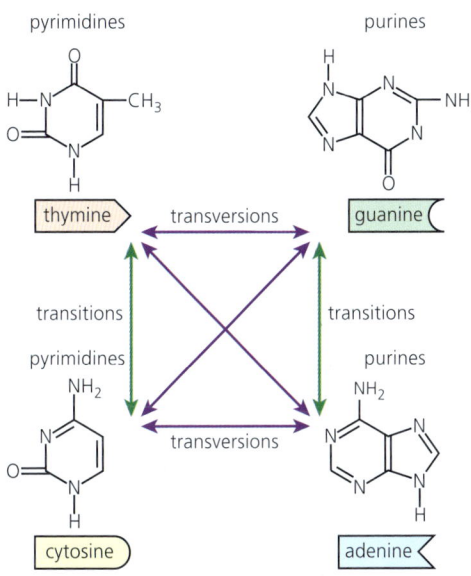

■ **Figure D1.3.3** Different types of substitution mutations: transversions are between different types of base (purines and pyrimidines), while transitions are between the same type of base

Link

For HL students, methylation as an example of epigenetic tagging is discussed in Chapter D2.2, page 673.

5 Explain why some bases have a higher probability of mutating than others.

Top tip!

Environmentally induced mutations due to exposure to a mutational stimulus are biased because of the relative likelihood of exposure to specific stimuli, so they are not all equally likely.

Mutations are not randomly distributed in genomes as may be expected. There are 'hot spots' of mutation. Possible reasons for this include:
- variations in the **superstructure** of DNA ('DNA superstructure' refers to large-scale organizational structures within chromosomes, such as segments of code that can be tens of millions of nucleotides long)
- regions where accessory proteins are less common, leading to an increased localised error rate that exceeds the correction rate.

D1.3 Mutation and gene editing

Consequences of mutation in germ cells and somatic cells

◆ **Germ cell**: a reproductive cell of the body.

Mutations occurring in ovaries or testes (or anthers or embryo sacs of flowering plants) are called **germ line mutations**. Since these mutations occur in **germ cells** that give rise to gametes, they may be passed to the offspring. These can result in inherited diseases, for example Huntington's disease (page 638).

Meanwhile, mutations that occur in body cells of multicellular organisms (**somatic mutations**) are only passed on to the immediate descendants of those cells, and they disappear when the organism dies. New mutations acquired during the lifetime of an individual in somatic cells are not heritable. Such mutations can lead, however, to **cancers** (see Chapter D2.1, page 667).

Link
For HL students, tumours and cancer are discussed in Chapter D2.1, page 667.

6 **Distinguish** between the effects of mutations in germ cells and somatic cells.

Mutation as a source of genetic variation

In Chapter A3.1, we saw that variation between organisms is a defining feature of life, with no two individuals identical in all their traits. The original source of all genetic variation is gene mutation. Alleles, which are different versions of the same gene (Chapter A1.2, page 21), are created by random mutations in genes. Many mutations, if they affect part of the non-coding region of a chromosome, are neutral and have no phenotypic effect. Other mutations may be harmful (e.g. sickle cell anaemia, page 623). However, mutations are essential in the long term for evolution by natural selection (Chapter D4.1). Because all individuals in a population are different, changes to the environment can lead to some having a selective advantage, with increased chance of survival, leading ultimately to changes in the gene pool of a species. If reproductive isolation occurs (Chapter A4.1, page 149), speciation can take place.

Link
Variation between organisms as a defining feature of life is covered in Chapter A3.1, page 102; evolution as change in the heritable characteristics of a population is covered in Chapter A4.1, page 138.

Link
Natural selection is covered in detail in Chapter D4.1, page 779.

7 **State** the original source of all genetic variation.

Concept: Change
Mutation leads to variation within a population. This variation is the basis for selection.

■ **Figure D1.3.4** Variation in the colour and patterning of ladybird beetles – the result of mutations over many generations

TOK

How might the context in which knowledge is presented influence whether it is accepted or rejected?

A genetics test prescribed by a doctor may be more acceptable to an individual or society than one bought over the counter. Why would this be the case, if the genetic tests are identical?

Nature of science: Global impact of science

Predictive genetic tests can provide consumers with knowledge of their health status and potential future health and disease risk, as well as providing commercial opportunities. 'Over-the-counter' and mail-order genetic tests are now available. However, this information could be problematic without expert interpretation. Issues associated with these tests include:
- Regulation of testing companies is currently lacking.
- This type of genetic testing cannot tell definitively whether an individual will or will not get a particular disease. Results often need to be confirmed with genetic tests carried out by a doctor or nurse.
- Tests may not be available for the health conditions that are of interest to an individual.
- They only test for a subset of variants within genes, so disease-causing variants can be missed.
- Unexpected information about the health of an individual may be stressful or upsetting.
- Test are carried out outside of a healthcare clinic, and so individuals are often not provided with genetic counselling.
- Individuals may make important decisions about disease treatment or prevention based on inaccurate, incomplete or misunderstood information from test results.

To address some of these issues, the sale of genetic tests to individuals can be made subject to ethical guidelines, for example to protect vulnerable groups.

Gene knockout

◆ **Gene knockout (KO)**: technique that produces a genetically engineered organism with one non-functional gene, allowing researchers to investigate the function of that gene.

◆ **Mutant**: organism with altered genetic material (abruptly altered by a mutation).

Gene knockout (KO) produces an organism with one non-functional gene, allowing researchers to investigate the function of that gene. The technique is used on **model organisms**. A model organism is a non-human organism that can be used to understand simple and complex biological functions. One model organism used for gene knockout is the mouse (*Mus musculus*). Researchers knock out or destroy the function of a particular gene (usually by introducing a deletion) and observe the effect on the phenotype.

The gene knockout technique has been used to create the p53 KO mouse, which has a mutated version of the *p53* gene. Normally, the *p53* gene directs the synthesis of a protein that regulates the cell cycle. When the protein is absent, tumour cells are created, increasing the risk of bone and other cancers in mammals.

Another remarkable example of this technique is the Methuselah KO mouse; increased longevity is shown in these **mutants**, allowing researchers to study the complexity of aging and potentially leading to prevention of human aging diseases and better treatments. As shown in Figure D1.3.5, the obese KO mouse can give insights into the metabolic disorders that are involved in obesity in humans.

Following the development of the gene knockout technique, the 2007 Nobel Prize in Physiology or Medicine was awarded to three scientists: Mario R. Capecchi (Italy), Martin J. Evans and Oliver Smithies (both UK). There is now a library of knockout organisms for some species that are used as models in research (see next section).

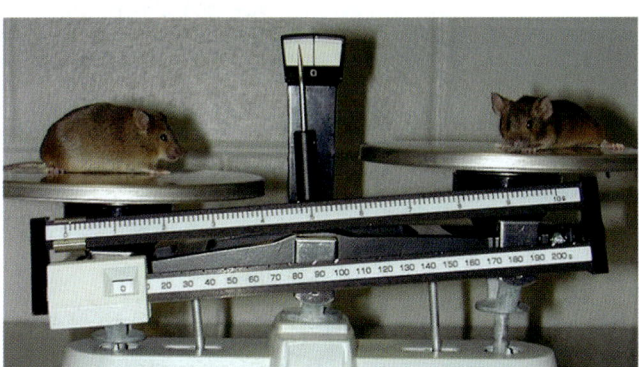

■ **Figure D1.3.5** A mouse that has a knockout gene (left) compared with a normal mouse (right); the mouse with the knocked out gene showed an increase in body fat and developed metabolic-related disorders, leading to obesity

8 **Define** the term *gene knockout*.

Model organisms are useful for studying gene function

The way in which a gene controls a function can be studied using model organisms, enabling researchers to then apply the findings to other organisms, such as humans. These model organisms can range from bacteria, such as *Escherichia coli* (*E. coli*), to mammals, such as the mouse (*Mus musculus*). Some plant models, for example *Arabidopsis thaliana*, are widely used for investigating crops and other plants.

It is the universality of the genetic code and commonality of the basal biological processes performed by almost all living organisms, such as oxidative phosphorylation during aerobic respiration, that make the use of model organisms possible. Studies, performed in vivo using model organisms, such as **gene knockout**, are extremely valuable as they can predict similar effects in closely related organisms.

■ Table D1.3.2 Some of the most common model organisms used

Name and type of organism		Example of its use
Caenorhabditis elegans (*C. elegans*)	nematode (soil roundworm)	understanding organ development and programmed cell death (apoptosis)
Drosophila melanogaster	insect (fruit fly)	a high homology with human cells means that the fruit fly is being used to study Parkinson's disease and Alzheimer's disease in humans
Danio rerio	vertebrate (zebra fish)	embryos are transparent, allowing non-invasive in vivo imagery to obtain data about the circulatory system; also useful in the understanding of the process of metastasis
Arabidopsis thaliana	angiospermophyte (flowering plant)	first plant to have had its genome sequenced; used in a variety of studies, such as understanding floral production and the effects of environmental stress in plants

> **Common mistake**
>
> Many people have an idea of why model organisms are used, but few can outline the benefits clearly. Make sure you know why each model organism is used in scientific investigations (see Table D1.3.2).

Inquiry 1: Exploring and designing

Exploring; designing

Research a model organism and how it is used in experimental research. Ensure you consult a variety of sources and select sufficient and relevant sources of information.

Design your own theoretical investigation using the organism. Which organism have you selected and why? In which ways does your organism acts as a model?

Formulate a research question and hypothesis. State and explain your predictions using your scientific understanding.

Use of the CRISPR sequences and the enzyme Cas9 in gene editing

◆ **CRISPR sequences**: sequences in the genomes of some prokaryotes that act as a genomic record of previous viral attack.

◆ **Cas9 enzyme**: a bacterial endonuclease that forms a double-strand break in DNA at a specific target site within a larger recognition sequence.

Prokaryotic cells have a mechanism to defend themselves against infection comparable to, but very different from, the immune response found in eukaryotic cells. In a prokaryotic cell, the response to a 'foreign' invasion is based on a region of DNA sequences called CRISPR, which stands for **c**lustered **r**egularly **i**nterspaced **s**hort **p**alindromic **r**epeats. These **CRISPR sequences** are used for an immune response that protects the prokaryote from bacteriophages (a type of virus that attacks bacterial cells). Along with the CRISPR-associated **Cas9 enzyme**, bacteria use the sequences to recognize and deactivate future invading viruses.

- When bacteria are infected by a virus, they use their CRISPR system to cut up the invading viral DNA and insert pieces of it (**spacers**) into their own genome as a 'memory' of the infection.

- Bacteria transcribe the spacers into RNA, which can form a complex with the Cas9 enzyme. These complexes monitor the cell for any DNA sequence complementary to the RNA. When DNA sequences complementary to the DNA are identified, the Cas9 enzyme is activated.
- If matching (viral) DNA is encountered, the spacer RNA–Cas9 complex binds to it and cuts the viral DNA at specific locations to prevent it from replicating. This halts the viral infection.

Scientists Emmanuelle Charpentier and Jennifer Doudna were awarded the Nobel Prize for Chemistry in 2020 for their collaborative work with the CRISPR-Cas9 'genetic scissors', developing the precise genome-editing technology. This was the first time a Nobel Prize had been won by two women alone. Scientists can use this technology to edit the human genome (Figure D1.3.6).

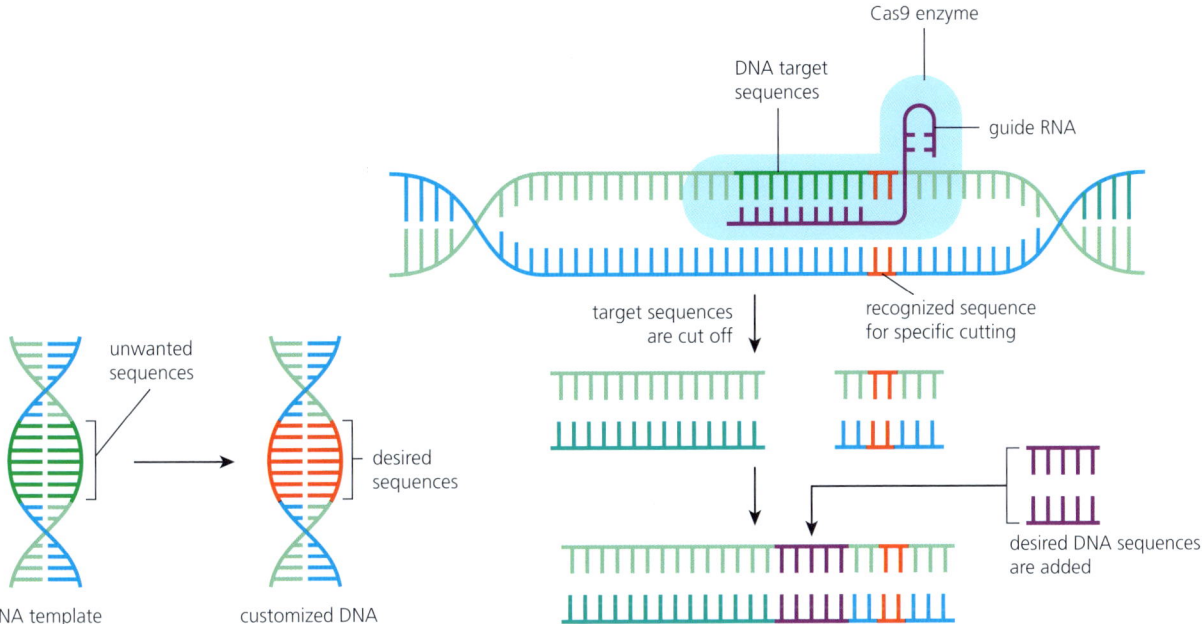

■ **Figure D1.3.6** CRISPR and its application in gene editing

ATL D1.3A

Find out about how CRISPR technology can be used to treat genetic diseases. Use this weblink as a starting point:

www.bio-rad.com/en-uk/applications-technologies/crispr-cas-gene-editing-teaching-resources?ID=Q58I0DWDLBV5

Summarize, in a poster or PowerPoint, how CRISPR technology can be used to treat genetic diseases.

The precise specificity of the CRISPR system offers many opportunities for use in modifying any gene of interest with significantly reduced risk of accidentally affecting other genes. Researchers are looking to CRISPR as a technique for editing out genetic defects that result in cystic fibrosis, sickle cell disease, haemophilia and muscular dystrophy, for example, as well as for developing more targeted and effective cancer treatments. One study showed that adult rats genetically engineered to have a genetic form of blindness could be treated using CRISPR gene therapy, which means that their offspring would not have the genes that cause genetic blindness. The goal of human gene editing is to remove diseased cells, 'fixed' (i.e. the genetic mutation removed and corrected for a code that is functional) using CRISPR technology, and then returned to patients to treat the genetic condition.

◆ **Genome editing**: the manipulation of DNA by deleting, replacing or inserting a DNA sequence, with the aim of, for example, correcting a genetic disorder.

Genome editing requires bio-molecular tools to specifically recognize a target sequence on the genome and cut it in a precise way. Cas9 recognizes the target DNA via a guide RNA. Multiple types of Cas enzymes are found in nature, but Cas9 is commonly used in the laboratory. Research is being carried out to identify and manufacture guide RNA for specific genetic diseases. Scientists have used CRISPR technology to insert particular selected sequences of a genome into the Cas9 protein. This allows for adding, replacing and removing genetic sequences from genes, opening many opportunities for research.

Concept: Change

Genome editing changes the genetic code of an organism. This can be used to treat genetic disorders.

Successful use of CRISPR technology: treating sickle cell anaemia

CRISPR is revolutionizing many aspects of biotechnology, scientific research and genetic disorder treatment. Sickle cell disease is a blood disorder that can be difficult to treat effectively (page 623). In an experimental treatment, scientists used CRISPR to edit a gene from a patient's bone marrow so the cells would produce foetal haemoglobin – a form of haemoglobin that stops being made shortly after birth. By restarting production of foetal haemoglobin, scientists hoped this form of haemoglobin could compensate for the sickle-shaped haemoglobin. In one patient, blood tests showed that approximately 46% of the haemoglobin was foetal haemoglobin and it remained present in 99.7% of red blood cells. Further samples of bone marrow found more than 81% of the cells contained the genetic change needed to produce foetal haemoglobin. This showed the edited cells were continuing to survive and function in the patient's body for a sustained period.

ATL D1.3B

Find out more about CRISPR technology and how it is being used to treat sickle cell anaemia. Use this site as a starting point: www.synthego.com/crispr-sickle-cell-disease#:~:text=Victoria%20Gray%20was%20the%20first,without%20any%20major%20side%20effects

Produce a poster to summarize the key points. What is sickle cell anaemia and what causes this genetic disease? What current treatments exist? How can CRISPR technology be used to treat the disease?

TOK

That scientists work under regulatory systems is a prime example of how knowledge communities do not exist in isolation. Social, political and financial concerns can be present in the development of scientific knowledge and can impact on its development. The peer review of scientific research also means that scientists work within a framework that ensures collaboration and mutual support.

Nature of science: Global impact of science

Despite the potential of CRISPR-Cas9 in treating many genetic diseases, the use of CRISPR to change genetic traits in humans has raised serious concerns within the scientific community. Possible unintended effects include:
- potential off-target mutations (unintended mutations) in the genome; such mutations may be deleterious
- the cost of germ line-editing technology is very high, to the extent that only people from rich countries might afford it.

The ease of applying CRISPR has caused worry about the potential misuse of the technology. Certain potential uses of CRISPR raise ethical issues that must be addressed before implementation, including:
- genome editing in human embryos, which could have unpredictable effects on the future generation
- the use of the technology for non-therapeutic modifications, leading to loss of human diversity and eugenics.

Scientists across the world are subject to different regulatory systems. For this reason, there is an international effort to harmonize the regulation of the application of genome-editing technologies such as CRISPR.

> **ATL D1.3C**
>
> Write an essay on one of the following two titles:
> - **What measures can be taken to regulate the application of genome editing?**
>   (Propose specific measures to help regulate the application of genome editing. In explaining these measures, include details of the CRISPR-Cas9 technology, its applications, and ethical issues relating to its use.)
> - **To what extent should CRISPR technology be encouraged, restricted or prohibited?**
>   (On a continuous scale from widespread use to total prohibition, choose and define a position. In defending that position, use and explain details of the CRISPR-Cas9 technology, its applications, and ethical issues relating to its use.)

Hypotheses for conserved or highly conserved sequences in genes

In Chapter A2.1, conserved genetic sequences were identified as a way of providing evidence about the structure and function of the first life on Earth (LUCA). **Conserved sequences** are identical or similar sequences in nucleic acids (DNA and RNA) across species or a group of species. Highly conserved genetic sequences are identical or similar over long periods of evolution.

Examples of highly conserved sequences include proteins associated with ribosomes, which are organelles present in all domains of life. There are two hypotheses that explain the mechanism for conserved sequences:

- functional requirements for the gene products; highly conserved sequences are usually required for basic cellular stability, function and reproduction, and so must be maintained from one generation to the next and on throughout evolutionary history
- slower rates of mutation than the background mutation rate.

Research has shown that mutation rate is linked to the level of gene expression. Highly transcribed genes generally show lower mutation rates that less expressed genes. It has also been noted that the transcribed strand of a gene shows a lower mutation rate than the non-transcribed strand. The impact of gene expression on mutation rate may be due to enhanced proofreading and, therefore, repair mechanisms for these sections of DNA.

In addition to protein coding sequences, the genome contains a significant amount of regulatory DNA (page 626). These non-coding regions contain highly conserved sequences. In vertebrates, these sequences are in and around genes that act as developmental regulators. Highly conserved regions for transcription factors related to embryonic and subsequent development is further evidence for a common origin of all vertebrates (see also Chapter A2.1, page 47).

◆ **Conserved sequences**: identical or similar sequences in nucleic acids (DNA and RNA), across species or a group of species, that have remained essentially unchanged throughout evolution.

Link

Conserved genetic sequences and the evidence for the evolution of the last universal common ancestor is discussed in Chapter A2.1, page 47.

9 **Suggest** why some genes are conserved.

> **LINKING QUESTIONS**
> 1. How can natural selection lead to both a reduction in variation and an increase in biological diversity?
> 2. How does variation in subunit composition of polymers contribute to function?

D2.1 Cell and nuclear division

Guiding questions

- How can large numbers of genetically identical cells be produced?
- How do eukaryotes produce genetically varied cells that can develop into gametes?

SYLLABUS CONTENT

This chapter covers the following syllabus content:
- ▶ D2.1.1 Generation of new cells in living organisms by cell division
- ▶ D2.1.2 Cytokinesis as splitting of cytoplasm in a parent cell between daughter cells
- ▶ D2.1.3 Equal and unequal cytokinesis
- ▶ D2.1.4 Roles of mitosis and meiosis in eukaryotes
- ▶ D2.1.5 DNA replication as a prerequisite for both mitosis and meiosis
- ▶ D2.1.6 Condensation and movement of chromosomes as shared features of mitosis and meiosis
- ▶ D2.1.7 Phases of mitosis
- ▶ D2.1.8 Identification of phases of mitosis
- ▶ D2.1.9 Meiosis as a reduction division
- ▶ D2.1.10 Down syndrome and non-disjunction
- ▶ D2.1.11 Meiosis as a source of variation
- ▶ D2.1.12 Cell proliferation for growth, cell replacement and tissue repair (HL only)
- ▶ D2.1.13 Phases of the cell cycle (HL only)
- ▶ D2.1.14 Cell growth during interphase (HL only)
- ▶ D2.1.15 Control of the cell cycle using cyclins (HL only)
- ▶ D2.1.16 Consequences of mutations in genes that control the cell cycle (HL only)
- ▶ D2.1.17 Differences between tumours in rates of cell division and growth, and in the capacity for metastasis and invasion of neighbouring tissue (HL only)

◆ **Parent cell**: cell that divides to form daughter cells by mitosis or meiosis.

◆ **Daughter cells**: cells produced when a cell divides by mitosis or meiosis.

Concept: Continuity

All cells arise from pre-existing cells via cell division. This process allows the genetic code, which determines the structure and function of the organism, to be passed on to new cells.

Generation of new cells in living organisms by cell division

Cell theory states that all cells arise from pre-existing cells (Chapter A2.2, page 50). Multicellular organisms begin life as a single cell, which grows and divides. During growth, this cycle is repeated, forming many cells. Initially, stem cells can become any other type of cell, but during embryonic development cell specialization occurs (see Chapter B2.3). It is these cells that eventually make up the adult organism. In all living organisms, a **parent cell** – often referred to as a 'mother' cell – divides to produce two **daughter cells**.

There are two types of cell division – one produces genetically identical daughter cells and the other genetically different (varied) cells. Both types of cell division will be discussed in this section.

ATL D2.1A

Do cells divide a fixed number of times, or can they do this continually with no set number of divisions? Find out about the 'Hayflick limit', which is the number of times a normal group of differentiated (i.e. specialized) human body cells will divide before cell division stops. Are there exceptions to this 'limit'?

Theme D: Continuity and change – Cells

♦ **Cytokinesis**: splitting of cytoplasm in a parent cell between daughter cells, following the division of the nucleus.

Cytokinesis

In cell division, the nucleus and cytoplasm will divide. Division of the cytoplasm is known as **cytokinesis** and follows the final stage of nuclear division. During division, cell organelles such as mitochondria and chloroplasts become distributed evenly between the cells. In animal cells, division is by in-tucking of the plasma membrane at the equator of the cell, 'pinching' the cytoplasm in half (Figure D2.1.1). A ring of actin, myosin and other proteins form a 'contractile ring' that pinches the cell membrane together to split the cytoplasm.

In plant cells, the Golgi apparatus forms vesicles of new cell wall materials, which collect along the line of the equator of the spindle, known as the cell plate. Here the vesicles combine to form the new plasma membranes and cell walls between the two cells (Figure D2.1.2).

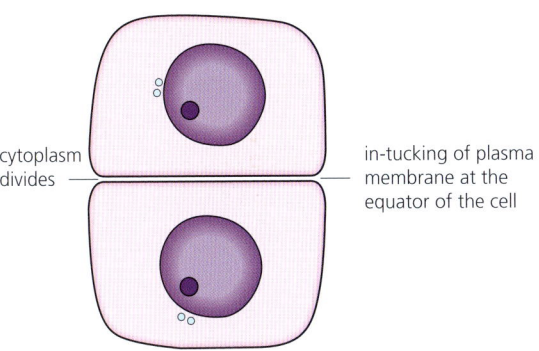

■ Figure D2.1.1 Cytokinesis: the division of the cytoplasm

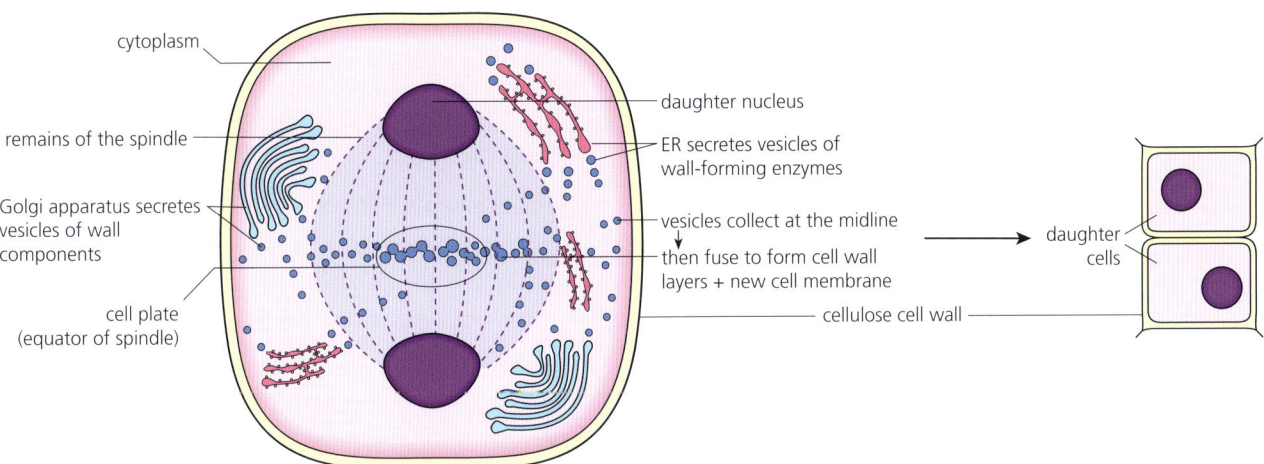

■ Figure D2.1.2 Cytokinesis in a plant cell

> ● **Top tip!**
>
> Cytokinesis differs between animal and plant cells. Plant cells cannot proceed with a cleavage furrow because they have a cell wall and animal cells do not. Plants must first form a cell plate, which divides the old cell into two new ones. Vesicles assemble sections of membrane and cell wall to achieve splitting. In an animal cell, a ring of contractile actin and myosin proteins pinches the cell membrane together to split the cytoplasm.

Equal and unequal cytokinesis

When cells divide to form new cells in a body tissue, the division of cytoplasm is usually equal, so that two daughter cells of equal size are produced (Figure D2.1.2). Both daughter cells must receive at least one mitochondrion and, in plants, a chloroplast. These are the only organelles that can be made by dividing a pre-existing structure.

However, in some cases the division of cytoplasm is not equal. For example, in egg production in humans, a process known as **oogenesis**, one large egg (ovum) is formed as well as several small cells that do not go forward for fertilization. Oogenesis begins in the ovaries of the foetus before birth, but

Link
Oogenesis is covered in detail in Chapter D3.1, page 710.

D2.1 Cell and nuclear division

1 **Explain** why some cell divisions produce cells that are unequal in size.

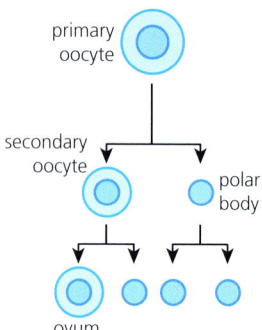

■ **Figure D2.1.3** The unequal division of the cytoplasm in human oogenesis

the final development of oocytes is only completed in adult life. An oocyte is an immature egg cell (ovum). When a primary oocyte divides, the cytoplasmic division that follows is unequal, forming a tiny polar body and a secondary oocyte. Similarly, when the secondary oocyte divides, another polar body is formed (Figure D2.1.3).

The polar bodies degenerate, and their cytoplasm and cellular contents are absorbed by the final ovum. This means that one large egg is formed rather than four smaller structures. The ovum needs a large cytoplasm to provide nutrients and energy for the growing embryo.

ATL D2.1B

Yeast, a single-celled fungus, also divides by unequal division of the cytoplasm. Cell division in yeast is known as **budding**. Find out about budding in yeast, and the reasons why cytokinesis forms daughter cells that are unequal in size.

Roles of mitosis and meiosis in eukaryotes

Divisions of the nucleus occur by a very precise process, ensuring the correct distribution of chromosomes between the new cells (daughter cells). Nuclear division must happen before cell division to avoid production of **anucleate** cells (cells without a nucleus). If nuclear division does not take place, one daughter cell would have a nucleus and the other would not.

There are two types of nuclear division: **mitosis** and **meiosis**. Both produce daughter cells from a parent cell. The number of cells produced and the form of genetic material varies between mitosis and meiosis, as will be discussed later in this chapter. Mitosis maintains the chromosome number and genome of cells, whereas meiosis halves the chromosome number and generates genetic diversity.

In **mitosis**, the daughter cells produced have the same number of chromosomes as the parent cell, typically two of each type, known as the **diploid** ($2n$) state. Mitosis is the nuclear division that occurs when an organism grows, when old cells are replaced, and when an organism reproduces asexually. Mitosis is explained on page 653 in this chapter.

♦ **Mitosis**: nuclear division in which the daughter nuclei have the same number of chromosomes as the parent cell.

♦ **Meiosis**: nuclear division with daughter cells (gametes) containing half the number of chromosomes of the parent cell.

♦ **Diploid**: cells with nuclei containing two sets of chromosomes.

♦ **Haploid**: cells with nuclei containing one set of chromosomes.

In **meiosis**, the daughter cells contain half the number of chromosomes of the parent cell. That is, one chromosome of each type is present in the nuclei formed; this is known as the **haploid** (n) state. Meiosis is the nuclear division that occurs when sexual reproduction occurs, normally during the formation of the gametes.

The differences between mitosis and meiosis are summarized in Figure D2.1.4.

 **Top tip!**

Mitosis and meiosis only occur in eukaryotes. Prokaryotes divide by a process called binary fission.

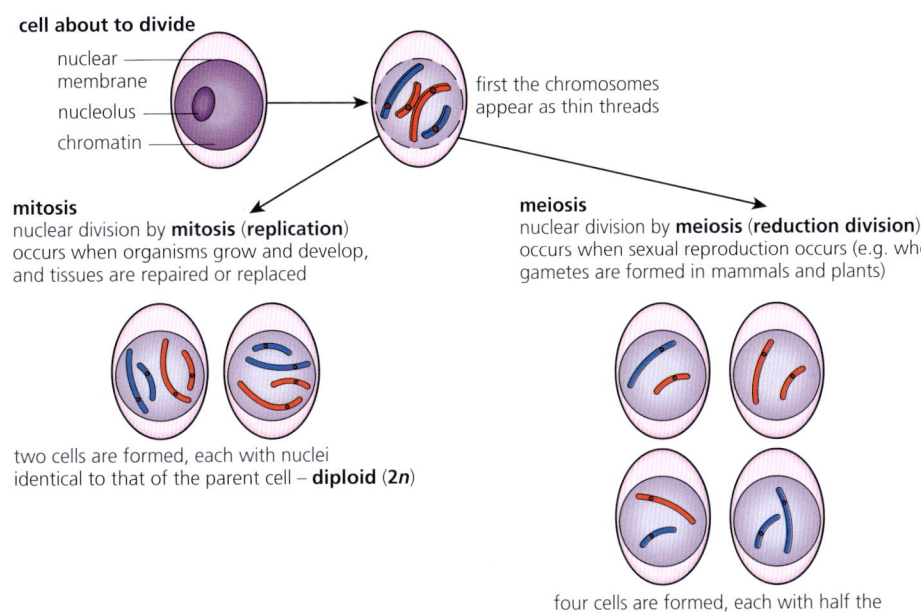

■ Figure D2.1.4 Mitosis and meiosis – the significant differences

2 **Suggest** why it is essential that nuclear division is a precise process.

◆ **Interphase**: the period between nuclear divisions when the nucleus controls and directs the activity of the cell and replicates chromosomes.

◆ **Chromatid**: one of two copies of a chromosome after it has replicated, joined together at a centromere.

◆ **Centromere**: constriction of the chromosome, the region that becomes attached to the spindle fibres during nuclear division.

Link
For HL students, the details of interphase are on page 665.

Concept: Continuity

For new cells to form, DNA must first replicate so that copies of the genome can be sent into each new cell. In this way, the genetic code of an organism continues into new cells.

DNA replication as a prerequisite for both mitosis and meiosis

DNA replication occurs in a stage of the cell cycle called **interphase**. The process produces double-stranded chromosomes. Each strand is called a **chromatid**. Chromatids are held together by a **centromere**: a specialized DNA sequence that can be seen as the constricted region of a chromosome (Figure D2.1.5). During nuclear division, each of the sister chromatids separate and move into different cells.

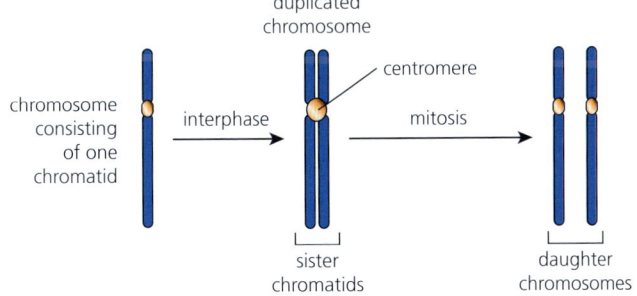

■ Figure D2.1.5 Changes in chromosomes during interphase and mitosis (nuclear division)

● Common mistake

To avoid confusion in terminology, you should refer to the two parts of a chromosome as sister chromatids while they are attached to each other by a centromere in the early stages of mitosis. Once the chromatids have been separated into separate cells, they can be referred to as distinct chromosomes.

D2.1 Cell and nuclear division

Condensation and movement of chromosomes as shared features of mitosis and meiosis

The total length of the DNA in all 46 human chromosomes is over 2 m. Each chromosome contains one very long DNA molecule. A typical chromosome of 5 μm length contains DNA that is approximately 50 mm long, which means that about 50 000 μm of DNA is packed into 5 μm of chromosome. Today we know that, while some of the proteins of the chromosome are enzymes involved in the copying and repair reactions of DNA, the bulk of chromosome protein has a support and packaging role for DNA.

DNA in the nucleus is packaged with proteins in a complex known as **chromatin**. Here, the much-coiled DNA double helix of each chromosome is looped around **histone** protein beads (**nucleosomes**, Figure D2.1.6). Histones are one sort of packaging protein. Histones are a basic (positively charged) protein containing a high concentration of amino acid molecules with additional basic functional groups ($-NH_2$), such as lysine and arginine. These histones occur clumped together, and provide support to the lengths of the DNA double helix that occur wrapped around them, giving the appearance of beads on a thread. The 'bead thread' is itself coiled up, forming the chromatin fibre. The chromatin fibre is again coiled, and the coils are looped around a 'scaffold' protein fibre, made of a non-histone protein. This whole structure is folded again (**supercoiled**) into the much-condensed metaphase chromosome (Figure D2.1.7).

◆ **Histone**: protein (rich in the amino acids arginine and lysine) that forms the scaffolding of chromosomes and is used in chromosome condensation to form nucleosomes.

◆ **Nucleosome**: the basic unit of eukaryotic chromosome structure consisting of a ball of eight histone molecules wrapped about by two coils of about 220 base pairs of DNA.

◆ **Supercoiling**: form of DNA in which the double helix is further twisted about itself within nucleosomes, forming a tightly coiled structure.

Link
For HL students, nucleosome structure is covered in Chapter A1.2, page 26.

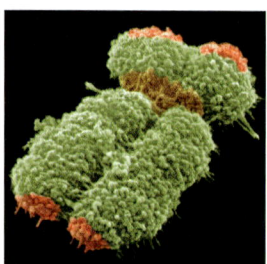

■ **Figure D2.1.7** The packaging of DNA in the chromosomes

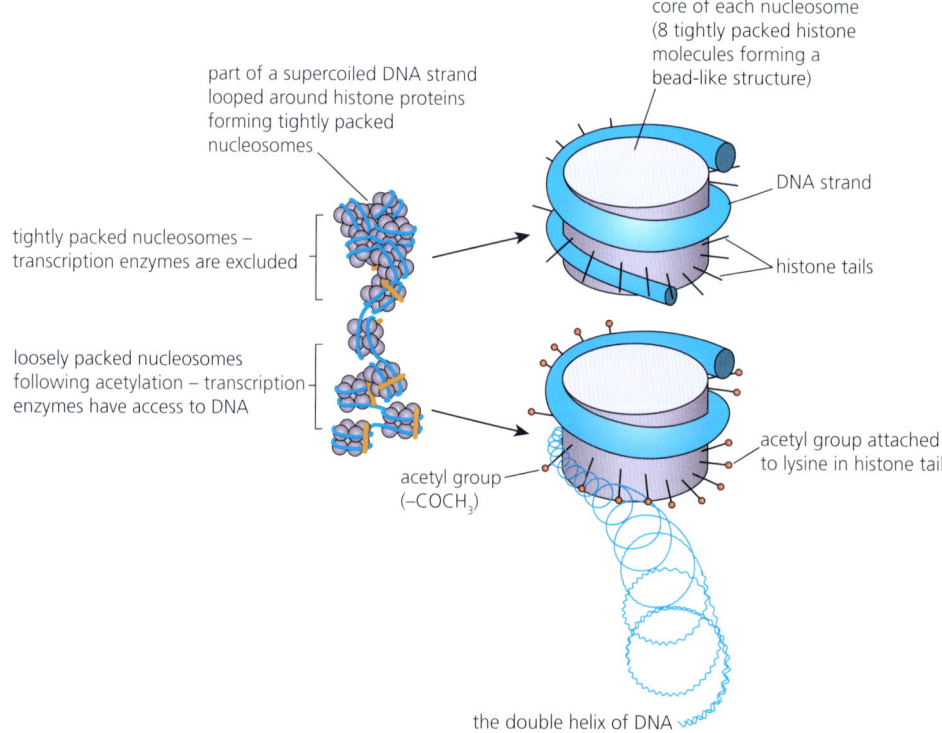

■ **Figure D2.1.6** Histone proteins forming tightly packed nucleosomes

Chromosomes are moved during cell division using **microtubules** and **microtubule motors**. Microtubules are cytoskeletal fibres that are able to lengthen and shorten by polymerization and depolymerization of tubulin (Chapter A2.2, page 72). The movement of chromosomes during cell division is achieved by the lengthening and shortening of microtubules. The movement of chromosomes is facilitated by motor proteins (Chapter A2.2, Figure A2.2.25, page 73) carrying them along the microtubules to the middle of the cell, ready for nuclear division.

> In 1922 the number of chromosomes counted in a human cell was 48. This remained the established number for over 30 years, until 1956 when Joe Hin Tjio and Albert Levan published a paper in which they proposed that the number of human chromosomes was 46, not 48. Why are we slow to change our beliefs? How can we judge when evidence is adequate?

3 **Suggest** a main advantage of chromosomes being 'supercoiled' during metaphase of mitosis.

Phases of mitosis

When cell division occurs, the nucleus divides first. In mitosis, the chromosomes, present as the chromatids formed during the interphase part of the cell cycle (see page 665, HL only), are separated, and accurately and precisely distributed to two daughter nuclei. Each of the daughter cells are identical to each other and to the parent cell.

Here, mitosis is presented and explained as a process in four phases (Figure D2.1.8), but remember this is for convenience of description only. Mitosis is a continuous process with no breaks between the phases.

You can follow the events of mitosis in Figure D2.1.8.

- In **prophase**, the chromosomes become visible as long thin threads. Now, they increasingly shorten and thicken by a process of supercoiling. You can see an electron micrograph of a supercoiled chromosome in Figure D2.1.7. It is only possible to see at the end of prophase that chromosomes consist of two chromatids held together at the centromere. At the same time, the nucleolus gradually disappears and the nuclear membrane breaks down.
- In **metaphase**, the centrioles move to opposite ends of the cell. Microtubules in the cytoplasm start to form into a **spindle**, radiating out from the centrioles (Figure D2.1.8). Microtubules attach to the centromeres of each pair of chromatids, and these are arranged at the equator of the spindle. (Note that, in plant cells, a spindle of exactly the same structure is formed, but without the presence of the centrioles.)
- In **anaphase**, the centromeres divide, the spindle fibres shorten and the chromatids are pulled by their centromeres to opposite poles. Once separated, the chromatids are referred to as chromosomes.
- In **telophase**, a nuclear membrane reforms around both groups of chromosomes at opposite ends of the cell. The chromosomes decondense by uncoiling, becoming chromatin again. The nucleolus reforms in each nucleus. Interphase follows division of the cytoplasm.

Top tip!

You need to know the names of the phases of mitosis and how the process as a whole produces two genetically identical daughter cells.

- **Prophase**: first stage in nuclear division, mitotic or meiotic, where the chromosomes condense.
- **Metaphase**: stage in nuclear division (mitosis and meiosis) in which chromosomes become arranged at the equator of the spindle.
- **Anaphase**: stage in nuclear division where chromosomes move away from one another to opposite poles of the cell.
- **Telophase**: a phase in nuclear division when the nuclear membrane reforms around daughter cell nuclear material.

Common mistake

To avoid confusion in terminology, you should refer to the two parts of a chromosome as sister chromatids while they are attached to each other by a centromere in the early stages of mitosis. Once the chromatids have been separated into separate cells, they can be referred to as distinct chromosomes.

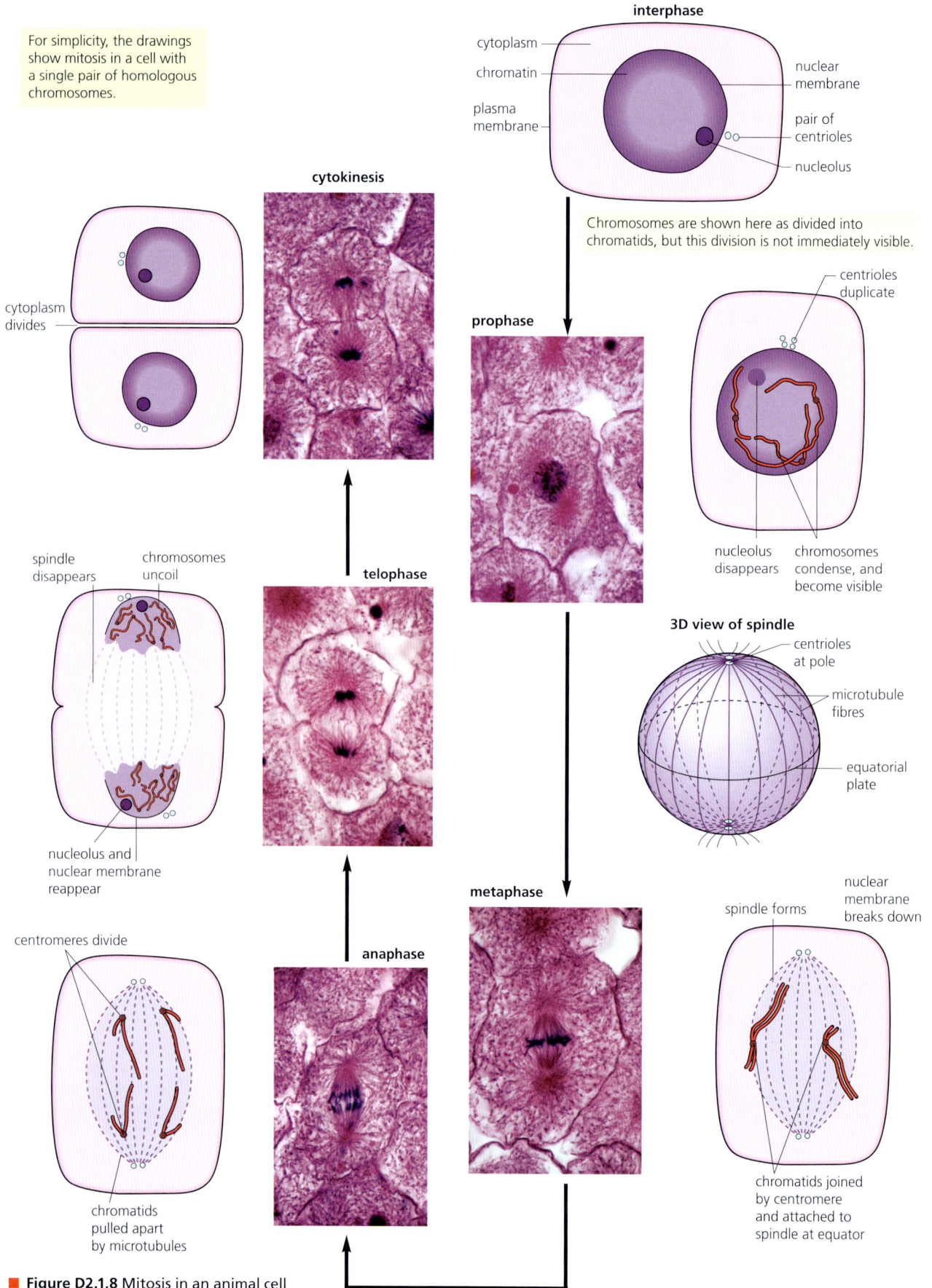

■ Figure D2.1.8 Mitosis in an animal cell

Identification of phases of mitosis

Tool 1: Experimental techniques

Preparation of temporary mounts

Observing chromosomes during mitosis

After looking at the photos and diagrams in Figure D2.1.8, apply your knowledge by identifying the phases of mitosis in cells viewed with a microscope.

Actively dividing cells, such as those at the growing points of the root tips of plants, include many cells undergoing mitosis. This tissue can be isolated, stained with an orcein ethanoic (acetic orcein) stain, squashed, and then examined under the high-power lens of a microscope. Nuclei at interphase appear red–purple with almost colourless cytoplasm, but the chromosomes in cells undergoing mitosis will be visible, rather as they appear in the photomicrographs in Figure D2.1.8. The procedure is summarized in the flow diagram in Figure D2.1.9. From the resulting temporary slides, the proportion of cells with nuclei at any stage of mitosis can be observed.

Produce a stained slide of a root tip squash, or other growing region of a plant. Identify cells that are undergoing mitosis and draw the position of the chromosomes in these cells, identifying which stage of mitosis you can see.

You could also go on to try identifying the phases of mitosis in diagrams or micrographs of cells.

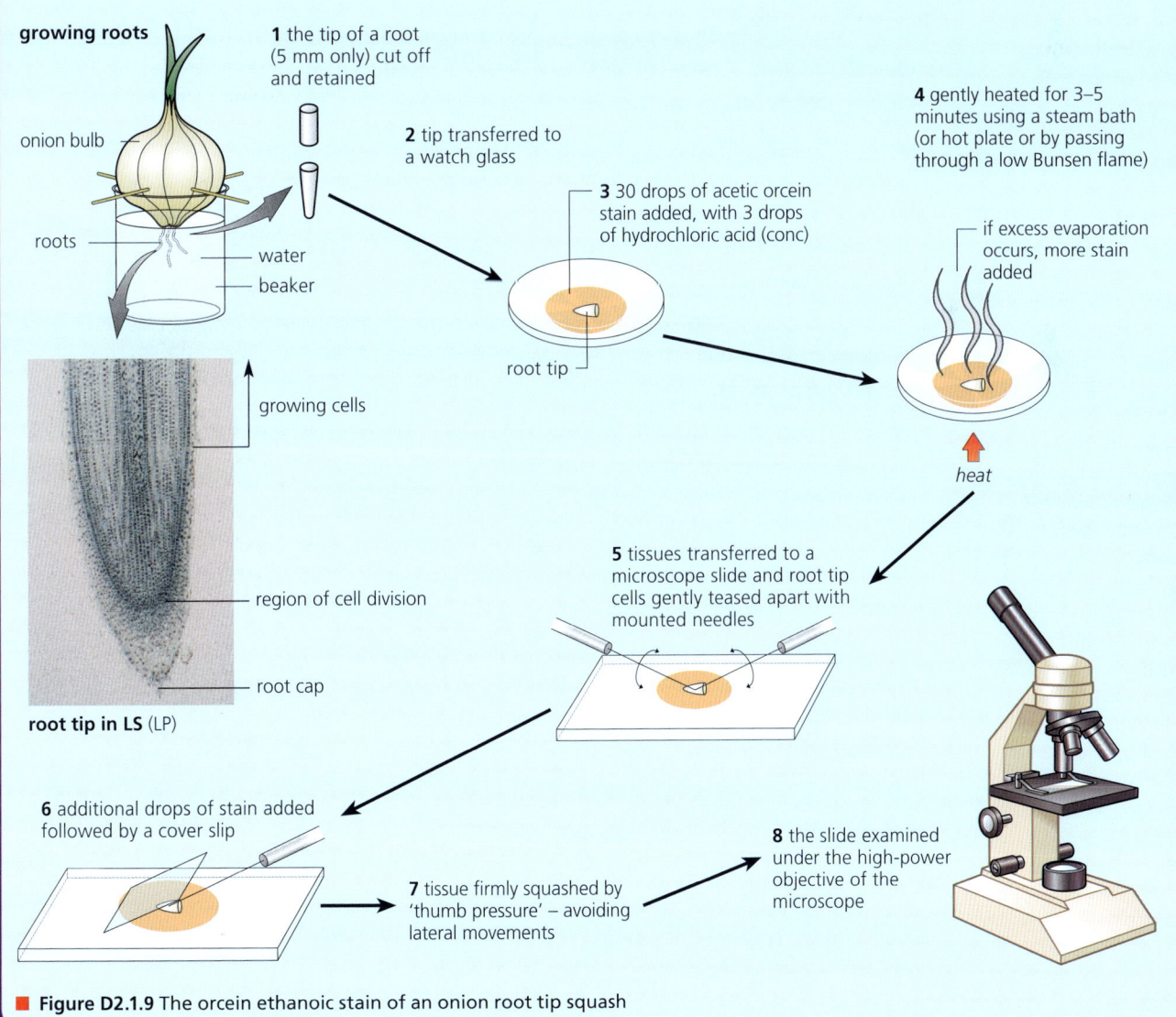

■ Figure D2.1.9 The orcein ethanoic stain of an onion root tip squash

D2.1 Cell and nuclear division

> **Top tip!**
>
> When drawing the stages of mitosis, make sure you show a membrane (intact or disappearing) in prophase, the number of chromosomes changing during the different stages, and the movement of chromosomes (not chromatids) in anaphase. Make sure diagrams are large enough so that their structures are distinct and easily labelled.

4 Using slides they had prepared to observe chromosomes during mitosis in a plant root tip (Figure D2.1.9), five students observed and recorded the number of nuclei at each stage in mitosis in 100 cells as shown in the table below.

Number of nuclei counted					
Stage of mitosis	Student 1	Student 2	Student 3	Student 4	Student 5
prophase	64	70	75	68	73
metaphase	13	10	7	11	9
anaphase	5	5	2	8	5
telophase	18	15	16	13	13

a **Calculate** the mean percentage of dividing cells at each stage of mitosis and present your results as a pie chart.

b Assuming that mitosis takes about 60 minutes to complete in this species of plant, **deduce** what these results imply about the lengths of the four steps.

Meiosis as a reduction division

Meiosis involves **two divisions** of the nucleus, known as **meiosis I** and **meiosis II**. The parent cell is **diploid**, and meiosis produces daughter cells that are **haploid**. Figure D2.1.10 shows the stages of meiosis – refer to this as you read through the text below.

As in mitosis, chromosomes replicate to form chromatids during interphase before meiosis occurs. Then, early in meiosis I, **homologous chromosomes pair up**. By the end of meiosis I, homologous chromosomes have separated again, but the chromatids they consist of do not separate until meiosis II. Thus, meiosis consists of two nuclear divisions but only **one replication of the chromosomes**.

In the interphase (page 665) that precedes meiosis, the chromosomes are replicated as chromatids, but between meiosis I and II there is no further interphase, so no replication of the chromosomes occurs during meiosis.

As meiosis begins, the chromosomes become visible. The parent cell is diploid, containing homologous pairs of chromosomes. The homologous chromosomes pair up. (Remember, in a **diploid** cell each chromosome has a partner that is the same length and shape and with the same linear sequence of genes. It is these partner chromosomes that pair.)

When the homologous chromosomes have paired up closely, each pair is called a **bivalent**. Members of the bivalent continue to shorten – a process known as condensation.

During the coiling and shortening process within the bivalent, the chromatids frequently break. Broken ends rejoin more or less immediately. When non-sister chromatids from homologous chromosomes break and rejoin, they do so at exactly corresponding sites, so that a cross-shaped structure called a **chiasma** is formed at one or more places along a bivalent. The event is known as a **crossing over** because lengths of genes have been exchanged between chromatids. This generates variation (see page 102).

Next, the spindle forms. Members of the bivalents become attached by their centromeres to the fibres of the spindle at the equatorial plate of the cell. Spindle fibres pull the homologous chromosomes apart to opposite poles, but the individual chromatids remain attached by their centromeres.

5 **Distinguish** between the terms *haploid* and *diploid*.

◆ **Bivalent**: a pair of duplicated homologous chromosomes, held together by chiasmata during meiosis.

◆ **Chiasma (plural, chiasmata)**: site of crossing over (exchange) of segments of DNA between homologous chromosomes during meiosis.

◆ **Crossing over**: exchange of genetic material between two homologous chromosomes during meiosis.

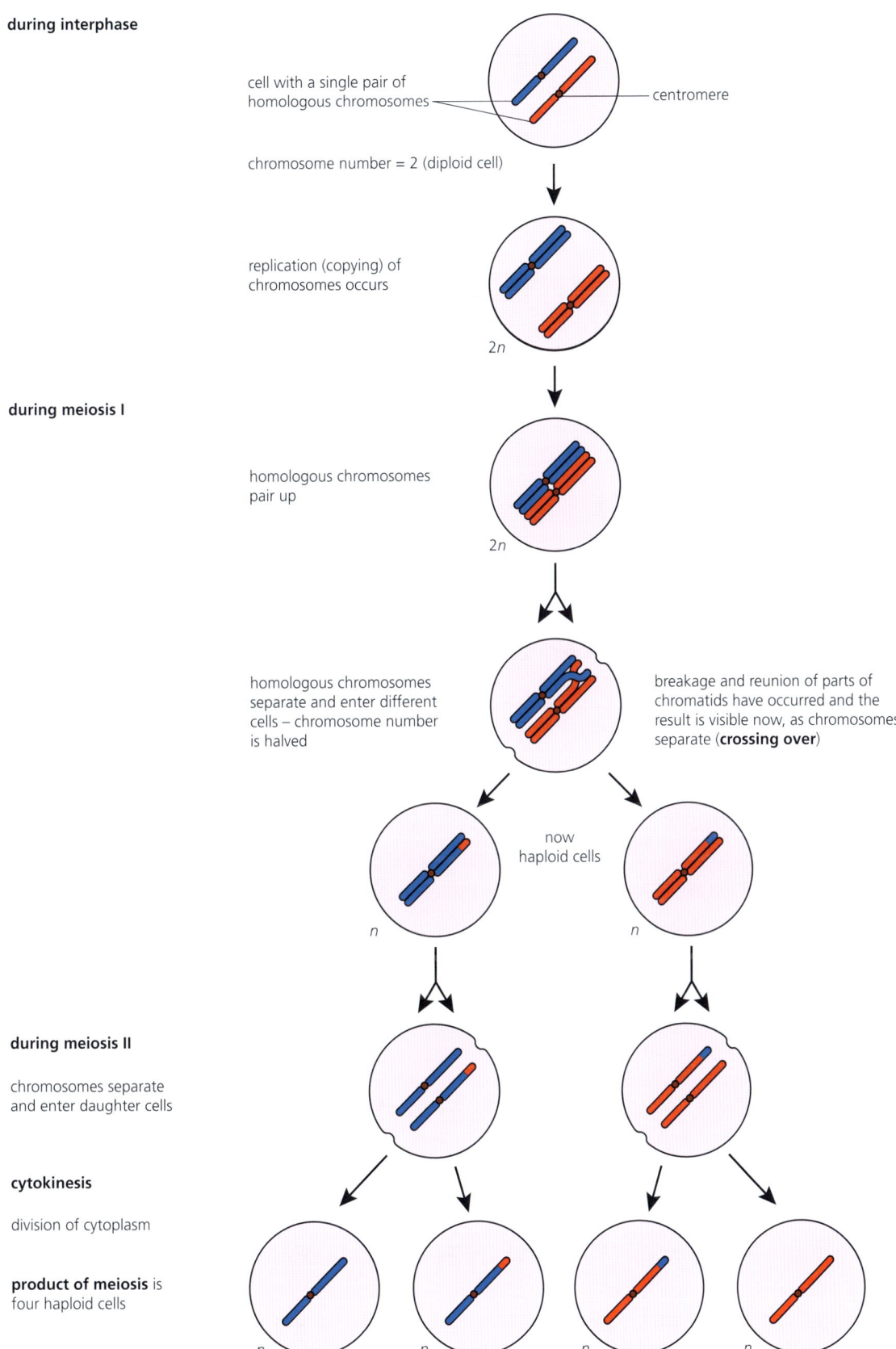

■ **Figure D2.1.10** What happens to chromosomes during meiosis?

> **Common mistake**
>
> Limited use of accurate and effective terminology is sometimes lacking in descriptions of meiosis, which restricts descriptions concerning, for example, the reduction division of meiosis. Accurate use of the words 'homologous', 'maternal', 'paternal', 'diploid (2n)', 'haploid (n)', and 'random' will help describe movement of chromosomes and how this results in genetic variety during reduction division and ensures better answers.

♦ **Non-disjunction**: the failure of homologous chromosomes or sister chromatids to separate properly during meiosis.

> **Common mistake**
>
> A common misunderstanding is that non-disjunction can only happen if additional chromosomes are present rather than if one were missing. Non-disjunction can involve either the addition or reduction in the standard number of chromosomes.

Meiosis I ends with two cells each containing a single set of chromosomes made of two chromatids. These cells do not go into interphase, but rather continue smoothly into meiosis II. This takes place at right angles to meiosis I, and is exactly like mitosis. Centromeres of the chromosomes divide and individual chromatids now move to opposite poles. Following division of the cytoplasm, there are four cells – each with half the chromosome number of the original parent cell. (The four cells are said to be **haploid**.)

> **ATL D2.1C**
>
> This animation shows the process of meiosis: https://javalab.org/en/meiosis_en
>
> Compare (look at similarities) and contrast (consider the differences) the details in this animation to those in Figure D2.1.10. Which details does the animation not include, and how would you change the animation to more accurately reflect the events of meiosis?
>
> Why is the first series of events in meiosis known as the 'reduction division'? Look again at the animation and write a summary of what is happening to the number of chromosomes during meiosis.

> **Concept: Continuity**
>
> Somatic body cells are diploid, containing pairs of homologous chromosomes. By producing haploid gametes, meiosis allows the number of chromosomes to be maintained from one generation to the next. A haploid male gamete fuses with a haploid female gamete at fertilization, forming a diploid zygote. Without the reduction division of meiosis, the number of chromosomes would double at each fertilization.

Nature of science: Observations

The discovery of meiosis

Meiosis was discovered by careful microscope examination of dividing germ-line cells. This was possible after the discovery of dyes that, when applied to tissues, specifically stained the contents of the nucleus. First, chromosomes were observed, described and named. Further careful studies then revealed the steps of mitosis and meiosis. The unravelling of the complexities of meiosis followed the observation of the doubling of the chromosome number at fertilization. Appreciation of a need for a reductive division preceded its discovery.

Germ-line cells are found in the gonads (testes and ovaries). Meiosis is part of the life cycle of every organism that reproduces sexually. In meiosis, four daughter cells are produced – each having half the number of chromosomes of the parent cell. Halving of the chromosome number of gametes is essential because at fertilization the number is doubled.

Down syndrome and non-disjunction

Very rarely, errors occur in the precisely controlled movements of the chromosomes during meiosis. The outcome is an alteration to part of the chromosome set. For example, chromosomes that should separate and move to opposite poles during the nuclear division of gamete formation fail to do so. Instead, a pair of chromosomes can move to the same pole. This malfunction event is referred to as **non-disjunction**. It results in some gametes with more than and some with less than the haploid number of chromosomes.

◆ **Down syndrome**: a congenital condition in which a person has an extra chromosome 21 as a result of non-disjunction.

For example, people with **Down syndrome** have an extra chromosome 21, giving them a total of 47 chromosomes. How this non-disjunction arises is illustrated in Figure D2.1.11. The symptoms of Down syndrome are variable but, when severe, they include congenital (hereditary) heart and eye defects. The incidence of all forms of chromosomal abnormalities increases significantly with age. People over the age of 40 who become pregnant are advised to have the chromosomes of the foetus assessed by screening.

Non-disjunction, resulting in trisomy (a condition in which an extra copy of a chromosome is present), does not only occur in chromosome 21. It can occur in chromosomes 18 and 13, as well as occasionally in other pairs of chromosomes.

An extra chromosome causes Down syndrome. The extra one comes from a meiosis error. The two chromatids of chromosome 21 fail to separate, and both go into the daughter cell that forms the secondary oocyte.

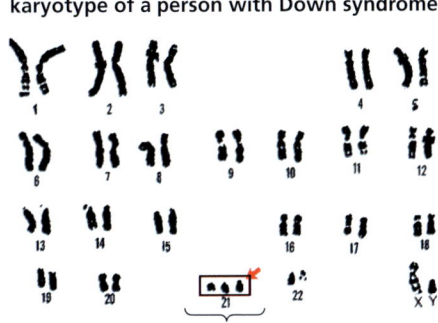

karyotype of a person with Down syndrome

an extra chromosome 21

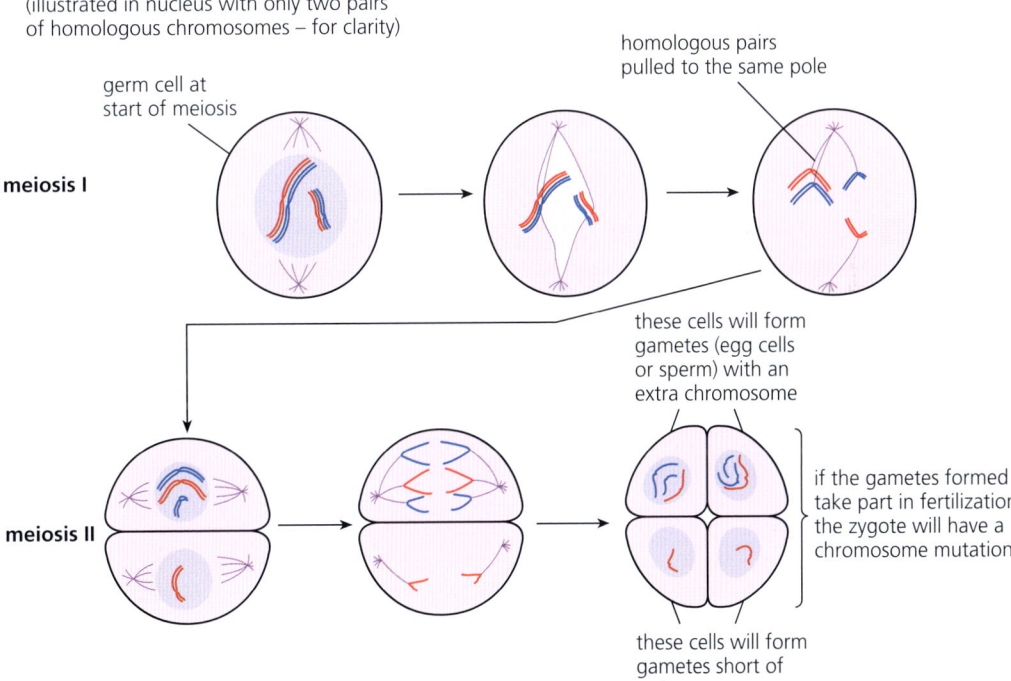

■ **Figure D2.1.11** Down syndrome, an example of a non-disjunction

ATL D2.1D

Find out about other examples of congenital conditions resulting from trisomy in other chromosomes. Research and present your findings to the rest of your class. These sites are useful starting points:
www.betterhealth.vic.gov.au/health/conditionsandtreatments/trisomy-disorders
www.nhs.uk/conditions/pataus-syndrome

> **Going further**
>
> Chromosome mutations are detected by karyotyping the chromosome set (see Chapter A3.1, page 111) of a foetus. The two methods for obtaining foetal cells are known as amniocentesis and chorionic villus sampling. Both techniques carry a slight risk of miscarriages of the foetus, so they are only recommended in cases where there is significant likelihood of genetic defect. Parents must consent to the procedure.

◆ **Random orientation**: refers to the way chromosomes line up in the centre (equator) of the cell during meiosis, with each chromosome of a given pair behaving independently to the chromosomes in other pairs. It results in each sperm and each egg having different combinations of chromosomes from the mother and father.

The four haploid cells produced by meiosis differ genetically from each other for two reasons:
- There is **crossing over** of segments of individual maternal and paternal homologous chromosomes. These events result in new combinations of genes on the chromosomes of the haploid cells produced.
- There is **random orientation** (**independent assortment**) of maternal and paternal homologous chromosomes. This happens because the way the **bivalents** line up at the equator of the spindle in meiosis I is entirely random. Which chromosome of a given pair goes to which pole is unaffected by (independent of) the behaviour of the chromosomes in other pairs.

Crossing over

As the chromosomes thicken in the first stage of meiosis, homologous chromosomes come together in specific pairs all along their length. The product of each pairing is called a **bivalent**.

The homologous chromosomes of the bivalents continue to shorten and thicken. Later in prophase, the individual chromosomes can be seen to be double-stranded, as the sister chromatids (of which each consists) become visible.

Within the bivalent, during the coiling and shortening process, breakages of the chromatids occur frequently. Breaks are common in non-sister chromatids, at the same points along their lengths. Broken ends rejoin more or less immediately but, where these 'repairs' are between non-sister chromatids, swapping of pieces of the chromatids occurs, hence the term 'crossing over'.

Once crossing over is complete, the non-sister chromatids continue to adhere at that point, called a **chiasma** (plural, **chiasmata**). The chiasma stabilizes the bivalent.

Virtually every pair of homologous chromosomes forms at least one chiasma at this time, and to have two or more chiasmata in the same bivalent is very common (Figure D2.1.12).

Chiasmata increase genetic variability because the process results in the exchange of DNA between maternal and paternal chromosomes. Remember, crossing over can occur many times and between different chromatids within each bivalent. So, crossing over can produce new combinations of alleles on the chromosomes of the haploid cells that are finally formed by meiosis, followed by cytokinesis.

● **Top tip!**

Non-disjunction, resulting in **trisomy** (a condition in which an extra copy of a chromosome is present), does not only occur in chromosome 21. It can also occur in chromosome 13 and chromosome 18.

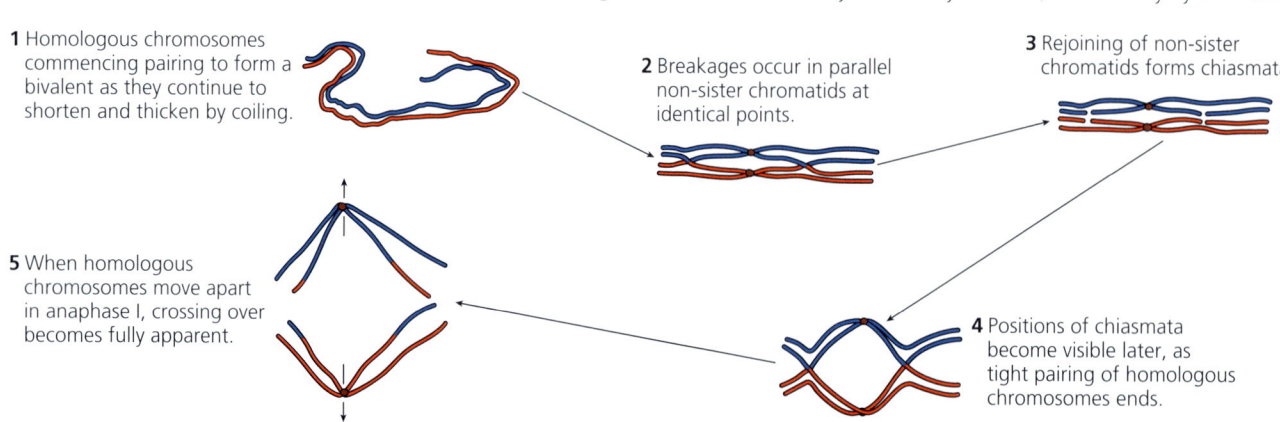

1. Homologous chromosomes commencing pairing to form a bivalent as they continue to shorten and thicken by coiling.
2. Breakages occur in parallel non-sister chromatids at identical points.
3. Rejoining of non-sister chromatids forms chiasmata.
4. Positions of chiasmata become visible later, as tight pairing of homologous chromosomes ends.
5. When homologous chromosomes move apart in anaphase I, crossing over becomes fully apparent.

■ **Figure D2.1.12** Formation of chiasmata

Random orientation

The random orientation of chromosomes is illustrated in Figure D2.1.13 in a parent cell with a diploid number of four chromosomes. In human cells, the number of pairs of chromosomes is 23; the number of possible combinations of chromosomes that can be formed by random orientation during meiosis is 2^{23}, which is over 8 million. We see that random orientation alone generates a huge amount of variation in the coded information carried by different gametes into the fertilization stage.

Random orientation is illustrated in a parent cell with two pairs of homologous chromosomes (four bivalents). The more bivalents there are, the more variation is possible. In humans, for example, there are 23 pairs of chromosomes giving over 8 million combinations.

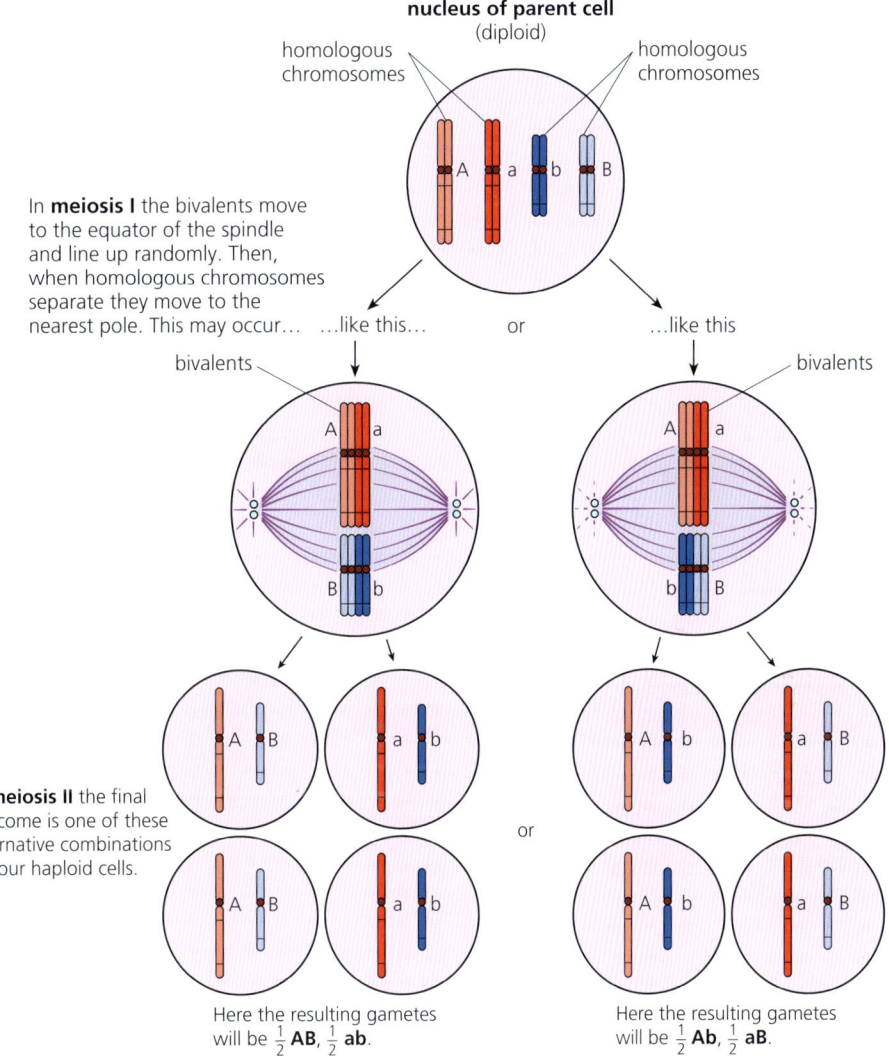

6 **Explain** how variation is generated during meiosis.

7 **Compare and contrast** mitotic and meiotic cell division.

Concept: Change

Crossing over and random orientation during meiosis generates variation. Each gamete is different from every other gamete. These differences are a source of variation for evolution by natural selection.

■ Figure D2.1.13 Genetic variation due to random orientation

In fertilization, the fusion of gametes from different parents then promotes genetic variation as well.

● Top tip!

The second stage of meiosis also results in an increase of variety, not just the first stage. Crossing over in meiosis I has led to non-identical chromatids in meiosis II. During meiosis II, the sister chromatids separate and are randomly distributed to the gametes. The outcome of which chromatid will go into which gamete is random, so that each gamete has a potentially unique combination of genetic material.

Cell proliferation for growth, cell replacement and tissue repair

Both animals and plants grow from a single cell, the zygote, by repeated cell divisions to form an embryo. **Proliferation** for growth then occurs (i.e. rapid increase in number), which happens in animal early-stage embryos and in plant meristem tissue.

Plant meristems

Once a plant has grown past the early embryo stage, all later growth of the plant occurs at restricted points in the plant, called meristems. A **meristem** is a group of cells that retain the ability to divide by mitosis. These cells are small, with thin cellulose walls and dense cytoplasmic contents. Vacuoles in the cytoplasm are mostly absent, marking them apart from typical mature plant cells (which have large, fluid-filled vacuoles). Meristems occur either at terminal growing points of stems and roots, or they are found laterally. In Figure D2.1.14, both types of meristem can be identified.

◆ **Proliferation**: increase in the number of cells as a result of cell growth and cell division.

◆ **Meristem**: a group of cells in plants that retains the ability to divide by mitosis and form new cells and tissues throughout the life of a plant.

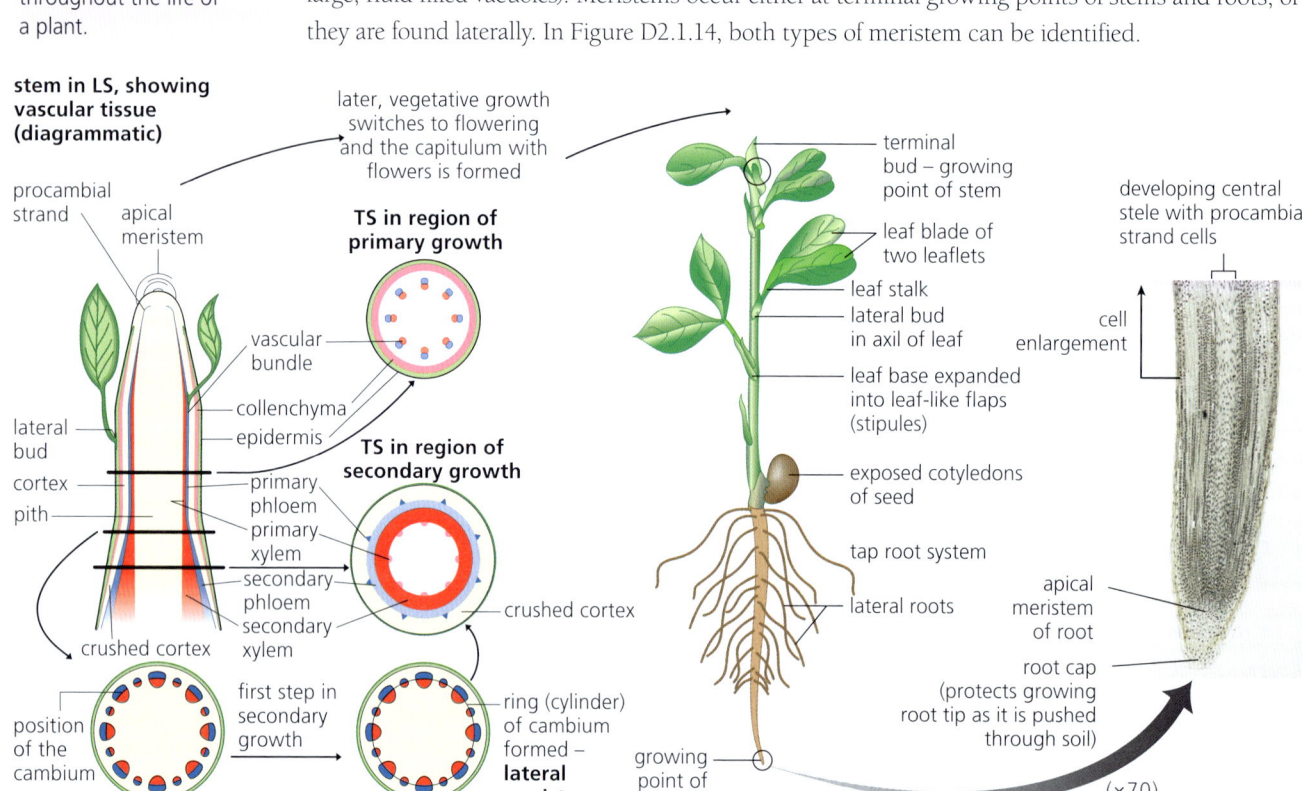

■ **Figure D2.1.14** The roles of apical and lateral meristems in the growth of stems and roots

Apical meristems occur at the tips of the stem and root, and are responsible for their primary growth (Figure D2.1.14). Cell division and the subsequent growth of the cells produced here leads to the formation of stem (and root) tissue. First, the new cells formed by division rapidly increase in size. This cell enlargement phase is then followed by cell differentiation.

Lateral meristems (Figure D2.1.14) form from the **cambium** cells (meristematic tissue – undifferentiated cells that are capable of indefinite division) in the centre of vascular bundles, between the (outer) phloem tissue and the (inner) xylem tissue. When the lateral meristem forms and grows, it causes the secondary growth of the plant, resulting in an increase in the girth of the stem. Growth of the lateral meristem increases the circumference and also the strength of the stem.

Table D2.1.1 compares the growth due to apical and lateral meristems.

■ Table D2.1.1 Comparison of growth due to apical and lateral meristems

Growth due to apical meristem		Growth due to lateral meristem
occurs at tip of stems and roots	position of meristem	occurs laterally, between primary phloem and primary xylem
product of embryonic cells	origin	cambium – meristematic cells left over from primary growth
produces initial tissues of actively growing plant from the outset	timing of activity	functions in older stems (and roots), and in woody plants from the early development
produces growth in length and height of plant	outcome for stem	produces growth in girth of stem, plus strengthening of stem

8 **Define** the term *meristem*.

Link
The development of unspecialized cells following fertilization is covered in Chapter B2.3, page 256.

Early-stage animal embryos

Following fertilization, the **zygote** divides rapidly by mitosis to form a ball of cells (Figure D2.1.15). After approximately six days, a **blastocyst** has formed, from which the placenta develops and the main body of the organism. At this stage, all cells are undifferentiated, with the potential to become any type of cell and tissue in the body.

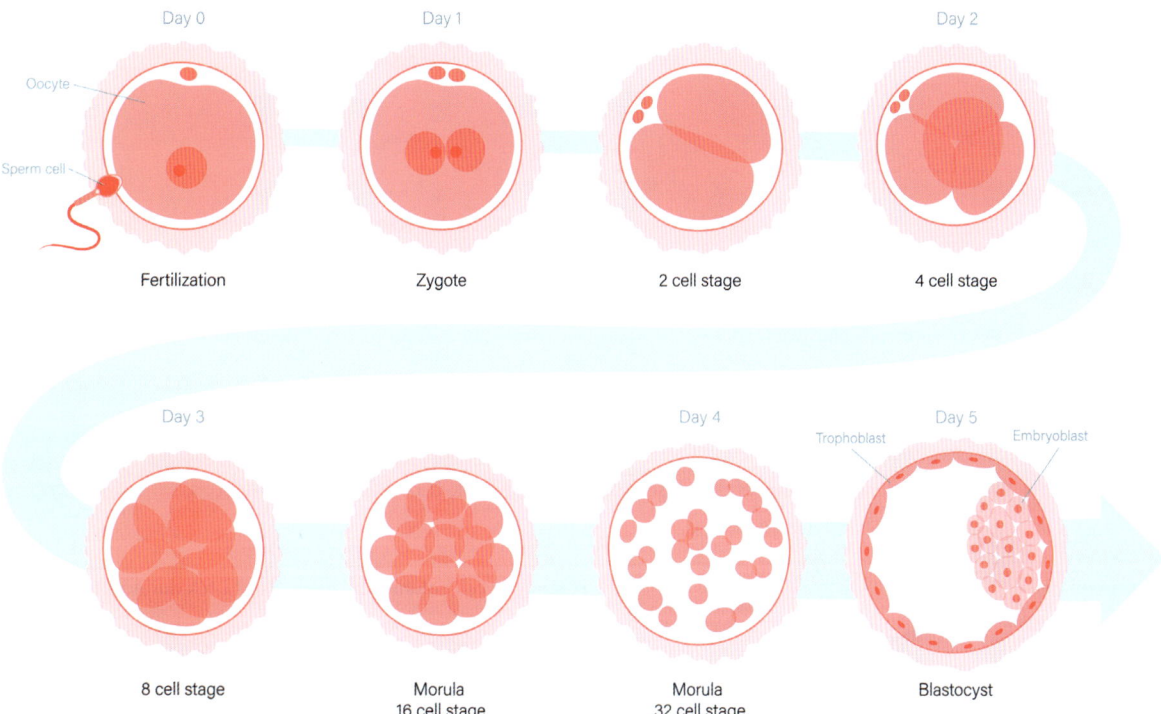

■ Figure D2.1.15 Cell proliferation in the early-stage embryo of an animal

Cell proliferation during routine cell replacement and wound healing

Cells in the body are constantly being worn out and need replacing. The average adult human loses approximately 500 million skin cells per day, and red blood cells at a rate of 2 million per second. Cell replacement is a routine process for all cell types (except those which cannot carry out mitosis, such as nerve cells, see page 467).

Cell proliferation is the process of generating an increased number of cells through cell division (mitosis). Because cells are regularly lost from the skin, routine cell replacement is needed. The outer layers of the skin epidermis are replaced many thousands of times during a lifetime. Stems cells in the basal (bottom) layer of the epidermis remain undifferentiated and continue dividing throughout life.

◆ **Cell proliferation**: the process of generating an increased number of cells through cell division (mitosis).

D2.1 Cell and nuclear division

Daughter cells of these cells differentiate and leave the basal layer. The process can be maintained because the basal cells are self-renewing.

Wound healing in the skin also involves cell proliferation. Following blood clotting (see Chapter C3.2, page 520), inflammation causes increased blood flow to the wound, enabling white blood cells (macrophages) and **fibroblasts** to travel rapidly to the damaged skin. The macrophages remove infection and the fibroblasts produce proteins that help wound closure (Figure D2.1.16).

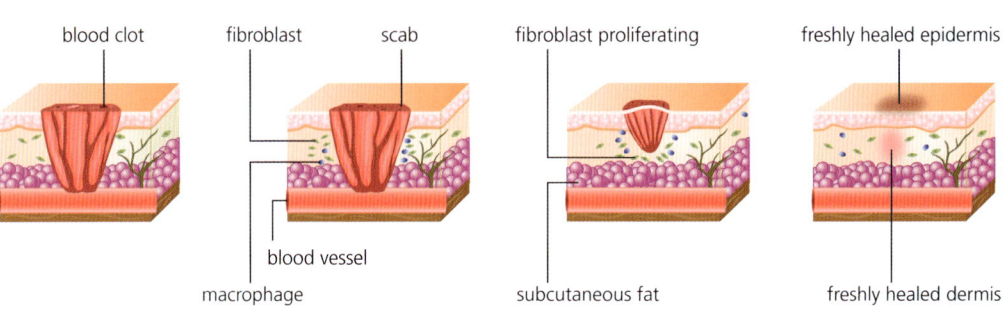

■ Figure D2.1.16 The process of wound healing

9 **Outline** how skin repairs itself to heal a wound.

Fibroblasts proliferate to enable wound healing to occur rapidly. They break down the fibrin clot and create new extracellular matrix (ECM; made from collagen), that supports the other cells associated with effective wound healing as well as narrowing the wound.

Phases of the cell cycle

◆ **Cell cycle**: repeating process of an orderly sequence of events where cells arise by the division of existing cells, grow, and then divide.

The cycle of growth and division of cells is called the **cell cycle**. Cell proliferation is achieved using the cell cycle. This cycle has three main stages (see Figure D2.1.17):

- **interphase** which includes G_1, S and G_2 phases (see Figure D2.1.17 for details of each)
- **division of the nucleus** by a process (mitosis) that results in two nuclei, each with an identical set of chromosomes
- **division of the cytoplasm** and whole cell (known as **cytokinesis**).

In fact, in each stage of the cell cycle particular events occur. These events are summarized in Figure D2.1.17, and they are also discussed below. *Look at the subdivision of interphase now – distinctive features are identified in each stage.*

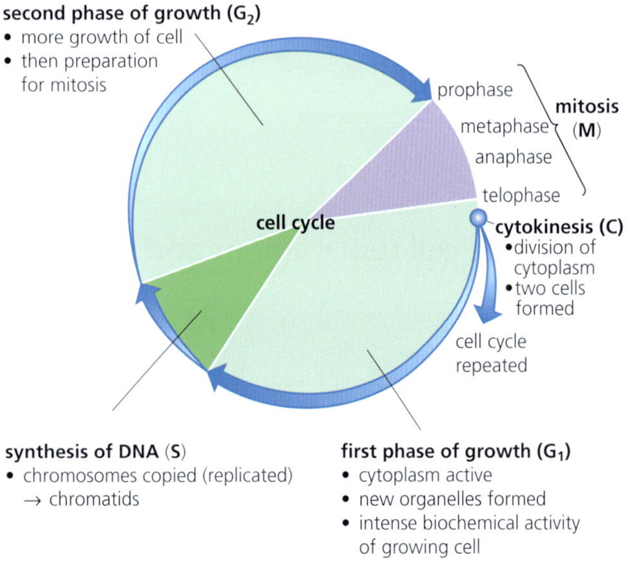

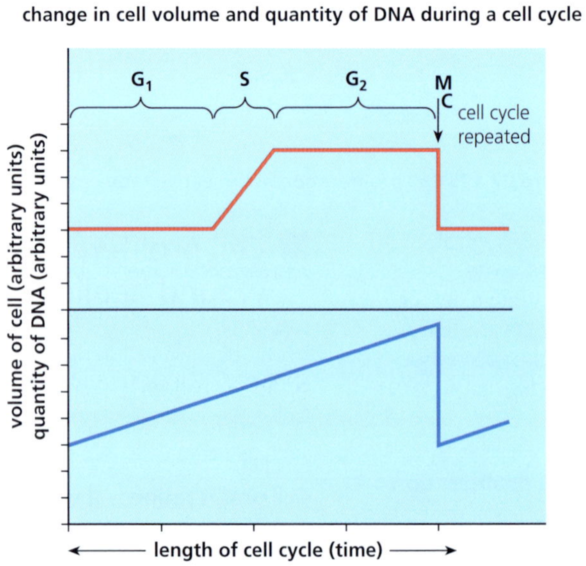

■ Figure D2.1.17 The stages of the cell cycle

Theme D: Continuity and change – Cells

■ The steps of interphase

During the first phase of growth (G_1), the synthesis of new organelles takes place in the cytoplasm. This is also a time of intense biochemical activity in the cytoplasm and organelles, and there is an accumulation of energy stores before nuclear division occurs again.

Next is a period of synthesis of DNA (S), when each chromosome makes a copy of itself. It is said to **replicate**. The two identical structures formed are called **chromatids**. The chromatids remain attached until they divide during mitosis.

Finally, there is a second phase of growth (G_2), which is a continuation of the earlier time of intense biochemical activity and increase in amount of cytoplasm.

> 10 **List** the stages of the cell cycle.

Cell growth during interphase

Interphase is always the longest part of the cell cycle, but it is of extremely variable length. When growth is fast, as in a developing human embryo and in the growing point of a young stem, interphase may last about 24 hours or less. On the other hand, in mature cells that divide infrequently, it lasts a very long period, for example in liver, bone, kidney and lung cells. Some cells, once they have differentiated, rarely or never divide again, such as brain cells. Here, the nucleus remains at interphase permanently.

■ An overview of interphase

When the nucleus of a living cell at interphase is observed by light microscopy, the nucleus appears to be 'resting'. This is not the case. Interphase is a metabolically active period and that growth involves the synthesis of cell components including proteins and DNA. The numbers of mitochondria and chloroplasts are also increased by growth and division of these organelles. During interphase, the chromosomes are actively involved in protein synthesis. From the chromosomes, copies of the information of particular genes or groups of genes (in the form of mRNA, page 616) are taken for use in the cytoplasm. It is in the ribosomes of the cytoplasm that proteins are assembled from amino acids, combined in sequences dictated by the information from the gene and relayed in the form of mRNA.

> 11 **State** what structures of the interphase nucleus can be seen by electron microscopy.

The distinctively compact chromosomes, visible during mitosis (Figure D2.1.8), become dispersed in interphase. They are now referred to as **chromatin**. Among the chromatin can be seen one or more dark-staining structures, known as **nucleoli** (singular, nucleolus). Chemically, the nucleoli consist of protein and RNA, and they are the site of synthesis of the ribosomes. These tiny organelles then migrate out into the cytoplasm.

Control of the cell cycle using cyclins

Look back at the stages of the cell cycle (Figure D2.1.17). Note that it consists of distinct phases, represented in shorthand as G_1, S, G_2, M and C.

> ◆ **Cyclin**: regulatory protein whose concentration rises and falls at specific times during the eukaryotic cell cycle.

The cell cycle is regulated by a **molecular control system**. The key points of this system are outlined below, best understood together with Figure D2.1.18.
- In the cell cycle there are key **checkpoints** where signals operate. These are stop points which have to be overridden.
- Three checkpoints are recognized – at G_1, G_2 and in M.
- At the G_2 checkpoint, if the 'go-ahead' signal is received here, the cell goes through to M then C, for example.
- The molecular control signal substance in the cytoplasm of cells are proteins known as **kinases** and **cyclins**.

Concept: Continuity

Each cell divides into two, which then themselves divide. The cell cycle is a repeating and continuous process that ensures the continuity of cells within an organism and between generations.

- Kinases are enzymes that either activate or inactivate other proteins. Kinases are present in the cytoplasm all the time, though sometimes in an inactive state.
- Kinases are activated by specific cyclins, so they are referred to as cyclin-dependent kinases (CDKs).
- Cyclin concentrations in the cytoplasm change constantly. As the concentration of cyclins increases, they combine with CDK molecules to form a complex that functions as a mitosis-promoting factor (MPF) (Figure D2.1.18).
- As MPF accumulates, it triggers chromosome condensation, fragmentation of the nuclear membrane and, finally, spindle formation – that is, mitosis is switched on.
- By anaphase of mitosis, destruction of cyclins commences (but CDKs persist in the cytoplasm).

The concentration of different cyclins increases and decreases during the cell cycle, and there is a threshold level of a specific cyclin required to pass each checkpoint in the cycle. Figure D2.1.18 shows how the action of cyclin and CDK control cell division at the G_2 checkpoint.

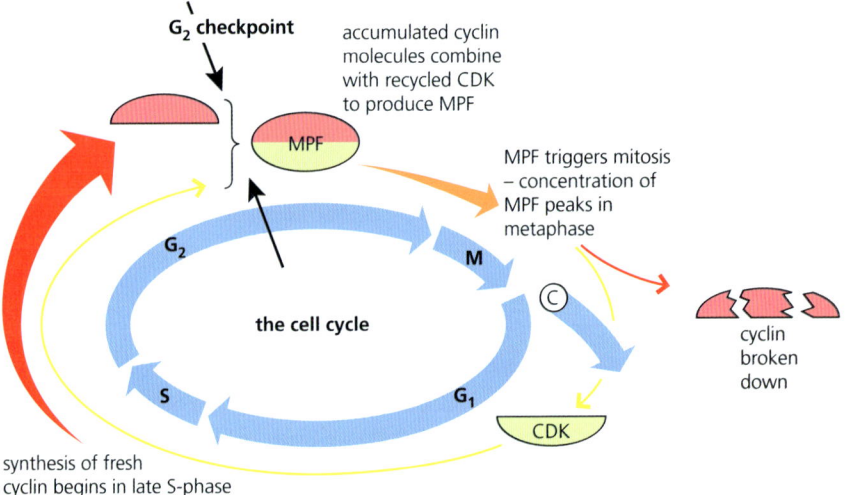

■ Figure D2.1.18 The molecular control system of the cell cycle: the role of kinase and cyclins in controlling mitosis at the G_2 checkpoint

Nature of science: Observations

The accidental discovery of cyclins

The discovery of the proteins that control the cell cycle came partly from work by Tim Hunt, as his team investigated protein synthesis more generally in the eggs of sea urchins. While the synthesis of most new proteins proceeded steadily, as anticipated, a minority of others went through short, abrupt cycles of increasing and decreasing concentration. High threshold levels of these individual proteins were found to correlate with changes in the cell cycle. By this means, and with the contributions of others (including from work on yeasts) the roles of four different proteins in the cell cycle were discovered. They were named cyclins at this stage.

In 2001, Paul Nurse, Tim Hunt and Leland Hartwell were awarded the Nobel Prize in Physiology or Medicine for their contributions to the discovery of the control of the cell cycle. Tim Hunt makes clear that his discovery of cyclins was accidental in his Nobel Prize lecture, which you can access as a paper, or watch the video made at the ceremony:
www.nobelprize.org/prizes/medicine/2001/hunt/lecture

The contributing studies of Paul Nurse were made on yeasts. The background of this scientist, a former President of The Royal Society, makes an interesting contrast with that of the others, and helpfully (and encouragingly) establishes how diverse the paths to a distinguished career in research can be: www.crick.ac.uk/research/find-a-researcher/paul-nurse

> **Link**
> G proteins and cascade of protein kinases are discussed in Chapter C2.1 (HL only), page 459.

◆ **Proto-oncogene**: gene that codes for proteins that stimulate the cell cycle and promote cell growth and proliferation.

◆ **Oncogene**: a gene that, when activated, can potentially make a cell cancerous. Typically a mutant form of a normal gene (proto-oncogene) involved in the control of cell growth or division.

◆ **Tumour-suppressor gene**: a gene that in a normal tissue cell inhibits cancerous behaviour. Loss or inactivation of both copies of such a gene from a diploid cell can cause it to behave as a cancer cell.

Consequences of mutations in genes that control the cell cycle

In a healthy cell, the cell cycle is regulated by a molecular control system in which cyclins are involved (page 665). When mutations occur in genes that control the cell cycle, cancer can result because the normal controls of the cell cycle are absent.

Whatever the cause of a cancer, two types of genes play a part in initiating a cancer if they mutate:

- **Proto-oncogenes** are genes that encode proteins that stimulate normal cell division. Mutations in proto-oncogenes converts them to **oncogenes**, which leads to increase in the amount of a proto-oncogene's protein product or permanently activated proteins. This results in uncontrolled cell division, possibly leading to cancer.

    For example, the *Ras* gene codes for Ras protein, a G protein that relays a signal from a growth factor receptor on the cell surface membrane to a cascade of protein kinases, which leads to normal cell division. The *Ras* oncogene codes for a permanently activated Ras protein that triggers the kinase cascade in the absence of growth factor, resulting in uncontrolled cell division. The mutation results in a dominant allele, and the effect of the normal allele is masked by the mutated allele. Only one allele of a gene needs to be mutated to have an effect.

- **Tumour-suppressor genes** encode for proteins that inhibit cell division or promote apoptosis (controlled cell death) if damaged DNA is being copied. Mutation of tumour-suppressor genes leads to no protein product, a decrease in the amount of protein product or permanently deactivated proteins. This will result in uncontrolled cell division, possibly leading to cancer.

    For example, the *p53* gene codes for p53 protein, a transcription factor that promotes the synthesis of protein that triggers cell cycle arrest or promotes apoptosis when DNA damage is detected. The mutated *p53* tumour-suppressor genes result in no p53 being produced. Cells with damaged DNA that are not supposed to proliferate will continue to divide, resulting in uncontrolled cell division.

    The mutation results in a recessive allele as the normal dominant allele encodes functional protein. Two alleles of a gene need to be mutated to have an effect.

> **Top tip!**
> When a proto-oncogene mutates it becomes a cancer-causing oncogene. The role of tumour-suppressor genes is to prevent cancer – when they mutate they can lead to tumour formation.

Differences between tumours in rates of cell division and growth

 ## Cancer – diseases of uncontrolled cell division

There are many different forms of cancer, affecting different tissues of the body. Cancer is not thought of as a single disease. Today, in more economically developed countries, one in three people will suffer from cancer at some point in their life, and approximately one in four will die from it. In these regions, the most common cancers are of the lung in men and of the breast in women. However, in many parts of the world, cancer rates are different – often they are significantly lower. Biologists in laboratories throughout the world are carrying out research into the causes and treatment of cancer. You can see the range of common cancers and their incidences worldwide at the Global Cancer Observatory: **https://gco.iarc.fr**.

HL ONLY

12 Describe *three* environmental conditions that may cause normal cells to become cancerous cells.

13 Describe how the behaviour of cancerous cells differs from that of normal cells.

◆ **Benign tumour**: a tumour that tends to grow slowly and does not spread to other parts of the body.

◆ **Malignant tumour**: a tumour that grows rapidly, invades and destroys nearby normal tissues, and spreads throughout the body.

◆ **Primary tumour**: a cancer growing at the site where the abnormal growth first occurred.

◆ **Secondary tumour**: formed when cancerous cells detach from the primary tumour, penetrate the walls of lymph or blood vessels, and circulate around the body, causing tumours elsewhere.

◆ **Metastasis**: the movement of cells from a primary tumour to set up secondary tumours.

14 Distinguish between a tumour and cancer.

◆ **Mitotic index**: the number of cells undergoing mitosis divided by the total number of cells visible, which can also be shown as a percentage.

In all cancers, cells start to divide repeatedly by mitosis, without control or regulation. Where this occurs in the body, the rate of cell multiplication is much faster than the rate of cell death. An irregular mass of cells is formed, called a **tumour** (Figure D2.1.19). Some tumours are **benign**: these grow slowly and do not pass to other parts of the body. Other tumours are **malignant**, growing rapidly, destroying nearby normal tissues and spreading throughout the body. Benign tumours are non-cancerous whereas malignant tumours are cancerous.

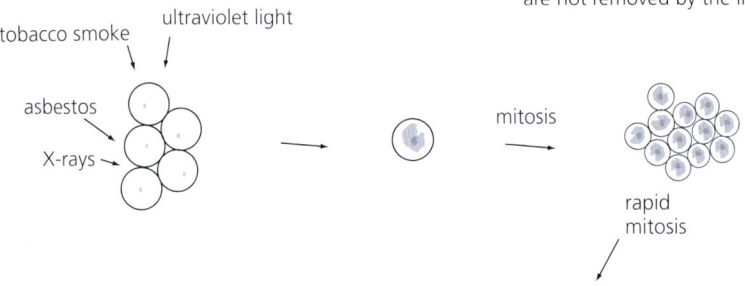

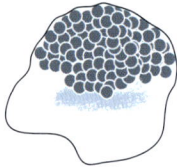

■ **Figure D2.1.19** Steps in the development of a malignant tumour

Sometimes tumour cells break away from the **primary tumour** and are carried to other parts of the body, where they form a **secondary tumour**. This process is known as a **metastasis**. Unchecked, cancerous cells ultimately take over the body at the expense of the surrounding healthy cells, leading to malfunction and death.

Calculating the mitotic index

When observing a population of cells, we can determine its **mitotic index**. The mitotic index measures how many cells in a sample are in mitosis compared to the total number of cells. It is calculated in the following way:

(number of cells in mitosis ÷ total number of all cells) × 100

For example, Figure D2.1.20 shows a micrograph with cells, some of which are in mitosis.

Theme D: Continuity and change – Cells

> ### Tool 1: Experimental techniques
>
> #### Counts
>
> Micrographs can be used to perform counts, e.g. of cells in stages of mitosis. Repeated counts of objects in the field of view at high power illustrate the variability of biological material and the need for replicate trials.

● **Common mistake**

Sometimes the word 'tumour' is used as an equivalent term to 'cancer'. This is not the case. Tumours are growths caused by uncontrolled cell division, but benign tumours do not cause cancer. Only malignant tumours are cancerous.

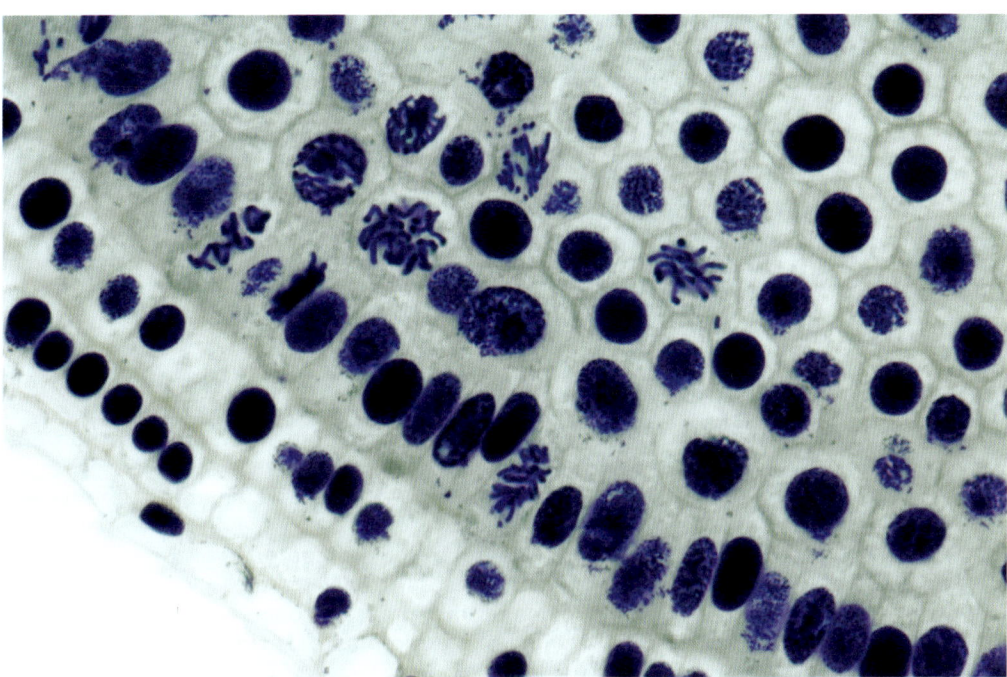

■ **Figure D2.1.20** Cells in different stages of the cell cycle, including mitosis – this can be used to calculate the mitotic index

> ### Tool 3: Mathematics
>
> #### Using basic arithmetic
>
> To calculate the mitotic index, the number of cells visible in the micrograph is counted, and the number of cells undergoing mitosis. In the case of Figure D2.1.20:
>
> - number of cells visible = 95
> - number of cells undergoing mitosis = 10
>
> So, the mitotic index = $\left(\frac{10}{95}\right) \times 100 = 0.105 \times 100 = 10.5\%$

For greatest accuracy, the mean of three or more samples of 100 cells should be used. This is important because the mitotic index is used to differentiate benign from malignant tumours (benign tumours have a lower mitotic index than malignant tumours) – a critical distinction. Tissue with a high mitotic index indicates a rapidly dividing cell mass – a possible indicator of tumour formation. The mitotic index can also be used to investigate the response to chemotherapy in most types of cancer (i.e. a reduction in the mitotic index indicates that treatment has been successful in reducing the cancer).

> **LINKING QUESTIONS**
>
> 1. What processes support the growth of organisms?
> 2. How does the variation produced by sexual reproduction contribute to evolution?

D2.1 Cell and nuclear division

D2.2 Gene expression

Guiding questions

- How is gene expression changed in a cell?
- How can patterns of gene expression be conserved through inheritance?

SYLLABUS CONTENT

This chapter covers the following syllabus content:
- ▶ D2.2.1 Gene expression as the mechanism by which information in genes has effects on the phenotype (HL only)
- ▶ D2.2.2 Regulation of transcription by proteins that bind to specific base sequences in DNA (HL only)
- ▶ D2.2.3 Control of the degradation of mRNA as a means of regulating translation (HL only)
- ▶ D2.2.4 Epigenesis as the development of patterns of differentiation in the cells of a multicellular organism (HL only)
- ▶ D2.2.5 Differences between the genome, transcriptome and proteome of individual cells (HL only)
- ▶ D2.2.6 Methylation of the promoter and histones in nucleosomes as examples of epigenetic tags (HL only)
- ▶ D2.2.7 Epigenetic inheritance through heritable changes to gene expression (HL only)
- ▶ D2.2.8 Examples of environmental effects on gene expression in cells and organisms (HL only)
- ▶ D2.2.9 Consequences of removal of most but not all epigenetic tags from the human ovum and sperm (HL only)
- ▶ D2.2.10 Monozygotic twin studies (HL only)
- ▶ D2.2.11 External factors impacting the pattern of gene expression (HL only)

♦ **Phenotype**: observable traits of an organism resulting from genotype and environmental factors.

Common mistake

Do not confuse the terms genotype and phenotype. Genotype refers to the 'genetic makeup' of an individual (the genetic information in the cell), whereas phenotype is the outward effect of the genotype on the body.

Gene expression and its effect on the phenotype

Gene expression is the mechanism by which information in genes has effects on the **phenotype**. Information contained in the genetic code is transcribed into mRNA, which is in turn translated into proteins at ribosomes. The proteins produced by an organism determine its phenotype.

Gene expression is explored throughout this IB Biology course:
- Transcription is the first stage of gene expression and is a key stage where the expression of a gene can be switched on and off (Chapter D1.2).
- The existence of non-coding sequences in DNA that do not code for polypeptides, and that these include regulators of gene expression, was introduced in page 626.
- Different patterns of gene expression are the basis for cell differentiation, often triggered by changes in the environment (Chapter A2.2).
- Biochemical gradients affect gene expression within an early-stage embryo (Chapter B2.3).

> **Common mistake**
>
> The relationship between gene expression and cell differentiation is sometimes not fully appreciated: genes are selected for expression to produce specific proteins needed by the cell.

- Intracellular receptors affect gene expression when a signalling chemical binds to it, causing the activated receptor to bind to specific DNA sequences to promote gene transcription (Chapter C2.1).
- Auxin affects gene expression mechanisms operating in the nucleus, regulating growth and development in plants (Chapter C3.1).

To summarize, the most common stages in gene expression are transcription, translation and the function of a protein product, such as an enzyme.

> **Concept: Continuity**
>
> The mechanism of gene expression ensures that the genetic code is transcribed into mRNA, which is in turn translated into proteins at ribosomes. The proteins produced by an organism determine its phenotype. There is therefore a continuity between genetic code and phenotype, maintained by gene expression.

Regulation of transcription by proteins that bind to specific base sequences in DNA

♦ **Transcription factor**: a protein that binds to specific DNA sequences to control the transcription of mRNA.

♦ **Enhancer**: regulatory sequences on the DNA which increase the rate of transcription when activator proteins bind to them.

Transcription factors are proteins that bind to specific sequences of DNA to control transcription. Each transcription factor has a DNA-binding site that gives it the ability to bind to specific sequences of DNA called **enhancer** or **promoter** sequences. Enhancer sequences are **regulatory sequences** that, when bound by transcription factors, enhance the transcription of an associated gene. Regulatory sequences can be thousands of base pairs and are generally upstream from (i.e. before) the gene being transcribed, although they can also be downstream. Regulation of transcription is the most common form of gene control. The action of transcription factors allows for unique expression of each gene in different cell types and during development.

The definition and role of promoters in regulating transcription was discussed in Chapter D1.2 (page 625). Promoter regions occur on DNA strands just before the start of a gene's sequence of bases. Only when this transcription complex of proteins (enzyme plus factors) has been assembled, can transcription of the template strand of the gene begin.

> **Concept: Change**
>
> Regulation by transcription factors determines which genes are encoded into proteins within the cell. In this way, the proteome of the cell can change according to the needs of the cell and its function.

The rate of transcription may be increased (or decreased) by the binding of specific transcription factors on the enhancer site for the gene (Figure D1.1.7, page 612). This is at some distance prior to the promoter and gene sequence. However, when activator proteins bind to this enhancer site, a new complex is formed and makes contact with the polymerase–transcription factor complex. Then the rate of gene expression is increased.

Control of the degradation of mRNA as a means of regulating translation

mRNA molecules are synthesized during the process of transcription. From there, the mRNA is examined, modified and transported to ribosomes, before eventually being translated into proteins. mRNA is eventually broken down (degraded) by nucleases. In human cells, mRNA may exist for minutes or up to days. The 3′ poly-A tail (Chapter D1.2, page 628) must be degraded before the mRNA is broken down. When the mRNA is no longer being used for translation, poly-A tail shortening is one of the key steps initiating degradation of the mRNA. Degradation of mRNA ensures that proteins are synthesized only when they are needed, and that mRNA is removed once its job is done.

◆ **Epigenesis**: the development of patterns of differentiation in the cells of a multicellular organism.

◆ **Epigenetics**: the study of heritable changes in gene activity that are not caused by changes in the DNA base sequences. Mechanisms that produce such changes are DNA methylation and histone modifications.

◆ **Genotype**: the combination of alleles inherited by an organism.

1 **Outline** how mRNA is degraded and why this is necessary.

2 **Describe** how gene expression is controlled in eukaryotic organisms.

Top tip!

Proteomics is the study of the structure and function of the entire set of proteins of organisms.

Link

Cell differentiation and the development of specialized tissues in multicellular organisms (HL only) is covered in Chapter A2.2, page 85.

Epigenesis

Epigenesis refers to the development of an organism by differentiation from an undifferentiated zygote. It is the development of patterns of differentiation in the cells of a multicellular organism. As we have seen, this is determined by the genome of an organism and gene expression. Research has shown, however, that environmental factors can change the activity of the genes, and these changes can be passed on to offspring, thereby affecting their development.

Epigenetics is the study of heritable changes in gene activity that are *not* caused by changes in the DNA base sequences (*epi* = outside). Examples of mechanisms that produce such changes are DNA methylation and histone modifications (see page 673, this chapter). Note that these modifications affect the way cells 'read' the genes, rather than alter the base sequences of DNA itself (which we would describe as a mutation).

The term **phenotype** refers to the characteristics or appearance of an organism. The phenotype is determined by the **genotype** (the genetic code of an organism). DNA base sequences are not altered by epigenetic changes, so the phenotype, but not the genotype, is altered. Later in this chapter (page 674), we will examine how epigenetic changes take place.

Differences between the genome, transcriptome and proteome of individual cells

As we saw in Chapter A2.2, the **proteome** of an organism refers to all the proteins expressed within a cell, tissue or organism at a certain time. Because the **genome**, the whole of the genetic information of an organism, is unique to each individual then the proteome it causes to be expressed is also unique.

The genome is the same in all the cells of an organism because all cells in a multicellular organism are ultimately derived from one original cell by cell division. The genome instructs the expression of proteins. Because individual tissues have specific structures relating to their function, different proteins are expressed in each specific tissue cell type. For example, in the gut, specific organs synthesize digestive enzymes, whereas in bone and other skeletal structures, collagen is produced. No cell expresses all its genes; the process of gene expression determines which genes are switched on or off at any one time. It requires energy to express genes, so it is more efficient to turn on only the genes that are needed. It would waste energy to have all genes expressed at the same time when they are not required. Specific genes are therefore expressed (turned on or off) in different cells according to a required function.

As the proteome is all the proteins produced by a cell, the proteome varies with the function, location or environmental conditions of the cell through the process of cell differentiation. This means that while the genome stays the same, the proteome varies and is dynamic due to gene expression.

genome
(~22 000 genes)

alternative promoters, alternative splicing, mRNA editing

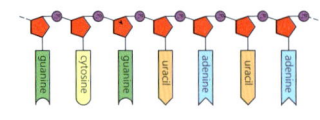
transcriptome
(~200 000 transcripts)

post-translational modifications

proteome
(>1 million proteins)

■ **Figure D2.2.1** The relationship between genome, transcriptome and proteome

◆ **Transcriptome**: the range of mRNA transcripts produced in a specific cell or tissue type at a particular time.

The **transcriptome** is the range of mRNA transcripts produced in a specific cell or tissue type at a particular time. This depends of the differential expression of genes within a cell. As we saw in Chapter A2.2, the pattern of gene expression in a cell determines how it differentiates. The transcriptome will determine which proteins are synthesized within a cell, which in turn depends on which parts of the genome are activated via gene expression (see Figure D2.2.1).

3 **Distinguish** between genome, transcriptome and proteome.

Top tip!

Every individual has a unique proteome. Most, but not all, organisms assemble their proteins from the same amino acids, but the proteome of an organism is unique. This is because the genome, the whole of the genetic information of an organism, is unique to each individual (except in the example of monozygotic twins) and so the proteome it causes to be expressed is also unique.

Concept: Continuity and change

In contrast with the genome, which is characterized by its stability and continuity, the transcriptome actively changes. An organism's transcriptome varies depending on many factors, including stage of development and environmental conditions.

Methylation of the promoter and histones in nucleosomes

Nucleosomes are stable protein–DNA complexes (page 26), but they are not static. For example, they can inhibit or facilitate transcription. **Methylation** is the reversible addition of a methyl group (–CH$_3$) within the chromatin, sometimes to histone tails but usually to the DNA molecule itself at the **promoter** region. These are known as **epigenetic tags**. Enzymes bring about this addition to the base cytosine (Figure D2.2.2). The addition occurs while the DNA is wrapped around histone proteins. The effect is to change the activity of the gene – usually extensive methylation inactivates a gene, repressing transcription and therefore expression of the gene downstream. Removal of methyl groups may turn genes back on again. Methylation of amino acids in histones either enables or disables the recruitment of regulatory proteins to the chromatin, causing transcription to be repressed or activated.

◆ **Methylation**: the reversible addition of a methyl group (–CH$_3$) within chromatin.

◆ **Epigenetic tag**: chemical tags (such as methyl or acetyl groups) that are added directly to DNA or on to histone proteins to regulate gene expression, blocking or allowing access to a gene's 'on' switch.

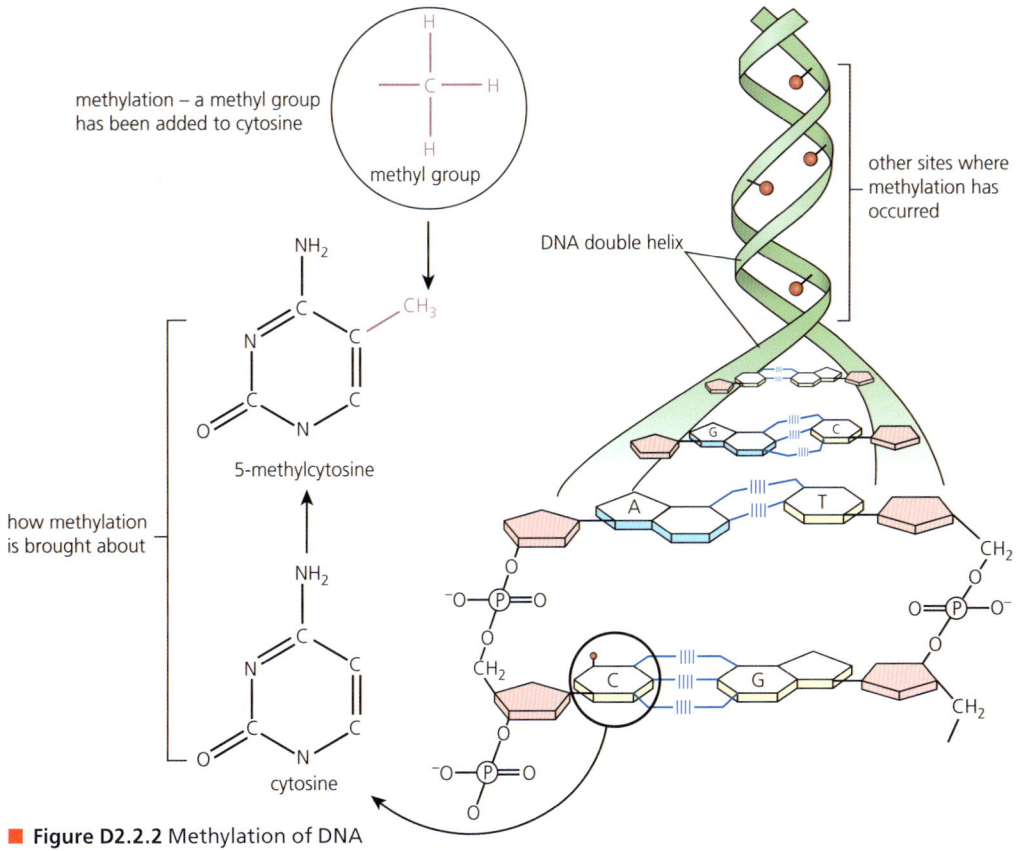

■ Figure D2.2.2 Methylation of DNA

D2.2 Gene expression

Once a gene has been methylated, it may remain in this condition. The methyl groups persist in situ from cell division to cell division. Furthermore, external conditions adverse to the cell or organism can leave their mark on our DNA through extensive methylations, leading to switched-off genes.

> **Top tip!**
>
> The promoter is an example of a length of non-coding DNA with a special function.

> **Top tip!**
>
> Cell differentiation and the regulation of gene expression allow proteins (e.g. insulin) to be produced in only certain types of body cell. When explaining these processes, chemical modification of DNA needs to be discussed, as well as the role of non-coding sequences in the regulation of gene expression.

Epigenetic inheritance through heritable changes to gene expression

Epigenetic inheritance is the acquisition of characteristics gained by a parent in their lifetime and passed on to their offspring. It comes about by modification of chromatin via epigenetic tags, such as DNA methylation or histone modification. Phenotypic changes in a cell or organism can therefore be passed on to daughter cells or offspring without changes in the nucleotide sequence of DNA. It can occur if epigenetic tags remain in place during mitosis or meiosis. If the tags are removed during cell division, epigenetic inheritance will not take place.

◆ **Epigenetic inheritance**: an inheritance pattern in which a modification to chromatin alters gene expression in an organism; a parent's experiences, in the form of epigenetic tags, can be passed down to offspring.

> **Nature of science: Patterns and trends**
>
> It now appears that, not only does methylation last a lifetime, it can be, and often is, transmitted to offspring – and sometimes to further generations. The environment of a cell and an organism may have an impact on gene expression for generations.
>
> In 1944, communities in the western Netherlands were deprived of food supplies as a result of wartime hostilities for over six months. Many people died of extreme starvation, but among the survivors were pregnant mothers who gave birth. Although these babies born after the famine were underweight initially and were fed normally after the war was over, they ended up heavier than average as adults and suffered from higher rates of obesity. It seems the parents' starvation was imprinted on their children's DNA, where as a survival strategy children developed a heavier mass than normal.
>
> From the time Mendel deduced the existence of 'factors' (page 726), through to Crick and Watson's discovery of the nature of the gene in chemical terms (page 20), genes and alleles were seen as unchangeable by external factors. The effects of environmental change were believed to be restricted to which genes (alleles) survived and contributed to the gene pool of future generations. The environment's impact was held to be on selection, not gene performance. Today, it seems poor diets (for example) can interfere with the performance of genes in succeeding generations, presumably as a result of 'markers' attached to parent DNA in earlier times.
>
> *Information based on www.nytimes.com/2018/01/31/science/dutch-famine-genes. html and https://theanalyticalscientist.com/fields-applications/a-lasting-legacy*

Examples of environmental effects on gene expression in cells and organisms

Many external factors can affect gene expression, such as diet, oxygen levels, humidity, light cycles, drugs, temperature and exposure to mutagens. Here, we will explore the effect of air pollution on gene expression.

Air pollution caused by traffic includes particulate matter (small particles produced by diesel vehicles that can enter lung tissue and cause damage), ozone (O_3), carbon monoxide (CO) and nitrogen oxides (NO_x). Air pollution has numerous harmful effects on health and contributes to the development of cardiovascular disease and a number of lung conditions, including asthma and chronic obstructive pulmonary disease (COPD). Recent research indicates that exposure to air pollution can modify DNA and histone methylation patterns, and that these changes might in turn affect inflammation and the development of diseases.

Exposure to air pollution has been correlated with long-term negative respiratory health outcomes, including the development of lung diseases. Interstitial lung diseases (ILDs) is an overall term for diseases that cause inflammation of tissues within the lung, including the bronchioles, alveoli and the capillaries surrounding the alveoli, leading to scarring and thickening of tissue. In the alveoli, thickening of tissue can lead to decreased rates of oxygen diffusion into the blood. Inflammation of the bronchioles causes asthma. Hypersensitivity pneumonitis (HP), one of the most common forms of ILD, is triggered by the inhalation of organic and inorganic substances from, for example, air pollution. These pollutants influence HP development through epigenetic modifications.

Treatments, such as exercise and B vitamins, have been suggested as solutions to reduce the impact of air pollution on methylation patterns and health.

> **Link**
> The adaptations of mammalian lungs for gas exchange is covered in Chapter B3.1, page 277.

4 Outline the role of methylation in epigenetic inheritance.

♦ **Imprinting**: process by which only one copy of a gene in an individual (either from their mother or their father) is expressed, while the other copy is suppressed. Gene expression is silenced by the epigenetic addition of chemical tags to the DNA during egg or sperm formation.

5 Explain why most epigenetic tags are removed from human ovum and sperm.

> ● **Common mistake**
> Do not confuse genetic imprinting with a different type of imprinting you may have read about, for example animal imprinting, where young birds follow their mother.

Consequences of removal of most but not all epigenetic tags from the human ovum and sperm

Many organisms retain and pass on epigenetic marks from environmental exposure via gametes. This means that the epigenetic tags can be passed on to the next generation. In humans and other mammals, egg and sperm cells undergo extensive reprogramming during development, which removes nearly all epigenetic marks (the epigenetic tags are said to be 'silenced'). Any tag that is not removed may affect the development and function of subsequent generations, as we have seen earlier (page 674). Research has suggested that one of the important purposes of epigenetic reprogramming in human sperm and egg cells is to remove epigenetic changes that might be caused by the environment. The mechanism evolved to prevent changes in methylation patterns caused by environmental disturbances from being transmitted across generations. For a small minority of genes (approximately 1%), however, epigenetic tags survive this process and pass unchanged from parent to offspring, through a process known as **imprinting**. Imprinting is a process where the DNA in sperm and eggs are modified during gametogenesis (i.e. during the formation of gametes by meiosis). This involved the addition of epigenetic tags that can switch 'on' parts of the DNA and switch 'off' other parts.

Different genes are epigenetically silenced in eggs and sperm. For most genes, two working copies are inherited, one from each parent. With imprinted genes, only one working copy is inherited, due to the differential silencing of epigenetic tags in sperm and egg development. Depending on the gene, either the copy from mother or the copy from father is epigenetically silenced. In sperm development, maternal tags are silenced and in egg development, paternal tags are silenced.

■ Imprinting in tigons and ligers

Tigons and ligers are both lion–tiger hybrids produced by interbreeding lions and tigers. Crossing a male tiger with a female lion produces a **tigon**, which is about the same size or smaller than either parent. However, the cross between a female tiger and a male lion produces a **liger**, which is known for its great size as it is bigger than both lions and tigers.

The differences can be accounted for by genetic imprinting: tigers and lions imprint their DNA differently. It is possible these differences have evolved because of the different reproductive habitats and lifestyles of the two species. Female lions may mate with multiple males, so a male lion passes on genes that encourage growth in his own cubs in that litter. In contrast, the female lion has evolved imprinted genes that are anti-growth (Figure D2.2.3), because she is equally related to all the potential cubs and so wants to equally distribute resources to maximize the number who might survive.

Tigers have a different social structure to lions and are not subject to the same pressures. A female tiger mates with one male, who is equally related to all the cubs. His genes would not want to 'encourage' growth because there is no competition between cubs from the male's perspective. The female tiger does not therefore need to evolve anti-growth imprinting 'defences'. If a female tiger, with no imprinted defences against paternally inherited genes encouraging growth in her offspring, is crossed (i.e. bred) with a male lion, who will contribute genes that encourage growth, a liger is produced (Figure D2.2.3). Growth in the liger is therefore a consequence of different patterns of epigenetic and genetic inheritance in lions and tigers. This shows how the phenotypic differences in tigons and ligers have an epigenetic origin, although the exact genes that are imprinted in lions or tigers, or which of these genes make the biggest differences in terms of growth, are not currently known.

> **Top tip!**
> You can read more about tigons and ligers, and other examples of genetic imprinting, here: www.thetech.org/ask-a-geneticist/why-are-ligers-so-big

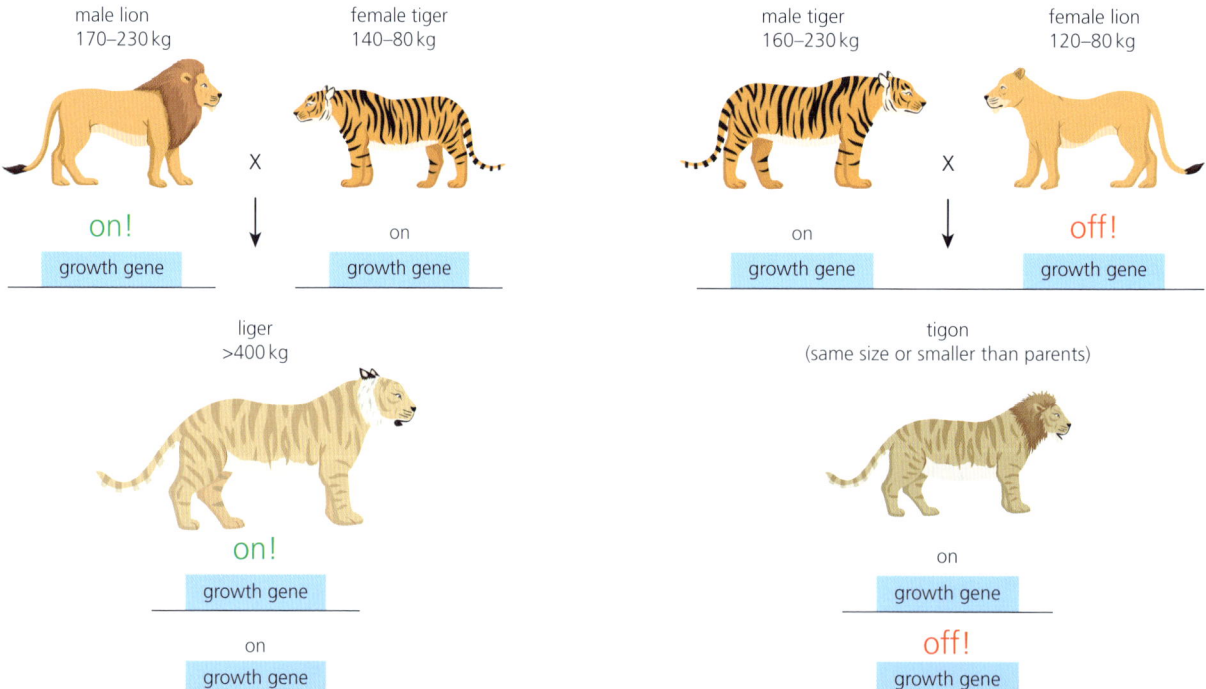

■ Figure D2.2.3 Different patterns of epigenetic inheritance in lions and tigers explains different phenotypes in ligers and tigons

6 **Define** the term *epigenetic inheritance*.

7 **Outline** the effect of epigenetic tags remaining in sperm and eggs.

> **ATL D2.2A**
>
> Read more about how epigenetic tags can be inherited, and the role of imprinting, using these websites:
> https://learn.genetics.utah.edu/content/epigenetics/inheritance
> https://learn.genetics.utah.edu/content/epigenetics/imprinting
>
> Write an essay explaining how hybrid animals can be used to understand the role of epigenetic inheritance in the determination of phenotypic traits.

Monozygotic twin studies

Within families there are remarkable similarities between parents and their offspring, but no two members of a family are identical, apart from identical twins.

Differences can be due to:
- **genetic factors**: some differences may be controlled by genes – such as human blood groups
- **environmental factors**: other differences between individuals may be due to the effect of our environment, such as a suntan acquired from exposure to the Sun
- a **combination of genetic and environmental factors**: other differences between individuals may be due to both genetics and environment, such as our body height and weight.

You may have heard the question 'nature or nurture?' This refers to the origin of phenotypic variation: whether it is due to the genetic code of an individual ('nature') or the environment they were brought up in ('nurture'). Studies using **monozygotic twins** attempt to answer this question and are designed to measure the contribution of genetics, as opposed to the environment, for a given trait.

Identical twins are **monozygotic**, meaning that they come from the same dividing cell, so are genetically identical. They cannot be distinguished by DNA fingerprinting. This contrasts with **dizygotic** twins, which are non-identical (i.e. may be a male and a female) because they develop from two separate fertilized eggs (Figure D2.2.4). Because non-identical twins come from two different egg cells, each one with a different combination of alleles, fingerprinting techniques are able to find differences between them.

◆ **Monozygotic twins**: identical twins result from the fertilization of a single egg by a single sperm, with the fertilized egg then splitting into two. Identical twins share the same genomes and are always of the same sex.

◆ **Monozygotic**: derived from a single ovum, and so genetically identical.

> **Common mistake**
>
> Do not just refer to 'twins' when referring to siblings born at the same time – refer to the specific type of twin, i.e., either monozygotic (developed from the same zygote) or dizygotic (developed from two different zygotes) twins.

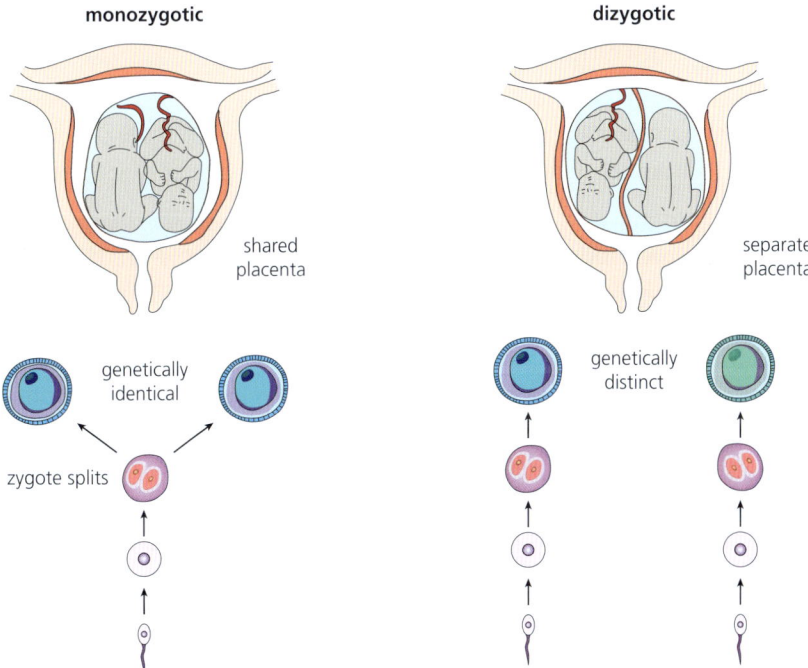

■ Figure D2.2.4 Comparing monozygotic (identical) and dizygotic (non-identical) twins

Studies can compare twins that have been brought up together or separately. Measurements of specific phenotypic traits can be measured and variation attributed to genotype or the environment (or possibly a combination of both). If separated twins show similar characteristics then the cause of the trait is probably genetic, whereas if there are significant differences then it is probably due to the environment.

8 **Describe** the role of monozygotic twin studies.

HL ONLY

> **ATL D2.2B**
>
> Identical genes in monozygotic twins may be expressed differently due to epigenetic changes – differences in methylation patterns influenced by the environment. Read the following research paper, which investigates changes in levels of methylation between 3-year-old twins and 50-year-old twins: https://www.pnas.org/doi/10.1073/pnas.0500398102 Summarize the findings of the study. What conclusions do you deduce from the methylation patterns seen in the two sets of twins?

In twin studies, results for monozygotic twins can be compared to those for dizygotic twins. In these investigations, many pairs of twins are studied to gather an accurate estimate of the relative importance of genes and the environment. Monozygotic twins share all their genes and so differences between twins can therefore be due to non-shared environmental influences. If both monozygotic and dizygotic twins resemble each other closely, this must be due to shared environmental influences. If monozygotic twins resemble each other more closely than dizygotic twins, then it suggests that genetic factors may play a role. Although monozygotic twins share the same genotype, most monozygotic twins are not identical. They may have differences in physical features and predisposition to disease, for example. One explanation for these observations are epigenetic changes, due to differences in DNA methylation and histone acetylation over the lifetime of the twins. Methylation levels can be expected to increase with age.

One twin study, carried out by geneticist Claude Bouchard in 1990 and published in *The New England Journal of Medicine*, examined the importance of genes for body-fat storage and the development of obesity. Bouchard gave male monozygotic twins excess calories (an additional 1000 calories a day) for three months. Every participant was heavier by the end of the experiment, although weight gain within pairs of twins was much more similar than weight gain between different twin pairs. Twins in each pair tended to gain weight in the same parts of the body, suggesting that genetic factors are involved. You can read more about this study here: https://pubmed.ncbi.nlm.nih.gov/2336074

Inquiry 1: Exploring and designing

Exploring; designing

If you were designing a monozygotic twin study, which phenotypic aspect would you study? What ethical issues would you have to address? Consult a variety of sources and from these select sufficient and relevant sources of information.

Formulate a research question and hypothesis for the monozygotic twin study you have proposed. Be sure to state and explain your predictions using your scientific understanding, and identify and justify your dependent, independent and control variables. Justify the range and quantity of measurements in your investigation. Refer back to the investigations of twin studies you have read about in research papers. How does your hypothesis, prediction and methodology compare to these studies?

External factors impacting the pattern of gene expression

Factors from outside the cell can affect gene expression. For example, both hormones and biochemical factors can impact the pattern of gene expression.

The effect of a hormone

We have seen in Chapter C2.1 how steroid hormones oestradiol, progesterone and testosterone can affect gene expression (page 465). In each case, the hormone binds to receptors in the cytoplasm, which act as ligand-gated transcription factors that in turn bind to DNA, with the help of co-activators and co-repressors, to alter DNA transcription.

The effect of lactose in bacteria

In bacteria, control of gene expression occurs most commonly at the level of transcription, which means that bacteria regulate how much mRNA is made from most genes. A second way for bacteria to regulate gene expression is to control the rate at which mRNA is translated into protein, that is, translational level of control.

In prokaryotes, **operons** are regions of DNA that contain a group of closely related genes consisting of structural genes and regulating elements. The regulatory sequences (non-coding DNA regions) include a promoter, **operator** and terminator. Structural genes encode the enzymes of a metabolic pathway. The operator position is important as its regulation enables or prevents the transcription of genes into mRNA.

The bacterium *E. coli* can be found in the human gut. Lactose is available to *E. coli* when we drink milk. Lactose metabolism begins with its hydrolysis into monosaccharides. By producing the appropriate enzymes only when the nutrient is available, the *E. coli* cell avoids wasting energy and resources making proteins that are not needed. It is the **lac operon** (lactose operon) that ensures that the *E. coli* cell only produces the enzymes required to metabolize lactose when lactose is present. The process is illustrated in Figure D2.2.5.

This mechanism involves a **repressor** molecule (coded for by a regulator gene), an **operator** gene situated close to the gene that is being regulated (a structural gene, in this case coding for the lactose-metabolizing enzyme), and a **promoter** gene. When there is no lactose, the repressor molecule binds to the operator and prevents transcription of the structural gene that codes for the lactose-metabolizing enzyme. If lactose is present, the lactose molecule reacts with the regulator protein, preventing it from binding with the operator gene. The lactose-metabolizing gene can then be transcribed and lactose metabolized by the cell. When all the lactose has been metabolized and it is no longer present, the repressor molecule blocks transcription again.

◆ **Operon**: sequence of adjacent bacterial genes all under the transcriptional control of the same operator.

◆ **Operator**: a DNA region at one end of an operon that acts as the binding site for a repressor protein. When the operator is complexed with the repressor, transcription is prevented.

◆ **Lac operon**: an inducible operon including three loci involved in the uptake and breakdown of lactose in the bacterium *E. coli*.

◆ **Repressor**: the protein product of a regulator gene that acts to control transcription of an operon. The repressor binds to the operator and prevents transcription by RNA polymerase of an operon.

◆ **Inducer**: a molecule that is capable of activating the transcription of a gene by combining with and inactivating a genetic repressor.

In the presence of the substrate lactose:

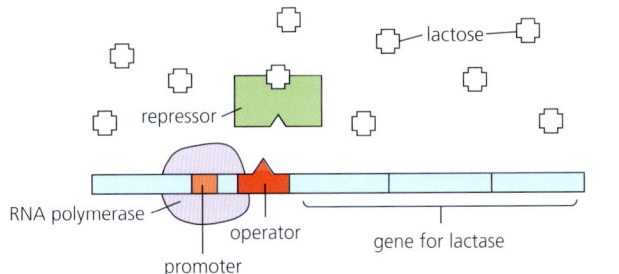

1 Lactose has combined with the repressor and inhibited its action.

2 RNA polymerase has bound to the promoter and is about to express the structural gene for lactase. mRNA for lactase synthesis will be transcribed and pass straight to ribosomes in the surrounding cytoplasm, where lactase is synthesized.

3 Eventually, all the lactose will have been metabolized, and the repressor will be free to bind to the operator again.

In the absence of the substrate lactose:

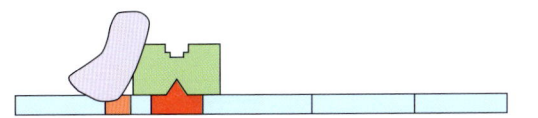

The gene is turned off.

1 There is no lactose to combine with the repressor.

2 The repressor binds with the operator.

3 The RNA polymerase is obstructed from binding to the promoter.

■ **Figure D2.2.5** The function of the lac operon in bacteria

Top tip!

Operons are only found in prokaryotes (bacteria and cyanobacteria). The regulator-gene mechanisms in eukaryotes are more complicated.

Link

The definition of allosteric regulators and more details about them is in Chapter C1.1, page 399.

The lac operon is an **inducible** operon. Transcription is turned on by the presence of a small effector molecule called the **inducer**, in this case, lactose. The ability of the repressor to bind the operator and inhibit transcription depends on the protein's conformation, which is **allosterically regulated** by an inducer. Thus, the concentration of the inducer determines the activity of the operon.

9 **Suggest** why operons are necessary in bacteria.

> **Top tip!**
> The lac operon is an inducible operon. Its transcription is usually switched off but can be turned on when a specific small inducer molecule (lactose) binds allosterically and inactivates the repressor. In the absence of lactose or when lactose concentration is low:
> - no lactose binds to the lac repressor
> - the lac repressor binds to the operator site and blocks RNA polymerase from transcribing the structural genes, thus inhibiting transcription.

The effect of tryptophan in bacteria

E. coli synthesizes tryptophan (an amino acid) from a precursor molecule in a series of metabolic steps, each catalysed by a specific enzyme. The five coding regions for the tryptophan enzymes that synthesize tryptophan are arranged sequentially (starting with trpE) on the chromosome in the operon (Figure D2.2.6). When *E. coli* needs to make tryptophan for itself because it lacks the amino acid, all the enzymes in the metabolic pathway are synthesized at one time.

The **trp operon** is a **repressible** operon because its transcription is usually turned on but can be inhibited (repressed) when a specific small molecule (tryptophan) binds allosterically (see Chapter C1.1) to a regulatory protein.

- The trp repressor is the product of a regulatory gene called trpR, which is located some distance away from the operon it controls, and it has its own promoter.
- The trp repressor is synthesised in an inactive form with little affinity for the trp operator.
- Only if tryptophan binds to the trp repressor does the repressor protein change to the active form that can bind to the operator, inhibiting transcription of the structural genes.

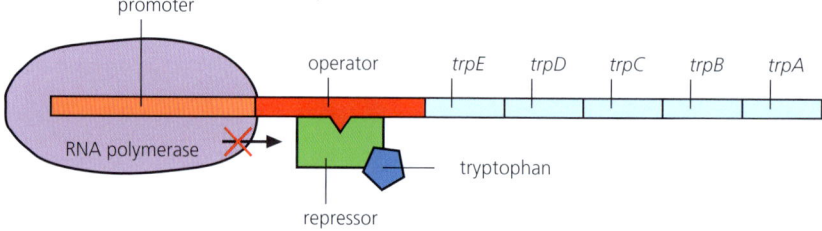

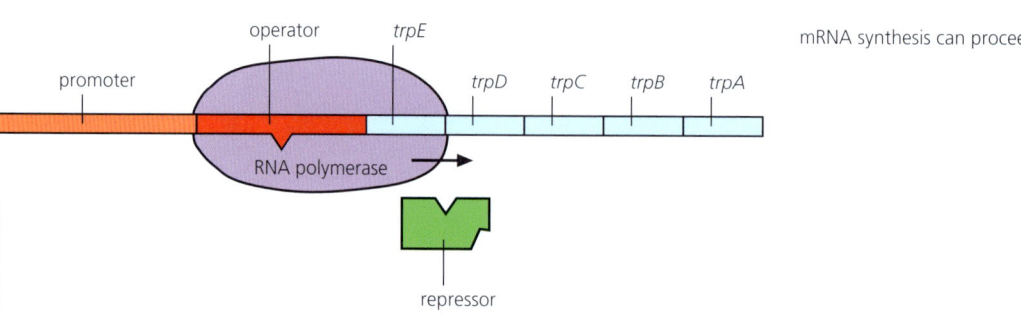

■ Figure D2.2.6 The function of the trp operon

Tryptophan functions as a co-repressor (as opposed to lactose acting as the inducer in lac operon) that cooperates with a repressor protein to switch an operon off. As more tryptophan accumulates, more tryptophan molecules can then bind to the trp repressor, which can then bind to the trp operator and inhibit the synthesis enzymes involved in the tryptophan biosynthetic pathway.

LINKING QUESTIONS

1. What mechanisms are there for inhibition in biological systems?
2. In what ways does the environment stimulate diversification?

D2.3 Water potential

Guiding questions

- What factors affect the movement of water into or out of the cell?
- How do plant and animal cells differ in their regulation of water movement?

SYLLABUS CONTENT

This chapter covers the following syllabus content:
- ▶ D2.3.1 Solvation with water as the solvent
- ▶ D2.3.2 Water movement from less concentrated to more concentrated solutions
- ▶ D2.3.3 Water movement by osmosis into or out of cells
- ▶ D2.3.4 Changes due to water movement in plant tissue bathed in hypotonic and hypertonic solutions
- ▶ D2.3.5 Effects of water movement on cells that lack a cell wall
- ▶ D2.3.6 Effects of water movement on cells with a cell wall
- ▶ D2.3.7 Medical applications of isotonic solutions
- ▶ D2.3.8 Water potential as the potential energy of water per unit volume (HL only)
- ▶ D2.3.9 Movement of water from higher to lower water potential (HL only)
- ▶ D2.3.10 Contributions of solute potential and pressure potential to the water potential of cells with walls (HL only)
- ▶ D2.3.11 Water potential and water movements in plant tissue (HL only)

Solvation with water as the solvent

◆ **Solvation**: the interaction of a solvent with dissolved molecules or ions.

Solvation is the interaction of a solvent with the molecules and ions that dissolve in it (the solute). As seen in Chapter A1.1, water is an ideal solvent because of its polarity, where the hydrogen ends of the V-shaped molecule are positive and the oxygen end is negative. Water molecules can therefore interact with the positive and negative charges of the solute and form hydrogen bonds, or ion-dipole forces (if ions), between the two. Solvent polarity is the most important factor in determining how well the solvent solvates a particular solute. Polar solvent molecules can solvate polar solutes and ions because they can orient the appropriate partially charged portion of the molecule towards the solute through hydrogen bonding or ion-dipole forces (Figure D2.3.1). This stabilizes the system and creates a hydration shell around each particle of solute (Figure A1.1.9, page 8).

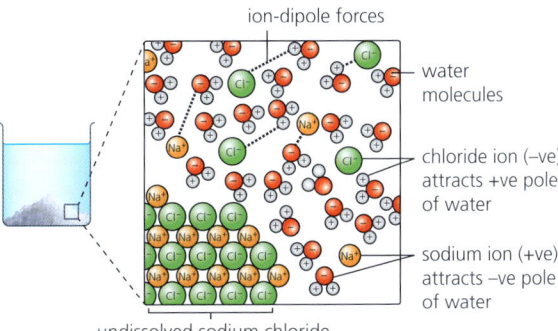

■ **Figure D2.3.1** Solvation of a salt solute (sodium chloride) in a solvent (water)

Link

The definition of osmosis is on page 227. The movement of water molecules across membranes by osmosis and the role of aquaporins starts on Chapter B2.1, page 228.

◆ **Hypotonic**: when the external solution is less concentrated than the cell cytoplasm and there is a net inflow of water into the cell by osmosis.

◆ **Hypertonic**: when the external solution is more concentrated than the cell cytoplasm and there is a net outflow of water from the cell by osmosis.

◆ **Isotonic**: when the external solution is the same concentration as the cell cytoplasm and there is no net entry or exit of water from the cell by osmosis.

● Common mistake

It is incorrect to refer to water as 'osmosising'. The correct terminology is to say 'water moves by osmosis'.

Water movement from less concentrated to more concentrated solutions

Osmosis is a special case of diffusion involving water molecules. Because water is an important component of cells, the diffusion of water is referred to specifically as 'osmosis'. Water, although polar, is small enough to move through the phospholipids of the bilayer, although it also moves through special pore proteins called aquaporins (page 228). Osmosis is the net (i.e. overall) movement of water molecules from a **less concentrated** solution (i.e. more dilute) to a region of **more concentrated** solution, across a selectively permeable membrane. Osmosis is therefore the passive movement of water molecules across a partially permeable membrane, from a region of lower solute concentration to a region of higher solute concentration (Figure D2.3.2).

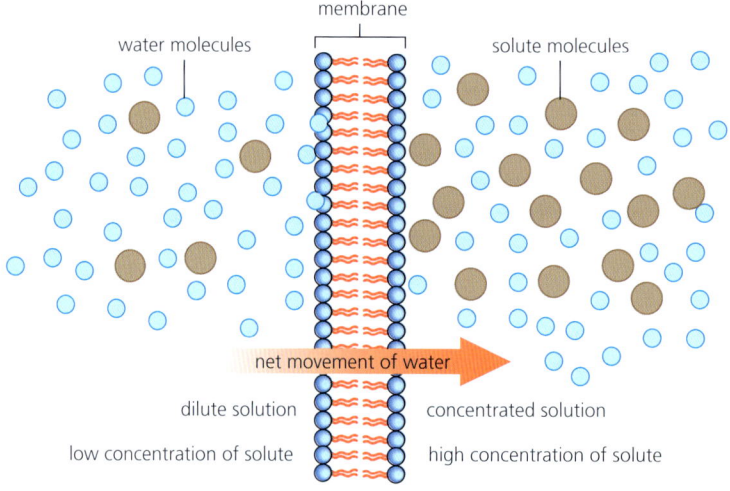

■ **Figure D2.3.2** Water moves by osmosis from a less concentrated to a more concentrated solution

Figure D2.3.3 shows three different concentrations of a solution outside a cell. If the solution outside the cell (i.e. below the membrane in Figure D2.2.3) is more dilute than the cytoplasm inside the cell, it is referred to as **hypotonic**. This means that the external solution is less concentrated than the cell cytoplasm. In these conditions, there will be a net inflow of water into the cell by osmosis. If the solution outside the cell is more concentrated than the cytoplasm inside the cell, the solution is said to be **hypertonic**. This means there will be a net outflow of water from the cell by osmosis. If the solute concentration is the same on both sides of a membrane (i.e. inside the cell and in the external solution), the external solution is said to be **isotonic**, and there is no net entry or exit of water from the cell by osmosis.

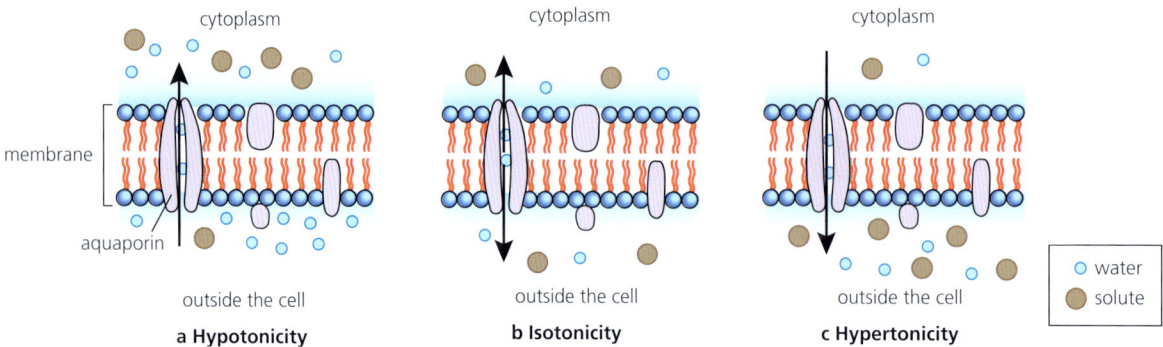

■ **Figure D2.3.3** Flow of water into or out from a cell depends on the osmotic concentration

> ● **Top tip!**
>
> A way to remember the difference between hypertonic and hypotonic is that in a <u>hypo</u>tonic solution, the solute concentration is <u>low</u>er than the cell cytoplasm, whereas in a <u>hyper</u>tonic solution the solute has a <u>higher</u> solute concentration than the cell cytoplasm; i.e. hypo = low (in terms of solute concentration) and hyper = high.

● **Common mistake**

It is incorrect to say that there is *no* movement of water in an isotonic solution. There is movement of water across the membrane, but the movement out from the cell is balanced by movement into the cell, i.e. there is dynamic equilibrium, and so no *net* osmosis.

Water movement by osmosis into or out of cells

The direction of the net movement of water can be predicted depending on whether the external solution surrounding a cell is hypotonic or hypertonic. In Figure D2.3.3, the area below the plasma membrane represents the solution outside the cell, the area above the membrane is the cytoplasm. In an isotonic solution, there is no net movement of water because solute concentration is the same on both sides of the membrane. The condition in an isotonic solution is said to be in **dynamic equilibrium**, which means that although water moves across the membrane (i.e. the situation is dynamic) there is no overall change in water concentration on either side of the membrane (i.e. the situation is at equilibrium).

> **Concept: Continuity**
>
> Water moves into and out of cells by osmosis. In this way cells can maintain their osmotic concentration. This is important in animal cells, where the internal environment needs to be the same as the external environment to avoid cell lysis (bursting) or shrinkage.

Changes in plant tissue bathed in hypotonic and hypertonic solutions

■ Estimation of osmotic concentration of plant tissue

When plant cells are bathed in a solution that is isotonic with the cell cytoplasm, there is no net entry or exit of water from the cells. The tissue retains the same dimensions and mass. Alternatively, if similar plant tissue is placed in a hypertonic solution, the tissue decreases in dimensions and mass. When placed in a hypotonic solution, it increases in dimensions and mass. This observation is the basis of the experiment illustrated in Figure D2.3.4. Here the aim is to discover the concentration of the bathing solution isotonic with the cells of potato tuber.

Study the sequence of steps involved in the experiment. Note the importance of:
- accurate weighing out of the solute (sucrose in this case)
- accurate pipetting of solution from tube to tube in the process of serial dilution
- the use of replicate samples of tissue in each tube
- the accurate measurement of the length of the tissue strips at the end of the experiment.

Examine the graph to identify what molar concentration of sucrose is isotonic with the cytosol of the potato tissue used in this experiment.

1 **State** the significance of the use of three tissue strips in each of the tubes in Figure D2.3.4.

1 Preparing different concentrations of sucrose solution, 0.8 mol dm⁻³ → 0.2 mol dm⁻³

100 cm³ of 1 mol dm⁻³ solution was taken and the following dilutions carried out:

Volume of distilled water (cm³)	Volume of 1 mol dm⁻³ sucrose (cm³)	Concentration of sucrose (mol dm⁻³)
2	8	0.8
4	6	0.6
6	4	0.4
8	2	0.2

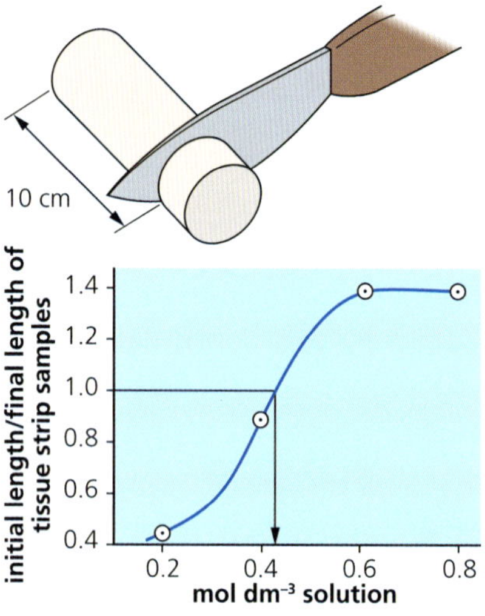

2 Preparing the tissue strips and the setting up of the experiment

replicate strip (10 × 1 × 0.5 cm) or cylinders (10 × 1 cm cork borer) cut from a large potato tube
tissue strips/cylinders washed and three placed in each tube

3 Measuring the final lengths of the tissue strips

After a period of 30 minutes the tissue strips were retrieved, blotted dry, and their lengths measured accurately. The mean change in length of the three strips in each tube was calculated.

4 Graphing the results and estimating the osmotic concentration of potato tuber tissue

A graph of the results was plotted: The relationship between the molar concentration of sucrose solutions and the change in length of tissue strips

■ Figure D2.3.4 The investigation of the osmotic concentration of potato tuber tissue

Estimation of osmotic concentration of tissues by bathing samples in hypotonic and hypertonic solutions

◆ **Osmotic concentration**: the measure of solute concentration, defined as the number of osmoles (Osm) of solute per litre (L).

Changes in tissue length and mass can be measured and analyzed to deduce isotonic solute concentration. The aim of this experiment is to determine the solute concentration, or **osmotic concentration**, of a plant tissue. Potatoes are ideal for the experiment, although other root vegetables can be used. Potato tissue is put in a range of sucrose solutions of different osmotic concentration to see how they change in mass and length. Osmotic concentration of potato tissue is estimated by finding the concentration of sucrose that results in no change in the potato tissue mass or length.

We can use standard deviation and standard error to help in our analysis of the data. These measurements indicate variation around the mean. Standard deviation and standard error could be determined for the results of this experiment if there are repeats for each concentration, which would allow the reliability of length and mass measurements to be compared. You are not required to memorize formulae for calculating these statistics.

Inquiry 2: Collecting and processing data

Collecting data

For the investigation you will need the following equipment:

- 1.00 mol dm⁻³ sucrose solution
- cork borer/chip-maker
- distilled water
- pipettes
- measuring cylinders
- boiling tubes
- electronic balances
- stop clock
- ruler or calipers
- balances

Safety

Take care cutting the potato chips. Use a white tile to cut the potato on and do not cut towards the body.

Procedure

1. Make up six sucrose solutions of 1.00, 0.80, 0.60, 0.40, 0.20 and 0.00 mol dm^{-3} (see Table D2.3.1).
2. Use the cork borer to prepare 30 chips of potato, each 30 mm in length.

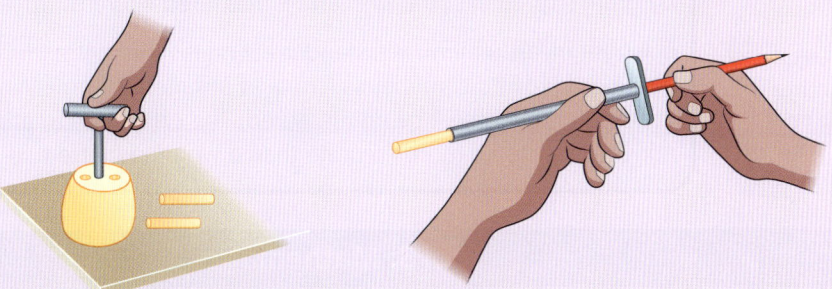

■ **Figure D2.3.5** Preparing potato chips using a cork borer

3. Weigh and measure each chip and record its mass (each length should be 30 mm).
4. Put one chip in each of the solutions. Repeat the test five times with each sucrose concentration (i.e. have five boiling tubes containing the solution at each concentration, with a chip of known mass and length in each tube).
5. After 40 minutes, remove the chips and reweigh and remeasure them – taking care to remove any excess solution first. (Why do you do this?)

Processing data

6. Calculate the percentage change in mass and the percentage change in length for each chip. This is calculated by working out the change in mass or length, dividing it by the original mass or length, and multiplying by 100 (to produce a percentage) (see Chapter C3.2, page 546).
7. Plot a graph of percentage change (*y*-axis) against sucrose concentration (*x*-axis) for both length and mass.
8. Estimate the concentration of the potato tissue (its solute potential). This is the point when there is no change in mass/length (i.e. no net osmosis, because the solute potential is the same in the solution as the cell cytoplasm). *Is it the same for both length and mass?*

■ **Table D2.3.1** Preparing different concentrations of sucrose solution, 0.80 mol dm^{-3} to 0.20 mol dm^{-3}

Volume of distilled water/cm³	Volume of 1.00 mol dm^{-3} sucrose/cm³	Concentration of sucrose/mol dm^{-3}
2.00	8.00	0.80
4.00	6.00	0.60
6.00	4.00	0.40
8.00	2.00	0.20

> ● **Top tip!**
>
> The most important part of this experiment is uniform blotting (i.e., drying of potato tissue). Ideally, this should take into account a uniform pressure applied by the blotting/tissue paper when drying a chip, for a known time.

Tool 1: Experimental techniques

Accurately measuring mass and length

Accurate measurements are needed in scientific experiments. The term 'accuracy' relates to how close your results are to the true value. Accuracy can be improved by carefully measuring mass and length in the experiment on page 685. Limitations in your equipment and method will reduce the accuracy of the results. The mass of a substance can be measured in the laboratory with an electronic balance. The SI unit for mass is the kilogram (kg), smaller masses can be measured in grams (g), and milligrams (mg).

In length measurement, the standard SI unit is the metre (m). However, a metre rule will be subdivided into centimetres (cm) and millimetres (mm). Which of these units is used depends on the object being measured. The calibration (markings) of a metal or plastic rule are likely to be more accurate than a wooden metre rule.

◆ **Standard error (SE)**: an estimate of the reliability of the mean of a population sample. A small standard error indicates that the mean value is close to the actual mean of the population.

Tool 3: Mathematics

Applying general mathematics

Standard deviation

Calculate the **standard deviation** for each chip. The equation and calculation method are shown in Chapter C4.1, page 552.

Standard error

The **standard error (SE)** represents how well the sample mean approximates to the population mean. The larger the sample, the smaller the standard error, and the closer the sample mean approximates to the population mean. The standard error is obtained by dividing the standard deviation, SD, by the square root of n, the sample size.

$$\text{Standard error (SE)} = \frac{\text{standard deviation (SD)}}{\sqrt{n}}$$

Processing uncertainties

When graphs are presented showing mean values, the standard error can be shown graphically as error bars. Error bars are added to each plotted value to demonstrate the deviation of the sample from the true population mean. Error bars (±1SE) extend above and below the points plotted on a graph to show this variability (Figure D2.3.6). Non-overlapping error bars demonstrate that the difference between mean values is significant (as shown in Figure D2.3.6), whereas overlapping error bars suggest a non-significant difference.

Data must be expressed to an appropriate number of decimal places. Raw and processed data should be recorded to a number of decimal places appropriate to the sensitivity of the apparatus or instrument. All raw data of the same type should be recorded to the same number of decimal places. In Table D2.3.2, percentage change was recorded to two decimal places and so the mean values should be expressed to the same level of precision.

Data were collected using the methodology outlined in the Inquiry 2 above. This is shown in Table D2.3.2.

■ **Table D2.3.2** Results from investigation to establish the osmotic concentration of potato tissue

Sucrose concentration/ mol dm^{-3}	Percentage change in the mass of the potato chip							
	Repeat 1	Repeat 2	Repeat 3	Repeat 4	Repeat 5	Mean	SD	SE
0.0	18.43	15.65	21.55	16.53	19.98	18.43	2.42	1.08
0.2	3.88	6.44	2.22	5.23	1.05	3.76	2.18	0.98
0.4	−6.55	−3.05	−6.55	−4.88	−4.12	−5.03	1.53	0.69
0.6	−11.09	−9.55	−10.05	−15.53	−9.55	−11.15	2.53	1.13
0.8	−15.40	−18.55	−17.55	−13.76	−13.42	−15.74	2.27	1.01
1.0	−16.58	−17.23	−21.65	−14.23	−15.22	−16.98	2.86	1.28

Standard deviation (SD) was calculated using the STDEV function in Excel. Standard error was then calculated using the formula: =STDEV(B3:F3)/SQRT(COUNT(B3:F3)). This shows the calculation for data between cells B3 and F3: the equation was copied and applied to other cells.

A graph of the data was plotted (Figure D2.3.6). Error bars were added by clicking on the line of best fit, then selecting 'error bars – custom – specify values' and then highlighting the SE calculations from Excel (for both positive and negative values).

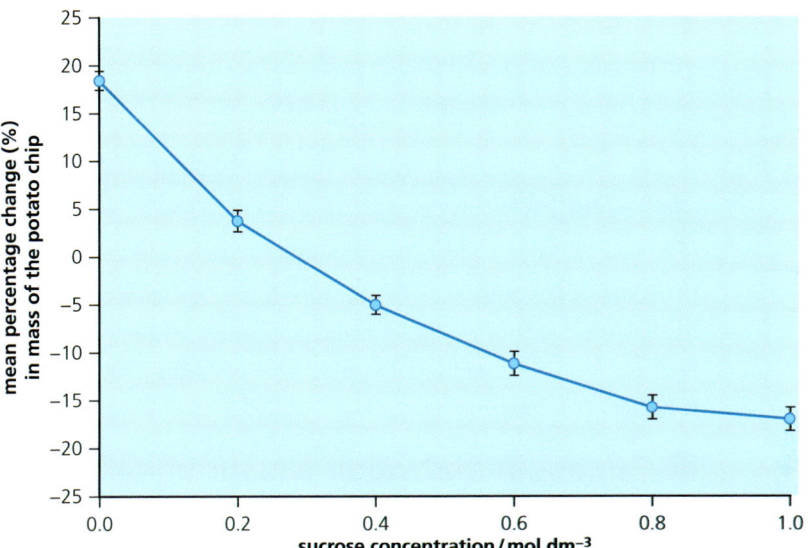

■ **Figure D2.3.6** Graph showing the relationship between solute concentration and mean percentage change in mass; error bars indicate variation of data around mean values

The results show that the chip does not change mass at a sucrose concentration of $0.3\,mol\,dm^{-3}$. This is the isotonic point, where there is no net osmosis (i.e. no overall movement of water into or out of the chip). This gives the osmotic concentration of the chip. The SE for mean values between 0.0 and $0.6\,mol\,dm^{-3}$ are non-overlapping, which suggests that these differences are real and significant. The SE bars for 0.8 and $1.0\,mol\,dm^{-3}$ do overlap, however, which suggests that the percentage change of chips at these two sucrose concentrations are not significantly different.

Inquiry 3: Concluding and evaluating

Concluding

Describe and explain your results from the Inquiry on page 684. Interpret your processed data and analysis to justify your conclusions. Compare the outcomes of your investigation to the accepted scientific context discussed in this chapter. The point where the line between the data points crosses the x-axis is the point at which there is no change in mass. This means that there is no net osmosis. (*Why is it incorrect to say that there is 'no osmosis'?*) This indicates the concentration of the cell solution (i.e. it is the same as the external solution), that is, the osmotic concentration.

Relate the outcomes of your investigation to your stated research question or hypothesis. What was the impact of uncertainties on your conclusions?

Evaluating

1. What were the limitations of the experiment? How might they have affected the accuracy of the result? How could you improve the experiment to reduce these limitations and improve accuracy?
2. Comment on reliability and identify any anomalous results. What is standard deviation and standard error, and what does this tell you about your results? (*Hint: variation around the mean.*) Identify and discuss the sources and impacts of random and systematic errors.
3. Which results provided the more accurate results – the percentage change in length or the percentage change in mass? *Why is this?*
4. Explain any realistic and relevant improvements you could make to the investigation.

Effects of water movement on cells that lack a cell wall

The effects of osmosis on plant and animal cells can be quite different.

Can you suggest why?

In an animal cell, the absence of a protective cellulose wall generates a serious problem in terms of water relations. A typical animal cell – for example a red blood cell – when placed in pure water or a **hypotonic** solution will quickly swell and break open from the pressure generated by the entry of an excessive amount of water by osmosis (Figure D2.3.7). Notice that the same cells, when placed in a **hypertonic** solution, shrink in size due to net water loss from the cytoplasm. Cells in hypertonic solutions are described as **crenated**. These cells are less likely to move smoothly through capillaries, which can lead to blood clots. Isotonic tissue fluid (the solution that surrounds cells) therefore needs to be maintained in multicellular organisms to prevent harmful changes.

In mammals and other animals, the osmotic concentration of body fluids (blood plasma and tissue fluid) is very carefully regulated, maintaining the same osmotic concentration inside and outside body cells (isotonic conditions) to avoid such problems. This process is an aspect of osmoregulation (Chapter D3.3, page 768).

Osmosis in freshwater aquatic unicellular animals

Many unicellular animals survive in freshwater aquatic environments where the external medium is normally hypotonic to their cell solution. These organisms experience a continuous net inflow of water by osmosis and are in danger of disruption of the plasma membrane by high internal pressure.

The protozoan *Amoeba* is one example. In fact, the cytoplasm of amoebae contains a tiny water pump, known as a **contractile vacuole**, which works continuously to pump out excess water. The importance of the contractile vacuole to these organisms is fatally demonstrated if the cytoplasm is temporarily anaesthetized. The protist quickly bursts (Figure D2.3.8).

When the external solution is less concentrated than the cell contents (hypotonic)
cell swells and bursts

net inflow of water

When the external solution is at the same concentration as the cell contents (isotonic)
no net water movement
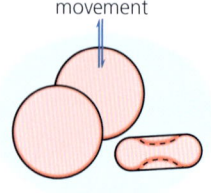

When the external solution is more concentrated than the cell contents (hypertonic)
cell outline becomes crinkled
net outflow of water
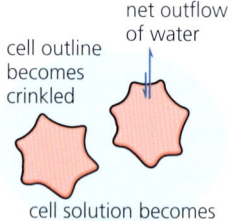
cell solution becomes more concentrated

■ Figure D2.3.7 Osmosis in animal cells

◆ **Contractile vacuole**: a small vesicle in the cytoplasm of many freshwater protists that expels excess water.

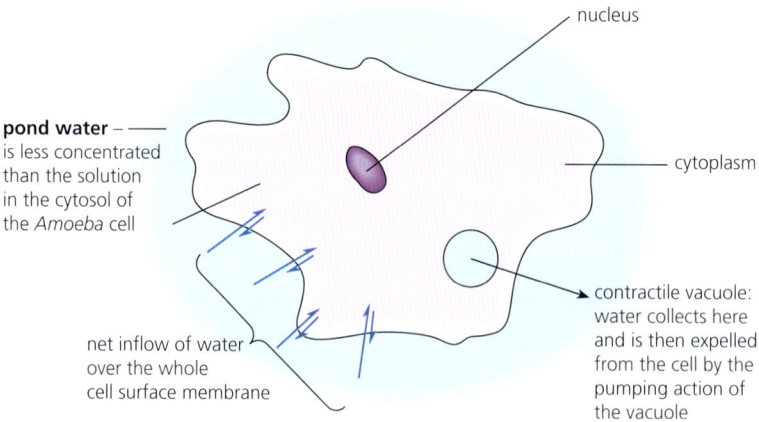

■ Figure D2.3.8 *Amoeba* – the role of the contractile vacuole

Concept: Continuity and change

Single-celled organisms without a cell wall are adapted to change by using contractile vacuoles to remove water from the cell. This maintains the osmotic concentration of the cytoplasm and functioning of the organism.

◆ **Turgor pressure**: hydrostatic pressure in a cell that pushes the plasma membrane against the cell wall.

◆ **Turgid**: condition where the vacuole of a plant cell is full of water, pushing the cell membrane against the cell wall.

◆ **Flaccid**: plant cell that has become soft and less rigid than normal because the cytoplasm within its cells has shrunk and contracted away from the cell walls through loss of water.

◆ **Plasmolysis**: condition when plant cells are in hypertonic solutions when the cell membrane pulls away from the cell wall due to reduced cell volume and pressure.

Effects of water movement on cells with a cell wall

Osmosis in an individual plant cell, with its cellulose cell wall (Figure D2.3.9), produces different outcomes to those of animal cells and protists that do not have a cell wall. Whether the net direction of water movement is into or out of a plant cell depends on whether the concentration of the cell solution is more or less concentrated than the external solution.

When the external solution is less concentrated than the cell solution (**hypotonic**), there is a net inflow of water into the cell by osmosis and the cell solution becomes diluted. Then the permanent vacuole swells due to water uptake, and the membrane of the vacuole pushes against the cytoplasm, which in turn presses hard against the cell wall, increasing **turgor pressure**. If this happens the cell is described as **turgid**. The pressure that develops (due to the stretching of the cell wall) eventually becomes so great that it prevents further uptake of water. The cell wall has protected the delicate cell contents from damage due to osmosis, but the tissue may be quite rigid due to the internal pressure.

When the external solution is more concentrated than the cell solution (**hypertonic**), there is a net flow of water out of the cell by osmosis and the cell solution becomes more concentrated. As the volume of cell solution (cytoplasm and permanent vacuole) decreases, the cytoplasm pulls away from parts of the cell wall (contact with the cell wall is maintained at points where there are cytoplasmic connections between cells). The cell becomes **flaccid** and it is said to be **plasmolysed** (from plasmo = cytoplasm, lysis = splitting).

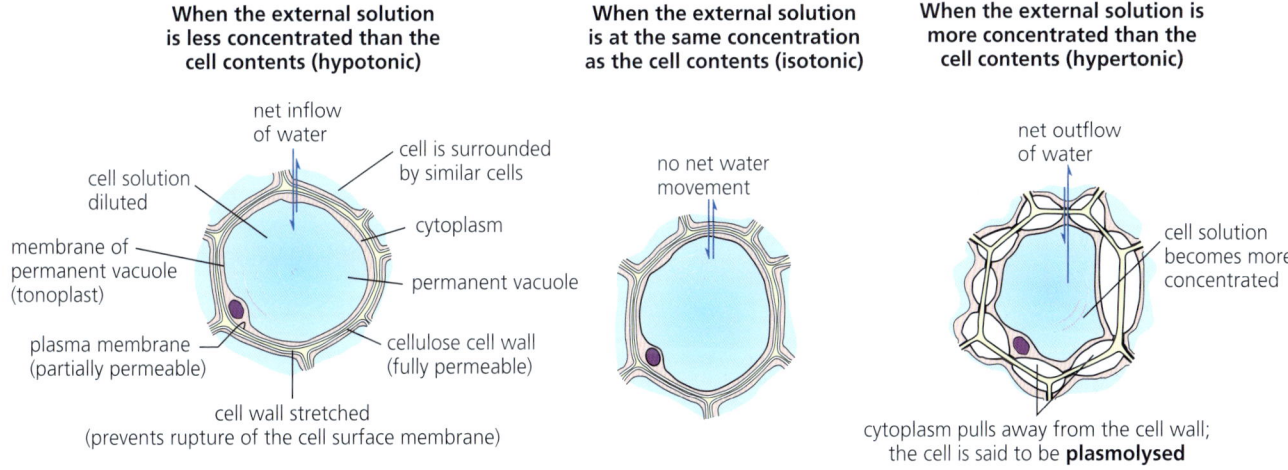

■ Figure D2.3.9 Osmosis in plant cells

In the case of hypotonic solutions, the cell wall stops the cell bursting and allows the cell to develop turgor pressure, which helps to support plant tissues (in the absence of supportive structures such as skeletons or shells that animals have). In hypertonic solutions, plant tissue can survive under plasmolysed conditions, although not for extensive periods of time. The movement of the plasma membrane away from the cell wall enables the solute to enter the space between the wall and membrane. Over time this can lead to damage to the membrane.

ATL D2.3A

Why are plants rare in sea water? What issues do they face? Use your knowledge of osmosis to explain the problems plants encounter in salt water.

Some plants *do* exist around coasts and spend some or much of their time in sea water, such as mangrove forests. Seagrasses (of which around 60 species exist) are the only flowering plants that grow underwater in marine environments. What adaptations do marine plant species have that enable them to exist in this environment? (Look back at Chapter B4.1, page 342, to help you answer.)

Medical applications of isotonic solutions

An intravenous (IV) drip is a piece of medical equipment that delivers fluids directly into a patient's circulatory system. The fluid is contained in a small bag, held on a stand above the patient so liquid can drip down easily. It is used to treat dehydration (i.e. to replenish lost salts and water to the body) and to deliver medicine.

A sterile saline solution of 0.9% NaCl is safe to use because it has the same solute concentration as body tissues and the cells in the blood. If a hypotonic solution is introduced to the body it would cause a decrease in the solute concentration of blood plasma, leading to red blood cells swelling and bursting, resulting in decreased delivery of oxygen to cells. A high concentration of salt would cause water to be lost from red blood cells, causing them to move less freely in capillaries, with the potential of blood clots forming.

In the same way that drips need to be isotonic in relation to body tissues and blood plasma, when human organs are donated for transplant surgery, they must be maintained in a saline solution that is isotonic with the cells of the tissues and organs. This is to prevent damage to cells due to water uptake or loss during transit to the recipient patient.

> **Concept: Continuity**
>
> The osmotic concentration of a drip, or a solution used to bathe transplant organs, must be the same as that of the blood plasma and tissue fluid (i.e. isotonic). This maintains the osmotic concentration of the blood plasma and body tissue, to ensure continuity in body functions.

2 **Explain** why the osmotic concentration of the solution used to bathe the organ needs to be determined.

Water potential as the potential energy of water per unit volume

It is useful to be able to quantify the potential of water to move from one area to another, in order to predict the direction and degree of movement. **Water potential** is used to quantify the tendency of water to move from dilute to concentrated solutions due to osmosis. Water potential is the potential energy of water per unit volume. As the movement of water within cells generates pressure against the membrane, water potential is measured in kilopascals (kPa), which is an SI unit. It is represented by the Greek letter ψ_w (psi).

It is not possible to measure the absolute quantity of potential energy of water, and so measures of water potential are standardized relative to pure water at standard atmospheric pressure and 20 °C.

◆ **Water potential**: the potential energy of water per unit volume, relative to pure water, denoted by the symbol ψ_w (psi) and expressed in units of pressure (kPa).

3 **Define** the term *water potential*.

Movement of water from higher to lower water potential

Water with a low concentration of solute, and therefore a high water potential, will have a high potential energy for movement, whereas a solution that has a high concentration of solute and therefore low water potential, will have a lower potential energy for the movement of water. In solutions containing a solute, the water molecules form hydrogen bonds with the solute (Figure D2.3.1, page 681), restricting the freedom of movement of water and lowering the potential energy. Water containing a solute will have a lower water potential than pure water. Water therefore moves from a region where it has a higher water potential (i.e. lower solute concentration and higher potential energy) to a region where it has a lower water potential (i.e. higher solute concentration and lower potential energy).

- 4 **Explain** why water moves from a higher to lower water potential.
- 5 **Describe** the effect of adding salt to a solution that is bathing animal tissue.

Contributions of solute potential and pressure potential to the water potential of cells with walls

As we have seen, the concentration of a solute determines the water potential of a solution. The pressure within a cell also determines the movement of water – a higher pressure will result in greater potential energy and, therefore, water potential.

Dissolving any solute in pure water lowers the water potential (i.e. leads to a more negative value). The highest water potential is generally taken to be zero, which is found in pure water (Figure D2.3.10).

Solution	Water potential (ψ_w)	
pure water	0 (zero)	highest water potential
dilute solution (of solute)	negative e.g. –50 kPa	
concentrated solution (of solute)	more negative e.g. –100 kPa	lower water potential

■ **Figure D2.3.10** Water potential becomes more negative as solute is added to a solution

Water potential can be calculated using the following equation:

$$\psi_w = \psi_s + \psi_p$$

where ψ_w = water potential

ψ_s = solute potential

ψ_p = pressure potential

Both solute and pressure potentials are measured in terms of their effect on water potential, and so the unit in both cases is kPa. Because adding a solute to a solution lowers the water potential, solute potentials can range from zero downwards (Figure D2.3.10).

HL ONLY

6 **Distinguish** between solute potential and pressure potential.

7 **Describe** how solute potential and pressure potential are used to determine water potential.

8 **Deduce** the changes to the pressure and solute potential of a plant tissue placed in a hypotonic solution.

Concept: Change

The water potential of a cell will change if placed in a hypertonic or hypotonic solution, although the effects on solute and pressure potentials are different in each case.

Pressure potentials, because of the movement of water molecules pushing against the cell membrane and the cell wall of plant, fungi and prokaryote cells, are generally positive inside cells. Negative pressure potentials can occur, however, inside xylem vessels where sap (mainly water) is being drawn up the plant under tension (see Chapter B3.2, page 303).

Water potential and water movements in plant tissue

As we saw in the experiment to determine the osmotic concentration of plant tissue (page 683), plant cells gain water in a hypotonic solution and lose water in a hypertonic solution by osmosis. The following changes occur when plant tissue is bathed in these solutions.

In a **hypotonic** solution:

- As water moves into the cell, **pressure potential increases**. This is because there are a greater number of water molecules in the cell. The increased potential energy of water means that there is increased potential for water molecules to move and exert pressure on the plasma membrane.
- The **solute concentration decreases** because the number of solute molecules relative to water molecules decreases.
- Overall, the water potential of the cell increases (due to a decreasingly negative solute concentration and an increasingly positive pressure potential).

In a **hypertonic** solution:

- As water moves out from the cell, **pressure potential decreases**. This is because there are fewer water molecules in the cell. The decreased potential energy of water means that there is a decreased potential for water molecules to move and exert pressure on the plasma membrane.
- The **solute concentration increases** because the number of solute molecules relative to water molecules increases.
- Overall, the water potential of the cell decreases (due to an increasingly negative solute concentration and a decreasing pressure potential).

LINKING QUESTIONS

1 What variables influence the direction of movement of materials in tissues?
2 What are the implications of solubility differences between chemical substances for living organisms?

Reproduction

> **Guiding questions**
> - How does asexual or sexual reproduction exemplify themes of change or continuity?
> - What changes within organisms are required for reproduction?

SYLLABUS CONTENT

This chapter covers the following syllabus content:
- ▶ D3.1.1 Differences between sexual and asexual reproduction
- ▶ D3.1.2 Role of meiosis and fusion of gametes in the sexual life cycle
- ▶ D3.1.3 Differences between male and female sexes in sexual reproduction
- ▶ D3.1.4 Anatomy of the human male and female reproductive systems
- ▶ D3.1.5 Changes during the ovarian and uterine cycles and their hormonal regulation
- ▶ D3.1.6 Fertilization in humans
- ▶ D3.1.7 Use of hormones in in vitro fertilization (IVF) treatment
- ▶ D3.1.8 Sexual reproduction in flowering plants
- ▶ D3.1.9 Features of an insect-pollinated flower
- ▶ D3.1.10 Methods of promoting cross-pollination
- ▶ D3.1.11 Self-incompatibility mechanisms to increase genetic variation within a species
- ▶ D3.1.12 Dispersal and germination of seeds
- ▶ D3.1.13 Control of the developmental changes of puberty by gonadotropin-releasing hormone and steroid sex hormones (HL only)
- ▶ D3.1.14 Spermatogenesis and oogenesis in humans (HL only)
- ▶ D3.1.15 Mechanisms to prevent polyspermy (HL only)
- ▶ D3.1.16 Development of a blastocyst and implantation in the endometrium (HL only)
- ▶ D3.1.17 Pregnancy testing by detection of human chorionic gonadotropin secretion (HL only)
- ▶ D3.1.18 Role of the placenta in foetal development inside the uterus (HL only)
- ▶ D3.1.19 Hormonal control of pregnancy and childbirth (HL only)
- ▶ D3.1.20 Hormone replacement therapy and the risk of coronary heart disease (HL only)

Differences between sexual and asexual reproduction

◆ **Asexual reproduction**: reproduction not involving gametes and fertilization.

◆ **Sexual reproduction**: reproduction involving the production and fusion of gametes.

There are two types of reproduction: **asexual** and **sexual**. In asexual reproduction there are no gametes (e.g. eggs and sperm) and no fertilization. Eukaryotic cells undergo mitosis (see Chapter D2.1) and a structure which breaks away from the main body is formed: this structure grows into a new creature. All offspring produced by asexual reproduction are genetically identical.

Single-celled organisms tend to reproduce asexually as they cannot have specialized reproductive organs. A yeast cell, for example, simply divides in two to reproduce (Figure D3.1.1). Single-celled protists also divide asexually, such as *Amoeba*.

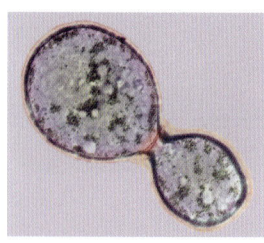

■ **Figure D3.1.1** A yeast cell reproducing asexually – a process known as budding

■ **Figure D3.1.2** Asexual reproduction in hydra

Examples of asexual reproduction in plants include the production of runners (for example, in strawberries) that grow into clones of the parent plant, bulbs that can divide and produce clones, and tubers (such as the potato) that can split to produce several genetically identical new plants.

Some animals can reproduce asexually, for example many insects such as greenfly, and aquatic hydra. Hydra reproduce asexually by budding off a daughter polyp (as seen in Figure D3.1.2). The mouth of the hydra (upper centre) is surrounded by tentacles that it uses to capture small items of food that float past on the current.

Sexual reproduction is the production of offspring from two parents using gametes (e.g. eggs and sperm). The cells of the offspring have two sets of chromosomes (one from each parent). Sexual reproduction involves two stages:

- meiosis – the special cell division that makes gametes with half the number of chromosomes (one set) (Chapter D2.1)
- fertilization – the fusion of male and female gametes to form a zygote (back to two sets of chromosomes).

In sexual reproduction, DNA from one generation is passed to the next by gametes. Sexual reproduction is a source of genetic variation, when DNA from a mother is 'shuffled' with DNA from a father. It involves the random fusion of gametes, which also leads to variation (any sperm can fertilize the egg). Variation is therefore produced by combining genes in different combinations in the gametes and by random fusion of gametes.

Sexual reproduction is a much more complex process than asexual reproduction. Some animals reproduce through external fertilization, where males and females release their gametes into the water. Many species of fish do this – the process is called spawning. Most animals on land reproduce through internal fertilization, where males ejaculate sperm into the bodies of the females.

By reproducing sexually, organisms generate variation in a population, which is an advantage if the environment is changing and they need to adapt to these changes. Or they can reproduce asexually when the conditions are constant and all offspring can be identical. For example, aphids (greenfly) reproduce asexually in spring and summer, when conditions are predictable and stable, and sexually in autumn, creating variation in offspring to allow populations to adapt to changing conditions.

> ● **Top tip!**
>
> While reproduction ensures continuity of a species though time, with sexual reproduction the genetic code of a species is passed on to the next generation with modification derived from meiosis and random fertilization, which produces variation. Variation allows a population to adapt to a changing or changed environment. In asexual reproduction, offspring are genetically identical to their parent, which allows a population to grow rapidly in an unchanging environment.

◆ **Gamete**: a haploid male or female sex cell that is able to unite with another of the other sex to form a zygote.

◆ **Fertilization**: the fusion of male and female gametes to form a zygote.

Link

The process of meiosis is covered in Chapter D2.1, page 656.

Role of meiosis and fusion of gametes in the sexual life cycle

In sexual reproduction, a male and a female **gamete** (specialized sex cells) fuse to form a **zygote**, which then grows into a new individual. In the process of gamete formation, a nuclear division by **meiosis** (page 656) halves the normal chromosome number. Gametes are **haploid** and **fertilization** restores the **diploid** number of chromosomes (Figure D3.1.3).

Without the reductive nuclear division in the process of meiosis, the chromosome number would double in each generation. Remember, the offspring produced by sexual reproduction are unique, in complete contrast with offspring formed by asexual reproduction.

Common mistake

The chromosome numbers shown in Figure D3.1.3 are only for *Homo sapiens*. Other species have different numbers of chromosomes. Defining haploid cells as having 23 chromosomes is incorrect outside the context of human reproduction.

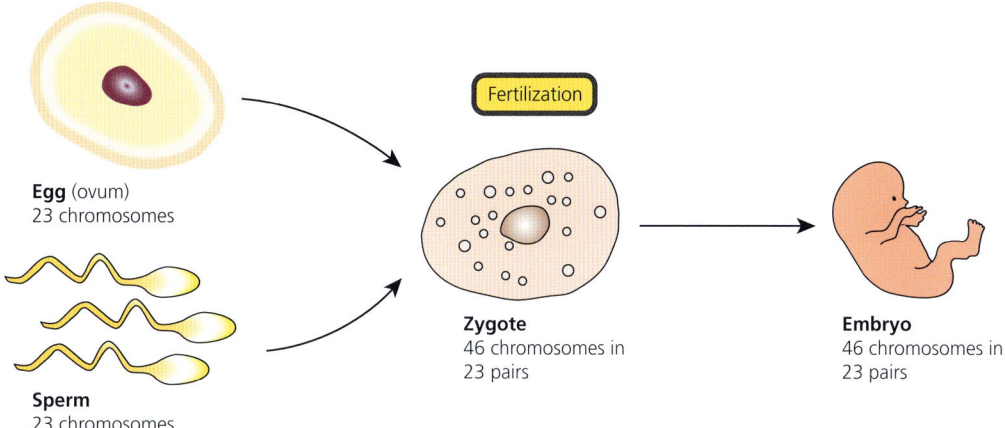

■ **Figure D3.1.3** Meiosis maintains the number of chromosomes in the cells of offspring, preventing the doubling of the number of chromosomes that would occur otherwise

During meiosis, paternal and maternal chromosomes are randomly assorted into new haploid cells (page 110). In this way, meiosis breaks up parental combinations of alleles. Gametes, due to random assortment and crossing over, have a unique sequence of alleles that were not present in either parent (see page 660).

The fusion of gametes is random – any male gamete can fertilize the egg, and the egg itself has a unique set of alleles not present in any other egg. The fusion of gametes is known as fertilization. This process produces new combinations of alleles that are unique to the new organism.

Differences between male and female sexes in sexual reproduction

Male and female gametes (sex cells) are adapted to their specific roles in reproduction (Chapter B2.3, page 271). The male gamete travels to the female gamete, so it is smaller, with fewer energy reserves than the egg. For at least one male gamete to reach the egg, large numbers must be produced – another reason for its small size. Female gametes are fewer and larger in size to provide nutrients and energy for the growing embryo.

In the animal kingdom, because females generally produce relatively few eggs compared to sperm, they tend to choose their mate carefully and not have multiple partners. In mammals, once pregnant, the female feeds and nurtures the growing offspring; there is no benefit of sexual intercourse with other males, and there are periods when she cannot carry out further reproduction. The reproductive strategy of a male is different, because of the large number of sperm produced and the ability to fertilize eggs from multiple females. Males can have several mates and reproduce continually throughout life.

Top tip!

Make sure you can draw diagrams of the male-typical and female-typical reproductive systems, and annotate them with names of structures and functions.

Anatomy of the human male and female reproductive systems

Organisms that carry out sexual reproduction have highly specialized reproductive organs to ensure the gametes from the male can reach the gametes of the female. In mammals, as in most of the vertebrates, the sexes are separate (unisexual individuals). In the body, the reproductive and urinary (excretory) systems are closely bound together, especially in the male, so biologists refer to these as the urinogenital system. Here, just the human reproductive systems are considered (Figures D3.1.4 and D3.1.5).

D3.1 Reproduction

◆ **Testis (plural, testes)**: site of sperm production; produces testosterone.

◆ **Sperm (spermatozoa)**: motile male gamete.

◆ **Sperm duct**: tube carrying sperm from the epididymis to the urethra.

◆ **Seminal fluid (semen)**: male reproductive fluid, containing spermatozoa in a liquid that supports the sperm.

◆ **Prostate gland**: releases alkaline secretion that neutralises any urine left in the urethra and aids sperm mobility.

◆ **Urethra**: tube that runs from the bladder to the outside of the body; in male vertebrates also conveys semen.

◆ **Penis**: an external male organ used for urination and sexual intercourse.

Male reproductive system

The male reproductive system has the following anatomy:
- Two **testes** (singular, testis) are situated in the scrotal sac (**scrotum**), hanging outside the main body cavity; this allows the testes to be at the optimum temperature for sperm production, 2–3 °C lower than the normal body temperature. As well as producing the male gametes, **spermatozoa** (singular, spermatozoon) or **sperm**, the testes also produce the male sex hormone, **testosterone**; the testes are, therefore, also endocrine glands (page 493).
- The **epididymis** stores the sperm and the **sperm ducts** carry them in a fluid, called **seminal fluid (semen)**, to the outside of the body during a process called ejaculation.
- Ducted or exocrine glands secrete the nutritive seminal fluid (made of alkali, proteins and fructose) in which the sperm are transported; these glands include the **seminal vesicles** and **prostate gland**. Seminal fluid (semen) therefore contains sperm as well as fluid from the seminal vesicles and prostate gland.
- The **urethra** is a duct that carries semen during an ejaculation (and urine during urination) to the outside.
- The **penis** contains spongy erectile tissue that can fill with blood when the male is sexually stimulated. This causes the penis to enlarge, lengthen and become rigid, in a condition known as an **erection**. The erect penis penetrates the vagina in sexual intercourse.

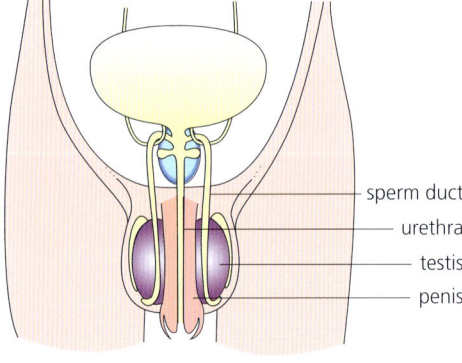

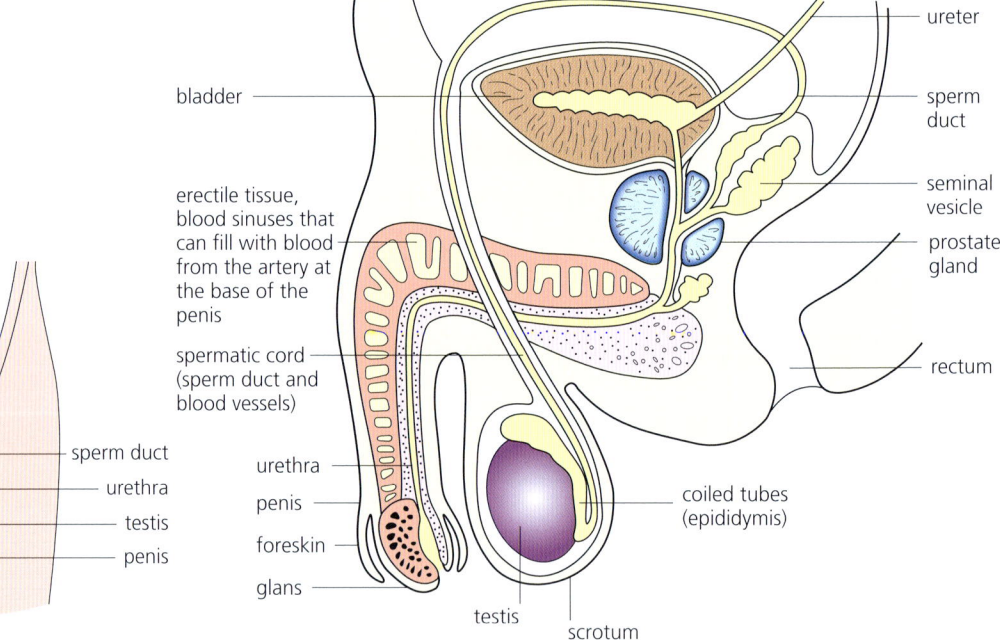

■ Figure D3.1.4 The male urinogenital system

◆ **Ovary**: female reproductive organ in which the female gametes are formed; gland which secretes oestradiol and progesterone.

◆ **Oviduct**: tube connecting ovary to uterus; site of fertilization.

Female reproductive system

The female reproductive system has the following anatomy:
- The **ovaries** are held near the base of the abdominal cavity. As well as producing the female gametes, **ova** or **egg cells**, the ovaries are also endocrine glands, secreting the female sex hormones **oestradiol** (which used to be known as oestrogen) and **progesterone**.
- A pair of **oviducts** (also known as fallopian tubes) extend from the uterus and open as funnels close to the ovaries. The oviducts transport egg cells and are the site of fertilization.

- **Uterus**: the organ in which the embryo develops in female mammals.
- **Endometrium**: the lining of the uterus.
- **Menstruation**: shedding of the endometrium from the uterus.
- **Vagina**: muscular canal leading from uterus to outside of body.
- **Cervix**: ring of muscles at neck of uterus.

- The **uterus**, which is about the size and shape of an inverted pear, has a thick muscular wall and an inner lining of mucous membrane that is richly supplied with arterioles. This lining, called the **endometrium**, undergoes regular change in, on average, a 28-day cycle. The lining is built up each month in preparation for implantation and early nutrition of a developing embryo, should fertilization occur. If it does not occur, the endometrium disintegrates and **menstruation** begins.
- The **vagina** is a muscular tube that can enlarge to allow entry of the penis and exit of a baby at birth. The vagina is connected to the uterus at the **cervix** and it opens to the exterior at the **vulva**.

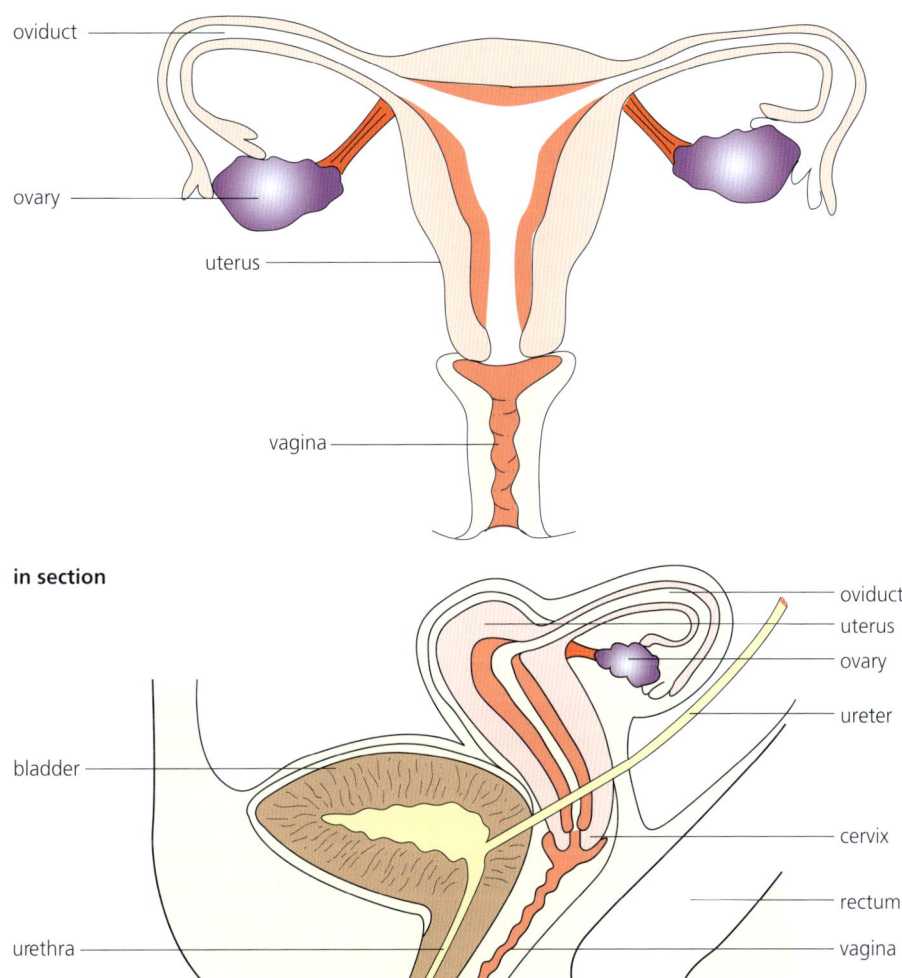

■ Figure D3.1.5 The female urinogenital system

Top tip!
Diagrams drawn as a side view tend to be better in terms of proportions and relative positions of the different structures.

Common mistake
When drawing the reproductive system, it is important to ensure that connections between the parts of the reproductive system are correct. In the male reproductive system, the distances between and connections of the sperm duct, prostate gland and urethra need to be drawn correctly, for example. In the female reproductive system, oviducts need to lead to the space within the uterus, not into the wall of the uterus.

Changes during the ovarian and uterine cycles and their hormonal regulation

> **Concept: Change**
>
> Changes in the blood concentration of four hormones, controlled by feedback loops, coordinate the menstrual cycle.

◆ **Follicle-stimulating hormone (FSH)**: hormone that stimulates the growth of ovarian follicles in the ovary.

◆ **Luteinizing hormone (LH)**: hormone that stimulates ovulation and corpus luteum formation.

◆ **Menstrual cycle**: monthly cycle of ovulation and menstruation in human females.

◆ **Ovarian cycle**: the monthly changes that occur to ovarian follicles leading to ovulation and the formation of a corpus luteum.

◆ **Uterine cycle**: cycle of changes to the uterus lining (approximately 28 days).

In the female reproductive system:
- The secretion of **oestradiol** and **progesterone** is cyclical, rather than at a steady rate. Together with **follicle-stimulating hormone (FSH)** and **luteinizing hormone (LH)**, the changing concentrations of all four hormones bring about a repeating cycle of changes that we call the menstrual cycle.
- The **menstrual cycle** consists of two cycles: the **ovarian cycle** taking place in the **ovaries** and the **uterine cycle** in the **uterine lining**. The ovarian cycle is concerned with the monthly preparation and shedding of an egg cell from an ovary, while the uterine cycle is concerned with the build-up of the lining of the uterus. 'Menstrual' means 'monthly'; the combined cycles take, on average, 28 days (Figure D3.1.6).

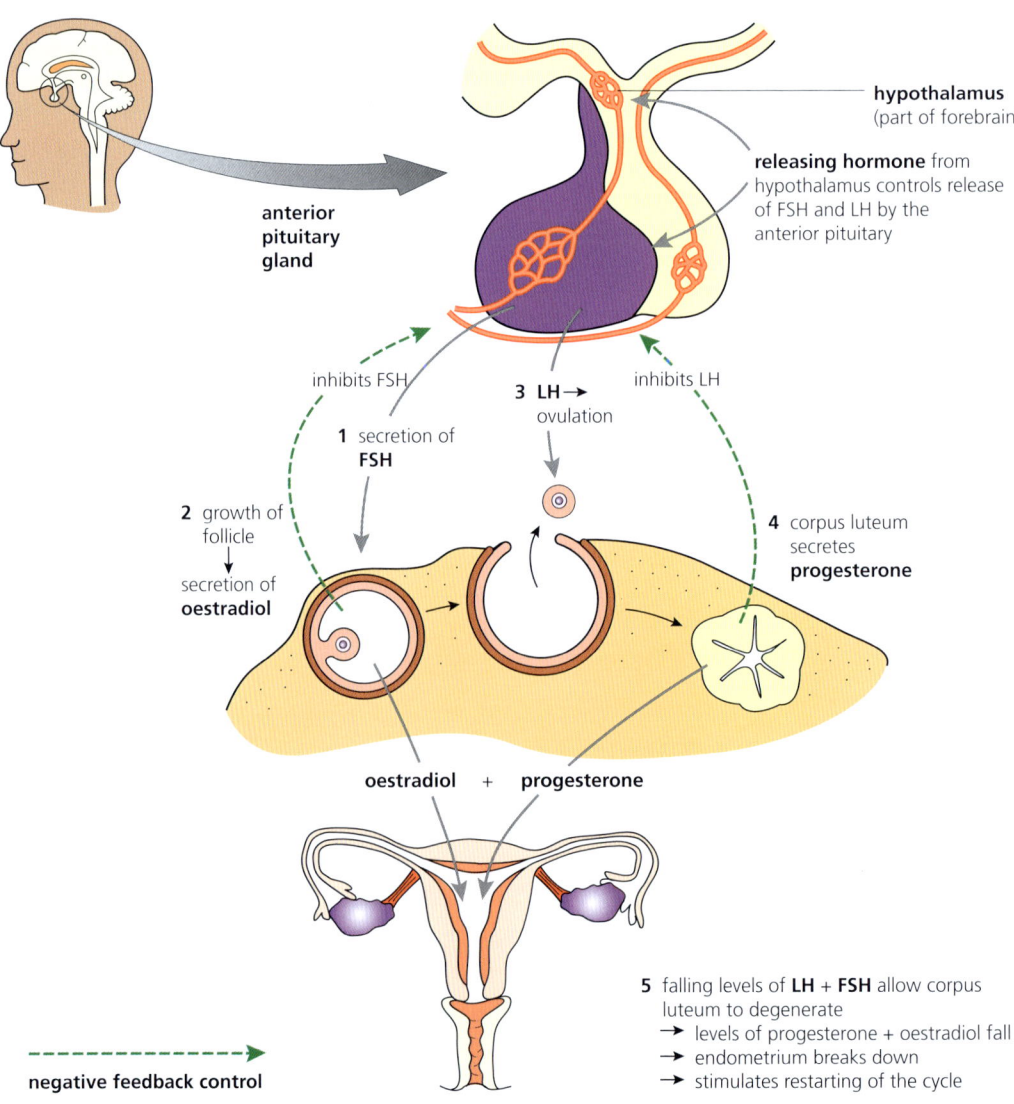

■ Figure D3.1.6 Hormone regulation of the menstrual cycle

Link
Negative feedback is also covered in the following chapters: C2.1 (HL only), page 465; C4.1, page 557, and D3.3, page 761. Positive feedback is also covered in the following chapters: C3.1, page 516; D3.1, page 723, and D4.3, page 825.

◆ **Ovarian follicle**: a fluid-filled spherical sac that contains and nourishes an immature egg, or oocyte.

◆ **Ovulation**: release of oocyte (egg) from ovary.

◆ **Corpus luteum**: a hormone-secreting structure that develops from an ovarian follicle after an oocyte has been discharged. It degenerates after a few days unless pregnancy has begun.

Top tip!
The graph in Figure D3.1.7 on page 700, shows the various hormone changes during the menstrual cycle. Can you explain what each hormone does, what feedback mechanisms (negative and positive) are involved and how they operate?

The menstrual cycle is controlled by **negative** and **positive feedback** mechanisms involving ovarian and pituitary hormones. The changing concentrations of all four hormones bring about a repeating cycle of changes.

By convention, the start of the cycle is taken as the first day of menstruation (bleeding), which is the shedding of the endometrium lining of the uterus. The steps, also summarized in Figure D3.1.7, are as follows.

1. **FSH** is secreted by the pituitary gland and stimulates the development of several immature egg cells (in primary follicles) in the ovary. Only one will complete development into a mature egg cell (now in the **ovarian follicle**).

2. The developing follicle then secretes **oestradiol**. Oestradiol has two targets:
   a. In the uterus it stimulates the build-up of the endometrium, the lining, for a possible implantation of an embryo should fertilization take place.
   b. Leads to an increase in FSH receptors in the follicles, increasing oestradiol production further (this is an example of **positive feedback**).

3. The concentration of oestradiol continues to increase to a peak value just before the mid-point of the cycle. When oestradiol reaches its highest level, it inhibits further secretion of FSH from the pituitary gland. This prevents the possibility of further follicles being stimulated to develop (an example of **negative feedback**).

4. The high level of oestradiol stimulates the secretion of **LH** by the pituitary gland. LH stimulates **ovulation** (the shedding of the mature egg cell from the ovarian follicle and it being released from the ovary) on day 14 of the cycle.

   As soon as the ovarian follicle has discharged its egg, LH also stimulates the conversion of the vacant follicle into an additional temporary gland, called a **corpus luteum** (plural, corpora lutea).

5. The corpus luteum secretes **progesterone** and, to a lesser extent, oestradiol. Progesterone has two targets:
   a. In the uterus, it continues the build-up of the endometrium, further preparing for a possible implantation of an embryo should fertilization take place.
   b. In the pituitary gland, it inhibits further secretion of LH, and also of FSH; this is a second example of **negative feedback control**.

6. The levels of FSH and LH in the bloodstream now rapidly decrease. Low levels of FSH and LH allow the corpus luteum to degenerate. As a consequence, the levels of progesterone and oestradiol also fall. Soon the levels of these hormones are so low that the extra lining of the uterus is no longer maintained. The **endometrium breaks down** and is lost through the vagina in the first five days or so of the new cycle. Falling levels of progesterone again cause the secretion of FSH by the pituitary.

7. A new cycle is under way.

8. **If the egg is fertilized** (the start of a pregnancy), then the developing embryo itself immediately becomes an endocrine gland, secreting a hormone that circulates in the blood and maintains the corpus luteum as an endocrine gland for at least 16 weeks of pregnancy. When, eventually, the corpus luteum does break down, the **placenta** takes over as an endocrine gland, secreting oestradiol and progesterone. These hormones continue to prevent ovulation and maintain the endometrium.

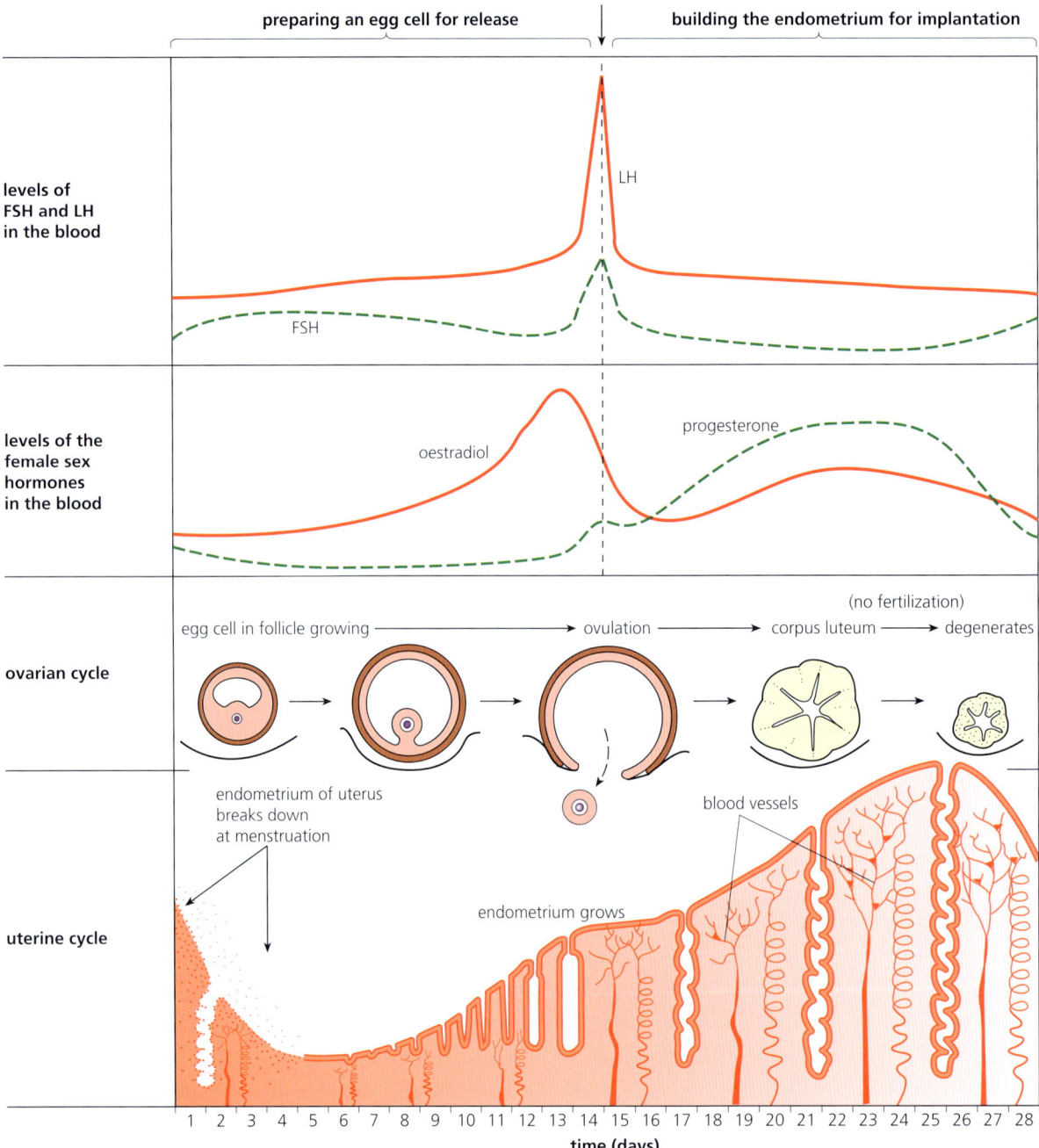

■ Figure D3.1.7 Changing levels of hormones in the menstrual cycle

1 **Identify** the critical hormone changes that respectively trigger ovulation and cause degeneration of the corpus luteum.

Common mistake

The roles of FSH and LH are sometimes confused. Make sure you know the differences between the two. The two pituitary hormones have distinct roles and each of the different effects needs to be discussed and explained. For example, LH promotes secretion of oestradiol by cells in the developing follicle and stimulates ovulation. Follicle development itself is stimulated by FSH only. The LH surge is a good predictor of ovulation for someone wanting to conceive because LH stimulates ovulation.

ATL D3.1A

The change in the mean diameter of follicles and the subsequent corpora lutea (see Figure D3.1.23, page 716) in the ovary over a 40-day period is recorded in the table below.

Days	0	2	4	6	8	10	12	14	16	18	20	22	24	26	28	30	32	34	36	38	40
Follicles/mm																					
	1.5	1.8	2.0	2.5	3.0	3.5	4.0	4.1	4.3	8.0	11.0	6.0									
Corpora lutea/mm																					
												6.0	8.0	9.0	10.0	9.5	9.0	8.4	7.2	6.4	5.0
Next batch of follicles/mm																					
									1.5		2.0			10.0		4.0		4.3			11.0

1. Using an Excel spreadsheet, plot a graph to show the change in mean diameter in follicles and corpora lutea over the 40-day cycle.
2. How do we know, from the data given, that fertilization did not occur during these 40 days?
3. When was fertilization most likely to happen, if insemination had occurred?
4. Draw a labelled diagram of the follicle at the time of fertilization.
5. What is the role of the corpora lutea?
6. a What hormones trigger the growth and development of follicles on days 0 and 20?
   b Where are these hormones released from?

Fertilization in humans

At fertilization, the sperm's cell membrane fuses with an egg cell membrane (see Figure B2.3.5, page 261). The sperm nucleus enters the egg but the tail and mitochondria remain outside the egg and are ultimately destroyed. All the mitochondria in a cell therefore derive from the egg.

Following the fusion of the two cells, the nuclear membranes of sperm and egg nuclei break down. The newly joined homologous pairs of chromosomes, one set from the egg and one from the sperm, condense and participate in a joint mitosis to produce two diploid nuclei.

In mammals, including humans, fertilization – the fusion of male and female gametes to form a **zygote** – is internal. It occurs in the upper part of the oviduct.

Use of hormones in IVF treatment

Natural fertilization and the production of a zygote requires both partners to be fertile, i.e. able to produce gametes and, in the case of the male, deliver them to the female. Not all partners, however, are fertile. Either the male or female, or both, may be infertile, due to a number of different causes (Table D3.1.1).

■ **Table D3.1.1** Causes of infertility

In males, infertility may be due to:	In females, infertility may be due to:
failure to achieve or maintain an erect penis	conditions of the cervix that cause death of sperm
structurally abnormal sperm	conditions in uterus that prevent implantation of embryo
sperm with poor mobility	eggs that fail to mature or be released
short-lived sperm	blocked or damaged oviducts, preventing egg from reaching sperm
too few sperm	
a blocked sperm duct, preventing semen from containing sperm	

D3.1 Reproduction

◆ **In vitro fertilization (IVF)**: medical procedure in which an egg is fertilized by sperm in a test tube (in vitro = in glass).

In some cases, a couple's infertility may be overcome by the process of fertilization of eggs outside the body – **in vitro fertilization (IVF)**. The key step in IVF is the successful removal of sufficient eggs from the ovaries. To achieve this, normal menstrual activity is temporarily suspended with hormone-based drugs. Then, the ovaries are induced to produce a large number of eggs simultaneously, at a time controlled by doctors. In this way, the optimum moment to collect the eggs can be known accurately.

The steps of in vitro fertilization are as follows:

1 **Down-regulation** is the first step in IVF, shutting down of the menstrual cycle by stopping secretion of the pituitary and ovarian hormones. The process takes about two weeks and allows better control of superovulation. Down-regulation is done with a drug, commonly in the form of a nasal spray.
2 The next step is **superovulation**. High doses of FSH are injected over approximately a 10-day period to stimulate the development of multiple follicles.
3 When follicles reach 15–20 mm in diameter, an injection of hCG (human chorionic gonadotropin hormone, normally secreted by the developing embryo and which maintains the corpus luteum) is given to start the **maturation** process; after approximately 36 hours, follicles (typically 8–12) are collected from the ovaries under a general anaesthetic.
4 Prepared eggs (i.e. removed from the follicles) are combined with sperm in sterile conditions.
5 Successfully fertilized eggs are then incubated before implantation.
6 For approximately two weeks before implantation the woman takes progesterone (which maintains the endometrium) to aid implantation. This treatment is continued until pregnancy testing and, if positive, until 12 weeks of gestation.
7 Embryos at the eight-cell stage may be placed in the uterus. If one (or more) embed there, then a normal pregnancy may follow.

2 **Explain** what in vitro and in vivo mean.

Because the natural success rate of implantation is around 40%, usually two or three blastocysts (growing fertilized eggs) are implanted: as a consequence, the chance of IVF treatment leading to multiple pregnancies is high.

The first 'test-tube baby' was born in 1978. Today, the procedure is regarded as a routine one.

Sexual reproduction in flowering plants

Flowering plants contain their reproductive organs in the flower. Flowers are often **hermaphrodite** structures, carrying both male and female parts. Note: reproduction in flowering plants is sexual, even if a plant species is hermaphroditic.

◆ **Ovule**: in the flowering plant flower, the structure in an ovary which, after fertilization, grows into the seed.

◆ **Pollen**: microspore produced in anthers (and male cones) containing male gamete(s).

◆ **Pollination**: transfer of pollen from anther to stigma.

◆ **Pollen tube**: grows out of a pollen grain attached to a stigma, and down through the style tissue to the embryo sac.

The reproductive structures of plants are very different to those of animals. The ovum is contained within a structure called an **ovule**, which also contains a store of nutrients. The ovule, once the ovum has been fertilized, becomes the seed of the plant. The male reproductive cells are **pollen**, which travel to the female reproductive part in the same plant, or from plant to plant, either carried by insects (e.g. bees) or on the wind.

Pollination is the transfer of pollen from a mature anther to a receptive stigma. When the pollen travels to the female organs of the flower, pollination occurs. The pollen travels down a structure in the flower to allow it access to the ovule and, ultimately, the ovum, where fertilization occurs to produce an embryo.

Fertilization (fusion of male and female gametes to form a zygote) in flowering plants can occur only after an appropriate pollen grain has landed on the **stigma** and germinated. The pollen grain produces a **pollen tube**, which grows down between the cells of the style into the ovule (Figure D3.1.8).

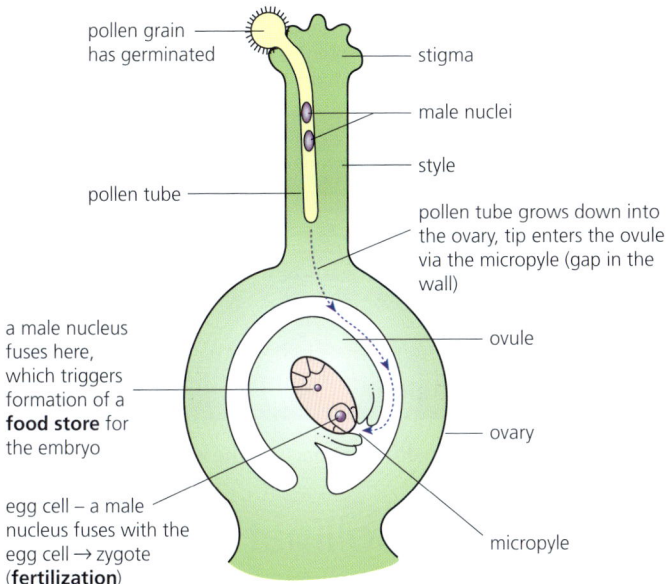

■ Figure D3.1.8 Fertilization in a flowering plant

Labels (clockwise): pollen grain has germinated; stigma; male nuclei; style; pollen tube grows down into the ovary, tip enters the ovule via the micropyle (gap in the wall); ovule; ovary; micropyle; egg cell – a male nucleus fuses with the egg cell → zygote (**fertilization**); a male nucleus fuses here, which triggers formation of a **food store** for the embryo; pollen tube.

Incidentally, the pollen tube delivers *two* male nuclei. One of these male nuclei fuses with the egg nucleus in the embryo sac, forming a diploid zygote. The other fuses with another nucleus, triggering formation of the food store for the developing embryo. This 'double fertilization' is unique to flowering plants.

Features of an insect-pollinated flower

The parts of flowers occur in rings or whorls, attached to the swollen tip of the flower stalk, called the receptacle. The **sepals** (collectively, the calyx) enclose the flower in the bud, and are usually small, green and leaf-like. The **petals** (collectively, the corolla) are often coloured and conspicuous, and may attract insects or other small animals. The **stamens** are the male parts of the flower, and consist of **anthers** (housing pollen grains) and the **filaments** (stalk). The **carpels** are the female part of the flower. There may be one or many, free-standing or fused together. Each carpel consists of an **ovary** (containing ovules), a **stigma** (a surface for receiving pollen) and a connecting **style**.

The buttercup flower is shown in Figure D3.1.9, and other flowers that are common in different parts of the world are shown in Figure D3.1.10.

3 Explain the differences between pollination and fertilization in the flowering plant.

◆ **Sepal**: green structures that cover the flower while it is developing.

◆ **Petal**: large, colourful structure that attracts pollinators to the flower.

◆ **Stamen**: the male part of the flowers, consisting of the anther and filament.

◆ **Anther**: produces pollen, which contains the male gamete.

◆ **Filament**: structure that holds up the anther, presenting it to visiting animals.

◆ **Carpel**: the female part of the flower, made from the stigma, style and ovary.

◆ **Stigma**: the part of the female reproductive organs where the pollen lands.

◆ **Style**: holds up the stigma and is the path down which the pollen tube grows, carrying the pollen nuclei to the ovum.

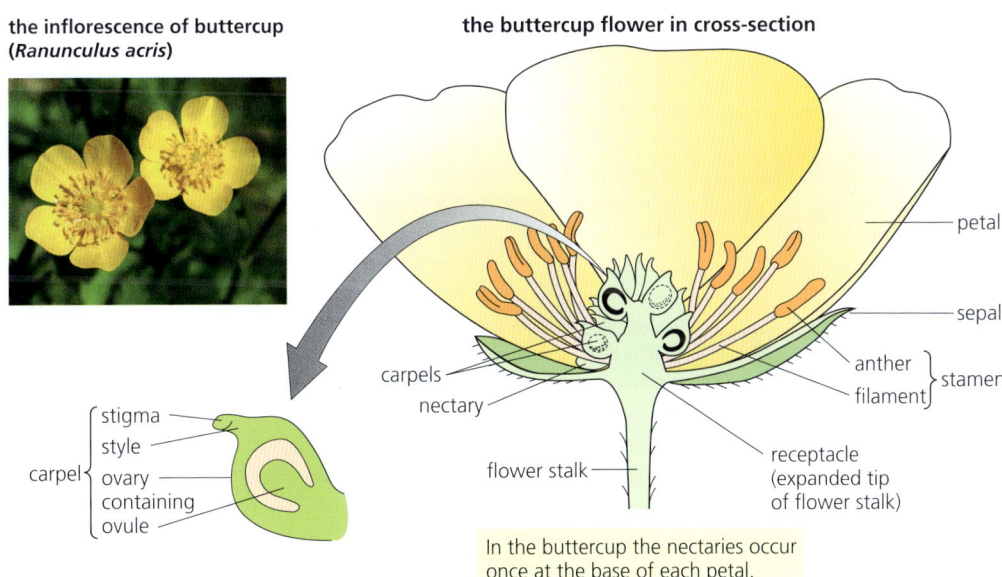

■ Figure D3.1.9 The buttercup (*Ranunculus*) flower

In the buttercup the nectaries occur once at the base of each petal.

> ## Tool 1: Experimental techniques
>
> ### Drawing an annotated diagram from observation
>
> Select a flower and create a half-flower drawing of your specimen. You will have to cut the flower in half, in a longitudinal section (i.e. down the longest length of the flower), to do this. Annotate your drawing and add the magnification.

D3.1 Reproduction

ATL D3.1B

Choose an insect-pollinated flower that is available near your school or college. Study its structure and then work out its pollination mechanism. What insects can visit and benefit?

Bougainvillea rosenka Native of tropical and sub-tropical South America. The flowers are small but surrounded by brightly coloured leaves (bracts).

Hibiscus syriacus Native of warm temperate, tropical and sub-tropical regions throughout the world. The large, trumpet-shaped flowers have five petals.

■ Figure D3.1.10 Other animal-pollinated flowers common in different parts of the world

Methods of promoting cross-pollination

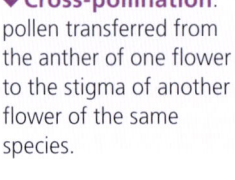

◆ **Cross-pollination**: pollen transferred from the anther of one flower to the stigma of another flower of the same species.

Pollen may come from flowers on a different plant of the same species, which is referred to as **cross-pollination**. Transfer of pollen is often brought about by animals (Figure D3.1.11). Pollinators include insects, such as butterflies or bees. In other flowers, it may be bird or bat visitors that unwittingly carry out pollination. The pollinator is typically attracted by colour or scent (or both), and is rewarded by a sugar solution, called nectar, and pollen, which usually form a key part of the diet. In return, they accidentally transfer pollen between flowers and between plants. Thus, there is a mutualistic relationship between pollinator and plant in plant sexual reproduction. Alternatively, pollen may be transferred by the wind or, occasionally, by running water.

■ Figure D3.1.11 Pollinators at work

A species of plant may want to encourage cross-fertilization, i.e. fertilization with another flower rather than with itself (self-pollination). Cross-pollination is achieved by:
- the stamen and stigma maturing at different times (Figure D3.1.12)
- the stigma and anthers at different heights in same flower (Figure D3.1.13)
- separate male and female flowers.

Theme D: Continuity and change – Organisms

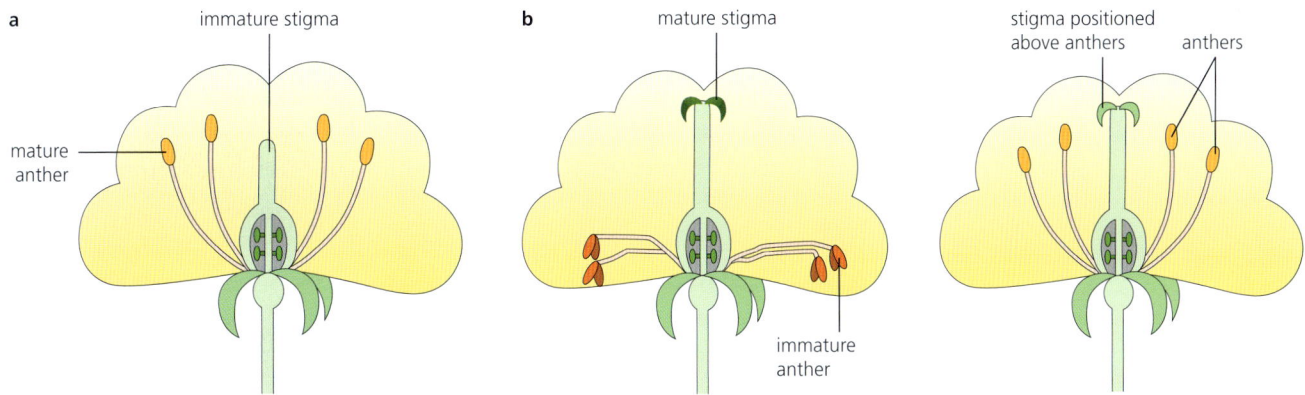

■ **Figure D3.1.12** Anthers and stigma can mature at different times to avoid self-pollination: a) anthers are mature but not the stigma; b) stigma is mature but not the anthers

■ **Figure D3.1.13** Stigma and anthers at different heights to avoid self-pollination

Cross-pollination results in outbreeding.
- Advantage: variation and some genomes more successful than others – good if environment changes (i.e. evolutionary significance).
- Disadvantage: pollen need to find the stigma of other flowers – element of chance (especially in wind-pollinated plants).

> **Common mistake**
>
> It is incorrect to say that pollination delivers pollen to the ovary. Pollination delivers pollen grains to the stigma, from which a pollen tube grows, which delivers pollen nuclei down the style and into the ovary.

> **ATL D3.1C**
>
> The figures in this section, e.g. Figure D3.1.11, show animal-pollinated plants. Find out about wind-pollinated plants. How does the structure of their flowers differ from animal-pollinated plants? Why is this? Draw a diagram of a wind-pollinated flower and compare it with an animal-pollinated flower.

Self-incompatibility mechanisms to increase genetic variation within a species

The male and female reproductive organs of plants are found in the same flower, which can lead to gametes from the same individual coming together at fertilization. This is known as **self-pollination**. Self-pollination results in inbreeding.
- Advantages: preserves good genomes suited to a stable environment.
- Disadvantages: reduction in variation; greater chance of two undesirable recessive alleles coming together; decreases genetic diversity.

◆ **Self-pollination**: when pollen is transferred to the stigma in the same flower.

Genetic mechanisms in many plant species ensure that the male and female gametes fusing during fertilization are from different plants. **Self-incompatibility** refers to the recognition and rejection of pollen by the carpel of the same flower. This mechanism is genetically controlled and inhibits the growth of the pollen tube down the style if the pollen has been produced in the same flower as the style, preventing pollen from fertilizing the ovum. Self-incompatibility prevents inbreeding and promotes outcrossing.

D3.1 Reproduction

Dispersal and germination of seeds

The seed develops from the fertilized ovule and contains an embryo plant and a food store. After fertilization, the following occur:
- The zygote grows by repeated **mitotic** division to produce cells that form an **embryonic plant**, consisting of an embryo root, an embryo stem and either a single cotyledon (seed leaf) or two cotyledons.
- Formation of **stored food reserves** is triggered. In many seeds, the developing food store is absorbed into the cotyledons, rather than remaining as a separate store that is packed round the embryonic plant. For example, this is the case in peas and beans (Figure D3.1.14). Note that the formation of food reserves can only occur if fertilization occurs – in the absence of fertilization, food reserves are not moved into the unfertilized ovule.

As the seed matures, the outer layers of the ovule become the protective seed coat or **testa**, and the whole ovary develops into the **fruit**. Next, the water content of the seed decreases and the seed moves into a dormancy period. In a mature, fully dormant seed, water makes up only 10–15% of seed weight.

ATL D3.1D

The drawing of a broad bean seed in Figure D3.1.14 shows how structure can be recorded, once the seed has been examined. Try this with other seeds, such as a sunflower seed. Does the structure of the seed vary with species or size of plant?

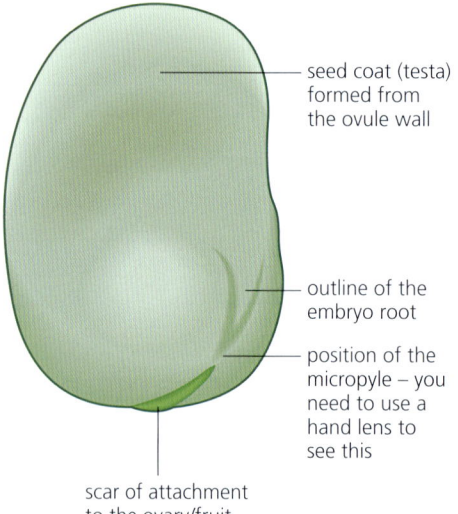 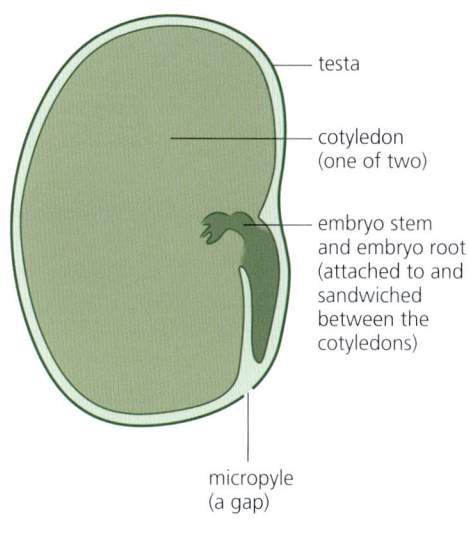

■ Figure D3.1.14 The structure of a broad bean seed

Dispersal of seeds

The seed is a form in which the flowering plant may be dispersed. **Seed dispersal** is the carrying of the seed away from the parent plant. It is needed to ensure that the offspring of a parent plant eventually germinate some distance away. Seed dispersal is a different process to pollination. Pollination is the transfer of pollen grains from the anther to the stigma of flowers, whereas seed dispersal is a mechanism by which a plant's seeds (fertilized ovules) are moved to new areas, away from the parent plant, ready for germination and the establishment of new plants.

The plant structures to aid dispersal that have evolved exploit air currents (wind), passing animals or flowing water to transport seeds (Figure D3.1.15a, b and d). In a few plants, seeds are flung away from the ripening fruit by an explosive mechanism (Figure D3.1.15c). The force that generates the explosion is from turgor pressure within the fruit or due to internal tensions within the fruit. Other plants (such as coconuts) use water for dispersal. All seeds are compact, nutritious and relatively lightweight – in effect, they are food packages to a hungry animal. Many seeds taken for food are dropped and lost, or stored and forgotten. In this way, some seeds are successfully dispersed.

◆ **Seed dispersal**: the carrying of the seed away from the parent plant.

● Top tip!

Seeds need to be dispersed away from the parent plant in order to reduce competition for space, light, nutrients and water.

4 **Distinguish** between seed dispersal and pollination.

■ **Figure D3.1.15** Methods of seed dispersal: a) wind – seed of the South East Asian hardwood species *Dipterocarpus obtusifolius* being dispersed by wind; b) water – coco de mer or sea coconut (*Lodoicea maldivica*), a rare species of palm tree native to the Seychelles archipelago in the Indian Ocean, is distributed by water; it forms the largest seed on Earth; c) mechanical – high-speed photo of the seeds of Himalayan balsam (*Impatiens glandulifera*) being dispersed by an explosive mechanism in the fruit; d) animal – a cedar waxwing (*Bombycilla cedrorum*) eating a mulberry fruit in North America

■ Germination of seeds

◆ **Germination**: the resumption of growth by an embryonic plant in seed or fruit, at the expense of stored food.

Many seeds do not **germinate** as soon as they are formed and dispersed. Such seeds are said to have a dormant period and germinate only when this has elapsed. Dormancy may be imposed within the seed, due to:

- **incomplete seed development** that causes the embryo to be immature, and which is overcome in time
- the **presence of a plant growth regulator** – abscisic acid, for example – that inhibits development, and which only disappears from the seed tissues with time
- an **impervious seed coat** that is eventually made permeable – for example, by abrasion with coarse soil or by the action of micro-organisms
- a **requirement for pre-chilling** under moist conditions before the seed can germinate; some seeds need to be held at or below 5°C for up to 50 days (possibly the equivalent of winter in temperate climates).

Once dormancy is overcome, germination occurs if the following essential external conditions are met (Table D3.1.2):

- **water** uptake has occurred, so that the seed is fully hydrated and the embryo can be physiologically active
- **oxygen** is present at a high enough partial pressure to sustain aerobic respiration; growth demands a continuous supply of metabolic energy in the form of ATP that is best generated by aerobic cell respiration in all the cells
- **a suitable temperature** exists, one that is close to the optimum temperature for the enzymes involved in the mobilization of stored food reserves, the translocation of organic solutes in the phloem, and the synthesis of intermediates for cell growth and development. For example, wheat seeds germinate in the range 1–35°C, and maize in the range 5–45°C.

■ **Table D3.1.2** Conditions for germination

External	Internal
water uptake – hydration of the cytoplasm of cells of the embryo	overcoming of dormancy
ambient **temperature** – within optimum range for enzyme action	production of plant growth regulator(s) by embryo cells to initiate biochemical changes of germination, leading to production of hydrolytic enzymes for mobilization of stored food
oxygen – to sustain aerobic cell respiration	respiration provides ATP for growth and metabolic processes

D3.1 Reproduction

5 **Define** the term *germination*.

6 **Explain** the factors needed for germination.

In germination, food reserves are mobilized (i.e. stored food such as starch is made available to the growing embryo). The steps of germination are summarized in Figure D3.1.16. Note that a particular plant growth substance (known as gibberellic acid, GA) is produced by the cells of the embryo. This growth-promoting substance passes to the food stored in the cotyledons. Here, protein reserves are converted to hydrolytic enzymes that mobilize the stored food reserves. The main event is the production of the enzyme amylase, which hydrolyses starch to maltose. This disaccharide is then hydrolysed to glucose. The resulting soluble sugar (and other compounds) sustain **respiration** and also provide the building blocks for synthesis of the intermediates essential for new cells.

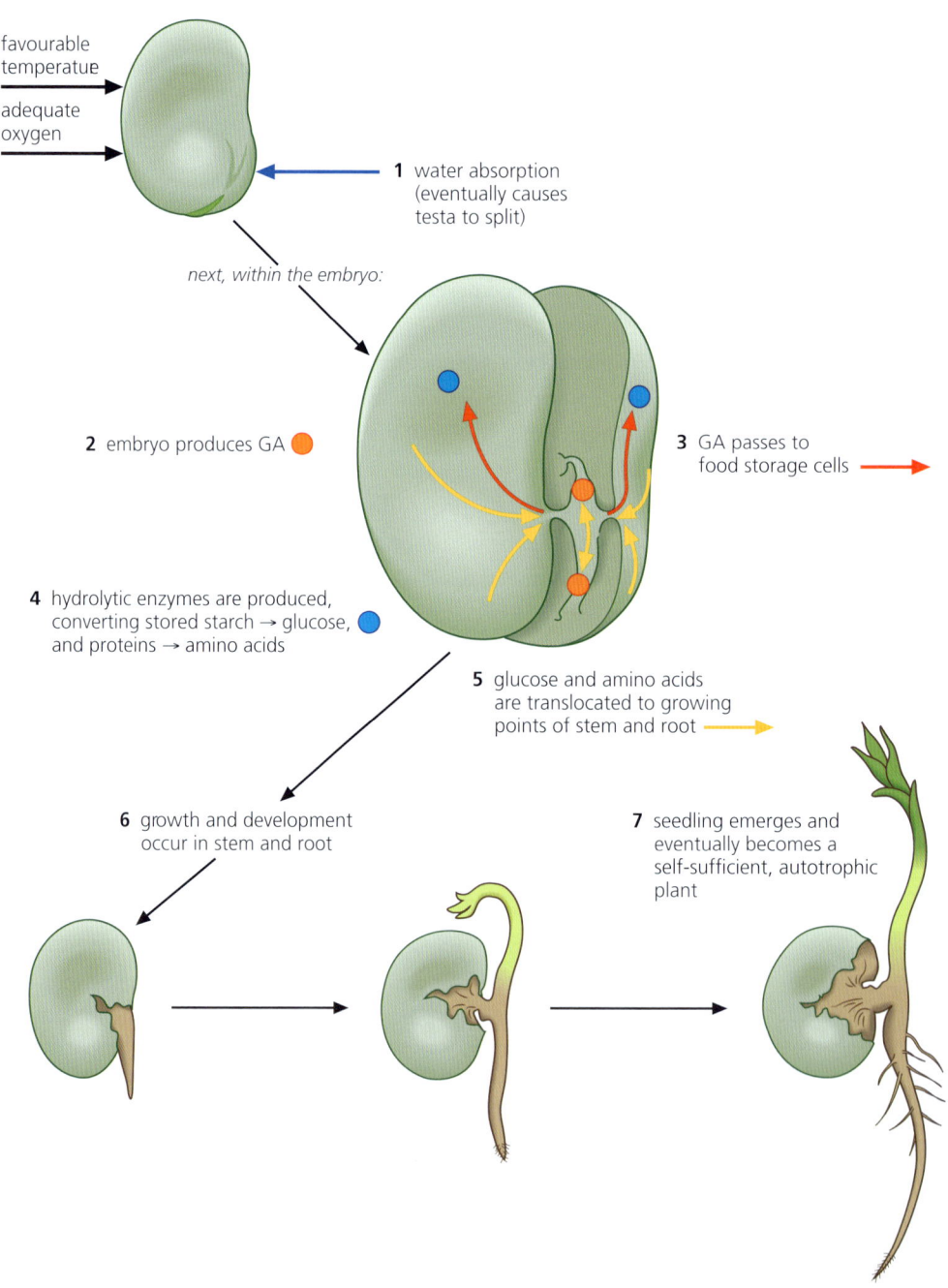

■ Figure D3.1.16 Metabolic events of germination in a starchy seed

Inquiry 1: Exploring and designing

Designing

Design an experiment to investigate the effect of an independent variable on the germination of seeds (the dependent variable). Identify and justify the choice of dependent and control variables.

Samples of seeds can be set to germinate on damp filter paper in a Petri dish, but how many seeds per dish would make an appropriate sample, and how many different species might you test? Given this simple apparatus, you would be able to investigate the effect of light (presence or absence), and perhaps temperature (low and room temperatures – unless you also have access to temperature-controlled cabinets).

More ambitious investigations might look at the effects of intense cold treatment on newly formed seeds, or the effects of brief and prolonged pre-soaking (in effect, the degree of hydration).

Which variables would you keep the same (control variables)? These are variables that could affect your dependent variable. How would you control these variables? For example, keeping the seedlings in a temperature-controlled greenhouse would enable you to keep the temperature the same.

Whatever is investigated, there is the issue of the percentage of seeds in a sample that might be long-term dormant or non-viable.

---

7 An experiment was carried out to investigate the factors needed for germination. Figure D3.1.17 shows the results of the experiment.

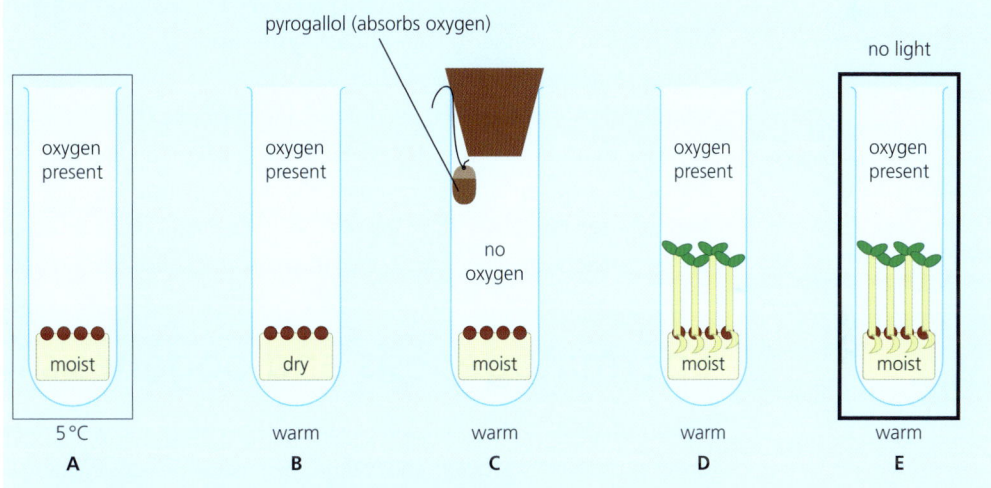

■ **Figure D3.1.17** Experiment investigating the factors needed for germination

   a Analyse the results shown by the experiment.
   b Explain why germination has occurred in tubes D and E but not the others.

Control of the developmental changes of puberty by GnRH and steroid sex hormones

◆ **Puberty**: the stage in life when an individual becomes sexually mature, marking the transition from childhood to adulthood.

During the first 10 years of life, the human reproductive systems remain in a juvenile state. Then, at **puberty**, the secondary sexual characteristics begin to develop. This is a time of growth and development at the beginning of sexual maturation. Now, there is a significant increase in the production of the sex hormones by the gonads – testosterone in cells of the testes, and oestradiol and progesterone in cells of the ovaries. These hormones are chemically very similar molecules; they are manufactured from the steroid cholesterol that is both made in the liver and absorbed as part of the diet. (Steroids are a form of lipid, page 204.)

D3.1 Reproduction

Perhaps the most noticeable effect of the increased secretion of the sex hormones is the stimulation they cause in muscle protein formation and bone growth. Because of this effect, testosterone, oestradiol and progesterone are known as **anabolic steroids** (anabolic means 'build up'). The effects of the female sex hormones are less marked in this respect than those of testosterone in the male. The onset of puberty occurs, on average, about two years earlier in females.

■ Table D3.1.3 Secondary sexual characteristics

Secondary sexual characteristics of a human female	Secondary sexual characteristics of a human male
• maturation of the ovaries and enlargement of the vagina and uterus • development of the breasts • widening of the pelvis • growth of pubic hair and underarm hair • monthly ovulation and menstruation • changes in behaviour associated with a sex drive	• development and enlargement of the testes, scrotum, penis and glands of the reproductive tract • increased skeletal muscle development; enlargement of the larynx, deepening of the voice • growth of pubic hair, underarm hair and body hair • continuous production of sperm and, in the absence of sexual intercourse, occasional erections and the discharge of seminal fluid • changes in behaviour associated with a sex drive

♦ **Gonadotropin-releasing hormone (GnRH)**: hormone that causes the pituitary gland to secrete luteinizing hormone (LH) and follicle-stimulating hormone (FSH).

♦ **Gametogenesis**: the production of sex cells (gametes).

The onset of puberty is triggered by a part of the brain called the **hypothalamus**. Here, production and secretion of **gonadotropin-releasing hormone (GnRH)** causes the nearby pituitary gland (the 'master' endocrine gland) to produce and release two hormones into the blood circulation: **follicle-stimulating hormone (FSH)** and **luteinizing hormone (LH)**. They were named because their roles in sexual development were discovered in the female, although they do operate in both sexes. Their first effects are to enhance secretions of the sex hormones. Then, in the presence of FSH, LH and the respective sex hormones, there follows the development of the secondary sexual characteristics and the preparation of the body for its role in sexual reproduction.

Spermatogenesis and oogenesis in humans

Gametogenesis is the formation of gametes by mitosis first of all, to proliferate cells that will then undergo meiosis. In order to understand the process of gametogenesis in males and females, further detail of the meiotic divisions need to be understood.

■ The process of meiosis

Once started, meiosis proceeds steadily as a continuous process of nuclear division. The steps of meiosis are explained in four distinct phases (**prophase**, **metaphase**, **anaphase** and **telophase**), but this is just for convenience of analysis and description – there are no breaks between the phases in nuclear division. There are two divisions: meiosis I and meiosis II (see Figure D3.1.18). Both follow the same stages. The arrival of homologous chromosomes at opposite poles signals the end of meiosis I. Meiosis II is remarkably similar to mitosis.

The behaviour of the chromosomes in the phases of meiosis is shown in Figure D3.1.18. For clarity, the drawings show a cell with a single pair of homologous chromosomes.

MEIOSIS I

prophase I (early)
Chromosomes begin to condense (shorten and thicken) and become visible with a light microscope.

prophase I (mid)
Homologous chromosomes pair up (becoming bivalents) as they continue to shorten and thicken. Centrioles duplicate.

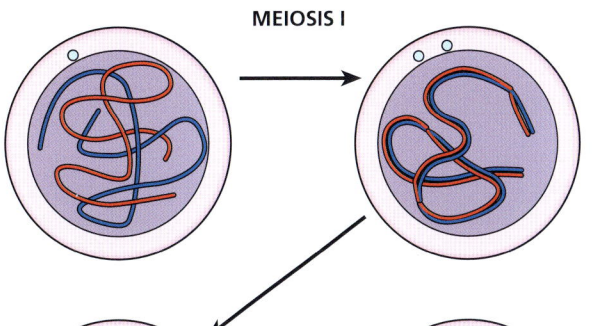

prophase I (late)
Chromosomes can now be seen to consist of chromatids. Sites where chromatids have broken and rejoined, causing crossing-over, are visible as chiasmata.

metaphase I
Nuclear membrane breaks down. Spindle forms. **Bivalents** line up at the equator, attached by centromeres. At the end of metaphase I, members of the bivalents start to repel each other and separate, but are still held together by one or more chiasmata.

anaphase I
Homologous chromosomes separate. Whole chromosomes are pulled towards opposite poles of the spindle, centromere first (dragging along the chromatids). Individual chromatids remain attached by centromeres. Meiosis I has separated homologous pairs of chromosomes, but not sister chromatids.

telophase I
Nuclear membrane re-forms around the daughter nuclei. The chromosome number has been halved. The chromosomes start to decondense. Spindle breaks down. These two cells do not enter interphase, but continue into meiosis II.

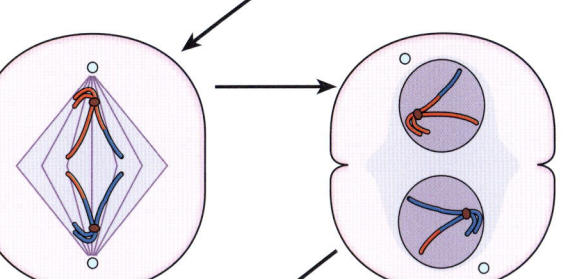

there is no interphase between **MEIOSIS I** and **MEIOSIS II**

MEIOSIS II

prophase II
The chromosomes condense (shorten and re-thicken by coiling). Centrioles duplicate and move to the poles.

metaphase II
The nuclear membrane breaks down and the spindle apparatus reforms (at right angles to the original spindle). The chromosomes attach by their centromere to spindle fibres at the equator of the spindle.

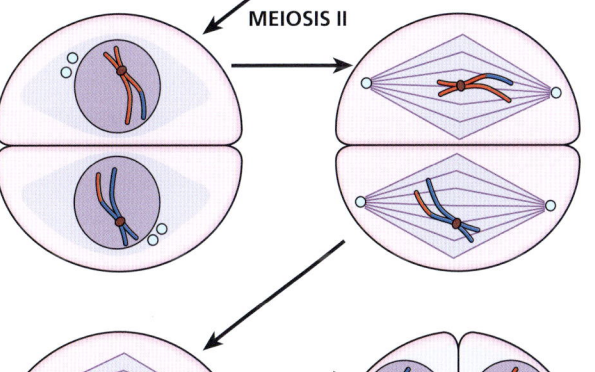

anaphase II
The centromeres divide and the chromatids are pulled to opposite poles of the spindle, centromeres first.

telophase II
The chromatids (now called chromosomes) decondense. The nuclear membrane re-forms. Nucleoli reform. There are now four cells, each with half the chromosome number of the original parent cell.

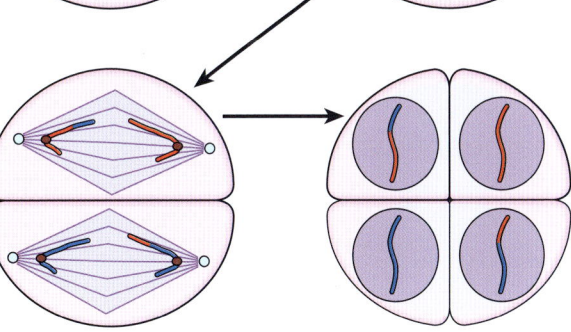

■ **Figure D3.1.18** The stages of meiosis

Gametogenesis

In **gametogenesis**, many gametes are produced, although relatively few of them are ever used in reproduction. The processes of gamete formation in testes and ovaries have a common sequence of phases.

First, there is a **multiplication phase** in which the gamete mother cells divide by mitotic cell division (Figure D2.1.8, page 654). This division is then repeated to produce many cells with the potential to become gametes. Secondly, each developing sex cell undergoes a **growth phase**. Third, and finally, comes the **maturation phase**. This involves meiosis and results in the formation of the haploid gametes.

The products of meiosis I are secondary **spermatocytes** and secondary **oocytes**, and the products of meiosis II are **spermatids** and **ova**. The steps of meiosis are described in Figure D3.1.18.

These phases in sperm and ova production are summarized in Figure D3.1.19.

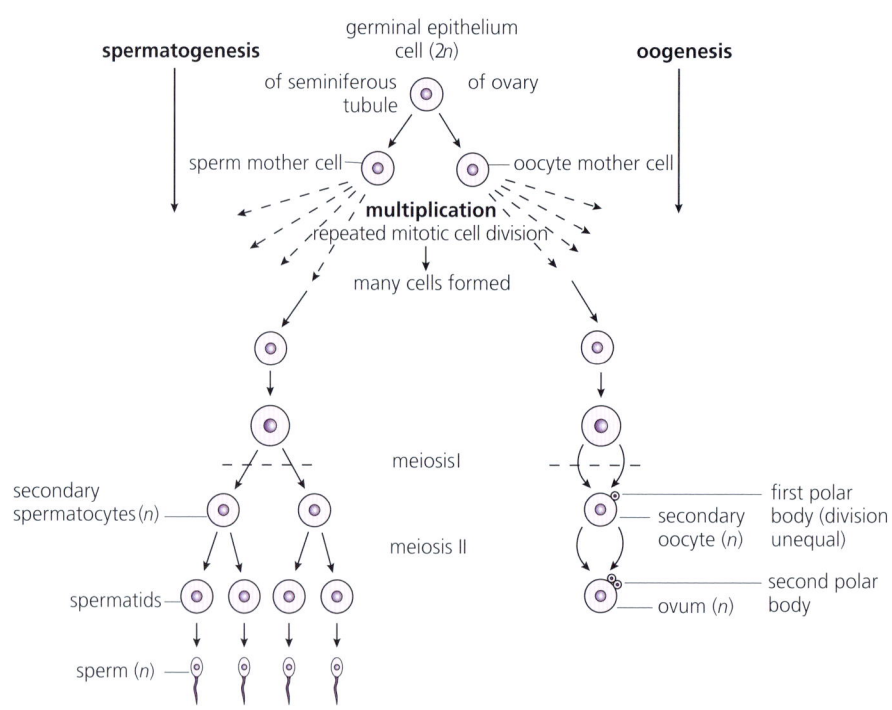

■ **Figure D3.1.19** The phases and changes during gametogenesis

Link
Examples of unequal division of cytoplasm between daughter cells is covered in Chapter D2.1, page 649.

◆ **Spermatogenesis**: the production of sperm in the testes.

◆ **Seminiferous tubule**: elongated tubes in the testes, the site of sperm production.

Figure D3.1.19 shows how gametogenesis results in different numbers of sperm and eggs, and different amounts of cytoplasm in the oocytes, ovum and polar bodies (see also Chapter D2.1, page 649).

Spermatogenesis

The structure and functioning of the testis

In the human foetus, testes develop high on the posterior abdominal wall and migrate to the scrotum in about the seventh month of pregnancy. In the scrotum, the paired testes are held at a temperature 2–3 °C below body temperature after birth. This lower temperature is eventually necessary for sperm production – testes that fail to migrate do not later produce sperm.

Spermatogenesis begins in the testes at puberty and continues throughout life. Each testis consists of many **seminiferous tubules**. These are lined by germinal epithelial cells that divide repeatedly. Tubules drain into a system of channels leading to the epididymis, a much-coiled tube that leads to the **sperm duct**. Between the individual seminiferous tubules is connective tissue containing blood capillaries, together with groups of **interstitial cells**. These latter cells are hormone secreting (the testis is also an endocrine gland). Testes are suspended by a spermatic cord containing the sperm duct and blood vessels (Figure D3.1.4).

♦ **Spermatogonia**: male germ cells (stem cells) which make up the inner layer of the lining of the seminiferous tubules, and give rise to spermatocytes by mitosis.

♦ **Spermatocyte**: cell formed in seminiferous tubules of testes; develops into sperm.

♦ **Spermatid**: cells derived from secondary spermatocytes in the second meiotic division, which then differentiate into mature sperm.

In the seminiferous tubules, the **germinal epithelial cells** are attached to the basement membrane, along with the **nutritive cells**. Cells from the subsequent steps of sperm production (**spermatogonia**, primary **spermatocytes**, secondary **spermatocytes** and **spermatids** – see Table D3.1.4 below) occur lodged in the surface of the **Sertoli cells**. Sertoli cells are large cells in the seminiferous lining that nourish the developing sperm and on which they are dependent until they mature into spermatozoa (Figures D3.1.20 and D3.1.21). Figure D3.1.19 and Table D3.1.4 show the stages typical of nuclear division and the different stages that lead to the development of mature **sperm**.

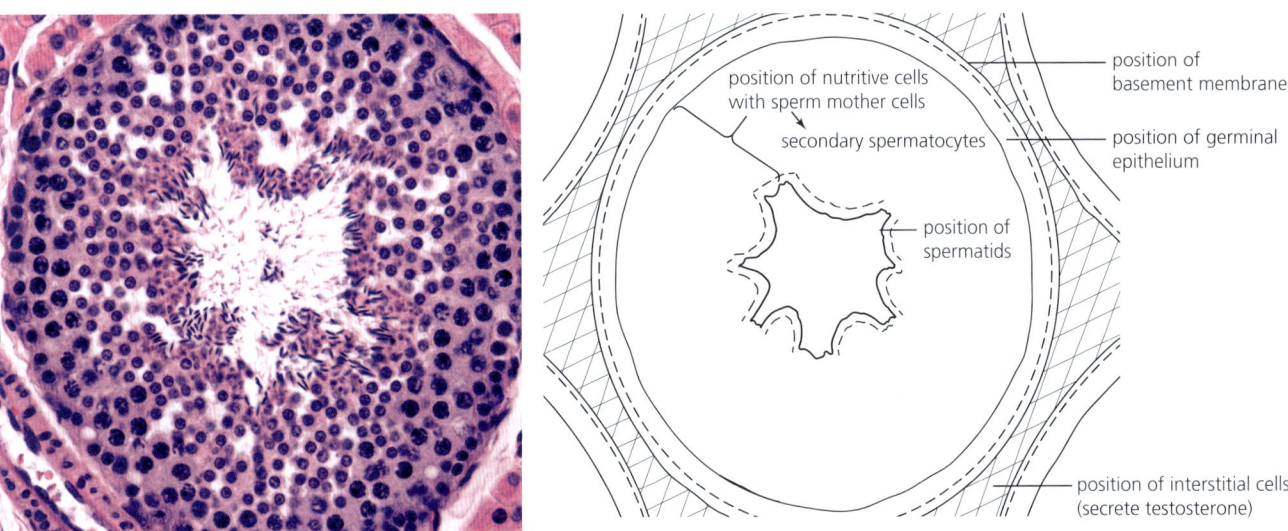

■ **Figure D3.1.20** Photomicrograph of testis tissue in section, and interpretive drawing

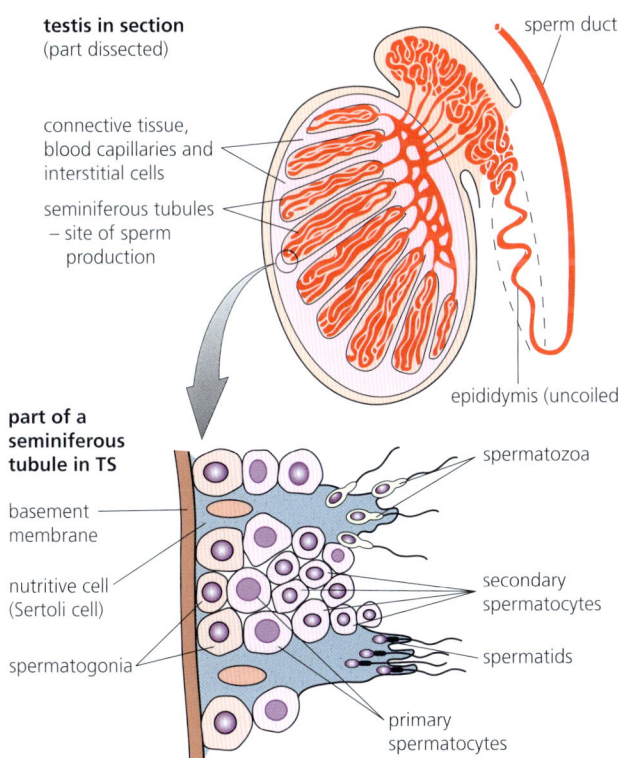

■ **Figure D3.1.21** Structure of a seminiferous tubule – site of sperm production

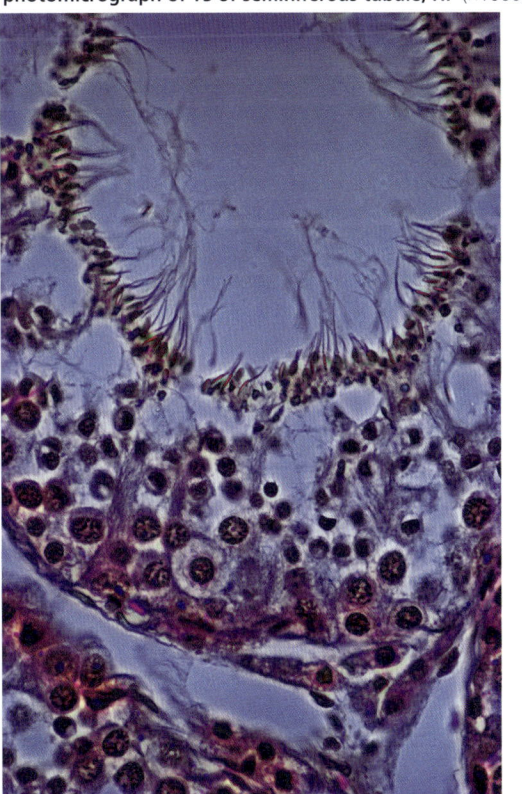

D3.1 Reproduction

HL ONLY

> ⊙ **Common mistake**
>
> The words spermatogonia, spermatocyte, and spermatid, and the different stages of meiosis they relate to, are sometimes not used when describing spermatogenesis – they should be!

Link

The structure of a mature sperm is shown in Chapter B2.3, page 271.

8 **Draw and annotate** a diagram of a mature sperm cell.

Table D3.1.4 summarizes the stages of sperm development, including the type and stage of cell division.

■ **Table D3.1.4** The stages of sperm development

Cells	Type and stage of cell division
spermatogonium	$2n$ (diploid) cells made by mitosis from cells lining the seminiferous tubules
primary spermatocyte	$2n$ cells produced by mitosis of above
	undergoes meiosis to make secondary spermatocyte
secondary spermatocyte	n (haploid) cells made from first meiotic division
spermatid	n cells made from second meiotic division
spermatozoon/sperm	n spermatids differentiate into mature spermatozoa

The mature sperm, and the production of semen

The sperm are immobile when first formed. From the seminiferous tubules, they pass into the much-coiled epididymis, where maturation is completed and storage occurs. During an ejaculation, the sperm are moved by waves of contraction in the muscular walls of the sperm ducts. Sperm are transported in a nutritive fluid that is secreted by glands, mainly the seminal vesicles and prostate gland. These glands add their secretions just at the point where the sperm ducts join with the urethra, below the base of the penis (Figure D3.1.4, page 696). As well as providing nutrients for the sperm, semen is a slightly alkaline fluid, the significance of which we will return to later. During an ejaculation, the sphincter muscle at the base of the bladder is closed.

> ⊙ **Common mistake**
>
> Do not confuse the production of semen with spermatogenesis. Semen is sperm and the fluid from the prostate and seminal vesicles that the sperm travel in, while spermatogenesis is the process of sperm formation.

The roles of hormones in spermatogenesis

In male puberty, a hormone from the hypothalamus triggers the secretion of FSH and LH by the anterior lobe of the pituitary. The first effect of FSH is to initiate sperm production in the testes. LH stimulates the endocrine cells of the testes to secrete testosterone. Subsequently, testosterone and FSH together maintain continued sperm production and growth of the essential Sertoli cells that support sperm with nutrients as they grow and develop in the testes.

Subsequently, secretion of testosterone continues throughout life. Over-secretion of testosterone is regulated by **negative feedback control**, as an excessively high level of testosterone in the blood inhibits secretion of LH. Only when the concentration of LH in the blood has fallen significantly, will testosterone production recommence. (Similarly, over-activity of the nutritive cells inhibits secretion of FSH for a while.)

■ Oogenesis

In the female, the ovaries are about 3 cm long and 1.5 cm thick. These paired structures are suspended by ligaments near the base of the abdominal cavity. As well as producing egg cells, the ovaries are endocrine glands. They secrete the female sex hormones **oestradiol** and **progesterone**. A pair of oviducts extend from the uterus and open as funnels close to the ovaries. The oviducts transport oocytes and are the site of fertilization. In the event of fertilization, development of the foetus will occur in the uterus.

The steps of oogenesis occur in the ovary. **Ovulation**, the process by which an egg is released to the oviduct, occurs at the secondary **oocyte** stage. Development of a secondary oocyte into an ovum is triggered in the oviduct, if fertilization occurs. Consequently, a thin section through a mature ovary, examined by light microscopy, shows the developing oocytes at differing stages (Figure D3.1.22).

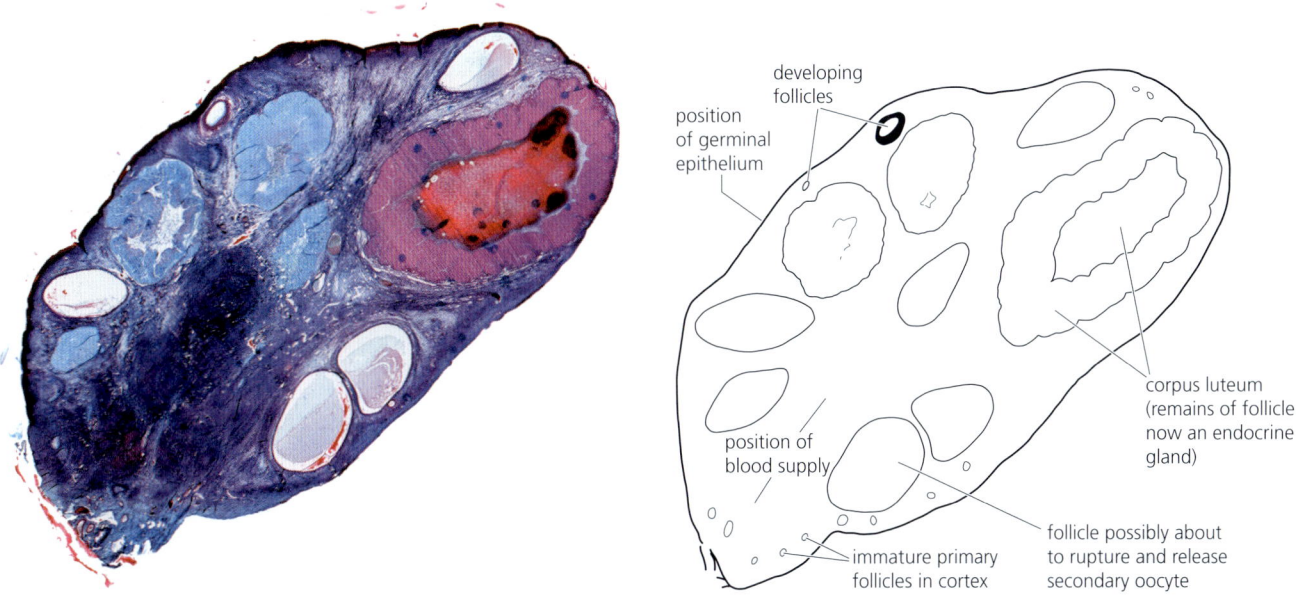

■ **Figure D3.1.22** Photomicrograph of an ovary in section, and interpretive drawing

The structure of the ovary and the steps of oogenesis

◆ **Oogenesis**: the development of an ovum.

◆ **Oogonia**: immature female sex cell in the process of developing into an oocyte.

◆ **Oocyte**: a female sex cell in the process of a meiotic division to become an ovum.

Oogenesis begins in the ovaries of the foetus before birth, but the final development of oocytes is only completed in adult life (Figure D3.1.23). The germinal epithelium, which lines the outer surface of the ovary, divides by mitotic cell division (page 653) to form numerous **oogonia**. These cells migrate into the connective tissue of the ovary, where they grow and enlarge to form **oocytes**. Each oocyte becomes surrounded by layers of follicle cells, and the whole structure is called a **primary follicle**.

By mid-pregnancy, production of oogonia in the foetus ceases – by this stage there are several million in each ovary. Very many degenerate, a process that continues throughout life. At the onset of puberty, the number of primary oocytes remaining is about 250 000. Fewer than 1% of these follicles will complete their development; the remainder never become secondary oocytes or ova.

Between puberty in females (at about 11 years old) and the cessation of ovulation at menopause (typically at about 55 years old), primary follicles begin to develop further. Several start growth each month, but usually only one matures. Development involves progressive enlargement and, at the same time, the follicles move to the outer part of the ovary. The primary follicle then undergoes **meiosis I** (page 656), but the cytoplasmic division that follows is unequal, forming a tiny polar body and a **secondary oocyte** (Figure B2.3.18, page 272). The second meiotic division, **meiosis II**, then begins, but it does not go to completion. In this condition the **egg cell** (it is still a secondary oocyte) is released from the ovary (ovulation), by rupture of the follicle wall (Figure D3.1.23).

HL ONLY

summary of changes from oogonium to ovum
– steps in the growth and maturation phases of gametogenesis in the ovary

diagrammatic representation of the sequence of events in the formation of a secondary oocyte for release and the subsequent changes in the ovary

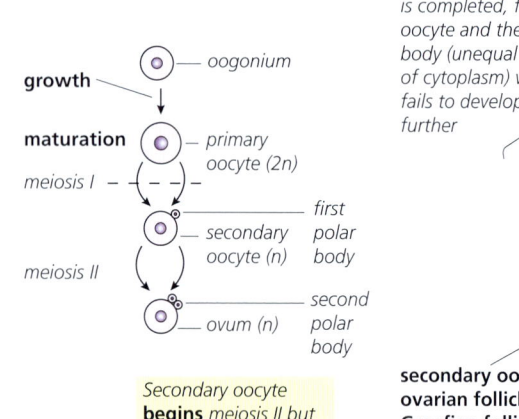

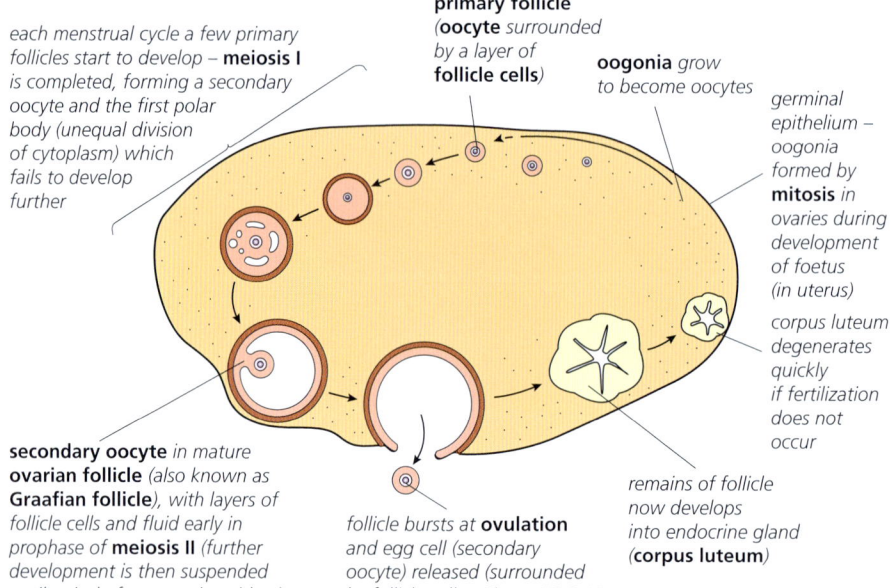

each menstrual cycle a few primary follicles start to develop – **meiosis I** is completed, forming a secondary oocyte and the first polar body (unequal division of cytoplasm) which fails to develop further

primary follicle (**oocyte** surrounded by a layer of **follicle cells**)

oogonia grow to become oocytes

germinal epithelium – oogonia formed by **mitosis** in ovaries during development of foetus (in uterus)

corpus luteum degenerates quickly if fertilization does not occur

secondary oocyte in mature ovarian follicle (also known as Graafian follicle), with layers of follicle cells and fluid early in prophase of **meiosis II** (further development is then suspended until arrival of sperm – in oviduct)

follicle bursts at **ovulation** and egg cell (secondary oocyte) released (surrounded by follicle cells – Figure B2.3.18

remains of follicle now develops into endocrine gland (**corpus luteum**)

Secondary oocyte **begins** meiosis II but this does not complete until sperm nucleus penetrates cytoplasm of oocyte.

■ Figure D3.1.23 The ovary, and stages in oogenesis

9 Distinguish between spermatogenesis and oogenesis.

Ovulation occurs from one of the two ovaries about once every 28 days. Meanwhile, the remains of the primary follicle immediately develops into the yellow body, the corpus luteum (page 699). This is an additional but temporary endocrine gland, with a role to play if fertilization occurs (see below).

Mechanisms to prevent polyspermy

◆ **Polyspermy**: fertilization of an egg by many sperm.

◆ **Zona pellucida**: protective glycoprotein layer; jelly coat of egg. After entry of one sperm it becomes stronger and no more sperm can enter.

◆ **Acrosome**: sac at head of sperm containing protease enzymes.

◆ **Cortical reaction**: prevents polyspermy during fertilization.

One or more of the few sperm that reach a secondary oocyte pass between the follicle cells surrounding the oocyte. The entry of more than one sperm into the oocyte is known as **polyspermy**. Fertilization involves a mechanism that *prevents* polyspermy. Look at the steps to fertilization in Figure D3.1.24.

First, the coat that surrounds the oocyte, which is made of glycoprotein and is called the **zona pellucida**, must be crossed. At the tip of the sperm is a membrane-bound sac of hydrolytic enzymes called the **acrosome**. When in contact with the zona pellucida, these enzymes are released and digest a pathway for the sperm to the oocyte membrane – a process known as the **acrosome reaction**.

Next the plasma membrane around the head of a sperm fuses with the plasma membrane of the oocyte. The male nucleus enters the oocyte. As this happens, **cortical granules** in the outer cytoplasm of the oocyte release their contents outside the oocyte by exocytosis. This is known as the **cortical reaction**. The result is that the oocyte plasma membrane cannot be crossed by another sperm. This prevents the possibility of fusion of more than one male nucleus with the oocyte nucleus.

As the sperm nucleus enters the oocyte, completion of meiosis II is triggered and the second polar body is released. The male and female haploid nuclei come together to form the diploid nucleus of the zygote. Fertilization is completed.

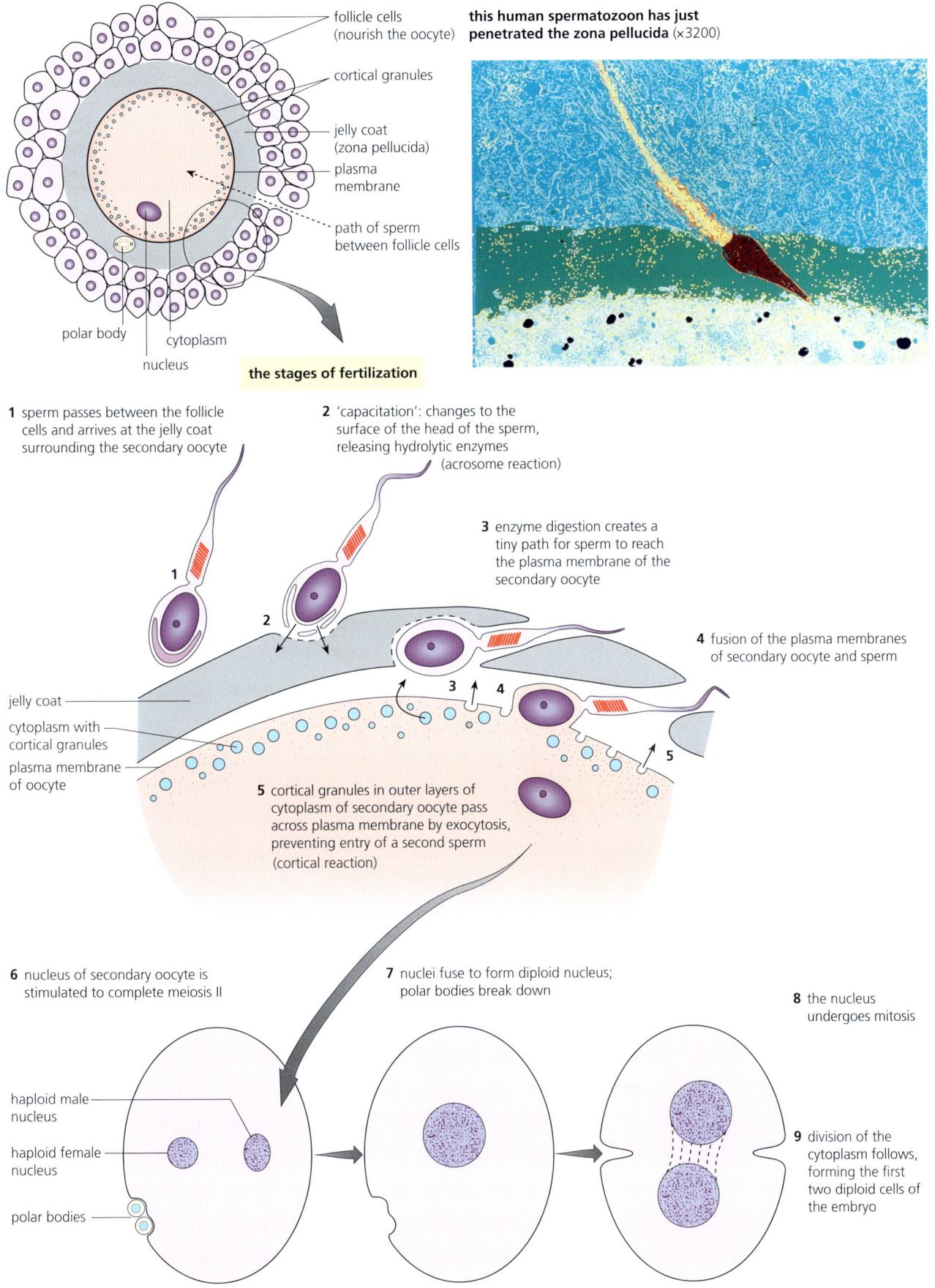

■ **Figure D3.1.24** Fertilization of a human secondary oocyte

D3.1 Reproduction

Development of a blastocyst and implantation in the endometrium

Link
The term blastocyst is defined and covered in Chapter B2.3, page 257.

◆ **Implantation**: embedding of the blastocyst (developed from the fertilized ovum) in the uterus wall.

Fertilization occurs in the upper oviduct. As the zygote is transported down the oviduct by ciliary action, mitosis and cell division commence. The process of the division of the zygote into a mass of daughter cells is known as **cleavage**. This is the first stage in the growth and development of a new individual. The embryo does not increase in mass at this stage. By the time the embryo has reached the uterus, it is a solid ball of tiny cells. Division continues and the cells organize themselves into a fluid-filled ball, the **blastocyst**.

In humans, by day 7, the blastocyst consists of about 100 cells. It now starts to become embedded in the endometrium, a process known as **implantation**. Implantation takes from day 7 to day 14, approximately. At this stage, some of the blastocyst appears grouped as the **inner cell mass** and these cells will eventually become the foetus. Other parts of the blastocyst become the placenta. Once implanted, the embryo starts to receive nutrients directly from the endometrium of the uterus wall (Figure D3.1.25).

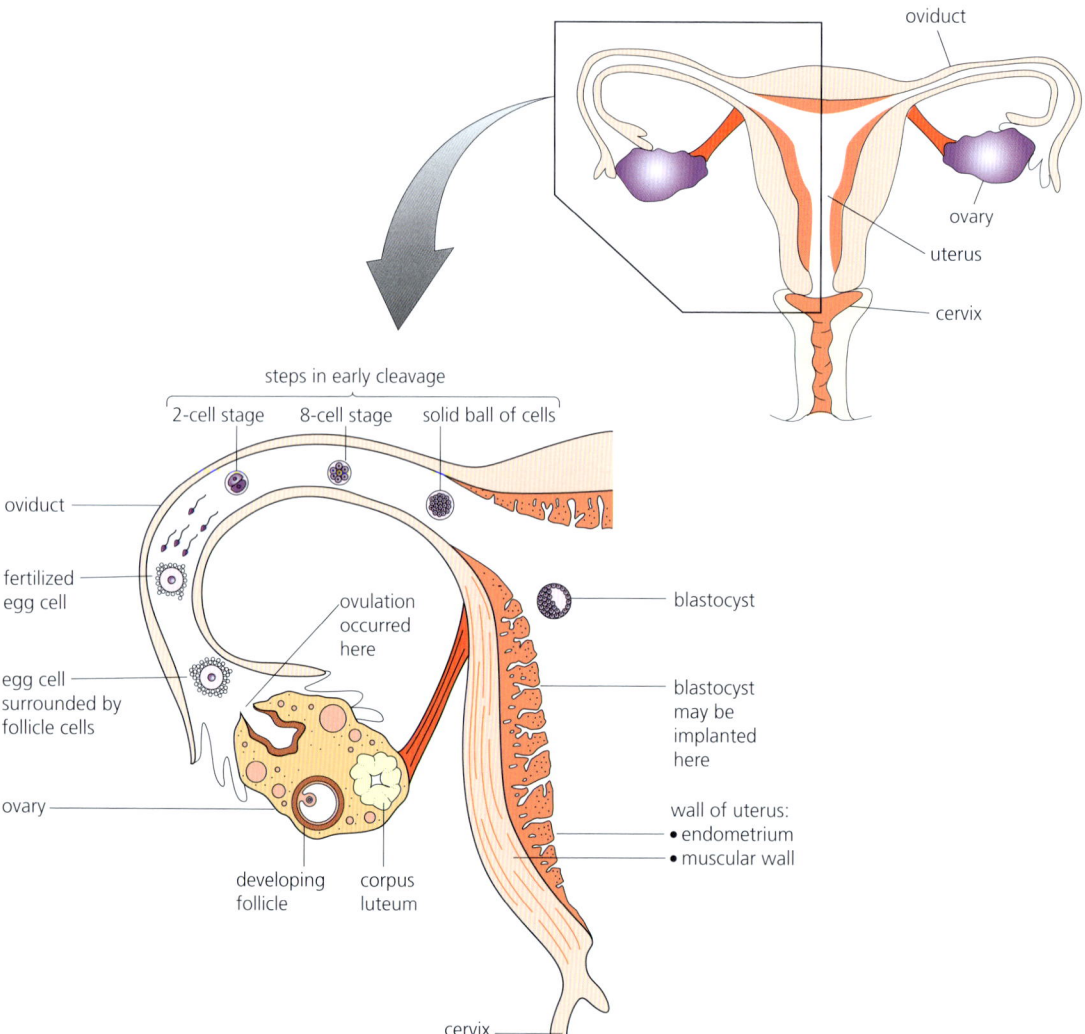

■ **Figure D3.1.25** The site of fertilization and early stages of development

> ### ● Common mistake
>
> When discussing the stages of fertilization, make sure you focus on the relevant detail. The events of fertilization are considered to start with the arrival of sperm at the surface of the oocyte, including the acrosome reaction and the cortical reaction.

Pregnancy testing by detection of human chorionic gonadotropin secretion

If the egg is fertilized (the start of a pregnancy) the developing embryo secretes a hormone, **hCG (human chorionic gonadotropin)**, that circulates in the blood and maintains the corpus luteum (Figure D3.1.23) as an endocrine gland for at least 16 weeks of pregnancy. When the corpus luteum eventually breaks down, the placenta takes over as an endocrine gland, secreting oestradiol, progesterone and hCG. These hormones continue to prevent ovulation and maintain the endometrium.

■ Monoclonal antibodies

White blood cells in the blood provide our main defences against invasion of the body by harmful micro-organisms (Chapter C3.2, page 522). Antibodies are effective in the destruction of antigens within the body, but the plasma cells (the product of a B-lymphocyte, Figure C3.2.7, page 527) that secrete them have an extremely short lifespan and cannot, themselves, divide. As a consequence, antibodies cannot be used outside the body.

◆ **Monoclonal antibody**: antibody produced by a single clone of B-lymphocytes; it consists of a population of identical antibody molecules.

Monoclonal antibodies are the product through which antibodies are made available in the long term, for applications in entirely new circumstances. A monoclonal antibody is a single antibody that is stable and that can be used over a period of time. Each specific antibody is made by one particular type of B-cell. The problem of the normally brief existence of a plasma cell is overcome by fusing the specific lymphocyte with a cancer cell – which, unlike other body cells, goes on dividing indefinitely. The resulting hybrid cell, known as a **hybridoma cell**, divides to form a clone of cells that persists and which conveniently goes on secreting the antibody in significant quantities. Hybridoma cells are virtually immortal, provided they are kept in a suitable environment.

■ Pregnancy testing

Monoclonal antibodies are used in pregnancy testing. A pregnant person has a significant concentration of the hormone hCG in their urine, whereas a non-pregnant person has a negligible amount. Monoclonal antibodies to hCG have been engineered to carry (become attached to) coloured granules, so that in a simple test kit the appearance of a coloured strip in one compartment provides immediate and visual confirmation of pregnancy. How this works is illustrated in Figure D3.1.26.

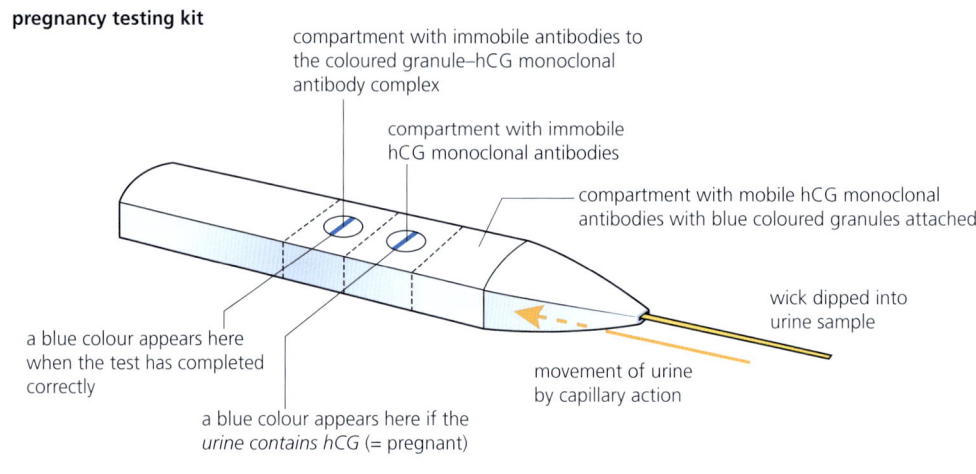
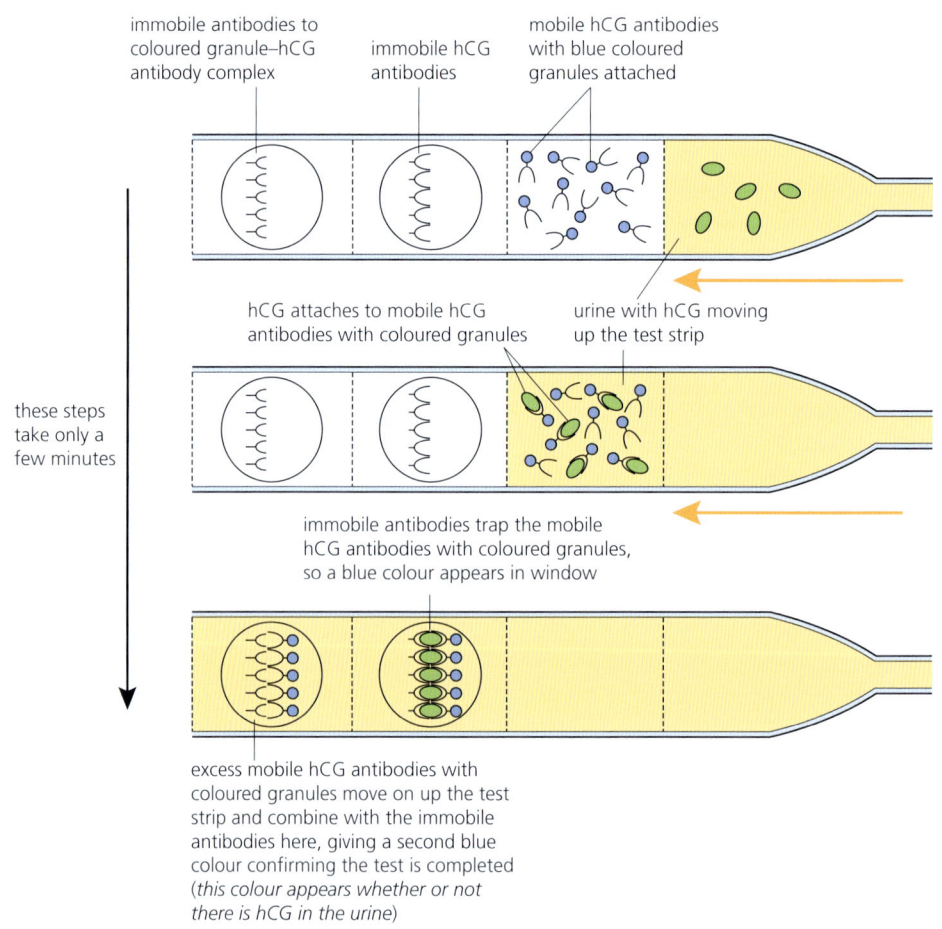

■ Figure D3.1.26 Detecting pregnancy using monoclonal antibodies

◆ **Placenta**: a temporary organ that joins the mother and foetus, transferring oxygen and nutrients from the mother to the foetus, with carbon dioxide and other waste material transported from foetus to mother.

◆ **Gestation**: the period of development in the mother's body, lasting from conception to birth.

Role of the placenta in foetal development inside the uterus

From early in the development of the embryo, this tiny and delicate structure is contained, supported and protected by a membranous, fluid-filled sac. It is the outer layers of the tissues of the embryo that grow and give rise to the membranes and that also form the **placenta** (see below). By the end of two months' development, the beginnings of the principal adult organs can be detected within the embryo and the placenta is operational. During the rest of **gestation**, the developing offspring is called a **foetus**. The placenta allows the foetus to be retained in the uterus to a later stage of development than in mammals that do not develop a placenta (such as marsupials, for example the kangaroo and wombat).

Top tip!

The distinction between embryo and foetus is made based on the amount of time spent in the uterus. An embryo is the early stage of human development in which organs and other critical body structures are formed. An embryo is termed a foetus from the 11th week of pregnancy, i.e. the 9th week of development after fertilization of the egg.

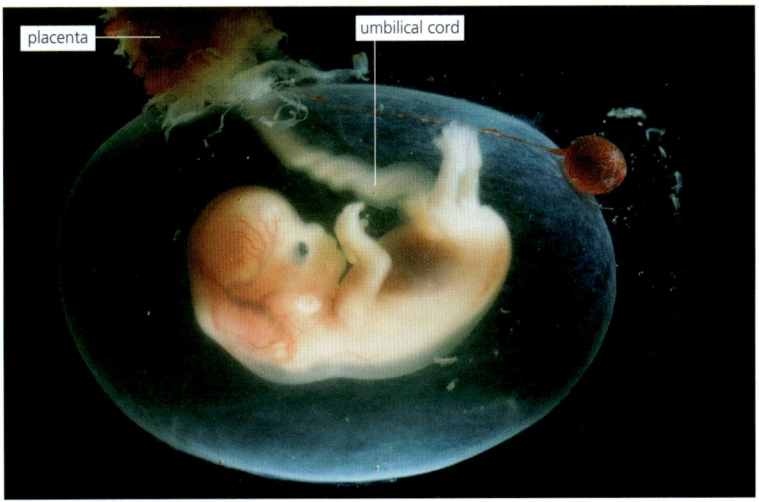

■ Figure D3.1.27 Human embryo at the six-week stage

The placenta – structure and function

The **placenta** is a disc-shaped structure composed of maternal (endometrial) and foetal membrane tissues. Here the maternal and foetal blood circulations are brought very close together over a huge surface area, but they do not mix. Placenta and foetus are connected by arteries and a vein in the umbilical cord (Figure D3.1.28).

Exchange in the placenta is by diffusion and active transport. Movements across the placenta involve:
- respiratory gases, which are exchanged; oxygen diffuses across the placenta from the maternal haemoglobin to the foetal haemoglobin, and carbon dioxide diffuses in the opposite direction
- water, which crosses the placenta by osmosis; glucose, which crosses by facilitated diffusion; and ions and amino acids, which are transported actively
- excretory products, including urea, leaving the foetus
- antibodies present in the mother's blood, which freely cross the placenta, so the foetus is initially protected from the same diseases as the mother (passive immunity, page 541).

The placenta is a barrier to bacteria, although some viruses can cross it.

Top tip!

The placenta is adapted for diffusion in much the same way as other exchange organs.
- It has a large surface area (with lots of villi-like projections).
- It is only a few cells thick, so there is a short diffusion pathway.
- It has a constant blood supply that maintains the concentration gradient between the mother's blood and the blood of the foetus.

D3.1 Reproduction

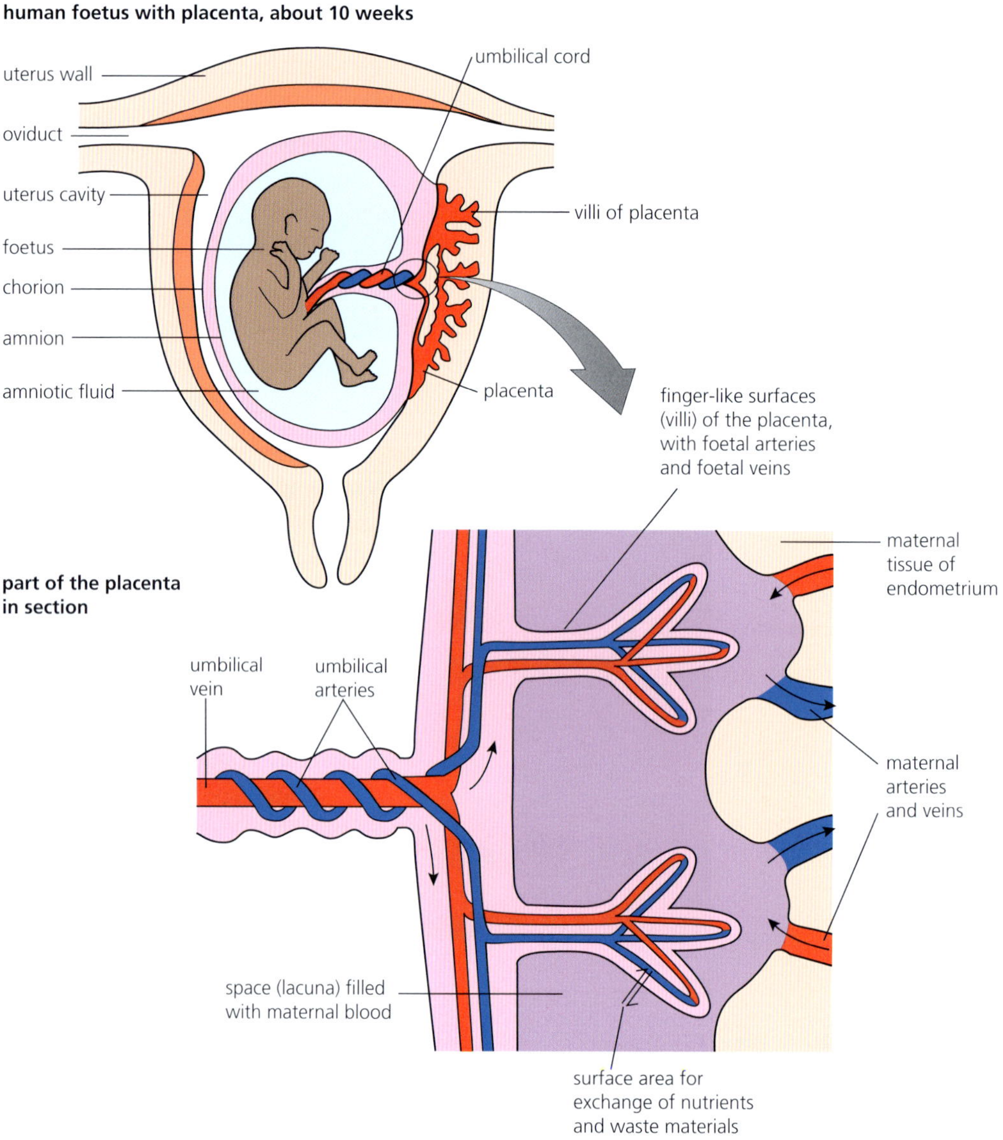

human foetus with placenta, about 10 weeks

■ Figure D3.1.28 The placenta – site of exchange between maternal and foetal circulations

10 **Suggest** why it is so important that the blood of the mother and offspring do not mix together in the placenta.

11 **List** the structural features of the placenta that contribute to efficient exchange and **explain** why each is important.

Hormonal control of pregnancy and childbirth

Pregnancy

The placenta is also an endocrine gland, initially producing an additional sex hormone **human chorionic gonadotrophin** (**hCG**). hCG appears in the urine from about seven days after conception. We have already noted that it is the presence of hCG in a sample of urine that is detected using monoclonal antibodies in a pregnancy-testing kit (Figure D3.1.26).

hCG is initially secreted by the cells of the blastocyst, but later it comes entirely from the placenta. The role of hCG is to maintain the corpus luteum as an endocrine gland (secreting progesterone) for the first 16 weeks of pregnancy. When the corpus luteum eventually does break down, the placenta secretes oestradiol and progesterone (Figure D3.1.29). Progesterone maintains the lining of the uterus, and therefore maintains the continuation of the pregnancy. Without maintenance of these hormone levels, conditions favourable to a foetus are not maintained in the uterus and a spontaneous abortion results.

Theme D: Continuity and change – Organisms

◆ **Oxytocin**: hormone released by the pituitary gland that causes increased contraction of the uterus during childbirth.

Link

Other examples of positive feedback loops are in Chapters C3.1, page 516, and Chapter D4.3, page 825.

Childbirth

Immediately before birth, the level of **progesterone** declines sharply. As a result, progesterone-driven inhibition of contraction of the muscle of the uterus wall is removed.

At the same time, the posterior pituitary begins to release a hormone, **oxytocin** (Figure D3.1.29). This relaxes the elastic fibres that join the bones of the pelvic girdle, especially at the front, and thus aids dilation of the cervix for the head (the widest part of the offspring) to pass through. Oxytocin also stimulates rhythmic contractions of the muscles of the uterus wall. Subsequently, control of contractions during birth occurs via a **positive feedback loop** (Figure D3.1.30). The resulting powerful, intermittent waves of contraction of the muscles of the uterus wall start at the top of the uterus and move towards the cervix. Progressively during this process (known as labour), the rate and strength of the contractions increase, until they expel the offspring.

Finally, less powerful uterine contractions separate the placenta from the endometrium and cause the discharge of the placenta and remains of the umbilicus (called the afterbirth).

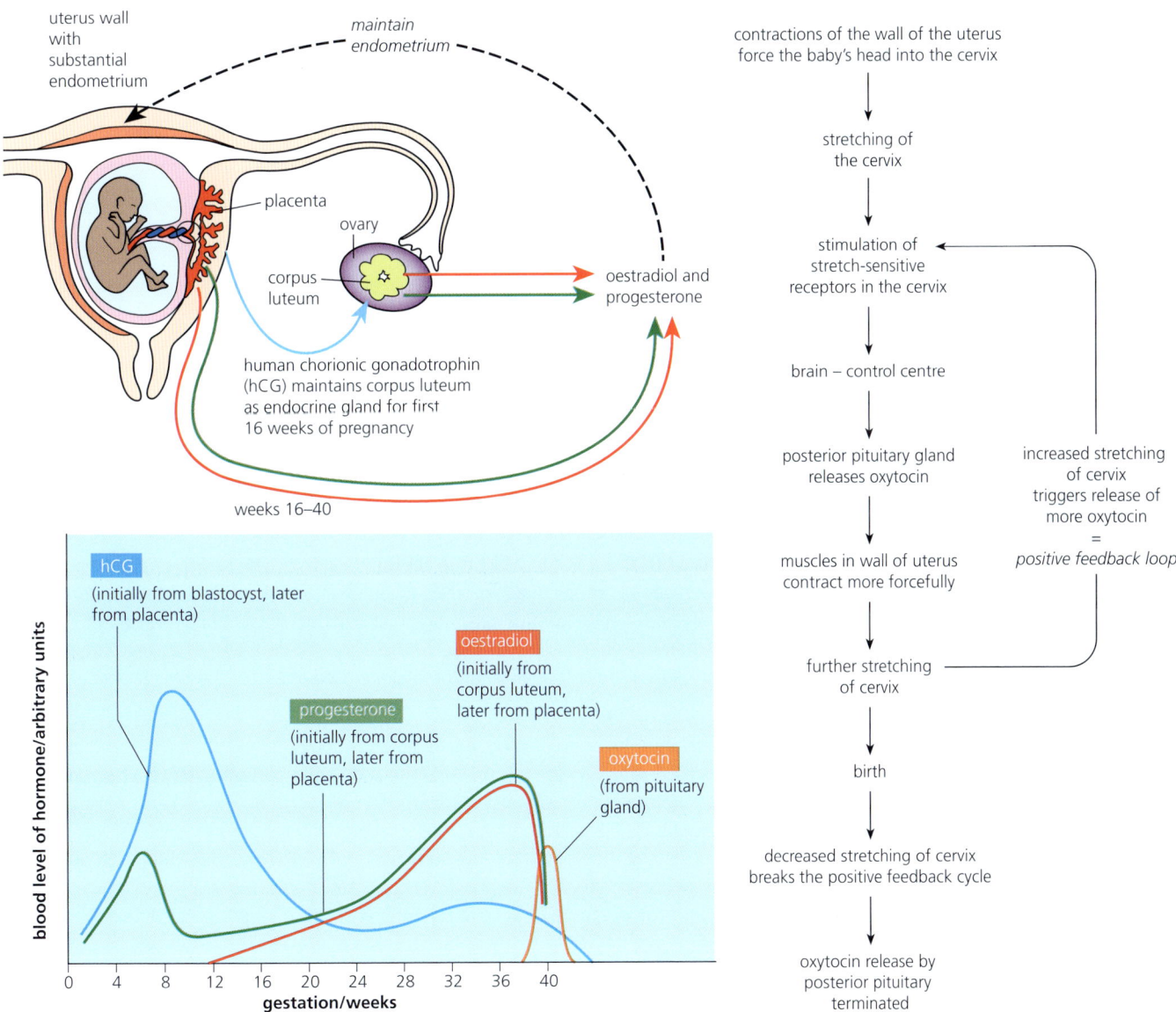

■ **Figure D3.1.29** Blood levels of sex hormones during gestation

■ **Figure D3.1.30** The positive feedback loop in the control of childbirth

D3.1 Reproduction

Top tip!

The continuity of pregnancy is maintained by progesterone secretion initially from the corpus luteum and then from the placenta, whereas the changes during childbirth are triggered by a decrease in progesterone levels, allowing increases in oxytocin secretion due to positive feedback.

12 Suggest why the immediate production of hCG by the embryo while it still consists of relatively few cells is significant in a successful outcome to gestation.

13 Distinguish between negative and positive feedback processes.

◆ **Menopause**: the period in life (typically between the ages of 45 and 55) when menstruation ceases.

◆ **Hormone replacement therapy (HRT)**: medication containing female hormones that is taken to replace the oestradiol and progesterone that the body stops making during menopause.

Hormone replacement therapy and the risk of coronary heart disease

Menstruation typically ceases between the ages of 45–55. A person is said to be in **menopause** when they have not had a period for 12 months in a row. Menopause results from the ovaries reducing the production of oestradiol and progesterone, which regulate the menstrual cycle, and fertility declines. After the menopause, the ovaries no longer respond to stimulation by FSH and LH (the pituitary gland hormones) and stop releasing eggs and oestradiol. Due to lower oestradiol and progesterone levels, there is a loss of negative feedback to the pituitary gland.

Oestradiol affects many parts of the body, including the blood vessels, heart, bone, uterus, urinary system, skin and brain. Loss of oestradiol is believed to be the cause of many of the symptoms associated with menopause, which include brief, periodic increases in body temperature, an increased heart rate, sudden perspiration as the body tries to reduce its temperature, joint and muscle stiffness, and changes in concentration levels.

Hormone replacement therapy (HRT) is a treatment used to relieve the symptoms of menopause by replacing the female hormones oestradiol and progesterone. HRT can be taken as tablets, patches, creams or gels, under advice from a doctor.

Nature of science: Patterns and trends

There are marked variations in the prevalence of different health problems in societies around the world. Claims are based on:
- epidemiological studies – these provide circumstantial evidence of health risks; they suggest connections but do not establish a cause or biochemical connection
- clinical studies of individual patients with health problems, which attempt to show causal relations between diseases and diets, for example (however, it is not possible to carry out 'controlled' experiments for ethical reasons).

Epidemiological studies have been used to monitor the outcomes of hormone replacement therapy (HRT) on the incidence of coronary heart disease (CHD). In early epidemiological studies, it was argued that women undergoing HRT had reduced incidence of CHD and this was deemed to be a cause-and-effect relationship.

A **randomized control trial** is a test in which subjects are randomly assigned to one of two groups: one (the experimental group), which receives the intervention that is being tested, and the other (the comparison group or control) receiving an alternative (conventional) treatment. The two groups are then subsequently tested to see if there are any differences between them in outcome. The results and subsequent analysis of the trial are used to assess the effectiveness of the treatment.

Randomized controlled trials followed on from earlier epidemiological studies and showed that the use of HRT led to a small increase in the risk of CHD. The correlation between HRT and decreased incidence of CHD is therefore not actually a cause-and-effect relationship. HRT patients have a higher socioeconomic status (SES), and this status has a causal relationship with lower risk of CHD. The link between higher household income and uptake of HRT is uncertain, although one hypothesis is that women with high SES use health services more comprehensively and regularly than those from lower income families, and so are more likely to take up preventative health measures such as HRT.

LINKING QUESTIONS

1. How can interspecific relationships assist in the reproductive strategies of living organisms?
2. What are the roles of barriers in living systems?

D3.2 Inheritance

> **Guiding questions**
> - What patterns of inheritance exist in plants and animals?
> - What is the molecular basis of inheritance patterns?

SYLLABUS CONTENT

This chapter covers the following syllabus content:
- ▶ D3.2.1 Production of haploid gametes in parents and their fusion to form a diploid zygote as the means of inheritance
- ▶ D3.2.2 Methods for conducting genetic crosses in flowering plants
- ▶ D3.2.3 Genotype as the combination of alleles inherited by an organism
- ▶ D3.2.4 Phenotype as the observable traits of an organism resulting from genotype and environmental factors
- ▶ D3.2.5 Effects of dominant and recessive alleles on phenotype
- ▶ D3.2.6 Phenotypic plasticity is the capacity to develop traits suited to the environment experienced by an organism, by varying patterns of gene expression
- ▶ D3.2.7 Phenylketonuria as an example of a human disease due to a recessive allele
- ▶ D3.2.8 Single-nucleotide polymorphisms and multiple alleles in gene pools
- ▶ D3.2.9 ABO blood groups as an example of multiple alleles
- ▶ D3.2.10 Incomplete dominance and codominance
- ▶ D3.2.11 Sex determination in humans and inheritance of genes on sex chromosomes
- ▶ D3.2.12 Haemophilia as an example of a sex-linked genetic disorder
- ▶ D3.2.13 Pedigree charts to deduce patterns of inheritance of genetic disorders
- ▶ D3.2.14 Continuous variation due to polygenic inheritance and/or environmental factors
- ▶ D3.2.15 Box-and-whisker plots to represent data for a continuous variable such as student height
- ▶ D3.2.16 Segregation and independent assortment of unlinked genes in meiosis (HL only)
- ▶ D3.2.17 Punnett grids for predicting genotypic and phenotypic ratios in dihybrid crosses involving pairs of unlinked autosomal genes (HL only)
- ▶ D3.2.18 Loci of human genes and their polypeptide products (HL only)
- ▶ D3.2.19 Autosomal gene linkage (HL only)
- ▶ D3.2.20 Recombinants in crosses involving two linked or unlinked genes (HL only)
- ▶ D3.2.21 Use of a chi-squared test on data from dihybrid crosses (HL only)

Production of gametes as the means of inheritance

Chapter D3.1 shows how haploid gametes are produced in meiosis and that this process is essential in maintaining the number of chromosomes across generations. At fertilization, haploid gametes from the male and female fuse, restoring the diploid number of chromosomes (Figure D3.2.1). This pattern of inheritance is common to all eukaryotes with a sexual life cycle.

There are two types of chromosomes: **autosomal** and **sex chromosomes**. Autosomal chromosomes are the numbered chromosomes in a karyogram (see page 111), ranged from smallest to largest.

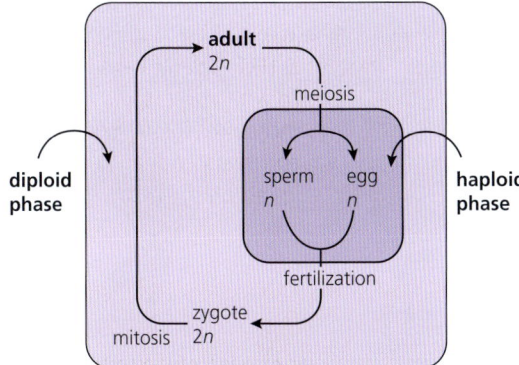

■ Figure D3.2.1 Meiosis and the diploid life cycle

◆ **Autosomal gene**: gene located on one of the numbered, or non-sex, chromosomes.

Sex chromosomes contain genes that determine the gender of the offspring. In diploid cells, chromosomes are in homologous pairs, so a diploid cell has two copies of each **autosomal gene**.

Methods for conducting genetic crosses in flowering plants

Concept: Continuity and change

Gametogenesis changes diploid cells into haploid gametes, enabling the sexual life cycle to continue.

How do characteristics get passed down from one generation to the next? The mechanism of inheritance was successfully investigated before chromosomes had been observed or genes were detected. A monk called Gregor Mendel (Figure D3.2.2) made the first discovery of the fundamental laws of heredity. Mendel cross-bred different types of pea plant in the gardens of the monastery. In total he planted 30 000 plants over eight years. He crossed rounded seed pods with wrinkled ones, yellow peas with green peas, tall stems with dwarf stems, and so on.

● Nature of science: Experiments

Gregor Mendel was born in Moravia (now in the Czech Republic) in 1822, the son of a peasant farmer. As a young boy, he worked to support himself through schooling, but at the age of 21 he was offered a place in the monastery at Brno. The monastery was a centre of research in natural sciences and agriculture, as well as in the humanities. Mendel was successful there. Later, he became abbot.

Mendel discovered the principles of heredity by studying the inheritance of seven contrasting characteristics of the garden pea plant. These did not 'blend' on crossing, but retained their identities, and were inherited in fixed mathematical ratios. He concluded that hereditary factors (we now call them genes) determine these characteristics, that these factors occur in duplicate in parents, and that the two copies of the factors segregate from each other in the formation of gametes.

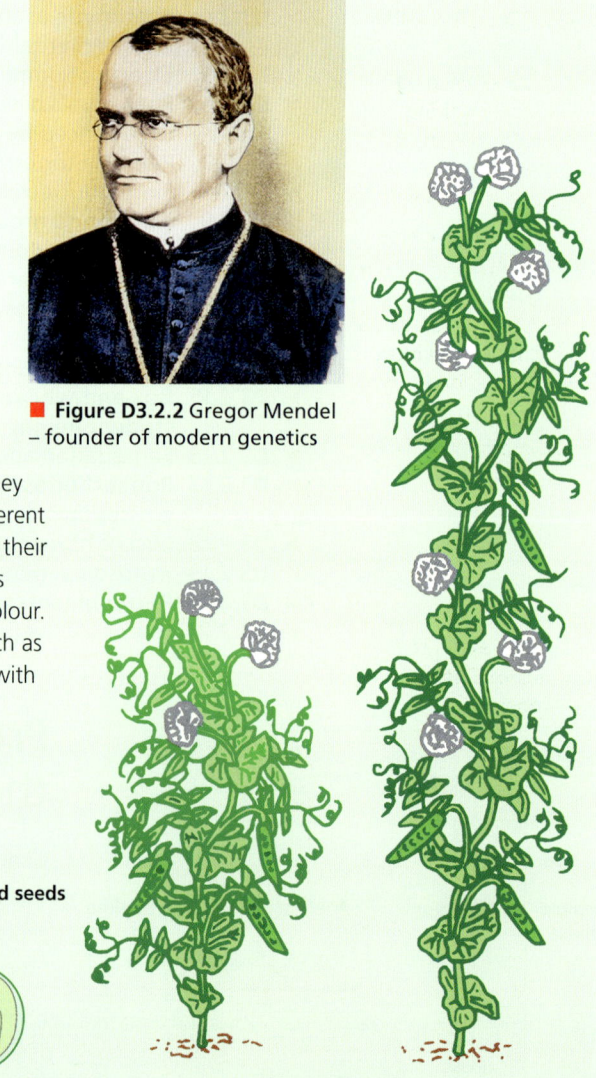

■ **Figure D3.2.2** Gregor Mendel – founder of modern genetics

Before Mendel, scientists did not know how inheritance worked but they thought that it must work by 'blending' (a bit like mixing paints of different colours). For example, if a tall person has children with a short person, their genes will be 'blended' and the children will be of medium height. This seemed to work with several characteristics, such as height and skin colour. However, it was an inadequate explanation of other characteristics, such as eye colour (how can you have one sibling with blue eyes and another with brown eyes?).

Features of the garden pea

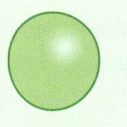  round v. wrinkled seeds

green v. yellow cotyledons (seed leaves)

dwarf v. tall plants

Today, we often refer to Mendel's laws of heredity, but Mendel's results were not presented as laws – which may help to explain the difficulty others had in seeing the significance of his work at the time. Mendel was successful because:
- his experiments were carefully planned, and used large samples
- he carefully recorded the numbers of plants of each type but expressed his results as ratios
- in the pea, contrasting characteristics are easily recognized
- by chance, each of these characteristics was controlled by a single factor (gene)* rather than by many genes, as most human characteristics are
- pairs of contrasting characters that he worked on were controlled by factors (genes) on separate chromosomes*
- in interpreting results, Mendel made use of the mathematics he had learnt.

* Genes and chromosomes were not known then.

◆ **Parental generation (P)**: the first set of parents in a genetic cross. The parents' genotypes are used to predict the genotype of their offspring (F1 generation).

◆ **Monohybrid cross**: cross (breeding experiment) involving one pair of contrasting characters exhibited by homozygous parents.

◆ **F1 generation**: first filial generation – arises by crossing parents (P) and, when selfed or crossed via sibling crosses, produces the F2 generation.

◆ **F2 generation**: offspring produced by the F1 generation.

ATL D3.2A

Figure D3.2.3 shows pea pods from Mendel's experiments. Mendel noted that **parental generations** with either yellow (A) or green peas (B) produced offspring that were yellow in colour (C). When these offspring were bred together, the offspring were a mixture of yellow and green peas (D).

Read the information in the Nature of science box above and look at the results of one of his experiments (Figure D3.2.3). What do you conclude about the mechanism of inheritance? How does it disprove the 'blending' hypothesis? Discuss with a partner and present your conclusions to the rest of the class.

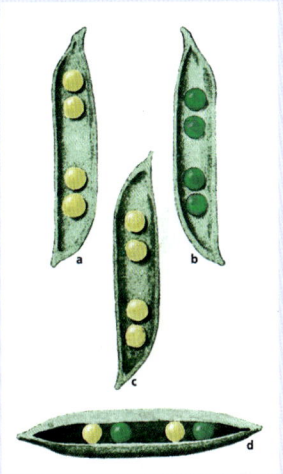

■ **Figure D3.2.3** Mendel's peas. Historical artwork of the peas (*Pisum* sp.) used by Gregor Mendel (1822–84) in his experiments

Link
Self-pollination is covered in Chapter D3.1, page 705.

Initially, Mendel investigated the inheritance of a single contrasting characteristic, known as a **monohybrid cross**. Mendel had noticed that the garden pea plant was either tall or dwarf. *How was this contrasting characteristic controlled?*

Mendel's investigation of the inheritance of height in pea plants began with plants that always 'bred true' (Figure D3.2.4). This means that the tall plants produced progeny that were all tall and the dwarf plants produced progeny that were all dwarf when each was allowed to self-fertilize. Pollen contains male gametes and female gametes are in the ovary, so pollination is needed to carry out a cross (the breeding of two parents with different alleles that produce offspring with characteristics of both parents). Plants such as peas produce both male and female gametes on the same plant, allowing self-pollination and, therefore, self-fertilization.

Mendel crossed true-breeding (i.e. both alleles the same) tall and dwarf plants and found the progeny, the **F1 generation**, were all tall. The offspring were allowed to self-pollinate (and so self-fertilize) to produce the **F2 generation**. The progeny of this cross consisted of both tall and dwarf plants in the ratio of 3 tall:1 dwarf.

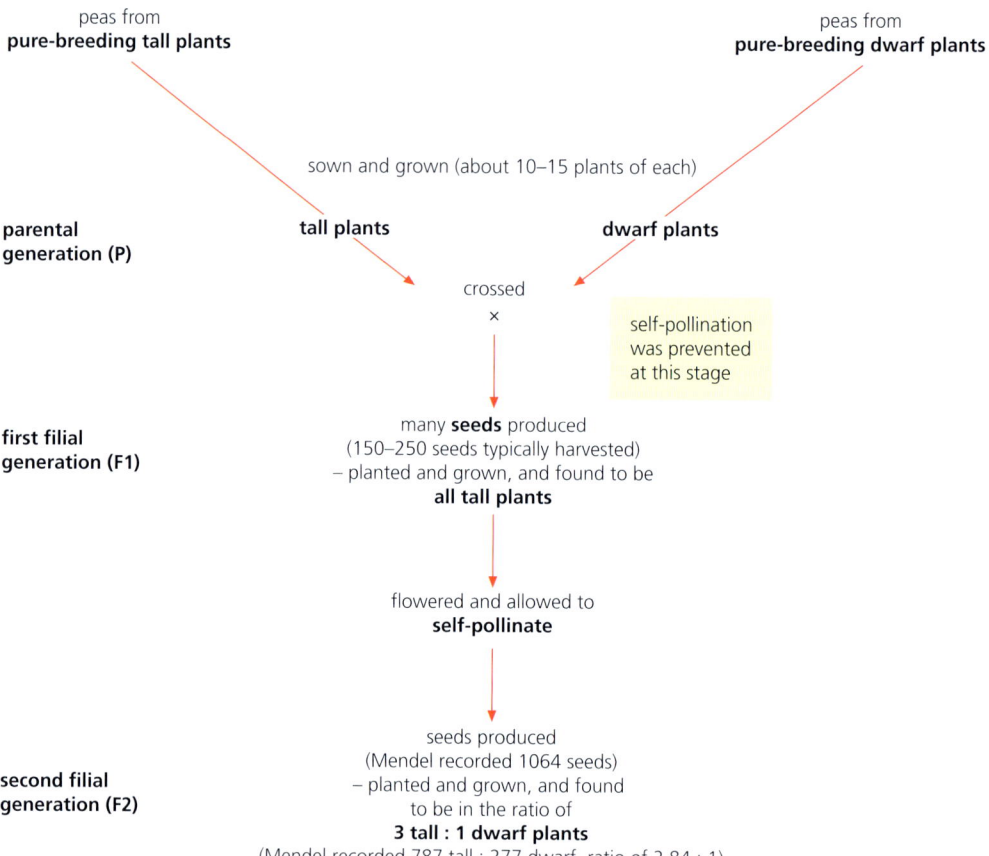

■ **Figure D3.2.4** The steps of Mendel's monohybrid cross

♦ **Dominant allele**: an allele that has the same effect on the phenotype whether it is present in the homozygous or heterozygous state.

♦ **Recessive allele**: an allele that has an effect on the phenotype only when present in the homozygous state.

♦ **Punnett grid**: diagram used to show and calculate all the combinations and frequencies of different genotypes and phenotypes among the offspring of a genetic cross.

 **Top tip!**

Punnett grids show the combining process of the gametes of two parents. The diagrams can be used to calculate the probabilities that particular gametes and zygotes will occur. Punnett grids can either be shown in a diamond conformation, as shown opposite, or as a square. They may also be referred to as 'Punnett squares'.

In Mendel's interpretation of the monohybrid cross he argued that, because the dwarf characteristic had apparently disappeared in the F1 generation and reappeared in the F2 generation, there must be a factor controlling dwarfness that remained intact from one generation to another. However, this factor for 'dwarf' did not express itself in the presence of a similar factor for tallness. In other words, as characteristics, tallness is **dominant** and dwarfness is **recessive**; the **dominant allele** totally masks the effects of a **recessive** (non-dominant) **allele**. Logically, there must be two independent factors for height, one received from each parent. A sex cell (gamete) must contain only one of these factors.

Mendel saw that a 3:1 ratio could be the product of randomly combining two pairs of unlike factors (**T** and **t**, for example). This can be shown using a grid, now known as a **Punnett grid**, after the mathematician who first used it (Figure D3.2.5).

Mendel's conclusions from the monohybrid cross were that:
- within an organism there are breeding factors controlling characteristics such as 'tall' and 'dwarf'
- there are two factors in each cell
- one factor comes from each parent
- the factors separate in reproduction and either one can be passed on to an offspring
- the factor for 'tall' is an alternative form of the factor for 'dwarf'
- the factor for 'tall' is dominant over the factor for 'dwarf'.

Top tip!

Alleles are represented by letters, which are (usually) derived from the characteristic being studied, e.g. 'T' or 't' for tall or short plants. The capital letter (upper case) represents the dominant allele and the lower case the recessive. Make sure that the letters used to show the genotype alleles are sufficiently different in lower case and upper case, for example T and t, or A and a, but not L and l or S and s, which are difficult to distinguish visually.

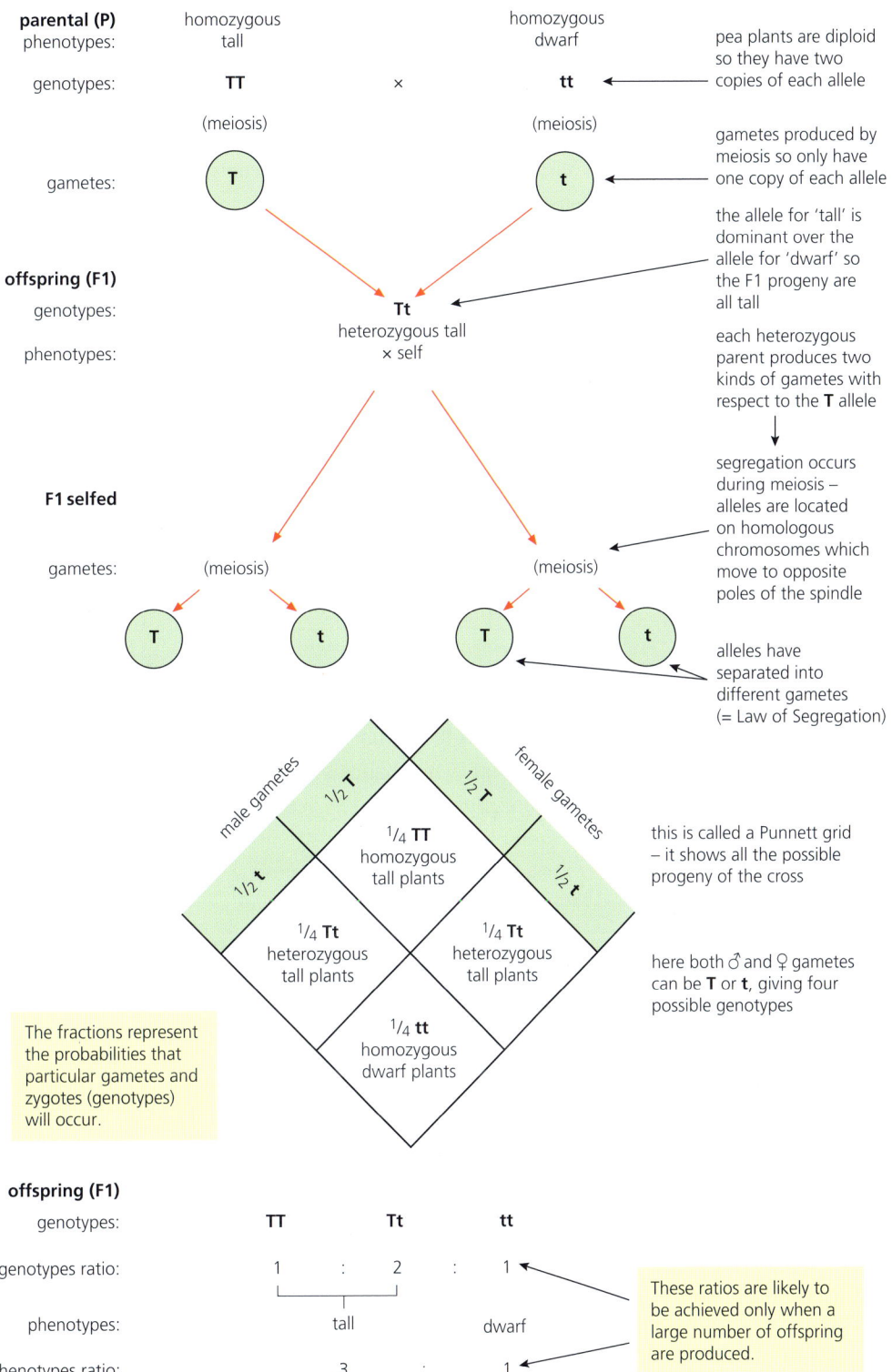

■ **Figure D3.2.5** Genetic diagram showing the behaviour of alleles in Mendel's monohybrid cross

Notice the use of 'P', 'F1' and 'F2' for the three generations involved. Look carefully at the use of a Punnett grid to represent the process and products of the F1 cross. This device shows clearly the combining process of the gametes. The fractions represent the probabilities that particular gametes and zygotes will occur.

D3.2 Inheritance

Concept: Continuity and change

Punnett grids show both continuity and change: continuity in the delivery of alleles to new generations, and change in the arrangement of alleles in offspring.

● **Common mistake**

In Punnett grids, parental genotypes are often missing and gametes on the Punnett grid are usually shown but not labelled as gametes. Make sure that parental genotypes are shown and that the gametes are labelled.

◆ **Genotype**: the combination of alleles inherited by an organism.

◆ **Homozygous**: having two identical alleles of a gene.

◆ **Heterozygous**: having two different alleles of a gene.

Link

Genes and alleles are also covered in Chapter A1.2, page 21.

Mendel never stated his discoveries as laws, but it would have been helpful if he had done so. For example, he might have said: 'Each characteristic of an organism is determined by a pair of factors of which only one can be present in each gamete.'

Today we call a similar statement Mendel's first law, the **Law of segregation**:

The characteristics of an organism are controlled by pairs of alleles which separate in equal numbers into different gametes as a result of meiosis.

Mendel used pea plants to develop his laws of inheritance. Today, genetic crosses are widely used to breed new varieties of crop or ornamental plants.

1. The actual results that Mendel obtained in three of his later investigations of inheritance in the garden pea plant are as shown in Table D3.2.1. These experiments followed on from Mendel's cross shown in Figure D3.2.5 and developed from the conclusions he made as a result of the first monohybrid cross.

■ Table D3.2.1 Mendel's later experimental results

Character	Cross	Number of F2 counted	Number showing dominant character	Number showing recessive character
Position of flowers	axial × terminal	858	651 axial	207 terminal
Colour of seed coat	grey × white	929	705 grey	224 white
Colour of cotyledons	yellow × green	8023	6022 yellow	2001 green

a **State** what ratio of offspring he would have predicted from these crosses.
b **Calculate** the actual ratios obtained in each of these crosses.
c **Suggest** what chance events may influence the actual ratios of offspring obtained in breeding experiments like these. (We will return to this issue at a later point. For example, the statistical test known as the chi-squared test can be applied to the results of genetic crosses to test whether the actual results are close enough to the predicted results to be significant.)

Genotype

The alleles that an organism carries (present in every cell) make up the **genotype** of that organism. You will recall from Chapter A1.2 that **allele** is a term that refers to a different version of the same **gene**. A genotype in which the two alleles of a gene are the same is **homozygous** for that gene. In Figure D3.2.5, the parent pea plants (P generation) were either homozygous tall or homozygous dwarf.

If the alleles are different, the organism is **heterozygous** for that gene. In Figure D3.2.5, the progeny (F1 generation) were heterozygous tall.

2. **State** whether a person with sickle-cell trait (page 623) is homozygous or heterozygous for sickle-cell haemoglobin.

So, the **genotype** is the genetic constitution of an organism. Alleles interact in various ways and with environmental factors. The outcome is the **phenotype** (see below).

Phenotype

♦ **Phenotype**: observable traits of an organism resulting from genotype and environmental factors.

The **phenotype** is the way in which the genotype of the organism is expressed – including the appearance of the organism. In Mendel's monohybrid cross (Figure D3.2.5) the heights of the plants were their phenotypes.

Sometimes organisms may have the same phenotype but different genotypes. For example, plants with genotypes TT and Tt will have the same phenotype (tall) because they both have a dominant allele. Plants like these can only be distinguished by the offspring they produce in a particular cross. When the tall heterozygous plants (**Tt**) are crossed with the homozygous recessive plants (**tt**), the cross yields 50% tall and 50% dwarf plants (Figure D3.2.6). This type of cross has become known as a test cross. If the offspring are all tall, then we know that the tall plants under test are homozygous plants (**TT**). Of course, sufficient plants have to be used to obtain these distinctive ratios.

Phenotype can be determined by genotype, environmental factors, or due to interaction between genotype and environment (as explored with monozygotic twin experiments). Eye colour is determined by genotype only, as is blood type. Changes due to exposure to the Sun, e.g. developing a suntan, is solely environmental. Height and weight are a combination of genotype and phenotype.

Link
Variation in monozygotic twins is discussed in Chapter D2.2 (HL only), page 677.

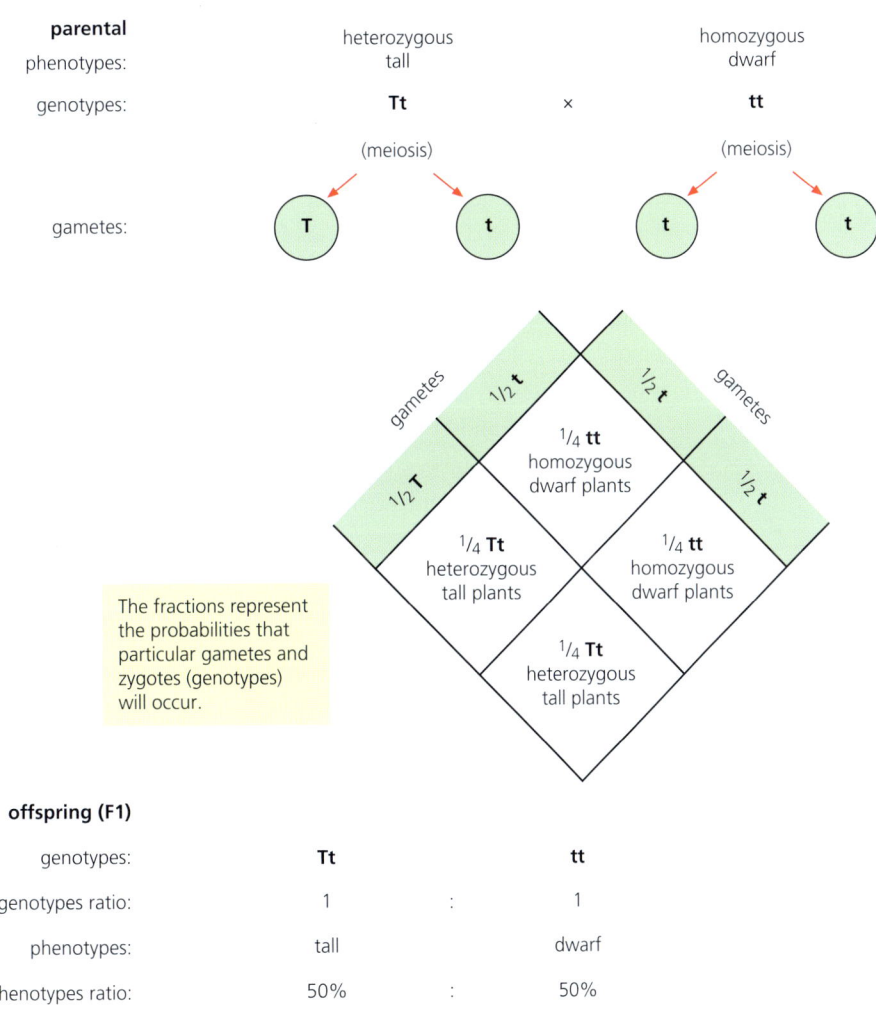

■ **Figure D3.2.6** Genetic diagram of Mendel's test cross. Plants with the phenotype 'tall' can either be homozygous or heterozygous. Test crosses can be used to determine the genotype – in this diagram, the tall plants are demonstrated to be Tt (heterozygous) not TT (homozygous)

3 **Construct** a table to explain the relationship between Mendel's Law of Segregation and meiosis.

● TOK

Mendel's theories were not accepted by the scientific community for a long time. What factors would encourage the acceptance of new ideas by the scientific community? In what ways do our values affect our acquisition of knowledge?

Inquiry 2: Collecting and processing data

Interpreting results

Access the following website: https://javalab.org/en/test_cross_en

In the simulation, you can breed heterozygous pea plants that have green and yellow peas. You can breed pure-bred plants (i.e. that are homozygous for a trait) by removing either yellow or green peas from the breeding programme.

Carry out three investigations: one with general breeding, one where yellow breeding occurs only, and another where only green breeding occurs.

Compare the outcomes from the three experiments. Identify, describe and explain patterns, trends and relationships. What do your results show about the mechanism for inheritance?

■ Effects of dominant and recessive alleles on phenotype

Mendel's experiments on peas showed an important feature of inheritance – that some alleles can be **dominant** and others **recessive**. In the experiment, for example, where yellow pea plants produced a mixture of green and yellow peas, the outcome can be explained by a dominant/recessive relationship between the two colours. This is because the colour of the peas is controlled by an allele in which yellow colour is dominant. An allele is a variant of a gene, with different alleles having a different effect on the same characteristic, for instance controlling pea colour. By breeding the F1 generation peas together (Figure D3.2.3, page 727), Mendel found that the next generation had a mixture of pea colours. The ratio of yellow to green peas was 3:1. His work formed the basis of genetic theory.

If an organism shows a recessive characteristic in its phenotype (like the dwarf pea) it must have a homozygous genotype (**tt**). But if it shows the dominant characteristic (like the tall pea) then it may be either homozygous for a dominant allele (**TT**) or heterozygous for the dominant allele (**Tt**). In other words, **TT and Tt look alike**; they have the same phenotype but different genotypes. A homozygous-dominant genotype and heterozygous genotype for a particular trait will therefore produce the same phenotype because, in both cases, there is a dominant allele that shows through in the phenotype. The only way that a recessive characteristic can show through is in the homozygous condition.

● Common mistake

It is incorrect to say that all dominant alleles give an advantage to an individual and result in the most common phenotype. Not all dominant alleles confer an advantage. For example, Huntington's disease is caused by a dominant allele and results in progressive mental deterioration and involuntary muscle movements (page 638). Six fingers, as another example, is another dominant trait that is not the most common phenotype.

◆ **Phenotypic plasticity**: the capacity to develop traits suited to the environment experienced by an organism, by varying patterns of gene expression.

■ Phenotypic plasticity

Traits can be developed that suit the environment experienced by an organism. This is known as **phenotypic plasticity**. It is generated by varying patterns of gene expression. Phenotypic plasticity is not due to changes in genotype, and the changes in traits may be reversible during the lifetime of an individual. For example, the results of exercise and/or dieting affect human morphology and physiology during the lifetime of an individual.

◆ **Phenylketonuria (PKU)**: a recessive genetic condition caused by mutation in an autosomal gene that codes for the enzyme needed to convert phenylalanine to tyrosine.

◆ **Carrier**: an individual that has one copy of a recessive allele that causes a genetic disease in individuals that are homozygous for this allele.

Phenylketonuria as an example of a human disease due to a recessive allele

Phenylketonuria (PKU) is a rare inherited disorder caused by a mutation in a gene (PAH) on chromosome 12 that codes for phenylalanine hydroxylase, an enzyme needed to convert the amino acid phenylalanine to a different amino acid, tyrosine. As phenylalanine hydroxylase is not synthesized, phenylalanine builds up in the body.

Untreated infants with PKU tend to have unusually light eye, skin and hair colour due to the high phenylalanine levels interfering with the production of melanin, a substance that causes pigmentation in the skin, eyes and hair. High blood phenylalanine levels can cause disruptions in neurotransmitters such as serotonin and dopamine in the brain, which are important for mood, learning, memory and motivation. According to the National Organization of Rare Disorders (NORD), neurological symptoms include seizures, abnormal muscle movements, tight muscles, involuntary movements or tremors.

Without treatment, most people with PKU develop severe brain damage. To prevent this, treatment consists of a carefully controlled, phenylalanine-restricted diet beginning during the first days or weeks of life.

Phenylketonuria is caused by a mutation on an allele that is recessive. This means that two copies of the faulty gene are needed to develop the disease. If a person inherits a copy of the faulty gene from one parent but a working gene from the other, they will not develop the disease. Such people are said to be **carriers** because they carry the faulty allele but do not show it in their phenotype.

The inheritance of genetic diseases can be shown using the diagram in Figure D3.2.7.

Common mistake

Do not confuse the terms 'carrier' and 'affected'. A carrier is a person who has one copy of a faulty recessive allele but does not show this in the phenotype as it is masked by a dominant allele. In the case of phenylketonuria, a carrier does not have the disease. An affected person is someone who has two copies of the recessive faulty allele and so has the symptoms of the disease.

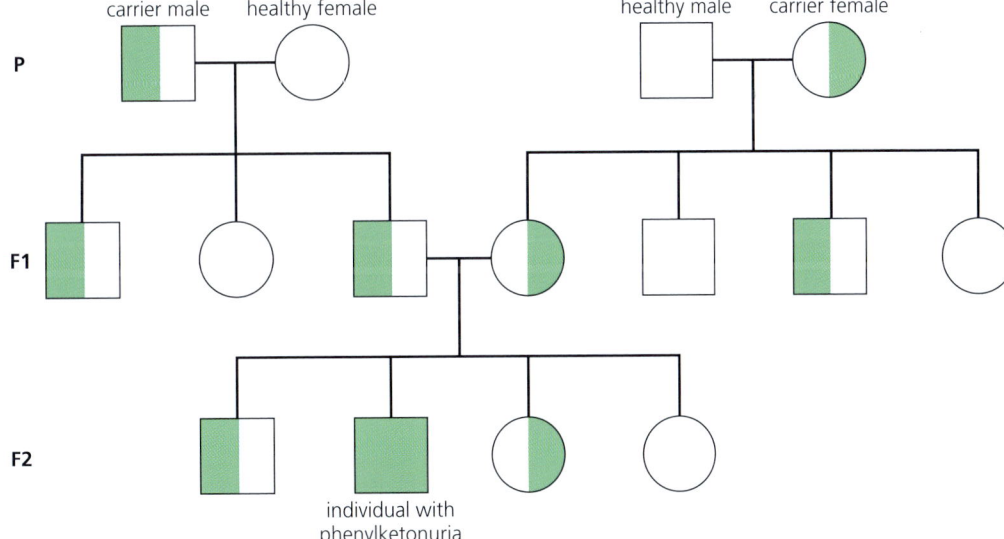

■ **Figure D3.2.7** The inheritance of phenylketonuria

Figure D3.2.7 shows a **pedigree chart**. These are diagrams used to deduce patterns of inheritance of genetic disorders. We will return to this on page 741, later in this chapter.

> ### Concept: Change
>
> Mutation can lead to a change in the genetic code of a gene, which affects the expression of the gene in the phenotype, which can result in a genetic disease.

Single-nucleotide polymorphisms and multiple alleles in gene pools

Link

Single-nucleotide polymorphisms (SNPs) have also been discussed in Chapter A3.1, page 115, and Chapter D1.3, page 638.

A **single-nucleotide polymorphism (SNP)** represents a difference in a single nucleotide. For example, an SNP may replace the nucleotide guanine (C) with the nucleotide adenine (A). This can create an alternative version of a gene (an allele). Many different versions of a gene can exist in a gene pool, depending on how many SNPs have occurred over time and whether these have affected the gene in question. Because only one copy of an allele can exist on each chromosome, an individual can only inherit two alleles (one from each parent) rather than the full number available in the gene pool. An example of a gene with multiple alleles, determining blood type, is discussed below.

ABO blood groups as an example of multiple alleles

The genes introduced so far exist in two forms (two alleles); for example, the height gene of the garden pea exists as tall and dwarf alleles. This means that in genetic diagrams we can represent alleles with a single letter (here, **T** or **t**) according to whether they are dominant or recessive.

For simplicity we began by considering inheritance of a gene for which there are just two alleles. However, we now know that not all genes are like this. In fact, most genes have more than two alleles, and these are cases of genes with **multiple alleles**.

With multiple alleles, we choose a single capital letter to represent the locus at which the alleles may occur, and the individual alleles are then represented by an additional single letter (usually capital) in a superscript position – as with codominant alleles. An excellent example of multiple alleles is found in the genetic control of the ABO blood group system in humans. Human blood belongs to one of four groups: A, B, AB or O. Table D3.2.2 lists the possible phenotypes and the genotypes that may be responsible for each blood group.

■ Table D3.2.2 The ABO blood groups – phenotypes and genotypes

Phenotype	Genotypes
A	$I^A I^A$ or $I^A i$
B	$I^B I^B$ or $I^B i$
AB	$I^A I^B$
O	

Top tip!

You should use the following to denote the different blood alleles: I^A, I^B and i (the latter denotes the allele for blood type O).

Common mistake

Do not confuse the terms 'blood group' and blood 'allele'. A person's blood group (A, B, AB or O) is determined by which combination of three alternative alleles they have (I^A, I^B or i).

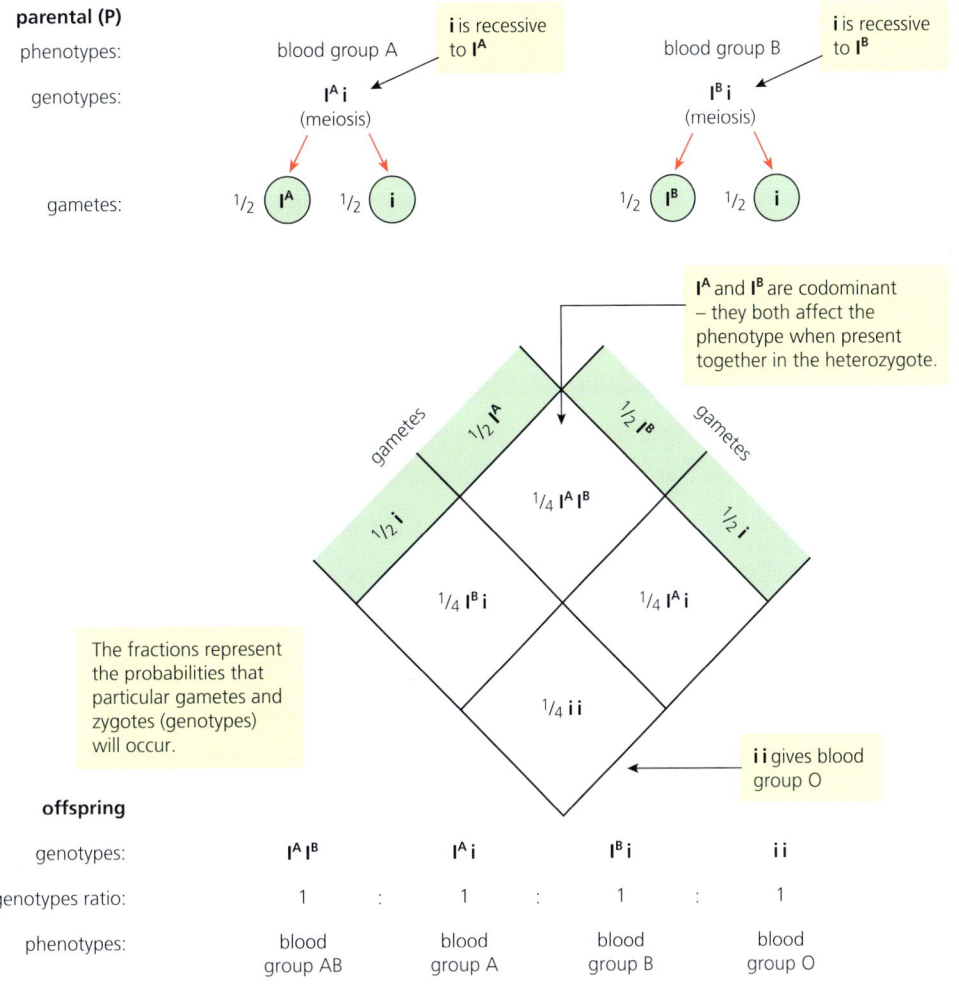

Figure D3.2.8 Inheritance of blood groupings A, B, AB and O

Link

Blood groups have also been discussed in Chapter B1.1, page 195.

Top tip!

A person with blood type A may have a genotype of I^AI^A or I^Ai. Similarly, someone with blood type B may have genotype I^BI^B or I^Bi. These alternative genotypes must be considered when working out the outcomes of genetic crosses involving the ABO blood system.

So, the ABO blood group system is determined by combinations of alternative alleles. In each individual, only two of the three alleles exist, but they are inherited as if they were alternative alleles of a pair. However, I^A and I^B are codominant alleles, and both I^A and I^B are dominant to the recessive i. Figure D3.2.8 shows the way in which the alternative blood groups may be inherited.

Inquiry 1: Exploring and designing

Exploring

Access the following website: **https://javalab.org/en/abo_blood_type_en**

Select two different blood groups by clicking on the relevant symbols for the male and female. Formulate a hypothesis and predict the outcome of the cross. Draw the cross on paper and explain your prediction using your scientific understanding. Now slide the red button along the slider to reveal the outcome. Did you predict correctly? Repeat the procedure for other blood types.

> ### Common mistake
>
> Although the blood type O does not show through in the phenotype, for example if the genotype is $I^A i$ the person has blood type A, it is incorrect to say that i is a recessive allele. It appears to have no phenotypic effect because this allele does not code for carbohydrates in the cell membrane, and so the red blood cells do not have a surface antigen.

Incomplete dominance and codominance

In certain types of monohybrid cross, the 3:1 ratio is not obtained. Two of these situations are illustrated next.

■ Codominance

In **codominance**, heterozygotes have a dual phenotype. For example, the AB blood type $I^A I^B$ is an example of codominance (see page 735), because both allele I^A and allele I^B show through equally in the phenotype.

■ Incomplete dominance

In the case of some genes, both alleles may be expressed simultaneously, rather than one being dominant and the other recessive. In these cases, the dominant allele does not completely mask the effects of a recessive allele, resulting in heterozygotes having an intermediate phenotype. This is known as **incomplete dominance**.

Mirabilis jalapa (the marvel of Peru, or four o'clock flower) is a commonly grown decorative plant. *Mirabilis jalapa* has two types of pure-breeding plants: red flowered and white flowered. When red-flowered plants are crossed with white-flowered plants, the F1 plants have pink flowers (Figure D3.2.9). When pink-flowered *Mirabilis jalapa* plants are crossed, the F2 offspring are found to be red, pink and white in the ratio 1:2:1, respectively.

Pink colouration of the petals occurs because both alleles are expressed in the heterozygote – both a red and a white pigment system are present. Red and white show incomplete dominance. In genetic diagrams, alleles that show incomplete dominance are represented by a capital letter for the gene, and different superscript capital letters for the two alleles, in recognition of the fact that the dominant allele does not completely mask the effects of a recessive allele, resulting in a new phenotype. (Figure D3.2.9).

◆ **Codominance**: both alleles are expressed in a phenotype.

◆ **Incomplete dominance**: where a dominant allele does not completely mask the effects of a recessive allele, resulting in heterozygotes having an intermediate phenotype.

> ### Common mistake
>
> Do not confuse codominance with incomplete dominance. In codominance, both alleles show through in the phenotype equally, because no allele can block or mask the expression of the other allele. Incomplete dominance is a condition where a dominant allele does not completely mask the effects of a recessive allele, resulting in a new phenotype.

> ### Top tip!
>
> Unlike monohybrid crosses with a dominant–recessive relationship for alleles, which produce a 3:1 ratio if both parents are heterozygous, a genetic cross where there is incomplete dominance produces a 1:2:1 ratio (i.e. there is an extra phenotype for the combined alleles).

> **Concept: Change**
>
> In incomplete dominance, a combination of two different alleles for the same trait leads to a third, different phenotype. The outcome of a genetic cross of two heterozygous individuals therefore does not result in a 3:1 phenotypic ratio of dominant to recessive phenotype, but a changed ratio of 1:2:1, with 50% of offspring showing the intermediate phenotype.

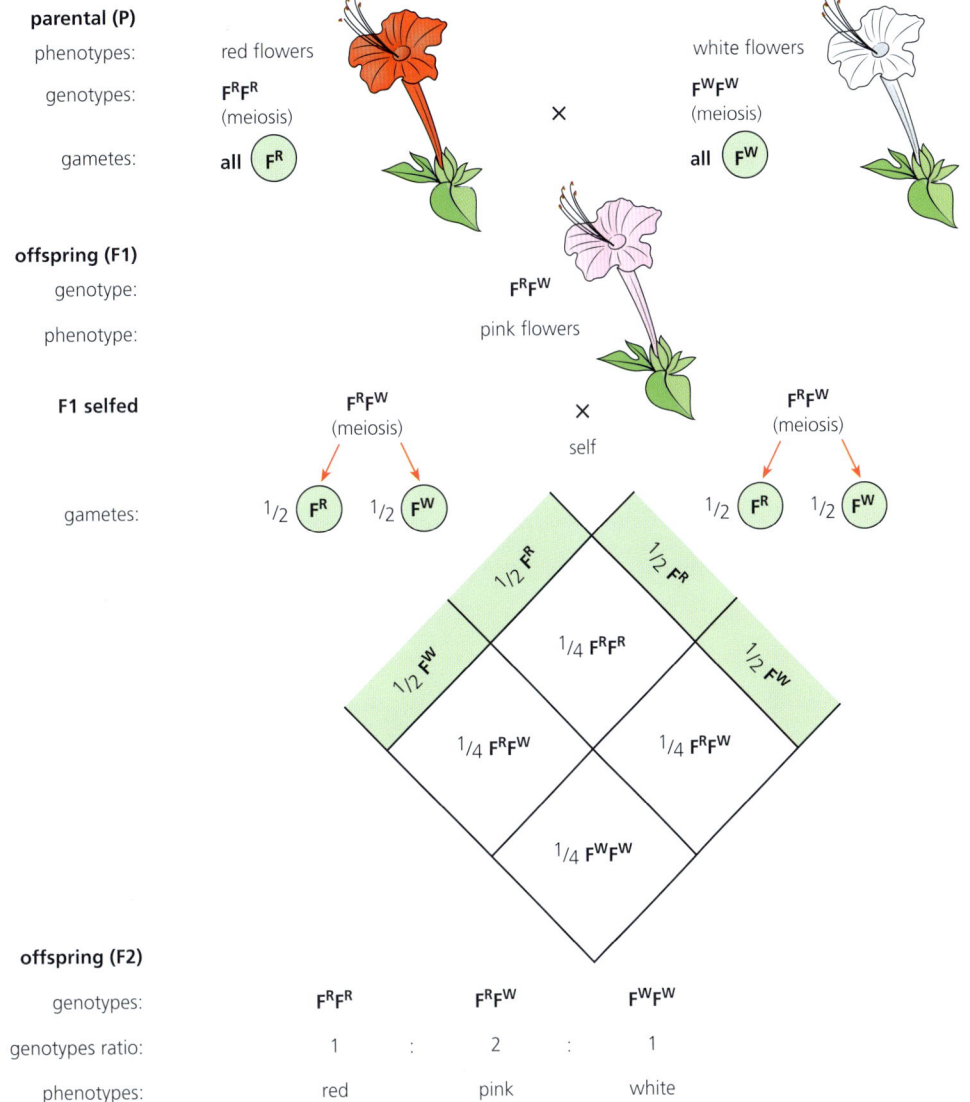

■ Figure D3.2.9 Incomplete dominance in the ornamental flower, *Mirabilis jalapa*

> 4 a **Construct** (using pencil and paper) a monohybrid cross between cattle that have a gene for coat colour with red and white alleles that shows incomplete dominance. Homozygous red and homozygous white parents cross to produce roan offspring (red and white hairs together).
>   b **Predict** what offspring you would expect and in what proportions, when a sibling cross (equivalent to selfing in plants) occurs between roan offspring.

Sex determination in humans and inheritance of genes on sex chromosomes

Chromosomes can be arranged in order of size, shown as a **karyogram** (Figure A3.1.10, page 112). In the karyogram, the final pair of chromosomes is not numbered. Rather, they are labelled X and Y. These are known as the **sex chromosomes**; they decide the sex of the individual. All the other chromosomes (pairs numbered 1 to 22) are autosomes. The X chromosome is longer than the Y chromosome, so it can carry far more genes.

> **Link**
>
> Karyotyping and karyograms are covered in Chapter A3.1, page 111.

Egg cells produced by meiosis all carry an X chromosome, but 50% of sperm carry an X chromosome and 50% carry a Y chromosome. At fertilization, an egg cell may fuse with a sperm carrying an X chromosome, leading to a female offspring. Alternatively, the egg cell may fuse with a sperm carrying a Y chromosome, leading to a male offspring. So, the sex of offspring in humans (and all mammals) is determined by the male partner. We would expect equal numbers of male and female offspring to be produced over time by a breeding population (Figure D3.2.10).

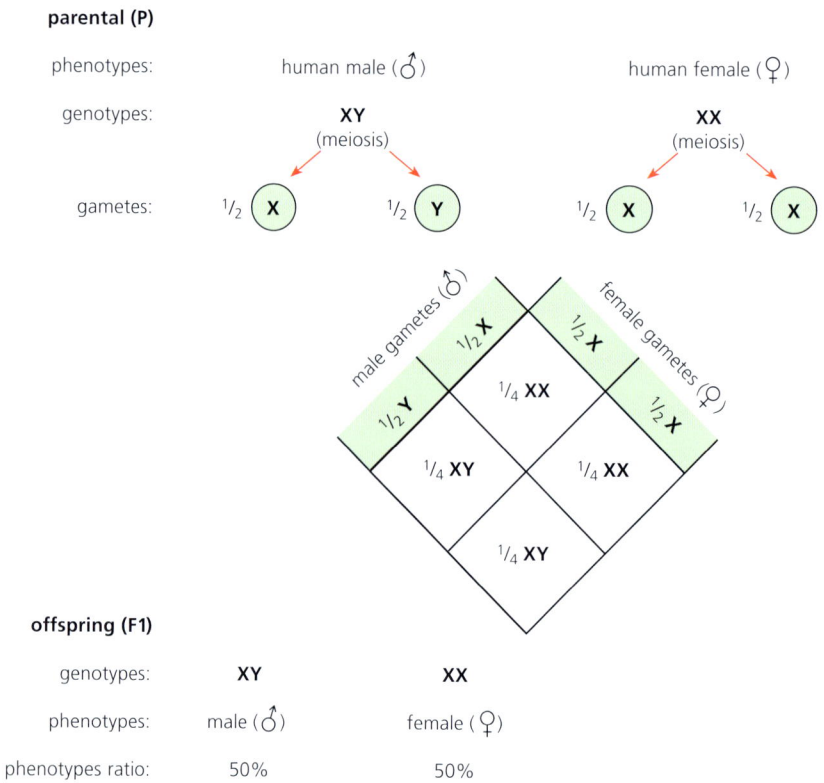

■ **Figure D3.2.10** X and Y chromosomes and the determination of sex

Human X and Y chromosomes and the control of sex

Initially, male and female embryos develop identically in the uterus. At the seventh week of pregnancy, however, a cascade of developmental events is triggered, leading to the growth of male genitalia if a Y chromosome is present in the embryonic cells.

On the Y chromosome is the **prime male-determining gene**. This gene codes for a protein called **testis-determining factor (TDF)**. TDF functions as a molecular switch; on reaching the embryonic gonad tissues, TDF initiates the production of a relatively low level of testosterone. The effect of this hormone at this stage is to inhibit the development of female genitalia, and to cause the embryonic genital tissues to form testes, scrotum and penis.

In the absence of a Y chromosome, the embryonic gonad tissue forms an ovary. Then, partly under the influence of hormones from the ovary, the female reproductive structures develop. It is therefore the sex chromosome in sperm which determines whether a zygote develops certain male-typical or female-typical physical characteristics.

Common mistake

The condition for determining a female is the absence of the Y chromosome rather than the presence of the X chromosome.

Haemophilia as an example of a sex-linked genetic disorder

◆ **Sex linkage**: a special case of linkage occurring when a gene is located on a sex chromosome (usually the X chromosome).

Genes present on the sex chromosomes are inherited with the sex of the individual. They are said to be **sex-linked** characteristics. Sex linkage is a special case of linkage occurring when a gene is located on a sex chromosome (usually the X chromosome).

The inheritance of these sex-linked genes is different from the inheritance of genes on the autosomal chromosomes. This is because the X chromosome is much longer than the Y chromosome (since many of the genes on the X chromosome are absent from the Y chromosome). In a male (XY), most alleles on the X chromosome lack a corresponding allele on the Y and will be apparent in the phenotype **even if they are recessive**. The area of the X chromosome that does not have a corresponding sequence of genes on the Y chromosome is called the **non-homologous region** (Figure D3.2.11).

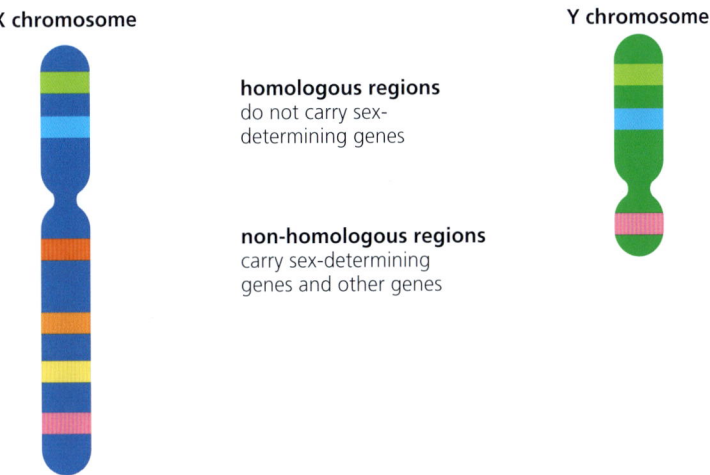

■ Figure D3.2.11 Comparing an X and Y chromosome

In a female (XX), a single recessive gene on one X chromosome may be masked by a dominant allele on the other X and would not be expressed. A human female can be homozygous or heterozygous with respect to sex-linked characteristics, whereas males have only one allele.

A heterozygous individual with a recessive allele of a gene that does not have an effect on their phenotype is known as a **carrier**; they carry the allele but it is not expressed. So, female carriers are heterozygous for sex-linked recessive characteristics. Of course, the unpaired alleles of the Y chromosome are all expressed in the male. However, the alleles on the (short) Y chromosome are mostly concerned with male structures and male functions. Examples of recessive conditions controlled by genes on the X chromosome are red–green colour blindness and haemophilia.

If a break occurs in the circulatory system of a mammal, there is a risk of uncontrolled bleeding. Usually, this risk is averted by the blood-clotting mechanism (Chapter C3.2, page 520). Haemophilia is a rare, genetically determined condition in which the blood does not clot normally. The result is frequent and excessive bleeding.

There are two forms of haemophilia, known as haemophilia A and haemophilia B. They are due to a failure to produce adequate amounts of particular blood proteins that are essential to the complex blood-clotting mechanism. Today, haemophilia is effectively treated by the administration of the clotting factor that the patient lacks.

> ● **Top tip!**
>
> In a female (XX), a single recessive gene on one X chromosome may be masked by a dominant allele on the other X and would not be expressed. A human female can be homozygous or heterozygous with respect to sex-linked characteristics, whereas males have only one allele.

> ● **Top tip!**
>
> In examples of sex linkage, you should show alleles carried on X chromosomes as superscript letters on an uppercase X.

Haemophilia is a sex-linked condition because the genes controlling production of these blood proteins are located on the X chromosome. Haemophilia is caused by a recessive allele. As a result, haemophilia is largely a disease of the male – since a single X chromosome carrying the defective allele (X^hY) will result in disease. For a female to have the disease, she must be homozygous for the recessive gene (X^hX^h), but this condition is usually fatal in the uterus, typically resulting in a natural abortion.

A female with only one X chromosome with the recessive allele (X^HX^h) is a carrier. She has a normal blood-clotting mechanism. When a carrier is partnered by a normal male, there is a 50% chance of the daughters being carriers and a 50% chance of the sons having haemophilia (Figure D3.2.12).

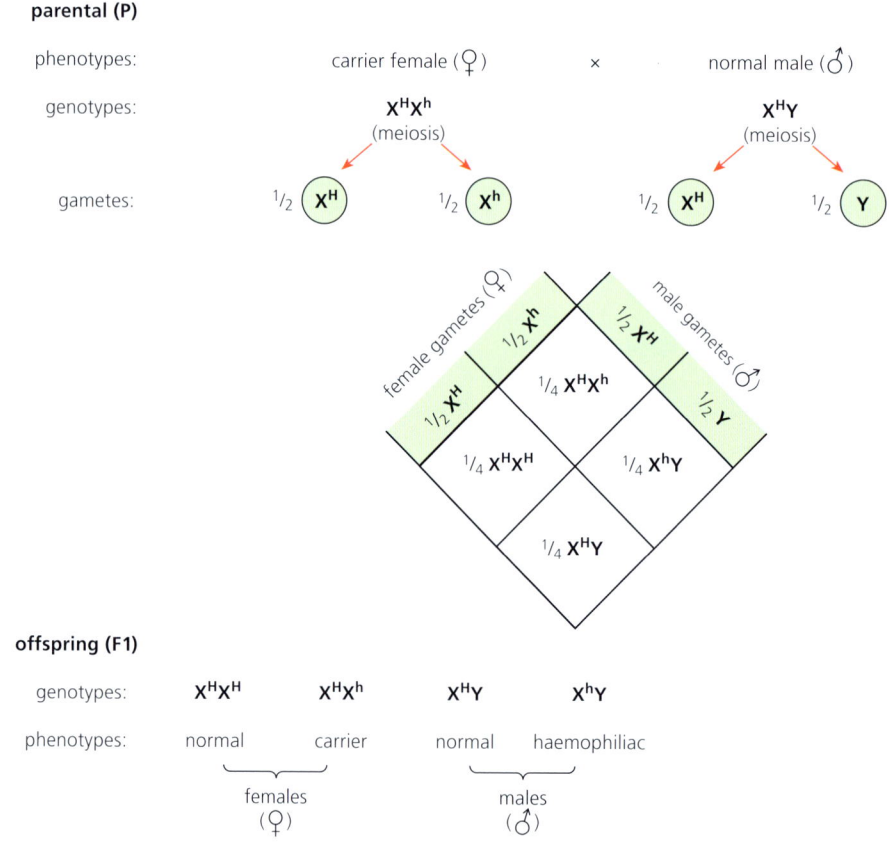

■ Figure D3.2.12 The inheritance of haemophilia

5 Haemophilia results from a sex-linked gene. The disease is most common in males, but the haemophilia allele is on the X chromosome. **Explain** this apparent anomaly.

● **Common mistake**

The appropriate sex-linkage allele symbols must be used when showing genetic crosses, with upper-case and lower-case letters shown as superscripts to indicate genotype on the X chromosome only. Y chromosomes are not associated with sex-linked genes as these are invariably linked to the X chromosome, not the Y.

Pedigree charts to deduce patterns of inheritance of genetic disorders

Tool 3: Mathematics

Representing familial genetic relationships using pedigree charts

Studying human inheritance by experimental crosses (with selected parents, sibling crosses, and the production of large numbers of progeny) is out of the question. Instead, we may investigate the pattern of inheritance of a particular characteristic by researching a family pedigree, where appropriate records of their ancestors exist. A human pedigree chart uses a set of rules. These are identified in Figure D3.2.13.

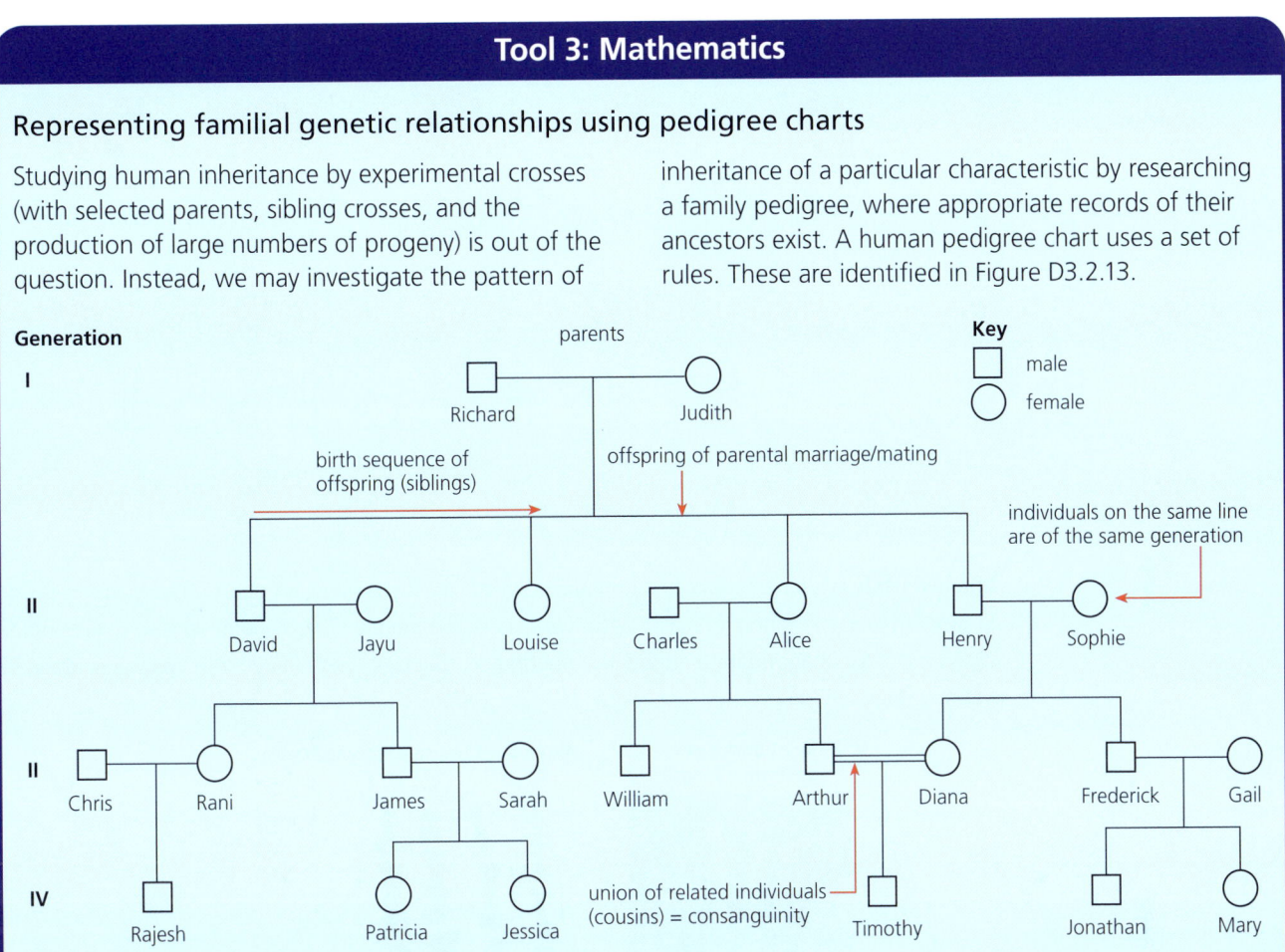

■ Figure D3.2.13 An example of a human pedigree chart

6 In the human pedigree chart in Figure D3.2.13 **state**:
   a who are the female grandchildren of Richard and Judith
   b who are Rajesh's (i) grandparents and (ii) uncles
   c how many people in the chart have parents unknown to us
   d the names of two offspring who are cousins.

■ Analysis of pedigree charts to deduce the pattern of inheritance

We can use a pedigree chart to detect conditions that are due to dominant and recessive alleles. In the case of a characteristic due to a dominant allele, the characteristic tends to occur in one or more members of the family in every generation. On the other hand, a recessive characteristic is seen infrequently, often skipping many generations.

For example, albinism is a rare inherited condition of humans and other mammals in which the individual has a block in the biochemical pathway by which the pigment melanin is formed. Albinos have white hair, very light-coloured skin and pink eyes. Albinism shows a pattern of recessive monohybrid inheritance in humans. In the chart shown of a family with albino members, albinism occurs infrequently, skipping two generations altogether (Figure D3.2.14).

People with albinism must be homozygous for the recessive albino allele (**pp**).
People with normal skin pigmentation may be homozygous normal (**PP**) or carriers (**Pp**).

Key: ☐ ♂ / ○ ♀ normal skin colour; ■ / ● people with albinism

Generation I — 1, 2

Generation II — carrier, carrier, carrier

Generation III — 1, 2, 3, 4, 5, 6, 7 (X), 8

Generation IV

■ **Figure D3.2.14** Pedigree chart of a family with members with albinism

This is a typical family tree for inheritance of a characteristic controlled by a recessive allele. In generation I, individual 2 must be **pp** (with albinism). In generation II, all offspring must be carriers (**Pp**) because they have received a recessive allele (**p**) from their mother. In generation III, individual 8 must be a carrier and her partner **X** must also be a carrier, since their offspring include a son with albinism.

Brachydactyly is a rare condition of humans in which the fingers are very short. Brachydactyly is due to a mutation in the gene for finger length. Unusually, the mutant allele is dominant, so the condition shows a pattern of dominant monohybrid inheritance; that is, it tends to occur in every generation in a family (Figure D3.2.15).

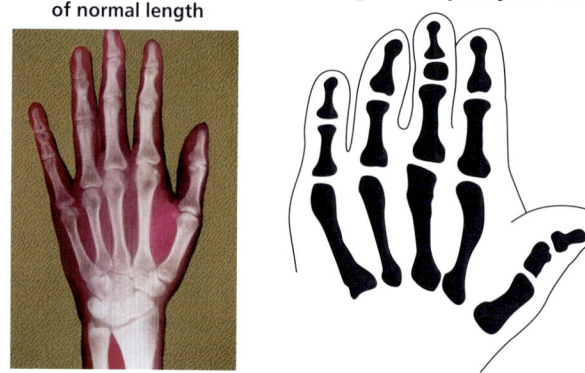

X-ray of bones of hand of normal length

Drawing of brachydactylous hand

7 If a homozygous normal-handed parent (**nn**) had a child with a heterozygous brachydactylous parent (**Nn**), **calculate**, using a Punnett grid, the probability of an offspring with brachydactylous hands. **Construct** a genetic diagram to show your workings.

■ **Figure D3.2.15** Brachydactyly, and pedigree chart of a family with brachydactylous genes

Pedigree chart of family with brachydactylous alleles

Key: ☐ ♂ / ○ ♀ normal; ■ / ● brachydactylous individuals

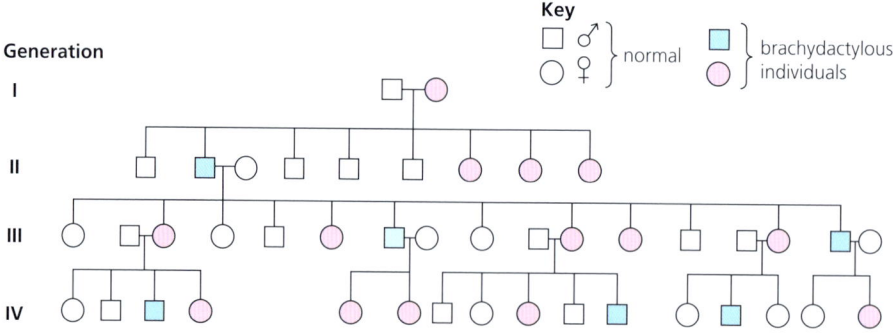

In **generation I** the parents are assumed to be normal male (**nn**) and brachydactylous female (heterozygous **Nn**), as the ratio of offspring is similar to that of a test cross.

In **each subsequent generation** about half the offspring are brachydactylous (i.e. **Nn** or **NN**) and half are normal (**nn**).

Theme D: Continuity and change – Organisms

> **Top tip!**
>
> Inbreeding increases the chance of homozygous recessive offspring for autosomal genetic disorders.

Many societies prohibit marriage between close relatives – as well as for societal or religious reasons, there is also a genetic basis for such laws. In closely related individuals there is a reduction in variation due to meiosis because of the smaller number of near ancestors. The overall gene pool is therefore smaller than in the overall society. If two siblings have a child, the child has two rather than four grandparents. The probability is therefore increased of the child inheriting two copies of a harmful recessive allele rather than only one, which is more likely to have harmful effects.

> **Nature of science: Patterns and trends**
>
> Scientists draw general conclusions by inductive reasoning when they base a theory on observations of some but not all cases. A pattern of inheritance may be deduced from parts of a pedigree chart and this theory may then allow genotypes of specific individuals in the pedigree to be deduced. Inductive and deductive reasoning are covered in more detail in the Nature of science in A1.2, page 32.

> **TOK**
>
> The explanatory power of genetic inheritance when considering many aspects of human variation is very strong. However, since Mendel's initial work and throughout the nineteenth and twentieth centuries, inheritance has been used in attempts to explain all sorts of human variation that have little to do with genes, or in cases where genes are only a part of a full explanation. 'Social Darwinism' was a prominent theory in early twentieth century America used to explain political, social and economic views, and suggested that Darwin's theory that genetic mutations and environmental 'fitness' meant that people in power (social, political or economic) were somehow a genetically better fit with those positions. Many people at the margins of society were deemed to be unfit and were forced to undergo sterilization in the expectation that a 'better' gene pool would alleviate many social issues.
>
> This highlights the potential damage that can be done when facts and theories within the scope of one Area of Knowledge are used to try to solve problems in another. Here the truths of genetic inheritance were used uncritically to try to explain and solve problems in the human sciences and the result was great human injustice.

> **Link**
>
> Continuous and discontinuous variation are defined and covered in Chapter A3.1, page 103.

◆ **Polygenic inheritance**: inheritance of phenotypic characters (such as height and eye colour in humans) that are determined by the collective effects of several different genes.

> **Common mistake**
>
> Do not confuse multiple genes with multiple alleles. 'Multiple alleles' refers to several different versions of the same gene, resulting in different phenotypic properties. 'Multiple genes' refers to several different genes that relate to polygenic inheritance.

Continuous variation due to polygenic inheritance and/or environmental factors

Variation is a feature of life, and there are two types of variation: **continuous** and **discontinuous**.

The clear-cut difference in an inherited characteristic, such as blood type, is an example of **discontinuous** or **discrete variation** – there is no intermediate form and no overlap between the two phenotypes.

In fact, very few characteristics of organisms are controlled by a single gene. Mostly, characteristics of organisms are controlled by a number of genes. Groups of genes which together determine a characteristic result in **polygenic inheritance**.

The genes that may result in polygenic inheritance are often (but not necessarily always) located on different chromosomes. Any one of these genes has a very small effect on the phenotype, but the combined effect of all the genes of the polygene is to produce infinite variety among the offspring. This variety we refer to as **continuous variation**.

Many features of humans are controlled by polygenes, including body weight and height (see Figure A3.1.2, page 103).

Human skin colour

The colour of human skin is due to the amount of the pigment called **melanin** produced in the skin. Melanin synthesis is genetically controlled. It seems that three, four or more separately inherited genes control melanin production. The outcome is an almost continuous distribution of skin colour from very pale (no alleles coding for melanin production) to very dark brown (all alleles for skin colour code for melanin production).

In Figure D3.2.16, polygenic inheritance of human skin colour involving only two independent genes is illustrated. This is because dealing with all four genes is unwieldy and the principle can be demonstrated sufficiently clearly using just two genes.

It should be noted, too, that both human height and skin colour are characteristics that may be influenced by environmental factors.

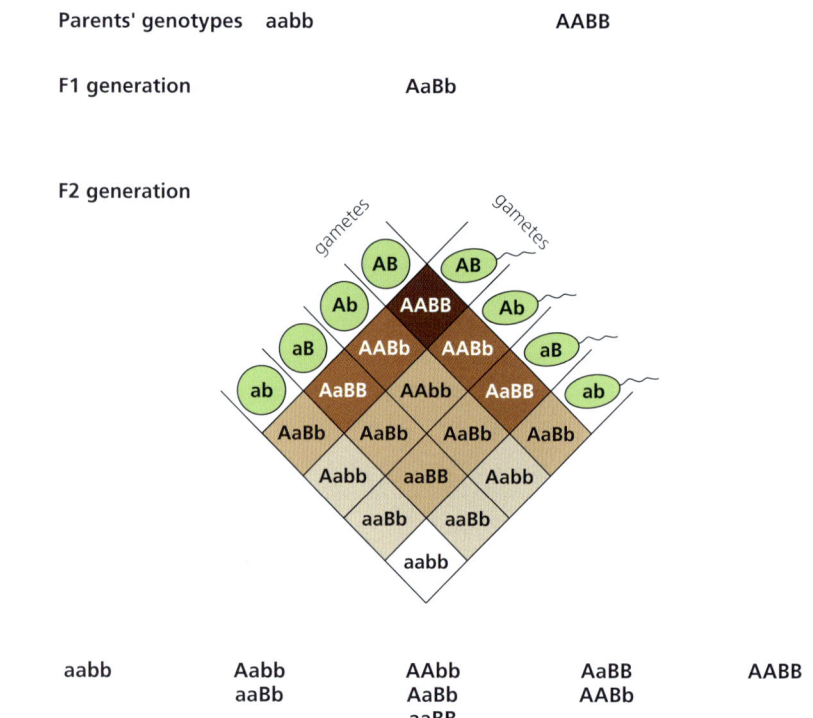

■ **Figure D3.2.16** Human skin colour as a characteristic controlled by two independent genes – an illustration of polygenic inheritance

> ● **Top tip!**
> You will only need to be able to draw and interpret dihybrid crosses, as shown in Figure D3.2.16, at HL not SL. The figure here is to indicate the complexity of patterns in the inheritance of skin colour. Dihybrid crosses are explored in the HL section of this chapter.

There are many other examples of polygenic inheritance. In all cases, the number of genes controlling a characteristic does not have to be large before the variation in a phenotype becomes more or less continuous. The outcome is that characteristics controlled by polygenes show continuous variation. Nevertheless, the individual genes concerned are inherited in accordance with the principles established above. However, there are so many intermediate combinations of alleles that the discrete ratios are not observed.

■ Distinguishing between continuous and discrete variables

You need to understand the distinction between continuous variables such as skin colour and discrete variables such as ABO blood group (see Chapter A3.1, page 103). You also need to be able to apply measures of central tendency such as mean, median and mode. We will look at this now.

> **Top tip!**
> You can use the range for data that are not normally distributed.

Data obtained from biological experiments may show a 'normal distribution': this means that when the frequency of particular classes of measurement is plotted against the classes of measurement, a symmetrical bell-shaped curve is obtained (Figure D3.2.17). Explanations of mode, median and mean are found below.

Normal distribution curve

Most biological data show variability, but with values grouped symmetrically around a central value.

Here the mode, median and mean coincide.

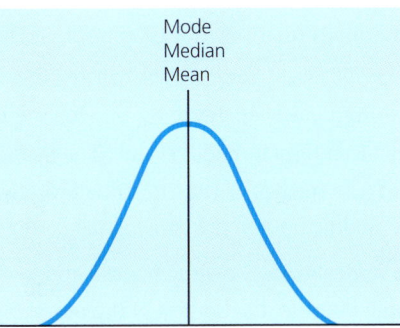

Skewed distribution

Values reduce in frequency more rapidly on one side of the most frequently obtained value than on the other.

Here the difference between the mean and mode is a measurement of 'skewness' of the data.

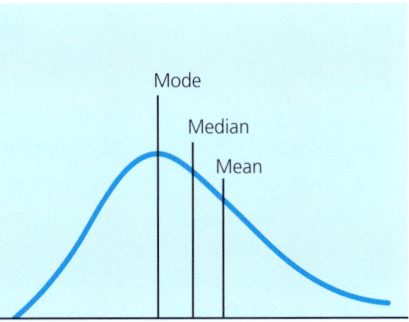

■ **Figure D3.2.17** Frequency distributions of normal and skewed (non-normal) data

Tool 3: Mathematics

Calculating measures of central tendency: mean, median and mode

The normal distribution shows symmetrical distribution of data around the central tendency (a central value for a probability distribution). There are three different ways of calculating the central tendency:

- the **mode**: the most frequent value in a set of values
- the **median**: the middle value in a set of values arranged in ascending order. If there is an even number of items in a data set then the median is found by taking the average (mean) of the two middle numbers.
- the average or arithmetic **mean**: calculated by dividing the sum of the individual values by the number of values obtained. The formula for the arithmetic mean is:

$$\bar{x} = \frac{\sum x}{n}$$

where $\bar{x}$ = arithmetic mean

$\sum x$ = the sum of all the measurements

n = the total number of measurements.

Simple mean and range

The purpose of averaging data (provided it follows a normal distribution) is to reduce the effect of random errors.

For example, the mean, median and mode can be calculated for the data concerning the frequency of blood group A shown in Table D3.2.3.

■ **Table D3.2.3** The top 12 countries of the world whose populations have the highest concentration of blood type A

Country	Brazil	Spain	Norway	Australia	Netherlands	Belgium
Population (%)	8.0	8.0	7.2	7.0	7.0	7.0
Country	UK	France	Denmark	Sweden	Austria	Portugal
Population (%)	7.0	7.0	7.0	7.0	7.0	6.6

Mean = (8 + 8 + 7.2 + 7 + 7 + 7 + 7 + 7 + 7 + 7 + 7 + 6.6) / 12 = 6.98

Mode = 7.00

Median = 7.00

D3.2 Inheritance

Box-and-whisker plots to represent data for a continuous variable

A **box-and-whisker plot** is used to show differences in the mean and range (that is, the difference between the maximum value and the minimum value) of a series of data.

For data tables with data that are not normally distributed, the appropriate descriptive statistics are **medians** and **quartiles**, and the appropriate graph is a box-and-whisker plot (Figure D3.2.18).

Quartiles divide a distribution, ordered from low to high values, into four equal parts. Quartiles are determined by splitting the data in half around their median value. The median of the 'lower half' of the data will give you your first quartile. The median of the 'upper half' of the split data will give you your third quartile. Approximately, the middle 50% of the data fall inside the box.

The ticks at the top and bottom of the vertical lines show the highest and lowest values in the set of data. The top of the box shows the upper quartile, the bottom of the box shows the lower quartile, and the horizontal line within the box represents the median. The central rectangle spans the interquartile range (IQR), i.e. the first quartile to the third quartile. The IQR shows the spread of data in the middle half of the distribution. A data point is categorized as an outlier if it is more than $1.5 \times$ IQR above the third quartile or below the first quartile.

> **Top tip!**
>
> The mode is not typically used as a measure of central tendency in biology, but it can be useful in describing a bimodal distribution, which has two peaks or modes. This type of distribution can be caused by disruptive selection.

> **Top tip!**
>
> Box-and-whisker plots can either be shown vertically or horizontally (Figure D3.2.18).

Tool 3: Mathematics

Constructing box-and-whisker plots

Consider the following set of data:

1, 1, 2, 2, 4, 6, 6, 7, 8, 8, 9, 10, 10, 12

The first quartile is 2 (the median of the first half of the data), the median is 6.5 (because there are 14 data points, the median is calculated by taking the mean of the middle two values i.e. $(6 + 7) \div 2 = 6.5$), and the third quartile is 9 (the median value of the upper half of data). The smallest value is 1, and the largest value is 12. The following image shows the constructed box plot.

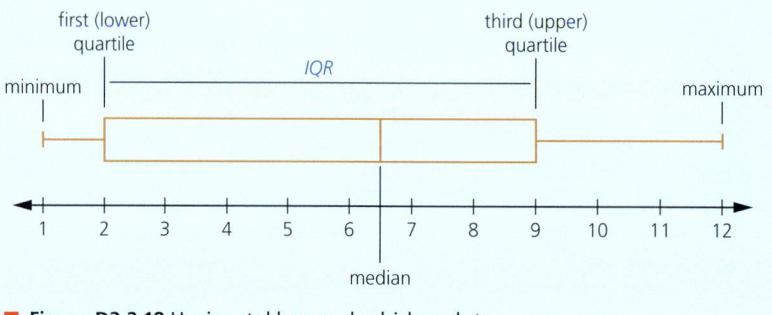

■ **Figure D3.2.18** Horizontal box-and-whisker plot

Inquiry 2: Collecting and processing data

Collect data for the height of each member of your class. You need to decide how best to record accurate data.

Process the data by calculating the median and IQR range for your data. Plot the results using a box-and-whisker graph. Identify and justify the removal or inclusion of outliers in data.

Identify, describe and explain the patterns shown in your data.

Constructing histograms to plot data for continuous variables

Tool 3: Mathematics

Constructing histograms

Histograms can be used to show the distribution of continuous data. A study of fruit length in a commercial crop was undertaken on a sample size of 350 fruits. Each fruit had a different length, so the classes of size were artificially defined. This type of data is best plotted as a histogram.

Length/mm	Frequency of fruits
40–9	0
50–9	2
60–9	9
70–9	21
80–9	29
90–9	43
100–9	69
110–19	66
120–9	48
130–9	32
140–9	19
150–9	11
160–9	1
170–9	0

Construct a **histogram** of these data, using an Excel spreadsheet.

- To do this, the frequency classes are entered in column 1 and the numbers at each class in column 2.
- Highlight the two columns, and then in the 'insert' menu click 'chart'. Choose the column option.
- Your histogram requires a title, and each axis needs labelling.
- The spaces between the bars must be removed because the data are continuous. To do this, click on one of the 12 bars and hold down until the bars are highlighted. Using the 'format' menu select 'selected data point'. Now use the 'options' tab and set the gap width to zero.

Applying general mathematics

Data obtained from biological experiments like this frequently show a 'normal distribution' (i.e. a symmetrical bell-shaped curve is obtained when the frequency of particular classes of measurements are plotted against the classes of measurements).

We can now examine the data you used to plot the histogram.

1 **Calculate** how clustered (closely spaced) the readings are. This can be expressed as:
   - the **average** or arithmetic mean (the sum of the individual values divided by the number of values)
   - the **mode**, the most frequent value in a set of values
   - the **median**, the middle value in a set of values arranged in ascending order.
2 **Deduce** how widely distributed (spread out) the readings are. This can be expressed as the **standard deviation** (SD) of the mean. This is a measure of the variation from the mean of a set of values. A low SD indicates that the values differ very little from the mean.

Segregation and independent assortment of unlinked genes in meiosis

Mendel crossed pure-breeding pea plants (P generation) from round seeds with yellow cotyledons (seed leaves) with pure-breeding plants from wrinkled seeds with green cotyledons. All the progeny (F1 generation) were round, yellow peas.

When plants grown from these seeds were allowed to self-fertilize the following season, the resulting seeds (F2 generation) – of which there were more than 500 to be classified and counted – were of the following four phenotypes, and they were present in the ratio shown in Table D3.2.4.

■ Table D3.2.4 Mendel's F2 generation

Phenotypes	round seed with yellow cotyledons	round seed with green cotyledons	wrinkled seed with yellow cotyledons	wrinkled seed with green cotyledons
Ratio	9	3	3	1

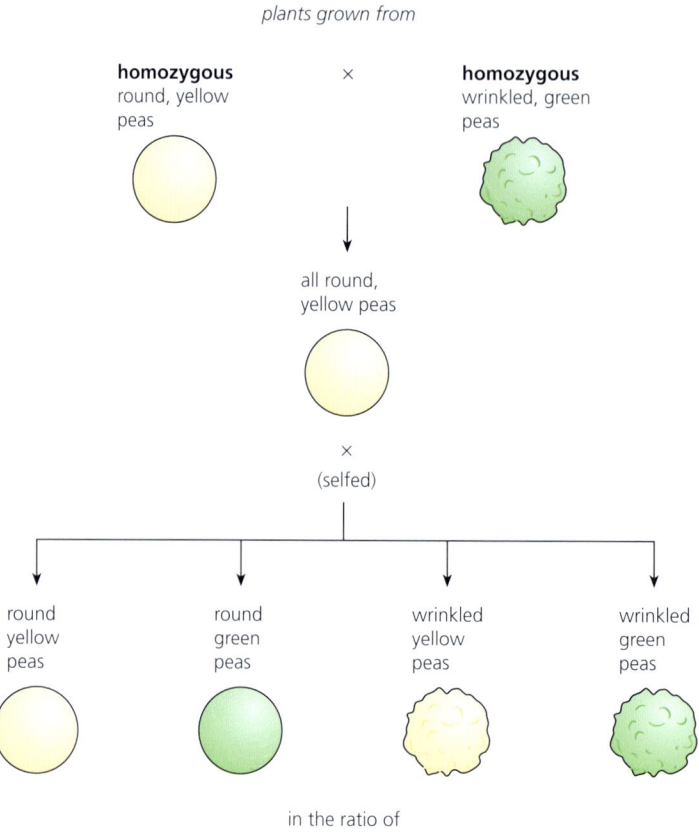

■ Figure D3.2.19 Mendel's dihybrid cross

8 In Figure D3.2.19, **identify** the progeny that are
   a heterozygous for both traits
   b homozygous for both traits.

Mendel noticed that two new combinations, not represented in the parents (i.e. **recombinations**), appeared in the progeny; both round and wrinkled seeds appear with either green or yellow cotyledons. This result shows that two pairs of factors were inherited independently and, therefore, were on separate chromosomes. Mendel noticed that either one of a pair of contrasting characters could be passed to the next generation. This meant that a heterozygous plant must produce four types of gametes in equal numbers (Figures D3.2.19 and D3.2.20).

Dihybrid crosses involving pairs of unlinked autosomal genes

Mendel did not express the outcome of the dihybrid cross as a succinct law. However, today we call Mendel's second law the **Law of independent assortment**. It is stated as:

Two or more pairs of alleles segregate independently of each other as a result of meiosis, provided the genes concerned are not linked by being on the same chromosome.

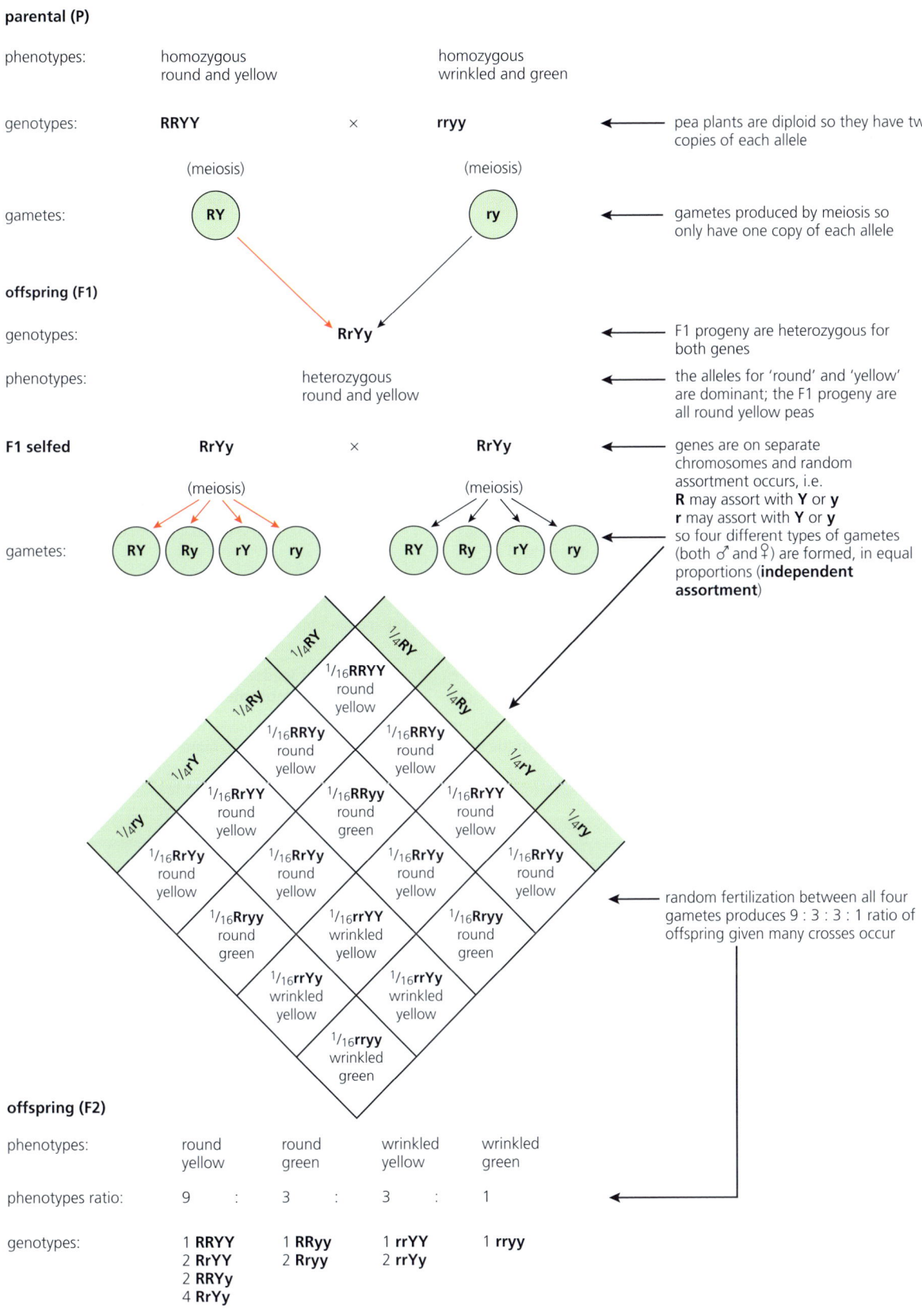

■ **Figure D3.2.20** Genetic diagram showing the behaviour of alleles in Mendel's dihybrid cross

> **HL ONLY**
>
> **9 Calculate** the phenotype ratio of offspring in a dihybrid cross when one parent is heterozygous for both genes and the other is homozygous recessive.

> ### ● Common mistake
>
> In a dihybrid cross, the gametes from each parent should include one allele of each gene, e.g. ry, rY, Ry or RY. It is incorrect to show one parent with the pair of alleles of one gene and the second parent with the alleles of the other gene.

> **10 Deduce** the positions of the genes for round/wrinkled and for yellow/green cotyledons within the nucleus of the pea plant. **Explain** the significance of their location (which was unknown to Mendel).

The phenotype ratio of offspring when both parents are heterozygous for both genes is therefore 9:3:3:1.

Notice the use of a Punnett grid diagram to predict the outcome of a breeding investigation in which independent assortment of alleles is occurring (Figure D3.2.20). By this device, every possible combination of maternal and paternal gametes – the product of random fertilization – is made. These are shown with as many rows/columns as there are unique male and unique female gametes. Each fraction represents the probability that a particular gamete or zygote will occur.

The relationship between Mendel's Law of independent assortment and meiosis is detailed in Table D3.2.5.

■ **Table D3.2.5** How the Law of independent assortment relates to meiosis

Mendel's dihybrid cross	Feature of meiosis
Within an organism exist 'breeding factors' that control characteristics like round or wrinkled seeds, and yellow or green cotyledons. These factors remain intact from generation to generation.	Each chromosome holds a linear sequence of genes. A particular gene always occurs on the same chromosome in the same position (locus) after each nuclear division.
There are two factors for each characteristic in each cell. One factor comes from each parent. (A recessive factor is not expressed in the presence of a dominant factor.)	The chromosomes of a cell occur in pairs, called homologous pairs. One of each pair came originally from each parent.
Factors separate in reproduction; either can be passed to an offspring. Only one of the factors can be in any gamete.	At the end of meiosis, each cell (gamete) contains a single member of each of the homologous pairs of chromosomes present in the parent cell.
The factors for seed shape and seed colour segregate independently of each other as a result of meiosis.	The genes for seed shape and seed colour are **on separate chromosomes.**
The 9:3:3:1 ratio shows that all four types of gamete are equally common. The inheritance of the two characteristics is separate.	The arrangement of bivalents at the equatorial plate of the spindle is random; maternal and paternal homologous chromosomes are independently assorted. In a large number of matings, all possible combinations of chromosomes will occur in equal numbers.

● Nature of science: Theories

9:3:3:1 and 1:1:1:1 ratios for dihybrid crosses are based on what has been called Mendel's second law. This law only applies if genes are on different chromosomes or are far apart enough on one chromosome for recombination rates to reach 50%. If genes are located closer together, these ratios may not be shown. This shows that there are exceptions to biological 'laws' under certain conditions.

A scientific law predicts the results of certain initial conditions, whereas a theory provides an explanation for these results. Predictions can be generated by theories, and if these are tested thoroughly by repetition and remain uncontested, they can become a law, such as demonstrated by Mendel's laws of inheritance.

Loci of human genes and their polypeptide products

You should explore genes and their polypeptide products in databases. You should find pairs of genes with loci on different chromosomes and also in close proximity on the same chromosome.

Genes are located on chromosomes. Each gene occupies a specific position on a chromosome; therefore each eukaryotic chromosome is a linear series of genes. Furthermore, the gene for a particular characteristic is always found at the same position or **locus** (plural, loci) on a particular chromosome (see Chapter A1.2, page 21).

Online databases can be used to locate the locus of specific genes in the human genome. The databases also provide information about the function of the gene (e.g. its polypeptide product).

ATL D3.2B

Access the following site: www.ncbi.nlm.nih.gov/gene

To identify a specific gene locus:
1. Type in the name of the gene of interest, such as testis-determining factor (TDF), which switches foetal development to 'male', or chloride channel protein (CFTR) – a mutant allele that causes cystic fibrosis.
2. Choose the species of interest (*Homo sapiens*) from the drop-down menu.
3. Click on the link for the gene.
4. Scroll to the 'Genomic context' section on the right-hand side of the screen to determine the specific position of the gene locus.

To identify the polypeptide products of the gene, carry out steps 1–3 above: the polypeptide product should be identified within the 'Summary' section.

To search for genes on the same chromosome, access this site: www.ncbi.nlm.nih.gov/genome/gdv/?org=homo-sapiens
1. In the box 'Search in genome' type in the gene of interest.

2. Access 'Ideogram view' on the left-hand side – this may appear automatically.
3. The highlighted chromosome shows you the location of the gene. Click on the image of the chromosome.
4. At the top of the screen, the chromosome is shown horizontally, with the gene highlighted with a horizontal purple line across the chromosome showing its location, or alternatively as small green arrows if the polypeptide product is the result of genes at more than one locus.
5. Click to the left of the gene and drag the cursor to the right to zoom in on this region of the chromosome. The top line of horizontal green lines shows the genes on this part of the chromosome. Click on a gene to identify its function.

Find pairs of genes in close proximity on the same chromosome and others with loci on different chromosomes. The relevance of whether genes are located on the same chromosome (i.e. linked) or on separate chromosomes has implications for how characteristics are inherited, which will be discussed in the next section.

Autosomal gene linkage

After the rediscovery of Mendel's work in the early 1900s, geneticists investigated other dihybrid crosses to confirm his results. For example, William Bateson and Reginald Punnett (who had devised the 'Punnett grid') crossed pure-breeding sweet pea plants with purple flowers and long pollen grains with plants having red flowers and round pollen grains. All the F1 plants were purple-flowered with long pollen grains. This shows that the allele for purple flower is dominant over the allele for red flower, and the allele for long pollen is dominant over that for round pollen.

When the F1 were self-crossed, however, most of the offspring resembled the parental phenotypes, but with a small number of **recombinants**. The actual results obtained are shown in Figure D3.2.21.

◆ **Recombinant**: a chromosome in which the genetic information has been rearranged.

parents	homozygous purple flowers, long pollen	×	homozygous red flowers, round pollen
genotypes:	F E / F E		f e / f e
gametes:			

offspring (F1): heterozygous purple flowers, long pollen

F E / f e

× self

offspring (F2):

flowers of sweet pea (*Lathyrus odoratus*)

Outcome	Phenotypes			
	Purple flower long pollen	Purple flower round pollen	Red flower long pollen	Red flower round pollen
expected	240	80	80	27
ratio	9	3	3	1
actual	296	19	27	85

■ **Figure D3.2.21** An example of linked genes in the sweet pea plant

Top tip!

In crosses involving linkage, the symbols used to denote alleles should be shown alongside vertical lines representing homologous chromosomes.

The Mendelian ratio of 9:3:3:1 was not obtained. Since most of the F2 offspring resembled the parental phenotypes (with a small number of recombinants), it seemed reasonable to conclude that the genes for flower colour and pollen shape were present on the same chromosome. If so, these genes were linked – they did not segregate in meiosis but were inherited together.

Notice that in crosses involving linkage, the alleles are typically shown as vertical pairs, as shown on the left, rather than as FfEe, for example.

Concept: Change

Linked genes change the expected genotype and phenotype of the F2 generation in a dihybrid cross.

Common mistake

Do not confuse 'gene linkage' with 'sex linkage'. Linkage refers to genes present together on the same chromosomes, whereas sex linkage refers to genes specifically associated with one of the sex chromosomes (usually the X chromosome) and not present on the other.

ATL D3.2C

The closer together the genes are on the same chromosome, the more frequently they will be inherited together.

Read further about gene linkage and how it can be studied using this site:
https://learn.genetics.utah.edu/content/pigeons/geneticlinkage

Produce a poster that shows how linked genes are inherited and how this differs from unlinked genes.

Recombinants in crosses involving two linked or unlinked genes

What caused the **recombinants** in the sweet pea experiment? If the genes concerned were on the same chromosome, when the F1 plants were crossed the appearance of the F2 offspring depended on whether a chiasma formed between these alleles, or elsewhere along the chromosome, during meiosis in gamete formation. *Look at Figure D3.2.22 now.* Here, the consequence of a chance chiasma – and its location – is made clear.

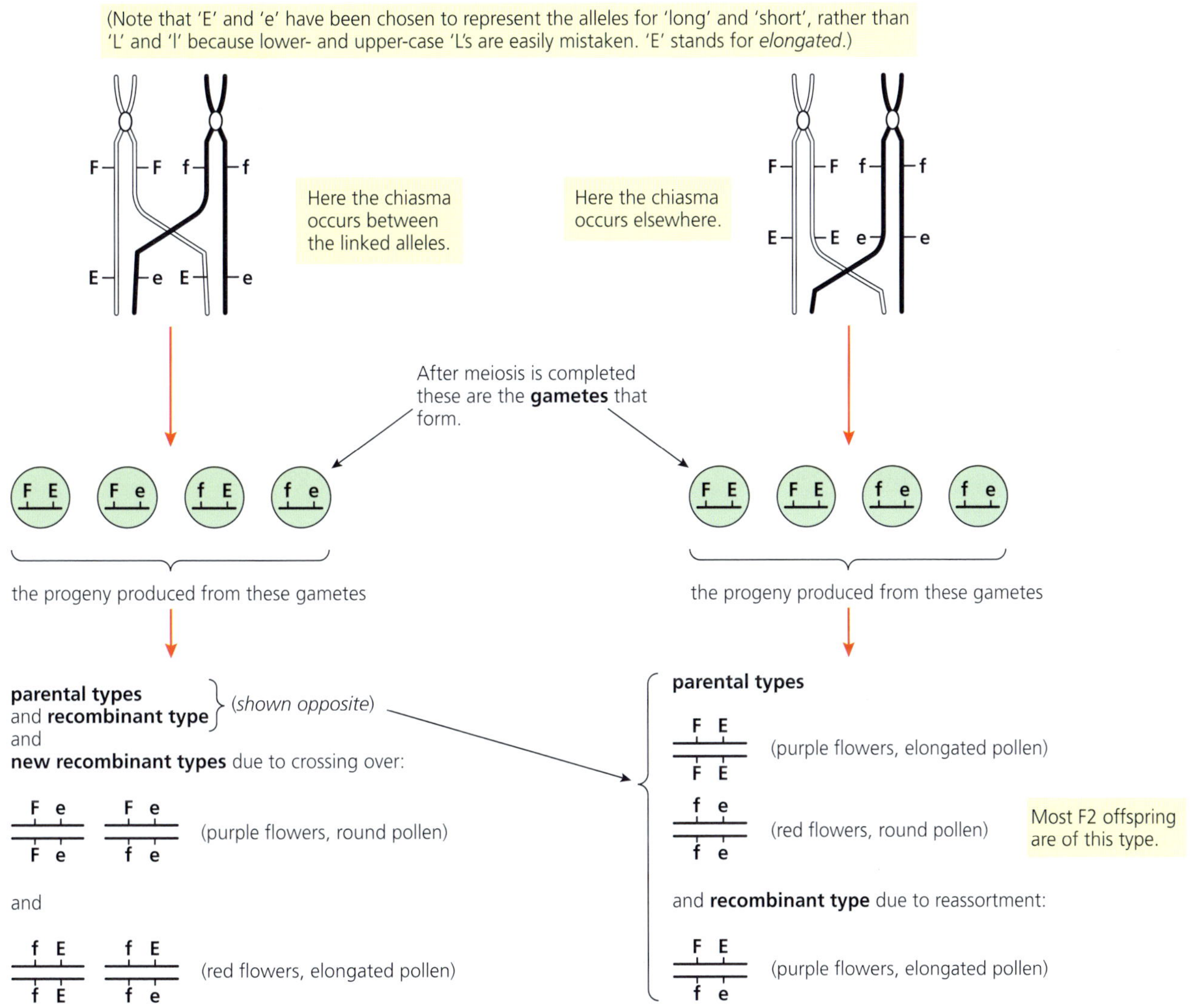

(Note that 'E' and 'e' have been chosen to represent the alleles for 'long' and 'short', rather than 'L' and 'l' because lower- and upper-case 'L's are easily mistaken. 'E' stands for *elongated*.)

Here the chiasma occurs between the linked alleles.

Here the chiasma occurs elsewhere.

After meiosis is completed these are the **gametes** that form.

the progeny produced from these gametes

the progeny produced from these gametes

parental types
and **recombinant type** (shown opposite)
and
new recombinant types due to crossing over:

$$\frac{F\ e}{F\ e} \quad \frac{F\ e}{f\ e}$$ (purple flowers, round pollen)

and

$$\frac{f\ E}{f\ E} \quad \frac{f\ E}{f\ e}$$ (red flowers, elongated pollen)

parental types

$$\frac{F\ E}{F\ E}$$ (purple flowers, elongated pollen)

$$\frac{f\ e}{f\ e}$$ (red flowers, round pollen)

Most F2 offspring are of this type.

and **recombinant type** due to reassortment:

$$\frac{F\ E}{f\ e}$$ (purple flowers, elongated pollen)

■ **Figure D3.2.22** Chiasmata and the origin of recombinants

In this example, both individuals are heterozygous for both genes. Recombinants can be identified in genotypes of offspring and in phenotypes of offspring. In Figure D3.2.25, page 756, the outcomes of a cross between an individual heterozygous for both genes and an individual homozygous recessive for both genes is demonstrated, and also how the recombinants, genotypes and phenotypes can be identified.

Drosophila and the work of Thomas Morgan

Drosophila melanogaster (the fruit fly) was first selected in 1908 by an American geneticist, Thomas Morgan, as an experimental organism for his series of investigations of Mendelian genetics, in this case in an animal. The remarkable breakthroughs in understanding that Morgan achieved resulted in the award of a Nobel Prize in 1933. His experimental work:

- showed that non-Mendelian ratios are commonly obtained in breeding experiments with *Drosophila* – but not always
- established that Mendel's 'factors' are linear sequences of genes on chromosomes (this is now called the **Chromosome Theory of Inheritance**)
- discovered sex chromosomes and sex linkage (page 739)
- demonstrated crossing over and the exchange of alleles between chromosomes, resulting from the chiasmata that form during meiosis.

Drosophila is an organism that commonly occurs around rotting vegetable material, existing in a form called a 'wild type' (a non-mutant form) and in various naturally occurring mutant forms (Figure D3.2.23). This animal rapidly became a useful experimental animal in the study of genetics because:

- *Drosophila* has only four pairs of chromosomes (Figure D3.2.23).
- From mating to emergence of adult flies (generation time) takes about 10 days at 25 °C.
- A single female fly produces hundreds of offspring.
- The flies are relatively easily handled, cultured on sterilized artificial medium in glass bottles (they can be temporarily anaesthetized for setting up cultures and sorting progeny).

11 Define the term *mutant*.

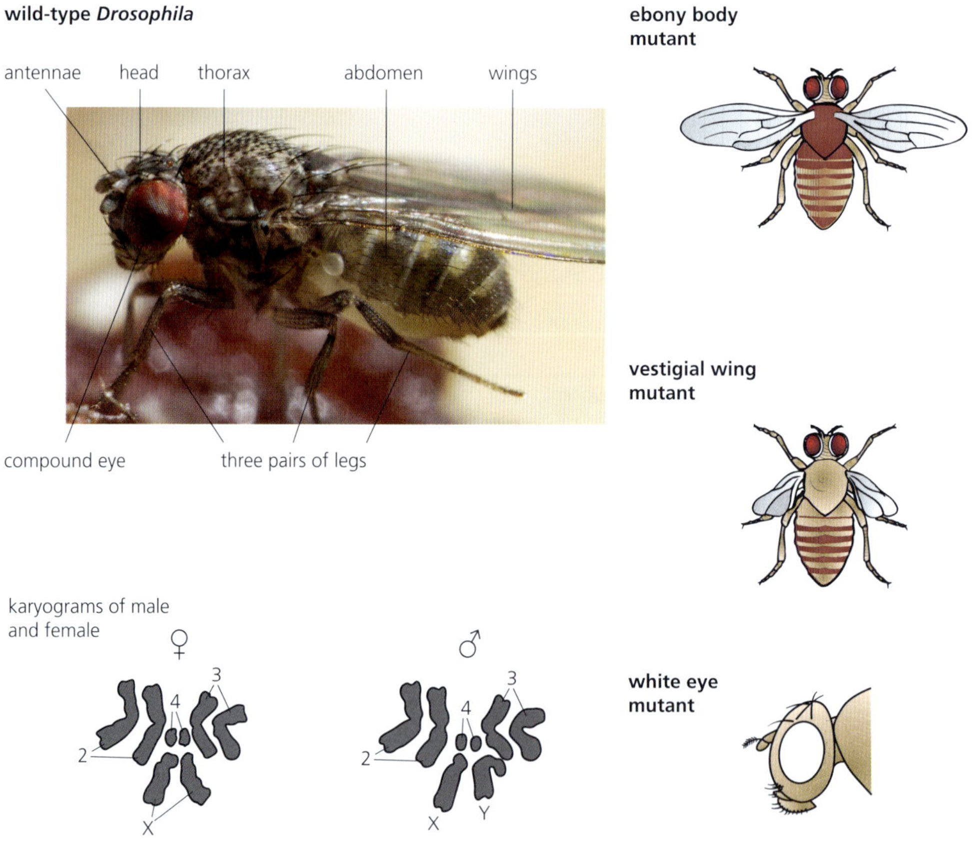

■ Figure D3.2.23 Wild-type *Drosophila* and some common mutants

Nature of science: Patterns and trends

Non-Mendelian ratios in *Drosophila*

Thomas Morgan, by careful observation and record keeping, made the discovery of linkage in *Drosophila*. Like others, he was aware of Mendel's discoveries. The genetic crosses he conducted gave results at odds with the expected Mendelian ratios. The cross in Figure D3.2.25, page 756, is one example. There, a fly homozygous for ebony body and curled wing was crossed with a fly heterozygous for normal body and straight wing. Note that the characteristic 'curled wing' is different from the characteristic 'vestigial wing' shown in Figure D3.2.24 (a cross concerned with contrasting characteristics controlled by genes on separate chromosomes).

This experiment, and many others, confirmed the existence of linked genes in *Drosophila*.

Recombinants in crosses involving unlinked genes

First, we will consider a dihybrid cross in *Drosophila* involving unlinked genes. In this experiment we are crossing normal flies (wild type) with flies that are homozygous for vestigial wing and ebony body (Figure D3.2.24). These characteristics are controlled by genes that are located on separate **autosomal chromosomes** (chromosomes other than the sex chromosomes, page 725).

The prediction of genotype and phenotype ratios in a dihybrid cross involving unlinked autosomal genes

After you have examined Figure D3.2.24, respond to question 12 that follows. This requires you to determine the genotype and phenotype of the F2 generation raised in Figure D3.2.24.

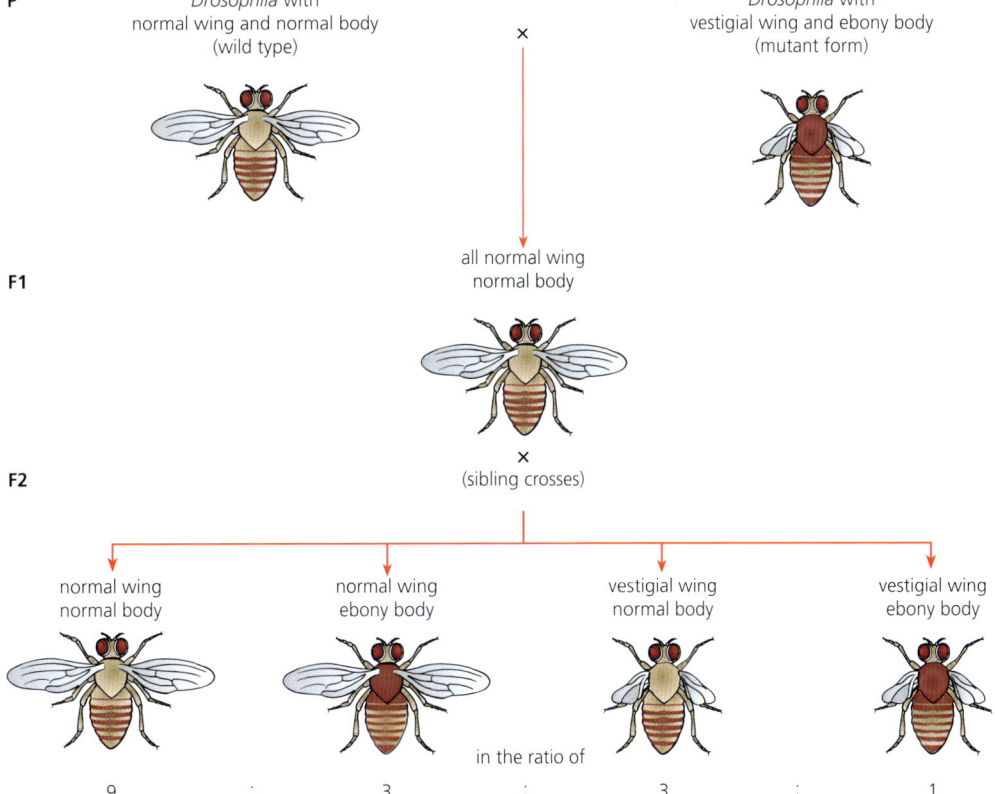

■ **Figure D3.2.24** A dihybrid cross in *Drosophila* – a summary

12 **Construct** a genetic diagram for the dihybrid cross shown in Figure D3.2.24, using the layout given in Figure D3.2.20. **Determine** the genotype and phenotype ratios of the F2 generation.

Recombinants in crosses involving linked genes

We have seen that for some traits, genes for two different characteristics are located on the same chromosomes, i.e. they are linked. Figure D3.2.25 shows two linked genes in *Drosophila* – body colour (ebony or normal) and wing shape (curled or straight). One of the flies in the cross is homozygous for both characteristics, and the other heterozygous. The heterozygous fly (GgSs) has dominant alleles for both traits on one homologous chromosome (GS) and recessive alleles on the other (gs). Without crossing over, the only gametes produced by this fly would either be GS or gs. Crossing over allows recombinants to be formed, i.e. gametes that contain Gs or gS: this means that all possible combination of the alleles for both traits can be produced (Figure D3.2.25).

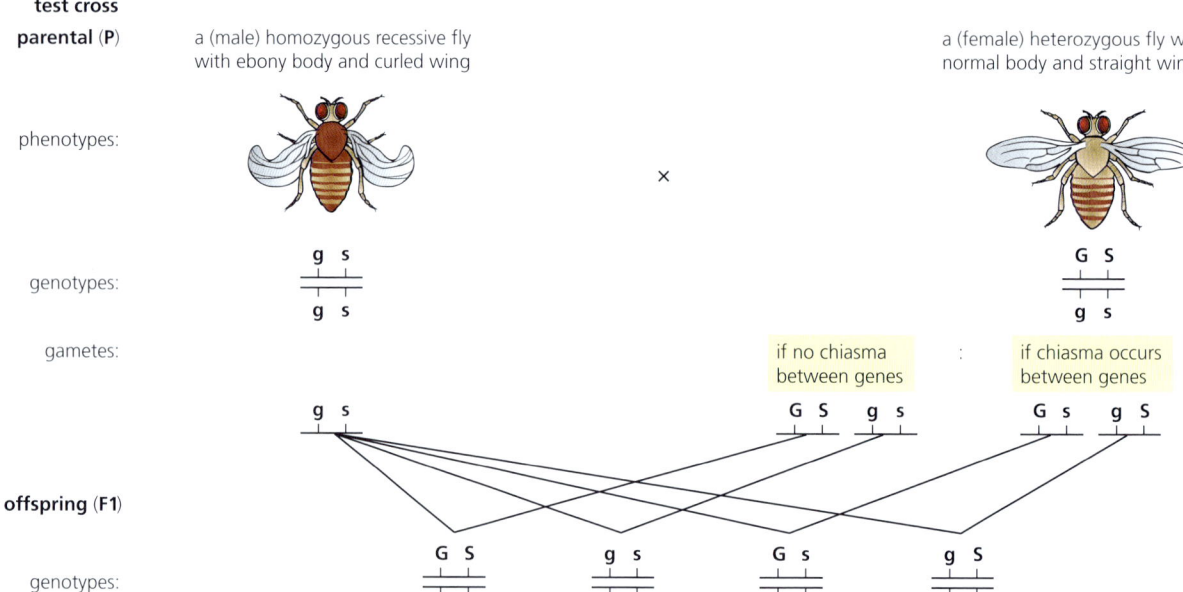

Notes:

1 If these genes were on separate chromosomes we would expect these offspring in the ratio 1:1:1:1, but these genes are linked;

2 If no crossing over occurred we would expect parental types only, in the ratio 1:1;

The outcome of this experiment was:

Offspring	Phenotypes	Genotypes	Numbers obtained
parental types	normal body straight wing	G S / g s	536
	ebony body curled wing	g s / g s	481
recombinants	normal body curled wing	G s / g s	101
	ebony body straight wing	g S / g s	152

■ Figure D3.2.25 A non-Mendelian ratio in a *Drosophila* cross

3 The majority of offspring were parental types, so we can conclude that few chiasmata occurred *between* the gene loci of these linked genes.

13 **Deduce** what recombinants may be formed in a cross involving the linked genes:

TOK

The law of independent assortment was soon found to have exceptions, such as when looking at linked genes. What is the difference between a 'law' and a 'theory' in science? Do these terms have the same meaning in other Areas of Knowledge? In what ways does confusion about how terms are used differently in other Areas of Knowledge lead to challenges in constructing knowledge?

Use of a chi-squared test on data from dihybrid crosses

Statistical tests

Statistical tests, sometimes called inferential statistics, involve calculations that allow differences to be compared between experimental treatments or samples to see if they are likely to have occurred as a result of variation in the data or if they are a treatment effect. If the statistics show a treatment effect, then the results are said to be **statistically significant**. A statistically significant result is one where there is a less than 5% probability (which can be denoted as $p = 0.05$) that it has occurred by chance alone.

All statistical techniques involve hypothesis-testing. They test a statement called the **null hypothesis**. Statistical analyses test whether data match the null hypothesis or significantly vary from it. The **alternative hypothesis** is the 'opposite' of the null hypothesis, that is, that there is a difference or association shown by the data.

Because of the complexity of the data, scientists can never be 100% certain whether their results are true or not. Statistical tests allow them to be 95% certain that any associations or correlations found in the data are real and not due to chance (that is, there is a 5% chance that they are due to chance). The outcome of a statistical test is therefore a probability that the null hypothesis is true. A probability (known as the p value) varies from 0 (impossible) to 1 (certain). Since the p values are small, they are given as a percentage (0 to 100%) to avoid possible confusion with small numbers. The lower the probability, the less likely it is that the null hypothesis is true.

◆ **Statistical significance**: a calculated value that is used to establish the probability that an observed trend or difference represents a true difference that is not due to chance alone.

◆ **Alternative hypothesis**: there is a statistically significant difference between two variables.

Link
The null hypothesis is defined in Chapter C4.1, page 572.

Applying the chi-squared test

We know the expected ratio of offspring of a dihybrid cross is 9:3:3:1. Actually, the offspring produced in many dihybrid-cross experiments do not *exactly* agree with the expected ratio. This is illustrated by the results of the experiment described in Figure D3.2.24 with mutant forms of *Drosophila*, shown in Table D3.2.6.

■ **Table D3.2.6** Observed and expected offspring

Offspring in F2 generation	normal wing, normal body = 315	normal wing, ebony body = 108	vestigial wing, normal body = 101	vestigial wing, ebony body = 32	total = 556
Predicted ratio	9	3	3	1	
Expected numbers of offspring	313	104	104	35	

Clearly, these results are fairly close to (but not precisely in) the predicted ratio 9:3:3:1.

What, if anything, went wrong?

D3.2 Inheritance

Well, we can expect precisely this ratio among the progeny *only* if three conditions are met:
- Fertilization is entirely random.
- There are equal opportunities for survival among the offspring.
- Very large numbers of offspring are produced.

In the experiment with *Drosophila*, the exact ratio may not be obtained because, for example:
- More male flies of one type may have succeeded in fertilizing the females on this occasion.
- More females of one type may have died before reaching egg-laying condition.
- Fewer eggs of one type may have completed their development.

Similarly, in breeding experiments with plants, such as the pea plant, exact ratios may not be obtained. This could, perhaps, be due to parasite damage to some seeds or plants, to the action of browsing predators on the anthers or ovaries in some flowers, or because some pollen types fail to be transported by pollinating insects as successfully as others. Such accidental events are quite common.

The chi-squared test on data from a dihybrid cross

Experimental geneticists often ask the question:

Do the observed values differ significantly from the expected outcome?

This question is resolved by a simple statistical test, known as the **chi-squared (χ^2) test**. This is used to estimate the probability that any difference between the observed results and the expected results is due to chance. If it is not due to chance, it may be due to an entirely different explanation and the phenomenon needs further investigation.

The chi-squared (χ^2) test

$$\chi^2 = \sum \frac{(O - E)^2}{E}$$

where:

O = observed result, E = expected result and $\sum$ = the sum of the values.

Chi-squared applied

We can test whether the observed values obtained from the dihybrid cross between *Drosophila* of normal flies (wild type) with flies homozygous for vestigial wing and ebony body differ significantly from the expected outcome.

First we calculate χ^2 (Table D3.2.7).

■ Table D3.2.7 Calculating χ^2

Category	Predicted	O	E	O – E	(O – E)²	$\frac{(O - E)^2}{E}$
normal wing, normal body	9	315	312.75	2.25	5.062	0.016
normal wing, ebony body	3	108	104.25	3.75	14.062	0.135
vestigial wing, normal body	3	101	104.25	–3.25	10.562	0.101
vestigial wing, ebony body	1	32	34.75	–2.75	7.562	0.218
Total		556		$\Sigma(\chi^2)$		0.47

In this example we have calculated χ^2 to be 0.47.

To see if this value of chi-squared represents a significant difference between the observed and expected results, we now consult a table, such as that shown in Table D3.2.8, of the **distribution of χ^2**. We can find out the **probability (*p*)** of obtaining by chance alone a deviation as large as (or larger than) the one we have observed.

Note that the table takes account of the number of independent comparisons involved in our test. In our example, there were four categories and, therefore, three comparisons were made – we call this 'three **degrees of freedom (df)**'. (Another way of putting this is that for any one condition there are three alternatives.)

So, we look along the row 'df = 3' to see whether 0.47 lies to the left or to the right of the 0.05 level of probability (shown in red).

■ **Table D3.2.8** Table of χ^2 distribution

Degrees of freedom	Probability greater than							
	0.99	0.95	0.90	0.50	0.10	0.05	0.01	0.001
df = 1	0.00016	0.004	0.016	0.455	2.71	3.84	6.63	10.83
df = 2	0.0201	0.103	0.21	1.386	4.60	5.99	9.21	13.82
df = 3	0.115	0.35	0.58	1.39	6.25	7.81	11.34	16.27

Using the χ^2 distribution table, we can resolve whether the difference between the result we expected and the result we actually observed is due to chance – or whether the difference is, in fact, significant.

- If the value of χ^2 is *bigger* than the critical value highlighted in red (a probability of 0.05) then we can be at least 95% confident that the difference between the observed and expected results is *significant*.
- If the value of χ^2 is *smaller* than the critical value highlighted in red (a probability of 0.05) then we can be confident that the difference between the observed and expected results is *due to chance*.

In biological experiments we take a probability of 0.05 or larger to indicate that the difference between the observed (O) and expected (E) results is not significant. We can say it is due to chance.

In this example, the value (0.47) lies between a probability of 0.95 and 0.90. This means that a deviation of this size can be expected 90–95% of the times the experiment is carried out (due to chance). So, there is clearly *no* significant deviation between the observed (O) and the expected (E) results; the data conform to a Mendelian ratio.

The chi-squared test is similarly applicable to the results of other test crosses.

In any chi-squared test that produces a significant deviation of observed from expected results (it does *not* confirm that the results conform to the anticipated values) by giving a value for χ^2 that is *bigger* than the critical value and a probability that is *smaller* than 0.05, we must reconsider our experimental hypothesis. In this outcome, the statistical test gives no clue as to the true location or behaviour of the alleles. Further genetic investigations are required.

● **Nature of science: Measurement**

Statistical testing often involves using a sample to represent a population. In this case the sample is the F2 generation. In many experiments the sample is the replicated or repeated measurements.

14 In the dihybrid test cross between homozygous dwarf pea plants with terminal flowers (ttaa) and heterozygous tall pea plants with axial flowers (TtAa), the progeny were:
- tall, axial = 55 peas
- tall, terminal = 51 peas
- dwarf, axial = 49 peas
- dwarf, terminal = 53 peas.

Use the χ^2 test to **determine** whether or not the difference between these observed results and the expected results is significant.

LINKING QUESTIONS

1. What are the principles of effective sampling in biological research?
2. What biological processes involve doubling and halving?

D3.3 Homeostasis

Guiding questions

- How are constant internal conditions maintained in humans?
- What are the benefits to organisms of maintaining constant internal conditions?

SYLLABUS CONTENT

This chapter covers the following syllabus content:
- ▶ D3.3.1 Homeostasis as maintenance of the internal environment of an organism
- ▶ D3.3.2 Negative feedback loops in homeostasis
- ▶ D3.3.3 Regulation of blood glucose as an example of the role of hormones in homeostasis
- ▶ D3.3.4 Physiological changes that form the basis of type 1 and type 2 diabetes
- ▶ D3.3.5 Thermoregulation as an example of negative feedback control
- ▶ D3.3.6 Thermoregulation mechanisms in humans
- ▶ D3.3.7 Role of the kidney in osmoregulation and excretion (HL only)
- ▶ D3.3.8 Role of the glomerulus, Bowman's capsule and proximal convoluted tubule in excretion (HL only)
- ▶ D3.3.9 Role of the loop of Henle (HL only)
- ▶ D3.3.10 Osmoregulation by water reabsorption in the collecting ducts (HL only)
- ▶ D3.3.11 Changes in blood supply to organs in response to changes in activity (HL only)

Homeostasis

Living organisms face changing and sometimes hostile environments; some external conditions change slowly, others dramatically. For example, temperature changes quickly on land that is exposed to direct sunlight, but the temperature of water exposed to sunlight changes very slowly (Chapter A1.1, page 9). *How do organisms respond to environmental changes?*

An animal that can maintain a constant internal environment, enabling it to continue normal activities more or less whatever the external conditions, is known as a **regulator**. For example, mammals and birds maintain a high and almost constant body temperature over a very wide range of external temperatures. Their bodies are at or about the optimum temperature for most of the enzymes that drive their metabolism. Their muscles contract efficiently and the nervous system coordinates responses precisely, even when external conditions are unfavourable. Regulators have greater freedom in choosing where to live.

Homeostasis is the ability to maintain a constant internal environment. Homeostasis means 'staying the same'. The internal environment consists of the blood circulating in the body and the fluid that circulates among cells (tissue fluid that forms from blood plasma), delivering nutrients and removing waste products while bathing the cells. Mammals are excellent examples of animals that maintain remarkably constant internal conditions. They successfully regulate their body temperature, blood pH, blood glucose concentration and blood osmotic concentration at constant levels or within narrow limits (Figure D3.3.1). These homeostatic variables are kept within pre-set limits, despite fluctuations in the external environment.

Link
Homeostasis is defined and covered in Chapter A2.2, page 73.

Negative feedback loops in homeostasis

Negative feedback is the type of control in which conditions are brought back to a set value as soon as it is detected that they have deviated from it (Figure D3.3.2). Negative feedback loops are used in homeostasis, rather than positive feedback loops, because they involve a return to the original set point. Positive feedback moves conditions further away from equilibrium (the set point) and so would not restore the original conditions. Negative feedback returns homeostatic variables to the set point from values above and below the set point, for example temperature above and below the optimum temperature for the body.

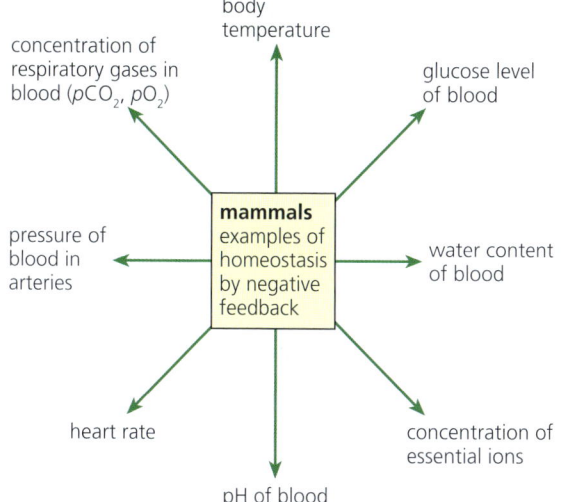

■ Figure D3.3.1 Homeostasis in mammals

> ## Concept: Continuity and change
>
> Organisms experience changes in their internal and external environment that cause deviation away from equilibrium. Negative feedback mechanisms return body systems to their original set point. Homeostatic mechanisms allow for continuity in body systems. For example, control of body temperature allows enzymes to work at their optimum, and water balance maintains osmotic concentrations within safe limits.

> ## Common mistake
>
> Homeostasis is control of the internal environment and does not involve a person controlling their external environment. Do not confuse homeostasis with responses to external stimuli, such as touching a hot object.

In mammals, regulation of body temperature, blood sugar level and the amounts of water and ions in blood and tissue fluid (osmoregulation) are regulated by negative feedback. The detectors are specialized cells in either the brain or other organs, such as the pancreas. The effectors are organs such as the skin, liver and kidneys. Information passes between these cells and organs via the nerves of the nervous system or via hormones (the endocrine system), or both. The outcome is an incredibly precisely regulated internal environment.

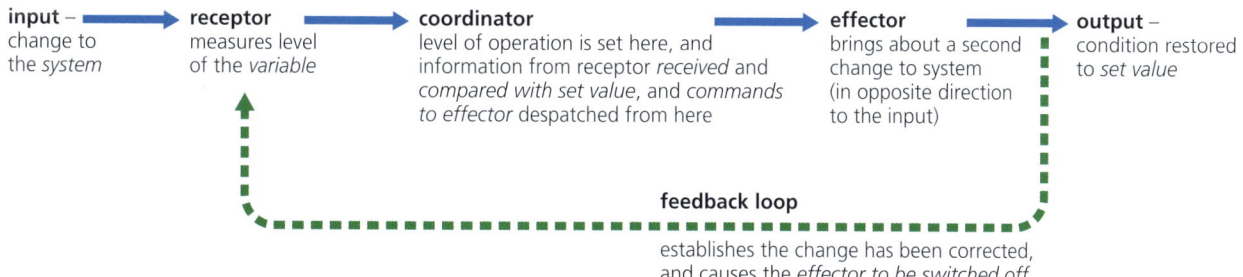

■ Figure D3.3.2 Negative feedback, the mechanism

> **Link**
>
> Negative feedback is defined in Chapter C4.1, page 557; for HL students it is also covered in Chapter C2.1, page 465.

Regulation of blood glucose as an example of the role of hormones in homeostasis

Transport of glucose to all cells is a key function of the blood circulation. In humans, the **normal level of blood glucose** is about 90 mg of glucose in every 100 cm^3 of blood, but it can vary. For example, during an extended period without food, or after prolonged physical activity, blood glucose may fall to as low as 70 mg. After a meal rich in carbohydrate has been digested, blood glucose may rise to 150 mg.

Respiration is a continuous process in all living cells. To maintain their metabolism, cells need a regular supply of glucose, which can be quickly absorbed across the cell membrane. Glucose is the main respiratory substrate for many tissues. Most cells (including muscle cells) hold reserves in the form of glycogen, which is quickly converted to glucose during prolonged physical activity. However, glycogen reserves may be used up quickly. In the brain, glucose is the only substrate the cells can use and, here, there is no glycogen store held in reserve.

The maintenance of a constant level of this monosaccharide in the blood plasma is important for two reasons:

- If our blood glucose falls below 60 mg per 100 cm^3, we have a condition called **hypoglycaemia**. If this is not quickly reversed, we may faint. If the body and brain continue to be deprived of adequate glucose levels, convulsions and coma follow.
- An abnormally high concentration of blood glucose, known as **hyperglycaemia**, is also a problem. Since a high concentration of any soluble metabolite lowers the water potential of the blood plasma, water is drawn from the cells and tissue fluid by osmosis, back into the blood. As the volume of blood increases, water is excreted by the kidney to maintain the correct concentration of blood. As a result, the body tends to become dehydrated and the circulatory system is deprived of fluid. Ultimately, blood pressure cannot be maintained.

For these reasons, it is critically important that blood glucose is held within set limits.

■ Mechanism for regulation of blood glucose

After the digestion of carbohydrates in the gut, glucose is absorbed across the epithelial cells of the villi in the small intestine and into the hepatic portal vein. The blood carrying the glucose reaches the liver first. If the glucose level is too high, glucose is withdrawn from the blood and stored as glycogen. Despite this action, blood circulating in the body immediately after a meal does have a raised level of glucose. At the pancreas, the presence of an excess of blood glucose is detected in patches of cells known as the **islets of Langerhans** (Figure D3.3.3). These islets are hormone-secreting glands (endocrine glands); they have a rich capillary network, but no ducts that would carry secretions away. Instead, their hormones are transported all over the body by the blood. The islets of Langerhans contain two types of cell: **alpha (α) cells** and **beta (β) cells**.

◆ **Hypoglycaemia**: condition when blood sugar levels are too low.

◆ **Hyperglycaemia**: condition when blood sugar levels are too high.

◆ **Islets of Langerhans**: groups of endocrine cells located in the pancreas.

◆ **Alpha (α) cell**: glucagon-secreting cell of the islets of Langerhans in the pancreas.

◆ **Beta (β) cell**: insulin-secreting cell of the islets of Langerhans in the pancreas.

● Top tip!

In biology, if the prefix to a word is 'hypo' then some factor is too low; if the prefix is 'hyper' then it means a factor is too high. For example, 'hypoglycaemia' means that blood sugar levels are too low and 'hyperglycaemia' that blood sugar is too high.

TS of pancreatic gland showing an islet of Langerhans

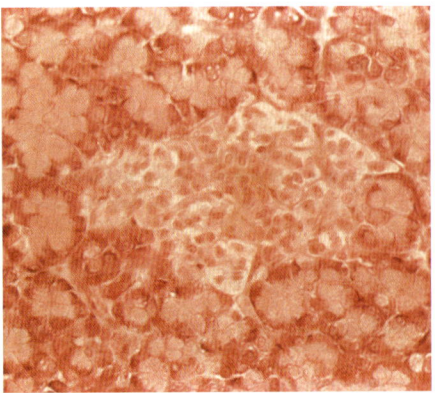

drawing of part of pancreatic gland

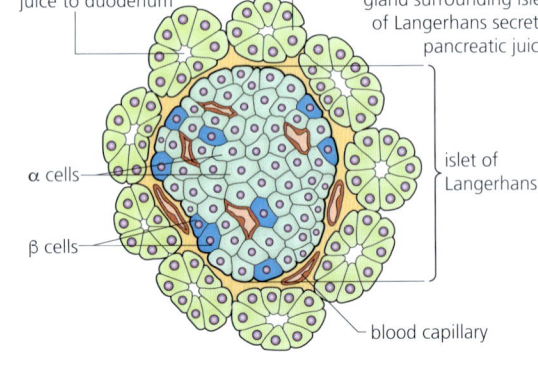

■ Figure D3.3.3 Islet of Langerhans in the pancreas

1 **Predict** what type of organelle you would expect to see most frequently when a liver cell is examined by electron microscopy.

A **raised blood glucose level** stimulates the β **cells**, and they secrete the hormone **insulin** into the capillary network. Insulin causes the uptake of glucose into cells all over the body, but especially by the liver and the skeletal muscle fibres. It also increases the rate at which glucose is used in respiration, in preference to alternative substrates (such as fat). Another effect of insulin is to trigger conversion of glucose to glycogen in cells (**glycogenesis**), and of glucose to fatty acids and fats, and finally the deposition of fat around the body.

As the blood glucose level reverts to normal, this is detected in the islets of Langerhans, and the β cells stop insulin secretion. Meanwhile, the hormone is excreted by the kidney tubules and the blood insulin level falls.

When the **blood glucose level falls below normal**, the α **cells** of the pancreas secrete a hormone called **glucagon**. This hormone activates the enzymes in cells that convert glycogen to glucose (glycogenolysis) and amino acids to glucose (**gluconeogenesis**). Glucagon also reduces the rate of respiration (Figures D3.3.4).

As the blood glucose level reverts to normal, glucagon production ceases and this hormone, in turn, is removed from the blood in the kidney tubules.

♦ **Glucagon**: hormone made in the pancreas that promotes the breakdown of glycogen to glucose in the liver and muscle cells.

● Common mistake

Do not confuse glucagon with glycogen. Glucagon is a hormone and glycogen a storage product of glucose. Make sure these terms are spelt correctly to avoid confusion.

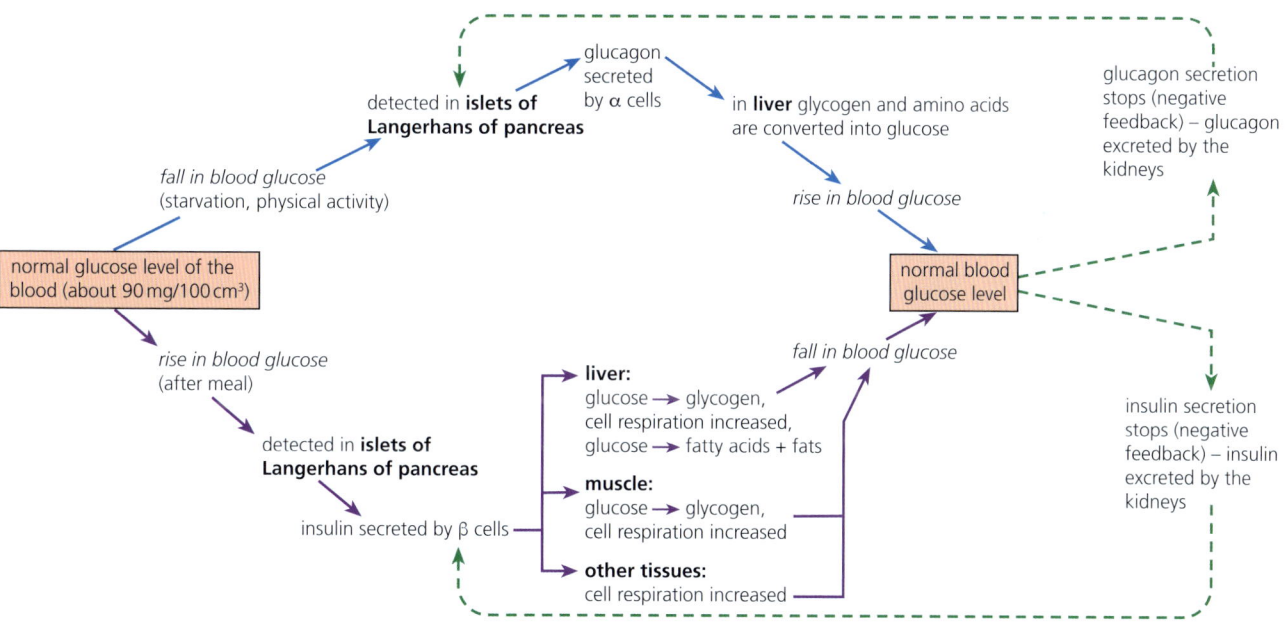

■ Figure D3.3.4 Glucose regulation by negative feedback

● Common mistake

Glucagon is a hormone, not an enzyme, so it is incorrect to state that glycogen is broken down into glucose by glucagon. It is correct to say that glucagon causes cells to breakdown glycogen.

● Common mistake

The β cells in the pancreatic islets monitor blood glucose concentration directly – the hypothalamus is not involved. It is incorrect to state that the hypothalamus monitors blood glucose concentration and, when the concentration is high, sends messages to the pancreas to stimulate insulin secretion.

Physiological changes that form the basis of type 1 and type 2 diabetes

◆ **Diabetes**: failure to regulate blood glucose levels.

◆ **Type 1 diabetes**: the result of a failure of insulin production by the β cells.

◆ **Type 2 diabetes**: failure of the insulin receptor proteins on the cell membranes of target cells.

Diabetes is where the body fails to regulate blood glucose levels correctly. There are two types: **type 1 diabetes** is the result of a failure of insulin production by the β cells and **type 2 diabetes** (diabetes mellitus) is a failure of the insulin receptor proteins on the cell membranes of target cells (Table D3.3.1). As a result, blood glucose level is more erratic and, generally, permanently raised. Glucose is also regularly excreted in the urine. If this condition is not diagnosed and treated, it carries an increased risk of circulatory disorders, renal failure, blindness, strokes or heart attacks.

■ **Table D3.3.1** Diabetes, causes and treatment

Type 1 diabetes 'early onset diabetes'	Type 2 diabetes 'late onset diabetes'
relatively rare	the common form (90% of all cases of diabetes are of this type)
	this form of diabetes is having an increasing effect on human societies around the world, including young people and even children in economically developed countries, seemingly because of poor diet
affects young people below the age of 20 years	common in people over 40 years, especially if overweight
due to the destruction of the β cells of the islets of Langerhans by the body's own immune system	β cells continue functioning but body stops responding to insulin
symptoms: • constant thirst • undiminished hunger • excessive urination	symptoms: • mild – people with type 2 diabetes usually have sufficient blood insulin, but insulin receptors on cells have become defective
treatment: • injection of insulin into the bloodstream daily • regular measurement of blood glucose level	treatment: • largely by diet alone

Risk factors for type 2 diabetes include: being overweight or obese, age 40 or older, having a family history of diabetes, having high blood pressure, having a low level of HDL ('good' cholesterol) or a high level of triglycerides, or being inactive. Scientists think type I diabetes is caused by genes and environmental factors, such as viruses, that might trigger the disease (that is, cause an autoimmune response where the immune system attacks and destroys the insulin-producing β cells of the pancreas). Having a parent or sibling with the disease may increase the chance of developing type 1 diabetes, as may exposure to specific viruses.

Type 2 diabetes can be controlled by diet, to reduce weight. Type 1 diabetes is controlled by insulin injections and insulin pumps: blood sugar levels are monitored throughout the day and insulin administered when there is a danger of blood sugar becoming too high.

● **Top tip!**

Rather than stating that type 2 diabetes is caused by having a high-sugar diet, it is more correct to state that there is a link and that such diets are one of a number of risk factors.

Inquiry 2: Collecting and processing data

Interpreting results

The glucose tolerance test involves giving sugar to individuals and then sampling blood to determine how quickly glucose is cleared from their blood. Typically, a standard dose of glucose (75 g for adults) is given orally – all to be consumed within 5 minutes. Blood glucose levels are monitored subsequently.

In an investigation to find out how quickly glucose was removed from the blood, two subjects were treated in this way, and their blood glucose was measured at intervals of 30 minutes for a period of 5 hours. One subject had normal glucose metabolism and one subject had diabetes. The results are given in Table D3.3.2.

■ Table D3.3.2 Measuring blood glucose concentrations in unaffected and diabetic individuals

Time after glucose ingestion (hours)	Unaffected individual blood glucose (mg/100cm³)	Diabetic blood glucose (mg/100cm³)
0.0	75	150
0.5	150	250
1.0	160	300
1.5	140	325
2.0	110	375
2.5	100	325
3.0	90	300
3.5	85	275
4.0	80	250
4.5	75	225
5.0	75	200

Using the data in Table D3.3.2, answer the following questions:
1. For normal diabetes screening in many countries, a sample of blood is taken for analysis at time 0 and again at 2 hours after the glucose was ingested, only. **Comment** on why these times may have been selected.
2. **Describe** where the excess glucose in blood plasma is detected in the body, in a way that leads to activation of the glucose level adjustment mechanism.
3. **Explain** how and where in the body the excess glucose is metabolized.
4. **Outline**, using a table, the ways in which an excess of blood glucose is harmful to body organs and tissues.

Tool 3: Mathematics

Constructing graphs for raw and processed data

Draw a line graph of the results shown in the Inquiry box above. Annotate the line graph to identify the maximum glucose levels and the lengths of time taken for the levels to return to those at the start.

Nature of science: Theories

Scientists have developed general explanations (i.e. theories) to explain the occurrence of diseases such as type 1 diabetes, based on observed patterns or tested hypotheses. Type 1 diabetes is considered to be an **autoimmune disease**. Autoimmune diseases are not always well understood in the medical community. Doctors know that autoimmune diseases are caused when our own immune system (hence 'auto'-immune) recognizes our own body cells as foreign cells or pathogens and begins to damage and destroy them, just as the immune system normally does to actual pathogenic cells. However, the reason why an individual's immune system starts fighting against their body cells is not completely understood. For that reason, it is often difficult to predict who will develop an autoimmune disease and how to manage it.

Thermoregulation as an example of negative feedback control

◆ **Thermoregulation**: regulation of body temperature.

◆ **Thyroid gland**: an endocrine gland found in the neck of vertebrates, site of production of thyroxin and other hormones influencing the rate of metabolism.

Negative feedback is when changes away from a set-point in the body are corrected and conditions returned to equilibrium. Regulation of our body temperature, known as **thermoregulation**, involves controlling both heat loss across the surface of the body and heat production within the body. It is an example of a negative feedback control.

Significant changes in temperature derive from the environment. The body needs various mechanisms to respond to these changes. White adipose tissue forms an insulating barrier around the outside of the body and is therefore an important means of regulating heat loss. There is also another form of adipose tissue – brown adipose tissue – that can be used to release heat when there is a decrease in temperature.

Thyroxin, an iodine-containing hormone produced in the **thyroid gland**, also plays a part in the control of temperature regulation. The **hypothalamus** secretes thyrotropin-releasing hormone which, in turn, stimulates the **pituitary gland** to produce thyroid-stimulating hormone. This hormone stimulates the production of thyroxin. The presence of thyroxin in the blood circulation stimulates oxygen consumption and increases the basal metabolic rate of the body organs. Variations in secretion of thyroxin help the control of body temperature.

Temperature changes are detected by peripheral thermoreceptors in the skin and hypothalamus. These pass on messages to the effectors, as we explore below.

Thermoregulation by physiological and behavioural means

Mammals and birds maintain a high and relatively constant body temperature by generating heat within the body from the reactions of metabolism and by being able to regulate this heat production. These are called physiological controls. Mammals also control the loss of heat through the skin (see below). An animal with this form of thermoregulation is called an **endotherm**, meaning 'inside heat'.

Mammals and birds also regulate their body temperature by modifying their behaviour: birds open up their wings and spread their feathers to allow circulating air to reach their skin and lower their body temperature. Mammals can seek cooler areas of their habitat if they begin to overheat.

■ Thermoregulation mechanisms in humans

Humans hold their inner body temperature (core temperature) just below 37 °C. In fact, human core temperature only varies between about 35.5 and 37.0 °C within a 24-hour period, when we are in good health. Heat is lost to the environment by convection, radiation and conduction. The body also loses heat by evaporation.

Maintaining homeostasis through **negative feedback** requires a stimulus, receptor, control centre and effector.

◆ **Vasoconstriction**: the narrowing (constriction) of blood vessels by small muscles in their walls, restricting blood flow.

◆ **Vasodilation**: the widening (dilation) of blood vessels by small muscles in their walls, increasing blood flow.

The human body's temperature regulatory control centre is in the **hypothalamus** of the brain. The hypothalamus receives electrical impulses from temperature receptors in the skin and brain that the body temperature is either higher or lower than the set point, and sends signals to effectors to respond to the change in temperature. Effectors include the skin, liver and muscles. They regulate blood flow from the arteries into the capillaries through **vasoconstriction** (when the arterioles contract and narrow to reduce blood flow) and **vasodilation** (when the arterioles relax and widen to allow greater blood flow).

Theme D: Continuity and change – Organisms

Increase in temperature

Physiological changes if there is an increase in temperature include:
- **Vasodilation**: arterioles supplying blood to the skin dilate, allowing increased blood flow to the capillaries in the skin. This leads to increased radiation of heat away from the body.
- Muscles attached to hairs in the skin relax, so **hairs lie flat** on the surface of the skin, allowing radiation of heat from the body.
- Sweat glands are activated to produce **increased sweat**, which evaporates from the skin and cools the body.

Decrease in temperature

Physiological changes if there is a decrease in temperature include:
- **Vasoconstriction**: arterioles supplying blood to the skin constrict, causing decreased blood flow to the capillaries in the skin. This leads to decreased radiation of heat away from the body.
- Muscles attached to hairs in the skin contract, causing **hairs to be erect** on the surface of the skin, trapping heat.
- Sweat glands **decrease sweat production**.
- Skeletal muscles contract and relax rapidly causing **shivering**, which generates heat.
- **Metabolic activity in the liver increases**, heating the body.

In mammals, brown adipose tissue is used to maintain the body temperature of newly born offspring. Brown adipose tissue metabolizes triglycerides in mitochondria to emit heat. Usually, mitochondria generate ATP (they are said to be 'coupled' to ATP production), which is then used to support life processes such as movement. The function of the mitochondria can be 'uncoupled' (i.e. separated from their usual function), however, and the energy that would have been stored in ATP is used to release heat instead. This is known as **uncoupled respiration**. Newborns have a large surface area compared to their size, and so lose heat rapidly. Uncoupled respiration in brown adipose tissue enables them to generate sufficient heat to maintain body temperature.

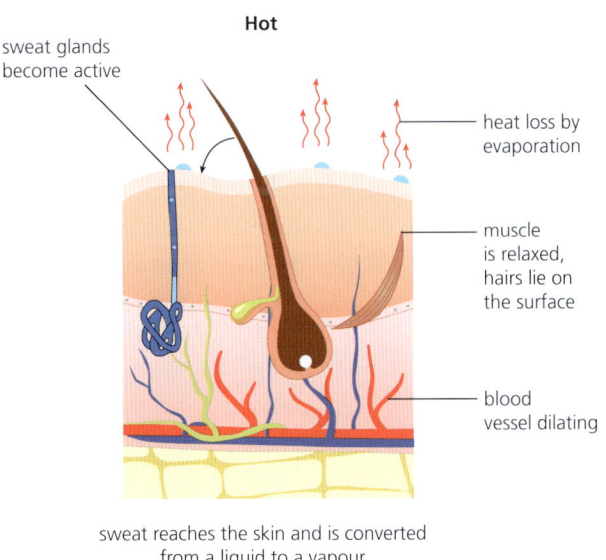
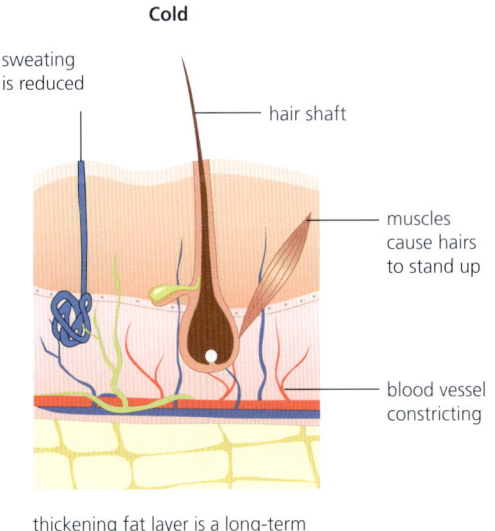

■ **Figure D3.3.5** Thermoregulation in humans

The negative feedback mechanisms involved in thermoregulation are shown in Figure D3.3.6.

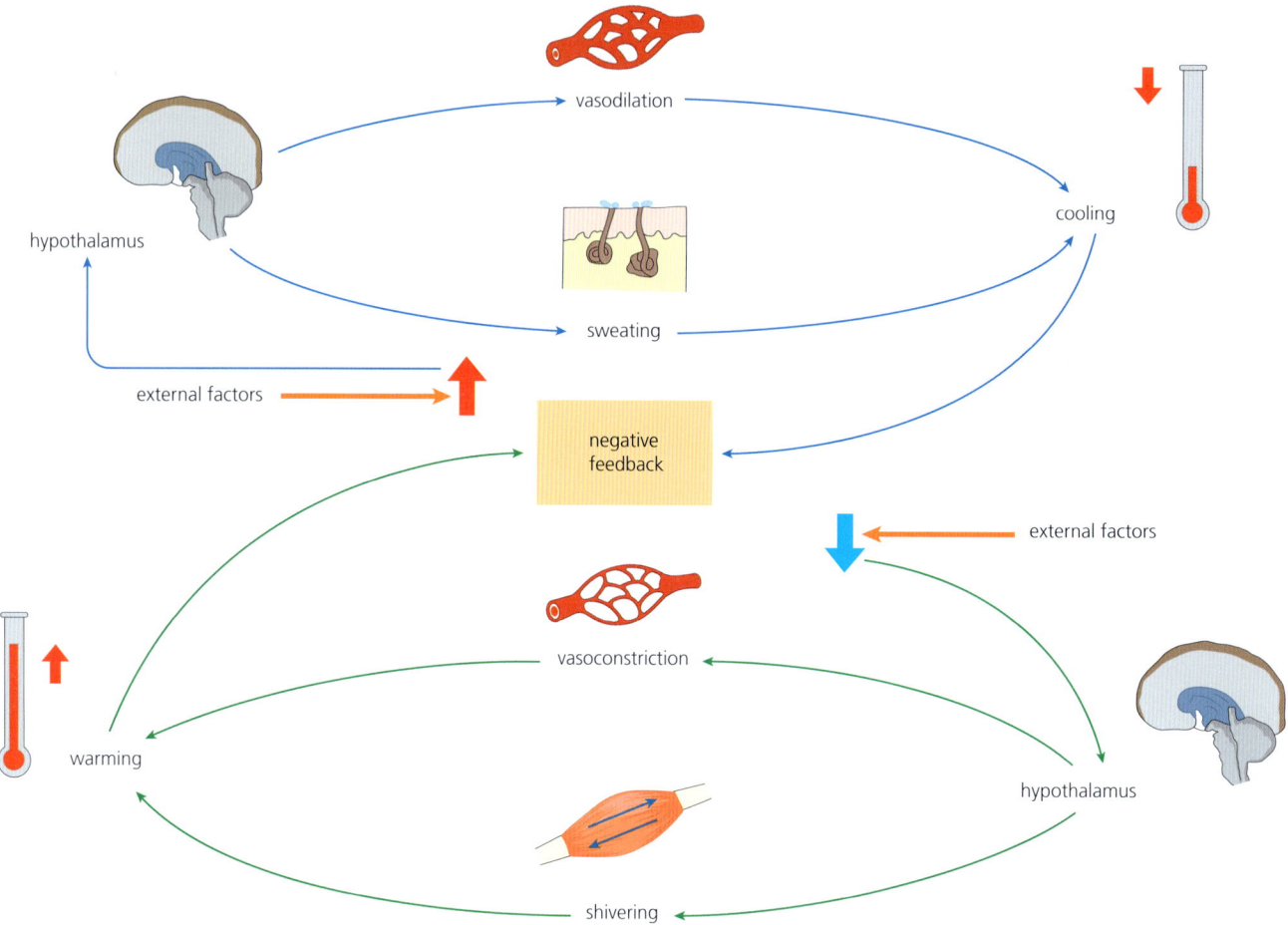

■ Figure D3.3.6 Negative feedback mechanisms in thermoregulation

◆ **Excretion**: the removal from the body of the waste products of cell metabolism.

◆ **Osmoregulation**: the maintenance of a proper balance of water and dissolved substances in the organism; regulation of osmotic concentration.

◆ **Osmotic concentration**: the measure of solute concentration, defined as the number of osmoles (osmol) of solute per litre (L).

Role of the kidney in osmoregulation and excretion

The chemical reactions of metabolism give by-products. Some of these would be toxic if they were allowed to accumulate in the organism. **Excretion** is the removal from the body of the waste products of metabolism. It is a characteristic activity of all living things. Metabolites that are present in excessive concentrations are also excreted.

In mammals, excretion plays an important part in the process by which the internal environment is regulated to maintain more or less constant conditions (**homeostasis**).

Associated with excretion, and very much part of homeostasis, is the process of **osmoregulation**. Osmoregulation is the maintenance of a proper balance of water and dissolved substances in the organism. The kidney maintains the **osmotic concentration** of tissue fluid and blood plasma.

Link

Osmosis is covered in detail in Chapter D2.3, page 682.

2 Distinguish between excretion and osmoregulation by means of both definitions and examples.

All animals need to balance water uptake and loss from the body. We know that water enters and exits cells by osmosis (page 682). If water uptake into animal cells is excessive, the hydrostatic pressure that quickly develops stretches the plasma membrane to the point of bursting. Alternatively, if water loss is excessive, a cell will shrink and die, because water is an essential and major component of the cytoplasm of all living organisms – it makes up about 90% of cell volume.

> **Top tip!**
>
> Osmotic concentration is measured in osmoles per litre (osmol L^{-1}). It used to be known as osmolarity – older websites and books may still use this term.

The human kidney – an organ of excretion and osmoregulation

The role of our kidneys is to regulate the body's internal environment by constantly adjusting the composition of the blood. Waste products of metabolism are transported from the metabolizing cells by the blood circulation, removed from the blood in the kidneys and excreted in a solution called urine. At the same time, the concentrations of inorganic ions, such as Na$^+$ and Cl$^-$, and of water in the body are also regulated in the kidneys. The functioning of the kidney is another example of **homeostasis** by **negative feedback**.

The position of the kidneys in humans is shown in Figure D3.3.7. Each kidney is served by a renal artery and drained by a renal vein. Urine from the kidney is carried to the bladder by the **ureter**, and from the bladder to the exterior by the **urethra**, when the bladder sphincter muscle is relaxed. Together these structures are known as the urinary system.

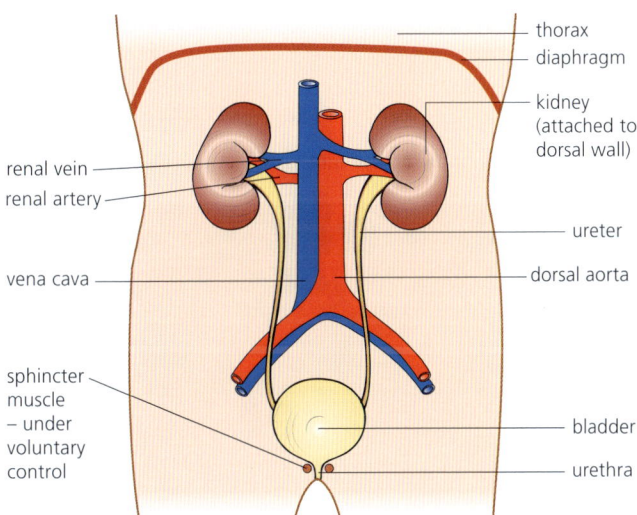

■ **Figure D3.3.7** The human urinary system

In section, a kidney can be seen to consist of an outer **cortex** and inner **medulla**, and these are made up of a million or more tiny tubules called **nephrons**. The shape of a nephron and its arrangement in the kidney are shown in Figure D3.3.8.

Blood vessels are closely associated with each of the distinctly shaped regions of the nephrons. For example, the first part of the nephron is formed into the cup-shaped **Bowman's capsule**, and the capillary network here is known as the **glomerulus**. These occur in the cortex. The **convoluted tubules** occur partly in the cortex and partly in the medulla, but notice that the extended **loops of Henle** and **collecting ducts** occur largely in the medulla.

Each region of the nephron has a specific role to play in the work of the kidney, and the capillary network serving the nephron plays a key part, too, as we shall now see.

◆ **Bowman's (renal) capsule**: the cup-shaped closed end of a nephron which contains the glomerulus.

◆ **Glomerulus**: network of capillaries which are surrounded by the Bowman's capsule.

◆ **Loop of Henle**: loop of mammalian kidney tubule, passing from cortex to medulla and back, important in the process of concentration of urine.

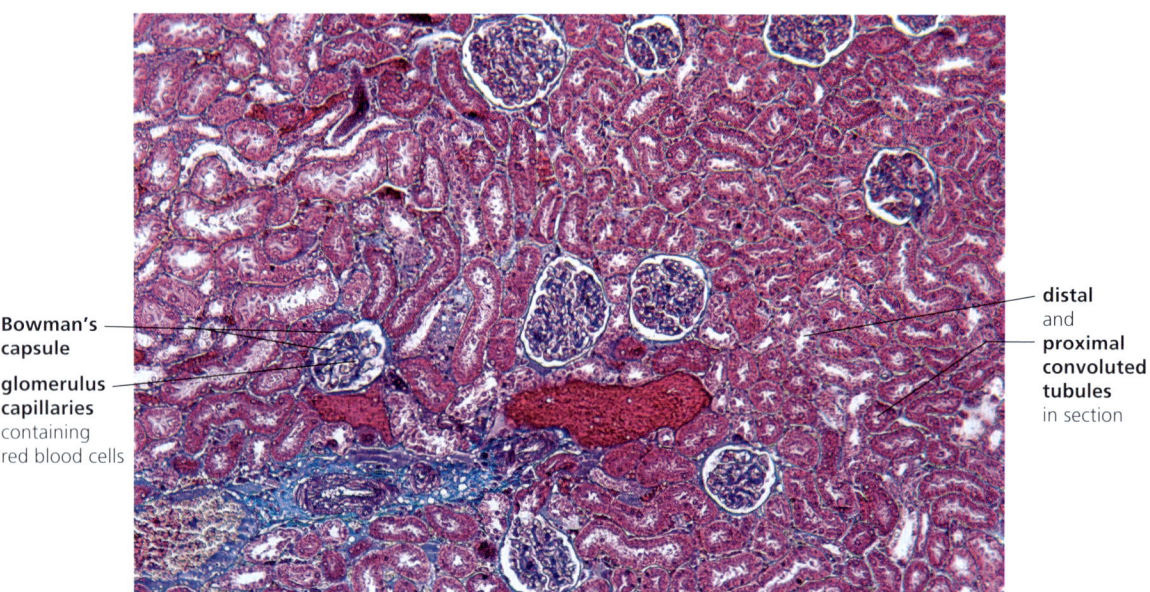

■ Figure D3.3.8 The kidney and its nephrons – structure and roles

Role of the glomerulus, Bowman's capsule and proximal convoluted tubule in excretion

The formation of urine

In humans, about 1.0–1.5 L of urine is formed each day, typically containing about 40–50 g of solutes, of which urea (about 30 g) and sodium chloride (up to 15 g) make up the bulk. The nephron produces urine in a continuous process, which we can conveniently divide into several steps to show how the blood composition is so precisely regulated.

Step 1: Ultrafiltration in the renal capsule

◆ **Ultrafiltration**: occurs through the tiny pores in the capillaries of the glomerulus, under pressure from the blood.

In the glomerulus, much of the water and many relatively small molecules present in the blood plasma, including useful ions, glucose and amino acids, are forced out of the capillaries, along with urea, into the lumen of the capsule (Figure D3.3.9). This fluid is called the **glomerular filtrate** and the process is described as **ultrafiltration**, because it is powered by the pressure of the blood (**hydrostatic pressure**). The blood pressure here is high enough for ultrafiltration because the input capillary (afferent arteriole) is significantly wider than the output capillary (efferent arteriole).

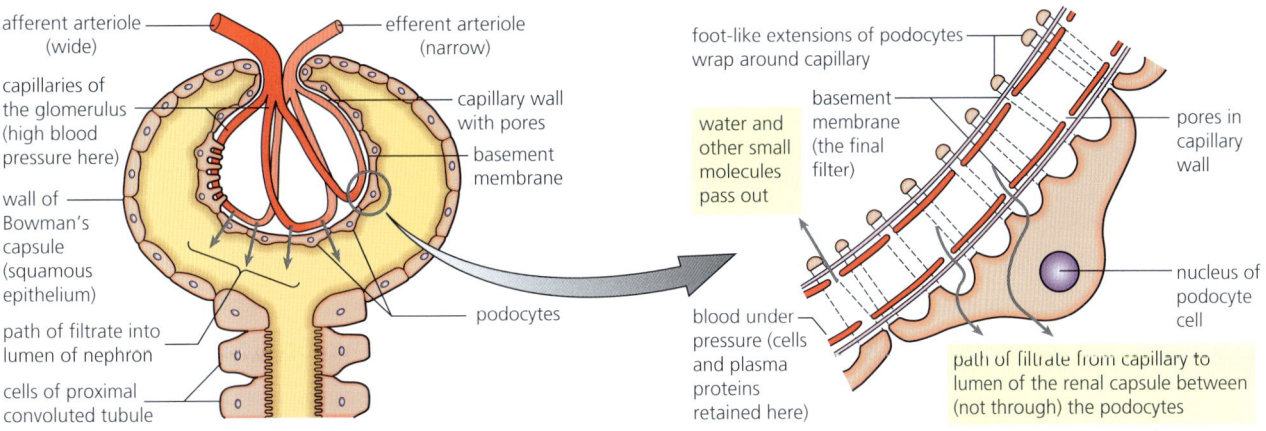

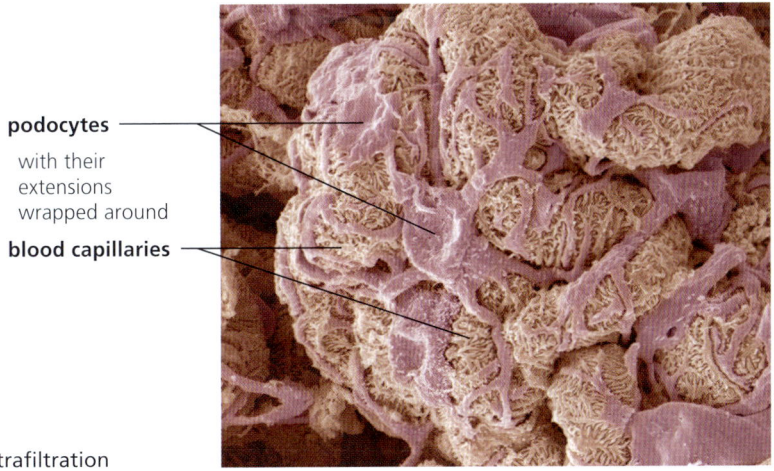

■ **Figure D3.3.9** The site of ultrafiltration

HL ONLY

3 **State** the source of energy for ultrafiltration in the glomerulus.

The barrier between the blood plasma and the lumen of the **Bowman's capsule** functions as a filter or 'sieve' through which ultrafiltration occurs. This sieve is made of two layers of cells (the endothelium of the capillaries of the glomerulus and the epithelium of the capsule wall), between which is a basement membrane.

You can see this arrangement in Figure D3.3.9.

The cells of the inner wall of the capsule are called podocytes because they have feet-like extensions. These wrap around the capillaries of the glomerulus, leaving a network of slits between the extensions.

Similarly, the endothelium of the capillaries has pores. These are large enough for fluid to pass through, but not large enough for the passage of blood cells. This detail has only become known because of studies using the electron microscope; these filtration gaps are very small indeed.

Finally, there is the basement membrane. This is a layer that surrounds and supports the capillary walls. This structure consists of a mesh-work of glycoproteins that allows the filtrate to pass, but which retains almost all of the plasma proteins.

The fluid that has filtered through into the renal capsule is very similar to blood plasma, but it has a significant difference (Table D3.3.3). Not only are blood cells retained in the plasma, but the majority of blood proteins and polypeptides also remain there.

■ **Table D3.3.3** A comparison of chief components of the blood plasma and the glomerular filtrate (units: *mol dm^{-3}/**mg dm^{-3})

	Blood plasma	**Filtrate**
urea*	5	5
glucose*	5	5
sodium ions*	150	145
chloride ion*	110	115
proteins**	740	3–4

Step 2: Selective reabsorption in the proximal convoluted tubule

The proximal convoluted tubule is the longest section of the nephron and it is here that that a large part of the filtrate is reabsorbed into the capillary network. The walls of the tubule are one cell thick and their cells are packed with mitochondria; we expect this if **active transport** is a key part of the mechanism for reabsorption. The cell membranes of the cells of the tubule wall (in contact with the filtrate) all have a 'brush border' of microvilli. These increase the surface area where reabsorption occurs enormously (see Chapter B2.3, page 266, Figure B2.3.10). The individual mechanisms of transport are given in Table D3.3.4.

4 The cells of the walls of the proximal convoluted tubule have a brush border. **Describe** what this means and **explain** how it helps in tubule function.

■ **Table D3.3.4** Mechanisms of selective reabsorption in the proximal convoluted tubule

Component of filtrate	Mechanism of reabsorption
glucose	**potassium pumps** maintain a gradient of sodium between the lumen of the proximal convoluted tubule and the inside of the epithelial cells lining the tubule, and sodium-dependent glucose cotransporters are used to transport glucose into the cell along with sodium ions; glucose is absorbed by facilitated diffusion at the bottom of the epithelial cells, into blood capillaries
amino acids and other essential metabolites	by **active transport** into the epithelial cells, and by facilitated diffusion at the bottom of the epithelial cells, into blood capillaries
ions (Na$^+$, Cl$^-$ and others)	by a combination of active transport, facilitated diffusion and some exchange of ions
urea	diffusion
proteins	pinocytosis
water	osmosis

Link

The adaptations of the proximal convoluted tubule in the nephron are covered in Chapter B2.3, page 265.

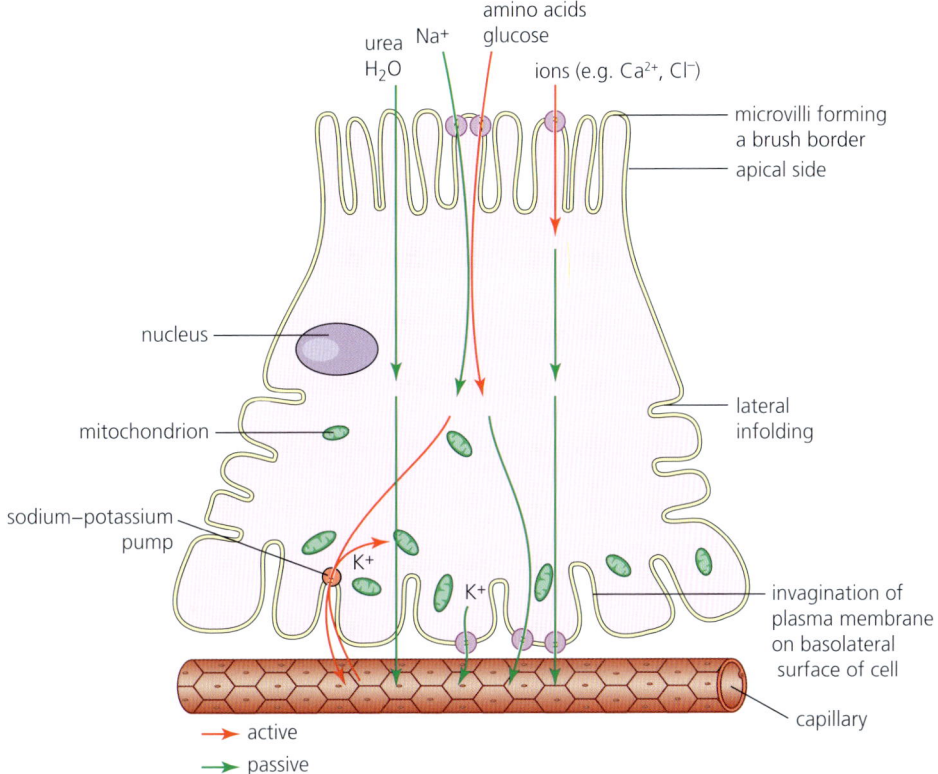

■ **Figure D3.3.10** Reabsorption in the proximal convoluted tubule

The basolateral area of the cell contains many mitochondria, which provide energy for sodium–potassium pumps (see page 241) located in the plasma membrane. The ion gradient created by these pumps in turn supports ion and water reabsorption on the apical side of the cell. By pumping three sodium ions out of the cell for reabsorption into the bloodstream and pumping two potassium ions back into the cell, a gradient is set up so that sodium ions can move by **facilitated diffusion** into the cell at the apical surface (Figure D3.3.10). Sodium enters the cell down its electrochemical gradient into tubule epithelial cells. At the same time, glucose is carried into the cell by sodium-dependent glucose cotransporters (for further information about these membrane proteins, see page 242).

Ions, such as sodium, and molecules (glucose and amino acids) are absorbed by facilitated diffusion at the basolateral side of the cell, into blood capillaries. The invaginations increase the surface area for absorption and for the activity of the sodium–potassium pumps.

Role of the loop of Henle

The function of the **loop of Henle** is to enable the kidneys to conserve water. Since urea is expelled from the body in solution, some water loss in excretion is inevitable. There is a potential problem here. Water is a major component of the body and it can be a scarce resource for terrestrial organisms.

The structure of the loop of Henle with its descending and ascending limbs, together with a parallel blood supply, the vasa recta, is shown in Figure D3.3.11. The vasa recta is part of the same capillary network that surrounds a nephron.

The gradient across the medulla is from a less concentrated salt solution near the cortex to the most concentrated salt solution at the tips of the pyramid of the medulla (Figure D3.3.8). The pyramid region of the medulla consists mostly of the collecting ducts. Thus, the loop of Henle maintains hypertonic conditions around the collecting ducts. The osmotic gradient allows water to be withdrawn from the collecting ducts if circumstances require it.

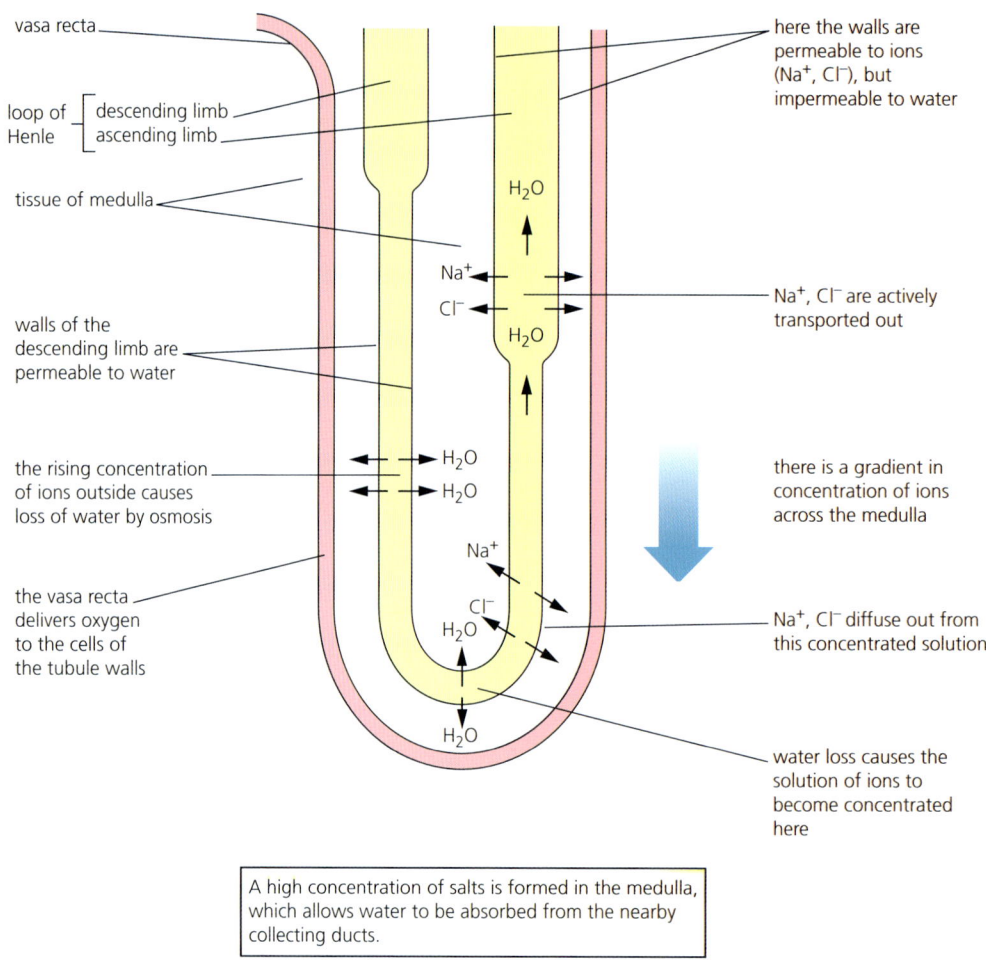

■ **Figure D3.3.11** The functioning loop of Henle

Look first at the second half of the loop, the **ascending limb**. In the upper (thick-walled) part of the ascending limb, sodium and chloride ions are pumped out of the filtrate into the fluid between the cells of the medulla, called the interstitial fluid. The energy to pump these ions is transferred from ATP. In the lower (thin-walled) part of the ascending limb, sodium and chloride ions diffuse out into the interstitial fluid. This movement of sodium chloride out of the tubule helps maintain the osmotic concentration of the interstitial fluid in the medulla. All along the ascending limbs, the walls are unusual in being impermeable to water. So, water in the ascending limb is retained in the filtrate as salt is pumped out.

Opposite is the *first* half of the loop, the **descending limb**. This limb is fully permeable to water, but is of very low permeability to solutes generally. Here, water passes out into the interstitial fluid by osmosis, due to the salt concentration in the medulla.

At each level in the loop, the salt concentration in the descending limb is slightly higher than the salt concentration in the adjacent ascending limb. As the filtrate flows, the concentrating effect is multiplied and so the fluid in and around the hairpin bend of the loops of Henle is saltiest.

The function of the vasa recta is to deliver oxygen to and remove carbon dioxide from the metabolically active cells of the loop of Henle. The vasa recta also absorbs water that has passed into the medulla at the collecting ducts.

Osmoregulation by water reabsorption in the collecting ducts

Osmoregulation, the control of the water content of the blood (and therefore of the whole body), is a part of homeostasis – another example of regulation by negative feedback. The composition of the blood is continuously monitored here by **osmoreceptors**, as it circulates through the capillary networks of the **hypothalamus**, and from sensory receptors located in certain organs in the body. All these inputs enable the hypothalamus to control accurately the activity of the pituitary gland.

The **pituitary gland** is situated below the hypothalamus, but is connected to it (Figure C3.1.13, page 505). In the process of osmoregulation, it is the posterior part of the **pituitary** that stores and releases **antidiuretic hormone** (ADH), among other hormones.

When nerve impulses from the hypothalamus trigger release of ADH into the capillary networks in the posterior pituitary, ADH circulates in the bloodstream. However, the targets of this hormone are the walls of the collecting ducts of the kidney tubules.

When the water content of the blood is low, ADH is secreted from the posterior pituitary gland. When the water content of the blood is high, little or no ADH is secreted.

■ How does ADH change the permeability of the walls of the collecting ducts?

The cell plasma membranes of the cells that form the walls of the collecting ducts contain a high proportion of channel proteins (**aquaporins**). Aquaporins are stored in the membrane of vesicles within the cells, and move from these intracellular vesicles to the cell membranes of the collecting ducts when ADH levels increase.

When there is an excess of ADH in the blood circulating past the kidney tubules, this hormone binds to receptor molecules in the collecting-duct membrane, causing the protein channels in the membranes to open. As a result, much water diffuses out into the medulla and very little diffuses from the medulla into the collecting ducts (Figure D3.3.12).

The water entering the medulla is taken up and redistributed in the body by the blood circulation. Only a small amount of very concentrated urine is formed. Meanwhile, the action of the liver continually removes and inactivates ADH. This means that the presence of freshly released ADH has a regulatory effect.

When ADH is absent from the blood circulating past the kidney tubules, the protein channels in the collecting-duct plasma membranes are closed. The amount of water that is retained by the medulla tissue is now minimal. The urine becomes copious and dilute.

> **Top tip!**
>
> When the intake of water exceeds the body's normal needs, the urine produced is copious and dilute. We notice this after drinking a lot of water. On the other hand, if we have taken in very little water, have been sweating heavily (part of our temperature regulation mechanism) or if we have eaten very salty food, perhaps, then a small volume of concentrated urine is formed.

> **Link**
>
> The structure of a fluid mosaic membrane and its channel proteins are discussed in Chapter B2.1, page 234.

5 **Explain** why small quantities of concentrated urine are produced when you are dehydrated.

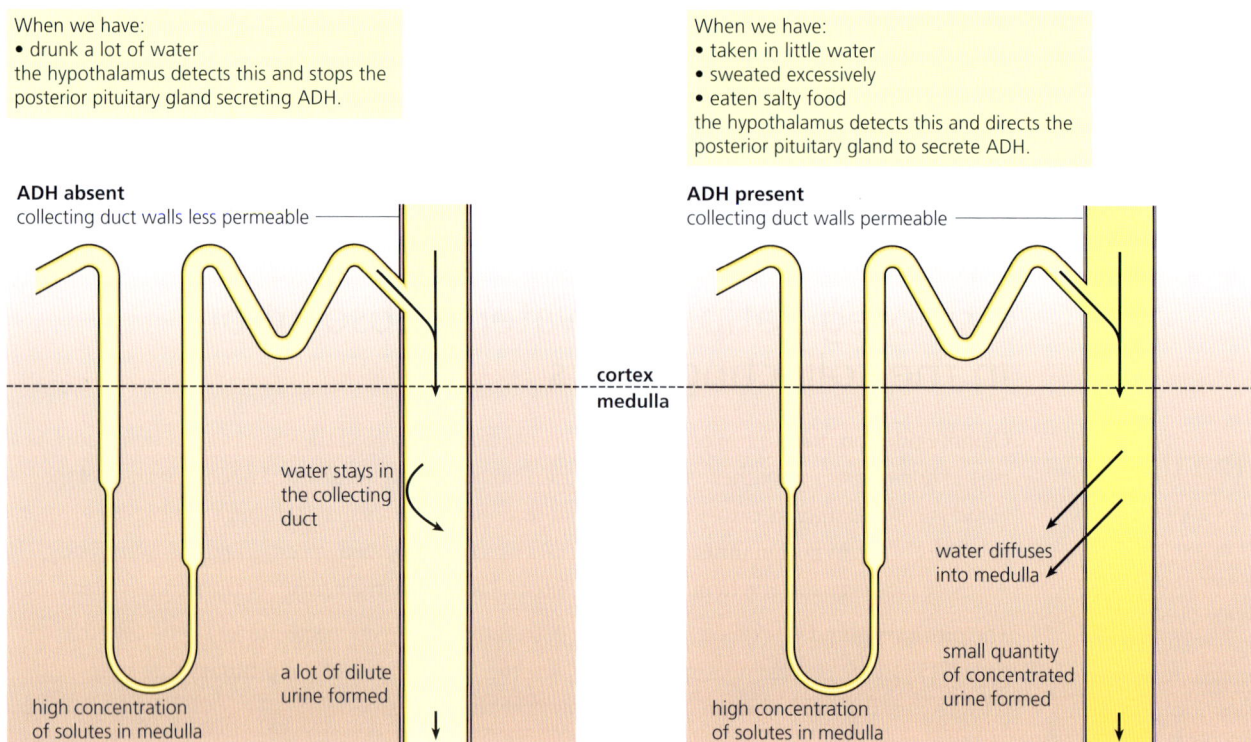

■ **Figure D3.3.12** Water reabsorption in the collecting ducts

A summary of osmoregulation by the kidneys is shown in Figure D3.3.13.

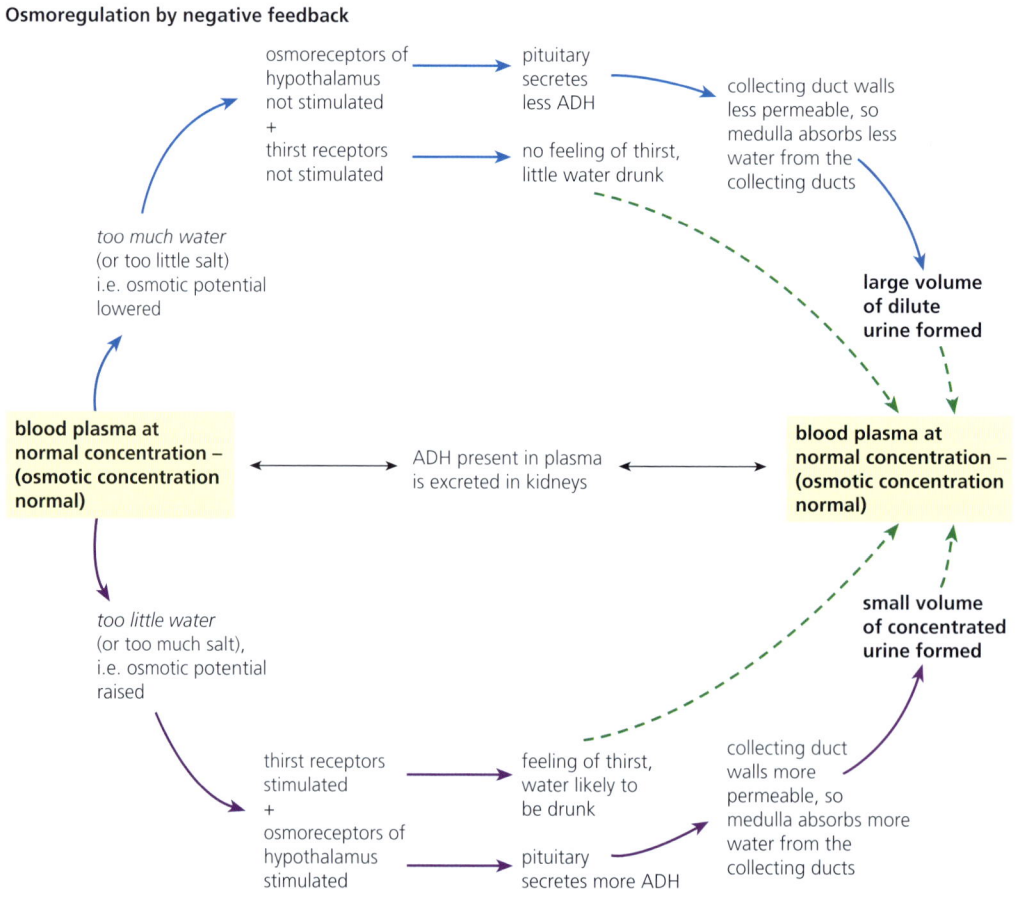

■ **Figure D3.3.13** Homeostasis by osmoregulation in the kidneys – a summary

> **ATL D3.3A**
>
> Construct a diagram of an individual nephron, extending from the Bowman's capsule to the collecting duct, with its associated blood supply.
>
> Use the knowledge you have gained to annotate the following regions and structures, highlighting the features of each and recording its function in the production of urine: Bowman's capsule, proximal convoluted tubule, loop of Henle, collecting duct.
>
> Also include the blood vessels: afferent arteriole, glomerulus, efferent arteriole, capillary bed surrounding the convoluted tubules, vasa recta, venules.

■ Differences in composition of blood plasma, glomerular filtrate and urine

The composition of urine that is excreted from the body is variable – depending on diet (including salt intake and amount of protein consumed), water intake, degree of physical activity, environmental conditions, water loss by other routes (particularly in sweat) and on our state of health.

On the other hand, the composition of the blood is held more or less constant. This is due to the efficiency of the homeostatic mechanisms of the body, particular those operating at the kidney tubules.

Similarly, the composition of the filtrate passing into the renal capsule is remarkably constant. This is the result of ultrafiltration in the glomerulus, the pressure of the blood, and the sizes of the blood proteins and polypeptides dissolved in the plasma – they are mostly too large to be filtered.

In Table D3.3.5, we have evidence of the power of the ultrafiltration mechanism and the scale of selective reabsorption. We see the largest divergence in concentration when the composition of the filtrate is compared with the composition of urine formed in one day.

6 The composition of the blood in the renal veins leaving the kidneys differs significantly from that in the renal arteries. **Identify** and **list** the differences you would anticipate.

■ **Table D3.3.5** Composition of blood, glomerular filtrate and urine

Substance	Total amount in the plasma	Filtered (per day)	Reabsorbed (per day)	Urine (excreted per day)
water	3 litres	180 litres	178–179 litres	1–2 litres
glucose	2.7 g	162.0 g	162.0 g	0.0 g
urea	0.9 g	54.0 g	27.0 g	27.0 g*
proteins and polypeptides	200.0 g	2.0 g	1.9 g	0.1 g
Na$^+$ ions	9.7 g	579.0 g	575.0 g	4.0 g
Cl$^-$ ions	10.7 g	640.0 g	633.7 g	6.3 g

*some is also secreted by the cells of the tubule into the urine

Changes in blood supply to organs in response to changes in activity

The blood supplies nutrients and oxygen to body tissue and removes waste products. Metabolically active tissues, such as muscles during exercise, require an increased supply of blood to support increased rates of respiration. Depending on the levels of activity, blood flow changes around the body. Vasoconstriction and vasodilation of arterioles controls the rate of blood flow to tissues.

During sleep, although there is little overall change in total brain blood flow, regional differences within the brain may occur. For example, during rapid eye movement (REM) sleep, blood flow to the hypothalamus and brainstem may be significantly higher than during wakefulness.

On the other hand, blood flow to most skeletal muscle groups, and particularly to the diaphragm, is lower during sleep than during wakefulness.

During vigorous exercise, blood is redirected away from organs, such as the gut, not directly involved in the activity, and towards skeletal muscles. Redistribution of blood also occurs during a fight or flight response, when epinephrine is released from the adrenal gland. This causes blood flow to be increased to skeletal muscle and away from organs not required by the response, such as the gut. Epinephrine achieves this by causing constriction in many networks of arterioles but dilation of blood vessels to the skeletal muscles. During wakeful rest, less blood will be needed in the muscles. If a meal has been eaten recently, blood flow to the gut will be increased to help in the digestion and absorption of food.

The blood flow to the kidneys (renal blood flow) is greater when a person is lying down than when standing, and reduced by prolonged vigorous exercise that constricts arterioles leading to the kidney and diverts blood to other organs. Overall, blood flow to the kidneys must remain fairly constant to maintain osmoregulation and the excretion of waste.

> **LINKING QUESTIONS**
> 1. For what reasons do organisms need to distribute materials and energy?
> 2. What biological systems are sensitive to temperature changes?

D4.1 Natural selection

Guiding questions
- What processes can cause changes in allele frequencies within a population?
- What is the role of reproduction in the process of natural selection?

SYLLABUS CONTENT

This chapter covers the following syllabus content:
- ▶ D4.1.1 Natural selection as the mechanism driving evolutionary change
- ▶ D4.1.2 Roles of mutation and sexual reproduction in generating the variation on which natural selection acts
- ▶ D4.1.3 Overproduction of offspring and competition for resources as factors that promote natural selection
- ▶ D4.1.4 Abiotic factors as selection pressures
- ▶ D4.1.5 Differences between individuals in adaptation, survival and reproduction as the basis for natural selection
- ▶ D4.1.6 Requirement that traits are heritable for evolutionary change to occur
- ▶ D4.1.7 Sexual selection as a selection pressure in animal species
- ▶ D4.1.8 Modelling of sexual and natural selection based on experimental control of selection pressures
- ▶ D4.1.9 Concept of the gene pool (HL only)
- ▶ D4.1.10 Allele frequencies of geographically isolated populations (HL only)
- ▶ D4.1.11 Changes in allele frequency in the gene pool as a consequence of natural selection between individuals according to differences in their heritable traits (HL only)
- ▶ D4.1.12 Differences between directional, disruptive and stabilizing selection (HL only)
- ▶ D4.1.13 Hardy–Weinberg equation and calculations of allele or genotype frequencies (HL only)
- ▶ D4.1.14 Hardy–Weinberg conditions that must be maintained for a population to be in genetic equilibrium (HL only)
- ▶ D4.1.15 Artificial selection by deliberate choice of traits (HL only)

Natural selection as the mechanism driving evolutionary change

Link
Evolution is also covered in detail in Chapter A4.1, page 141.

◆ **Natural selection**: process where organisms better adapted to their environment survive and produce more offspring than competitors; the mechanism through which evolution occurs.

Evolution is the development of life from its earliest beginnings to the diversity of organisms we know about today, both living and extinct. Evidence for evolution can be found from a variety of sources, including fossils (Chapter A2.1, page 47), base sequences in DNA or RNA and amino acid sequences in proteins, homologous structures, and from the selective breeding of domesticated animals and crop plants (Chapter A4.1, page 141).

Natural selection is the mechanism that explains how evolution occurs. Natural selection operates continuously and over billions of years, resulting in the biodiversity of life on Earth. It can also happen over much more rapid timescales, for example in such cases as the evolution of viruses and antibiotic-resistant bacteria (Chapter A2.3, page 98, and Chapter C3.2, page 534).

The theory of evolution by natural selection was developed by Charles Darwin. Darwin (1809–82) was a careful observer and naturalist who made many discoveries in biology (Figure D4.1.1). After attempting to become a doctor at Edinburgh University and then a clergyman at Cambridge University, he became the unpaid naturalist on an Admiralty-commissioned expedition to the southern hemisphere, on a ship called the *HMS Beagle* (Chapter A3.1, page 102). During Darwin's lifetime, the age of the Earth was increasingly recognized as being millions of years old, rather than thousands. His observations of living and fossilized species made during the voyage made him question the Genesis creation myth and the immutability of species.

On this five-year expedition around the world, and in his later investigations and reading, Darwin developed the idea of **organic evolution by natural selection**. Figure D4.1.2 shows a page from one of Darwin's notebooks, where he first developed his ideas for evolution.

■ Figure D4.1.1 Portrait of Charles Darwin, at an age when he travelled on the *HMS Beagle* and first developed his ideas of evolution by natural selection

Concept: Change

Natural selection is the process that drives evolutionary change.

ATL D4.1A

Look at the sketch in Charles Darwin's 'B' notebook (Figure D4.1.2), one of his notebooks used to record his thoughts and ideas (a reflection diary). The diagram shows a tree-like structure. What do you think Darwin was trying to represent here? Why is this sketch significant?

Darwin always remained very anxious about how the idea of evolution might be received, and he made no moves to publish it until the same idea was presented to him in a letter by another biologist and traveller, **Alfred Russel Wallace**. Only then (1859) was *On the Origin of Species by Natural Selection* completed and published. The arguments and ideas in the *Origin of Species* are summarized in Table D4.1.1.

■ Figure D4.1.2 Charles Darwin's vision of 'descent with modification' from his 'B' notebook

■ Table D4.1.1 Charles Darwin's ideas about the origin of species, summarized in four statements (S) and three deductions (D) from these statements

	Statements and deductions
S1	Organisms produce a far greater number of progeny than ever give rise to mature individuals.
S2	The number of individuals in species remains more or less constant.
D1	Therefore, there must be a high mortality rate.
S3	The individuals in a species are not all identical, but show variations in their characteristics.
D2	Therefore, some variants will succeed better than others in the competition for survival. So the parents for the next generation will be selected from those members of the species better adapted to the conditions of the environment.
S4	Hereditary resemblance between parents and offspring is a fact.
D3	Therefore, subsequent generations will maintain and improve on the degree of adaptation of their parents, by gradual change.

Top tip!

Make sure that you really understand the concept of natural selection so you can apply it to unfamiliar situations. Do not just rote learn it.

1 **Outline** the theory of evolution by natural selection.

TOK

In 1858, Charles Darwin unexpectedly received a letter from a young naturalist, Alfred Russel Wallace. Wallace outlined a remarkably similar theory of natural selection to Darwin's own. Wallace had come up with the idea while travelling in South East Asia. The men had developed the same theory independently. Why was this possible? Common experiences seem to have been crucial, and both had read similar books. For two individuals to arrive at one of the most important theories in science independently and at the same time is remarkable.

Nature of science: Theories

A paradigm shift is a significant and fundamental change in how the world is viewed. In Charles Darwin's time, it was widely understood that species evolved, but the mechanism was not clear. Darwin's theory provided a convincing mechanism and replaced Lamarckism. Paradigm shifts occur when scientific research contradicts previously held assumptions: Darwin's (and Wallace's) concept of natural selection replaced previously held views about the origin of species and is therefore an important example of a paradigm shift in scientific thinking.

While Darwin initially may have had to rely on strong – but limited – evidence, and while his initial argument may have included much speculation, the strength of nearly 200 years of genetic research continues to support Darwin's initial ideas. The fact that evolution by natural selection is called a scientific 'theory' underscores the highly grounded and evidenced nature of the ideas, even if they are prone to adjustment as further data are offered. The 'theory' is not speculation or even very speculative – all the available evidence, properly understood, supports Darwin's ground-breaking ideas.

TOK

What is the role of paradigm shifts in the progression of scientific knowledge?

A scientific theory provides a framework within which to carry out research (investigations) and determines the types of research questions that a biologist thinks is appropriate to formulate and test. A new theory in biology replaces an old theory when the new theory is better supported by current data and when it gives a better explanation of biological phenomena.

A theory may be considered better because:
- it is more effective in giving a framework (tools and processes) for problem solving and making predictions
- it simply works better and is more useful (i.e. the results make further progress possible) – this is the 'pragmatic theory of truth'
- it fits more consistently with other existing biological theories – this is the 'coherence theory of truth'
- it seems, in terms of evidence, to more accurately reflect what is really there (in an independent external reality) – this is the 'correspondence theory of truth'.

When an accepted scientific theory fails to explain increasingly anomalous data that do not fit in with the predictions of the theory, a scientific evolution occurs and a completely new hypothesis, theory or model is developed, which requires that old data and the anomalous data be interpreted in a totally new way. American philosopher Thomas Kühn (1922–96) called this sudden shift from the old theory to the new theory a **paradigm shift**. The most important paradigm shift in biology was probably the development and later acceptance of Darwin's concept of evolution by natural selection.

Darwin's publication of the *Origin of Species* in 1859 provided a simple explanation for the great diversity of life and showed that all living organisms, including humans (*Homo sapiens*), are related by descent from a common evolutionary ancestor. The view was 'simpler' than the creationist view in the sense that it did not need to postulate a divine being as the explanation for the diversity of life, a being which would then itself need explaining. His explanation of evolution via natural selection is the basis of all of biology and, in the words of the eminent geneticist Theodosius Dobzhansky (1900–75), 'nothing in biology makes sense except in the light of evolution'.

● Common mistake

When explaining the role of sexual reproduction in evolution, it is not enough to state that sexual reproduction produces variation – *how* the variation occurs also needs to be explained. It is not necessary to say what evolution is, but rather to explain how sexual reproduction facilitates evolution. The role of mutation does not need to be discussed, because this is not part of sexual reproduction.

2 **Describe** how mutation and sexual reproduction lead to variation in populations.

The roles of mutation and sexual reproduction in generating variation

Natural selection can only occur if there is variation among members of the same species. The individuals in a species are not all identical but show variations in their characteristics. Today, modern genetics has shown us that there are several ways by which **genetic variations** arise.

1 Variation arises in **sexual reproduction** during both **meiosis** in gamete formation and fertilization.

   We have seen that genetic variations arise via:

   - **random assortment** of paternal and maternal chromosomes in meiosis (in the process of gamete formation, page 660)
   - **crossing over** of segments of individual maternal and paternal homologous chromosomes (resulting in new combinations of genes on the chromosomes of the haploid gametes produced by meiosis, page 660)
   - the **random fusion** of male and female gametes during fertilization.

   These processes operate each time meiosis occurs and are followed by fertilization. The results are **new combinations** of existing characteristics that may favour individuals in their lifetime – they may affect survival and opportunities to reproduce. If so, a particular individual's success in reproduction will result in certain alleles being passed on to the next generation in greater proportions than other alleles.

2 Variations arise as the product of mutation (page 637), giving entirely **new alleles**. We have seen that mutations are changes in DNA. Mutations occurring in ovaries or testes (or anthers or embryo sacs of flowering plants) are called **germ line mutations**. Since these mutations occur in the cells that give rise to gametes, they may be passed to the offspring. Meanwhile, mutations that occur in body cells of multicellular organisms (**somatic mutations**) are only passed on to the immediate descendants of those cells, and they disappear when the organism dies. So, new characteristics acquired during the lifetime of an individual are not heritable.

As a result of all these factors, the individual offspring of parents are not identical. Rather, they show variations in their characteristics.

Overproduction of offspring and competition for resources

■ **Table D4.1.2** Numbers of offspring produced

Organism	No. of eggs/seeds/young per brood or season
rabbit	8–12
great tit	10
poppy	6000
honey bee (queen)	120 000
cod	2–20 million

Link
Carrying capacity is covered in Chapter C4.1, page 557.

Darwin did not coin the phrase 'struggle for existence', but it does sum up the point that the overproduction of offspring in the wild leads naturally to their competition for resources. Table D4.1.2 is a list of the normal rate of production of offspring in some common species.

How many of these offspring survive to breed themselves?

In fact, in a stable population on average, a breeding pair gives rise to a single breeding pair of offspring. All their other offspring are casualties of the 'struggle'; many organisms die before they can reproduce.

So, populations do not show rapidly increasing numbers in most habitats or, at least, not for long. Population size is naturally limited by restraints that we call **environmental factors**. These include space, light and the availability of food. These factors limit the **carrying capacity** of a species' population. The never-ending competition for resources means that the majority of organisms fail to survive and reproduce. In effect, the environment can only support a certain number of organisms,

Theme D: Continuity and change – Ecosystems

Link

Examples of temperature as a density-independent factor are given in Chapter B4.1, pages 344 and 354.

Link

Abiotic factors as selection pressures is covered more in Chapter A4.2, page 164.

3 **List** three abiotic factors that can act as selection pressures.

and the number of individuals in a species remains more or less constant over a period of time. Overproduction of offspring and competition for resources are therefore factors that promote natural selection.

Abiotic factors as selection pressures

Abiotic factors are the non-living components of an ecosystem. These can act as density-independent factors that may affect the survival of individuals in a population, causing numbers of individuals in a population to fluctuate (Chapter C4.1). An example of a density-independent factor is high or low temperatures.

Common mistake

It is incorrect to refer to species '...adapted by developing new characteristics according to the different environments'. Instead, different environments will have different selective pressures on isolated populations of the same species, leading to different adaptations to the local environment.

Differences between individuals as the basis for natural selection

When genetic variation arises in organisms:
- the favourable characteristics are expressed in the phenotypes of some of the offspring
- these offspring may be better able to survive and reproduce in a particular environment; others will be less able to compete successfully to survive and reproduce.

Thus, natural selection operates to determine the survivors and the genes that are perpetuated in future progeny and results in offspring with favourable characteristics. In time, this selection process may lead to new varieties and new species. The operation of natural selection is sometimes summarized in the phrase **survival of the fittest**, although these were not words that Darwin used, at least not initially. Differences between individuals in adaptation, survival and reproduction are the basis for natural selection.

4 **Deduce** the importance of modern genetics to the theory of the origin of species by natural selection.

♦ **Fitness**: an organism's ability to pass its genetic material to its offspring, i.e. its reproductive success.

Fitness is a term used to refer to the survival value and reproductive potential of a genotype. To avoid the criticism that 'survival of the fittest' is a circular phrase (how can fitness be judged except in terms of survival?), the term 'fittest' is understood in a particular context. For example, the fittest of the wildebeest (hunted herbivores) of the African savannah may be those with the most acute senses, quickest reflexes and strongest leg muscles for efficient escape from predators. These adaptations relate to the genotype of members of the population – those with the alleles that confer a selective advantage are more likely to breed and pass on these alleles to future generations. By natural selection for these characteristics, the health and survival of wildebeests is assured.

Link

Intraspecific competition is covered in Chapter C4.1, page 563. Niches are covered in Chapter B4.2, page 363.

Natural selection is the result of **intraspecific competition** – competition between individuals of the same species. Individuals will usually interact with more members of their own species than of other species. Individuals of the same species share the same **niche** and so intraspecific competition plays a stronger role in evolution than interspecific competition (competition between individuals of different species) because in intraspecific competition, individuals compete for the same resources, and have the same biotic and abiotic interactions that influence the growth, survival and reproduction of the species. Any mutation that gives individuals in a population a selective advantage improves their chance of surviving intraspecific competition.

When the environment changes, some individuals present may be at a disadvantage. If so, natural selection is likely to operate on individuals and cause changes to gene pools. We have already noted that individuals possessing a particular allele, or combination of alleles, may be more likely to survive, breed and pass on their alleles than other less well-adapted individuals. This process is also referred to as 'differential mortality'.

Requirement that traits are heritable for evolutionary change to occur

Genes are the heritable unit through which evolution occurs. Mutation and remixing of alleles following sexual reproduction produces new arrangements of the genetic code that are passed down through the generations. Characteristics acquired during an individual's life due to environmental factors are not encoded in the base sequence of genes and so are not heritable.

As discussed in Chapter A4.1, Jean-Baptiste Lamarck put forward an explanation for change in species over time (page 139). His theory proposed that all the physical changes occurring in an individual during its lifetime can be inherited by its offspring (i.e. the inheritance of acquired traits). The theory influenced evolutionary thought throughout the first half of the nineteenth century. This was subsequently replaced by Darwin's theory of evolution by natural selection, which is compatible with our current knowledge of genetics.

> ## Common mistake
> 'Competition between individuals within a species' should be referred to rather than 'competition between species'. Similarly, you should refer to fitter individuals within a population, which survive and pass on adaptive alleles, and not 'survival of the fittest species'.

> ## Nature of science: Falsification
> All theories are tentative and can be falsified. For example, the evidence Darwin gathered while on the Galápagos Islands during his journey on the *Beagle* falsified Lamarck's earlier theory of evolution by inheritance of acquired traits. The theory of evolution by natural selection, in contrast to Lamarck's theory, allows for prediction and explanation. The term 'theory' (as opposed to 'hypothesis' or 'conjecture') indicates that a scientific community generally feels that the ideas provided by the theory are the best and most full explanation of the data available.

> ### ATL D4.1B
> Why have organisms not evolved following the mechanism proposed by Lamarck (what is known as Lamarckism)? Read the following article: www.americanscientist.org/article/experimental-lamarckism
>
> Summarize the main ideas of the article in 400 words. Use your knowledge of DNA and the molecular basis of inheritance to explain the ideas and concepts in the article.

> ## Top tip!
> If you are taking HL, you will be aware that, in mammals, imprinted genes are not all stripped of their epigenetic tags, and so changes due to environmental factors are still heritable, even though they are not causing changes to the base sequence of genes.

> ## Link
> Evolution as change in the heritable characteristics of a population is covered in Chapter A4.1, page 138.

■ Natural selection in action: the peppered moth (*Biston betularia*)

During the Industrial Revolution in Britain, in the early part of the nineteenth century, air pollution by gases (such as sulfur dioxide) and solid matter (mainly soot) was distributed over the industrial towns, cities and surrounding countryside. Here, lichens and mosses on brickwork and tree trunks were killed off and these surfaces were blackened. The numbers of dark varieties of some 80 species of moth increased in these habitats during this period. This rise in proportion of darkened forms is known as industrial melanism.

The dark-coloured (melanic) form of the peppered moth, *Biston betularia*, followed this trend in industrialised areas, but their numbers were low in unpolluted countryside, where pale, speckled forms of the moths were far more common (Figure D4.1.3). In sooty areas, the melanic form was effectively camouflaged from predation by insectivorous birds and became the dominant species.

Theme D: Continuity and change – Ecosystems

Biston betularia

pale form observed in non-polluted habitats

melanic form observed in industrially polluted habitats

experimental evidence

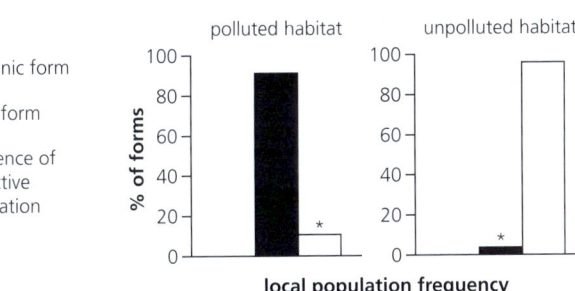

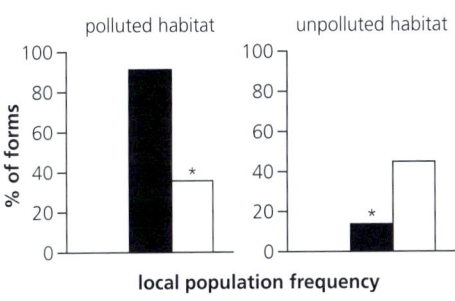

Key:
- melanic form
- pale form
- * evidence of selective predation

■ **Figure D4.1.3** The peppered moth and industrial melanism

 **TOK**

In the study of evolutionary history, do experiments have any part to play in establishing knowledge? If an experimental approach has a limited role, is the study of evolution a 'science'?

Tool 2: Technology

Generating data from models and simulations

Although it is not possible to study the selection of the pale and melanic varieties of peppered moths in the wild, there are simulations available online that allow you to do this, e.g.:

https://askabiologist.asu.edu/peppered-moths-game/play.html

You can select either a dark, polluted forest (with trees covered by soot and no lichen) or a light, unpolluted forest (with trees covered in lichen). Run the programme on one type of forest – how many melanic and pale moths remain at the end of the experiment? Why is this? What does the experiment tell you about natural selection?

Repeat the experiment using the second forest type.

D4.1 Natural selection

Inquiry 1: Exploring and designing

Designing

Design an investigation using the peppered moth simulation. *Which variable will you alter?* Identify and justify your choice of dependent, independent and control variables. Justify the range and quantity of measurements you would take and explain your methodology.

For example, you could adjust the brightness of your computer screen to investigate how this affects the levels of predation, and selection pressure, on peppered moth populations. You need to ensure the control of variables that may affect your dependent variables, such as ensuring background lighting in the room is constant, and allowing the same amount of time for each experiment.

The investigation should be replicated and mean results calculated. A control experiment could be carried out by running the simulation under full light intensity. Include screenshots from the simulation in the method and analysis, to clearly explain your methodology and results.

◆ **Sexual dimorphism**: differences in appearance between males and females of the same species, such as in colour, shape, size and structure.

◆ **Sexual selection**: selection arising through preference by one sex for certain characteristics in individuals of the other sex.

Sexual selection as a selection pressure in animal species

We have seen how alleles that confer an advantage in the phenotype of a species lead to the survival of individuals and the passing on of these adaptive alleles to subsequent generations. This can be shown in the evolution of the plumage of birds of paradise. Birds of paradise are found in the rainforests of Papua New Guinea in South East Asia. Alfred Russel Wallace was one of the first naturalists to observe these and other species of birds of paradise in the wild. Figure D4.1.4 shows the Raggiana bird of paradise. The male bird is brightly coloured but the female is plain.

Differences in physical appearance and courtship display in birds of paradise have led to the evolution of numerous species within the same forest. Male birds of paradise have bright and colourful feathers, which they use to attract females, and different species also have different dancing displays. Changes in the appearance or behaviour of populations may result in males and females of those populations no longer being attracted to each other and therefore not breeding together (see Chapter A4.1, page 147).

■ **Figure D4.1.4** A male Raggiana bird of paradise (*Paradisaea raggiana*) displaying to a female in Varirata National Park, Papua New Guinea

In addition to acting as an isolating mechanism, this form of selection can result in **sexual dimorphism**, i.e. distinct difference in size or appearance between the sexes of an animal. In the majority of bird of paradise species, the male birds are brightly coloured and often perform elaborate courtship displays, whereas the female's plumage is greys and browns (Figure D4.1.4). The differences between males and females is the result of **sexual selection**, which results in the development of physical and behavioural traits, usually in the male, which can be used as a sign of overall fitness. Females will choose the most impressive and attractive males based on their displays.

Bright feathers and an energetic dance are used to signal fitness of an individual to a female – those with the most impressive plumage and courtship display are more likely to attract a female and mate with her. Over time, the genes associated with these physical and behavioural traits are passed down and the characteristics become more prominent within the species. In this way, sexual selection can affect success in attracting a mate, and so drive the evolution of an animal population.

> ### ● Top tip!
>
> Natural selection is the result of competition for resources, but sexual selection is the result of competition for mates. On an island with abundant resources and few predators, such as Papua New Guinea, females are the primary selection pressure in determining how males evolve. Sexual selection enhances mating success, while natural selection tends to produce individuals well-adapted to their environment. Sexual selection does not adapt the individuals to their environment.

> ### ● Nature of science: Theories
>
> Charles Darwin developed the concept of sexual selection, and it can be seen as his second great insight. He defined it as depending on the 'advantage which certain individuals have over other individuals of the same sex and species solely in respect of reproduction'. Sexual selection can therefore be defined as intraspecific *reproductive* competition. Darwin made a clear distinction between natural selection and sexual selection – he realized that he needed to explain the existence of characteristics that were apparently not favoured by natural selection. Darwin said that sexual selection 'depends not on the struggle for existence, but on the struggle between males for possession of females'.

Peacocks are the male of the species of peafowl. There are only three species of peafowl: Indian (*Pavo cristatus*), Green (*Pavo muticus*) and Congo (*Afropavo congensis*). Peacocks have bright, iridescent tail feathers with eye-spot patterning that are presented during courtship displays to a female (peahen) – see Figure D4.1.5.

■ **Figure D4.1.5** Peacock (*Pavo cristatus*) displaying to a female

Sexual selection has resulted in the tail feathers becoming increasingly long and more elaborately patterned over time. The tail reduces manoeuvrability, powers of flight and makes the bird more conspicuous – these traits make the bird more prone to predation and reduce the likelihood of survival. This demonstrates a key difference between natural selection and sexual selection: natural selection produces individuals well-adapted to their environment whereas sexual selection does not. Sexual selection enhances the chance of breeding success. Sexual selection can be seen as a stronger evolutionary force than natural selection because variation in mating success can amplify selection and maintain new genetic variation among individuals, which ultimately leads to rapid evolutionary change.

> **Nature of science: Experiments**
>
> **Testing sexual selection**
>
> The two sexes are similar in the barn swallow (*Hirundo rustica*), except for the outermost tail feather, which is about 16% longer in the male than the female. The French researcher Anders Møller tested whether females open-endedly prefer males with longer tails by experimentally shortening the tails of some males. He cut them off with a pair of scissors, and lengthened the tails of others by sticking those cut tail feathers on to other intact males with very fast drying superglue. He then measured how long it took the different males to find a female mate. Males with longer tails mated faster, resulting in higher reproductive success.

5 **Distinguish** between natural selection and sexual selection.

Modelling of sexual and natural selection based on experimental control of selection pressures

■ Figure D4.1.6 Guppies (*Poecilia reticulata*) show a variety of colours and patterns

Guppies (*Poecilia reticulata*) are a fish species native to the mountain forest streams of north-east Venezuela, Margarita, Trinidad and Tobago. They show complex variation in colouration and patterning (Figure D4.1.6), and sexual dimorphism (with brightly coloured males and drab females). Scientist John Endler carried out field work in forest streams in Trinidad and found that colour patterning varied with predation pressure. Following these observations, Endler designed and carried out both lab and field research on guppy populations to investigate the effects of natural and sexual selection on the evolution of the fish.

6 **Suggest** why males and females of some species are different colours.

Endler noted that guppies derive most of their camouflage from their spots. Spots, often quite large ones, resemble the gravel bottoms of their native streams. Some streams have coarser, more pebbly gravel and others finer, sandier gravel. Changing the environments and substrates affects survival because certain environments will provide more protection and hiding spots from predators. Hypotheses for varying levels of male colouration include:

1 When there are predators present, the substrate type of the stream affects survival, leading to spot brightness on the guppies changing

2 Spot brightness on male guppies increases when there is a lack of predators or a low predation rate, due to sexual selection.

3 High spot brightness attracts predators, so as predation increases, male guppy spot brightness decreases.

In his laboratory experiment, Endler used a greenhouse to resemble the tropical environment of the guppies. He set up ten ponds inside the greenhouse and put gravel on the bottom of the ponds: five ponds with coarse gravel and five with fine gravel. Endler predicted that when guppies were exposed to strong predation, the populations should diverge from each other, each evolving to match its own background. Where there was less predation, male guppies should tend to become more conspicuous to attract females.

At the start of the experiment, guppies were randomly assigned to the ten tanks and allowed to breed for six months with no predation. Endler then used three levels of predation: two ponds (one with fine and one with coarse gravel) had no predation, four ponds (two with coarse and two with fine gravel) contained a dangerous predator, the pike cichlid (*Crenicichla alta*), and in the remaining four ponds he introduced a weak predator, the killifish (*Rivulus hartii*). The different number of dangerous predators and weak predators was used to reflect the density of predators in the wild. After introducing the predators, the experiment ran for five months. Endler then counted and measured the spots on all the guppies. After nine more months, further data were recorded.

Before the experiment, the guppies showed a wide range of spot size and number, reflecting the range of streams and predator content they were taken from. In the six months with no predation, the number of spots increased, presumably because of selection by females. At the point when the predators were introduced, there were significant changes:
- in the four ponds with the dangerous predator, the mean number of spots decreased
- in the ponds with no or weak predation, the number of spots continued to increase.

In the presence of either weak or strong predation, coarse gravel favoured larger spots whereas finer gravel favoured smaller spots. This can be interpreted as spot size mimicking gravel size. In the ponds with no predation, Endler found the reverse: fine gravel favoured large spots on male guppies and coarse gravel favoured small spots. He interpreted the results as males being more conspicuous if they do not match the background, which is good for attracting females.

In his field experiment, Endler transferred dull-coloured male guppies from an area with a dangerous predator (*Crenicichla alta*) to a region with a weak predator (*Rivulus hartii*) and left them for 15 guppy generations (two years). He then returned to study the effect of the transplants. In the stream with introduced fish, the males had evolved more colourful patterning, due to sexual selection and the absence of strong predation.

Endler's experiments showed that the evolution in the guppy populations is a dynamic process of natural and sexual selection. Natural selection can occur in guppies as a response to the ability of finding food and avoiding predation. Predators can more easily see brightly coloured males, reducing their chance of survival, leading to selection of less highly coloured/spotted individuals. Sexual selection can also occur: traits beneficial for sexual selection (e.g. bright colouration) will confer a reproductive advantage. Males with greater brightness/spot colouration and size are more likely to mate, reproduce and pass on the alleles that code for these ornate characteristics. A trade-off between the two processes can be expected – in areas with high predation, it is less likely that brightly coloured males will survive, irrespective of their reproductive advantage.

Top tip!

You can read more about Endler's experiments on guppies here: https://evolution.berkeley.edu/lines-of-evidence/experiments

Interpreting the data

Data from Endler's field and laboratory experiment are shown in Figure D4.1.7. The shaded bars represent the guppies in the greenhouse, while the unshaded bars represent the guppies from the field experiment.

In both studies, the colour of the guppies was scored by counting the colour, brightness and number of spots after 15 generations.

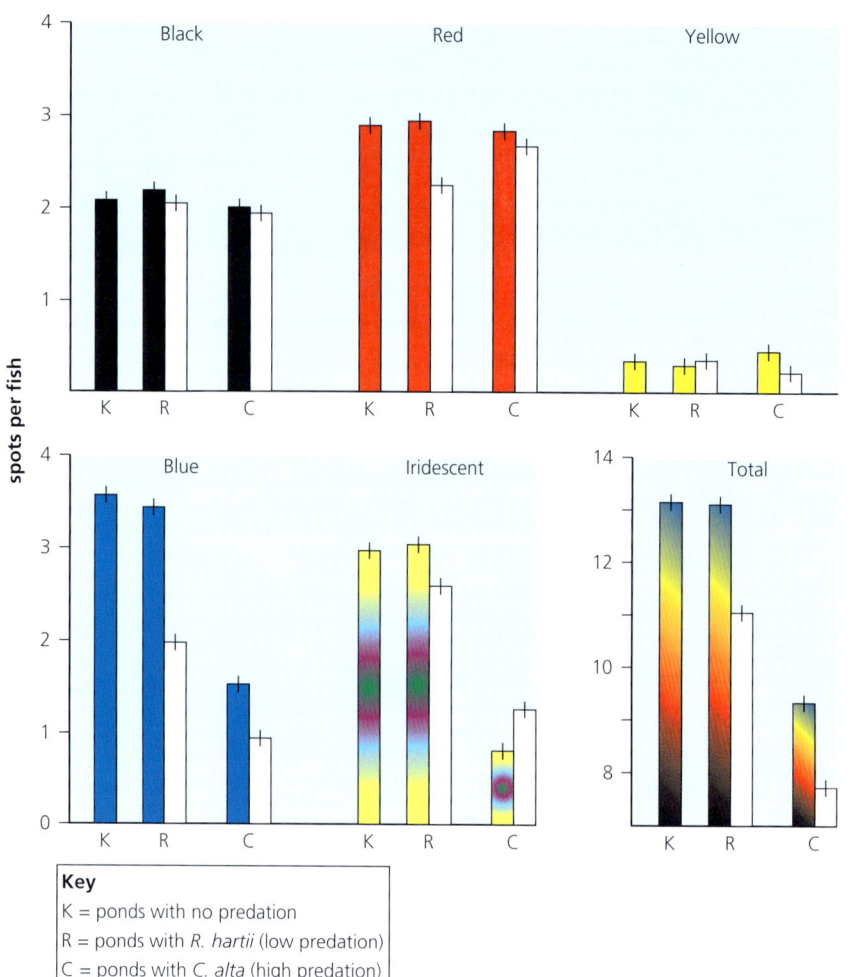

■ **Figure D4.1.7** Guppy experimental data: number of spots per fish in the greenhouse and in the field. Small vertical lines are two standard errors

Analyse the data shown in the charts. Compare the data from the field and data from the greenhouse. Your summary should include a specific statement regarding how predation influences colouration.

ATL D4.1C

Endler also examined colour patterns, spot patterns and sizes of spots. Read Endler's full paper 'Natural selection on color patterns in *Poecilia reticulata*' here:
https://onlinelibrary.wiley.com/doi/pdf/10.1111/j.1558-5646.1980.tb04790.x

Look at Figures 1 (page 80) and 4 (page 85) from Endler's paper and **analyse** the data shown. Summarize the data and **explain** the results. What is indicated by the convergence of x and r, and the divergence from c, in Figure 4?

Concept of the gene pool

◆ **Gene pool**: all the genes (and their alleles) present in a breeding population.

Population genetics is the study of genes in populations (breeding groups). The total of all the genes and their different alleles present in a population make up a **gene pool**. A sample of the genes of the gene pool will form the **genomes** (gene sets of individuals) of the next generation, and so on, from generation to generation.

When the breeding group is a large one, and all the individuals of the population have an equal opportunity of contributing gametes, random matings will maintain the original proportions of alleles in the population. In *these* circumstances, the allele frequency will not change.

We can demonstrate this by setting two alleles of a gene at some arbitrary frequency in an infinitely large gene pool, and then following the frequency of possible genotypes these will produce by random matings over several generations. The example of two alleles, **A** and **a**, present in the population at frequencies of 0.8 and 0.2 is illustrated in Figure D4.1.8. Look at the single mating cycle illustrated there. You can see why the frequency of genes will not change after one cycle. In fact, this frequency will never change however many matings occur *unless* there are what geneticists call 'disturbing factors' at work.

When there is no change in a gene pool, we can say the population is not evolving.

Link
Gene pools are also covered in Chapter A3.1, page 108, Chapter A4.1, page 148, and Chapter D3.2, page 734.

7 Predict the factors that may cause the composition of a gene pool to change. (*Think about the changes that may take place in a population and between its members.*)

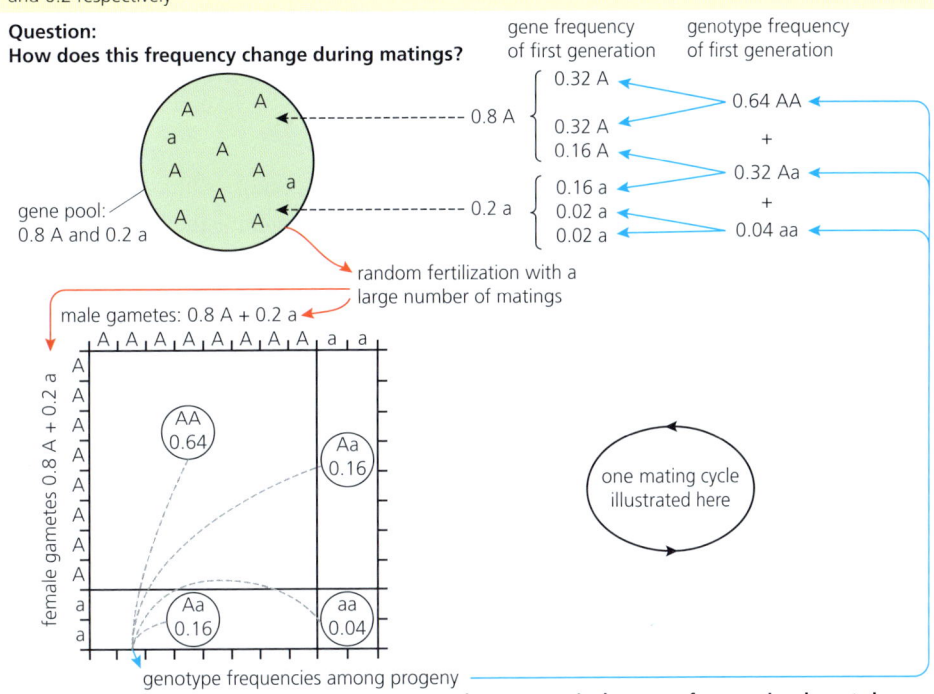

■ Figure D4.1.8 The fate of genes in a gene pool

Allele frequencies of geographically isolated populations

In Chapter A3.1, we saw how the composition of gene pools can change (Figure A3.1.8, page 108), and in Chapter A4.1, we saw how changes to gene pools in isolated populations can lead to speciation (page 147).

The composition of a gene pool can change due to a range of factors that may alter the proportions of some alleles. With one or more 'disturbing factors' operating, allele frequencies are likely to change from generation to generation. For example, some alleles may increase in frequency because of the advantage they give to the individuals that carry them. Because of these alleles, an organism may be more successful – producing more offspring, for example. If we can detect change in a gene pool (that is, if the proportions of particular alleles are altered) we may be seeing evolution before a new species is observed.

Using databases to search for allele frequencies

Biological databases have been created to store data, such as genetic databases. These allow scientists to browse, share and compare data while doing research studies. In this section, we will look some databases to research allele frequencies.

NCBI Short Genetic Variation database

The NCBI Short Genetic Variation database (dbSNP) catalogues short variations in nucleotide sequences in humans. These variations include single nucleotide variations, including insertions, deletions and short tandem repeats less than 50 nucleotides in length.

www.ncbi.nlm.nih.gov/snp

The rsID number is a unique label ('rs' followed by a number) used by researchers and databases to identify a specific SNP (single-nucleotide polymorphism). It stands for Reference SNP cluster ID and is the naming convention used for most SNPs. To find the rsID number for a specific gene, go to:
www.ensembl.org/index.html

Using the Ensembl website, change 'all species' to 'human' in the search box, and underneath type in the gene of interest. For example, if researching the LPL gene, which provides instructions for making an enzyme called lipoprotein lipase, type 'LPL' in the search box under 'human'. On the page that you are transferred to, select 'variant' on the left – this will show you variant SNPs for the gene. Each variant has a different rs number. Some variants produce disease or other clinical effects: type the rs number into an internet search engine to find out the phenotypic effect of the variant, if known. Select a variant of interest, for example rs268 for a variant of the LPL gene.

Now access the dbSNP for the gene of your choice by typing the web address followed by the rs number; for example, for the LPL gene this is www.ncbi.nlm.nih.gov/snp/rs268. The database shows you a table of a selection of worldwide locations and frequencies of the allele. 'Ref' is the allele included in the search, and 'Alt' is any other allele found at that locus on the chromosome. The Frequency tab lists the allele frequency data from major studies, broken down by subpopulation if available. This provides a way to evaluate the impact of a variant. If the gene has more than two alleles, the most common is shown. For example, the frequency of LPL globally is $A = 0.983727$ and that of the alternate allele is $G = 0.016273$. These numbers refer to the frequency of the allele in the population, for example 0.5 would be 50% incidence. This is the notation used in the Hardy–Weinberg equation (see page 796).

> **Link**
> For HL only students, the Hardy–Weinberg equation is covered later in Chapter D4.1, page 796.

> **Top tip!**
> Neo-Darwinism is a modified theory of Darwinism which explains the origin of species on a genetic basis. The main difference between them is that Darwinism describes favourable and inheritable phenotypic variations as the driving force of speciation, whereas neo-Darwinism states that it is only inheritable genetic variations that are the driving force of speciation.

> **Top tip!**
> This website has further information and a video about the database:
> www.ncbi.nlm.nih.gov/snp/docs/gsr/alfa
>
> This website has a factsheet about the database:
> https://ftp.ncbi.nlm.nih.gov/pub/factsheets/Factsheet_SNP.pdf

Allele Frequency Net database

The following database contains allele frequencies for genes related to the immune system, such as the human leukocyte antigen (HLA) system: www.allelefrequencies.net

The HLA system is a complex of genes on chromosome 6 in humans that encodes cell-surface proteins (antigens) responsible for the regulation of the immune system.

Click on the 'populations' tab at the top of the screen and 'browse populations by geographic regions'. This enables you to select specific areas of the world and examine allele frequencies for the HLA system.

Concept: Change

Changes in allele frequency in the gene pool occur as a result of natural selection and other processes such as genetic drift.

Changes in allele frequency in the gene pool as a consequence of natural selection

When Darwin developed the theory of evolution by natural selection, he had no knowledge of DNA or Mendelian genetics. Despite this, his theory explains the evolution of life on Earth that is still used today. Following Darwin's work, scientists have integrated genetics with natural selection: this is known as **neo-Darwinism**. This has allowed an explanation of how species evolve at the molecular level, i.e. changes in allele frequency in the gene pool are a consequence of natural selection between individuals according to differences in their heritable traits.

ATL D4.1D

Evolutionary biologist Richard Dawkins (Figure D4.1.9) has written about neo-Darwinism in a variety of his books and essays. Two of his best-known works are *The Selfish Gene* and *The Blind Watchmaker*. These explore a gene-centred view of evolution.

Which other scientists can you find who have developed a modern synthesis of evolution? Select one of the books they have written and write a review for a school magazine or journal.

■ **Figure D4.1.9** Professor Richard Dawkins – evolutionary biologist and science author

■ **Figure D4.1.10** Three different types of selection

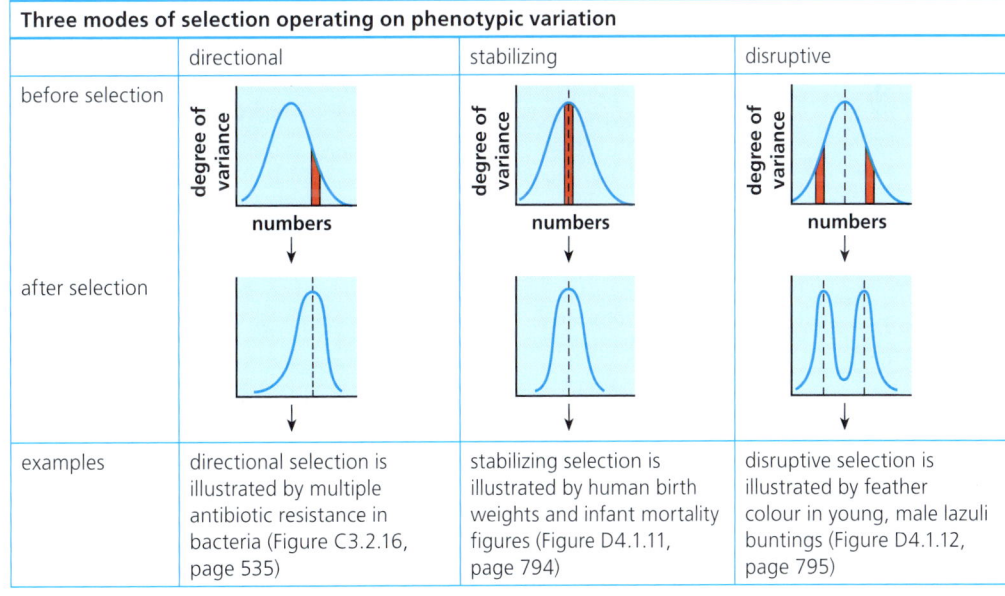

Differences between directional, disruptive and stabilizing selection

Natural selection operates to change the composition of gene pools, but the outcomes of this vary. We can recognize three different types of selection, and these are explained in Figure D4.1.10. In each case, selection results in a change in allele frequency.

Stabilizing selection occurs when there is an increase in frequency of an average trait. It is a mechanism that maintains a favourable characteristic and the **alleles** responsible for it, and eliminates variants and abnormalities that are useless or harmful. It determines the frequency and values for the variation seen in a population. Most populations probably undergo stabilizing selection. Our example (in Figure D4.1.11) comes from human birth records on babies born between 1935 and 1946, in London. It shows there was an optimum birth weight for babies, where 'optimum' refers to the middle birth weight, and those with birth weights heavier or lighter were at a selective disadvantage. The data show how environmental factors can affect continuous variation of a trait in a population, and the mechanism which determines the range of values.

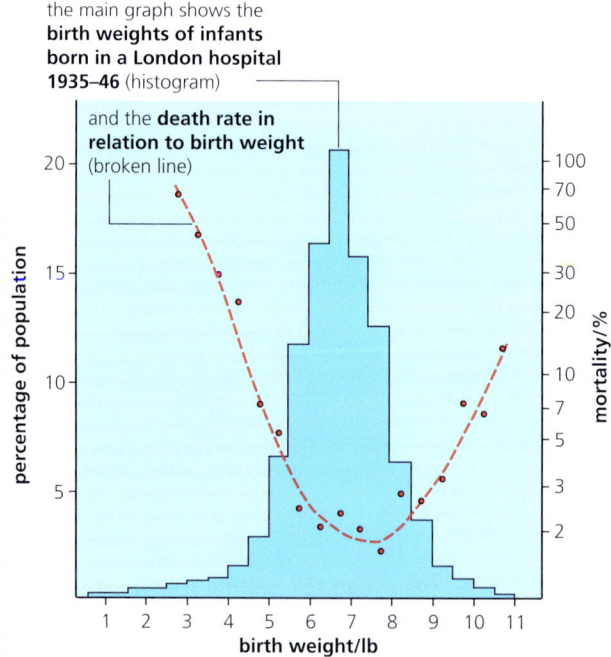

The birth weight of humans is influenced by **environmental factors** (e.g. maternal nutrition, smoking habits, etc.) and by **inheritance** (about 50%).

When more babies than average die at very low and very high birth weights, this obviously affects the gene pool because it tends to eliminate genes for low and high birth weights.

The data provide an example of continuous variation. The 'middleness' or central tendency of this type of data is expressed in three ways:

1 **mode** (modal value) – the most frequent value in a set of values
2 **median** – the middle value of a set of values where these are arranged in ascending order
3 **mean** (average) – the sum of the individual values, divided by the number of values

■ **Figure D4.1.11** Birth weight and infant mortality, a case of stabilizing selection

Directional selection may result from changing environmental conditions. In these situations, the majority form of an organism may become unsuited to the environment because of change. Some other alternative phenotypes may have a selective advantage.

An example of directional selection is the development of resistance to an antibiotic by bacteria (Figure C3.2.16, page 535). Certain bacteria cause disease, and patients with bacterial infections are frequently treated with an antibiotic to help them overcome the infection. However, in a large population of a species of bacterium, some may carry a gene for resistance to the antibiotic in question. A 'resistant' bacterium has no selective advantage in the absence of the antibiotic and must compete for resources with non-resistant bacteria. But, when the antibiotic is present, most bacteria in the population are killed. Resistant bacteria remain and create the future population, all of which now carry the gene for resistance to the antibiotic. The genome has been changed abruptly.

Disruptive selection occurs when particular environmental conditions favour the extremes of a phenotypic range over intermediate phenotypes. Here, individuals of a species that show the extreme phenotypes have the highest fitness and those individuals with intermediate values are at a fitness disadvantage. To be 'disruptive' there are intermediate phenotypes, but individuals in between the extremes are less able to adapt to the environment and become less frequent over time. This phenomenon is shown by feather colour in young (yearling, i.e. one year or less in age) male lazuli buntings (*Passerina amoena*), native to North America. Adult birds show bright plumage (Figure D4.1.12), but yearling male birds take a year or two to develop full adult colouration.

The delayed maturation of colour means that yearling birds show a greater variation in colour than adults, with colouration ranging from dull brown to bright blue individuals (as bright as adult males). A study published in *Nature* in 2008 scored colouration from 14 for extremely dull males to 36 for extremely bright males (Figure D4.1.12).

The researchers showed that the dullest and brightest yearling males generally obtained better territories. Intermediate yearlings (i.e. with colour in between the two extremes) did not gain as good territories. Male birds in high-quality territories generally paired with females more frequently than males in poorer territories. This led to a bimodal pattern in pairing success because females preferentially paired with males in high-quality territories (i.e. with brown or brightly coloured males). In this case, data suggested that territorial quality was a key component in the disruptive selection shown here.

8 **Distinguish** between stabilizing, directional and disruptive selection.

■ Figure D4.1.12 Colourful male lazuli bunting with brown female

ATL D4.1E

The research in this example of disruptive selection came from a 2008 article in *Nature*, 'Disruptive sexual selection for plumage coloration in a Passerine bird'. Read the article and see photographs of different individual feather colours here:

www.researchgate.net/publication/12258797_Disruptive_sexual_selection_for_plumage_coloration_in_a_Passerine_bird

Can you find other examples of variation in a population that leads to selection for phenotypic extremes, and where individuals with intermediate values are at a fitness disadvantage?

So, why do bright and dull yearlings obtain good territories, while intermediate males do not? Adult males may see dull, brown-coloured yearlings as non-threatening, possibly because they resemble females, and so allow them to become resident in their area. Adult males also leave brightly coloured yearlings alone, perhaps seeing them as a threat. Therefore, both bright and dull yearling males can establish territories and attract females. However, yearling males with intermediate plumage are more often attacked by adult males and fail to obtain territories and a mate. This results in disruptive selection, favouring the colour extremes.

Hardy–Weinberg equation and calculations of allele or genotype frequencies

> **Top tip!**
>
> The proportion of alleles in a population is called the allele frequency. The frequency of an allele can vary from 0 to 1. The frequencies of two alleles of any given gene must add up to 1, e.g. if the frequency of the allele for blue eyes is 0.25 then the frequency of the allele for brown eyes must be 0.75.

We have noted that in any population:
- The total of the alleles of the genes located in the reproductive cells of the individuals makes up a **gene pool**. A sample of the alleles of the gene pool will contribute to form the genomes (gene sets of individuals) of the next generation, and so on, from generation to generation.
- When the gene pool of a population remains more or less unchanged, that population is not evolving. If the gene pool of a population is changing (the proportions of particular allele pairs is altered), then evolution may be happening.

How can we detect change or constancy in gene pools?

The answer is by a mathematical formula called the **Hardy–Weinberg equation**.

The general formula to represent the frequency of dominant and recessive alleles is:

$$p + q = 1$$

where p = frequency of the dominant alleles

and q = frequency of the recessive alleles.

The problem in estimating gene frequencies is that it is not possible to distinguish between homozygous dominants and heterozygotes based on their appearance or phenotype, as has been noted previously (page 732). The Hardy–Weinberg equation overcomes this problem:

$$p^2 + 2pq + q^2 = 1$$

frequency of dominant homozygous individuals + frequency of heterozygous individuals + frequency of homozygous recessive individuals = TOTAL

> **Concept: Continuity**
>
> In population genetics, if allele frequencies remain unchanged and match Hardy–Weinberg conditions, then this indicates stability and continuity in the genotype.

Using the Hardy–Weinberg formula

We can use the Hardy–Weinberg formula to find the frequency of a gene in cases of dominance in which we are unable to distinguish between the homozygous dominants and the heterozygotes on the basis of phenotype.

In humans, the ability to taste the chemical phenylthiocarbamide (PTC) is conferred by the dominant allele T. Both the dominant homozygotes (**TT**) and the heterozygotes (**Tt**) are 'tasters'. The non-tasters are the recessive homozygotes (**tt**).

In a sample of a local population in Western Europe of 200 people, 130 (65%) were tasters and 70 (35%) were non-tasters (Table D4.1.3).

> **Top tip!**
>
> Allele frequencies are likely to be very different in geographically isolated populations compared to the original populations, due to natural selection, mutation and random genetic drift.

■ Table D4.1.3 Tasters and non-tasters

Phenotypes	Tasters	Non-tasters
Genotypes	TT + Tt	tt
Frequency	0.65	0.35

Applying these data to the Hardy–Weinberg equation, we know the value of q^2 to be 0.35.

Taking the square root, the value of $q = 0.59$.

So, the frequency of the non-tasting alleles (**t**) in this European population was 0.59.

Theme D: Continuity and change – Ecosystems

> **Top tip!**
>
> Remember to use p and q to denote the two allele frequencies. It is important to understand that $p + q = 1$ so genotype frequencies are predicted by the Hardy–Weinberg equation: $p^2 + 2pq + q^2 = 1$. If one of the genotype frequencies is known, the allele frequencies can be calculated using the same equations.

9 In two small isolated populations the incidence of PTC non-tasters was as shown in the table. **Calculate** the frequency of the non-tasting alleles in each of the two sample populations.

Population	Sample size	Number of non-tasters
Indigenous Australians	500	245
Indigenous North Americans	500	20

Hardy–Weinberg conditions that must be maintained for a population to be in genetic equilibrium

If genotype frequencies in a population do not fit the Hardy–Weinberg equation, this indicates that one or more of the conditions is not being met, for example mating is non-random or survival rates vary between genotypes.

When the gene pool of a population remains more or less unchanged, then this indicates that the population is not evolving. If the gene pool of a population is changing (the proportions of particular allele pairs is altered), then evolution may be happening.

Inquiry 2: Collecting and processing data

Collecting data; interpreting results

Look back at Endler's experiments on guppies on page 788. Then access the following guppy evolution simulation, which lets you explore the causes of evolution and Hardy–Weinberg equilibrium: **www.labxchange.org/library/items/lb:LabXchange:d884c485:lx_simulation:1**

Click 'start simulation' and then read the background information section (select the tab at top of screen). You can alter one of several different variables to investigate the effect of this on two alleles:

p = wt allele, ornamental (wild-type) allele

q = gol allele, drab (golden) allele

Select the 'Hardy–Weinberg equilibrium' tab. The simulation tracks the population dynamics of a wild-type allele and a golden allele. In the wild-type allele, guppies produce visible ornamentation in their skin. The golden allele has a mutation that leads to less pigmented skin. Pigmentation seems to drive sexual selection in wild populations.

Work in pairs to explore the effects of different variables on allele frequency. Variables you can manipulate include level of predation, the salinity of the water, the mutation rate (which will affect the rate of evolution of the guppy populations) and migration rate (movement into the population from external areas). You can also change habitat.

Change one factor and then run the simulation to see the effect on the frequency of the two alleles. Then select a different factor. Did the outcomes match your predictions? Identify, describe and explain patterns, trends and relationships from your investigations.

> **Links**
>
> Artificial selection (selective breeding) is covered in Chapter A4.1, page 141. Bacterial resistance to antibiotics is covered in Chapter C3.2, page 534.

Artificial selection by deliberate choice of traits

Artificial selection is carried out in crop plants and domesticated animals by choosing individuals for breeding that have desirable traits. Unintended consequences of human actions, such as the evolution of resistance in bacteria when an antibiotic is used (page 534), are due to natural rather than artificial selection.

10 **Explain** the key difference between natural and artificial selection.

> **LINKING QUESTIONS**
>
> 1 How do intraspecific interactions differ from interspecific interactions?
> 2 What mechanisms minimize competition?

D4.2 Stability and change

Guiding questions

- What features of ecosystems allow stability over unlimited time periods?
- What changes caused by humans threaten the stability of ecosystems?

SYLLABUS CONTENT

This chapter covers the following syllabus content:
▶ D4.2.1 Stability as a property of natural ecosystems
▶ D4.2.2 Requirements for stability in ecosystems
▶ D4.2.3 Deforestation of Amazon rainforest as an example of a possible tipping point in ecosystem stability
▶ D4.2.4 Role of keystone species in the stability of ecosystems
▶ D4.2.5 Assessing sustainability of resource harvesting from natural ecosystems
▶ D4.2.6 Factors affecting the sustainability of agriculture
▶ D4.2.7 Eutrophication of aquatic and marine ecosystems due to leaching
▶ D4.2.8 Biomagnification of pollutants in natural ecosystems
▶ D4.2.9 Effects of microplastic and macroplastic pollution of the oceans
▶ D4.2.10 Restoration of natural processes in ecosystems by rewilding
▶ D4.2.11 Ecological succession and its causes (HL only)
▶ D4.2.12 Changes occurring during primary succession (HL only)
▶ D4.2.13 Cyclical succession in ecosystems (HL only)
▶ D4.2.14 Climax communities and arrested succession (HL only)

Stability as a property of natural ecosystems

Ecosystems are open systems that are supported by input of energy and matter. Sunlight or energy from chemical reactions ensure that producers synthesize glucose, which supports the whole food chain. Inputs of matter, from organisms entering the ecosystem or the input of inorganic material (e.g. carbon dioxide), provide the elements from which biological molecules are created, such as carbohydrates, lipids and proteins. Given an ongoing supply of energy and matter, ecosystems are sustainable. The term **stability** has many definitions, but in terms of ecosystems it can be defined as the ability to maintain or support systems and processes continuously over time. There is evidence for some ecosystems on Earth persisting for millions of years, such as the rainforests of South East Asia, which are estimated to be at least 70 million years old (Figure D4.2.1). Steady conditions, with high inputs of rainfall and sunlight throughout the year, combined with warm temperatures, have ensured the stability of these ecosystems, at least until the advent of humanity.

◆ **Stability**: the ability to maintain or support systems and processes continuously over time.

Link
Ecosystems as open systems is covered in Chapter C4.2, page 579.

Concept: Continuity

Stability represents continuity within ecosystems. Inputs of energy and matter, and nutrient cycling, support and maintain ecosystems.

■ **Figure D4.2.1** Rainforest at Danum Valley, Malaysia

> **ATL D4.2A**
>
> Find out about the age of ecosystems in the area where you live. What are the oldest ecosystems? Why have they survived unchanged for such long periods of time? What are the characteristics that determine the longevity of natural systems?

Requirements for stability in ecosystems

For an ecosystem to remain stable it requires a continual supply of energy, nutrient recycling, a diversity of organisms and a climate that remains within tolerance levels for the species living there.

In Chapter C4.2 we saw how transfers of energy and matter are essential in the maintenance of ecosystems. We also saw how the recycling of all chemical elements is needed to support living organisms in ecosystems. Energy and the recycling of nutrients contribute to the stability of ecosystems. Ecosystems require constant inputs of energy and matter, and the recycling of nutrients within an ecosystem adds additional stability.

Genetic diversity refers to the range of genetic material present in a gene pool or population of a species. Although the term normally refers to the diversity within one species, it can also be used to refer to the diversity of genes in all species within an area. Ecosystems that are complex (supported by a large biomass of producers) lead to an increased number of niches, which increases species and genetic diversity, resulting in greater stability. Ecosystems that contain species with small population sizes can lead to low genetic diversity, which leaves species more prone to disease.

Biome distribution is determined by climatic variables that need to remain within tolerance levels for the plant species that are found there. These climatic variables include temperature, precipitation and insolation (page 354). The type of stable ecosystem that will emerge in an area is predictable based on climate. If conditions vary from tolerance levels, the survival of the ecosystem can be threatened (as we will see in the case of the Amazon rainforest, below).

Link
Transfers of energy and matter in ecosystems is covered in Chapter C4.2, page 578; the recycling of chemical elements in ecosystems is on page 601.

Link
Biodiversity as the variety of life in all its forms is covered in Chapter A4.2, page 156.

Link
Abiotic factors as the determinants of terrestrial biome distribution is covered in Chapter B4.1, page 354.

Deforestation of Amazon rainforest as an example of a possible tipping point in ecosystem stability

◆ **Tipping point**: a critical threshold when even a small change can have dramatic effects and cause a disproportionately large response in the overall system.

Tipping points occur when there is a dramatic change in the ecological state away from stability. They represent points beyond which irreversible change or damage occurs. **Equilibrium** refers to a state of balance among the components of a system. **Positive feedback** loops tend to amplify changes and drive the system towards a tipping point where a new equilibrium is adopted (Figures D4.2.2 and D4.2.3b). Such changes are caused by human population growth and associated factors, such as:

- resource consumption
- habitat transformation and fragmentation
- energy production and consumption
- climate change.

All these factors exceed, in both rate and magnitude, the changes seen in the most recent global-scale shift in equilibrium at the end of the last ice age. Most projected tipping points are linked to climate change (Chapter D4.3). Increases in carbon dioxide levels above a certain value (450 ppm) would lead to increased global mean temperature, causing melting of the ice sheets and permafrost.

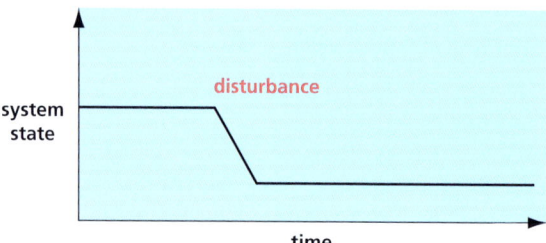

■ **Figure D4.2.2** Unstable equilibrium: the system moves to a new equilibrium following disturbance

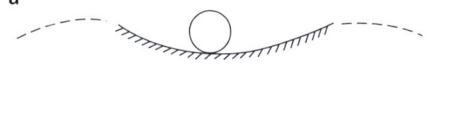

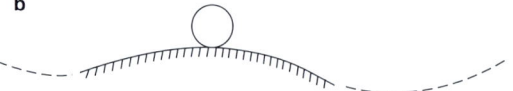

■ **Figure D4.2.3** Diagrams showing the difference between (a) stable and (b) unstable equilibrium

Concept: Change

Disturbance to ecosystems and positive feedback mechanisms can move them away from stability and towards tipping points. Change to ecological systems can lead to ecosystem collapse.

Deforestation of Amazon rainforest

The Amazon is a rainforest biome that spans nine countries in South America: Brazil, Bolivia, Peru, Ecuador, Colombia, Venezuela, Guyana, Suriname and French Guiana. The rainforest covers 6 million km² and contains half of the planet's remaining tropical forests.

Economic, ecological and climatic systems in Amazonia may now be interacting in ways that will move the forest beyond a tipping point. Chapter C4.2 discusses how fires in the Amazon are caused by increased frequency of drought conditions (page 598). Burning and deforestation of the Amazon rainforest to make space for grazing land or housing, and to provide timber, leads to loss of large areas of rainforest. Illegal mining operations also lead to loss of forest habitat. Remaining areas can become fragmented, forming islands of habitat. Continued burning and clearance, and the establishment of grasslands, prevents re-establishment of rainforest.

A large area of rainforest is needed for the generation of atmospheric water vapour by continuous transpiration, with consequent cooling, air flows and rainfall. There is scientific uncertainty over the minimum area of rainforest that is sufficient to maintain these processes.

Climate change further reinforces many of these processes by increasing air temperatures, dry season severity and the frequency of extreme weather events. One projection from a 2008 paper in the journal *Philosophical Transactions of the Royal Society B: Biological Sciences* suggests that by 2030, 24% of the Amazon rainforest will be damaged by drought or logging and 31% of the Amazon's closed-canopy forest will be deforested (Figure D4.2.5). If 'business as usual' continues, logging will further reduce the Amazonian forest carbon store size by 15%; drought damage will cause another 10% reduction in forest biomass. More recent studies, such as a January 2022 World Wide Fund for Nature (WWF) briefing at the Westminster Hall debate on deforestation in the Amazon, suggest 40% of the Amazon rainforest will be lost by 2050. The effects of more frequent fires could release an additional 20% of forest carbon to the atmosphere. Thus, some predicted 15–26 billion tonnes of carbon contained in the Amazon will be released to the atmosphere in less than 30 years as a result of large-scale dieback.

> **Link**
> The release of carbon dioxide into the atmosphere during combustion of biomass, peat, coal, oil and natural gas is covered in Chapter C4.2, page 597.

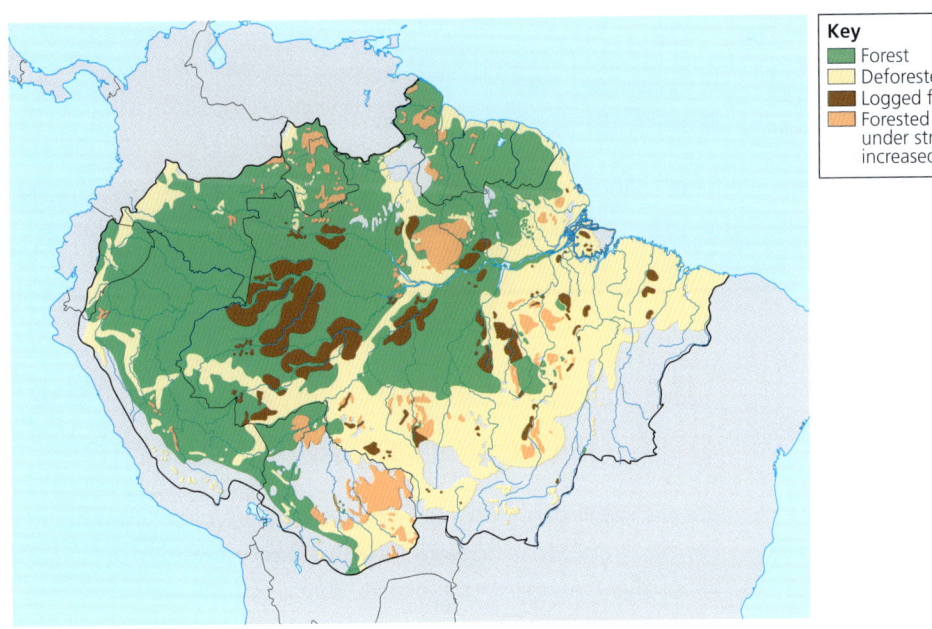

■ **Figure D4.2.4** Projected status of the Amazon rainforest, 2030, showing drought-damaged, logged and cleared forest assuming 'business as usual'

Tool 3: Mathematics

Calculating and interpreting percentage change

To assess the extent of deforestation in the Amazon rainforest, we can calculate the percentage change from the original area of forest. Figure D4.2.6 shows deforestation of the Amazon rainforest. Sequences of such photos are available that show the loss of rainforest over time.

These photos can be used to estimate forest cover at different points in time, and these data used to calculate percentage change using the following formula:

$$\text{percentage change} = \frac{\text{change in forest cover}}{\text{initial forest cover}} \times 100$$

Where change in forest cover = final forest cover − initial forest cover.

The following site has data you can use: https://rainforests.mongabay.com/amazon/deforestation_calculations.html

For example, in 2017 estimated natural forest cover in the Brazilian Amazon (Amazonia) was $3\,399\,308\,km^2$, and in 2018 it was $3\,390\,835\,km^2$.

The percentage change in forest cover therefore:

$$= \frac{3\,390\,835 - 3\,399\,308}{3\,399\,308} \times 100$$

$$= -\frac{8473}{3\,399\,308} \times 100$$

$$= -0.25\%$$

A negative sign in percentage change calculations represents a loss.

■ **Figure D4.2.5** Satellite photo showing deforestation in the Amazon rainforest

This is a yearly change in forest cover. In terms of overall change from the original area of forest cover, the total percentage change is:

Original estimated forest cover (pre-1970)
= $4\,100\,000\,km^2$

Area of natural forest cover in 2018 = $3\,390\,835\,km^2$.

The percentage change in forest cover is therefore

$$= \frac{3\,390\,835 - 4\,100\,000}{4\,100\,000} \times 100$$

$$= -\frac{709\,165}{4\,100\,000} \times 100$$

$$= -17.30\%$$

This calculation shows that 17.3% of Amazonia natural forest cover has been lost since 1970. A tipping point may be reached at 20% to 25% of deforestation.

Use of a model to investigate the effect of variables on ecosystem stability

◆ **Mesocosm**: enclosed environments that allow a small part of a natural environment to be observed under controlled conditions.

The stability of an ecosystem may change when an external 'disturbing' factor that disrupts the natural balance is applied. An investigation of this may be attempted in natural habitats or in experimental, enclosed systems called a **mesocosm**. Both approaches have advantages and drawbacks (Table D4.2.1).

> **Concept: Continuity and change**
>
> Mesocosms can be used to study the factors that maintain stability in ecosystems, and also how altering biotic or abiotic factors can lead to change.

■ **Table D4.2.1** Alternative approaches to investigating ecosystem stability

	A natural ecosystem, for example an entire pond or lake	A small-scale laboratory model aquatic system (a mesocosm)
Advantages	realistic – actual environmental conditions are experienced	able to control variables – opportunity to measure the degree of stability or extent of change in a community, and to investigate the precise impact of a disturbing factor
Disadvantages	variable conditions – minimum or non-existent control over 'controlled variables'	unrealistic – possibly of disputed relevance and applicability to natural ecosystems

Mesocosms are enclosed experimental areas that are set up to explore ecological relationships. Because they are contained experimental areas, they can be closely controlled and variables monitored. Studying natural ecosystems can be difficult because there are so many variables that cannot be controlled – mesocosms enable all variables other than the independent and dependent variable to be kept constant, so as to ensure a fair test.

Sealed glass vessels can be used for mesocosms because entry and exit of matter can be prevented, but light can enter and heat can leave. Both aquatic and terrestrial systems can be used.

The stability of an ecosystem may change when an external 'disturbing' factor that disrupts the natural balance is applied. Investigation of this may be attempted in natural habitats or in experimental, enclosed systems. Mesocosms can be used as an enclosed system to investigate the effect of altering one variable on the stability of an ecosystem. Both approaches have advantages and drawbacks, as detailed in the table above.

Nature of science: Experiments

Care and maintenance of the mesocosms should meet with the standards of the IB ethical experiment policy. Make sure you recognize and address any relevant safety, ethical and environmental issues when you are investigating mesocosms.

Inquiry 1: Exploring and designing

Designing

Set up a mesocosm to investigate the effect of changing one variable on the stability of the system. For example, in an aquatic system you could set up one mesocosm with fish and the other without fish to investigate the effect of fish on the aquatic ecosystem.

When setting up the mesocosm you need to consider:

- What variables will you be controlling (keeping the same) and why? What effect might these variables have on the system if they are changed (i.e. why do you need to control them)?
- For terrestrial mesocosms, large glass jars can be used, although plastic containers can be just as effective. Which will you use, and why? Should the sides of the container be transparent or opaque?
- Which groups of organisms will you need to include in the mesocosm? Think of the organisms that would be present in a natural ecosystem (e.g. autotrophs, consumers and decomposers).
- Because the mesocosm will be sealed, you need to consider how the organisms will obtain fresh sources of oxygen. Photosynthetic organisms will ensure a supply of oxygen, but how will you ensure these are kept alive?
- Because the mesocosm is a closed system there is a danger that the organisms in it will suffer from lack of food, competition, excess heat, and so on. How will you ensure the well-being of the organisms in your mesocosm?

D4.2 Stability and change

◆ **Eutrophication**: the natural or artificial enrichment of a body of water, particularly with respect to nitrates and phosphates. There are both natural and human causes of eutrophication.

Lakes and ponds where there is an excess of nutrients can undergo a process called **eutrophication**. Excess nutrients may come from fertilizer run-off from surrounding land, for example. The excess nitrates and phosphates provide high-levels of nutrition for algae, which undergo rapid population growth (an algal bloom occurs). The algae block light to underwater plants which die, providing detritus for bacteria to feed on – these bacteria undergo rapid growth and remove oxygen from the water, causing fish and other aquatic animals to die. A few organisms can survive in these conditions and prosper, but the death of many aquatic organisms results.

Could a mesocosm be set up to investigate eutrophication, so avoiding the destruction of a natural ecosystem?

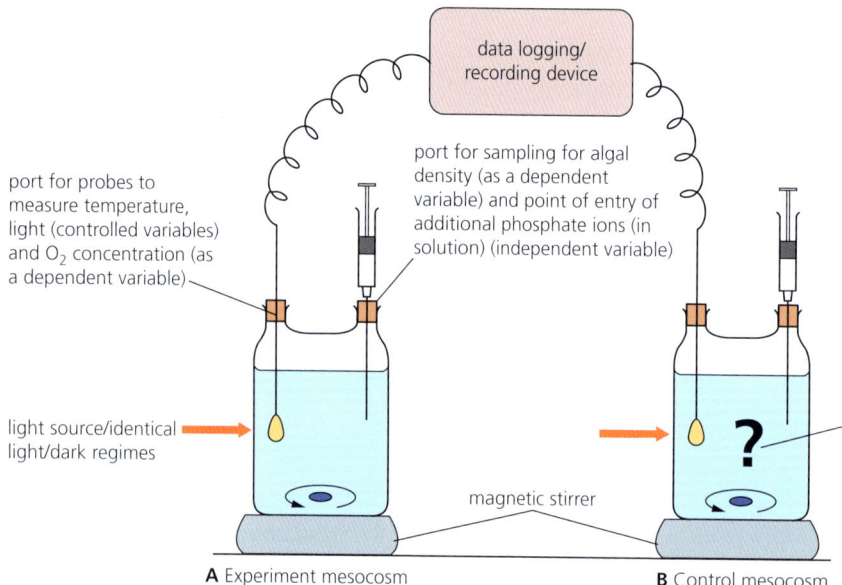

Possible steps to the investigation – what 'control' flask is required?:

1 Set up of mesocosms A (experiment) and B (control) with identical cultures of algal suspensions in pond water. Allowed to stabilize, and give evidence of normal algal growth

2 Addition of a quantity of concentrated phosphate solution to A

What would the control flask require?

3 Regular monitoring of change in algal cell density and O_2 concentration in A and B mesocosms
Issues: Does an algal bloom develop? How do the patterns of algal cell densities and O_2 concentration change with time?

■ Figure D4.2.6 An experimental mesocosm investigating the effect of nutrient enrichment

Inquiry 3: Concluding and evaluating

Concluding; evaluating

Carry out the investigation laid out in Figure D4.2.7, investigating the effect of nutrient enrichment:

- Is the mesocosm an appropriate practical investigation of eutrophication under controlled laboratory conditions? What changes might be made?
- Two dependent variables have been proposed in this experiment rather than just one. Why?
- If the additional phosphate ions added to mesocosm (A) resulted in an algal bloom, how would the control (B) be arranged to establish that phosphate ions influx caused it?
- How would you expect the oxygen concentrations in both mesocosms to change over an extended period?
- Compare the outcomes of the investigation to the accepted scientific context.
- Evaluate the implications of methodological weaknesses, limitations and assumptions on your conclusions.
- Explain what realistic and relevant improvements you could make to the investigation.

◆ **Keystone species**: a species which has a disproportionately large effect on community structure relative to its abundance.

Role of keystone species in the stability of ecosystems

In architecture, a keystone is a central stone at the apex (top) of an arch. The keystone supports the whole structure – without it, the arch would collapse (see Figure D4.2.8).

■ **Figure D4.2.7** An arch from the Temple of Hadrian, Ephesus, Turkey. The keystone (the central stone) supports the arch – without it the arch would collapse

■ **Figure D4.2.8** An agouti (*Dasyprocta* sp.) – a keystone species

The concept of a 'keystone' has been applied to ecology. In communities, certain species are vital for the continuing function of the ecosystem. Without them, a fundamental shift in the community takes place; it effectively 'collapses'. Such organisms are, therefore, 'keystones' for the ecosystem. Just as the keystone is not under the greatest pressure in the arch (this is focused on stones lower down), a **keystone species** may be only a relatively small part of the ecosystem in terms of biomass or productivity, but it still plays a fundamental role in the community. These species have a disproportionate impact on community structure and there is a risk of ecosystem collapse if they are removed.

An example of a keystone species is the agouti of tropical South and Central America (see Figure D4.2.9), which feeds on the nuts of the Brazil nut tree. The Brazil nut tree (*Bertholletia excelsa*) is a hardwood species that is found from eastern Peru and northern Bolivia across the Brazilian Amazon, and some specimens are among the oldest and tallest (they grow up to 50 m) trees in the Amazon. The agouti is a large forest rodent, and the only animal with teeth strong enough to open the Brazil nut tree's tough seed pods, to access the nuts inside.

The agouti buries many of the nuts around the forest floor for times when the Brazil nuts are less abundant. Inevitably, the agouti does not remember to dig up and eat all these buried seeds, and the remaining ones are able to germinate and grow into adult plants. Without the agouti, the Brazil nut tree would not be able to distribute its seeds and the species would eventually die out. Without the Brazil nut tree, other animals and plants that depend on it would be affected, such as harpy eagles that use them for nesting sites. Brazil nuts are one of the most valuable non-timber products found in the Amazon: they are used as a protein-rich food source, and their extracted oils are a popular ingredient in many cosmetic products. The collection and sale of Brazil nuts provides an important source of income for many local communities.

● Top tip!

Using the analogy of a keystone in architecture is a good way of remembering the importance of keystone species in ecosystems – if the stone/species is removed, the arch/ecosystem collapses.

1 **Outline** the role of a keystone species in an ecosystem.

ATL D4.2B

Find out about a keystone species in an ecosystem from your local area or region. What is the effect of the keystone species on the ecosystem? What would happen the species was removed?

Concept: Continuity

Keystone species are essential species within a community; they maintain the continuity and stability of ecosystems.

D4.2 Stability and change

◆ **Sustainability**: the responsible maintenance of ecological systems so that there is no reduction of conditions for future generations, ensuring the long-term viability of a system.

Assessing sustainability of resource harvesting from natural ecosystems

Renewable resources are ones that continue to exist despite being consumed or ones that can replenish themselves naturally over a period of time even as they are used. Humans use natural resources for food and materials. **Sustainability** depends on the rate of harvesting being lower than the rate of replacement. If used sustainably, renewable resources can be used indefinitely.

Terrestrial plant species

Trees are a renewable natural resource, providing they are replanted at the same rate as they are harvested. Trees are used to produce a large variety of different materials such as paper, cardboard and furniture. Tree chemicals are used to produce food, medicine and rubber.

Finland has promoted the multi-functionality of the forests as an important part of forest management and sustainability. Over 70% of the country is forested and the majority (61%) of Finland's forests are privately owned. Finland's sustainable forest management is based on balancing economic, social and environmental aspects of the forest. The Scots pine (*Pinus sylvestris*) is one of the four main logged trees in Finland (Figure D4.2.10). The sustainability of harvesting trees can be assessed by surveying soil disturbance and forest structure in logged and unlogged sites. To achieve sustainable harvesting the amount of replanting needs to exceed the number of trees cut down. The volume of timber extracted can be measured and new trees planted which, once fully grown, will replace the extracted timber. The long-term life of the forest is prioritized over short-term profits. In 2019, the President of the Republic of Finland, Sauli Niinistö, said that Finland is committed to achieving carbon neutrality by 2035 and become carbon negative soon after that. Forests provide an important storage of carbon (page 595) so, for forests to remain sustainable, the volume of timber removed needs to be replaced by new growth.

● **Top tip!**

Sustainability depends on the rate of harvesting being lower than the rate of replacement. If removal exceeds the rate of replenishment of the resources, for example forests harvested at a faster rate than they can regrow, then resource use is unsustainable.

■ **Figure D4.2.9** Scots pine (*Pinus sylvestris*) plantation and harvest

Marine fish

Fish are an important source of food for many human populations. Most fish populations that are exploited commercially as food sources are marine species, and it is commonly the case that the populations are shared among a few harvesting countries. Fishing vessels from many nations take their harvest from common waters and overfishing is a global issue. Chapter A4.2 discusses how overharvesting has impacted wild populations of animals (page 170). Overfishing of the Atlantic (or North Sea) cod (*Gadus morhua*) led to a collapse of fish stocks (see Figure A4.2.11, page 170).

If populations of fish are persistently overfished (as they regularly are), stocks will be rapidly depleted to the point where they collapse and can no longer support a commercial fishery. The extent to which species are fished should be based on population dynamics that consider their reproductive and growth rates. Population size and age structure can be used to inform practices associated with sustainable fishing.

The effect of population size, age and reproductive status on sustainable fishing practices

Sustainable fishing practices are those that use methods of catching fish that do not diminish the fish stock. For example, this can be done to establish the sustainability of Atlantic cod. The **maximum sustainable yield (MSY)** of a stock represents the maximum average catch that a stock can sustain over a long period of time. This catch corresponds to the optimum balance between the reproductive rate and growth rate of the stock and the deaths due to harvesting and natural mortality. Typically, harvesting at MSY requires much lower fishing rates than occur in many fisheries. If fish were harvested at the MSY then fishing would be able to continue in a sustainable way. A growing fish population is indicated by having a relatively larger number of younger fish, whereas a declining population would have a larger number of older fish. Such population age structures can, along with population size, indicate whether fishing methods are sustainable or not.

Calculation of MSY requires good knowledge of the relationship between the size of the stock and the number of juveniles produced each year. The natural variation in the number of juveniles produced annually makes establishing this relationship difficult. It usually requires a very long series of data – at least 20 years' worth. Nevertheless, there are reliable methods that provide an adequate approximation to MSY. This method relies on the cooperation of the fishing community – because estimates of fish stocks rely on accurate measurement of age structure. Any operators who work outside the regulations and discard restricted fish at sea before returning to port create skewed data that can lead to a miscalculation of stock size.

■ Figure D4.2.10 Atlantic cod (*Gadus morhua*)

Factors affecting the sustainability of agriculture

Humans started to grow food using agriculture around 10 000 years ago. Today, it provides the majority of food for human societies.

Various physical factors can affect agriculture, such as:

- precipitation: type, frequency, intensity, amount
- temperature: growing season (>6 °C), ground frozen (0 °C), range of temperatures
- soil: fertility, nutrient status, structure, texture, depth
- pests: vermin, locusts, disease
- location: slope gradient, relief, altitude, aspect (shady or sunny).

Environmental impacts, such as pollution, habitat loss, reduction in biodiversity and soil erosion, determine whether agriculture is sustainable or not.

When forests or other ecosystems are cleared for agriculture, soil is exposed and this can lead to **soil erosion**. Fertile soil can be considered a non-renewable resource because, once depleted, it can take significant time to restore its fertility. In some cases it may never recover.

◆ **Soil erosion**: process that occurs when the impact of water or wind detaches and removes soil particles, causing the soil to deteriorate.

◆ **Leaching**: the loss of water-soluble nutrients from the soil.

◆ **Agrochemical**: a chemical used in agriculture, such as a pesticide or a fertilizer.

◆ **Carbon footprint**: the amount of carbon dioxide released into the atmosphere because of the activities of a particular individual, organization or community.

Soil erosion and degradation reduce the quality of soil. This makes the amount of land available per person much lower now than in the past, and therefore has a potential impact on food production.

The exposure of soil can also lead to the **leaching** of nutrients from the soil. When it rains, minerals in the soil are washed away and so are not available for plant growth. For example, nitrates are essential for the synthesis of amino acids in plants, and thus loss of nitrates leads to a reduction in protein synthesis and growth in plants.

Fertilizers are added to soils to increase productivity. For example, nitrates and phosphates can be applied, which are used by plants to synthesize amino acids, DNA and other essential biological molecules. However, the over-application of fertilizers can lead to **eutrophication** (see page 808). Fertilizers are an example of an **agrochemical**. Other chemicals used to enhance crop yield are pesticides. These kill pests that either eat the crop (for example, insects) or compete with it (for example, weeds), reducing crop growth. Targeted pesticides that are biodegradable (i.e. target specific pests and can break down naturally) are used to minimize environmental damage. The use of non-biodegradable pesticides has a serious and detrimental effect on food chains, leading to **biomagnification** (see page 810), where increased levels of toxins accumulate in the tissues of consumers in higher trophic levels.

■ **Figure D4.2.11** Soil erosion on a farm after heavy rain

Another issue to consider is the distance that agricultural crops are transported. Transportation uses energy generated from fossil fuels, which adds to the amount of carbon dioxide in the atmosphere. This addition of carbon dioxide, a greenhouse gas, is leading to global warming (see page 824). Ideally, transport should be minimized so that the **carbon footprint** of the crop is reduced. The carbon footprint refers to the amount of carbon dioxide (and also methane, – CH_4) released into the atmosphere as a result of a particular activity. Growing crops locally to where they are consumed minimizes their carbon footprint. The demand for unseasonal foods (such as strawberries in Europe during the winter months) has increased the global transport of food, which has increased the carbon footprint of food. Consumption of seasonal crops, grown locally, will reduce the impact of agriculture on the environment.

2 **List** the factors that affect the sustainability of agriculture.

Eutrophication of aquatic and marine ecosystems due to leaching

Eutrophication is the natural or artificial enrichment of a body of water, particularly with respect to nitrates and phosphates. There are both natural and human causes of eutrophication. Natural effects include nutrients being added from decomposing biomass and run-off from areas surrounding the body of water. Human causes include run-off of fertilizers or manure from agricultural land, domestic waste water containing phosphates from detergents, and non-treated sewage. Leaching of mineral nutrients from agricultural land into rivers can also cause eutrophication.

The addition of extra nutrients causes algae living in the ion-enriched water to multiply rapidly, causing an algal bloom. Raised water temperatures in the summer months add to the algal growth rates. The process results in depletion of oxygen levels in the water. Run-off and leaching from agriculture can also end up in the sea, where it can cause eutrophication.

The chain of events leading to eutrophication can be summarized as follows:
- as more nutrients are added to the system, through leaching of nutrients from surrounding ground, the biomass of algae increases due to the availability of nutrients
- the growth of algae gives lower light penetration, causing underwater plants to die and create more nutrients as they decompose
- more nutrients leads to further growth of algae
- the increased death of algae and underwater plants leads to an increase in dead organic matter (DOM)
- the increase in DOM leads to an increase in bacteria that feed on the dead biomass, causing it to decompose
- bacterial respiration leads to increased BOD (see below), which causes a lowered oxygen content of water (hypoxia)
- oxygen-dependent organisms, such as fish, die due to a lack of oxygen. The fish community can become dominated by surface-dwelling coarse fish, such as roach and rudd.

■ **Figure D4.2.12** Eutrophication in the Ijssel River in the Netherlands

> ● **Common mistake**
>
> In explanations for anoxic water following eutrophication, the involvement of bacteria is often omitted. Bacteria feed on dead plants and remove oxygen from the water, causing low oxygen conditions. This then leads to the death of animals in the ecosystem affected.

Further effects of eutrophication include:
- net primary productivity is usually higher compared with unpolluted water, and may be indicated by extensive algal or bacterial blooms
- diversity of primary producers changes and finally decreases; the dominant species change
- the length of the food chain decreases as algae lock up the nutrients and block sunlight from reaching the riverbed
- with less sunlight penetrating the water, macrophytes (submerged aquatic plants) disappear because they are unable to photosynthesize
- as eutrophication proceeds, early algal blooms give way to cyanobacteria (blue-green algae) – these are toxic to wild animals and humans
- fish populations are adversely affected by reduced oxygen availability, and the fish community becomes dominated by surface-dwelling coarse fish, such as roach and rudd.

3 Define the term *eutrophication*.

Biochemical oxygen demand

◆ **Biochemical oxygen demand (BOD):** a measure of the amount of dissolved oxygen required to break down the organic material in a given volume of water through aerobic biological activity.

Biochemical oxygen demand (BOD) is a measure of the amount of dissolved oxygen required to break down the organic material in a given volume of water through aerobic biological activity. BOD increases in water that has organic pollution, such as sewage, or that has been nutrient-enriched by fertilizers (leading to eutrophication). Aerobic organisms use oxygen in respiration. When there are more organisms and faster respiration, more oxygen will be used. Thus, the BOD at any point is affected by:
- the number of aerobic organisms
- their rate of respiration.

BOD is an indirect method used to assess pollution levels in water. The presence of an organic pollutant causes an increase in the population of organisms that feed on and break down the pollutant, such as bacteria. Organic pollution causes a high BOD.

- **4 Explain** the relationship between BOD and eutrophication.
- **5 Explain** why eutrophication reduces the length of an aquatic food chain.

Both chemical testing and indicator species can be used to test the quality of the water. BOD can be measured directly:

1 Take a sample of water of measured volume.
2 Measure the oxygen level using an oxygen probe.
3 Place the sample in a dark place at 20 °C for five days. (Lack of light prevents photosynthesis, which would release oxygen and give an artificially low BOD.)
4 After five days, remeasure the oxygen level.
5 BOD is the difference between the two measurements.

> **Concept: Change**
>
> Pollution can change the abiotic factors within an ecosystem, which in turn impacts the biotic (living) components. These changes have a detrimental effect on the ecosystem.

Biomagnification of pollutants in natural ecosystems

The use of non-biodegradable toxins (such as DDT or mercury) can affect food chains, especially top predators. Increased concentrations occur because consumers feed on a number of organisms, so accumulate the toxins present in all of them. Organisms higher up the food chain live longer and so have more time to accumulate the toxin. Top carnivores are therefore at risk from poisoning from the toxin.

Wild plants and animals have evolved alongside their predators and parasites, and most live in balance with them. In contrast, cultivated crops and herds have been artificially selected to be especially productive and high-yielding, but typically have limited resistance to local parasites and predators. Crops grown as monocultures are especially prone to local predators. **Pesticides** are chemicals that are used to control harmful organisms which are a danger to crops or herds. Pesticides have improved productivity in agriculture enormously, but their use has generated problems in the environment.

The revolution in the chemical control of insect pests came when a particular substance, known as dichlorodiphenyltrichloroethane (DDT), was found to be a very effective insecticide (substance used to kill insects), causing rapid death even when applied in low concentrations.

DDT is fat soluble and is selectively retained in the fatty tissues of animals, rather than circulating in their blood to be excreted by the kidneys. **Bioaccumulation** refers to the build-up of non-biodegradable or slowly biodegradable chemicals in the body. DDT is stored in fat tissues because it is not recognized as a toxin and is not excreted.

As a result, at each stage of the food chain, DDT becomes concentrated. The process by which chemical substances become concentrated in the tissues of organisms at higher trophic levels is called **biomagnification** – the result can be that a top predator may have an accumulation that is several thousand times greater than that of a primary producer. In vertebrates, fish and birds, for example, DDT concentrations sometimes reached toxic levels (Figure D4.2.13). It began to be concentrated in top carnivores with devastating consequences.

♦ **Pesticide**: chemical that is used to control organisms that are a danger to crops or herds.

♦ **Bioaccumulation**: the build-up of non-biodegradable or slowly biodegradable chemicals in the body.

♦ **Biomagnification**: the process by which chemical substances become concentrated in the tissues of organisms at higher trophic levels.

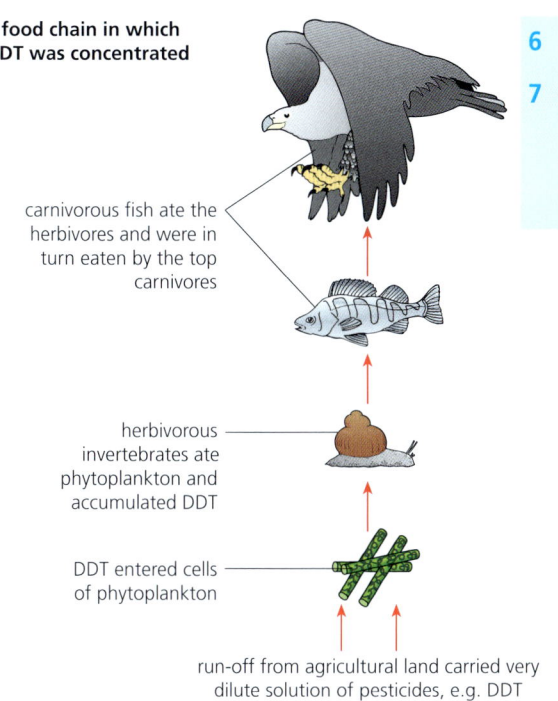

a food chain in which DDT was concentrated

carnivorous fish ate the herbivores and were in turn eaten by the top carnivores

herbivorous invertebrates ate phytoplankton and accumulated DDT

DDT entered cells of phytoplankton

run-off from agricultural land carried very dilute solution of pesticides, e.g. DDT

6 **Distinguish** between bioaccumulation and biomagnification.

7 **Analyse** the steps by which pesticides, applied at apparently safe concentrations, may cause the death of organisms seemingly unrelated to the crops or herds for which the pesticides are designed.

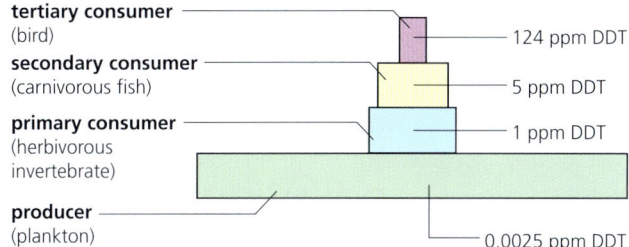

pyramid of biomass of the food chain, showing typical concentrations of DDT in ppm/organism

- tertiary consumer (bird) — 124 ppm DDT
- secondary consumer (carnivorous fish) — 5 ppm DDT
- primary consumer (herbivorous invertebrate) — 1 ppm DDT
- producer (plankton) — 0.0025 ppm DDT

■ Figure D4.2.13 Biomagnification of DDT

Common mistake

Do not confuse bioaccumulation with biomagnification. Bioaccumulation is the build-up of toxins in the biomass (usually lipids) of organisms. Biomagnification is the increasing concentration of the toxin per gram of biomass as the chemicals pass along the food chain.

The causes and consequences of biomagnification

Examine the following figure (Figure D4.2.15). What do the data show you about the causes and consequences of biomagnification?

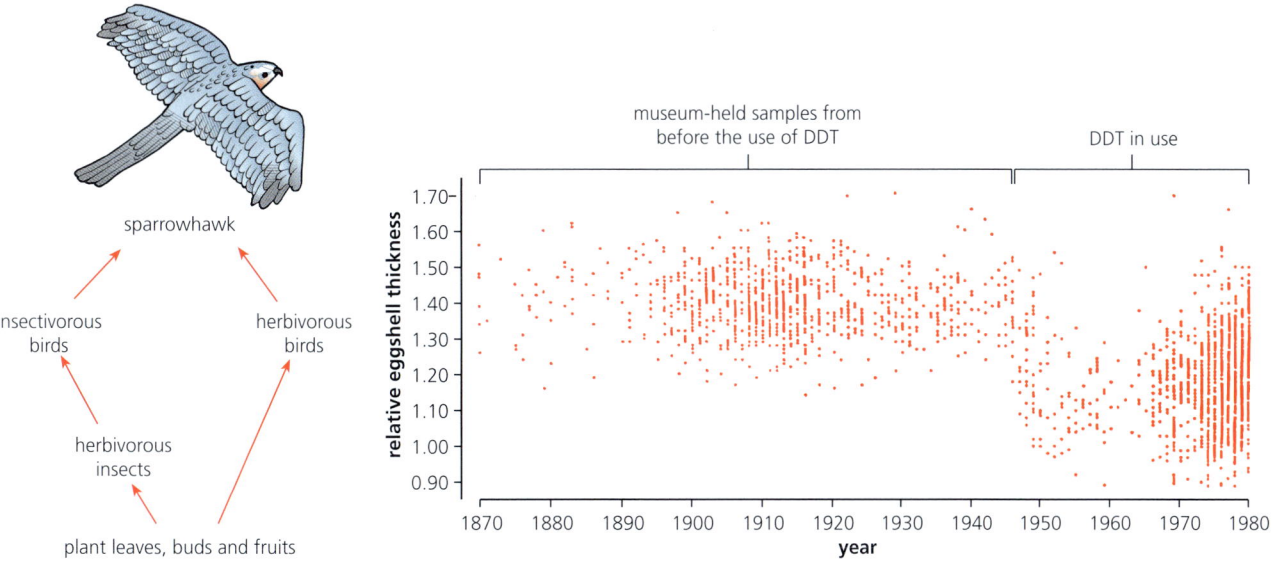

■ Figure D4.2.14 A food web of a sparrowhawk and a graph showing eggshell thickness studies in sparrowhawks

D4.2 Stability and change

◆ **Macroplastic**: relatively large, easily visible plastic debris found especially in the marine environment (typically more than 5 mm in size), such as bottles, plastic bags, rubbish and other materials that have not degraded.

◆ **Microplastic**: extremely small pieces of plastic debris, less than 5 mm in size, resulting from the breakdown of consumer products and industrial waste.

Figure D4.2.14 shows the food web for a sparrowhawk. When DDT was used as an insecticide, it was probable that this pesticide would bioaccumulate at each trophic level in the food chain where it was used and biomagnify in the top carnivore. In the eggshell thickness study of British sparrowhawks 1870–1980 (Figure D4.2.14) more than 2000 clutches of eggs were measured, each dot representing the mean shell thickness of a clutch (typically five eggs).

Although DDT is not a nerve poison in birds and mammals, in breeding birds it does inhibit the deposition of calcium in the eggshell. Affected birds lay thin-shelled eggs that crack easily. There was a rapid decline in numbers of birds of prey in areas where DDT had become widely used in agriculture. When this harmful effect of DDT was discovered, this quality of chemical 'stability' was renamed 'persistence'. Once the wider effects of organochlorine pesticides on wildlife were recognized, a ban was imposed in economically developed countries.

Effects of microplastic and macroplastic pollution of the oceans

■ **Figure D4.2.15** Plastic garbage contamination of the sea and beaches

Plastics are used in an increasing number of disposable consumer products. These plastics are persistent in the natural environment due to their non-biodegradability. **Macroplastics** are large debris that is easily visible, such as bottles, plastic bags, rubbish and other material that has not degraded (Figure D4.2.15). The physical and chemical breakdown of macroplastic into smaller, less-visible **microplastic** causes its own problems.

Figure D4.2.16 shows how macroplastics enter the marine environment, and what happens to them subsequently.

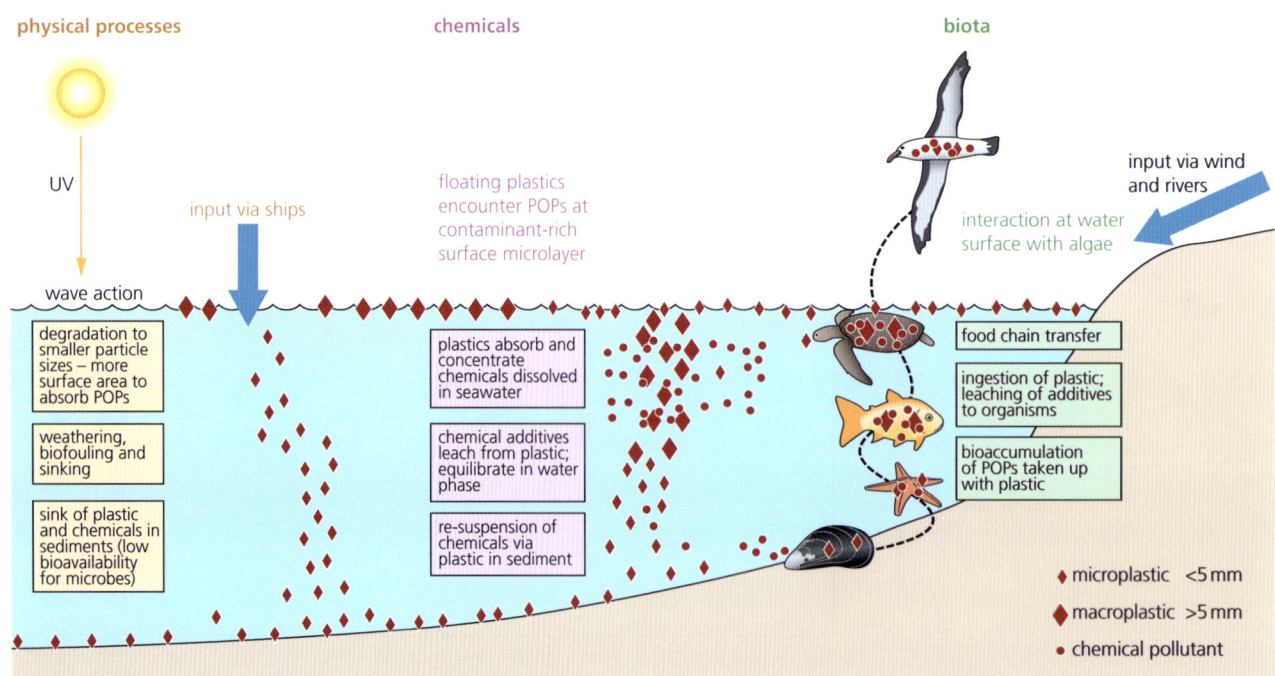

■ **Figure D4.2.16** Sources of marine macroplastics and microplastics, and the various abiotic (physical and chemical) and biotic processes affecting them in the marine environment. POPs are persistent organic pollutants that are resistant to environmental degradation: they can therefore accumulate through food chains

Macroplastic and microplastic debris has accumulated in marine environments. Macroplastics can be confused with food by marine predators. There is evidence that the degradation of marine plastics releases non-biodegradable contaminants that can **bioaccumulate** in the food chain. Once chemicals enter food chains, the top predators are often at extra risk because of the **biomagnification** effects of some chemicals. Studies of Southern fur seals have shown that seal scats (faeces) can contain plastic particles from the night-feeding myctophids (lanternfish), which are active near the sea surface and which are consumed by the seals.

The impact of marine plastic debris on the Laysan albatross

The Laysan albatross (*Phoebastria immutabilis*) ranges across the North Pacific, with 99.7% of the population breeding on the north-western Hawaiian Islands. They are large birds, with a wingspan of about 2 m, and require great quantities of fish to survive. One island on which they nest, Midway Atoll, is in the North Pacific gyre, and large volumes of plastic are washed on to its beaches. Many birds accidentally eat plastic and other marine debris floating in the ocean, mistaking it for food. The problem is made worse by the way in which the Laysan albatrosses catch fish: they catch squid and other seafood by skimming the surface of the water with their beak. By catching food in this way they also accidentally pick up a lot of floating plastic, which they then feed to their chicks. Chicks, unlike adults, cannot regurgitate plastic that they have swallowed, and so the plastic can fill their stomachs, potentially leading to starvation.

The impact of marine plastics on sea turtles

Plastic garbage harms and kills approximately 100 000 sea turtles and other marine animals each year. Sea turtles mistake plastic bags for jelly fish (one of their favourite foods) and so eat them – the consequences of ingesting plastic can be fatal; if the plastic debris gets lodged in the mouth, the turtle can have problems feeding, eventually leading to starvation.

> ### Nature of science: Global impact of science
>
> Scientists can influence the actions of citizens if they provide clear information about their research findings. Popular media coverage of the effects of plastic pollution on marine life has changed public perception globally, which has driven measures to address this problem. Groups such as The Ocean Cleanup (https://theoceancleanup.com) have information about the extent of plastic pollution and strategies for reducing it. The following website lists many groups active in reducing plastic pollution and raising public awareness of the issues:
> https://innovations-oceans-sans-plastique.com/en/ngo-from-a-z

8 **Outline** how micro- and macroplastics can affect food chains.

Restoration of natural processes in ecosystems by rewilding

Chapter A4.2 discusses how different methods could be used to conserve species, habitats and ecosystems. One technique is rewilding (see page 176). Rewilding aims to restore ecosystems and reverse declines in biodiversity by allowing wildlife and natural processes to reclaim areas no longer under human management. Rewilding reintroduces lost animal species to natural environments, such as top (apex) predators that have a significant effect on the food web and trophic levels in an ecosystem. Keystone species are also reintroduced, which are essential for the functioning of the ecosystem (see page 805). The establishment of connectivity of habitats over large areas, using corridors of habitat that connect different areas, allows for movement of animals between fragments and gives larger animals the area they need to feed and breed. Agriculture and other resource harvesting are no longer allowed and natural ecosystems are allowed to recover through ecological management.

The Hinewai Reserve is an ecological restoration project on Banks Peninsula, New Zealand (Figure D4.2.17).

■ **Figure D4.2.17** Otanerito Bay from Hinewai Reserve, Banks Peninsula, New Zealand

Parts of the reserve were cleared by human settlers, initially by Polynesian settlers (from about 700 years ago) and then by European settlers (from around 1850 onwards) who completed the clearance. The area covers 1250 hectares and is managed to ensure the natural regeneration of native vegetation and wildlife. The ecological management involves minimal intervention, allowing succession to occur resulting in ecosystems (nearly all forest) similar to those that existed before forest clearance. Alien species are removed, such as exotic trees and vines, allowing the endemic flora and fauna to be re-established.

ATL D4.2C

Read more about the Hinewai Reserve in New Zealand here: www.hinewai.org.nz

Biologist Dr Hugh Wilson has been reforesting the Hinewai Reserve since 1987. Read more about him and his work here: https://brightvibes.com/2048/en/meet-the-kiwi-whos-spent-over-30-years-reforesting-a-corner-of-new-zealand

Produce a poster summarizing how communities and habitats in the reserve are being restored. Use your knowledge of stability of ecosystems to include detail of how the continuity of the ecosystem can be assured.

Link
The need for several approaches to conservation of biodiversity is covered in Chapter A4.2, page 174.

◆ **Succession**: the orderly process of change over time in a community.
◆ **Pioneer community**: the first stage of an ecological succession that contains hardy species able to withstand difficult conditions.
◆ **Climax community**: a community of organisms that is more or less stable and in equilibrium with natural environmental conditions such as climate. It is the end point of ecological succession.

Ecological succession and its causes

All ecosystems develop over time from the point of colonization of a pristine area to the maturation of a stable final community. **Succession** is the term for the orderly process of change over time in a community. Succession can be triggered by changes in both an abiotic environment and in biotic factors. Changes in the community of organisms frequently cause changes in the physical environment, allowing another community to become established and eventually replace the former through competition. Often, but not inevitably, the later communities in such a sequence (or sere) are more complex than those that appear earlier. The first communities are called **pioneer communities**, while the final stage of the succession is the **climax community**. Two examples of succession, from pioneer community to climax community, could be:

- Succession in freshwater:
  aquatic plants (e.g. water lilies) → reeds → low woodland species (e.g. willow)
- Succession in an abandoned quarry:
  mosses and lichens → grasses and herbs → shrubs (e.g. birch) → woodland

Changes occurring during primary succession

Primary succession

> **Concept: Change**
>
> Succession is the orderly process of change over time in a community. Succession can be triggered by changes in both an abiotic environment and in biotic factors.

The formation of an ecosystem in an environment devoid of vegetation and lacking soil, for example bare rock, is called a **xerosere** (Figure D4.2.18):

- pioneer species arrive (for example, lichens, mosses and bacteria)
- as pioneers die, soil is created
- new species of plants arrive that need soil to survive – these displace pioneer species
- growth of plants causes changes in the environment (factors such as light, wind and moisture)
- growth of roots enables soil to be retained; nutrients and water in the soil increase
- nitrogen-fixing plants arrive, adding nitrates to the soil
- soil depth increases further, allowing shrubs and other taller plants to arrive
- animal species arrive as species of plant they rely on become established
- a climax community is established.

A **xerosere** = succession under dry, exposed conditions where water supply is an abiotic factor limiting growth of plants, at least initially.

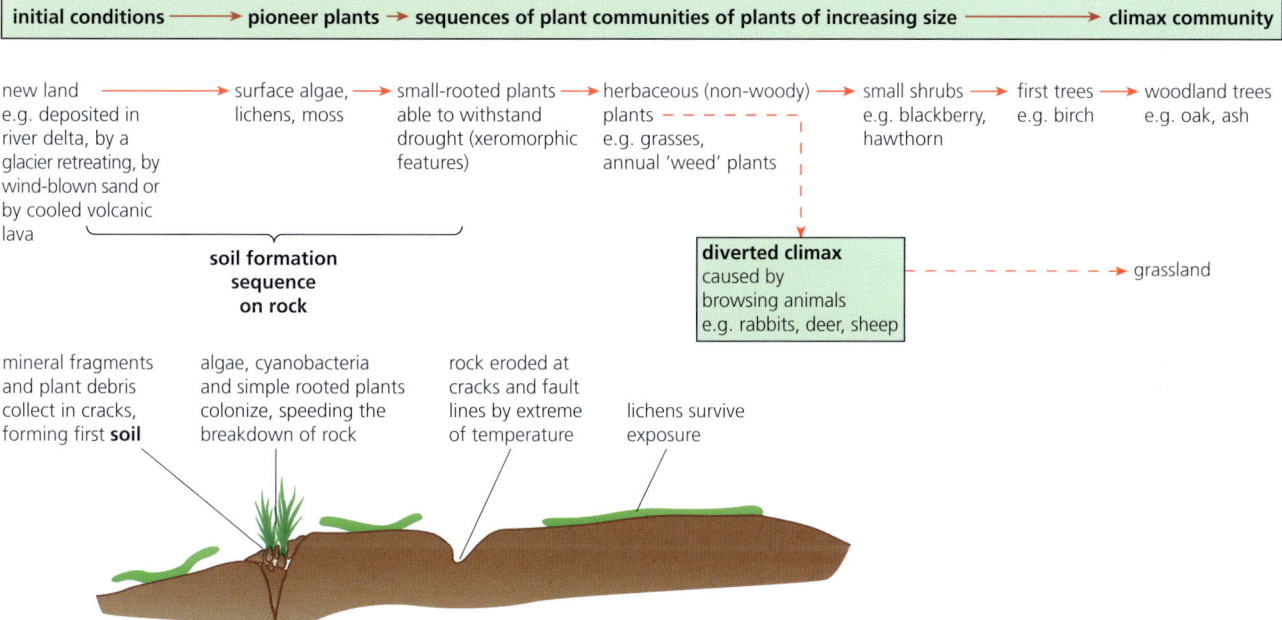

This succession sequence is not a rigid ecological process, but an example of what may happen. It is influenced by factors such as:
a how quickly humus builds up and soil forms
b rainfall or drought, and the natural drainage that occurs
c invasions of the habitat by animals and seeds of plants.

■ Figure D4.2.18 A primary succession on dry land – a xerosere

D4.2 Stability and change

HL ONLY

Inquiry 2: Collecting and processing data

Processing data; interpreting results

The table below shows a study of the vegetation of a system of developing sand dunes. Note how both the species diversity and physical parameters change as the dunes age. Analyse the data, then process the data through the following questions.

Calculate the Simpson's reciprocal diversity indices (see Chapter A4.2, page 157) of the fore dune and semi-fixed dune (using a spreadsheet). **Comment** on the results.

State which species are pioneers and which form the climax community.

■ Figure D4.2.19 Sand dune community development – a student field study exercise

Study of a sand dune succession from embryo state to fixed dune				
	Site 1	Site 2	Site 3	Site 4
Stage in succession	Embryo dune	Fore dune	Semi-fixed dune	Fixed dune
Number of pins dropped	220	220	220	220
sea couch grass	169	9	0	0
marram grass	1	123	19	0
fescue grass	0	0	126	182
spear-leaved orache	4	0	0	0
prickly saltwort	2	0	0	0
bindweed	1	44	0	0
bird's foot trefoil	0		0	6
biting stonecrop	0	0	0	1
buttercup	0	0	0	1
cat's ear	0	0	5	2
clover	0	0	0	68
common stork's bill	0	0	0	11
daisy	0	0	2	8
dandelion (common)	0	0	5	1
eyebright	0	0	0	4
hawkbit	0	0	20	1
ladies bedstraw	0	0	86	35
medick	0	0	0	62
mouse-ear chickweed	0	0	0	2
ragwort	0	0	8	1
restharrow	0	0	25	15
ribwort plantain	0	0	0	37
sand sedge	0	0	32	17
stagshorn plantain	0	0	0	36
tufted moss	0	0	0	119
yellow clover	0	0	0	5
Yorkshire fog	0	0	2	0
wild thyme	0	0	0	82
Total live hits	177	176	330	706
Bare ground hits	48	68	6	0
% bare ground	27	38	3	0
soil moisture (%)	5.0	8.5	16.0	14.3
soil density (g cm^{-3})	1.6	0.9	0.6	0.5
soil pH	8.0	7.5	7.0	7.0
wind speed (m s^{-1})	10.0	9.3	2.4	1.3

■ Changes in energy, matter and complexity

Both energy and matter change through a succession. Productivity changes through a succession in the following way:

- In pioneer communities, gross productivity is low due to the initial conditions and the low density of producers. The proportion of energy lost through community respiration is also relatively low. **Net primary productivity (NPP)** is therefore high: the system grows and biomass accumulates.
- **Gross primary productivity** increases due to the increase in biomass of the community. Sunlight is available for all plants as there is little shading.
- In later stages there is an increased consumer community, with increased levels of respiration, leading to a reduction in NPP. The rate of accumulation of biomass begins to slow. In climax communities gross productivity may be high, but this is balanced by community respiration and so overall NPP may be lower than earlier in the succession.

The size of plants increases through a succession, and the complexity of food webs also increases. Nutrients initially come from outside the succession in the pioneer stages, but through the succession, with increases in biomass, soil depth and complexity, nutrient cycling within the ecosystem becomes increasingly important. With increased complexity of the ecosystem during the succession, there are increased opportunities for a variety of different niches and so species diversity increases.

Nutrient availability changes from poor initial stores through to gradual accumulation of nutrients as biomass develops and soil fertility increases.

> ### Going further
>
> A secondary succession occurs in an environment after an ecosystem has been disrupted or destroyed due to a disturbance that reduces the population of the original inhabitants. Soil already exists in this type of succession. Recolonization of woodland after fire is an example of secondary succession.

Link
Net primary productivity and gross primary productivity are defined and discussed in Chapter C4.2, page 592.

9 Distinguish between pioneer and climax communities in a succession.

● Common mistake
It is not enough to say that 'productivity increases through a succession'. GPP increases but NPP decreases as community respiration increases.

10 Explain the interrelationship between abiotic and biotic factors during a succession.

Cyclical succession in ecosystems

In some ecosystems there is a cycle of communities rather than a single, unchanging climax community. For example, wood pasture cycles from pasture to scrub to woodland and back to pasture.

Wood pasture may be defined as tree-land on which farm animals or deer are systematically grazed. Wood pastures were formed and maintained initially by large wild herbivores, although following the extinction of many large mammalian species, livestock and deer are the grazers that maintain this ecosystem. The grazing regimes of wood pasture systems prevent the mass regeneration of trees, and the reduced competition for light allows sufficient tree regeneration for replacement of the tree population over time. Grazing prevents mass regeneration that would develop into high forest.

Wood pastures are mosaic habitats valued for their trees, especially veteran and ancient trees, and the plants and animals that they support. Grazing animals (either domestic livestock or deer) are fundamental to the existence of this ecosystem along with flowers providing nectar sources and open grassland or heathland ground vegetation.

Nature of science: Science as a shared endeavour

■ Figure D4.2.20 Old oak trees on a wood pasture in southern Transylvania, Romania

Wood pastures are being studied in the Transylvania region of Romania, as well as in adjacent areas in other parts of Europe, such as Germany. This region contains the highest number of wood pastures from central and eastern Europe as well as from Romania. This example of international cooperation is giving scientists a better understanding of wood pastures, and how they can best be managed and conserved.

Read more about this project here:
https://transylvanian-wood-pastures.eu/en

Climax communities and arrested succession

Given any specific environmental conditions, ecological succession tends to lead to a particular type of climax community. However, human influences can prevent this from developing.

Climatic and edaphic (soil) factors determine the nature of a climax community. Climax communities are more stable than earlier seral stages because:
- they contain more complex food webs – this provides more stability because, if one organism becomes extinct, it can be replaced by another
- **negative feedback mechanisms** lead to steady-state equilibrium
- each stage in the succession helps to create a deeper and more nutrient-rich soil, so allowing larger plants to grow
- a climax community is more productive, providing more energy to support consumers and decomposers
- increased biomass leads to an increased number of niches, which increases species and genetic diversity, resulting in greater stability.

Disturbance can stop the process of succession so that the climax community is not reached. Interrupted succession is known as **arrested succession** or **plagioclimax**. Human activity can have various effects on climax communities:
- decrease in productivity through the removal of primary producers
- reduction in producers, leading to reduced habitat diversity and fewer niches, which threatens more specialized species
- deterioration in abiotic factors, leading to harsher conditions that fewer species can adapt to
- species extinction, leading to shorter food webs
- a less complex community, leading to decreased stability.

These activities divert the progression of succession to an alternative stable state so that the original climax community is not reached.

◆ **Arrested succession (plagioclimax)**: an area where human activity has prevented the ecosystem developing a climax community.

■ **Figure D4.2.21** Interrupting a succession: grazing prevents a climax community from forming and maintains heather moorland

■ **Figure D4.2.22** Drainage network establishment for peat extraction field, next to the natural peat bog wetland

Arrested succession: grazing by farm livestock

Large parts of the UK were once covered by deciduous woodland. Some heather would have been present in the north, but relatively little. From 1300–1500 (the Middle Ages) and onwards, forests were cleared to supply timber for fuel, housing, construction of ships (especially oak) and to clear land for agriculture. As a result, soil deteriorated and heather came to dominate the plant community. Sheep grazing and associated burning has prevented the regrowth of woodland by destroying young saplings (Figure D4.2.21).

Controlled burning of heather also prevents the re-establishment of deciduous woodland. The heather is burned after 15 years, before it becomes mature. If the heather matured, it would allow colonization of the area by other plants. The ash adds to the soil fertility and the new heather growth that results increases the productivity of the ecosystem.

Arrested succession: drainage of wetlands

Drainage of wetlands is another example of arrested succession. Wetlands include a variety of different ecosystems and are found across the world, including deltas and estuaries, mudflats, floodplains and peat bogs. Humans and wildlife have relied on wetlands for thousands of years. According to the Wildfowl and Wetlands Trust (WWT), approximately 87% of the world's wetlands were lost in the last 300 years, as hundreds of thousands of hectares of wetlands were drained to provide land for housing and industry.

In many areas, soils that formed as wetlands have been drained for agricultural use, such as the drainage of more than 1 million acres in South Florida for agriculture (sugarcane, vegetables and other crops). In other areas, wetlands are drained so that peat can be extracted (Figure D4.2.22).

> ### Top tip!
> Various factors can divert the progression of succession to an alternative stable state by modifying the ecosystem, such as the use of fire in an ecosystem, agriculture, grazing pressure or resource use (such as deforestation). This diversion may be more or less permanent depending upon the resilience of the ecosystem.

LINKING QUESTIONS

1. What is the distinction between artificial and natural processes?
2. Over what timescales do things change in different biological systems?

D4.3 Climate change

Guiding questions

- What are the drivers of climate change?
- What are the impacts of climate change on ecosystems?

SYLLABUS CONTENT

This chapter covers the following syllabus content:
- D4.3.1 Anthropogenic causes of climate change
- D4.3.2 Positive feedback cycles in global warming
- D4.3.3 Change from net carbon accumulation to net loss in boreal forests as an example of a tipping point
- D4.3.4 Melting of landfast ice and sea ice as examples of polar habitat change
- D4.3.5 Changes in ocean currents altering the timing and extent of nutrient upwelling
- D4.3.6 Poleward and upslope range shifts of temperate species
- D4.3.7 Threats to coral reefs as an example of potential ecosystem collapse
- D4.3.8 Afforestation, forest regeneration and restoration of peat-forming wetlands as approaches to carbon sequestration
- D4.3.9 Phenology as research into the timing of biological events (HL only)
- D4.3.10 Disruption to the synchrony of phenological events by climate change (HL only)
- D4.3.11 Increases to the number of insect life cycles within a year due to climate change (HL only)
- D4.3.12 Evolution as a consequence of climate change (HL only)

Anthropogenic causes of climate change

◆ **Climate**: a long-term average of the weather, over 20–30 years.

◆ **Climate change**: a long-term change in global or regional climate patterns, caused by natural or human factors, such as increased levels of atmospheric carbon dioxide produced by the use of fossil fuels.

◆ **Greenhouse effect**: process in which greenhouse gases trap outgoing long-wave radiation from the Earth, causing the planet to be warmer than it would otherwise be.

Climate is the long-term average of the weather, over 20–30 years, not year-to-year variations. Climate change is a long-term change in global or regional climate patterns, caused by natural or human factors. **Climate change** since the mid- to late-twentieth century onwards is attributed primarily to the increased levels of atmospheric greenhouse gases produced by the use of fossil fuels and land-use change (the latter, i.e. deforestation and cattle, account for almost 40% of historic change). To understand the causes of climate change, we first need to understand the greenhouse effect.

■ The greenhouse effect

The radiant energy reaching the Earth from the Sun is mainly visible light with some ultraviolet and shorter-wavelength infrared radiation (together termed shortwave radiation). It is this shortwave radiation that is largely transmitted through the atmosphere to directly warm up the sea and the land, although some is reflected by clouds before it reaches the surface. As the surface warms, the Earth itself radiates more infrared radiation at longer wavelengths back towards space. However, much of this heat does not escape from our atmosphere. This radiation is absorbed by clouds and by gases in the atmosphere, which are warmed. In this respect, the atmosphere traps heat somewhat like a greenhouse, which is why this phenomenon is called the **greenhouse effect** (Figure D4.3.1).

The atmosphere lets through much of the sunlight but absorbs much of the outgoing infrared radiation. We must recognize that the greenhouse effect is very important to life on the Earth – without it, surface temperatures would be altogether too cold for life (see Chapter A2.1 (HL only), page 34).

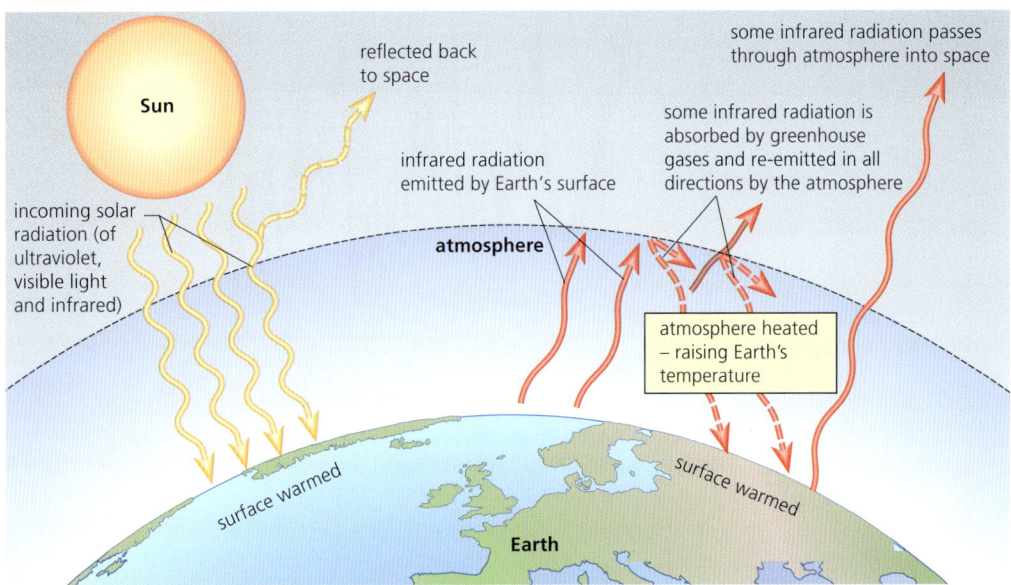

■ Figure D4.3.1 The greenhouse effect

◆ **Greenhouse gas**: a gas that contributes to the greenhouse effect by absorbing infrared radiation.

The gases in the atmosphere that absorb infrared radiation are referred to as **greenhouse gases**. **Carbon dioxide** and **water vapour** are the most significant greenhouse gases. Other gases, including **methane** and **nitrogen oxides**, have less impact.

The contribution of each gas to the greenhouse effect is largely a product of the properties of the gas and of how abundant it is at any time. For example, methane is considerably more powerful as a greenhouse gas than the same mass of carbon dioxide. However, methane is present in lower concentrations than carbon dioxide. Also, it is a relatively short-lived component of the atmosphere – when exposed to (sun)light, methane molecules are steadily oxidized to carbon dioxide and water. Table D4.3.1 lists figures for the typical contributions of the chief greenhouse gases.

■ Table D4.3.1 The direct consequence of greenhouse gases to the greenhouse effect

Compound	Approximate contribution to greenhouse effect / %
water vapour (+ clouds)	36–72
carbon dioxide	9–26
methane + nitrous oxide	4–9

The **long-term records** of changing levels of greenhouse gases (and associated climate change) are based on evidence obtained from ice cores drilled in the Antarctic and Greenland ice sheets. The ice there has formed from accumulation of layer upon layer of frozen snow, deposited and compacted over thousands of years. Gases from the surrounding atmosphere were trapped as the layers built up. Data on the composition of the bubbles of gas obtained from different layers of these cores from the Vostok (East Antarctic) Ice Sheet provide a record of how the carbon dioxide and methane concentrations have varied over a period of 400 000 years of Earth's history. Similarly, variations in the concentration of oxygen isotopes from the same source indicate how temperature has changed during the same period (Figure D4.3.2).

D4.3 Climate change

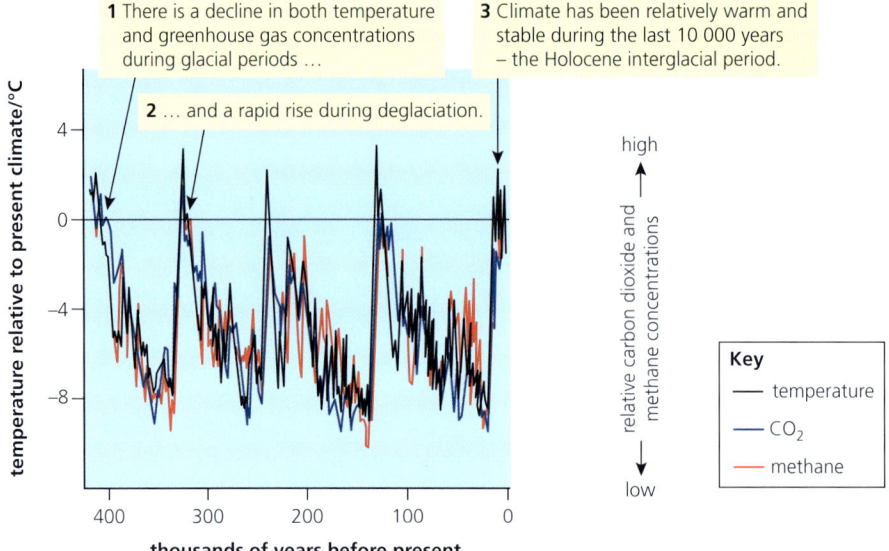

■ **Figure D4.3.2** Three types of data recovered from the Vostok ice cores over 400 000 years of Earth history

1 Using data from Figure D4.3.3, **estimate** the length of interglacials (warm periods) and their frequency.

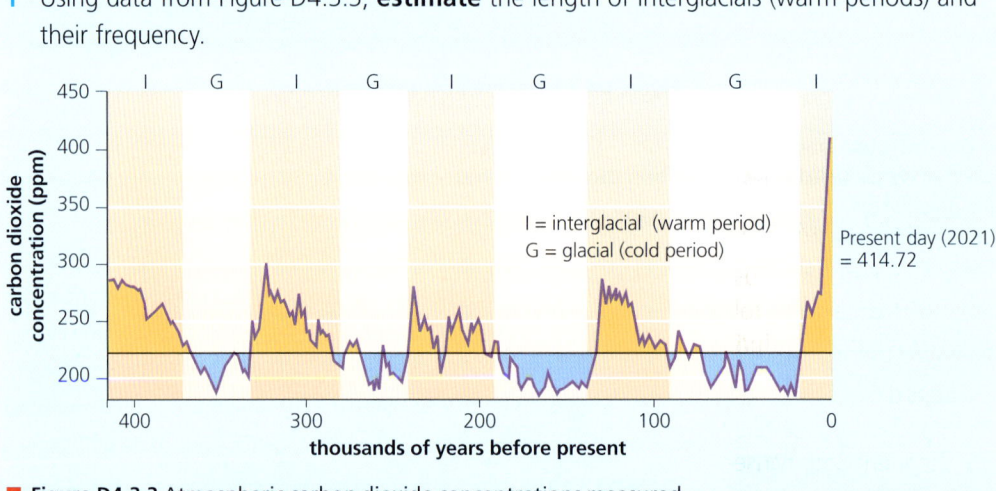

■ **Figure D4.3.3** Atmospheric carbon dioxide concentrations measured from Vostok ice cores and recent Mauna Loa data

2 **Explain** why the current concentration of carbon dioxide is different from those of previous warm periods.

Nature of science: Patterns and trends

From the graphs in Figures D4.3.2 and D4.3.3, we can see that the levels of atmospheric carbon dioxide have varied quite markedly. Note that many millions of years ago, there were periods in Earth's history when it was especially raised. Earth was a very different place then, with continental drift, volcanic emissions and weathering of chalk and limestone playing a role over very long timescales. Despite variation, data from Antarctic ice cores show a positive correlation between global temperatures and atmospheric carbon dioxide concentrations over hundreds of thousands of years (Figure D4.3.2). Current carbon dioxide concentrations are well above the natural variability in the last 800 000 years, and other evidence shows that these changes are caused by human activities that are driving current climate change.

> **Top tip!**
>
> The greenhouse effect is a natural phenomenon but there has been an anthropogenic increase in carbon dioxide concentrations that is positively correlated with global warming. Most climate scientists accept that there is a causal link between increased greenhouse gas emissions due to human activities and climate change.

Link

Analysis of the Keeling curve is covered in Chapter C4.2, page 599.

In June 2022, carbon dioxide was present in the atmosphere at a concentration varying between 413 and 422 parts per million (ppm) (see Figure C4.2.19, page 600). We might expect the amount of atmospheric carbon dioxide to be maintained by a balance between the fixation of this gas during photosynthesis and release of carbon dioxide into the atmosphere by respiration, combustion and decay by micro-organisms – an interrelationship illustrated in the carbon cycle (Figure C4.2.16, page 595). In fact, photosynthesis does withdraw almost as much carbon dioxide during daylight hours as is released into the air by all the other processes (including human-caused factors), day and night – but not quite as much. As a result, the level of atmospheric carbon dioxide is now rising (Figure D4.3.3) as shown by the Keeling curve (a daily measure of global atmospheric carbon dioxide concentration) in Figure C4.2.18, page 599.

> **TOK**
>
> The discussions around climate change have become quite political and might, for some, seem an odd inclusion in a biology textbook. However, just because a topic is political does not mean that it is beyond the scope of science. Political claims about the world must be consistent with how the world actually is, which is best determined through the methods developed by the scientific community over hundreds of years. Scientific findings that are politically inconvenient do not necessary mean they are untrue. Furthermore, just because there might be disagreement at the political level over certain claims about climate change does not mean that there is genuine disagreement within the scientific community about those claims.

Tool 2: Technology

Generating data from models and simulations

How can models be used to predict future climate change? Access the following site:
https://scied.ucar.edu/interactive/simple-climate-model

The model explores how the rate of carbon dioxide emissions affects the amount of carbon dioxide in the atmosphere and, consequently, the Earth's temperature.

Select an emissions rate for carbon dioxide using the sliding scale. Press 'play' – the simulation will show increases in carbon dioxide and temperature linked to the emission rate you have selected. The graph also shows the carbon emissions – these will level off at the rate you have set.

Repeat the exercise at a range of different emissions rates. Plot a graph of emissions rate against temperature increase between the current date and 2100. In 2021, global energy-related carbon dioxide emissions were 31.5 Gt according to the International Energy Agency's 2021 Global Energy Review. Extrapolate, using your graph, to estimate the increase in temperature if this rate of emissions is maintained.

ATL D4.3A

A novel way of showing increasing global temperatures was developed by Professor Ed Hawkins at the University of Reading in 2018. Read about his initiative here:
www.reading.ac.uk/planet/climate-resources/climate-stripes

The graphic does not use words, graphs or numbers, just a series of vertical, coloured bars. What is the advantage of this approach compared to more traditional methods of showing data?

Create a presentation or display for your school or college using the climate stripes. Perhaps you could create a mural using the climate stripes as a theme, using recycled waste material? Put your display in a prominent area that is commonly used in the school, to raise awareness about the issue of climate change.

◆ **Global warming**: an increase in the global average temperature of the Earth's surface and atmosphere.

■ An enhanced greenhouse effect resulting in global warming

The increase in greenhouse gases such as carbon dioxide and methane have anthropogenic causes (i.e. are caused by humans) and has resulted in an enhanced greenhouse effect. This has resulted in the global temperature increasing – a process known as **global warming**.

Carbon dioxide

The enhanced greenhouse effect is due to the build-up of certain greenhouse gases because of human activity. Carbon dioxide levels have risen from about 315 ppm in 1950 to approximately 418 ppm in 2022, and could reach 600 ppm by 2050 depending on policy decisions by governments. The increase is due to human activities, particularly the burning of fossil fuels (coal, oil and natural gas), deforestation and agricultural emissions.

Since the advent of industrial revolutions (the past 200 years or so), there has been a sharp and accelerating rise in the level of this greenhouse gas. This is mainly attributed to the burning of coal and oil; these 'fossil fuels' were mostly laid down in the Carboniferous Period. As a result, we are now adding to our atmosphere carbon that had been locked away for about 350 million years. This is an entirely new development in geological history.

Deforestation of the tropical rainforests has a double impact – not only does it increase atmospheric carbon dioxide levels, it also removes the trees that convert carbon dioxide into glucose, producing oxygen as a by-product. Deforestation accounts for about 12–15% of global carbon dioxide emissions, with emissions coming from the burning of trees. The resulting graph of increasing carbon dioxide concentrations (see Figure D4.3.3 and Figure C4.2.18, page 599) is a key piece of evidence for anthropogenic changes to the global carbon budget.

■ **Table D4.3.2** Changing levels of atmospheric carbon dioxide – recorded and predicted

Time period	CO_2 levels/ppm
pre-Industrial Revolution level (before 1850)	280 (± 10)
by mid 1970s	330
by 1990	360
by 2007	380
by 2013	400
by 2050 (**if current rate of anthropogenic emission is maintained**)	600

ATL D4.3B

Will higher carbon dioxide levels boost plant productivity? If so, what impact will this have on ecosystems and the atmosphere?

https://allianceforscience.cornell.edu/blog/2018/04/will-rising-carbon-dioxide-levels-really-boost-plant-growth

Discuss the issues with members of your class. The outcomes may be more complex than first appear and depend on the type of photosynthesis carried out by plants (see Going further).

Concept: Change

Human activities are increasing the greenhouse gases in the atmosphere, such as carbon dioxide and methane. This has led to the enhanced greenhouse effect and climate change.

Going further

Plants can be categorized based on the way they photosynthesize. Most plants are C3 plants because their first photosynthetic product is a three-carbon compound. Examples of C3 plants include barley, oats, potato, rice and wheat, which are commonly grown in temperate regions. On the other hand, C4 plants produce a four-carbon compound as their first photosynthetic product. Examples of C4 plants are common grass crops of tropical regions, such as maize, millet, sorghum and sugarcane.

Methane

Methane (CH_4) is the second-largest contributor to global warming and is increasing at a rate of between 0.5 and 2% per annum. Cattle alone release between 65 and 85 million tonnes of methane per year. Natural wetlands and paddy fields are another important source – paddy fields emit up to 150 million tonnes of methane annually. As global warming increases, bogs trapped in permafrost will thaw and could release vast quantities of methane (see pages 596 and 828).

> ● **Common mistake**
>
> The greenhouse effect and global warming are often considered to be the same. They are not, but they are related. The greenhouse effect is a natural process that traps some outgoing longwave (infrared) radiation and enables life on Earth. The enhanced greenhouse effect is the process in which human activities have led to an increase in the amount of greenhouse gases in the atmosphere, and an increased trapping of infrared radiation by these greenhouse gases, leading to an increase in global surface temperatures, i.e. global warming.

3 Distinguish between the greenhouse effect and the enhanced greenhouse effect.

Positive feedback cycles in global warming

There are many examples of **feedback** in the process of global warming. **Positive feedback** mechanisms involve, for example, increasing temperatures, thawing permafrost and the release of methane. As methane is a greenhouse gas, it has the potential to increase temperatures, thereby reinforcing the rise in temperature.

■ Methane feedback

Around one-quarter of the Earth's surface is affected by continuous or sporadic permafrost, including tundra, polar and mountain regions. Globally, permafrost covers 23 million square kilometres (mostly in the Earth's northern hemisphere). It formed during past cold glacial periods and has persisted through warmer interglacial periods, including the Holocene (the last 10 000 years).

Reports indicate that methane-storing permafrost is now shrinking at an alarming rate. According to scientists at NASA, temperatures in Newtok, Alaska, have risen by 4 °C since the 1960s, and by as much as 10 °C in winter months. The effects of permafrost thawing are potentially magnified over time by positive feedback loops.

1 As the atmosphere warms, more permafrost is expected to thaw.
2 This will release large amounts of methane (some researchers estimate that the volume of this gas stored in permafrost equates to more than double the amount of carbon currently in the atmosphere).
3 The atmosphere will warm up even more quickly.
4 This will cause (i.e. feedback) even more methane to be released by more melting permafrost.

Climate-related drying of wetlands would increase emissions of methane and carbon dioxide in peatlands that have been deposited over millions of years and would increase the potential for peat fires, which are difficult to extinguish.

Research has shown that increased mean global temperatures increases the rate of decomposition. If humidity remains stable, an increase of the atmospheric temperature of 1 °C leads to an increase in the rate of decomposition of up to 10%, according to a 2019 paper on the effects of climate change on decomposition process in the Andean Paramo ecosystem. In terms of peat, decomposition rates increase, which leads to increased production of methane and carbon dioxide, which in turn increases the global temperature – another example of positive feedback.

■ Terrestrial and marine carbon feedback

Figure D4.3.4 shows another positive feedback model. Here, combined terrestrial and marine feedback loops are shown which both result from, and accelerate further, global temperature rises. The elements of this model include:

- increasing water vapour in the atmosphere; because water vapour is also a greenhouse gas, this could lead to further temperature rises
- terrestrial permafrost thawing and methane release
- a reduction in seawater's ability to absorb surplus carbon dioxide from the atmosphere: warmer water is less effective at absorbing carbon dioxide than colder water, and so may begin to release, rather than absorb, carbon dioxide from deep ocean waters if temperatures continue to rise. However, life in the oceans absorbs more carbon dioxide and so the overall outcome is not so clear.

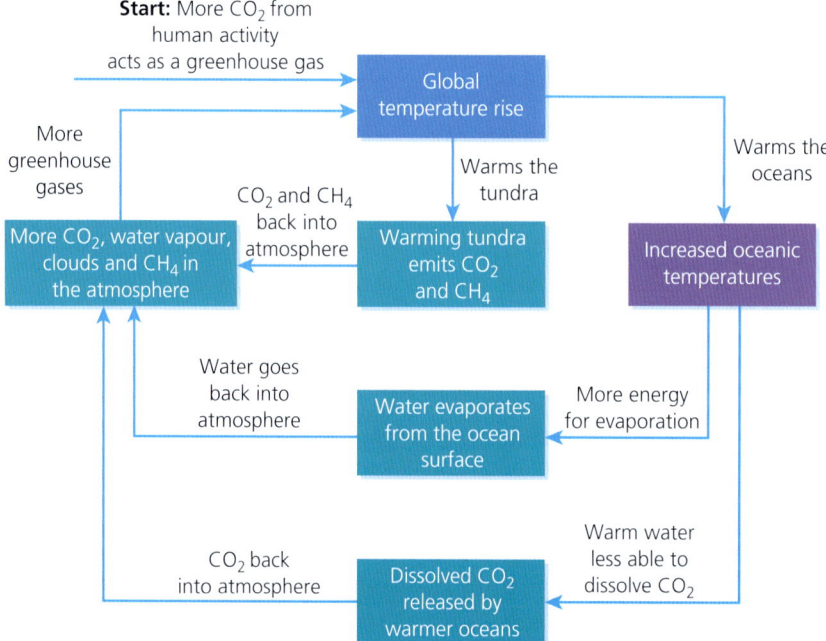

■ **Figure D4.3.4** A system diagram showing how interrelations between climate change, water flows and carbon flows could create positive feedback loops

Carbon dioxide dissolved in the surface of the ocean can be transferred to the deep ocean in areas where cold, dense surface waters sink. This down-welling carries carbon molecules to great depths where they may remain for centuries. Carbon is also carried to the floor of the ocean in the bodies of small organisms. Ocean phytoplankton absorb carbon dioxide through photosynthesis. These organisms form the bottom of the marine food web, and they live in the oceans' surface layer where sunlight penetrates. Phytoplankton are consumed by other marine organisms and carbon is subsequently transferred along food chains by fish and larger sea animals as they consume one another. Organic carbon may eventually be transferred to the deep ocean when dead organisms sink towards the ocean floor. Increasing temperatures in the deep ocean releases carbon dioxide through decreased solubility of carbon dioxide and increased decomposition of dead organisms on the ocean floor.

Increased temperature through global warming melts more of the ice in polar ice caps, glaciers and sea ice (Figures D4.3.5 and D4.3.6), leading to a decrease in the Earth's reflectivity (**albedo**). Thus the Earth absorbs more of the Sun's energy, which makes the temperature increase even more, melting more ice.

◆ **Albedo**: the fraction of solar radiation reflected by a surface or object, often expressed as a percentage.

> **Common mistake**
>
> The word 'positive' usually means something that is good or useful. In terms of feedback, the word 'positive' does not mean that the feedback loop has a constructive effect on the environment. Positive feedback increases change in a system, moving it further away from steady-state equilibrium.

■ Figure D4.3.5 Arctic ice melting, leading to positive feedback through decreased planetary albedo

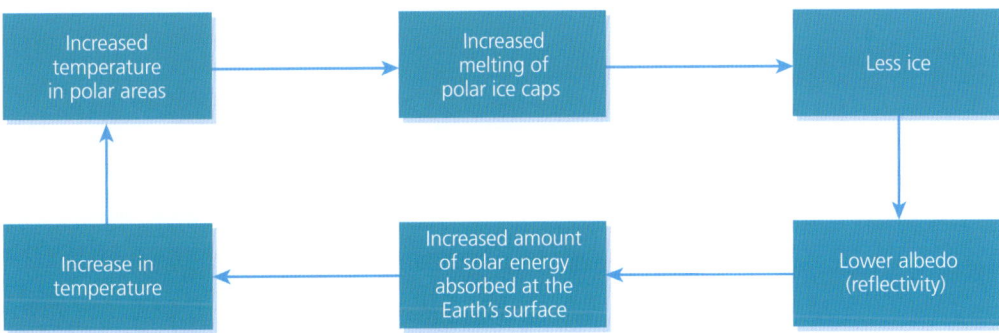

■ Figure D4.3.6 Positive feedback in global warming

Droughts and forest fires

A further example of positive feedback linked to climate change is the increased incidence of droughts and forest fires. With increased global temperatures, there is greater chance that droughts will occur and, subsequently, increase the chance of fire. Combustion of vegetation releases carbon dioxide into the atmosphere, which increases global warming still further. High temperatures, drought and high winds in Europe in the summer of 2022 made fighting the higher numbers of wildfires in France, Portugal, Spain and Greece difficult. These fires released large quantities of carbon dioxide into the atmosphere and removed important carbon sinks.

Change from net carbon accumulation to net loss in boreal forests as an example of a tipping point

Boreal (coniferous) forests (also known as taiga) store more carbon than tropical rainforests globally because they are distributed across a greater area, including much of northern Russia and North America. In boreal forests and wetlands, frozen soils contain large reserves of carbon (in the form of carbon compounds).

The Siberia Integrated Regional Study (SIRS) aims to investigate environmental change in Siberia (Figure D4.3.7) due to current global changes, and the potential impact on Earth system dynamics. The actual and projected consequences of global warming are well documented for Siberia. Temperatures have already increased, particularly during the winter in Eastern Siberia (0.5°C per decade), and the number of frost days and growing season length have also increased (roughly one day per year). Decreased winter snowfall has led to increased incidence of drought and reductions in primary production in taiga. Future climate change threats include the shift of permafrost boundaries northward, and dramatic changes in land cover.

According to SIRS, there are three main regional challenges that are very important for the global carbon cycle:

- Permafrost degradation, especially its border shift, might form a significant carbon and methane source to the atmosphere (while also seriously threatening human infrastructure). Climate-related drying would alter methane emissions in peatlands that have been deposited over millennia and would increase the potential for peat fires that cannot be extinguished.
- Temperature and precipitation changes, which increase forest browning and cause increases in the frequency and intensity of forest fires, thus significantly changing the carbon cycle of the region.
- Shifts in ecosystem borders northwards. The taiga zones (currently two-thirds of Siberia) will move northward and be reduced to 40% of the present area (Figure D4.3.7).

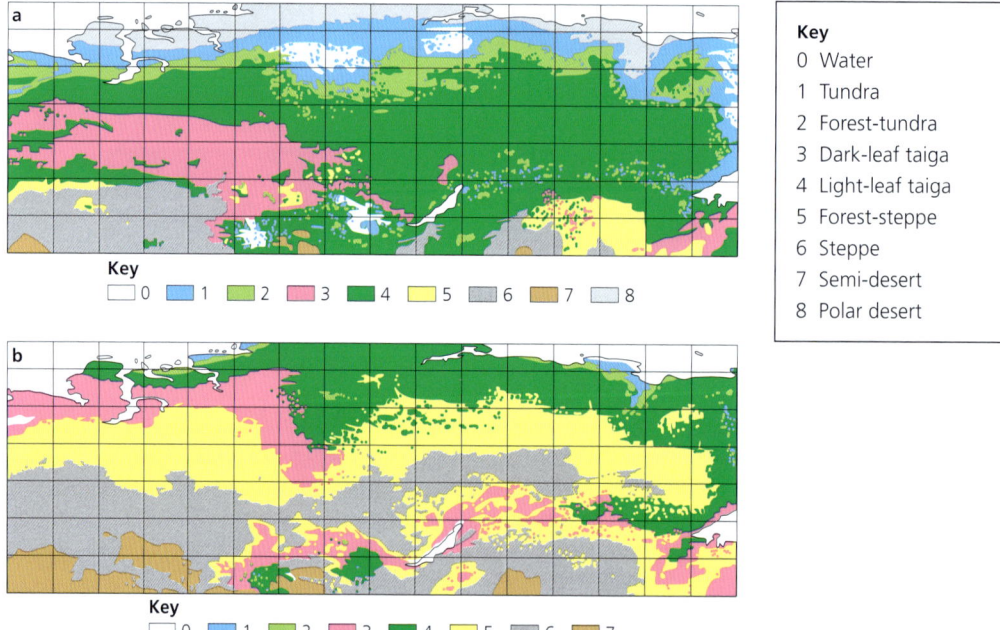

■ **Figure D4.3.7** Vegetation over Siberia: a) 2009, and b) predicted for 2090

4 **Outline** how positive feedback mechanisms result in climate change.

The southern borderline of taiga in Siberia will be more affected by forest fires. In this region, conditions favourable to fire have become unusually frequent during the past 20 years. Fire and the thawing of permafrost are considered to be the principal mechanisms that will trigger vegetation changes across the Siberian landscape and lead to increased release of carbon into the atmosphere in the form of carbon dioxide.

Melting of landfast ice and sea ice as examples of polar habitat change

The Arctic is a highly sensitive region – ice cover varies naturally according to the season. However, since 1979, the decline in the September Arctic sea ice is 12.6% per decade according to NASA, compared to its average extent during the period of 1981 to 2010. At this rate, there may be no ice by the summer of 2060. The associated Greenland ice is similarly in decline. Loss of sea ice has impacts for animals that use the ice to feed and breed. For example, the loss of sea ice habitat has affected walruses (*Odobenus rosmarus*) in the Arctic. Walruses cannot swim continuously so use sea ice as a place for rest between dives to the sea floor where they feed on clams and mussels. Sea ice floes are especially important for walrus calves (Figure D4.3.8).

■ **Figure D4.3.8** Adult female Atlantic walrus (*Odobenus rosmarus*) and juvenile resting on pack ice, Svalbard Archipelago, Arctic Norway

In Antarctica the picture is more complex. According to NASA, Antarctica is losing approximately 150 billion tonnes of ice mass a year due to melting. Also, major sections of the Antarctic ice shelf have broken off. Meanwhile, at times, the interior ice has become cooler and thicker, as circular winds around the land mass have prevented warmer air reaching the interior. Warmer seas may be eroding the ice from underneath, but the UN's Intergovernmental Panel on Climate Change (IPCC) predicts that the Antarctic's contribution to rising sea levels will be small.

◆ **Landfast ice**: sea ice that is 'fastened' (attached) to the coastline, to the sea floor along shoals or to grounded icebergs.

Landfast ice is sea ice that is 'fastened' to the coastline, to the sea floor along shoals or to grounded icebergs. As temperatures increase, these areas of sea ice break away from their attachment ('breakout'). As the planet warms through global warming, breakout of landfast ice is happening earlier in the year. Early breakout of landfast ice in the Antarctic has implications for animals that live and breed in these regions, such as the emperor penguin (*Aptenodytes forsteri*) – the largest penguin on Earth. These penguins use landfast ice, such as that in the Weddell Sea, to live and breed, and so early melting of the sea ice leads to loss of their breeding grounds. This could severely impact on the species' ongoing survival.

Changes in ocean currents

On a global scale, the oceanic conveyor belt (Figure D.4.3.9) transfers energy around, and links, the world's great oceans. Waters that flow from the Equator are generally warm, whereas those that flow from the high latitudes are cold. At the Earth's poles, cold, salty water sinks and is replaced by surface water that is warmed in the tropics. Now, with increasing temperatures at the poles, more melting ice decreases ocean salinity, which then slows the great ocean currents. Ocean currents convey heat energy from warmer to colder regions through their pattern of convection. Thus, for example, as the Gulf Stream (which, to date, has kept temperatures in Europe relatively warmer than in Canada) slows down, more heat is retained in the Gulf of Mexico. Here, hurricanes get their energy from warmer water, and they are becoming more frequent and more severe.

◆ **Upwelling**: process where deep, cold water rises toward the surface of the ocean.

Upwelling is a process where deep, cold water rises toward the surface and it occurs in the north Pacific Ocean and the Indian Ocean (Figure D.4.3.9). Upwelling draws nutrients from the deep ocean to the surface, causing an increase in phytoplankton abundance, which in turn leads to an increase in zooplankton. Some marine animals, such as whales, migrate to these areas of high primary production. If climate change affects the oceanic conveyor belt, and the ocean primary production decreases in these areas, there will be a knock-on effect for marine food chains.

D4.3 Climate change

The changes in ocean currents affect the availability of nutrients in the ocean. Warmer surface water can prevent nutrient upwelling to the surface, which decreases ocean primary production and energy flow through marine food chains.

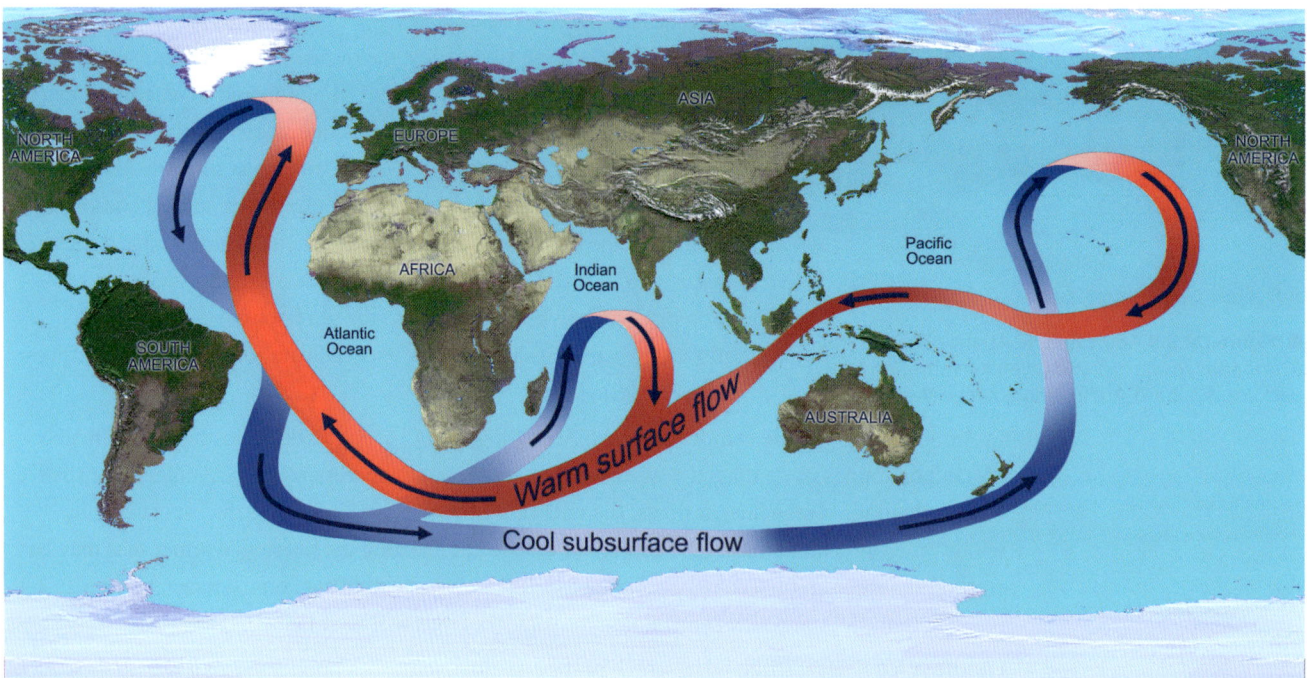

■ Figure D4.3.9 The Earth's ocean currents

Poleward and upslope range shifts of temperate species

With increasing mean global temperatures, effects can be expected on the distribution of habitats, communities and ecosystems.

Research has shown that the distributions of many terrestrial organisms are currently shifting in latitude or elevation in response to changing climate. Range changes in elevation include species that move upwards in mountain ranges as warming conditions cause habitat ranges to shift higher. Species have zones of tolerance (page 345) and so need to shift higher, to colder areas, as the areas they live in warm. As the conditions required by their niche change, so too does the distribution of the species. These shifting geographic distributions have been recorded in temperate species and correlated to anthropogenic temperature increases. Recent studies have examined the effect of climate change on tropical species. In Papua New Guinea, studies have shown that montane bird species (i.e. species that live in mountainous regions) have moved higher in the mountains in which they live, a process known as **upslope range shift**.

◆ **Upslope range shift**: process where montane species move higher up the mountains in response to recent temperature increases.

Data were collected from two mountains in Papua New Guinea in 2012–13 – Mt Karimui and Karkar Island – and compared with historical data from the 1960s (Figure D4.3.10). On both mountains, bird species shifted their upper limits upslope significantly, on Mt Karimui by 113 m, and Karkar Island by 152 m. Upslope shifts also occurred at species' lower limits on Mt Karimui by 95 m.

The researchers found that few species were restricted to montane elevations on Karkar Island, so upslope shifts in species' lower limits were not statistically significant there. In addition, the birds' upslope shifts significantly outnumbered their downslope shifts in both mountains at the upper limits (Mt Karimui: 87 upslope, 36 downslope; Karkar Island: 17 upslope, 5 downslope) and lower limits (Mt Karimui: 39 upslope, 14 downslope).

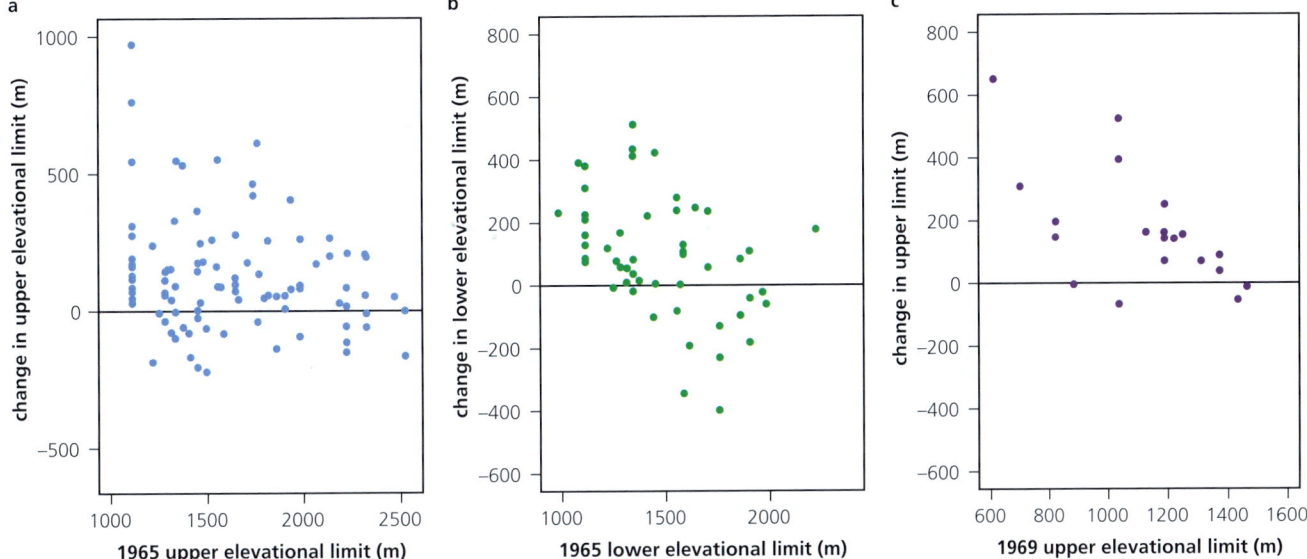

■ **Figure D4.3.10** Graphs showing a) the changes in species' elevational limits for Mt Karimui upper elevational limits, b) Mt Karimui lower elevational limits, and c) Karkar Island upper elevational limits. Changes in species' elevational limits between historical and modern resurveys are plotted against historical elevational limits measured in the 1960s. Points on the solid zero-change lines represent species with unchanged elevational limits

ATL D4.3C

Can you find other examples of similar outcomes to those recorded in Papua New Guinea? What affect may upslope shifts in animal species have on biodiversity? What may happen to species adapted to high elevation when their niche requirements move further upslope? Apart from 'global action' to reduce warming, are there 'local solutions' communities might try? Discuss your findings and thoughts with someone else in your class.

These articles are a useful starting point for your research and discussions:
www.pnas.org/doi/10.1073/pnas.0809320106
www.sciencedaily.com/releases/2009/01/090121091239.htm
https://phys.org/news/2022-04-lost-golden-toad-heralds-climate.html
www.environment.vic.gov.au/__data/assets/pdf_file/0018/32355/Mountain_Pygmy-possum_Burramys_parvus.pdf

The distribution of biomes is dependent on climate (page 354). Changes in patterns of rainfall associated with climate change, together with alterations in land and sea surface temperatures, will lead to shifts in the distribution of biomes.

The global biome pattern can be expected to change, for example, in response to rising global mean surface temperature. Research has shown a range contraction and northward spread in North American tree species. One study sampled 92 species in 43 000 plots across eastern North America. Latitudinal differences were compared with twentieth-century temperature change. More than half of the tree species showed indications that their ranges were shrinking at both their northern and southern habitat boundaries. Slightly more species showed their ranges were shifting just north, and only a few species indicated an increase at both north and south limits of their range.

Another study in Quebec, Canada, used forest inventory data of 16 temperate species from 30 years. This study also showed that tree species' ranges had shifted predominantly northward, although the responses for different species varied. The shifts in range were greater for tree saplings (0.34 km yr^{-1}) than for full-grown trees (0.13 km yr^{-1}).

The data above are based on the following papers: 'Evidence for range contraction of eastern North American trees', from 96th ESA Annual Convention 2011, and 'Tree range expansion in eastern North America fails to keep pace with climate warming at northern range limits', from Global Change Biology, *23(8) pages 3292–3301.*

Threats to coral reefs as an example of potential ecosystem collapse

Link
Conditions required for coral reef formation are covered in Chapter B4.1, page 353; mutualism between coral and zooxanthellae is in Chapter C4.1, page 567.

Corals are colonies of small animals embedded in a calcium carbonate shell that they secrete. They form underwater structures, known as coral reefs, in warm, shallow water where sunlight penetrates. Specific microscopic photosynthetic algae (zooxanthellae) live sheltered and protected in the cells of tropical corals. Coral reefs are the most diverse of ecosystems known and, although they cover less than 0.1% of the surface of the oceans, these reefs are home to about 25% of all marine species.

■ **Figure D4.3.11** Coral reefs – a crisis situation; the photo on the left shows healthy coral reef and, on the right, bleached coral as the result of climate change

> ● **Top tip!**
>
> Climate change does not just affect individual species – it also affects the interactions between species, such as the mutualistic relationship between coral polyps and zooxanthellae.

Link
A case study on the loss of the Great Barrier Reef is in Chapter A4.2, page 166.

Increased release of carbon dioxide (see Table D4.3.2, page 824), ocean currents changing (see Figure D4.3.9, page 830) and oceans absorbing more heat as greenhouse gases trap more heat, means that ocean temperatures are rising. The corals have a very sensitive range of tolerance: when under environmental stress (such as high water temperature), the algae inside the coral are expelled, which causes the coral to turn white (bleaching). Mass bleaching events occurred in the Great Barrier Reef in 1998, 2002, 2016, 2017 and 2020. Multiple bleaching events, or bleaching and other damage, kill the coral.

The coral skeleton is made from calcium carbonate in a process called calcification (see page 353). Increased carbon dioxide concentrations cause ocean acidification, however, and the suppression of calcification in corals. Once the coral has been lost, other species lose their niches and the coral reef loses its biodiversity. Today, coral reefs are dying all around the world, and this loss of corals causes the collapse of reef ecosystems. The effects of thermal stress are likely to be exacerbated under future climate scenarios.

Afforestation, forest regeneration and restoration of peat-forming wetlands as approaches to carbon sequestration

Afforestation

◆ **Afforestation**: the establishment of forests in an area where there was no previous tree cover.

Common mistake

Do not confuse afforestation and reforestation. Afforestation is the addition of forests to land that did not originally have them; reforestation is the restoration of forests to lands where they once existed.

Afforestation involves planting trees in deforested areas or places that have never been forested. New trees act as carbon sinks and can therefore help with climate change mitigation.

The UN-REDD Programme, launched in 2008, is the United Nations Collaborative Programme on Reducing Emissions from Deforestation and Forest Degradation in Developing Countries. UN-REDD provides incentives for developing countries to conserve their rainforests by placing a monetary value on forest conservation. This is an important example of successful global governance. UN-REDD stresses the role of conservation, the sustainable management of forests and the increase of forest carbon stocks. By 2020, the total funding had reached almost US$325 million, with contributions from Norway, the EU, Denmark, Spain, Japan, Luxembourg and Switzerland.

China has the largest afforested area in the world, with an average gain of 19 370 km² of forest per year. Between 2020 and 2025, China plans to plant 36 000 km² of new forest a year, which is more than the total area of Belgium.

As well as acting as carbon storage and helping the country offset carbon emissions, the tree-planting campaign also aims to stop the spread of deserts (for example of the northern borders of China, adjacent to the Gobi Desert), stabilize water resources, and provide local people with sustainably sourced resources such as timber and other forest products, according to the World Forest Organisation.

■ **Figure D4.3.12** Farmers use machines to drill tree holes for afforestation activities during the Spring Festival in northern China

■ **Figure D4.3.13** Hard tree ferns (*Blechnum* sp.) growing in burnt forest after 2019–20 bushfires in Australia

■ Agroforestry and forest regeneration

Agroforestry combines agriculture with forestry, allowing the farmer to continue cropping while using trees for animal food, fuel and building timber: trees protect, shade and fertilize the soil, decreasing rates of decomposition and related rates of respiration and increasing photosynthesis. Such practices improve **carbon sequestration** in agricultural soils and above-ground biomass through a range of soil, crop and livestock management practices, and protect existing carbon in the system by slowing decomposition of organic matter and reducing erosion.

Forest regeneration is the process by which new trees and forest species become established after forest trees have been harvested or have died from fire, insects or disease. In the same way that afforestation increases carbon sequestration, forest regeneration ensures that the storage of carbon in trees increases and, as a consequence, reduces carbon dioxide in the atmosphere through the process of photosynthesis.

◆ **Carbon sequestration**: the capture and storage of carbon dioxide from the atmosphere by physical or biological processes such as photosynthesis.

■ Restoration of peat-forming wetlands

In acidic and anaerobic conditions found in waterlogged soils, dead organic matter is not fully decomposed but accumulates as peat. The largest terrestrial carbon store is found in peatlands. From 1640 onwards, large areas of wetland, such as the East Anglian Fens in the UK, were drained for farming, which degraded the peat. No longer waterlogged, the peat shrank, decomposed and became eroded by the wind, which increased the flux of carbon dioxide to the atmosphere. Restoring rich peatlands that were lost during agricultural development would enhance the carbon sequestration of these wetland ecosystems. Restoring these wetlands increases the land carbon store and decreases the atmospheric store. Degraded and drained peatlands emit more than 2 gigatonnes of carbon dioxide (GtC) annually: through rewetting of peatlands it is possible to restore carbon levels in peat soils that have already been degraded.

There is active scientific debate over whether plantations of non-native tree species or rewilding with native species offer the best approach to carbon sequestration. Peat formation occurs naturally in waterlogged soils in temperate and boreal zones, and also very rapidly in some tropical ecosystems.

5 **Define** the term *carbon sequestration*.

6 **Explain** how afforestation can be used to mitigate the effects of climate change.

Phenology as research into the timing of biological events

The timing of biological events that depend on seasonal cycles, such as flowering, budburst and bud set, bird migration and nesting, can be affected by variables such as **photoperiod** (day length) and temperature patterns.

◆ **Photoperiod**: the period of time each day during which an organism receives light, i.e. day length.

Flowering needs to coincide with the availability of pollinators, and so timing needs to be carefully controlled. Day length determines whether flowering occurs or not. Some species require long days (and short nights) to flower, whereas others need shorter days and longer nights.

Link
The black-throated loon is discussed in more detail in Chapter A1.1, page 10.

Many animals leave the area where they spend spring and summer to escape the colder winter months (Chapter B3.3, page 338). For example, the black-throated loon (*Gavia arctica*) migrates south to areas around the Black Sea and the Mediterranean Sea, and to north-east Atlantic coasts and the eastern and western Pacific Ocean. The bird returns to its breeding grounds in early April when sea ice in those areas has melted. For birds to rest, temperatures must be warm enough to allow a sufficient supply of food for young, such as insects.

The study of the timing of biological events is known as **phenology**. Research has shown that increases in average global temperatures are affecting the timing of biological events.

◆ **Phenology**: research into the timing of seasonal or cyclical biological events, such as flowering, budburst and bud set, bird migration and nesting.

Disruption to the synchrony of phenological events by climate change

Climate change can lead to a disruption of the synchrony of phenological events. *Cerastium arcticum* (Arctic mouse-ear) is a shrub native to Greenland, northern Canada, Svalbard (Norway) and western Siberia. Spring growth of this shrub is advancing (i.e. coming earlier in the year) due to climate change. Warmer climate provides an increase in the length of the growing season for the species, allowing the plants more time for establishing, reproducing and spreading.

Many migratory animals, such as the caribou (*Rangifer tarandus*), also known as reindeer (Figure D4.3.14), track the growth of spring vegetation during a synchronized spring migration. Studies have shown that a combination of snow and temperature has a strong influence on autumn migratory movements. Earlier availability of ground vegetation can be expected to advance caribou migration.

■ **Figure D4.3.14** Caribou (*Rangifer tarandus*) migrating in Alaska, USA

Inquiry 1: Exploring and designing

Exploring

This activity will enable you to demonstrate independent thinking, initiative or insight. The following site includes a time-lapse image of patterns of migration of arctic animals, and includes a link to the Arctic Animal Movement Archive (AAMA):

www.nasa.gov/feature/goddard/2020/arctic-animals-movement-patterns-are-shifting-in-different-ways-as-the-climate-changes

Discuss in groups:
1. What are some of the similarities and differences in the patterns of movement by each of the three animal groups represented on the map?
2. Why do you think the animals were grouped in this way? Formulate a hypothesis.
3. What differences are suggested in the future adaptability of their movements in response to climate change? State and explain your predictions using scientific understanding.
4. Why do you think the technical information (data sets and sensors used) about individual studies is hidden behind a link (AAMA)? Who (mostly) are the people who have provided these data?
5. How would you change this presentation to make its message more accessible to (a) an ecologist, and (b) a younger audience?
6. Why do you think the National Aeronautics and Space Administration (NASA) has an interest in sharing this research?

D4.3 Climate change

■ Figure D4.3.15 Great tit (*Parus major*) with a caterpillar to feed to its young

The timing of reproduction for animals has consequences on the numbers of offspring that will survive and reproduce themselves (**fitness**). The timing of breeding in insectivorous bird species (species that feed on insects), for example, needs to coincide with an abundance of their food species. A study on great tits (*Parus major*), published in 2006 in the journal *Global Change Ecology*, measured insect biomass over a 20-year period. The researchers showed that the annual peak date in biomass correlated with temperatures from 8 March to 17 May. Egg-laying dates also correlate with temperatures, but over an earlier period (16 March–20 April). The food peak also correlates with these temperatures. The major difference is that due to climate change, with the synchrony between offspring needs and the caterpillar biomass, this cycle was disrupted in the recent warm decades. This study indicates that changes in phenology may have severe consequences for both the number of fledglings as well as their fledging mass, which is affected by this synchrony of insect biomass. Models were used for both the caterpillar biomass peak and for the great tit laying dates, in order to predict shifts in caterpillar and bird phenology between 2005–2100, using a model of predicted global warming from the IPCC. The model predicts that great tits will start breeding earlier and this is predicted to be at the same time period as the earlier food peak.

Increases to the number of insect life cycles within a year due to climate change

Beetles (the order Coleoptera) are the most diverse and abundant group of insects on Earth. They inhabit every ecosystem except for the oceans. While most beetles provide vital roles in ecosystem dynamics, some species have become pests. One group, the bark beetles, feed on tree plantations, damaging and killing trees. There are more than 2000 species of bark beetles.

♦ **Life cycle**: the series of changes in the life of an organism, including reproduction.

Several species of bark beetle attack spruce and pine forests. The great spruce bark beetle (*Dendroctonus micans*) is a non-native (alien) pest of spruce and pine trees in Europe, while the Eurasian spruce bark beetle (*Ips typographus*) is one of the most destructive forest insects in Europe, where it has killed more than 50 million m³ of Norway spruce (*Picea abies*) since the 1940s.

Insect species follow one of two different types of **life cycle**. One life cycle has the development of nymphs from eggs, followed by instars (intermediate developmental stages) and finally adults. Insects such as locusts and grasshoppers follow this life cycle. Other insects produce larvae from eggs (e.g. caterpillars) which then form a pupa (where the developing organism is usually enclosed in a cocoon or protective covering). In the pupa, the organism undergoes internal changes (metamorphosis) where larval structures are replaced by those of the adult. Beetles follow the latter life cycle (see Figure D4.3.16).

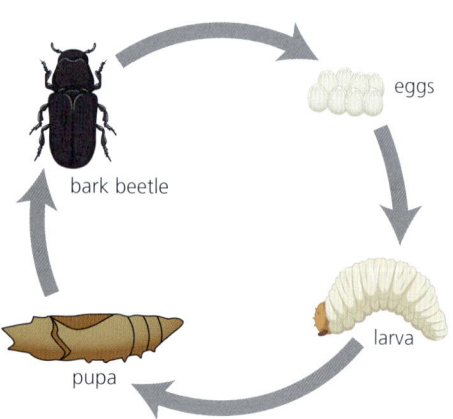

■ Figure D4.3.16 Life cycle of a bark beetle

Bark beetle females bore into the bark of the tree and tunnel through to the phloem where they lay eggs, underneath the bark layer (bark is the outer, protective layer of a tree). When the larvae emerge, they feed on the sap of the phloem as well as the tissue itself. After pupation, new adults make new holes through the bark as they emerge.

Phloem transport sucrose and other organic compounds around the plant (see page 323). The larvae damage this tissue, which weakens the tree and can lead to its death. Large areas of forest have been affected by outbreaks of bark beetle (Figure D4.3.17).

Link
The adaptations of phloem sieve tubes and companion cells for translocation of sap is covered in Chapter B3.2, page 323.

Insects are sensitive to temperature, have short life cycles and are very mobile. Changes in temperature due to climate change therefore can be expected to affect their rate of development and geographical distribution. Warming summer and winter temperatures have a major effect on beetle populations and

Figure D4.3.17 Trees killed by spruce bark beetle, Colorado, USA

increase the chance of outbreaks across the USA and Europe. Temperatures also lead to range expansion in some species (for example, beetles spreading to new areas as temperatures warm). Climate change may increase the risk of bark beetle outbreaks, particularly in old forests in northern Europe. Research in Norway has shown that higher temperatures will favour the beetles and cause a transition from one to two generations per year in northern latitudes. With two generations per year, there will also be two attack periods on spruce, one in the spring and one in July/August. The increased number of generations of beetles per year will mean that there will be a greater chance that the critical number of beetles needed to kill a tree will be reached. In addition, a second generation of bark beetles can more easily kill trees, as fewer beetles are needed to overwhelm less-resistant trees.

> **Top tip!**
>
> Review your knowledge of evolution by natural selection before reading this section.

Evolution as a consequence of climate change

We have seen how changes in the environment can lead to changes in the selective pressure on species, ultimately resulting in the evolution of new species. Can the changes caused by climate change result in the evolution of new species?

Evidence suggests that tawny owls (also known as the brown owl, *Strix aluco*) in southern Finland are evolving because of milder winters over many decades. The milder winters have resulted in less snow, and so birds with lighter feathers are at a selective disadvantage as they are more visible to their prey and so catch less food. Research has shown there has been an increase in dark brown tawny owls in a population that was usually dominated by pale grey owls. Darker-coloured brown owls, which used to form only 30% of the tawny owl population in Finland, now form 50%. In years when winter weather is more severe, there is a higher mortality rate in the brown owl population: this could be because brown owls are more visible to predators when there is thick snow cover. This research has provided the first evidence for a population evolving in response to climate change.

ATL D4.3D

Which other pest species are being affected by climate change, and what impact is this having on the species on which they feed and reproduce? Research one species from your local area and produce a fact sheet about the organism and its impact. Use your knowledge of biology to explain why the incidence of the pest is increasing and how it affects other species. What solutions are there to reduce the impact of the pest?

> **Top tip!**
>
> We have seen how industrialization led to the increase in the melanic form of the peppered moth (see page 784). Changes in the colour variants of tawny owls is another example of human-induced change to the environment affecting the fitness of individuals within a population. In both cases, the colour variants are still within the same species but, given time, natural selection could lead to the evolution of new species.

Figure D4.3.18 Tawny owls have two plumage colours, brown and grey

LINKING QUESTIONS

1. What are the impacts of climate change at each level of biological organization?
2. What processes determine the distribution of organisms on Earth?

D4.3 Climate change

Acknowledgements

Authors' acknowledgements

This book has been extensively reviewed by experts in their field, and we would like to thank them for their thorough and helpful comments, which have ensured that the information throughout is both accurate and relevant. Sections were reviewed by: Professor Tim Flowers, University of Sussex (D2.3), Emeritus Professor Andrew J. Easton, University of Warwick (A2.3 and D1.3), Professor Mike Williamson, University of Sheffield (C2.1 and C1.1), Professor Susan Ward, University of Leeds (B3.1), Professor Roger Butlin, University of Leeds (A4.1), Dr Anthony Poole, University of Auckland (B2.2), Dr Tom Williams (A2.1), University of Bristol, Professor Jon Cooper, University College London (B1.2 and C1.1), Professor Tod Lowary, University of Alberta (B1.1), Emeritus Professor Nancy Curtin, Imperial College, London (B3.3), Professor David G. Nicholls, Buck Institute for Research on Aging, California (C1.2 and C1.3), Professor Hans Degens, Manchester Metropolitan University (B3.3), Annie Termatt, (D3.2), Professor Richard Allan, University of Reading (D4.3), Professor Piers Forster, University of Leeds (D4.3)

We would also like to thank Marcela Rodriguez, who reviewed the book for the IB. Her attention to detail and constructive comments have ensured that the book closely matches the aims and aspirations of the Guide and IB Diploma, and we are very grateful to her for the extensive work she has put into this project.

Finally, we are indebted to the publishing team at Hodder Education who have coordinated and guided this book through to completion: Catherine Perks, who expertly and with great attention to detail edited the book and provided advice throughout, Phillipa Allum, who was project manager and ensured the smooth coordination of what was a very complex operation, and Sophie Clark who copy edited the manuscript with great expertise: their skill and patience have brought together the text and illustrations as we have wished, and we are most grateful to them.

Publisher's acknowledgements

The Publishers would like to thank the following for permission to reproduce copyright material. Every effort has been made to trace and acknowledge ownership of copyright. The publishers will be glad to make suitable arrangements with any copyright holders whom it has not been possible to contact.

Photo credits

p5 © Biophoto Associates/Science Photo Library; p6 © Rossi Paolo – Fotolia.com; p10 l © Mps197/stock.adobe.com; r © Joe McDonald/Shutterstock.com; p20 © Science Photo Library; p21 © Science Photo Library; p26 © A.Barrington Brown, Gonville & Caius College/Science Photo Library; p28 © Biophoto Associates/Science Photo Library; p30 © Biozentrum, University of Basel/Science Photo Library; p47 © Photo by Matthew Dodd and Dominic Papineau; p48 Image courtesy of Submarine Ring of Fire 2014 – Ironman, NOAA/PMEL, NSF; p56 © Power and Syred/Science Photo Library; p59 l © J.C.Revy, ISM/Science Photo Library; r © Biophoto Associates/Science Photo Library; p60 l © Sinclair Stammers/Science Photo Library; r © Power and Syred/Science Photo Library; p61 © CEITEC Masaryk University; p62 tl © Dennis Kunkel Microscopy/Science Photo Library; tr © Biophoto Associates/Science Photo Library; b © Drimafilm/stock.adobe.com; p64 l © J.C.Revy, ISM/Science Photo Library; r © Gene Cox; p67 © Kwangshin Kim/Science Photo Library; p69 © CNRI/Science Photo Library; p70 t © Medimage/Science Photo Library; b © Omikron/Science Photo Library; p71 © Carolina Biological Supply Company/Science Photo Library; p73 © Dr_Microbe/stock.adobe.com; p75 © Don W. Fawcett/Science Photo Library; p76 © Dr Kari Lounatmaa/Science Photo Library; p79 © Gene Cox; p80 tl © David T. Moran, J. Carter Rowley, Visual Histology, Wolters Kluwer, 1988; tr © Dr. Jeremy Burgess/Science Photo Library; b © Steve Gschmeissner/Science Photo Library; p81 © Kevin Mackenzie; p82 © Steve Gschmeissner/Science Photo Library; p85 © CDC; p90 l © James Cavallini/Science Photo Library; c © Siddhesh/stock.adobe.com; r © Dr M. Wurtz/Biozentrum, University Of Basel/Science Photo Library; p91 © Dennis Kunkel Microscopy/Science Photo Library; p92 tl © A. Dowsett, Health Protection Agency/Science Photo Library; bl © Nigel Cattlin/Alamy Stock Photo; br © Nigel Cattlin/Alamy Stock Photo; p101 From https://commons.wikimedia.org/wiki/File:SIR-Modell.svg; p102 © Dr Keith Wheeler/Science Photo Library; p105 tl © Sheila Terry/Science Photo Library; bl © Natural History Museum, London/Science Photo Library; r © Creativemarc/Fotolia; p106 tl © DDniki – Fotolia.com; tr © Picture.jacker/stock.adobe.com; bl © Marek R. Swadzba/stock.adobe.com; bc © Lip Kee (lipkee@gmail.com); br © Guy Bryant/stock.adobe.com; p112 l © Leonard Lessin/Science Photo Library; p117 © Tek Image/Science Photo Library; p136 tl © Steve Taylor ARPS/Alamy Stock Photo; tr © Top-Pics TBK/Alamy Stock Photo; cl © C.J Clegg; cr © Frank Teigler/Premium Stock Photography GmbH/Alamy; bl © Chris Howes/Wild Places Photography/Alamy Stock Photo; br © C.J Clegg; p142 t © Dave Watts/Alamy; bl © Holly Grogan/stock.adobe.com; br © David Daniel/stock.adobe.com; p143 b © Bezvershenko/stock.adobe.com; p147 l © Slowmotiongli/stock.adobe.com; r © Joern/stock.adobe.com; p151 b © Eric Gevaert/stock.adobe.com; t © Papa Bravo/stock.adobe.com; p152 © Konrad Wothe/Minden Pictures/Alamy Stock Photo; p155 © PT pictures/stock.adobe.com; p161 © Dewessa/stock.adobe.com; p163 © Naoki Nishio/stock.adobe.com; p164 t © M2/stock.adobe.com; b © Paul D Stewart/Science Photo Library; p.165 l © Andrew Davis; r © Andrew Davis; p166 Jacques Descloitres, Modis Rapid Response Team, NASA/GSFC; p171 l © Rich Carey/Shutterstock.com; r © Oliver Sved/123rf; p173 t © Crisod/stock.adobe.com; b © Cavan/stock.adobe.com; p177 t © Jozef Durok/iStock/Getty Images; b © Ken Griffiths/stock.adobe.com; p178 © Adisak Banpot/stock.adobe.com; p193 © Biophoto Associates/Science Photo Library; p201 © Steve Gschmeissner/Science Photo Library; p213 © Courtesy of J.B.Cooper, UCL; p214 l © Courtesy of J.B.Cooper, UCL; r © Courtesy of J.B.Cooper, UCL; p221 © Steve Gschmeissner/Science Photo Library; p235 © Don W. Fawcett/Science Photo Library; p236 © Don W. Fawcett/Science Photo Library; p249 © CNRI/Science Photo Library; p250 © Dr Kenneth R. Miller/Science Photo Library; p261 t © Thierry Berrod, Mona Lisa Production/Science Photo Library; b © Biophoto Associates/Science Photo Library; p265 © Aldona/stock.adobe.com; p266 © Steve Gschmeissner/Science Photo Library; p268 © Gene Cox; p269 © Gene Cox; p270 © P. Navarro, R. Bick, B. Poindexter, Ut Medical School/Science Photo Library; p278 © Gene Cox; p284 © Gene Cox; p285 t © Gene Cox; b © Gene Cox; p286 © BioFoto/stock.adobe.com; p296 © CNRI/Science Photo Library; p300 © Image Source/Photodisc/Getty Images; p304 © Dr David Furness, Keele University/Science Photo Library; p308 © Gene Cox; p309 © Gene Cox; p323 © Biophoto Associates/Science Photo Library; p326 © Stuporter/stock.adobe.com; p327 t © RLS Photo/

stock.adobe.com; *b* © Josehidalgo87/stock.adobe.com; **p328** *t* © Mark Rothery (http://www.mrothery.co.uk/); *b* © Mark Rothery (http://www.mrothery.co.uk/); **p337** © Robert Kneschke/stock.adobe.com; **p338** *t* © Ian Dyball/stock.adobe.com; *tc* © Tom & Pat Leeson/Science Photo Library; *bc* © Stephane Bidouze/stock.adobe.com; *b* © Justin/stock.adobe.com; **p339** © Rocky Grimes/stock.adobe.com; **p342** © Joe Trentacosti/stock.adobe.com; **p343** *tl* © Steve Lowry/Science Photo Library; *tr* © Steve Gschmeissner/Science Photo Library; *bl* © Tonguy324/stock.adobe.com; *br* © Rus/stock.adobe.com; **p344** © Kwanchaichaiudom/stock.adobe.com; **p346** © Dr. Craig Cary, PhD. University of Delaware, USA and University of Waikato, New Zealand; **p349** © Melissa Douglas, UCSC; **p356** © Stéphane Bidouze/stock.adobe.com; **p357** *t* © Yare yare/stock.adobe.com; *b* © PiLensPhoto/stock.adobe.com; **p358** *t* © Kjekol/stock.adobe.com; *b* © Troutnut/Shutterstock.com; **p360** *t* © Buteo/Shutterstock.com; *c* © EcoView/stock.adobe.com; *b* © Stan/stock.adobe.com; **p361** *t* © Classic/stock.adobe.com; *c* © Thanagon/stock.adobe.com; *b* © Sdbower/stock.adobe.com; **p362** *l* © Ahen Hendry/Shutterstock.com; *r* © Eric Isselée/stock.adobe.com; **p367** © Biophoto Associates/Science Photo Library; **p370** © DaveCarlson/CarlsonStockArt.com; **p372** *t* © Clouds Hill Imaging Ltd/Science Photo Library; *c* © Duelune/stock.adobe.com; *b* © Chrispo/stock.adobe.com; **p373** *t* © Greg Mailaender/stock.adobe.com; *bl* © Albert Beukhof/stock.adobe.com; *br* © Photomic/stock.adobe.com; **p374** *t* © Michael Ireland/stock.adobe.com; *cl* © Piotr Krzeslak/stock.adobe.com; *cr* © DS light photography/stock.adobe.com; *bl* © Sacha Lubow/stock.adobe.com; *br* © Mattjeppson/stock.adobe.com; **p375** *t* © Starsphinx/stock.adobe.com; *c* © Kikkerdirk/stock.adobe.com; *b* © Jayvee18/stock.adobe.com; **p376** *t* © Joppi/stock.adobe.com; *b* © RealityImages/stock.adobe.com; **p377** *tl* © Alexey Stiop/stock.adobe.com; *tr* © Uwe Bergwitz/stock.adobe.com; *b* © Ksenia Ragozina/Shutterstock.com; **p428** © Frank Fox/Science Photo Library; **p438** © Duke University; **p453** © Diveivanov/stock.adobe.com; **p464** © Alila Medical Media/stock.adobe.com; **p476** © Prof S. Cinti/Science Photo Library; **p485** © Omikron/Science Photo Library; **p488** © Signal Photos/Alamy Stock Photo; **p511** © Abet/stock.adobe.com; **p521** © CNRI/Science Photo Library; **p528** © Steve Gschmeissner/Science Photo Library; **p529** © Thomas Deerinck, NCMIR/Science Photo Library; **p533** © St Mary's Hospital Medical School/Science Photo Library; **p538** *t* © Cavallini James/BSIP SA/Alamy Stock Photo; *c* © James Cavallini/Science Photo Library; *b* © Nobeastsofierce/stock.adobe.com; **p539** © Boomerang11/stock.adobe.com; **p548** © Ernst August/stock.adobe.com; **p550** © Andrew Davis; **p562** *t* © Ongushi/stock.adobe.com; *b* © R.M. Nunes/stock.adobe.com; **p563** *t* © EcoView/stock.adobe.com; *b* © Oscar/stock.adobe.com; **p.564** *t* © Vera Kuttelvaserova/stock.adobe.com; *b* © Zsuzsanna Bird/stock.adobe.com; **p565** *t* © C.J. Clegg; *b* © Dante Fenolio/Science Photo Library; **p566** © Dr Jeremy Burgess/Science Photo Library; **p567** *l* © Eye Of Science/Science Photo Library; *r* © Garlezki/stock.adobe.com; **p569** *l* © Menno Schaefer/stock.adobe.com; *r* © Taviphoto/stock.adobe.com; **p571** *l* © Alexisaj/Alamy Stock Photo; *r* © WildPictures/Alamy Stock Photo; **p573** *t* C.J. Clegg; *bl* © Richard Becker/Alamy Stock Photo; *br* D. Hurst/Alamy Stock Photo; **p577** *l* © Barmalini/stock.adobe.com; *r* © Vodolej/stock.adobe.com; **p584** © Paroli Galperti/REDA & CO srl/Alamy Stock Photo; **p600** © NOAA; **p610** © David Parker/Science Photo Library; **p621** © Dee-sign/stock.adobe.com; **p639** © Rumruay/stock.adobe.com; **p642** © Alexey Protasov/stock.adobe.com; **p643** Courtesy of Dcirovic/Wikimedia Commons; **p652** © Science Photo Library; **p654** *tr* © Michael Abbey/Science Photo Library; *br* © Michael Abbey/Science Photo Library; *bl* © Michael Abbey/Science Photo Library; *cl* © Michael Abbey/Science Photo Library; *tl* © Michael Abbey/Science Photo

Library; **p655** © Gene Cox; **p.659** © U.S. Department of Energy Human Genome Program; **p662** © Gene Cox; **p663** © Pikovit/stock.adobe.com; **p669** © Steve Gschmeissner/Science Photo Library; **p694** *t* © London School of Hygiene & Tropical Medicine/Science Photo Library; *b* © Science Pictures Limited/Science Photo Library; **p703** © Patrick Hermans/stock.adobe.com; **p704** *tl* © Bill Ross/The Image Bank/Getty Images; *tr* © Jane Sugarman/Science Photo Library; *bl* © Daniel Prudek/stock.adobe.com; *bc* © Vaclav/stock.adobe.com; *br* © David/stock.adobe.com; **p707** *l* © BJ.Photo/stock.adobe.com; *cl* © Aleh Varanishcha/stock.adobe.com; *cr* © Dr Keith Wheeler/Science Photo Library; *r* © Nd700/stock.adobe.com; **p713** *l* © Jose Calvo/Science Photo Library; *r* © Gene Cox; **p715** © Jean Claude Revy-A.Goujeon, ISM/Science Photo Library; **p717** © Edelmann/Science Photo Library; **p721** © Dr G Moscoso/Science Photo Library; **p726** © Science Photo Library; **p727** © Sheila Terry/Science Photo Library; **p742** © Chris Bjornberg/Science Photo Library; **p752** © Roger de Montfort/stock.adobe.com; **p754** © Stephen Dalton/Avalon.Red; **p762** © Gene Cox; **p770** © J M Barres/Agefotostock/Alamy Stock Photo; **p771** © Steve Gschmeissner/Science Photo Library/Getty Images; **p780** *t* © Photostock-Israel/Science Photo Library; *b* © Mario Tama/Getty Images; **p786** © Feathercollector/stock.adobe.com; **p787** © Rebius/stock.adobe.com; **p788** © Vera Kuttelvaserova/stock.adobe.com; **p793** © Max Alexander/Science Photo Library; **p795** © Eric/stock.adobe.com; **p799** © Kim/stock.adobe.com; **p802** © NASA's Earth Observatory http://earthobservatory.nasa.gov/Features/WorldOfChange/deforestation.php; **p805** *l* © Pasta Design/stock.adobe.com; *r* © Duelune/stock.adobe.com; **p806** *l* © Kletr/stock.adobe.com; *r* © Martin/stock.adobe.com; **p807** © Elena/stock.adobe.com; **p808** © Meryll/stock.adobe.com; **p809** © Kruwt/stock.adobe.com; **p812** © Somchairakin/stock.adobe.com; **p814** © Photo by Michal Klajban; **p818** © Frank Wagener (https://transylvanian-wood-pastures.eu/); **p819** *t* © Eddie Cloud/stock.adobe.com; *b* © Mati Kose/stock.adobe.com; **p827** © Christinezenino http://creativecommons.org/licenses/by/2.0/http://www.flickr.com/photos/chrissy575/3977247173_130424; **p829** © Incredible Arctic/stock.adobe.com; **p830** © NASA; **p832** *l* © Crisod – Fotolia; *r* © Suzanne Long/Alamy Stock Photo; **p833** © Zhang yongxin/stock.adobe.com; **p834** © Doug Gimesy/Nature Picture Library/Science Photo Library; **p835** © David W Shaw/stock.adobe.com; **p836** © Milan/stock.adobe.com; **p837** *tl* © Jim West/Science Photo Library; *bl* © Photo by Katja Koskenpato; *br* © Photo by Katja Koskenpato

b = bottom, *c* = centre, *l* = left, *r* = right, *t* = top

Text and artwork credits

p97 Figures A.2.3.9 and A.2.3.10 © Sarah Ajee, Where Did Viruses Come From?, STEM in Context, August 05, 2020 © Let's Talk Science; **p113** Figure A3.1.12 © Chris Mooney, "You Share 98.7 Percent of Your DNA With This Sex-Obsessed Ape", February 26, 2014, © Mother Jones and the Foundation for National Progress; **p116** Figure A3.1.15 © EMBL-EBI; **p170** Figure A4.2.12 © C. J. Clegg, *Biology for the IB Diploma 2nd Edition*, Hodder Education Group, 2014; **p175** Figure A4.2.20 © Paul Guinness and Garett Nagle, *Advanced Geography Concepts and Cases*, Hodder Education Group, 1999; **p248** Figure B2.2.3 © GrahamColm; **p252** Figure B2.2.6 © Martin, W. (2005). Archaebacteria (Archaea) and the origin of the eukaryotic nucleus. Current opinion in microbiology, 8(6), 630-637; **p252** Figure B2.2.7 © Geoffrey M. Cooper, Robert E. Hausman, "The Cell A Molecular Approach", 2004, © ASM Press; **p290** Figure B3.1.15 © Lígia T. Bertolino, Robert S. Caine and Julie E. Gray, Impact of Stomatal Density and Morphology on

Water-Use Efficiency in a Changing World, Frontiers Media S.A; **p301** Figure B3.2.7 Adapted from *Report of the WHO Study Group*, Andrew Edmonson and David Druce, Oxford University Press, 1996; **p346** Figure B4.1.8 © Ravaux J, Hamel G, Zbinden M, Tasiemski AA, Boutet I, Léger N, et al. (2013) Thermal Limit for Metazoan Life in Question: In Vivo Heat Tolerance of the Pompeii Worm. PLoS ONE 8(5): e64074; **p348** Figure B4.1.9 © Munns, R., & Tester, M. (2008). Mechanisms of salinity tolerance. Annual review of plant biology, 59, 651; **p355** Figure B4.1.12 © Robert Harding Whittaker, "Communities and Ecosystems", Macmillan Publishers, 1975; **p356** Figure B4.1.13 © Anne Davis and Garrett Nagle, *A-level Geography Topic Master: The Water and Carbon Cycles*, Hodder Education, 2018; **p401** Figure C1.1.20 © Denise R. Ferrier, "Biochemistry", 2014, Wolters Kluwer Health; **p404** Figure C.1.1.12 © Mariya Lobanovska, Giulia Pilla, "Penicillin's Discovery and Antibiotic Resistance: Lessons for the Future?", Yale J Biol Med. 2017 Mar; 90(1): 135–145. © Yale Journal of Biology and Medicine; **p441** Figure C1.3.18 © The Light-Dependent Reactions of Photosynthesis, Rice University; **p463** Figure C2.1.12 © Du, Z., & Lovly, C. M. (2018). Mechanisms of receptor tyrosine kinase activation in cancer. In Molecular Cancer (Vol. 17, Issue 1). Springer Science and Business Media LLC; **p473** Figure C2.2.5 © Bruce M. Koeppen, Bruce A. Stanton, "Berne & Levy Physiology, Updated Edition E-Book", 11 December 2009. © Elsevier Health Sciences; **p474** © De Hert, S., De Baerdemaeker, L., & De Maeseneer, M. (2013). What the phlebologist should know about local anesthetics. In Phlebology: The Journal of Venous Disease (Vol. 29, Issue 7, pp. 428–441). SAGE Publications; **p536** Figure C3.2.17 © Crown copyright; **p537** Figure C3.2.3 © Antibiotic Resistance Threats in the United States, 2013, CDC, U.S. Department of Health and Human Services; **p543** Figure C3.2.21 Adapted from 'Tubercolosis: the global challenge', *Biological Science Review* 8, S.J.G. Kavanagh and D.W. Denning, Hodder Arnold, 1981; **p545** Figure C3.2.22 © U.S. Department of Agriculture; **p547** © Baden, L. R., et al (2021). Efficacy and Safety of the mRNA-1273 SARS-CoV-2 Vaccine. In New England Journal of Medicine (Vol. 384, Issue 5, pp. 403–416). Massachusetts Medical Society; **p560** Figure C4.1.11 © Simpson, M. J., Browning, A. P., Warne, D. J., Maclaren, O. J., & Baker, R. E. (2022). Parameter identifiability and model selection for sigmoid population growth models. In Journal of Theoretical Biology (Vol. 535, p. 110998). Elsevier BV.; **p571** Figure C4.1.27 Data from Joseph Connell's experiment showing the effect of Semibalanus removal from the upper shore and middle/lower shore, from South China Normal University; **p599** Figure C4.2.18 © Scripps Institution of Oceanography at UC San Diego; **p626** Figure D1.2.10 © National Human Genome Research Institute; **p630** Figure D1.2.13 © Pearson Education, "Biology Campbell", Inc, 5 January 2011; **p636** Figure D1.2.18 © John Hines and Craig M. Crews, "The structure and function of the cell's protein-degrading machine", Apr 30, 2017. © The Scientist; **p640** Figure D1.3.2 © Ezria Copper, "A Cure for HIV? Past Research of the CCR5 Delta 32 Mutation", Jul 22, 2022, Owlcation. © The Arena Media Brands; **p676** Figure D2.2.3 © Trisha Chong, "Lions are big. Tigers are bigger. Lion-tiger hybrids are biggest. Why?"; **p745** Figure D3.2.3 © Soo T. Tan, "Finite Mathematics for the Managerial, Life, and Social Sciences", 2014. © Cengage Learning; **p790** Figure D4.1.7 © Endler, J. A. (1980). Natural Selection on Color Patterns in Poecilia reticulata. In Evolution (Vol. 34, Issue 1, p. 76). JSTOR; **p801** Figure D4.2.4 © Nepstad, D. C., Stickler, C. M., Filho, B. S.-, & Merry, F. (2008). Interactions among Amazon land use, forests and climate. In Philosophical Transactions of the Royal Society B: Biological Sciences (Vol. 363, Issue 1498, pp. 1737–1746); **p812** Figure D4.2.16 © Frank Kleissen, Andre Dick Vethaak, "Microplastic Litter in the Dutch Marine Environment Providing facts and analysis for Dutch policymakers concerned with marine microplastic litter"; **p825** © Andrew Davis, Garrett Nagle, *Environmental Systems and Societies for the IB Diploma Study and Revision Guide*, Hodder Education Group, 2017; **p828** Figure D4.3.7 © Vygodskaya, N. N., et al (2007). Ecosystems and climate interactions in the boreal zone of northern Eurasia. In Environmental Research Letters (Vol. 2, Issue 4, p. 045033). IOP Publishing; **p829** Data on arctic sea extents from: https://climate.nasa.gov/vital-signs/arctic-sea-ice/; data on Antarctica ice loss from https://climate.nasa.gov/vital-signs/ice-sheets/; **p830** Figure D4.3.9 © NASA/JPL; **p831** Figure D4.3.10 © Freeman, B. G., & Class Freeman, A. M. (2014). Rapid upslope shifts…, In Proceedings of the National Academy of Sciences (Vol. 111, Issue 12, pp. 4490–4494); **p824** Figure D4.3.2 © Scripps Institution of Oceanography at UC San Diego for the data 1970-2013.

Index

abiotic environment 164
abiotic factors 341, 344–8, 579
    effect on species distribution 344–53
    effect on terrestrial biome distribution 354–5
    role in natural selection 783
ABO blood groups 195–6, 734–5
abscisic acid (ABA) 513
absorption spectra 431–2, 433
accessory pigments 427
accuracy 40, 510
acetylcholine (ACh) 240, 456, 477
    transmembrane receptors 459
acrosome reaction (capacitation) 716, 717
actin 327–30
action potentials 471
    depolarization and repolarization 479–80
    frequency of 481
    oscilloscope traces 482
    propagation of 482
    propagation velocity 472–4
    saltatory conduction 483
    threshold potential 481
action spectra 432–3
activation energy 395–6
active immunity 541
active sites 383–5
active transport 230–2
    indirect 242
adaptations
    blood vessels 296, 297–8, 299
    foetal and adult haemoglobin 293
    to habitats 342–4, 359–62
    leaves 285–6, 290
    lungs 277
    to modes of nutrition 372–7
    specialized cells 265–72
    for swimming 339–40
    xylem 308
adaptive immune system 522, 523–5
adaptive radiation 144–5, 152–3
adenine 15–16
adenylyl cyclase /adenylate cyclase 462
adhesion, water 7–8, 303
adipose tissue 201–2
    brown 767
ADP (adenosine diphosphate) 231–2, 407
adrenal glands 494
adrenaline *see* epinephrine
adrenergic receptors 461–2
adult stem cells 257
aerobic respiration 365, 408
    see also respiration

afforestation 833
agglutination 196
agouti (*Dasyptocta* species) 805
agriculture
    sustainability 807–8
    see also crops
agrochemicals 808
agroforestry 834
AIDS (acquired immunodeficiency syndrome) 529, 531, 539
    see also HIV
air pollution 675
albedo 826
albinism 741–2
alcoholic fermentation 416–18
alcohols 184
aldehydes 184
algae 428
alien species 172, 173, 568–9
alleles 21–2, 103, 111, 732
    consequences of natural selection 793
    dominant and recessive 728
    incomplete dominance and codominance 736–7
    multiple 734–5
allelopathy 576–7
allopatric speciation 152
allosteric proteins 293
allosteric regulators 399–401
a cells, pancreatic islets 762–3
a-helices 213, 215
alternative hypothesis 757
alternative RNA splicing 629–30
altruism 564
alveoli 266–7, 276–9
Amazon 598, 801–2
amino acids 205–6
    activation of 630–2
    chain formation 207
    chemical diversity 211
    effect on protein structure 216
    enantiomers 206, 213
    essential and non-essential 208–9
    isoleucine 403
    recycling 636
Amoeba 50, 51, 688
amphipathic molecules 40, 203
amylase 187, 387
amylopectin 191
amylose 191
anabolism 382–3
anaerobic respiration 364–5, 408, 416–18
analogous structures 145

anaphase
    meiosis 711
    mitosis 653, 654
animal cells 63–5, 68
    effects of osmosis 688
    structure 67–73, 75–6, 80
antagonistic muscle pairs 331–2, 334–5, 337–8
anther 703
anthropogenic species extinctions 162–4
antibiotic resistance 404, 534–7
antibiotics 532, 577
    discovery of penicillin 532–3
    mechanisms of action 533–4
    quest for new drugs 536–7
antibodies 195–6, 524–5, 527, 528
    monoclonal 719
anticodons 619
antidiuretic hormone (ADH) 775–6
antigenic drift 99
antigenic shift 99
antigen presentation 526
antigens 195–6, 525
antigen-specific B-cells 526
antiviral drugs 542
aorta 297
apical meristems 662–3
apoplast pathway 305, 306, 307
apoptosis 71, 459
aquaporins 228–9, 234, 775
archaea 137, 368–9
arrested succession 818–19
arteries 296–7
    adaptations 297–8
    atherosclerosis 300
arterioles 298
artificial classification 145
artificial selection 141–4, 797
asexual reproduction 693–4
assimilation 495
asteroids, water content 11
atherosclerosis 300–2, 402
atmosphere
    carbon dioxide concentration 597–601, 822–4, 826
    of early Earth 34–5
    importance of photosynthesis 426–7
ATP (adenosine triphosphate) 231–2, 397, 406–7
    amino acid activation 631, 632
    production in thylakoids 442
    synthesis in the mitochondrion 421–2
    yield from aerobic respiration 414–15, 419, 420–1

ATP synthase (ATPase) 248, 421–3
atria (singular: atrium) 315, 316
atrioventricular node 320–1
atrioventricular valves 316
autoimmune diseases 765
autoinducers 453
autolysis 71
autonomic nervous system (ANS) 508
autosomal gene linkage 751–2
autosomal genes 726
autotrophs 48, 366, 584–5
auxin (indoleacetic acid, IAA) 513, 513–14
    actions of 515, 516
    movement and control 514–15
axons 468, 469
B-lymphocytes (B-cells) 523, 524–8
bacteria 50, 51
    antibiotic resistance 404, 534–7
    cell structure 65–7
    comparison with viruses 89
    CRISPR sequences 644–6
    gene expression 678–80
    genetic transfer 118–20
    quorum sensing 452–3
    Rhizobium 566
    species concept 120, 162
bacteriophage lambda 94–5, 96
bacteriophages 29, 89, 90, 91, 92, 119
    determination of particle concentration 95–6
    Hershey and Chase experiment 30–1
ball-and-socket joints 335, 336
bark beetles, impact of climate change 836–7
barnacles, competitive exclusion 569–71
baroreceptors 506, 507
base substitutions 638
behavioural adaptations 376
behavioural isolation 152
belt transects 349
Benedict's test 192
Bengal tiger (*Panthera tigris tigris*) 178
benign tumours 668
b cells, pancreatic islets 762–3
b-sheets 213, 215
bias 552
bicuspid valve 316
binary fission 118
binomial system 105–6
bioaccumulation 810, 813
biochemical oxygen demand (BOD) 809–10
biodiversity 156–7, 799
    conservation approaches 174–8
    current and past levels 160–1
    ecosystem loss 165–7
    international conventions 178–9
    Simpson's reciprocal index 157–9
biodiversity crisis 168–73
biofilms 453
bioinformatics 118

biological species concept 106–7
bioluminescence 453
biomagnification 808, 810–12, 813
biomes 356–9
    influence of climate 354–5, 830–1
biotic factors 341, 344, 579
bivalents 656
black-throated loon (*Gavia arctica*) 10, 11
bladder 769
blastocyst 257, 663, 718
blood, transport of materials 495
blood-clotting mechanism 520–1
blood glucose regulation 466, 761–3
    diabetes 764–5
blood groups 195–6
blood plasma 312, 522, 772, 777
blood pressure 297
blood supply to organs, regulation of 777–8
blowholes 339–40
blubber 11, 201, 339
body temperature regulation 766–8
Bohr shift 294
bone marrow 258
boreal forest (taiga) 357
    carbon storage 827–8
bottom-up control 576
Bowman's capsule 769, 770, 771–2
box-and-whisker plots 746
brachydactyly 742
brain 487–9, 496–7, 498
    cerebellum 502
    hypothalamus 494, 504–5, 698, 710, 766, 775, 776
    medulla oblongata 506, 507, 508
    respiratory centre 507, 508
Brazil nut tree (*Bertholletia excelsa*) 805
bread making 417
brewing 417
bronchi and bronchioles 276
broomrape (*Orobanche* species) 584
brown adipose tissue 767
brush border 265–6
bulk transport 238–9
buoyancy 9, 11
cacti 360
calcium ions
    role in muscle contraction 329–30
    role in signalling 454
Calvin, Melvin 448
Calvin cycle 250, 445–6
cambium 309, 662
camels 360
camouflage 374
cancer
    development of 668
    role of gene mutations 667
    types of 667
capillaries 296, 310–12
capillary action 8

capillary tubes 8
capsaicin 486–7
capsid 92
capsomeres 92
captive breeding programmes 177, 178
capture–mark–release–recapture method 555–6
carbohydrates 189
    monosaccharides 189–91
    see also glucose
    polysaccharides 191–4
    see also cellulose; glycogen; starch
    as respiratory substrates 407, 424
carbon, chemical properties 181–4
carbon compounds, origin of 38–40
carbon cycle 594–7
carbon dioxide
    effect on oxygen dissociation curve 293, 294
    as a greenhouse gas 821, 824
    Keeling curve 599–601
    ocean acidification 353
    release during combustion 597–8
    solubility in water 8
    use in photosynthesis 426, 427–8
carbon dioxide, atmospheric 597–601, 822–4
    positive feedback cycles 826
carbon dioxide concentration, effect on photosynthesis rate 436
carbon dioxide enrichment experiments 438
carbon fluxes 596–7, 601
carbon footprints 808
carbon sequestration 833–4
carbon sinks and carbon sources 595–6
carbon storage, boreal forest (taiga) 827–8
carboxylic acids 184
cardiac cycle 318–19
cardiac muscle 268–9, 315
Caribbean monk seal (*Neomonachus tropicalis*) 163–4
carnivores 366
carpel 703
carriers 733
carrying capacity 557, 558, 564–5
cartilage 335
Cas9 enzyme 644, 645, 646
Casparian strip 306, 307, 310
catabolism 382–3
catalase 391–3, 399–401
catalysts 380
CCR5 gene 639–40
cell adaptations
    cardiac and striated muscle fibres 268–70
    to increase surface area-to-volume ratio 265–6
    in lungs 266–7
    sperm and egg cells 271–2
cell-adhesion molecules (CAMs) 243
CellCams 56
cell–cell recognition 195

cell cycle   256, 664
   control of   665–7
   interphase   665
cell differentiation   85–6
cell division   252, 648
   condensation and movement of chromosomes   652–3
   cytokinesis   649–50
   interphase   651
   meiosis   656–8
   mitosis   653–6
   roles of mitosis and meiosis   650–1
cell junctions   243
cell membrane *see* plasma membrane
cell proliferation   663–4
cell respiration *see* respiration
cells   36, 49
   animal and plant   63–5
   common features   37
   microscopic examination   52–6
   origin of   37
cell size   51, 260–2
   calculation of   58
   measurement   54–5
   surface area-to-volume ratios   262–4
cell structure   63
   atypical structures   78–9
   drawing annotated diagrams   80–2
   eukaryotic cells   67–73
   fungi   77
   organelles not found in animal cells   76–7
   organelles not found in plant cells   75–6
   prokaryotic cells   65–7
cell theory   50
cellulose   8, 64, 77, 193–4
cell ultrastructure   60
cell walls   64, 77, 689
   bacteria   67
central dogma   42, 100
central nervous system (CNS)   467, 468, 487–9, 496–9, 504–5
   see also brain
central tendency, measures of   745, 794
centrifugation   245
centrioles   51, 68, 75, 653, 654
centromeres   22, 110, 111, 651
centrosome   75
cerebellum   502
cerebral hemispheres   487–8, 498
cervix   697
channel proteins   229, 234
Chargaff, Erwin   31, 33
Chase, Martha   29–31
checkpoints   665–6
chemical defences   375
chemical signalling   450–2
   in animals   453–4
   cellular response   458–9
   intracellular receptors   456, 457–8, 464
   ligand–receptor interaction   458

   local and long-distance   455–6
   quorum sensing   452–3
   regulation   465–6
   signal transduction   458
   transmembrane receptors   456–7, 459–64
   see also hormones; neurotransmitters
chemiosmosis   421–2, 442
chemoautotrophs   368, 585
chemoreceptors   506–7
chemosynthesis   47, 368, 369
chiasmata (singular:chiasma)   656, 660
   recombinants   753
childbed fever (puerperal septicaemia)   519
childbirth   723
chi-squared (c^2) test   572–4, 757–9
chitin   77, 334
Chlamydomonas   50, 51
Chlorella   74
chlorophyll   250, 425, 427
   absorption spectrum   431–2, 433
   action spectrum   432–3
   photoactivation   440–1
chloroplasts   51, 59, 64, 68, 76, 249–50
   DNA   84, 85
   endosymbiotic theory   83–5
cholera   518
cholesterol   203
   and atherosclerosis   301, 402
   and membrane fluidity   237–8
chromatids   111, 651
chromatin   68, 652, 665
chromatograms   429–31
chromatography   429–31
chromosome numbers   110, 120
chromosome replication   111
chromosomes   21–2, 68, 110–11, 725–6
   condensation and movement of   652–3
   crossing over   656, 657, 660
   humans compared with chimpanzees   113
   karyotypes   111–12
   nucleosome structure   26, 28–9
   random orientation in meiosis   660, 661
cilia (singular: cilium)   76, 520
circadian rhythms   502–3
circulatory systems   295, 313–14
   arteries   296–8
   atherosclerosis   300
   capillaries   296
   human   315
   veins   296–7, 299
cisternae   254
citizen science   168
citric acid cycle *see* Krebs cycle
cladistics   128–9
   figwort family reclassification   135–6
   limitations   129–30
   sequence differences   130
cladograms   128, 129
   construction of   131–3

   interpretation   133–4
classes   126
classification   104–6, 125–7, 162
   artificial   145
   cladistics   128–36
   domains   137
   natural (phylogenetic)   144
   of viruses   89, 90
clathrin   255
climate   820
climate change   799, 820, 823
   boreal forest carbon storage   827–8
   carbon sequestration approaches   833–4
   effect on coral reefs   832
   evolution as a consequence   837
   impact on insect life cycles   836–7
   ocean current changes   829–30
   phenological effects   835–6
   polar ice cover   829
   positive feedback cycles   825–7
   upslope range shifts   830–1
climax communities   814, 818
climographs   354–5
closed circulation   314
closed systems   579
cocaine   484
coding strand, DNA   616–17
codominance   736
codons   17, 617, 620
   deduction of   632
   mRNA   621
coefficient of determination (R^2)   474–6
coenzymes   413
cohesion, water   5–7, 303
cohesion–tension theory   303
collagen   219–20, 221
collecting ducts   769, 770
   water reabsorption   775–6
colorimetry   232–3
communities   164, 341, 562
   climax communities   814, 818
   interspecific relationships   564–8
   intraspecific relationships   563–4
   predator–prey relationships   574–5
companion cells   323, 324
comparative serology   131–2
competition   563, 564
   allelopathy   576–7
   between endemic and invasive species   568–9
   interspecific   569, 569–70
   intraspecific   783–4
   role in natural selection   782–3
competitive exclusion   378–9, 569–71
competitive inhibitors   398–401, 401
complementary base pairing   23, 617
   tRNA and mRNA   619
computer modelling   4, 101
   see also simulations

Index

concentration gradients 275
   and transpiration 288
condensation reactions 15–16, 18, 186–7, 198, 207
confirmation bias 4
conjugated proteins 217–18
conjugation 119
Connell, Joseph 569–70
consciousness 488–9
conservation
   associated issues 180
   ex situ 177–8
   in situ 174–7
conserved sequences 24–-5, 647
consumers 366, 581, 586
   energy flow 589
continuous variation 103, 743
contractile vacuoles 688
convergent evolution 145–6
cooperation 563–4
cooperative binding 292
coral reefs 166–7, 353, 567–8, 832
coronary arteries 315, 316
coronary heart disease 300–2, 724
coronaviruses 91
   see also SARS-CoV-2
corpus luteum 698, 699, 700, 715, 716, 719
correlation 301–2, 474–6
cortical reaction 716
cotransporters 242
covalent bonds 3, 182
COVID-19 14, 100, 539–40, 546, 547
   see also SARS-CoV-2
crenation 688
Crick, Francis 20, 23, 25–6
CRISPR sequences 644–6
cristae 69, 248, 249
crops
   land use 171–2
   polyploidy 154–5
   salt tolerance 347–8
   selective breeding 143–4
crossing over 656, 657, 660, 782
   recombinants 753
cross-pollination 704–5
cryogenic electron microscopy 61, 220–1
cyanobacteria 428
cyclical metabolic pathways 397–8
cyclical succession 817
cyclic AMP (cyclic adenosine monophosphate) 461
cyclic photophosphorylation 443
cyclins 665–6
cystic fibrosis 453
cytokines 454
cytokinesis 649–50, 654
cytokinin 516
cytoplasm 63
cytosine 15–16
cytoskeleton 72–3

Darwin, Charles 102, 128, 139, 141–2, 780–1, 787
databases 114, 115, 116, 792
daughter cells 648
Dawkins, Richard 793
DDT (dichlorodiphenyltrichloroethane) 810–12
decarboxylation 418
decay 583–4
decomposers 582–3, 601
deductive reasoning 51–2
deforestation 170–1, 801–2, 824
degeneracy, genetic code 620, 638
degrees of freedom 759
deletion mutations 637
   CCR5 gene 639–40
denaturation of proteins 210, 387
dendrites 467, 469
dendrons 468, 469
density 9
   water and ice 10
density-dependent factors 557–8, 575
dentition 369–72
depolarization 479–80
deserts 359–60
detrivores 583–4
diabetes 764–5
diaphragm 276, 280–1
diastole 317, 318–19
dichotomous keys 121–3
diet 201
   amino acid requirements 208–9
   and atherosclerosis 301
   lipid content 402
   relationship to dentition 369–72
   vegetarian 592
differentiation 63
diffusion 7, 225–6
   determinants of rate 274
   facilitated 229
digital microscopy 82
dihybrid crosses 748–50
   in Drosophila 755–6
   statistical tests 757–9
dimerisation 463
diploid cells 110, 650, 651, 726
dipterocarp forests 165–6
directional selection 793, 794–5
disaccharides 186, 187
   test for 192
discontinuous variation 103
diseases 172, 517–19
disruptive selection 793, 795
distal convoluted tubule (DCT) 770
disulfide bridges (S–S bridges) 215, 219
divergent evolution 146
division of labour 86
dizygotic twins 677
DNA (deoxyribonucleic acid) 14–15, 51
   base sequence stability 618
   coding and template strands 616–17

comparison with RNA 22
complementary base pairing 23
data storage capacity 23–4
directionality 25, 27
and evolution 140
gel electrophoresis 606–8
genetic code 17, 23, 24
Hershey and Chase experiment 29–31
methylation 673–4
in mitochondria and chloroplasts 84, 85
non-coding sequences 626–7
nucleosome structure 26, 28–9
polymerase chain reaction 605–6
promoters 625–6
purine to pyrimidine bonding 25–6, 27
DNA barcoding 124
DNA hybridization 133
DNA ligase 612, 613
DNA polymerase 604, 613, 640
   directionality 611
DNA polymerase III 611, 613
DNA primers 606
DNA profiling (genetic fingerprinting) 608–10
DNA proofreading 614, 640, 641
DNA replication 602–3, 651
   enzyme functions 613
   helicase and DNA polymerase 604
   leading and lagging strands 611–13
DNA structure
   base pairing 23, 31
   discovery of 20–1, 23, 25–6
   double helix 19
   nucleotides 15–16
   polynucleotides 16–17
DNA viruses 88, 90–1
domains 127, 137
domestication of animals 142–3
dominant alleles 728, 732
dopamine 484
double bonds 183–4
   in lipids 199–200
double circulation 314, 315
double helix, DNA 19
Down syndrome 659
Drosophila melanogaster (fruit fly) 754–6
droughts 827
duckweed (Lemna species) 562
early Earth, conditions on 34–5
Earth Summit 179
ecological species concept 107
ecosystem loss 170
   dipterocarp forests 165–6
   Great Barrier Reef 166–7
ecosystem restoration 176–7
ecosystems 164, 579
   pollution 809–13
   rewilding 813–14
   succession 814–19
   sunlight as energy source 579–80

ecosystem stability 798
    keystone species 805
    mesocosm investigations 802–4
    requirements for 799
    sustainability 806–8
    tipping points 799–800
EDGE species 179
effector proteins 451, 452
egg cells 261, 271–2
    epigenetic reprogramming 675
    fertilization 716–17
electrochemical gradients 479
electron carriers 413
electron microscopy 59, 60–2, 80
    cryogenic 220–1
    drawing annotated diagrams 80–1
electron transport chain (ETC) 412, 420–1, 423
electrophoresis 605, 606–8
embryonic stem cells 257
embryo 256, 663, 721
emergent properties 491
empirical data 59
enantiomers 206
endangered species 177–8, 179
endergonic reactions 396–7
Endler, John 788–90
endocrine glands 494
endocrine system 493–5
    control of 504–5
    see also hormones
endocytosis 238–9, 255
endodermis 306, 307, 310
endometrium 697, 700
endoplasmic reticulum 68, 70, 81, 254
endoskeletons 334
endosymbiotic theory, eukaryotic cells 83–5
endothelins 297
endothelium 296
end-product inhibition 402–3
energy flow 580–1, 588, 589
energy pyramids 586–8
energy release 585
energy sources 579–80
enhanced greenhouse effect 824
enhancers 671
enteric nervous system (ENS) 508–9
envelopes, viral 93
environment 341
    effects on gene expression 674–5, 732
enzyme inhibitors 398–9
    allosteric regulators 399–401
    competitive 401
    end-product inhibition 402–3
    mechanism-based inhibition 403–4
enzymes 9, 70, 380–1
    denaturation 387
    effect of pH 390–1
    effect of substrate concentration 391–3
    effect of temperature 387–9

effect on activation energy 395
immobilization techniques 386
induced-fit binding 384–5
intracellular and extracellular 396
role in metabolism 381
specificity 382, 384
structure 383
enzyme–substrate (ES) complex 383
epidemiological studies 724
epidemiology 542
epidermis, plants 283, 284, 309, 310
epididymis 712, 713, 714
epigenesis 672
epigenetic inheritance 140, 674–6
epigenetic tags 673–5
epinephrine (adrenaline) 461–3, 503–4
epiphytes 376–7
equilibrium
    ecosystems 799
    genetic 797
error bars 686–7
errors 549
Escherichia coli 50, 51, 66–7
essential amino acids 208–9, 403
ester bonds 198
esters 184
ethanol, production from pyruvate 416–17
ethical issues
    conservation 180
    diet 592
    experiments and research 410, 411, 497
    genetics 20–1, 646–7
    stem cells 258
ethylene 466, 516
eubacteria 137
eukarya 137
eukaryotic cells 65
    autogenous theories 84
    endosymbiotic theory 83–5
    structure 67–73
eutrophication 804, 808–9
evaluation 40, 302
evolution 37, 41, 109, 138–40, 779–80
    of antibiotic resistance 534–5
    convergent 145–7
    directional, disruptive and stabilizing selection 793–5
    evidence for 141–5
    guppy populations 788–90
    last universal common ancestor 43–8
    of multicellularity 87
    natural selection 782–5
    in response to climate change 837
    sexual selection 786–8
    viruses 98–100
    see also speciation
evolutionary species concept 107
exchange transporters 241–2
excretion 768

exercise, redistribution of blood flow 777–8
exergonic reactions 396–7
exocrine glands 494
exocytosis 238–9
exons 246
exoskeletons 334
experimental techniques ix
expiration 280
expiratory reserve 282
exponential growth 558–9, 560
ex situ conservation 177–8
extinctions 139, 149
    anthropogenic 162–4
extracellular enzymes 396
extracellular matrix (ECM) 243
extraterrestrial life 12–13, 36
extremophiles 47, 48
eyes, position of 373
F1 and F2 generations 727, 728
FACE experiments 438
facilitated diffusion 229
facultative anaerobes 365
FAD (flavin adenine dinucleotide) 413
Falkland Islands wolf (*Dusicyon australis*) 164
falsification 32–3, 100, 114
families 126
fats *see* lipids
fatty acids 197, 198
    and membrane fluidity 237
    omega-3 201
    saturated and unsaturated 199–201
feedback 402–3, 465–6, 575
    see also negative feedback; positive feedback
feeding, carbon cycle 594–5
female reproductive system 696–7
    menstrual cycle 698–701
fertilization 120, 271, 694, 695
    flowering plants 702–3
    humans 701, 716–17
fertilizer use 808
fibroblasts 664
fibrous proteins 221
field experiments 438
figwort family (Scrophulariaceae) 135–6
filament 703
fishing, sustainability 806–7
fitness 783
flaccidity 284
flagella (singular: flagellum) 66, 76
Fleming, Alexander 532
flowering plants
    cross-pollination 704–5
    insect-pollinated 703–4
    reproductive structures 702–3
    seed dispersal 706–7
    seed germination 707–9
    seeds 706
    self-incompatibility 705

Index

fluid mosaic model 216, 234, 236
fluorescence microscopy 62–3
flying lizards (*Draco* species) 361
foetal haemoglobin 293
foetus 721
follicle-stimulating hormone (FSH) 698, 699, 700, 710
    role in spermatogenesis 714
food chains 580–1, 586
    biomagnification 810–12
    energy transfers 589–91
    heat loss 590–1
    number of trophic levels 591, 593
    and world hunger 591–2
food webs 581–2, 586
forensic investigations 610
forest fires 827
forest regeneration 834
forests 356–7
    see also rainforests; boreal forest
fossil fuels 597–8, 824
fossils 46, 47, 161
founder effect 109
frameshift mutations 637
Franklin, Rosalind 21
freeze etching 62
fruit ripening 466, 516
functional groups 184
fundamental niches 378
fungi, antibiotic production 577
Galápagos finches 152–3
gametes 120, 694, 725
    cell size 261
gametogenesis 710–12
    oogenesis 714–16
    spermatogenesis 712–14
gap junctions 243, 269, 456
gas exchange 273–4
    concentration gradients 275
    lungs 275–83
    plants 283–6
gas-exchange surfaces 274
gated ion channels 239–41, 479–80
gel electrophoresis 605, 606–8
gene editing 20
gene expression 464, 618, 670–3
    environmental effects 674–5
    epigenetic inheritance 674–6
    external factors 678–80
    imprinting 675–6
    methylation 673–4
    monozygotic twin studies 677–8
gene knockout (KO) technique 643
gene linkage 751–2
gene mutations 637–8
    see also mutations
gene pools 108, 148, 743, 790–1
    consequences of natural selection 793
    Hardy–Weinberg equation 796–7

gene probes 608
genes 21–2, 24, 86, 111, 615, 626
    conserved sequences 24–5, 647
    loci of 751
genetic code 14, 17, 23, 44, 209, 616–17, 620
    conserved sequences 24–5
    deduction of first codon 632
    mRNA codons 621
genetic databases 114, 115, 116, 792
genetic disorders
    albinism 741–2
    brachydactyly 742
    cystic fibrosis 453
    Down syndrome 659
    phenylketonuria 733
    sex-linked conditions 739–40
    sickle cell anaemia 622–3, 638, 646
genetic diversity 799
genetic drift 109
genetic fingerprinting 20
genetic testing 643
genome editing 646
genomes 86, 115, 672
genome sequencing 118
    Human Genome Project 117, 246
genome sizes 24, 116
genotype 672, 730
genus 105–6, 126
geographical isolation 150–1
germination 707–9
germ line mutations 642, 782
germ plasm conservation 177
germ theory 519
gestation 721
    see also pregnancy
gibberellins (GAs) 513, 708
gibbons (*Hylobates* species) 361
global warming 824
    positive feedback cycles 825–7
globular proteins 221–2
    enzymes 383
glomerular filtrate 771, 772, 777
glomeruli (singular: glomerulus) 769, 770
    ultrafiltration 771–2
glucagon 763
gluconeogenesis 463
glucose 187, 189–91
    in aerobic respiration 408, 412
    glycolysis 414–15
    in photosynthesis 426, 427–8, 446
    regulation of blood levels 466, 761–3
glycerate-3-phosphate 445, 446
glycerol 197
glycocalyx 195, 234
glycogen 192–3, 763
glycogenesis 464
glycolipids 234
glycolysis 248, 412, 414–15

glycophytes 347
glycoproteins 77, 93, 195, 234, 525
glycosidic linkage 186
golden lion tamarin (*Leontopithecus rosalia*) 177
'Goldilocks zone' 12
Golgi apparatus 68, 71, 72, 81, 254
gonadotropin-releasing hormone (GnRH) 710
goniometers 337
G protein 459–60
G protein-coupled receptor (GPCR) 460–1
grana (singular: granum) 76, 249–50
graphs x
    logarithmic 561
grasslands 358
grazing 819
Great Barrier Reef 166–7
greenhouse effect 34, 820–2
    enhanced 824
greenhouse gases 34, 821, 824–5
gross primary productivity (GPP) 592
gross secondary productivity (GSP) 593
GTP (guanosine triphosphate) 459–60
guanine 15–16
guard cells 284, 285
guppies (*Poecilia reticulata*) 788–90
habitat loss 170–1
habitats 341–2
    abiotic factors 344–8
    adaptations to 342–4
    upslope rage shifts 830–1
haemoglobin 217–18, 275, 292
    oxygen dissociation curves 292–4
    sickle cell anaemia 622–3
haemophilia 739–40
hair follicles 258–9
halophytes 347
haploid cells 110, 650, 651
    production in meiosis 656–8
Hardy–Weinberg equation 796–7
heart 315–17
    cardiac cycle 318–19
heart attack (myocardial infarction) 300
heartbeat, origin and control 320–1
heart rate control 506–7
heather burning 819
heat loss 590–1
helicase 604, 613
helper T-cells 526, 527
herbivores 366
    adaptations 372
herbivory 564
    plant resistance to 372–3
herd immunity 544–5
heredity *see* inheritance
heritable traits 784
Hershey, Alfred 29–31
heterotrophs 366–7, 585
heterozygous condition 111, 730–2

hierarchy of subsystems 492–3
hinge joints 335, 336
histograms 747
histones 26, 28, 652
HIV (human immnodeficiency virus) 88, 99–100, 529–31, 539
    and CCR5 mutation 639–40
HLA system 792
holozoic nutrition 366–7
homeopathy 4
homeostasis 504, 760
    negative feedback loops 761–3
    osmoregulation 768–9, 775–6
    thermoregulation 766–8
homeotic mutations 103
Hominidae, diet and dentition 369–72
homologous chromosomes 22, 110
homologous structures 127, 144–5
homozygous condition 111, 730–2
hormone replacement therapy (HRT) 724
hormones 453, 454, 456, 493–5
    antidiuretic hormone 775–6
    blood glucose regulation 762–3
    in childbirth 723
    effect on gene expression 678
    epinephrine 461–3, 503–4
    from hypothalamus and pituitary 504–5
    insulin 218–19, 221–2, 463–4
    IVF treatment 702
    melatonin 502
    and menstrual cycle 698–701
    phytohormones 512–16
    in pregnancy 719, 722, 723
    puberty 709–10
    role in spermatogenesis 714
    sex hormones 204, 464–5
    thyroxin 766
hot deserts 359–60
HTT gene 638–9
human chorionic gonadotropin (hCG) 719, 722, 723
Human Genome Project 117, 246
human population growth 169, 171, 591
Huntington's disease 638–9, 732
hybridization 120
    barriers to 153–4
    polyploidy 154–5
hydrogen bonds 3–4, 5
    in ice 10
    in proteins 213, 215, 216
    role in transcription 617
    and surface tension 5–6
hydrolysis reactions 186, 187–8
hydrophilic substances 7, 8–9
hydrophobic substances 8
hydrothermal vents 47–8, 346, 580
hyperglycaemia 762
hypertonic solutions 682–3
hyphae (singular: hypha) 78
hypoglycaemia 762

hypothalamus 494, 504–5, 698, 710
    osmoregulation 775, 776
    thermoregulation 766
hypotheses 37, 45–6, 114, 132, 572
hypotonic solutions 682–3
ice
    density 10
    thermal conductivity 10
ice caps 826–7, 829
ice core studies 821–2
immunity 522, 526–8, 529
    active and passive 541
    adaptive 523–5
    herd immunity 544–5
    innate 522–3
    recognition of 'self' and 'non-self' 525
immunization see vaccines
immunofluorescence 62, 63
immunoglobulins 524
implantation 718
imprinting 675–6
inbreeding 743
incomplete dominance 736–7
independent assortment, Law of 748–50
indirect active transport 242
induced-fit binding 384–5
inducers 679
inductive reasoning 32, 51–2
infections 517
infectious (communicable) diseases 517–19
    zoonoses 537–40
infertility 701
influenza (flu) 98–9, 100
inheritance
    autosomal gene linkage 751–2
    dihybrid crosses 748–50
    in Drosophila 754–6
    epigenetic 674–6
    genotype 730
    heritable traits 784
    incomplete dominance and codominance 736–7
    loci of human genes 751
    Mendel's experiments 726–30, 747–8
    Mendel's laws of 730, 748–50
    multiple alleles 734–5
    pedigree charts 741–2
    phenotype 731
    polygenic 743–4
    recombinants 751–3
    sex chromosomes 737–9
    sex-linked conditions 739–40
inhibitory synapses 484–5
innate immune system 522–3
inorganic compounds 185
inquiry process 1
insect-pollinated flowers 703–4
insertion mutations 637
    HTT gene 638–9

in situ conservation 174–6
inspiration 280
inspiratory reserve 282
insulin 218–19, 221–2, 466, 494, 635, 763
    and diabetes 764
insulin signalling 463–4
integral membrane proteins 216, 226, 227, 234
integration of body systems 491–3
    central nervous system 496–9
    endocrine system 493–5
    heart rate control 506–7
    hormones 503–5
    transport in blood system 495
    ventilation rate control 507–8
interaction 381
intercalated discs 268, 269
intercostal muscles 276, 280–1, 337–8
interdependence 381
interphase 651, 654, 665
interpolation 390
interquartile range (IQR) 746
interspecific relationships 563, 564–5
    competition 568–71
    mutualism 566–8
intracellular enzymes 396
intracellular receptors 456, 457–8, 464
intraspecific relationships 563–4
    competition 783–4
intravenous drips 690
introns 246, 626, 627, 628
invagination 255, 265–6
invasive species 172–3, 568–9
island biogeography theory 174–5
islets of Langerhans 762–3
isolated populations 149–52
    allele frequencies 791
isoleucine 403
isomers 190–1
isotonic solutions 682
    medical applications 690
IVF (in vitro fertilization) 702
Japanese encephalitis 539
Jenner, Edward 89
joints 335–7
kangaroo rats (Dipodomys species) 360
karyotypes 111–12, 660, 737
Keeling curve 599–601
ketones 184
keystone species 805
kidneys 769–70
    osmoregulation 775–6
    urine formation 771–3
    water conservation, loop of Henle 773–5
kinases 462, 665–6
kingdoms 126, 127
Krebs cycle (citric acid cycle) 249, 412, 418–19
lac operon 679
lactic acid fermentation 408, 416–18

Index

lactose intolerance 386
lagging strand, DNA replication 612–13
Lamarck, Jean-Baptiste 139, 784
landfast ice 829
land use 171–2
last universal common ancestor (LUCA) 43–8
lateral meristems 662–3
leaching 808
leading strand, DNA replication 612
leafcutter ants (*Atta* species) 564
leaves
    gas exchange 285–6
    stomatal density 290
    structure 283–6
lever systems 335
life, definition of 36
life, origin of 34–5
    estimated date 46
    formation of cells 37
    hypotheses 45–6
    last universal common ancestor 43–8
    origin of carbon compounds 38–40
    RNA world hypothesis 41–3
    spontaneous formation of vesicles 40–1
    theories 38
    timeline 43
life cycles, impact of climate change 836–7
life processes 73
    unicellular organisms 74
ligaments 335
ligand–receptor interaction 458
ligands 450–2, 453–4
    receptor proteins 456–8
light 427
light-dependent reactions, photosynthesis 439–44
light-harvesting complexes (LHCs) 440
light-independent reactions, photosynthesis 439, 445–8
light intensity, effect on photosynthesis rate 436
lignin 304, 369
limiting factors 344–8, 557
limits of tolerance 345
Lincoln index 555–6
linear metabolic pathways 397–8
line transects 349
link reaction 412, 418
Linnaeus, Carl 104, 105
lionfish (*Pterois volitans*) 173
lipids 197–9
    adipose tissue 201–2
    in diet 201
    as respiratory substrates 420, 424
    saturated and unsaturated fatty acids 199–201
    see also phospholipids
Lister, Joseph 519
local chemical signalling 455–6

loci (singular: locus) 21–2, 111
    human genes 751
lock-and-key model, enzymes 385
locomotion 338
    reasons for 338
    swimming 339–40
logarithmic graphs 561
logging 806
logistic sigmoid function 561
London forces 215
long-distance chemical signalling 456
loops of Henle 769, 770, 773–5
lungs 266–7
    adaptations 277
    efficiency 279
    gas exchange 277–8
    structure 275–6, 278
    ventilation of 279–81
lung volumes 281–3
luteinizing hormone (LH) 698, 699, 700, 710
    role in spermatogenesis 714
lymphatic system 312–13
lymph nodes 312, 313
lymphocytes 522, 523–5
    immune response 526–8
lysogenic viral life cycle 95
lysosomes 51, 68, 71–2, 81, 247
lytic viral life cycle 94
macromolecules 181
macrophages 267, 526
magnification 57–8
major histocompatibility complex (MHC) 525
malaria 118
male reproductive system 696
malignant tumours 668
maltose 187
mangrove swamps 343–4
Margulis, Lynn 83, 84
marram grass 342–3
mass extinctions 139, 161, 162
mass flow 295
mathematical tools x
matrix, mitochondrial 69
maximum sustainable yield (MSY) 807
mean 553, 745, 794
measurements 59, 291
    microscopy 54–5
mechanism-based inhibition 403–4
median 745, 794
medulla oblongata 506, 507, 508
meiosis 252, 650–1, 656–8, 710–11
    completion at fertilization 716, 717
    crossing over 660
    discovery of 658
    and Law of independent assortment 750
    non-disjunction 658–9, 660
    oogenesis 712, 715–16
    random orientation 661
    role in sexual reproduction 694–5

    sources of genetic variation 660–1, 782
    spermatogenesis 712, 714
melatonin 502–3
membrane polarization 470
membrane potential 470
memory cells 524, 526, 527, 529
Mendel, Gregor 726–30, 747–8
menopause 724
menstrual cycle 465, 698–701
meristems 662–3
mesocosms 802–4
mesophyll tissue 283, 284, 285
messenger RNA (mRNA) 18, 616, 619
    alternative splicing 629–30
    codons 621
    control of degradation 671
    post-transcriptional modification 627–8
metabolic pathways 397–8
    end-product inhibition 402–3
    mechanism-based inhibition 403–4
metabolic water 202, 424
metabolism 67, 381
    anabolic and catabolic reactions 382–3
    heat generation 396–7
metaphase
    meiosis 711
    mitosis 653, 654
metastasis 668
methane 821, 825
methylation 673–4
microfilaments 51, 73
microhabitats 342
micro-organisms 89
    modes of nutrition 366, 368–9
    modes of respiration 365
    see also bacteria; fungi; protists
microscopy 52–3
    CellCams 56
    digital 82
    electron microscopes 59, 60–2, 80, 220–1
    eyepiece graticule 54–5
    fluorescence microscopes 62–3
    magnification 57–8
    of mitosis 655–6
    resolution 59
    temporary mounts 54
microspheres 40–1
microtubules 51, 72–3, 653
microvilli 80, 265–6
Miller–Urey experiment 39, 40
mimicry 361–2, 374
Mirabilis jalapa 736–7
Mitchell, Peter 421
mitochondria 51, 68, 69, 81, 248–9
    aerobic respiration 415–16
    ATP synthesis 421–2
    DNA 84, 85
    endosymbiotic theory 83–5

mitosis 252, 650–1, 653–4
    microscopy 655–6
mitotic index 669
mixotrophs 367
moa (*Dinornis novaezealandiae*) 163
mode 745, 794
model organisms 644
models 236, 265, 561
molecular clock 46, 130
molecular visualization software 28–9
monoclonal antibodies 719
monosaccharides 186, 187, 189–91
    test for 192
    see also glucose
monohybrid crosses, Mendel's experiments 727–30
monozygotic twin studies 677–8
Morgan, Thomas 754–5
morphogens 256, 257
morphological species concept 104, 107
motility 326, 327
motor neurons 332–3, 468, 469, 499
motor proteins 72, 73, 653
MRI (magnetic resonance imaging) 488–9
MRSA (methicillin-resistant *Staphylococcus aureus*) 535, 536
Mucor 78
mucous membranes 520
multicellular organisms 50, 87
multinucleate cells 78
multiple alleles 734
multipotent stem cells 259–60
muscle contraction 327–31
mutagens 640–1
mutation rates 647
mutations 24, 103, 622–3, 637–8
    base substitutions 638
    as cause of cancer 667
    causes 640–1
    in germ cells and in somatic cells 642
    insertions and deletions 638–40
    randomness 641
    as a source of genetic variation 642, 782
    in viruses 98, 99
mutualism 565
    corals and zooxanthellae 567–8
    mycorrhizae 567
    root nodules 566
mycorrhizae 567
myelin sheath 332, 333, 468, 469, 471, 473–4, 483
myofibrils 268, 327–8
myosin 327–30
NAD (nicotinamide adenine dinucleotide) 413
'naked' DNA 28, 66
natural classification 144
natural selection 102, 109, 779–80
    abiotic factors 783
    changes in allele frequencies 793
    competition for resources 782–3
    directional, disruptive and stabilizing 793–5
    guppy populations 788–90
    mechanisms of variation 782
    overproduction of offspring 782–3
    peppered moth 782–5
    role of genetic variation 783–4
nature reserves 175
negative feedback 402–3, 465–6, 575, 714, 761
    control of population size 557–8
    menstrual cycle 699
    osmoregulation 775–6
    thermoregulation 766–8
negative-sense RNA 91
neo-Darwinism 792, 793
neonicotinoids 483
nephrons 265, 769–70
nerves 500
nervous system 467, 495, 496–9
    CNS 496–9
    enteric nervous system 508
    nerves 500
    pain reflex arcs 500–1
net primary productivity (NPP) 592
    in succession 817
net secondary productivity (NSP) 593–4
neural signalling
    action potential 471, 479–82
    all-or-nothing principle 481
    conduction speeds 472–4
    effects of exogenous chemicals 483–4
    inhibitory synapses 484
    pain perception 486–7
    refractory period 481
    resting potential 469–70
    saltatory conduction 483
    stimulus intensity 481
    summation 485
    synapses 476–8
neuromuscular junctions 332–3, 477
neurons 262, 467, 498–9
    gated ion channels 239–41
    motor 332–3
    structure 468–9
neurotransmitter-gated ion channels 240–1
neurotransmitters 453, 455–6, 476–8
    transmembrane receptors 459
niches 363–4, 569–71
    competitive exclusion 378–9
    fundamental and realized 377–8
    modes of nutrition 365–9
    modes of respiration 364–5
nicotine 483
nicotinic receptors 240
nitric oxide (NO) 455
nitrogen fixation 566
nitrogenous bases 15–16, 17, 31
    complementary pairing 23
    purine to pyrimidine bonding 25–6, 27
nitrogen oxides 821
nodes of Ranvier 332, 468, 469, 471, 473, 483
non-competitive inhibitors 398–9
non-conjugated proteins 218–20
non-cyclic photophosphorylation 443
non-disjunction 658–9, 660
non-histone chromosomal protein 28
normal distribution 552, 745
nuclear envelope 68
nuclear lamina 252
nuclear membrane 246–7, 251–2
    origin of 253
nuclear pores 68, 246–7, 251, 252
nuclear receptors 464
nucleic acids 14
    digestion 188
nucleocapsid 93
nucleoid 67
nucleoli (singular: nucleolus) 69, 81, 665
nucleosomes 28–9, 652
    methylation 673–4
nucleosome structure 26
nucleotides 15–16
    sugar–phosphate bonding 16–17
nucleus 51, 63, 64, 65, 68–9, 81
null hypothesis 572, 573, 757
nutrient cycles 583–4, 601
    carbon cycle 594–601
nutrients 208
nutrition, modes of 365–9
obligate aerobes and anaerobes 365
Occam's razor 132
ocean currents 829–30
oestradiol 204, 465, 710
    menopause and HRT 724
    menstrual cycle 698–701
    pregnancy 722, 723
Okazaki fragments 612, 613
omega-3 fatty acids 201
oncogenes 667
oocytes 271–2, 715–16
oogenesis 649–50, 714–16
oogonia 715
open systems 579
operators 679
operons 679–80
optical isomers 190–1
    amino acids 206
orchid mantis (*Hymenopus coronatus*) 361–2
orders 126
organelles 61, 64, 244
    advantages of compartmentalization 247–50
    see also chloroplasts; Golgi apparatus; mitochondria; nucleus; ribosomes
organic compounds 182–5
organs 85, 492
organ systems 493
oscilloscopes 482

Index

osmoregulation 768–9, 775–6
osmosis 227–8, 682
    effects on animal cells 688
    effects on plant cells 689
osmotic concentration of plant tissue 683–7
ovarian cycle 698–700
ovarian follicles 465, 715
    menstrual cycle 698–700
ovaries 494, 696, 697, 714
    oogenesis 715–16
overharvesting 170, 806–7
oviducts (fallopian tubes) 696, 697, 714
ovulation 699, 716
ovule 702, 703
oxidation 412–13
oxygen
    in aerobic respiration 423
    in photosynthesis 427, 428, 441–2
    solubility in water 9
oxygen dissociation curves 292–3
    Bohr shift 294
    foetal and adult haemoglobin 293
oxytocin 723
ozone 34, 35, 427
p values 757, 759
pacemaker (sinoatrial node) 316, 320–1, 506
pain perception 486–7
pain reflex arcs 500–1
palisade mesophyll 283, 284, 285
pancreas 80, 494, 762–3
paradigm shifts 128, 781
Paramecium 74, 379
parasites 88, 565
parasympathetic nervous system 506
parental generations 727, 728
parent cells 648
parsimony 132
partial pressures 291
passive immunity 541
Pasteur, Louis 89, 519
pathogenicity 565
pathogens 517–18
    primary defences against 519–20
peat 597
peatland restoration 834
pedigree charts 733, 741–2
peer review 168, 546
penicillin 403–4, 532–3, 577
penis 696
pentose sugars 15–16
peppered moth (*Biston betularia*) 782–5
peptide linkages 207
percentage change 546, 547, 802
percentage difference 546, 547
pericardium 315
peripheral membrane proteins 226, 234
peripheral nervous system (PNS) 467, 468, 496
peristalsis, control of 508–9

pesticides 810–12
petals 703
pH
    effect on enzyme activity 390–1
    effect on proteins 210
phagocytes 522–3
phagocytosis 239, 247–8
phenology 834–5
    impact of climate change 835–6
phenotype 670, 731
    effects of dominant and recessive alleles 732
phenotypic plasticity 732
phenylketonuria (PKU) 733
phloem 309, 310, 323–4
    pressure–flow hypothesis 324–5
    structure 78–9
phosphodiesterase 462
phosphodiester bonds 18
    hydrolysis 188
phospholipid bilayers 202–4, 223–4
phospholipids 8, 137, 198–9, 203, 223
phosphorylases 462
phosphorylation 414
phosphorylation cascades 458
photoautotrophs 366, 585
photolysis 427, 428
photoperiods 834
photophosphorylation 439
    cyclic and non-cyclic 443
photosynthesis 249–50, 425–6
    C3 and C4 plants 824
    Calvin's experiments 448
    carbon cycle 594–5
    carbon dioxide enrichment experiments 438
    interdependence of reactions 449
    light-dependent reactions 439–44
    light-independent reactions 439, 445–8
    product synthesis steps 447–8
    as source of oxygen 427, 428
    steps of 427–8
    wider significance of 426–7
photosynthesis rate
    influencing factors 436–7
    measurement 434–5
photosynthetic pigments 427
    absorption spectra 431–2, 433
    action spectra 432–3
    chromatography 429–31
photosystem I 440
    NADP reduction 443–4
photosystem II 440
    photolysis of water 441–2
photosystems 439–40
    structured array 441
phototropism 510–12
    role of auxin 514
phyla (singular: phylum) 126
phylogenetic classification 144

phylogenetic diversity 179
phylo-morphological species concept 130
phytohormones 512–13
    auxins 513–16
    cytokinin 516
    ethylene 466, 516
pili 66, 67
pineal gland 494, 502, 504
pinocytosis 239
pioneer communities 814
pitcher plants 361
pituitary gland 494, 505, 698, 766
placenta 699, 721–2
plagioclimax (arrested succession) 818–19
plan diagrams 308–10
plant cells 63–5, 68
    effects of osmosis 689
    structure 67–73, 76–7, 80
plant receptors 466
plants
    adaptations for harvesting light 376–7
    allelopathy 576–7
    gas exchange 283–6
    meristems 662–3
    reproduction in flowering plants 702–9
    resisting herbivory 372–3
    transverse sections 308–10
plant tissue
    osmotic concentration 683–7
    water movements 692
plasma cells 524, 527, 528
plasma membrane 63, 64, 67, 72, 137, 195, 216, 223
    Davson–Danielli model 235–6
    effect of temperature 233
    evolution 41
    fluidity 237–8
    fluid mosaic model 234, 236
    integral and peripheral proteins 226–7
    mechanisms of movements across 225–32, 238–42
    movements across 224
    phospholipid bilayers 202–4
    receptor proteins 456–7, 459–64
    selective permeability 232
plasmids 66, 67
plasmodesmata 79, 243, 306, 323, 456
plasmolysis 689
plastic pollution 812–13
plastids 76
platelets 520–1
pleural membranes 276
pluripotent stem cells 259–60
pneumocytes 266–7
podocytes 771, 772
point mutations 622–3, 637
polar ice cover 826–7, 829
polarity of water 3–4
pollen 702
pollen tube 702–3

pollination 702, 704–5
pollution 172
    and biochemical oxygen demand 809–10
    biomagnification 810–12
    effects on gene expression 675
    plastics 812–13
polyadenylation 627–8
polygenic inheritance 103, 743–4
polymerase chain reaction (PCR) 605–6
polymerization 39
polymers 187
polynucleotides 16–17
polypeptides 205
    digestion 188
    formation 187
    infinite variety of 209
polyploidy 154–5
polysaccharides 191–4
    formation and digestion 187
    see also cellulose; glycogen; starch
polysomes 253, 622
polyunsaturated fatty acids 200
Pompeii worm 346–7
pond skaters 6
Popper, Karl 32, 45, 51
population growth curves
    exponential 558–9, 560
    sigmoid 559–62
populations 107–8, 341, 548
    density-dependent factors 557–8
    predator–prey interactions 575
    top-down and bottom-up control 575–6
population size estimation 549
    capture–mark–release–recapture method 555–6
    quadrats 550–1
positive feedback 465–6
    childbirth 723
    fruit ripening 516
    in global warming 825–7
    menstrual cycle 699
    tipping points 799–802
positive phototropism 510–12
    role of auxin 514
positive-sense RNA 91
postsynaptic neurons 476
post-transcriptional modification 246, 627–8
post-translational modification 635
potometers 287–8
precision 40, 510
predation 564
predator–prey relationships 574–5
predators, adaptations 373–4, 376
pregnancy 699
    childbirth 723
    embryo and foetus 721
    hormonal control 722
    implantation of blastocyst 718
    placenta 721–2

pregnancy testing 719–20
premature babies 279
pressure–flow hypothesis 324–5
presynaptic neurons 476
prey, adaptations 373–6
primary consumers 586
primary production 592–3
primary structure of proteins 212, 215
primary succession 815
primary tumours 668
primase 612
producers 580
    energy flow 589
productivity 592–4
    changes in succession 817
progesterone 465, 710
    menstrual cycle 698–701
    pregnancy 722, 723
programmed cell death (PCD) 256, 459
prokaryotic cells 65, 82
    CRISPR sequences 644–6
    structure 65–7
promoters 625–6, 673, 674
proofreading, DNA 614, 640, 641
prophase
    meiosis 711
    mitosis 653, 654
prostate gland 696
proteasomes 636
proteins 205
    amino acid chain formation 207
    amino acid structure 206
    conjugated 217–18
    denaturation 210
    digestion 188
    fibrous and globular 221–2
    infinite variety of 209
    non-conjugated 218–20
    in plasma membrane 226–7, 229, 230–1, 234, 235
    processing in Golgi apparatus 254
    translation termination 634–5
protein structure 212
    primary 212, 215
    quaternary 217–20
    secondary 213
    super-secondary 214
    tertiary 214–16
protein synthesis
    alternative splicing 629–30
    polypeptide elongation 633–4
    post-transcriptional modification 627–8
    post-translational modification 635
    ribosome structure and function 632–3
    transcription 615–18
    transcription initiation 625–6
    translation 618–22
    translation initiation 630–2, 633
proteolysis 636

proteomes 86, 209, 672
protists 50
protocells 38, 40–1, 45
proto-oncogenes 667
proximal convoluted tubule (PCT) 265–6, 770
    selective reabsorption 772–3
pseudoscience 4
puberty 709–10
pulmonary circulation 314, 315
pulse rate 298
pump proteins 230–1
Punnett grids 728, 729, 730
    for dihybrid crosses 749, 750
purines 16
    relative amounts 31
purine to pyrimidine bonding 25–6, 27
pyramids of energy 586–8
pyrimidines 16
    relative amounts 31
pyruvate 248
    conversion to lactate 416–17
    formation in glycolysis 414–15
quadrats 349, 550–1
qualitative data 59
quantitative data 59, 291
quartiles 746
quaternary structure of proteins 212, 217–20
quorum sensing 452–3
rabies 538
rainforests 165–6, 170–1, 356
    adaptations for harvesting light 376–7
    adaptations to 360–2
    deforestation 801–2, 824
randomized controlled trials 724
random sampling 549–50
    quadrats 550–1
range of data 552
range of tolerance 345
rate of reaction 394, 399–401
realized niches 378
receptor proteins 450–2
    epinephrine receptors 461–2
    intracellular 456, 457–8, 464
    transmembrane 456–7, 459–64
receptor tyrosine kinases (RTKs) 463–4
recessive alleles 728, 732
    phenylketonuria 733
recombinants 751–3
    in *Drosophila* 755–6
recombination 99
red blood cells (erythrocytes) 60, 69, 78, 261, 265
redox reactions 412–13, 585
Red Queen hypothesis 149
reducing sugars 192
reduction 412–13
reflex arcs 500–1
reflexes 498
refractory period 481

regression lines 474–5
regulation of technology 646
regulatory DNA 626, 647
relay neurons 468, 469
relay proteins 463
repolarization 479–80
repressors 679
reproduction 693–4
   see also asexual reproduction; sexual reproduction
reproductive isolation 150, 153–4
reproductive systems
   female 696–701
   male 696
resolution of an image 59
resource competition 568–71
respiration 248–9, 273, 405–6, 585
   aerobic 408
   anaerobic 408, 416–18
   carbon cycle 594–5
   coenzymes 413
   electron transport chain 420–1, 423
   glycolysis 414–15
   Krebs cycle 418–19
   link reaction 418
   modes of 364–5
   redox reactions 412–13
   sites of 415–16
   stages of 412
respiration rate
   determination of 409–10
   influencing factors 409
respiratory centre 507
respiratory substrates 407–8, 420, 424, 762
respirometer 409–10
resting potentials 469–70
   oscilloscope traces 482
retroviruses 99–100, 530
reverse transcriptase 99, 100
rewilding 176, 813–14
R_f values 430, 431
R-groups 184
   amino acids 211
Rhizobium 566
ribosomal RNA (rRNA) 18
ribosomes 51, 67, 68, 69–70, 81, 253–4, 616, 619, 632–3
   polypeptide formation 622
ribozymes 42
ribulose bisphosphate 446
ringed seal (*Pusa hispida*) 10
RNA (ribonucleic acid) 14–15
   comparison with DNA 22
   directionality 25
   mRNA 616, 619, 627–9, 671
   tRNA 619, 630–3
RNA polymerase 616
RNA polymers 17–18
RNA primers 611, 612
RNA splicing 246, 627–8

RNA viruses 88, 91
RNA world hypothesis 41–3
rocky shores, transect studies 349–51
root hairs 305
root nodules 566
root pressure 321–2
roots 304–5
   transverse sections 310
rough endoplasmic reticulum (RER) 68, 70, 81, 254
rubisco (ribulose bisphosphate carboxylase) 398, 445
S-population curves 559
   case study 560–1
   modelling 562
saltatory conduction 483
sampling error 549
sampling methods 549–50
   capture–mark–release–recapture method 555–6
   quadrats 550–1
saprotrophs 367–8, 583, 584
sarcolemma 268, 270
sarcomeres 327–8
sarcoplasmic reticulum 270
SARS-CoV-2 14, 90, 91, 100, 539–40
   see also COVID-19
saturated fatty acids 197, 199, 200–1
scaffold proteins 451, 452
scale bars 57–8
Schwann cells 332, 333, 468, 469, 471, 473
scientific method 4, 32, 51–2
scorpions 360
secondary consumers 586
secondary production 593–4
secondary structure of proteins 212, 213–14, 215
secondary tumours 668
second messengers 451, 452, 461
seeds 706
   dispersal 706
   germination 707–9
segregation, Law of 730
selective breeding 141–4
selective permeability 232
self-incompatibility 705
self-pollination 705
semi-conservative DNA replication 603
semilunar valves 315, 316
seminal fluid (semen) 696, 714
seminiferous tubules 712–13
Semmelweis, Ignaz 519
sense organs 373
sensors 352–3
sensory neurons 468, 469, 498–9
sepals 703
sessile organisms 326–7
sex chromosomes 112, 737–40
sex determination 738

sex hormones 464–5
   menstrual cycle 698–701
   puberty 709–10
   role in spermatogenesis 714
sex-linked conditions 739–40
sexual dimorphism 104, 786
sexual reproduction 693, 694
   fertilization 701, 716–17
   flowering plants 702–9
   gametogenesis 710–16
   generation of variation 782
   human reproductive systems 695–701
   male and female sexes 695
   pregnancy 718–22
   role of meiosis 694–5
sexual selection 786–8
   guppy populations 788–90
short tandem repeats (STRs) 610
sickle cell anaemia 622–3, 638
   genome editing 646
sieve plates 323
sieve tubes 69, 78–9, 323
sigmoid (S-shaped) growth curves 559
   case study 560–1
   modelling 562
signal cascades 458
signalling see chemical signalling; neural signalling
signal transduction 450–1, 458
Simpson's reciprocal index 157–9
simulations 437, 472, 785, 797
single circulation 314
single-nucleotide polymorphisms (SNPs) 115, 638, 641, 734
SIR model 101
SI units 185
skeletal (striated) muscle 262, 269–70
   antagonistic muscle pairs 331–2, 334–5
   contraction 327–31
   neuromuscular junctions 332–3, 477
   structure 78
skeletons 334–5
skewed distribution 745
skills viii–x
skin
   barrier function 519–20
   cell proliferation 663–4
   thermoregulation 767
skin colour, polygenic inheritance 744
sliding-filament hypothesis 327–30
small intestine, cell structure 80
smallpox eradication 544
smallpox vaccine 89, 543
smooth endoplasmic reticulum (SER) 68, 70
Snow, John 518
social insects 564
sodium chloride, solubility in water 8
sodium–potassium pumps 241–2
soil, capillary tubes 8
soil erosion 807–8

solutes 8
solvation 681
solvent properties of water 8–9
solvents 8
somatic mutations 642
Southern blotting 609
specialized cells 85–6, 265–72
speciation 108–9, 147–9
    abrupt 154–5
    adaptive radiation 152–3
    cladograms 133, 134
    reproductive and geographical isolation 150–1
    reproductive isolation 153–4
    sympatric and allopatric 152
species 102, 103–4, 126, 341
    chromosome number 120
    estimated number of 160
species concept 104, 106–7, 130
    bacteria and asexually reproducing species 120
species identification
    dichotomous keys 121–3
    DNA barcoding 124
species richness and diversity 157
specific heat capacity 9
    water 10
spermatids 713–14
spermatocytes 713–14
spermatogenesis 712–14
spermatogonia 713–14
sperm cells (spermatozoa) 261, 271, 696
    epigenetic reprogramming 675
    fertilization 716–17
sperm duct 696, 712
spinal cord 498–9
spindle 75, 653, 654, 656
spirometers 281–3
spliceosomes 627
spongy mesophyll 283, 284, 285
    cell structure 80
square-cube law 264
squirrels (*Sciurus* species), resource competition 568–9
stabilizing selection 793, 794
stamen 703
standard deviation (SD) 552–3
standard error (SE) 686–7
starch 191, 193
    comparison with cellulose 194
    hydrolysis 187, 387–9
    test for 192
statins 401–2
statistical significance 572–4, 757–9
stem, transverse sections 308–9
stem cells 256, 257
    niches 258–9
    totipotent, pluripotent and multipotent 259–60
steroid receptors 457–8

steroids 203, 204
stigma 703
stomata (singular: stoma) 284–5, 289
stomatal density 290
stop codons 634–5
storage compounds 193, 202
    see also glycogen; starch
stratified sampling 550
striated muscle *see* skeletal muscle
stroma, chloroplasts 76
style 703
substitution mutations 637, 641
substrate concentration, effect on enzyme activity 391–3
substrates 383
succession 814
    arrested 818–19
    cyclical 817
    data processing activity 816
    primary 815
    productivity changes 817
sugar–phosphate bonding 16–17
supercoiling of chromosomes 652
superoxide dismutase 216
surface area-to-volume ratios 262–4
    and cell adaptations 265
    and gas exchange 274
surface tension 5–7
surfactant 267, 277
survival of the fittest 783–4
sustainability 806–8
swimming 339–40
symbiosis 353, 566–8
sympathetic nervous system 506
sympatric speciation 152
symplast pathway 306
synapses 240, 455, 476–8
    effects of exogenous chemicals 483–4
    inhibitory 484–5
synovial joints 335
systematic sampling 550
systemic circulation 314, 315
systole 317, 318–19
T-lymphocytes (T-cells) 523, 524, 528
    immune response 526–8
taiga (boreal forest) 357
    carbon storage 827–8
Taq polymerase 605, 606
taxonomy 104–6, 126–7
    viruses 89, 90
technological tools ix
teeth (dentition) 369–72
telomeres 626, 627
telophase
    meiosis 711
    mitosis 653, 654
temperate forests 357
temperature
    effect on enzyme activity 387–9

    effect on photosynthesis rate 437
    effect on proteins 210
template strand, DNA 616–17
temporal isolation 152
tendons 335
tertiary consumers 586
tertiary structure of proteins 212, 214–16
test crosses 731
testes (singular: testis) 494, 696
    spermatogenesis 712–14
testis-determining factor (TDF) 112
testosterone 204, 464, 710, 714
theories 51, 84–5, 140
    paradigm shifts 781
thermal conductivity 9
    water 10
thermophilic species 346–7
thermoregulation 766–8
thin-layer chromatography (TLC) 430–1
threshold potential 481
thylakoids 76, 249–50, 440, 441, 444
    ATP production 442
thymine 15–16
thymus gland 494, 523, 524
thyroid gland 494, 766
thyroxin 766
tidal volume 282
tipping points 799–800
    boreal forest carbon balance 827–8
    deforestation 801–2
tissue fluid
    comparison with blood plasma 312
    release and reuptake 310–12
tissues 49, 85, 492
titin 331–2
tolerance 344–8
    Pompeii worm 346–7
    to salt 347–8
tolerance ranges 349–51
tonoplast 76
tools ix–x
top-down control 575–6
totipotent stem cells 259–60
toxicity 375
trachea 276
transcription 615–18
    directionality 625
    initiation 625–6
transcription factors 671
transcriptome 673
transduction 119
transects 349–52
transfer RNA (tRNA) 18, 619
    amino acid activation 630–2
    sites of action 632–3
transformation 119
translation 618–22
    directionality 625
    initiation 630–2, 633
    termination 634–5

Index

translocation 324
transmembrane receptors 456–7
    epinephrine receptors 461–2
    G protein activation 459–60
    insulin signalling 463–4
    for neurotransmitters 459
    plants 466
    tyrosine kinase activity 463
transpiration 5, 286–9, 303
    advantages 304
transpiration stream 303, 307
tree of life 44–5
tricuspid valve 316
triglycerides 197–8, 199, 203
trinucleotide repeat sequences 638–9
trophic levels 575–6, 582, 58
    energy loss between 589–91
    energy pyramids 586–8
    restriction on number 591, 593
tropical rainforests *see* rainforests
tropic responses (tropisms) 509
    positive phototropism 510–12
tropomyosin 328
troponin 329
trp operon 680
tuberculosis (TB) 537–8, 542–3
tumours 668, 669
tumour-suppressor genes 667
tundra 358
turgid cells 689
turgidity 284
turgor pressure 689
twin studies 677–8
Twort, Frederick 89
type 1 and type 2 diabetes 764
tyrosine kinases 463
ubiquitin 636
ultrafiltration 771
umbilical cord 721, 722
uncertainties x, 389, 686–7
uncoupled respiration 767
unicellular organisms 50
    effects of osmosis 688
    life processes 74
unsaturated compounds 183
unsaturated fatty acids 197, 199, 200–1
upslope range shifts 830–1
upwelling 830
uracil 17, 18
urbanization 170
ureters 769
urethra 696, 769
urinary system 769
    see also kidneys
urine
    composition of 777
    formation of 771–3
uterine cycle 698–700
uterus 697

blastocyst implantation and development 718
    childbirth 723
    foetal development 721–2
vaccination 540–2
    effectiveness 542–3
    herd immunity 544–5
vaccine efficacy 547
vaccine hesitancy 546
vacuolar pathway 306
vacuoles 64, 68, 688
    non-permanent 71, 72
    permanent 76–7
vagina 697
valves
    in the heart 316–17, 318
    in veins 299
Van Valen, Leigh 149
variable number tandem repeats (VNTRs) 608, 626
variation 102
    continuous and discontinuous 103, 743
    human 743
    mechanisms of 782
    mutation as a source 642
    role in natural selection 783–4
    role of meiosis 660–1
    single-nucleotide polymorphisms 115
variation of data values 552
vasa recta 773–5
vascular bundles (veins) 283, 284, 285, 309
vasoconstriction 766–7
vasodilation 766–7
veins 296–7, 299
ventilation of the lungs 279–81
ventilation rate 283
    control of 507–8
ventricles 315, 316
venules 299
vesicles 81, 255
    spontaneous formation 40–1
viral genomes 90–1
viral vaccines 541–2
virions 93
virulence 94, 539
viruses 14, 36, 88, 518
    antiviral drugs 542
    classification 89, 90
    comparison with bacteria 89
    determination of particle concentration 95–6
    discovery of 89
    evolutionary origins 96–8
    HIV 529–31
    infectivity 93
    rapid evolution 98–100
    reproductive life cycles 93–5
    zoonoses 538–40
virus structure 89–93

viscosity, water 7, 11
vital capacity (VC) 283
voltage-gated channels 241, 479–80
Wallace, Alfred Russel 780
water 2
    adhesion 7–8, 303
    buoyancy 11
    cohesion 5–7, 303
    density 10
    extraplanetary origin 11–12
    hydrogen bonds 3–4, 5–6
    metabolic 202, 424
    movement by osmosis 682–3
    physical properties 9–11
    search for extraterrestrial life 12–13
    solvent properties 8–9
    specific heat capacity 10
    surface tension 5–7
    thermal conductivity 10
    transport in plants 5, 7–8, 304–8, 321–2
    viscosity 7, 11
water hyacinth (*Eichhornia crassipes*) 173
water potential 690–2
    and water movements in plant tissue 692
water vapour 821
Watson, James 20, 23, 25–6
wetlands 819
white blood cells (leucocytes) 261
    lymphocytes 523–5
    phagocytes 522–3
    see also B-lymphocytes; T-lymphocytes
Woese, Carl 127, 137
World Health Organization (WHO) 518
world hunger 591–2
wound healing 664
X chromosome 112, 737–9
X-ray crystallography 20, 61, 220
xerophytes 342
xeroseres 815
xylem 287, 303, 304, 309, 310
    adaptations 308
    cohesive and adhesive forces 7
    generation of root pressure 321–2
Y chromosome 112, 737–9
yeast cells 77, 650
    anaerobic respiration 417–18
Z lines 327–8
zona pellucida 716, 717
zoonoses 537–40
zooxanthellae 567–8
zygote 256, 663, 701